# International Seminar on Chemical Engineering Soehadi Reksowardojo (STKSR 2020)

IOP Conference Series: Materials Science and Engineering
Volume 1143

Online
16-17 November 2020

ISBN: 978-1-7138-4619-2
ISSN: 1757-8981

**Printed from e-media with permission by:**

Curran Associates, Inc.
57 Morehouse Lane
Red Hook, NY 12571



Institute of Physics
Dirac House, Temple Back
Bristol BS1 6BE UK

Phone: 44 1 17 929 7481
Fax: 44 1 17 920 0979

techtracking@iop.org

**Additional copies of this publication are available from:**

Curran Associates, Inc.
57 Morehouse Lane
Red Hook, NY 12571 USA
Phone: 845-758-0400
Fax: 845-758-2633
Email: curran@proceedings.com
Web: www.proceedings.com

# TABLE OF CONTENTS

IOP Conf. Series: Materials Science and Engineering 1143 (2021) 011001 doi:10.1088/1757-899X/1143/1/011001

## Preface

International Seminar on Chemical Engineering Soehadi Reksowardojo or STKSR is an annual seminar held by the Department of Chemical Engineering Institut Teknologi Bandung in honor of Prof. Soehadi Reksowardojo's contribution to the early developments of chemical engineering higher education in Indonesia. This year, STKSR 2020 will be held in conjunction with Symposium on phocatalyst and Photocatalysis (Symhosis) 2020 by virtual conference, starting from November 16$^{th}$ to November 17th, 2020. Bringing forward the theme "Circular Economy for a Better World", we would like to invite scholars and practitioners from all aroundthe world to contribute to these seminars.

Symposium on Photocatalyst and Photocatalysis (Symphosis) 2020 is a platform to disseminate and discuss recent advances in the development and application of photocatalysts. It will be the very first typical symposium on this specific subject to be held in Indonesia. This year's symposium will come up with energy and environment-related issues: Water treatment, $CO_2$ reduction and $H_2$ production. In addition to the research discussion involving researchers and academics, this event enables the industrial practitioners to observe any prospects for technology incubation in the field of photocatalyst and photocatalysis. Supported but the Photocatalyst Industry Association of Japan and Ikatan Alumni Habibie, this event features internationally recognized researchers and industrial practitioners as the invited speakers as well as the introduction and official launching of the Indonesia Photocatalyst Society.

This seminar is made because of the importance of natural resources in life. Natural resources are really important but they are limited. This limitation triggers awareness to use the natural resources and products made wisely without reducing the value and to optimize the use of renewable resources. With the theme of "Circular Economy for A Better World", this event is made as a platform to exchange information between researchers, industry and anyone with importance.

## Scientific Committee

• Prof. Tjandra Setiadi

Institut Teknologi Bandung, Indonesia

• Prof. Ashok Pandey

CSIR-Indian Institute of Toxicology Research, India

• Prof. Ramaraj Boopathy

Nicholls State University, USA

• Prof. S. Vigneswaran

University of Technology Sydney, Australia

• Prof. Ir. Dwiwahju Sasongko, Ph.D.

Institut Teknologi Bandung, Indonesia

• Prof. Danu Ariono

Institut Teknologi Bandung, Indonesia

• Prof Lienda Aliwarga

Institut Teknologi Bandung, Indonesia

• Prof. Yazid Bindar, Ph.D.

Institut Teknologi Bandung, Indonesia

• Prof. Ir. I. Gede Wenten

Institut Teknologi Bandung, Indonesia

• Prof. Mikrajuddin Abdullah

Institut Teknologi Bandung, Indonesia

• Prof. Takashi Ogi

Hiroshima University, Japan

• Dr. Kakeru Fujiwara

Yamagata University, Japan

• Dr. Ferry Iskandar

Institut Teknologi Bandung, Indonesia

• Dr. Pramujo Widiatmoko

Institut Teknologi Bandung, Indonesia

• Dr. Aditya Farhan Arif

PT Rekayasa Industri

• Dr. Hendri Widiyandari

Universitas Sebelas Maret, Indonesia

• Dr. Osi Arutanti

Lembaga Ilmu Pengetahuan Indonesia

**Organizing Committee**

**STKSR – SYMPHOSIS Chairman:**

Prof. Tjandra Setiadi, Institut Teknologi Bandung, Indonesia

**Co-Chairman:**

Dr. Eng. Aditya Farhan Arif (PT Rekayasa Industri)

Ardiyan Harimawan, Ph.D. (ITB)

**Team Members:**

Vita Wonoputri, Ph.D.

Dian Shofinita, Ph.D.

Hary Devianto, Ph.D.

Made Tri Ari Penia Kresnowati, Ph.D.

Anggit Raksajati, Ph.D.

Wibawa Hendra Saputera, Ph.D.

Helen Julian, Ph.D.

IOP Conf. Series: Materials Science and Engineering 1143 (2021) 011001 doi:10.1088/1757-899X/1143/1/011001

Antonius Indarto, Ph.D.

Dr. Eng. Jenny Rizkiana

Guntur Adisurya, M.T.

Owen Khosashi

Fathur Rahman Wibisono

Sendy Susanto

Priscilla Angelica

Claritta Sukmalovelina

Jonathan Sasmito

Reynaldo Jonatan

Vincent Augusta Primayudha

Eprianto Pasaribu

**Sub-theme**

1. Advanced Science – Material
2. Bioprocess Technology
3. Chemurgy – Biobased Material
4. Bioenergy – Alternative Energy
5. Food Technology
6. Industrial Application
7. Separation Technology
8. Process Simulation
9. Reaction – Control Engineering
10. Symphosis
11. Kurita

Sponsored by:

PT. EMI (Persero)
Energy Management Indonesia

KURITA

Rekind
Integrated EPC and Industrial Solutions

Supported by:

PERSATUAN INSINYUR INDONESIA
PII

IATK-ITB
IKATAN ALUMNI TEKNIK KIMIA INSTITUT TEKNOLOGI BANDUNG

HIMPUNAN MAHASISWA
HIMATEK - ITB
TEKNIK KIMIA - ITB

Organized by:

IOP Conf. Series: Materials Science and Engineering 1143 (2021) 011002 doi:10.1088/1757-899X/1143/1/011002

# Peer review declaration

All conference organisers/editors are required to declare details about their peer review. Therefore, please provide the following information:

- **Type of peer review: Single-blind**
- **Conference submission management system: Manual by Email**
- **Number of submissions received: 89 manuscript**
- **Number of submissions sent for review: 89 manuscript**
- **Number of submissions accepted: 77 manuscript**
- **Acceptance Rate (Number of Submissions Accepted / Number of Submissions Received X 100): 86.52**
- **Average number of reviews per paper: 1**
- **Total number of reviewers involved: 20**
- **Any additional info on review process:**

  A review process was conducted in four main steps: (i) A minimum of one reviewer was assigned to referee the manuscript. (ii) suggestions/comments from reviewers were mailed to the corresponding authors. There will be 21 days period for the authors to revise the manuscript (iii) once the manuscript was declared accepted by the reviewers, the committee looked over the format of the manuscript and conduct the similarity check using Turnitin software. (iv) if the manuscript has a similarity higher than 40%, the manuscript was sent back again to the corresponding authors and ask for another revision. On the other hand, if the format and the similarity check has fulfilled the criteria, the manuscript was ready to be submitted. The main criteria when assessing the paper are originality, quality and clarity of contents, suitability of references used, contribution to the related fields, depth of research, and quality of English. The reviewer was asked to fill in the comments in the provided tables including general comment and comments on introduction, methodology, results and discussion, bibliography/references, other comments, and final notes. There are five referee's decisions on review process including: (i) can be published without revision (ii) can be published with minor revision (iii) can be published with major revision (iv) shall be returned to the reviewer after revision (v) rejected.

- **Contact person for queries: Wibawa Hendra Saputera (hendra@che.itb.ac.id)**

Please submit this form along with the rest of your files on the submission date written in your publishing agreement.

The information you provide will be published as part of your proceedings.

IOP Conf. Series: Materials Science and Engineering 1143 (2021) 012001 doi:10.1088/1757-899X/1143/1/012001

# Effect of Bubble Size on Electrochemical Reduction of Carbon dioxide to Formic Acid

**Pramujo Widiatmoko*, Wibawa Hendra Saputera, Hary Devianto, Isdiriayani Nurdin, Esperanza Rivana and Albert Angkasa**

Department of Chemical Engineering, Institut Teknologi Bandung, Bandung 40132 Indonesia

*pramujo@che.itb.ac.id

**Abstract.** A variety of technologies to diminish greenhouse gases $CO_2$ emission have been extensively explored in recent years. One promising technology called electrochemical reduction of $CO_2$ have been developed to convert $CO_2$ into value-added chemicals, such as methanol, ethanol, ethylene, and formic acid. One of important parameter that need to be considered is the size of $CO_2$ bubbles. In general, decreasing bubbles size can increase retention time of carbon dioxide in an electrolyte. In this research, effect of the microbubble size on the effectiveness of electrochemical reduction of carbon dioxide to formic acid was studied. Bubbles with average diameter 0.321 and 0.952 mm were generated through perforated glass apparatus. The FTIR analysis showed peak at 1634 $cm^{-1}$ which attributed to C=O bonding of formic acid. Smaller bubble size showed relatively more stable and higher electric current during electrolysis, indicating better retention time of carbon dioxide in the electrolyte. *Keywords*: Microbubbles; Electrochemical reduction of $CO_2$; Formic Acid; Retention time; Size effect; Electrolyte composition

## 1. Introduction

Concerning on the global warming issue, recycling of $CO_2$ into valuable chemicals through electrolysis recently become attractive [1]. The electroreduction of $CO_2$ at various metal electrodes yields many kinds of organic substances, e.g. CO, $CH_4$, $C_2H_6$, and Ethanol [2]. Transition metals such as Au, Ag and Zn produce CO while p-block metals e.g. Sn, Pb and In mainly produce formate. In formic acid production, the Pb electrode shows high selectivity while the Sn electrode shows high current efficiency [3].

The electrolyte used for the electrocatalytic reduction of $CO_2$, both aqueous and non-aqueous, also have been summarized [4]. Common catholyte used in the process is $KHCO_3$ as aqueous solution [5]. Low solubility of $CO_2$ in water under standard conditions (about 34 mM), however, is a key challenge to improve efficiency of the electrocatalytic reduction of $CO_2$ [6]. Usage of organic electrolytes such as acetonitrile, dimethyl formamide, and methanol increase solubility of $CO_2$ up to eight times [7]. Despite the solubility is higher when using organic solvent, toxicity of the organic solvent and safety hazards are need to be considered. It is also important to ensure that the products come from the $CO_2$ reactant instead of the electrolytes.

Another interesting method to overcome the solubility problem is using smaller bubble size of $CO_2$. Smaller bubble size decreases rising velocity, retain the $CO_2$ in the electrolyte longer. In example, the 0.4 mm bubble diameter has theoretical rising velocity of 0.05 m/s meanwhile 0.1 mm bubble has almost

IOP Conf. Series: Materials Science and Engineering 1143 (2021) 012001 doi:10.1088/1757-899X/1143/1/012001

one tenth at about 0.003 m/s [8]. Moreover, the microbubble, with 10-50 μm in diameter, tend to gradually decrease the size by time and dissolute the gas into surrounding water [9, 10]. While $CO_2$ concentration on the cathode surface decrease due to formation of formic acid, high concentration of $CO_2$ in catholyte is maintained by dissolution of $CO_2$ gas from the bubble. This condition ensures high mass transfer during the reaction.

In this work, effect of bubble size to performance of $CO_2$ electrochemical reduction to formic acid was studied. Alloy of Pb-Sn electrode was used to optimize both selectivity and current efficiency. The result show that smaller bubble size increases performance of the electrolysis. It is also shown that diameter ratio of electrode ring to reactor affects the performance.

## 2. Methodology

This work was carried out by bubbling $CO_2$ gas using sparger into reactor during electrolysis, with equipment's arrangement shown in Figure 1. Two spargers types were used i.e. the $CO_2$ atomizer and bubble stone. Arise bubble was photographed using microscope and analyzed using ImageJ software to measure cross-sectional diameter. The Sn-Pb alloy wire with 90 cm length was used as cathode in a spiral shaped with various diameter, while Pt-Ir wire was used as anode. Solution of $KHCO_3$ 0.5 M was used as a catholyte solution and $H_2SO_4$ 0.1 M as anolyte. Prior to electrolysis, the catholyte was saturated by $CO_2$ bubbling for 1 hour. The electrolysis was carried out at overpotential of 1 V for 3 hours. Electric current was recorded every 10 minutes. Theoretical formic acid formed was calculated based on data of electric current vs. time using the Faraday's law of electrolysis. Reaction products was analyzed using Fourier-Transform Infrared Spectroscopy (FTIR).

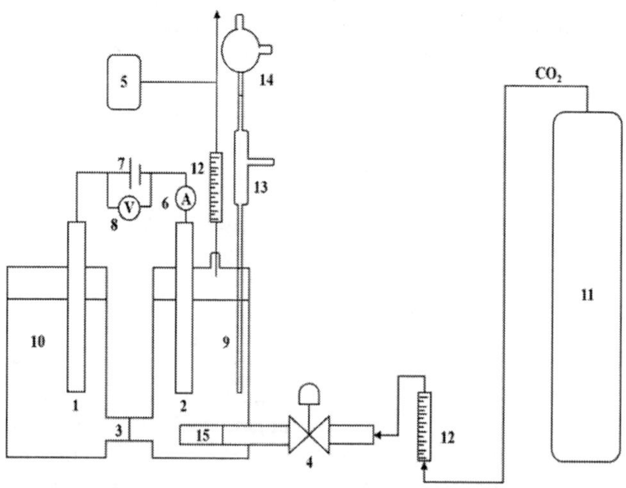


1. Anode
2. Cathode
3. Membrane separator
4. Gate valve
5. Gas bag
6. Amperemeter
7. DC Power supply
8. Voltmeter
9. Catholyte
10. Anolyte
11. $CO_2$
12. Flowmeter
13. Sampler
14. Filler

**Figure 1.** Schematic diagram of experiment.

## 3. Result and Discussion

### *3.1. Bubble generation*

The spargers produce different size of bubble, as shown in Figure 2. The $CO_2$ atomizer and the bubble stone sparger produce bubble diameter of 0.35 mm and 1.96 mm in average at diameter ratio of electrode spiral to reactor ($D_e / D_t$) of 0.25. With increasing the $D_e/D_t$ ratio, secondary peak of bigger bubble size rises at 0.48 mm and 2.65 mm, respectively. In this case, the diameter of reactor ($D_t$), was kept constant.

Terminal rising velocity ($U_t$) of the bubble at low Reynold number can be predicted using Stokes equation as shown in Equation (1), where $\Delta\rho$, $r$, $g$, and $\mu$ are density difference between the bubble and water, radius of bubble, gravitational acceleration, and water dynamic viscosity, respectively [11].

$$U_{t(ST)} = \frac{2r^2 \Delta\rho g}{9\mu} \tag{1}$$

Average terminal rising velocity for the bubbles produced using the $CO_2$ atomizer and the bubble stone sparger are shown in Table 1. The terminal rising velocity in this work was calculated based on properties under 30°C and 1 atm condition. The value of $\Delta\rho$, $g$, and $\mu$ are 993.886 kg/m$^3$, 9.806 m/s$^2$, and 0.0007972 kg/(m.s), respectively.

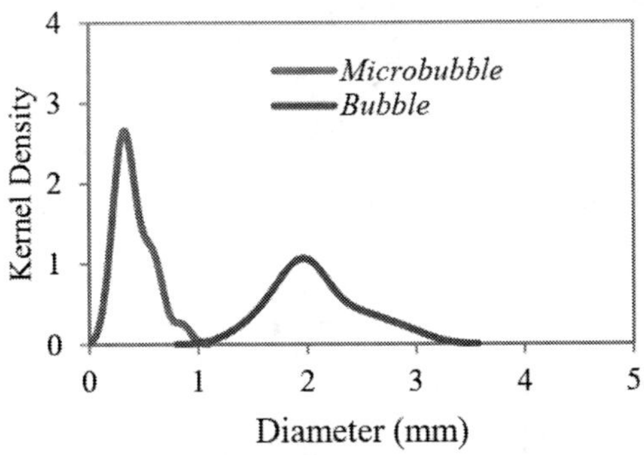


(a)

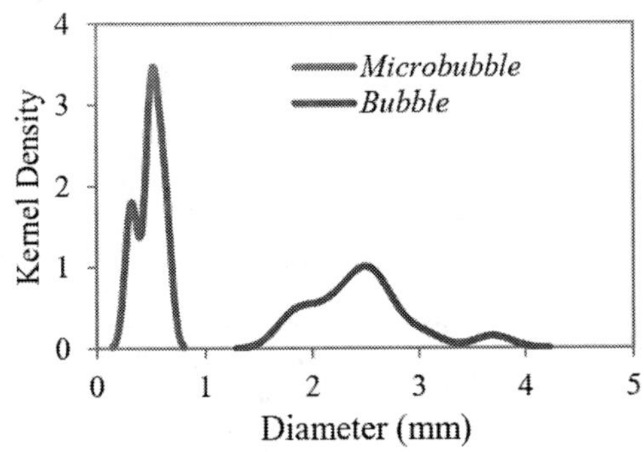


(b)

**Figure 2.** Distribution of bubbles generated in reactor with $D_c / D_t$ ratio of (a) 0.25 and (b) 0.5.

**Table 1.** Terminal rising velocity of $CO_2$ bubbles.

| Type of sparger | Bubble size (mm) | Terminal rising velocity (m/s) |
|---|---|---|
| Atomizer | 0.35 | 0.083 |
| | 0.48 | 0.156 |
| Bubble stone | 1.96 | 2.609 |
| | 2.65 | 4.769 |

Replacing the bubble stone with the atomizer reduces significantly the terminal rising velocity up to 30 times slower. In our 20 cm length reactor, retention time of the bubbles therefore will increase from 0.07 s to 2.40 s. Ideally, microbubbles size at 10 – 50 μm in diameter, which has terminal rising velocity of 4.075 mm/min, will provide much longer retention time at 49 min. To be noted, experiment conducted by Parkinson et al. shows higher terminal velocity of $CO_2$ bubble in pure water than ones predicted by the Stokes equation [12]. The effect is attributed to formation of boundary layer around the small bubble due to enhanced $CO_2$ solubility.

### *3.2. Effect of bubble size on electrolysis performance*

Figure 3 shows effect of the bubble size and $D_c/D_t$ ratio to electrical current on the electrolysis system. The electrical current increases with decreasing of bubble size. As we know, the reaction rate is proportional to the current. It is supposed that increasing $CO_2$ gas in the electrolyte enhance the diffusion rate, move the reaction out from mass control region. Increasing the $D_c/D_t$ ratio means increase in electrode spiral diameter. As the bubble generator was located in the bottom centre of the spiral, we supposed that wider diameter accommodates better distribution of bubble. With decreasing the spiral diameter, there are more bubble move to outside of the electrode spiral. It will reduce the contact between the bubble and electrode as well as the reaction rate.

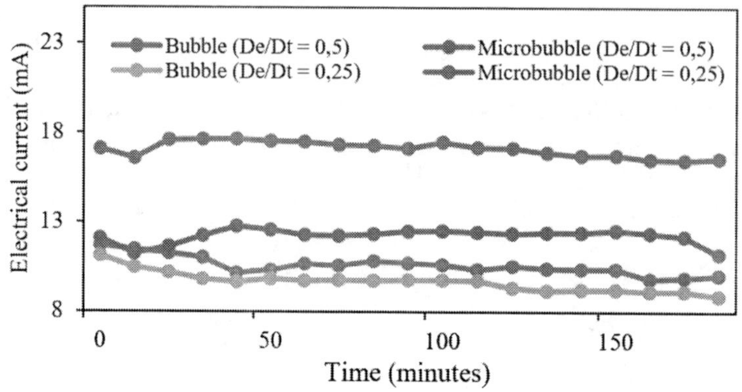


**Figure 3.** Effect of Bubble size and $D_e/D_t$ ratio on performance of electrolysis.

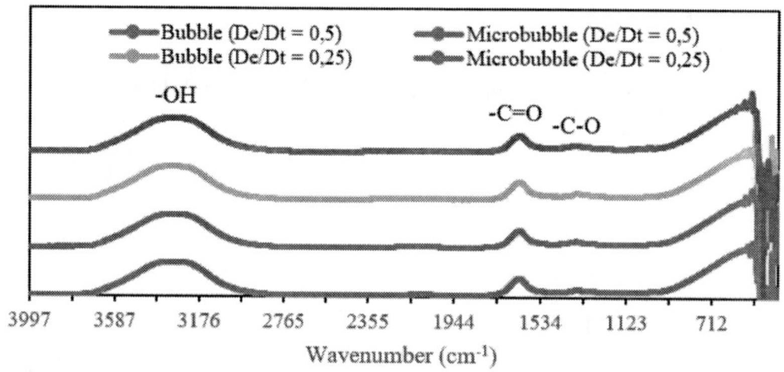


**Figure 4.** Infrared spectra of electrolysis products using Sn-Pb cathode with various bubble size and $D_e/D_t$ ratio.

Figure 4 shows the IR spectra of the catholyte after 3 hours electrolysis. Peaks at the wavenumber of 1364, 1634 and 3328 $cm^{-1}$ indicating the presence of -C-O, -C = O, and -OH groups, respectively. The spectra results are compared with the IR spectra of the possible by-products such as methanol and CO. The presence of the –C = O group at the wavenumber 1634 $cm^{-1}$ indicates formic acid in the product.

Reduction of carbon dioxide into formic acid follows reaction in Equation 2 [13]. On the Pb-Sn electrode, the reduction reaction is intermediated by formation of $CO_2^{*-}$ radical [14, 15]. Theoretical amount of produced formic acid is then calculated using the Faraday's law of electrolysis as shown in Equation 3.

$$CO_2 + 2\,H^+ + 2e^- \rightarrow HCOOH \qquad E^0 = -0.12\,V/RHE \tag{2}$$

$$m = \frac{\int Idt}{nF}\eta \tag{3}$$

Using electric current ($I$) – total electrolysis time ($t$) data shown in Figure 3, Faraday constant ($F$) of 96485 C/mol, number of involved electron ($n$) of 2, and assuming 100% efficiency ($\eta$), the theoretical amount of produced formic acid ($m$) within 3 h electrolysis is about 9 x $10^{-3}$ to 1,5 x $10^{-2}$ mmol. In average, reducing the bubble size by 0,17 times increases amount of formic acid by 1.4 times. Further, a quantitative analysis is required to determine the actual electrolysis efficiency.

## 4. Conclusion and Future Works

Performance of $CO_2$ electrochemical reduction to formic acid is enhanced by decreasing of $CO_2$ bubble size. Infrared spectra indicate formic acid presence in the electrolysis product. Future work on quantitative analysis using High Performance Liquid Chromatography is being conducted to determine the current efficiency of electrolysis.

## Acknowledgments

This work was funded by research, community service, and innovation (P3MI) program 2020, Institut Teknologi Bandung.

## References

[1] Nitopi S, Bertheussen E, Scott S B, Liu X, Engstfeld A K, Horch S, Seger B, Stephens I E L, Chan K, Hahn C, Nørskov J K, Jaramillo T F, and Chorkendorff I 2019 *Chem. Rev.* **119** 761
[2] Jitaru M 2007 *J. the University of Chem. Tech. Met.* **42** 333
[3] Hori Y 2008 *Modern Aspects of Electrochemistry* vol 42. (New York: Springer) p 89
[4] de Salles Pupo M M and Kortlever R 2019 *Chem. Phys. Chem.* **20** 2926
[5] Wu J., Risalvato, F and Zhou X D 2012 *ECS Trans.* **41** 49
[6] Konig M, Vaes J, Klemm E, and Pant D 2019 *iScience* **19** 135
[7] Kaneco S, Iiba K, Katsumata H, Suzuki T, and Ohta K 2007 *J. Solid State Electrochem.* **11** 490
[8] Zimmerman W B, Tesar V, Butler S, Bandulasena H C H 2008 *Microbubble Generation, Recent Patents on Engineering* Vol. 2 (Sharjah: Bentham Science Publishers) p 1
[9] Agarwal A, Ng W J, and Liu Y 2011 *Chemosphere* **84** 1175
[10] Takahashi, M.,Chiba K, and Li P 2007 *J. Ohys. Chem. B* **111** 11443
[11] Stokes G G 1851 *Cambridge Philos. Trans.* **9** 8
[12] Parkinson L, Sedev R, Fornasiero D, and Ralston J 2008 *J. Colloid and Interface Sci.* **322** 168
[13] Nitopi S, Bertheussen E, Scott S B, Liu X, Engstfeld A K, Horch S, Seger B, Stephens I E L, Chan K, Hahn C, Nørskov J K, Jaramillo T F, and Chorkendorff I 2019 *Chem. Rev.* **119** 7610
[14] Sun Z, Ma T, Tao H, Fan Q, and Han B 2017 *Chem* **3** 560
[15] Innocent B, Pasquier D, Ropital F, Hahn F, Leger J M, and Kokoh K B 2010 *Appl. Catal. B* **94** 219

IOP Conf. Series: Materials Science and Engineering 1143 (2021) 012002 doi:10.1088/1757-899X/1143/1/012002

# An overview of synthetic polymer-based membrane modified with chitosan for direct methanol fuel cell application

**S R Adiyar[1*], A Satriyatama[2], A N Azjuba[1], N K A K Sari[1]**

Chemical Engineering Department, Universitas Diponegoro, Jl. Prof. Soedarto SH, Semarang 50275, Indonesia
Chemical Engineering Department, Institut Teknologi Bandung, Jl. Ganesha 10, Bandung 40132, Indonesia

sabrinarahmiad@gmail.com

**Abstract**. The materials used for fuel cell membrane must have a high proton conductivity, a strong enough wall to block the reactant flow rate and be chemically or mechanically stable in the environment around the fuel cell. To improve the effectiveness of fuel cell membranes and reduce production costs, several synthetic polymer membranes have been developed, including polyethersulfone, polysulfone, polyvinyl alcohol, and polystyrene. Membranes from this polymer have the advantage of being cheap, commercially available, and allowing its structure to store moisture so it can operate at higher temperatures, yet it has low hydrophilic property. Chitosan, as a biopolymer that has strong hydrophilicity property resulted from numerous hydrophilic groups (e.g. –OH, $-NH_2$ and $-NR_3$ þ), can be used for various chemical modifications including to increase mechanical and chemical stability and modification to the possibility of producing ion exchange and increasing ionic conductivity which is a requirement for fuel cell membrane. The purpose of this study is to review the use of chitosan as synthetic polymer-based membrane modification from its structure and properties. Recent achievements and prospect of its applications have also been included.

## 1. Introduction

The rapid growth of population and industry contributes to the demand and availability of energy which is increasingly depleted and has led the world to an energy crisis. New and renewable energy sources in the future will increasingly have a critical role in meeting energy needs. This is because the use of fossil fuels for conventional power plants in the long term will deplete the sources of oil, gas and coal whose reserves are getting decreasing over time [1]. Fuel cells are one of potential energy sources generating electricity through chemical pathways, as for the recent decades many incredible researches and developments of compact fuel cells have been conducted for various stationary and mobile applications [2], [3]. Polymer electrolyte membrane fuel cells have gained huge consideration, particularly within the automobile and portable power source industries, because of their high efficiency, power density, fast on-off, and low pollution [4]. DMFC is a type of proton exchange membrane fuel cell (PEMFC) which uses methanol as fuel which is fed directly to the anode compartment [5]. Polymer electrolyte membrane (PEM) is one of the essential components in DMFC, which is expected to have high proton conductivity and low methanol permeability [6], [7]. This membrane in the DMFC functions as a transport agent for hydrogen ions ($H^+$) that generated from anode to cathode and also as a separator between anode and cathode compartment [8]. The membrane that is commercially used today is a perfluorinated sulfonic acid membrane known by its trade name Nafion. However, Nafion membrane

IOP Conf. Series: Materials Science and Engineering 1143 (2021) 012002 doi:10.1088/1757-899X/1143/1/012002

still has several disadvantages including high cost, high methanol permeability which results in reduced cell performance, methanol crossover, and loss of conductivity above 80 °C in DMFC [9], [10]. Therefore, development and research on membranes for DMFC applications will always continue so that global problems regard to energy can be resolved with appropriate and environmentally friendly solutions through the DMFC application [5]. To overcome the deficiency of Nafion, membranes made from synthetic polymers with various modifications have been created. There are numerous advantages to polymeric membranes, such as an excellent mechanism for pore-forming control, good mechanical strength, reasonable cost, and more flexible [11]. Some examples of modified synthetic polymers are poly (vinylalcohol) (PVA) [12], poly (ether ketone) [13], poly (ether ether ketone) (PEEK) [14], poly (ether sulfone) (PES) [15], poly (vinylidene fluoride) (PVDF) [16][17], polyetherimide (PEI) [18] and many have been studied further for their potential as membranes in DMFC application [19].

## 2. Direct Methanol Fuel Cell

One type of fuel cell based on the type of electrolyte used, which is being studied extensively is the methanol fuel cell, so it is widely known as the Direct Methanol Fuel Cell (DMFC)[20]. One of the crucial components of DMFC is the membrane, which functions as a proton conductor from the anode to the cathode and as a separator between anode and cathode space[21]. DMFC offers a simple plan design and possibly higher in general productivity than the reformate-fed fuel cells by removing the reformer. Significant signs of progress in $H_2$ or air polymer electrolyte fuel cells and high power density have been noticed in the application of DMFC[22].

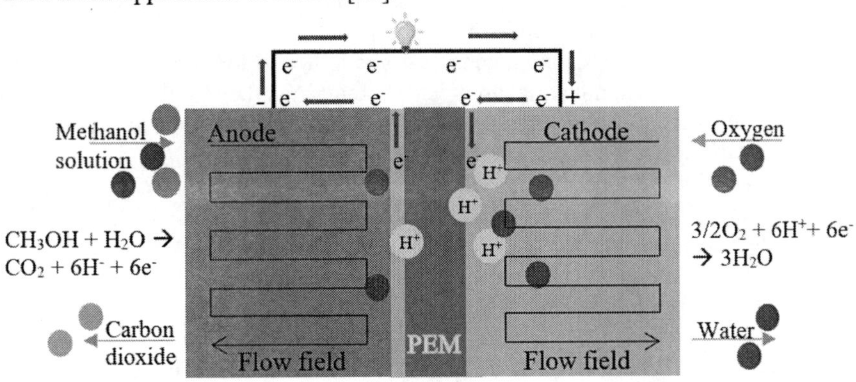


**Figure 1.** Basic illustration of the direct methanol fuel cell (DMFC) system [23].

In DMFC, methanol is oxidized at the anode to produce protons and electrons. Protons arrive at the cathode through the electrolyte and meet oxygen at the cathode to produce water, thus providing the electricity[24]. The proton transfer from anode to cathode is ascribed to the proton-exchange membrane, also maintaining methanol and air (oxygen) separated in the two electrode compartments. The primary problem of DMFCs are their efficiency and performance because of the methanol crossover through the membrane, with consequent efficiency losses[25]. Therefore, it is needed to develop functionalized membranes that can diminish the methanol crossover and at the same time to ensure an effective proton move to the cathode compartment is important [26]. The electrochemical reactions taking place in a DMFC as follows:

$$\textit{Anode: } CH_3OH + H_2O \rightarrow CO_2 + 6H^+ + 6e^- \quad [27]$$
$$\textit{Cathode: } 3/2O_2 + 6H^+ + 6e- \rightarrow 3H_2O \quad [28]$$
$$\textit{Overall: } CH_3OH + 3/2O_2 \rightarrow CO2 + 2H_2O \quad [27]$$

The free energy change, ΔG, also known as free enthalpy is a thermodynamic potential which can be used to calculate the maximum reversible work that a thermodynamic system can perform at constant temperature and pressure [29].

$$\Delta G = - nFE \quad (1)$$

where n is expressed as the number of electrons present in the chemical reaction (n = 6 electrons per mole of methanol), F is the Faraday constant (96,487 coulombs per mole), and E is the equilibrium (reversible) cell potential [30].

## 3. Synthetic Polymer Based Membrane

Making process of membranes from natural polymers has been studied in recent years [31]. In general, the membrane that is often used for membrane is made of cellulose and chitin and their derivatives, which two of them are the most well-known because of their accessibility and their existence is abundant in nature[32]. In Direct Methanol Fuel Cell (DMFC), membrane functions as a transport medium for hydrogen ($H^+$) ions resulting from the oxidation reaction at the anode and a barrier between the two electrodes. The electrolyte membrane requirements for DMFC applications are chemical, electrochemical, mechanical, acidic, good proton conductor, water adsorption, and as small as possible methanol permeability. The proton transport in the electrolyte membrane is determined by the negatively charged groups (usually sulfonates, $SO_3H$). The larger the sulfonate groups, the greater the conductivity. In addition to the presence of sulfonate groups in the membrane, water is also needed to facilitate proton transport, which can increase proton conductivity [33].

Polymer electrolyte membrane (PEM) is an essential part of a DMFC. A variety of PEMs have been developed and studied for DMFC applications. PEM offer verified long-term performance (60,000 h), high proton conductivity, remarkable mechanical strength, and long-term stability. Nevertheless, they are more expensive. Besides, high methanol permeation limits their application and creates a barricade for the DMFC technology commercialization. These factors have added additional motivation for preparing other varieties of PEMs[32].

The most desired properties of polymer electrolyte membranes are:

- High perm selectivity
- Low electrical resistance
- Good mechanical and form stability and
- High chemical and thermal stability

However, the properties depend on numerous factors determined by the parameters such as the membrane material, polymer network density, hydrophilic and hydrophobic nature of the polymer matrix, type, and concentration of ions. The polymeric membrane is the vital part of the fuel cell. It uses as the main component for the conduction of protons and effective separation of the electrodes and reactants. In general, properties such as (a) heat and chemical resistance, (b) fire safety, (c) lightweight, (d) high strength and accuracy, and (e) The feasibility of manufacturing in high volumes with low processing costs make them well suited for a broad range of applications [34]. Some of the modified synthetic polymers that have been studied as potential membranes in DMFC include poly(vinyl alcohol) (PVA)[12], poly (ether ketone) [13], poly (ether ether ketone) (PEEK)[14], poly (ether sulfone) (PES)[15], poly (vinylidene fluoride) (PVDF) [16][17], polyetherimide (PEI)[18], and many more. The common modification made to the making of the proton exchange membrane is the modification using a sulfonation reaction and consolidation of inorganic fillers [19].

## 4. Synthetic polymer-based membrane modification with chitosan

One form of polymer membrane composite membrane, which is widely used in DMFC is electrolyte polymer membrane. The type of membrane that is usually used is polyethylene oxide (PEO) with added KOH because it shows good voltage-current density characteristics. However, due to the high degree of crystallinity of the PEO membrane, the resulting fuel cell current density cannot be increased [35]. Chitosan is a natural carbohydrate biopolymer derived from the deacetylation process of chitin, a significant component of shrimp shells. Chitosan has biocompatible, biodegradable, and non-toxic properties. Chitosan is also an alkaline polyelectrolyte and has good chemical, mechanical and thermal stability. The addition of chitosan which is a biopolymer that has an amine (-$NH_2$) functional group allows it to be modified physically and chemically, where it can improve the properties of the electrolyte membrane such as proton conductivity and reduce methanol permeability[36]. However, suppose the chitosan membrane is used as the fuel cell. In that case, the resulting current density is still low because the ion delivery of the chitosan membrane is still relatively low. Besides, the crystallinity of the chitosan membrane influences the ionic conductivity of the membrane. To increase the ionic conductivity of the membrane and particularly to trigger the rate of ion migration through the membrane, it is necessary to modify the chitosan membrane [37]. Several studies have been carried out on the manufacture of DMFC application membranes by combining synthetic polymers with chitosan. Most studies combined chitosan with other materials that function as fillers to increase overall conductivity because in actual conditions, chitosan has a low conductivity due to its hydrogen atoms being tightly bound to its monomers, so it cannot be immobilized under the influence of electric fields [38]. Alam [39], made a membrane using chitosan biopolymer extracted from waste shrimp shell as a matrix and montmorillonite as a filler. Chitosan or montmorillonite composite membranes have been successfully prepared by modifying the silane coupling agent where the increase in the proton conductivity of the membrane, along with the increase in silane concentration and the operating temperature was found. Ari et al. [40] have synthesized a membrane made from chitosan by adding silica modified by cetyl trimethyl ammonium bromide. Modification is made by adding hydrophobic organic filler such as silica. Nur Fatin et al. [41] have conducted a study to make chitosan bio composites with a mixture of cellulose with various amounts of sulfosuccinic acid (SSA). The membranes were prepared using the casting method to improve their properties as well as the conductivity of the membrane. In this study, the results show that the membranes have better hydrophilic characteristic when compared to the Nafion membrane. This is due to the existence of three polar functional groups where chitosan is also highly capable of forming a hydrogen bond with water. In addition, summary of the chitosan-based membrane for DMFC is discussed in Table 1.

**Table 1.** Summary of the chitosan-based membrane for DMFC applications.

| Membrane Material | Methanol Concentration (mol/L) | Ionic conductivity (mS/cm) | Methanol Permeability ($cm^2/s$) | Reference |
|---|---|---|---|---|
| Quaternized Chitosan | 3 | 0.17 | $2.1 \times 10^{-7}$ | [42] |
| Chitosan/SPVDF | 2 | 2.85 | $5.7 \times 10^{-7}$ | [43] |
| Quaternized Chitosan/PVDF | 2 | 4.1 | $12.6 \times 10^{-7}$ | [44] |
| Zeolite/Chitosan | 5 | 2 | $1.162 \times 10^{-6}$ | [45] |
| $H_2SO_4$ modified chitosan | 1 | - | $1.4 \times 10^{-6}$ | [46] |
| Sorbitol-plasticized chitosan/zeolite | 12 | - | $4.9 \times 10^{-7}$ | [47] |

*4.1. Membrane fabrication*
There are 2 points that need to be considered in the membrane fabrication; these are the original properties of the material and the intended morphology in the membrane. There are several techniques used in the manufacture of polymeric membranes: stretching, phase inversion, solution coating, track-etching and sintering [48]. In the DMFC application, membrane fabrication mostly uses phase inversion techniques [19].

IOP Conf. Series: Materials Science and Engineering 1143 (2021) 012002 doi:10.1088/1757-899X/1143/1/012002

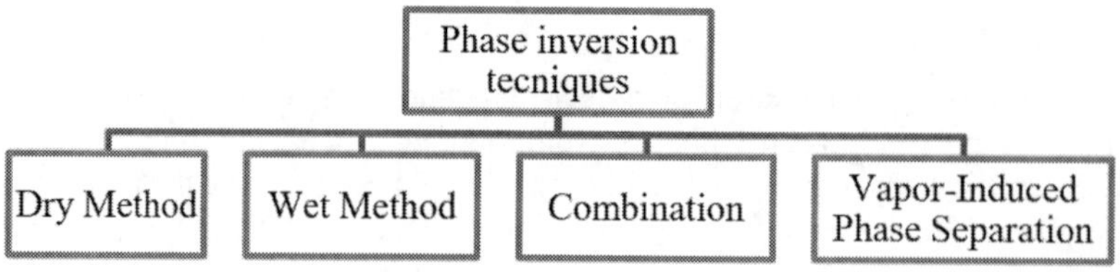


**Figure 2**. Different routes for membrane fabrications via phase inversion techniques [19].

The process of modifying chitosan on polymer membranes is carried out in several ways. First is with sulfonation. Sulfonated chitosan is obtained through the incorporation of sulphonate groups into chitosan matrix [49]. On a membrane with a sulfonate group, a sulfonation process is carried out, which is a process of attaching the sulfonate group ($HSO^{3-}$) to chitosan. The sulfonation method for chitosan solution was done by adding acid reagents [50]. Also, the process of modifying chitosan on polymer membranes can be done by cross-linking. Cross-linking is a commonly used method of chemical modification. The purpose of cross-linking is to ensure good chemical and mechanical stability of chitosan and insoluble chitosan in the aqueous medium [49]. Cross-linking occurs when the sulfuric acid content in the chitosan membrane has been combined with protonation [18]. Vijayalekshmi and Khastgir [4] prepared chitosan-sulfonated poly(vinylidene fluoride) blend membranes by solution casting technique followed by cross-linking with sulfuric acid. In Enggita and Santoso [51], also and Ambili et al. [52], the addition of the poly vinyl alcohol membrane was by dissolving chitosan in an acetic acid solution and then adding it to the PVA solution by stirring until a homogeneous solution was formed.

## 5. Characterization and performance of membrane

### *5.1. Methanol permeability*

If the membrane more resistive, the permeability value will decrease[53]. Permeability was measured using a glass cell with two identical spaces. The membrane commonly used in DMFC, Nafion, has good conductivity, with a range of 0.1 $Scm^{-1}$. However, this is due to an increase in methanol crossover and conductivity in hydrophilic membranes with the membrane hydration level (λ), it is necessary to improve the properties of this property in order to have better final performance results[54]. The equations used to determine methanol permeability ($mS.cm^{-1}$) is:

$$P = \frac{C_{B(t)} x \, V_B \, x \, l}{C_A \, x \, A \, x \, t} \qquad (2)$$

Where $C_a$ and $C_b$ denote the concentration of methanol in each compartments A and B, $l$ is membrane's thickness, $V_b$ is volume of liquid of compartment B, $t$ as time, and $A$ represents the area of membrane [55].

**Table 2.** Methanol permeability for Nafion 117 and MMW chitosan membranes.

| Membrane type | Permeability ($10^{-6}$ $cm^2/s$) | mol/L | Temperature (°C) |
|---|---|---|---|
| Nafion 117 | 4.65 | 12 | 50 |
| Nafion 117 | 2.38 | 18.5 | 20 |
| MMWC | 4.25 | 12 | 50 |
| MMWC | 0.56 | 18.5 | 20 |

Based on the experimental table 2, at a temperature of 20° C each chitosan membranes (fabricated from medium molecular weight (MMW) chitosan) at various concentrations of methanol has lower methanol permeability than Nafion (117). However, if the experiment was conducted at high temperatures, the mechanical strength of the chitosan membrane will weaken, which will increase methanol permeability[56]. From the experiment table, it can also be concluded that in the chitosan membrane, an increase in methanol concentration will cause the methanol permeability to decrease, which is inversely proportional to what happens to membrane Nafion. Sulfonated PVDF (poly (vinylidene fluoride)) with chitosan blends as polymer electrolyte membranes showed good performance due to the ability to hydrate ion groups higher. Also, this membrane excludes organic solvents, which is one of the crucial things in polyelectrolyte membranes [4]. Permana found that the methanol permeability of composite membranes from chitosan and PVA with hematite still had a lower methanol permeability than Nafion [57]. A study by Hidayati et al. (2019) that the membrane printed from sPEEK/chitosan has a methanol permeability that is two times lower than the pure chitosan membrane at room temperature, but higher than the pure sPEEK membrane. It was comparable to that of Nafion 115, although the thickness of Nafion 115 was doubled [58].

### 5.2. *Proton conductivity*

Proton conductivity (σ) is one of the most important properties for proton-conducting membranes used in a fuel cell [59]. The two mechanisms that determine proton conductivity are "proton leap" (Grotthus) and migration of hydrated protons (Vehicle). The Grotthus mechanism of proton transfer occurs through hydrogen bonds, whereas the Proton transfer mechanism occurs with the help of water[60]. If the proton conductivity value is bigger than the quality of the proton-conducting membrane will be better because the proton conductivity shows the efficiency of the cell reactions that occur in the fuel cell system. The equations used to determine proton conductivity ($mS.cm^{-1}$) is:

$$\sigma = \frac{l}{A \, x \, R} \tag{3}$$

With $l$ is distance between electrodes, $A$ denote cross-sectional area, and $R$ is membrane resistance [61]. Nafion, membrane usually used in DMFCs, already has high proton conductivity that is 4.7 x $10^{-2}$ S/cm below 100°C, high thermal stability (280°C), and low percentage of water absorption. However, Nafion membrane still has weakness, where it has high methanol permeability, amounting to 27.6 x$10^{-8}$ $cm^2/s$ [62]. Membranes made of chitosan components require a combination with other materials as a filler to increase the conductivity of the membrane because chitosan has low conductivity [38]. Kharisma et al. in their study made a membrane proton electrolyte made from a combination of polyvinyl alcohol (PVA) and chitosan and by adding clay with silica as a filler. The highest conductivity was obtained with a value of 6.96282 x $10^{-7}$ S / cm, where it was known that the proton conductivity increases with increasing clay composition added to the membrane. This showed that the addition of silica into the polymer matrix affects the amount of proton conductivity [63]. A study conducted by Julian and Santoso [64] to determine the optimal ratio of chitosan/PVA composition with the addition of graphene oxide cross-linked with sulfuric acid showed that membranes with more chitosan composition required a longer immersion time in the manufacturing process to have bigger proton conductivity value.

### 5.3. *Swelling degree*

Water swelling test is used to determine the absorption capacity of the membrane to water. The water absorbed in the composite membrane does function as a proton transfer medium, but if swelling degree of the membrane is too high, the membrane will tend to be brittle. Swelling degree is determined based on the difference in membrane length in wet and dry conditions. Wet weight or length is determined after immersion of the membrane in water [65].

$$SD\ (by\ length) = \frac{l_{wet} - l_{dry}}{l_{dry}} x100\% \tag{4}$$

$$SD\ (by\ thickness) = \frac{d_{wet} - d_{dry}}{d_{dry}} x100\% \tag{5}$$

Where $l_{wet}$ is the length of the wet membrane, $l_{dry}$ is the length of the dry membrane, $d_{wet}$ is the thickness of the wet membrane, and $d_{dry}$ is the thickness of the dry membrane. By knowing the swelling degree of the membrane, it can give an overview of methanol's crossover properties and looseness of the membrane structure. If the membrane has low swelling degree, it indicates that the membrane is denser so that it can hold the rate of methanol from anode to cathode better [65]. Hidayati [66] found that in the membrane of chitosan and acrylonitrile butadiene styrene (ABS), ABS has the lowest value on both swelling degree and water absorption; on the contrary chitosan shows the opposite performance. In membranes made of chitosan and sPEEK, pure sPEEK membranes had the lowest water absorption while membranes made of pure chitosan had the highest water absorption. Water uptake of sPEEK/chitosan membrane was in between pure membranes [58]. Some factors influence water uptake from the sPEEK/chitosan membrane; one of them is the ionic interaction between -$SO^3$ and -$NH^3$ groups in chitosan [67].

## 6. Conclusion and Future Outlook

Based on several literatures reviewed, chitosan is a suitable material for DMFC because it has thermal and chemical properties which are stable and cheap. However, mixing chitosan alone with synthetic polymer-based membranes has not provided a much more significant value than Nafion in several aspects, such as methanol permeability. It is necessary to add other materials as a filler of chitosan. The addition of material to the membrane can improve some aspects of membrane morphology, but on the other side it can also reduce other aspects. Further studies are needed to obtain the type of material and optimum composition to enhance properties and increase the performance of fuel cells by optimizing membrane structure and fabrication techniques with excellent proton and methanol barrier properties.

## References

[1] Ramadhan A I, Diniardi E, Mukti S H 2016 Analisis desain sistem pembangkit listrik tenaga surya kapasitas 50 WP *Jurnal Teknik* **37** (2) 59-63

[2] Gong X, Dai Y, Yan X, Wu X, Wang Q, Zhen D, He G 2018 Electrospun imidazolium functionalized multiwalled carbon nanotube/polysulfone inorganic-organic nanofibers for reinforced anion exchange membranes *International Journal of Hydrogen Energy* **43** (46) 21547-21559

[3] Rambabu K, Show P L, Bharath G, Banat F, Naushad M, Chang J S 2020 Enhanced biohydrogen production from date seeds by Clostridium thermocellum ATCC 27405 *International Journal of Hydrogen Energy* **45** (42) 22271-22280

[4] Vijayakumar V, Khastgir D 2020 Polymeric composite membranes for anion exchange membrane fuel cells. *In Synthetic Polymeric Membranes for Advanced Water Treatment, Gas Separation, and Energy Sustainability* 365-381 Elsevier

[5] Hidayati N, Mujiburohman M, Abdillah H, Harmoko T, Arimurti R D 2017 Sintesis dan Karakteristik Membran Komposit Akrilonitril Butadiena Stirena (ABS)-Kitosan Tersulfonasi untuk Direct Metanol Fuel Cell (DMFC) *Jurnal Matematika dan Sains* **22** (1) 20-23

[6] Imannurya P C and Hidayati N 2020 Pengaruh Penambahan Silika Carbon Nanotube dalam Membran sPEEK PVA terhadap Karakteristik Membran pada Sel Bahan Bakar Metanol *In Seminar Nasional Teknik Kimia Kejuangan* 8

[7] Yang T 2008 Preliminary study of SPEEK/PVA blend membranes for DMFC applications *International Journal of Hydrogen Energy* **33** (22) 6772-6779

[8] Handayani S, Dewi E L, Wahyu W, and Soemantojo R W 2007 Sel Bahan Bakar Metanol Langsung Suhu Tinggi **8** (3) 192–197

[9] Sharaf O Z and Orhan M F 2014 An overview of fuel cell technology: Fundamentals and applications *Renewable and sustainable energy reviews* **32** 810-853

[10] Xue S, Yin G, Cai K, Shao Y 2007 Permeabilities of methanol, ethanol and dimethyl ether in new composite membranes: A comparison with Nafion membranes *Journal of membrane science* **289** (1-2) 51-57

[11] Tofighi A and Rahimnejad M 2020 Synthetic polymer-based membranes for microbial fuel cells *In Synthetic Polymeric Membranes for Advanced Water Treatment, Gas Separation, and Energy Sustainability* 309-335 Elsevier

[12] Xu D, Xu J, Wang X, and Wang Z 2016 Excellent performance of resistance methanol of a novel

sulfonated poly (aryl ether ketone sulfone)/poly (vinylalcohol) composite membrane for direct methanol fuel cell applications *International Journal of Hydrogen Energy* **41** (45) 20536-20548

[13] Liu D, Dong B, Zhang H, Xie Y, Pang J, and Jiang Z 2021 High methanol resistant polyelectrolyte membrane based on semi-crystalline Poly (ether ketone) with densely sulfonated side chain for direct methanol fuel cell *Journal of Power Sources* **482** 228982

[14] Zhang C, Yue X, Luan J, Li P, Zhang S, Liu J, Wang G 2020 Remarkable reinforcement effect of pore-filled semi-crystalline poly (ether ether ketone) membranes for high concentration direct methanol fuel cells *Electrochimica Acta* **363** 137242

[15] Hasani-Sadrabadi M M, Dashtimoghadam E, Ghaffarian S R, Sadrabadi M H H, Heidari M, Moaddel H 2010 Novel high-performance nanocomposite proton exchange membranes based on poly (ether sulfone) *Renewable Energy* **35** (1) 226-231

[16] Prakash G S, Smart M C, Wang Q J, Atti A, Pleynet V, Yang B, McGrath K, Olah G A, Narayanan S R, Chun W, Valdez T, Surampudi S 2004 High efficiency direct methanol fuel cell based on poly (styrenesulfonic) acid (PSSA)–poly (vinylidene fluoride)(PVDF) composite membranes *Journal of fluorine chemistry* **125** (8) 1217-1230

[17] Liu G, Tsen W C, Jang S C, Hu F, Zhong F, Liu H, Gong C 2019 Mechanically robust and highly methanol-resistant sulfonated poly (ether ether ketone)/poly (vinylidene fluoride) nanofiber composite membranes for direct methanol fuel cells *Journal of Membrane Science* **591** 117321

[18] Yang Z, Peng H, Wang W, Liu T 2010 Crystallization behavior of poly (ε-caprolactone)/layered double hydroxide nanocomposites *Journal of applied polymer science* **116** (5) 2658-2667

[19] Junoh H, Jaafar J, Nordin N A H M, Ismail A F, Othman M H D, Rahman M A, Yusof N 2020 Synthetic polymer-based membranes for direct methanol fuel cell (DMFC) applications *In Synthetic Polymeric Membranes for Advanced Water Treatment, Gas Separation, and Energy Sustainability* 337-363 Elsevier

[20] Suka I G, Simanjuntak W, Dewi E L 2010 Pembuatan Membran Polimer Elektrolit Berbasis Polistiren Akrilonitril (SAN) untuk Aplikasi Direct Methanol Fuel Cell *Jurnal Natur Indonesia* **13** (1) 1-6

[21] Dewi E L, Ismujanto T, Chandrasa G T 2008 *Pengembangan dan aplikasi fuel cell* In Prosiding Seminar Nasional Teknoin

[22] Zaidi S M J 2002 *Technology of Direct Methanol Fuel Cell* **2** December

[23] Junoh H, Jaafar J, Nordin N A H M, Ismail A F, Othman M H D, Rahman M A, Yusof N 2020 Performance of polymer electrolyte membrane for direct methanol fuel cell application: Perspective on morphological structure *Membranes* **10** (3) 34

[24] Vietanti F, Susanti D, Purwaningsih H, Kurniawan F 2019 Pengaruh Reduktor Zink pada Sintesis Graphene terhadap Performa PdAu/Graphene sebagai Material Elektrokatalis DMFC *In Prosiding Seminar Nasional Sains dan Teknologi Terapan* **1** (1) 379-384

[25] Sarkar J and Bhattacharyya S 2012 Application of graphene and graphene-based materials in clean energy-related devices Minghui *Arch. Thermodyn* **33** (4) 23-40

[26] Jiang R and Chu D 2003 Comparative studies of methanol crossover and cell performance for a DMFC *Journal of the Electrochemical Society* **151** (1) A69

[27] Kotrbacek P, Long N V, Thi C M, Nogami M, and Physical T 2012 *Chapter 3* Nguyen Viet Long 3, May 2014

[28] Sánchez-Sánchez C M, Solla-Gullón J, Vidal-Iglesias F J, Aldaz A, Montiel V, Herrero E 2010 Imaging structure sensitive catalysis on different shape-controlled platinum nanoparticles *Journal of the American Chemical Society* **132** (16) 5622-5624

[29] Chan C C 2007 The state of the art of electric, hybrid, and fuel cell vehicles *Proceedings of the IEEE* **95** (4) 704-718

[30] Joghee P, Malik J N, Pylypenko S, O'Hayre R 2015 A review on direct methanol fuel cells–In the perspective of energy and sustainability *MRS Energy & Sustainability* **2**

[31] Kusumawati N, and Tania S 2012 Pembuatan dan uji kemampuan membran kitosan sebagai

membran ultrafiltrasi untuk pemisahan zat warna Rhodamin B *Molekul* **7** (1) 43-52

[32] Mansoori S, Davarnejad R, Matsuura T, Ismail A F 2020 Membranes based on non-synthetic (natural) polymers for wastewater treatment *Polymer Testing* **84** 106381

[33] Altaf F, Gill R, Batool R, Majeed H, Abbas G, Jacob K 2020 Synthesis and applicability study of novel poly (dopamine)-modified carbon nanotubes based polymer electrolyte membranes for direct methanol fuel cell *Journal of Environmental Chemical Engineering* **8** (5) 104118

[34] Larminie J, Dicks A, McDonald M S 2003 *Fuel cell systems explained* 2207-225 Chichester, UK: J. Wiley

[35] Agel E, Bouet J, and Fauvarque J F 2001 Characterization and use of anionic membranes for alkaline fuel cells *Journal of Power Sources* **101** (2) 267-274

[36] Osifo P O, and Masala A 2010 Characterization of direct methanol fuel cell (DMFC) applications with H2SO4 modified chitosan membrane *Journal of Power Sources* **195** (15) 4915-4922

[37] Zulfikar M, Wahyuningrum D, and Berghuis D N T 2009 Pengaruh konsentrasi kitosan terhadap sifat membran komposit kitosan-silika untuk sel bahan bakar *In Prosiding Seminar Kimia Bersama UKM-ITB VIII* **9** 11

[38] Mohamed N S, Subban R H Y, and Arof A K 1995 Polymer batteries fabricated from lithium complexed acetylated chitosan *Journal of Power Sources* **56** (2) 153-156

[39] Alam M N 2016 Modifikasi Membran Komposit Kitosan dengan Filler Montmorillonit Untuk Aplikasi Sel Bahan Bakar *Dinamika* **07** (2) 269–283

[40] Wulandari A V, Kusumastuti E, and Sulistyaningsih T 2017 Pengaruh Penambahan Abu Layang Termodifikasi terhadap Karakteristik Membran Elektrolit Berbahan Dasar Kitosan *Indonesian Journal of Chemical Science* **6** (2) 104-109

[41] Rahman N F A, Loh K S, Mohamad A B, Kadhum A A H, and Lim K L 2016 Synthesis and characterisation of chitosan-cellulose biocomposite membrane for fuel cell applications *Malaysian Journal of Analytical Sciences* **20** (4) 885-891

[42] Ryu J, Seo J Y, Choi B N, Kim W J, Chung C H 2019 Quaternized chitosan-based anion exchange membrane for alkaline direct methanol fuel cells *Journal of Industrial and Engineering Chemistry* **73** 254-259

[43] Vijayalekshmi V, and Khastgir D 2018 Chitosan/partially sulfonated poly (vinylidene fluoride) blends as polymer electrolyte membranes for direct methanol fuel cell applications *Cellulose* **25** (1) 661-681

[44] Liu G, Tsen W C, Jang S C, Hu F, Zhong F, Zhang B, Gong C 2020 Composite membranes from quaternized chitosan reinforced with surface-functionalized PVDF electrospun nanofibers for alkaline direct methanol fuel cells *Journal of Membrane Science* **611** 118242

[45] Wang J, Zheng X, Wu H, Zheng B, Jiang Z, Hao X, Wang B 2008 Effect of zeolites on chitosan/zeolite hybrid membranes for direct methanol fuel cell *Journal of Power Sources* **178** (1) 9-19

[46] Osifo P O, and Masala A 2010 Characterization of direct methanol fuel cell (DMFC) applications with H2SO4 modified chitosan membrane *Journal of Power Sources* **195** (15) 4915-4922

[47] Yuan W, Wu H, Zheng B, Zheng X, Jiang Z, Hao X, and Wang B 2007 Sorbitol-plasticized chitosan/zeolite hybrid membrane for direct methanol fuel cell *Journal of power sources* **172** (2) 604-612

[48] Drioli E 2016 *Polyamide Membrane*

[49] Rosli N A H, Loh K S, Wong W Y, Yunus R M, Lee T K, Ahmad A, and Chong S T 2020 Review of chitosan-based polymers as proton exchange membranes and roles of chitosan-supported ionic liquids *International journal of molecular sciences* **21** (2) 632

[50] Riniati R, and Chamidy H N 2013 Pembuatan Membran Kitosan Sulfonat Untuk Aplikasi Direct Ethanol Fuel Cell *In Prosiding Industrial Research Workshop and National Seminar* **4** 286-289

[51] Enggita A P 2015 *Pengaruh Komposisi terhadap Perilaku Membran Komposit PVA/Kitosan/Grafin Oksida yang Terikat Silang Trisodium Sitrat* (Doctoral dissertation,

Institut Teknologi Sepuluh Nopember)

[52] Ambili K, Joseph S, and Sidharthan A 2019 A polyvinyl alcohol/chitosan blend proton exchange membrane for direct methanol fuel cell *In AIP Conference Proceedings* **2162** (1) AIP Publishing LLC.

[53] Altaf F, Gill R, Batool R, Drexler M, Alamgir F, Abbas G, and Jacob K 2019 Proton conductivity and methanol permeability study of polymer electrolyte membranes with range of functionalized clay content for fuel cell application *European Polymer Journal* **110** 155-167

[54] Aricò A S, Sebastian D, Schuster M, Bauer B, D'Urso C, Lufrano F, and Baglio V 2015 Selectivity of direct methanol fuel cell membranes *Membranes* **5** (4) 793-809

[55] Wootthikanokkhan J, and Seeponkai N 2006 Methanol permeability and properties of DMFC membranes based on sulfonated PEEK/PVDF blends *Journal of applied polymer science* **102** (6) 5941-5947

[56] Mukoma P, Jooste B R, and Vosloo H C M 2004 A comparison of methanol permeability in Chitosan and Nafion 117 membranes at high to medium methanol concentrations *Journal of Membrane Science* **243** (1-2) 293-299

[57] Permana D, Ilimu E, Faariu N M, Setyawati A, Kadidae L O, and Ramadhan L O A N 2020 Synthesis and characterization of chitosan-polyvinyl alcohol-Fe2O3 composite membrane for DMFC application *Makara Journal of Science* **24** (1) 1

[58] Hidayati N, Harmoko T, Mujiburohman M, and Purnama H 2019 Characterization of sPEEK/chitosan membrane for the direct methanol fuel cell *In AIP Conference Proceedings* **2114** (1) 060008 AIP Publishing LLC.

[59] Fang J H 2018 Polyimide proton exchange membranes *In Advanced Polyimide Materials* 323-383 Elsevier

[60] Li W 2009 *Development of New Membranes Based on Aromatic Polymers and Heterocycles for Fuel Cells* Philosophy

[61] Nugraeni C D, Hayati N, Santoso M, Kusumawati Y, and Atmaja L 2018 Preparation of proton exchange membrane based on chitosan phthaloyl and chitosan blend for direct methanol fuel cell application *In AIP Conference Proceedings* **2049** (1) 020077 AIP Publishing LLC

[62] Smitha B, Sridhar S, and Khan A A 2005 Chitosan–sodium alginate polyion complexes as fuel cell membranes *European Polymer Journal* **41** (8) 1859-1866

[63] Kharisma T, Ariesta N, and Arrisujaya D 2020 Karakteristik Membran Komposit Berbasis Kitosan/Pva Termodifikasi Lempung Dari Babakan Madang Bogor *Jurnal Sains Natural* **10** (1) 33-42

[64] Matematika F 2015 *Pengaruh Komposisi Terhadap Perilaku Membran Komposit PVA / Kitosan / Grafin Oksida yang Terikat Silang Asam Sulfat* **4** (2) 2–6

[65] Hidayati N 2015 Karakteristik Membran Komposit Poli Eter Eter Keton Tersulfonasi untuk Direct Methanol Fuel Cell *In Seminar Nasional Teknik Kimia Kejuangan* 5-1

[66] Hidayati N U R, Mujiburohman M, Purnama H, Dwilaksita A, and Zubaida F R 2019 Preliminary study of ABS/chitosan blend polymer for DMFC membranes In *Materials Science Forum* **961** 23-29 Trans Tech Publications Ltd

[67] Aini N A, Yahya M Z A, Lepit A, Jaafar N K, Harun M K, and Ali A M M 2012 Preparation and characterization of UV irradiated SPEEK/chitosan membranes *Int. J. Electrochem. Sci.* **7** 8226-8235.

IOP Conf. Series: Materials Science and Engineering 1143 (2021) 012003 doi:10.1088/1757-899X/1143/1/012003

# The synthesis of magnetic molecularly imprinted polymer against di-(2-ethylhexyl)phthalate

**Asyifa Rizqi Utami*, Muhammad Ali Zulfikar and Deana Wahyuningrum**

Department of Chemistry, Institut Teknologi Bandung, Bandung 40132 Indonesia

*asyifarizqi@gmail.com

**Abstract**. Magnetic molecularly imprinted polymer (MMIP) was prepared against di-(2-ethylhexyl)phthalate, using methacrylic acid as a functional monomer, ethylene glycol dimethacrylate as a crosslinker and benzoyl peroxide as an initiator. MMIP was characterized using Scanning Electron Microscope and infrared spectrophotometer. IR spectrum shows successful preparation of MMIP. The peak at 586 $cm^{-1}$ is a typical vibration of Fe-O. The strong peaks at 1728 $cm^{-1}$, 1261 $cm^{-1}$, and 1153 $cm^{-1}$ are stretching vibration of C=O and C-O of methacrylic acid and DEHP. The peak at 1462 $cm^{-1}$ due to C-C aromatic vibration of DEHP. A broad absorption band at 3468 $cm^{-1}$ due to the stretching vibration of hydroxyl bonds between target molecule and functional monomer. The MMIP has a spherical morphology and pores that important for adsorption process. Experimental results show the potential of magnetic molecularly imprinted polymer for recognition of di-(2-ethylhexyl) phthalate.

## 1. Introduction

Di(2-ethylhexyl)phthalate (DEHP) or dioctylphthalate is specified as phthalate esters group. This compound used as a plasticizers, an additive compound that added to improve the flexibility of plastic, especially for polyvinyl chloride. This compound also used in many products such as medical device, plastic furniture, baby clothes, vinyl flooring, some toys, shoes, shower curtains and packaging material [1]. DEHP is physically dispersed on the surface of polymer so it will leach out to the environment easily [2]. DEHP has harmful effects on animal and human health [3]. Earlier study by the US Environmental Protection Agency (USEPA) showed that this compound classified as a pollutant that act as endocrine disruptors compound (EDG) [4],[5]. DEHP also causes endometriosis in woman, and also gives effects on fetus, heart, kidneys, lungs, and liver [6].

Gas chromatography and liquid chromatography are the common method to analyze DEHP. Sometimes, another separation method is needed before chromatography analysis to concentrate the DEHP from a complex sample [7]. Adsorption is one of the common separation method for organic compounds because of its easy operation [8],[9],[10]. Some adsorbents such as chitosan [11], nanoporous gold polyethyleneimine [12], and graphene [13] were used to adsorbed DEHP. Unfortunately, these adsorption methods has poor selectivity for recognition of DEHP [2].

Recently, researchers were used molecularly imprinted polymer to improve the selectivity of the adsorbent. This polymer has a specific recognition sites towards the template [14]. The monomer could interact with the target molecule through covalent or non covalent interaction, resulting prepolymerization complexes[15]. Polymerization took place with the presence of initiator and crosslinker. The template (target molecule) were extracted from the polymers to form a recognition sites [16]. Bulk polymerization is a common synthetic method. However, MIPs synthesized by these methods have an irregular shape, the polymer have heterogeneous binding sites, and the template were hard to completely removed [17]. These problems could be minimalized by surface molecularly imprinted polymer.

Magnetic molecularly imprinted polymers (MMIPs) is one of the surface imprinting technology. In this method, polymerization took place on the surface of magnetit. The MMIPs have advantages such as fast binding and separation due to the magnetic properties and have uniform shape so that the polymers doesn't need to be ground [18]. In conventional method, polymer separated by filtration or centrifugation. In MMIPs, polymer separated from the solution easily using a magnetic field [19].

In this paper, a new magnetic molecularly imprinted polymers (MMIPs) were prepared using DEHP as a template, $Fe_3O_4$ particles as a core, ethylene glycol dimethacrylate as a crosslinker, methacrylic acid as a functional monomer, and benzoyl peroxide as an initiator. The MMIPs were characterized using Fourier transform infrared (FTIR) and scanning electron microscopy (SEM).

## 2. Materials and Methods

### 2.1. Chemicals

di(2-ethylhexyl) phthalate (DEHP) (>99.5%) were obtained from Sigma Aldrich. Methacrylic acid (MAA), benzoyl peroxide (BPO), acetonitrile, ethylene glycol dimethacrylate (EGDMA), iron (III) chloride ($FeCl_3.6H_2O$), iron (II) chloride ($FeCl_2.4H_2O$), polyvinylpyrrolidone (PVP), oleic acid, methanol, sodium hydroxide, acetic acid glacial, ethanol, , were analytical grade and obtained from Merck.

### 2.2. Synthesis of MMIPs

In the first step, the $Fe_3O_4$ particles was synthesized by the co-precipitation method [20]. Iron (II) and iron (III) were dissolved in 100 ml water with the comparation 1:2. The solution were then stirred with the presence of nitrogen gas in a hotplate at 80°C. Sodium hydroxide (2 M) were added dropwise to the mixture resulting a change of color solution from orange to black. The black precipitant is $Fe_3O_4$ particle that could separated from the solvent using external magnetic field. The $Fe_3O_4$ then washed with water and ethanol several times. The $Fe_3O_4$ were dried in the oven at 60°C. After that, 1 g of $Fe_3O_4$ mixed with 1 mL of oleic acid, then the mixture were sonicated for 15 min.

MMIPs were synthesized by mixing DEHP (1 mmol) and methacrylic acid (4 mmol) in acetonitrile (30 mL). In another glass, 20 mmol EGDMA mixed with 1 g $Fe_3O_4$- oleic acid. The complex solution of DEHP and MAA was added to this mixture and then sonicated for 30 minutes. This was a pre-polymerization solution. In another glass, 0.4 g PVP was dissolved in 100 ml ethanol (80%) then stirred with the presence of nitrogen gas at 400 rpm. This solution was added to the pre-polymerization solution. After that, BPO (150 mg) was added to the solution. The nitrogen gas was added to the solution for 10 min. Polymerization took place in an oil bath (70°C) with continuous stirring for 1 hour, resulting MMIPs that could separated from the solution using magnetic field. The template were removed by washing the polymer with a methanol:acetic acid solution (9:1, v/v) several times until no DEHP detected. Then, the MMIPs were dried in the oven at 60°C. The magnetic non-molecularly imprinted polymers (MNIPs) were prepared with the same procedure without the presence of DEHP.

### 2.3. Characterization

The morphology of MMIPs were observed by Scanning Electron Microscope (SEM JSM-6510A). FTIR spectra of MMIPs, MNIPs and $Fe_3O_4$ were obtained by Jasco FTIR-4200.

## 3. Results and Discussion

### *3.1. Synthesis of MMIPs*

The MMIPs and MNIPs were synthesized by surface imprinting method. The polymerization took place on the surface of $Fe_3O_4$. $Fe_3O_4$ were obtained by reacting iron (II) and iron (III) in an alkaline conditions. The surface of $Fe_3O_4$ were modified using oleic acid. The hydroxy group on the surface of $Fe_3O_4$ would react with carboxylic group of oleic acid. The next step is polymerization using DEHP, MAA, and EGDMA with the mol ratio 1:4:20. The specific sites of the polymer were due to the non covalent interaction between DEHP and methacrylic acid. DEHP has four oxygen atoms that could interact with hydroxy functional group in methacrylic acid through hydrogen bond. Moreover, there could be a van der Waals forces beetwen long alkyl chain of DEHP with the polymers. The DEHP were extracted from the polymers by eluting the MMIPs with MeOH: acetic acid (9:1, v/v) and dried in oven at 60°C. In this synthesis method, MMIPs doesn't need to be grounded and sieved because the polymer was fabricated on the surface of magnetite and it expected to have more uniform particular size. This is one of the advantage of surface polimerization method over conventional method that time consuming. The separation of MMIPs from the solution was done by using external magnetic field while in conventional method polymers should be filtrated or centrifugated. **Figure 1** show the synthesis scheme of MMIPs.

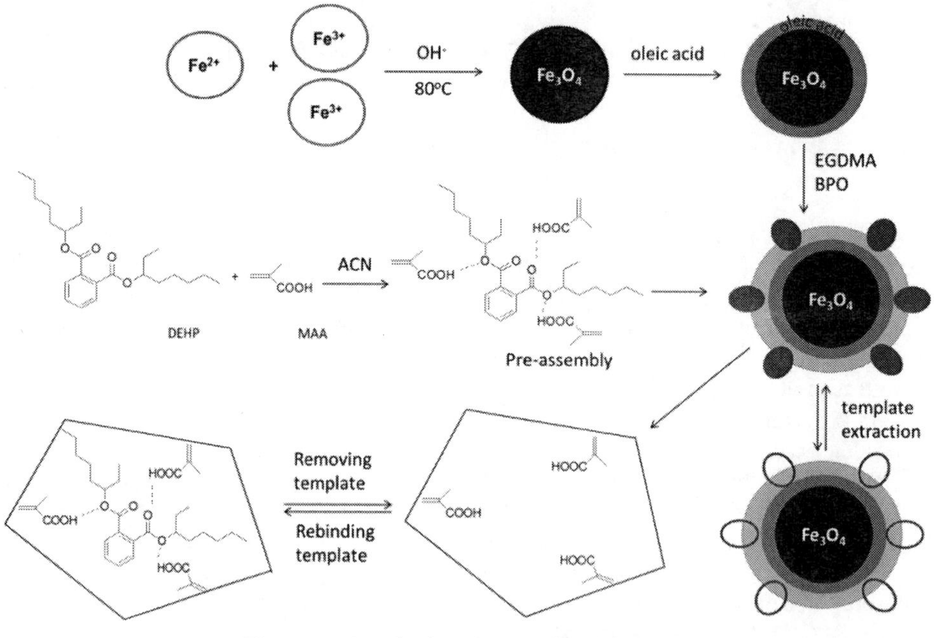


**Figure 1.** Synthetis scheme of MMIPs.

### *3.2. SEM analysis*

The morphology of MMIPs were shown in **Figure 2**. The MMIPs have a pores and the particles were spherical. The pores on the surface of the polymers was useful in the adsorption process. In the synthesis process, DEHP interact with methacrylic acid by hydrogen bonding and van der Waals forces. After template removal, it will produce cavities due to the shape, size and functional group of DEHP. However, further study was needed to ensure the formation of a spesific cavities of MMIP towards DEHP.

**Figure 2.** SEM image of MMIPs, taken under 20.000x magnification.

### *3.3. FTIR analysis*

The prepared $Fe_3O_4$-oleic acid, MMIPs and MNIPs were characterized by FTIR. The IR spectra was shown in **Figure 3**. The presence of $Fe_3O_4$-oleic acid was confirmed from the peak 586 $cm^{-1}$ that was an absorption band of Fe-O bond. The Methacrylic acid as a monomer could be seen at peak 1728 $cm^{-1}$ confirmed the presenc of C=O from MAA. The peak at 1153 $cm^{-1}$ and 1261 $cm^{-1}$ were vibration of C-O in MAA. The C-H vibration at 2954 $cm^{-1}$ was atributed to oleic acid. The presence of DEHP was confirmed at 709 $cm^{-1}$. The presence of C=C aromatic shown at 1600 $cm^{-1}$ and 1462 $cm^{-1}$. The band at 3468 $cm^{-1}$ in the curve (c) atributed to the hidroxy bond interaction between DEHP and MAA.

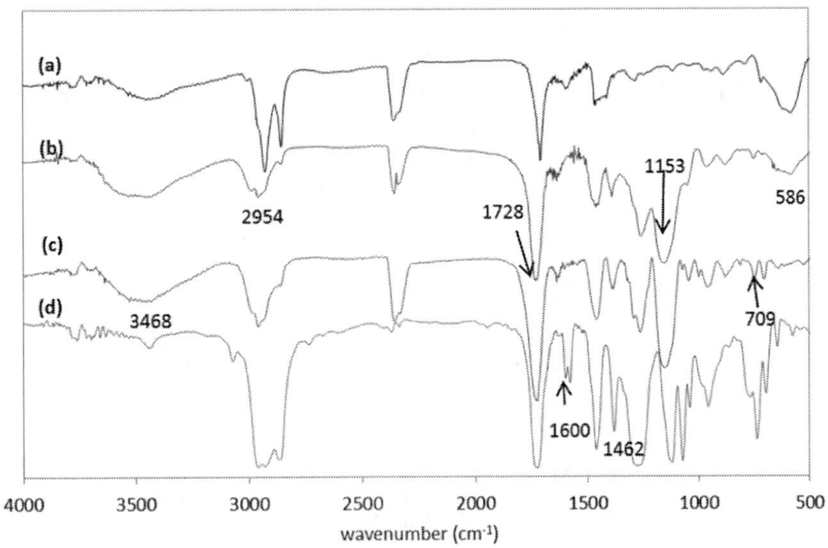


**Figure 3.** Infrared spectra of oleic acid-coated $Fe_3O_4$ (a), MNIPs (b), MMIPs (c) and DEHP (d).

Nevertheless, there is still a challenge to ensure the formation of a specific pores of MMIP in molecular level. Further characterization such as Brunauer Emmet Teller could provide the differences of surface area beetwen MMIP and MNIP. Furthermore, selectivity study using another compound that similar to DEHP in an adsorption process was needed to proves the succesful fabrication of specific pores of MMIP.

## 4. Conclusions

A new magnetic molecularly imprinted polymers were prepared against DEHP. The morphology and IR spectra of MMIPs were characterized. SEM image show that MMIPs have pores and a spherical morphology. The infrared peaks at 1153 $cm^{-1}$ , 586 $cm^{-1}$, 1462 $cm^{-1}$, 2358 cm-1, 709 $cm^{-1}$, 3468 $cm^{-1}$, 2954 $cm^{-1}$ and 1728 $cm^{-1}$ confirmed the presence of EGDMA co-MAA polymers, respectively. However, further study was needed to investigate the adsorption performance of MMIP towards DEHP.

## References

[1] Agency for Toxic Substances and Disease Registry (ATSDR) 2002 *Toxicological profile for di(2-ethylhexyl)phthalate (DEHP)* Atlanta, GA: U.S. Department of Health and Human Services, Public Health Service

[2] Li C, Ma X, Zhang X, Wang R, Li X and Liu Q 2017 Preparation of magnetic molecularly imprinted polymer nanoparticles by surface imprinting by a sol–gel process for the selective and rapid removal of di-(2- ethylhexyl) phthalate from aqueous solution *Journal of Separation Science* **40** 1429–1630

[3] Svechnikova I, Svechnikov K and Söder O 2007 The influence of di-(2-ethylhexyl) phthalate on steroidogenesis by the ovarian granulosa cells of immature female rats *J. Endocrinol.* 194(3) 603–9

[4] Environmental Protection Agency (EPA) 1991 *Di-2-ethylhexyl phthalate, Integrated risk information* U.S

[5] Sayles D G 2002 Environmental engineering and endocrine disrupting chemicals *Journal of Environmental Engineering* 128 1-2

[6] Tickner J A, Schettler T, Guidotti T, Mccally M and Rossi M M 2001 Health risks posed by use of di-2-ethylhexyl phthalate (DEHP) in PVC medical devices: a critical review *Am. J. Ind. Med.* **39** 100–11

[7] Heise S and Litz N 2004 *Deskstudy Phthalates* German Federal Environmental Agency, Berlin. pp. 20

[8] Morais E C, Correa G G, Brambilla R, dos Santos J H Z and Fisch A G 2013 Selective silica-based sorbent materials synthesized by molecular imprinting for adsorption of pharmaceuticals in aqueous matrices *J. Sep. Sci.* **36** 636–43

[9] Altmann J, Zietzschmann F, Geiling E L, Ruhl A S, Sperlich A and Jekel M 2015 Impacts of coagulation on the adsorption of organic micropollutants onto powdered activated carbon in treated domestic wastewater *Chemosphere* 125 198–204

[10] Abdul G, Wang P, Zhang D and Li H 2015 Adsorption of diethyl phthalate on carbon nanotubes: pH dependence and thermodynamics *Environ. Eng. Sci.* **32** 103–10

[11] Chen C and Chung Y 2006 Removal of phthalate esters from aqueous solutions by chitosan bead *J. Environ. Sci. Heal.* A. **41** 235–48

[12] Zhang C, Zeng G, Huang D, Lai C, Huang C, Li N and He X 2014 *Combined removal of di (2-ethylhexyl) phthalate (DEHP) and Pb (II) by using a cutinase loaded nanoporous gold-polyethyleneimine adsorbent* RSC Adv. **4** 55511–8

[13] Yang G C and Tang P L 2016 Removal of phthalates and pharmaceuticals from municipal wastewater by graphene adsorption process *Water Sci. Technol.* **73** 2268–74

[14] Alexander C, Andersson H S, Andersson L I, Ansell R J, Kirsch N, Nicholls I A, O'Mahony J, Whitcombe M J 2006 Molecular imprinting science and technology: a survey of the literature for the years up to and including 2003 *J. Mol. Recogn.* **19** 106

[15] Turiel E and Martín-Esteban A 2010 Molecularly imprinted polymers for sample preparation: a review *Anal. Chim. Acta.* **668**

[16] Ansari S and Karimi M 2017 Recent progress, challenges and trends in trace determination of drug analysis using molecularly imprinted solid-phase microextraction technology *Talanta* **164** 612-25

[17] Chen L X, Xu S F and Li J H 2011 Recent advances in molecular imprinting technology: current status, challenges and highlighted applications *Chem. Soc. Rev.* **40**

[18] Zhang Z M, Tan W, Hu Y L and Li G K 2011 Simultaneous determination of trace sterols in complicated biological samples by gas chromatographyemass spectrometry coupled with extraction using b-sitosterol magnetic molecularly imprinted polymer beads *J. Chromatogr. A* **1218**

[19] Kubo T and Otsuka K 2016 Recent progress in molecularly imprinted media by new preparation concepts and methodological approaches for selective separation of targeting compounds *TrAC Trends Anal. Chem.* **81**

[20] He D, Zhang X P, Gao B, Wang L, Zhao Q, Chen H Y, Wang H and Zhao C 2014 Preparation of magnetic molecularly imprinted polymer for the extraction of melamine from milk followed by liquid chromatography-tandem mass spectrometry *Food Control* **36** 36-41

**Acknowledgments.** This study were supported by the Indonesia Endowment Fund for Education (LPDP).

IOP Conf. Series: Materials Science and Engineering 1143 (2021) 012004 doi:10.1088/1757-899X/1143/1/012004

# Production of activated carbon from palm empty fruit bunch as supercapacitor electrode material

**N N Wulandari[1*], H Rustamaji[1], W U Fibarzy[1], T Prakoso[1], J Rizkiana[1], H Devianto[1], P Widiatmoko[1], and I Nurdin[1]**

[1]Department of Chemical Engineering, Institut Teknologi Bandung, Bandung 40132 Indonesia

*E-mail: ninakusnadi@students.itb.ac.id

**Abstract.** Activated carbon is prepared by carbonization out by hydrothermal carbonization. Hydrother-mal carbonization is a process that carried out with aqueous media in a pressurized reactor at a lower temperatur that the pyrolysis process. This research focused on production of activated carbon from empty fruit bunch by hydrothermal carbonization process using $CaCl_2$ as activat-ing agents. Ratio of raw material/activating agent/water was 1:2:3. Hydrothermal carbonization was carried out at 275°C for an hour. Furthermore, hydrochar was activated at a temperatur of 800°C in presence of $CO_2$ for 2 hours. Activated carbon was modified with 1 M nitric acid ($HNO_3$). A self-supporting and flexible activated carbon/graphene/carbon nanotube (AC/GP/CNT) electrode has been rationally designed for constructing high-performance supercapacitor. The CNT is beneficial for improving the electronic conductivity of the electrode, while AC particles could effectively suppress the aggregation of GP and CNT due to their blocking effect. The synergistic effect among the AC-Ca, GP and CNT validate the AC-Ca/GP/CNT as a promising electrode for supercapacitor, exibiting greatly enhanced electrochemical performances in comparison with the pure GP, pure CNT and AC-Ca electrode. The AC-Ca/GP/CNT electrode delivers a high specific capacitance of 53.51 F $g^{-1}$.

## 1. Introduction

Electrical energy is a major requirement that cannot be eliminated in this era. Technology which has a greater energy and power density as well as shorter charging times are needed to the future technology. The new energy storage device is a supercapacitor, with several advantages such as wide range of operating temperature , high power density and long cycling life. Supercapacitor devices are usually assembled in a traditional two electrode configuration using (Ni) foam or aluminum (Al) foil as current collectors [1]. Supercapacitors used a surface electrode and a thin dielectric electrolyte solution to achieve a larger cpacitance than conventional capacitors.

Activated carbon has a hydrophobic surface that can decrease interaction between electrode and electrolyte ions. Therefore, surface modification is required to increase the accessible surface of activated carbon and can be improved through surface functionalization such as oxidation [2]. The surface oxidation of activated carbon produces a more hydrophilic surface with a large number of oxygenated surface groups. These functional groups increase the hydrophilicity of the activated carbon surface area, thus increasing the adsorption of electrolyte ions on the surface of the electrodes and increasing the effective surface area for the formation of an electric double-layer capacitance of electrode. Various reagents that have been used as oxidizers are concentrated nitric or sulfuric acid, sodium hypochlorite, permanganate, bichromate, hydrogen peroxide, transition metals, and ozone-based gas mixtures [3-4].

Cheng and Teng (2003) observed surface modification of carbon using 2 N nitric acid in an ultrasonic mixer can increase the specific capacitance [5]. Another study confirmed that immersed the carbon with 4 M nitric acid can increase specific capacitance [6]. Shim et al (2001) confirmed that modification of carbon fiber with 1 M nitric acid using reflux can decrease the surface area and increase surface oxygen group [4]. Therefore, this study intends to investigate the effect of nitric acid concentration on surface modification using reflux on the physical characteristic and electrochemical behavior of activated carbon. scanning electron microscopy (SEM) can study the structural properties of activated carbon, $N_2$ adsorption-desorption isotherms, and FTIR (Fourier transform-infrared). The electrochemical analysis of activated carbon-based supercapacitor electrode was studied by cyclic voltammetry (CV), galvanostatic charge–discharge, and electrochemical impedance spectroscopy (EIS).

## 2. Materials and methods

### *2.1. Preparation*

Electrode of supercapacitor required additional materials other than activated carbon, namely graphene and CNT (commersial). Activated carbon was produced from Empty fruit bunch (EFB) wich was collected from PT Perkebunan Nusantara VII in Bogor. EFB was washed and dried. Figure 1, dried EFB was then crushed using a disc mill and sieved into coarse granules (35-80 mesh) [7]. The precursor was leached with 1 M hydrofluoric acid (reagent grade, 1:25) for two hours [8]. Furthermore, the sample was washed with demineralized water until a pH of 6 was obtained for the rinse. The sample was dried at 105°C for 24 hours. Activated carbon was prepared by physicochemical activation of the sample with $CaCl_2$ (SmartLab, analytical grade) and $CO_2$. Nitric acid (reagent grade) was used to surface modification of activated carbon. Hydrochloric acid (Merck, 37%) was used to wash the activated carbon, KOH (Merck, analytical grade) and ethanol (Merck, analytical grade) for the supercapacitor electrode preparation [9].

### *2.2. Activated carbon production*

Hydrothermal reactor was charged with 10 g sample (EFB after leaching with HF) and $CaCl_2$ solution in mass ratio of EFB:$CaCl_2$:water=1:2:3. The samples were held at 275°C for an hour. The obtained hydrochar was filtered and dried at 105°C for 24 hours [7-9].

The hydrochar was loaded into ceramic boat and positioned inside the tubular furnace for activated. The furnace temperature was ramped to 800°C at a rate of 10°C/min in the presence of $N_2$ flow. After the temperature was reached, $N_2$ was replaced with $CO_2$ flow for two hours. Afterward, the furnace was chilled to room temperature in the presence of $N_2$ flow. The products were stirred for 30 min in 150 mL hydrochloric acid (0.1 M) and washed with demineralized water until a pH of 6 was obtained for the rinse. Then, activated carbon was dried at 105°C for 24 hours [7-9].

### *2.3. Activated carbon production*

Activated carbon was modified by oxidation treatments using nitric acid with varying concentrations. 1 g of activated carbon was suspended in 150 mL of 1 M nitric acid solution at 75°C under refluxing for 4 hours. The activated carbon was washed with demineralized water until a pH of 6 was obtained for the rinse and dried at 105°C for 24 hours [9].

### *2.4. Activated carbon production*

Scanning electron microscopy (SEM) using SEM Hitachi SU3500 conducted in Pusat Penelitian Nanosains dan Nanoteknologi Institut Teknologi Bandung was carries out to see the physical characteristics of activated carbon. Surface area and pore diameter of activated carbon were analyzed by a gas sorption Quantochrome Instruments Nova 3200e in Laboratorium Instrumentasi Teknik Kimia ITB. The Crystal structure of CNT was obtained from a powder X-ray diffraction pattern by using a Bruker D8 advanced diffractometer with monochromatized Cu K radiation [9].

CV tests on potentiostat/ galvanostat (Gamry V3000) use to analyzed the electrochemical properties of the supercapacitors and electrochemical impedance spectroscopy (EIS) with a frequency response detector at room temperature. Symmetrical two-electrode configurations were used to evaluate the capacitive performance of produced activated carbon. Nanocarbon (AC-Ca:GP:CNT) was wetted by ethanol, then mixed with polyvinylidene fluoride (PVDF) in a mass ratio of 9:1. Some variations in the characterization of AC-Ca:GP:CNT was 90:10:0, 90:0:10, and 90:5:5. The materials were pressed and dried at 105°C. Sandwich-type supercapacitor was assembled with two activated carbon electrodes separated by Nafion 212 in a button cell. The electrochemical behavior of samples was characterizations using cyclic voltammetry (CV), galvanostatic charge–discharge (GCD), and electrochemical impedance spectroscopy (EIS).

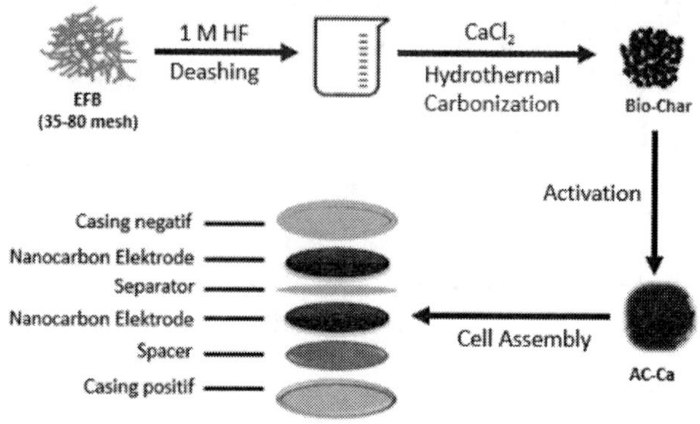


**Figure 1.** Process diagram of activated carbon electrode.

## 3. Results and discussion

Activated carbon was produced through hydrothermal carbonization and activation processes. In this study, two-stage activation processes consisting of a chemical activation step ($CaCl_2$ as an activating agent during hydrothermal carbonization at 275°C) followed by physical activation ($CO_2$ activation at 800°C) [9]. Physicochemical activation has also been used to enhance porosity development and uniformity the pore size distribution of activated carbon [10]. The mass yield of activated carbon is 3.54 g from 10 g of EFB. $CaCl_2$ supports the polymerization reactions which occur among the aromatic hydrocarbons and tar-forming compounds. These reactions result in the formation of larger molecules in the structure of activated carbon products and an increase in the carbon yield [11].

**Figure 2.** SEM images of (a) AC-Ca, (b) Graphene commercial and (c) CNT commercial.

Morphologies of raw materials of activated carbon with $CaCl_2$ as an activating agent (AC-Ca), GP and CNT presented in Figure 2, can be observed that each material electrodes present the self-supporting, mechanically and physical characteristics. Figure 2a indicates that AC-Ca has more pore on the surface. A large surface area is more advantageous for applied as a supercapacitor electrode. Figure 2b, the GP exhibits highly structured properties on the surface and layer structure on the inside of the electrode, this phenomenon is called the typical restacking of the graphene sheets, usualy driven by the layer-to-layer van der Waals force. However, the stacked surface and oriented layer structure are unfavorable for the fast charge/discharge processes of the suitable supercapacitor, resulting in a small enough specific capacitance due to the little utilization of the specific surface of GP sheet [12]. Randomly tangled and interwined structure as displyed at Figure 2c is pure CNT. In addition with the CNT, the falue of conductivity of electrode could also be increased.

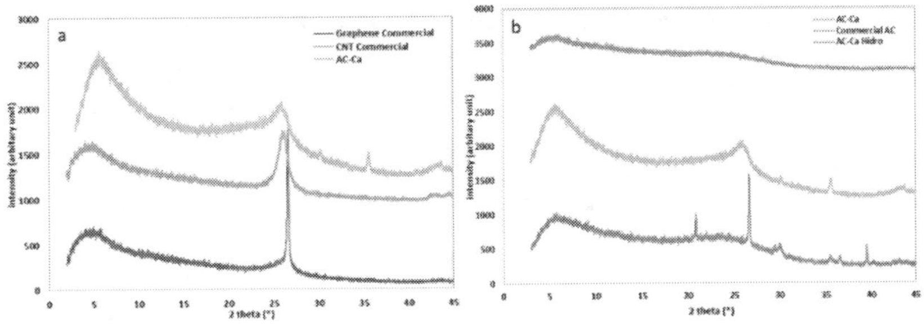

**Figure 3.** XRD curve of (a) AC-Ca, GP and CNT (b) AC-Ca, AC-Ca Hidro and AC commercial

The XRD curve of AC-Ca, GP, CNT, AC commercial and AC-Ca hidrofilization shown in Figure 3. Shows that only the GP, CNT and AC commercial shows a diffraction peak at 26.0°, which the meaning its high graphic characteristic and good conductivity. This fixed value of 5% is the best percentage weight for CNT and GP in the mix electrode of Ac-Ca/GP/CNT, whisch is good for efficient electron transportation during charge/discharge processes without sacrificing much of the total specific capacitance [13]. Figure 3. shows the AC-Ca shows a weak long range ordering characteristic. The pattern of AC-Ca Hydrofilization from XRD is not too prominent because the activated carbon has been oxidized.

Figure 4a show the CV curves of the electrode supercapacitors using the pure AC-Ca electrode at different scan rate of 2, 4, 6, 8 and 10 mV $s^{-1}$. The shape of the CV curve resembles a rectangles according to the electrostatic adsorption capacitance [14]. All the electrodes show the CV curves of the defect rectangles, which indicate the presence of electrochemical adsorption capacitance and electrostatic adsorption as shown in Figure 4. The electrostatic capacitance from the AC, GP or CNT, while the electrochemical adsorption capacitance is adsorption and desorption occuring on the OFG (oxygen containing functional group) (Figure 4c). It is found that the AC-Ca/GP/CNT electrode shows the maximum CV curve area, which showed the largest specific capacitance achieved by AC-Ca/GP/CNT at 90:5:5. Compared to the AC-Ca/GP/CNT 90:10:0 electrode indicates the smallest curve area (Figure 4b). Figure 4c, AC/GP/CNT 90:5:5 electrode CV curve is shown at different scan rate at 2, 4, 6, 8, and 10 mV $s^{-1}$. The CV curves shows a curve shape similar to a deformed rectangle although at high scan rate it can be seen, which are a good enough power capability of the AC-Ca/GP/CNT 90:5:5 electrode. Comparison on Figure 4d of AC-Ca before and after hydrophilization modify.

Table 1. shows the modification of activated carbon has the smallest surface area and average pore size. This is caused by a decrease in micropore volume due to pore blocking by surface oxide groups obtained from nitric acid oxidation [4]. The reason could be an activated carbon absorbs the molecule or ions of nitric acid, which occupies some pore volume. Furthermore, the produced functional groups on the carbon surface reduce pore size, even blocking partial micropores. Besides, increasing the nitric acid concentration leads to a decrease in surface area and average pore size. These results indicate that high concentrations of nitric acid cause the collapse of partial pores and oxygen-containing groups can block some micropores [3]. Surface area and pore size are very influential on activated carbon performance when applied as a supercapacitor electrode material. High surface area coupled with the presence of mesopores of appropriate sizes in activated carbon can improve the charge storage performance of supercapacitors [15].

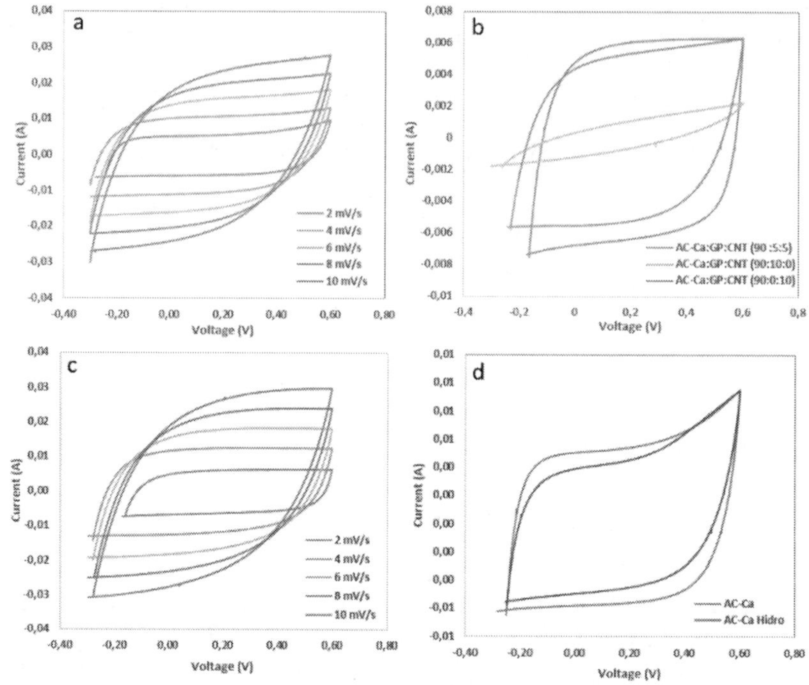


**Figure 4.** CV curves of (a) pure AC-Ca electrode at different scan rate; (b) variation of AC-Ca/GP/CNT electrode at scan rate 2 mV $s^{-1}$; (c) electrode of AC-Ca/GP/CNT 90:5:5at different scan rate; and (d) AC-Ca hydrophilization.

**Table 1.** Spesific capacitance of activated carbon electrodes

| Sample | Spesific capacitance (F/g) |
|---|---|
| AC-Ca | 29.46 |
| AC-Ca-hydro | 24.85 |
| AC-Ca/GP/CNT (90:5:5) | 53.51 |
| AC-Ca/GP/CNT (90:10:0) | 5.36 |
| AC-Ca/GP/CNT (90:0:10) | 24.19 |

The EIS measurement has been carried out to investigate the electrochemical properties. The EIS curve of the assembled electrode double-layer supercapacitors as shown in Figure 5. The impedance spectrum shows the semicircle located at high to medium frequency region that is associated with the charge transfer impedance ($R_{ct}$), and an oblique line located at low frequency region representing the Warbug impedance (W) [4]. The intecept at high frequency with real axis is the electrolyte resistance ($R_e$). The EIS results are shown in Figure 5 it can be observed that the AC-Ca hydro (Figure 5b) electrode smallest resistance (0.10 ohm) while the AC-Ca before hydrophilization (Figure 5a) shows the largest resistance (22 ohm). In comparison of the AC-Ca/GP/CNT 90:10:0 (Figure 5d) electrode smallest resistance (-20 ohm) while the AC-Ca/GP/CNT 90:0:10 (Figure 5c) shows the largest resistance (17 ohm).

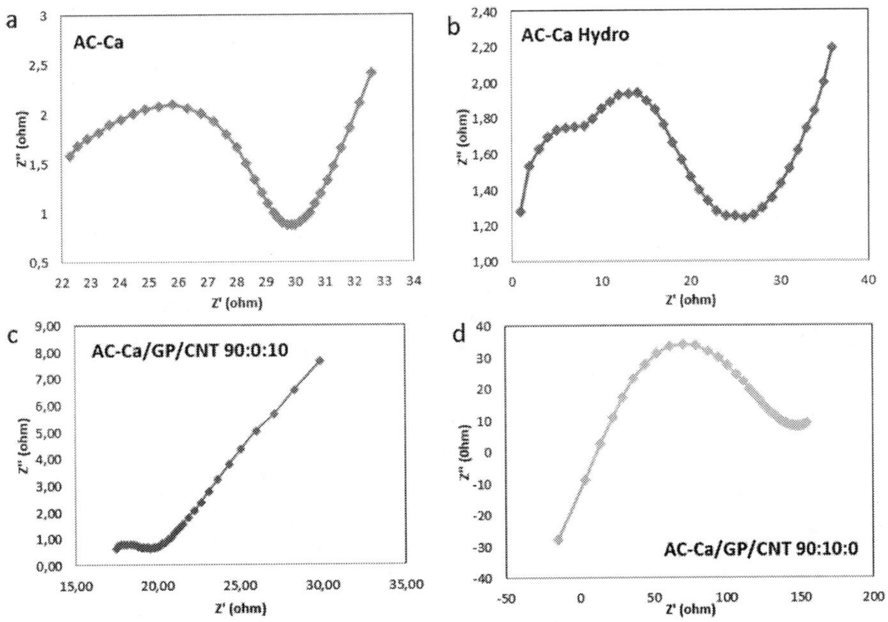


**Figure 5.** EIS of (a) AC-Ca elecrode; (b) AC-Ca hydrophilization with 1 M $HNO_3$; (c) AC-Ca/GP/CNT 90:0:10 electrode; and (d) AC-Ca/GP/CNT 90:10:0 electrode.

The galvanostatic charge-discharge curve of all activated carbon supercapacitor cells have a triangle and symmetric shape which indicates the supercapacitor cells have the same charging and discharging time [17]. GCD analysis for each supercapacitor cell was performed at a constant current density of 1 mA/cm$^2$. The GCD curve of AC-Ca has the lowest cell voltage. The GCD curve of AC-Ca hydro (Figure 6c) have higher cell voltages that cause a decrease in specific capacitance. The specific capacitance of GCD results in accordance with the results of cyclic voltammetry (CV) analysis.

The charging and discharging curve (GCD) of the electrodes for different materials can be seen in Figure 6. This analysis ar carried out at a constant current and measures the magnitude of the voltage as a function of charging and discharging times. It can be seen in the GCD analysis image of each sample in the form of a linear and symmetrical curve. This indicates that the supercapacitor made has good capacitive performance. Because the results of the electrode GCD curve with carbon material will generally be isosceles triangle. It can be observed in Figure 4a, similar to the CV curve, it also shows that electrostatic adsorption/desorption mainly occurs at the low potential stage. The pure AC-Ca electrode in Figure 6a has almost no adsorption capacitance, which is only around -0.069 V this results loss of the specific area. In Figur 6c, it can be compared that shows a higher specific capacitance of charge/discharge than the AC-Ca electrode during the longer cycle (Figure 6ca). Occurs that the AC-Ca/GP/CNT electrodes show higher capacitance retention than the AC-Ca electrodes, which is considered a synergistic effect of the AC, GP and CNT compound.

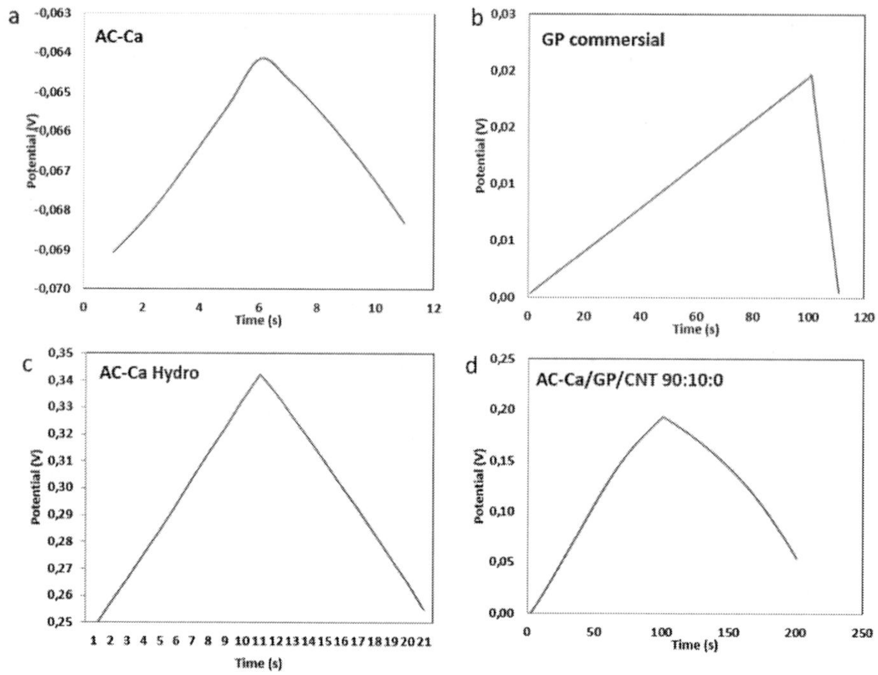


**Figure 6.** long-term cycling performances of the (a) AC-Ca; (b) GP; (c) AC-Ca hydro; and (d) AC-Ca/GP/CNT 90:10:0

## 4. Conclusions

Activated carbon material as an electrode material of supercapacitor can be produced from oil palm empty fruit bunch through the hydrothermal carbonization process and physicochemical activation. The specific capacitance value of supercapacitor from unmodified activated carbon (AC-Ca) is 29.46 F $g^{-1}$. Increasing the nitric acid concentration on surface modification leads to an increase in hydrophilicity of activated carbon, a decrease surface area, a decrease pore diameter and a decrease in the specific capacitance. Modified activated carbon with 1 M nitric acid has a specific capacitance value of 24.85 F $g^{-1}$. The self-supporting AC-Ca/GP/CNT electrode has been successfully synthesized for supercapacitor and work very well. The surrounding CNT could be enhanced the electronic conductivity value of the AC-Ca and GP. With the AC-Ca/GP/CNT 90:5:5 electrode is able to deliver a high specific capacitance of 53.51 F $g^{-1}$.

## Acknowledgments

This research is supported by Crude Palm Oil Fund of the Ministry of Finance, funding year of 2018.

## References

[1] Xing Li, Yao Tang, Junhua Song, Wei Yang, Mingshan Wang, Cheng Zhu, Wengao Zhao, Jianming Zheng, Yuehe Lin. *Self-supporting activated carbon/carbon nanotube/reduce graphene ooxide flexible electrode for high performance supercapacitor*. 2018 *Carbon*
[2] Davies A and Yu A 2011 *Can. J. Chem. Eng.*
[3] Liu Y, Hu Z, Xu K, Zheng X and Gao Q 2008 *Acta Phys-Chim Sin.*
[4] Wang H, Yuan X, Zeng G, Wu Y, Liu Y, Jiang Q, Gu S *Three dimensional graphene based materials: synthesis and applications from energy storage and conversion to electrochemical sensor and environmental remediation.*
[5] Cheng P and Teng H 2003 *Carbon*

[6] Jin H, Wang X, Gu Z and Polin J 2013 *J. Power Sources*
[7] Hendriansyah R, Prakoso T, Widiatmoko P, Nurdin I and Devianto H. *manufacturing carbon material by carbonization of cellulosic palm oil waste for supercapacitor material.* 2018 *MATEC Web Conf.*
[8] Larasati T D, Prakoso T, Rizkiana J, Devianto H, Widiatmoko P and Nurdin I. *Nano carbon produced by advanced mild hydrothermal process of oil palm biomass for supercapacitor* material. 2019 *IOP Conf. Ser. Mater. Sci. Eng.*
[9] Gumulya C, Prakoso T. Rizkiana J, Devianto H, Widiatmoko P and Nurdin I 2019 *IOP Conf. Ser. Master Sci. Eng*
[10] Sevilla M and Mokaya R 2014 *Energy Environ. Sci.*
[11] Ahmadpour A and Do D D 1997 *Carbon*
[12] Chen H, Muller M B, Gilmore K J, Wallace G G, Li D Mechanically strong, electrically conductive, and biocompatible graphene paper *Adv. Mater.*
[13] Huang H, Zhang W, Fu Y, Wang X. Controlled growth of nanostructured MnO2 on carbon nanotubes for high-performance electrochemical capacitors *Electrochim.*
[14] Xiong C, Li T, Zhao T, Dang A, Li H, Ji X, Jin W, Jiao S, Shang Y, Zhang Y, and conversion to electrochemical sensor and environmental remediation, Adv.
[15] Jain A, Xu C, Jayaraman S and Balasubramanian R 2015 *Microporous Mesoporous Mater.*
[16] Yu A, Chabot V and Zhang J 2013 *Electrochemical Supercapacitors for Energy Storage and Delivery* (Boca Raton:CRC Press)
[17] Abioye A M, Noorden Z A and Ani F N 2017 *Electrochim. Acta*
[18] Farma R, Deraman M, Awitdrus A, Talib I A, Omar R, Manjunatha J G, Ishak M M, Basri N H and Dolah B N M 2013 *Int. J. Electrochem. Sci.*

# Wettability improvement of carbon nanotube for supercapacitor electrode

**H Devianto, N Luthfiana, P Widiatmoko, I Nurdin and T Prakoso**

Department of Chemical Engineering, Bandung Institute of Technology, 40132, Bandung, Indonesia

E-mail : hardev@che.itb.ac.id

**Abstract.** Carbon Nanotube (CNT) has outstanding properties such as electrical conductivity, specific surface area, charge transport capability, mesoporosity, and very high electrolyte accessibility. Based on these properties, CNTs are very suitable to be used as high-performance supercapacitor electrodes which can be seen from the capacitance. One of the factors that can lead to increase capacitance is the high interface interaction between CNT and the electrolyte. The interface interaction of CNT and electrolyte can be improved by increasing the CNT wettability using hydrophilization. CNT is synthesized by pyrolysis with Palm Oil Mill Effluent (POME) as the main raw material. In this study, hydrophilization was carried out using chemical activation $HNO_3$ with various concentrations and activation times (1M for 1, 3, 6 hours, and 13M for 1 hour). The results of CNT structure were characterized using Scanning Electron Microscopy (SEM), Transmission Electron Microscope (TEM), and X-Ray Diffraction (XRD), as well as electrochemically characterized using cyclic voltammetry (CV), Electrochemical Impedance Spectroscopy (EIS) dan galvanostatic charge-discharge (GCD), as supported data to evaluate which concentration and activation time with the highest efficiency to improve CNT performance as a supercapacitor. Based on CV characterization, the capacitance of CNT before hydrophilization, and after hydrophilization using $HNO_3$ (1M for 1 hour, 3 hours, 6 hours, and 13M for 1 hour) were 10 F/g, 13.30 F/g, 14.28 F/g, 26.95 F/g and 24.5 F/g respectively. When CNT hydrophilized at 13M for 1 hour, capacitance decreased. Due to surface damage of CNT, therefore the performance of the supercapacitor decreased. This was supported by the results of the GCD characterization, where some charges could not completely release.

## 1. Introduction

Indonesia is one of the largest country that produces palm oil in the world. Palm oil then processed to make crude palm oil (CPO) with a large amount for each year, in 2020 CPO had been produced about 40 million tons [1]. From the CPO process then produce biomass waste such as solid waste and liquid waste. POME (Palm Oil Mill Effluent) is one of the liquid wastes that had a 70% bigger amount than solid waste [2]. Utilization of POME such as biogas, compost, and carbon nanotube are the attempts to make use of it that are not wasted. Production of biogas and compost needs a large field and longer time to decompose, while the production of carbon nanotube did not need a large field, and POME can be used directly without decomposition. On the other side, POME had carbon content that made it possible to produce carbon nanotube which has higher sales value than others. CNT is a carbon allotrope with a nanometer size of the cylindrical model. CNT has outstanding properties such as the biggest electrical conductivity than copper with 3000 W/K, tensile strength 11-63 GPa, modulus young 270-950 GPa, and small resistance with $0.34\text{-}10^{-4}$ ohm.cm [3]. From that characteristic, CNT can be used for the supercapacitor electrode because of the carbon characteristic which had corrosion resistance, stable at high temperature, high surface area, and low operating cost. KPEE-ITB Laboratory succeeded to produce CNT from POME by pyrolysis. However, the capacitance of supercapacitor with the CNT electrode still low because of the hydrophobicity of CNT. This study investigated the effect concentration and activation time of nitric acid on CNT morphology and capacitance of supercapacitor by performing hydrophilization on CNT.

## 2. Methods

One of the synthesized processes of CNT by using the pyrolysis method. Pyrolysis is one of the thermodynamical processes which material degraded become smaller component at high temperature without the presence of oxygen. Pyrolysis is the easiest process that had many advantages such as storage, transportation, and flexibility for example turbines, boiler combustion equipment, engine, etc [4]. Here are the materials and methods that needed to synthesize CNT and hydrophilization.

### 2.1. *Materials*

The main material used for synthesized of CNTs is POME that was collected from the palm oil mill of PT. Perkebunan Nusantara VIII, in Cikasungka Bogor. The POME is wastewater with rich organic carbon with a colloidal suspension of 95 – 96% water, 0.6 – 0.7% oil, and 4 – 5% total solids, including 2 – 4% suspended solids (Kumar, 2016). Formaldehyde (Merck, purity 37%) used to polymerize POME and ammonium hydroxide (Merck, purity 28%) as a catalyst. The resin that produces by polymerization method used for synthesized CNT by pyrolysis method. Ferrocene (purity 98%) used as a catalyst in nitrogen flow. Ferrocene is frequently used as a precursor to prepare carbon nanostructures because it cannot only act as a carbon source but also give rise to small metal clusters as catalysts. For CNT hydrophilization by using $HNO_3$ (Merck, purity 63%).

### 2.2. *CNT Production*

The resin that has been ready to mix with 30% ferrocene (w/w) and ethanol mixed in order to homogeneous between resin and ferrocene. Once it was homogeneous, the ethanol evaporated and after that, the mixture put into the furnace to be pyrolyzed. The pyrolysis process is carried out by adjusting the increase rate of the temperature of 10 °C/minute under the nitrogen atmosphere. During the process, the temperature held at 200 °C for 1 hour for the carbonization process. Then the temperature was raised until 900 °C and held for 1.5 hours for the formation of the CNT.

### 2.3. *CNT Hydrophilization*

The CNT that has been formed then hydrophilized to improve the wettability of CNT. The chemically activation process was using $HNO_3$ with several variations of concentration and duration of activation. CNT used for hydrophilization was 0.25 gr for each variation and the temperature was set close to the boiling point of the solution with $HNO_3$ concentration 1M for 1 hour, 3 hours, and 6 hours and 13M for 1 hour.

### 2.4. *Supercapacitor Cell*

Electrode preparation of 0.06 gr of CNT that had been measured contact angle wettability, mixed with polyvinylidene fluoride (PVDF), and dissolved with ethanol until it shaped like a slurry. SS mesh used as a current collector. The upper surface of the ss mesh is coated by 1 $cm^2$ slurry CNT then dried in the oven for 24 hours at temperature 80 °C. After slurry CNT dried then weighed and stored as electrodes. The electrodes dripped with 5 drops of KOH solution as electrolyte with 6M concentration for each electrode by using microsyringe and bounded by nafion 212.

### 2.5. *Characterization*

The morphology of CNT was characterized by using scanning electron microscopy (SEM) (Hitachi SU3500, Japan) and transmission electron microscope (TEM) (Hitachi H9500, Japan). The crystal structure of CNT was obtained from a powder X-ray diffraction pattern by using a Bruker D8 advanced diffractometer with monochromatized Cu K radiation.

The electrochemical characterization of supercapacitor were studied by cyclic voltammetry (CV), galvanostatic charge-discharge (GCD), and electrochemical impedance spectroscopy (EIS) methods. The electrochemical measurements were carried out using Gamry V3000 potentiostat-galvanostat. The data obtained were needed to evaluate the highest efficiency of the hydrophilization method with the optimum concentration and activation time to improve CNT performance as a supercapacitor. The specific capacitances of supercapacitors were computed from CV curves by using the following equation (1) [5].

$$C_{sp} = \frac{I\Delta t}{m\Delta V} \quad (1)$$

Where '$C_{sp}$' is the specific capacitance ($F.g^{-1}$), 'I' is the current (A), '$\Delta t$' is the discharge time (s), m is the mass for one electrode material (g) and '$\Delta V$' is the change in voltage during the discharge process (V).

## 3. Result and Discussion

Here is the result of the research based on structure characterization and electrochemical characterization of CNT as a supercapacitor electrode.

### 3.1. *CNT Structure Characterization*

The samples analysed consisted of commercial CNTs, CNT before hydrophilization, and CNT after hydrophilization (CNTH). Before that, the contact angle of CNT was measured first to review the effects before and after chemical activation performed on CNT wettability.

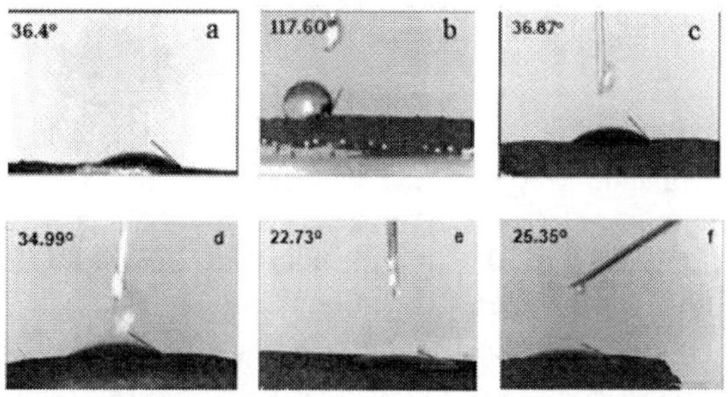


**Figure 1.** Contact angle measurement (a) commercial CNT (b) lab synthesized CNT (c) CNTH 1M 1 hour (d) CNTH 1M 3 hours (e) CNTH 1M 6 hours (f) CNTH 13M 1 hour.

Based on the results of contact angle measurements shown in Figures 1 (a) and (b) indicates that the contact angle on commercial CNT was lower (hydrophilic) than lab synthesized CNT. Lab synthesized CNTs are still hydrophobic, so to make CNTs become hydrophilic the surface of CNTs need functionalization by hydrophilization using nitric acid. In figures 1 (c) to (f) it can be seen that the contact angle decreases, indicating that the CNT is hydrophilic. The lower the contact angle, the greater the wettability. That way, the ions contained in the electrolyte easily enter the CNT pore and stick to the inner surface so that it can increase the capacitance [6].

IOP Conf. Series: Materials Science and Engineering 1143 (2021) 012005 doi:10.1088/1757-899X/1143/1/012005

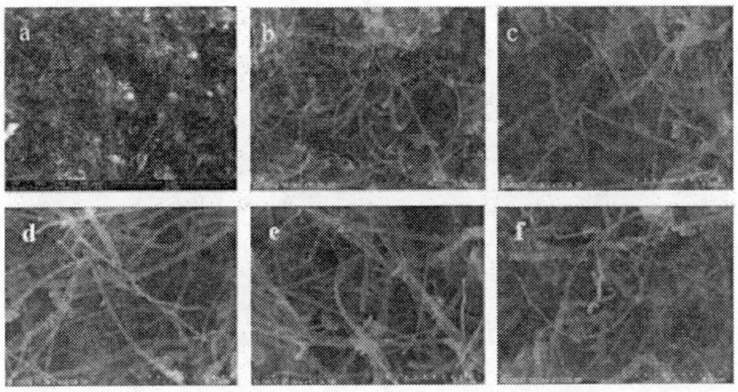


**Figure 2.** CNT SEM characterization (a) commercial CNT (b) lab synthesized CNT (c) CNTH 1M 1 hour (d) CNTH 1M 3 hours (e) CNTH 1M 6 hours (f) CNTH 13M 1 hour.

Based on the SEM results above in Figure 2 (b), it shows the formation of a CNT that almost resembles a commercial CNT. Then comparing Figures 2 (c) to (f), it shows that there is a change in the structure on the CNT surface. The nitric acid that used for the hydrophilization process can reduce impurities on the CNT, but also can change the structure of the CNT. In this figure, it can be seen that at the end of the CNT found a white spot that looks like it was broken into several parts. This is due to the side effect of nitric acid which made the CNT structure truncated and shorter. On the other hand, the concentration of nitric acid that been used and the duration of activation also greatly influence the changes in the CNT structure [7].

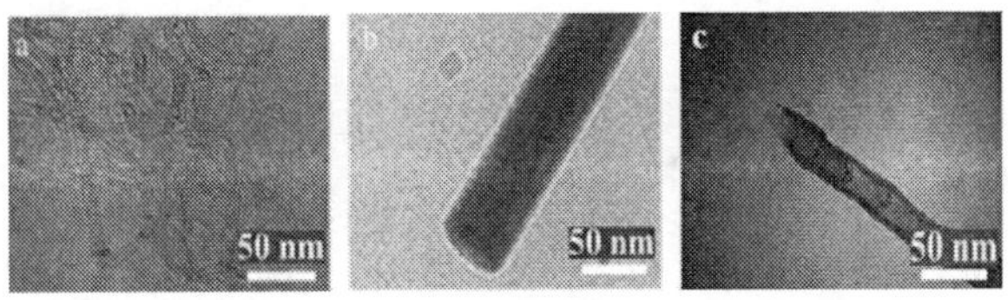


**Figure 3.** TEM characterization (a) commercial CNT (b) lab synthesized CNT by previous researcher (c) current lab synthesized CNT.

TEM characterization was carried out to see the innermost part of the CNT where the nanocarbon has tubular space so the internal and external diameters can be measured. In Figure 3 (a) the CNT has a diameter of 5-10 nm, while Figure 3 (b) has a diameter of 14 nm. Compared to commercial CNT, the CNT diameter between lab synthesized by the previous researchers and commercial CNT had little difference. While in the current lab synthesized CNT had a bigger diameter, approximately 35.7 nm that shows in Figure 3 (c). This was caused when synthesizing CNT by pyrolysis, the reaction temperature increases and there was agglomeration between metal particles which causes the size of the metal particles to increase, resulting in a larger CNT diameter [8].

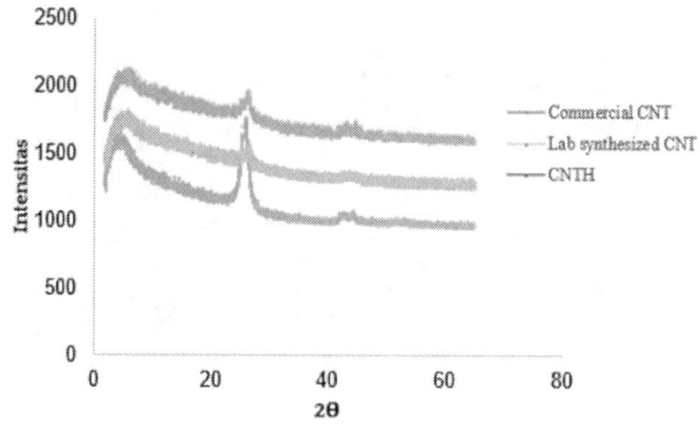


**Figure 4.** XRD characterization of commercial CNT, lab synthesized CNT and CNTH.

Based on the XRD characterization above, the graphite C (002) structure was at the vertex of 2Θ with a value of about 26.97°. the peak of the graphite structure on the CNT before and after hydrophilization tends to be low compared with commercial CNT. The vertex of the 2Θ angle at a value of 44.14° (100) was a reflection of the CNT multi-walled graphite structure [9].

### 3.2. *CNT Electrochemical Characterization*

Characterization of CV aims to determine the specific capacitance of supercapacitor cells before and after hydrophilization with various concentrations and activation time. The result of CV characterization is shown below.

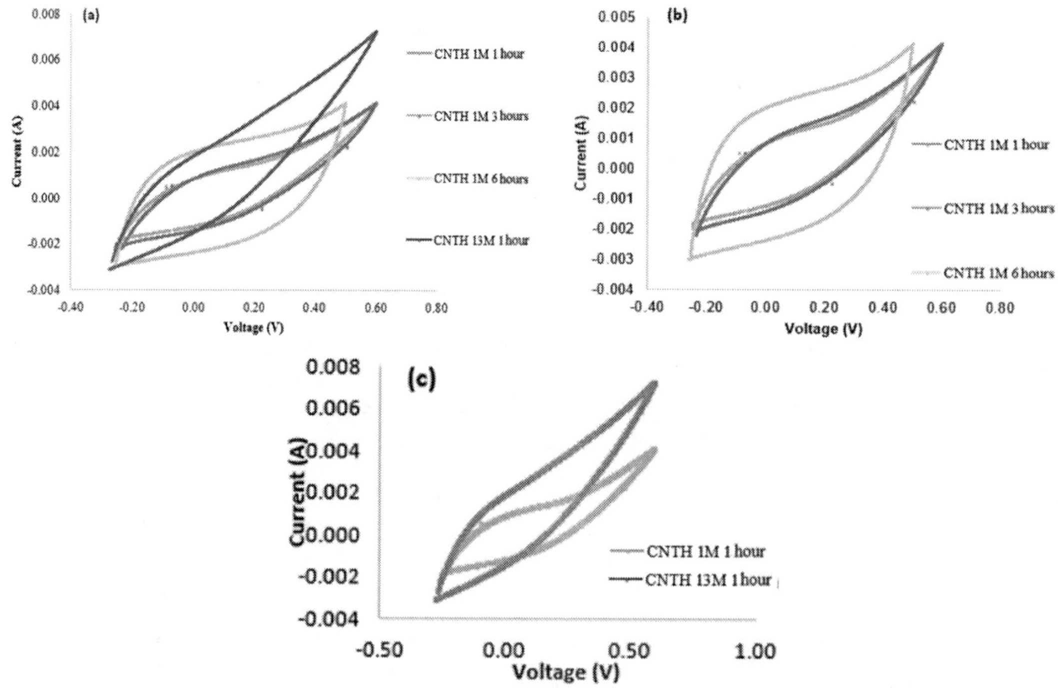


**Figure 5.** CV characterization with scan rate 2 mV/s on (a) lab synthesized CNT (b) duration of chemical activation (c) concentration of nitric acid.

IOP Conf. Series: Materials Science and Engineering 1143 (2021) 012005 doi:10.1088/1757-899X/1143/1/012005

From the curve above it can be seen that the longer activation, the greater the capacitance that obtained. Hydrophilized CNT based on the different concentration in Figure 5 (c) shows that the higher the concentration, the capacitance also increase. The capacitance of hydrophilized CNT with a concentration of 13M was 24.5 F/g that higher compared to hydrophilized CNT with concentration 1M which only had a capacitance of 13.4 F/g. The specific capacitance for each CNTH can be seen in the table below.

**Table 1.** Specific capacitance for each hydrophilized CNT.

| Sample | Capacitance (F/g) |
|---|---|
| CNTH 1M 1 hour | 13.4 |
| CNTH 1M 3 hours | 14.28 |
| CNTH 1M 6 hours | 26.95 |
| CNTH 13M 1 hour | 24.5 |

Figure 5 (a) is a combination of CNT that had been hydrophilized. To see how the effect of activation time on fixed nitric acid concentration is shown in Figure 5 (b) and the effect of different concentrations on fixed activation time is shown in Figure 5 (c). For the effect of activation time, the CNT that has the largest capacitance was when 6 hours of activation time with 26.95 F/g.

GCD characterization aims to see how much voltage of the supercapacitor cell when it was tested at a constant current. The curve formed on GCD characterization was isosceles triangle which shows the time required for the charging process was the same as the time for discharge of the charge. The supercapacitor cells that had been tested were CNT electrode commercial CNT, lab synthesized CNT, and hydrophilized CNT.

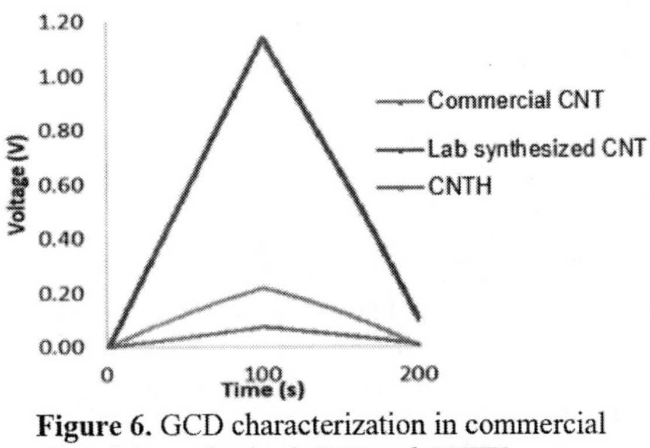


**Figure 6.** GCD characterization in commercial CNT, lab synthesized CNT and CNTH.

In Figure 6 (a) above, the lab synthesized CNT was not discharged completely. In the 200$^{th}$ second the curve should touch axis 0, but the curve stops at point 0.01 second. GCD curve shows the Faradaic process which in the supercapacitor cell there was a reaction of oxygen adsorbed on the electrode surface with water to form $OH^-$ or $HO_2^-$. Because of this reaction, the charge that should had been released during the discharge process was trapped until the discharge time had been completed [10].

Furthermore, EIS characterization was carried out which aims to provide information on the interaction between electrodes and electrolytes from the supercapacitor cells. The setup for EIS measurement was made at a frequency of 100,000 Hz to 0.001 Hz. The result was a Nyquist curve that could be seen as the amount of resistance in the supercapacitor cells. The following are the results of EIS characterization on commercial CNT, lab synthesized CNT, and CNTH.

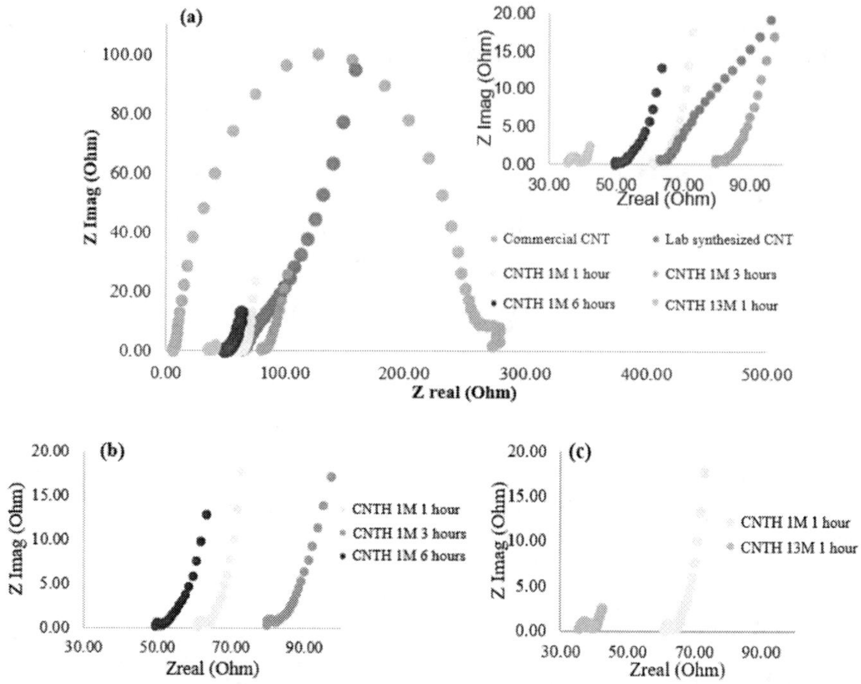


**Figure 7.** Supercapacitor cell's Nyquist plot (a) commercial CNT, lab synthesized CNT & CNTH (b) based on activation time (c) based on the concentration of nitric acid.

The Nyquist plot for supercapacitor cells generally consists of 3 parts, such as semicircles at high frequencies (>100,000 Hz), straight lines with a slope of 45°, and vertical lines with an angle of 90° at medium to low frequencies [11]. The starting point of the curve shows the resistance from the electrolyte ($R_s$), while the semicircles show the resistance of charge transfer at the electrical double layer interface of the electrode with the electrolyte. The value of $R_s$ did not affect the capacitance of the supercapacitor electrode ($Z_{imaginer}$) and was real resistance ($Z_{real}$).

The frequency that used for this EIS measurement was 100,000 Hz-10 Hz, which the semicircles results shown in Figure 7 (a). When the semicircles were compared between lab synthesized CNT with commercial CNT, the semicircle in lab synthesized CNT was smaller than those of commercial CNT. This semicircle represents a voltage as shown in Figure 7 (a) requiring a high voltage to store the charge. The higher the voltage, the smaller the capacitance. One of the factors of low semicircles size in the research was due to the increased wettability of the electrode surface that reducing the repulsive force between the electrode and the electrolyte [12].

## 4. Conclusions

From the results that had been researched, it can be concluded that the nitric acid that used for hydrophilization of the CNT greatly affects the CNT structure. When the CNT was hydrophilized, the tip of the CNT had a white spot which indicates that the CNT had broken into several pieces and became shorter. On the other hand, CNTH had the lowest contact angle 22.73° with a concentration of nitric acid 1M and an activation time of 6 hours obtained the specific capacitance 26.95 F/g.

## 5. Acknowledgment

This research is supported by Crude Palm Oil Fund of the Ministry of Finance, funding year of 2019.

**References**

[1] Ministry of Industry 2020 *Indonesian Palm Oil Production*, (Indonesia : Kemenperin)
[2] Mangoensoekarjo S Semangun H 2005 *Gajah Mada University Press*. Yogyakarta
[3] Saeed K Ibrahim 2013 *Carbon Letters*. **14** 131
[4] Zaman C Z *et al* 2017 *London : Intech Open Science*. 35
[5] Zhou Y *et al* 2018 *Nature*.
[6] Liu T *et al* 2019 *Green Energy & Environment*.
[7] Pumera M *et al* 2012 *Chemical Science*. **3** 3347
[8] Ming H *et al* 2016 *Journal of Nanomaterials*. 8106845
[9] Huang Y *et al* 2018 *Journal Polymer English*.
[10] Yeager E 1984 *Electrochimia Acta*. **29** 1527
[11] Farma R *et al* 2013 *Bioresource Technology*. **132** 254
[12] Cheng Q Chen Z 2013 *Journal Electrochemical Science*. **8** 8282

# Effective Deagglomeration in Biosynthesized Nanoparticles: A Mini Review

**S Khairunnisa, V Wonoputri and T W Samadhi***
Department of Chemical Engineering
Institut Teknologi Bandung, Bandung 40132 Indonesia

*Corresponding Author's E-mail: twsamadhi@che.itb.ac.id

**Abstract.** Materials with nanoscale particle size have different properties from its bulk phase, which allows for wider application of the material. There are various methods to synthesize nanoparticles, namely physical, chemical, and biological method. Nowadays, nanoparticle synthesis method is focused on biological method because of its advantages, such as environmentally friendly, relatively simple procedures, and lower production costs. Biosynthesis by co-precipitation method using extracts from biological agents is considered the most efficient among other biological methods. Biochemical compound in the extract have a dual role in synthesis, they act as a reducing agent which reduces metal salt to metal ion, and as a capping agent which stabilizes the nanoparticle. Biosynthesis has been shown to result in nanoparticles as good as physical and chemical method. However, several studies report that the synthesized nanoparticles have low stability regardless of the presence of their capping agent, resulting in agglomeration of nanoparticles, which reduces its efficiency. Until now, studies on particle deagglomeration especially during nanoparticle biosynthesis have not been widely carried out. This mini review will explain the phenomenon of agglomeration during biosynthesis. Moreover, deagglomeration treatment using physical and chemical approaches will be examined. Each approach is considered to be able to deagglomerate nanoparticles well, and the combination of the two is projected to be able to provide better results.

## 1. Introduction

Nanoparticles are defined as materials with particle sizes ranging from of 1 to 100 nm. Consequently, nanoparticles have large surface area, allowing for high number of atoms to be exposed on the surface resulting in higher reactivity than that of bulk form. [1,2]. With this advantage, the nanoparticle has wide applications, such as antimicrobe, catalyst, remediating agent for pollutant and oil waste, sensor, etc. Selection of the appropriate synthesis method is very important as the slightest difference in synthesis routes can have a big influence on the final characteristics [3]. There are various synthesis methods that have been studied, both chemical, physical, and biological. Biological synthesis is a method that has recently been developed as a safe synthesis alternative.

Biosynthesis, also known as green synthesis, utilizes living cells, extracts of biomolecules from living cells, or cell free supernatants to drive nanoparticle formation reactions [3]. Biosynthesis using biomolecular extracts by precipitation method is the most studied due to the ease of the synthesis procedure. Although the synthesis route is different, the characteristics of the biosynthesized nanoparticles are close to those of the chemical and physical methods. In fact, the organic components attached to the surface of the nanoparticles are able to provide additional functions when the nanoparticles are applied, one of which is to increase the antimicrobial effect [4].

During the synthesis process takes place, the nanoparticles with high reactivity will be susceptible to agglomeration. This resulted in a decrease of nanoparticle toxicity to microbes [4], decrease in pigment degradation photocatalytic activity [5], and poor interfacial/interphase properties and tensile strength of nanocomposite [6], etc. Agglomeration occurs spontaneously and cannot be avoided immediately, so it can becomes a major problem in nanoparticle synthesis [7,8]. Agglomeration does not only occur when the nanoparticles are dispersed in the media, but especially during the synthesis process where the nanoparticles experience strong adhesion between their particles so they are easily agglomerated [1]. For biosynthesis method, it is said that the biological components which are involved in the synthesis process can have a role to prevent agglomeration by acting as a capping agent. Capping agents are amphiphilic compounds capable of covering the surface of the nanoparticles and preventing two or more nanoparticles from approaching each other and join to form large particles [9]. However, several studies stated that these components are still unable to withstand adhesion strength, making agglomeration to be unavoidable [10–12]. So far, there have not been many studies on agglomeration during biosynthesis. Likewise, research of nanoparticles deagglomeration during biosynthesis is limited, especially compared to similar studies performed for chemical synthesis methods. This review will examine how the phenomenon of agglomeration in biosynthesis can occur and also the methods that can be applied to the biosynthesis of nanoparticles to prevent and disperse nanoparticle agglomeration.

## 2. Nanoparticles Synthesis

Nanoparticles can be produced by reducing the size of the material from a micro or larger scale to a nanoscale called the top-down approach, or the opposite way through the formation of atoms or molecules into nanoparticles, which is referred to as a bottom-up approach. Compared with the top-down, bottom-up approach is more advantageous as it can produce nanoparticles with a size range of 1 to 50 nm with a narrow nanoparticle size distribution [13–15]. Physical and chemical bottom-up methods have been studied for a long time. The physical synthesis requires expensive and complex equipment with the use of high temperatures and pressures. Meanwhile, chemical synthesis frequently involves chemicals that are hazardous to the environment and people [16]. As an alternative, biosynthesis has become a popular bottom-up method currently being developed. Compared to physical and chemical methods, biosynthesis uses a simple set up tools so that it is economical and energy efficient. The reaction for the formation of nanoparticles is fast and the reaction parameters are easy to control [13,17]. Biosynthesis is also often referred to as green synthesis because it avoids the production of synthetic waste, reduces pollution and uses safe materials, and uses renewable feed stocks [18].

Through biosynthesis, nanoparticles are formed by the help of the living organisms activities, including bacteria, yeast, fungi, and algae, and plants [16]. Biological components are reductive, so generally, biosynthesis is carried out for the production of metal or metal oxide-based nanoparticles. This biosynthesis can be carried out intracellularly, extracellularly, or using cell extracts. Intracellular biosynthesis takes advantage of the metabolic system in cells to reduce metal salts by reductase[14], while in extracellular biosynthesis, ionized metal salts adhere to and are reduced by the membrane on the cell surface[19]. The challenge using these biological agents is the need to provide rich grown medium and to be carried in specific environmental conditions during the synthesis process. In addition, the process can take a long time ranging from hours to days [3]. In synthesis using cell extracts, it is the organic compound content that reduces metal ions and stabilizes the formed nanoparticles [20]. The advantages of using cell extracts are that they are able to complete the synthesis in minutes to hours and the synthesis process are simpler than using intra or extracellular [3]. The most common method of biosynthesis using cell extracts is precipitation. In addition, there is synthesis method using microwaves that emerging in recent years because it can be done only in few minutes. However, the microwave method makes it difficult to observe the particle formation reaction at each synthesis time, nor can it be scaled-up for large-scale production [21]. The precipitation method involves more synthesis parameters that are easily modified so that it is more interesting to study in depth. Therefore, biosynthesis in this paper is focused on the precipitation method.

*2.1 Nanoparticle Biosynthesis by Precipitation Method*

One of the most studied biological nanoparticle synthesis methods is precipitation. There are several advantages of the precipitation method among other biosynthetic methods, which include ease of preparation, low cost, and high product yield [22]. The principle of the precipitation method is to add the extract of the biological agent to the metal precursor solution gradually until nanoparticles are formed in the form of precipitates. Generally, the extract with precursors is mixed at room temperature for several hours with constant stirring. Synthesis can also be continued by aging at room temperature for 24 hours. The nanoparticle precipitates are then centrifuged and dried with hot air at a temperature of 70-80°C [4,10]. As for metal oxide-based nanoparticles, the dry precipitates are calcined at 350-400°C for 1-3 hours.

In the biosynthesis of nanoparticles, biological extracts act as reducing agents and capping agents. Metal precursors that dissolved in water are ionized into positively charged metal ions, which are further reduced by hydroxyl groups. Zero charged metal atoms undergo nucleation and grow into a larger trench with the attachment of metal ions that have not or have been reduced on the surface of the nanoparticles. Afterwards, the biological component that has a role as a capping agent will cover the nanoparticles through polar interactions with the nanoparticle surface and non-polar interactions with the environment. Capping agent prevents particle overgrowth so that it can keep the size of the nanoparticles small and uniform. Capping agents can also increase the electrostatic and steric repulsive forces of the nanoparticles, where the two repulsive forces can prevent adjacent particles from coming into contact, so that the stability of the nanoparticles in suspension can be maintained properly [9].

## 3. Agglomeration in Nanoparticle Biosynthesis

As previously mentioned, there is always a chance for the nanoparticles to undergo agglomeration during synthesis. This agglomeration phenomenon is closely related to the size and reactivity of the particles. The very small-sized particle makes the number of atoms that come into contact and react to be high. Therefore, the formed nanoparticles have high reactivity. Nanoparticles will interact with their environment to make them stable, one of which is through van der Waals interactions. The van der Waals force is the interaction of attraction between two or more adjacent nanoparticles through the London dispersive force, Debye inductive force, and Keesom orientational force [23]. Nanoparticles that are reactive and move freely in solution will easily experience collisions, where the collision of particles induces van der Waals force. The nanoparticle clusters will be solidified into agglomerates through chemical bridge between clusters. Next, crystal growth occurs which causes the agglomerates to bind through chemical bonds. Deagglomeration can still be done by shear stress of stirring if the bonds to the agglomerates are still weak [24]. However, as the bonds get stronger, the agglomerates will be more difficult to break even if they are given a larger physical dispersion force. It is very important to pay attention to the phenomenon of agglomeration formation during biosynthesis of nanoparticle.

The bioactive compounds are believed to act as capping agents that can prevent agglomeration. However, several studies have reported that plant extracts actually induce agglomeration. For instance, biomolecules from *Phoenix dactylifera* leaf extract produces sphere-shaped $Fe_3O_4$ particles that agglomerate into non-uniform flower shapes [12]. The same thing happened in the synthesis of $TiO_2$ synthesized with *Phyllanthus niruri* leaf extract, where the agglomeration due to the electrostatic forces between the biomolecules on the nanoparticle surface [10]. ZnO synthesized using the pericarp extract of *Garcinia mangostana* fruit undergoes agglomeration due to the strong polarity and electrostatic attraction between the nanoparticles [11]. In the biosynthesis of nanoparticles using fungi, agglomeration can occur due to the production of enzymes and proteins by the fungi which then accumulate in suspension as happened in extracellular Ag synthesis using *Trichoderma longibrachtium* [25]. The same thing happened in the study of Vetchinkina et al [26] who carried out biosynthesis using extracellular from *Pleurotus ostreatus, Lentinus edodes, Ganoderma lucidum,* and *Grifola frondosa.*

The resulting Ag nanoparticles are large, irregular in shape, and agglomerated (Figure 1). The potential zeta value of the Ag suspension is -9 to -12 mV, where this figure shows that the repulsion force between the particles is weak. In addition to the influence of capping agents from several types of biological agents, the agglomeration of nanoparticles on biosynthesis is also related to the shape of the particles. Biosynthesis method is advantageous because it is able to control the shape of the nanoparticles into a uniform sphere [27,28]. However, sphere-shaped nanoparticles tend to agglomerate more easily because of the ease of contact between the particles, while nanoparticles with irregular shapes need to adjust their position to suit other particles [29].

Another cause of agglomeration in nanoparticle biosynthesis is the acidity of the biological agent. Most cases of agglomeration occur when biosynthesis is carried out with acidic extracts, where acidic conditions reduce steric repulsion and increase interparticle hydrogen bonding. Biosynthesis of Ag nanoparticles using *Mangifera indica* peel extract at pH 3 caused the deprotonation of functional group to decrease, resulting in non-uniform and agglomerated nanoparticles [29]. In the synthesis of ZnO using *Averrhoa bilimbi* fruit extract, the pH which was originally around 3.5-4 decreased to a pH of 1.5-2 due to heating. This extreme acidic condition not only facilitates agglomeration due to decreased capping agent ability, but also slows down ZnO production. In the application of ZnO nanoparticles synthesized with *A. bilimbi* fruit extract, the formation of nanoparticle agglomerates causes the nanoparticles to settle so that their antimicrobial ability decreases [4]. When the nanoparticle suspension is to be investigated further, filtering using a syringe filter can be performed to eliminate the agglomerates and produce a monodisperse suspension of nanoparticles [30]. However, the effect of filtering was considered insignificant when viewed from the measurement results of particle distribution based on TEM and DLS. For this reason, an additional treatment to prevent agglomeration or to aid the breakdown of the formed nanoparticle agglomerates is needed.

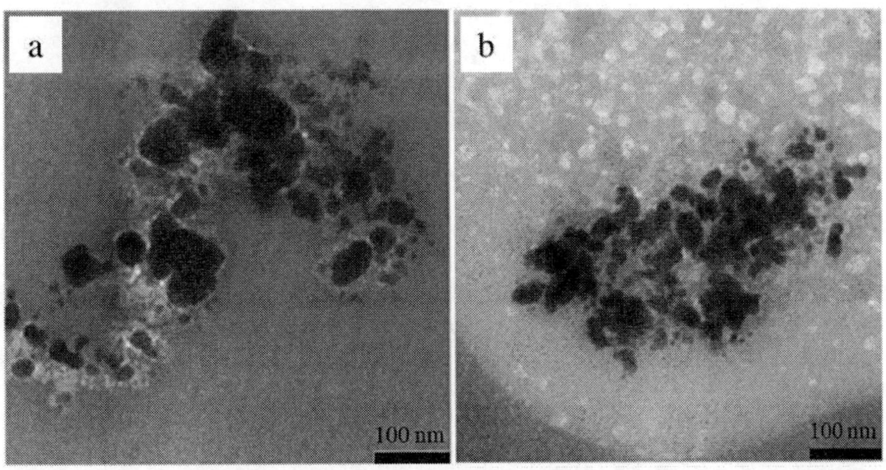


**Figure 1.** Agglomeration of synthesized Ag nanoparticles using extracellular of (a) *L. edodes* and (b) *G. frondose* (From "Green synthesis of nanoparticles with extracellular and intracellular extracts of basidiomycetes" by E. Vetchinkina et al., 2015, *PeerJ*, *6.* CC BY 4.0.).

## 4. Deagglomeration in Nanoparticle Biosynthesis

Various studies have been carried out to disperse the nanoparticle agglomerates formed during synthesis, most of which have been investigated on chemical synthesis. In this section, we will discuss deagglomeration methods that can potentially be applied during the biosynthesis process of nanoparticles, physically and chemically.

### *4.1 Physical Deagglomeration*

Physical deagglomeration aims to break down the agglomerates that are formed. There are various physical deagglomeration that can be done during synthesis and post-synthesis. In this review, the methods discussed are methods that can be applied during the biosynthesis process, namely mechanical stirring and ultrasonication.

*4.1.1 Mechanical stirring.* Generally, biosynthesis by the precipitation method is accompanied by mechanical stirring. Besides aiming to homogenize the solution, stirring also provides a deagglomeration effect of nanoparticles during synthesis. The effectiveness of stirring for deagglomeration can be increased with the speed of rotation. One study observed the effect of stirring at a speed of 500-900 nm during the synthesis of $Fe_3O_4$ nanoparticles. $Fe_3O_4$ crystal size decreased from 24 nm to 22.5 nm by increasing the stirring speed from 500 rpm to 600 rpm. However, TEM observations indicated that the nanoparticles generated at a speed of 600 rpm were agglomerated quite severely [31]. The smaller crystal size is thought to produce a greater magnetic attraction, thereby facilitating agglomeration. On the other hand, the crystal diameter increased to 25 nm when the speed was increased to 900 rpm. At such high speed, the collision between metal ions increases, causing larger crystal growth.

Khan et al. [30] observed the effect of stirring at speeds of 500, 1000, 1500, and 2000 rpm on the growth of ZnO particles and agglomerates. At the highest velocity, ZnO agglomerates are less than that at the lowest velocity. The amount of agglomeration correlated to the lethality of the nanoparticles toward *Bacillus subtilis*, *Eschericia coli*, and *Candida albicans*, where the synthesis at 2000 rpm (lowest agglomeration) had the best inhibition ability of the three microbes. A similar thing was found in a study by Gulrajani et al. [31], where the lowest stirring speed (700 rpm) produced the largest average size of nanoparticles (510 nm), whereas the highest stirring speed (2500 rpm) produced the lowest particle size (370 nm) with lowest agglomeration too. The high stirring speed will prevent the growth of particles and induce new particles to produce nanoparticles with small diameter and low agglomeration. In addition, high velocity can also produce a more regular shape of the particles. Based on these studies, deagglomeration in laboratory-scale nanoparticle biosynthesis carried out by the mechanical stirring method will be effective when the stirring speed is at 2000-2500 rpm.

*4.1.2 Ultrasonication.* Ultrasonication can break down soft-agglomerates effectively and it can even break down hard agglomerates that are difficult to do by stirring or dispersion method with ball milling [32,33].The principle of ultrasonication is to pass the waves on a liquid medium which can cause rarefaction and compression cycle. Rarefaction occurs when the media pressure is low where air bubbles are formed, which is called cavitation. Furthermore, compression causes stress cavitation due to higher pressure conditions. Rarefaction and compression alternately make the cavitation larger and unstable until it breaks by generating mechanical and thermal energy. This mechanical energy will crush and erode the nanoparticle agglomerates [34,35].

Even though it requires more energy, ultrasonication is able to disperse nanoparticles much more effectively than the stirring method. Kawashima et al. [36] proved that the synthesis of $CaCO_3$ nanoparticles by stirring, produced aggregates with a size of more than 2 μm, while synthesis process that used ultrasonication did not resulted in any aggregates. The same thing was reported by Estrada-Monje et al. [37] who observed the effect of the deagglomeration method on the dispersion of $TiO_2$ nanoparticles in ethylene-vinyl acetate (EVA) solution. STEM observations clearly show that the $TiO_2$ produced by mechanical stirring forms large agglomerates. Meanwhile, deagglomeration by ultrasonication with 100% amplitude, 80% pulse, for 10 minutes resulted in much smaller agglomerate. In its application, EVA film with $TiO_2$ which was deagglomerated by ultrasonication was able to inhibit *Escherichia coli* colonies more than deagglomeration by mechanical stirring.

Parameters that need to be considered when applying ultrasonication as a deagglomeration method include the type of ultrasonicator, time, temperature, pulse, power, and frequency. The probe type ultrasonicator has higher deagglomeration ability than the bath type because it is able to transmit more energy to the nanoparticle system. However, treatment with the sonicator probe can increase the temperature of the nanoparticle suspension rapidly, which then might induce agglomeration. In addition, sonication time that are too long can also lead to nanoparticle re-agglomeration. The phenomenon of agglomeration in the sonication treatment is related to excess gasification and intense cavitation friction, which changes the charge on the surface of the nanoparticles [32]. Nanoparticle biosynthesis accompanied by sonication deagglomeration can be carried out at room temperature or high temperature up to 90°C, with a time of 10 to 180 minutes. Synthesis time can be shortened by changing other parameters. For example, biosynthesis at 50 W ultrasonication power can be completed in 40 minutes and when the power is increased to 100, 150, and 200 W, the synthesis can be shortened to 20, 16, and 10 minutes, respectively [38].

*4.2 Chemical Deagglomeration*

The key to nanoparticle agglomeration is the high reactivity on the surface which causes the attraction between the particles to increase. Therefore, it is necessary to control the repulsive force on the surface of the nanoparticles which can be done by chemical deagglomeration. This method involves adding a chemical that can modify the electrostatic and steric stability of the nanoparticles to maintain the stability of the nanoparticles in suspension.

*4.2.1 Electrostatic Stabilization.* Electrostatic stabilization aims to increase the repulsive force by the charged layer on the surface of the nanoparticles called the Electric Double Layer (EDL). EDL consists of unreduced metal atoms and ions from biological extracts that are attracted to the surface of the nanoparticles. It is this collection of charged ions that produces an electrostatic repulsion so that when the nanoparticles are close together and the EDL of each particle overlaps, the ions experience an osmotic repulsive interaction which then cause them to drifts away from one another [39]. Increasing the electrostatic repulsive force can be achieved by adjusting the pH during biosynthesis. Nanoparticle agglomeration occurs when the pH of the synthesis approaches its isoelectric pH, which is pH where the electric potential value is 0. At this pH, the attraction between the particles reaches a maximum, so it is easy to form hard agglomerates [40]. pH also affects the characteristics of the extract. Changes in pH will change the ability of biomolecules to chelate and reduce metal ions, which in turn changes the shape and yield of the resulting nanoparticles [20]. Biological extracts, especially plants, tend to be acidic. The level of acidity can affect the ability of phytochemical compounds to bind and reduce metal ions. In addition, the more acidic the extract pH, the slower reduction and nucleation processes of nanoparticles will be, so that agglomerates and aggregates are easier to form [41].

In addition to pH modification, electrostatic stability control can also be done by adding a surfactant type dispersing agent that can be ionized in the water. Espitia et al. [42] tried to add $Na_4P_2O_7$ dispersing agent to increase the stability of ZnO nanoparticles. $Na_4P_2O_7$ ionizes to form $Na^+$ and $P_2O_7^{2-}$ ions, likewise, some of the Zn atoms from the nanoparticles are also ionized in the water. Positive Zn atoms will be attracted to the surface of the ZnO nanoparticles to form a stern layer. Pyrophosphate which is negatively charged will cover the outer side to form a diffuse layer, and the counter ion from pyrophosphate, namely natrium, will inhabit the outer side. The higher the pyrophosphate anion that covers ZnO, the greater the electrical potential, so the greater its electrostatic repulsive force also. The level of electrostatic stability can be calculated by the zeta potential, where good stability that prevents agglomeration can be achieved when the value is more than positive or negative 25 mV [43]. Moreover, addition of dispersing agents that have the potential to be used to increase the electrostatic stabilization of nanoparticles can also be done. Some examples of dispersing agents that can be used in the biosynthesis of metal nanoparticles are sodium dodecyl sulphate, cetyltrimethyl ammonium bromide, tertiary butyl ammonium, sulphonium, phosphonium, and imidazolium moieties [9,44]. However, the interaction of these chemicals with the biological extract has yet to be determined.

*4.2.2 Steric Stabilization.* To maintain their stability, nanoparticles can attract complex components such as polymers and alcohols to form an additional protective layer apart from EDL, that is steric layer [39,44]. The repulsion force caused by stern layer maintains the position of the nanoparticles with a minimum distance so there is no interaction that led to agglomeration. The components involved in the formation of this layer are capping agents, and in biosynthesis, capping agents can be polyphenols, citric acid, vitamins, biodegradable polymers, silica, and etc. [9]. This capping agent provides steric stabilization on the nanoparticle suspension to prevent agglomeration. The role of steric stabilization may be weak if the effectiveness and concentration of the biochemicals that act as capping agents are low, or if the biochemicals are damaged by drying and calcination as they cannot withstand high temperatures [28].

Steric stabilization can be increased by the addition of a complex dispersing agent. Thus, the protective layer on the nanoparticles will be thicker so that the steric repulsive force is able to prevent agglomeration and maintain the stability of the nanoparticles during synthesis and also after drying [44]. Polymer materials such as poly(ethylene glycol), poly(vinyl alcohol), and poly (vinyl pyrrolidine) are commonly used to increase steric repulsive force [9]. It should be noted that in order to produce maximum steric stability, the determination of the type and concentration of the dispersing agent must be accurate. Qiang et al. [45] tried four types of dispersing agents to stabilize CuO nanoparticles. It is said that each type of dispersing agent has a different effect, even if it is added with the same concentration. This is related to the absorption of dispersing agent on the surface of the particles that change the characteristics and reactivity of the nanoparticles [46]. For CuO dispersed in water with a concentration of 2% wt, sodium polyacrylate with a concentration of 0.4 to 0.8% (w/w) under conditions of pH 10 gave the best stability. At a concentration that is too low, the amount of dispersant is not yet equivalent to CuO resulting in sedimentation of the nanoparticles. On the other hand, a too much dispersing agent causes flocculation which is also undesirable. The use of dispersing agent with polymer types is considered to be able to maintain the stability of the nanoparticle solution for a long time, as evidenced by the Ag nanoparticles resulting from biosynthesis with PVA addition that remains stable for 3 months with room temperature storage [47]. In its potential use in biosynthesis method, a dispersing agent should be added to the metal precursor solution before mixing it with the biological extract.

### 4.3 Combined Deagglomeration Treatments

Some researchers combine two deagglomeration methods simultaneously to maximize the dispersal of the nanoparticles. It has been mentioned that ultrasonication is known to be able to break down agglomerations well but does not eliminate the potential for re-agglomeration after the sonication process is complete. To maintain the deagglomeration effect longer, a component that can increases its repulsive or steric stability can be added [32,33]. Pradhan et al. [46] conducted a research to determine optimum dispersing method for Mn and Cu nanoparticles. They proved that sonication for 15 minutes with the addition of a dispersing agent of Bovine Serum Albumin (BSA) solution in 0.05 vol% resulted in the higher percentage of the Mn and Cn nanoparticles measured at <20 nm than the treatment without the addition of BSA. Another study by Kawashima et al. [36] showing that $CaCO_3$ nanoparticles dispersed by the probe ultrasonicator settles and are not affected by large variations in sonication intensity and hydrophobic properties of $CaCO_3$. However, when combined with a polycarboxylate-based superplasticizer dispersing agent, the stability of the deagglomerated nanoparticles is maintained. The increase in sonication time and amplitude also has a positive effect on $CaCO_3$ dispersion in solution, where the best stability is produced at 3 hours and an amplitude of 40%. Cunniffe et al. [48] also states that the synthesis of high concentrations of nanoparticles accompanied by sonication deagglomeration produces micron-sized particles, while additional deagglomeration using a dispersing agent can reduce the particle size to the nanoscale. In their research, sonication for 4 minutes produced nanoparticles with a size of less than 100 nm. This combination of deagglomeration methods can be referred to for application in nanoparticle biosynthesis.

## Conclusion

This review has discussed the biosynthesis of nanoparticles and the accompanying agglomeration issues, along with the methods that can solve these problems. The environmental friendliness and easy procedures make biosynthesis a widely explored method of producing nanoparticles. However, nanoparticles resulting from biosynthesis method is prone to agglomeration. Therefore, some potential deagglomeration is proposed in this study. Physical deagglomeration can be done by mechanical stirring at a speed of 2000-2500 rpm or by ultrasonication using a bath type ultrasonicator. Chemical deagglomeration can be carried out by increasing electrostatic stability through pH modification and addition of ionic surfactants or increasing steric stability by adding a polymer type dispersing agent. The combination of these methods also has the opportunity to improve the deagglomeration and stability of the nanoparticles. The discussion in this review is limited to the principle of deagglomeration, therefore it is necessary to conduct direct research to further prove that it is true that the deagglomeration method that has been mentioned can be adapted to the biosynthetic process.

**Reference**

[1] Yokoyama T, Masuda H, Suzuki M, Ehara K, Nogi K, Fuji M, Fukui T, Suzuki H, Tatami J, Hayashi K and Toda K 2012 Basic properties and measuring methods of nanoparticles *Nanoparticle Tehcnology Handbook* ed M Hisokawa, K Nogi, M Naito and T Yokoyama (Amsterdam: Elsevier Ltd.) pp 5–48

[2] Yuwono A H, Sofyan N, Hamdani V, Sholehah A and Arief M 2014 The effect of precursor mixing temperature during precipitation process on the size of ZnO nanoparticles and the dispersion of ZnO@SiO2 core-shell nanostructure *Appl. Mech. Mater.* **525** 108–16

[3] Rana A, Yadav K and Jagadevan S 2020 A comprehensive review on green synthesis of nature-inspired metal nanoparticles: Mechanism, application and toxicity *J. Clean. Prod.* 122880

[4] Rahayu E, Wonoputri V and Samadhi T W 2020 Plant extract-assisted biosynthesis of zinc oxide nanoparticles and their antibacterial application *IOP Conf. Ser. Mater. Sci. Eng.* **823** 012036

[5] Li G, Lv L, Fan H, Ma J, Li Y, Wan Y and Zhao X S 2010 Effect of the agglomeration of TiO2 nanoparticles on their photocatalytic performance in the aqueous phase *J. Colloid Interface Sci.* **348** 342–7

[6] Ashraf M A, Peng W, Zare Y and Rhee K Y 2018 Effects of size and aggregation/agglomeration of nanoparticles on the interfacial/interphase properties and tensile strength of polymer nanocomposites *Nanoscale Res. Lett.* **13** 214

[7] Rouxel D, Cherni and Binet S 2017 Synthesis of nanoparticles of titanium dioxide TiO2 for toxicological studies: A short review *Rev. Artic.* **3** 18–22

[8] Halamoda-Kenzaoui B, Ceridono M, Urbán P, Bogni A, Ponti J, Gioria S and Kinsner-Ovaskainen A 2017 The agglomeration state of nanoparticles can influence the mechanism of their cellular internalisation *J. Nanobiotechnology* **15** 48

[9] Gulati S, Sachdeva M and Bhasin K K 2018 Capping agents in nanoparticle synthesis: Surfactant and solvent system *AIP Conf. Proc.* **1953** 030214

[10] Shanavas S, Priyadharsan A, Karthikeyan S, Dharmaboopathi K, Ragavan I and Vidya C 2020 Green synthesis of titanium dioxide nanoparticles using Phyllanthus niruri leaf extract and study on its structural, optical and morphological properties *Mater. Today Proc.* **26** 3531–4

[11] Aminuzzaman M, Ying L P, Goh W S and Watanabe A 2018 Green synthesis of zinc oxide nanoparticles using aqueous extract of Garcinia mangostana fruit pericarp and their photocatalytic activity *Bull. Mater. Sci.* **41** 50

[12] Ahmed J A, Salah L and Abderrhmane B 2020 Green synthesis and characterization of iron oxide nanoparticles by pheonix dactylifera leaf extract and evaluation of their antioxidant activity *Sustain. Chem. Pharm.* **17** 100280

[13] Jamkhande P G, Ghule N W, Bamer A H and Kalaskar M G 2019 Metal nanoparticles synthesis: An overview on methods of preparation, advantages and disadvantages, and applications *J. Drug Deliv. Sci. Technol.* **53** 101174

[14] Khanna P, Kaur A and Goyal D 2019 Algae-based metallic nanoparticles: Synthesis, characterization and applications *J. Microbiol. Methods* **163** 105656

[15] Sweet M J, Chesser A and Singleton I 2012 Review: Metal-based nanoparticles; size, function, and areas for advancement in applied microbiology *Advances in Applied Microbiology* vol 80,

ed S Sariaslani and G M Gadd (Cambridge: Academic Press) pp 113–42

[16] Agarwal H, Kumar S V and Rajeshkumar S 2017 A review on green synthesis of zinc oxide nanoparticles – An eco-friendly approach *Resour. Technol.* **3** 406–13

[17] Ramzan M, Obodo R M, Mukhtar S, Ilyas S Z, Aziz F and Thovhogi N 2020 Green synthesis of copper oxide nanoparticles using Cedrus deodara aqueous extract for antibacterial activity *Mater. Today Proc.* https://doi.org/10.1016/j.matpr.2020.05.472

[18] Singh J, Dutta T, Kim K H, Rawat M, Samddar P and Kumar P 2018 "Green" synthesis of metals and their oxide nanoparticles: Applications for environmental remediation *J. Nanobiotechnology* **16** 84

[19] Malik P, Shankar R, Malik V, Sharma N and Mukherjee T K 2014 Green chemistry based benign routes for nanoparticle synthesis *J. Nanoparticles* **2014** 302429

[20] Sajjad S, Leghari S A K, Ryma N-U-A and Farooqi S A 2018 Green synthesis of metal-based nanoparticles and their applications *Green Metal Nanoparticles: Synthesis, Characterization and their Applications* ed S Kanchi and S Ahmed (New Jersey: Wiley) pp 23–77

[21] Kooti M and Naghdi Sedeh A 2013 Microwave-assisted combustion synthesis of ZnO nanoparticles *J. Chem.* **2013** 562028

[22] Kim D-W, Kim D-S, Kim Y-G, Kim Y-C and Oh S-G 2006 Preparation of hard agglomerates free and weakly agglomerated antimony doped tin oxide (ATO) nanoparticles by coprecipitation reaction in methanol reaction medium *Mater. Chem. Phys.* **97** 452–7

[23] Park S J and Seo M K 2011 Intermolecular Force *Interface Science and Technology* vol 18 (Oxford: Elsevier Ltd.) pp 1–57

[24] Nore P and Mersmann A 1993 Batch precipitation of barium carbonate *Chem. Eng. Sci.* **48** 3083–8

[25] Elamawi R M, Al-harbi R E and Hendi A A 2018 Biosynthesis and characterization of silver nanoparticles using Trichoderma longibrachiatum and their effect on phytopathogenic fungi *Egypt. J. Biol. Pest Control* **28** 1–11

[26] Vetchinkina E, Loshchinina E, Kupryashina M, Burov A, Pylaev T and Nikitina V 2018 Green synthesis of nanoparticles with extracellular and intracellular extracts of basidiomycetes *PeerJ* **6** e5237

[27] Gautam S, Dubey P and Gupta M N 2013 A facile and green ultrasonic-assisted synthesis of BSA conjugated silver nanoparticles *Colloids Surfaces B Biointerfaces* **102** 879–83

[28] Castro L, Li J, González F, Muñoz J A and Blázquez M L 2020 Green synthesis of tellurium nanoparticles by tellurate and tellurite reduction using Aeromonas hydrophila under different aeration conditions *Hydrometallurgy* **196** 105415

[29] Lu J and Wang J K 2006 Agglomeration, breakage, population balance, and crystallization kinetics of reactive precipitation process *Chem. Eng. Commun.* **193** 891–902

[30] Khan M F, Ansari A H, Hameedullah M, Ahmad E, Husain F M, Zia Q, Baig U, Zaheer M R, Alam M M, Khan A M, Alothman Z A, Ahmad I, Ashraf G M and Aliev G 2016 Sol-gel synthesis of thorn-like ZnO nanoparticles endorsing mechanical stirring effect and their antimicrobial activities: Potential role as nano-Antibiotics *Sci. Rep.* **6** 27689

[31] Gulrajani M L, Gupta D, Periyasamy S and Muthu S G 2008 Preparation and application of silver nanoparticles on silk for imparting antimicrobial properties *J. Appl. Polym. Sci.* **108** 614–23

[32] Yao S Y and Xie Z H 2007 Deagglomeration treatment in the synthesis of doped-ceria nanoparticles via coprecipitation route *J. Mater. Process. Technol.* **186** 54–9

[33] Xiao F, Ye J, Wang Y and Rao P 2005 Deagglomeration of HA during the precipitation synthesis *J. Mater. Sci.* **40** 5439–42

[34] Leong T, Ashokkumar M and Sandra K 2011 The fundamentals of power ultrasound - A review *Acoust. Aust.* **39** 54–63

[35] Taurozzi J S, Hackley V A and Wiesner M R 2011 Ultrasonic dispersion of nanoparticles for environmental, health and safety assessment issues and recommendations *Nanotoxicology* **5** 711–29

[36] Kawashima S, Seo J W T, Corr D, Hersam M C and Shah S P 2014 Dispersion of CaCO3 nanoparticles by sonication and surfactant treatment for application in fly ash-cement systems *Mater. Struct. Constr.* **47** 1011–23
[37] Estrada-Monje A, Zitzumbo-Guzmán R, Bañuelos-Díaz J A and Zaragoza-Contreras E A 2019 Ultrasonic dispersion and activation of TiO2 nanoparticles and its effect on bacterial inhibition in EVA films *Mater. Chem. Phys.* **235** 121760
[38] Nouri A, Tavakkoli Yaraki M, Lajevardi A, Rezaei Z, Ghorbanpour M and Tanzifi M 2020 Ultrasonic-assisted green synthesis of silver nanoparticles using Mentha aquatica leaf extract for enhanced antibacterial properties and catalytic activity *Colloids Interface Sci. Commun.* **35** 100252
[39] Polte J 2015 Fundamental growth principles of colloidal metal nanoparticles - a new perspective *CrystEngComm* **17** 6809–30
[40] Park S, Lee W J, Park S, Choi D, Kim S and Park N 2019 Reversibly pH-responsive gold nanoparticles and their applications for photothermal cancer therapy *Sci. Rep.* **9** 20180
[41] Makarov V V, Love A J, Sinitsyna O V, Makarova S S and Yaminsky I V 2014 " Green" nanotechnologies: synthesis of metal nanoparticles using plants *Acta Naturae* **6** 35–44
[42] Espitia P J P, Soares N D F F, Teófilo R F, Vitor D M, Coimbra J S D R, De Andrade N J, De Sousa F B, Sinisterra R D and Medeiros E A A 2013 Optimized dispersion of ZnO nanoparticles and antimicrobial activity against foodborne pathogens and spoilage microorganisms *J. Nanoparticle Res.* **15** 1324
[43] Singh A K 2016 *Engineered Nanoparticles: Structure, Properties, and Mechanism of Toxicity* (Cambridge: Cambridge University Press)
[44] Manojkumar K, Sivaramakrishna A and Vijayakrishna K 2016 A short review on stable metal nanoparticles using ionic liquids, supported ionic liquids, and poly(ionic liquids) *J. Nanoparticle Res.* **18** 103
[45] Qiang A H, Zhao L M, Xu C J and Zhou M 2007 Effect of dispersant on the colloidal stability of nano-sized CuO suspension *J. Dispers. Sci. Technol.* **28** 1004–7
[46] Pradhan S, Hedberg J, Blomberg E, Wold S and Odnevall Wallinder I 2016 Effect of sonication on particle dispersion, administered dose and metal release of non-functionalized, non-inert metal nanoparticles *J. Nanoparticle Res.* **18** 285
[47] Chandran S, Ravichandran V, Chandran S, Chemmanda J and Chandarshekar B 2016 Biosynthesis of PVA encapsulated silver nanoparticles *J. Appl. Res. Technol.* **14** 319–24
[48] Cunniffe G M, O'Brien F J, Partap S, Levingstone T J, Stanton K T and Dickson G R 2010 The synthesis and characterization of nanophase hydroxyapatite using a novel dispersant-aided precipitation method *J. Biomed. Mater. Res.* **95** 1142–9

# Effects of Fe-Doped Electrolyte and Feed Flow Rate Evaluation in Home Made Solid Oxide Fuel Cell

**Hary Devianto*, Isdiriyani Nurdin, Pramujo Widiatmoko, Kafi Adi Prasetya, Basil Pradipta**

Department of Chemical Engineering Institut Teknologi Bandung, Bandung 40132 Indonesia

*hardev@che.itb.ac.id

**Abstract**. Solid Oxide Fuel Cell (SOFC) is one type of fuel cell that has several advantages over other types of fuel cells. To achieve low cost SOFC, the materials used are nickel oxide (NiO) for the anode, Calcia Stabilized Zirconia (SCZ) for the electrolyte and Calcia Cobalt Oxide Zinc Oxide (CCZO) as cathode. $Fe_2O_3$ is a doping agent to form a denser electrolyte. The cell with and without Fe doping was tested using hydrogen and ambient air. Additionally, the effect of the amount of fuel flow rate is also investigated using Aspen Plus Modelling. Based on physical characterization, it is determined that 3%-mole Fe doping on the electrolyte increases the electrolyte density from 83% to 90%. Using electrochemical impedance spectroscopy, it can be concluded that SOFC with Fe doped electrolyte has a lower resistance than SOFC without Fe doping. On the other hand, using a correction factor, an equation of power correlated to fuel flow for the SOFC is generated to show increasing fuel flow enhanced SOFC performance.

## 1. Introduction

Energy is becoming one of the basic human needs, along with clothing, food, and shelter. Advances in technology increases the global needs of energy. With the increasing global needs of energy, clean energy improvement has to be made. Fuel cell is one of the new and alternative energy conversion technology that converts chemical energy directly into electrial energy. This technology is highly efficient, it has a wide range of fuels, and relatively environmentally friendly. Fuel cell's unique and favorable characteristics compared to regular heat engine to produce electricity attracts the world's attention. Research development by institutes and commercialization effort about fuel cell has been done. For fuel cell to be a commercial power generator, fuel cell has to produce power as needed while also being inexpensive. This can be done by stacking individual cells into a stacked fuel cell system [1, 2].

Solid oxide fuel cell (SOFC) is a breakthrough in clean energy convertion technology that can uses fuel that is widely available in Indonesia. The performance of SOFC is directly effected by cell fabrication methods, and one of the most important part of SOFC that requires fabrication is its electrolyte. SOFC performance will increase along with the increasing of electrolyte density and ionic conductivity[3, 4]. One of the more widely used electrolyte for SOFC is calcia-stabilized zirconia (CSZ) ceramic [5, 6]. To fabricate CSZ, calcia and zirconia powder is sintered at its eutectic temperature at 1300 °C. It is known that metal oxide doping during sintering process can lower the eutectic temperature and increase the ionic conductivity of CSZ thus increasing the overall performance of SOFC [7]. Aim of this research is to determine the effect of Fe foping for CSZ and the evaluation of flow rate variation for single sel SOFC fueled by hydrogen.

## 2. Methodology

In general, this research is divided into three main sections, starting with material and equipment preparation, cell fabrication along with physical test of cell's micro and macrostructure, and on-off cell cycle evaluation. Physical characterization of cell is carried out for cell fabrication evaluation, while the cell characteristic test is carried out using electrochemical and physical characterization. Then proceed with modeling with SOFC operation simulation using Aspen Plus simulation software.

### *2.1. SOFC Cell Fabrication*

Anode supported SOFC fabrication starts with anode fabrication. SOFC anode used in this research was manufactured by means of dry casting. The main material of the anode used in this research was a ceramic powder made of Nickel Oxide (NiO) purchased from Changsha Easchem Co., Ltd (technical grade, Changsa, China) and zirconia purchased from Pingxiang Ball-Tec New Materials Co., Ltd (technical grade, Jianxi, China) in a rasio 65:35, into the powder polyvynil alcohol (PVA) 1 wt% and maizena 10 wt%. To make the powder homogen, the powder was put into a ball mill for 1 hour. The mixed powder is then formed using dry casting method at 60 kN/m$^2$. The casted powder is sintered in a furnace at 1100 °C and atmospheric pressure for 3 h to make a stable ceramic. After sintering process, the physical structure of the anode was tested.

After anode fabrication, the electrolyte is made. Electrolyte used in this research is a zirconia based electrolyte with CaO purchased from Bratachem (technical grade, Bandung, Indonesia) as a stabilizer, and $FeO_{1.5}$ purchased from Bratachem (technical grade, Bandung, Indonesia) of 3 mole% as a doping. The powder mixture is then dispersed in isopropyl alcohol (IPA). Using automated air brush sprayer, the mixture is then sprayed on the anode. The electrolyte and anode is then sintered in a furnace at 1100 °C and atmospheric pressure for 3 h. After the sintering process the physical structure sintered electrolyte and anode was tested. Cathode fabrication is the last step of SOFC fabrication. The cathode was made by using solid state reaction of calcium oxide, cobalt, and zinc. The oxide mixture was measured stoichiometricly and polyvinyl alcohol (PVA) is added into the mixture. The wet cathode mixture was then sprayed on the electrolyte using automated air brush sprayer. After the cathode was sprayed onto the electrolyte, the cell was sintered in a furnace at 900 °C and atmospheric pressure for 5 h.

### *2.2. SOFC Performance Test*

SOFC cells are attached to main equipment set in the tubular furnace. To operate the SOFC cell, a start-up procedure is required from the cell, which is heating to a temperature of 800 °C followed by hydrogen flow into the cell to perform reduction reaction to the anode to be ready to operate. As SOFC cell is ready to operate, the electrochemical performance of the SOFC cell is checked. The SOFC cell performance testing was carried out using Gamry 3000 Potentiostat and Gamry software. Voltage and current tests on SOFCs are carried out using potentiodynamic and electrochemical impedance spectroscopy (EIS) on the potentiostat. After an hour of cell electrochemical testing, the furnace was turned off and the fuel feed flow was stopped for a period of the shutdown cycle. After the cycle off, SOFC cell is started up again, then tested for another electrochemical properties. After having tested the second time, the cells were turned off and the feed flow was stopped. After the temperature reaches ambient temperature, SOFC are removed from the furnace and tested physically using SEM to evaluate its microstructure and ASTM C373-88, standard test method for bulk density of fired whiteware products, to determine its porosity, respectively.

### *2.3. SOFC Simulation using Aspen Plus*

SOFC was simulated using Aspen Plus software to generate voltage, current, and power production data. The simulation was adapted from a study from Doherty [8]. The flowsheet for the simulation is shown in Figure 1. In this simulation, the main blocks are prereforming, anode, cathode, and post combustion block. For a hydrogen fueled SOFC, the prereforming and post combustion blocks is not used. To simulate an anode in Aspen Plus, RGibbs block was used. The use of RGibbs makes the fuel and oxidizing agent react according to Gibbs free energy. In the anode block, the fuel utilization is set to 85%. To simulate the cathode, separator block was used. In this block, the oxygen used in anode reaction is separated with the excess oxygen and nitrogen from air. The voltage, current, and power is calculated using Aspen Plus calculator.

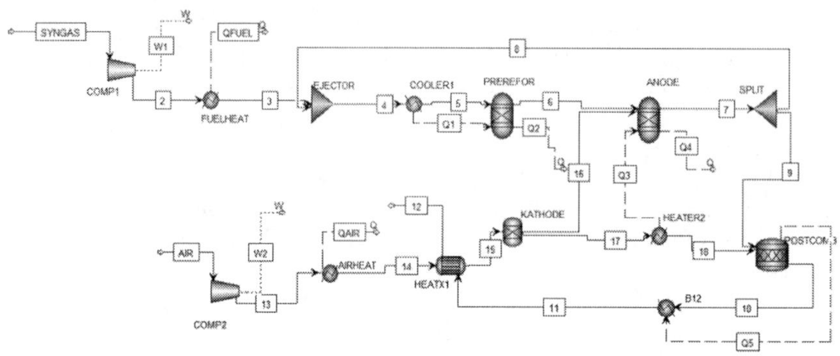


**Figure 1.** SOFC simulation process flow diagram.

## 3. Results and Discussion

### *3.1. SOFC Cell Performance*

The experimental result of the SOFC cell is shown with the IVP and EIS curve the cell produced. The IVP of the cell is shown in the Figure 2. From the IVP figure, it can be seen that with the same flow rate, the fuel cell with $FeO_{1.5}$ doping has a higher maximum power than the fuel cell without $FeO_{1.5}$ doping. The increase of maximum power happens because the Fe doping decreases resistances in the electrolyte. From the figure, it can be seen that with the increasing rate of hydrogen fuel, the maximum power also increases. This happens because the increasing rate of fuel increases the current in the SOFC system, thus increases the produced power [9].

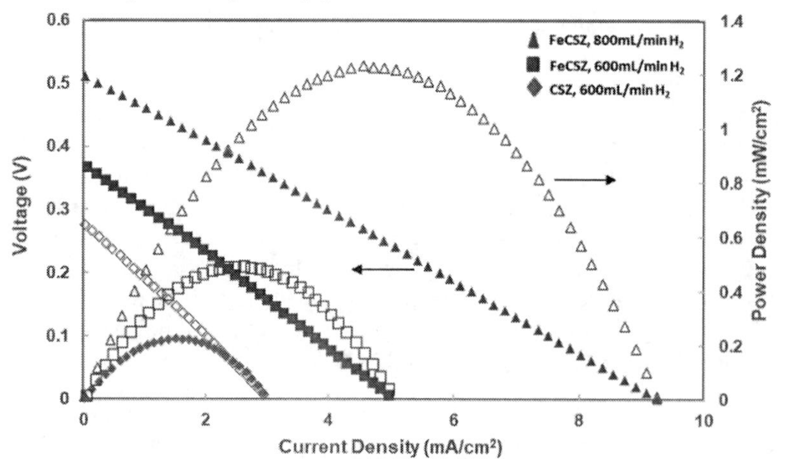


**Figure 2.** Experimental IVP result at 800 °C.

Both electrolytes were tested by using EIS method. The EIS result, equivalent circuit model and resistance values can be seen on the Figure 3 and Table 1. Ohmic resistance (Rohm) is the bulk resistance of the electrolyte, grain boundary resistance (R1) is the ionic resistivity, double layer resistance (R2) is the electronic resistivity of the cell. On the whole, the resistivity of the $FeO_{1.5}$ doped electrolyte is lower than the electrolyte without $FeO_{1.5}$ doping. The addition of $FeO_{1.5}$ doping also increases the oxygen vacancy thus increasing the potential of oxygen migration and lowering its grain boundary resistance [7].

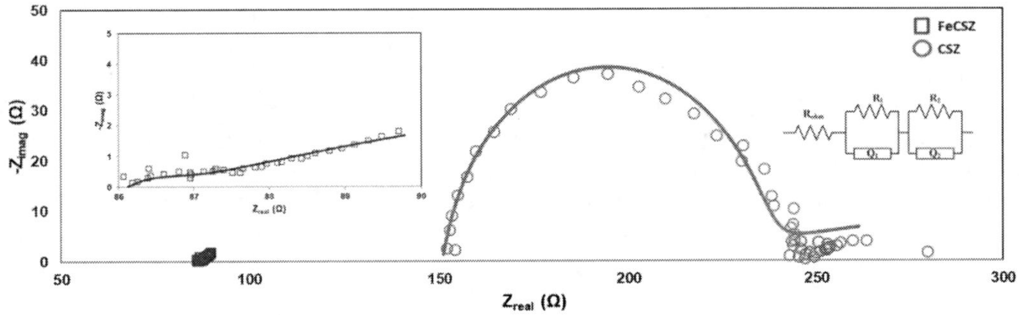


**Figure 3.** Impedance spectra of CSZ and FeCSZ at 800 °C.

**Table 1.** Resistance values from equivalent circuit model.

| | Rohm | R1 | Y1 | a1 | R2 | Y2 | a2 |
|---|---|---|---|---|---|---|---|
| CSZ | 139.60 | 53.84 | $8.84\ 10^{-3}$ | $1.47\ 10^{-3}$ | 83.25 | $5.93\ 10^{-7}$ | $9.42\ 10^{-1}$ |
| CSZ + $FeO_{1.5}$ | 84.31 | 2.51 | $1.47\ 10^{-3}$ | $4.68\ 10^{-1}$ | 19.63 | $1.24\ 10^{-1}$ | $3.84\ 10^{-1}$ |

The porosity of the cell is tested using ASTM C373-88 method. The porosity for the electrolyte without $FeO_{1.5}$ doping is 17.18% and the porosity for the electrolyte with $FeO_{1.5}$ doping is 10.12%. This indicates the increase of the electrolyte density. Both electrolytes were further tested using scanning electron microscope (SEM). The cross section of the cell can be seen on the Figure 4. From the SEM result, it can be seen that the electrolyte with $FeO_{1.5}$ doping is denser than the electrolyte without the addition of $FeO_{1.5}$ doping.

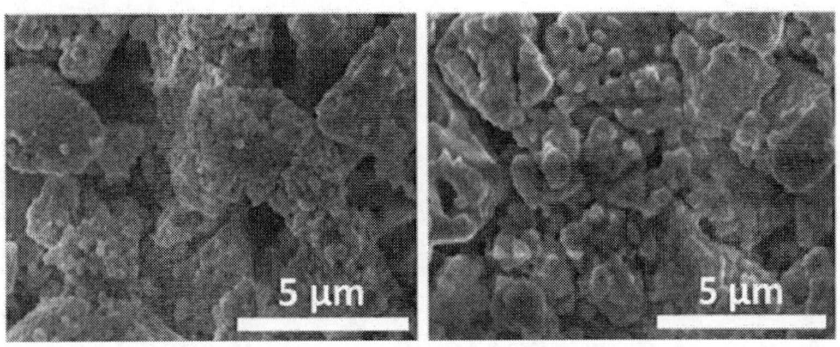


**Figure 4.** Cross section SEM of the electrolyte without doping (left) and with Fe doping (right).

*3.2. SOFC Simulation using Aspen Plus*

Aspen plus simulation was used to determine the effect of different varieties of fuel. The simulation is done by adapting a simulation that was done by Doherty. The simulation was then modified until it has the same operating condition as the one from the experiment. The first comparison was the effect of recycle on the fuel cell performance. The cell performance can be seen on the Table 2.

**Table 2.** Comparison of using a recycled flow.

| Variation | Inlet Flowrate (kmol/h) | OCV (V) | J ($A/m^2$) | Resistance (Volt) | | | Voltage (V) | Power Density ($W/m^2$) |
|---|---|---|---|---|---|---|---|---|
| | | | | Ohmic | Activation | Concentration | | |
| 1 | 8.54 | 1.11 | 2035.98 | 0.20 | 0.14 | 0.11 | 0.68 | 1392.61 |
| 2 | 4.78 | 1.14 | 1884.03 | 0.20 | 0.13 | 0.15 | 0.70 | 1316.94 |

From the table, it can be seen that the cell equipped with a recycling stream has a lower voltage, higher current, and higher power. The recycling stream increases the amount of fuel in the inlet side of the anode, thus according to Faraday Law, the current will increase accordingly. The decrease of voltage is because of Nernst Law, the impurity on the inlet side of the anode will decrease the OCV. The next comparison was the effect of temperature on the fuel cell performance. The increase of temperature increases the concentration resistivity of the cell, but the ionic and activation resistivity will decrease significantly. The rapid decrease of activation resistivity is because the nature of SOFC material which are semiconductors that activates at a high temperature. The effect of temperature on the resistance can be seen on the Figure 5.

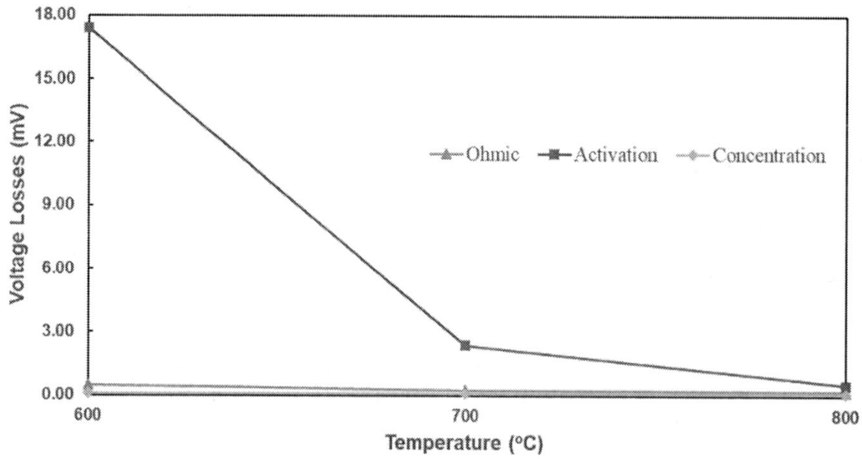


**Figure 5.** The effect of temperature on the resistance of the cell.

The next comparison is fuel flowrate. According to Nernst equation, variation on the fuel flow will affect the OCV of the cell. The increase of fuel flow will increase the current and current density according to Faraday Law. With the increase of current density there will be an increase on the ohmic, activation, and concentration resistivity of the cell. This resistivity will decrease the voltage of the cell. The change of current is more significant than the change of current, thus the power produces increases along with the current.

### *3.3. Correction Factor of Experiment Results with Simulation Results*

The data generated from simulations and experiments have significant discrepancy. Table 3 showed the comparison of simulation and experimental results with variations in hydrogen flow rate.

**Table 3.** Comparison of simulation and experimental results.

| $H_2$ inlet flowrate(mL/min) | P simulation (mW/cm$^2$) | P experiment (mW/cm$^2$) | V simulation (V) | V experiment (V) | I simulation (mA/cm$^2$) | I experiment (mA/cm$^2$) |
|---|---|---|---|---|---|---|
| 500 | 59.24 | - | 0.90 | - | 55.00 | - |
| 600 | 68.71 | 0.39 | 0.87 | 0.27 | 66.00 | 1.46 |
| 700 | 77.12 | - | 0.84 | - | 77.00 | - |
| 800 | 84.65 | 0.98 | 0.80 | 0.38 | 88.00 | 2.60 |
| 900 | 91.08 | - | 0.77 | - | 99.00 | - |
| 1000 | 96.45 | - | 0.73 | - | 110.00 | - |

The table shows current has a large deviation value from the experiment. It affects the value of the activation and concentration voltage loss. Therefore, it is necessary to have a correction factor that could adjust simulation data to experimental data. The correction factor is used to assess any deviation in cell performance characteristics. In this experiment, the correction factor is considered a gradient value that manipulates the graph line equation of the current value against the flow rate. This is justified by two things, namely the Faraday Equation which shows the current value is directly proportional to the amount of fuel and the findings that show a linear trend of changes in current to flow rate with the same fuel utilization value [10]. Based on Eq (1), the current is affected by the mole of $H_2$ entering and fuel utilization. In this case, the experimental fuel utilization value is held constant. Using Equation 1, the value of correction factor could be determined. K value then modeled against the current data generated through the simulation to see the results of the correction.

$$K = \frac{f(I_{percobaan})}{f(I_{simulasi})} = \frac{0.003\ nH_{2,in}}{0.1319\ nH_{2,in}} = 0.022744 \tag{1}$$

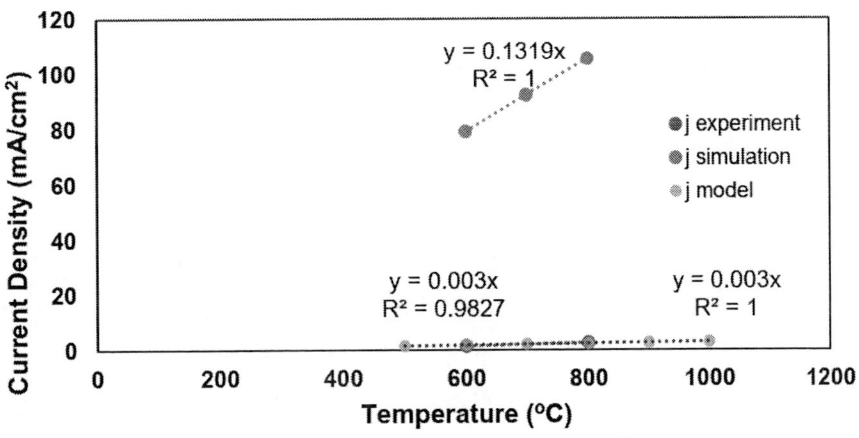


**Figure 6.** The effect of temperature on the performance of the cell.

Through the model line equation in Figure 6, the equation of the power density relationship with hydrogen flow rate is determined as shown in Eq (2). Using this power density value, the simulation model can show results that are closer to the experimental results.

$$P = -1 \times 10^{-6} \left(nH_{2,in}\right)^2 + 0.0033\ nH_{2,in} \tag{2}$$

## 4. Conclusion

$FeO_{1.5}$ doping on the electrolyte will lower its internal resistivity due to oxygen vacancy formation mechanism. It also increases its density because of its transformation mechanism during sintering, thus minimizing fuel cross over. Because the doping lowers the overall resistivity, it also increases the power of the SOFC. By using Aspen Plus simulation a mathematical model was made for its power generation when using 500 ml/min up to 1000 ml/min hydrogen as fuel.

## References

[1] Sharaf O.Z and Orhan M.F 2014 *Renew. Sustain. Energy Rev* **32** 810-853
[2] Akinyele D, Olabode E and Amole A 2020 *Inventions* **5** 1-35
[3] Timurkutluk B, Timurkutluk C, Mat M.D and Kaplan Y 2016 *Renew. Sustain. Energy Rev* **56** 1101–1121
[4] Zakaria Z, Mat Z.A, Hassan S.H.A and Kar Y.B 2020 *Intl. J. Energy Res* 1-18
[5] Zhou M & Ahmad A 2006 *Mater. Res. Bull* **41** 690–696
[6] Muccillo R, Netto R.C and Muccillo E.N 2001 *Mater. Lett* **49** 197–201
[7] Gao H, Liu J, Chen H, Li S, He T, Ji Y and Zhang J 2008 *Solid State Ion* **179** 1620-1624
[8] Doherty W, Reynolds A and Kennedy D 2010 *Energy* **35** 4545-4555
[9] Larminie J and Dicks A 2018 *Fuel Cell System Explained* (West Sussex: John Wiley & Sons Ltd)
[10] Chakraborty U.K 2019 *Appl. Sci* **9** 1-24

IOP Conf. Series: Materials Science and Engineering 1143 (2021) 012008 doi:10.1088/1757-899X/1143/1/012008

# Techno-economic and Environmental Analyses on Reduced Graphene Oxide modified Kapok Fiber as Oil Absorbent

**Graecia Lugito[1*], Tjokorde Walmiki Samadhi[1], Tirto Prakoso[2], Muhammad Lauda[2], and Vincent Laurent Tanujaya[2]**

[1]Department of Chemical Engineering, Institut Teknologi Bandung, Bandung 40132 Indonesia.
[2]Department of Bioenergy and Chemurgy Engineering, Institut Teknologi Bandung, Bandung 40132 Indonesia.

graecia@che.itb.ac.id

**Abstract.** The industrialization of synthetic fiber has resulted in an economic depression for both domestic and global kapok fiber industries as well as the farmers. On the other hand, the development of ssthe oleochemical industry in Indonesia demands a separation process that is less cost- and energy-intensive. Supporting the sustainable development goals (SDGs) number 1, 8, 9, 12, and 13, in this study, the under-tapped kapok fiber has been utilized to show its potential value as an oil sorbent material. Despite having high oil sorption capacity and being biodegradable, the raw kapok fiber has low oil selectivity, low oil retention, and a fragile structure. The kapok fiber has been modified with reduced graphene oxide (rGO) to obtain hydrophobic-oleophilic absorbent with high absorption capacity, high oil selectivity, high oil retention, and strong structure to maximize the recovery yield. Modification has been done by soaking kapok fiber in GO supernatant and reducing them with ascorbic acid, then freeze-drying the fiber to get a robust superhydrophobic sponge. Techno-economic and life-cycle analyses have been conducted and the results indicate that rGO-KF aerogel is economically viable with a potential GPM greater than 36,27%, and environmentally beneficial as it can potentially decrease product carbon emissions by 32.8 tons/ton recovered oil.

## 1. Introduction

Kapok (*Ceiba pentandra*) is one of the Indonesian commodities which market has been declining and dormant for more than half a century. In 1936/1937, Indonesia was the world's largest kapok fiber exporter accommodating about 85% of the global demand or about 28,400 tons. The industrialization of synthetic fibers and the global market competitions have turned down the economy of the kapok industry and farmers. In recent years, however, kapok fiber with its unique features attracts increasing attention, particularly in its utilization as an oil sorbent material [1]. Kapok fiber is hydrophobic, oleophilic, lightweight, and has a lumen (air cavity) in the middle of its fiber wall which makes it superior organic absorbent. Each gram of pristine kapok fiber can absorb approximately 40 grams of oil [2,3].

There are several interesting chemical dan physical modifications made on Kapok fiber. Chemical treatment with alkali/acid and acetylation as well as the simple blending with synthetic polymer (such as polyethylene) are among the oldest methods [2]. With technological advances and comprehensive knowledge of the chemical reaction mechanism, methods such as sol-gel mineralization [4,5], rGO-modified [6] as well as the use of ultrasonication in the process promote the advancement of the methods.

IOP Conf. Series: Materials Science and Engineering 1143 (2021) 012008 doi:10.1088/1757-899X/1143/1/012008

**Table 1** summarizes the oil sorption capacity and reusability of some kapok fiber-based composite materials.

**Table 1.** Oil sorption capacity and reusability of some kapok fiber-based composite materials.

| Composite Material | Oil sorption capacity | | Reusability | | Ref. |
|---|---|---|---|---|---|
| | Type of oil | (g/g) | No. Cycle | Capacity reduction | |
| Pristine kapok | Diesel | 32 | 1 | | [3] |
| | Soybean | 50 | | | |
| Organo-silane | Diesel | 46.9 | 8 | ~ 20% | [4] |
| | Soybean | 58.8 | | | |
| | Palm oil | 48.6 | 10 | 37.4% | [5] |
| rGO | Diesel | 34 | 8 | ~ 6% | [6] |
| | Soybean | 42 | | | |

On the other hand, the growth of vegetable and essential oil industries in Indonesia increases the need for an effective and economical technology that can improve the quality as well as the yield of the produced oil and easily applied with low energy consumption levels without generating concerns over the possibility of becoming waste in the future. Palm oil is a key vegetable industry in Indonesia; crude palm oil (CPO) production in 2019 reached 42.9 million tons. About 0.6%wt. CPO has lost as palm oil mill effluent (POME) [7]. Reclaiming the oil content in POME enhances the yield production of the palm oil industry as well as reduces the BOD of the effluent. Oil recovery from POME using membrane faces a major drawback of fouling. Therefore, a pre-separation process that is less cost- and energy-intensive is desirable; and absorption material utilizing the long-forgotten commodity such as kapok fiber could be an option to address the needs.

Modification with rGO potentially strengthens the structural integrity, enhancing oil sorption capacity, reusability, and oil recovery of the kapok fiber without losing its affinity to oil and water repellency. In this study, rGO modified kapok fiber (rGO-KF) modification procedures have been formulated with considerations to the techno-economic feasibility and the environmental impacts through comprehensive literature reviews and simulations. Expectantly, the market of the long-forgotten commodity could revive with the positive impacts of this research.

## 2. Methodology

This research focuses on the techno-economic analysis and life-cycle assessment of the rGO-KF aerogel. **Figure 1** shows the framework of this research started from literature study at stage 1 to the process innovation at stage 2, case study at stage 3, and techno-economic analyses and life cycle assessment at stage 4. The techno-economic analyses have been carried out based on the implementation scenario of crude palm oil (CPO) reclamation from palm oil mill effluent (POME). Process material and energy requirements are simulated using Aspen Hysys V.8. Assumptions taken in the rGO-KF aerogel production process simulation and assessment are summarized in **Table 2**. Life cycle assessment (LCA) has been conducted based on the acid and carbon emissions of rGO-KF aerogel in comparison to the synthetic polypropylene absorbent.

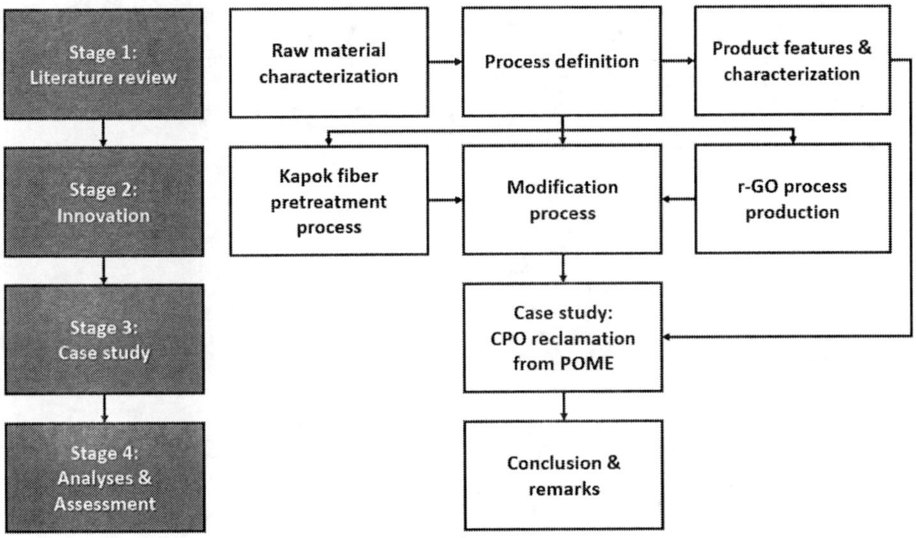


**Figure 1.** Research framework.

**Table 2.** Assumptions taken for the rGO-KF aerogel production process simulation and assessment.

| Assumptions | Value |
|---|---|
| Raw kapok fiber | 1000 ton/year |
| Production rate rGO-KF aerogel | 1133 ton/year |
| **Product specifications** | |
| rGO:kapok fiber (mass) | 3:20 |
| Oil sorption capacity | min. 46 g oil/g absorbent |
| Number of cycles | 5 |
| Capacity reduction | max. 10% |
| **Operational cost (Aspen Hysys simulation)** | |
| Raw materials and chemicals | Price (million IDR) |
| Kapok fiber (per year) | 2,705 |
| Chemicals (per year) | 122,820 |
| Energy (per year) | 21,317 |
| Utility (per year) | 7,313 |
| Recovered oil (per ton) | 7.5 |

## 3. Results and Discussion

Prior to the techno-economic analyses, the process flow diagram of rGO-KF aerogel production should be determined. **Figure 2** displays the process flow diagram of rGO-KF aerogel production. Prior to the modification processes, the raw kapok fiber was separated from seeds and cleaned from dust by aeration then oxidized by sodium chlorite (5%wt. $NaClO_2$) under acid condition (pH 4.5, temperature 80°C) to eliminate the waxy layer from the kapok surface exposing the hydroxyl group on the cellulosic material [5].

The Marcano method can be utilized to oxidize the graphite into graphite oxide. This method uses a mixture of sulfuric acid ($H_2SO_4$) and phosphoric acid ($H_3PO_4$) without using sodium nitrate ($NaNO_3$); thus, the production of acid waste can be reduced and the toxic gases (i.e., $NO_2$ and $N_2O_4$) can be eliminated from the synthesis process [8]. The graphite oxide could be exfoliated into graphene oxide through ultrasonication [9]. Kapok fiber bundle is added into the GO solution (1g/L) with a weight ratio of 20:3; the modification has been carried out for 2 hours with stirring at room temperature. Reduction takes place at 95°C for 1 hour with 1 M sorbic acid (1/10 v/v GO solution) instead of hydrazine to reduce the formation of unwanted chemical waste [10]. After reduction, the rGO is neutralized with aqua demin at room condition. Lastly, the rGO-KF aerogel is made from freeze-drying 2 %wt. suspension rGO-KF in the aqua demin at 0.1 mbar, -55°C for 72 hours [6].

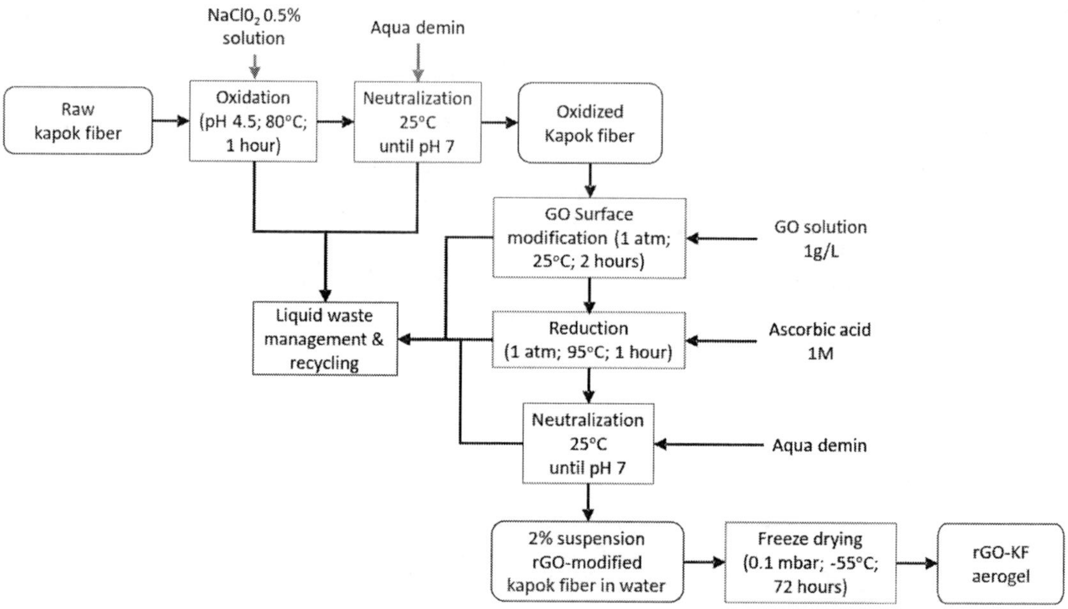


**Figure 2.** Process flow diagram of rGO-KF aerogel production.

Techno-economic and life cycle assessments have been carried out to evaluate the feasibility of rGO-KF process production as well as its utilization for CPO reclamation from POME. The economic analysis carried out in this report is limited to the gross profit margin (GPM) of the product while the revenue comes from the sales of the reclaimed oil. Conventional palm oil mill lost about 0.6%wt. of its CPO production during condensate disposal of boiling stations and clarification stations [11]. Taking Indonesian CPO production in 2019, the amount of prospectively reclaimable CPO is about 264 thousand tons/year. Therefore, the reclamation potentially achieves revenue of IDR 1.98 trillion/year. **Figure 3** displays the GPM assessment against the value of %-oil recovery from the utilization of rGO-KF aerogel. Based on CSI Market 2020, the average GPM value of the chemical industry in Q1 2020 is 36.27%; this value is used as a benchmark for the economic feasibility of rGO-KF aerogel utilization for CPO reclamation from POME. Based on the graph, with only 17%-oil recovery, the sorbent is economically viable with a potential GPM greater than 36,27%.

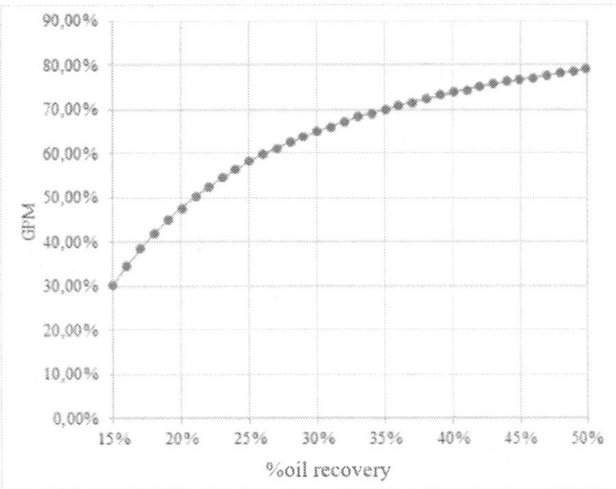


**Figure 3.** GPM against %-oil recovery value from the utilization of rGO-KF aerogel.

The utilization of rGO-KF aerogel in oil reclamation from POME is expected to support sustainable development goals (SDGs) number 1, 8, 9, 12, and 13. To assess whether the rGO-KF aerogel meets the expectations as environmentally friendly oil absorbents, LCA has been carried out. There are four stages in the LCA manufacturing process as stipulated in ISO 14040 and ISO 14044: identification of the definition of purpose and scope; inventory analysis; LCIA (life cycle impact analysis); and data interpretation. In the context of products from natural biomass, LCA is used to compare the environmental effects between biomass products and products from synthetic fibers [12]. The approach used is cradle-to-grave, which is the analysis of products from the production of raw materials to the end of product life. The inventory analysis of the rGO-KF aerogel production is summarized in **Table 3**.

**Tabel 3.** Inventory analysis of the rGO-KF aerogel production.

| Activity | Input | | Output | |
|---|---|---|---|---|
| | **Type (unit)** | **Amount** | **Type (unit)** | **Amount** |
| **Kapok plantation** | Water (kg) | 41,43 | Kapok fiber (kg) | 1 |
| | Fertilizers (kg) | $9{,}19\times10^{-2}$ | $CO_{2\ eq.}$ (kg) | 0,12 |
| | Energy (MJ) | 0,30 | | |
| | $CO_2$ (kg) | 1624,83 | | |
| **Transportation** | Diesel (L) | $2{,}30\times10^{-2}$ | $CO_{2\ eq.}$ (kg) | $8{,}04\times10^{-4}$ |
| **Graphite production** | Energy (MJ) | 163 | Graphite (kg) | 1 |
| | | | $CO_{2\ eq.}$ (kg) | 14,74 |
| **Graphene oxide production** | Energi (MJ) | 1,70 | GO (kg) | 1,90 |
| | $H_2SO_4$ (kg) | 12,67 | $SO_{2\ eq.}$ (kg) | 12,93 |
| | $H_3PO_4$ (kg) | 1,29 | $CO_{2\ eq.}$ (kg) | 0,33 |
| | Graphite (kg) | 1 | | |
| **rGO-modified Kapok fiber production** | Kapok Mentah (kg) | 0,95 | Absorbent (kg) | 1 |
| | rGO (kg) | 0,05 | $CO_{2\ eq.}$ (kg) | 14,93 |
| | Energy (MJ) | 76,19 | | |
| **Waste burning** | Absorbent (kg) | 1 | Energy (MJ) | 12,08 |
| | | | $CO_{2\ eq.}$ (kg) | 1,09 |

The LCA in this report aims to compare the environmental impacts of two types of oil absorbents: rGO-KF aerogel and polypropylene oil boom [13]. The scope of comparison is the greenhouse gas and acid emissions in units of kg of $CO_2$ equivalent and kg of $SO_2$ equivalent, respectively. **Table 4** summarizes the comparative LCA between rGO-KF aerogel and synthetic PP oil boom. Based on the value of carbon gas emissions from LCA, it shows that the utilization of the rGO-KF aerogel is beneficial for the environment because it reduces $CO_2$ gas emissions by 32.8 tons/ton absorbed oil.

**Table 4.** Comparative LCA between rGO-KF aerogel and synthetic PP oil boom.

| Comparative factors | rGO-KF aerogel | Synthetic PP [13] |
|---|---|---|
| **Oil sorption capacity (kg)** | 1000 | 1000 |
| **Required absorbent (kg)** | 21,46 | 68,83 |
| **Emission (kg eq. $CO_2$)** | -32814,89 | 536,37 |
| **Energy consumption (MJ)** | 1575,41 | 1863,49 |
| **Acidification (kg eq. $SO_2$)** | 13,89 | 3,34 |

## 4. Conclusion

rGO strengthens the structural integrity of the kapok fibers, enhancing its oil selectivity, retention, and recovery of the fiber while kapok fiber reduces the acid and other unwanted chemical wastes generation caused by the production of pure rGO aerogel. rGO is chemically compatible with the surface of oxidized kapok fiber; thus, to achieve homogeneous coverage, pretreatment of kapok fiber with a strong oxidation agent (e.g., $NaClO_2$) is necessary. Modified Hummer method (Marcano method) to oxidize graphite and ascorbic acid to reduce graphene oxide potentially reduce the acid waste and eliminate the toxic gases (i.e., $NO_2$ and $N_2O_4$) as well as hydrazine footprint from the rGO synthesized process. With 17% oil recovery, the rGO-modified kapok fiber is economically viable with a potential GPM greater than 36,27%. Furthermore, the use of rGO-KF aerogel can reduce CO2 gas emissions in the environment by 32.81 tonnes / 1 ton of absorbed oil.

## References

[1] Zheng Y, Wang J, Zhu Y, Wang A 2015 Research and application of kapok fiber as an absorbing material: A mini review *J. Environ. Sci.* **27** 21-32

[2] Abdullah M A, Rahmah A U, Man Z 2010 Physicochemical and sorption characteristics of Malaysian Ceiba pentandra (L.) Gaertn. as a natural oil sorbent *J. Hazard. Mater.* **177** 683-91

[3] Wang J T, Zheng Y A, Kang Y R, Wang A Q 2013 Investigation of oil sorption capability of PBMA/SiO2 coated kapok fibre. *Chem. Eng. J.* **223** 632-7

[4] Wang J T, Zheng Y A, Wang A Q 2012 Superhidrofobik kapok fibre oil-absorbent: Preparation and high oil absorbency. *Chem. Eng. J.* **213** 1-7

[5] Lugito G, Kustiana A, Martuani R, Wenten I G 2020 Kapok fibre as potential oil-absorbing material: Modification mechanism and performance evaluation *IOP Conf. Series: Materials Science and Engineering* **823** 012033

[6] Luo Z, Li D, Tan S, Huang L 2020 Preparation and oil-water separation of 3D kapok fiber-reduced graphene oxide aerogel *J. Chem. Tech. & Biotech.* **95** 639-48

[7] Ahmad A L, Chan C Y, Abd Shukor S R, Mashitah M D 2008 Recovery of oil and carotenes from palm oil mill effluent (POME) *Chem. Eng. J.* **141** 383–6

[8] Husnah M, Fakhri H A, Rohman F, Aimon A H, Iskandar F 2017 A modified Marcano method for improving electrical properties of reduced graphene oxide (rGO) *Mater. Res. Express* **4** 064001

[9] Arvidsson R, Kushnir D, Sanden B A, Molander S 2014 Prospective Life Cycle Assessment of Graphene Production by Ultrasonication and Chemical Reduction *Environ. Sci. Technol.* **48** 4529-36

[10] Andrijanto E, Shoelarta S, Subiyanto G, Rifki, S 2016 Facile synthesis of graphene from graphite using ascorbic acid as reducing agent *AIP Conference Proceedings* **1725** 020003

[11] Yusmartato Y, Parinduri L, Hasibuan A, Ramayulis, Sulaiman O K 2018 Optimize effort oil collection in fat pit palm oil factory tanjung seumantoh *Inter. J. Eng. Tech.* **7** 41-3

[12] Mansor M R, Salit M S, Zainudin E S, Aziz N A, Ariff H *Life Cycle Assessment of Natural Fiber Polymer Composites* In: Hakeem K, Jawaid M, Alothman OY. (eds) Agricultural Biomass Based Potential Materials. Springer, Cham. 121-41

[13] Harding K G, Dennis J S, Blottnitz H V, Harrison S T L 2017 Environmental analysis of plastic production processes: Comparing petroleum-based polypropylene and polyethylene with biologically-based poly-β-hydroxybutyric acid using life cycle analysis *J. Biotech.* **130** 57-66

**Acknowledgement**
This research has been financially supported by Osaka Gas Foundation 2019-2020 as well as Research Programs, Community Services, and Innovation, Institut Teknologi Bandung (P3MI ITB).

# Lipase immobilization onto Cellulose Nanocrystals (CNC) for catalyzing lipolysis of triglycerides

**Elvi Restiawaty[1,2], Fauzatu Arabica Yatasya[2], Ellys[2], Neng Tresna Umi Culsum[1], Akhmaloka[3], Yogi Wibisono Budhi[1,]***

[1]Department of Chemical Engineering, Faculty of Industrial Technology, Institut Teknologi Bandung
[2]Department of Bioenergy Engineering and Chemurgy, Faculty of Industrial Technology, Institut Teknologi Bandung
[3]Department of Chemistry, Faculty of Mathematics and Natural Sciences, Institut Teknologi Bandung
[4]Biosciences and Biotechnology Research Center, Institut Teknologi Bandung
[5]Research Center for Nanosciences and Nanotechnology, Institut Teknologi Bandung

*Corresponding author's e-mail: y.wibisono@che.itb.ac.id

**Abstract.** Over the past decade, the technology of enzyme immobilization has been developed because it is able to produce reusable immobilized enzymes to reduce production costs. This research aims to synthesize cellulose nanocrystals (CNC) from sugarcane bagasse. The CNC was then used as a matrix in immobilization of lipase to be applied in triglyceride lipolysis. Lipase was immobilized onto CNC through covalent bonding method at 25°C dan pH 7 with immobilization period variation of 0.5–4.5 hours. The immobilized lipase will be utilized to catalyze lipolysis reaction of triglyceride from palm oil, with lipolysis period 2–10 hours, reaction temperature of 25–60°C, and pH of 6.25–11.25. Lipase was successfully immobilized onto CNC so it can be utilized in lipolysis of palm oil with degree of hydrolysis about 10%. The optimum immobilization time and reaction time of lipolysis was 1.5 hours and 6–8 hours, respectively. Both free lipase and immobilized lipase had the optimum temperature at 40°C. In contrast to the optimum temperature, the optimum pH of the lipolysis reaction using free lipase is different from that of lipolysis using immobilized lipase. The optimum pH of lipolysis using free lipase is pH 7.25, while the optimum pH for reaction using immobilized lipase is shifted to pH 8.25.

## 1. Introduction

Free fatty acids (FFA) are aliphatic monocarboxylic acids with 4 to 28 carbon atoms. FFA are intermediate compounds that are widely used in industries such as the soap, cosmetics, pharmaceutical and bioenergy industries. During this time, FFA are produced from the hydrolysis of vegetable oils by the Colgate-Emery process which requires large amounts of energy (250°C, 4.82 MPa) [1]. To overcome these problems, enzymatic hydrolysis technology (lipolysis) was developed. The use of lipase in hydrolysis reactions can take place at low temperatures and pressures. However, the use of enzyme on an industrial scale tends to require high production costs due to expensive enzyme prices. This is because the enzyme purification process is difficult and tends to require expensive operating costs [2].

Over the past decade, the development of enzyme immobilization has been a great research topic. Immobilized enzymes exhibit high stability in high range of operation conditions, such as temperature, and pH. In addition, the immobilized enzymes can be reused because they can be easily separated from the product. Moreover, the reusability of immobilized enzymes might reduce the production costs in enzymatic process industry. One of the important factors in the immobilization of this enzyme is a matrix to support the enzyme. Most matrices tend to be difficult to decompose in nature, so their use is not environmentally friendly. There are even several types of matrices that exhibit high toxicity [3]. Choosing the right matrix material is important to be considered. Organic materials that are more environmentally friendly and low in toxicity are attractive to be used as a matrix. One of the potential organic materials to be used as a matrix or enzyme support is cellulose nanocrystallines (CNC) [4]. CNC exhibits some superior material properties, such as low density, high mechanical properties, high surface area, and high biocompability [5–7].

The development of current immobilization technology research tends to lead to the immobilization of enzymes by the covalent bonding method in inorganic supports. One of the study is conducted by Xie and Ma (2009), immobilization of the lipase from *Thermomyces lanuginose* by the covalent bonding method on $Fe_3O_4$ nanoparticles matrix. The results from the study is yield of immobilization obtained by 84% [8]. Along with the development of renewable-based materials, the researchers developed immobilization technology on organic supports. Research conducted by Kim et al. (2012), immobilization of the lipase from *Candida rugosa* by entrapment method on a 5% cellulose-biopolymer composite hydrogel buffer resulting in an immobilization yield of 24% [9]. Based on these two studies, it shows that immobilization of enzymes using the covalent bonding method is very potential. However, immobilization of the lipase in organic matrix by the covalent bonding method is still rarely performed. Therefore, this study was conducted to examine the covalent bonding method in the process of immobilizing lipase onto CNC for triglyceride lipolysis.

## 2. Methodology

### *2.1. Materials*

The main raw material used in CNC synthesis is sugarcane bagasse. CNC was synthesized on a laboratory scale using the TEMPO ((2,2,6,6-Tetramethylpiperidin-1-yl) oxyl) oxidation method [10]. In TEMPO oxidation, the oxidant used was NaOCl to oxidize the alcohol group into carboxylic group. Lipase type VII originally obtained from *Candida rugosa*, N-hydrosuccinimide (NHS) and 1-Ethyl-3-(3-dimethylaminopropyl) carbodiimide (EDC) were from Sigma Aldrich. Other materials used were materials for enzyme activity analysis and lipolysis. The lipolysis substrate used was Refined Bleached Deodorized Palm Oil (RBDPO) (Bimoli, produced by PT Salim Ivomas Pratama Tbk.). For the titration analysis, potassium hydroxide in ethanol was used. Universal indicators used were phenolphthalein indicators (lipolysis) and thymolphthalein indicators (enzyme activity).

### *2.2. Preparation of activated cellulose nanocrystals*

Before being used to immobilize the lipases by covalent binding, the CNC needed to be activated. The inactivated CNC as amount of 2.5 mg was added into 9 mL of 0.1 M phosphate buffer (pH 7.0) and dispersed under digital ultrasonic for 60 min. The suspension was mixed with 2.3 mL of NHS (or equivalent to 50 mg/mL) and 1.2 mL of EDC (or equivalent to 10 mg/mL) and then the mixture was stirred for 30 min. pH was maintained in the range 7.5–8.0 by dropping 1.5–2.0 mL of 0.5 M NaOH/HCl. After centrifugation for 10 min at 5,000 rpm, the supernatant was removed. The product was washed with distilled water three times.

### *2.3. Immobilization of lipase on cellulose nanocrystals*

For the preparation of lipase solution, 1 mg of lipase powder was added into 0.125 mL of 0.1 M phosphate buffer (pH 7.0). Activated CNC was mixed with 9.9 mL of 0.1 M phosphate buffer (pH 7.0) and dispersed under digital ultrasonic for 60 min. The suspension was mixed with lipase solution at 320 rpm in the shaker for certain time (0.5 h, 1.5 h, 2.5 h, 3.5 h, and 4.5 h). The mixture was centrifuged at 2,000 rpm for 5 min and then the supernatant was removed to remove unbound enzyme. Additionally, the celluloses were washed with 0.1 M phosphate buffer (pH 7) two times. Prepared immobilized lipase were stored at 4°C until use.

### *2.4. Lipolysis triglycerides*

Immobilized lipase was added with 0.5 mL of Ammonia-$NHCl_4$ buffer (pH 8.25). The solution was mixed with 0.5 g of palm oil. The reaction mixture was agitated for 6 h at 45°C and 200 rpm in incubator shaker. For analysis, lipolysis product was settled for 5 min and two layers are formed. Sample was pipetted from upper layer for ± 1 g. Sample was solved in 0.5 mL of ethanol and 0.5 mL of chloroform. Sample was titrated with 0.1 N ethanolic KOH. Degree of hydrolysis was determined with Equation (1) [11].

$$Degree\ of\ hydrolysis(\%) = \frac{V_{KOH} \times 10^{-3} \times N_{KOH} \times MM}{m \times f} \times 100 \quad (1)$$

$V_{KOH}$ is volume of ethanolic KOH needed for titration (mL), $N_{KOH}$ is normality of ethanolic KOH (N), *MM* is molar mass of fatty acid in palm oil (g/mol), $m$ is sample mass (g) and $f$ is fraction of fatty acid in palm oil.

## 3. Result and discussion

### *3.1. The effect of mixing time in lipase immobilization through EDC/NHS crosslink method*

The mixing time between the enzymes dissolved in the phosphate buffer and the CNC during the lipase immobilization process can affect the amount of enzymes bound in the CNC [12]. The amount of bound enzymes will affect the enzyme activity. The effect of mixing time between soluble enzymes and CNC during the lipase immobilization on the enzyme activity was investigated. The enzymatic activities of lipase immobilized in CNC that was observed at each different mixing time are shown in Figure 1.

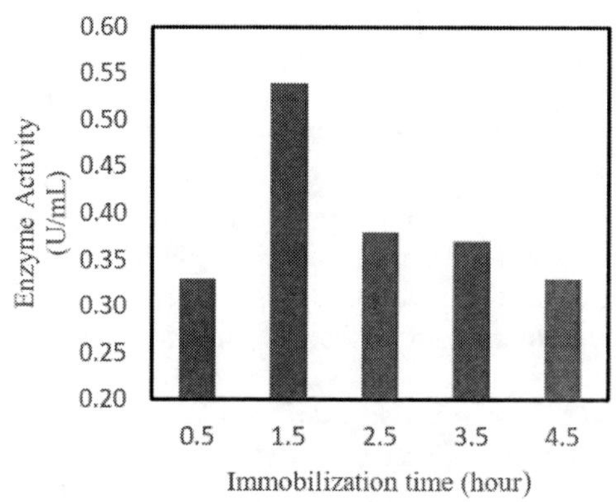


**Figure 1.** The effect of immobilization time to the lipase activity.

The mixing time of 0.5 hours between lipase and CNC during the immobilization process indicated the presence of enzymes attached to the CNC or in other words, enzyme immobilization occurred. It was proved that immobilized lipase activity was detected at 0.33 U/mL for the enzyme that had been immobilized for 0.5 hours (see Figure 1). A longer immobilization time can result in more enzyme binding to CNC. Immobilization time of 1.5 hours showed higher enzyme activity (i.e. 0.54 U/mL) than immobilization time of 0.5 hours. There was an increase in enzyme activity by 64% with an increase in time for one hour. However, an increase in immobilization time longer than 1.5 hours did not indicate an increase in enzyme activity. This is because the longer the immobilization time, the more lipases are bound to CNC, but the lipases that are bound close to each other cause steric hindrance on the active site of the lipase [10]. Steric hindrance on the lipase active site may cause some of the substrate has no direct contact with the lipase active site so that the interfacial activation does not occur [13]. According to the results, it is concluded that the optimum immobilization time is obtained at a duration of 1.5 hours. This is consistent with the results reported by Chen et al., (2016). They reported that a favorable immobilization time for lipases derived from Candida rugosa was between 1 hour and 2 hours [14].

*3.2. The effect of lipolysis condition using immobilized lipase*

The results of experiments on the effect of lipolysis condition can be seen in Figure 2. Constant variable used in this study were 40°C, pH 8.25 and 6 hours lipolysis time. Generally, the reaction rate will increase with increasing the reaction temperature. However, reactions that utilize enzymes as biocatalysts may have limited activity in high operating temperature [15]. Based on Figure 2 (a), free lipase and immobilized lipase had the optimum temperature at 40°C. Babaei et al. (2014) reported that the optimum temperature for immobilized lipase from *Candida sp.* in magnetic nanoparticles and free lipase were at 40°C [16]. At a temperature of 50°C, the decrease in activity on free lipase is greater than immobilized lipase, which is 9.53%, while the decrease in activity on immobilized lipase is 1.69%. Meanwhile the reaction temperature at 60°C also showed an indication of decreased in the degree of hydrolysis of the free lipase, which was 8.63% and 3.92% for immobilized lipase. This is because there is a CNC matrix that prevents changes in protein structure or reduces the flexibility of enzyme conformation so that the denaturation process will only occurs when the enzyme receives sufficient energy such as heating treatment [17]. But some of the active sites of the lipase were blocked or covered by the CNC matrix and the substrate did not reach the active sites of enzyme which is covered by CNC matrix. Thus, the activity for immobilized lipase is lower than the free lipase.

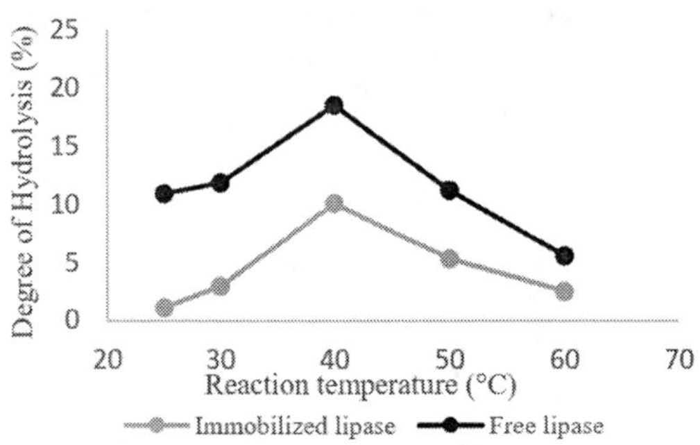


(a)

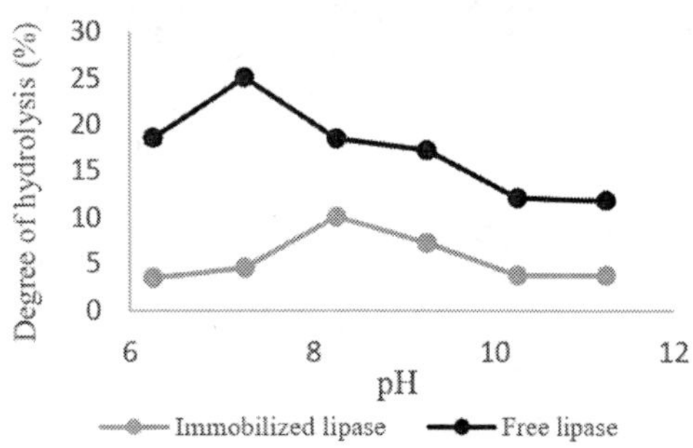


(b)

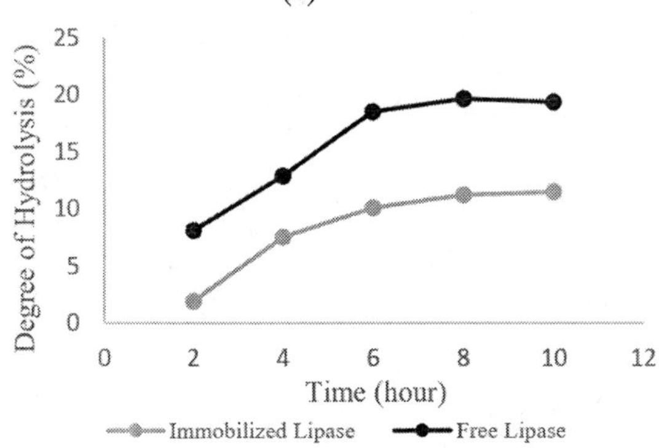


(c)

**Figure 2.** The effect of (a) reaction temperature, (b) pH, (c) reaction time to degree of hydrolysis of free lipase and immobilized lipase.

The general profile of the pH dependency of free lipase and immobilized lipase was very similar above pH 8.25 based on observations shown in Figure 2 (b). At pH 8.25, the degree of hydrolysis using free enzymes is 1.83 times higher than immobilized enzymes. Free lipase was optimal at pH 7.25 while immobilized lipase was optimal at pH 8.25. This is consistent with the results reported by Atiroglu (2018) that free lipase was optimal at pH 7 while immobilized lipase was optimal at pH 8 [18]. The shifted optimum pH for immobilized lipase was probably caused by covalent bonds between the enzyme and matrix [19]. Covalent bonds cause ionic interactions between the anion compounds from CNC and cation compounds from lipase. Meanwhile carboxyl anion compounds from CNC are more dominant at basic pH [20]. Based on Figure 2 (b), the degree of hydrolysis of lipase immobilized at CNC is more stable than free lipase at acidic and basic pH. So, the advantage of lipase immobilized onto CNC is that it can be used in variety of chemical reactions that require a variety of pH conditions [21].

The effect of reaction (lipolysis) time was analyzed by measuring the degree of hydrolysis in various time and it was illustrated in Figure 2 (c). Based on Figure 2 (c), the degree of hydrolysis produced using free lipase was higher than immobilized lipase. This is probably caused by the differences in the amount of enzymes that react in lipolysis and due to the modification of amino acids from lipase which is not preferred [22]. Based on the results in Figure 2 (c), the longer the reaction time, the greater the degree of hydrolysis produced. The optimum degree of hydrolysis is showed after 6–8 hours lipolysis time. This optimum time range of lipolysis was also reported by Fadiloglu & Soylemez (1998) and Knezevic et al. (2002) [23, 24]. After 8 hours, the degree of hydrolysis of free lipase and immobilized lipase tend to be constant. This may due to chemical equilibrium between the substrate and the product or the accumulation of glycerol on the CNC surface, thereby reducing the active phase of the lipase. In addition, this can also be caused by the low rate of hydrolysis of monoglycerides by the lipase [25].

## 4. Conclusion

Lipase was successfully immobilized onto CNC so it can be utilized to obtain degree of hydrolysis of 10.1%. Optimum immobilized lipase activity was obtained in 1.5 hours for immobilization at pH 7. The proportion of arginine was lower than lysine in the amino acid of *Candida rugosa* lipase, which causes the lipase tend to be more active and more stable at low temperatures. There is an optimum pH shift from pH 7.25 for free lipase to pH 8.25 for immobilized lipase. This was because the carboxylic anion from CNC is more dominant at basic pH. The chemical balance between the substrate and the product in the lipolysis reaction was obtained after 6–8 hours.

## 5. Acknowledgments

Authors wish to acknowledge Program Riset ITB 2019 for financial support and the Asahi Glass Foundation, Japan year 2019–2020.

## 6. References

[1] Barnebey H L and Brown, A C 1948 *J. American Oil Chemists' Societ.* **25** 95-99
[2] Raja S, Murty V R, Thivarahan V, Rajasekar V and Ramesh V 2012 *Sci. Technol.* **1** 7–16
[3] Chiang C L, Sung C Sm Wu T F, Chen C Y and Hsu C Y 2005 *J. Chromatogr. B. Anal. Technol. Biomed. Life Sci.* 822 54–60
[4] Budhi Y W, Fakhrudin M, Culsum N T U, Suendo V and Iskandar F 2018 *IOP Conf. Ser.: Earth Environ. Sci.* **105** 012063
[5] Wibowo A, Madani H, Judawisastra H, Restiawaty E, Lazarus C and Budhi Y W 2018 *IOP Conf. Ser.: Earth Environ. Sci.* **105** 012059
[6] Brinchi L, Cotana F, Fortunati E and Kenny J 2013 *Carbohyd. Polym.* **94** 154–69
[7] Li B, Xu W, Kronlund W, Maattanen A, Smatt J H, Peltonen J, Willfor S, Mu X and Xu C 2015 *Carbohyd. Polym.* **133** 605–12
[8] Xie W and Ma N 2009 *Energy fuels* **23** 1347–51
[9] Kim M H, An S, Won K, Kim H J and Lee S H 2012 *J. Mol. Catal. B* **75** 68–72

[10] Zhang D H, Yuwen L X, Li C and Li Y Q 2012 *Bioresour. Technol.* **124** 233–6
[11] Rooney D and Weatherley L R 2001 *Process Biochem.* **36** 947–53
[12] Vashist S K 2012 *J Mol Diagn.* **2** 23-33
[13] Ortiz C, Ferreira M L, Barbosa O, Santos J C S, Rodrigues R C, Murcia A B, Briand L E and lafuente R F 2019 *Catal. Sci. Technol.* **9** 2380–20
[14] Chen Y W, Lee H V and Juan J C 2016 *Carbohydr. Polym.* **151** 1210–19
[15] Ho J, Bong S and Woo S 2011 *Bioresour. Technol* **102** 2105–08
[16] Babaei M, Karimi A and Hejazi M A 2014 *Chem. Ind. Chem. Eng. Q.* **20** 371–8
[17] Parikh D T, Lanjekar K J and Rathod V K 2019 *Biotechnol. Lett.* **10** 1163–75
[18] Atiroglu V 2018 *J. Bio. Macro.* **145** 456–65
[19] Goswami D, Basu J K and De S 2009 *Biotechnol. Bioprocess Eng.* **14** 220–4
[20] Zhu Q, Wang Y, Li M, Liu K, Hu C, Yan K, Sun G and Wang D 2017 *Sep. Purif. Technol.* **186** 70–77
[21] Akova A and Ustun G 2000 *Biotechnol. Lett.* **22** 355–9
[22] Montero S, Blanco A, Virto M D, Landeta L C, Atud I, Solozabal R, Lascaray J M and Renobales M 1993 *Enzyme Microb. Technol* **15** 239-47
[23] Fadiloglu S and Soylemez Z 1998 *J. Agric. Food Chem.* **46** 3411–14
[24] Knezevic Z, Bobic S, Milutinovic A, Obradovic B, Mojovic L and Bugarski B 2002 *Process Biochem.* **38** 313–18
[25] Kang S T and Rhee J S 1989 *Biotechnol. Bioeng* **33** 1469–76

# Utilization of Kaolin as a Filling Material for Rubber Solid Tire Compounds for Two-wheeled Electric Scooters

**Nasruddin[1], Sri Agustini[2], Muhammad Sholeh[3]**

[1,2]Palembang Institute for Industrial Research and Standardization, Ministry of Industry. Jl. Prindustrian II. No. 12 K. 9. Palembang, Indonesia.
[3]Center for Leather, Rubber and Plastics, Jl. Sokonandi No. 9 Yogyakarta, Indonesia

*Corresponding author. e-mail: nas.bppi@gmail.com

**Abstract.** Kaolin is a mining material widely used as a filler in the manufacture of various types of finished goods made from natural rubber. Kaolin in this study was used as a special filler for solid tires for two-wheeled electric scooters commonly used by children. The purpose of this study was to determine the effect of the kaolin loading on the curing properties of the rubber compound, the mechanical properties of the vulcanizate, and the thermal properties of the rubber vulcanizate. Kaolin loading was varied from 22 to 42 per hundred rubber. The compounds were mixed on a two-roll mill. The vulcanization process took place at 140 °C for 17 minutes. The test results showed that kaolin loading had a strong effect on the maximum torque, scorch time, optimum vulcanization time, and mechanical properties of the vulcanizate. Thermo gravimetric analysis showed that the thermal stability of the vulcanizate was influenced by kaolin loading.

## 1. Introduction

Natural rubber (NR) is a polymer widely used for various purposes including tires. It is a non-polar polymer with double crosslinks, which has high flexibility but not resistant to non-polar solvents, ozone and oil [1-2]. The mechanical properties of NR can be improved by adding synthetic rubbers and additives. The advantages of synthetic rubber are oil, ozone, and acid resistance, but not elasticity. The combination of NR with synthetic rubber at a certain loading can improve the mechanical properties by producing a better product.

Solid tires are a product that has been developed for a long time. The solid tires for special purposes such as forklifts are generally made of NR which is composite with synthetic rubber. On the other hands the solid tires for children's bicycles, patient wheelchairs, two-wheeled electric scooters for children's, conveyances, are generally made of synthetic rubber such as styrene-butadiene rubber (SBR), butadiene nitrile rubber (NBR), polyurethane. The weakness of solid tires from synthetic rubber is stiff, and it is easy to suffer physical damage. This is because synthetic rubber is not elastic like NR, when it crosses the road the friction will result in a vibration. By looking at the mechanical properties of NR, this study used NR combined with SBR, by using fillers consisted of kaolin, $CaCO_3$ and silica.

The development of solid tires using NR with various types of fillers has been carried out. Nasruddin and Aprillena [3] have researched to make solid tire vulcanizates for forklifts from NR materials with SBR, butadiene rubber (BR), chloroprene rubber (CR), and ethylene propylene diene monomer (EPDM). Nasruddin [4] has made solid tires for wheelchairs using NR without synthetic rubber with fillers of silica, fly ash and kaolin. Rukkur et al [5] developed a heat-resistant solid tire from NR with carbon-N-330 and silica (VN3) fillers.

NR as the material for two-wheeled solid tires for electric scooters used by children in this study was combined with styrene-butadiene (SBR) rubber. The combination of NR with SBR can improve the mechanical properties of solid tires. Several studies revealed that SBR can improve abrasion, heat, oil, and crack resistance due to ozone attack [6]. In the case of amorphous elastomer, it does not undergo strain-induced crystallization, increases processability, and improves product mechanical properties [7].

The filler material is one of the ingredients that determine the mechanical properties and cost of the finished rubber product. The loading of Kaolin as the object of research for the two-wheeled motorcycle was varied. While for $CaCO_3$, silica and other additives, the loading is conditioned. Determination of the loading of kaolin intended to obtain better elasticity, abrasion resistance and mechanical properties.

Kaolin is a clay mineral that contains kaolinite which consists of hydrated aluminosilicates. According to Hassan and Abdu [8] kaolin is white, soft, crystalline mineral with porous cell structure, irregular shape, containing $SiO_2$, muscovite, quartz, feldspar, and anaphase. Yahaya et al [9] reported that the particle size of kaolin was between 25-35 μm, and 0.4–0.75 μm. Widodo et al [10] stated that kaolin consisted of 65.78% $SiO_2$, 19.55% $Al_2O_3$, 0.90% $Fe_2O_3$ and 8.29% loss of ignition (LOI).

Kaolin with the formula $Al_2O_7Si_2$ or $Al_2Si_2O_5(OH)_4$ [11-12] can be used as a functional filler [13]. The use of kaolin for paint mixtures, ceramic materials has been widely used [14]. Kaolin in the manufacture of rubber finished goods is used as a non-reinforcing filler which does not crystallize [15]. Fillers are used to reducing rubber content, reduce costs, strengthen or lighten compounds, increase the number of chains, modify the physical properties of rubber, and to improve the mechanical properties of rubber [16-17].

The use of kaolin as a filler for rubber compounds has been widely used by previous researchers. Nasruddin and Susanto [18] used kaolin as a filler in NR, NBR and EPDM composites. According to Surya et al, [19] the addition of 5 per hundred rubber (phr) kaolin with alkanolamide can increase the cross-link density of NR compounds. Wu and Tian [20] reported that kaolin with a plate-like structure can reduce rubber heat, as a substitute for silica for certain rubber products.

Several researchers reported the effect of kaolin on the mechanical properties of rubber vulcanizates. It can be used as a reinforcing filler in NR and SBR composites [21-24]. Jincheng et al [25] reported kaolin has better tensile strength than carbon black. According to Peter et al [26] kaolin and silica modified with prepolymer phosphorylated cashew shell aqueous solutions improved the tensile properties and thermal stability of natural rubber.

Kaolin and $CaCO_3$ are used as volume-enhancing fillers and silica as reinforcing fillers. $CaCO_3$ as a filler is widely used for the manufacture of rubber compounds. It can increase the modulus of elasticity, compressive strength, high tensile strength and polyethylene density [27]. $CaCO_3$, type of polar filler, is easier to spread in polar rubber than non-polar rubber [28]. The addition of filler in the elastomeric matrix to improve physical properties and simplify the production process.

Silica as an inorganic reinforcing filler has been widely used [29-35]. The combination of reinforcing fillers with fillers that increase volume, in addition to reducing production costs, can simultaneously improve the mechanical properties of the vulcanizate. To increase filler activity with the NR-SBR matrix, silane-89 was added as a linking agent [30-31, 36-37]. The coupling agent can enhance polymer-polymer and polymer-filler interactions. Carbon black together with titanium dioxide at predetermined loading forms the grey color of solid tires [4].

By looking at the molecular structure, shape and mechanical properties of kaolin, $CaCO_3$, silica as a filler, NR, and SBR as described above, rubber vulcanizate can be used for solid tires for two-wheeled electric scooters for children. Thus, the objectives of this study were to determine the effect of the kaolin loading on the curing properties of the rubber compound, the mechanical properties of the vulcanizate, and mass degradation.

## 2. Material and methods

### *2.1. Material*

The material used in this research is natural rubber (NR) according to the Indonesian Rubber Standard (SIR-20) from one of the crumb rubber factories in Palembang City. Styrene-butadiene rubber (SBR-1500) with mechanical properties of viscosity Mooney ML (1 + 4) 100 between 47-56, elongation at break, 550-750%, reflection elasticity 39% min, tensile strength 27.9 Mpa min. Titanium dioxide ($TiO_2$ - R105), density 4.10 g/cm$^3$, content 95%, the average particle size is 0.3 μm. Black carbon (CB-N330) ash content 0.56%, sulfur content 0.88%. Zinc oxide (ZnO) molecular weight 81.408 g/mol, density 5.606 g/cm$^3$, water-solubility 0.16 mg/100 ml (30 °C). Stearic acid ($C_{18}H_{36}O_2$) einecs no: 166-828-5, cas no: 57-11-4, molecular formula (Mf): $C_{18}H_{36}O_2$, purity 99%, freezing point 53~59 °C. Kaolin 400 mesh size from Tanjung Pandan Belitung Indonesia. Calcium carbonate ($CaCO_3$) cas no. 471-34-1, purity 99%, humidity 0.33%. Silicon dioxide ($SiO_2$) einecs No. 215-683-2, purity 99.8%. Silane-89. Coumarone resin (CR) cas no. 63393-89-5, purity 99%. Paraffinic oil-60. Paraffin wax melting point 136-140 °F, oil content 2%. Butylated hydroxytoluene (BHT) mf: $C_{15}H_{24}O$, purity 99%. 2-mercaptobenzothiazole (MBTS-CAS: 120-78-5). Tetramethyl thiuram disulfide (TMTD) cas no. 137-26-8, mf $C_6H_{12}N_2S_4$, melting point $\leq$ 140 °C, ash (max) 0.30%. PVI (anti scorching agent) cas no. 17796-82-6, mf: $C_{14}H_{15}O_2SN$, purity 98%. Sulphur CAS No: 7704-34-9, purity: 99%.

### *2.2. Methods*

*2.2.1. Rubber compounding.* Five formulation of NR-SBR composites with various kaolin loadings (22, 27, 32, 37, and 42 phr) were prepared (Table 1).

**Table 1.** Compound formulations of solid tire.

| Material | Amount, phr | | | | |
|---|---|---|---|---|---|
| | L-01* | L-02* | L-03* | L-04* | L-05* |
| NR | 92.75 | 92.75 | 92.75 | 92.75 | 92.75 |
| SBR | 7.25 | 7.25 | 7.25 | 7.25 | 7.25 |
| $TiO_2$ | 5.25 | 5.25 | 5.25 | 5.25 | 5.25 |
| CB | 0.15 | 0.15 | 0.15 | 0.15 | 0.15 |
| ZnO | 4.75 | 4.75 | 4.75 | 4.75 | 4.75 |
| Stearic acid | 2.25 | 2.25 | 2.25 | 2.25 | 2.25 |
| Kaolin | 22 | 27 | 32 | 37 | 42 |
| $CaCO_3$ | 25 | 25 | 25 | 25 | 25 |
| Silica | 47 | 47 | 47 | 47 | 47 |
| Silane-89 | 3.25 | 3.25 | 3.25 | 3.25 | 3.25 |
| Coumarone resin | 7 | 7 | 7 | 7 | 7 |
| Paraffinic oil | 8 | 8 | 8 | 8 | 8 |
| Paraffin wax | 2.5 | 2.5 | 2.5 | 2.5 | 2.5 |
| BHT | 1.75 | 1.75 | 1.75 | 1.75 | 1.75 |
| MBTS | 1.25 | 1.25 | 1.25 | 1.25 | 1.25 |
| TMTD | 0.75 | 0.75 | 0.75 | 0.75 | 0.75 |
| PVI | 0.25 | 0.25 | 0.25 | 0.25 | 0.25 |
| Sulphur | 2.85 | 2.85 | 2.85 | 2.85 | 2.85 |

Note: *Ratio of material in per hundred rubber (phr)

Firstly, NR was milled using an open mill until soft, then the SBR was added until homogeneous. $TiO_2$ was added until the color became white, then the CB was added until the color became grey. Next, ZnO and stearic acid were incorporated until homogeneous. Finally, kaolin, $CaCO_3$, silica, silane-89, coumarone resin, paraffin oil, paraffin wax, BHT, PVI, MBTS, TMTD and Sulphur were added subsequently.

*2.2.2. Test method.* The resulting rubber compounds were tested by a moving die rheometer (UR-2010, U-CAN Dynatex Inc, Taiwan) at 150 °C. In order to prepare the vulcanized rubber sheets, the rubber compound of the five treatments were compression molded using a hydraulic hot-press.

Testing of mechanical properties of rubber vulcanization includes: specific gravity (ASTM D.297-15), hardness testing before ageing (ASTM D.2240-15), hardness after ageing ASTM D. 573-04 (2015), compression set, defl 25%, 70 °C, 22 hours (ASTM D.395-16e1), tensile strength before and after ageing (ASTM D.412-16). Mass degradation testing using thermogravimetric analyzer (TGA).

## 3. Results and discussion

### *3.1. Cure Characteristics*

The results of the rubber compound curing characteristics shown in Table 2 provide information on the maturation profile of the compound described by delta torque $\Delta_s$ (Max – Min), curing time ($t_{90}$), and scorch time ($ts_2$). The test results of rubber compounds at 22 phr kaolin loading (L-01) shows the highest $\Delta_s$ (Table 2). The increase in kaolin loading from 27 phr (L-02) to 42 phr (L-05) cause $\Delta_s$ to decreased from 12.57 kg-cm (L-02) to 10.76 kg-cm (L- 05). The $\Delta_s$ illustrates the crosslinks density formed during vulcanization. The fillers dispersed in the polymer matrix affect the interaction between polymer and filler [38], thus affect the compound cure characteristic (Table 2).

**Table 2.** Curing characteristics of NR-SBR composite.

| Curing Characteristic | L-01 | L-02 | L-03 | L-04 | L-05 |
|---|---|---|---|---|---|
| S* Max – Min, kg-cm ($\Delta_s$) | 12.79 | 12.57 | 10.08 | 11.60 | 10.76 |
| S* Maximum, kg-cm | 13.38 | 13.27 | 12.28 | 13.01 | 13.07 |
| S* Minimum, kg-cm | 0.59 | 0.70 | 2.20 | 1.41 | 2.31 |
| Opt cure time ($t_{90}$), min: sec | 8:10 | 8:48 | 0:56 | 1:29 | 1:03 |
| Scorch time ($ts_2$), min: sec | 0:40 | 0:52 | 0:24 | 0:40 | 0:30 |

The cure time ($t_{90}$) illustrates the compound maturation time. The shorter the maturation time of the compound, the cheaper the vulcanization costs. The cure time is the time to form crosslinks during vulcanization [39]. The difference in curing time in Table 2 is influenced by the kaolin loading in the NR-SBR matrix. The curing time is found to be the longest at 27 phr kaolin loading (L-02).

The formation of crosslinking is caused by the reaction of sulphurs with the polymer matrix which is accelerated by activators, co-activators and accelerators. Kaolin, $CaCO_3$, and silica with the help of silane-89 during vulcanization inserted between the NR and SBR molecules contribute to improving the mechanical properties of the vulcanizate. Crosslinking affects hardness, density, compression and other mechanical properties.

The test results of the scorch time ($ts_2$) for formula L-01 (kaolin 22 phr) was 0:40 sec. This value is the same as the formula L-04 (37 phr). At a kaolin loading of 32 phr (L-03) a scorch time (ts2) 0:24 sec is the shortest time among treatments. Considering the kaolin loading (Table 1) and the results of testing the cure characteristic compounds (Table 2), one can conclude that the increase in the loading of kaolin in the NR-SBR compound matrix did not reflect a linear increase or decrease in scorch time ($ts_2$). The effect of filler distribution and crosslinking reaction speed in each treatment affects the difference in scorch time ($ts_2$). The distribution of the filler into the NR-SBR compound was influenced by the rate at which silane-89 softened the polymer matrix. In addition, stearic acid as a coactivator added to the rubber matrix acts as a plasticizer to distribute the filler into the NR-SBR polymer chain.

### *3.2. Mechanical properties of a rubber vulcanizate*

The test results on the mechanical properties of solid tire vulcanization from five formula have been investigated as shown in Figures 1 to 4. The effect of kaolin loading combined with $CaCO_3$ and silica on the NR and SBR matrices gave different results. Kaolin as a filler in the NR and SBR matrices affects the vulcanization process, compound maturation, curing time, and scorch time (Table 2).

In the vulcanization process, kaolin affects the formation of crosslink density which is illustrated by the results of the specific gravity test. An increase in the kaolin loading (Table 1) in NR and SBR vulcanizate affected the curing time and scorch time (Table 2), followed by an increase in specific gravity (Figure 1).

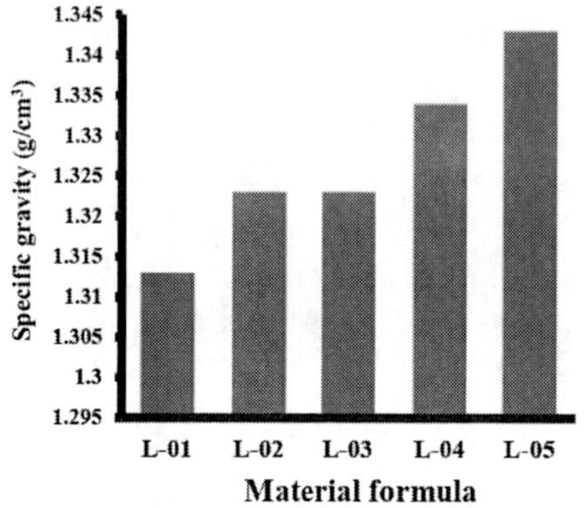


**Figure 1.** Effect of kaolin loading on the specific gravity of solid tires.

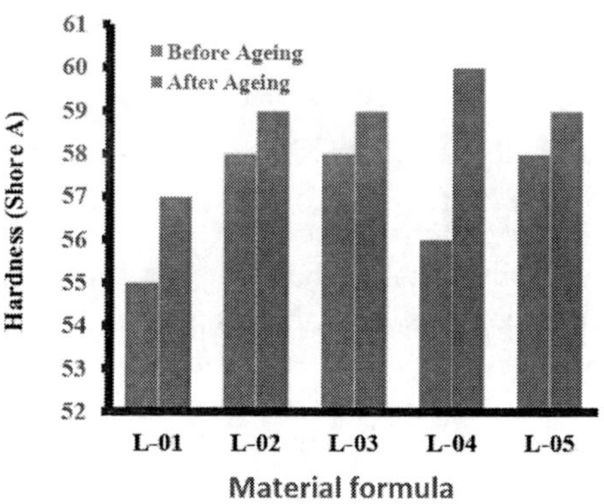


**Figure 2.** Effect of kaolin loading on hardness before and after ageing in solid tires.

NR and SBR matrices with kaolin filler with a loading of 22 phr (L-01) had a lower specific gravity value (1.313 g/cm$^3$) than other formulas. Increasing the loading of kaolin to 27 phr (L-02), and 32 phr (L-03) had the same specific gravity (1.323 g/cm$^3$). At 37 phr (L-04) kaolin loading there was an increase in density (1.334 g/cm$^3$), an increase in the loading of 42 phr (L-05) kaolin in NR and SBR compounds also an increase in density up to 1.343 g/cm$^3$.

The distribution of fillers in rubber composites affects the mechanical properties of rubber vulcanizates [40-41]. This is related to the density of the material linked to the NR and SBR vulcanizates. The higher the specific gravity value, the increase in material density. The change in the value of specific gravity is theoretically followed by a change in the value of hardness which affects the change in the value of the compression set.

The hardness test results for kaolin loading of 22 phr (L-01) before ageing 55 Shore A, after ageing increased (57 Shore A). Increasing the kaolin loading of 27 phr (L-02), increasing the hardness before ageing (58 Shore A), and after ageing (59 Shore A). At the loading of kaolin 32 phr (L-03), 42 phr (L-05) hardness before ageing is the same as hardness L-02, likewise with hardness after ageing L-03 and L-05 the value is 59 shore A equal to hardness before ageing L-02.

For the kaolin loading of 37 phr, the hardness before ageing 56 shore A was lower than the hardness of L-02, L-03 and L-05. In contrary, the hardness after ageing for the loading kaolin of 37 phr was 60 shore A, higher than L-01, L-02, L-03 and L-05. The increase in hardness after ageing in all formulas was due to the shrinkage of the rubber vulcanizate volume during heating. The vulcanization was heated in an oven at a temperature of 72 °C for 72 hours. At this temperature and time, the low boiling point evaporates easily, at the same time there is a narrowing of the rubber vulcanizate volume which causes an increase in hardness. When viewed from the test result data, the facts show that an increase in the loading of kaolin is not necessarily followed by an increase in hardness values (Figure 2). This is probably due to the distribution of the filler, the softener material used, and the imperfect cross-linking reaction. The high hardness value of the two-wheeled electric scooter solid tire for children affects the user's comfort. The solid tires that are too hard will cause shocks when used on uneven roads.

Changes in the value of specific gravity and hardness affect the compression set. If the specific gravity and hardness values are high, the solid tire does not return to its original state when compressed. Theoretically, the filler added to the rubber matrix in a certain loading can fill the space between the rubber molecules. Compression set test data for the loading of kaolin 22 phr (L-01) to 27 phr (L-02) showed an increase from 23.42% to 27.11% following the increase in specific gravity (Figure 1 and Figure 2). At a kaolin loading of 27 phr (L-02) to 32 phr (L-03) followed by an increase in the compression set of 30.82%. At the loading of kaolin 37 phr (L-04) up to 42 phr (L-05) the increase in the compression set value was not significant from 31.01% to be 31.14%. The value of the compression set cannot be separated from the influence of the distribution of fillers (kaolin, $CaCO_3$ and silica) into the NR-SBR rubber vulcanizate. The distribution of fillers in the NR-SBR matrix in this study was accelerated by paraffin oil as a softener, stearic acid, coumarone resin, paraffin wax, and xilane-89 as a coupling agent.

The abrasion resistance is an important parameter for determining the service life of a solid tire, where when a solid tire is susceptible to abrasion its service life is shorter. The bond strength among the solid tire molecules determines the abrasion resistance. Abrasion cannot be separated from the influence of the bonds among molecules that form the rubber vulcanizate structure. In addition, the specific gravity and hardness values describe the density and hardness of the rubber vulcanized structure which affects the abrasion resistance. The higher the specific gravity value, theoretically, the amount of material in space is denser/denser and the level of hardness is high. The high density of a material in space describes the stronger the bonds among the molecules of the material which have an impact on hardness and abrasion resistance.

The results of the abrasion resistance test for each formula (Table 1) increased (Figure 4) along with the increase in the kaolin loading. At the loading of kaolin 22 phr (L-01) there was an abrasion of 201.4 $mm^3$, at the loading of kaolin 27 phr (L-02) there was an increase in the abrasion of 211.5 $mm^3$. Increasing the loading of kaolin to 32 phr (L-03), the abrasion increased to 225.8 $mm^3$, the abrasion increased to 239.6 $mm^3$ at the loading of kaolin to 37 phr (L-04), and the loading of kaolin to 42 phr (L-05) there was an abrasion of 236.5 $mm^3$.

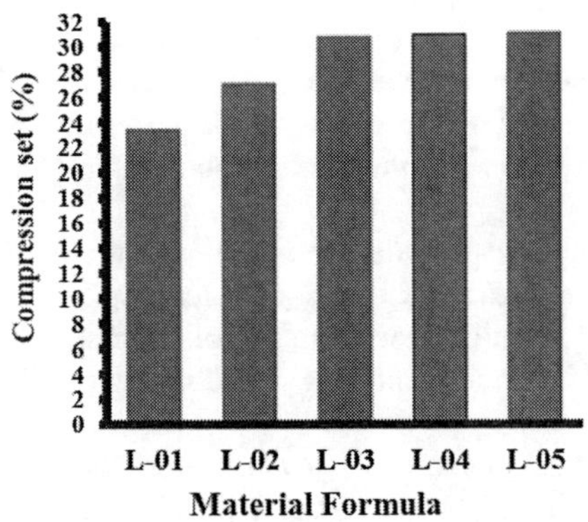


**Figure 3.** Effect of kaolin loading on solid tire compression set.

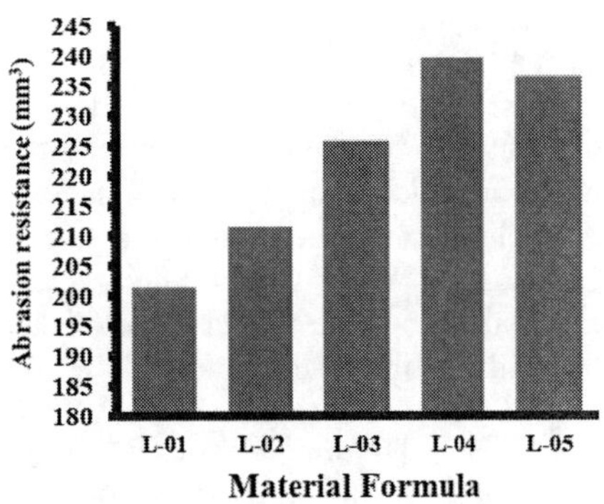


**Figure 4.** Effect of kaolin loading on the abrasion resistance of solid tires.

The increase in abrasion can be caused by weak bonds between the material molecules, and there is excess space between the rubber composite molecules and the filler. Therefore, friction with other objects to experience abrasion quickly. Facts showed the addition of fillers (Table 1) to the rubber matrix must be balanced with the strength of the bonds between molecules (adhesion and cohesion) that form it. One of the ways to increase the bond strength among the material molecules in the polymer matrix can be done by increasing the cross-link reaction with the addition of additives to strengthen the solid structure of the tire.

Determination of tensile strength and tear strength in the rubber vulcanizate is to see the ability of solid tires when experiencing pull and tear from other objects. The tensile strength, tear strength, and abrasion strength are related to the strength of the bonds between molecules, cohesion, and adhesion that form them. Where if there is friction or pulling, it can survive until it breaks and tears. The bond among the molecules of the material is influenced by the homogeneity of the material which is evenly distributed into the rubber matrix and the vulcanization process which forms cross-links.

Tensile strength and tear strength test data as in Figure 5 and Figure 6, are not linear with the increase in the kaolin loading. The results of the tensile strength test for the loading of kaolin 22 phr (L-01) were higher (9.1 MPa) compared to the loading of kaolin 27 phr (L-02) 6.9 MPa, 32 phr (L-03) 7.5 MPa, and 42 phr (L-05) 6.5 MPa. The addition of kaolin with a loading of 37 phr (L-04) was the highest (10.4 MPa) than all formulas (Figure 5).

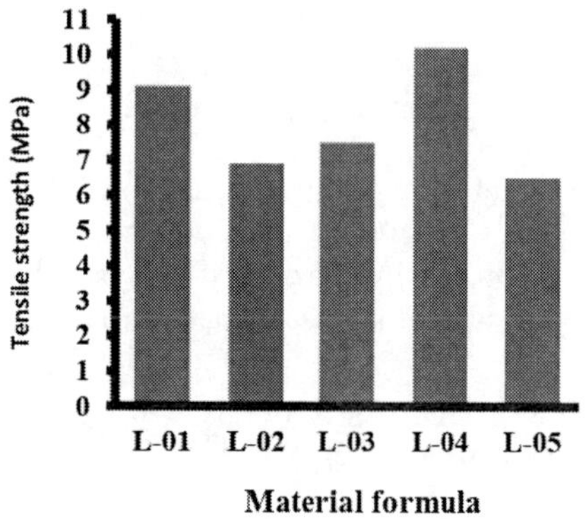


**Figure 5**. Effect of kaolin loading on the tensile strength of solid tires.

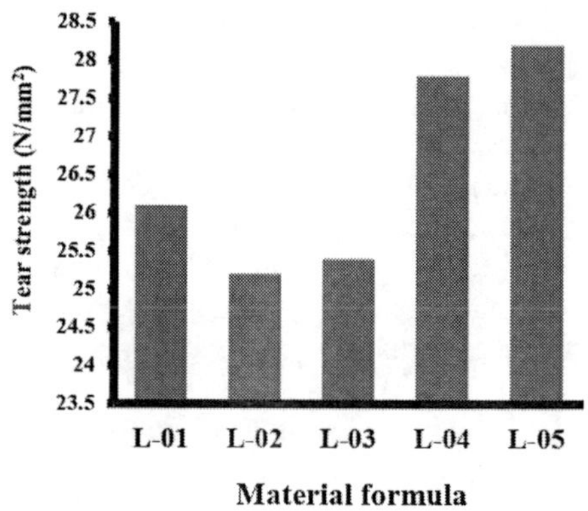


**Figure 6.** Effect of kaolin loading on the tear strength of solid tires.

The changes that occur are caused by changes in the crosslink structure, and cutting the main chain [42]. In addition, the changes in tensile strength are caused by chain scission and an increase in cross-link density [43]. Increasing the loading of kaolin in a mixture of $CaCO_3$ and silica in the NR-SBR matrix causes sideloading which affects the cross-link reaction and the bonds among the molecules of the material which ultimately affect the tensile strength.

As well as the tensile strength test result, the tear strength test results are not linear with the increase in the loading of kaolin (Table 1 and Figure 6). Solid tires vulcanizate with kaolin loading of 22 phr (L-01), tear strength value of 26.1 N/mm$^2$ is higher than that of vulcanizate with a kaolin loading of 27 phr (L-02) and a kaolin loading of 32 phr (L-03), however, it is lower than the kaolin loading of 37 phr (L-04) and 42 phr (L-05) (Figure 6).

Kaolin with the formula $2H_2O.\ Al_2O_3.2SiO_2$ with the SiO-Si composition at different loadings in the rubber matrix affects the space for the polymer and the bonds between molecules. The difference in space is caused by differences in the loading of kaolin that occupies the available space. The more the kaolin loading reduces the space for the polymer and affects the value of the tear strength.

Based on the data from the tensile strength and tear strength test results as described above, the kaolin distributed in the NR-SBR matrix may not be evenly distributed when compounding at the open mill. The spread of kaolin and additives into the NR-SBR matrix during compounding in an open mill affects the tensile strength and tear strength values.

*3.3. Mass degradation analysis*

IOP Conf. Series: Materials Science and Engineering 1143 (2021) 012010 doi:10.1088/1757-899X/1143/1/012010

Mass degradation can be caused by breaking bonds between molecules, induced by other particles causing molecules to spontaneously polar. The polarity of a compound is due to the difference in electronegativity between the atom and the shape of the molecule. Solid tires consist of fillers and additives in the NR-SBR matrix (Table 1). When the bonds among molecules are weak, the mass will naturally be degraded by environmental conditions. The mass of solid tires can be reduced due to abrasion when there is friction with other objects and the influence of heat that damages the surface, slowly towards the center point. The results of the TGA analysis are shown in Figures 7(a)-(d).

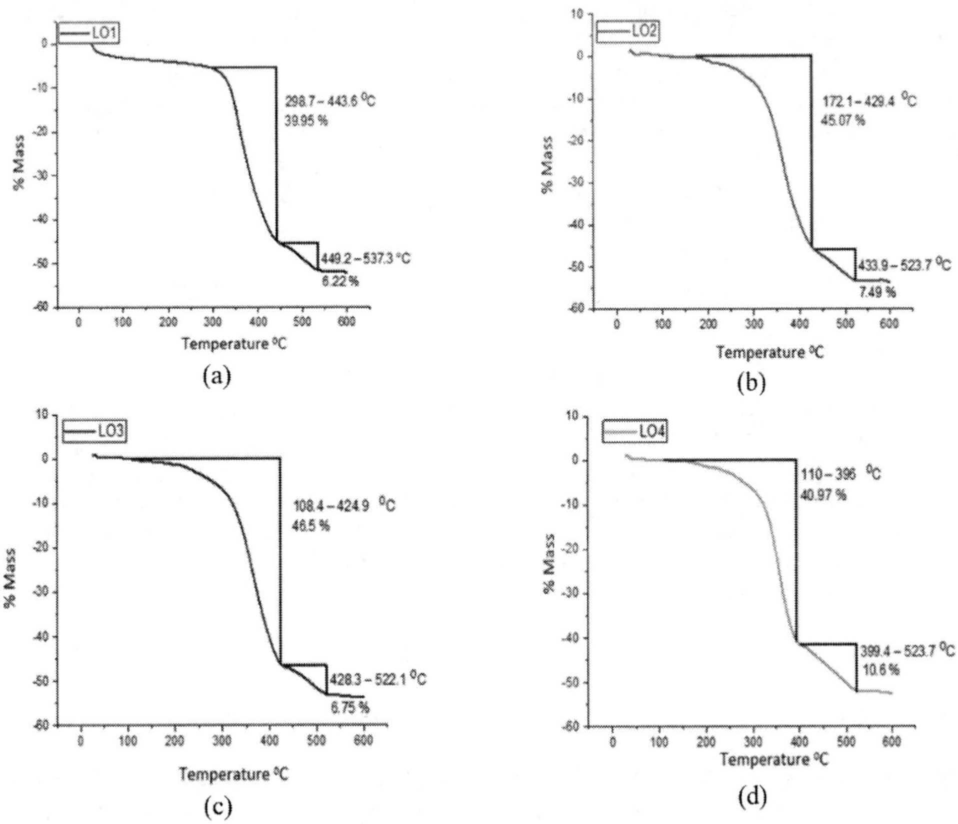


**Figure 7**. Material mass reduction thermogram with (a) 22 phr, (b) 27 phr, (c) 32 phr, and (d) 37 phr of kaolin filler on solid tire vulcanizate.

To determine the mass loss, it was analyzed using the thermo gravimetric method (TGA). The TGA method is used to determine the thermal stability of the volatile component fractions by calculating the change in mass with temperature changes. The measurement of mass degradation using thermogravimetric analysis has been widely used [44-46]. The mass reduction analysis of four solid tire vulcanization formula used the thermo gravimetric method. The test data for the four samples are based on the temperature differences that occur in two stages. Each degradation stage was caused by a different loading of kaolin added to the composite matrix. Theoretically, of all the materials used to make rubber compounds into solid tire vulcanizate, NR-SBR and additives are conditioned except for kaolin. The difference in each stage is influenced by the amount of kaolin that is distributed in the mixture of ingredients.

The results of TGA analysis for sample L-01 (Figure 7(a)) showed that the mass reduction in the first stage occurred in the temperature range of 298.7 °C to 443.6 °C. In this temperature range, there was a massive decrease in mass, namely 39.95%. The mass loss in the first stage comes from volatile elements with a low boiling point below 443.6 °C. The second stage of degradation occurred at a temperature range of 449.2 °C to 537.3 °C with a mass decrease of 6.22%. The mass that is degraded in the second stage is the mass which has a high boiling point, possibly from zinc and titanium.

The decrease in the mass of the L-02 sample (Figure 7(b)) in the first stage occurred at the temperature range of 172.1 to 429.4 °C with a mass decrease of 45.07%. The second stage of degradation occurred at a temperature range of 433.9 - 523.7 °C with a mass of 7.49%The decrease in sample mass with code L-03 (Figure 7(c)) in the first stage occurred at the temperature range of 108.4 °C to 424.9 °C with the mass decrease of 46.5%, while the second degradation occurred at a temperature of 428.3 °C - 522.1 °C with a decrease in mass of 6.75%. The degradation of the sample with the L-O4 code in the first stage occurred at a temperature of 110 - 396 °C with a mass reduction of 40.97%, while the second stage degradation at a temperature of 399.4 - 523.7 °C with a mass decrease of 10.6%.

Based on the mass of degradation of each stage, for the first stage, the four samples were derived from volatile elements such as Silane, paraffin oil, NR, SBR, stearic acid, paraffin wax, coumarone resin. The second stage of the mass reduction comes from the titanium element's, Zn, silica, kaolin, $CaCO_3$, and other additives such as MBTS, TMTD, PVI. Loss of mass due to the decomposition process is characterized by breaking chemical bonds, evaporation, increasing temperature, reducing the interaction between substances with reducing agents and desorption. The TGA curve occurs based on the degradation stage which is limited by the temperature range. Changes in temperature are converted into changes in energy and mass during heating and then plotted against temperature and time.

## Conclusion

The mechanical properties of solid tires are influenced by the constituent materials such as natural rubber, synthetic rubber, additives, fillers and material loadings. Kaolin as a volume increasing filler at different loadings affected the mechanical properties of solid tires. The hardness, compression set, elasticity and abrasion were important parameters to assess the safety and comfort of using solid tires. Test results of curing characteristic for a compound with 27 phr kaolin loading were as follows: delta torque of 12.57 kg-cm, cure time of 8:48 seconds, and scorch time of 0.52 second. The results of mechanical properties testing were as follows: hardness before and after ageing of 58 and 59 Shore A, respectively, compression set test of 25%, and abrasion resistance of 211.5 $mm^3$. The mass degradation for each sample occurs in two stages, each of which is differentiated based on temperature. The difference in the loading of kaolin affects the degradation temperature and mass degradation.

## Acknowledgements

We wish to thank the Head of the Palembang Institute for Industrial Research and Standardization, and the technicians who have helped in this research activity. The authors also thank Director of PT. Shima Prima Utama regarding research collaboration and information on the use of solid tires.

## References

[1] Limhengha S and Sriseubsai W 2018 *Int. J. GEOMATE* **15** 189–96
[2] Zhong B, Jia Z, Hu D, Luo Y, Jia D and Liu F 2017 *Part A Appl. Sci. Manuf.* **96** 129–36
[3] Nasruddin and Bondan A T 2019 *J. Phys. Conf. Ser.* **1282**
[4] Nassruddin 2017 *J. Din. Penelit. Ind.* **28** 20–1
[5] Rukkur S, Dechwayukul C, Thongruang W and Patarapaiboolchai O 2014 *Adv. Mater. Res.* **844** 445–9
[6] Noriman N Z and Ismail H 2013 *Int. J. Polym. Mater. Polym. Biomater.* **62** 252–9
[7] Mathew G, Huh M Y, Rhee J M, Lee M H and Nah C 2004 *Polym. Adv. Technol.* **15** 400–8
[8] Hassan U J and Abdu S G 2014 *Sci. World J.* **9** 33–7
[9] Yahaya S, Jikan S S, Badarulzaman N A and Adamu A D 2017 *Path Sci.* **3** 4001–5
[10] Widodo., Subari and Bagus D 2016 *Indones. Min. J.* **19** 89–96
[11] Aleanizy F S, Alqahtani F, Al Gohary O, El Tahir E and Al Shalabi R 2015 *Saudi Pharm. J.* **23** 167–76

[12] Yahaya S, Jikan S S, Badarulzaman N A and Adamu A D 2017 *Path Sci.* **3** 1001–4
[13] Zhang Y, Liu Q, Zhang S, Zhang Y, Zhang Y and Liang P 2016 *Appl. Clay Sci.* **124–125** 167–74
[14] Dewi R, Agusnar H, Alfian Z and Tamrin 2018 *J. Phys. Conf. Ser.* **1116**
[15] Sheikh S H, Yin X, Ansarifar A and Yendall K 2017 *J. Reinf. Plast. Compos.* **36** 1132–45
[16] Azrem A A, Noriman N Z and Razif M N 2014 *Key Eng. Mater.* **594–595** 867–71
[17] El-Wakil A A and Abd El-Megeed A A 2011 *ARPN J. Eng. Appl. Sci.* **6** 24–9
[18] Nasruddin and Susanto T 2020 *Indones. J. Chem.* **20** 967–78
[19] Surya I, Maulina S and Ismail H 2018 *Malaysian J. Anal. Sci.* **22** 1014–22
[20] Wu W and Tian L 2013 *Appl. Clay Sci.* **80–81** 93–7
[21] Zoromba M S, Belal A A M, Ali A E M, Helaly F M, El-Hakim A A A and Badran A S 2007 *Polym. - Plast. Technol. Eng.* **46** 529–35
[22] Zhang Q, Liu Q, Zhang Y, Cheng H and Lu Y 2012 *Appl. Clay Sci.* **65–66** 134–8
[23] Ahmed N M, El-Nashar D E and Abd El-Messieh S L 2011 *Mater. Des.* **32** 170–82
[24] Zhang Y, Liu Q, Zhang S, Zhang Y and Cheng H 2015 *Appl. Clay Sci.* **111** 37–43
[25] Jincheng W, Yuehui C and Jihu W 2005 *J. Elastomers Plast.* **37** 169–80
[26] Peter R, Vijay V R, Ramakrishnan S, Sukumar R and Menon A R R 2015 *Appl. Clay Sci.* **105–106** 186–91
[27] Jong L 2020 *J. Compos. Sci.* **4** 113
[28] Surya I, Hayeemasae N and Ginting M 2018 *IOP Conf. Ser. Mater. Sci. Eng.* **343** 1–7
[29] Xiong L, Liang H, Wang R and Chen L 2011 *J. Polym. Res.* **18** 1017–21
[30] Hualong Peng, Lan Liu, Yuanfang Luo, Haoqun Hong D J 2009 *J. Appl. Polym. Sci.* **112** 1967–73
[31] Yang S, Liu L, Jia Z, Jia D and Luo Y 2011 *Polymer (Guildf).* **52** 2701–10
[32] Ahmed K, Nizami S S, Raza N Z and Habib F 2013 *J. King Saud Univ. - Sci.* **25** 331–9
[33] Liu Y, Li Z and Du B 2014 *Appl. Phys. Lett.* **105**
[34] ten Brinke J W, Debnath S C, Reuvekamp L A E M and Noordermeer J W M 2003 *Compos. Sci. Technol.* **63** 1165–74
[35] Ismail H and Haw F S 2010 *J. Reinf. Plast. Compos.* **29** 105–11
[36] Kapgate B P, Das C, Basu D, Das A and Heinrich G 2015 *J. Elastomers Plast.* **47** 248–61
[37] Pongdong W, Nakason C, Kummerlöwe C and Vennemann N 2015 *J. Chem.* Article ID 796459, 15 pages
[38] Bandyopadhyay S, De P P, Tripathy D K and De S K 1995 *Polymer (Guildf).* **36** 1979–84
[39] Ogbebor O J, Okieimen F E, Ogbeifun D E and Okwu U N 2015 *Chem. Ind. Chem. Eng. Q.* **21** 477–84
[40] Sui G, Zhong W H, Yang X P and Yu Y H 2008 *Mater. Sci. Eng. A* **485** 524–31
[41] Basfar A A, Abdel-Aziz M M and Mofti S 2002 *Radiat. Phys. Chem.* **63** 81–7
[42] Nakason C, Wannavilai P and Kaesaman A 2006 *Polym. Test.* **25** 34–41
[43] Fan R L, Zhang Y, Li F, Zhang Y X, Sun K and Fan Y Z 2001 *Polym. Test.* **20** 925–36
[44] Pimolsiriphol V, Saeoui P and Sirisinha C 2007 *Polym. - Plast. Technol. Eng.* **46** 113–21
[45] Rohana Yahya Y S, Azura A R and Ahmad Z 2011 *J. Phys. Sci.* **22**(2), 1–14
[46] Rostek E and Biernat K 2013 *J. Sustain. Dev. Energy, Water Environ. Syst.* **1** 163–71

IOP Conf. Series: Materials Science and Engineering 1143 (2021) 012011 doi:10.1088/1757-899X/1143/1/012011

# The effect of nanostructured silica synthesis temperature on the characteristics of silica filled natural rubber composite

**Muhammad Sholeh[1,2,*], Rochmadi Rochmadi[1], Hary Sulistyo[1], Budhijanto Budhijanto[1], Martin Doloksaribu[3]**

[1]Department of Chemical Engineering, Faculty of Engineering, Universitas Gadjah Mada, Yogyakarta, Indonesia
[2]Center for Leather, Rubber and Plastics, Jl. Sokonandi No. 9 Yogyakarta, Indonesia
[3]Metal Industries Development Center, Jl. Sangkuriang No. 12 Kota Bandung, Indonesia

*Corresponding author. E-mail: muhammad-sholeh@kemenperin.go.id

**Abstract.** The processing, curing, mechanical, and morphological characteristics of natural rubber (NR) filled with nanostructured silica (NS) were discussed in this paper. NS was synthesized at 60-90°C using bagasse ash as raw material. NR was mixed with NS using Haake Rheomix equipped with roller rotors. Curing kinetics of NR compounds was studied using the data obtained from a moving die rheometer. Lower compounding temperature and lower compound viscosity were observed with increasing silica synthesis temperature. Highest bound rubber content and modulus at break were found at natural rubber filled with silica synthesized at 70°C. The state-of-mix of the filler was found to increase with increasing silica synthesis temperature. The curing curve obtained showed that the vulcanization kinetics follow autocatalytic model.

## 1. Introduction

About 60% of world rubber production is used in tire manufacturing. Silica addition into tire compound is widely known to improve the tire rolling resistance [1]. Silica is made through three synthesis routes: precipitation of an aqueous sodium silicate solution (precipitated silica), high-temperature hydrolysis of silicon tetrachloride (fumed silica), and hydrolysis of tetraethylorthosilicate [2,3]. The primary particle diameters of precipitated silica and fumed silica range between 10-80 nm and 5-20 nm, respectively. Smaller primary particle size of fumed silica provides a higher level of reinforcement, but its expensive price limits its use only for silicone rubber and other specialty synthetic rubber. Sodium silicate produce more refined and uniform particle size along with the higher silica concentration than tetraethylorthosilicate [4]. As a reinforcing filler for natural rubber (NR), 15 phr (parts per hundred rubber) silica gives the composite a tensile strength of 27 MPa [2].

Many researchers have tried to use biomass-based silica as a rubber filler. Huabcharoen et al. [5] studied the effect of acid treatment of bagasse ash on the mechanical properties of NR composite. The authors reported that the optimum bagasse ash concentration was 15 phr. Higher bagasse ash decreased the mechanical properties of rubber composites. Zeng et al. [6] modified rice husk ash using a rare earth coupling agent. The modified ash was used as a NR filler by the latex process. The highest tensile strength of 25.96 MPa was obtained at 4 wt% ash content and then dropped to 18.42 MPa at 10 wt% ash content. Other researchers also found that purification and/or surface modification only gave biomass ash a role as a semi-reinforcing filler [7–9].

Boonme et al. [4] synthesized silica nanoparticle from bagasse ash by the sol-gel method. The average diameter of the particles obtained was 90 nm. The silica obtained has a high filler-filler interaction and non-homogeneous dispersion in the NR matrix thus gave a negative effect on the mechanical properties of composites at a concentration of 10 phr. Jembere & Fanta [10] applied a xerogel from rice husk ash as a replacement of commercial precipitated silica. At a xerogel level of 12 phr, the increase in tensile strength obtained was only 17.5% compared to the unreinforced rubber matrix. Filler agglomeration was the main problem of the small optimum silica concentration and/or mechanical properties in some of the studies above. Commercial rubber products require a silica concentration of 15-100 phr to fulfill the targeted mechanical properties. Agglomeration at high silica levels needs to be minimized by using silica with sufficient reinforcing properties and the addition of a coupling agent [2].

According to our previous work, nanostructured silica (NS) from bagasse ash has the potential as a rubber reinforcing filler at high silica loading. We found lower water content (as silanol) and higher primary particle size of the nanostructured silica with increasing synthesis temperature [11,12]. Filler polarity affects filler dispersion and filler primary particle size influences the surface area that interacts with the rubber. This study was aimed to extend our previous findings on the use of nanostructured silica from bagasse ash as a reinforcing filler in NR. To the best of our knowledge, no report has yet described in detail the influence of silica synthesis temperature on the processing, curing, mechanical, and morphological characteristics of NR/NS composite.

## 2. Materials and methods

NS was synthesized using bagasse ash as raw material at 60, 70, 80, and 90°C. NR was mixed with NS using Haake Rheomix equipped with roller rotors. NS was synthesized and mixed with NR according to previously published methods as can be seen from Figure 1 [12].

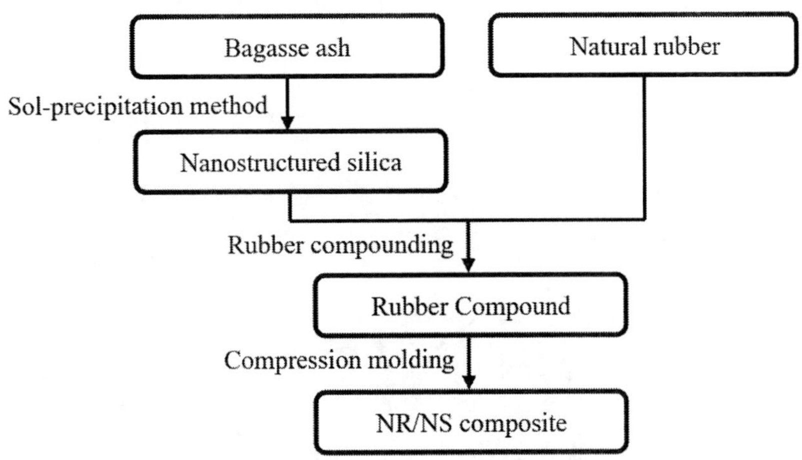


**Figure 1.** Steps of the silica synthesis and mixing with natural rubber.

Vulcanization kinetics was studied at 140, 150, and 160°C on a Gotech 3000A moving die rheometer (MDR) according to ASTM D2084-01. Mechanical properties of the vulcanizate were determined using samples with a thickness of 2 mm obtained by compression molding at a temperature of 150°C under a pressure of 14.7 MPa. Mooney viscosity of the compound was measured by a Mooney shearing disk viscometer (MonTech MV3000) according to ASTM D 1646. Mooney viscosity was reported as ML (1+ 4), 100°C. Bound rubber content (BRC) was calculated using the relation [13]

$$\%BRC = \frac{W_{fg} - W[\frac{m_f}{m_f+m_p}]}{W[\frac{m_p}{m_f+m_p}]} \times 100 \quad (1)$$

where $W$ and $W_{fg}$ were the weight of the sample before and after immersion in toluene, respectively. $m_f$ and $m_p$ were the weight of silica and rubber in the uncured mix, respectively. Morphology of the vulcanizate was characterized using an optical microscope (Arcs Video Measuring System) with a 112.5 magnification factor. The fraction of agglomerates area was determined using imageJ software. The stress-strain behavior of the vulcanizate was monitored according to ASTM D 412-06 with dumbbell-shaped samples using a universal testing machine (Tinius Olsen H25K).

## 3. Results and discussion

The incorporation of NS into NR was performed in a Haake 3000P Rheomix. The torque and compounding temperature in the internal mixer during mixing process was recorded. From Figure 2, it can be seen that the first torque peak curve represents the natural rubber mastication process. The parts for the fast torque increase to the second and third peaks value represent the incorporation of nanostructured silica into the natural rubber. Higher silica synthesis temperature produces shorter incorporation time due to lower surface area and water content [12]. Afterward, a slow decrease in torque values represents the dispersion of the silica [14]. Lower torque and lower compounding temperature are observed with increasing silica synthesis temperature. The torque is proportional to the viscosity of the blends [15]. Lower viscosity could be attributed to the higher dispersion state of filler in rubber [16]. According to Sholeh et al. [12], higher synthesis temperature produces silica with lower surface area, thus resulting in the decrease of the shear force and friction, which results in a substantial decrease in compounding temperature [14].

Fluctuating trend of the torque is attributed to difficulty in incorporating the second portion of the filler (7th min). At synthesis temperature of 60°C, the second portion of the filler was pushed upward to about three-quarter the height of the hopper's neck. This can be explained as follows. The lower the synthesis temperature, the greater the water content (more polar) [12]. Friction between the silica loaded with the rubber became smaller, resulting in slippage due to differences in polarity between the filler and the elastomer.

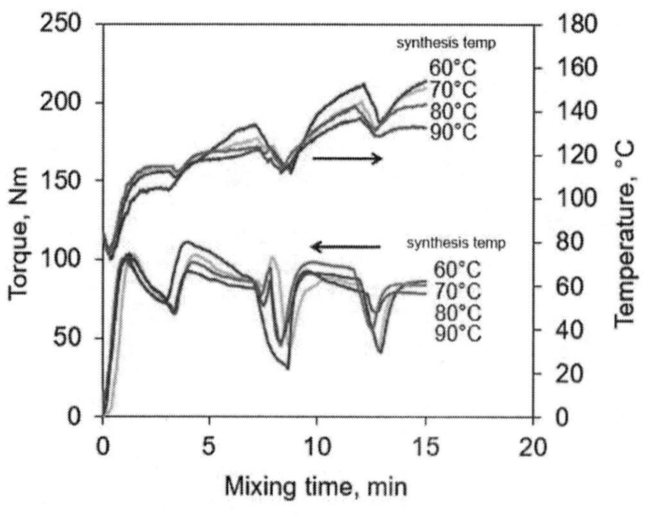


**Figure 2.** Torque and compound temperature during mixing.

Table 1 shows the effect of silica synthesis temperature on Mooney viscosity. It can be seen that Mooney viscosity decreases with increasing synthesis temperature. A lower viscosity means higher processability [17]. According to Ryu et al. [16], viscosity reduction is due to better dispersion. Sholeh et al. [12] reported that higher synthesis temperature produces silica with lower water content, thus resulting in a decrease of the surface polarity. Silica with lower surface polarity gives a better rubber-filler interaction, which leads to an improvement in the dispersion state.

The results in Table 1 reveal that bound rubber content increase with increasing synthesis temperature from 60 to 70°C and then decrease from 70°C to 90°C. Although bound rubber content is proportional to the level of filler dispersion, the result doesn't show the opposite trend to Mooney viscosity. It seems that from 70°C to 90°C, the decrease of bound rubber content with increasing synthesis temperature is attributed to the decreasing surface area of the silica [12].

**Table 1.** Mooney viscosity and bound rubber content of the compound.

| Synthesis temperature, °C | Mooney viscosity, MU | Bound rubber content, % |
|---|---|---|
| 60 | 30.5 | 33.05 |
| 70 | 29.4 | 36.95 |
| 80 | 20.8 | 31.75 |
| 90 | 16.7 | 27.20 |

Figure 3 shows images results from the optical microscope. NS agglomerates appear as luminous spots, while the area of well-dispersed silica is shown in black color. It is clear from Figure 3 that the number of visible agglomerates become much smaller with the higher synthesis temperature, thus means better silica macrodispersion. This result corroborate the Rheomix and Mooney viscosity data. The image in Figure 3 was transformed into a black and white image with a given threshold using ImageJ software, then the fraction of agglomerates area was obtained by dividing the total area of agglomerate with a diameter larger than 5 µm by the image area [18]. The calculation result is depicted in Figure 4. According to the classification by Waddell et al. [19], excellent silica dispersion is achieved at agglomerate area less than 2%.

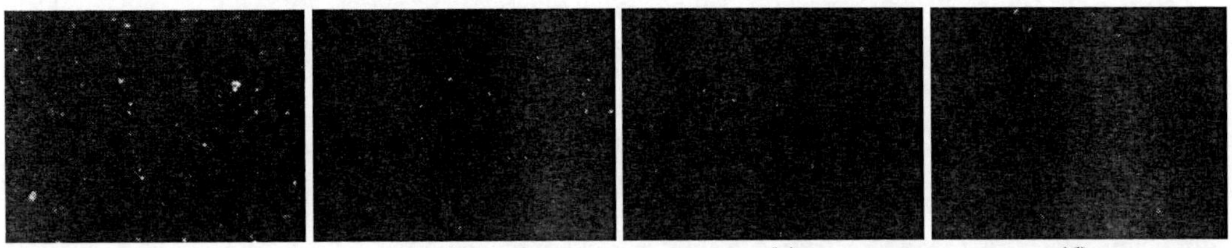

(a) (b) (c) (d)

**Figure 3**. Optical images (1395 mm x 1052 mm) obtained by Arcs video measuring system. NS synthesis temperature (a) 60°C, (b) 70°C, (c) 80°C, and (d) 90°C.

To explore the curing behavior of NR/NS compound, the curing characteristics were studied at different vulcanization temperatures. Figure 5 shows curing curves of NR/NS composite at 140, 150, and 160°C. It is clear that vulcanizing curves exhibited three distinct regions. The first region is induction period that provided a time for the compound to easily flow to fill the mold or through a die. Uncrosslinked compound in this region provides the lowest torque on the rheograph. The second region is the curing reaction period during which the crosslinking of rubber is formed. The torque rises towards the maximum value in this region. The third region is the overcure region that happens as an equilibrium or reversion cure behaviors [20]. Equilibrium behavior is shown by a relatively constant torque while reversion cure behavior is shown by a slow decrease in torque values. Shorter induction and curing reaction region are observed with increasing vulcanization temperature. Reversion begins to appear at a vulcanization temperature of 150°C for compounds filled with silica synthesized at 60 and 70°C (Figure 4(a) and (b)). At 160°C, significant reversion is detected for all compounds. It is well known that NR is susceptible to degradation at high temperature due to its chemical structure.

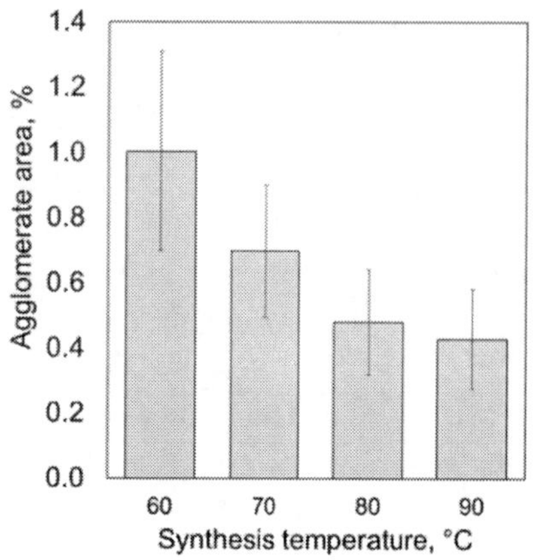


**Figure 4**. The fraction of agglomerate area calculated using ImageJ.

The percentage of reversion ($\%R$) at 600 min is defined as [21]:

$$\%R = \frac{M_H - M_{600}}{M_H - M_L} \times 100 \quad (2)$$

where $M_H$, $M_L$, and $M_{600}$ are the maximum torque, minimum torque, and torque at 600 s, respectively. The calculation result of $\%R$ is shown in Table 2. It can be seen from Table 2 that $\%R$ is found to increase with decreasing synthesis temperature. According to Kok [22], silanol group on silica surface can react with zinc oxide (activator) to form zinc silicate. These reactions can reduce the efficiency of the formation of the zinc-accelerator complex, believed to be responsible for the desulphuration reaction. Lower synthesis temperature produces silica with higher surface area and silanol content [12], thus could resulting in an increase of the zinc silicate formation, which results in an increase in $\%R$.

Detailed curing characteristics at a vulcanization temperature of 160°C are given in Table 2. It shows that scorch time ($t_{s2}$) increases with increasing synthesis temperature from 60 to 80°C and then decreases from 80°C to 90°C. Both state-of-mix of the filler and silanol content affect on scorch time [2,16]. Based on the silica dispersion result above, scorch time should increase from synthesis temperature 60 to 90°C. However, it is clear from Figure 4 that silica synthesized at 90°C only gives a slightly better silica macrodispersion than silica synthesized at 80°C. Thus it is believed that scorch time at synthesis temperature 80°C is longer than 90°C because the effect of higher silanol content is more dominant. The curing time ($t_{90}$) shows a similar trend as scorch time.

The dependence of minimum torque and maximum torque on silica synthesis temperature is also illustrated in Table 2. Compounds having higher silica synthesis temperature give lower minimum torque and higher maximum torque. The result of minimum torque is in line with the result from Mooney viscosity test. The trend of $M_H$ and $M_H$-$M_L$ show that higher silica synthesis temperature gives stiffer vulcanizate and higher crosslink density, respectively.

Vulcanization kinetics model is used to predict the degree of cure at any time-temperature history. The torque obtained from the rheometer is believed to be proportional to the crosslink density of the rubber. Thus the degree of curing ($\alpha$) can be obtained from the torque values obtained from rheometer using the relation [20]:

$$\alpha = \frac{M_t - M_L}{M_H - M_L} \quad (3)$$

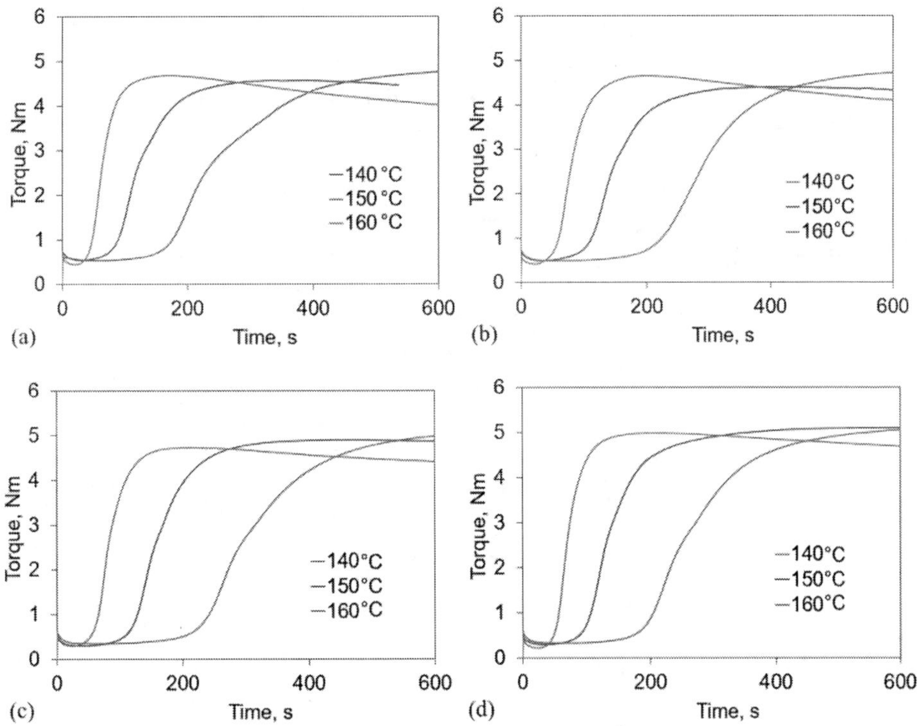


**Figure 5**. Curing curve of NR/NS composite at different curing temperatures: NS synthesis temperature (a) 60°C, (b) 70°C, (c) 80°C, and (d) 90°C.

**Table 2.** Curing characteristics of NR/NS composite.

| Synthesis temp, °C | $t_{s2}$, s | $t_{90}$, s | $M_L$, Nm | $M_H$, Nm | $M_H$-$M_L$, Nm | $M_{600}$, Nm | %$R$ |
|---|---|---|---|---|---|---|---|
| 60 | 38 | 106 | 0.47 | 4.38 | 3.91 | 3.92 | 11.6 |
| 70 | 49 | 117 | 0.44 | 4.46 | 4.02 | 4.11 | 8.69 |
| 80 | 53 | 120 | 0.29 | 4.74 | 4.46 | 4.40 | 7.68 |
| 90 | 49 | 109 | 0.26 | 4.87 | 4.61 | 4.60 | 5.92 |

where $M_t$ are the torque values at time $t$. An appropriate kinetic model to describe vulcanization reaction is the autocatalytic reaction model. The follow relationship is assumed [20,23]:

$$\frac{d\alpha}{dt} = k\,\alpha^m(1-\alpha)^n \quad (4)$$

Where $k$, $m$ and $n$ denote the rate constant, reaction orders of autocatalytic reaction, and reaction orders of non-autocatalytic reaction, respectively. The parameters $k$, $m$ and $n$ were determined using nonlinear regression analysis, and the results are shown in Table 3. The fitted curves are illustrated in Figure 6. A comparison between the experimental and the model curves was made. Autocatalytic reaction is indicated by the increasing of the reaction rate with the degree of curing and passes through a maximum [23]. The experimental conversion rate values were closely fitted with the model for all samples. It is clear that the rate constant value is increasing with the increase in curing temperature and silica synthesis temperature. Higher synthesis temperature produces silica with lower surface area and silanol content [12], thus could resulting in a decrease of the zinc silicate formation, which results in an increase in $k$. The rate constant follows the Arrhenius equation below:

$$k = A\, e^{-E/RT} \tag{5}$$

where $A$ and $E$ are pre-exponential factor and the activation energy, respectively. The calculation result is shown in Table 3.

**Table 3.** Kinetic parameters of NR/NS composites.

| $T_{synthesis}$, °C | $T_{curing}$, °C | $k$, $s^{-1}$ | $m$ | $n$ | $E$, J/mol | $A$, $s^{-1}$ | $R^2$ |
|---|---|---|---|---|---|---|---|
| 60 | 140 | 0.0350 | 0.836 | 1.93 | $5.66.10^4$ | $5.27.10^5$ | 0.937 |
| | 150 | 0.0611 | 0.870 | 1.56 | | | |
| | 160 | 0.0746 | 0.832 | 1.25 | | | |
| 70 | 140 | 0.0318 | 0.946 | 1.48 | $7.68.10^4$ | $1.73.10^8$ | 0.959 |
| | 150 | 0.0648 | 1.00 | 1.56 | | | |
| | 160 | 0.0891 | 0.899 | 1.36 | | | |
| 80 | 140 | 0.0326 | 0.876 | 1.74 | $7.20.10^4$ | $4.40.10^7$ | 0.956 |
| | 150 | 0.0639 | 0.968 | 1.67 | | | |
| | 160 | 0.0856 | 0.860 | 1.30 | | | |
| 90 | 140 | 0.0347 | 0.800 | 1.65 | $8.06.10^4$ | $5.69.10^8$ | 0.965 |
| | 150 | 0.0720 | 0.934 | 1.77 | | | |
| | 160 | 0.102 | 0.844 | 1.36 | | | |

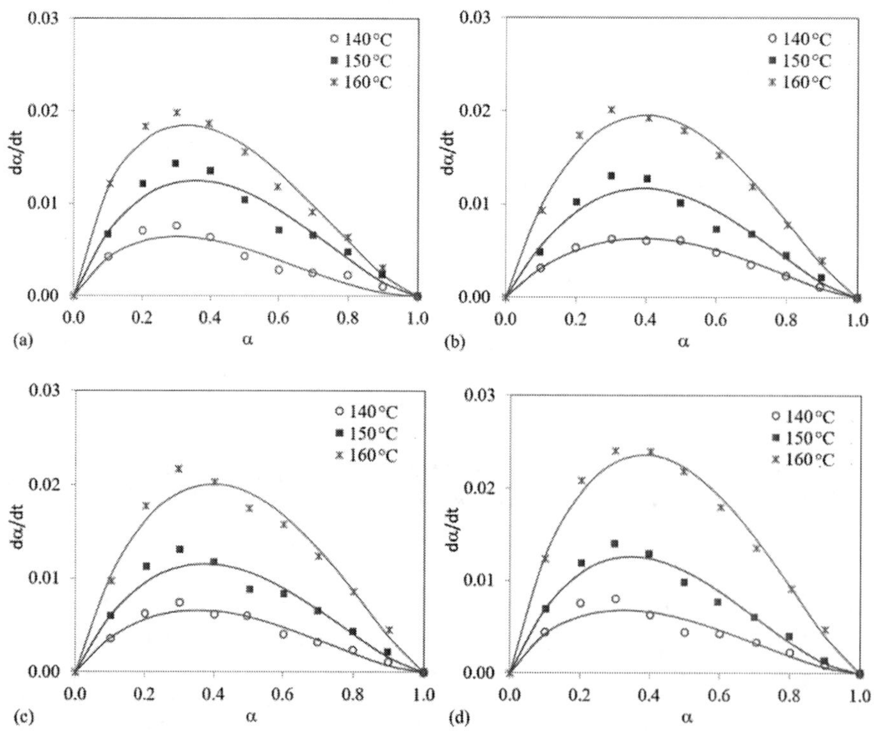


**Figure 6**. Reaction rate versus degree of curing: NS synthesis temperature (a) 60°C, (b) 70°C, (c) 80°C, and (d) 90°C.

Uniaxial tension experiments on a NR/NS composite was conducted to obtain the stress-strain curve. It can be seen from Figure 6 that higher synthesis temperature produces stiffer vulcanizate. This result agrees well with maximum torque obtained by rheometer. However, the modulus at break doesn't follow the same trend with $M_H$-$M_L$. It is obvious that modulus at break increases with increasing synthesis temperature from 60 to 70°C and then decreases from 70°C to 90°C. The detrimental effect on modulus at break at higher synthesis temperature may be attributed to the higher primary particle diameter of the nanostructured silica [11].

The experimental stress-strain curve was fitted to the Ogden constitutive model as follow [24]:

$$\sigma_1 = \frac{2\mu}{\mathrm{A}}\left[\lambda^{\mathrm{A}-1} - \lambda^{-1-\left(\frac{\mathrm{A}}{2}\right)}\right] \qquad (6)$$

where $\sigma_1$, $\lambda$, $\mu$, and A are the uniaxial stress, stretch in the loading direction, shear modulus, and strain hardening exponent, respectively. Material constants can be determined using a nonlinear regression analysis and the results is shown in Table 4. Figure 7 illustrates that the experimental stress-strain values were closely fitted with Ogden model for all samples.

**Table 4.** Ogden constants evaluated from uniaxial tensile stress versus strain response of NR/NS composite.

| | Synthesis temperature, °C | | | |
|---|---|---|---|---|
| parameter | 60 | 70 | 80 | 90 |
| $\mu$, MPa | 0.976 | 1.13 | 1.27 | 1.21 |
| A | 2.88 | 2.81 | 2.77 | 2.89 |

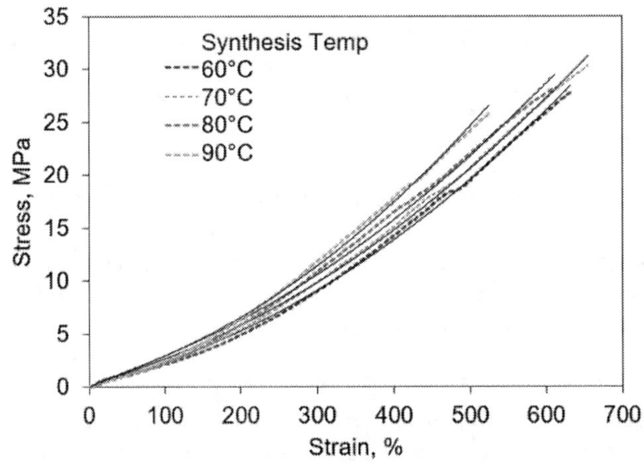


**Figure 7.** Stress vs strain for the NR/NS composite. Data fitting by Ogden equation represented by a solid line.

## 4. Conclusion

Compounding of NR with NS synthesized at four different temperatures was conducted in an internal mixer. Higher silica synthesis temperature produced shorter incorporation time, lower torque, lower compounding temperature, and lower Mooney viscosity, thus higher processability. Highest bound rubber content was found at NR filled with NS synthesized at 70°C. Images results from the optical microscope revealed that all vulcanizates had excellent NS dispersion with the best state-of-mix was found at NS synthesized at 90°C. Rheometer data showed that scorch time increased with increasing synthesis temperature from 60 to 80°C and then decreased from 80°C to 90°C. Higher silica synthesis temperature gave stiffer vulcanizate, higher crosslink density, and lower percent reversion. However, modulus at break decreased from 70°C to 90°C attributed to the higher primary particle diameter of the nanostructured silica. The experimental curing and stress-strain curve value were closely fitted with autocatalytic and Ogden model, respectively.

## References

[1] Olthuis M 2020 *Relevance and development of new rubber technology competences for a sustainable automotive industry* (University of Twente)
[2] Hewitt N 2007 *Compounding Precipitated Silica in Elastomers* (William Andrew Inc.)
[3] Leksomboon A and Jongsomjit B 2011 Synthesis of spherical silica by sol-gel method and its application as catalyst support *ASEAN J. Chem. Eng.* **10** 25
[4] Boonmee A and Jarukumjorn K 2020 Preparation and characterization of silica nanoparticles from sugarcane bagasse ash for using as a filler in natural rubber composites *Polym. Bull.* **77** 3457–72
[5] Huabcharoen P, Wimolmala E, Markpin T and Sombatsompop N 2017 Purification and characterization of silica from sugarcane bagasse ash as a reinforcing filler in natural rubber composites *BioResource* **12** 1228–45
[6] Zeng Z, Li Y, Zhao P and Yu H 2020 Fabrication of Rice Husk Ash/Natural Rubber Composites by the Latex Process *J. Wuhan Univ. Technol. Sci. Ed.* **35** 42–6
[7] Ooi Z X, Ismail H and Abu Bakar A 2014 Thermal properties and aging characteristics of chemically modified oil palm ash-filled natural rubber composites *Iran. Polym. J. (English Ed.* **23** 723–30
[8] Ooi Z X, Ismail H and Abu Bakar A 2014 Study on the ageing characteristics of oil palm ash reinforced natural rubber composites by introducing a liquid epoxidized natural rubber coating technique *Polym. Test.* **37** 156–62
[9] Pangamol P, Malee W, Yujaroen R, Sae-Oui P and Siriwong C 2018 Utilization of Bagasse Ash as a Filler in Natural Rubber and Styrene–Butadiene Rubber Composites *Arab. J. Sci. Eng.* **43** 221–7
[10] Jembere A L and Fanta S W 2017 Studies on the Synthesis of Silica Powder from Rice Husk Ash as Reinforcement Filler in Rubber Tire Tread Part : Replacement of Commercial Precipitated Silica *Int. J. Mater. Sci. Appl.* **6** 37–44
[11] Sholeh M, Rochmadi R, Sulistyo H and Budhijanto B 2020 Synthesis of precipitated silica from bagasse ash as reinforcing filler in rubber *IOP Conf. Ser. Mater. Sci. Eng.* **778** 012012
[12] Sholeh M, Rochmadi R, Sulistyo H and Budhijanto B 2020 Nanostructured silica from bagasse ash: the effect of synthesis temperature and pH on its properties *J. Sol-Gel Sci. Technol.*
[13] Jang S H, Kim W S, Kang Y G, Han M H and Chang S M 2013 Study on Mixing Condition of the Rubber Composite Containing Functionalized S-SBR, Silica and Silane: II. Effect of Mixing Temperature and Time. *Elastomers Compos.* **48** 103–13
[14] Sui G, Zhong W, Yang X and Zhao S 2007 Processing and material characteristics of a carbon-nanotube-reinforced natural rubber *Macromol. Mater. Eng.* **292** 1020–6
[15] Bondan F, Soares M R F and Bianchi O 2014 Effect of dynamic cross-linking on phase morphology and dynamic mechanical properties of polyamide 12/ethylene vinyl acetate copolymer blends *Polym. Bull.* **71** 151–66
[16] Ryu C, Yang J, Park W, Kim S J, Kim D Il and Seo G 2020 Silica-filled NR compounds prepared by dry and wet masterbatches with different mixing times *J. Appl. Polym. Sci.* 49548
[17] Sattayanurak S, Sahakaro K, Kaewsakul W, Dierkes W K, Reuvekamp L A E M, Blume A and Noordermeer J W M 2020 Synergistic effect by high specific surface area carbon black as secondary filler in silica reinforced natural rubber tire tread compounds *Polym. Test.* **81**
[18] Morozov I A, Solodko V N and Yu A 2015 Quantitative study of fi lled rubber microstructure by optical and atomic force microscopy *Polym. Test.* **44** 197–207
[19] Waddell W H, Evans L R, Lee S, Lin T Y and Wu C S 2018 Quantitative silica-filled tread

dispersion rating system *Rubber Plast. News* 16–21

[20] Alwaan I M 2018 Rheological characterization and modeling of vulcanization kinetics of natural rubber/starch blends *J. Appl. Polym. Sci.* **135** 46347

[21] Boonkerd K, Deeprasertkul C and Boonsomwong K 2016 Effect of sulfur to accelerator ratio on crosslink structure, reversion, and strength in natural rubber *Rubber Chem. Technol.* **89** 450–64

[22] Kok C M 1987 The effects of compounding variables on the reversion process in the sulphur vulcanization of natural rubber *Eur. Polym. J.* **23** 611–5

[23] Zachariah A K, Chandra A K, Mohammed P K and Thomas S 2019 Vulcanization kinetics and mechanical properties of organically modified nanoclay incorporated natural and chlorobutyl rubber nanocomposites *Polym. Test.* **76** 154–65

[24] Mahmoud W E, Mansour S A, Hafez M and Salam M A 2007 On the degradation and stability of high abrasion furnace black (HAF)/acrylonitrile butadiene rubber (NBR) and high abrasion furnace black (HAF)/graphite/acrylonitrile butadiene rubber (NBR) under cyclic stress-strain *Polym. Degrad. Stab.* **92** 2011–5

**Acknowledgments**

We wish to thank the LPDP Ministry of Finance, Indonesia for the financial support to this project. M. Sholeh wishes to thank Mr. Parsono and Mr. Andri (CLRP), Indonesia for the compounding prosess and preparation of the vulcanizates.

# Numerical Investigation of Double Chamber Acetate-Fed Microbial Fuel Cell in Unsteady-State Condition

**I Subadri[1], A Satriyatama[1], I D M Budi[1], A Harimawan[1*]**

[1]Department of Chemical Engineering, Faculty of Industrial Technology, Institut Teknologi Bandung, Bandung 40134, Indonesia

*Email: intan_subadri@students.itb.ac.id

**Abstract.** Microbial fuel cells (MFCs) are devices that utilize the work of microorganisms to oxidize organic substrate involving biochemical pathways. Several studies have been done based on experiments while simulation and modelling remain unexplored. Basically, MFCs have a lot of similarities to chemical fuel cell systems, which modelling and simulation have been widely developed. Hence, a study should be done to develop the model in order to widen the implementation of MFCs. In order to evaluate MFCs performance with less cost and time, numerical modelling might be an effective approach. Models could also be easily developed or modified for various operation conditions and configurations to generate experimental data on MFCs. A number of papers on simulation and modelling focused on cell voltage as function of both cell current density and chemicals concentration. In this paper, a double chamber acetate MFCs under continuous operation and unsteady state condition would be investigated. MFCs model based are developed by calculating biochemical reactions, Butler-Volmer equation, and electrochemical equations using MATLAB 2018a software. The parameters and constants data reported from recent literature are used. Results show that periodic flow rate of fuel could improve the power production. This result also gives the prediction of cell voltage and current density. Nevertheless, models with various conditions or configurations could be developed to scale-up or create more efficient MFCs using simple methods.

## 1. Introduction

The energy crisis in recent year, including electricity has become a global issue in this world. It is happened because of the massive use of fossil fuels, that is equal to 50.66%, reported by Dudley in 2018. In fact, fossil fuels are not a sustainable energy source since fossil fuel cannot be renewed and is decreasing year by year. Moreover, the use of fossil fuels actually contributes to the land degradation, $CO_2$ emission in the atmosphere, and also water pollution. These phenomena have led to the development of sustainable and eco-friendly energy.

Microbial fuel cell (MFC) is a electrochemical cell that could convert chemical energy into electrical energy via biochemical reactions directly. MFC utilizes energy from organic substrates by using exoelectrogenic microorganisms [1]. Regarding MFC applications, the ability of microorganisms to utilize organic materials, can be found in wastewater, biosensors, and chemostat assays [2]. Moreover, MFC can also be used in desalination, hydrogen production, pollution remediation, and remote power source [3]. MFC could be operated in the ambient temperature, neutral pH, and atmospheric pressure. Power production from organic substrate is one of most interest areas in MFC. The MFC technology may become great promise since effective and wide scale MFC are not available yet.

Multiple chemical, biological, and physical indicators have an important role to determine MFC work performance [4]. The analysis of a MFC system requires many discipline knowledges to achieve better understanding. Efforts have been made to analyze the aspects related to microorganisms such as molecular analysis, biochemical rate, and internal electron transfer [4, 5]. Problems faced by MFC such as stability, reliability, and electricity production efficiency must be fixed before putting into larger scale operation.

Researches have been developed to investigate the molecular reaction of the electron transfer and microbial community while recent studies have been focusing on engineering and electricity calculation. The variations of the organic compound's concentration, biological environment, load disturbance, and the number of microbes is directly affected by the current and voltage produced by the MFC. Therefore, it is important to analyse the complexity of the MFC to provide complete understanding of the high-bioelectrochemical performance to maintain the price, reproducibility, and process control of the MFC.

Pinto et al. [6] proposed the two populations model to describe the competition of anodophilic and methanogenic microbes. Later, models by Zeng et al. [5] were developed, explained the MFC model as function of biochemical, Butler-Volmer, and mass balances expression. Picoreanu et al. [1] developed the model predictive control for MFC performance improvement. Capodaglio et al. [7] published the mathematical model related to MFC application in wastewater treatment. Then, Oliveira et al. [8] reported an MFC model, focused on parameters such as substrate concentration, current density, and cell temperature.

In in this paper, the MFC studies are generally divided into three main sections, which are biochemical reaction process, thermodynamic studies, and electrochemical process. With the development of mathematical modelling, the non-linear and hysteresis characteristic of the MFC can be solved using MATLAB. As a result, the optimum condition of the MFC to provide constant current and voltage output is obtained.

## 2. Methods

In this research, simulation platform is established mainly based on Zeng et al. [5] MFC model. Zeng and teams developed a model close to chemical-based fuel cells. Biochemical reactions, mass/charge balances, and Butler-Volmer expression are integrated on two-chamber configuration to simulates unsteady behavior of MFC. Modelling methods described are applied for making some modification [9, 10].

Considering acetate as substrate on anode compartment, the mass balance equation is considered in the equation (1) [11]. Since the anode compartment is operated under anaerobic conditions, the acetate reaction could be modelled with Monod-type equation. In electrochemical cell, the bioelectrochemical reactions are controlled by the electrical potential. Butler-Volmer expression (2) can be used to mathematise the reaction rate [12].

$$(CH_2O)_2 + 2H_2O \rightarrow 2CO_2 + 8H^+ + 8e^- \quad (1)$$

$$r_1 = k_1^0 \exp\left(\frac{\alpha F}{RT}\eta\right)\frac{C_{ac}}{K_{ac}+C_{ac}}X \quad (2)$$

In cathode compartment, the reduction of dissolved oxygen described in Eq. (3). The Monod-type behavior of dissolved oxygen shows that oxygen reduction is ignorable [13]. The rate of reaction formula in cathode compartment explained in Eq. 4.

$$O_2 + 4e^- + 2H_2O \rightarrow 4OH^- \quad (3)$$

$$r_2 = -k_2^0 \frac{C_{O_2}}{K_{O_2}+C_{O_2}} \exp\left[(\beta - 1)\frac{F}{RT}\eta_c\right] \quad (4)$$

The continuously stirred tank reactor (CSTR) is assumed in anode and cathode compartment. Hence, mass transport, phase mixture, and redox reactions is assumed to be fast processes. In addition, acetate and carbon dioxide are assumed not diffuse to membrane, transportation loss is neglected, and fluid density in anode and cathode compartment is equal. The chemical components mass balances in anode compartment, which are acetate, dissolved $CO_2$, biomass, and $H^+$ are described in following Eq. (5), (6), (7), and (8). The subscript 'a' denotes anode and 'c' denotes cathode.

$$V_a \frac{dC_{Ac}}{dt} = Q_a\left(C_{Ac}^{in} - C_{Ac}\right) + A_m r_1 \tag{5}$$

$$V_a \frac{dC_{CO_2}}{dt} = Q_a\left(C_{CO_2}^{in} - C_{CO_2}\right) + 2A_m r_1 \tag{6}$$

$$V_a \frac{dC_H}{dt} = Q_a\left(C_H^{in} - C_H\right) + 8A_m r_1 \tag{7}$$

$$V_a \frac{dX}{dt} = Q_a \frac{(X^{in} - X)}{f_X} + A_m Y_{ac} r_1 - V_a K_{dec} X \tag{8}$$

Meanwhile, in the cathode compartments, the mass balances of dissolved O2, OH, and M+ are written in following Eq. (9), (10), (11).

$$V_c \frac{dC_{O_2}}{dt} = Q_c\left(C_{O_2}^{in} - C_{O_2}\right) + A_m r_2 \tag{9}$$

$$V_c \frac{dC_{OH}}{dt} = Q_c\left(C_{OH}^{in} - C_{OH}\right) - 4A_m r_2 \tag{10}$$

$$V_c \frac{dC_M}{dt} = Q_c\left(C_M^{in} - C_M\right) + N_m A_m \tag{11}$$

The cell current density and the flux of ions relation can be described in Eq. 12.

$$i_{cell} = F \sum_i z_i N_i \tag{12}$$

$N_i$ is the superficial flux, $i_{cell}$ denotes the cell current density, and $z_i$ is the charge number. It gives the charge balances at the anode and cathode as described below.

$$C_a \frac{d\gamma_a}{dt} = 3600 i_{cell} - 8F r_1 \tag{13}$$

$$C_c \frac{d\gamma_c}{dt} = 3600 i_{cell} - 8F r_2 \tag{14}$$

It has been assumed that we neglect the electric connections and the ohmic drops in the current-collectors. The total potential or the cell voltage $U_{cell}$ is calculated using following equation.

$$U_{cell} = U^0 - \gamma a + \gamma c - \left(\frac{d^m}{k^m} + \frac{d_{cell}}{k^{aq}}\right) i_{cell} \tag{15}$$

$d_{cell}$ and $d^m$ are distance of the electrode and membrane thickness. $k$ represents the conductivities characteristic. To determine the overpotential of the MFC, the approximation of Butler-Volmer equation, Tafel equation is used. Tafel equation can be written as below.

$$i = nFKC\, exp(\pm \alpha F \left(\frac{eta}{RT}\right)) \tag{16}$$

The microorganism biofilm formation in this model is observed. The chemotaxis equations below us used to determine the production rate of the sensing molecules from attached and suspended bacteria

$$\frac{\partial \mathrm{E}}{\partial \mathrm{t}} = \mathrm{D_E} \frac{\partial^2 \mathrm{E}}{\partial \mathrm{Z}^2} - \lambda \mathrm{E} + \alpha_\mathrm{S} \mathrm{X_S} \tag{17}$$

E is the sensing molecule concentration and $D_E$ is the coefficient of it. λ is the sensing molecule degradation rate, while $\alpha_s$ is the sensing molecule production rate. The boundary conditions are defined as following.

$$D_E \frac{\partial E}{\partial z}(z=0) = \alpha X \tag{18}$$

$$\frac{\partial E}{\partial z}(z=l) = 0 \tag{19}$$

The first boundary condition is defined for the biofilm formed on the surface of the anode. The second boundary condition is defined for the membrane. The chemotaxis equation of suspended microorganism is described below.

$$\frac{\partial x_s}{\partial t} = \mu_{SE0}\frac{\partial^2 x_s}{\partial z^2} - X\frac{\partial}{\partial z}\left(x_s\frac{\partial E}{\partial z}\right) + \gamma x_s q_s \tag{20}$$

The boundary condition for the condition is described as below.

$$\frac{\partial x_s}{\partial z}(Z=l) = 0 \tag{21}$$

The bacteria is sticking on the anode surface so that the bacteria density is assumed to be zero.

$$x_s(Z=0) = 0 \tag{22}$$

The main of the microbial fuel cells are modeled and simulated using MATLAB 2018a software of two-chamber. The model also can be used to simulate the running of MFC in various conditions. All parameters, constant, and assumptions are given in Table 1.

**Table 1.** All parameters, constants, and conditions used on acetate-fed MFC were developed from Zeng et al. (2010).

| Symbol | Description | Unit | Value |
|---|---|---|---|
| $F$ | Faraday's constant | Coulombs $mol^{-1}$ | 96485.4 |
| $T$ | Temperature | K | 303 |
| $R$ | Gas constant | J $mol^{-1}$ $K^{-1}$ | 8.314 |
| $A_m$ | Area of membrane | $m^2$ | $5 \times 10^{-4}$ |
| $Y_{ac}$ | Bacterial yield | Dimensionless | 0.05 |
| $K_{dec}$ | Decay constant for acetate | $h^{-1}$ | $8.33\times10^{-4}$ |
| $f_x$ | Reciprocal of wash-out fraction | Dimensionless | 10 |
| $V_a$ | Volume of anode compartment | $m^3$ | $5.5 \times 10^{-5}$ |
| $V_c$ | Volume of cathode compartment | $m^3$ | $5.5 \times 10^{-5}$ |
| $Q_a$ | Flow rate of fuel feed to anode | $m^3$ $h^{-1}$ | $2.25 \times 10^{-5}$ |
| $Q_c$ | Flow rate feeding to cathode compartment | $m^3$ $h^{-1}$ | $1.11 \times 10^{-3}$ |
| $C_{Ac}^{in}$ | Concentration of influent acetate in anode compartment | mol $m^{-3}$ | 1.56 |
| $C_{CO_2}^{in}$ | Concentration of influent $CO_2$ in anode compartment | mol $m^{-3}$ | 0 |
| $X^{in}$ | Concentration of influent bacteria in anode compartment | mol $m^{-3}$ | 0 |
| $C_H^{in}$ | Concentration of influent $H^+$ in anode compartment | mol $m^{-3}$ | 0 |
| $C_{O_2}^{in}$ | Concentration of dissolved $O_2$ in the influent of cathode compartment | mol $m^{-3}$ | 0.3125 |
| $C_M^{in}$ | Concentration of influent $M^+$ in cathode compartment | mol $m^{-3}$ | 0 |

IOP Conf. Series: Materials Science and Engineering 1143 (2021) 012012 doi:10.1088/1757-899X/1143/1/012012

| | | | |
|---|---|---|---|
| $C_{OH}^{in}$ | Concentration of influent $OH^-$ in cathode compartment | mol m$^{-3}$ | 0 |
| $k_1^0$ | Rate constant of anode reaction at standard condition | mol m$^{-2}$ h$^{-1}$ | 0.207 |
| $k_2^0$ | Rate constant of cathode reaction at standard condition | m$^{12}$ mol$^{-4}$ h$^{-1}$ | $3.288 \times 10^{-5}$ |
| $K_{ac}$ | Rate constant for acetate | mol m$^{-3}$ | 0.592 |
| | Rate constant for dissolved oxygen | mol m$^{-3}$ | 0.004 |
| $\alpha$ | Charge transfer constant of anode | Dimensionless | 0.051 |
| $\beta$ | Charge transfer constant of cathode | Dimensionless | 0.663 |
| $U^0$ | Cell open circuit potential | volt | 0.77 |
| $x_s$ | suspended microorganisms concentration in liquid bulk | kg$_{CODX}$ m$^3$ | 1.6 |
| $m_{SE0}$ | Mobility coefficient | cm$^2$ day$^{-1}$ | 1.296 |
| $\gamma$ | chemotactic coefficients of sensing molecules | kg$_{CODX}$ kg$_{CODS}$$^{-1}$ | 0.243 |
| $q_{s,max}$ | Substrate consumption in the anolyte | kg$_{CODS}$ kg$_{CODX}$$^{-1}$day$^{-1}$ | 52.86 |

## 3. Results and Discussion

The predicted results on substrate concentration change, electrochemical analysis, and work efficiency in double chamber acetate-fed microbial fuel cell are described below.

### *3.1 Reactions and Kinetic Study*

For MFCs mass balances, the changes of all substrates and chemicals are considered. Meanwhile, the internal diffusion between substrate and biofilm are neglected. All the primary components in MFCs are represented in mass and charge balances, especially chemical and biomass. The predicted results of reaction rates and mass balances calculations expressed on Fig. 1 below.

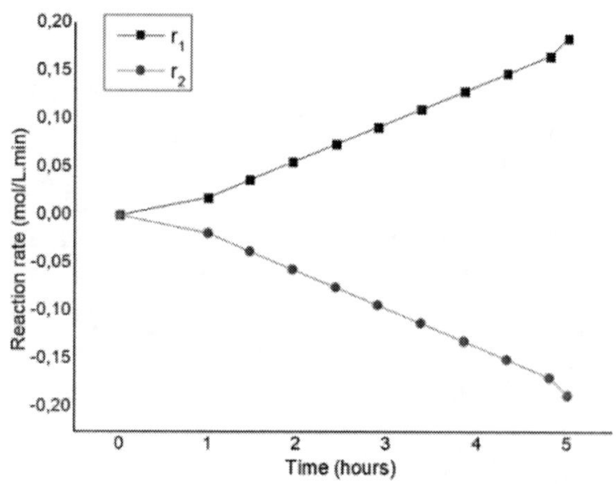

IOP Conf. Series: Materials Science and Engineering 1143 (2021) 012012 doi:10.1088/1757-899X/1143/1/012012

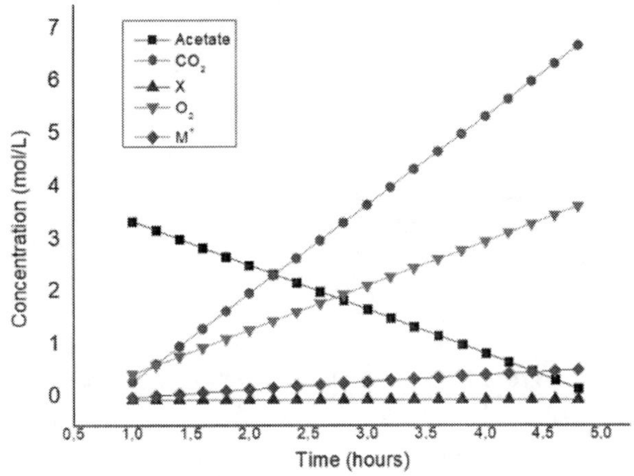


**Figure 1.** Simulation results of acetate MFC a) reaction rates; b) Concentrations of chemical substances.

The component concentration of substrate and reaction rates are evaluated using all parameters listed in **Table 1. Figure. 1a** shows that the reaction rate of oxygen reduction in cathode chamber and acetate oxidation in anode chamber increase proportionally with cell current density. This results could be explained by **Eq.** (12) and (13) integration.

Figure. 1b explains the concentration changes of acetate, $CO_2$, biomass, $M^+$, and dissolved oxygen during operation. From the plot, it can be seen that the decrease of acetate followed by the increase of $CO_2$ and biomass. The reactions also represent that the higher current density, the higher acetate will be consumed and $CO_2$ will be produced in anode chamber. Similar results also happened in cathode chamber, where dissolved oxygen decreased influenced by higher current density. This results could be confirmed based from Picoreanu et al. [14], where the acetate consumption was limited by the both electron acceptor and donor.

### *3.2. Electrochemical Model*

The rate of an electrochemical reaction to the overpotential can be calculated using the Tafel equation. The electrode' concentrations are assumed to be practically equal to the bulk electrolyte' concentrations, so that the current can be expressed as a function of potential only [15]. Each data point is gained by simulating the Tafel equation in MATLAB. From anode and cathode potential, the output voltage of MFCs could be determined. The polarisation effect can cause the voltage losses since it can lead to the overpotential.

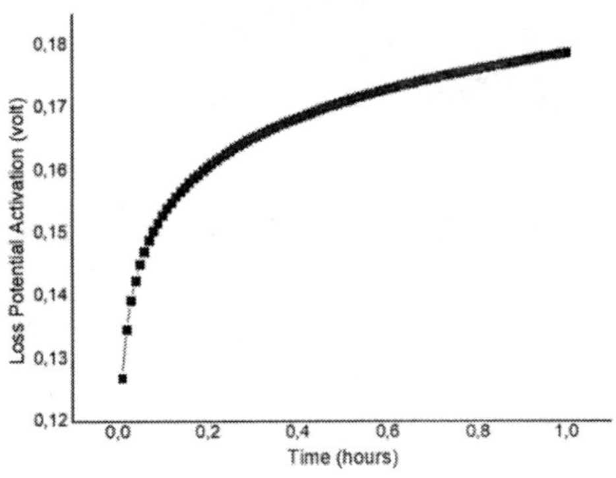

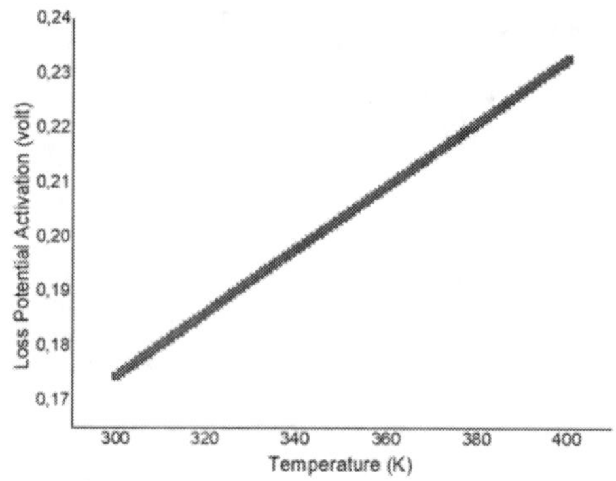


**Figure 2.** Results of activated potential loss as the function of a) current density and b) temperature.

Based on the graph above, the activated potential rises due to the increase of the current density. The raised of anode potential also influenced from the increased pH caused by acidified process at anode. The growth of the bacteria, that consists of 4 main points, affects the total current density. As the number of the bacteria grow, both the rate of acetate oxidation and oxygen reduction are faster and the cell current density. The kinetic reaction is faster when the temperature is increased. Therefore, the reaction will be directly shift to the right, then the anode product increases the activated potential loss.

*3.3. Bacteria Distribution*

The distribution of bacteria in anolyte of MFC is shown in Fig. 3. The trend of spatial distribution of suspended bacteria indicating the adsorption of suspended bacteria to biofilm. The variation of suspended bacteria indicates the two characteristics of bacteria movement, including bacteria diffusion and the bacteria movement as function of concentration gradient.

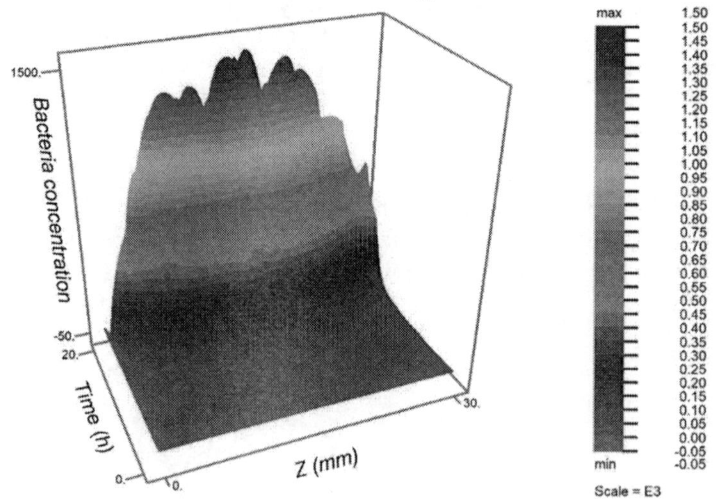


**Figure 3.** Biofilm concentration distribution.

The concentration of attached bacteria is increased by increasing the distance from anode that attributed to accessibility of nutrient in surface of biofilm which the net growth rate of these bacteria would be higher.

## 4. Conclusion

The mathematical models for MFCs work prediction are evaluated in this paper. The representative models from previous literature are investigated and applied. Mass and charges balance with bio-electrochemical reactions integration is used as the basis method. These results explain the trends in the simulated data in unsteady-state condition and the parameter's influence for power production. Activated potential rises due to the increase of the current density and temperature. Overall, these mathematical models could provide a convenient tool for process optimization and process control further.

## References

[1] Picioreanu C, van Loosdrecht M C, Curtis T P, Scott K 2010 Model based evaluation of the effect of pH and electrode geometry on microbial fuel cell performance *Bioelectrochemistry* **78** 8-24

[2] Choi S, Chae J 2012 An array of microliter-sized microbial fuel cells generating 100 μW of power *Sensors Actuators A Phys.* **177** 10–15

[3] Pandey P, Shinde V N, Deopurkar R L, Kale S P, Patil S A, Pant D 2016 Recent advancesin the use of different substrates in microbial fuel cells toward wastewater treat-ment and simultaneous energy recovery *Appl. Energy* **168** 706–723

[4] Picioreanu C, Head I M, Katuri K P, van Loosdrecht M C, Scott K 2007 A computational model for biofilm-based microbial fuel cells *Water Res.* **41** 2921–2940

[5] Zeng Y, Choo Y F, Kim B H, Wu P 2010 Modelling and simulation of two-chamber mi- crobial fuel cell *J. Power Sources* **195** 79–89

[6] Pinto R P, Srinivasan B, Manuel M F, Tartakovsky B 2010 A two-population bio-electrochemical model of a microbial fuel cell *Bioresour. Technol.* **101** 5256–5265

[7] Capodaglio A G, Hlavínek P, Raboni M 2016 Advances in wastewater nitrogen removal by biological processes: State of the art review. Ambient. e Agua—An Interdiscip. *J. Appl. Sci.* **11** 250.

[8] Oliveira V B, Simões M, Melo L F, Pinto A M F R A 2013 1D mathematical model for a microbial fuel cell *Energy* **61** 463–471

[9] Pinto R P, Tartakovsky B, Srinivasan B 2012 Optimizing energy productivity of microbial electrochemical cells' *J. Process Control* **22** (6) 1079–1086

[10] Corrêa J M, Farret F A, Canha L N, Simoes M G 2004 An electrochemical-based fuel-cell model suitable for electrical engineering automation approach *IEEE Transactions on industrial electronics* **51** (5) 1103-12

[11] Rozendal R A, Hamelers H V, Rabaey K, Keller J, and Buisman C J 2008 Towards practical implementation of bioelectrochemical wastewater treatment *Trends in biotechnology* **26**(8) 450-459

[12] Kato Marcus, Andrew, César I Torres, and Bruce E Rittmann 2007 Conduction-based modeling of the biofilm anode of a microbial fuel cell *Biotechnology and bioengineering* **98.6** 1171-1182

[13] Logan B E, Hamelers B, Rozendal R, Schröder U, Keller J, Freguia S, Aelterman P, Verstraete W and Rabaey K 2006 Microbial fuel cells: methodology and technology *Environmental science & technology* **40** (17) 5181-5192

[14] Picioreanu C, Katuri K P, Head I M, Van Loosdrecht M C M, Scott K 2008 Mathematical model for microbial fuel cells with anodic biofilms and anaerobic digestion *Water. Sci. Technol* **57** 965–971

[15] Kazemi S, Barazandegan M, Mohseni M 2016 Systematic study of separators in air-breathing flat-plate microbial fuel cells – part 2: numerical modeling *Energies* **9** (2) 79

# Evaluation of Heat Distribution And Aeration On Xylanase Production From Oil Palm Empty Fruit Bunches Using Tray Bioreactor

**Diah Meilany[1,2*], M T A P Kresnowati[1], Tjandra Setiadi[1,3]**

[1]Chemical Engineering, Institut Teknologi Bandung, Jl. Ganesha no 10, Bandung 40132, Indonesia
[2]Chemical Engineering, Politeknik Negeri Malang, Jl. Sukarno Hatta No 9, Malang 65141, Indonesia
[3]Center for Environmental Studies (PSLH), Institut Teknologi Bandung, Jl. Sangkuriang 42A, Bandung 40135, Indonesia

*diah.meilany@polinema.ac.id

**Abstract**. Indonesia imports ±99% of its industrial enzyme supply, mostly from China, India, Japan, and Europe. Xylanase is one of the most widely used enzymes in industries, mainly pulp and paper, animal feed, and bakery. Previous research has shown that *Aspergillus fumigatus* ITBCCL170 can produce xylanase enzyme using the Solid State Fermentation (SSF) process in a tray bioreactor using OPEFB. The tray bioreactor model for solid-state fermentation was developed and further used to simulate the effects of aeration and heat distribution on biomass growth and enzyme production. The results showed that solid-state fermentation in a tray bioreactor is indeed far from ideal as the model simulated. Yet, parallel aeration can increase cell and enzyme production, which needed further validation.

## 1. Introduction

The essential parameters in the SSF system are the distribution of heat and mass transfer [1,2]. Several things influence the heat and air distribution in the tray bioreactor, namely, the water content in the bed ($W_b$) and the bioreactor ($W_i$). Furthermore, the bioreactor's oxygen content (*Oks*) influenced the number of cells that grow ($X$). The heat of metabolism is generated by microorganisms and the heat of evaporation that occurs during incubation. Therefore we need a model that can describe the changes in the tray bioreactor system.

Microorganisms release some metabolism heat ($Q_{met}$) along with their growth. This metabolic heat can increase the bed temperature ($T_b$) and further reduce water in the bed. Both may harm the growth of microorganisms, which can only occur at a limited range of temperature and above a minimum level of humidity. To overcome this, the air supplied into the system (the aeration) can lower bed temperature. The airflow pattern inside the bioreactor contributes to the heat distribution as it may create a proper environment for the microbes to grow. However, the introduction of air into the system should not cause the evaporation of free water in the bed. The effect of airflow, humidity, and mixing on the growth of microorganisms in the two-tray bioreactor modelling reported that the use of saturated air of 0.1 LPM of air flow rate and mixing once a day gave the highest xylanase activity of 236.3 U/g [3].

Rajagopalan and Modak [4] introduced simultaneous modelling of heat and mass balance in the growth of *Aspergillus niger* in a tray bioreactor. The pan's height influenced the profile of oxygen

IOP Conf. Series: Materials Science and Engineering 1143 (2021) 012013 doi:10.1088/1757-899X/1143/1/012013

concentration in the tray bioreactor [5]. Montero [6] has reported the modelling of forced aeration on a tray bioreactor. However, since the model build was for a single tray, he could not simulate the effect of aeration distribution. Another model for the tray bioreactor is the growth profile in the aeration variation of *Aspergillus oryzae* and *Aspergillus awamori* in a circular tray bioreactor [7].

This research studied the effects of aeration and heat distribution on cell and enzyme production during a solid-state fermentation for xylanase production using *Aspergillus fumigatus* on Oil Palm Empty Fruit Bunches (OPEFB) as the primary substrate.

## 2. Materials and Method

The model was built as an ideal system due to the airflow pattern. The specific growth rate as a function of bed temperature was fitted empirically to approximate the experimental cells' values and enzymes. It was assumed that the system was homogenous, and the bed was initially saturated with Oxygen. The water vapour transferred into the bioreactor headspace was assumed to occur only from the surface's thin layer. The mixture of OPEFB and water in the bed was considered a pseudohomogenous system and will only be represented by OPEFB. The system was set as an isothermal system because the bioreactor chamber temperature was controlled at 37°C. The saturated air was supplied from the bottom of the bed at a constant flow rate of 0.006 $m^3$/h and 28°C. As for the bioreactor, it is assumed that the distance between the tray does not affect heat transfer.

The model is divided into three parts: the basic model in which the heat and aeration are assumed to be homogeneously mixed inside the bioreactor (one chamber bioreactor). In this model, the influence of the presence and absence of air is simulated. The model was built as combinations of smaller homogenously mixed chambers in series or parallel: the two-chamber and four-chamber bioreactors.

### 2.1. *Building a model*

The model built consists of a mass balance and a heat balance to obtain a complete picture of the process. Water mass balance is in the form of water content in the bed due to the water transfer from the bed to the bioreactor. While the bioreactor water content was due to mass transfer from the saturated air into the bioreactor. The heat balance is in the form of sensible heat generation, metabolic heat, and the release of evaporative heat from the bed to the bioreactor. Besides, the growth rate of microorganisms and product formation is the main focus of modelling.

### 2.2. *Basic model*

The one-chamber model means placing 1000 grams of EFB in one pan with a dimension of 26x26x7 cm, and the bioreactor consists of one tray in a 0.12 $m^3$ bioreactor room. Figure 1. shows the mass and heat transfer process in a one-chamber bioreactor.

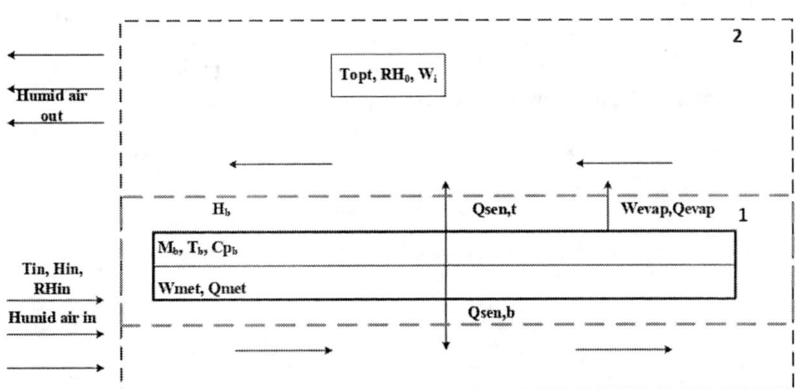


**Figure 1.** Heat and mass transfer scheme in a one-chamber bioreactor.

Figure 1 shows that the inlet air will absorb some heat generated in the bed ($Q_{met}$) and evaporation heat ($Q_{evap}$). The incubation took place at a constant $T_{opt}$ with $RH_0$ 85%. Some metabolic water ($W_{met}$) is produced, and some water in bed evaporated ($W_{evap}$), which moves into the flowing air. Because there is a temperature gradient between the bed and the bioreactor, a significant amount of sensible heat is released ($Q_{sen,t}$ and $Q_{sen,b}$). Changes in bed temperature in subsystem two (2) will affect subsystem one (1).

### *2.2.1. Growth kinetic*

Several model equations describe the growth of microorganisms in solid-phase bioreactors, as presented in Table 1. [8]

**Table 1.** Growth Kinetic Equation.

| | | |
|---|---|---|
| Linear | $r_X = \frac{dX}{dt} = K$ | (1) |
| Exponential | $r_X = \frac{dX}{dt} = \mu X$ | (2) |
| Logistic | $r_X = \frac{dX}{dt} = \mu X \left(1 - \frac{X}{X_{max}}\right)$ | (3) |

The model used an exponential growth kinetic equation, a first-order kinetic equation, with μ value was empirically constructed following polynomial regression as a function of bed temperature in a range *Aspergillus fumigatus* growth temperature. The specific growth rate was used as a constraint in the iteration of the model's equations simultaneously.

### *2.2.2. Product formation kinetic*

Similar difficulties in modelling products' formation, such as the rate of nutrient utilization, make the model equations more complex [9]. The empirical equations can present a proportional relationship between cell mass and product [10]. This study assumed the $m_p$ value to 0, which had been reported before [11].

$$r_P = \frac{dP}{dt} = Y_{P/X} \frac{dX}{dt} + m_P X \quad (4)$$

(5)

### *2.2.3. Oxygen balance*

The Oxygen balance in the tray bioreactor describes the change in the mole fraction of Oxygen at any time due to the intake of airflow, the outflow of air, and microorganisms' Oxygen consumption. The equation which states the relationship between the flow of Oxygen is,

$$\frac{dOks}{dt} = F_g\, yo_{2,in} - F_g\, yo_{2,out} + OUR \quad (6)$$

Whereas the rate of oxygen consumption (OUR) correlates with the growth of microorganisms [9].

$$r_{Oks} = OUR = Y_{O/X}\, r_X + m_O X \quad (7)$$

### *2.2.4. Heat balance in the bioreactor*

Heat transfer in the tray bioreactor occurs in a thin layer of the bed surface due to airflow at the top and bottom of the bed. Besides, there is a certain amount of heat generated from the metabolism of microorganisms. The heat transferred to the top of the bed is evaporation heat, which causes some of the bed water to evaporate. As microbes grow, a temperature gradient has formed accordingly. This temperature gradient produces sensible heat transferred from the bed into the air at the top of the bed. In general, the equation is written,

$$m_b C p_b \frac{dT_b}{dt} = Q_{met} - Q_{sen} - Q_{evap} \tag{8}$$

The metabolic heat produced because of microorganism growth was defined and written as,

$$r_Q = Q_{met} = Y_{Q/X} r_X + m_O X \tag{9}$$

$$Q_{met} = (\frac{460.000}{Y_{O/X}})\, r_X \tag{10}$$

The formation of metabolic heat can increase the bed temperature; in the tempeh production process, it has been reported that an increase in temperature reaches 3°C/cm with a bed height of 5.1 cm [12]. Even the centre's bed temperature can reach 47°C to 49.6°C under saturated air entering at a rate of 0.47 Lpm. Thus the metabolic heat has a significant impact on the increase in bed temperature. One way to reduce the bed's temperature is to provide saturated air with relative humidity above 95%. If the saturated air temperature is lower than the bed's temperature, there is a sensible heat transfer. The sensible heat transfer is expressed in the equation;

$$Q_{sen} = -U_t A_b (T_b - T_{in}) - U_b A_b (T_b - T_{in}) \tag{11}$$

Because the conductivity of aluminium (pan material) is much greater than that of the bed, then Ut = Ub so that

$$Q_{sen} = -2\, U_t A_b (T_b - T_{in}) \tag{12}$$

As long as the bed temperature is higher than the bioreactor temperature, the bed's cooling by the release of sensible heat is still quite significant. However, if the bed temperature is the same as the bioreactor temperature, the sensible heat release is no longer effective. Thus, cooling the bed needs to be assisted by introducing saturated air so that the bed's temperature reduction is more effective [9].

In the evaporation process, a diffusion process requires a mass transfer coefficient in the model's equation. Because this study was lack of experiment data, the model uses a correlation equation for the mass transfer coefficient [13]. The humidity gradient between the bed and air becomes the driving force of the evaporation process so that the heat equation for evaporation becomes,

$$Q_{evap} = \lambda\, k_m A_b (\mathcal{H}_b - \mathcal{H}_{in}) \tag{13}$$

*2.2.5. Water balance*

The system needs humidity to dissipate heat metabolites because water's heat capacity is greater than air, making it easier to absorb heat. If the heat can be controlled, water evaporation from the bed can be prevented to maintain productivity. The water can be free water available outside of microorganisms, metabolized water evaporated into the macro environment, and water becomes a nutrient solvent. The water balance is divided into two: the water balance in the bed and the bioreactor.

Because the bed is a closed system, it only involves changes that occur in the bed, namely the formation of water metabolites and loss of evaporated water.

$$r_{W_b} = r_{W_{met}} - r_{W_{evap}} \tag{14}$$

Metabolite water is water produced when microorganisms grow, so the water metabolite equation is analogous to the enzyme production equation, namely,

$$r_{W_{met}} = Y_{W/X}\, r_X \tag{15}$$

Evaporation water is water that is evaporated from the bed because there is a humidity gradient or water activity between the bed and the bioreactor [9]. The rate of evaporation was shown in equation (16) and needed adjustments so that each component of the equation has the same unit,

$$r_{W_{evap}} = k_m A_b(\mathcal{H}_b - \mathcal{H}_{in}) \tag{16}$$

There were some adjustments made on the model to maintain each term had the same unit, as $k_m$ was in m/h while $r_{W_{evap}}$ was in g water/g OPEFB-h.

### 2.3. Simulation

The model simulation was performed using MATLAB and ode45 solver according to the model that has been built, one-chamber bioreactor, serial airflow in two-chamber, and four-chamber bioreactor and parallel airflow in the four-chamber bioreactor. The coefficients used are tabulated in Table 2.

The experiments that have been performed by using 1 kg of OPEFB in 2 trays [3] compared to the model built.

**Table 2.** Coefficients and parameters used in the model.

| SYMBOL | NAME | VALUE | UNIT (SOURCE) |
|---|---|---|---|
| $A_i$ | Cross-sectional area of bioreactor | 0,2 | $m^2$ (calculated) |
| $A_b$ | Bed surface area | 0,2704 | $m^2$ (calculated) |
| $C_{p,b}$ | Specific heat capacity of OPEFB | 1,4827 | J/g-K [14] |
| $E$ | Enzyme produced | Simulated | gram |
| $F_g$ | The volumetric flow rate of air | 0,006 | $m^3$/hr |
| $\mathcal{H}_b$ | Bioreactor humidity | Simulated | g water/g dry air |
| $\mathcal{H}_{in}$ | Air inlet humidity | Calculated | g water/g dry air |
| $h_c$ | Convective heat transfer coefficient | 61,5 | W/$m^2$-K [15] |
| $k_m$ | Water mass transfer coefficient | Correlation | [13] |
| $L$ | Pan length | 0,52 | m |
| $m_b$ | OPEFB mass | 1000 | gram |
| $Oks$ | $O_2$ mole fraction | Simulated | - |
| $P_{tot}$ | The total pressure of the system | 1 | atm |
| $p^{sat}$ | Saturated pressure | Calculated | atm |
| $p_a$ | The partial pressure of water vapour | Calculated | atm |
| $R$ | Ideal gas constant | 0,000082 | m3-atm/mol-K |
| $T_b$ | Bed temperature | Simulated | K |
| $T_{in}$ | Inlet air temperature | 301,15/310,15 | K |
| $V_i$ | Bioreactor volume | 0,12 | $m^3$ (calculated) |
| $V_b$ | Bed volume | 0,0189 | $m^3$ (calculated) |
| $y_{O_{2,in}}$ | $O_2$ mole fraction of inlet air | Simulated | - |
| $y_{O_{2,out}}$ | $O_2$ mole fraction of outlet air | Simulated | - |
| $y_{air,in}$ | Water mole fraction of inlet air | Simulated | - |
| $y_{air,out}$ | Water mole fraction of outlet air | Simulated | - |
| $Y_{E/X}$ | Yield coefficient of enzyme produced per grown cell | 1,4-gram/gram cell | Data |
| $Y_{Q/O}$ | Yield coefficient of heat produced per $O_2$ consumed | 460 | kJ/mol [16] |

**Table 3.** Coefficients and parameters used in the model (continued).

| SYMBOL | NAME | VALUE | UNIT (SOURCE) |
|---|---|---|---|
| $Y_{O/X}$ | Yield coefficient of consumed $O_2$ per grown cell | 0,7 | [17] |
| $U_t$ | Global heat transfer coefficient | 4524,589 | J/jam-m$^2$-K (calculated) |
| $\lambda$ | Heat of evaporation | 2502 | J/g water [18] |
| $X$ | Cell mass | Simulated | gram |
| $\mu$ | The specific growth rate coefficient | Calculated | 1hr |

## 3. Results

The simulation for studying the effect of aeration is presented in figure 2. The simulation was performed on one-chamber bioreactor model.

### *3.1. Effect of absence and presence of aeration on one-chamber bioreactor*

Figure 2a showed that aeration caused a sudden decrease in the bed temperature (solid line), while a non-aerated process can maintain the bed temperature at ±310.15K for about 45 hours (dashed-line). The sudden decrease in bed temperature affected cell growth, as can be seen in figure 2b. As cell growth was decreased, the Oxygen was consumed at a lower rate, and it did not show any changes up to 120 hours of fermentation time. While the water balance showed that there were only slight changes in water in bed, and bioreactor humidity did not show any changes. These changes will not occur in the experiment because as microorganisms grow, there will be enough metabolic heat produced and increase bed temperature [12]. The expected decrease in bed temperature was too fast, presumably due to relatively low growth, while the air rate was too large. For this reason, changes in the rate of air are also simulated at a value of 0.003 m3 / hour and 0.012 m3 / hour. The results are presented in Table 3.

As shown in Table 3, the higher the aeration flow rate, the bed temperature decreases started from early simulation time and becomes lower than the optimal temperature (310.15K) at the end of simulation time. The difference between those three is the rate of the bed temperature increase. The model explains the effect of the airflow rate of air on changes in bed temperature. However, the decrease in bed temperature due to aeration appears extreme because it occurs as soon as the bed comes into contact with air.

The model simulation showed different results compared to real experimental data. This difference occurs because the air and heat were distributed homogeneously in the model. The real condition indicated that the bioreactor was not homogeneously mixed. Furthermore, the experiment can not validate the model because the bioreactor was not as ideal as the model, and the comparisons can be seen in Table 4.

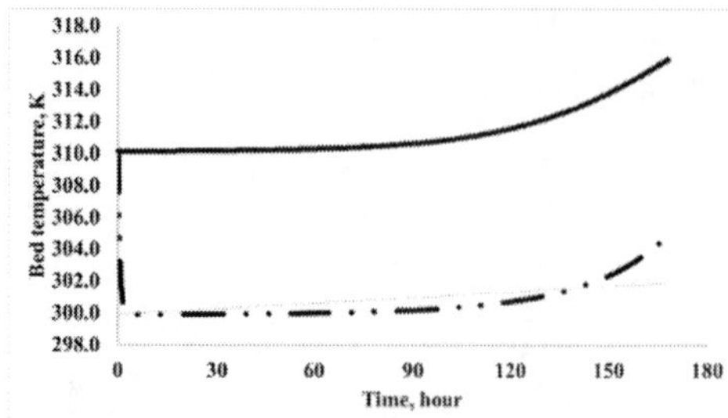


(a) Bed temperature, K

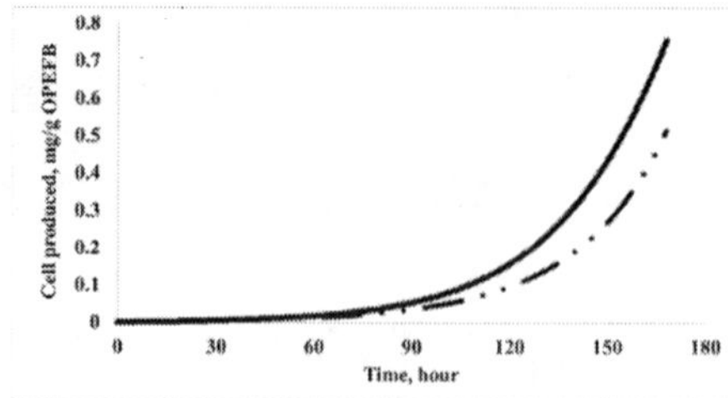


(b) The cell produced mg/mg OPEFB

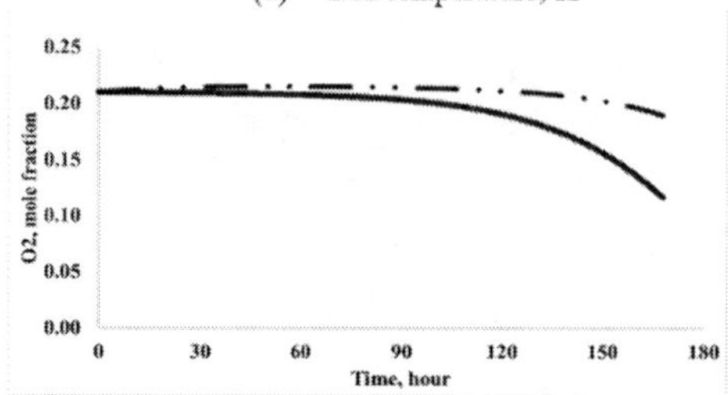


(c) O2 mole fraction

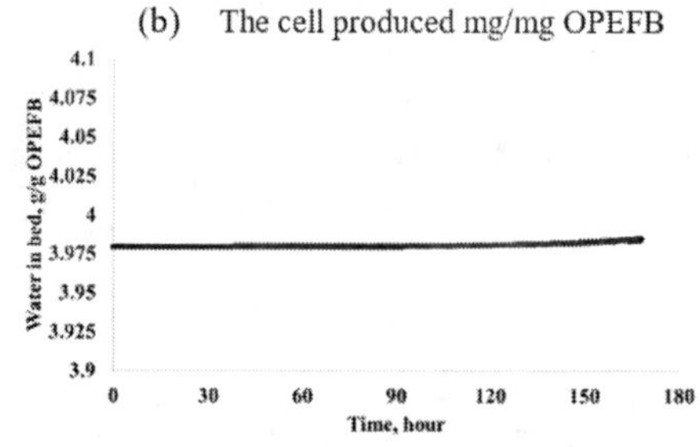


(d) Water in bed, g/g OPEFB

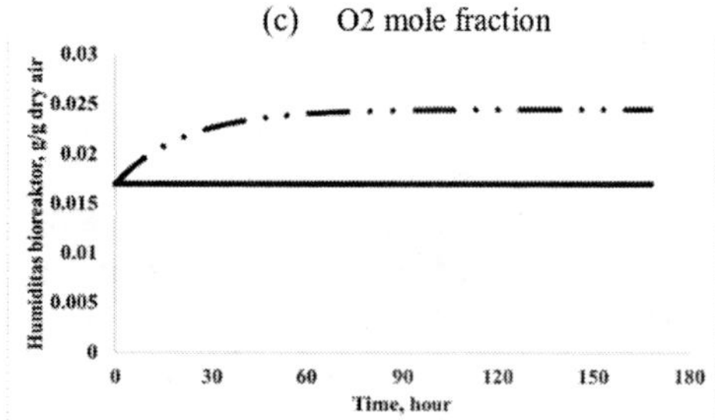


(e) Bioreactor humidity, g/g dry air

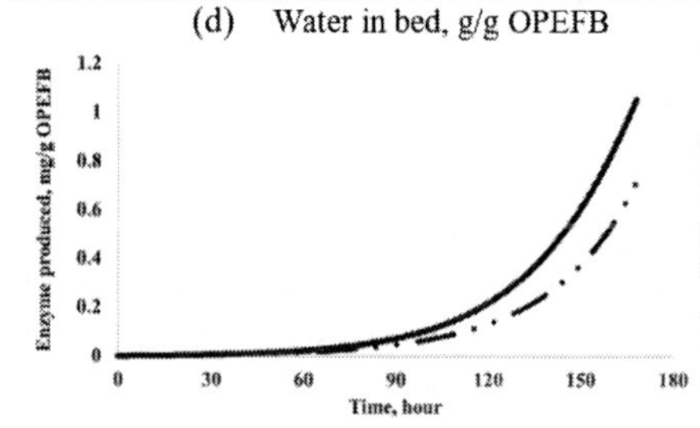


(f) The enzyme produced, mg/g OPEFB

**Figure 2.** The aeration effect on one-chamber bioreactor, solid line for no aeration and dashed line for aeration at 0.1 $LM^{-1}$.

**Table 4.** Aeration flow rate effect on bed temperature.

| Airflow rate, m3/jam | Tb at 24 h (K) | Tb at 100 h (K) | Tb at 168 h (K) |
|---|---|---|---|
| 0 | 310.2 | 310.9 | 315.9 |
| 0.003 | 300.6 | 301 | 306 |
| 0.006 | 299 | 300 | 305 |
| 0.012 | 299 | 299 | 302 |

**Table 5.** Comparison between experimental data and model.

| Aeration | Experiment | | Model | |
|---|---|---|---|---|
| | Cell, mg/g OPEFB | The enzyme mg/g OPEFB | Cell, mg/g OPEFB | The enzyme mg/g OPEFB |
| Without | 0.495 | 0.632 | 0.755 | 1,054 |
| With | 0.796 | 0.946 | 0.516 | 0.719 |

### *3.2. The effect of the division of the bioreactor chamber in serial airflow pattern*

As has been reported, increasing the tray bioreactor's production capacity due to an increase in the bioreactor space in the form of the number of trays [19]. The simulation was tabulated in table 5.

From this simulation, it can be concluded temporarily that the serial flow pattern in two and four-chamber causes the system to be not homogeneous. It can be seen from the gap in the number of cell products and enzymes in each bioreactor chamber. The division of the bioreactor is identical to the division of the mass of the media. In contrast, the division of the flow pattern in series causes Oxygen's supply to the second chamber, third and fourth to decrease because it has been consumed. Therefore, with the same airflow rate of 0.006 m3 / hour, the four chambers oxygen intake is higher than the two-chamber. This condition does not cause a decrease the bioreactor's total productivity because the cell and enzyme production is accumulated. Following what was stated by Mitchell et al. (2006), increasing the chambers used will increase fermentation production in a tray bioreactor. Although each chamber's production is different, the total production of cells and enzymes increases with increasing bioreactor chambers.

**Table 6.** Resume of simulation results.

| Chamber | Cell, mg cell/g OPEFB | | | The enzyme, mg xylanase/g OPEFB | | |
|---|---|---|---|---|---|---|
| | One-Chamber | Two-Chamber | Four-Chamber | One-Chamber | Two-Chamber | Four-Chamber |
| **Chamber 1** | 0.52 | 0.85 | 1.90 | 0.98 | 1.60 | 3.60 |
| **Chamber 2** | | 0.03 | 0.03 | | 0.05 | 0.03 |
| **Chamber 3** | | | 0.02 | | | 0.02 |
| **Chamber 4** | | | 0.02 | | | 0.03 |
| **Total** | | 0.44 | 0.50 | 0.98 | 0.82 | 0.92 |

## 3.3. *The effect of parallel airflow pattern*

Airflow patterns simulations were carried out by dividing the compressor's airflow, fixed at 0.006 m3 / hr, into four bioreactor chambers. The results obtained are presented in table 6.

**Table 7.** Resume of simulation.

| Chamber | Cell, mg/g OPEFB | | | The enzyme mg/g OPEFB | | |
|---|---|---|---|---|---|---|
| | One-Chamber | 4-Paralel | 4-Serial | One-Chamber | 4-Paralel | 4-Serial |
| Chamber 1 | 0.52 | 2.43 | 1.90 | 0.98 | 4.60 | 3.60 |
| Chamber 2 | | 2.43 | 0.03 | | 4.60 | 0.03 |
| Chamber 3 | | 2.43 | 0.02 | | 4.0 | 0.02 |
| Chamber 4 | | 2.43 | 0.02 | | 4.60 | 0.03 |
| Total | 0.52 | 2.43 | 0.50 | 0.98 | 4.60 | 0.92 |

Based on the tabulation results in Table 6, it can be seen that a parallel bioreactor, resulting in higher total cell and enzyme production when compared to the series flow. There is an increase of approximately five times the yield in the four-chambers bioreactor. The parallel bioreactor ensures the availability of fresh air supply so that cell growth is maximized and uniform even though the first chamber of the bed in the serial bioreactor produces more cells. The first chamber's serial bioreactor produces more cells and enzymes than the first chamber of a parallel bioreactor. This condition was due to the more intense heat transfer that occurs in the one-chamber bioreactor. Even though the total intake air flow rate is the same, the air's linear velocity in the bioreactor is different. The linear velocity of air in the one-chamber bioreactor is higher than in the four-chamber parallel bioreactor. This condition allows for better heat transfer to occur in parallel bioreactors. Heat transfer takes place more slowly so that the decrease in layer temperature is slighter and can even return to the optimal temperature for microbial growth. As a result, microbes' oxygen consumption occurs slower in a one-chamber bioreactor because the bed temperature is lower than the optimal temperature. The bed water in the one-chamber bioreactor appears to increase faster than in the parallel bioreactor. All of these changes result in better productivity of the parallel bioreactor than the one-chamber bioreactor.

This result follows Mitchell et al. [19], who stated that to increase production capacity using a tray bioreactor, namely by increasing the number of trays or trays. Thus it is important to apply parallel bioreactor arrangement so that homogeneous conditions are maintained, and the tray bioreactor can increase productivity. The use of four lamps in the experiment was an attempt to distribute heat. Thus in the model, four lamps will be modelled as a variation of the four-chamber bioreactor. The results of the comparison are presented in Table 7.

Model simulation results and experimental results show a similar phenomenon; cell production increases when the process occurs in more chambers. However, there was an anomaly in the production of enzymes experimentally. But it was believed this was due to the inaccurate protein analysis process because it cannot distinguish between cells and fungal spores. The enzyme production simulated by the series bioreactor arrangement (0.92 mg / g OPEFB) was close to experimental results (0.78 mg / g OPEFB). However, in the experiment using four lamps, data was not obtained for each pan, so it was impossible to determine the system conditions. Therefore, model validation is needed to ensure the correctness of the model.

**Table 8.** Comparison of four-chamber in Experiment and Simulation.

| Chamber | Experiment | | Simulation | |
|---|---|---|---|---|
| | Cell, mg/g OPEFB | The enzyme mg/g OPEFB | Cell, mg/g OPEFB | The enzyme mg/g OPEFB |
| One-Chamber | 1.18 | 0.82 | 0.52 | 0.98 |
| Four-Chamber | 1.49 | 0.78 | | |
| Four-Chamber serial | - | - | 0.50 | 0.92 |
| Four-Chamber parallel | | | 2.43 | 4.60 |

## 4. Conclusion

A model on tray bioreactor for solid-state fermentation has been developed. An ideal, homogeneously mixed heat and aeration (one chamber) can not explain the phenomena: aeration improves growth and enzyme productivity. This phenomenon showed, in reality, the fermentation is far from a homogenously mixed system. The phenomenon of the presence or absence of aeration was explained well in 2 or 4 bioreactor chambers. What happens to the bioreactor can be predicted with a partially mixed system. The airflow pattern in the bioreactor tray used can be changed to parallel to increase productivity.

## References

[1] Cerda A, Martínez M E, Soto C, Zúñiga M E, and Poirrier P 2016 *Renew. Energy* **87** (1) 326
[2] Chen H 2013 *Modern Solid State Fermentation: Theory and Practice* Springer Science & Business Media
[3] Meilany D, Anugeraheni D, Aziz A, Kresnowati M T A P, and Setiadi T 2020 *Reaktor* 20, 32
[4] Rajagopalan S and Modak J M 1994 *Chem. Eng. Sci.* **49** 2187
[5] Raghava Rao K S M S, Gowthaman M K, Ghildyal N P, and Karanth N G 1993 *Bioprocess Eng.* **8** 255
[6] Figueroa-Montero A, Esparza-Isunza T, Saucedo-Castañeda G, Huerta-Ochoa S, Gutiérrez-Rojas M, and Favela-Torres E 2011 *J. Chem. Technol. Biotechnol.* **86** 1321
[7] Manan M A and Webb C 2018 *Biotechnol. Biotechnol. Equip.* **32** 1455
[8] Mitchell D A, von Meien O F, Krieger N, and Dalsenter F D H 2004 *Biochem. Eng. J.* **17** 15
[9] Mitchell D A, Berovič M, and Krieger N, 2006 *Solid-State Ferment. Bioreact. Fundam. Des. Oper.*, edited by Mitchell D A, Berovič M, and Krieger N 2006 *Springer Berlin Heidelberg, Berlin, Heidelberg* 1–12
[10] Luedeking R and Piret E L 1959 *J. Biochem. Microbiol. Technol. Eng.* **1** 393
[11] Ooijkaas L P, Buitelaar R M, Tramper J, and Rinzema A 2000 *Biotechnol. Bioeng.* **69** 292
[12] Rathbun B L and Shuler M L 1983 *Biotechnol. Bioeng.* **25** 929
[13] Cussler E L and Cussler E L 1997 *Diffusion: Mass Transfer in Fluid Systems* Cambridge University Press
[14] Kaban H, Penelit J 2009 *Sains* **12** 168252

[15] Rodrigues L G G, Parisotto E I B, Carciofi B a M, Laurindo J B, Rodrigues L G G, Parisotto E I B, Carciofi B a M, and Laurindo J B, Braz 2017 *J. Chem. Eng.* **34** 171

[16] Nielsen J and Villadsen J 1994 *Bioreaction Engineering Principles* Plenum Press, New York, NY

[17] Pandey A 2003 *Solid-State Ferment.* **13** 81

[18] Perry R H, Green D W, and Maloney J O 1997 *Perry's Chemical Engineers' Handbook* McGraw-Hill

[19] Mitchell D A, von Meien O F, Luz L F L, and Berovič M, 2006 *Solid-State Ferment. Bioreact. Fundam. Des. Oper.*, edited by Mitchell D A, Berovič M, and Krieger N *Springer, Berlin, Heidelberg* 57–64

# Syngas fermentation for production of ethanol

**N A Istiqomah [1*], M T A P Kresnowati [1,2] and T Setiadi [1,3]**

[1] Department of Chemical Engineering, Faculty of Industrial Technology Institut Teknologi Bandung, Jl. Ganesa 10, Bandung 40132, Indonesia
[2] Department of Food Engineering, Institut Teknologi Bandung, Jl. Ganesha 10, 40132 Bandung, Indonesia
[3] Centre for Environmental Studies (PSLH), Institut Teknologi Bandung, Jl. Sangkuriang No. 42A, 40135 Bandung, Indonesia

noviani105@students.itb.ac.id

**Abstract**. The global demand for fossil fuels has increased sharply in the past 50 years. Plant biomass is one of the fourth largest non-fossil renewable energy sources after geothermal, solar, and wind energy. Biomass can be first converted into another energy form, such as ethanol, to provide a more efficient energy source. Syngas fermentation enables converting all biomass components, including lignin, into ethanol, thereby increasing the biomass quantity. The development of the syngas fermentation process will be reviewed in this article, which focuses on the types of microorganisms used, the effect of syngas composition, and the gas-liquid mass transfer to be used as a reference for optimizing the syngas fermentation process.

## 1. Introduction

Syngas conversion into bioenergy through fermentation has been carried out since 1983, and this research spike occurred in 2010 until now. Syngas is a gas mix containing CO, $H_2$, and $CO_2$ produced from gasification of raw material lignocellulose and other carbon raw materials [1].

The conversion of synthetic gas to liquid biofuel and biochemistry can be carried out in two ways, first, using a metal catalyst through Fischer-Tropsch Synthesis (FT) and second, utilize a microbial catalyst known as syngas fermentation [2]. The weaknesses in FT Synthesis are the high cost of the metal catalysts, Inert gases and contaminants can be potentially toxic when reacting with catalysts, operating using high temperatures and pressures, the ratio and quality of CO and $H_2$ gases must be consistent [3].

(Homo) acetogen is the name for a microbial catalyst that can ferment syngas. There are more than 20 genera and more than 100 different species of acetogens [4]. Acetogens can produce various types of chemicals and biofuels and have advantages over FT synthesis, including a higher specificity, a more flexible ratio of CO and $H_2$, not using expensive metal catalysts so that the potential for metal poisoning is small, operations are carried out in ambient conditions [3]. However, fermented syngas still has challenges that need a solution, namely $H_2$ and CO's low solubility in the water phase and low product conversion.

## 2. Syngas Characteristics

The operation process and the type of raw material from the gasification determine the gas's composition. The gas composition can be a mixture of $CO_2$, CO, $H_2$, $N_2$, $CH_4$, and a small amount of

NOx, SOx, tar, char, and hydrocarbons [34]. The composition of the gas affects the results of syngas fermentation. If pure syngas is used, the resulting product will be high.

Syngas impurities can be nitrogen oxide (NO), tar, $H_2S$, and carbonyl sulfide (COS). Nitrogen oxide above 40 ppm will affect cell growth and hydrogenase activity. Cell growth can be inhibited by the presence of tar [35]. However, acetogen tolerance to toxic impurities, $H_2S$ (<2%), and COS are still better than inorganic catalysts [36].

## 3. Microbiology and Metabolic Pathways

Syngas fermentation carried out by acetogens can produce various useful products, besides anaerobic conditions can reduce the risk of contamination. The fixation of $CO_2$ to acetyl-CoA in acetogen takes place effectively through the Wood Ljungdahl Pathway (WLP) so that WLP is known as acetyl-CoA reductive pathway [5].

The fermentation stage is generally divided into two stages: the acidogenesis and solventogenesis stages [6]. Acidogenesis stage was a growth stage when cells would grow, accompanied by acetate and butyrate as products. This process causes the pH of the medium to drop. As more acetate is formed, the cell growth rate will continue to decline until it eventually becomes zero. In this time, the acetogens respond by consuming the acetic acid. Once acetate starts to be toxic to cells, it is converted to solvent, which in this case, is ethanol. This process causes the pH of the solution to rising again. The process will continue until the medium's nutrients run out, or the acid crash phenomenon occurs. In this phenomenon, cells are already inactive due to too much acid and have poisoned the cells.

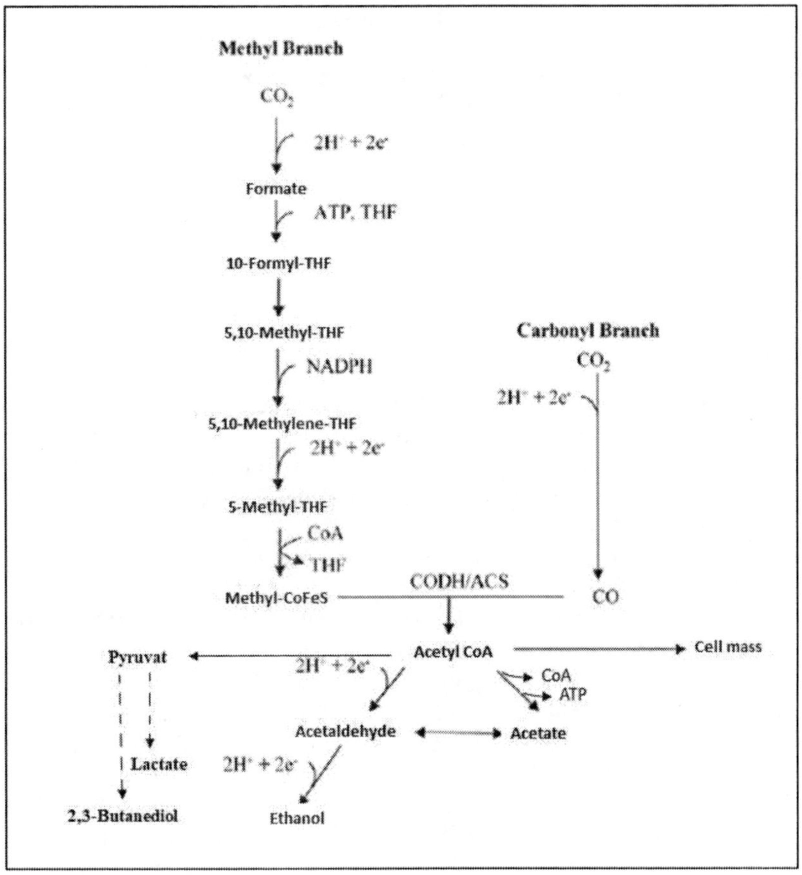


**Figure 1**. Wood-Ljungdahl Pathway (WLP), adapted from [7,8].

Energy for the sustainability of WLP comes from ATP from the formation of acetate from acetyl-CoA, in figure 1. Production ATP takes place in the cell growth phase. When the cells enter the stationary phase, acetyl-CoA is reduced to ethanol. Apart from ethanol, acetyl-CoA can form other products, such as isopropanol, butyrate, butanol, hexanol, and butanediol [9]. The difference in the resulting product depends on the type of acetogen used. Information regarding the types of acetogens, fermentation conditions (pH and temperature optimal), and the resulting products with CO, $H_2$, and $CO_2$ as the substrates can be seen in table 1.

**Table 1.** Acetogen for syngas fermentation to produce ethanol.

| Organism | $pH_{opt}$ | $T_{opt}$ (°C) | Products | References |
|---|---|---|---|---|
| *Clostridium drakei* | 3,6-6,8 | 25-30 | Ethanol, acetate, butyrate | [10–14] |
| *Clostridium scatologenes* | 5,4-7,5 | 37-40 | Ethanol, acetate, butyrate | [10,11] |
| *Clostridium autoethanogenum* | 5,8-6 | 37 | Ethanol, acetate, lactate, 2,3-butanediol | [15–18] |
| *Butyribacterium methylotropicum* | 6 | 37 | Ethanol, acetate, butyrate, butanol | [19–21] |
| *Clostridium ljungdahlii* | 6 | 37 | Ethanol, acetate, lactate, 2,3-butanediol | [22–25] |
| *Clostridium ragsdalei*/P11 | 6,3 | 37 | Ethanol, acetate, lactate, 2,3-butanediol | [16,26] |
| *Clostridium coskatii* | 5,8-6,5 | 37 | Ethanol, acetate | [27] |
| *Alkalibaculum bacchi* | 8,0-8,5 | 37 | Ethanol, acetate | [28,29] |
| *Clostridium carboxidivorans*/ P7 | 6,2 | 38 | Ethanol, acetate, butyrate, butanol, lactate | [11,30] |
| *Clostridium difficile* | 6,5-7,0 | 35-40 | Ethanol, acetate, butyrate | [31–33] |

## 4. Factors that Affect Syngas Fermentation

### *4.1 Temperature and pH*

The temperature factor affects microorganisms' growth and the mass transfer of gas to liquid fermentation syngas medium. Each microorganism used has an optimum temperature for its respective growth. However, operation at optimum growth time does not guarantee that the expected product is formed. The temperature used in syngas fermentation depends on the type of microbe and the product expected.

For example, *C. carboxidivorans* P7 grown at an optimum temperature of 37 °C, the product produced in high quantities is acetic acid, while ethanol is obtained in low amounts. The condition occurs because the production of acetic acid is linear with cell growth. When fermentation operation at a temperature of 25 °C, cell growth is not fast, but the ethanol produced is much more [37].

Another effect of temperature is its effect on gas transfer to the liquid medium of fermentation. The solubility of gases in the liquid will decrease the higher temperature. However, increasing the temperature will decrease the solution's viscosity, which reduces in the size of the bubbles so that the area of contact between the gas-liquid phases will increase so that the kLa (gas-liquid transfer coefficient) value will increase. The use of temperature in syngas fermentation is determined based on the microorganisms used. Mesophilic microorganisms grow in a temperature range of 37 to 40 °C, while thermophilic microorganisms are at a temperature of 55 °C and extreme-thermophilic at a temperature of 80 °C [1].

Like temperature, each microorganism has an optimum pH value. However, in general, a decrease in pH triggers solventogenesis (formation of ethanol and butanol). The use of different fermenters can be done to get optimal results. The first reactor aims to support growth and acid formation so that it is maintained at pH 6, then flowed to the second fermenter, which is maintained at pH 5 so that acid and cells from the first fermenter are used to form ethanol [38].

### *4.2 Growth media culture*

The growth medium must meet microbes' needs to grow optimally, including vitamins, minerals, trace elements, and reducing agents. The growth medium used, pH, and temperature are determined by the microorganisms used and the product expected. Special media are readily available at the American-Type Culture Collection (ATCC).

### *4.3 Mass transfer*

In syngas fermentation, the gas substrate acts as a source of carbon and energy to produce by-products. The gas transfer steps to microbial cells include gas diffusion from the bulk gas to the gas-liquid interface, then movement across the gas-liquid interface, transportation to the bulk liquid around the microbial cell, and diffusive via the liquid-solid frontier [34].

The syngas fermentation process's biggest problem is the transfer of mass gas to the liquid fermentation medium. The mass transfer problem is more significant than aerobic processes because syngases' solubility is much less than oxygen. CO Solubility is only about 76% of oxygen, while $H_2$ is only about 55% [39], and low substrate solubility results in low product productivity.

The mass transfer rate is quantified in the following equation:

$$\frac{dC}{dt} = k_L a\,(C^* - C_L) \qquad (1)$$

In that equation, dC/dt is the amount of gas that moves from the gas to liquid phase per unit time, $k_L$ ($ms^{-1}$) is the mass transfer coefficient, an ($s^{-1}$) is the specific mass transfer area, C* is the concentration of saturated gas in liquids. $C_L$ is the concentration of the gas dissolved in the liquid, which is considered to be in equilibrium with the gas phase concentration according to Henry's law. Based on these equations, the components that can be modified to increase the mass transfer rate increase the area of mass transfer or enlarging the driving force. If using gas with low solubility, the $k_La$ depends on the size and number of bubbles by controlling the agitation speed, the gas and liquid medium's flow rate, and the reactor configuration [40].

A simple approach to increasing the area of mass transfer is to increase the propeller's rotational speed, but this is less applicable because a massive rotational speed will damage cells in the fermenter, high energy costs, and decreased productivity [41]. Minimizing the driving force's influence can be done by increasing the gas's residence time through an increase in the gas flow rate, especially CO, as the primary carbon source. However, CO can inhibit cell hydrogenase enzyme activity, so another alternative is needed to solve mass transfer [4].

### *4.4 Reactor configuration*

The mass transfer effectiveness is also related to the bioreactor configuration. Parameters in designing bioreactor efficiency include high mass transfer rate, bioreactor size and ease of scale-up, high cell density, and low budget operation and maintenance [1].

A mass transfer without agitation can be done to minimize power, including by modifying the bubble flow pattern in the Bubble Column Reactor (BCR) [44], using packaging media in the Trickle Bed

Reactor (TBR) [45], and the use of microbubble generators in Hollow Fiber Membrane Reactors (HFMR). HFMR has the highest mass transfer effectiveness to date, with a $k_La$ CO value of 1096.2 $h^{-1}$ [43]. The kLa values for various reactors can be seen in table 2.

## 5. Conclusion

Syngas fermentation can be carried out by acetogens consisting of various species and can produce various useful products through the Wood-Ljungdahl Pathway (WLP). The growth medium used, pH, and temperature are determined by the microorganisms used and the product expected. Bottleneck from syngas fermentation lies in the low mass transfer of gas to the liquid fermentation medium. Transfer mass gas-liquid depends on the size and number of bubbles by controlling the agitation speed, the gas and liquid medium's flow rate, and the reactor configuration. HFMR is the right choice for syngas fermentation because it has the highest mass transfer. The use of HMFR for a large scale still has several challenges because it is predicted to require high costs. The use of membranes from reusable materials to reduce costs is an efficient alternative to commercial membranes.

Further research on alternative materials based on sustainability, cost, and durability needs to be done. The efficiency of the HFMR can still be improved by making internal adjustments. Application features such as height to diameter ratio and the use of micro sparger can be studied further.

IOP Conf. Series: Materials Science and Engineering 1143 (2021) 012014 doi:10.1088/1757-899X/1143/1/012014

**Tabel 2.** Ethanol production from syngas fermentation in various reactor

| Type of Reactor | Reactor Scale | Microbe | Substrat $CO:H_2:CO_2:N_2$ | Gas Flow | T (°C) | pH | Cell Density (g/L) | $K_La$ ($h^{-1}$) | Yield Ethanol (g/L) | Reference |
|---|---|---|---|---|---|---|---|---|---|---|
| Batch | 125 mL | *C. ljungdahlii* | 55:20:10:Ar 15 | 200 uL | 37 | | 1,15 | 135 | 0,55 | [46] |
| CSTR | 250 mL | *C. ljungdahlii and C. autoethanogenum* | 20:10:20:50 | 5,75 and 10 mL/min | 37 | 5,5 and 6,8 | 0,5 | NM | 0,17-0,29 | [47] |
| Fed-batch | 5 L | *C. ljungdahlii* | 80:0:20 | NA | 37 | 5,4 | 1,5 | NM | 0,9 | [48] |
| CSTR | 3,3 L | *A. bacchi* | 20:5:15:60 | 200 sccm | 37 | 8 | 0,4-5,05 | NA | 0,2-6,39 | [49] |
| Batch | 250 mL | *C. ljungdahlii* | 20:5:20:55 | | 37 | 6,8 | | NA | 0,306 | [50] |
| CSTR | 1,2 L | *C. autoethanogenum* | CO 100% | NA | 30 | 4,75 and 5,75 | 0,15 | NM | 0,35 | [51] |
| Trickle bed | 500 mL | *C. ragsdalei* | 38:28,5:28,5:5 | | 37 | | | NM | 5,7; 12,3 | [42] |
| CSTR | 2 L | *Clostridium ljungdahlii* | 12:35:5:0 | 5-15 mL/min | 37 | 4,5 | 2,4 | | 0,8-3,75 | [52] |
| CSTR | 1,2 L | *C. carboxidivorans* | 30:20:10:40 | | 33 | varies | OD 600=1 | | 0,2-1,2 | [38] |
| STR-external HFM | 8 L | *C. carboxidivorans* | 20:5:15:60 | 300 mL/min | 37 | 4,5-5,5 | | 1096,2 | 23,93 | [43] |
| STR-external HFM | 800 mL | *C. ljungdahlii* | 25:15:20:40 | 80 mL/min | 37 | 6 | | 300,5 | 1,09 | [25] |

**References**

[1] Yasin M, Cha M, Chang I S, Atiyeh H K, Munasinghe P and Khanal S K 2019 *Syngas fermentation into biofuels and biochemicals* (Elsevier Inc.)

[2] Munasinghe P C and Khanal S K 2011 Biomass-derived syngas fermentation into biofuels *Biofuels* (Elsevier Inc.) pp 79–98

[3] Bredwell M D, Srivastava P and Worden R M 1999 Reactor Design Issues for Synthesis-Gas Fermentations *Biotechnol. Prog.* **15** 834–44

[4] Bertsch J and Müller V 2015 CO metabolism in the acetogen Acetobacterium woodii *Appl. Environ. Microbiol.* **81** 5949–56

[5] Drake H L, Gößner A S and Daniel S L 2008 Old acetogens, new light *Ann. N. Y. Acad. Sci.* **1125** 100–28

[6] Jiang Y, Guo D, Lu J, Dürre P, Dong W, Yan W, Zhang W, Ma J, Jiang M and Xin F 2018 Consolidated bioprocessing of butanol production from xylan by a thermophilic and butanologenic Thermoanaerobacterium sp. M5 *Biotechnol. Biofuels* **11** 1–14

[7] Phillips J R, Huhnke R L and Atiyeh H K 2017 Syngas fermentation: A microbial conversion process of gaseous substrates to various products *Fermentation* **3**

[8] Liew F M, Martin M E, Tappel R C, Heijstra B D, Mihalcea C and Köpke M 2016 Gas Fermentation-A flexible platform for commercial scale production of low-carbon-fuels and chemicals from waste and renewable feedstocks *Front. Microbiol.* **7**

[9] Schoelmerich M C and Müller V 2019 Energy conservation by a hydrogenase-dependent chemiosmotic mechanism in an ancient metabolic pathway *Proc. Natl. Acad. Sci. U. S. A.* **116** 6329–34

[10] Küsel K, Dorsch T, Acker G, Stackebrandt E and Drake H L 2000 Clostridium scatologenes strain SL1 isolated as an acetogenic bacterium from acidic sediments *Int. J. Syst. Evol. Microbiol.* **50** 537–46

[11] Liou J S C, Balkwill D L, Drake G R and Tanner R S 2005 Clostridium carboxidivorans sp. nov., a solvent-producing clostridium isolated from an agricultural settling lagoon, and reclassification of the acetogen Clostridium scatologenes strain SL1 as Clostridium drakei sp. nov *Int. J. Syst. Evol. Microbiol.* **55** 2085–91

[12] Gößner A S, Picardal F, Tanner R S and Drake H L 2008 Carbon metabolism of the moderately acid-tolerant acetogen Clostridium drakei isolated from peat *FEMS Microbiol. Lett.* **287** 236–42

[13] Jeong Y, Song Y, Shin H S and Cho B K 2014 Draft genome sequence of acid-tolerant Clostridium drakei SL1T, a potential chemical producer through syngas fermentation *Genome Announc.* **2**

[14] Bengelsdorf F R, Poehlein A, Schiel-Bengelsdorf B, Daniel R and Dürre P 2015 Genome sequence of the acetogenic bacterium Oxobacter pfennigii DSM 3222T *Genome Announc.* **3**

[15] Abrini J, Naveau H and Nyns E J 1994 Clostridium autoethanogenum, sp. nov., an anaerobic bacterium that produces ethanol from carbon monoxide *Arch. Microbiol.* **161** 345–51

[16] Köpke M, Mihalcea C, Liew F M, Tizard J H, Ali M S, Conolly J J, Al-Sinawi B and Simpson S D 2011 2,3-Butanediol production by acetogenic bacteria, an alternative route to chemical synthesis, using industrial waste gas *Appl. Environ. Microbiol.* **77** 5467–75

[17] Brown S D, Nagaraju S, Utturkar S, De Tissera S, Segovia S, Mitchell W, Land M L, Dassanayake A and Köpke M 2014 Comparison of single-molecule sequencing and hybrid approaches for finishing the genome of Clostridium autoethanogenum and analysis of CRISPR systems in industrial relevant Clostridia *Biotechnol. Biofuels* **7** 40

[18] Marcellin E, Behrendorff J B, Nagaraju S, Detissera S, Segovia S, Palfreyman R W, Daniell J, Licona-Cassani C, Quek L E, Speight R, Hodson M P, Simpson S D, Mitchell W P, Köpke M and Nielsen L K 2016 Low carbon fuels and commodity chemicals from waste gases-systematic approach to understand energy metabolism in a model acetogen *Green Chem.* **18**

3020–8
[19] Zeikus J G, Lynd L H, Thompson T E, Krzycki J A, Weimer P J and Hegge P W 1980 Isolation and characterization of a new, methylotrophic, acidogenic anaerobe, the marburg strain *Curr. Microbiol.* **3** 381–6
[20] Lynd L, Kerby R and Zeikus J G 1982 Carbon monoxide metabolism of the methylotrophic acidogen Butyribacterium methylotrophicum. *J. Bacteriol.* **149** 255–63
[21] Grethlein A J, Worden R M, Jain M K and Datta R 1991 Evidence for production of n-butanol from carbon monoxide by Butyribacterium methylotrophicum *J. Ferment. Bioeng.* **72** 58–60
[22] Tanner R S, Miller L M and Yang D 1993 Clostridium ljungdahlii sp. nov., an acetogenic species in clostridial rRNA homology group I *Int. J. Syst. Bacteriol.* **43** 232–6
[23] Köpke M, Held C, Hujer S, Liesegang H, Wiezer A, Wollherr A, Ehrenreich A, Liebl W, Gottschalk G and Dürre P 2010 Clostridium ljungdahlii represents a microbial production platform based on syngas *Proc. Natl. Acad. Sci. U. S. A.* **107** 13087–92
[24] Nagarajan H, Sahin M, Nogales J, Latif H, Lovley D R, Ebrahim A and Zengler K 2013 Characterizing acetogenic metabolism using a genome-scale metabolic reconstruction of Clostridium ljungdahlii *Microb. Cell Fact.* **12** 118
[25] Anggraini I D, Keryanti, Kresnowati M T A P, Purwadi R, Noda R, Watanabe T and Setiadi T 2019 Bioethanol production via syngas fermentation of clostridium ljungdahlii in a hollow fiber membrane supported bioreactor *Int. J. Technol.* **10** 481–90
[26] Huhnke R L, Lewis R S and Tanner R S 2010 Isolation and characterization of novel clostridial species
[27] Zahn J A and Saxena J 2012 *US20120156747A1 - Novel Ethanologenic Clostridium species, Clostridium coskatii - Google Patents*
[28] Allen T D, Caldwell M E, Lawson P A, Huhnke R L and Tanner R S 2010 Alkalibaculum bacchi gen. nov., sp. nov., a CO-oxidizing, ethanol-producing acetogen isolated from livestock-impacted soil *Int. J. Syst. Evol. Microbiol.* **60** 2483–9
[29] Liu K, Atiyeh H K, Tanner R S, Wilkins M R and Huhnke R L 2012 Fermentative production of ethanol from syngas using novel moderately alkaliphilic strains of Alkalibaculum bacchi *Bioresour. Technol.* **104** 336–41
[30] Bruant G, Lévesque M J, Peter C, Guiot S R and Masson L 2010 Genomic analysis of carbon monoxide utilization and butanol production by clostridium carboxidivorans strain P7T ed N Ahmed *PLoS One* **5** e13033
[31] Rieu-Lesme F, Dauga C, Fonty G and Dore J 1998 Isolation from the rumen of a new acetogenic bacterium phylogenetically closely related to Clostridium difficile *Anaerobe* **4** 89–94
[32] Monot M, Boursaux-Eude C, Thibonnier M, Vallenet D, Moszer I, Medigue C, Martin-Verstraete I and Dupuy B 2011 Reannotation of the genome sequence of Clostridium difficile strain 630 *J. Med. Microbiol.* **60** 1193–9
[33] Köpke M, Straub M and Dürre P 2013 Clostridium difficile Is an Autotrophic Bacterial Pathogen ed M M Heimesaat *PLoS One* **8** e62157
[34] Klasson K T, Ackerson C M D, Clausen E C and Gaddy J L 1992 Time ( hrs ) Time ( hrs ) **17** 281–8
[35] Ahmed A, Cateni B G, Huhnke R L and Lewis R S 2006 Effects of biomass-generated producer gas constituents on cell growth, product distribution and hydrogenase activity of Clostridium carboxidivorans P7T *Biomass and Bioenergy* **30** 665–72
[36] Worden R M, Grethlein A J, Jain M K and Datta R 1991 Production of butanol and ethanol from synthesis gas via fermentation *Fuel* **70** 615–9
[37] Ramió-Pujol S, Ganigué R, Bañeras L and Colprim J 2015 Incubation at 25°C prevents acid crash and enhances alcohol production in Clostridium carboxidivorans P7 *Bioresour. Technol.* **192** 296–303
[38] Abubackar H N, Veiga M C and Kennes C 2018 Production of acids and alcohols from syngas

in a two-stage continuous fermentation process. *Bioresour. Technol.* **253** 227–34
[39] Gevantman L H 2000 *CRC Handbook of Chemistry and Physics* (Boca Raton: CRC)
[40] Munasinghe P C and Khanal S K 2010 Biomass-derived syngas fermentation into biofuels: Opportunities and challenges *Bioresour. Technol.* **101** 5013–22
[41] Yasin M, Jeong Y, Park S, Jeong J, Lee E Y, Lovitt R W, Kim B H, Lee J and Chang I S 2015 Microbial synthesis gas utilization and ways to resolve kinetic and mass-transfer limitations *Bioresour. Technol.* **177** 361–74
[42] Devarapalli M, Atiyeh H K, Phillips J R, Lewis R S and Huhnke R L 2016 Ethanol production during semi-continuous syngas fermentation in a trickle bed reactor using Clostridium ragsdalei *Bioresour. Technol.* **209** 56–65
[43] Shen Y, Brown R and Wen Z 2014 Syngas fermentation of Clostridium carboxidivoran P7 in a hollow fiber membrane biofilm reactor: Evaluating the mass transfer coefficient and ethanol production performance *Biochem. Eng. J.* **85** 21–9
[44] Shen Y, Brown R and Wen Z 2014 Enhancing mass transfer and ethanol production in syngas fermentation of Clostridium carboxidivorans P7 through a monolithic biofilm reactor *Appl. Energy* **136** 68–76
[45] Shen Y, Brown R, Energy Z W-A and 2017 U 2017 Syngas fermentation by Clostridium carboxidivorans P7 in a horizontal rotating packed bed biofilm reactor with enhanced ethanol production *Elsevier*
[46] Younesi H, Najafpour G and Mohamed A R 2006 Liquid fuel production from synthesis gas via fermentation process in a continuous tank bioreactor (CSTBR) using Clostridium ljungdahlii *Iran. J. Biotechnol.* **4** 45–53
[47] Cotter J L, Chinn M S and Grunden A M 2009 Ethanol and acetate production by Clostridium ljungdahlii and Clostridium autoethanogenum using resting cells *Bioprocess Biosyst. Eng.* **32** 369–80
[48] Xie B T, Liu Z Y, Tian L, Li F L and Chen X H 2015 Physiological response of Clostridium ljungdahlii DSM 13528 of ethanol production under different fermentation conditions *Bioresour. Technol.* **177** 302–7
[49] Liu K, Atiyeh H K, Tanner R S, Wilkins M R and Huhnke R L 2012 Fermentative production of ethanol from syngas using novel moderately alkaliphilic strains of Alkalibaculum bacchi *Bioresour. Technol.* **104** 336–41
[50] Kim Y K, Park S E, Lee H and Yun J Y 2014 Enhancement of bioethanol production in syngas fermentation with Clostridium ljungdahlii using nanoparticles *Bioresour. Technol.* **159** 446–50
[51] Kennes D, Abubackar H N, Diaz M, Veiga M C and Kennes C 2016 Bioethanol production from biomass: Carbohydrate vs syngas fermentation *J. Chem. Technol. Biotechnol.* **91** 304–17
[52] Acharya B, Roy P and Dutta A 2014 Review of syngas fermentation processes for bioethanol *Biofuels* **5** 551–64

# Phenolic Compound, Antioxidant and Antibacterial properties of Electrospun PVP Nanofiber loaded with *Bassela rubra linn* extract and Alginate from *Sargassum sp.*

**Eka Lutfi Septiani[1*], Okky Putri Prastuti[1], Siti Machmudah[2], Sugeng Winardi[2], Wahyudiono[3], Hideki Kanda[3] and Motonobu Goto[3]**

[1]Department of Chemical Engineering, Universitas Internasional Semen Indonesia, Gresik 61122 Indonesia
[2]Department of Chemical Engineering, Institut Teknologi Sepuluh Nopember, Surabaya 60111 Indonesia
[3]Department of Material process Engineering, Nagoya University, Nagoya 464-8603 Japan

[*]Corresponding Author's E-mail: eka.septiani@uisi.ac.id

**Abstract.** An alternative wound dressing based on nanofiber mats have been developed recently. The antioxidant and antibacterial activity play an important role in wound healing process. This study aims to combine the properties of *Bassela rubra linn extract* (BRLE) and Alginate from brown macroalgae into Polivynil Pyrrolidone (PVP) nanofiber using electrospinning method. Firstly, the optimization step was conducted to obtain the best operation voltage and distance. Afterwards, the nanofiber composites of PVP/BRLE/Alginate were injected into electrospinning tool in various Alginate concentrations of 1%, 2%, and 3% (w/w) respectively. The morphology of PVP nanofiber observed by Scanning Electron Microscopy (SEM) conveys that the closest distance and the highest voltage were the best. Meanwhile, the in vitro analysis through 2,2-diphenyl-1-picrylhydrazyl (DPPH) assay and antibacterial activity to *S. aureus* and *E. choli* bacteria show a quite strong radical scavenging ability around 60% and adequate antibacterial property as good as reference a well-known material consecutively.
**Keywords:** Nanofiber, Electrospinning, Total Phenolic Compound, Antioxidant, Antibacterial.

## 1. Introduction

A serious wound generally appears in the skin because of surgery, accident, and disease such as diabetes. This wound may alter to a chronic wound once the healing mechanism experiences a failure in which the wound cannot heal completely in 12 weeks [1]. Regarding to this issue, wound dressing become an importance part in a proper healing process by maintaining moist environment at the wound area. As a result, the closed wound would contribute re-ephytelialization, collagen synthesis, and angiogenesis faster than the open ones [2]. A cotton based wound dressing is widely used since a long time ago due to its advantages in cost and absorbing exudate from the wound. However, it is not only give a small supporting in bacterial prevention and oxygen exchange, but also it should be removed oftentimes which lead to a more painful feeling, sometimes even cause re-injury [3]. Therefore, a cheaper and more adequate wound dressing development is required.

Nanofiber through electrospinning method is one of promising wound dressing materials with many benefits, particularly, in its simply, versatile, cheap cost production and its properties [4]. In general, polymer nanofiber is used as drugs carrier while it is applied as a wound dressing. Several research reveal superiority of polymer nanofiber combined with antiinflammation compound from plants extract. A good enhancement in healing process was represented by *Lawsonia Inermis*/gelatin/starch by means of in vivo test [5]. Garcinia mangostana [6], Green tea extract [7], the other plant extracts in nanofiber mats may speed up healing process [8]. Meanwhile, *Basella rubra linn* (as known as Malabar spinach, Gendola, red Binahong) was confirmed has an anti-inflammatory compound [9] that is believed to accelerate an injury recovery. Polyninyl Pirrolydone/Basella rubra linn extract (PVP/BRLE) was evaluated in this previous research through in vitro analysis [10]. Nevertheless, its antioxidant property is relatively low, below 50%. Hence, recent research needs to propose Alginate as an added material to raise the antioxidant activity which is related to healing process acceleration potency. Lavender-Alginate nanofiber indicates alleviation of inflammatory response to wound burn on animal body [11]. The electrospun carboxymethylcellulose/sodium alginate/CaCl2/polyethylene oxide polymeric solution also provide significant percentage of oxidant inhibition compared to the mats without alginate [12]. A comparison study between extract spirulina-Polycaprolactone (Spi-PCL) and extract spirulina-Alginate-Polycaprolactone (Spi-Alg-PCL) was conducted by Choi et. al. (2017). The result reveals that the alginate presence promotes eminent wound recovery in a rat model, especially in 5 days observation [13]. Accordingly, this research is purposing a nanofiber composite of PVP/BRLE/Alginate in order to understand its potency as a wound dressing via free radical scavenging activity using 2,2-Diphenyl-1-picrylhydrazyl (DPPH) assay and antibacterial tests towards *Staphylococcus aureus* and *Escherichia coli*. This study carried several operation parameters to gain prominent nanofiber size.

## 2. Experimental Section

### 2.1. Material

Dried *Bassela rubra linn* (Binahong) leaves with Premium Herb brand was acquired from Madiun City, East Java, Indonesia. The leaves, then, were leached by Soxhlet extractor to get its phenolic and alkaloid compounds. Further, the *Bassela rubra linn* extract (BRLE) solution was concentrated using vacuum evaporation process. Fresh *Sargassum sp.* (brown macroalgae) was purchased from local market in Lampung City, Sumatera island, Indonesia. Alginate compound from the macroalgae was obtained from extraction process by hydrothermal auto-clave reactor with the ratio of the macroalgae mass to water 1:20 w/v [14]. 1,300,000 g/mol MW PolyvynilPyrrolidone (PVP) and 2,2-Diphenyl-1- picrylhydrazyl (DPPH) were produced by Aldrich Chemistry. Methanol, ethanol, and sodim carbonate were manufactured by Wako Pure Chemical Corporation. Folin-Ciocalteu's agent was supplied from Sima-alorich. Deionized water was used in the experiment. Commercial Povidone-iodine (Betadine) was bought from a local market in gresik City, East Java, Indonesia.

### 2.2. Experimetal setup and method

At the first, the 8% (w/w) of PVP in the ethanol solvent was synthesized by electrospinning method as can be found somewhere [10] with the tip-to-collector distances of 10 cm and 12 cm. Various voltage of 8, 10, 12 kV by a power supply (Matsusada) were applied to obtain the best process parameter. Furthermore, the BRLE and alginate solution were mixed into 8% (w/w) PVP solution to get a precursor solution that was changed into nanofiber composite of PVP/BRLE/Alginate using optimized parameters. The composition of BRLE was fixed in the smallest percentage of previous study [10], 2% (w/w). Meanwhile, the percentage of Alginate solution was adjusted in the variation of 1, 2, and 3 % (w/w) of the total solution. The different PVP/BRLE/Alginate solutions were injected from syringe into electrospinning chamber by mean of tiny capillary pipe to produce the nanofiber mats. To set the rate at 1 mL/hour, a syringe pump (HARVARD APARATUS) was employed. As the solution move out from the tip of capillary spinneret, the solution was spinning to generate nanofiber sheet.

Scanning Electron Microscopy (SEM, Hitachi S-4300, Japan) analysis with carbon coating was conducted to determine the best distance and voltage of the electrospinning process of PVP through the nanofiber morphology. Moreover, in vitro analysis was conducted by Total Phenolic Compound, DPPH, and antibacterial assay. The total content of phenolic compounds (TPC) of the PVP/BRLE/Alginate nanofiber composite was analyzed by Folin-Ciocalteu method in 1 mL of the sample solutions [15], generated from 50 ppm of PVP/BRLE/Alginate nanofiber mat, were mixed with Folin-Ciocalteu solution and left for 3 minutes. Afterwards, 1 ml of 7.5% (w/w) sodium carbonate and 7 ml pure water were added into the the solution. The total phenolic compound of the sample solutions was measured by UV-VIS Spectrophotometer (V-550, Jasco, Japan) using wavelength of 750 nm after 2 hours incubation. Besides that, the radical scavenging ability was specified by using DPPH assay. The absorbance of the mixture from 50 ppm DPPH solution and 50 ppm sample solution in the ratio of 1:1 (v/v) was measured as A sample while the absorbance of 25 ppm DPPH solution was measured as A control, thus, the antioxidant activity can be calculated by equation (1) [16]

$$\%AA = (Acontrol - Asample)/Acontrol * 100 \quad (1)$$

The antibacterial activity of the PVP/BRLE/Alginate nanofiber composite was conducted using streak plate method. An agar solution was poured into a plate which is signed with four quadrants. Then, 6-mm in diameter of nanofiber was placed onto a streak of the strain drawn on each quadrant of the agar plate using metal loop, respectively. The inhibit zone, appear once the bacteria become susceptible, were observed in this study [17]. Povidone-iodine was used as reference in this assay.

## 3. Results

### *3.1. Optimization process*

The optimum parameter process is determined by 8% PVP nanofiber from SEM image as displayed as Figure 2. The morphology from 9000 magnificent of the pictures shows a nanofiber form with different range of diameter distribution in which determined by ImageJ software analysis. As can be seen, the longer distance shows the larger nanofiber mean diameter at all the voltages used. However, the general behavior is the increasing distance should cause the smaller the average diameter of nanofiber due to the orientation and the elasticity that preserved by the polymer jets [18]. This deviation may be affected by the weakening of the electric field at a certain distance [19] so that the force becomes unbalance and the jet becomes unstable [20]. On the other hand, the applied voltages in the electrospinning process also provide a prominent effect to the nanofiber mean diameter. According to the Figure 2, it conveys that the highest voltage produces the smallest mean diameter and standard deviation of 372 ±132 nm and 407 ± 161 nm for the tip-to-collector distance of 10 cm and 12 cm successively. In term of 10 cm distance, however, the average diameter of nanofiber elevates from 428 ±121 nm at 8 kV to 463 ±169 nm at 10 kV before decrease at 12 kV. The lowest voltage maintains the polymer jet elongation due to oscillation until reaching the collector whereas the middle voltage (12 kV) encourages a rapid spinning movement in the process [21]. Hence, the polymer solution under 10 kV has longer drying time and stretching range than those of the solution under 12 kV [22]. Furthermore, the increase of voltage results a multiple jet that causes similar nanofiber diameter even though the spinning process runs promptly [23]. Regarding another distance, the higher the employed voltage, the smaller nanofiber average diameter. It can be led by sufficient stretching of the solution during electrospinning process [24]. In the wound dressing application, the smaller and the more homogeneous mean diameter of nanofiber are important because these characteristics can enhance the proliferation process during wound healing mechanism [25] and hasten the adessive process [26]. Thus, the best parameter condition for the experiment is applying the tip-to-collector distance at 10 cm and the voltage of 12 kV.

### *3.2 In vitro analysis*

As a result of the optimizing the operation condition, the electrospinning process of PVP/BRLE/Alginate nanofiber was conducted at the tip-to-collector distance at 10 cm as well as the voltage of 12 kV since the little amount addition of the organic extract may affect inconsiderably [7]. According to the Table 1, the evaluated phenolic compound of the nanofiber composite of PVP/BRLE/Alginate shows a fluctuating result while the antioxidant activity result describes an escalating tendency as the increase of Alginate concentration. The data reveals that both total phenolic content and antioxidant activity cannot be parallelized. The radical scavenging ability levels up by the addition of Alginate concentration. Nevertheless, the highest total phenolic compound is not indicating the highest antioxidant activity as well. It may be caused by prooxidant activity from the brown macro algae [27] which allows the maximum percentage of antioxidant activity at 60.7 ± 0.06 with the increase of phenolic compound.

Table 1. Total Phenolic Content and Antioxidant Activity of PVP/BRLE/Alginate nanofiber mats.

| Sample | Total Phenolic Content (mg GAE/mL precursor) | Antioxidant activity (%) |
|---|---|---|
| 8/2/1 PVP/BRLE/Alginate | 32± 0.74 | 58.6 ± 0.31 |
| 8/2/2 PVP/BRLE/Alginate | 34 ± 1.19 | 60.7 ± 1.34 |
| 8/2/3 PVP/BRLE/Alginate | 32 ± 0.37 | 60.7 ± 0.06 |

The antibacterial test against S. aureus as shown as the Figure 3 conveys that the raising of the Alginate percentage from 1% to 2% in the nanofiber composite can broaden the inhibit zone, yet the more addition cannot extend the inhibit zone. This data may figure out the relevance with the antioxidant activity of the nanofiber mats.

Table 2 represents the antibacterial zone from the reference material and the PVP/BRLE/Alginate nanofiber composite from S. aureus and E. coli. All materials show better antibacterial activity to resist E. coli (gram-negative) growth than S. aureus ones (gram-positive). It occur because the gram-negative bacteria have a thin layer of cell wall no more than 10 nm although they have three layers of wall whilst the gram-positive bacteria have a thick layer, peptidoglycan, between 20 and 80 nm [28].

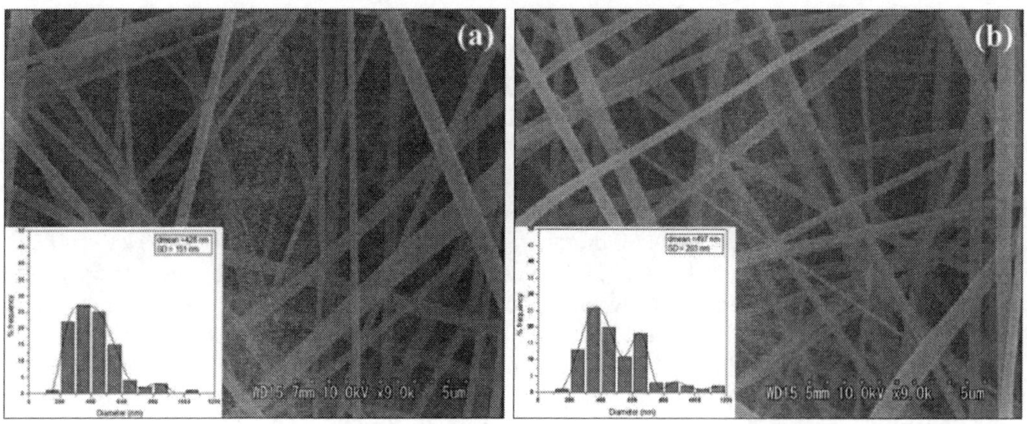

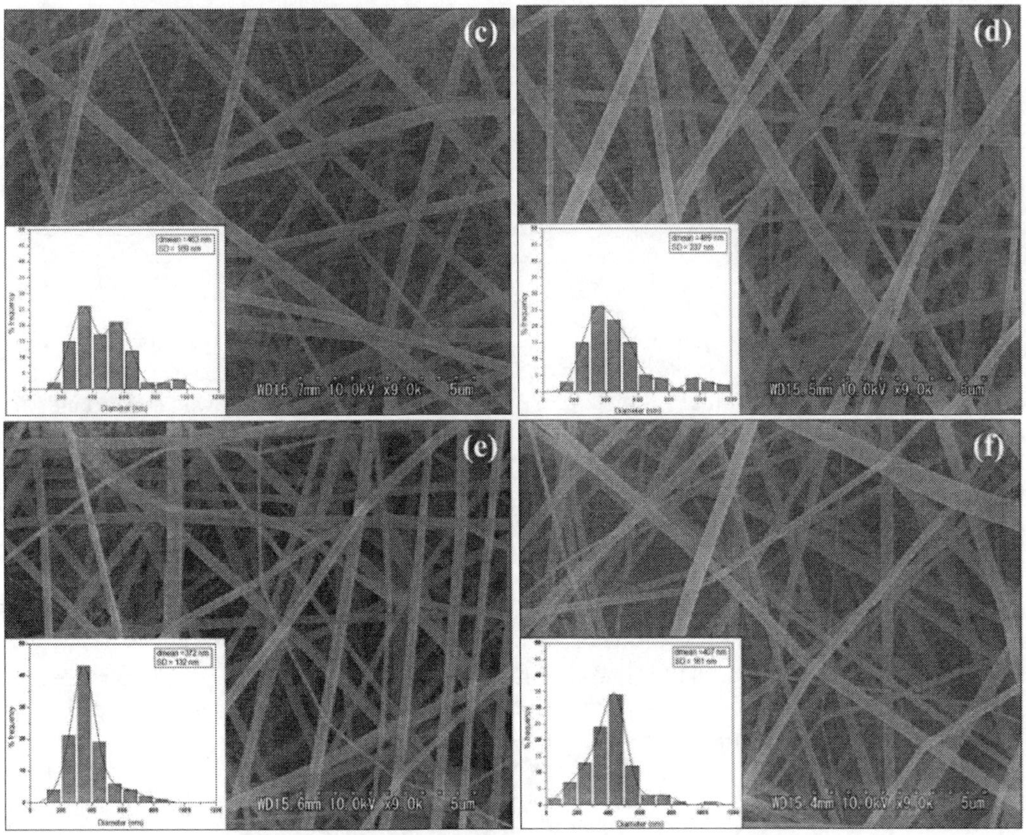


**Figure 2.** Morphology of PVP nanofiber (a),(c),(e) 8 kV, 10kV, 12 kV at the distance of 10 cm. (b),(d),(f) 8 kV, 10kV, 12 kV at the distance of 12 cm.

Based on the antibacterial susceptibility, the nanofiber composite of PVP/BRLE/Alginate exhibits an adequate property as good as a commercial medicine even better, especially in S. aureus prevention.

**Figure 3.** Resistance zone of PVP/BRLE/Alginate towards S. aureus in various Alginate composition (a) 1%, (b) 2%, (c) 3% w/w.

**Table 2.** Inhibit zone of reference material and PVP/BRLE/Alginate nanofiber.

| Sample | Inhibit zone (mm) | |
|---|---|---|
| | E. coli | S. aureus |
| Commercial Povidone-iodine | 6 | 2 |
| 8/2/1 PVP/BRLE/Alginate | 5 | 3 |
| 8/2/2 PVP/BRLE/Alginate | 5 | 4 |
| 8/2/3 PVP/BRLE/Alginate | 6 | 4 |

## 4. Conclusion

Synthesis of the PVP/BRLE/Alginate nanofiber composite was conducted at the superlative operation condition, achieved from different distances and voltages, tip-to-collector distance at 10 cm and the voltage of 12 kV. The variation of Alginate proportion in the composites was evaluated through in vitro analysis, Total Phenolic Compound, antioxidant ability, antibacterial susceptibility. The results show that the phenolic compound and antioxidant properties are not analogous. However, the antioxidant activity can be related to the antibacterial activity which reveal that the higher antioxidant activity in percentage, the wider resistance zone. The highest antioxidant activity and the largest area of bacterial resistance are 60.7 ± 0.06% and 6 mm in turn.

## Reference

[1] Sheehan P, Jones P, Caselli A, Giurini J M, Veves A 2003 Percent change in wound area of diabetic foot ulcers over a 4-week period is a robust predictor of complete healing in a 12-week prospective trial *Diabetes care* **26**(6) 1879-1882

[2] Dhivya S, Padma V V, Santhini E 2015 Wound dressings–a review *BioMedicine* **5**(4) 24–28

[3] Ovington L G 2001 *There ' s More to Dressings Than Gauze Wet-to-Dry and Wet-to-Moist — What ' s Gauze Dressings : Patient Issues* **19** (8) 477–484

[4] Agarwal S, Wendorff J H, Greiner A 2008 Use of electrospinning technique for biomedical applications *Polymer (Guildf)* **49** (26) 5603–5621

[5] Hadisi Z, Nourmohammadi J, Nassiri S M 2018 The antibacterial and anti-inflammatory investigation of Lawsonia Inermis-gelatin-starch nano-fibrous dressing in burn wound *International journal of biological macromolecules 107* 2008-2019

[6] Charernsriwilaiwat N, Rojanarata T, Ngawhirunpat T, Sukma M, Opanasopit P 2013 Electrospun chitosan-based nanofiber mats loaded with Garcinia mangostana extracts *International journal of pharmaceutics* **452** (1-2) 333-343

[7] Pusporini P, Edikresnha D, Sriyanti I, Suciati T, Munir M M, Khairurrijal K 2018 Electrospun polyvinylpyrrolidone (PVP)/green tea extract composite nanofiber mats and their antioxidant activities *Materials Research Express* **5**(5) 054001

[8] Zhang W, Ronca S, Mele E. 2017 Electrospun nanofibres containing antimicrobial plant extracts *Nanomaterials* **7**(2) 42

[9] Singh M, Kumari R, Nandini D, Kotecha M 2016 Preliminary phytochemical screening of Basella rubra Linn *Journal of Pharmacognosy and Phytochemistry* **5**(4) 224

[10] Septiani E L, Gabriel A A, Prastuti O P, Indahsari D N, Ariyanti E D, Machmudah S, Goto M 2020 Correlation of Extract Composition on Antioxidant Activity of ELECTROSPUN PolyvinylPyrrolidone/Bassela rubra linn Leaf Extract Composite *In Key Engineering Materials* **851** 122-127 Trans Tech Publications Ltd.

[11] Summa M, Russo D, Armirotti A, Brunetti V 2016 *Materials Chemistry B*

[12] Peršin Z, Ravber M, Stana Kleinschek K, Knez Ž, Škerget M, Kurečič M 2017 Bio-nanofibrous mats as potential delivering systems of natural substances *Textile research journal* **87**(4)

444-459
[13] Choi J I, Kim M S, Chung G Y, Shin H S 2017 Spirulina extract-impregnated alginate-PCL nanofiber wound dressing for skin regeneration *Biotechnology and Bioprocess Engineering* **22** (6) 679-685
[14] Rajauria G, Jaiswal A K, Abu-Ghannam N, Gupta S 2010 Effect of hydrothermal processing on colour, antioxidant and free radical scavenging capacities of edible Irish brown seaweeds *International journal of food science & technology* **45**(12) 2485-2493
[15] Chhouk K, Uemori C, Kanda H, Goto M 2017 Extraction of phenolic compounds and antioxidant activity from garlic husk using carbon dioxide expanded ethanol *Chemical Engineering and Processing: Process Intensification* **117** 113-119
[16] Suwantong O, Pankongadisak P, Deachathai S, Supaphol P 2013 The potential of electrospun poly (L-lactic acid) fiber mats containing a crude Garcinia dulcis extract for use as wound dressings *Chiang Mai J Sci* **40** (3) 517-33
[17] Sanders E R 2012 *Aseptic Laboratory Techniques : Plating Methods 2 . Streak Plate Procedure : Isolation of Bacterial Colonies Using the Quadrant Method* May 1–18
[18] Doshi J and Reneker D H 1995 Electrospinning process and applications of electrospun fibers *Journal of electrostatics* **35** (2-3) 151-160
[19] Baumgarten P K 1971 Electrostatic spinning of acrylic microfibers *Journal of colloid and interface science* **36** (1) 71-79
[20] Ding W, Wei S, Zhu J, Chen X, Rutman D, Guo Z 2010 Manipulated electrospun PVA nanofibers with inexpensive salts *Macromolecular Materials and Engineering* **295** (10) 958-965
[21] Mazoochi T, Hamadanian M, Ahmadi M, Jabbari V 2012 Investigation on the morphological characteristics of nanofiberous membrane as electrospun in the different processing parameters *International Journal of Industrial Chemistry* **3**(1) 1-8
[22] Reneker D H and Yarin A L 2008 Electrospinning jets and polymer nanofibers *Polymer* **49**(10) 2387-2425
[23] Liu Z, Ju K, Wang Z, Li W, Ke H, He J 2019 Electrospun jets number and nanofiber morphology effected by voltage value: numerical simulation and experimental verification *Nanoscale research letters* **14** (1) 1-9
[24] Sharifi L, Assa F, Ajamein H, Mirhosseini S H 2017 Effect of Voltage and Distance on Synthesis of Boehmite Nanofibers with PVP by the Electrospinning Method **98** 63–74
[25] Hodgkinson T, Yuan X F, Bayat A 2014 Electrospun silk fibroin fiber diameter influences in vitro dermal fibroblast behavior and promotes healing of ex vivo wound models *Journal of tissue engineering* **2016**
[26] Chen S, Liu B, Carlson M A, Gombart A F, Reilly D A, Xie J 2017 Recent advances in electrospun nanofibers for wound healing *Nanomedicine* **12** (11) 1335-1352
[27] Kılıçgün H and Altıner D 2010 Correlation between antioxidant effect mechanisms and polyphenol content of Rosa canina *Pharmacognosy magazine* **6** (23) 238
[28] Mai-Prochnow A, Clauson M, Hong J, Murphy A B 2016 Gram positive and Gram negative bacteria differ in their sensitivity to cold plasma *Scientific reports* **6** (1) 1-11

IOP Conf. Series: Materials Science and Engineering 1143 (2021) 012016 doi:10.1088/1757-899X/1143/1/012016

# Estimation of the biomass yield and stoichiometric coefficient during bioproduct formation through thermodynamic approach: a case study of biosurfactant production

**R S Adiandri[1,3], R Purwadi[1,2], Hoerudin[3] and T Setiadi[1,4]**

[1]Department of Chemical Engineering, Institut Teknologi Bandung, Bandung 40135, Indonesia
[2] Food Engineering Department, Institut Teknologi Bandung, Jatinangor Campus, Sumedang 45363, Indonesia
[3]Indonesia Center for Agricultural Research and Development, Ministry of Agriculture of Indonesia, Bogor 16111; Indonesia
[4]Center for Environmental Studies (PSLH), Institut Teknologi Bandung, Bandung 40135, Indonesia

Corresponding author's e-mail address: resa.adiandri@gmail.com

**Abstract**. Microbial growth occurs on a wide variety of compounds. One of the critical parameters in biotechnological processes is biomass yield. Biomass yield for many different microbial systems extensively studied due to its primary importance. Besides biomass yield, a stoichiometric coefficient determination is also important because it can predict the number of reactants and products needed. This paper aims to explain how to determine the biomass yield on electron donor ($Y_{DX}$) and stoichiometric coefficient during bioproducts formation through a thermodynamic approach. The case study focused on the biosurfactant production process. The calculation shows that the electron donor and incubation temperature affect the electron donor's biomass yield ($Y_{DX}$). For biosurfactant production, glucose obtains a higher value of $Y_{DX}$ than propionate. By using the same electron donor, the higher the incubation temperature, the lower of $Y_{DX}$ value. $Y_{DX}$ value is useful for determining the stoichiometric coefficient of biomass growth during biosurfactants formation through elemental mass balance. The type of electron donor and temperature affect the stoichiometric coefficient of biomass growth during the biosurfactant production process.

## 1. Introduction

Biotechnology has a positive role in agriculture, health, and the environment. In agriculture, biotechnology is useful for producing bioproducts that can help reduce food crises, improve food quality, and increase agricultural production. The relevant variable of biotechnological processes' economic evaluation is the biomass yield on C-source, Yx / s [1]. The biomass yield in cell growth is one of the critical parameters in various scientific studies involving microbial culture because it can determine the final biomass or cell concentration. Biomass yield (Yx/s) is a stoichiometric coefficient that characterizes a particular microbial strain's growth efficiency. It can indicate how much biomass can be grown per amount of carbon or substrate energy consumed. It is essential to optimize yield in various biotechnological processes to produce a reasonable amount of biological material. Optimization of biomass yield is also a significant interest in the biotechnology industry for obtaining large quantities of products and synthesis rates, as they determine the economic viability of any bioprocess [2].

The biomass yield can vary from 0.01 to 1.0 C-mol biomass / C-mol substrate because it depends on microorganisms and their growth substrates [3]. The biomass yield can estimate oxygen consumption, carbon dioxide generation, and heat production. These parameters are of significant technological importance for large-scale bioreactors, particularly for $O_2$ transport, $CO_2$ removal, and cooling capacity [1].

One method to estimate biomass yield is a thermodynamic approach that expressed the biomass yield on electron donors ($Y_{DX}$). Estimating biomass yield from electron donors ($Y_{DX}$) is generally required in situations where the microorganism's biochemical properties are unknown or less well known concerning its biochemical properties. However, the carbon source, nitrogen source, donor electron, and acceptor electron are known. This information is generally referred to as a "black box" description. The biomass yield from electron donors ($Y_{DX}$) depends on several factors, namely: electron donors (through $\gamma_D$), electron acceptors (via $\Delta G_{AV}^{01}$), N-source (via $\gamma_X$), and temperature, growth rate, carbon source (via Gibbs energy dissipation (through $D_S^{01}/rx$) [4].

The parameter needed to estimate the C-source biomass yield must meet the requirements for black box information, general applications for chemotrophic growth systems, and their relationship with the Second Law of Thermodynamics. This parameter can be the Gibbs energy dissipation per C-mol produces measurable biomass from a well-established correlation [5]. The black box description provides a linear relationship between Gibbs principles of conservation and energy balance, which allows stoichiometric calculations and produces coefficients [3].

This paper attempts to explain how to estimate the biomass yield and stoichiometric coefficient during bioproducts formation through a thermodynamic approach. In this paper, a case study focused on the biosurfactant production process. Biosurfactant is a surfactant produced extracellularly by microorganisms such as bacteria, yeast, and fungi from various substrates such as sugars, oils, alkanes, and other substrates [6]. Biosurfactants have the potential to be applied commercially in the food and non-food industries. The advantages of biosurfactants are high biodegradability, effectiveness at extreme condition (pH, temperature, salinity), and can be able to produce using renewable substrates [7,8]

## 2. Methods

### *2.1. Research description used in this case study*

The data used in this case study refers to the research conducted by Sakthipriya et al. [9] and Heryani and Putra [10]. Description regarding carbon sources, microorganisms species, reactor type, and working volume used in those research shown in Table 1.

**Table 1**. Research description used in this case study.

| Research description | Sakthypriya et al.[11] | Heryani and Putra [10] |
|---|---|---|
| Condition process | Aerobic | aerobic |
| Carbon source | *waxy crude oil* | Glucose |
| Nitrogen source | Ammonium chloride ($NH_4CL$) | Ammonium nitrate ($NH_4NO_3$) |
| Microorganism | *Bacillus substilis* YB7 | *Bacillus sp.* BMN 14 |
| Reactor type | Batch | batch |
| Working volume | 50 mL in 250 mL Erlenmeyer flasks | 100 mL in 250 mL Erlenmeyer flasks |
| Temperature | 35, 50, 75°C | 37°C |
| Agitation | 180 rpm | 140 rpm |

### 2.2. *Black box description*

Heijnen et al. [4] defined a general system of chemotrophic microbial growth, including a list of compounds relevant to the microbial growth system. The compounds that are always present in the growth system include biomass, $H^+$, $HCO_3$-and $H_2O$. N-sources, C-sources, electron donors, and electron acceptors vary widely but are easy to identify. C-source is generally the reduced organic compounds (for heterotrophic growth) or CO2 (for autotrophic growth). The electron donor is a reduced organic (for organotrophic) and inorganic (for lithotropic) compounds. The electron acceptor is the oxidized compound.

The general system definition of microbial growth contains all the relevant compounds involved: electron donor, electron acceptor, N-source and C-source (always consumed); biomass, oxidized donor and reduced acceptor (always produced) and $HCO_3^-$, $H_2O$ and $H^+$ (can be consumed or produced dependent on the specific system). The Gibbs energy dissipation is always positive, according to the Second Law of Thermodynamics (Figure 1.) The black box description's basic feature represents the microbial growth by a macro chemical equation (Eq. 1)[4,5].

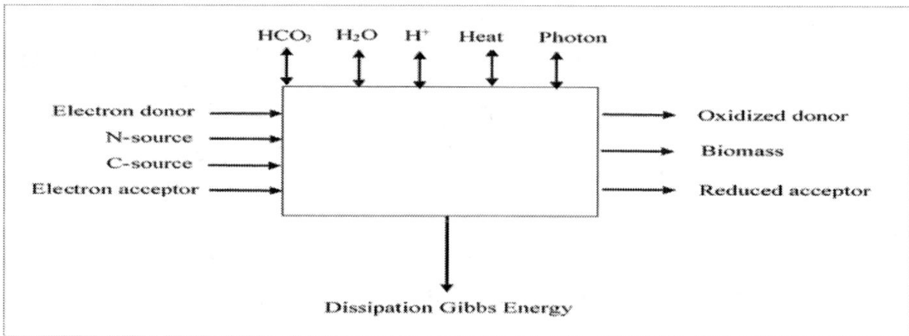


**Figure 1**. Generalized system definition of microbial growth [4,5].

$$(\ldots)C - source \pm (\ldots)donor + (\ldots)acceptor + (\ldots)N - source + (\ldots)H_20 + (\ldots)HCO_3^- + (\ldots)H^+ + 1CH_{1.8}O_{0.5}N_{0.2} + (\ldots)oxidized\ donor + (\ldots)reduced\ acceptor \quad (1)$$

### 2.3. *Biomass yield estimation.*

Chemotropic growth of microorganisms can occur on various chemicals within a wide range of temperatures and pH. The biomass yield ($Y_{DX}$) on electron donors can vary considerably. An alternative method used to predict biomass yield $Y_{DX}$ is the Gibbs energy dissipation. In microbial growth, part of the substrate is used for biomass production and partly for maintenance, so the equation for the total Gibbs energy dissipation ($Ds^{01}/rx$) becomes [5]:

$$\frac{D_S^{01}}{r_x} = \left[\frac{D_s{}^{01}}{r_x}\right]_{growth} + \left[\frac{m \in}{\mu}\right]_{maintenance} \quad (2)$$

For heterotrophic microbes, the above equation becomes :

$$\frac{D_S^{01}}{r_x} = 200 + 18(6-c)^{1,8} + exp[((3,8-\gamma_s)^2)^{0,16} \times (3,6+0,4c)] + \frac{4,5}{\mu} exp\left[-\frac{70000}{R}\left(\frac{1}{T} - \frac{1}{298}\right)\right] \quad (3)$$

In macro chemical equations, a general definition for microbial growth written as follows:

$$-\frac{1}{Y_{DX}}\,electron\ donor - (\ldots)N - source - \frac{1}{Y_{AX}}\,electron\ acceptor + 1C - mol\ biomass + (\ldots)H_2O + (\ldots)HCO_3^- + (\ldots)\,H^+ + \frac{Q}{r_x}\,heat + \frac{D_s^{01}}{r_x}\,Gibbs\ energy \quad (4)$$

From the macro chemical equations above (Eq. 3), there are three formulations of balance equations, i.e., the balance of degree of reduction, the Gibbs energy balance, and the enthalpy balance.

Balance of degree of reduction:

$$-\frac{\gamma_D}{Y_{DX}} - \frac{\gamma_A}{Y_{AX}} + \gamma_X = 0 \quad (5)$$

The Gibbs energy balance :

$$-\frac{\gamma_D\,\Delta G_{eD}^{01}}{Y_{DX}} - \frac{\gamma_A\,\Delta G_{eA}^{01}}{Y_{AX}} + \gamma_X\,\Delta G_{eX}^{01} + \frac{D_s^{01}}{r_x} = 0 \quad (6)$$

Enthalpy balance :

$$-\frac{\gamma_D\,\Delta H_{eD}^{01}}{Y_{DX}} - \frac{\gamma_A\,\Delta H_{eA}^{01}}{Y_{AX}} + \gamma_X\,\Delta H_{eX}^{01} + \frac{Q}{r_x} = 0 \quad (7)$$

From the above balances, three equations can predict the biomass yield and heat of the reaction with the following equation [5]:

Biomass yield on electron donor

$$Y_{DX} = \frac{\gamma_D\,\Delta G_{av}^{01}}{\frac{D_s^{01}}{r_x} + \gamma_X\,\Delta G_{av}^{01}} \quad (8)$$

Biomass yield on electron donor

$$Y_{AX} = \frac{(-\gamma_A)\,\Delta G_{av}^{01}}{\frac{D_s^{01}}{r_x}} \quad (9)$$

Mol acceptor/(C)-mol donor

$$Y_{DA} = -\frac{\gamma_D}{Y_{DX}} \times \frac{\frac{D_s^{01}}{r_x}}{\frac{D_s^{01}}{r_x} + \gamma_X\,\Delta G_{av}^{01}} \quad (10)$$

The heat production per C- mol produced biomass

$$\frac{Q}{r_x} = \frac{\Delta H_{av}^{01}}{\Delta G_{av}^{01}} \times \frac{D_s^{01}}{r_x} \quad (11)$$

The heat production per C-mol consumed substrate

$$\frac{Q}{r_S} = \left( \frac{\gamma_D \, \Delta H_{av}^{01}}{\frac{D_S^{01}}{r_x} + \gamma_X \; \Delta G_{av}^{01}} \right) \times \frac{D_S^{01}}{r_x} \tag{12}$$

### 2.4. *Stoichiometric coefficient*

The stoichiometric coefficient is useful for determining the ratio or ratio of moles between reactants and products. Determination of the stoichiometric coefficient of biomass growth during biosurfactant formation based on the $Y_{DX}$ value through the mass balance of elements. The biomass stoichiometric coefficients are assumed to be +1. The other coefficients (...) and yields must be measured or calculated from elemental (C, H, O, N) and electric charge conservation relations [5].

## 3. Results and Discussions

The steps for calculating the biomass yield on electron donor ($Y_{DX}$) and stoichiometric coefficient through a thermodynamic approach include 1) identifying the electron donor and other relevant compounds in the biosurfactant production process; 2) determining the Gibbs energy dissipation, the yield of biomass to the donor/acceptor and the heat of the reaction; 3) determining stoichiometric coefficient through elemental mass balances.

### 3.1. *Determining the electron donor and other relevant compounds*

The biosurfactant production process carried out by Sakthipriya *et al.* [11] used waxy crude oil as a carbon source. Waxy crude oil is a heterogeneous mixture of hydrocarbons consisting of paraffin, aromatics, resins, and asphaltene with a paraffin composition of more than 50%, and its paraffin molecular structure ranges from $C_{18}H_{38}$ to $C_{55}H_{112}$ [12]. The main fraction (> 50%) in crude oil is alkanes depending on the oil source. Alkanes are saturated hydrocarbons that are chemically very inert as nonpolar molecules [13,14].

Biomass is built from C4-C5 metabolite molecules known as central building blocks with a range of reduction degrees ($\gamma$=4). Suppose the carbon source is very different from the central building block, then several reactions (carbon-carbon coupling, oxidation, reduction) must be carried out in the primary pathway to convert the carbon source into a building block [4].

The substrate used in the research of the Sakthipriya et al.[11] was a long chain carbon ($C_{18}H_{38}$ to $C_{55}H_{112}$). It needs degradation or hydrolysis process to convert it into building blocks ($C_4$-$C_5$). Kumari et al. [15] stated that the essential elements that affect the hydrocarbon biodegradation process are biomass concentration, the population of microbes, enzyme activity, and the environmental parameters, such as pressure and temperature. The enzymes' high activity necessitates less energy to activate the hydrocarbon molecule to initiate biodegradation and speed up the process. Several enzymes responsible for the degradation of long-chain hydrocarbons are alkane monooxygenase, alcohol dehydrogenase, and aldehyde dehydrogenase. These enzymes cause various biochemical transformations of original oil substances and form a large number of intermediates.

The microorganism used in Sakthipriya *et al.* [11] is *Bacillus subtilis* YB7. This bacteria is one of the capable bacterias of producing enzymes that can help the biodegradation of crude oil under aerobic conditions. Aerobic degradation was initiated by the oxidation of a terminal methyl group to form primary alcohol. This activation has increased the solubility of hydrocarbon in water and introduced a reactive site for further reactions. The reaction's energy requirement was generated by the oxidation of a reduced biological intermediate NADH, which was re-oxidized by an electron acceptor. The primary alcohol was further oxidized by the enzyme (alcohol dehydrogenase) to form the corresponding aldehyde and converted into a fatty acid by aldehyde dehydrogenase [9].

The process of alkane degradation under aerobic conditions shown in more detail in Figure 3. The main pathway is the terminal oxidation of fatty acid. α - and ω - hydroxylation is catalyzed by the same set of enzymes.

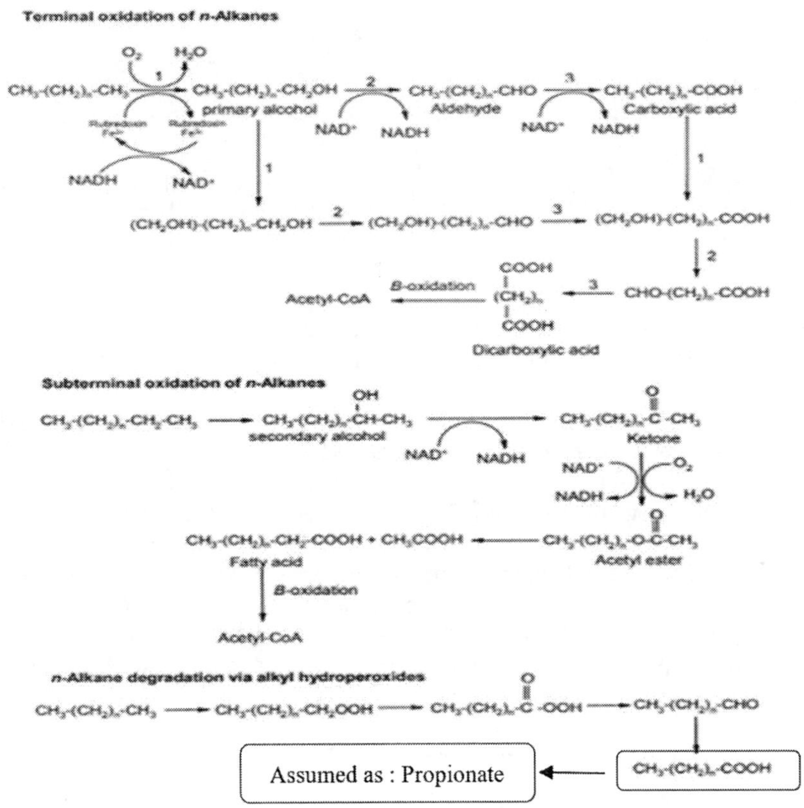


**Figure 2**. Peripheral aerobic pathways of n-alkane degradation [16].

Sakthipriya et al. [9] reported that the degradation rate of waxy crude oil reaches 70% within one day of incubation. The maximum extent of biodegradation (96%) was attained in about ten days. There was no significant increase in the percentage of degradation after ten days of incubation. Figure 4 shows the rate of degradation of waxy crude oil by *Bacillus subtilis* YB7.

The functional groups present in crude oil before and after degradation were confirmed using FTIR spectroscopy (Fig. 3). Figure 4(a) shows the FTIR spectrum of crude oil before and after microbial degradation. The peak obtained at 2984 $cm^{-1}$ was mainly due to the -C-H stretching mode of $CH_3$ and $CH_2$ in the alkyl chain, which confirms paraffin's presence. The peak in spectrum at 1736 $cm^{-1}$ stands for the C=O group, conforming to carboxylic acid production. Weak bonds at 1446 $cm^{-1}$ and 1372 $cm^{-1}$ resulted from the -C-$CH_2$ and -C-$CH_3$ groups. The stretch at 1232 cml indicated a C-O bond, which might be alcohol or ester, the band at 1097 cm-$^{1}$ means a bond of C-O-C, and an indentation at 786 $cm^{-1}$ indicated an alkene bond.

From the explanation of the degradation process above, it was assumed that fatty acid produced from the degradation process of waxy crude oil is propionate ($CH_3CH_2COOH$, $C_3H_6O_2$). Therefore, the carbon source used for estimating the yield of biomass yield on electron donor/acceptor, the heat of the reaction, and the stoichiometric coefficient is propionate. Propionate is a type of volatile fatty acid, can be used as a single carbon source, especially in activated sludge systems, and is a suitable substrate for EBPR (enhanced biological phosphorus removal) performance.

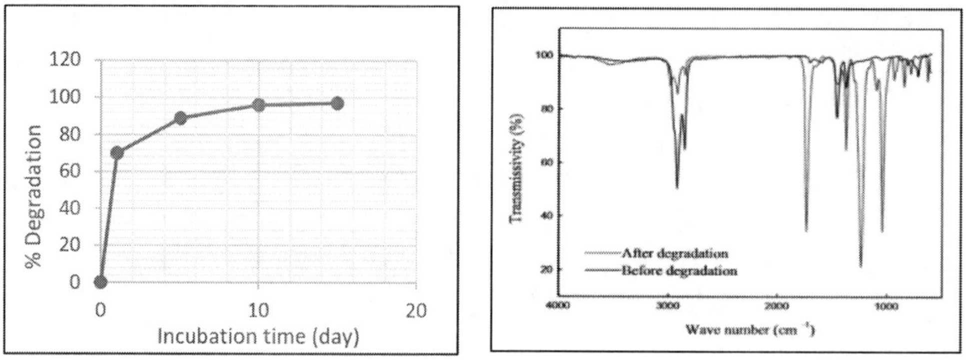


(a) (b)

**Figure 3.** a) percentage degradation of crude oil; b) Functional groups present before and after degradation using Bacillus subtilis at 50 °C and after 10 daysb

Meanwhile, Heryani and Putra [10] used glucose ($C_6H_{12}O_6$) as a carbon source in the biosurfactant production process, and the microorganism was *Bacillus subtilis* BMN 14. The biosurfactant production process also under aerobic conditions. With six C-atoms, glucose is readily substrate as a carbon source for forming biomass and metabolite products because the number of C-atoms is close to the central building block ($C_4$-$C_5$). Referring to the general definition of a microbial growth system that has been developed by Heijnen et al.[4] can then determine the relevant compounds involved in biomass growth during the biosurfactant production process as given in Table 2.

**Table 2.** Relevant compounds involved in biomass growth during the biosurfactant production*).

| Relevant compound | Sakthipriya et al.[11] Aerobic/Propionate | Heryani and Putra [10] Aerobic/Glucose |
|---|---|---|
| Biomassa | $CH_{1,8}O_{0,5}N_{0,2}$ | $CH_{1,8}O_{0,5}N_{0,2}$ |
| N-source | $NH_4^+$ | $NH_4^+$ |
| $H^+$ | $H^+$ | $H^+$ |
| $HCO_3^-$ | $HCO_3^-$ | $HCO_3^-$ |
| $H_2O$ | $H_2O$ | $H_2O$ |
| C-source | $C_3H_6O_2$ | $C_6H_{12}O_6$ |
| Electron donor | $C_3H_6O_2$ | $C_6H_{12}O_6$ |
| Oxidized donor | $HCO_3^-$ | $HCO_3^-$ |
| Electron acceptor | $O_2$ | $O_2$ |
| Reduced acceptor | $H_2O$ | $H_2O$ |

*) Referring to the general definition of the growth system developed by Heijnen et al., [4]

### *3.2. Determining the Gibbs energy dissipation and the biomass yield on electron donor ($Y_{DX}$)*

According to Heijnen et al. [4] , the substrate's growth process is used for biomass synthesis and maintenance. It means that Gibbs energy dissipation can distinguish between growth-related dissipation and maintenance related dissipation. Gibbs energy dissipation value can use to calculate the biomass yield and heat production that occurs during the biomass growth. The data required in calculating the Gibbs energy dissipation, biomass yield, and heat production shown in Table 3.

**Tabel 3**. The data used for the calculation of the Gibbs energy dissipation, biomass recovery, and reaction heat production during biomass growth in the biosurfactant production process.

| Data | Sakthipriya et al. [11] | Heryani dan Putra [10] | Electron acceptor ($O_2$) |
|---|---|---|---|
| Electron donor | Propionate ($C_3H_6O_2$) | Glucose ($C_6H_{12}O_6$) | |
| C-atoms | 3[a] | 6[a] | - |
| $\gamma$ | +4,67[a] | +4[a] | -4[a] |
| $\Delta G_{ef}^{01}$ (kJ/e-mol) | +26,939[a] | +39,744[a] | -78,719[a] |
| $\Delta H_{ef}^{01}$ (kJ/e-mol) | -33,80[a] | -25,75[a] | -143[a] |
| $\mu$ ($jam^{-1}$) | T : 35°C (308°K) = 0,075[b]<br>T : 50°C (323°K) = 0,084[b]<br>T : 75°C (348°K) = 0,071[b] | T : 37°C (310°K) = 0,097[b] | - |
| R (kJ/mol) | 8,314[a] | 8,314[a] | - |

[a]Data were taken from Table 4A (Value of degree of reduction and the electron Gibbs energies and enthalpies in the biological frame of references)
[b]Data were taken and processed from research data of Sakthipriya et al. [11]
[c]Data were taken and processed from research data of Heryani and Putra [10]

The data presented in Table 3 can calculate the Gibbs energy dissipation value during biomass growth and cell maintenance, which is shown in Table 4.

**Table 4**. Value of Gibbs energy dissipation for biomass growth and maintenance.

| Donor | $\gamma_s$ | C | $\mu$ | $D_S^{01}/r_X$ (kJ/C-mol) | | Total $D_S^{01}/r_X$ (kJ/C-mol) |
|---|---|---|---|---|---|---|
| | | | | Growth | Maintenance | |
| Propionate[a] | 4,67 | 3 | 0,075 (308°K)[a] | 428.616 | 150,155 | 578.771 |
| | | | 0,084 (323°K)[a] | 428.616 | 477,143 | 905.760 |
| | | | 0,071 (348°K)[a] | 428,616 | 3672,247 | 4100.864 |
| Glucose[b] | 4 | 6 | 0,097 (310°K)[b] | 236,051 | 138,491 | 374,542 |

[a]Data were taken and processed from research data of Sakthipriya et al.[11]
[b]Data were taken and processed from research data of Heryani dan Putra [10]

Table 4 shows that the carbon source/electron donor used in the biosurfactant production process affects the total Gibbs energy dissipation value. The production process using propionate produces a higher total Gibbs energy dissipation compared to a glucose substrate. In this case, one of the influencing factors was the degradation process of waxy crude oil. *Bacillus subtilis* YB7 performed substrate degradation process to form simple volatile fatty acids that can use as a carbon source; in this case, assume as propionate. The addition of a carbon-carbon coupling process or oxidation or reduction can increase energy dissipation [5]. Incubation temperature also affects the Gibbs energy dissipation value. Using the same carbon source (propionate), the Gibbs energy value of dissipation increases with increasing incubation temperature.

The values for biomass yield on electron donor ($Y_{DX}$), biomass yield on electron acceptor ($Y_{AX}$), mol acceptors to mol donors ($Y_{DA}$) shown in Table 5. Table 5 shows that the electron donor and incubation temperature affects the biomass gain against the electron donor ($Y_{DX}$). For biosurfactant production, glucose obtains a higher value of $Y_{DX}$ than propionate. By using the same electron donor, the higher the incubation temperature, the lower of $Y_{DX}$ value.

**Table 5**. The value of biomass yield on electron donor (*$Y_{DX}$*), biomass yield on electron acceptor (*$Y_{AX}$*), and mol acceptor/C-mol donor *($Y_{DA}$)*.

| Electron donor | T (°K) | Aerobic process: electron acceptor $O_2$ | | | | |
|---|---|---|---|---|---|---|
| | | $\Delta G_{av}^{01}$ (kJ/e-mol) | $D_S^{01}/r_x$ (kJ/C-mol) | $Y_{DX}$ (C-mol /C-mol) | $Y_{AX}$ (C-mol/mol) | $Y_{DA}$ (mol/C-mol) |
| Propionate[a] | 308 | 105,658 | 578,771 | 0,483 | 0,730 | 0,661 |
| | 323 | 105,658 | 905,760 | 0,366 | 0,466 | 0,783 |
| | 348 | 105,658 | 4100,864 | 0,109 | 0,103 | 1,053 |
| Glucose[b] | 310 | 118,463 | 374,542 | 0,543 | 1,265 | 0,429 |

[a] Taken and processed from research data of Sakthipriya et al. [11]
[b] Taken and processed from research data of Heryani and Putra [10]

The accuracy of the Gibbs energy dissipation approach has limitations because the calculations are based on the average microbiological behavior rather than the actual biochemistry of specific microorganisms. A more accurate value for biomass recovery obtains by calculation base on information about the generation/consumption of ATP or with the exact $Y_{DX}$ measurement. However, a thermodynamic approach via the Gibbs energy dissipation can be considered to estimate the initial biomass yield value on the electron donor ($Y_{DX}$) [4]

According to Heijnen et al. [4], in a microbial growth system, there is a heat conversion process which expressed in Q/rx (heat production per C-mol produced biomass) and Q/rs (heat production per C -mol consumed substrate). Heat (Q) use as one of the energetic characteristic parameters and can calculate using $\Delta G_{av}^{01}$ and $\Delta H_{av}^{01}$. $\Delta G_{av}^{01}$ and $\Delta H_{av}^{01}$ represent the energetic qualities of the catabolic (donor/acceptor) redox pair. The value of heat production per biomass produced (Q/rx) and heat production per substrate consumed (Q/rs) shown in Table 6. Table 6 shows that the heat production per produced biomass (Q/rx) and the heat production per substrate consumed (Q/rs) of propionate as an electron donor is higher than glucose. The temperature also influences the value of Q/rx and Q/rs. The higher the temperature of biosurfactant production, the higher the value of Q/rx and Q/rs. The high heat production is due to the maintenance process at a low specific growth rate ($\mu$), which causes an increase in heat production per C-mol of biomass (Q/rx) [4]. The Q/rx value obtained for both processes (using propionate and glucose) is positive, which means that both processes are exothermic. It means that the process releases heat from the system to the environment.

**Tabel 6**. The value of heat production per C-mol produced and heat production per C-mol consumed substrate.

| Electron Donor | T (°K) | Aerobic Process: Electron acceptor $O_2$ | | | | |
|---|---|---|---|---|---|---|
| | | $\Delta G_{av}^{01}$ (kJ/e-mol) | $D_S^{01}/r_x$ (kJ/C-mol) | $\Delta H_{av}^{01}$ (kJ/e-mol) | $Q/r_x$ (kJ/C-mol X) | $Q/r_s$ (kJ/C-mol S) |
| | 308 | 105,658 | 578,771 | 109,2 | 598,173 | 288,648 |
| Propionate[a] | 323 | 105,658 | 905,760 | 109,2 | 939,1236 | 342,273 |
| | 348 | 105,658 | 4100,864 | 109,2 | 4238,338 | 460,168 |
| Glucose[b] | 310 | 118,463 | 374,542 | 117,25 | 370,707 | 201,425 |

[a] Taken and processed from research data of Sakthipriya et al. [11]
[b] Taken and processed from research data of Heryani and Putra. [10]

### *3.3. Determining the stoichiometric coefficients*

The black box description can estimate the macro chemical equation's stoichiometric coefficients or the microbial growth description. The calculation is based on the elemental and electric charge conservation relations and an additional equation, which could be the experimental yield of biomass on C-source. The black box approach is beneficial when a preliminary estimation of the substrate's biomass yield is required or before knowing the microorganisms' properties.

$Y_{DX}$ value is useful for determining the stoichiometric coefficient of biomass growth during biosurfactants formation through elemental mass balance. $Y_{DX}$ use to assess the value of the coefficient 'a' (coefficient for electron donor). Determination of the electron donor coefficient (a) from the $Y_{DX}$ value is shown in Table 7. The stoichiometric equation for aerobic biomass growth with propionate and glucose as carbon sources shows in equation 13 and equation 14, respectively.

$$\mathbf{a\ C_3H_6O_2 + b\ O_2 + c\ NH_4^+ + d\ H_2O + e\ HCO_3^- + f\ H^+ + g\ CH_{1,8}O_{0,5}N_{0.2}} \quad (13)$$

Elemental mass balance for aerobic biomass growth with propionate as an electron donor as follow:

- C balance : $3a + e + g = 0$
- H balance : $6a + 4c + 2d + e + f + 1{,}8\ g = 0$
- O balance : $2a + 2b + d + 3e + 0{,}5\ g = 0$
- N balance : $c + 0{,}2g = 0$
- Charge balance : $c - e + f = 0$

$$\mathbf{a\ C_6H_{12}O_6 + b\ O_2 + c\ NH_4^+ + d\ H_2O + e\ HCO_3^- + f\ H^+ + g\ CH_{1,8}O_{0,5}N_{0.2}} \quad (14)$$

Elemental mass balance for aerobic biomass growth with glucose as an electron donor as follow:

- C balance : $6a + e + g = 0$
- H balance : $12a + 4c + 2d + e + f + 1{,}8\ g = 0$
- O balance : $2a + 2b + d + 3e + 0{,}5\ g = 0$
- N balance : $c + 0{,}2g = 0$
- Charge balance : $c - e + f = 0$

**Tabel 7.** The value of electron donor stoichiometric coefficient (a) from $Y_{DX}$ value.

| Electron Donor | T | number C-atoms | $Y_{DX}$ (C-mol biomass/C-mol electron donor) | $Y_{DX}$ (mol biomass/mol mol electron donor) | $a = -1/Y_{DX}$ |
|---|---|---|---|---|---|
| Propionate | 35 | 3 | 0.483 | 1.449 | -0.690 |
| | 50 | 3 | 0.366 | 1.098 | -0.911 |
| | 75 | 3 | 0.109 | 0.327 | -3.058 |
| Glucose | 37 | 6 | 0.543 | 3.258 | -0.307 |

The biomass stoichiometric coefficient is assumed to be +1. The other coefficients (...) and yields must be measured or calculated from elemental (C, H, O, N) and electric charge conservation relations [3]. Solving the equation 13 and 14 with the biomass stoichiometric coefficient (g) = 1 and the electron donor stoichiometric coefficient (a) is refer to Table 7, can then obtains other stoichiometric coefficients (b, c, d, e, and f) thus the stoichiometric equation can be formed.

The stoichiometric equation for aerobic biomass growth with propionate as an electron donor in biosurfactant production is as follows:

Propionate at T = 35°C

$$-0.690C_3H_6O_2\ -1.365O_2\ -0.2\ NH_4^+ + 0.4\ H_2O + 1.070\ HCO_3^- + 1.270\ H^+ + 1\ CH_{1,8}O_{0,5}N_{0.2} \quad (15)$$

Propionate at T = 50°C

$$-0.911C_3H_6O_2\ -2.138O_2\ -0.2\ NH_4^+ + 0.4\ H_2O + 1.733\ HCO_3^- + 1.933H^+ + 1\ CH_{1,8}O_{0,5}N_{0.2} \quad (16)$$

Propionate at T = 75°C

$$-3.058C_3H_6O_2\ -9.662O_2\ -0.2\ NH_4^+ + 0.418\ H_2O + 8.174\ HCO_3^- + 8.374H^+ + 1\ CH_{1,8}O_{0,5}N_{0.2} \quad (17)$$

The stoichiometric equation for aerobic biomass growth with glucose as an electron donor in biosurfactant production is as follows:

Glucose at T = 37°C

$$-0.307\ C_6H_{12}O_6\ -1{,}406\ O_2\ -0{,}2NH_4^+ + 0{,}4H_2O + 0.842HCO_3^- + 1.042\ H^+ + 1CH_{1,8}O_{0,5}N_{0.2} \quad (18)$$

## 4. Conclusion

The accuracy of estimating the yield value of biomass to the electron donor ($Y_{DX}$) through the thermodynamic approach (the Gibbs energy dissipation) has limitations, especially for specific polymeric carbon sources. However, a thermodynamic approach via the Gibbs energy dissipation can be considered to estimate the initial biomass gain value for the electron donor ($Y_{DX}$). The result of the calculation shows that the electron donor and incubation temperature affect the electron donor's biomass yield ($Y_{DX}$). For biosurfactant production, glucose obtains a higher value of $Y_{DX}$ than propionate. By using the same electron donor, the higher the incubation temperature, the lower of $Y_{DX}$ value. $Y_{DX}$ value is useful for determining the stoichiometric coefficient of biomass growth during biosurfactants formation through elemental mass balance. The type of electron donor and temperature affect the stoichiometric coefficient of biomass growth during the biosurfactant production process.

## References

[1] Krzystek L, Gluszcz P, Ledakowicz S 1996 Determination of yield and maintenance coefficients in citric acid production by Aspergillus niger *The Chemical Engineering Journal and the Biochemical Engineering Journal* **62** 215–22.

[2] von Stockar U, van der Wielen LAM 1997 Thermodynamics in biochemical engineering *Journal of Biotechnology* **59** 25–37

[3] Tobajas M, Garcia-Calvo E 1999 Determination of biomass yield for growth of Candida utilis on glucose: Black box and metabolic descriptions *World Journal of Microbiology and Biotechnology* **15** 431–8

[4] Hoijnen JJ, van Loosdrecht MC, Tijhuis L 1992 A black box mathematical model to calculate auto- and heterotrophic biomass yields based on Gibbs energy dissipation *Biotechnol Bioeng* **40** 1139–54

5] Heijnen J J, Dijken J P V 1992 In search of a thermodynamic description of biomass yields for the chemotrophic growth of microorganisms *Biotechnology and Bioengineering* **39** 833–58

[6] Mulligan C N 2005 Environmental applications for biosurfactants *Environmental Pollution* **133** 183–98.

[7] Desai J D, Banat I M 1997 Microbial production of surfactants and their commercial potential *Microbiol Mol Biol Rev* **61** 47–64

[8] Nitschke M, Costa S G V A O, Haddad R, Gonçalves L A G, Eberlin M N, Contiero J 2005 Oil Wastes as Unconventional Substrates for Rhamnolipid Biosurfactant Production by Pseudomonas aeruginosa LBI *Biotechnology Progress* **21** 1562–6

[9] Sakthipriya N, Doble M, Sangwai J S 2015 Fast degradation and viscosity reduction of waxy crude oil and model waxy crude oil using Bacillus subtilis *Journal of Petroleum Science and Engineering* **134** 158–66

[10] Heryani H, Putra M D 2017 Kinetic study and modeling of biosurfactant production using Bacillus sp *Electronic Journal of Biotechnology* **27** 49–54

[11] Sakthipriya N, Doble M, Sangwai J S 2018 Kinetic and thermodynamic behavior of the biodegradation of waxy crude oil using Bacillus subtilis *Journal of Petroleum Science and Engineering* **160** 412–21

[12] Bai C, Zhang J 2013 Thermal, macroscopic, and microscopic characteristics of wax deposits in field pipelines *Energy & Fuels* **27** 752–9

[13] Singh SN, Kumari B, Mishra S 2012 *Microbial Degradation of Alkanes*. In: Singh SN, editor. Microbial Degradation of Xenobiotics, Berlin, Heidelberg: Springer p. 439–69.

[14] Labinger J A, Bercaw J E 2002 Understanding and exploiting C–H bond activation *Nature* **417** 507–14

[15] Kumari B, Singh S N, Singh D P 2012 Characterization of two biosurfactant producing strains in crude oil degradation *Process Biochemistry* **47** 2463–71

[16] Olajire A A, Essien J P 2014 Aerobic degradation of petroleum components by microbial consortia *Journal of Petroleum & Environmental Biotechnology* **5** 1

# Cultivation of 'Botryococcus Braunii' Microalgae for Hydrocarbons Production and $CO_2$ Bio-fixation

**R G Dewi[1,2*], D Srikaton[1], S Sitorus[1], G N Sevie[1], P Bunga[1], and S Permata[1]**

[1] Chemical Engineering Study Program, Institut Teknologi Bandung, Indonesia
[2] Center for Research on Energy Policy, Institut Teknologi Bandung, Indonesia

[*]Email: gelangdewi@gmail.com , gelang@che.itb.ac.id

**Abstract** $CO_2$ fixation for climate change mitigation actions through photosynthesis can be carried out by growing intensively managed microalgae. The capability of microalgae in the $CO_2$ fixation is favourable compare to trees or crops, yielding three to five times more biomass per land area than typical plants. Another interesting feature of microalgae is microalgal farms can use arid, semi-arid land or highly saline water, which is practically not suitable for agriculture and other biomass production activities. In addition, algae biomass contains fibre, fatty acid, lipid, protein, polysaccharide, and hydrocarbon. These materials could be extracted for biofuels or biochemicals. Biofuels derived from microalgae are environmentally friendly and carbon neutral. However, microalgal farming seen only for producing biomass as sources of fuels and chemicals probably uneconomical. The yields of microalgal farm to produce fuels/chemicals are also low. Therefore, microalgal farm with the objective of solely producing fuels/chemicals is economically unfeasible. If it is designed for $CO_2$ mitigation measures and all at once for fuels/chemicals production, such endeavour could be economically justifiable. This paper presents results of a research on microalgae cultivation for producing biomass and hydrocarbons for biofuels that simultaneously fixing the $CO_2$ emissions from coal power plant effluent through photosynthesis process, which also for the implementation of circular economy concepts. A strain of 'Botryococcus Braunii' microalgae, which known can produce high hydrocarbons and lipids throughout its growth period has been used. The research identified and evaluated the effect of cultivation medium and $CO_2$ concentration to the growth rate, biomass production, lipid and hydrocarbon content and type. The microalgae were cultivated in bioreactor supported with airlift system for supplying $CO_2$/air and for mixing process. Results of the research show that modified Chu13 (MChu13) is appropriate cultivation medium for the growth, hydrocarbon/lipid production, and $CO_2$ fixation from B. braunii. The magnitude of biomass production in MChu13 is 0.145 g/L.9days, while in modified BG11 is 0,118 g/L.9days, Bold Bassal Medium (BBM) is 0.09 g/L.9days, and Bristol is 0.023 g/L.9days. This biomass production rate is in lined with the $CO_2$ fixation rate. In MChu13 the dry weight of B.brauni is 0.268 g/L.7days while in Bristol only 0.229 g/L.7days.
***Keywords:*** *biofixation, Botryococcus braunii, circular economy, $CO_2$ emissions, hydrocarbon, lipid, microalgae, mitigation measures, photosynthesis, uplift fermenter, vertical grow reactor*

## 1. Introduction

In recent years, atmospheric carbon dioxide level is increasing on global scale due to GHG emissions, which might be derived from increased activities in combustion of fossil fuels, industrial processes, petroleum productions, etc. Some measures undertaken for reducing these emissions, among of them are decreasing the fuel consumption by more efficient use of energy, increasing the utilization of low or no carbon emission fuels, end of pipe pollution control, $CO_2$ emissions fixation, etc. Regarding the $CO_2$ fixation, one of the options is photosynthesis by growing intensively managed microalgae. The capability of microalgae in $CO_2$ fixation is favourable compare to tree or crops, yielding three to five times more biomass per land area than typical crops or plants [1]. Other interesting feature of microalgae is that microalgal farm can use arid, semi-arid land, or highly saline water, which are practically not suitable for agriculture and other biomass production activities. In addition, algae biomass contains high concentration of important materials such as protein, polysaccharides, dyes or pigments, fibre, fatty acids, lipids, and bio-hydrocarbons. These materials could be extracted and converted to biofuels, biochemicals, or food [2]. Fuels derived from microalgae farming are environmentally friendly and can be considered as carbon neutral, in which the net of $CO_2$ emissions from such fuel is zero [3]. However, microalgal farming seen only as $CO_2$ abatement measure is probably uneconomical. The yields of microalgal farm to produce fuels/chemicals are low. Therefore, microalgal farm with the objective of solely producing fuels/chemicals is economically unfeasible. If it is designed both for $CO_2$ emissions abatement measure and for fuels and chemicals productions, such endeavour could be economically justifiable and considered as circular economy [4].

Research on microalgal farming for producing fuels/chemicals that simultaneously capturing and fixing the $CO_2$ emissions is still ongoing. The research focuses on searching of strains that appropriate for this specific objective (biofuels/biochemicals production as well as $CO_2$ fixation) and finding the optimum farming condition (medium characteristics, $CO_2$ concentration, and other related parameters to the growth rate of microalgae, biomass weight, hydrocarbon content and type). Previous research on chlorella microalgae farming was carried out in simulated saline water with $CO_2$ concentration of the air supply was similar to the flue gas of coal power plant. The research is intended to produce lipids that can be converted to biodiesel. However, results of the research show that the lipid content in the biomass of the chlorella is still low, i.e. less than 10% of the dry weight [5]. Therefore, a research that focusses on finding other strain, which contained high lipids or hydrocarbons are needed. A strain of 'Botryococcus Braunii' is used in the research with several reasons. Firstly, B. braunii is known as one of microalgae that can produce high hydrocarbons and lipids throughout its growth period. Secondly, B.braunii can grow in a high saline water while other microalgae usually grow in fresh water, and therefore B.braunii can grow solely for hydrocarbons/lipids production without any disturbance from other microalgae. Thirdly, capability of B.braunii as microalgae in the $CO_2$ bio-fixation is favourable compare to tree or crops, yielding three to five times more biomass per land area than typical plants. This paper presents results of research on B.braunii microalgal cultivation for producing hydrocarbons throughout its growth period that simultaneously producing biomass for $CO_2$ fixation [6].

## 2. Experimental procedures

### *2.1. Material, Equipment, and Procedures*

The experimental work was carried out using a set of fermenter unit named Airlift fermenter with 10 L volume made of transparent glassware (see **Error! Reference source not found.**). Transparent glassware was used due to allow incoming sunlight to fermenter for the photosynthesis process. In this research, florescent lamp was used to represent the sunlight. The illumination was supplied by 6 cool-white fluorescent-lights to keep the illumination of 150 $\mu$ E/m$^2$s where the power capacity of each lamp is 8/25 watt/lumen [6].

### *2.2. Experimental Design and Analysis of Data*

Experiment activities were designed into five stages, i.e. (i) isolation of monoculture B.braunii, (ii) preparation of culture media and stock culture of B.braunii, (iii) observation of B.braunii growth in several cultivation media, (iv) development of the growth curve and measurement of dry-cell weight of the B.braunii biomass in each cultivation medium at the same growing time, and (v) analysis of lipids production and hydrocarbons content in the B.braunii culture biomass at the same growing time.

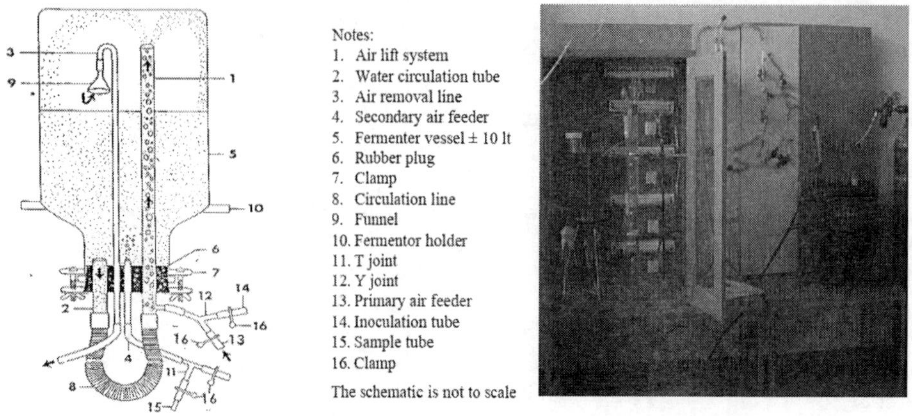


**Figure 1** Schematic of Airlift Fermenter System [6].

*2.2.1. Isolation and Activation of the Botryococcus braunii* B.braunii microalgae used in this study is originated strain of Indonesia. Search for this strain has been carried out in typical places where they might be found, namely swamps area, lakes, freshwater pond, rivers, ditch, and other slightly saline water (ephemeral lakes, brackish water, saltwater swamps, etc.). This search has been concentrated in the areas closed to oil and gas fields, in which about 40 samples have been collected from oil and gas field area (Bongas in West Java, East Kalimantan, East Java), oil and gas receiving terminal (Tanjung Priok Jakarta). Samples from freshwater lake (Situpatenggang and Kawah Putih) in West Java were also collected. From those 40 samples, B.braunii was found in samples from Bongas, Tuban, and Tanjung Priok. In this research, B.braunii from Bongas in West Java was used [6].

Isolated B.braunii appropriate for this research were obtained by separating B.braunii from a mix culture. Separation of B.braunii from other organisms was challenging, particularly from pediastrum (simplex/duplex), scenedesmus econis, staurastum anatinum, and anabaena sp. Floating microalgae colonies collected from the mix culture were diluted in serial dilutions (6 times dilution) on Bristol medium (see Table 1). Within two weeks, single colonies were separated from these cultures and isolated as single cell in a gelatine slants to obtain unialgal isolates.

**Table 1** Composition of Bristol (940 ml of distilled water + stock solution) [7,8].

| Ml | Stock solution | gram/400 ml $H_2O$ |
|---|---|---|
| 10 | $NaNO_3$ | 10 |
| 10 | $CaCl_2.2H_2O$ | 1 |
| 10 | $MgSO_4.7H_2O$ | 3 |
| 10 | $K_2HPO_4$ | 3 |
| 10 | $KH_2PO_4$ | 7 |
| 10 | NaCl | 1 |

Before commencing experiment stages, B.braunii culture needs to be cultivated on starter medium to activate the strain and make it more adaptable to the changes of nutrients, components of cultivation medium, $CO_2$ concertation, air supply flowrate, etc. The activation was carried out in 1 L glass bottle supported by airlift for supplying air (containing $CO_2$ and $O_2$). Selected unialgal culture with high B.braunii colonies was diluted ten times (10% isolates unialgal in 90% water). The adaptation process in the activation culture ordinarily takes place for 6-7 days until the growth phase of B.braunii has become stationary. Starter medium is used for B.braunii cultivation, in which four starter mediums are used, i.e. BBM, Bristol, BG-11, and MChu13. The MChu13 medium is used in this research with varied concentrations of potassium nitrate, potassium dihydrogen phosphate, magnesium sulphate and ferric citrate. The selection of appropriate medium was carried out base on the growth rate, biomass weight, and chemicals content (lipid or hydrocarbon).

*2.2.2. Preparation of Growth Culture Media and Cultivation of 'Botryococcus braunii'* Cultivation for B.braunii must be conditioned to meet intrinsic requirements of most organisms that may be physical (temperature, light, pH), chemicals (carbon, magnesium, nitrogen, phosphor, potassium, sulphur). These materials were used for structural and protoplasmic synthesis of the algae. The required concentration of these materials is presented in **Error! Reference source not found.**. Required nutrient compositions for algal cultivation are listed in the following section.

1) Algal biomass (50%) consists of carbon, sufficient carbon is needed for cultivation. In freshwater (low salinity) and neutral pH, carbon is supplied as inorganic substance (gaseous $CO_2$). In most photoautotrophic, carbon is supplied as bicarbonate or organic carbon (sugar or acetate).
2) Nitrogen is essential for algal growth (10% of algal biomass consists of nitrogen), which supplied as nitrate, ammonia, and urea. Changes in nitrogen content could influence metabolic pathways.
3) pH of algal cultivation is controlled under neutral or slightly acidic to avoid element precipitation.
4) Trace elements (iron, manganese, boron, cobalt, copper, molybdenum, and zinc) are essential trace elements, which are mostly supplied in small quantities. Sodium (in any form) is not essential and specific requirement for calcium is contradictory.

Other elements (potassium, magnesium, phosphate, sodium, sulphate), organic components, and growth-promoting substances (vitamins, nucleic acid) are required as nutrient for algal cultivation.

**Table 2** Required components of the B.braunii cultivation medium [9].

| Component | Required Concentration, mg/L | US Patent Plant 6, 169 Concentration, mg/L $H_2O$ |
|---|---|---|
| Ca $(NO_3).4H_2O$ | 50 – 250 | 26,5 |
| $NH_4Cl$ or $NH_4HCO_3$ | 10 – 150 | – |
| $MgSO_4.7H_2O$ | 10 – 100 | 25 |
| $K_2HPO_4$ or $KH_2PO_4$ | 0 – 50 | 10 |
| $H_3BO_3$ | 0 – 1 | 0.6 |
| $Na_2EDTA$ | 0 – 25 | 7.7 |
| $ZnCl_2$ | 0 – 10 | 0.624 |
| $CaCl_2.2H_2O$ | 0 – 10 | – |
| $NaMoO_4.2\ H_2O$ | 0 – 10 | 0.252 |
| $CoCl_2.6\ H_2O$ | 0 – 10 | 0.420 |
| $FeCl_3.6\ H_2O$ | 0 – 10 | 2.5 |
| $MnCl_2.4\ H_2O$ | 0 – 10 | 0.36 |
| $CuCl_2.2H_2O$ | 0 – 10 | 0.268 |
| MOPS Buffer | – | 3.14 |

*2.3. Experimental design and analysis of data*

Using 10 L batch airlift fermenter, B.braunii was cultivated in 6 L medium, in which isolated algal cultures were illuminated under 150 μ E/m-2s-1, in ambient condition (± 27- 30°C, 1 atm), and pH was adjusted (by acid or base) to the range from 6.5 to 7.5. Each culture has equal treatment, such as aeration using available air in the airlift fermenter, volume of initial cultivation (10% v/v) at the same time and place. Cultivation media for this experiment are the same as those in starter medium, i.e. BBM, Bristol, BG-11, and MChu13 with compositions such presented in **Error! Reference source not found.**. The aeration and agitation rate in the cultivation reactor must be maintained constantly using airlift system at 18 L/minutes gas rate. Reynolds number should exceed 3000 to remain turbulent flow to prevent sedimentation of microalgal biomass in the cultivation reactor while the flow velocity was kept around 0.3-0.4 m/s to avoid the cell damage. During cultivation, mix gas (air–$CO_2$) were introduced in the fermenter. The variation of the mix gas rate is discussed in Sub-chapter 3.2.

**Table 3** Composition of Botryococcus braunii cultivation medium [6].

| Composition | Bristol (gr/L) | MChu 13 | BG11 | BBM |
|---|---|---|---|---|
| $K^+$ | 0.168 | 0.28 | 0.08 | 0.165 |
| $NO_3^-$ | 0.25 | 0.2 | 1.5 | 0.25 |
| $PO_4^{3-}$ | 0.093 | 0.04 | 0.04 | 0.0915 |

| Composition | Bristol (gr/L) | MChu 13 | BG11 | BBM |
|---|---|---|---|---|
| $Na^+$ | 0.275 | - | 1.54 | 0.275 |
| $Mg^{2+}$ | 0.075 | 0.1 | 0.075 | 0.073 |
| $Fe^{3+}$ | 0.0005 | 0.01 | 0.006 | 0.005 |
| $SO_4^{2-}$ | 0.075 | 0.1 | 0.075 | 0.078 |
| $Cl^-$ | 0.0765 | 0.19 | 0.09 | 0,073 |
| $CO_3^{2-}$ | - | - | 0.02 | - |
| $Ca^{2+}$ | 0.025 | 0.08 | 0.036 | 0.024 |
| EDTA | - | - | 0.001 | 0.045 |
| Citric Acid | - | 0,1 | 0.006 | - |
| Ferric citrate | | 0.01 | | |

## 3. Results and Discussions

### *3.1. The Effect of Cultivation Medium to the Growth of Botryococcus braunii*

The research has identified the effect of cultivation media to the growth of B.braunii. The observation of the effects of each medium to the growth of B.braunii was carried out under the four growth phases, i.e. lag phase, exponential phase, stationary phase, and death phase. The growth does not occur in lag phase, therefore the growth rate is zero for the lag phase. Subsequently, the microalgae begin to be an active cell and the density will increase as a function of time during exponential phase. The growth rate in this phase is constant. After the exponential phase, the culture enters the stationary phase where the cell concentration remains constant at its maximum value and cell death. The growth phases of B.braunii in each medium composition is presented in Figure 2.

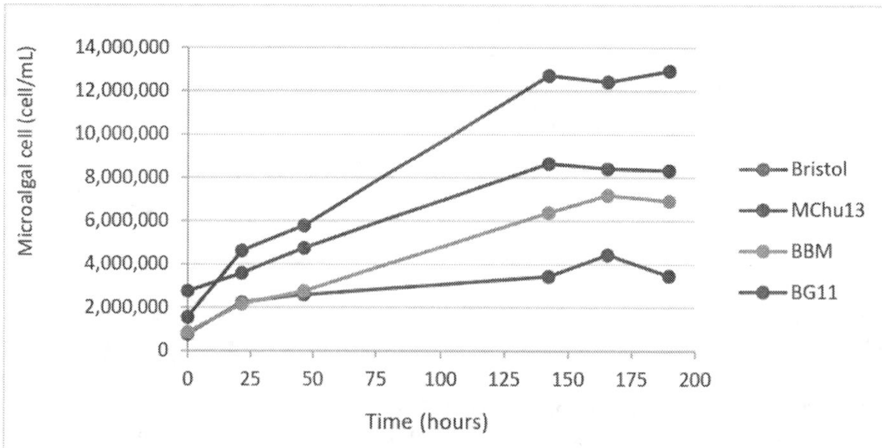


**Figure 2** The Growth curve of B.braunii.

In all microalgae growth phase, exponential phase begins after the first 160 hours. However, Figure 2 shows that the lag phase does not occur. It seems that B.braunii has adapted to the surrounding environment, i.e. medium, pH, temperature and lightning. The maximum growth rate of B.braunii for each medium are occurs when the cultivation carried out during 150 hours and the gradient of the curves reduces after 150 hours. The rapid growth shows a significant increase of cell density. After B.braunii reaches stationary phase, the curves are indicated constant and continued with the constant density. The density is still increasing or decreasing due to the balance of catabolism and anabolism cells. In addition, nutrient levels in microalgae are decreasing that will stimulate the cells to form a fatty acid which is the precursor for the formation of hydrocarbons.

It also can be seen in Figure 2 that MChu13 leads to the highest growth of microalgae cell. Refer to Table 3, each of cultivation media has specific compounds that can affect the growth of B.braunii. It can be seen that BG11, BBM and Bristol are dominated by $K^+$, $NO_3^-$, $PO_4^{3-}$, $Mg^{2+}$, $Fe^{3+}$, $SO_4^{2-}$, $Cl^-$, $Ca^{2+}$.

$NO_3^-$ and $Na^+$. There were additional essential components of EDTA in BBM and BG11, high content of $PO_4^{3-}$ in Bristol, and specific essential components in MChu13. The MChu13 does not contain sodium, which contributes to the transformation of nitrogen into ammonia in algal binding molecules. However, B.braunii does not require a source of nitrogen such as ammonia. The presence of ammonia will decrease photosynthetic rate, carbohydrate formation, and growth rate.

**Table 4** Biomass dry weight of B.braunii at various medium.

| Medium | Dry weight of B.braunii (g/L.9days) |
|---|---|
| MChu13 | 0.145 |
| Bristol | 0.023 |
| BG11 | 0.118 |
| Bold Bassal Medium (BBM) | 0.090 |

The growth of B.braunii were also represented as the growth of biomass dry weight. Table 4 shows the effect of cultivation medium to the dry weight of B.braunii. It can be seen from Table 4 that the highest biomass dry weight of B.Braunii was produced in MChu13. It is obviously that MChu13 medium is the best medium for cultivation medium of B.braunii. The highest growth of microalgae cell and biomass weight was obtained from MChu13 (Figure 2 and Table 4).

In this research, airlift fermenter is used to measure the growth rate of B.branuii and the ability of B.braunii to absorb $CO_2$ was identified by introducing more $CO_2$ in the cultivation reactor under the two variations of the growth medium, i.e. MChu13 and Bristol. Bristol was selected as comparison because this medium has been used for B.braunii cultivation in previous study so that required parameter for the maximum growth of B.braunii using the previous study. The growth of B.braunii cell for both medium, Bristol and MChu13 in airlift fermenter are presented in Figure 3.

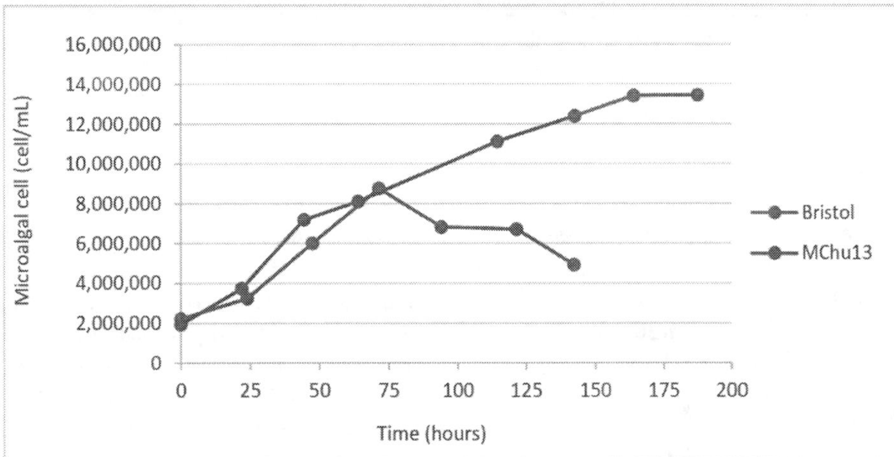


**Figure 3** The growth curve of B.braunii in airlift fermenter.

The results of this research show that the best growth is still provided by MChu13 medium. Bristol medium showed that the cell number of B.braunii increased in-line with increasing of dissolved $CO_2$ during the 75 hours cultivation duration with the same rate as MChu13. However, the Bristol medium indicates an earlier time of death phase. In comparison, the MChu13 shows that the cultivation is still continuous increasing until the death phase, i.e. 187.5 hours. The cell growth for B.braunii in Bristol was 0.019 cells/(mL.9hr) while in MChu13 was 0.22 cells/(mL.9hr). The MChu13 also results the best dry weight of B.brauni, i.e 0.268 g/L.7days while Bristol only 0.229 g/L.7days.

The cell growth and dry weight in MChu13 was higher than those in Bristol because the MChu13 has Fe and Mg, which is an essential component that supports the energy transfer for the formation of cell biomass [10]. However, Bristol contain $PO4^{3-}$ slightly higher compare to other medium. The $PO4^{3-}$ is essential to support biosynthesis of phospholipids, nucleic acids, and particular coenzyme, especially

IOP Conf. Series: Materials Science and Engineering 1143 (2021) 012017 doi:10.1088/1757-899X/1143/1/012017

nucleotide compounds (ATP, GTP, NAD, NADP, FAD) that are needed as electron carriers in energy transfer [11].

### *3.2. The effect of Dissolved $CO_2$ in Each Medium on Botryococcus braunii Growth*

In the growth phase, introducing $CO_2$ into cultivation culture determines the logarithmic phase of the growth but does not determine the lag phase. During cultivation, mix gas (air–$CO_2$) were introduced in the fermenter with variation of mix gas rate in this experiment is presented in Table 5.

**Table 5** Effect of dissolved $CO_2$ on the length of the B.braunii growth phase.

| Run | $CO_2$ rate, L/minute | $CO_2$ concentration, % | The Growth Phase | |
|---|---|---|---|---|
| | | | Lag | Logarithmic |
| 1 | 0 | 0.03 | 48 | 32 |
| 2 | 1.4 | 7.8 | 48 | 49 |
| 3 | 2.1 | 11.7 | 48 | 49 |
| 4 | 2.9 | 16.1 | 48 | 49 |

Notes: Total mix gas (air – $CO_2$) was constant at the rate of 18 L/minute

The determination of the supply rate of $CO_2$ has considered (i) typical $CO_2$ concentration in flue gas of fossil fuel combustions, which will be used as the source of $CO_2$ for supplying the B.braunii cultivation (ii) the ability of B.braunii in absorbing $CO_2$, and (iii) the solubility of $CO_2$ in the medium where B.braunii grow (freshwater). In each run of experiments, the algae were cultivated during 7 days until the growth achieved stationer phase. **Error! Reference source not found.** shows that introducing $CO_2$ into algal cultivation does not control microalgae adaptation, which was shown by the phase lag. The logarithmic phase's length shows that introducing $CO_2$ encounter photo-oxidation reaction. The B.braunii can still grow with longer time since $CO_2$ emissions increase carbon sources of the medium. The result of this experiment indicates that B.braunii can growth well by absorb $CO_2$ concentration of 0.03-16.1%. The increased of $CO_2$ rate the result obtained the algae was tolerant up to $CO_2$ rate of 2.9 L/minute depend on flue gas of fossil fuel combustion as source of $CO_2$.

### *3.3. Hydrocarbon Content of Botryococcus braunii*

B.braunii has known contains high hydrocarbon that could achieve 40-86% of its dry weight biomass although the typical range is 25-40%. An active colony produces unbranched olefins (C27:2, C29:2, C29:3, and C31:2) that could achieve 32% of its dry weight biomass. Resting-state colonies produce unusual branched olefins ($C_nH_{2n-10}$; n = 30 – 37) as terpenoid origin. These olefins are named as Botryococcenes components. The fraction of these components are 27-80% in natural cultivation.

This research shows that B.braunii contains unusual branched olefins, i.e. $C_nH_{2n-10}$; n = 32, 33, 34. Analysis of the hydrocarbon components is only for qualitative characteristics as the dry weight biomass produced from this experiment is too low (2–3.5-gram dry weight) for quantitative analysis. The B,braunii usually need a relatively long period of mass doubling time, which exceeds seven days. The measurement of hydrocarbon composition and production level would require a long experiment time. In this preliminary short time research, such measurement has not been performed.

### *3.4. Other Requirement Parameters for the Production System of Microalgae*

Generally, in addition to a carbon source, microalgae's growth depends on i) temperature: the temperature parameter affects the diffusion rate and the activity of the microalgae's inner molecules. The lower temperature causes the molecule movement and diffusion process to obstruct, while the higher temperature causes denaturation of proteins where these microalgae intracellular enzymes will be damaged; ii) pH: pH is an essential parameter in the cultivation process of Botryococcus braunii; the appropriate pH is maintained within range 7-9. Out of the optimum pH range will affect the equilibrium of the biomass formation reaction and form unwanted deposits; iii) circulation flow rate: the cultivation is also carried out by aeration of air to supply oxygen.

Oxygen is a terminal oxidizer in electron transport systems and the essential element in ATP formation for intracellular growth. Apart from that, aeration also plays a role in the homogenization of culture and the diffusion of $CO_2$. The more ratio of the surface area to the volume of air bubbles will increase the $CO_2$ diffusion rate into the culture; iv) Lighting: the experiment utilizes external lights to provide appropriate light intensity in the photosynthesis process of the microalgae. The light is used to form NADPH and ATP, which will be used in dark reactions. In the dark reaction, building blocks of glucose and other organic compounds will be formed.

It is also expected that the growth of Botryococcus braunii would be affected by the presence of other constituents in each medium where the microalgae live. The MChu13's predominance among the different autotrophic medium (BBM, Bristol, and BG11) is its high Fe dan Mg content [11]. Fe dan Mg is an essential component that supports the energy transfer events so that the formation of cell biomass was getting better.

## 4. Conclusions

In this research, B.braunii were successful cultivated in an Airlift Fermenter with several mediums of cultivation. The main objective of the cultivation is to produce biohydrocarbon and $CO_2$ biofixation. Among cultivation mediums used in this research, MChu13 was identified as the best for B.brauni cultivation with the cell growth of 0.22 cells/(mL.9hr) and high biomass productivity of 0.145 dry weight (g/L.9days) while in Bristol medium the cell growth is only 0.019 cells/(mL.9hr) and biomass productivity is only 0.023 dry weight (g/L.9days).

The optimum growth of other medium (Bristol) is only during the period of 75 hours after that period the microalgae will be in death phase. During the cultivation, cell population of the B.braunii increases in-line with the increasing of dissolved $CO_2$. After 75 hours cultivation period, the cell density decreases rapidly and the culture eventually collapses. As comparison, the microalgae growth optimum in MChu13 is in the period of 187.5 hours. The research shows that MChu13 is the best medium for $CO_2$ fixation. The dry weight biomass production is 0.268 g/L.7days in MChu13 medium while in Bristol is only 0.229 g/L.7days.

The $CO_2$ emission from fossil fuel-based power plant contributes significantly to the national GHG emissions of energy sector. The emission has been increased significantly during the past decade due to the increasing number of fossil fuels power plants in Indonesia. This paper presents results of a research on microalgae cultivation for producing biomass and hydrocarbons for biofuels that simultaneously fixing $CO_2$ emissions from coal power plant effluent through photosynthesis process, which also for the implementation of circular economy. B.braunii microalgae could be one of mitigation measures to reduce the $CO_2$ emissions due to the capability of microalgal to utilize $CO_2$ as a carbon source for their growth, which also at the same time serves as bio-fixation of $CO_2$ emissions. B.braunii can growth well by absorbing the $CO_2$ of about 0.03-16.1%. The increased of $CO_2$ rate the result obtained the algae was tolerant up to $CO_2$ rate of 2.9 L/minute depend on flue gas of fossil fuel combustion as source of $CO_2$. In addition, B.braunii microalgae could also produces alternative biofuels (gasoline and biodiesel).

B.braunii usually contain high hydrocarbon (40-86% of its dry weight biomass), an active colony produces unbranched olefins (C27:2, C29:2, C29:3, C31:2) that could achieve 32% of its dry weight biomass. A resting-state colony produce unusual branched olefins ($C_nH_{2n-10}$; n = 30-37) as terpenoid origin that could achieve 27-80% in natural cultivation. These olefins are Botryococcenes components. This research shows that B.braunii contains unusual branched olefins, i.e. $C_nH_{2n-10}$; n = 32, 33, 34.

**Acknowledgement**
We are grateful for receiving funding from P3MI (Program Penelitian Pengabdian kepada Masyarakat dan Inovasi) at Institut Teknologi Bandung (2020) and Osaka Gas Foundation (2003)

**References**

[1] Hauck J T, Olson G J, Scierka S J, Perry M B, and Ataai M M 1996 Effects of simulated flue gas on growth of microalgae *Abstracts of Papers. American Chemical Society, 212 Meeting, Pt.1, Fuel 118. Orlando, FL* 25–29 August 1996
[2] Hossein A A, Nahid R, Raul G, Abbas A, and Jose R M 2019 *Microbial Cell Factories* p18

[3] Cossarizka and Limawan 2005 *The Effect of Dissolved $CO_2$ in The B.Braunii Cultivation and $CO_2$ Bio-fixation* (Bandung, Indonesia: Institute Teknologi Bandung)
[4] Dewi RG 2003 *Proc. Int. Conf. Effect of Dissolved $CO_2$ on Growth and Hydrocarbon Content of Micro Algae 'Botryococcus Braunii'* Osaka Gas Foundation of International Cultural Exchange
[5] Jay M, Kawaroe M, Effendi H 2017 *Earth and Environmental Science*
[6] Dewi R G 2009 *Symp.,* on 16th Asean Regional Symposium on Chemical Engineering *Integrated Up Lift Vertical Grow Bioreactor for Microalgae Cultivation in the Context of Biofixation of $CO_2$ Emission and Production of Biofuel* December 1-2th, 2009, Manila Hotel, Philippines
[7] Droop M R 1967 *Br. Phycol. Bull* **3** 295–297
[8] Hegewald E 1985 *Cramer, Braunschweig* **41** 219-271
[9] Nomura A M and Calif D M 1986 (Berkely : The University of California)
[10] Khalid A, Mohamed T, Brian H M, Stella S, Andrew S B, and Eric M A 2015 *European Journal of Phycology* vol **51**
[11] Anindita R and Jayanthy M T 2010 *Studi Pertumbuhan Botryococcus braunii Pada Berbagai Medium Pertumbuhan dalam Reaktor Terintegrasi Airlift-Vertical Growth yang Ditunjukkan untuk Biofiksasi $CO_2$ dan Produksi Biofuel* (Bandung, Indonesia: Institute Teknologi Bandung)
[12] Wolf F R 1983 *Applied Biochemistry and Biotechnology* **8**
[13] Srikaton D 2009 *Integrated Up Lift Vertical Grow Bioreactor for Microalgae Cultivation in the Context of Biofixation of $CO_2$ Emission and Production Biofuel* (Bandung, Indonesia: Institute Teknologi Bandung)
[14] Vonshak A 1986 *Handbook for Algal Mass Culture,* Richmond A, CRC Press (Florida) 117- 146.
[15] Pengfei C, Yan W, Qiyong Y and Tianzhong L 2017 *International Journal of Agricultural and Biological Engineering* **10** 134–141

# Modelling and Simulation of Biobutanol Fermentation by *Clostridium saccharoperbutylacetonicum* N1-4

**Elvi Restiawaty**[a,b,e*], **Ardiyan Harimawan**[c,d], **Novaldio Rizki**[c], **Fauz Irfan Rafi**[c]

[a] Research Group of Chemical Engineering Process Design and Development, Faculty Industrial Technology, Institut Teknologi Bandung
[b] Deparment of Bioenergy Engineering and Chemurgy, Faculty Industrial Technology, Institut Teknologi Bandung
[c] Department of Chemical Engineering, Faculty Industrial Technology, Institut Teknologi Bandung
[d] Research Group of Chemical Engineering Product Design and Development, Faculty Industrial Technology, Institut Teknologi Bandung
[e] Biosciences and Biotechnology Research Center, Institut Teknologi Bandung

*Corresponding Author's E-mail: erestiawaty@che.itb.ac.id

**Abstract**. Biobutanol exhibits high octane number, non-hygroscopic, non-corrosive and can be mixed with gasoline without engine modification. Furthermore, the use of biobutanol as a fuel can reduce the negative impact of environmental due to the use of fossil-based fuel. Therefore, research on the development of biobutanol is still very much needed, including research on modelling and simulation of biobutanol production by fermentation. With modelling and simulation, we can predict various biobutanol production behaviour. This paper deals with the modelling and simulation of the production of biobutanol from glucose in batch bioreactor using *Clostridium saccharoperbutylacetonicum* N1-4. *C. saccharoperbutylacetonicum* N1-4 exhibits a good growth ability on simple or complex sugars, such as glucose, sucrose, or starch. This study concerned with the kinetics models of biobutanol production as developed by Shinto based on the *C. acetobutylicum* metabolic pathway which consists of the glycolysis, acidogenesis and solventogenesis pathways with several simplification of reactions. In this study, modelling was carried out by adapting the rate parameter value of $k_{13}$ indicating the rate constant of cell death. The adapted models were applied to determine the effect of the initial concentration of glucose and organic acids on the production of butanol and compared to the experimental data from previous research. Model simulations were performed by using MATLAB® software. The butanol yield was affected by the initial glucose concentration. The highest butanol yield of 77% was obtained at a glucose concentration of 22 g/L. The simulation results exhibit that the biobutanol yield decreases when the glucose concentration is higher than 22 g/L. These results are consistent with previous experimental data. The simulation results also show the effect of the concentration of acetic acid in the fermentation medium and can be optimized. From the simulation results, the presence of 7.5 g/L acetic acid can increase the biobutanol production by 20%. However, previous experimental data showed that a concentration of 2.5 g/L acetic acid was preferred. The simulation results also show the effect of the addition of butyric acid and lactic acid on the production of biobutanol.

## 1. Introduction

The issue of alternative energy is currently a hot topic of discussion due to the crisis of fossil-based fuels. Biofuel as an alternative energy can be produced from renewable resources. The use of renewable resources certainly has a positive impact on the environment and can produce environmentally friendly fuels. Global biofuel consumption is still quite low, i.e. about 3.8%, of which 60% is dominated by ethanol production in 2017 [1]. Although bioethanol is well developed, some of the properties of this fuel still have drawbacks. Furthermore, the development of biobutanol is currently being carried out considering that biobutanol has advantages over bioethanol. Biobutanol exhibits a higher energy density, quite similar to gasoline in octane number and air-fuel ratio, so that it is able to be used as a substitution of gasoline without engine modification [2].

In the bioprocess industry, biobutanol can be produced from fermentable sugars or biomass hydrolysate through acetone-butanol-ethanol (ABE) fermentation using *Clostridia* bacteria. In the early growth phase, *Clostridia* produces hydrogen, carbon dioxide, acetate, and butyrate which causes a decrease in the pH of the culture media in a batch reactor system so that this phase is referred to as the acidogenic phase. Then, cell metabolism undergoes a shift to the solventogenic phase after the culture enters the stationary growth phase. In this phase there is re-assimilation of acid, which occurs with continued consumption of carbohydrates, usually resulting in an increase in the pH of the culture medium. In this solventogenic phase, acetyl-CoA and butyryl-CoA are the main intermediaries in the production of ethanol and butanol, respectively [3]. There have been many studies using several solvent clostridial strains that can naturally produce ABE, such as *Clostridium acetobutylicum*, *C. beijerinckii*, *C. saccharobutylicum*, and *C. saccharoperbutylacetonicum* [4]-[6] and some use genetically engineered clostridial mutations for the production of biobutanol [7]-[9]. This study focuses on the biobutanol production from fermentable sugar using *C. saccharoperbutylacetonicum* N1-4. This strain is a mesophilic bacterium and can grow in an oxygen-free environment (anaerobic conditions). *C. saccharoperbutylacetonicum* N1-4 exhibits a good growth ability on both simple and complex sugars, such as glucose, sucrose, or starch [10].

The fermentation of biobutanol has been widely investigated both experimentally [11]-[16] and computer simulation [11]. Zhou *et al.* [15] studied the influence of acetate and lactate, meanwhile Al-Shorgani *et al.* [16] studied the influence of butyrate. Shinto *et al.* [11] have proposed mechanistic models of *C. saccharoperbutylacetonicum* N1-4 in mathematic equation forms that based on Michaelis-Menten's enzymatic reaction kinetics, except for the cell death reaction. Shinto *et al.* [11] varied the initial glucose concentration in their experiments to develop their mechanistic model of butanol fermentation. However, when the Shinto's model was applied to consider the laboratory work by Tanaka [13] and Ishizaki [14], in which high glucose concentration was used (90 g/L), it was not appropriate to validate the data. Those laboratory works indicated that glucose was still remained in certain concentration at the end of fermentation, meanwhile simulation of Shinto's model indicated that glucose was fully utilized. According to the Shinto model, the consumption rate of glucose correlates with the concentration of residual glucose, butanol, and biomass. In Shinto's mechanistic models, the biomass concentration is considered to be the most influencing one, since biomass concentration is consisted in every single kinetics rate equation. In other hand, the biomass concentration change is affected by the acetyl-CoA and butanol concentrations, and also death rate of the biomass. Furthermore, acetyl-CoA and butanol are intertwined with other compounds. It is leading to death rate as the most independent factor of biomass concentration, since it is only directly influenced by the biomass concentration itself.

The objective of this study is to investigate the production of biobutanol in a batch reactor by varying some process variables through modelling and simulations. The simulation is also able to estimate the performance of a system in several conditions [17]. In this study, modeling was carried out by adapting the Shinto's model. The adapted models were applied to determine the effect of the initial concentration of glucose ($S_0$) and organic acids on the production of butanol and compared to the experimental data from previous research.

## 2. Methodology

### 2.1. Simulation algorithm

The outline algorithm of this research consisted of literature studies, evaluating the models statistically, and applying the model to identify the effect of glucose and organic acids. This research began with a literature study on the metabolism mechanism of *C. saccharoperbutylacetonicum* N1-4 and kinetic models of the reaction mechanism by Shinto *et al.* [11], which is provided in Figure 1. In this study, the rate constant of biomass death rate (value of $k_{13}$ in Equation A.13) was tuned by several iterating. With the new value of $k_{13}$, the overall Shinto's model would be significantly improved.

The adapted mathematical models were validated with the previous laboratory experimental data investigated by Shinto *et al.* [11], Thang *et al.* [12], Tanaka *et al.* [13], and Ishizaki *et al.* [14]. Accuracy of the model was evaluated based on residual $R^2$ statistical method with the help of MATLAB® as the simulation tool. The statistical results are then used as an indicator of the adapted parameter to give more accurate results (higher $R^2$), with the same method. Once it is deemed accurate enough, the adapted model was applied to experiment variations, along with other experimental data.

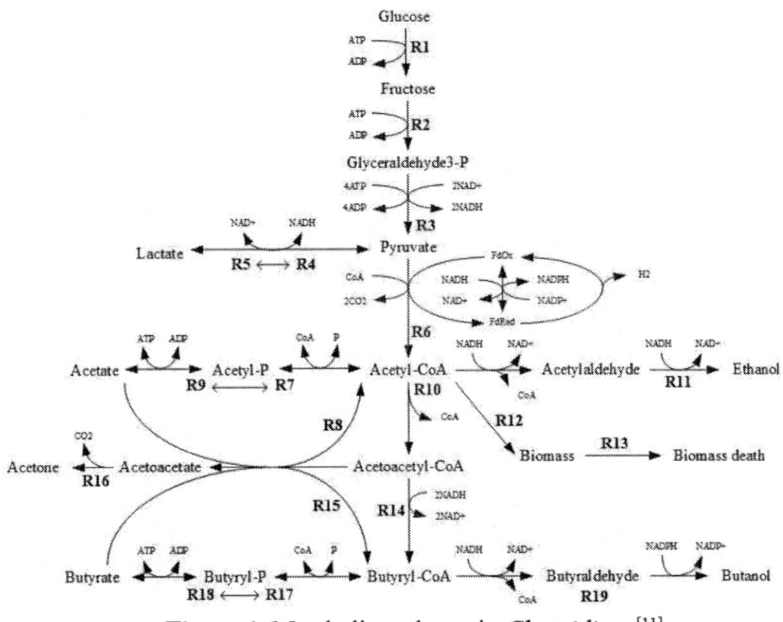


**Figure 1.** Metabolic pathway in *Clostridium* [11]

Mechanistic equations of reaction rates proposed by Shinto *et al.* [11] are presented below.

$$r_1 = \frac{V_{max,1}[Glucose][Biomass]}{K_{m1}\left(1+\frac{[Glucose]}{K_{is1}}\right)+[Glucose]\left(1+\frac{[Butanol]}{K_{ii1}}\right)} \quad (A.1)$$

$$r_2 = \frac{V_{max,2}[F6P][Biomass]}{K_{m2}+[F6P]}\times F \quad (A.2)$$

$$r_3 = \frac{V_{max,3}[G3P][Biomass]}{K_{m3}+[G3P]}\times F \quad (A.3)$$

$$\frac{d[Glucose]}{dt} = -r_1 \quad (A.20)$$

$$r_4 = \frac{V_{max,4}[Lactate][Biomass]}{K_{m4}+[Lactate]}\times F \quad (A.4)$$

$$\frac{d[F6P]}{dt} = r_1 - r_2 \quad (A.21)$$

$$r_5 = \frac{V_{max,5}[Pyruvate][Biomass]}{K_{m5}+[Pyruvate]}\times F \quad (A.5)$$

$$\frac{d[G3P]}{dt} = r_2 - r_3 \quad (A.22)$$

$$r_6 = \frac{V_{max,6}[Pyruvate][Biomass]}{K_{m6}+[Pyruvate]}\times F \quad (A.6)$$

$$\frac{d[Pyruvate]}{dt} = r_3 + r_4 - r_5 - r_6 \quad (A.23)$$

$$r_7 = \frac{V_{max,7}[Acetate][Biomass]}{K_{m7} + [Acetate]} \times F \quad \text{(A.7)}$$

$$r_8 = V_{max,8}\left(\frac{1}{1+\left(\frac{K_{m8A}}{[Acetate]}\right)}\right)\left(\frac{1}{1+\left(\frac{K_{m8B}}{[AACoA]}\right)}\right)[Biomass] \quad \text{(A.8)}$$

$$r_9 = \frac{V_{max,9}[ACoA][Biomass]}{K_{m9} + [ACoA]} \times F \quad \text{(A.9)}$$

$$r_{10} = \frac{V_{max,10}[ACoA][Biomass]}{K_{m10} + [ACoA]} \quad \text{(A.10)}$$

$$r_{11} = \frac{V_{max,11}[ACoA][Biomass]}{K_{m11} + [ACoA]} \times F \quad \text{(A.11)}$$

$$r_{12} = \frac{V_{max,12}[ACoA][Biomass]}{K_{m12}\left(1+\frac{[Butanol]}{K_{ii12}}\right) + [ACoA]\left(1+\frac{[Butanol]}{K_{ii12}}\right)} \quad \text{(A.12)}$$

$$r_{13} = k_{13}[Biomass] \quad \text{(A.13)}$$

$$r_{14} = \frac{V_{max,14}[AACoA][Biomass]}{K_{m14} + [AACoA]} \times F \quad \text{(A.14)}$$

$$r_{15} = V_{max,15}\left(\frac{1}{1+\left(\frac{K_{m15A}}{[Butyrate]}\right)}\right)\left(\frac{1}{1+\left(\frac{K_{m15B}}{[AACoA]}\right)}\right)[Biomass] \quad \text{(A.15)}$$

$$r_{16} = \frac{V_{max,16}[Acetoacetate][Biomass]}{K_{m16} + [Acetoacetate]} \quad \text{(A.16)}$$

$$r_{17} = \frac{V_{max,17}[Butyrate][Biomass]}{K_{m17}\left(1+\frac{K_{a17}}{[Butyrate]}\right) + [Butyrate]} \times F \quad \text{(A.17)}$$

$$r_{18} = \frac{V_{max,14}[BCoA][Biomass]}{K_{m18} + [BCoA]} \times F \quad \text{(A.18)}$$

$$r_{19} = \frac{V_{max,19}[BCoA][Biomass]}{K_{m19}\left(1+\frac{K_{a19}}{[Butyrate]}\right) + [BCoA]\left(1+\frac{[Butanol]}{K_{ii19}}\right)} \times F \quad \text{(A.19)}$$

$$\frac{d[Lactate]}{dt} = r_5 - r_4 \quad \text{(A.24)}$$

$$\frac{d[ACoA]}{dt} = r_6 + r_7 + r_8 - r_9 - r_{10} - r_{11} - r_{12} \quad \text{(A.25)}$$

$$\frac{d[Biomass]}{dt} = r_{12} - r_{13} \quad \text{(A.26)}$$

$$\frac{d[Acetate]}{dt} = r_9 - r_7 - r_8 \quad \text{(A.27)}$$

$$\frac{d[Ethanol]}{dt} = r_{11} \quad \text{(A.28)}$$

$$\frac{d[AACoA]}{dt} = r_{10} - r_8 - r_{14} - r_{15} \quad \text{(A.29)}$$

$$\frac{d[Acetoacetate]}{dt} = r_8 + r_{15} - r_{16} \quad \text{(A.30)}$$

$$\frac{d[BCoA]}{dt} = r_{14} + r_{15} + r_{17} - r_{18} - r_{19} \quad \text{(A.31)}$$

$$\frac{d[Butyrate]}{dt} = r_{18} - r_{15} - r_{17} \quad \text{(A.32)}$$

$$\frac{d[Acetone]}{dt} = r_{16} \quad \text{(A.33)}$$

$$\frac{d[CO_2]}{dt} = r_6 + r_{16} \quad \text{(A.34)}$$

$$\frac{d[Butanol]}{dt} = r_{19} \quad \text{(A.35)}$$

**Table 1.** Kinetic parameters estimated by Shinto *et al.* [11]

| Reaction | $k$ ($h^{-1}$) | $V_{max}$ ($h^{-1}$) | $K_m$ (mM) | $K_{is}$ (mM) | $K_{ii}$ (mM) | $K_a$ (mM) | $K_{mA}$ (mM) | $K_{mB}$ (mM) |
|---|---|---|---|---|---|---|---|---|
| $R_1$ | | 3.20 | 46.0 | 55.6 | 67.5 | | | |
| $R_2$ | | 40.0 | 10.0 | | | | | |
| $R_3$ | | 120 | 26.5 | | | | | |
| $R_4$ | | 7.50 | 177 | | | | | |
| $R_5$ | | 9.70 | 500 | | | | | |
| $R_6$ | | 180 | 1.50 | | | | | |
| $R_7$ | | 0.30 | 50.0 | | | | | |
| $R_8$ | | 19.0 | | | | | 40.0 | 70.0 |
| $R_9$ | | 26.5 | 51.0 | | | | | |
| $R_{10}$ | | 20.0 | 1.00 | | | | | |
| $R_{11}$ | | 7.45 | 30.0 | | | | | |
| $R_{12}$ | | 8.10 | 1.10 | | 23.0 | | | |
| $R_{13}$ | 0.017 | | | | | | | |
| $R_{14}$ | | 10.0 | 5.20 | | | | | |
| $R_{15}$ | | 80.0 | | | | | 15.0 | 50.0 |
| $R_{16}$ | | 12.0 | 10.0 | | | | | |
| $R_{17}$ | | 35.0 | 4.90 | | | 2.20 | | |
| $R_{18}$ | | 100 | 6.10 | | | | | |
| $R_{19}$ | | 3.15 | 5.00 | | 67.5 | 2.20 | | |

### 2.2. Statistical method

Shinto *et al.* [11] used Pearson correlation coefficient ($R^2_{Pearson}$) to measure the accuracy of their model on experimental data. $R^2_{Pearson}$ can be determined using a formula written in Equation (1), [18].

$$R^2_{Pearson} = \frac{\Sigma(y-\bar{y})(\hat{y}-\bar{\hat{y}})}{\sqrt{\Sigma(y-\bar{y})^2\,\Sigma\left(\hat{y}-\bar{\hat{y}}\right)^2}} \quad (1)$$

where $y$ is an experiment datum, $\hat{y}$ is a simulation datum corresponding to $y$, $\bar{y}$ is an average of the experiment data, and $\bar{\hat{y}}$ is an average of the simulation data [18]. In contrast to the method of determining accuracy by Shinto et al. [11], the model accuracy on experimental data in this study was examined using residual determination coefficients ($R^2$). The value of $R^2$ can be calculated using Equation (2) [18],

$$R^2 = 1 - \frac{\Sigma(y-\hat{y})^2}{\Sigma(y-\bar{y})^2} \quad (2)$$

The main difference of $R^2_{Pearson}$ and $R^2$ is the average simulation value ($\bar{\hat{y}}$) that $R^2_{Pearson}$ includes the $\bar{\hat{y}}$, while $R^2$ does not. Overall, $R^2$ illustrates how much total error between the simulation and the experimental results, compared to total variance of the experiment results itself. Variance states the extent of data distribution. A negative value of $R^2$ means that the simulation deviates too much from the distribution of the experimental data. For summarizing some values of $R^2_{Pearson}$ or $R^2$ in this study, average formula ($Avg$) is used as follows:

$$Avg = \frac{\sum_{i=1}^{n} R_i^2}{n} \quad (3)$$

with $R^2_i$ as the individual values of $R^2_{Pearson}$ or $R^2$, and n as number of $R^2_i$.

## 3. Results and Discussion

### 3.1. Model scripting evaluation

At the beginning of this study, an evaluation of script writing in MATLAB® from the models made by Shinto et al. [11] needed to be done. This was because Shinto et al. wrote a model script into different software, namely WinBEST-KIT. This evaluation was intended to ensure that the model script used in this study conforms to the model script written by Shinto et al. [11]. The evaluation was done by comparing the value of Pearson correlation coefficient ($R^2_{Pearson}$) obtained from Shinto et al. [11] and from the simulation results in this study. The experimental data are also in accordance with the data from Shinto et al. [11]. Table 2 shows the comparative values of $R^2_{Pearson}$ between the value obtained from Shinto et al. [11] using WinBEST-KIT and from this study using MATLAB®.

**Table 2.** Comparison of $R^2_{Pearson}$ values between literature and this study for changes in the concentration of compounds involved in butanol fermentation obtained by Shinto et al (2007).

| References | Pearson correlation coefficients on the concentration trend of compounds | | | | | | |
|---|---|---|---|---|---|---|---|
| | Glu.* | Ace.* | But.* | Act.* | Btr.* | Bms.* | $Avg$ * |
| **[11]** | 0.972 | 0.983 | 0.984 | 0.970 | 0.644 | 0.850 | 0.901 |
| **This study** | 0.975 | 0.984 | 0.986 | 0.976 | 0.679 | 0.831 | 0.905 |

*Abbreviation used in this study: Glu. stands for glucose, Ace. for acetone, But. for butanol, Act. for acetate, Btr. for butyrate, Bms. for biomass, and $Avg$ for average.

The difference in average error value between this study and the literature using Shinto's models is quite small, i.e. 0.04 (0.44%). The small error was accepted because the software used to simulate Shinto's models in this study was different with the Shinto's software. In addition, the data used in this study was data extracted from the visual graph provided by Shinto et al. [11], thus allowing the data to be read not completely accurate. It was concluded that the small error value indicates good-accuracy of the model-scripting, thus the simulation model used in this study can be considered the same as what Shinto *et al.* [11] have proposed.

### 3.2. Model adapting evaluation

As explained in model scripting evaluation, Shinto *et al.* [11] used Pearson correlation coefficient as a statistical method for justification of their model accuracy, but this method is considered inappropriate. Instead, a better and more common alternative is residual determination coefficient method ($R^2$) [18]. Re-evaluation of the model using $R^2$ have given negative values of $R^2$ on butyrate and biomass trend.

To achieve higher value of $R^2$, the model was adapted by tuning the parameter $k_{13}$ or cell death rate constant. This parameter is involved in a simpler first-order reaction kinetics equation, unlike other equations that follow the Michaelis-Menten kinetics which tends to be more complicated. In addition, this parameter is directly related to the biomass-concentration model. The formula of cell death kinetics rate is given in Equation A.13 with $k_{13}$ as a cell death rate constant ($h^{-1}$) and [*Biomass*] as concentration of cell biomass (g/L). Shinto *et al.* [11] obtained $k_{13} = 0.017$ $h^{-1}$.

In this study, tuning is done through trial-and-error by iterating $k_{13}$ in the range 0.035 – 0.07 $h^{-1}$. The results is depicted in figure 2 and show that the iteration with $k_{13}$ = 0.05 $h^{-1}$ is the best tune based on visual rating. Hereinafter, this adapted model is called 'Shinto bms2', while the reference model without tuning is 'Shinto original'.

### 3.3. Model evaluation using experiment results in other literatures

Evaluation of both Shinto original and Shinto bms2 models was carried out on other experimental data which have similar methodology, i.e., batch butanol fermentation with glucose as substrate using *C. saccharoperbutylacetonicum* N1-4, to enlarge the evaluation range. Those data were obtained from Thang *et al.* [12], Tanaka *et al.* [13], and Ishizaki *et al.* [14]. The results show that the Shinto bms2 model is more accurate than the Shinto original model based on negative $R^2$ frequency (see Table 3).

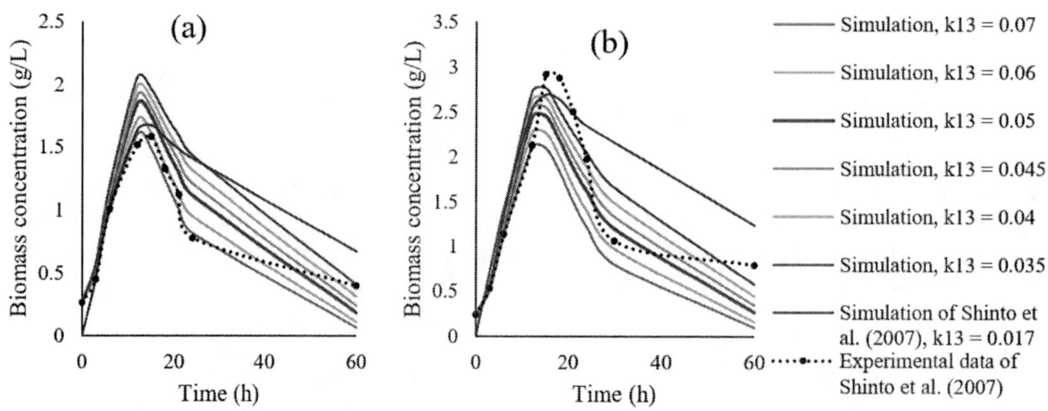


**Figure 2.** Tuning parameter evaluation for (a) $S_0$ = 12.7 g/L (a) and (b) $S_0$ = 22 g/L

Some $R^2$ values are negative, especially in the butyrate variables. The negative $R^2$ value is caused by not-fitted results between experimental data and the model simulation. By assumption the raw data is valid, there is two presumption causes of negative $R^2$ values. First, concentration of the related compounds during fermentation are low which resulting negative values after $R^2$ calculation. Second, the model itself is not suitable enough with actual condition, due to the model which is a simplification of complex metabolic mechanism. However, in general, the $R^2$ results after tuning (bms2) has more positive results counted than the original one, which is the new tuned-simulation more fit with the raw data. This indicates that the prior tuning indeed increases the accuracy of the model. Thus, Shinto bms2 was chosen as simulation model to determine the effect of initial concentration of glucose and organic acids to butanol production.

**Table 3.** Evaluation results of the models using other literatures

| Model | Data | $S_0$ (g/L) | $R^2$ (residual determination coefficients) | | | | | |
|---|---|---|---|---|---|---|---|---|
| | | | Glucose | Acetone | Butanol | Acetate | Butyrate | Biomass |
| Shinto original | Shinto *et al.* [11] | 6,5 – 53,1 | 0,923 | 0,811 | 0,918 | 0,788 | -0,997 | -0,890 |
| | Thang *et al.* [12] | 66 | 0,907 | 0,891 | 0,848 | 0,898 | -5,463 | -3,389 |
| | Tanaka *et al.* [13] | 90 | -0,517 | -0,187 | -0,032 | | | -0,020 |
| | Ishizaki *et al.* [14] | 90 | -0,745 | | -0,349 | | | 0,594 |
| Shinto bms2 | Shinto *et al.* [11] | 6,5 – 53,1 | 0,949 | 0,808 | 0,947 | 0,754 | -1,456 | 0,421 |
| | Thang *et al.* [12] | 66 | 0,954 | 0,880 | 0,924 | 0,724 | -7,338 | 0,128 |
| | Tanaka *et al.* [13] | 90 | 0,032 | 0,248 | 0,472 | | | -0,271 |
| | Ishizaki *et al.* [14] | 90 | 0,132 | | 0,450 | | | -0,502 |

### *3.4. Effect of initial glucose concentration*

Stoichiometrically, 1 mole of glucose produces 1 mole of butanol (0.411 g/g). However, the experiment shows the final butanol concentration never exceeds the theoretical concentration, because the metabolic reaction is more complex and produces other metabolites. Inhibition is another factor that also influences butanol production, especially when the butanol concentration does not increase when the initial glucose concentration is higher (figure 3). Higher initial glucose concentration causes higher final butanol concentration. But then, cell metabolism is increasingly inhibited by the butanol, reducing the butanol production itself. Regarding this phenomenon, Shinto bms2 model is only accurate for low initial glucose concentration, less than 66 g/L.

The same goes to butanol yield. The butanol yield also needs to be reviewed to determine the efficiency of glucose utilization by the bacteria. Experiments show that butanol yield never exceeds the stoichiometric yield (figure 3). The highest butanol yield of 77% was obtained at a glucose concentration of 22 g/L. The simulation results exhibit that the biobutanol yield decreases when the glucose concentration is higher than 22 g/L. These results are consistent with previous experimental data.

The butanol yield decreases with increasing initial glucose concentration. However, low initial glucose concentration is not an option to be applied to the industry because it produces low butanol concentration as well. Conversely, applying high initial glucose concentrations is also not appropriate because the butanol yield will be low and not economical well. Butanol productivity is the solution to determine the optimum initial glucose for butanol production. Productivity reviews the ratio of the final butanol concentration to the fermentation time when glucose runs out or reaches the end of utilization. Figure 3 shows the initial glucose concentration of 53 g/L could give the highest butanol productivity. This glucose concentration is in accordance with the use of sugar in the butanol industry, which uses molasses with sugar levels of 4-6% or approximately 40-60 g/L glucose [10].

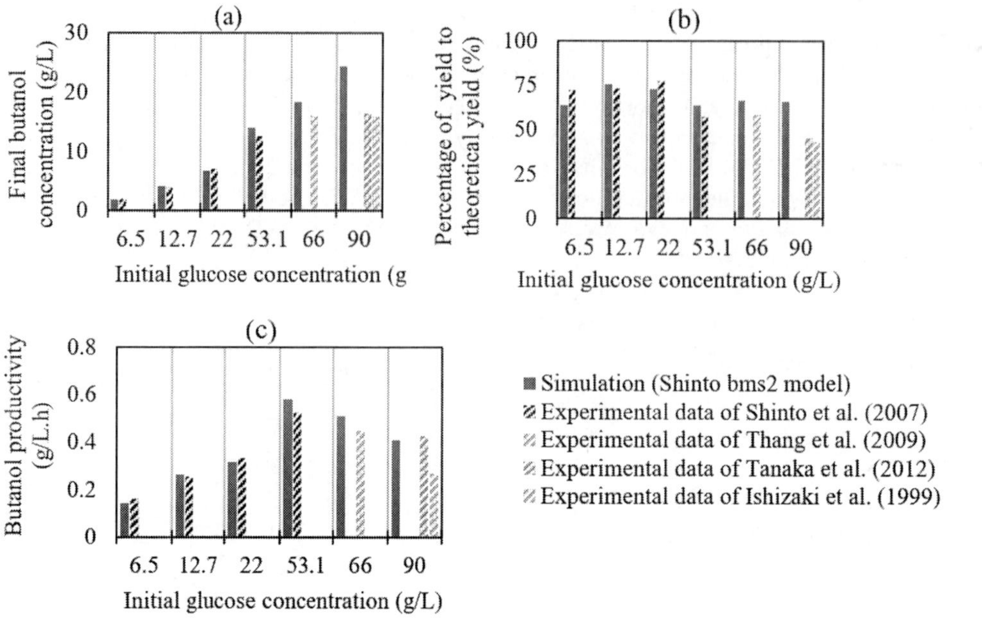


**Figure 3.** Effect of initial glucose concentration to butanol production: (a) final concentration, (b) yield, and (c) productivity

### 3.5. *Effect of initial organic acids concentration*

*3.5.1. Acetate.* Acetate is the primary metabolite in the *Clostridia* metabolic pathway, so its presence is thought to affect the production of intermediate compounds and secondary metabolites. However, Zhou *et al.* [15] have found the addition of acetate did not increase butanol production linearly. Butanol concentration tends to be constant with optimum acetate concentration of 2.5 g/L, in contrast to simulation which shows an increase (figure 4). This difference is thought to be caused by different types of fermentation medium used. Zhou *et al.* [15] used the nitrogen-free medium, whereas Shinto *et al.* [6] used the TYA (tryptone, yeast-extract, acetate) medium which contains nitrogen. Acidity is also another factor that influences butanol production. More acidic or basic fermentation conditions (compared to optimum condition) can cause inhibition of butanol production, so an increase of initial acetate concentration results in the decrease of butanol production. This effect cannot be described by the simulation. Also, the initial acetate concentration of 2.5 g/L is sufficient to produce butanol with the highest concentration. The value itself is the acetate concentration in the TYA medium.

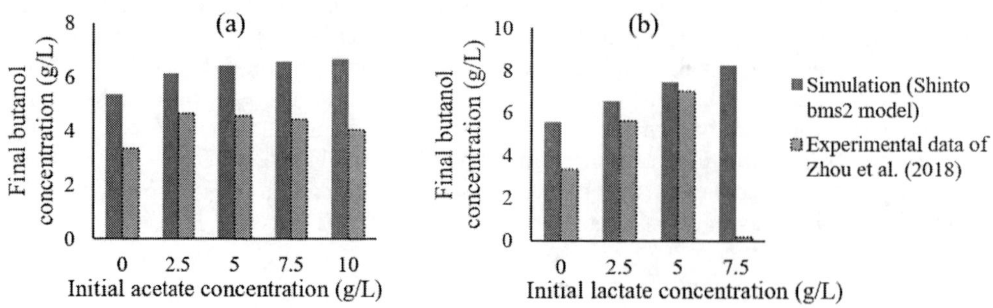


**Figure 4.** Effect of initial concentration of (a) acetate and (b) lactate to butanol final concentration

*3.5.2. Lactate.* Lactate is also an intermediate in the *Clostridia* metabolic pathway, which is directly related to pyruvate as the main precursor, so lactate is suspected to affect butanol production. Zhou et al. [15] have proved this, since their experiment results (figure 4) show the butanol production increases as the initial lactate concentration increases. At high initial lactate concentration, inhibition occurs because the fermentation medium is too acidic, so it inhibits the cell metabolism and therefore butanol production. The Shinto bms2 model cannot explain this phenomenon of inhibition, and the overall final butanol concentration from simulation is higher than the experimental data. This difference is thought due to differences in the type of medium used, as previously discussed.

*3.5.3. Butyrate.* Butyrate is also a primary metabolite. To produce butanol, butyrate is an important compound besides butyryl-CoA. Metabolic pathway from butyrate to butanol is relatively close, and butyrate is an inductor for butyryl-CoA, thus its effect is believed to be significant compared to other metabolites. Al-Shorgani *et al.* [16] reported that low initial glucose concentration (less than 15 g/L) can accommodate the addition of butyrate to increase butanol production, but not significant. When the initial glucose concentration is higher (more than 15 g/L), butanol production increases quite significantly, but only up to the initial butyrate-glucose ratio of 0.5 (figure 5). This shows that butyrate is also an inhibitor for butanol production, parallel to glucose.

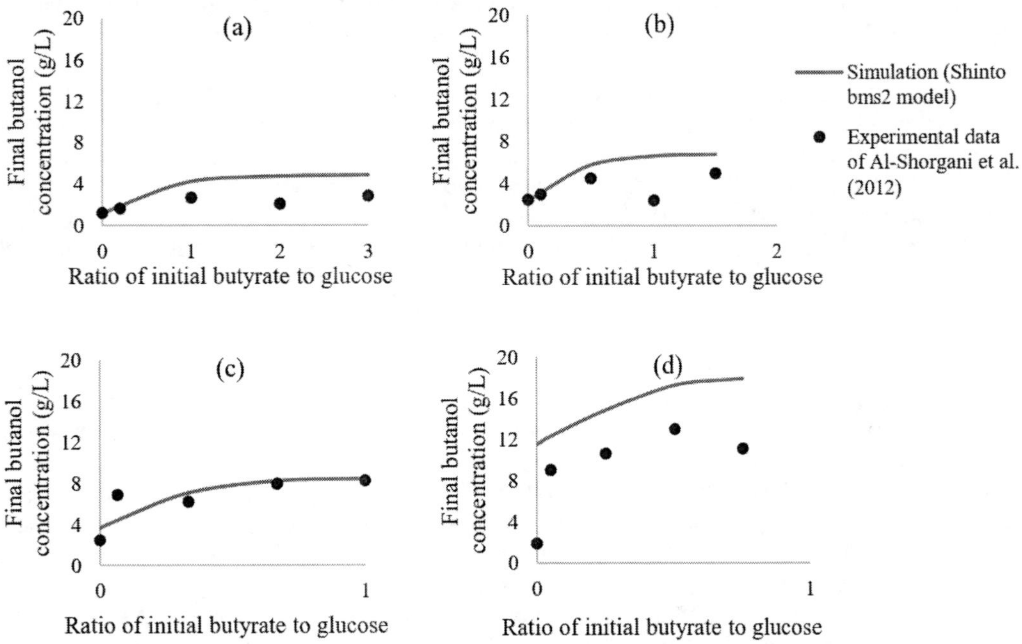


**Figure 5.** Effect of initial butyrate to glucose concentration ratio to final butanol concentration, with initial glucose concentration of (a) 5 g/L, (b) 10 g/L, (c) 15 g/L, and (d) 20 g/L

## 4. Conclusion

The adapted model (Shinto bms2) was obtained by tuning the parameter value of $k_{13}$ and more accurate than the original one. Furthermore, the simulation results show that the initial concentration of glucose and organic acids (acetic acid, butyric acid, and lactic acid) can increase butanol production up to a certain concentration value. However, if the concentration of glucose and these acids exceeds the limit concentration, there will be an inhibition. The optimum concentration of initial acetate is 2,5 g/L, as it was prescribed in TYA medium (common medium for *Clostridia*). Without butyrate and lactate addition, the optimum butanol yield of 77% and final concentration of 7 g/L was obtained at initial glucose concentration of 22 g/L. Meanwhile, the optimum concentration of initial lactate is 5 g/L, and the optimum ratio of initial butyrate to glucose is 0.5.

## References

[1] Lorne D and Bouter A 2019 Biofuels in the road transport sector Economic Outlook: Biofuels Dashboard. Available: https://www.ifpenergiesnouvelles.com/article/biofuels-dashboard-2019
[2] Dürre P 2007 Biobutanol: An attractive biofuel Biotechnol. J. 2:12 1525–1534
[3] Jones D T and Woods D R 1986 Acetone-butanol fermentation revisited Microbiol. Rev. 50:4 484 LP – 524
[4] Jones D T and Keis S 1995 Origins and relationships of industrial solvent-producing clostridial strains FEMS Microbiol. Rev. 17:3 223–232
[5] Restiawaty E and Grinanda D 2017 Ferrous Ion and Medium Composition Effects on Acidogenic Phase in Biobutanol Production from Molasses Journal of Physics: Conference Series 877:1
[6] Keis S, Shaheen R and Jones D T 2001 Emended descriptions of Clostridium acetobutylicum and Clostridium beijerinckii, and descriptions of Clostridium saccharoperbutylacetonicum sp. nov. and Clostridium saccharobutylicum sp. nov Int. J. Syst. Evol. Microbiol. 51:6 2095–2103
[7] Hillmann F, J Fischer R, Saint-Prix F, Girbal L and Bahl H 2008 PerR acts as a switch for oxygen tolerance in the strict anaerobe Clostridium acetobutylicum Mol. Microbiol. 68:4 848–860
[8] Riebe O, J Fischer R, Wampler D A, Kurtz D M and Bahl H 2009 Pathway for H2O2 and O2 detoxification in Clostridium acetobutylicum Microbiology 155:1 16–24
[9] Zhu L, Dong H, Zhang Y and Li Y 2011 Engineering the robustness of Clostridium acetobutylicum by introducing glutathione biosynthetic capability Metab. Eng. 13:4 426–434
[10] Keis S, Bennett C F, Ward V K and Jones D T 1995 Taxonomy and Phylogeny of Industrial Solvent-Producing Clostridia Int. J. Syst. Evol. Microbiol. 45:4 693–705
[11] Shinto H, Tashiro Y, Yamashita M, Kobayashi G, Sekiguchi T, Hanai T, Kuriya Y, Okamoto M and Sonomoto K 2007 Kinetic modeling and sensitivity analysis of acetone-butanol-ethanol production Journal of Biotechnology 131 45-56
[12] Thang V H, Kanda K and Kobayashi G 2010 Production of Acetone-Butanol-Ethanol (ABE) in Direct Fermentation of Cassava by Clostridium saccharoperbutylacetonicum N1-4 Applied Biochemistry and Biotechnology 161 157-170
[13] Tanaka S, Tashiro Y, Kobayashi G, Ikegami T, Negishi H, and Sakaki K 2012 Membrane-assisted extractive butanol fermentation by Clostridium saccharoperbutylacetonicum N1-4 with 1-dodecanol as the extractant Bioresource Technology 116 448-452
[14] Ishizaki A, Michiwaki S, Crabbe E, Kobayashi G, Sonomoto K and Yoshino S 1999 Extractive Acetone-Butanol-Ethanol Fermentation Using Methylated Crude Palm Oil as Extractant in Batch Culture of Clostridium saccharoperbutylacetonicum N1-4 (ATCC 13564) Journal of Bioscience and Bioengineering 87:3 352-356
[15] Zhou Q, Liu Y and Yuan W 2018 Kinetic modeling of lactic acid and acetic acid effects on butanol fermentation by Clostridium saccharoperbutylacetonicum Fuel 226 181-189
[16] Al-Shorgani N K N, Ali E, Kalil M S and Yusoff W M W 2012 Bioconversion of butyric acid to butanol by Clostridium saccharoperbutylacetonicum N1-4 (ATCC 13564) in a Limited Nutrient Medium BioEnergy Research 5 287-193
[17] Yuliandi D 2007 Analisis Model Pelayanan Antrian dengan Menggunakan Simulasi Kejadian Diskrit (Bandung: Universitas Komputer Indonesia)
[18] Spiess A N and Neumeyer N 2010 An evaluation of R2 as an inadequate measure for nonlinear models in pharmacological and biochemical research: a Monte Carlo approach BMC Pharmacology 10:6 1-11

IOP Conf. Series: Materials Science and Engineering 1143 (2021) 012019 doi:10.1088/1757-899X/1143/1/012019

# Medium Optimization for Production of *Monascus purpureus* Pigment through Solid-state Fermentation

**G A Ismail, A D Fitriana and U Sukandar**

Chemical Engineering Study Program, Institut Teknologi Bandung, Indonesia

E-mail: guntur@cheitb.id

**Abstract.** Colorants are component that is commonly added to many kinds of product, including food products. Regardless of its effectiveness and lower price, the effects of synthetic colorants towards human's health become a challenge for food colorant industries. Production of natural food colorant using microorganism as the producing agent is a promising prospect because microorganism has a high growth rate. *Monascus purpureus* can produce a set of natural pigments consist of yellow (monascin and ankaflavin), orange (rubropunctatin and monascorubrin) and red (rubropunctamine and monascorubramine) pigment. The objective of this research is to optimize the micronutrient composition in the medium for the production of Monascus pigment through solid-state fermentation using the job's tears (*Coix lacryma-jobi L*) as the substrate. Response surface method (RSM) was used to optimize the concentration of four substrate components: MSG, NaCl, $KH_2PO_4$, and $MgSO_4$. From the experiment, it was found that the maximum red pigment was produced in medium with additional of (w/w): MSG 1.496%, NaCl 1.0%, $KH_2PO_4$ 3.515% and MgSO4 0.206%. The yellow pigment was maximally produced in medium with additional of (w/w): MSG 1.5%, NaCl 1.0%, $KH_2PO_4$ 2% and $MgSO_4$ 0.2%. Biomass was maximally produced in the medium with additional of (w/w): MSG 1.41%, NaCl 1.01%, $KH_2PO_4$ 5.0% and $MgSO_4$ 0.2%. Among those nutrients, MSG gave the biggest impact on the increasing of pigment and biomass production.

## 1. Introduction

Food colorant is an important aspect in food as it is the one that first perceived by the consumer. Nowadays, the common coloring agent used is synthetic colorant. However, there are increasing concern in effect of synthetic coloring agent to human health as in potential of cancer and teratogenicity [1-2]. It is causing a shift in tendency to use natural food color over the synthetic one.

Fungi of *Monascus* genus has been well known to produce natural pigment. The pigment mainly consists of 3 colors: red (monascoruramine, rubropuntamine), yellow (monascin, ankaflavin), and orange (monascorubrin, rubropunctatin) [3]. *Monascus* pigment has been long used since ancient China as *angkak* to cure dengue fever. It also has good effect on promoting human health as it could lower cholesterol level, help to prevent osteoporosis and diabetes [4-6]. These pigments are secondary metabolite products, which are produced during stationary phase [7].

*Monascus* commonly utilized using rice as the substrate. There are already many researches to utilize *Monascus* using different substrates, such as jackfruit seed, soy bean waste, durian seed, and corn cob [8-11]. So far, there are no research to utilize Job's tears as substrate for pigment production yet.

Job's tears, which is also known as adlay, coix-seed, Chinese pearl barley, *hanjeli* (Indonesia) or *hato mugi* (Japan), is originated from China and now can be found worldwide. It is commonly mixed with rice and consumed to warm the body and stamina increasing. Job's tears consist of 16.2% protein,

4.65% fat and 79.17% carbohydrate. It also has some amino acid such as leucine, lysine, arginine and tyrosine [12]. It is also relatively easy to plant like paddy rice that don't need any special treatment.

In fermentation, micronutrients play an important role in microorganism growth and metabolite product formation. Micronutrients are nutrients that are added to fermentation medium in small amount (trace). Many of micronutrients are cofactor that play role in activating its respective enzyme, such as $Mg^{2+}$. The other examples of micronutrient are sulfur that is used as backbone for sulfur-containing amino acid and $Na^{+}$ that plays important role in regulating nutrient transport rate and osmotic pressure of the system [13-14]. Those micronutrients can support the microbial growth and production of metabolite product, but macronutrients such as carbon, nitrogen and oxygen are still essential for fermentation. Those macronutrients are the main backbone of cell composition. In case for pigment production, specific nitrogen sources also play role in promoting the growth and pigment production of the cell [15].

With the increasing trend of healthy lifestyle and concern about artificial substance, development of natural pigment sees an interesting prospect. Natural pigments can be obtained from plants and animals, but since both have long generation time, the quantity may be insufficient for industrial utilization, thus microbial-based pigment is more reliable and potential [16]. This study has main objective to utilize Job's tears as substrate to produce *Monascus* pigment via solid state fermentation and also optimize the composition of micronutrient usually added to grow *Monascus* fungi by using Response Surface Methodology (RSM) approach. RSM method is a model commonly used for optimization in biological process. RSM will yield mathematical model that can show the effect of each independent factor and its interaction with each other independent factors [17].

## 2. Material and Method

### 2.1. Culture

Culture of *Monascus purpureus* ITB CC L61 was obtained from Laboratory of Microbiology and Bioprocess Technology Culture Collection, Institut Teknologi Bandung. The culture was maintained in Potato Dextrose Agar (PDA) medium, preserved at 4°C and was re-cultured every month.

### 2.2. Inoculum preparation

*Monascus purpureus* was grown on PDA medium at 30°C until fully sporulated (10 days). The spore then scrapped by adding 5 ml of sterile water under aseptic condition. This suspension then used as inoculum.

### 2.3. Solid-state fermentation

Job's tears obtained from Punclut, West Java were used as substrate. Each variation used 35 grams of Job's tears. Job's tears were washed with clear water then filled into 250 mL Erlenmeyer flask. Micronutrients diluted in 105 mL of distilled water were then added into the medium with varied concentration. The medium was then cooked until the water content reached 65%. The flask was then covered with sterile fatty cotton and cooled until room temperature, and then inoculated with prepared inoculum. Fermentation took place at 30°C for 14 days.

### 2.4. Experimental design

Variation done in this research was the concentration of micronutrient and additional monosodium glutamate (MSG) as nitrogen source to find the optimal concentration. Optimization was done via RSM approach with full central composite design experiment, with detail shown in Table 1. In total, there were 31 runs done in this research.

**Table 1.** Experimental design for medium optimization.

| No. | MSG (A) (%w/w) | NaCl (B) (%w/w) | $KH_2PO_4$ (C) (%w/w) | $MgSO_4$ (D) (%w/w) |
|---|---|---|---|---|
| 1 | 0.5 | 1 | 5 | 0.2 |
| 2 | 0.5 | 1 | 5 | 1 |
| 3 | 0.5 | 1 | 2 | 1 |
| 4 | 0.5 | 1 | 2 | 0.2 |
| 5 | 0.5 | 3 | 5 | 0.2 |
| 6 | 0.5 | 3 | 2 | 1 |
| 7 | 0.5 | 3 | 5 | 1 |
| 8 | 0.5 | 3 | 2 | 0.2 |
| 9 | 0.5 | 2 | 3.5 | 0.6 |
| 10 | 1 | 1 | 3.5 | 0.6 |
| 11 | 1 | 2 | 2 | 0.6 |
| 12 | 1 | 2 | 3.5 | 1 |
| 13 | 1 | 3 | 3.5 | 0.6 |
| 14 | 1 | 2 | 3.5 | 0.6 |
| 15 | 1 | 2 | 3.5 | 0.6 |
| 16 | 1 | 2 | 5 | 0.6 |
| 17 | 1 | 2 | 3.5 | 0.6 |
| 18 | 1 | 2 | 3.5 | 0.6 |
| 19 | 1 | 2 | 3.5 | 0.6 |
| 20 | 1 | 2 | 3.5 | 0.6 |
| 21 | 1 | 2 | 3.5 | 0.6 |
| 22 | 1 | 2 | 3.5 | 0.2 |
| 23 | 1.5 | 1 | 5 | 0.2 |
| 24 | 1.5 | 2 | 3.5 | 0.6 |
| 25 | 1.5 | 3 | 2 | 0.2 |
| 26 | 1.5 | 3 | 5 | 1 |
| 27 | 1.5 | 1 | 2 | 1 |
| 28 | 1.5 | 3 | 5 | 0.2 |
| 29 | 1.5 | 1 | 5 | 1 |
| 30 | 1.5 | 3 | 2 | 1 |
| 31 | 1.5 | 1 | 2 | 0.2 |

### *2.5. Pigment analysis*
Pigment analysis was done by extracting pigment from 8 grams of fermented substrate. Pigment extraction was done by adding that fermented sample into a tube containing 40 mL ethanol 60%. The tube was then centrifuged with rotation speed of 200 rpm in room temperature. Extraction result was filtered through Whatman No. 1 filter paper and then analyzed by using spectrophotometry [8]. Blank used was the filtration result from unfermented substrate that undergone the same preparation. Measurement was done at wavelength of 510 nm to analyze the red pigment [15].

### *2.6. Biomass analysis*
Biomass analysis was done by measuring the N-acetyl-D-glucosamine concentration formed due to acid hydrolysis of chitin contained in the fungi cell wall [18]. A tube containing 0.5 gram of acid hydrolysis product was added with 1 mL of acetyl acetone and then heated at 100°C for 20 minutes. Sample was then cooled to 65°C, and then 6 mL of concentrated ethanol and 1 mL of Ehrlich Reagent was added and then incubated for 10 minutes. Sample was then analyzed by using spectrophotometer at 530 nm. Unfermented substrate was undergone the same preparation and used as blank for analysis. The optical

density (OD) obtained from spectrophotometer analysis was then converted into mass by using standard curve prepared by doing the same treatment to some certain concentrations of N-acetyl-D-glucosamine.

## 3. Result and discussion

### *3.1. Pigment production*

The result of analysis is shown in Table 2.

**Table 2.** Analysis result for each run.

| Variation | Pigment (OD) | Biomass (g/g) | Variation | Pigment (OD) | Biomass (g/g) |
|---|---|---|---|---|---|
| 1 | 5.6 | 0.532 | 17 | 5.0 | 0.250 |
| 2 | 8.6 | 0.622 | 18 | 5.7 | 0.272 |
| 3 | 12.2 | 0.394 | 19 | 6.7 | 0.262 |
| 4 | 8.0 | 0.304 | 20 | 7.7 | 0.224 |
| 5 | 3.6 | 0.528 | 21 | 6.6 | 0.275 |
| 6 | 1.6 | 0.158 | 22 | 8.7 | 0.258 |
| 7 | 9.5 | 0.192 | 23 | 3.7 | 0.374 |
| 8 | 7.9 | 0.116 | 24 | 4.4 | 0.452 |
| 9 | 2.6 | 0.278 | 25 | 8.0 | 0.452 |
| 10 | 6.9 | 0.258 | 26 | 7.9 | 0.452 |
| 11 | 4.6 | 0.128 | 27 | 7.8 | 0.482 |
| 12 | 4.3 | 0.082 | 28 | 8.1 | 0.344 |
| 13 | 8.0 | 0.19 | 29 | 8.2 | 0.452 |
| 14 | 8.8 | 0.258 | 30 | 11.0 | 0.43 |
| 15 | 1.6 | 0.258 | 31 | 13.2 | 0.472 |
| 16 | 8.0 | 0.498 | | | |

The data obtained in Table 2 was then analyzed using Minitab® in order to determine the optimal concentration for pigment production. Based on the optimization result, it was found that red pigment was maximally produced by adding 1.5 % MSG (A), 1.788 % NaCl (B), 2.909 % $KH_2PO_4$ (C), and 0.2 % $MgSO_4$ (D). Pigment production can be estimated by using equation (1), with effect of all additional component can be seen in Figure 1

$$Pigment = 7{,}70 + 2{,}29A - 1{,}05C - 0{,}56D + 2{,}1A^2 - 3{,}25B^2 - 2{,}35C^2 + 2{,}15D^2 - 0{,}68AD + 0{,}61BC + 1{,}04BD + 0{,}85CD \quad (1)$$

The equation was the simplified version that has already discarded insignificant coefficients that have p-value lower than 0.005. Based on the pareto diagram and the equation, MSG positively affecting pigment production. Monascus pigment contains amine substances which may be enhanced by the additional MSG in the medium. Close value between singular and quadratic constant of MSG from the equation (1) also means that the optimal value is within the variation used, in this case, the optimal value for MSG is 1.5 %. However, excessive MSG addition is not recommended since it may lead to inhibition of red pigment production [19].

The other significant factors in pigment production are $KH_2PO_4$ and $MgSO_4$ concentration, and both are affecting the pigment production negatively. It may be due to the nature of pigment as secondary metabolite product, which is produced during stationary state. Both components are essential in cell growth as cofactor. However, often that cell growth is not linear to secondary metabolite production, as

IOP Conf. Series: Materials Science and Engineering 1143 (2021) 012019 doi:10.1088/1757-899X/1143/1/012019

secondary metabolite product is usually produced as "emergency measure" when the growth is inhibited. If the growth is promoted, cell will find secondary metabolite production as unnecessary, thus reducing the amount. As shown in Figure 2, both $KH_2PO_4$ and $MgSO_4$ have optimal value between the minimum and maximum value, highlighting its importance for cell in general but may inhibit the product yield if given in abundant.

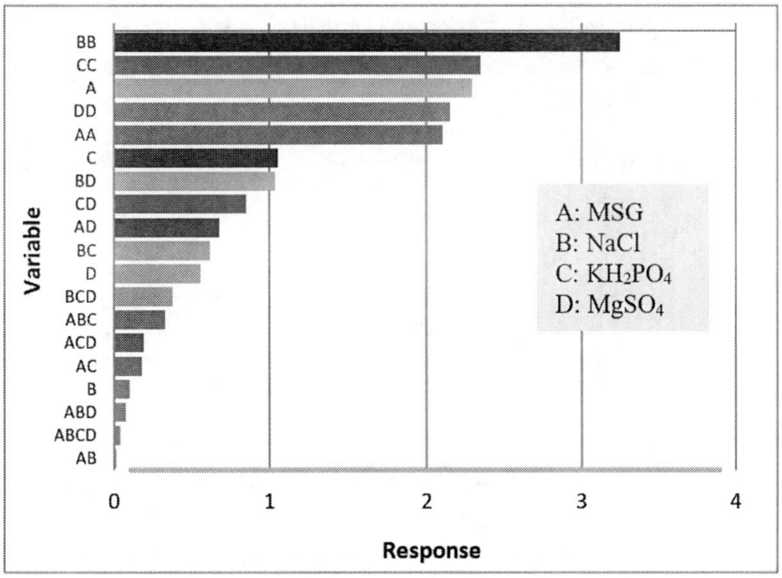


**Figure 1.** Pareto diagram for pigment production.

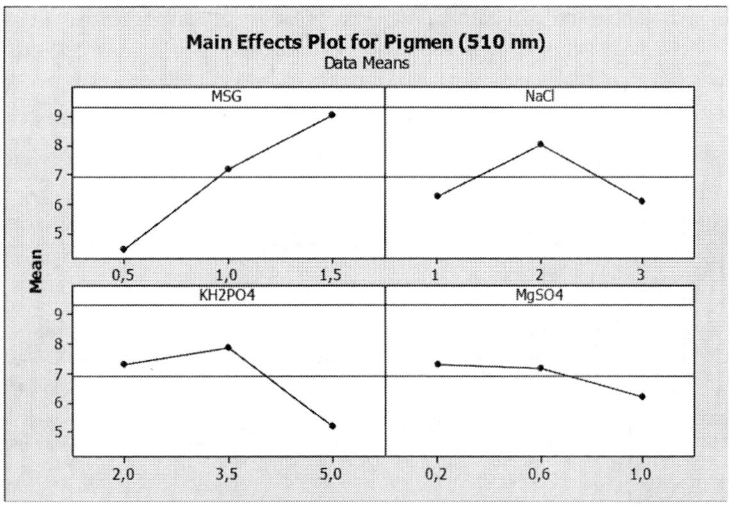


**Figure 2.** Main effects plot for pigment production.

### *3.2. Biomass production*

Similar to pigment analysis, data obtained in Table 2 was then analyzed using Minitab® in order to determine the optimal concentration for biomass growth. It was found that biomass was maximally produced by adding (w/w) 1.41 % MSG (A), 1.01 % NaCl (B), 5.0 % $KH_2PO_4$ (C), and 0.2 % $MgSO_4$ (D), and can be quantified by following equation (2), with the effect of all additional component is shown in Figure 3.

IOP Conf. Series: Materials Science and Engineering 1143 (2021) 012019 doi:10.1088/1757-899X/1143/1/012019

$$Biomass = 0{,}123 + 0{,}022\,A - 0{,}029\,B + 0{,}029\,C + 0{,}024\,AB - 0{,}035\,AC + 0{,}067A^2 \quad (2)$$

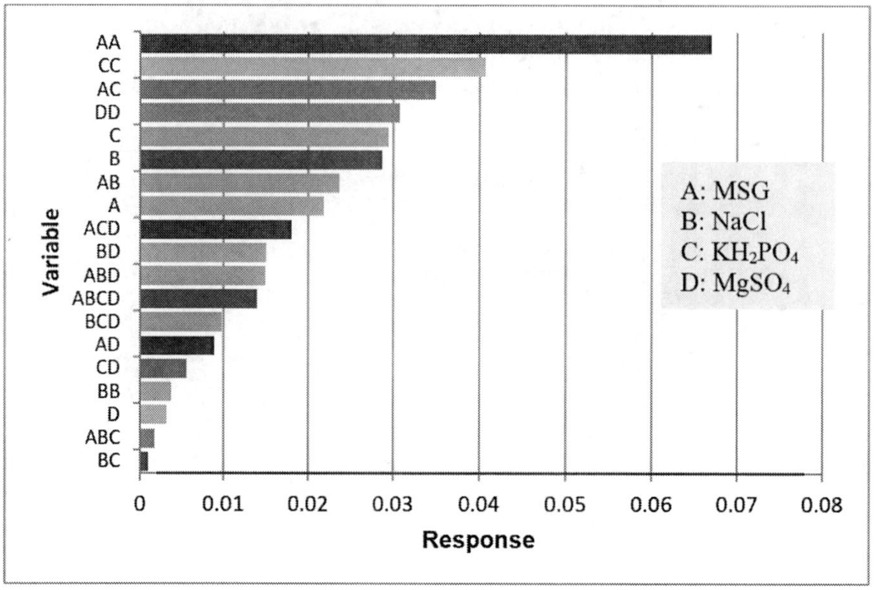


**Figure 3.** Pareto diagram for biomass production.

Based on the simplified equation, besides $MgSO_4$, additional nutrients have significant effect on biomass production, with only NaCl affect negatively towards cell growth. Positive effect from MSG and $KH_2PO_4$ can be understood as the additional of cofactor and nitrogen source will enhance the growth of microbial. Nitrogen is one of the cell's backbones, while increasing the activity of the enzyme included in main growth metabolism pathway will accordingly enhance the growth of the cell. As shown in Figure 4, excessive addition of each component will negatively affect the cell growth.

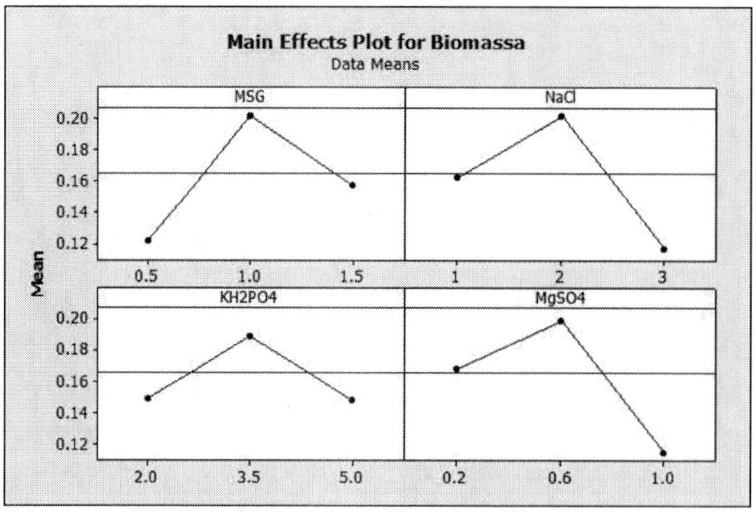


**Figure 4.** Main effects plot for biomass production.

The effect of salt is pretty interesting, since by itself, it will hinder the cell growth, but interaction between salt and MSG provide positive effect. It is well known that salt plays important rule in nutrient transport. However, excessive amount of salt will lead to hyperosmotic condition, in which, non-tolerant cell will be forced to adapt by changing the viscosity of cell membrane of the phospholipid composition of the cell wall [20]. However, the interaction between NaCl and MSG gives positive effect as additional nitrogen source will be transported into the cell more easily, thus somewhat offsetting the negative effect of salt as singular entity. However, further study may be required in order to confirm this hypothesis.

## 4. Conclusion

Adding MSG as additional nitrogen source can enhance the production of pigment and biomass for *Monascus purpureus*, however, excessive addition will instead inhibit the cell performance. Moreover, among all micronutrient added, only $MgSO_4$ shows no significant effect in both cell and pigment production. Based on the optimization result, maximum biomass that can be produced is 0.568 gr/gr by adding (w/w) MSG 1.41 %, NaCl 1.0 %, $KH_2PO_4$ 2.0 % and $MgSO_4$ 0.2 %. Maximum pigment that can be produced based on optimization result is 15.94 OD by adding (w/w) MSG 1.5 %, 1.788 %, 2.0 %, and 0.2 %. This study might be helpful for future development of Monascus fermentation, especially in the selection of additional micronutrient and helpful to avoid additional of components that may hinder the performance of each other.

## References

[1] Tsuda S, Murakami M, Matsusaka N, Kano K, Taniguchi K, and Sasaki Y F 2001 *Toxilogical Sciences* **61** 92-99
[2] Uysal O K and Aral E 1998 *J. of Medical Sciences* **28** 363-368
[3] Pattanagul P, Pinthong R, Phianmongkol A 2007 *Chiang Mai J. Sci* **34** 319-328
[4] Heber D, Yip I, Ashley J M, Elashoff D A, and Go V L W 1999 *The American Journal of Clinical Nutrition* **69** 231-236
[5] Wong R W K and Rabie B 2008 *Chinese Medicine* **3** 1-6
[6] Rajakaseran A, Kalaivani M and Sabitha R 2009 *Asian Journal of Scientific Research* **2** 180-189
[7] Carvalho J C, Oishi B O, Woiciechowski A L, Pandey A, Babitha S and Soccol C R 2007 *Indian Journal of Biotechnology* **6** 194-199
[8] Babitha S, Soccol C R and Pandey A 2006 *Food Technology and Biotechnology* **44** 465-471
[9] Japakaset J, Wongkkhalaung C and Leelawatcharamas V 2009 *Songklanakarin J. Sci. Technol* **31** 35-39
[10] Srianta I, Hendrawan B, Kusumawati N and Blanc P J 2012 *International Food Research Journal* **19** 941-945
[11] Velmurugan P, Hur H, Balachandar V, Kamala-Kannan S and Lee K-J 2011 *Journal of Bioscience and Bioengineering* **112** 590-594
[12] Hu S-Y 2005 *Food Plants of China* (Hong Kong: Chinese University Press)
[13] Patricia N, Winy M and Luc D V 2003 *International Journal of Food Microbiology* **88** 29-39
[14] Baker S B and Worthley L I G 2002 *Critical Care and Resuscitation* **4** 301-306
[15] Subhasree R S, Babu P D, Vidyalakshmi R and Mohan C 2011 *International Journal of Microbiology Research* **2** 184-187
[16] Suraiya S, Siddique M B, Lee J M, Kim E Y, Kim J M and Kong I S 2018 *J Appl Phycol* **30** 729-742
[17] Ajdari Z, Ebrahimpour A, Manan M A, Ajdari D, Abbasiliasi S, Hamid M, Mohammad R and Ariff A B 2013 *Malaysian Journal of Microbiology* **9** 68-83
[18] Sakurai Y, Lee T H and Shiota H 1977 *Agric. Biol. Chem* **41** 619-624
[19] Lee B K, Park N H, Piao H Y and Chung W J 2001 *Biotechnology and Bioprocess Engineering* **6** 341-346
[20] Smolyanuk E V, Bilaneko E N, Tereshina V M, Kachalkin A V and Kazolkina O V 2013 *Microbiology* **82** 600-608

IOP Conf. Series: Materials Science and Engineering 1143 (2021) 012020 doi:10.1088/1757-899X/1143/1/012020

# Study of the Effect of Glycerol Plasticizer on the Properties of PLA/Wheat Bran Polymer Blends

**A Satriyatama[1*], V A A Rochman[2], R E Adhi[3]**

[1]Department of Chemical Engineering, Faculty of Industrial Technology, Institut Teknologi Bandung, Bandung, Indonesia
[2]Faculty of Engineering, Universitas Diponegoro, Semarang, Indonesia
[3]Faculty of Business and Economics, Universitas Diponegoro, Semarang, Indonesia

Email: adhisatriyatama@students.itb.ac.id

**Abstract**. Polylactic Acid (PLA) is a natural polymer that can be used as a raw material for making plastics. PLA plastic manufacturing has developed rapidly since the crude oil supply is getting lower, besides PLA plastic is environmentally friendly. In this research, the development of PLA from starch and wheat bran was developed by using glycerol as a plasticizer. PLA film is made by lactic acid fermentation. After fermentation, PLA mixed with wheat bran with a ratio of 20:80 (w/w) and a variety of glycerol concentration of 0%; 1%; 3%; 5%; and 7%, then heating at 200 °C for 5 minutes. The film later printed in plastic size dimension of 15cm × 20cm and incubated at 50 °C for 5 hours, then dried in room temperature of ± 27 °C. The effect of glycerol addition as a plasticizer on PLA is analyzed using mechanical tests, swelling, solubility, and chemical resistance. Film with 1 wt% of glycerol composition is optimal composition of tensile strength value of 1.22 MPa, 0.014% elongation, swelling of 0.4, solubility of 0.0207%.

## 1. Introduction

The increasing of carbon dioxide due to fossil fuels combustion has focused the development of polymeric materials based on natural materials. PLA (Polylactic acid) is a nature-based polymer material that has the potential to be a petroleum-based polymer substituent. PLA is a polymer that is biocompatible, biodegradable, and comes from renewable resources [1]. This material can be obtained from lactic acid which comes from sugar, starch, cellulose, and glycerin [2]. L-lactic acid and D-lactic acid are lactic acid stereoisomers that can be used to form PLLA (poly (L-lactic acid)) and PDLA (poly (D-lactic acid)) [3].

Besides its good properties, PLA also has the weaknesses, such as low ductility, brittleness, and high hydrophobicity [4], [5]. To improve the mechanical properties, PLA can be modified by blending methods with other polymers. This method is classified as effective, simple, and versatile for developing new materials without synthesizing new polymers [6]. Several polymers are used, such as polyethylene glycol (PEG), polyhydroxybutyrate (PHB), polycaprolactone (PCL), polybutylene adipate coterepthalate (PBAT), chitosan, and starch [7]. The choice of polymer blending is very important because it greatly affects the thermal and mechanical stability during processing.

Wheat bran (WB) is produced by separating flour and the embryo from milled wheat. A total of quarter ton of WB is produced from one million tonnes of wheat. The current application of WB is classified as low value, because it is only used as animal feed [8]. The component distribution of WB consists of (dry-basis), 55-60% non-starch carbohydrate, 14-25% starch, 13-18% protein, 3-8%

minerals, and 3-4% fat [9]. The total fraction of non-starch carbohydrates is 52-70% is arabinoxylan. Arabinoxylan is a polysaccharide group that can be made into films with high strength and good lubrication [10]. Besides that, arabinoxylan is also a good hemicellulose for film formation with addition of plasticizers [11].

PLA and WB are biodegradable polymers from renewable resources. The mixture of PLA with WB can be one of the promising efforts because WB is an abundant and cheap biomaterial. The WB in the PLA/WB blend is used to increase the flexibility, hydrophilicity and strength of PLA. PLA/WB blend polymers can be produced by solution blending and melt-blending [4]. The blending of PLA and WB will improve the mechanical characteristics of the polymer blend indicated by a decrease in tensile strength and an increase in elongation so that the polymer has a greater stretch [12].

In this study, polymer blend PLA/WB plasticized with glycerol is made to obtain biomaterials that are safe to use for biomedical applications and food packaging. Glycerol is known to be a good plasticizer for PLA polymer blends, studied by some previous researchers [13], [14]. The purpose of this study is to determine the effect of plasticized glycerol on the polymer blend PLA/WB on its mechanical properties and chemical resistance.

## 2. Experimental

### *2.1. Materials*
The materials used in this study consist of lactic acid (90%, technical grade), distilled water, nitrogen gas, wheat bran (obtained from local farm), and glycerol (Merck). The glycerol concentration was varied from 0 to 5 wt%.

### *2.2. Preparation of Lactic Acid*
The lactic acid used in this study is prepared based on the method by Rahmayetty et al. (2017) [5]. Lactic acid of 85%-wt aqueous solution was added to the four-necked flask with hotplate magnetic stirrer and condenser. The reaction was held at 120 °C for 1 hour with a continuous flow of nitrogen gas. After that, polycondensation was carried out at 150 °C for 2 hours followed by 180 °C for 2 hours. The product in the form of solid PLA was cooled to room temperature for one day.

### *2.3. Preparation of polymer blend PLA/WB*
The procedure of making PLA/WB plasticized with glycerol solution was carried out based on study by Kanani et al. (2017) [15]. As much as 1 wt% WB was dissolved in distilled water and stirred at room temperature for 20 minutes. Glycerol was added to the solution and stirred continuously for 15 minutes at 70 °C. A total of 100 g of PLA was heated at 80 °C until it melted homogeneously. WB and glycerol solution were added and mixed for 20 minutes. Glycerol was varied as the composition of each material is shown in **Table 1**. The PLA/WB blend is cast by the casting method. Drying process was carried out for 6 hours at 70 °C.

**Table 1.** Material composition for PLA/WB preparation.

| Variation | PLA (%-wt) | Wheat Bran (%-wt) | Glycerol (%-wt) |
|---|---|---|---|
| PLA/WB-G1 | 80 | 20 | 0 |
| PLA/WB-G2 | 80 | 19 | 1 |
| PLA/WB-G3 | 80 | 17 | 3 |
| PLA/WB-G4 | 80 | 15 | 5 |
| PLA/WB-G5 | 80 | 13 | 7 |

### *2.4. Characterization*
Mechanical tests were carried out in this study using a tool called the Texture Analyzer. Tensile strength (σ) can be measured based on the maximum load (F) used to break the material divided by the initial

cross-sectional area (A0) shown in **Eq 1**.The elongation of a material (∈) can be determined by the ratio between the increase in length and the original length as in **Eq. 2.**

$$\sigma = \frac{F}{A} \tag{1}$$

$$\in (\%) = \frac{L_1 - L_0}{L_0} \times 100\% \tag{2}$$

Samples were also tested for swelling power and solubility in distilled water. The swelling power calculation is done based on the difference in the weight of the paste from the centrifuged supernatant from the water bath to the dry basis (**Eq 3**). Meanwhile, water solubility is calculated by dividing the weight of dry precipitate with the weight of the supernatant (**Eq 4**).

$$Swelling\ (\%) = \frac{Paste\ Weight}{Dry\ Weight} \times 100\% \tag{3}$$

$$Solubility\ (\%) = \frac{Dry\ Weight}{Supernatant\ Weight} \times 100\% \tag{4}$$

Chemical resistance test was carried out according to ASTM D1647-89 with the principle of damage, cracking, and loss of coating adhesion to the substrate by qualitative observation. The test was carried out for 3 days at room temperature. The sample was immersed in HCl solution (pH = 3) and NaOH solution (pH = 10). Sample assessment indicators are based on time (days).

## 3. Results and Discussion

### *3.1. Mechanical behavior of PLA/WB*

Tensile strength is the maximum stress a material can withstand when it is stretched prior to fracture. The effect of adding glycerol to PLA/WB on tensile strength and elongation is shown in Fig. 1. Based on the figure, it can be seen that the addition of glycerol plasticizer can reduce tensile strength, with the optimum value on PLA/WB-G1 of 1.22 MPa. While the optimum elongation value is found in the PLA/WB-G4 variable with a value of 0.017. elongation values below 15% indicate the material is brittle [7]. The addition of WB plasticized with glycerol can change the characteristics of the material. These results are consistent with research conducted by Yu et al. (2006), mixing PLA with wheat-derived products can increase tensile strength [12].

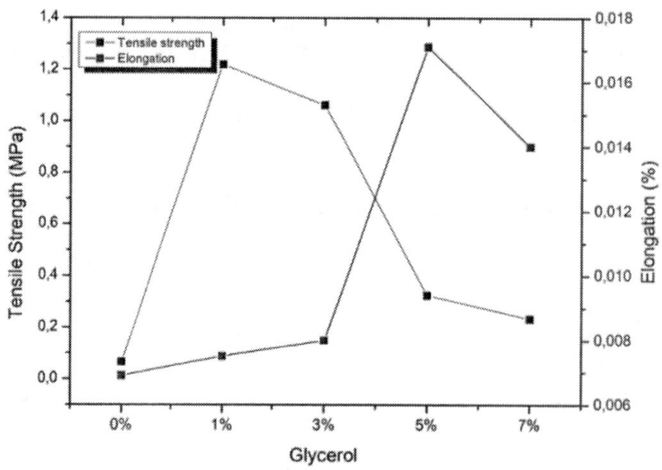


**Figure 1.** Effect of glycerol plasticizer with mechanical properties of PLA/WB.

### *3.3. Swelling ratio of PLA/WB*
The addition of glycerol plasticizer to PLA/WB can reduce the water absorption properties of the material, as shown in Fig. 2. Based on these results, it can be seen that the increase in glycerol concentration is directly proportional to the decrease in the swelling ratio. This decrease indicates the greater water resistance of polymer blends. The addition of glycerol as a plasticizer can cause flexibility, softening, and increase in resin elongation [16]. Besides that, glycerol is also known to keep polymer blends compact and there are no significant physical changes. The combination of hydrophilic and hydrophobic polymers is an effective way to improve degradation properties but does not reduce brittleness.

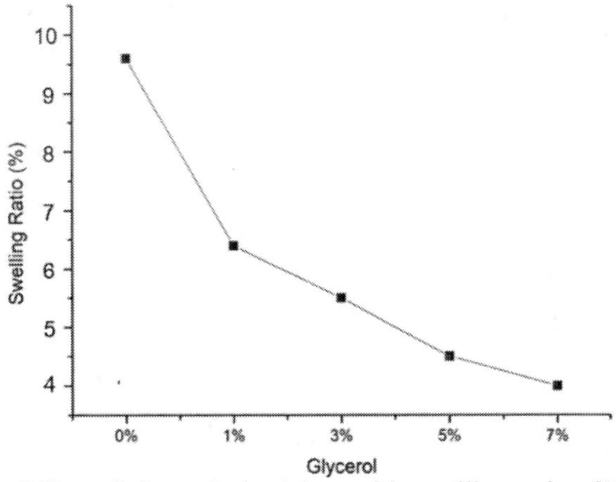


**Figure 2.** Effect of glycerol plasticizer with swelling ratio of PLA/WB.

### *3.4. Chemical resistance of PLA/WB*
Chemical resistance is a test used to evaluate the resistance of a material. Chemical resistance testing is carried out based on ASTM D1647-89. The test medium was an acid (HCl pH 3) and alkaline (NaOH pH 10) solution for 3 days. Failure to test indicated that the coating surface had lost its adhesion to the substrate and appeared to be peeling off. The test results are tabulated in Table 2.

**Table 2.** Effects of glycerol plasticizer with chemical resistance.

| Conditions | PLA/WB-G1 | PLA/WB-G2 | PLA/WB-G3 | PLA/WB-G4 | PLA/WB-G5 |
|---|---|---|---|---|---|
| Acid (pH 4) | Soluble | Soluble | Insoluble | Insoluble | Soluble |
| Neutral (pH 7) | Soluble | Insoluble | Insoluble | Soluble | Soluble |
| Base (pH 10) | Soluble | Soluble | Insoluble | Soluble | Soluble |

Based on the chemical resistance test results, it can be concluded that PLA/WB-G3 is the best layer. This can be seen from the presence of WB and glycerol which are strongly bonded to PLA so that they have good chemical resistance compared to other variations.

### *3.5. Solubility of PLA/WB*
The results of the solubility test on PLA/WB plasticized with glycerol showed quite varied results, as shown in Fig. 3. The results showed that the highest solubility value was found in PLA/WB-G5. This is due to the glycerol content, which in the preparation of plastics is very necessary for its function as plasticizers. Masutani (2014) reported that plastic is an organic solvent with a high boiling point which when added to a hard or stiff resin causes the accumulation of intermolecular forces on long chains to decrease [17]. Therefore, plasticization will affect the physical properties and mechanism of the film, including tensile strength, hardness elasticity and resistance to water.

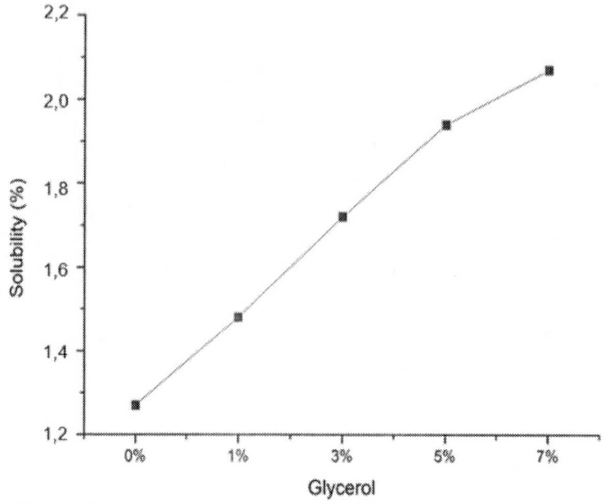


**Figure 3.** Effect of glycerol plasticizer with solubility properties of PLA/WB.

## 4. Conclusion

The addition of glycerol can improve the mechanical properties of the PLA/WB blend. The PLA/WB composition that has optimum tensile strength is found in PLA/WB-G2 with a value of 1.22 MPa and the highest elongation at break is found in PLA/WB-G3. The PLA/WB-G5 variable has the highest solubility and the lowest swelling compared to other variations. This research will further project the production of polymer packaging materials or biomedical applications with environmentally friendly biomaterials.

## References

[1] Mofokeng J P and Luyt A S 2015 Morphology and thermal degradation studies of melt-mixed poly(lactic acid) (PLA)/poly(ε-caprolactone) (PCL) biodegradable polymer blend nanocomposites with $TiO_2$ as filler *Polym. Test.* **45** 93–100

[2] Lasprilla A J R, Martinez G A R, Lunelli B H, Jardini A L, and Filho R M 2012 Poly-lactic acid synthesis for application in biomedical devices — A review *Biotechnol. Adv.* **30** (1) 321–328

[3] Lopes M S, Jardini A L, and Filho R M 2012 Poly (Lactic Acid) Production for Tissue Engineering Applications *Procedia Eng.* **42** 1402–1413

[4] Zuo Y, Gu J, Yang L, Qiao Z, Tan H, and Zhang Y 2014 Preparation and characterization of dry method esterified starch/polylactic acid composite materials *Int. J. Biol. Macromol.* **64** 174–180

[5] Rahmayetty, Sukirno, Prasetya B, and Gozan M 2017 Synthesis and characterization of L-lactide and polylactic acid (PLA) from L-lactic acid for biomedical applications

[6] Peesan M, Supaphol P, and Rujiravanit R 2005 Preparation and characterization of hexanoyl chitosan/polylactide blend films *Carbohydr. Polym.* **60** (3) 343–350

[7] El-hadi A M 2017 Increase the elongation at break of poly (lactic acid) composites for use in food packaging films *Sci. Rep.* 7 (1) 46767

[8] Prückler M *et al.* 2014 Wheat bran-based biorefinery 1: Composition of wheat bran and strategies of functionalization *LWT - Food Sci. Technol.* **56** (2) 211–221

[9] Hell J, Kneifel W, Rosenau T, and Böhmdorfer S 2014 Analytical techniques for the elucidation of wheat bran constituents and their structural features with emphasis on dietary fiber – A review *Trends Food Sci. Technol.* **35** (2) 102–113

[10] Whistler R O Y L 1993 *Chapter 1 - Introduction To Industrial Gums* R. O. Y. L. Whistler And J. N. B. T.-I. G. (Third E. Bemiller, Eds. London: Academic Press, 1993, pp. 1–19.

[11] Holloway W D, Tasman-Jones C, and Bell E 1980 The hemicellulose component of dietary fiber *Am. J. Clin. Nutr.* **33** (2) 260–263

[12] Yu L, Dean K, and Li L 2006 Polymer blends and composites from renewable resources *Prog. Polym. Sci.* **31** (6) 576–602

[13] Finkenstadt V L, Liu C-K, Cooke P H, Liu L S, and Willett J L 2008 Mechanical Property Characterization of Plasticized Sugar Beet Pulp and Poly(Lactic Acid) Green Composites Using Acoustic Emission and Confocal Microscopy *J. Polym. Environ.* **16** (1) 19–26

[14] Clarizio S C and Tatara R A 2012 Tensile Strength, Elongation, Hardness, and Tensile and Flexural Moduli of PLA Filled with Glycerol-Plasticized DDGS *J. Polym. Environ.* **20** (3) 638–646

[15] Kanani N, Wardalia W, Wardhono E, and Rusdi R 2017 Pengaruh Temperatur Pengeringan Terhadap Swelling dan Tensile Strength Edible Film Hasil Pemanfaatan Pati Limbah Kulit Singkong *J. Konversi* **6** (2) 75

[16] Koh J J, Zhang X, and He C 2018 Fully biodegradable Poly(lactic acid)/Starch blends: A review of toughening strategies *Int. J. Biol. Macromol.* **109** 99–113

[17] Masutani K and Kimura Y 2014 *Chapter 1. PLA Synthesis. From the Monomer to the Polymer* 1–36.

# Catalytic Process Development of Bio-BTX from Lignocellulose Derived Product: Preliminary Study Using Transition Metal Catalysts

**Haryo Pandu Winoto[1,2*], C B Rasrendra[1,2], Jenny Rizkiana[1,2], Johanes Kurniawan Leo[1], Andre Citawijaya[1]**

[1]Department of Chemical Engineering, Institut Teknologi Bandung, Bandung 40132 Indonesia
[2]Center for Catalysis and Reaction Engineering, Institut Teknologi Bandung, Bandung 40132 Indonesia

*Corresponding Author Email: haryo@che.itb.ac.id

**Abstract.** Lately, the increase in biofuel production has simultaneously led to lignocellulose material disposal activities. As a waste produced by the biofuel industry, lignocellulose materials are not utilized fully yet due to their complex polymeric structure. As one compound of lignocellulose, lignin possesses the lowest economic value due to its recalcitrant nature. In this work, guaiacol as one of monomeric substance originated from lignin is used as a representative molecule to be valorized due to the existence of both hydroxyl and methoxy moiety groups in it. One important reaction of lignin monomeric substance (phenolic compounds) valorization is oxygen removal. Aromatic substances such as Benzene and Toluene are produced through oxygen removal from the guaiacol molecule. Hydrodeoxygenation of guaiacol is aimed to remove oxygen atoms from it. Though conventionally hydrogen for the Hydrodeoxygenation process is supplied by external molecular hydrogen, in this work it is supplied from hydrogen atom abstraction of methylcyclohexane. Therefore, in a one-pot catalytic reaction system, methylcyclohexane dehydrogenation is the sole hydrogen provider for subsequent hydrodeoxygenation of guaiacol and this novel concept of a circular hydrogen economy has been proven its feasibility through heterogeneous catalytic reaction schemes conducted in this work. Metal supported on zeolite beta is selected as heterogeneous catalysts to evaluate the feasibility of one-pot hydrodeoxygenation and dehydrogenation reaction. Of every catalytic reaction attempt, multiple products consist of alkylated phenol, phenol itself, and toluene are observed proofing the feasibility of this concept. Possible interaction of catalyst surface acidity and metal contents are also probed through the distribution of byproducts. Overall, sequential dehydrogenation and hydrodeoxygenation have been proven through a catalytic reaction catalyzed by metal-supported zeolite beta catalyst and this work can potentially pave the way for further application.
**Keywords:** heterogeneous catalyst, Hydrodeoxygenation, dehydrogenation

## 1. Introduction

The race of biomass utilization has been affecting mankind dominantly. These days, the utilization of biomass-derived compounds such as sugar-based compounds, lignocellulose, and vegetable oils has attained significant attention from many governments and research groups[1]. There have been multiple attempts to use vegetable oils as substituents for fossil fuel. In this case, vegetable oils are being used as Fatty Acid Methyl Ester (FAME) and hydrodeoxygenation one[2, 3]. In particular, several tropical countries have been pushing forwards their utilization of vegetable oils as fuel substituents including Indonesia. Based on data from the Indonesia Bureau of Statistics report in 2018, as one of the biggest palm oil producers, Indonesia has been capable of producing around 35 million tons of crude palm oil per year and will still be increased in the future[4]. This huge production rate of course has also increased the burden on the environment due to empty fruit bunch disposal.

Palm oil plantation empty fruit bunch itself can be categorized as lignocellulose material[5]. As a polymeric material, it consists of cellulose, hemicellulose, and lignin as three dominant polymeric materials [6]. Among the three aforementioned polymeric materials, only cellulose and hemicellulose have been utilized widely these days. For lignin polymeric material, even though its utilization has been researched widely until this day, its utilization is still hindered by its recalcitrant nature [7]. Typically, among numbers attempt lignin utilization, ones always start it by depolymerization and ended with monomer utilization.

For monomer utilization, some previous research results have shown several options for phenolic monomers utilization by removing their oxygen atoms through a catalytic hydrodeoxygenation (HDO) reaction scheme [8]. HDO reactions of lignin-derived compounds are well catalyzed either by noble metals [9] and transition metals [10]. Both types of metals have been reported to be capable of removing oxygen atoms by using molecular hydrogen supplied externally. Even though this scheme has been well-proven, several improvements related to its hydrogen supply can still be discovered.

Lately, some alternative hydrogen sources such as formic acid and alcohols have been discovered as promising ones[11]. Their utilization might reduce the operating pressure of hydrogenation thus benign reaction conditions can be achieved. Formic acid utilization is reported to be well-catalyzed by transition and noble metals whilst alcohol ones are well catalyzed by Lewis acid-base pair catalysts [12]. However, using both formic acid and alcohols produce side products such as carbon dioxide and ketones. Thus, the further discovery of different hydrogen sources is still viable until this day.

In this work, another attempt of using in-situ transfer hydrogenation is attempted through dehydrogenating selected hydrocarbon. One promising option is to conduct methylcyclohexane (MCH) dehydrogenation in a one-pot reaction system simultaneously with hydrodeoxygenation of guaiacol. Guaiacol (GUA) itself is selected as a representative of monomers due to its hydroxyl and methoxyl functional groups. For methylcyclohexane dehydrogenation, several noble metals and transition metal catalysts have been proven as good options[13, 14]. In a typical paraffin dehydrogenation, metal active sites are known to abstract hydrogen from corresponding paraffin [15]. Subsequently, those hydrogen atoms must be detached from metal active sites to reactivate catalyst metal active sites. Hydrogen atoms are then released from the reaction mixture. By considering the requirements of both paraffin dehydrogenation and hydrodeoxygenation, the later reaction might utilize abstracted hydrogen from dehydrogenated paraffin. Putting it into a simpler line, this research is aimed to investigate the feasibility of one-pot methylcyclohexane dehydrogenation and guaiacol deoxygenation by utilizing abstracted hydrogen from methylcyclohexane to cleave oxygen from the later molecule. Effects of metal contents towards catalyst structure are probed through XRD and $N_2$ physisorption techniques whilst product distribution of each reactions is used to probe effects of catalyst composition towards their performances. Thus in the end, optimum catalyst composition and plausible mechanism can be investigated for one-pot MCH dehydrogenation and GUA deoxygenation reactions.

## 2. Material and Methods

### *2.1. Catalyst Preparation*

Catalyst preparation is started by zeolite beta (Si/Al = 25) impregnation with both Nickel Nitrate hexahydrate and Ammonium para molybdate as both Nickel and Molybdenum actives sites sources respectively. In a typical synthesis, controlled amounts of Nickel Nitrate hexahydrate and Ammonium para Molybdate are being dissolved in 50 ml of Deionized Water. Following precursors' dissolution, H-Beta Zeolite is suspended to the precursors solution and sonicated briefly to ensure complete dispersion of precursors. The suspension is then moved into a 100 ml Teflon-lined hydrothermal autoclave and heated at 150 °C for 24 hours for hydrothermal impregnation treatment. Subsequently, the cooled suspension is being dried in a drying oven at 100 °C for 24 hours to remove the water. As the final step, impregnated zeolite is then calcined at 500 °C for 3 hours in a muffle furnace to change the phase of metal precursors into their oxide forms. Prepared catalyst is then denoted as xNiyMo/H-Beta (x and y stand for %-wt composition of Nickel and Molybdenum respectively) and analyzed with X-Ray Diffraction techniques in Bruker D8 Advance XRD to check its crystal structure. Catalyst porosity is also being analyzed using Quantachrome NOVAtouch series $N_2$-physisorption equipment. One-pot Dehydrogenation-Hydrogenation Catalytic Reaction

### *2.2. Catalysts performance test*

In this work, MCH as molecular hydrogen source is mixed with guaiacol to form a homogenous solution. Into the previously mentioned solution, certain amount of Ni/Mo impregnated Beta is mixed so that Reactant/Catalyst ratio of 10 wt/wt is achieved. Into a 100 ml-autoclave reactor, reaction mixture is transferred and sealed to prevent it from escaping during elevated temperature reaction period. In order to remove air from sealed autoclave reactor, $N_2$ gas is flown into autoclave reactor, purged for at least three times, and used to pressurized the reactor up to 20 bar. Reaction initiation is then started by heating the autoclave reactor until desired temperature and mixing the reaction mixture by using magnetic stirrer vigorously. After certain time reaction has been reached, reaction mixture is then cooled, filtered, and analyzed by using Shimadzu GC-MS to determine product distribution of each batch. More detailed workflow and product distribution calculation are explained in supporting information.

## 3. Results and Discussion

### *3.1. Characterization of Prepared Catalyst*

Present crystalline phases of each constituent, either support and active sites, of prepared catalysts are being studied through wide-angle X-Ray Diffraction technique. From figure 1, it can clearly be observed that beta structures of zeolite beta are well preserved, proven by the appearance of double peaks at 7.9 and 22.5° 2ϴ diffraction angle. These results are well matched with numbers of previous works that use the same type of zeolite [16, 17]. Structure intactness has shown that during all steps of catalysts preparation, none to negligible change is applied to crystal structure of zeolite. As for active sites, NiO peaks at 50 and 70°[18]are not detected. From the same spectra, $MoO_3$ peak at 40° that can be used to distinguished $MoO_3$ [19]from beta structure is not detectable as well. Both NiO and $MoO_3$ peaks absence in X-Ray Diffractogram indicates that both active metal sites are well dispersed so that their crystal structure are not detected[20].

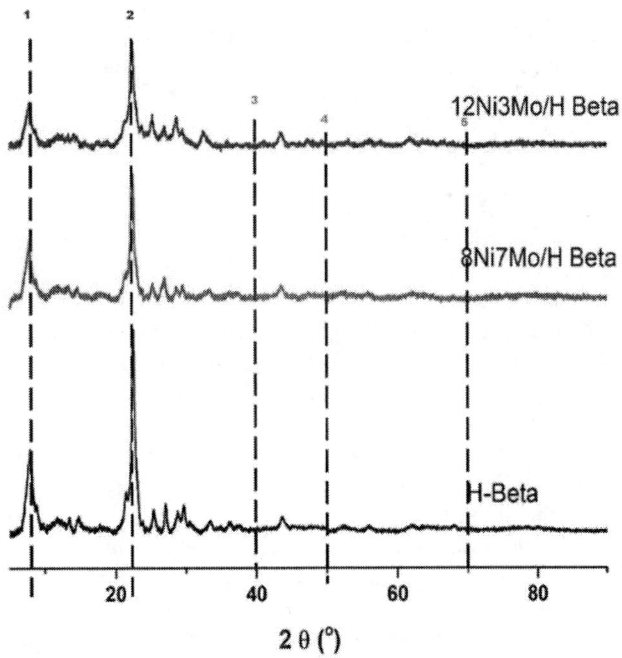


**Figure1.** X-Ray Diffractogram of prepared catalysts (1,2: attributed peaks for beta structure, 3: $MoO_3$ peak, 4,5: NiO peaks).

From figure 2, it can be observed that the addition of metal precursors has reduced both specific surface area and total pore volume of zeolite support significantly. This phenomenon can be related to pore filling by metal active sites[21, 22]. In term of the porosity abatement, both 8Ni7Mo and 12Ni3Mo have similar specific surface area and adsorption-desorption isotherms. Aforementioned similarity is mostly caused by similar loading amount of metal precursors, 15%-wt total loading. Thus, porosity analysis of prepared catalysts has proven the existence of impregnated substances, in this case metal active sites. Furthermore, since in this work the total metal loading is fixed at 15%-wt, almost negligible difference between each loading is observed.

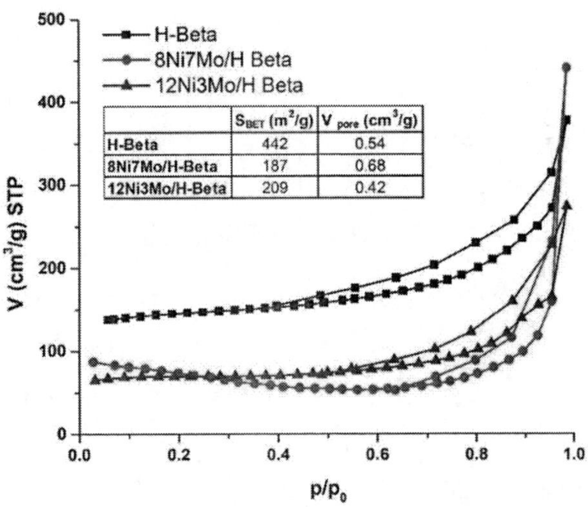


**Figure 2.** $N_2$ adsorption-desorption isotherm of prepared catalysts.

### *3.2. Catalytic Performance of Prepared Catalysts*

Catalytic conversion of both guaiacol (GUA) and methylcyclohexane (MCH) in a one-pot catalytic reaction system primarily consists of two subsequent reactions. The first reaction that happens in this scheme is hydrogen abstraction within the MCH ring so that aromatics compound can be obtained. Subsequently, abstracted hydrogen is used to cleavage either hydroxyl and methoxyl bonds within GUA. Thus, to present a clear and concise discussion, the one-pot reaction scheme is separated into the dehydrogenation aspect, oxygen atoms removal aspect, and synergistic aspect between hydrogenation and oxygen removal.

*3.2.1. Dehydrogenation aspect.* From the dehydrogenation side, it can be seen from figure 3 that mostly MCH molecules are being converted into methylated products, in this case, alkyl cyclopentane and methylated MCH. Dehydrogenated product, methyl cyclohex1-ene is only found in 12Ni3Mo/H-Beta. The dominance of methylated product can be related to the presence of acid sites from the support[23, 24]. For the case of NiO supported on metal oxides, NiO reducibility with zeolite support is not within favorable range for MCH dehydrogenation, proven by the work of Song's group in 2019[25]. In the previously mentioned work, it was observed that both Si and Al have poor metal-support interaction with NiO. Poor metal-support interaction causes NiO to have a high reducibility temperature, thus making it have a much slower turn over frequency in term of hydrogen abstraction due to the higher activation energy of NiO reduction[26]. This phenomenon explains the absence of dehydrogenated products in all catalysts except for 12Ni3Mo/H-Beta one. A considerable amount of NiO is required to catalyze the dehydrogenation of MCH. Oppositely when methylation and reforming reaction are being observed, pristine H-Beta has significantly higher conversion which leads to the MCH methylation and reformation. The existence of proton donors within the Beta zeolite framework in the aforementioned catalyst is not hindered by impregnated metal active sites, therefore acid catalyst has full capability of methylating and reforming MCH as cyclic paraffin. Even though methylation agent might come from cracked hydrocarbon, there is another resource that might also bring methylation agent. This aforementioned source is demethoxylated GUA discussed in the next section.

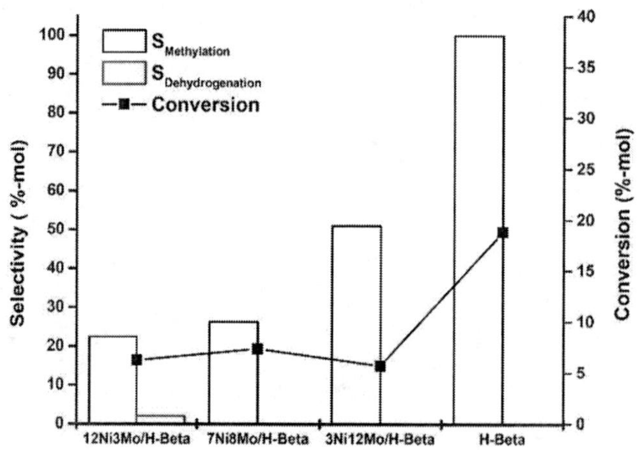


**Figure 3.** Product distribution of MCH dehydrogenation reaction (265 °C, 20 bar $N_2$, 4 h reaction time, R/C = 10, MCH/GUA = 10).

*3.2.2. Deoxygenation aspect* In the case of one-pot reactions conducted in this work, Molybdenum Oxide active site has the role of cleaving oxygen atoms from GUA molecules. In a normal cycle, Molybdenum Trioxide is reduced to Molybdenum Dioxide. Following the formation of Molybdenum Dioxide, oxygen atom from oxygenate substances is then being captured and cleaved leaving Molybdenum Trioxide as an inactive site. Reactivation of Molybdenum Trioxide to

IOP Conf. Series: Materials Science and Engineering 1143 (2021) 012021 doi:10.1088/1757-899X/1143/1/012021

Molybdenum Dioxide is done by hydrogenating it. The aforementioned catalytic cycle is called as reverse Mars Van Krevelen mechanism[27, 28]. In this work instead of supplying pressurized hydrogen for Molybdenum active sites reactivation, abstracted hydrogen from MCH is used instead.

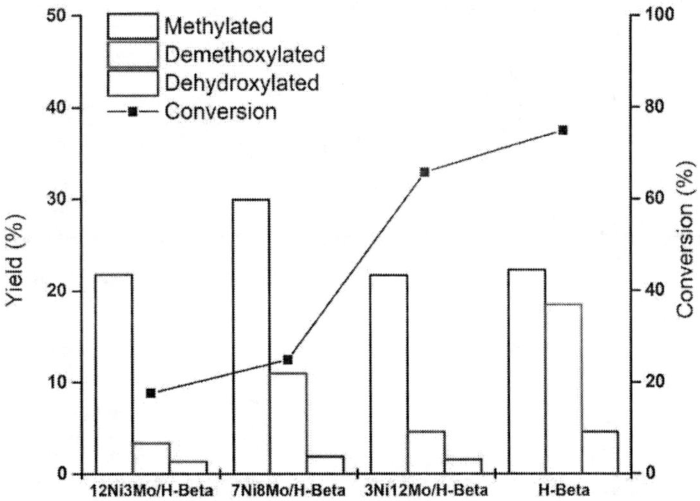


**Figure 4.** Product distribution of GUA deoxygenation reaction (265 °C, 20 bar $N_2$, 4 h reaction time, R/C = 10, MCH/GUA = 10)

From figure 4, it can be seen that demethoxylated products are more dominant compared to dehydroxylated ones. This phenomenon happens due to lower energy activation of methoxyl removal compared to hydroxyl[29]. Further discussing observed phenomena in figure 4, it is necessary for the discussion to be divided into impregnated and pristine H-beta catalyst since C-O cleavage undergoes different mechanism compared to reverse Mars van Krevelen[30]. For the metal impregnated ones, an increased amount of Molybdenum increases GUA conversion. Due to Molybdenum capability to abstract and cleave C-O bonds, the previous result is an apparent one due to Molybdenum active sites role for C-O cleavage. For the methylated ones, these products dominate in all catalysts and this phenomenon is mostly related to the existence of BrØnsted acid sites on the catalyst surface. As been reported from previous work by Svelle's group[31], C-O cleavage is possibly initiated by the protonation of aromatic rings of either MCH and GUA molecules to form carbenium ions. Subsequently, in the end, both methylated products and $H_2O$ molecules are produced. Thus, in all zeolite beta-supported catalysts, the methylation reaction rate has the highest reaction rate. Another interesting aspect observed from figure 4 is that pristine H-Beta outperforms in all aspects compared to metal impregnated catalysts. A possible explanation regarding this phenomenon can be related to reducibility aspects of both Ni and Mo metal active sites. Both aforementioned metals have a notably high temperature of reduction[32, 33] compared to reaction temperature at 265°C. Apparently, instead of adding catalyst activities, metal addition is just clogging Brønsted active sites used for C-O cleavage. Thus, complete plausible mechanism of existed reactions within one-pot catalytic conversion of MCH and GUA can be drawn as following:

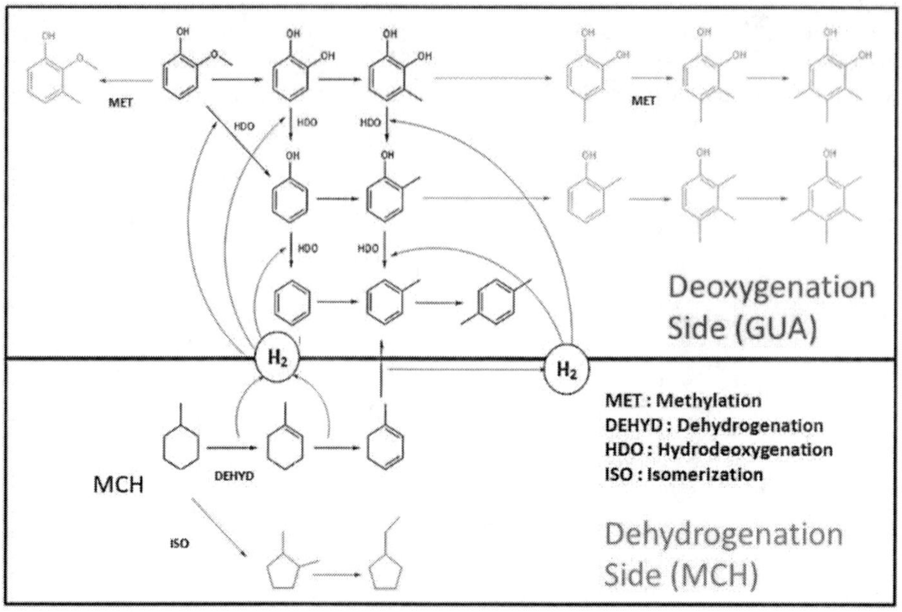


**Figure 5.** Complete proposed mechanism of one-pot GUA and MCH conversion.

## 4. Conclusion

Hydrodeoxygenation of Guaiacol by using hydrogen abstracted has been performed in a one-pot catalytic reaction system. Apparent lower performance of metal impregnated H-Beta can be related to bottleneck effect due to significantly lower rate of reduction of both metal active sites during reaction completion. Surprisingly, pristine H-Beta has better performance due to reactant accessibility to Brønsted active sites. Even though C-O cleavage undergoes different mechanism compared to Mo catalyzed one, during reaction completion time Brønsted acid site has proven its versatility during C-O cleavage. Nevertheless, at some certain amount of Ni active site, dehydrogenation of MCH is proven by product distribution analysis showing Nickel capability to abstract hydrogen from MCH and possibly with some catalyst modification, it can be subsequently used to cleavage C-O bond within oxygenates.

## 5. Acknowledgments

This research has been partially funded by P3MI FTI ITB year 2018 and KIST School Partnership Project year 2019.

## References

[1.] Balat M and Ayar G 2005 Biomass Energy in the World, Use of Biomass and Potential Trends *Energy Sources* **27**(10) 931-940

[2] Vargas E M, Ospina J L, Tarelho L A C, and Nunes M I 2020 FAME production from residual materials: Optimization of the process by Box–Behnken model *Energy Reports* **6** 347-352.

[3] Janampelli S and Darbha S 2019 Hydrodeoxygenation of Vegetable Oils and Fatty Acids over Different Group VIII Metal Catalysts for Producing Biofuels *Catalysis Surveys from Asia* **23**(2) 90-101

[4] Statistik BP 2017 *Statistik Kelapa Sawit Indonesia 2017*

[5] Sudiyani Y, Styarini D, Triwahyuni E, Sudiyarmanto, Sembiring K C, Aristiawan Y, Abimanyu H, and Han M H 2013 Utilization of Biomass Waste Empty Fruit Bunch Fiber of Palm Oil for Bioethanol Production Using Pilot–Scale Unit *Energy Procedia* **32** 31-38

[6] Potters G, Goethem D V, and Schutte F 2010 Promising Biofuel Resources: Lignocellulose and Algae *Nature Education* **3**(9) 14

[7] Zoghlami A and Paës G 2019 Lignocellulosic Biomass: Understanding Recalcitrance and Predicting Hydrolysis *Frontiers in Chemistry* **7**(874)

[8] Bu Q, Lei H, Zacher Alan H, Wang L, Ren S, Liang J, Wei Y, Liu Y, Tang J, Zhang Q, and Ruan R 2012 A review of catalytic hydrodeoxygenation of lignin-derived phenols from biomass pyrolysis *Bioresource Technology* **124** 470-477

[9] Gutierrez A, Kaila R K, Honkela M L, Slioor R, and Krause A O I 2009 Hydrodeoxygenation of guaiacol on noble metal catalysts *Catalysis Today* **147**(3) 239-246

[10] Boullosa-Eiras S, Lodeng R, Bergem H, Stocker M, Hannevold L, and Blekkan E A 2014 Catalytic hydrodeoxygenation (HDO) of phenol over supported molybdenum carbide, nitride, phosphide and oxide catalysts *Catalysis Today* **223** 44-53

[11] Gilkey M J and Xu B 2016 Heterogeneous Catalytic Transfer Hydrogenation as an Effective Pathway in Biomass Upgrading *ACS Catalysis* **6**(3) 1420-1436

[12] Chia M and Dumesic J A 2011 Liquid-phase catalytic transfer hydrogenation and cyclization of levulinic acid and its esters to γ-valerolactone over metal oxide catalysts *Chemical Communications* **47**(44) 12233-12235

[13] Choojun K, Worathanaseth A, Kuhatasanadeekul S, Kurato T, Ketinaruj S, Phichitsurathaworn P, Promchana P, Prkobtham K, Numwong N, Poo-arporn Y, and Sooknoi T 2019 Effect of cobalt complex precursors on reactivity of cationic cobalt catalysts: Cyclohexane dehydrogenation *Catalysis Communications* **125** 108-113

[14] Zhang X, Nie Y, Yuan Y, Lu F, and Geng Z 2020 Density functional theory investigation on the mechanism of dehydrogenation of cyclohexane catalyzed by heteronuclear NiTi+ *Computational and Theoretical Chemistry* **1184** 112820

[15] Zhao W, Chizalet C, Sautet P, and Raybaud P 2019 Dehydrogenation mechanisms of methyl-cyclohexane on γ-Al2O3 supported Pt13: Impact of cluster ductility *Journal of Catalysis* **370** 118-129

[16] Winoto H P, Ahn B S, and Jae J 2016 Production of γ-valerolactone from furfural by a single-step process using Sn-Al-Beta zeolites: Optimizing the catalyst acid properties and process conditions *Journal of Industrial and Engineering Chemistry* **40** 62-71

[17] Lu T, Yan W, and Xu R 2019 Chiral zeolite beta: structure, synthesis, and application *Inorganic Chemistry Frontiers* **6**(8) 1938-1951

[18] Yolcular S and Olgun Ö 2008 Ni/Al2O3 catalysts and their activity in dehydrogenation of methylcyclohexane for hydrogen production *Catalysis Today* **138**(3) 198-202

[19] Boufaden N, Akkari R, Pawelec B, Fierro J L G, Said Zina M, and Ghorbel A 2016 Dehydrogenation of methylcyclohexane to toluene over partially reduced silica-supported Pt-Mo catalysts *Journal of Molecular Catalysis A: Chemical* **420** 96-106

[20] Garbarino G, Riani P, Infantes-Molina A, Rodriguez-Castellon E, and Busca G 2016 On the detectability limits of nickel species on NiO/γ-Al2O3 catalytic materials *Applied Catalysis A: General* **525** 180-189

[21] Watanabe S, Sukino S, Miyasaka T, Sato K, Yashiro K, Kawada T, and Hashida T 2020 Influences of Ni content and porosity on mechanical properties of Ni–YSZ composites under solid oxide fuel cell operating conditions *Journal of Materials Science* **55**(20) 8679-8693

[22] Lewandowski M and Sarbak Z 1996 Structure and Texture of Nickel-Molybdenum Catalysts on Alumina Support Modified with Fluoride or Chloride Ions *Crystal Research and Technology* **31**(6) 773-781

[23] Wang C M, Wang Y D, and Xie Z K 2016 Methylation of olefins with ketene in zeotypes and its implications for the direct conversion of syngas to light olefins: a periodic DFT study *Catalysis Science & Technology* **6**(17) 6644-6649

[24] *Concepts and Analysis of Surface Acidity and Basicity*, in *Handbook of Heterogeneous Catalysis*. 1096-1122.

[25] Song Y, Lin W, Guo X, Dong L, Mu X, Tian H, and Wang L 2019 Aromatization and isomerization of methylcyclohexane over Ni catalysts supported on different supports *Green Energy & Environment* **4**(1) 75-82
[26] Cheng F, Dupont V, and Twigg M V 2016 Temperature-programmed reduction of nickel steam reforming catalyst with glucose *Applied Catalysis A: General* **527** 1-8
[27] Mironenko A V and Vlachos D G 2016 Conjugation-Driven "Reverse Mars–van Krevelen"-Type Radical Mechanism for Low-Temperature C–O Bond Activation *Journal of the American Chemical Society* **138**(26) 8104-8113
[28] Zacharopoulou V, Vasiliadou E S, and Lemonidou A A 2018 Exploring the Reaction Pathways of Bioglycerol Hydrodeoxygenation to Propene over Molybdena-Based Catalysts *ChemSusChem* **11**(1) 264-275
[29] Song W, Liu Y, Barath E, Zhao C, and Lercher J A 2015 Synergistic effects of Ni and acid sites for hydrogenation and C–O bond cleavage of substituted phenols *Green Chemistry* **17**(2) 1204-1218
[30] Huang J, Agrawal P K, and Jones C W 2009 Effects of Acidity on the Conversion of the Model Bio-oil Ketone Cyclopentanone on H−Y Zeolites *The Journal of Physical Chemistry C* **113**(38) 16702-16710
[31] Van der Mynsbrugge J, Visur M, Olsbye U, Beato P, Bjorgen M, Van Speybroeck V, and Svelle S 2012 Methylation of benzene by methanol: Single-site kinetics over H-ZSM-5 and H-beta zeolite catalysts *Journal of Catalysis* **292** 201-212
[32] Rajagopal S, Marini H. J. Marzari, J. A., and Miranda R 1994 Silica-Alumina-Supported Acidic Molybdenum Catalysts - TPR and XRD Characterization *Journal of Catalysis* **147**(2) 417-428
[33] Wang C B, Gau G Y, Gau S J, Tang C W, and Bi J L 2005 Preparation and characterization of nanosized nickel oxide *Catalysis Letters* **101**(3) 241-247

IOP Conf. Series: Materials Science and Engineering 1143 (2021) 012022 doi:10.1088/1757-899X/1143/1/012022

# Estimation of Xylose Recovery from Lignocellulosic Biomass

**M T A P Kresnowati[1,2*], D C Januardi[1] and S V Utomo[1]**

[1] Department of Chemical Engineering, Faculty of Industrial Technology, Bandung Institute of Technology, Indonesia
[2]Food and Biomass Processing Technology Research Group, Faculty of Industrial Technology, Bandung Institute of Technology, Indonesia

*Corresponding Author's E-mail: kresnowati@che.itb.ac.id

**Abstract**. Lignocellulosic materials are potential raw materials for (bio)chemical industries due to their abundance. Its hemicellulosic content, for example, can be hydrolysed to xylose and later converted to various valuable biochemical products, e.g. xylitol. Due to the variability in characteristics and composition of the lignocellulosic materials, however, thorough research is required before the utilization of each type of lignocellulosic materials. This paper presents the development of an empirical model to estimate the yield of xylose from various lignocellulosic materials. A comprehensive literature study was conducted to build lignocellulosic database, in which the yields of xylose from various lignocellulosic materials that were processed by using different pretreatment condition were mapped. An empirical model was developed to establish a correlation between the type of lignocellulosic materials as well as the pretreatment operation condition (severity factor) and the yield of xylose. Several correction factors, such as biomass composition, lignin structure, and the succeeding hydrolysis process have been proposed to improve its accuracy.

## 1. Introduction

As an agricultural country, Indonesian produce massive amount of agricultural and plantation of biomass waste. In 2010, Indonesia produces 145 MT of biomass waste every day [20]. Biomass is classified as a renewable energy source with low economic value. In addition, following the biorefinery concept, biomass can also be used as the raw materials to produce high value-added products. However, the utilization of biomass has not been maximized.

Lignocellulosic biomass is generally composed of starch, cellulose, hemicellulose, and lignin [1]. The potentials of cellulose or lignin components in biomass have been thoroughly studied. However, the potential of hemicellulose utilization has been overlooked. Xylan, the biopolymers backbone of hemicellulose, is known as the second most abundant biopolymer in nature [23]. Xylan can be processed to recover the monosaccharide xylose that can be further converted into xylitol.

The processing of lignocellulosic materials to recover monosaccharides includes pretreatment and hydrolysis. The pretreatment step is aimed to release the polysaccharides which bind closely to the lignin contained in the biomass. This first step can be conducted either chemically with acid or physically through hydrothermal whereas the hydrolysis which can be done using enzymes or acids [27]. To recover xylose from the lignocellulosic biomass, first the bonds between hemicellulose and lignin need to be broken and further the free hemicellulose needs to be extracted and hydrolysed. For xylitol production, the produced hydrolysate can be directly fermented by a suitable microbial platform to produce xylitol such as *Candida utilis* or *Debaryomyces hansenii* [9].

One of the critical factors in producing xylitol from biomass is the percentage of xylose that can be recovered from biomass. The later is affected by the operating conditions applied in the pretreatment and biomass hydrolysis [15,21]. However, the suitable pretreatment operating conditions may also vary with the type of lignocellulosic biomass used as the raw materials [28]. Thereby before using a particular biomass a thorough research needs to be conducted to obtain the optimum operation condition for the biomass processing.

The purpose of this study was to collect a solid database on lignocellulosic biomass processing to obtain xylose and to map those data to examine the effects of pretreatment type as well as the related operating conditions on xylose recovery. In particular the widely available data on biomass waste from corn processing, oil palm processing and sugar cane processing were used to build a model. A model would be development to enable further estimation of xylose recovery from any type of lignocellulosic biomass materials at a predesigned pretreatment condition. Thereby reducing the wet laboratory experiments that need to be conducted in an attempt for biomass utilization.

## 2. Methodology

This research used secondary data which was obtained from an extensive literature study on research papers on the processing and utilization of lignocellulosic biomass: corn fiber [22], corn cob [4,6,14,25], palm oil fruit bunch [5,18,19], corn stover [5], and sugarcane bagasse [2,26]. The research was focused on biomass waste from corn processing: corn cob, corn stover, and corn fiber; biomass waste from oil palm processing: oil palm empty fruit bunches (EFB); and biomass waste from sugar cane processing: bagasse; which were selected due to their high hemicellulose content and their high availability in Indonesia. Later, data on wheat straw was also used for model validation.

The data was processed to obtain the severity factor (SF) and the combined severity factor (CSF) of the applied lignocellulosic biomass pretreatment operating conditions, following eq. (1) and eq. (2). In particular, SF was applied for the datasets obtained from the hydrothermal pretreatment process whereas CSF was applied for the datasets obtained from other types of pretreatment process, for example weak acid hydrolysis.

$$SF = \mathrm{t} \times \mathrm{e}^{\frac{\mathrm{T}-100}{14{,}75}} \tag{1}$$

In which t is the duration of the pretreatment process [minute], and T is the temperature of the pretreatment (°C) [21]. Meanwhile,

$$\log CSF = \log SF - \mathrm{pH} \tag{2}$$

In which the acidity of the pretreatment solution is expressed as the pH [Ruiz].

Model development, that was model parameters estimation and model simulation, was conducted in MATLAB.

## 3. Results and Discussion

### *3.1. Mapping of the Effect of Operating Conditions on Xylose Recovery in Various Biomasses*

The mapping of the pretreatment operating conditions on xylose recovery from various kinds of lignocellulosic biomass: corn fiber, corn cobs, corn stover, palm oil EFB, and sugarcane bagasse, is presented in Figure 1.

A

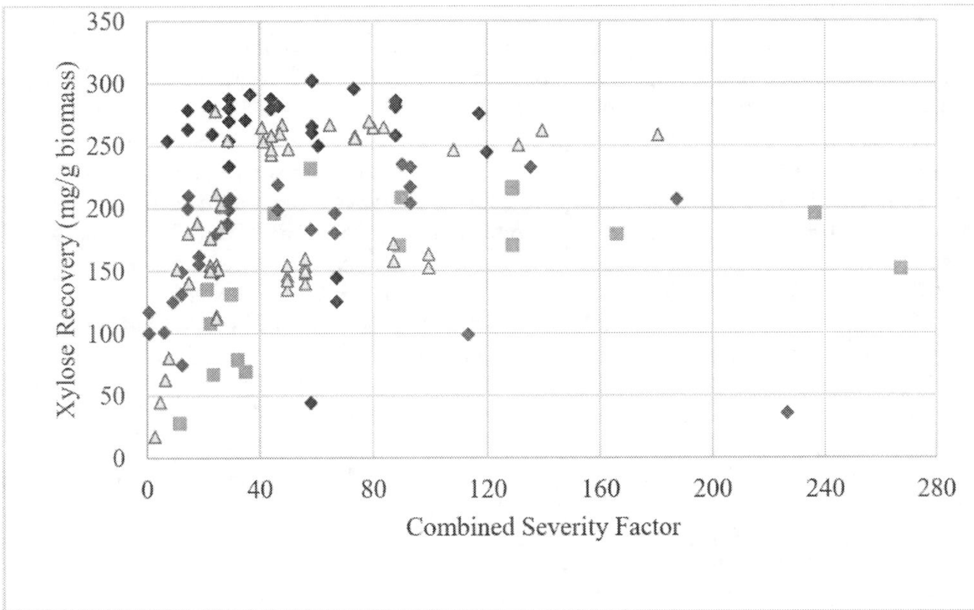


B

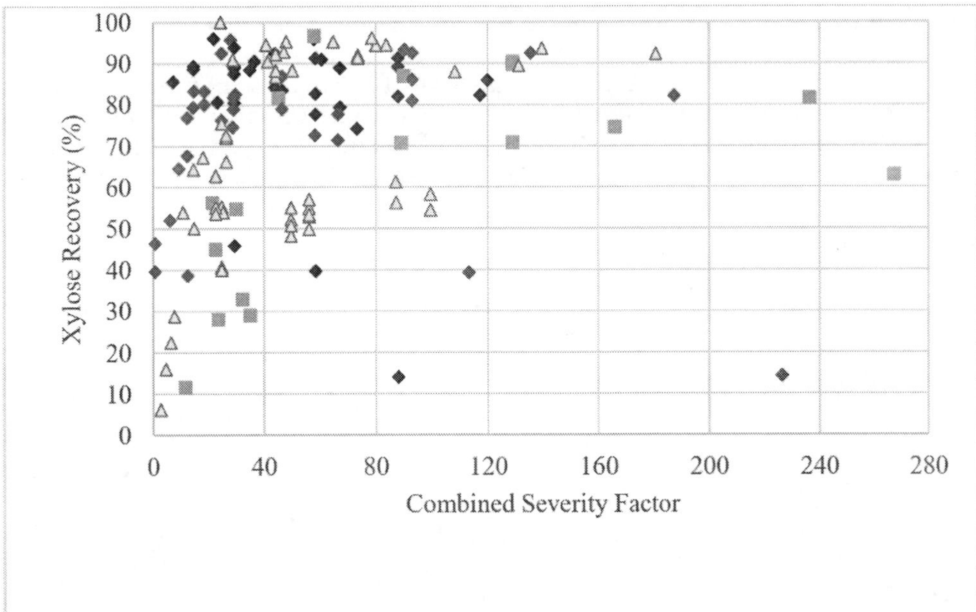


**Figure 1.** Effects of pretreatment operating conditions (CSF) on the xylose recovery from lignocellulosic biomass (a) the xylose recovery is expressed as [mg-xylose/g-dry biomass], (b) the xylose recovery is expressed as the percentage to its maximum/theoretical value, (blue diamonds: corn fiber, red diamonds: corn cob, purple diamonds: corn stover, green squares: oil palm EFB, orange triangles: sugarcane bagasse).

Literature reported moderate pretreatment condition for corn fiber [22]. The calculated CSF value was in the range of 25 – 30 and provided nearly the theoretical recovery of xylose, namely as much as 194 mg/g of corn fiber (Figure 1a-b).

Literatures reported higher pretreatment conditions for corn cob [4,25]. The calculated CSF value was in the range of 7 – 120 and provided the xylose recovery in the range of 70% - 96% of its theoretical value. The highest xylose recovery was obtained at CSF 58.7, giving 58.7% of theoretical xylose recovery which correspond to 302.3 mg/g corn cob (Figure 1a-b). The outlier data of corn cob, CSF 30-90, gave xylose recovery of 45-15% of its theoretical value [6].

Wider data set was obtained on corn stover, the reported pretreatment conditions were in the range of 0 – 230 [5]. At low CSF value that was lower than 100, an increasing trend of xylose recovery was observed along with the increase in the CSF value. Above higher CSF value, a decreasing trend of xylose recovery was observed (Figure 1a-b). The decrease in xylose recovery may be related to the partial xylose degradation at higher operation condition [28]. The maximum xylose recovery was observed in the CSF range of 10-100, giving xylose recovery in the range of 70-95% or 160-240 mg/g corn stover.

A wide range of pretreatment condition was reported for oil palm EFB, ranging from 0 – 267.43 [11,18]. However low xylose recoveries were mostly recorded, in the range of 30-80% or corresponded to 11.7 - 232 mg/g oil palm EFB. The highest xylose recovery of 96.7% was reported by literature, at the application of acid pretreatment method at temperature 120°C, 15 minutes duration, and 12,5% solid loading [19].

A large data set was obtained for sugarcane bagasse [2,26]. An increasing trend of xylose recovery was observed until the CSF of 40 (Figure 1). The maximum xylose recovery (90 – 100%) could be achieved starting at CSF value of 25. Interestingly, a constant trend of high xylose recovery was observed at higher CSF value.

Completely different dataset was obtained from the lignocellulosic biomass that was pretreated by the hydrothermal method. In general, no chemicals, nor acid nor base, is added for the hydrothermal pretreatment method and thus the pH of the system is neutral. Calculating the CSF for this dataset following eq. (2) gives negative values, and thus the pretreatment operation condition is parameterized by SF that is calculated following eq. (1). Similar trend of xylose recovery was observed. The xylose recovery increased along with the increase in severity factor until the maximum value of xylose recovery was achieved. Above that, a decreasing trend of xylose recovery was observed. The maximum xylose recovery was achieved at SF around 4920, giving xylose recovery of 348,87 mg/g hemicellulose which corresponded to xylose recovery of 52% of its theoretical value.

A

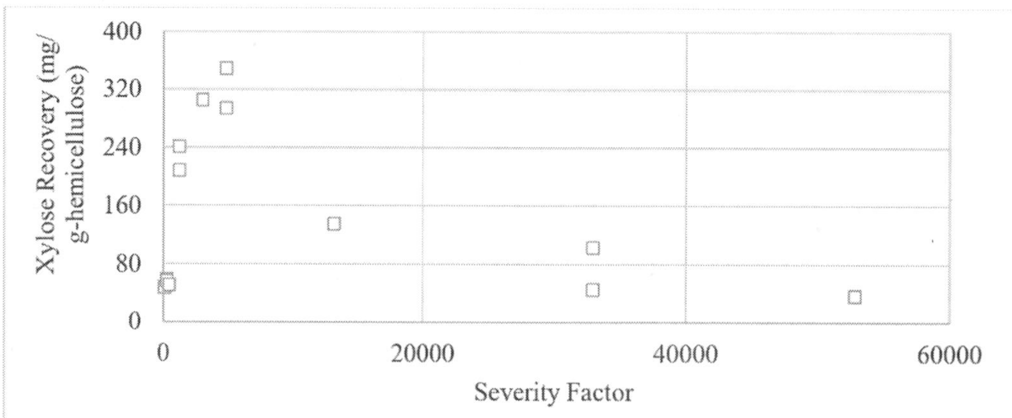

B

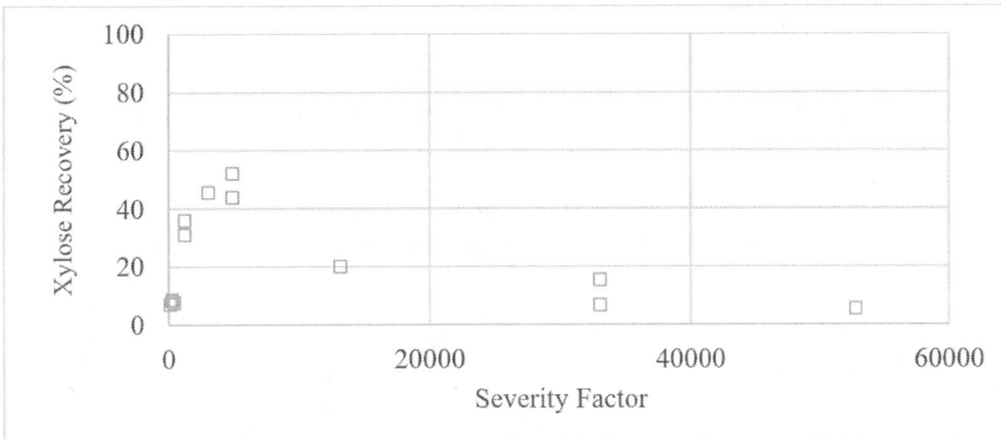


**Figure 2.** Effects of hydrothermal pretreatment operating conditions (SF) on the xylose recovery from lignocellulosic biomass oil palm EFB, (a) the xylose recovery is expressed as [mg-xylose/g-dry biomass], (b) the xylose recovery is expressed as the percentage to its maximum/theoretical value.

When the trends of the five lignocellulosic biomasses were compared (Figure 1), corn fiber is the biomass that requires the lowest CSF value compared to others, that is in the range of 0 – 30. This shows that corn fiber only needs a moderate pretreatment condition to achieve the maximum xylose recovery. The required mild pretreatment may be related to the presence of starch as well as low lignin content in the corn fiber [22]. Different trend was observed for corn cob, higher CSF value are required, at 0 – 120. Above a certain CSF value a decreasing trend of xylose recovery from corn cob was observed that may be related with the presence of lignin-hemicellulose bonds that cannot be broken by the pretreatment and the further degradation of xylose [28]. The similar trend observed between corn stover and corn cob (Figure 1) may be related to the similarity in the composition of cellulose and lignin in the two biomass. However, the lower xylose recovery that was obtained for the corn stover may be related with the tighter lignin-hemicellulose bonds on the corn stalk and leaf [8]. EFB biomass has data sets in 2 main zones, namely in the CSF value range 0 – 40, xylose recovery 10 – 60% and in the CSF value range 40 – 270, xylose recovery 60 – 100%. This trend indicates that in the first zone there are several lignin-hemicellulose bonds that cannot be removed. However, after passing the CSF value of 40, the bond was able to be released so that the pretreatment was able to effectively break down xylose in the biomass. In addition, what clearly distinguishes EFB from other biomass is the lower xylose recovery under the same pretreatment operating conditions. This may be caused by the higher lignin content of the oil palm EFB [7].

Above discussion indicated that lignin content in the biomass as well as lignin composition may affect the effectivity of the lignocellulosic biomass pretreatment process [24]. The structure of lignin in biomass is expressed by the composition of p-hydroxyphenyl (H), guaiacyl (G), and syringyl (S) contained in lignin biomass [12], and in particular the structure of lignin can be expressed by the ratio of syringyl and guaicyl [24]. Data on lignin content as well as lignin composition of the studied biomass is presented in Table 1.

**Table 1.** Lignin and hemicellulose factors of the five lignocellulosic biomass.

| Biomass | Total hemicellulose | Total lignin | Lignin composition[7],[8] | | | S/G | H/G |
|---|---|---|---|---|---|---|---|
| | | | p-hydroxyphenyl (H) | guaiacyl (G) | syringyl (S) | | |
| Corn fiber | 35%[18] | 8% | 18.5 | 35.5 | 46.0 | 1.30 | 0.52 |

| | | | | | | | |
|---|---|---|---|---|---|---|---|
| Corn cob | 35%[21] | 20%[7] | 17.0 | 33.0 | 50.0 | 1.52 | 0.52 |
| Corn stover | 26%[14] | 19%[7] | 4.0 | 35.0 | 61.0 | 1.74 | 0.11 |
| EFB | 26%[6] | 23%[6] | 3.0 | 43.0 | 54.0 | 1.26 | 0.07 |
| Sugarcane bagasse | 29%[2] | 20% | 2.0 | 38.0 | 60.0 | 1.58 | 0.05 |

The mapping of the effects of pretreatment operation conditions also shows that maximum xylose recovery can be achieved not only by using a single method. Figure 1 and Figure 2 show that the maximum xylose recovery from oil palm EFB can be achieved either by the weak acid pretreatment method or by the hydrothermal method, despite the different operating conditions required.

### *3.2. Model Development*

An empiric model was developed to determine the effects of pretreatment operating conditions on xylose recovery from lignocellulosic biomass. The data on five types of lignocellulosic biomass: corn fiber, corn cob, corn stover, oil palm EFB, and sugarcane bagasse, as is presented in Figure 1, was used for model development. The operating conditions that were expressed as log CSF, were set as the independent variable, whereas the xylose recovery [mg/g biomass] was set as the dependent variable. Only the increasing trend of xylose recovery was used in the curve fitting feature in the MATLAB, as the decreasing trend of xylose recovery at higher CSF was suspected to be related with further xylose degradation. The best model, however, only gave the R-square value of 0.305 (Table 2). The low R-square value implied that the model could not best represent the data. It indicated that the operating conditions are not the only parameter affecting xylose recovery. Series of model improvements were then conducted to obtain a model that could explain the data better.

**Table 2.** Development of Empirical Model on Xylose Recovery.

| Modeling | Equation | Information | |
|---|---|---|---|
| Basic model | $y = a + b \times \log(CSF) + c \times (\log(CSF))^2$<br>a= -159.6; b=197.7; c=-20.69 | R-*square* = 0.305 | (3) |
| incorporating of the effect of lignocellulosic biomass composition, i.e. hemicellulose content | $y = a + b \times \log(CSF) + c \times (\log(CSF))^2$<br>a= -515.5; b= 564; c= -61.78 | R-*square* = 0.359 | (4) |
| incorporating the effect of lignocellulosic biomass composition, i.e. lignin content | $y = a + b \times \log(CSF) + c \times (\%Lignin) + d \times (\log(CSF))^2 + e \times (\%Lignin)\log(CSF) + f \times (\%Lignin)^2$<br>a=-1694; b=444; c= 22940; d= -85,3 ; e= 87970; f=1254 | R-*square* = 0.5806 | (5) |
| incorporating the effect lignin composition, i.e. S/G Ratio | $y = a + b \times \log(CSF) + c \times \left(\frac{S}{G}\right) + d \times \log(CSF)^2 + e \times \left(\frac{S}{G}\right)\log CSF + f \times \left(\frac{S}{G}\right)^2$<br>a= -3115; b= 833; c= 2137; d= -78.81; e= 309,4; f= -122.9 | R-*square* = 0.5733 | (6) |

In the first attempt we changed the unit of xylose recovery to be mg/g of hemicellulose in the biomass (Table 1). This way, we corrected the model by incorporating the effect of lignocellulose biomass composition, in particular the hemicellulose content of the lignocellulose. A better R-square value was obtained, 0.359 (Table 2), which indicated that hemicellulose levels affected the correlation between CSF and xylose recovery. However, the obtained R-square value was still low.

In the second attempt we incorporated the effect of lignin content of the lignocellulosic biomass (Table 1). In this case, two independent variables were introduced to the model, that were the log CSF value and the lignin content of the biomass, whereas the dependent variable was the xylose recovery in .mg/g of hemicellulose in the biomass. A better R-square value was obtained, 0.5806 (Table 2). On the other hand, if it was the information on the lignin composition, i.e. the S/G value, that was incorporated into the model instead of the lignin content, a model with R-square value of 0.5733 was obtained (Table 2).

From these two models, it can be seen that the lignin content parameter has a greater correlation than the S/G parameter. However, the S/G equation has a smaller error value than the lignin content.Moreover, the R-square also states the goodness of fit. Therefore, this modeling equation was used to correct xylose recovery by entering the CSF and S/G values for each biomass into each experimental data. The mapping of the corrected xylose recovery results is presented in Figure 3.

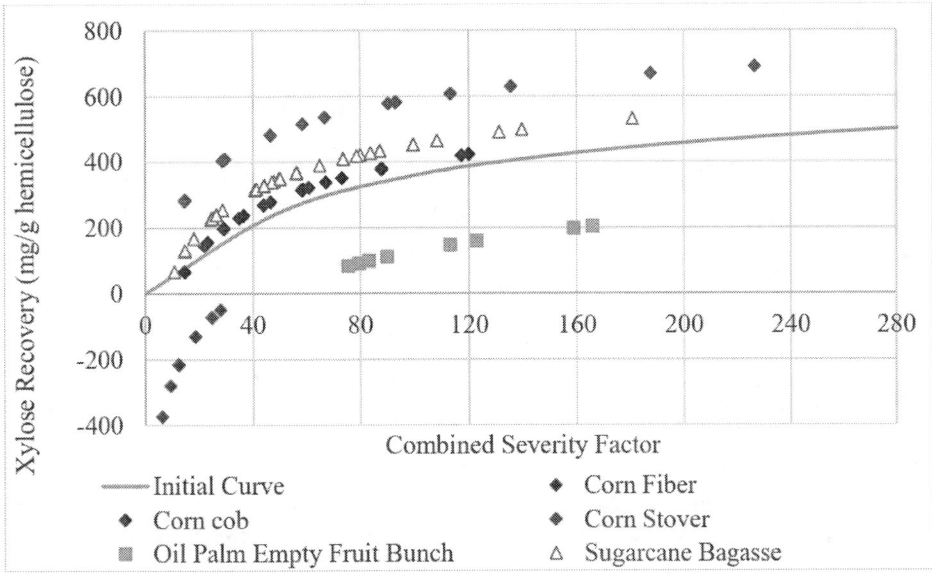


**Figure 3.** Comparison between data and xylose recovery estimated from model simulation.

The comparison between the data on the five lignocellulosic biomass and the xylose recovery estimated from model simulation is presented in Figure 3. At a similar CSF value, the xylose recovery would be higher on biomass with a higher S/G value. The estimated xylose recovery from corn cob, for example, is always higher than the xylose recovery from oil palm EFB for any value of CSF. However, for corn fiber biomass, the corrected xylose recovery gave negative value. That is, the obtained model can not estimate the xylose recovery correctly. This indicates that still other factor that should be taken into account in the model.

Another correction factor could also be introduced into the model to correct the estimate xylose recovery from lignocellulosic biomass pretreated by hydrothermal pretreatment method. Else, a specific model could also be developed for estimating the xylose recovery directly from the operating condition of the hydrothermal pretreatment method. However, due to data limitation this was not conducted here.

### 3.3. *Model Validation*

The developed model (eq. 6) was further validated by using it to estimate the xylose recovery from wheat straw, which data was not used in developing the model. The data on pretreatment operating condition as well as the related xylose recovery from wheat straw was obtained from literature [13].

Figure 4 shows that the developed model could estimate the xylose recovery from wheat straw on any given pretreatment condition quite well. The simulated model gave a similar trend to the data, although the estimated xylose recovery was mostly slightly lower than the data. This may be due to the S/G value of wheat straw was 1.3 [13], lower than the S/G values of the other biomass which were used in the model development. For example, the literature value for the xylose recovery from wheat straw that was obtained from a pretreatment at 170°C and pH 0.4, for 2 minutes, was 20 mg xylose/g of wheat straw. Meanwhile, the model simulation gave an estimation of xylose recovery of 19.4 mg/g wheat straw.

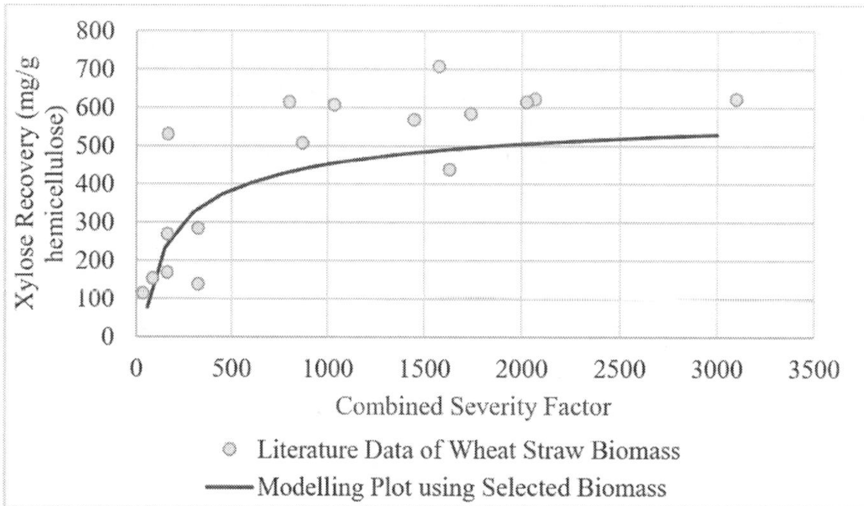


**Figure 4.** Model validation, simulation of the developed model to estimate the xylose recovery from wheat straw at any given operating condition (curve: model simulation; literature data [13] symbols: yellow circle).

## 4. Conclusion

The research has successfully mapped the effects of pretreatment operating conditions on xylose recovery from lignocellulosic materials, in particular on corn fiber, corn cob, corn stover, empty fruit bunches, and sugarcane bagasse. An increase in the pretreatment operating condition leads to an increase in the xylose recovery from lignocellulosic materials. However, above a certain condition, a decreasing trend of xylose recovery that may be related with further xylose degradation can be expected. Besides, the xylose recovery from lignocellulosic materials is also affected by biomass composition, lignin structure of the biomass, and the enzymatic hydrolysis process following to the hydrothermal pretreatment method. An empiric model has been developed to correlate the effect of pretreatment operating condition biomass, composition, lignin structure, and xylose recovery. The model has been successfully validated by using it to estimate the recovery of xylose from other lignocellulosic material, i.e. wheat straw. This model can be further used to estimate the expected xylose recovery from a specific lignocellulosic biomass material at any predesigned pretreatment condition.

**Acknowledgments**
Authors would like to acknowledge anyone who have contributed in completing the research, either directly or indirectly. In particular, Ms. Diah Meilany is acknowledge for sharing experimental data on EFB pretreatment.

**References**
[1] Das P K, Das B P and Dash P 2014 *Refining Biomass Residues for Sustainable Energy and Bioproducts* **2014** Elsevier Inc.
[2] De Paiva J E, Maldonade I R and Scamparini A R P 2009 *Revista Brasileira de Engenharia Agricola e Ambiental* **13(1)**, 75–80.
[3] del Río J C, Lino A G, Colodette J L, Lima C F, Gutiérrez A, Martínez Á T and Rencoret J 2015 *Biomass and Bioenergy* **81**, 322–338.
[4] Dominguez J M, Cao N, Gong C S and Tsao G T 1997 *Bioresource Technology* **61(1)**, 85–90.
[5] Feher A, Feher C, Rozbach M and Barta Z 2017 *Chemical and Biochemical Engineering Quarterly* **31(1)**, 77–87.
[6] Ghaffar A, Yameen M, Aslam N, Jalal F, Noreen R, Munir B, Mahmood Z, Saleem S, Rafiq N, Falak S, Tahir I M, Noman M, Farooq M U, Qasim S, and Latif F 2017 *Chemistry Central Journal* **11**, 97.
[7] Grande P M, Weidener D, Dietrich S, Dama M, Bellof M, Maas R and De María P D 2019 *ACS Omega*
[8] Guragain Y N, Herrera A I, Vadlani P V and Prakash O 2015 *Natural Product Communications* **10(1)**, 201–208.
[9] Harahap B M and Kresnowati M T A P 2018 *Biomass Conversion and Biorefinery* **8(2)**, 255–263.
[10] Kabel M A, Bos G, Zeevalking J, Voragen A G J and Schols H A 2007 *Bioresource Technology* **98(10)**, 2034–2042.
[11] Li Q, Ng W T, Puah S M, Bhaskar R V, Soh L S, Macbeath C and Wu J C 2014 *Biotechnology and Applied Biochemistry* **61(4)**, 426–431.
[12] Lourenço A and Pereira H 2018 *Lignin - Trends and Applications*
[13] Lupoi J S, Singh S, Parthasarathi R, Simmons B A and Henry R J 2015 *Renewable and Sustainable Energy Reviews* **49**, 871–906.
[14] Mardawati E, Andoyo R, Syukra K A, Kresnowati M T A P, and Bindar Y 2018 IOP Conference Series: Earth and Environmental Science **141**, 012019
[15] Meilany D, Kresnowati M T A P, Setiadi T, and Boopathy R 2020 *Applied Sciences (Switzerland)*, **10(4)**, 1391.
[16] Modenbach A A 2014 *Transactions of the ASABE*, 1187–1198.
[17] Nagata T, Lorz H and Widholm J M, 2009 *Biotechnology in Agriculture and Forestry* **February 2015**.
[18] Rahman S H A, Choudhury J P, Ahmad A L and Kamaruddin A H 2007 *Bioresource Technology* **98(3)**, 554–559.
[19] Rahman S H A, Choudhury J P, and Ahmad A L 2006 *Biochemical Engineering Journal* **30**, 97-103
[20] Romli M, Suprihatin, Indrasti N S and Aryanto A Y *Prosiding Seminar Tjipto Utomo Institut Tekhnologi Nasional* **October 2010**, B71-B7-9.
[21] Ruiz H A, Thomsen M H and Trajano H L 2017 *Hydrothermal Processing in Biorefineries: Production of Bioethanol and High Added-Value Compounds of Second and Third Generation Biomass*, 1–511.
[22] Saha B C and Bothast R J 1999 *Applied Biochemistry and Biotechnology - Part A Enzyme Engineering and Biotechnology* **76(2)**, 65–77.
[23] Scheller H V and Ulvskov P 2010 *Annual Review of Plant Biology* **61**, 263–289.
[24] Studer M H, DeMartini J D, Davis M F, Sykes R W, Davison B, Keller M and Wyman C E 2011

*Proceedings of the National Academy of Sciences of the United States of America* **108(15)**, 6300–6305.

[25] Tada K, Horiuchi J I, Kanno T and Kobayashi M 2004 *Journal of Bioscience and Bioengineering* **98(3)**, 228–230.

[26] Vargas, Betancur G J and Pereira N 2010 *Electronic Journal of Biotechnology* **13(3)**, 1–9.

[27] Venkateswar R L, Goli J K, Gentela J and Koti S 2015 *Bioresource Technology* **213** (February 2018), 299–310.

[28] Wyman C E, Yang B 2017 *Combined Severity Factor for Predicting Sugar Recovery in Acid-Catalyzed Pretreatment Followed by Enzymatic Hydrolysis.* 161-180.

# A Study of Producing Natural Red Color on Ikat Weaving Threads

**D P Dala Ngapa[1], Y Daud[1], and A C Sabuna[1]**

[1] Department of Biology Education, Universitas Kristen Artha Wacana, Kupang 85227 Indonesia

*Corresponding Author's E-mail: alan.sabuna@gmail.com

**Abstract**. The aims of this study are to find out what materials that used to produce red color on ikat weaving threads; what factors affect the making of red color on ikat weaving threads; the process of plants processing that used as a natural red colorant for the thread; and the kinds of red color produced from natural colorant plants. The samples were taken from the Kupang Regency area consists of three villages that are: Oesao, Sahraen, and Apraen village. The sample that took from three villages is Noni root, teak leaves, betel nut, and cherry fruit as the additional sample that was obtained from Dutalia Supermarket in Kupang city. The method that used in this study is the qualitative method with documenting technique and experiment. This study was done from January to March in the Biology Laboratorium of Universitas Kristen Artha Wacana, Kupang. The result of this study was found 11 red colors that produced from 4 kinds of plant that used for coloring the jark they are Noni root produced 4 colors (deep maroon, strawberry red, cardinal red, and terraccota. Cherry fruit produced 5 colors (coral pink, powder pink, hot pink, baby pink, and watermelon pink). Teak leaves produced the red-violet color and the last is Betel nut produced 2 colors (rust brown).

## 1. Introduction

The dye is an indispensable substance that gives additional artistic value and is used in varying a product (Jos et al, 2011). The dye is one of the main attractions that are more often used as the main colorant agent of the woven threads. Coloring of the weaving threads is usually done using synthetic dyes and natural dyes. Synthetic dyes are dyes derived from chemicals. While natural dyes are colors that can be produced from various types of plants that produce natural dyes that can be obtained from their parts such as leaves, bark, fruit skins, seeds, roots, and flowers that have gone through several processes, namely boiling, burning, pounding and directly used (Berlin et al., 2017).

Currently, the use of synthetic dyes began to replace the use of natural dyes, because the process was much easier and the resulting colors could be more diversified. The advantages of synthetic dyes when compared to natural dyes include fixed color composition, broader color selection, ease of use, brighter resulting color, available for all types of fibers and generally have better resistant to color fading (Pujilestari, 2015). According to Rymbai et al. (2011) colors derived from mineral derivatives (potassium dichromate and copper sulfate) can cause serious health problems and have a harmful effect on the environment. Considering the effect of synthetic dyes, it is very important to develop natural dyes, because they are more environmentally friendly and do not pollute the environment.

Most flora or plant species produce dyes that can be used in the textiles industry. According to Hariyanto (2017), there are 28 types of plants from 19 families that have been used as natural dyes. This shows that there are lots of plants that can be utilized as natural dyes in yarn, therefore, reducing the need to use synthetic dyes.

Some color-producing plants can produce various colors, one of which is the natural red color-producing plant for the weaving thread. So far, people have always thought that natural red dye only has limited and less varied colors, one of the colors that have a variety of colors is red. Red is one of the primary colors or base colors. Primary colors are colors that already exist by themselves, these colors can be combined and produce other derivative colors, made in various hues and shades of red by combining pure red with other colors (Meilani, 2013).

One example that shows how red can be combined with other colors is by adding white to red. However, if there is enough white color, it will turn to pink color. This is what proves that the color red can be varied, and there are several types of plants that can produce red. The absorption process of the dye in the fibers increases when mordant is added to the dye solution. Mordants also play an important role in the dyeing process. The mordants used in the coloring process here are Loba, alum, and lime. Mordant is also called a special substance that can increase the adhesion of various dyes to fabrics or threads. The mordant substance functions as a form of a chemical bridge between natural dyes and fibers so that the affinity (attraction) of the dye increases to the fiber, where metal salts will chemically bind the color carriers present in natural dyes and natural colors can be more easily react with the yarn fibers or it can facilitate the absorption of dye into the yarn fibers so that the resulting color can be more strongly absorbed and may produce different hues of colors.

The resulting color difference is influenced by the treatment of acids, alkalis or metal salts (Rosyida and Zulfiya, 2013). The alum mordant is used to produce threads that are light in color according to their original color, lime mordant produces threads with darker or darker colors. Meanwhile, Loba mordant is a mordant that is used as a special color extractor for noni roots. The aims of this study are a) to find out what materials that used to produced red color on ikat weaving threads. b) to find out what factors that effects the making of red color on ikat weaving threads. c) to find out the process of plants processing that used as natural red colorant for thread. d) to find out the kinds of red color that produced from natural colorant plants.

## 2. Research Method

The method used in this research is the qualitative method. The qualitative method is research on descriptive research and tends to rely on an analysis (Suryana, 2010). This research was conducted for 2 months starting from January 8 to March 5 2019 at the Biology Laboratory of Universitas Kristen Artha Wacana, Kupang. The work procedure in this study was carried out in several stages: 1) Sampling of plants used as the natural red dye was taken in the Kupang Regency which consists of 3 villages, namely Oesao, Sahraen, and Apren Village. 2) Preparation of tools and materials, before carrying out a color test at the laboratory, all equipment used in research must be clean. 3) The process of releasing the wax substance on the cotton yarn, the yarn is boiled for one hour so that the wax contained in the yarn fibers can be released and the yarn can absorb the dye evenly. 4) The dyeing process, which consists of two processes, the first stage of dyeing cotton threads using simultaneous mordant and the second stage of dyeing cotton threads using the preliminary mordant or pre-mordanting.

### *2.1. Data analysis*

The data analysis in this study used a qualitative descriptive analysis technique. There are three data collection techniques in this study, documentation, experiments and literature study.

*2.1.1. Documentation.* At the documentation stage, the researcher takes a picture of the sample that used in the yarn coloring, the dyeing process stage, to the resulting red color.

*2.1.2. Experiment.* Experiment is the process of taking the obtained plant specimens to the laboratory to have their color extracted and carry out color tests on the yarn.

IOP Conf. Series: Materials Science and Engineering 1143 (2021) 012023 doi:10.1088/1757-899X/1143/1/012023

*2.1.3. Literature study.* Methods of collecting data through printed and online media, such as books, color catalogs, magazines, and websites.

## 3. Results and discussion

First, the wax is removed from the cotton thread that will be used with natural dyes. This process is carried out so that the yarn could easily absorb the color in the dyeing process. The process of wax removal from the fabric / yarn is done by boiling it in water for about 1 hour so that the wax coating on the yarn fibers could come off and therefore enable the yarn fibers to easily absorb the natural dyes. In the process of dyeing the weaving threads using natural dyes derived from plants, each color produced is obtained from different methods and materials, be it the main ingredient (plant), the measurement of each main ingredient and, additional ingredients (solvent and mordant).

**Table 1.** Plant types, ingredients (plant) measurements and resulting red color in dyeing weaving thread.

| No | Plant | Ingredient (gram) | Parts | Water (ml) | Mordant | Boiling duration | Soaking duration | Resulting color | pH |
|---|---|---|---|---|---|---|---|---|---|
| 1 | Noni | Thread 7gram | Root 110 gram | Distilled water (ac) 250 ml | Loba 5 gram | 8 Minutes | 6 Days | Deep Maroon | 7 |
| 2 | Noni | Thread 7gram | Root 70 gram | Aquades 350 ml | Loba 5 gram | 10 Minutes | 4 Days | Strawberry red | 7 |
| 3 | Noni | Thread 7gram | Root 117 gram | Distilled water (ac) 500 ml | Loba 5 gram | 12 Minutes | 6 Days | Deep Maroon | 7 |
| 4 | Noni | Thread 7gram | Root 100gram | Aquades 250 ml | Alum 7 gram | 8 Minutes | 6 Days | Deep Maroon | 7 |
| 5 | Noni | Thread 7gram | Root 120gram | Distilled water (ac) 350 ml | Alum 7 gram | 10 Minutes | 6 Days | Cardinal Red | 7 |
| 6 | Noni | Thread 7gram | Root 105 gram | Distilled water (ac) 500 ml | Alum 7 gram | 12 Minutes | 4 Days | Terracotta | 7 |
| 7 | Noni | Thread 7gram | Root 100 gram | Aquades 250 ml | Lime 2 gram | 8 Minutes | 6 Days | Deep Maroon | 7 |
| 8 | Sweet Cherry | Thread 7gram | Fruit 34 gram | - | - | - | 3 Days | Coral pink (Pink with orange accent) | 3 |
| 9 | Sweet Cherry | Thread 7gram | Fruit 34 gram | Distilled water (ac) 50 ml | - | 5 Minutes | 3 Days | Coral pink (Pink with orange accent) | 3 |
| 10 | Sweet Cherry | Thread 7gram | Fruit 34 gram | Aquades 50 ml | Thread soaked in Alum Mordant | 4 Minutes | 3 Days | Powder pink | 3 |
| 11 | Sweet Cherry | Thread 7gram | Fruit 34 gram | Distilled water (ac) 70 ml | Thread soaked in Lime Mordant | 4 Minutes | 3 Days | Hot pink (Bright Pink) | 3 |
| 12 | Sweet Cherry | Thread7 gram | Fruit 69 gram | Aquades 10 ml | Alum 7 gram | - | 3 Days | Coral pink (Pink with orange accent) | 3 |
| 13 | Sweet Cherry | Thread 7 gram | Fruit 34 gram | Aquades 10 ml | Lime 2 gram | - | 3 Days | Baby pink (Bright Pink) | 3 |

| No | Plant | Ingredient (gram) | Parts | Water (ml) | Mordant | Boiling duration | Soaking duration | Resulting color | pH |
|---|---|---|---|---|---|---|---|---|---|
| 14 | Sweet Cherry | Thread 7 gram | Fruit 69 gram | Noni Root Extract 200 ml | - | 5 Minutes | 5 Days | Water melon | 7 |
| 15 | Teak | Thread 7 gram | Leaf 138 gram | Distilled water (ac) 1 liter | - | 15 Minutes | Soaked for 4 days in teak leaf extract, 1 week soaked in Noni extract | Red violet | 7 |
| 16 | Betel Nut and Betel | Thread 7 gram | Betel Nut 30 gram , and Betel 8 gram | Aquades 250 ml | Lime 2 gram | 8 Minutes | 6 Days | Rust brown (Brownish Red) | 7 |
| 17 | Betel Nut and Betel | Thread 7 gram | Betel Nut 70 gram , and betel 8 gram | Distilled water (ac) 500 ml | Lime 3 gram | 12 Minutes | 6 Days | Rust brown (Brownish Red) | 7 |

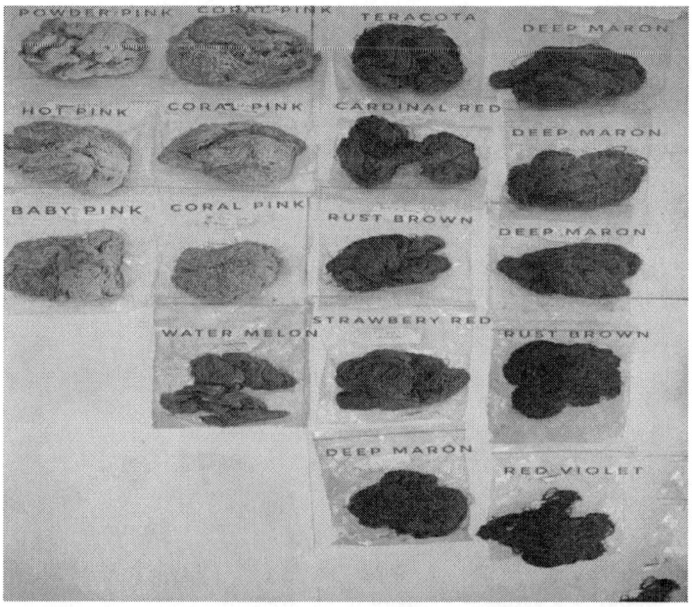


**Figure 1**. Different types of red colour (Doc. Dala Ngapa, 2019).

The materials used to produce a red color in the weaving thread or the carded cotton is generally used to make thick textured woven fabrics derived from natural fibers. According to Visalakshi and Jawaharlal (2013), water from taps or from nature contains a lot of minerals, such as iron and other substances that ultimately can affect the color extract. In order to avoid this color shift, distilled water and *Aquades* were used as both have been distilled and have only small amount of mineral content. This proves that distilled water and *Aquades* can be used as solvents in the coloring process. *Aquades* and distilled water were chosen as solvents in order to obtain a solution of color extract and can dye the yarn fibers well, detergent soap (Daia) is used to remove wax substances in the yarn fibers that were to be dyed, while mordant is used so that the dye can be absorbed more evenly into the yarn and can produce different colors. The resulting color difference is influenced by the treatment of acids, alkalis or metal salts (Rosyida and Zulfiya, 2013). The mordant used, (*Loba* skin, alum and lime) and the red color-producing plant organs used in the dyeing process, (noni root, teak leaves, betel nut and sweet cerry fruit). These plant organs are used as coloring agents in the dyeing process because the plant organs used contain anthocyanin pigments (red pigments).

The factors that influence the coloring process include the length of soaking time of the yarn in the color extract. According to Failisnur and Sofyan (2016), the longer the dyeing time of the fabric in the dye extract the darker (thick) the color density would be. This proves that the longer the soaking process, the darker or thicker the color will be. The yarn soaked for longer in the color extract, resulting in a darker color. pH is also one of the factors that can affect the color of the yarn. This is because anthocyanins are red dyes that are stable at low pH and neutral pH. Their stability will decrease if the pH is increased (Winarti and Firdaus, 2010). Therefore, pH paper is used to measure the acidity level of the color extract used to dye the yarn. Based on this theory, it was found that the pH contained in the color extract used in the coloring process, for the natural red color extract produced by noni roots, betel nut and teak leaves with the addition of *Loba* mordant, alum and lime have a pH value of 7 (neutral) and color extract from the fruit of cerry with the addition of mordant alum and lime has a pH value of 3, it can be concluded that the higher the pH value, the darker the red color and the lower the pH contained, the resulting color leads to a very light color or light pink.

## 4. Conclusion

Based on the results in the research, it can be concluded that:

1) Ingredients that were used to give red color to the yarn (carded cotton) are Aquades and distilled water as solvents. Detergent (Daia), Mordants (Loba skin, Alum and Lime) and natural red producing plants parts (noni root, teak leaves and sweet cherry fruit).
2) Factors that influenced the produced color is amount of pH and the soaking length.
3) The method of processing plants used as natural red dye in the weaving thread was done by these methods:
   a) Wax removal from the cotton thread
   b) Cotton thread coloring using the simultaneous mordant
   c) Cotton thread coloring using the preliminary mordant
4) Red color can be produced from naturally-producing red plants such as Noni root which can produce 4 different red color, deep maroon, strawberry red, terracotta, cardinal red. Cherry fruit can produce 5 red color, coral pink, powder pink, hot pink, baby pink, water melon. Whereas, teak leaves and betel nut can produce one red color, Red Violet and Rust Brown, respectively.

## References

[1] Jos B, Setyawan P E, and Satia Y 2011 Optimization of extraction and stability test of phycocyanin from microalgae (*Spirulina platensik*) *Teknik* **32** (3) 187-192
[2] Berlin S W, Linda R, and Mukarlina 2017 Utilization of plants as natural dyes by the dayak bidayuh tribe in kenaman village, sekayam district, sanggau regency *Protobiont* **6** (3) 303-309
[3] Pujilestari T 2015 Review : source and utilization of natural dyes for industrial use *Dinamika Kerajinan dan Batik* **32** (2) 93-106
[4] Rymbai H, Sharma R R, and Srivasta M 2011 Bio-colorants and its implications in health and food industry–a review *IJPRIF* **3** (4) 2228-44
[5] Hariyanto F 2017 *Study of plant diversity that has the potential as natural batik dyes in sidomulyo village, ampel district, boyolali regency* Thesis. Faculty of Teacher Training and Education. Universitas Muhammadiyah Surakarta.
[6] Meilani 2013 Color theory: the application of color circles in dress *Humaniora* **4** (1) 326-338
[7] Rosyida A, and Zulfiya A 2013 Coloring of textile materials using jackfruit wood extract and staining techniques to obtain optimal results *J. Rekayasa Proses* **7** (2) 52-58

[8] Suryana 2010 *Research methodology: a practical model of quantitative and qualitative research* Lecture textbooks. Universitas Pendidikan Indonesia.

[9] Visalakshi M, and Jawaharlal M 2013 Healthy hues- status and implication in industries- brief review *RRJAAS* **2** 42-51

[10] Failisnur and Sofyan 2016 Effect of temperature and time of dyeing cotton yarn on the natural coloring of gambier extract (*Incaria gambir* Roxb) *J. Litbang Industri* **6** (1) 25-37

[11] Winarti S, and Firdaus A 2010 Stability of red color rosella extract for food and beverage colorant *J. Teknologi Pertanian* **11** (2) 87-93

# Identification of Biomordant in Merbaun Village, West Amarasi District, Kupang Regency

**A K Taimenas[1], J Ngginak[1], and A C Sabuna[1]**

[1] Department of Biology Education, Universitas Kristen Artha Wacana, Kupang 85228 Indonesia

*Corresponding Author's E-mail: alan.sabuna@gmail.com

**Abstract**. East Nusa Tenggara (NTT) is a province that has cultural diversity. One of the cultures that are the pride of the people of East Nusa Tenggara is ikat weaving. In the observation stage related to this research, a small group of weaving craftsmen (*Runpah* and *Narwastu*) was found still use plants as natural dyes. The color produced from these plants is very distinctive, and besides using plant dyes, other ingredients are also added to prevent natural color fading. This study aims to determine the types of mordant plants, the used organs, the processing method, and the produced color by the craftsmen in Merbaun Village, West Amarasi District, Kupang Regency. The method used in this research is descriptive qualitative. The results of the research in Merbaun Village found 8 types of plants that were used as mordants, including Kepuh (*Sterculia foetida L*), Hibiscus (*Hibiscus rosa-sinensis L*), kesambi (*Schleichera oleosa L*), candlenut (*Aleurites moluccana (L.)* Willd, kapok randu alas (*Bombax ceiba L*)), pomegranate (*Punica granatum L.*), Loba (*Symplocos sp.*), and *Utaruna*, plant organs that are used, namely, fruit skins, leaves, bark, fruit and pulp. The processing process is burning, pounding, soaking, cooking and the resulting colors are yellow and red.

## 1. Introduction

East Nusa Tenggara (NTT) is a province that has cultural diversity. One of the cultures that are the pride of the people of East Nusa Tenggara is ikat weaving. Concerning the process of making woven fabrics, in principle, they use the available biological resources. Merbaun Village is one of the villages in West Amarasi Subdistrict, Kupang Regency, which so far still uses plant biological resources as the main ingredient in dyeing the weaving.

The observation phase related to this research found a small group of weaving craftsmen (*Runpah* and *Narwastu*) who still used plants as natural dyes. The plant parts used, such as noni root, tarum leaves, and turmeric rhizome, are used to produce natural colors. The use of dyed plants in the manufacture of woven fabrics means that the resulting color fades faster (Suirta et al. 2013). To maintain the color durability of the fabric, one alternative is to use mordant.

Mordant is a term for color enhancing or also as a special substance that can increase the adhesion of various dyes to fabrics. Mordant is divided into two, namely natural mordant and synthetic mordant. Plants that have the potential to become natural mordants are found in nature but are still underutilized. This type of mordant is harmless to the skin and is also environmentally friendly. To reduce the use of synthetic mordants, natural mordants are an alternative that can be used. According to Ramelawati *et al.* (2018), the use of natural mordant results in fairly bright colors, as well as good color evenness.

Craftsmen and the community already know about plants that can produce color and make use of mordant plants, but knowledge of Biomordant species has not been recorded properly. Starting from the description, it is necessary to research on: **"Identification of Biomordant in Merbaun Village, West Amarasi District, Kupang Regency".**

## 2. Research Method

The research was carried out from January to March 2019 in Merbaun Village, West Amarasi District, Kupang Regency, and the Biology Education Laboratory Universitas Kristen Artha Wacana. The tools used in the research include: the camera is used to take pictures of samples/types of plants obtained, knives/machetes are used to take plant organs to be used, stationery is used to record the results of interviews with respondents directly or field observations, mortal is used to smooth the samples, a hot plate is used to boil the sample, analytical scales are used to determine the appropriate sample weight required, a clock/stopwatch is used to control the time in each mordanting process, a stirrer/spatula is used in the mordanting process, a glass beaker is used for the mordanting process, sieve to filter the solution and stove to boil the sample. The materials used are cotton threads used for dyeing, water, and mordant plant organs taken from Merbaun Village, West Amarasi District, Kupang Regency.

The method used in this research is descriptive qualitative, the main objective of qualitative descriptive research is to describe systematically the facts and characteristics of the object and subject under study accurately. Sampling and data collection of potential mordant plants were carried out by exploration methods around the location of Merbaun Village, West Amarasi District, Kupang Regency, assisted by a group of craftsmen (*Runpah* and *Narwastu*) and people who have used plants as natural dyes and color generators (mordant). To find out the use of mordant plants, interviews were conducted. Interviews were intended to obtain data as information on plant species that have the potential for mordant. The data collected includes the name of the species, the part of the plant used.

## 3. Results and Discussion

### *3.1. The type of plant produced*

Based on the results of observations and interviews about mordant plants with informants in Merbaun Village, the observation results can be seen in the following table:

**Table 1.** *Biomordant* types / mordant plants.

| No. | Types of Species | | | |
|---|---|---|---|---|
| | General | Local | Scientific | Family |
| 1 | Kepuh | Nitas / nisfui | *Sterculia foetida* L. | Malvaceae |
| 2 | Hibiscus | Hibiscus | *Hibiscus rosa-sinensis* L. | Malvaceae |
| 3 | Kesambi | Kusambi / kulabe | *Schleichera oleosa* L. | Sapindaceae |
| 4 | Candlenut | Kemiri / fenu | *Aleurites moluccana* (L.) Willd | Euphorbiaceae |
| 5 | Kapok Randu alas | Kapok hutan / nekfui | *Bombax ceiba* L | Malvaceae and also Bombacaceae |
| 6 | White Pomegranate | Delima | *Punica granatum* L. | Punicaceae |
| 7 | Loba | Noba | *Symplocos* sp. | Symplocaceae |
| 8 | - | *Utaruna* | BT (Unidentified) | BT |

Based on table 1 above, there are 8 species found in Merbaun, namely from the family Malvaceae, Sapindaceae, Euphorbiaceae, Punicaceae, and Symplocaceae. In previous research in the East Sumba area, to be precise in Kaliuda Village, there were 3 types of plants used as mordants including Loba, Candlenut, and Dadap (Seran and Hana, 2018).

There is one type of flora that is found to enhance color, which has not been identified/has not been found the scientific name. This type of plant has characteristics: green leaves, serrated leaf edges, sharp leaf tips, stems reaching 2-5 meters, gray-white stems, growing freely in the forest. This plant is known to the public by the name *Utaruna*.

### 3.2. *Plant organs*

The plant organs used as mordant/*biomordant* obtained from interviews with craftsmen and communities in Merbaun Village are presented in Table 2.

**Table 2**. The organ of the mordant plant used.

| No. | Plant name | Scientific name | Organs used |
|---|---|---|---|
| 1 | Kepuh | *Sterculia foetida* L. | Rind |
| 2 | Hibiscus | *Hibiscus rosa-sinensis* L. | Leaf |
| 3 | Kesambi | *Schleichera oleosa* L. | Bark |
| 4 | Candlenut | *Aleurites moluccana* (L) Willd | Flesh of fruit |
| 5 | Kapok Randu alas | *Bombax ceiba* L. | Bark |
| 6 | White Pomegranate | *Punica granatum* L. | Fruit and leaves |
| 7 | Loba | *Symplocos* sp. | Bark |
| 8 | *Utaruna* | BT | Leaf |

Based on the table above, it can be seen that the organs used as mordants in kepuh plants are in the form of fruit skin/fruit shells. This kepuh fruit shell has a thick skin and a tapered tip. The level of maturity depends on the individual and the factors where it grows. Inside the kepuh fruit shell, there are shiny black seeds, which contain lots of oil so that they can be used to soften threads. There are also other benefits of kepuh seeds, namely as an industrial ingredient in the manufacture of shampoo, soap, paint, and plastics (Herdiana, 2005). The plant organ that is used from the porch is the bark. Kesambi can be used as a tanning agent, because according to the results of the study, in the skin of kesambi, 6.1-14.3% of tanning substances were found. Even in the past, the Balinese and Madurese used kesambi skin as a very effective skin medicine, especially against scurvy and other skin diseases (Bachli, 2007). However, Merbaun Village uses kesambi skin as a soaking thread. The plant organ used from candlenut is the fruit flesh. Candlenut pulp is generally used for cooking spices, medicines, raw materials for paint, soap, and cosmetics (Seran and Hana, 2018). In the handicraft of weaving in Merbaun Village, candlenut pulp is used as a mixture in coloring as a reinforcement and color preservative. In the village of Merbaun, craftsmen and the community use kapok skin as a soaking thread, to strengthen the color of the thread. The plant organs that are used from pomegranate are the fruit and leaves. Pomegranate leaf contains tannins because tannins are substances that are widely distributed in plants (Oci in Setiawati, 2014). The plant organs that are used from grease are dry bark. The inner grease of peppers (stems) contains a large amount of aluminum (up to 50% of the ash) which is a type of mordant substance that functions as a binding agent in dyeing yarn (Lemmens and Wulijarni-Soetjipto in Murniati and Takandjanji, 2015).

### 3.3. *Method of processing the mordant plant*

The methods of processing mordant plants are burning, pounding, soaking, and cooking. The flora found as a color enhancer by craftsmen and the Merbaun community, the initial processing method before entering the coloring stage as a whole is almost 100% done by pounding. Meanwhile, the dyeing process is carried out by soaking the threads and cooking / boiling the threads together with mordant and dye plants. Kuntorini (2005) states that traditional processing is one of the processes that is carried out by still maintaining its authenticity which has been passed down from generation to generation by the elders.

### *3.4. Color resulting from the the mordanting process*

**Table 3.** Colors resulting from the mordanting process.

| Types of color | Types of Dye Plants | Mordant | Mordanting stage | | | The resulting color |
|---|---|---|---|---|---|---|
| | | | 1 | 2 | 3 | |
| Yellow | Turmeric | Pomegranate and pomegranate leaves | | √ | | 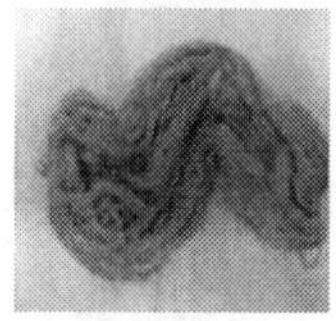 Turmeric + mordant pomegranate and pomegranate leaves |
| Red | Noni root | Loba, candlenut skin, Hibiscus flower, Utaruna, kesambi, randu alas, kepuh ash | √ | √ | | 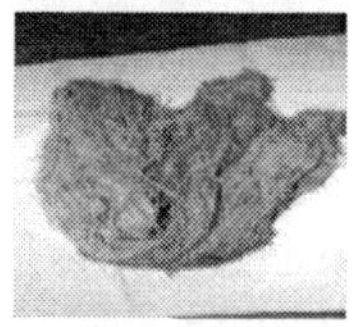 Soaking yarn with mordant (pre-mordanting) 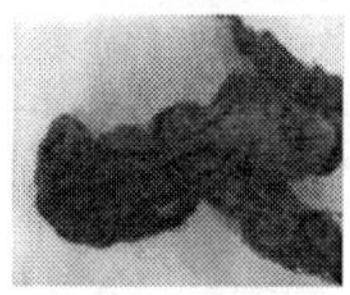 Yarn coloring (noni + loba) (meta-mordanting) |

The formation of the yellow color is required by types of plants, namely turmeric and pomegranate (leaves and fruit). The thread coloring stage is as follows: the crushed turmeric is weighed as much as 20 g, then the pomegranate is pounded as much as 12 g, and the ground pomegranate leaves weigh 12 g. After that, add 150 grams of water, then the crushed pomegranate leaves weigh 12 g. After that, add 150 g of water, then add the threads and cook together for 5 minutes, this process is called the meta-mordanting process, then remove the threads and then dry in the sun for 2 days. The resulting color is bright yellow. This is because turmeric contains chemical material curcumin, essential oils, resins, desmethoxycurcumin, oleoresin, and bidesmetoxicurcumin, resin, gum, fat, protein, calcium, phosphorus, and iron. However, turmeric contains active ingredients which are very useful in giving yellow color is curcumin (Shan and Iskandar, 2018), while pomegranate plants contain flavonoids. One of the flavonoid components from plants that can function as antioxidants is a natural dye called anthocyanin. The red color in pomegranates is caused by the high anthocyanin content in pomegranates. When turmeric and pomegranate are boiled together, the resulting color is bright yellow.

The formation of red color requires the types of mordant plants in the initial process of mordanting, namely kepuh ash, kesambi skin, randu alas skin, utaruna leaves, hibiscus leaves, candlenut pulp, and loba skin. The stages of coloring are as follows: dried kepuh fruit skin/shell is burned until the ash is left, then kepuh ash is weighed as much as 30 g, then the skin is cut into small pieces, then crushed using kepuh, then weighed as much as 12 g, skin kesambi cut into small pieces, then pounded using kepuh, then weighed as much as 12 g, then ground hibiscus leaves until smooth then weigh as much as 12 g, then the *utaruna* leaves that have been provided are pounded until smooth, then weighed 18 g. Next, separate the hazelnut fruit from the shell, then mash the hazelnut pulp to release the oil, then weigh 30 g, and don't forget to provide a thread that has weighed 7 g. After all the ingredients are ready, mix all the ingredients, then put the threads and add 400 ml of kepuh ash water. Soak the provided thread, then knead the thread so that the mordant plant mixture seeps into the thread. Leave the threads for 1 week, after 1 week removes the threads and dry them in the sun for 5 days. The color resulting from the soaking process is brownish. Then insert the thread and add 400 ml of kepuh ash water. Soak the provided thread, then knead the thread so that the mordant plant mixture seeps into the thread. Leave the threads for 1 week, after 1 week removes the threads and dry them in the sun for 5 days. The color resulting from the soaking process is brownish. Then insert the thread and add 400 ml of kepuh ash water. Soak the provided thread, then knead the thread so that the mordant plant mixture seeps into the thread. Leave the threads for 1 week, after 1 week removes the threads and dry them in the sun for 5 days. The color resulting from the soaking process is brownish.

These types of plants are used as soaking yarn, the soaking of yarn using all types of plants is intended so that before the yarn is colored the fibers are limp and open the pores so that they are easy to dye and weave (Rumeksa and Saftyaningsih, 2012). Subagiyo (2008) states that the mixed ingredients in coloring can increase the affinity of the dye molecules in the yarn fibers and chemically help diffuse molecules or red dyes to yarn fiber cells with the help of mordants.

The process of red coloring requires a type of noni root plant and greasy shell which acts as a mordant. The stages of staining the thread become red as follows: cut the noni root into small pieces, then mash until it is crushed, then weigh 300 g, then the dry grease skin is pounded until smooth, then weigh 20 g. After that, mix the noni root and loba bark into one, add 1000 ml of water, then insert the soaked thread into the mixture. Cook for 20 minutes, then remove the threads and dry in the sun for 1 week. The color that results from mixing dye and mordant plants is bright red.

Noni (*Morinda citrifolia*) contains morindin compounds (Lemmens and Wulijarni in Farida et al., 2015). Morindin is a disaccharide derivative of anthracenedione (anthraquinone) and has the molecular formula $C_{27}H_{30}O_{14}$ (molecular weight = 578). This compound produces a yellow color. While morindon, which is the result of hydrolysis of morindin glycosides and has a molecular formula of $C_{15}H_{10}O_8$ (molecular weight = 270), produces a red color (Hamid and Muhlis, 2005). Whereas for the grease skin containing large amounts of aluminum, this is the origin of *Symplocos'* work as a mordant (Nooteboom in Hanum, et al. 2012). The use of the *S. fasciculata* mordant is in conjunction with the coloring process, by pounding the dry skin then mixing it with a mixture of natural dyes.

Some things that need to be considered in its use as a natural color generator are when using noni (*Morinda citrifolia L*) for natural dyes without mordant *from S. fasciculata*, the red color will not appear. If the noni root is mixed with grease skin it will form the desired red color.

The mordant substance added to the dye plant functions to form a chemical bridge between natural dyes and fibers so that the affinity of the dye increases to the fiber. This shows that the mordant compound can bind colors so they don't fade easily (Fitrihana in Hanum, et al. 2012).

**Table 4.**Color Resulting from the Mordanting Process without Using Biomordants.

| No. | Dye Plants | Processing Method | Resulting Color |
|---|---|---|---|
| 1 | Turmeric | Pounded | 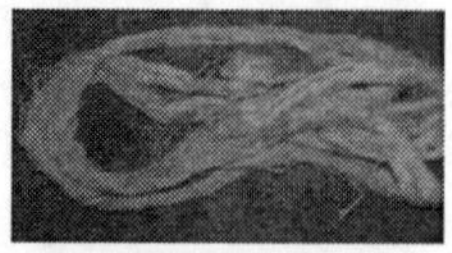 Pale yellow color |
| 2 | Noni | Pounded | 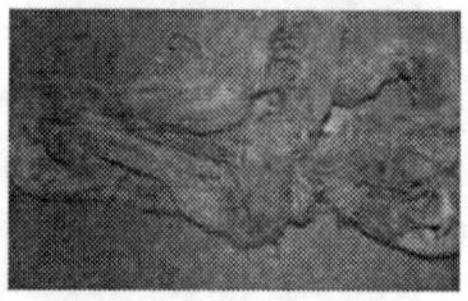 Pale red color |

Based on table 4 above, it can be explained that the color produced from the dyeing process without mordant mixture as shown in the table produces 2 types of colors, namely pale yellow and pale red.

*a.* *Pale Yellow Color*

The type of plant used to give a yellow color is turmeric. The stages of dyeing the threads are as follows: the turmeric that has been mashed is weighed as much as 12 g, then add 150 ml of water, then add the thread and boil it simultaneously with turmeric for 5 minutes, then remove the thread and dry in the sun for 1 day. The color produced after the yarn is dry is pale yellow and the texture of the thread is slightly coarse.

*b.* *Pale Red Color*

The coloring of the threads becomes red, using only the noni root, the resulting color is not good. The stages of dyeing the thread are as follows: cut the noni root bark into small pieces. Mash the noni roots using a mortar. Then weigh the noni root as much as 0.3 kg. Then add 1000 ml of water, then enter the provided thread and cook it together, for 20 minutes. Then remove the threads and dry the threads in the sun for 1 week, see the resulting color. The color resulting from the coloring process using the noni root is pale red.

## 4. Conclusion

*Biomordant* is a type of plant that has the potential as a reinforcement or color binder to the yarn. There are 8 types of biomordant found in Merbaun Village. Biomordant types include kepuh (*Sterculia foetida* L), hibiscus (*Hibiscus rosa-sinensis* L), kesambi (*Schleichera oleosa* L), candlenut (*Aleurites moluccana* (L.) Willd, kapok randu alas (*Bombax ceiba* L), pomegranate (*Punica granatum* L), loba (*Symplocos* sp.), and one plant species has not been identified, namely *utaruna*. The organs of mordant plants that are used are fruit skin/shells (kepuh), leaves (pomegranate, hibiscus, and *utaruna*), bark (kesambi, kapok randu alas, loba), and fruit (pomegranate). Biomordant type processing is burning, pounding, soaking, and cooking. The colors produced from the mordanting process are yellow and red.

## References

[1] Suirta I W, Widihati I A G, and Negara I P E S 2016 Utilization of merbau wood powder waste (*Instia spp.*) as coloring for cotton fabric with the addition of betel chalk *J. Kim* **10** (2) 175-80

[2] Ramelawati, Adriani, and Novrita S Z 2018 The effect of mordant alum and lime (*Citrus aurantifolia*) on the dyeing results of red onion extract (*Allium ascalonium* L.) in silk material *E-journal Home Economic and Tourism* **15** (2) 1-16

[3] Seran W and Hana Y W 2018 Identification of ikat weaving dye plants in kaliuda village, pahunga lodu district, east sumba regency *J. Agrikan* **11** (2) 1-8

[4] Herdiana N 2005 Cultivation potential of kepuh (*Sterculia foetida* Linn) *Proceedings of Research*

*Results of Baturaja Plantation Forests* 5 December 2005

[5] Bachli Y 2007 Kesambi plants and breeding fleas for prosperity *Agricultural Technology Research Center Bulletin* **1** (3)

[6] Setiawati R M 2014 *The effect of pomegranate (punica granatum linn.) composition variations on the physical properties of composite membranes to capture smoke-free radicals.* Thesis, Department of Physics, Faculty of Science and Technology, Universitas Islam Negeri Maulana Malik Ibrahim)

[7] Murniati M and Takandjandji M 2015 Utilization level of plant produce natural dye at cloth woven business in east sumba district *J. Penelitian Hutan Tanaman* **12** (1) 223-27

[8] Kuntorini 2005 Botani ekonomi suku zingiberaceae sebagai obat tradisional oleh masyarakat kotamadya banjarbaru *J. Bioscientiae* **2** (1) 25-36

[9] Shan C Y and Iskandar Y 2018 Study of chemical content and pharmacological activities of turmeric (*Curcuma longa* L.) *Farmaka* **16** (2)

[10] Rumeksa P N 2012 *Kapok fiber exploration (Ceiba pentandra) using ATBM and kempa weaving techniques* Thesis, Faculty of Art and Design. Institut Teknologi Bandung

[11] Subagiyo P Y 2008 *Traditional textiles, introduction to materials and techniques* Bekasi: Primastoria Studio

[12] Farida Atika V and Haerudin A 2015 The effect of pre mordant variations on batik coloring using noni roots (*Morinda citrifolia*) *Dinamika Kerajinan dan Batik* **32** (1) 1-8

[13] Hamid T S and Muhlis D 2005 Changes in physical and chemical properties of silk fabrics due to natural dyes of noni tree root bark *J. Teknologi* **19** (2) 163-170

[14] Hanum S F, Darma I D P and Sumerta I M d 2012 Utilization of loba trees (*Symplocos fasciculata* Zoll.) as natural color generators in weaving crafts in Pejeng Village, Tampak Siring, Gianyar, Bali *Berita Biologi* **11** (3) 367-372

# Cost-Benefit Analysis of Palm Kernel Shells as a Diesel Fuel Substitution for Hot-Mixed Asphalt

**Evi Gravitiani[1*], Nuri Resti Chayyani[1], Sunu Herwi Pranolo[2], Prabang Setyono[3], Ary Setyawan[4]**

[1]Postgraduate Program on Economic and Development Studies, Universitas Sebelas Maret, Surakarta, Indonesia
[2]Department of Chemical Engineering, Universitas Sebelas Maret, Surakarta, Indonesia
[3]Department of Civil Engineering, Universitas Sebelas Maret, Surakarta, Indonesia
[4]Department of Environmental Science, Universitas Sebelas Maret, Surakarta, Indonesia

*Corresponding Author's E-mail: evigravitiani_fe@staff.uns.ac.id

**Abstract**. Indonesia is one of the biggest oil palm producing countries. Solid waste from processing palm oil into crude palm oil (CPO) includes palm shells. This palm shell waste can be used as an alternative to diesel fuel for aggregate heating. This study aims to analyze the investment of the use of palm shells as an asphalt burner replacing diesel. Primary data were obtained from field and laboratory identification. In contrast, secondary data were obtained from relevant agencies, the Department of Public Development and Public Housing, PT Bara Energi Biomass and PT Damai Citra Mandiri. The method used is a quantitative descriptive analysis by calculating the annual costs and benefits over the machine's life, ten years. This study uses the Cost-Benefit Analysis. Four investment criteria are calculated: (1) Net Present Value (NPV), (2) Benefit-Cost Ratio (BCR), (3) Payback period (PbP), (4) Internal rate of return (IRR). The results found that palm shells' use as an asphalt burner for diesel substitutes showed high costs and top benefits. The calculation of economic feasibility with investment criteria shows that the NPV has a positive value, which means it is feasible. BCR shows a ratio of costs and benefits is more than one; it indicates that investment in asphalt production activities is feasible. The calculation of investment in replacing diesel fuel with palm shells shows a positive value of PbP since the first year. The internal rate of return shows that the investment value has met the minimum standard of general interest or replacement.
**Keywords**: palm shell, hot-mixed asphalt, aggregate heating, bioenergy, cost-benefit analysis

## 1. Introduction

There are two problems in utilizing natural resources: scarcity and dilemma in its utilization [4, 11]. The continuous use of non-renewable resources, such as fuel oil, can cause scarcity. Indonesia is a country that has excellent natural resource potential. Long-term energy management has become the government's concern, and various policies have been issued to optimize sustainable energy management [10]. One of these natural resources is infrastructure development, such as connecting roads between regions and provinces. So far, asphalt burners use non-renewable diesel fuel, so an alternative raw material for asphalt burners is needed. One alternative to substitute diesel fuel for asphalt burning machines is palm kernel shells. Indonesia is one of the largest palm oil-producing countries. Solid waste from processing palm oil into crude palm oil (CPO) includes palm kernel shells.

Indonesia has 1,688 companies engaged in the palm oil plantation industry. The larges palm oil plantations are found in 5 provinces, including Riau, West Kalimantan, North Sumatra, South Sumatra, and Central Kalimantan. The land area of Indonesia's oil palm plantations is 15.7 Million Hectares and is still growing. Based on the data, there is a potential palm shell of 10.4 million tons metric that can be utilized. During this time, palm kernel shells are used for energy and material mixtures [12]. As energy, palm kernel shells are used for dyer or broiler fuels. As a mixture of materials, palm shells are used as biomass because they contain ash important to the soil. The palm shell price is also still relatively low, which is around IDR 800,00per kg [5].

The abundance of palm kernel shells is then processed into fuel for burning asphalt [2,6]. Before undertaking an asphalt burner construction project, a cost-benefit analysis is necessary. The aim is to determine the number of costs required during the project and know the future benefits. Pranolo et al. [7] studied the use of palm shells for gas reduction emissions using a life-cycle analysis to reduce gas emissions instead of diesel fuel. Cost-benefit analysis has been done to calculate renewable energy, as in research [3], which measures electricity production's economic viability from small-scale renewable energy sources. Research on the cost-benefit analysis of using palm kernel shell as a replacement diesel fuel for asphalt burner has not been done much.

## 2. Methodology

Diesel is a fuel that cannot be renewed. Meanwhile, palm shells are fuel from nature and will be returned to nature to form sustainable natural management. The research uses a descriptive quantitative analysis method. The data used are primary data and secondary data. Primary data were obtained from field and laboratory observations, while secondary data were obtained from the Public Development and Public Housing Agency, PT Bara Energi, and related agencies. The analysis tool used is the cost-benefit analysis.

Cost-Benefit Analysis (CBA) calculates the overall costs and benefits used during economic life [9]. The economic age used in production activities using palm shells as an asphalt burner to replace diesel is 10 (ten) years. During the project's life, it is necessary to calculate the addition or reduction of the costs used.

The cost-benefit analysis uses four investment criteria. These investment criteria include Net Present Value (NPV), Benefit-cost ratio, Payback Period, and Internal rate of return.

### 2.1. *Net Present Value (NPV)*

Conducted to determine the difference between the present value of benefits and the flow of costs. The NPV calculation is

$$NPV = \sum_{t=1}^{n} \frac{B_t - C_t}{(1+i)^t} \tag{1}$$

Where

Bt : the benefits obtained per year
Ct : costs incurred each year
n : number of years
i : interest rate (discount)

Based on the NPV result, the investment eligibility criteria are if the NPV > 0 means that the project is declared profitable and feasible to be implemented. Suppose the NPV < 0; then the project does not produce the cost value used or is not feasible to implement. Moreover, if NPV = 0, then the project can return the initial capital, or the project will experience a balanced level of profit [1].

*2.2. Benefit-Cost Ratio*
To calculate this ratio, it is necessary first to calculate the net benefit each year. From this, we can see that there are years with positive net benefits (where the gross benefits are more significant than the costs), and there are years that have negative net benefits, where the gross benefits are smaller than the costs [1].

$$BCR = \frac{\Sigma DB}{\Sigma DC} \tag{2}$$

where
DB : Discounted benefit
Ct : Discounted cost

If the B/C ratio is > 1, then the project is feasible to run; the B/C ratio < 1, the project suffers a loss and is not feasible to run, and the B/C ratio is - 1, then the project is neither profitable nor loss [1].

*2.3. Payback Period*
The Payback Period is used to determine the payback period for an investment made.

$$Payback\ period = \frac{I}{A_b} \tag{3}$$

Where
I : the amount of investment required
Ab : the net benefit of the discount that can be obtained each year

A reasonable payback period is less than the defined project life.

*2.4. Internal Rate Of Return (IRR)*
The purpose of calculating the IRR is to determine whether an investment is profitable or whether it has met the minimum standard of interest or general return. Therefore it is often called the minimum attractive rate of return. The IRR method is also commonly called the "Economic Rate of Return" in economic analysis or the "Financial Rate of Return" in financial analysis. The IRR basis is which makes the Present Value of the net results as long as the project runs the same as zero. The method used is the trial and error method (trial and error system), namely by determining two or more interest rates, where the actual rate of interest is between the interest rates. By calculating the present worth (cash price), IRR can be calculated through interpolation [1].

$$IRR = i + \frac{NPV}{NPV - NPV'}(i' - i) \tag{4}$$

Where
i : Discount rate that results in a positive NPV
i' : Discount rate that results in a negative NPV
NPV : NPV is positive
NPV' : NPV is negative

## 3. Result and Discussion

### 3.1. Asphalt Mixing Plant Cost using Diesel Fuel

*3.1.1. Investment costs.* Investment costs are some costs incurred for business activities in goods with a certain benefit age. Investment costs used in asphalt burner and economy calculation activities with diesel fuel include 1 unit of a solar burner and ready-to-install solar tank worth IDR 220, 000, 000,-Tax for equipment purchased 10 percent worth IDR 22,000,000,-. The cost of making machine placement in the form of building a house of 10 M2 amounting to IDR10,000,000,-. The total investment cost of burning asphalt with diesel fuel is IDR 252,000,000,

*3.1.2. Production cost.* Production costs are the costs incurred to produce a product in the business. These costs include the cost of raw materials and fuel. The fuel used is a diesel with 16 litres/minute with a price per litter of IDR 9,000,-. The operating time is 8 hours a day, so the cost of Solar is IDR 69,120,000,-. The total cost of production is IDR 69,120,000.00

*3.1.3. Operational Cost.* Operational costs are associated with workmanship during businesses such as electricity and labour. Electricity needed for 25 Kwh with a price of IDR 1,467, - for 8 hours. The number of electricity costs IDR 293,400,- per day. Diesel fuel as an asphalt burner does not use human labour; it only uses tools. Total operating expenses are IDR 293,400,-

*3.1.4. Other costs.* Miserly or contingency costs must be added to the project costs, due to changes or errors in calculations. The total contingency cost stipulated is 5 percent of the total amount of IDR 16,070,670,-

**Table 1.** Cost summary of hot-mixed asphalt using diesel fuel.

| No | Cost | IDR | USD (1 USD = IDR 15,000) |
|---|---|---|---|
| 1 | Investment Costs | 252,000,000 | 16,800 |
| 2 | Production Costs | 69,120,000 | 4,608 |
| 3 | Operational Costs | 293,400 | 20 |
| 4 | Other Expenses | 16,070,670 | 1,071 |
| | Total | 337,463,400 | 22,499 |

Source: Primary data, 2020

Based on the calculation of these costs, to produce 384 Tons of HMA, it takes a fee of IDR 337,463,400.00. The production cost of 1-ton HMA using diesel fuel is IDR 878,810.9375 or USD 59.

### *3.2. Asphalt Mixing plant cost using Palm Kernell Shells*

The costs required for burning asphalt using palm kernel shells once consist of investment costs, production costs, operational costs, and taxes.

*3.2.1. Investment costs.* These costs include purchasing a gasifier BE-170 dryer heater for IDR 742,000,000, a 10 percent gasifier tax of IDR 74,200,000, and a single unit BE-76 asphalt boiler heater for IDR 247,000,000, along with a 10 percent tax of IDR 24,700,000. However, since the machine is assumed to be ten years old, the machine price and taxes are divided by ten so that the total cost of the device is IDR 108,790,000 per year. The casting of reinforced foundations, placing gasification machines and boiler heaters with a length of 16 meters, a width of 15 meters, a thickness of 15 $m^3$ and 5 $m^3$ of IDR 30,500,000, construction of a machine protective roof of IDR 25,000,000, a cooling water circulation tub of IDR 16,000,000, construction of a warehouse for storage of shells with an area of approximately 200 $m^2$, amounting to IDR 40,000,000, and delivery of machines from Surabaya to the Asphalt Mixing Plant location of IDR 30,000,000. The total investment cost of an asphalt burner using palm kernel shell fuel is IDR 1,229,400,000.

*3.2.2. Production costs.* Production costs include the cost of raw materials, namely palm kernel shells. The heating machine requires 50 kg/minute of palm shells in one production, which costs IDR 800 per kg. The determination of oil palm shells' price is influenced by the distance to the palm oil mill. The total production cost is IDR 19,200,000.

*3.2.3. Operational costs.* These costs consist of electricity, labour, water circulation, diesel fuel for the multitasking loader, maintenance of the shell storage, routine maintenance, and gearbox oil changes. Electricity for an asphalt burning machine with palm shell raw materials requires 45 KwH for one production day or 8 hours at the cost of IDR 528,120. Labour needs three people, with a production day cost of IDR 175,000. The workforce includes gasifier operators, boiler operators, and AMP operators. The total cost of labour is IDR 525,000. The water circulation in question is a cooling pool that requires the addition of 1 m3 of water. The water circulation fee is IDR 10,000 per hour. Engine maintenance costs include oil gearbox and minor damages estimated at IDR 48,000,000. The cost of shrinks and buildings is calculated using linear methods to vary from year to year. The total amount of operational and maintenance costs is IDR 1,484,648,800 per year.

*3.2.4. Other expenses.* Costs that must be added to project costs due to changes or errors in the calculation are 5 percent of the total investment costs, operating costs, and production costs IDR 16,257,156. The total contingency cost for oil palm shell fuel is IDR 463,863,078 in the first year.

Table 2 shows that the total investment costs, production costs, operational and maintenance costs, and taxes, an output of 384 tons per day of HMA, is IDR 287,400,276.00. Cost production for 1-*ton* HMA using palm kernel shells is IDR 748,438.21 or USD 50. The cost-benefit analysis calculation uses a cost per year data so that the figure is calculated to be per year, as shown in table 2.

**Table 2.** Cost summary of palm kernel shells use as fuel per day and per year.

| No | Cost | Cost per day (IDR) | USD (1 USD = IDR 15,000) | Cost Per Year |
|---|---|---|---|---|
| 1 | Investment Costs | 250,290,000 | 16,686 | 1,229,400,000 |
| 2 | Production Costs | 19,200,000 | 1,280 | 7,252,612,755 |
| 3 | Operational Costs | 1,653,120 | 110 | 1,484,648,800 |
| 4 | Other Expenses | 16,257,156 | 1,084 | 463,863,978 |
| | Total | **287,400,276.00** | 19,160 | **10,430,524,633** |

Source: Primary data, 2020

### *3.3. Benefits*

The analysis of the benefits of using palm kernel shells as an alternative to asphalt burners is divided into direct and indirect benefits. Direct benefits are benefits whose amounts are immediately known due to production activities. Meanwhile, indirect benefits are other benefits received outside of production activities.

*3.3.1. Direct benefit analysis.* Direct benefits due to palm kernel shells as raw material for asphalt burners are obtained from selling Hot Mix Asphalt (HMA) and cost savings due to changes in raw material costs from palm kernel shells.

- *Hot Mix Asphalt Sales.*
  The output produced by the gasifier engine during one shift or for eight operating hours is 384 tons of HMA. The selling price of 1 ton of HMA varies depending on the type, including Hot mix AC BC, Hot mix AC WC, Hot mix III Laston, Hot mix III Laston special, hot mix HRS, and Hot mix Sand Sheet. The average price for this type of hot mix is IDR 1,151,667. If the output produced is 384 tons, then the benefits received in one production shift for 8 hours are IDR 442,240,000 per day.
- *Fuel Savings.*
  Apart from the sale of hot mix, the direct benefit received from the change in fuel from diesel to palm kernel shells is fuel savings. The price of diesel is lower than the price of palm kernel shells. The savings made are the amount of diesel needed minus the price of the required palm shells. Total observations made are IDR 45,600,000 per day.

IOP Conf. Series: Materials Science and Engineering 1143 (2021) 012025 doi:10.1088/1757-899X/1143/1/012025

*3.3.2. Indirect benefit analysis.* The indirect benefits that are felt are reduced emissions and savings in foreign exchange.

- *Emission Reduction.*
  The indirect benefit that can be felt from using palm kernel shells for engine fuel is the resulting reduction in emissions [8]. The assumption used is that the asphalt burning machine operates for eight (8) hours in one day. When using diesel fuel, the resulting emission is 13.3632 tons CO2Eq per day, but using combustion biomass gasification using palm kernel shells, the resulting emissions are 7,968 tons of CO2Eq per day. The consequent reduction in emissions is the difference from this value. Combustion using palm kernel shell fuel has succeeded in reducing emissions by 5.3952 tons of CO2Eq per day. The assumed price of 1 ton CO2 Eq is the US $ 5, and the rupiah assumption is IDR 14,000. The emission reduction that has been done is IDR 70,000 per ton. Then the total emission reduction benefit is IDR 377,664 per day or IDR 90,639,360 per year.
- *Foreign Exchange.*
  The use of palm shells as a substitute for diesel fuel can save the country's foreign exchange because 80 % of the gasification equipment material is obtained from local materials. As much as 20 % is still imported. The equipment purchased was a Gasifier BE-170 Heating Dryer for IDR 742,000,000.00 and a BE-76 Asphalt Kettle Heater for IDR 247,000,000. The total purchase plus 10% VAT is IDR 1,087,900,000.00. Calculation by Domestic Component Level. Then there was a saving of IDR 870,320,000 in foreign exchange.

**Table 3.** Total Benefits per year.

| No | BENEFITS | PRODUCTION/ DAY | UNIT | IDR/TON | BENEFITS IDR/DAY | TOTAL BENEFITS IDR/YEAR |
|---|---|---|---|---|---|---|
| **A** | **Direct Benefits** | | | | | |
| 1 | Hot mix Sales | 384 | ton | 1,151,667 | 442,240,000 | 106,137,600,000 |
| 2 | Fuel Savings | | | | 49,920,000 | 11,980,800,000 |
| | Subtotal A | | | | | **118,118,400,000** |
| **B** | **Indirect Benefits** | | | | | |
| 3 | Emission Reduction | 5.3952 | ton CO2eq | 70,000 | 377,664 | 90,639,360 |
| 4 | Foreign Exchange | | | | | 870,320,000 |
| | Subtotal B | | | | | **960,959,360** |
| | **TOTAL BENEFIT A+B** | | | | | **119,079,359,360** |

Source: Primary data, 2020

Based on Table 3, the total benefits both directly and indirectly in hot mix production using palm kernel shell fuel for one year amounted to IDR 119,079,359,360.

*3.4. Cost-Benefit Analysis*

After obtaining the total costs and benefits for each year, the benefit costs are calculated. Table 3 shows the total costs and benefits.

**Table 4.** Costs and Benefits per year of palm kernel shells use as a substitute for diesel fuel.

| Year | Cost | Benefit |
|---|---|---|
| 1 | 10,430,524,633 | 119,079,359,360 |
| 2 | 9,059,895,133 | 107,595,279,360 |
| 3 | 8,961,415,633 | 96,981,519,360 |
| 4 | 8,831,436,133 | 86,367,759,360 |
| 5 | 8,732,956,633 | 75,753,999,360 |
| 6 | 8,602,977,133 | 65,140,239,360 |
| 7 | 8,504,497,633 | 54,526,479,360 |
| 8 | 8,374,518,133 | 43,912,719,360 |
| 9 | 8,260,288,633 | 33,298,959,360 |
| 10 | 8,146,059,133 | 22,685,199,360 |

Source: Primary data, 2020

In Table 4, in the cost column, it can be seen that there is a fluctuation in the total value per year because there are additional and subtraction costs for operational costs. The operating costs are in the form of machine maintenance every three years.

*3.4.1. Net Present Value.* An essential step in determining the NPV is that both the benefits and costs each year must be calculated in cash price, then compared. Each cost and benefit must be found in real value in the initial year (year 0) or discounted. A project is said to be a "go project" if the NPV value = 0, especially in economic analysis. Meanwhile, for financial analysis, NPV should be> 0.

NPV = $\sum$ (Net benefit × Discount Factor)
NPV = 429,243,328,916.56

The calculation result shows that the net present value is positive or NPV> 0. It indicates that the replacement of fuel from diesel fuel to palm kernel shells is feasible. NpV results are positive during the project period (10 years); the initial capital/investment invested can generate IDR429,243,328,916.56 making it very profitable.

*3.4.2. Benefit-Cost Ratio.* The next investment criterion is the Benefit-Cost Ratio (BCR), which aims to determine the cost and benefit ratio. Business activity is feasible if the gross B/C ratio of the project is ≥ 1.

$$BCR = \frac{\Sigma DB}{\Sigma DC}$$

$$BCR = \frac{484.171.363.572}{54{,}928{,}034{,}655}$$

BCR = 8.81 > 1

A BCR of more than 1 means that the project to replace diesel fuel with palm kernel shells is feasible to run. B-C ratio compares the current amount of value on benefit flows, and cost flows based on the cost of capital (OCC) opportunities. The OCC is the level of return if capital is invested in the best and easiest possible. B-C ratio is > 1, so the investment in the use of palm shells as a solar substitute for asphalt burners is feasible. The payback period is a calculation used to determine which year the NPV can show a positive value. The calculation of changing diesel fuel to palm kernel shells for asphalt burners has had a positive NPV since the first year.

*3.4.3. Payback period.* PbP analysis is used to determine the length of time it takes for the initial capital return and project time. Results show PBP 1 year or less of the project's life, which means that the project has been profitable since the first year the investment was run.

*3.4.4. Internal rate of return.* The internal rate of return is an indicator of the level of profit stated in the percentage. In this calculation, the IRR is more than 100% because NPV has a significant value.

## 4. Conclusion

Based on the calculations that have been done, the conclusions that can be obtained are as follows

a. The cost comparison of diesel fuel and palm shell is quite significant. Production costs for 1-ton HMA are USD 59 for diesel fuel and USD 50 for palm shell fuel. Thus, the difference is USD 9 or equivalent to IDR 120,000 – IDR 150,000.
b. The cost used to produce Hot Mix Asphalt in one year produces an output of 384 tons per day using oil palm shell as fuel is IDR 10,430,524,633. These costs consist of investment costs, production costs, operational and maintenance costs, and taxes. The benefits to be obtained either directly or indirectly in the hot mix production per year are IDR 118,042,559,360. The immediate benefits consist of selling hot mix and saving on raw materials. Meanwhile, indirect benefits consist of reduced emissions and savings in foreign exchange.
c. The benefit-cost analysis shows that NPV> 0, Benefit-Cost Ratio> 1, the payback period is less than the project life, and the IRR is less than the prevailing interest rate. It shows that the project using palm kernel shells as an asphalt burner to replace diesel fuel can be implemented in terms of benefits and costs.

## Acknowledgment

The authors would like to thank:

a. The Indonesian Oil Palm Plantation Fund Management Agency for funding this study (Contract Number: PRJ-89/DPKS/2018)
b. PT Bara Energi Biomas in Tangerang as an industrial partner for fabrication and installation of the gasification unit
c. PT Damai Citra Mandiri in Pontianak as hot mixed asphalt producer.

## References

[1] Rahardjo M 2010 *Studi Kelayakan Proyek* Cakra books, Surakarta
[2] Asadullah M, Adi A M, Suhada N, Malek N H, Saringat M I, and Azdarpour A 2014 Optimization of palm kernel shell torrefaction to produce energy densified bio-coal *Energy Conversion and Management* **88** 1086–1093
[3] Atănăsoae P, Pentiuc R D, Milici D L, Olariu E D, and Poienar M 2019 The Cost-Benefit Analysis of the Electricity Production from Small Scale Renewable Energy Sources in the Conditions of Romania *Procedia Manufacturing* **32** 385–389
[4] Barbier E B 2013 *Economics, Natural-Resource Scarcity, and Development (Routledge Revivals): Conventional and Alternative Views* Routledge
[5] Olutoge F A 2010 Investigations on Sawdust and Palm Kernel Shells As Aggregate Replacement *ARPN Journal of Engineering and Applied Sciences* **5** (4) 7–13
[6] Osei Bonsu B, Takase M, and Mantey J 2020 Preparation of charcoal briquette from palm kernel shells: a case study in Ghana *Heliyon* **6** (10) e05266
[7] Pranolo S H, Setyono P, and Fauzi M A 2020 Life cycle analysis of palm kernel shell gasification for supplying heat to an asphalt mixing plant *Waste Disposal & Sustainable Energy* **2** (1) 55–63
[8] Rashidi N A, and Yusup S 2017 Potential of palm kernel shells as activated carbon precursors through a single-stage activation technique for carbon dioxide adsorption *Journal of Cleaner Production* **168** 474–486

[9] Santoso L W, and Wirawan A 2015 *Analisis Investasi Sistem Informasi Akademik pada Universitas X dengan Metode Cost Benefit Analysis (CBA)* Petra Christian University
[10] Sugiyono A 2014 Permasalahan dan kebijakan energi saat ini *Prosiding Peluncuran Buku Outlook Energi Indonesia 2014 & Seminar Bersama BPPT Dan BKK-PI. Jakarta: Badan Pengkajian Dan Penerapan Teknologi* 9–16
[11] Suparmoko M, Hendarto E, Setyarko Y, and Widyantara G 2003 *Valuasi Ekonomi Sumberdaya Alam Kabupaten Sikka*
[12] Thiagarajan J, Srividhya P K, and Balasubramanian P 2018 Thermal behaviour and pyrolytic kinetics of palm kernel shells and Indian lignite coal at various blending ratios *Bioresource Technology Reports* 4 (July) 88–95

IOP Conf. Series: Materials Science and Engineering 1143 (2021) 012026 doi:10.1088/1757-899X/1143/1/012026

# Cogasification performance of deashed coal with various biomass

**Jenny Rizkiana[1,2,3]*, Sandy Fajar Maulana[1], Ghiffary Azka Nur Aulia[1], Nasywa Kamilah[1], Reyhan Fitri Ananda[1], Naafi Fauzan Mahendraputra Sastrohutomo[1], Edo Muhammad Nashirudin Latif[1], Winny Wulandari[1], Dwiwahju Sasongko[1]**

[1] Department of Chemical Engineering, Faculty of Industrial Technology, Institut Teknologi Bandung, Bandung 40132, Indonesia
[2] Department of Bioenergy Engineering and Chemurgy, Faculty of Industrial Technology, Institut Teknologi Bandung, Bandung 40132, Indonesia
[3] Center for Catalysis and Reaction Engineering, Institut Teknologi Bandung, Bandung 40132, Indonesia

*Email: jr@che.itb.ac.id

**Abstract.** Cogasification of lignite and EPFB is present as a production of energy sources that aim to get a synergy effect. However, not all cogasification gives better product quality than fuel parent gasification. In this study, demineralization acted as lignite pre-treatment stage is achieved through HF leaching at 70°C for 4 hours. The mixing ratio varied by lignite to empty palm fruit bunches (EPFB) are 1: 2, 1: 1, and 2: 1 both for non-demineralized lignite (BLM) and lignite demineralized (BLD). The negative synergy effect of two combination is caused by the high content of silica and alumina from the empty palm fruit bunches (EPFB). The lignite demineralization cogasification produces a lower $H_2$ yield than BLM because it also dissolves the mineral content which functions as a catalyst for cogasification synthesis.

## 1. Introduction

During 2006 to 2016, world energy consumption on average increased by 1.8% per year. World energy consumption reached 13,105 million tons in 2015 and exceeded 13,276.3 million tons in 2016 while the increase in Indonesia was 164.8 million tons and 175 million tons [1] respectively. Until 2015, world coal consumption showed a rising monotonous curve. However, in 2016 there was a decline in the use of coal fuel due to the enactment of rules on reducing $CO_2$ emissions to reduce the adverse effects of the environment. Of course, the use of environmentally friendly technology in processing coal into primary energy sources is very important and urgent to implement. This leads to the development of sustainable renewable energy technologies in realizing energy security and availability, one of which is biomass [2-3]. Indonesia has great potential renewable natural resources, such as palm oil that produces abundant waste, empty palm fruit bunches is the biggest one. Biomass is an environmentally friendly energy source that supplies 14% of the world's energy [4].

Gasification is a method of converting low rank coal that allows maximum energy extraction with a minimum amount of residue and high carbon conversion [2,5]. Lignite coal has an inhibition effect that is less desirable for processing even though its availability in nature is very large [6]. The disadvantages of biomass gasification are low energy density and high tar production so that the product needs to be treated. The shortcomings and advantages of coal and biomass gasification lead to the solution of cogasification of coal with biomass which combines high coal energy density and low biomass emissions. When biomass is involved in coal gasification, the rate of production of hydrogen gas signifies the quality of synthesis gas. This occurs due to the presence of alkali and alkaline earth (AAEM) [7]. The presence of AAEM has a synergistic effect on the cogasification process by increasing the overall gasification rate [8]. AAEM in biomass, especially potassium, is a good promoter in the gasification process [9]. Potassium in biomass ash acts as a catalyst to increase coal reactivity and suppress tar formation [10]. Cogasification studies of coal with various types of biomass have been studied. The effects of alkali and alkaline earth synergies in the cogasification of low rank coal were obtained from switch grass [11], straw [12], sawdust [7, 13, 14], corn stalks [15], pine powder [16], wheat powder, and wood waste [17].

The greater the AAEM content in biomass ash, the stronger the catalytic influence. Each type of biomass has different AAEM content. The AAEM content in biomass ash which improves the quality of coal gasification can be deactivated by silica [18]. Therefore, pre-treatment is carried out in the form of coal demineralization through acid leaching such as $HNO_3$, HF, and HCl. Karaca and Ceylan [19] examined that $H_2O_2$ solutions were able to reduce mineral content in coal by up to 70%. Ishaq et al. [20] found that demineralization process of coal with HCl and $HNO_3$ gave 87% mineral content decrease. Demineralization with HF and $HNO_3$ mixture, Kizgut et al. (2006) [21] explained that the acid leaching method was able to demineralize coal up to 0.12-0.41%-wt. The research conducted by Hacifazlioglu et al. [22], Steel et al. [23], and Vaccaro [24] state that in terms of decreasing mineral content in coal, HCl and HF solutions have very good performance. The dissolution of coal by HF at a temperature of 65°C has been shown to eliminate Al and Si so that it can reduce the yield of ash from 5.3% to 1.37%-wt. Unlike HF, HCl can only dissolve simple compounds such as carbonate and phosphate [23]. Based on those findings, it is interesting to further study the effect of acid leaching to the coal properties and and its gasification performance. The purpose of the present study is to examine the effect of pretreatment of HF leaching and mixing ratio on cogasification process of low rank coal with biomass.

## 2. Experiment

### *2.1. Materials*

Coal and biomass were crushed into sizes smaller than 232μm (48 mesh) and then dried in the oven at 110°C for 14 hours. Determination of ash content in biomass and coal was done through combustion in the furnace at 750°C for 2 hours. Elemental contents in ash were analyzed using EDS while the composition of coal and biomass was determined through proximate and ultimate analysis. The comparative analyses of non-demineralized lignite (BLM), lignite demineralized (BLD), and empty palm fruit bunches (EPFB) are shown in Table 1 and Table 2.

### *2.2. Coal demineralization by HF leaching method*

The 10g of dried coal was dissolved in 250 ml HF solution in Teflon glass. The solution was stirred for 4 hours at a temperature of 70°C. The solution was then neutralized by aqua dm. The filtered solid was dried in an oven for 14 hours at 110°C and then stored in a desiccator.

### *2.3. Lignit EPFB cogasification*

The cogasification process was performed in the fixed bed downdraft reactor for 80 minutes using steam at 90-92°C. The total mass of coal and biomass was 3 grams. The carrier gas (nitrogen) rate is 85 $cm^3$/minute. The furnace used is heated with a heating rate of 6°C/minute until it reached 700°C. The remainder solids after cogasification process was considered as the total mass of charcoal and ash. The gas product was then condensed with a temperature of 14-16°C cooling water. Some gases condensed to form tar were were collected in condenser bottle while the non-condensed gas would pass through and collected in a gas bag for every 10 minutes.

**Table 1.** Proximate and ultimate analysis of the samples.

| Parameter | BLM | BLD | EPFB |
|---|---|---|---|
| **Proximate** | | | |
| Ash | 9.54 | 1.63 | 8.97 |
| Volatile Matter | 47.38 | 51.53 | 72.18 |
| Fixed Carbon | 43.08 | 46.84 | 18.85 |
| **Ultimate** | | | |
| Sulfur | 1.07 | 1.07 | 0.14 |
| Carbon | 64.38 | 64.38 | 46.19 |
| Hydrogen | 3.74 | 3.74 | 5.52 |
| Nitrogen | 1.14 | 1.14 | 0.78 |
| Oxygen | 19.71 | 19.71 | 38.41 |

**Table 2.** Element content of the samples.

| Element (%-wt dry basis) | C | O | Na | Mg | Al | Si | S | K | Ca | P | Fe |
|---|---|---|---|---|---|---|---|---|---|---|---|
| **BLM** | 3.13 | 4.34 | 0.26 | 0.2 | 0.41 | 0.37 | 0.07 | 0.02 | 0.02 | 0.05 | 0.67 |
| **BLD** | 0.66 | 0.67 | 0.04 | 0.04 | 0.05 | 0.03 | 0.02 | 0.002 | 0.004 | 0.01 | 0.11 |
| **EPFB** | 3.34 | 3.8 | 0.24 | 0.27 | 0.27 | 0.25 | 0.07 | 0.04 | 0.03 | 0.05 | 0.59 |

### *2.4. Lignite EPFB cogasification*

The data obtained from the preparation stage, coal demineralization, and the cogasification process were processed, analyzed, and then presented in the graph. Ash content of raw materials was determined by the ASTM method D.3174 and synthetic gas composition was analyzed using gas chromatograph (GC) with a thermal conductivity detector (TCD) connected with 2 types of columns, Porapak Q and Molesieve 5A. Meanwhile, the surface area and size of charcoal were analyzed through isothermal adsorption and so did the proximate and ultimate analysis of three raw materials.

*2.4.1. Calculation of coal and biomass ash content.* Ash content in coal and biomass was obtained through the combustion process. Coal/biomass mass before and after combustion were used for Equation 1 calculation.

$$\boldsymbol{ash\ content} = \frac{\boldsymbol{mass\ ash}}{\boldsymbol{mass\ sample}}\ \boldsymbol{x\ 100\%} \tag{1}$$

*2.4.2. Theoretical calculation of cogasification.* The synthetic gas and charcoal yield from experimental were compared with predicted calculation. It was assumed there was no interaction between coal and biomass during cogasification [25]. The theoretical gas and charcoal yield were calculated by Equation 2 as follows,

$$Y_{predict} = X_{biomass} \times Y_{biomass} + (1 - X_{biomass}) \times Y_{coal} \tag{2}$$

with $X_{biomass}$ is the mass fraction of biomass, $Y_{biomass}$ and $Y_{coal}$ are gas yield of biomass and coal.

## 3. Results and Discussion

### *3.1. Effect of feed blend ratio on cogasification process*

Based on cogasification studies with variations of coal and biomass types, it was found that not all cogasification produced synergy effect. Krerkkaiwan et al. [12] stated that the interaction of biomass and coal in cogasification was determined by the type of biomass and type of coal. This study focused on the effect of adding oil palm empty bunches on $H_2$ yield from cogasification together with Indonesian lignite coal. To find out the synergy effect of cogasification, the experimental data were compared with the theoretical values calculated from the gasification of each feed: BLM, BLD, and EPFB. If there were differences between them, the synergy effect was indicated exist in cogasification [12]. If the experimental data produced a value greater than theoretical, the synergy effect would be positive whereas if the experimental data value was lower than the theoretical calculation, the synergy effect of cogasification would be negative. The synergy effect in cogasification occurred due to the catalytic effect of biomass ash which caused activation energy decreases [26].

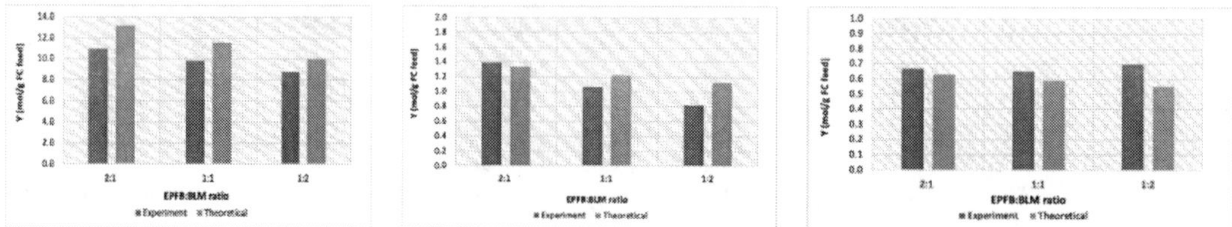


(a) (b) (c)

**Figure 1.** Product conversion of cogasification EPFB and BLM (a) Liquid, (b) Gas, and (c) Char.

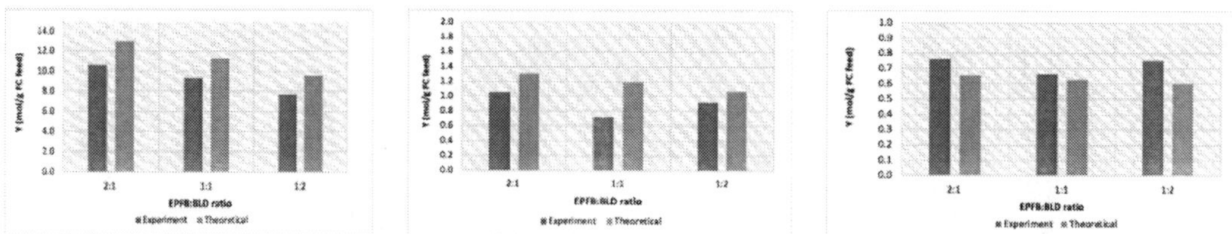


(a) (b) (c)

**Figure 2.** Synthesis gas conversion of cogasification EPFB and BLD (a) Liquid, (b) Gas, and (c) Char.

The main purpose of cogasification was to achieve the catalytic effect of biomass ash in increasing coal char pyrolysis conversion [27]. The pyrolysis char was converted into a gas product in the form of synthesis gas, liquid products consisted of volatile and tar condensates, and solid products as cogasified charcoal. This study focused on increasing the conversion of EPFB and lignite coal mixtures into synthesis gas with high $H_2$ gas yield. Based on this experiment, the addition of EPFB as a mixture of lignite coal feed, both BLM and BLD, increased percentage of cogasification gas products as shown in Figure 2. The same trend was produced by Patel and Narnaware [28] studies which examined six types of biomass with different mixing ratios to study the cogasification gas products. The six types of biomass provide the same behavior, increase in carbon conversion into gas products along with the increase percentage of biomass in the cogasification feed. This was caused by the synergy effect of biomass ash containing alkali and alkaline earth [27].

The results of the cogasification of EPFB-BLM and EPFB-BLD in Figure 3 show that the conversion of synthesis gas and liquid products produced a negative synergy effect for all mixing variations. Habibi et al. [28] and Krerkkaiwan et al. [12] explained the effect of negative synergy in cogasification due to the ratio of K/Al in feed and heating rate during the pyrolysis process. Habibi et al. [28] stated that the synergy effect occurs when the K/Al ratio in coal mixtures and biomass cogasification was more than 1. This comparison caused K bond Al and Si to form $KAlSi_3O_8$. The K/Al ratio for EPFB-BLM and EPFB-BLD in this experiment was 0.088 and 0.13, respectively. Although the content of Al and Si in coal has been minimized through leaching of HF acid, the cogasification mixture still had Si and Al content from EPFB with relatively high number. K and Al in carbon fuel were stored in the form of alumina ($Al_2O_3$) and silicate ($SiO_2$). Both of these compounds could react with alkali metals to form silica-aluminum compounds ($KAlSiO_4$, $KAlSi_3O_8$, $CaAl_2Si_2O_8$) which inhibited catalytic effect in cogasification. These compounds were formed irreversible reaction so that alkali and alkaline earth metals could not be reused in cogasification [10, 15, and 27]. In addition, effect of slow heating rates during pyrolysis process had caused no synergy effect in cogasification process [29-30].

Based on data from the composition measurement of cogasification synthesis gas EPFB: BLM every 10 minutes, sequence of rate formation of gas $H_2$, CO, and $CO_2$ stability respectively occurs in the ratio 1:2, 1:1, and 2:1. Those are showed on Figure 5 and Figure 6.

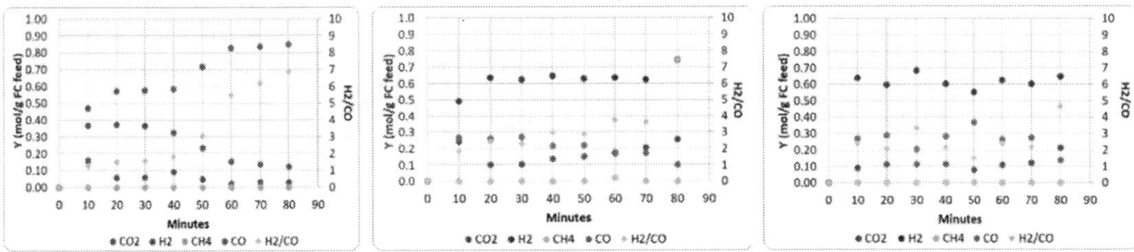


(a) (b) (c)

**Figure 3.** Profile of synthesis gas and $H_2/CO$ of EPFB:BLM mixing ratio at (a) 1:2, (b) 1:1, (c) and 2:1.

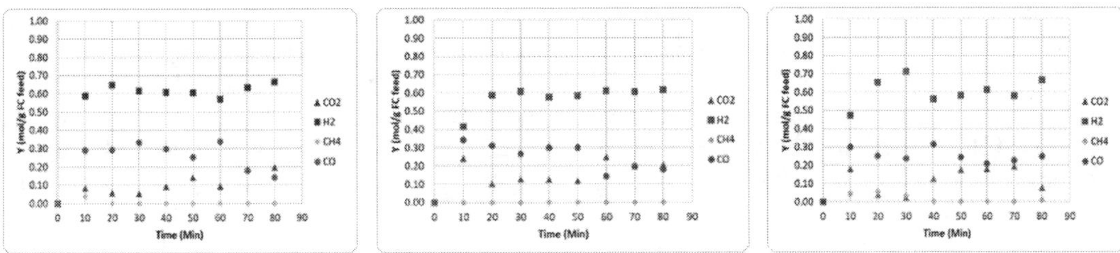


(a) (b) (c)

**Figure 4.** Profile of synthesis gas and $H_2/CO$ of EPFB:BLD mixing ratio at (a) 1:2, (b) 1:1, (c) and 2:1.

Cogasification of BLD had a better pattern of synthesis gas yield and $H_2/CO$ ratio than BLM. This was due to the alkali metal from biomass ash which was active on BLD-EPFB cogasification rather than BLM-EPFB which had high silica and alumina in lignite coal that could deactivate AAEM [15]. Cogasification of BLM-EPFB had high charcoal reactivity [31], so that both mixed cogasification provided an increasing and unstable yield profile. Low reactivity of charcoal BLD caused the mixture of cogasification of BLD-EPFB to have a stable gas yield [32]. The quality of synthesis gas in terms of synthesis gas yield was produced. Good synthesis gas had a large $H_2/CO$ ratio and low $CH_4/CO$. The more positive synergy effect of coal and biomass cogasification, the better the synthesis gas quality produced (high $H_2$ yield). All experimental results of $H_2$, $CO_2$, CO, and $CH_4$ were lower than theoretical values, indicating a negative synergy effect. This event was caused by the distribution of the mineral content in the cogasification feed mixture. Based on Table 2, the content of silica and alumina in EPFB was quite large so that it had strong strength in inhibiting the catalytic effect of AAEM from biomass. According to Habibi et al. [28] a low K/Al ratio will have a strong inhibitory effect in cogasification because the K element contained would react with Al to form non-catalytic compounds. The low K/Al ratio and the slow heating rate in the cogasification of the BLM-EPFB and BLD-EPFB cause both of them to have a negative synergy effect.

IOP Conf. Series: Materials Science and Engineering 1143 (2021) 012026 doi:10.1088/1757-899X/1143/1/012026

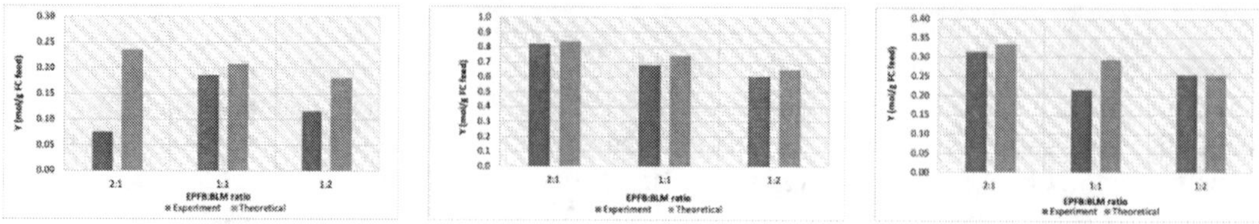


(a) (b) (c)

**Figure 7.** Profile of (a) $CO_2$, (b) $H_2$, (c) CO from EPFB:BLM cogasification at 1:2, 1:1, 2:1 mixing ratio.

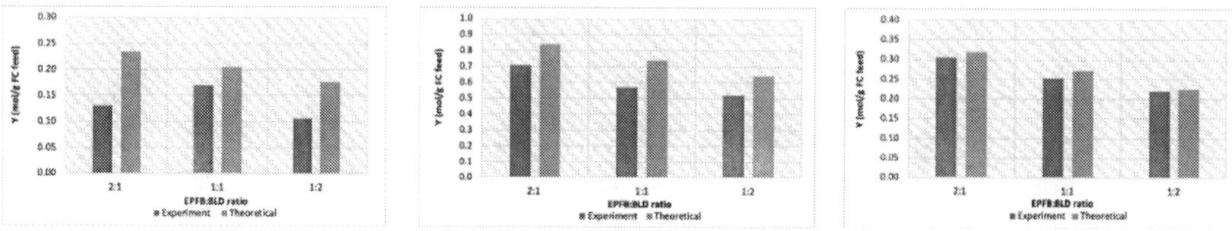


(a) (b) (c)

**Figure 8.** Profile of (a) $CO_2$, (b) $H_2$, (c) CO from EPFB:BLD cogasification at 1:2, 1:1, 2:1 mixing ratio.

In general, both the cogasification of EPFB-BLD and EPFB-BLM produced a negative synergy effect in producing $CO_2$ gas. Figure 7 and Figure 8 explained the most positive synergy effect of $CO_2$ gas yield from cogasification of lignite-EPFB occurring in mixing feed EPFB and lignite 1:1. The peak of $CO_2$ yield in the coal-biomass 1:1 mixing ratio was also obtained by the Krerkkaiwan et al. [12]. The increase in $CO_2$ gas yield when the percentage of biomass in the cogasification feed rose up due to the high oxygen content in the biomass which drove $CO_2$ oxidation. Mallick et al. [27] said there were two patterns of $H_2$ production when the percentage of biomass was added in the cogasification feed: increases or decreases. Based on the data of this study as illustrated in Figure 7 and Figure 8, it was agreed that $H_2$ increased when biomass was added to the cogasification feed. This was caused by an increase in biomass impact on higher H/C and O/C ratio in cogasification feed. The content of H and OH acted as a hydrogen donor species and encouraged the process of breaking aromatic components in coal [12]. The presence of AAEM in biomass which was mixed together with coal as a cogasification feed, increases $H_2$ and $CO_2$ with a decrease in CO formed. This occurs due to tendency of water shift gas reaction that converted CO to $H_2$ and $CO_2$ [33]. Biomass had more porous charcoal structure and higher charcoal reactivity. With this advantage, steam as gasification media easily diffused and reacted to produce special gas synthesis gases [34].

Although cogasification of raw and demineralization lignite resulted tendency to obtain $CO_2$ to the same percentage of biomass, the cogasification of lignite demineralization provides a higher theoretical experiment difference, indicating that the $H_2$ cogasification product of EPFB-BLD was not higher than of raw lignite $H_2$. Jenkins et al. [32]; Hippo and Walker [35] stated that Ca and Na in low rank coal had a significant effect on reactivity of charcoal while other minerals (K, Na and Fe) did not. A similar study also found that gasification reactivity originated from the content of K, Na, Ca, and coal char [36-38]. Therefore, demineralized coal (BLD) had a lower reactivity than BLM so that the yield of $H_2$ cogasification of BLD was lower than BLM.

The yield of methane gas in a variety of lignite-EPFB mixtures. Theoretically, the yield of methane was proportional to the addition of lignite in cogasification feed. However, according to the experiment, it was found that methane was only formed during cogasification involving non-demineralized lignite with 1:1 lignite-EPFB ratio while methane was formed at a ratio of 2:1 and 1:2 to the cogasification of lignite demineralization and EPFB. This indicated that methane only applied to a certain ratio for two different types of lignite. Figure 7 and Figure 8 showed that an increase in the amount of biomass leads to an increase in the yield of CO in synthetic gas. This was due to CO consumed in the water gas shift reaction. Yuan et al. [39] explained that the increase in CO along with the increase in the percentage of biomass caused by the supply of OH radicals was able to attack the aromatic coal chain and react with aliphatic species to form CO.

### *3.2. Effect of coal demineralization by HF leaching on gasification product*

Coal demineralization process was a pre-treatment cogasification phase of EPFB-lignite aimed at minimizing disadvantage minerals in coal ash, silica and alumina. The characteristics of the cogasification charcoal discussed consisted of surface area, pore diameter, and pore volume. Based on BET test with $N_2$ isothermal adsorption method, data obtained on charcoal cogasification characteristics of the mixture of BLM-EPFB and BLD-EPFB were shown in Table 3 below.

**Table 3.** Data characteristics of cogasification charcoal.

| Mixed Feed | Char Surface Area ($m^2/g$) | Pore Diameter (nm) | Pore Volume (ml/g) |
|---|---|---|---|
| **BLM : EPFB = 2 : 1** | 385.994 | 2.9054 | 0.2804 |
| **BLD : EPFB = 2 : 1** | 375.750 | 2.3668 | 0.2234 |

Based on Table 3, charcoal from the BLM-EPFB cogasification had larger surface area than BLD-EPFB. According to Nunes et al. (2017) [40] and Calahorro et al. [41], HF demineralization of coal was able to reduce the coal surface area since it reduced micropores and mesopores due to the displacement of insoluble inorganic substances from the coal carbon matrix. The transformation of micropores and mesopores into macropores causes lower surface area of BLD than BLM. Based on the experiments in this study, the BLM-EPFB cogasification charcoal had larger diameter and volume pore than the BLD-TKF. Reduction mesopores and micropores into macropores which resulted in coal pore volume decrease due to HF demineralization found by Nunes et al. [40] and Calahorro et al. [41]. According to Nunes et al. [40] and Calahorro et al. [41], demineralization of HF moved inorganic substances bound to carbon matrix then clogged the micropores and mesoporous' coal. Similar to pure sample gasification, the main product of lignite cogasification process along with oil palm empty bunches consists of synthesis gas ($H_2$, CO, $CO_2$ and $CH_4$), liquid products (volatile matter and condensable gases), and charcoal. Figures 11 and Figure 12 showed the distribution of cogasification products for EPFB-BLM and EPFB-BLD based on experiment and theoretical calculations. In general, the three variations in the mixing ratio provided a negative synergy effect on yield of liquid and gas products but not for BLD and BLM carbon conversion.

Based on the experiments, the cogasification conversion of EPFB-BLD and EPFB-BLM had significant differences for liquid, solid and gas products. In liquid products, the synergistic effect of cogasification of EPFB-BLD is more negative than EPFB-BLM, indicating that it is reducing volatile substances and condensed gases. HF is an acid solution with very strong oxidation properties and easily dissolves mineral substances in coal [23]. In case of cogasification that focuses on increasing the gas yield of synthesis, the demineralization process is not good enough because it has a more negative synergy effect than non-pre-treated lignite. Alkali metals that evaporate and stick to the coal surface hold a major effect of charcoal reactivity and catalytic effect in production of gas synthesis [10]. BLD produces a better synergy effect on char conversion because it has a silica content that can help the process of cutting large carbon molecules into smaller products [27]. This behavior occurs in all mixing ratios of EPFB-BLM and EPFB-BLD 2:1, 1:1, and 1:2.

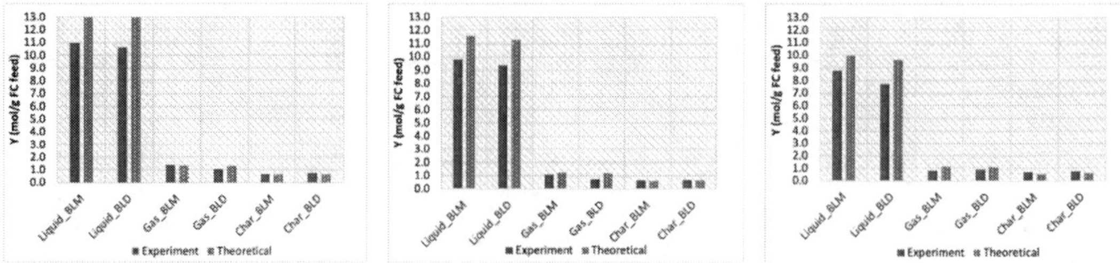


(a) (b) (c)

**Figure 9.** Cogasification products with EPFB:BLM and EPFB:BLD mixing ratios (a) 2:1, (b) 1:1, (c) 1:2.

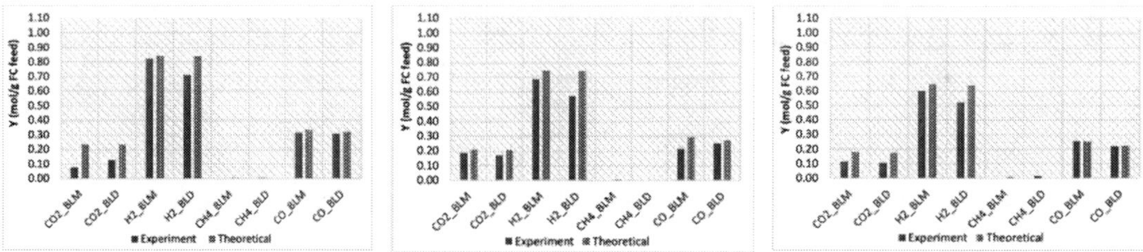


(a) (b) (c)

**Figure 10.** Cogasification gas yield with EPFB:BLM and EPFB:BLD mixing ratios (a) 2:1, (b) 1:1, (c) 1:2.

Figures 9 and Figure 10 show comparison of experimental data and theoretical calculations for cogasification synthesis gas EPFB-BLM and EPFB-BLD. In general, the three variations in the mixing ratio provide a negative synergy effect on the yield of $H_2$, $CO_2$, CO, and $CH_4$ for both BLD and BLM. Based on the experiment, the yield of CO and $CO_4$ gas between the EPFB-BLD and EPFB-BLM cogasification are alike. For $H_2$ concentration, negative synergy effect of EPFB-BLD cogasification was higher than EPFB-BLM, indicating that demineralization reduced $H_2$ gas yield. Therefore, demineralization process was not good enough for cogasification pretreatment which aimed to obtain high $H_2$ yield. Jenkins et al. [32] stated that demineralization of HF was able to reduce the amount of pyrolysis charcoal sevenfold which had an impact on reducing reactivity of lignite charcoal up to 34 times. Overall for the case of low rank coal, demineralization using HF acid solution caused a decrease in reactivity of charcoal [42].

## 4. Conclusion

Coal pre-treatment by demineralization through HF leaching reduces gasification performance in terms of a decrease in synthesis gas and $H_2$ yield. An increase in char yield is caused by a drop in coal reactivity significantly that affects the catalytic effect on the overall gasification process.

## References

[1] British Petroleum 2017 *BP Statistical Review of World Energy 2017* British Petroleum 1–52.
[2] Ahmed I I, Nipattummakul N and Gupta A K 2011 Characteristics of syngas from co-gasification of polyethylene and woodchips *Applied Energy* **88** 165–174.
[3] Fermoso J, Arias B, Plaza M G, Pevida C, Rubiera F, Pis J J, García-Peña F and Casero P 2009 High-pressure co-gasification of coal with biomass and petroleum coke *Fuel Processing Tech.* **90** 926–932.
[4] Saxena R C, Adhikari D K and Goyal H B 2009 Biomass-based energy fuel through biochemical routes: A review *Renewable and Sustainable Energy Rev.* **13** 167–178.

[5] Cormos C C 2012 Hydrogen and power co-generation based on coal and biomass/solid wastes co-gasification with carbon capture and storage *Intl. Journal of Hydrogen Energy* **37** 5637–48.

[6] André R N, Pinto F, Franco C, Dias M, Gulyurtlu I, Matos M. A. A and Cabrita I 2005 Fluidised bed co-gasification of coal and olive oil industry wastes *Fuel* **84** 1635–44.

[7] Xu C, Hu S, Xiang J, Zhang L, Sun L, Shuai C, Chen Q, He L and Edreis E M A 2014 Kinetic models comparison for steam gasification of coal/biomass blend chars *Bioresource Tech.* **171** 253–259.

[8] Ellis N, Masnadi M S, Roberts D G, Kochanek M A and Ilyushechkin A Y 2015 Mineral matter interactions during co-pyrolysis of coal and biomass and their impact on intrinsic char co-gasification reactivity *Chem. Eng. Journal* **279** 402–408.

[9] Hernández J J, Ballesteros R and Aranda G 2013 Characterisation of tars from biomass gasification: Effect of the operating conditions *Energy* **50** 333–342.

[10] Zhang Y, Zheng Y, Yang M and Song Y 2016 Effect of fuel origin on synergy during co-gasification of biomass and coal in $CO_2$ *Bioresource Tech.* **200** 789–794.

[11] Masnadi M S, Habibi R, Kopyscinski J, Hill J M, Bi X, Lim C J, Ellis N and Grace J R 2014 Fuel characterization and co-pyrolysis kinetics of biomass and fossil fuels *Fuel* **117** 1204–14.

[12] Krerkkaiwan S, Fushimi C, Tsutsumi A and Kuchonthara P 2013 Synergetic effect during co-pyrolysis/gasification of biomass and sub-bituminous coal *Fuel Processing Tech.* **115** 11–18.

[13] Xu C, Hu S, Xiang J, Zhang L, Sun L, Shuai C, Chen Q, He L and Edreis E M A 2014 Interaction and kinetic analysis for coal and biomass co-gasification by TG-FTIR *Bioresource Tech.* **154** 313–321.

[14] Yuan S, Dai Z H, Zhou Z J, Chen X L, Yu G S and Wang F C 2012 Rapid co-pyrolysis of rice straw and a bituminous coal in a high-frequency furnace and gasification of the residual char *Bioresource Tech.* **109** 188–197.

[15] Ding L, Zhang Y, Wang Z, Huang J and Fang Y 2014 Interaction and its induced inhibiting or synergistic effects during co-gasification of coal char and biomass char *Bioresource Tech.* **173** 11–20.

[16] Biagini E, Lippi F, Petarca L and Tognotti L 2002 Devolatilization rate of biomasses and coal-biomass blends: An experimental investigation *Fuel* **81** 1041–50.

[17] Vuthaluru H B 2004 Thermal behaviour of coal/biomass blends during co-pyrolysis *Fuel Processing Tech.* **85** 141–155.

[18] Mukherjee S and Borthakur P C 2001 Chemical demineralization/desulphurization of high sulphur coal using sodium hydroxide and acid solutions *Fuel* **80** 2037–40.

[19] Karaca H and Ceylan K 1997 Chemical cleaning of Turkish lignites by leaching with aqueous hydrogen peroxide *Fuel Processing Tech.* **50** 19–33.

[20] Ishaq M, Ahmad I, Shakirullah M, Bahader A and Taj N 2002 Characterization of Khushab coal (Punjab Pakistan) Journal of the Chemical Society of Pakistan **24** 240–245.

[21] Kizgut S, Baris K and Yilmaz S 2006 Effect of chemical demineralization on thermal behavior of bituminous coals *Journal of Thermal Analysis and Calorimetry* **86** 483–488.

[22] Hacifazlioglu H, Uckan D and Sahin M 2017 Environmental Effects Demineralization of Turkish Tosya lignite coal by boric acid leaching Demineralization of Turkish Tosya lignite coal by boric acid *Energy Sources, Part A: Recovery, Utilization, and Environmental Effects.* **38** 2777–81.

[23] Steel K M, Besida J, O'Donnell T A and Wood D G 2001 Production of Ultra Clean Coal - Part II - Ionic equilibria in solution when mineral matter from black coal is treated with aqueous hydrofluoric acid *Fuel Processing Tech.* **70** 193–219.

[24] Vaccaro S 2010 Demineralization and Desulfurization Process to Generate Clean Coal *Chem. Eng. Transactions* **21**.

[25] Rizkiana J, Guan G, Widayatno W B, Hao X, Huang W, Tsutsumi A and Abudula A 2014 Effect of biomass type on the performance of cogasification of low rank coal with biomass at relatively low temperatures *Fuel* **134** 414–419.

[26] McKee D W 1983 Mechanisms of the alkali metal catalyzed gasification of carbon *Fuel* **62** 170–

175.
[27] Mallick D, Mahanta P and Moholkar V S 2017 Co-gasification of coal and biomass blends: Chemistry and engineering *Fuel* **204** 106–128.
[28] Habibi R, Kopyscinski J, Masnadi M S, Lam J, Grace J R, Mims C A and Hill J M 2013 Co-gasification of biomass and non-biomass feedstocks: Synergistic and inhibition effects of switchgrass mixed with sub-bituminous coal and fluid coke during CO2 gasification *Energy & Fuels* **27** 494–500.
[29] Haykiri-Acma H and Yaman S 2007 Synergy in devolatilization characteristics of lignite and hazelnut shell during co-pyrolysis *Fuel* **86** 373–380.
[30] Jones J M, Kubacki M, Kubica K, Ross A B and Williams A 2005 Devolatilisation characteristics of coal and biomass blends *Journal of Analytical and Applied Pyrolysis* **74** 502–511.
[31] Nzihou A, Stanmore B and Sharrock P 2013 A review of catalysts for the gasification of biomass char, with some reference to coal *Energy* **58** 305–317.
[32] Jenkins R G, Nandi S P and Walker P L 1973 Reactivity of heat-treated coals in air at 500 °C *Fuel* **52** 288–293.
[33] Zhang Z, Pang S and Levi T 2017 Influence of AAEM species in coal and biomass on steam co-gasification of chars of blended coal and biomass *Renewable Energy* **101** 356–363.
[34] Xu Q, Pang S and Levi T 2011 Reaction kinetics and producer gas compositions of steam gasification of coal and biomass blend chars, part 2: Mathematical modelling and model validation *Chem. Eng. Science* **66** 2232–40.
[35] Hippo E and Walker P L 1975 Reactivity of heat-treated coals in carbon dioxide at 900 °C *Fuel* **54** 245–248.
[36] Fernández-Morales I, López-Garzon F J, López-Peinado A, Moreno-Castilla C and Rivera-Utrilla J 1985 Study of heat-treated Spanish lignites: Characteristics and behaviour in CO2 and O2 gasification reactions *Fuel* **64** 666–673.
[37] Fung D P C and Kim S D 1984 Chemical reactivity of Canadian coal-derived chars *Fuel* **63** 1197–1201.
[38] Miura K, Hashimoto K and Silveston P L 1989 Factors affecting the reactivity of coal chars during gasification, and indices representing reactivity *Fuel* **68** 1461– 75.
[39] Yuan S, Dai Z H, Zhou Z J, Chen X L, Yu G S and Wang F C 2012 Rapid co- pyrolysis of rice straw and a bituminous coal in a high-frequency furnace and gasification of the residual char *Bioresource Tech.* **109** 188–197.
[40] Nunes K G P, Engel K M, Osório E and Marcílio N R 2017 Demineralization of Brazilian coals for use in gasification and oxy-fuel combustion processes, aiming to reduce CO2 emissions *Energy and Fuels* **31** 8560–71.
[41] Calahorro C V, Cano T C and Serrano V G 1987 Effect of acid and heat treatments on surface area and porosity of a Spanish coal with high mineral matter content *Fuel* **66** 479–485.
[42] Laurendeau N M 1978 Heterogeneous kinetics of coal char gasification and combustion *Progress in Energy and Combustion Science* **4** 221–270.

IOP Conf. Series: Materials Science and Engineering 1143 (2021) 012027 doi:10.1088/1757-899X/1143/1/012027

# Mass Balance analysis of Bioethanol Production from Sweet Sorghum (Sorghum bicolor)

**Muhammad Lauda[1*], Nadiya Rahmawati[1], Wayda Rahma Putri Fajar[1], Aliya Ramadhani[1], Rahmah Amirah June[1], Meiti Pratiwi[1], Jenny Rizkiana[1,2]**

[1] Bioenergy Engineering and Chemurgy, Faculty of Industrial Technology, Bandung Institute of Technology, Indonesia
[2]Chemical Engineering, Faculty of Industrial Technology, Bandung Institute of Technology, Indonesia

Email: muhammadlauda@gmail.com

**Abstract.** Current fossil fuel reserves cannot keep up with the world's need for fuel, leading to a global energy crisis. The issue raises attention to renewable energy sources. Indonesia has committed to using 15% bioethanol in gasoline mixture by 2025, as outlined in Presidential Decree No. 5 of 2006. This article discusses past studies on sweet sorghum plants in their use as a raw material for bioethanol production from various aspects. The study shows that sweet sorghum juice has a high potential to be converted into bioethanol due to its high sugar content. Pretreated sweet sorghum seeds and bagasse also have great potential to be converted into bioethanol due to their rich oligomer and polymer sugar content. The main challenge of producing bioethanol from sweet sorghum is the low economic competitiveness of utilizing sweet sorghum as an energy crop compared to using sweet sorghum as a food crop. The present study focuses on the mass balance analysis of bioethanol production from sweet sorghum. It is expected that the results of the present study may give a preliminary overview of the bioethanol production potential from sweet sorghum.

## 1. Introduction

Bioethanol is a widely-used substitute for gasoline due to its compatibility with gasoline engines and high octane number, reaching 109. The mixture of bioethanol and gasoline has a higher octane number than pure gasoline and can prevent knocking in the engine [1]. Historically, bioethanol has been produced through the fermentation process from various substrates. The first generation of bioethanol uses sugar or starch from food crops, such as sugarcane and corn, as substrate. Brazil and the United States have successfully developed the first generation of bioethanol. Brazil uses sugarcane, while the United States uses corn as raw material for bioethanol production. Unlike the first generation, second-generation bioethanol uses lignocellulose as a fermentation substrate; the presence of a long carbon chain necessitates pretreatment of the substrate before fermenting the sugar in the initial material. Second-generation bioethanol can be made from biomass waste, which is easier to obtain than food crops [2].

Even though Indonesia has never been included in FAO's list of sorghum producers, exporters, and importers, it does not mean sorghum does not thrive in Indonesia. Sorghum was first planted in Indonesia before 1970 when Indonesia was still prone to food shortages and national food security had not been achieved yet [3]. Unfortunately, in Indonesia's status quo, sorghum has not been appropriately managed because it is seen as a low-grade crop among farmers [3]. Indonesia's highest sorghum producers in

2011 – based on harvested area and production yield – are Central Java, East Java, Yogyakarta, West Nusa Tenggara, and East Nusa Tenggara [4].

Sorghum production is closely related to sorghum productivity per hectare. Productivity is influenced by the sorghum variety as well as soil conditions and the environment in which sorghum grows. Sweet sorghum plants have a high level of productivity in Indonesia. In 2011, the production of sweet sorghum in Indonesia reached 7,695 tonnes [5].

Sweet sorghum has stems containing sugar-rich juice, bagasse waste containing lignocellulose, and seeds containing starch. The juice, bagasse, and sorghum seeds can all be used as raw material for bioethanol production [6]. The sugar content of sweet sorghum juice is approximately 11.8% while the lignocellulose content of bagasse waste sorghum is approximately 13% [7]. This study focuses on reviewing current knowledge and comparing the mass balance of ethanol production from various parts of the sweet sorghum plant to provide an overview of the potential of sweet sorghum to be used as a bioethanol feedstock.

## 2. Bioethanol production process review

Bioethanol can be derived from sweet sorghum juice, seed starch, and bagasse cellulosic material. Each raw material requires a different conversion process to be bioethanol.

### *2.1. Bioethanol production from sweet sorghum juice*

The processing of converting sweet sorghum juice into bioethanol consists of several steps that have been summarized into a flowchart in Figure 1.

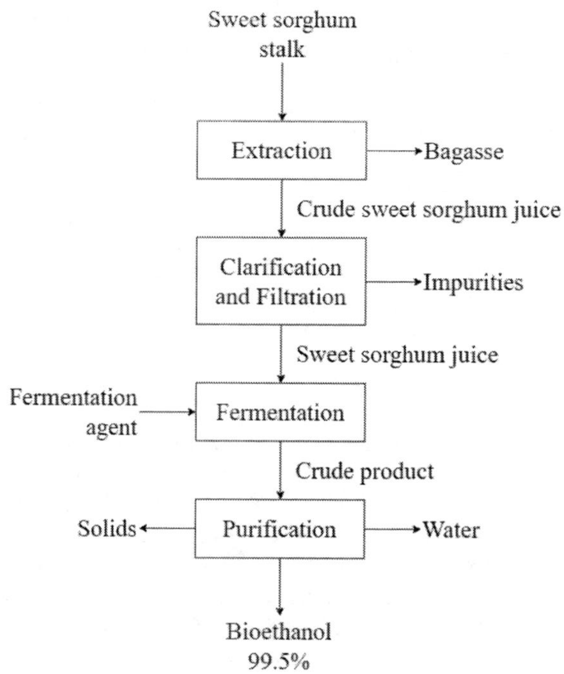


**Figure 1.** The block flow diagram of bioethanol production from sweet sorghum juice [8].

*2.1.1. Juice extraction.* Sweet sorghum juice extraction is done by crushing and squeezing the sweet sorghum stem using a crushing mill [9]. The process produces bagasse as a by-product. The extracted juice is separated from impurities through clarification and filtration processes. Sweet sorghum juice is very susceptible to bacterial contamination [10], so it must be protected from bacterial contamination by heating and concentrating the sugar solution. Heating is carried out to a temperature of 100°C to kill microbes [11]. Sugar concentrate is made by passing the solution through an evaporator. Then, the juice is fermented at 16-18° Brix. When not in use, it should be stored at a minimum of 65° Brix[9]. A preserving technique alternative is to keep the juice at -20°C [12].

*2.1.2. Fermentation.* In conventional industrial practices, fermentation is carried out with Saccharomyces cerevisiae's help due to its ability to produce high ethanol yields (93% theoretical yield) and its resistance to fermentation inhibitors [13]. The alternative fermentation agent commonly used is Zymomonas mobilis. Z. mobilis has a higher ethanol theoretical yield than S. cerevisiae, reaching a value of 97%. However, Z. mobilis is only useful in fermenting D-glucose. The performance of Z. mobilis in D-fructose and sucrose fermentation is not as good as D-glucose fermentation. Thus it is not very effective in fermenting molasses, sorghum, or starchy materials [14]. A different research reports that a 96-hour fermentation with Escherichia coli produced higher fermentation results (95% theoretical yield) than S. cerevisiae (90% theoretical yield). This trend has exceptions to varieties with juice that has high sucrose content due to E. coli's low effectiveness in sucrose fermentation [15].

*2.2. Bioethanol production from sweet sorghum seeds*

The starch content in sorghum seeds is higher, at 61%, than the sugar content in sweet sorghum juice, making the seeds' bioethanol potential higher than juice. Starch can be hydrolyzed and converted to fermentable sugar. The processes of producing bioethanol from sweet sorghum seeds include grinding, liquefaction, saccharification, fermentation, and purification. Solid waste from the fermentation process can be dried and used as animal feed. Figure 2 displays a block diagram of the bioethanol production process from sweet sorghum seeds.

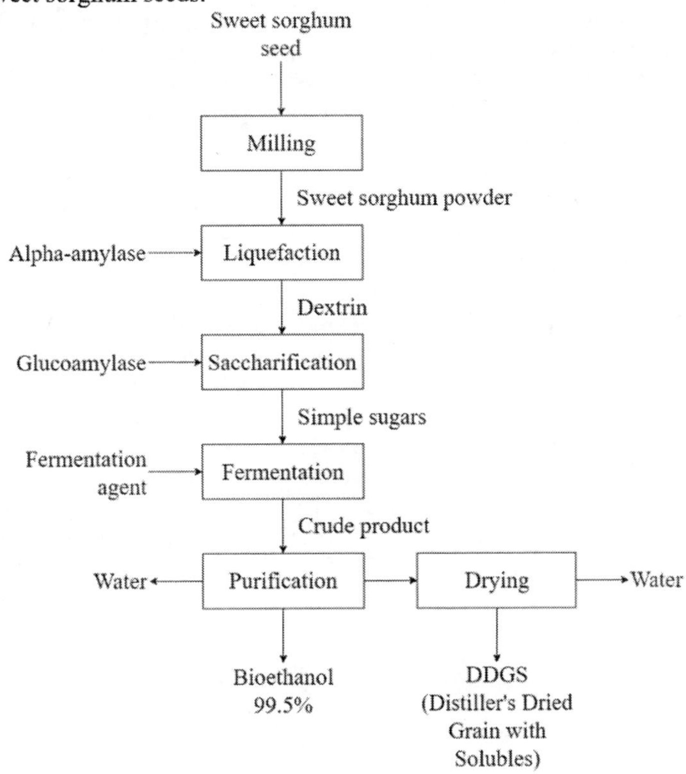


**Figure 2.** The block flow diagram of bioethanol production from sweet sorghum seeds [8].

*2.2.1. Grinding.* Sweet sorghum seeds consist of 60–77% starch [16]. Sweet sorghum seeds need to be ground to increase the contact area with water, enzymes, and microbes. The most common grinding method is dry milling – grinding without the addition of water – with a hammer mill. Wet milling – a grinding process in which sorghum seeds are soaked in water beforehand – can also reduce friction on the mill. In bioethanol production, dry milling is more commonly used because the equipment is simpler, cheaper, and the process requires less time [8]. The sweet sorghum particle size must be small enough to ensure water binding, heat transfer, and enzymatic reactions occur sufficiently. Research examining sweet sorghum seed particle size in the range of 0.3–0.7 mm shows that smaller particle size results in better bioethanol production [17].

*2.2.2. Liquefaction.* Ground sorghum seeds are soaked in water with a specific ratio of solids to liquids. The sorghum suspension is heated to 90°C. Then the acidity is adjusted before adding the enzyme α-amylase and/or β-amylase. Soaking and heating are done to dissolve the starch and to disrupt the structure of the seeds. This action can enhance the performance of enzymes starch hydrolysis (liquefaction). Liquefaction performance is measured by the sugar yield from starch hydrolysis. Around 86.4% of starch was successfully hydrolyzed using α-amylase enzyme at 90°C for 2 hours with a solid to liquid ratio of 1:7 g/ml. This study found that the higher the concentration of solids, the lower hydrolysis efficiency becomes, although a higher mass of sugar is obtained [17].

*2.2.3. Saccharification.* Saccharification is the process of converting dextrins obtained from liquefaction to simple sugar (glucose). Saccharification is generally done by adding the enzyme glucoamylase obtained from Aspergillus sp. to a starch solution. Factors affecting saccharification include temperature, saccharification duration, enzyme concentration, and dissolved solids concentration. A conversion rate of 90% can be achieved with saccharification using glucoamylase at 45°C and pH 5 for 24 hours [18]. Other studies report similar results, with 84.6% efficiency of liquefaction followed by saccharification using glucoamylase at pH 4.5 and temperature 55°C for 2 hours with a solid to liquid ratio of 1:7 [17]. Hydrolysis using amyloglucosidase enzyme obtained a 98% saccharification rate after 9 hours of processing time with an enzyme concentration of 0.4%, while lower enzyme concentrations took longer to achieve the same results [19]. These three studies show the correlation of reaction conditions with the yield of sugar. These studies suggest that the optimum saccharification temperature is in the range of 45–60°C, saccharification duration is generally directly proportional to the sugar yield, the concentration of dissolved solids is inversely proportional to the efficiency of saccharification, and the concentration of enzyme is proportional to saccharification rate.

*2.2.4. Fermentation.* Hydrolyzate (sugars obtained from starch hydrolysis) is converted into ethanol through the fermentation process. Sweet sorghum seed fermentation is similar to sweet sorghum juice fermentation.

### *2.3. Bioethanol production from sweet sorghum bagasse*

Sweet sorghum bagasse is a lignocellulosic by-product of sweet sorghum juice extraction. Before converting it into ethanol, bagasse components need to be separated. Cellulose in bagasse is hydrolyzed into sugar, which is used to produce bioethanol. The block flow diagram of the bioethanol production process from sweet sorghum bagasse is presented in Figure 3.

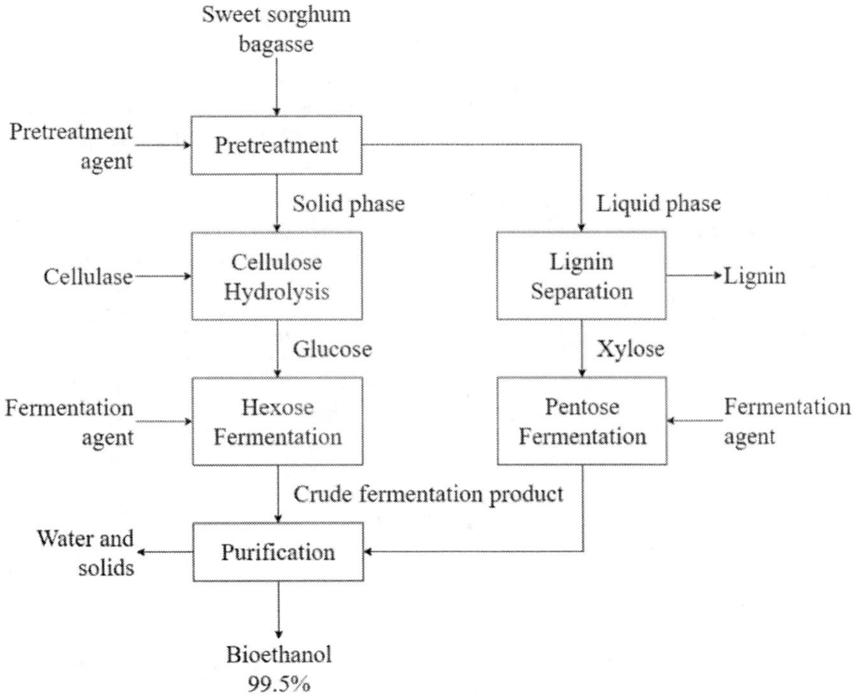


**Figure 3.** The block flow diagram of bioethanol production from sweet sorghum bagasse [8].

*2.3.1. Pretreatment.* Pretreatment of sweet sorghum bagasse can be divided into mechanical, physical, and chemical pretreatment. Physical and chemical pretreatment (or called pretreatment) makes biomass more easily digested by enzymes by separating lignin, hemicellulose, and cellulose. In contrast, mechanical pretreatment is done to increase the surface area so that enzymes can penetrate the fibers and reach the polymer and oligomer sugars in the fiber, such as cellulose, more easily[20]. Typically, bagasse chips 10–20 mm in size are ground to reach 0.2–2 mm in size [21]. Despite the benefits of decreasing bagasse particle size, this process has the disadvantage of higher energy consumption [22].

*2.3.2. Cellulose hydrolysis.* The pretreated bagasse is separated into liquid and solid phases. The liquid phase is a mixture of lignin and xylose. After lignin is separated, xylose can be fermented directly into bioethanol [23]. The solid phase (consists mainly of cellulose) must undergo a hydrolysis process first because cellulose cannot be fermented directly by the fermentation agents. Starch hydrolysis can be carried out with dilute acids or enzymes. The hydrolysis agent more commonly used is enzymes because, even though the unit price of enzymes is high, enzymes have high effectiveness and do not need detoxification step after the primary hydrolysis process. Acid hydrolysis results need to be detoxified with alkali pre-fermentation to accommodate the microbial agents, which have their optimum pH. Cellulase enzymes (which consist of endoglucanase, exoglucanase, and β-glucosidase) can be combined with xylanase for maximum hydrolysis results [24].

*2.3.3. Bagasse fermentation.* Sweet sorghum bagasse hydrolyzate contains glucose and xylose. Glucose can be fermented into bioethanol by biological agents S. cerevisiae and/or Z. mobilis. Xylose can be fermented by several wild type microbes such as Scheffersomyces stipitis and Thermoanaerobacter ethanolicus, as well as recombinant microbes such as E. coli ATCC 11303 (pLOI297). Genetic engineering is required to enable S. cerevisiae and Z. mobilis to digest xylose [8,25]. An alternative use for xylose is as a growth medium for cellulase-producing microbes.

### 2.4. *Bioethanol purification*
Crude sweet sorghum fermentation product usually has an ethanol concentration of 5–12%-v/v. Purification of bioethanol cannot be done by conventional distillation because water and ethanol form an azeotropic solution at 95% ethanol concentration. In contrast, ethanol for fuel purposes is required to have a purity of at least 99.5%. Some separation technologies such as azeotropic distillation, molecular sieve adsorption, and extractive distillation have been tested. All three of these technologies have been used commercially to produce anhydrous ethanol [12]. Extractive distillation with ethylene glycol solvent and reflux value of R = 0.5 can purify ethanol up to 99.5% [26]. Zeolite 3A molecular sieve technology can absorb water in ethanol, even in azeotropic solution [27]. Azeotropic distillation with benzene as a separating agent is able to produce ethanol with a purity of 99.3% [28].

### 2.5. *Sweet sorghum-based bioethanol production alternatives*
Research shows that sweet sorghum juice, seeds, and bagasse from the same plant can be used as bioethanol production raw material, although with different yields. Utilization of all parts of sweet sorghum plants can increase the yield of bioethanol per land area. Alternative uses of sweet sorghum plant parts are described in Figure 4.

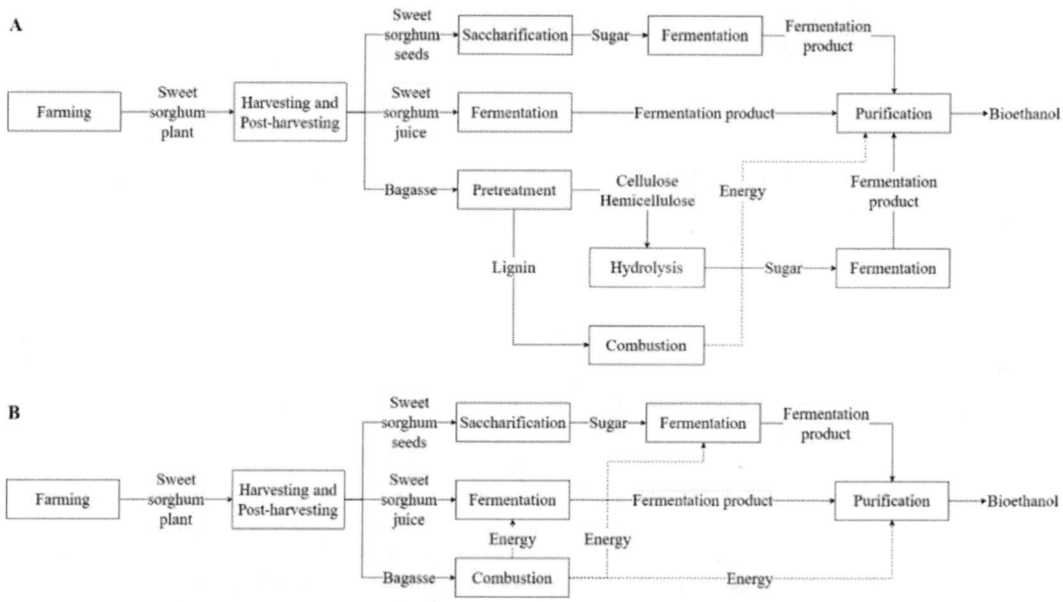


**Figure 4.** Sweet sorghum utilization as bioethanol production raw material routes: A) using all sweet sorghum plant components and B) using sweet sorghum juice and seeds.

## 3. Discussions

### 3.1. *Sweet sorghum bioethanol yield*
A comparative investigation of bioethanol yield from juice and bagasse of several sweet sorghum varieties has been conducted. GK-Coba, MN-4508, and SS-301 varieties stem juice was fermented with a mixture of *S. cerevisiae* and *Z. mobilis* cultures at 30°C for 48 hours. Bagasse from the same plant is hydrolyzed with dilute acid and fermented under similar conditions. SS-301 variety produces the highest bioethanol for both parts of the plant. However, bioethanol yield from bagasse is lower because the sugar concentration obtained from bagasse hydrolysis is lower than the juice [29]. The utilization of sweet sorghum bagasse and juice is summarized in the sorghum utilization chart in Figure 5.

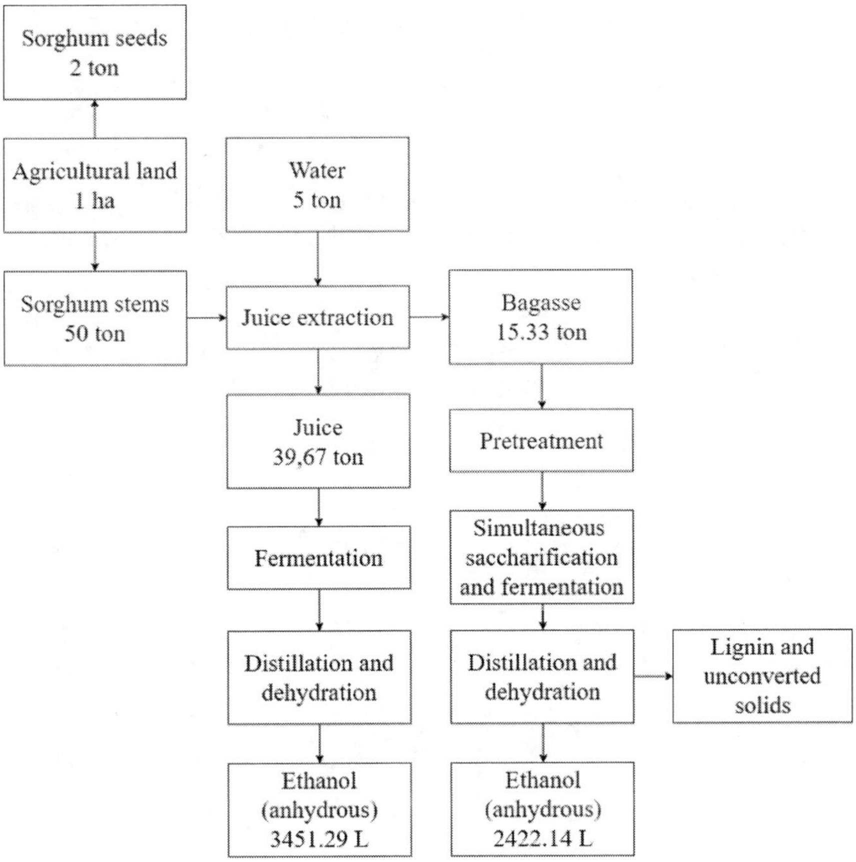


**Figure 5.** Sweet sorghum juice and bagasse utilization as bioethanol production raw material [30].

Ethanol can be derived from sweet sorghum juice, bagasse, and seeds. Each component of the sweet sorghum plant has different compositions. The ethanol yield of sweet sorghum seeds and juice data are available in a literature reference [31], but the ethanol yield from the sweet sorghum bagasse needs to be predicted with several assumptions: hydrolysis of glucans by cellulase produces 90% theoretical glucose yield, hydrolysis of xylan by xylanase has 75% theoretical xylose yield, and ethanol is produced from all sugars with the yield of 0.45 g ethanol/g sugar [31].

**Table 1.** Theoretical bioethanol yield from sweet sorghum (kg/ha, dry basis) [7].

| Variable | Sweet sorghum |
|---|---|
| Raw material | 1.000 |
| Juice | 719 |
| Fiber | 130 |
| **Sugars** | |
| Sugar from juice | 85 |
| Glucose from cellulose | 64 |
| Xylose from hemicellulose | 40 |
| **Ethanol** | |
| Ethanol from juice | 43 |
| Ethanol from cellulose | 33 |
| Ethanol from hemicellulose | 20 |
| Total ethanol | 97 |

**Table 2.** Predicted bioethanol yield, converted into post-distillation volume [32].

| Genotype | Predicted yield of bioethanol from juice (l/ha) | Predicted yield of bioethanol from bagasse (l/ha) | Predicted yield of bioethanol from seeds (l/ha) | Predicted yield of bioethanol from juice and bagasse (l/ha) | Predicted yield of bioethanol from juice and seeds (l/ha) | Predicted yield of bioethanol from seeds and bagasse (l/ha) | Predicted yield of bioethanol from juice, seeds, and bagasse (l/ha) |
|---|---|---|---|---|---|---|---|
| Watar Hammu Putih | 3,570 | 3,294 | 721 | 6,864 | 4,290 | 4,015 | 7,585 |
| 4-183A | 3,005 | 3,337 | 942 | 6,342 | 3,948 | 4,279 | 7,284 |
| 5-193C | 2,547 | 3,703 | 684 | 6,250 | 3,231 | 4,387 | 6,934 |
| 15011A | 2,989 | 4,583 | 1,020 | 7,572 | 4,009 | 5,603 | 8,592 |
| 15019B | 1,772 | 2,547 | 745 | 4,319 | 2,517 | 3,292 | 5,064 |
| 15011B | 3,033 | 3,486 | 649 | 6,520 | 3,682 | 4,135 | 7,168 |
| 15131B | 2,160 | 2,518 | 725 | 4,678 | 2,885 | 3,243 | 5,403 |
| 15020A | 1,566 | 2,028 | 1,288 | 3,594 | 2,854 | 3,316 | 4,882 |
| 1090A | 2,439 | 2,314 | 1,190 | 4,753 | 3,629 | 3,504 | 5,943 |
| 15105B | 2,631 | 2,926 | 914 | 5,557 | 3,545 | 3,840 | 6,471 |
| 15021A | 3,909 | 4,175 | 1,243 | 8,084 | 5,153 | 5,418 | 9,327 |
| Selayar hitam | 2,812 | 3,505 | 864 | 6,318 | 3,676 | 4,369 | 7,182 |
| Numbu | 2,272 | 3,311 | 697 | 5,583 | 2,969 | 4,008 | 6,280 |
| Sorghum hitam | 2,961 | 3,517 | 859 | 6,478 | 3,820 | 4,377 | 7,337 |
| **Average** | **2,690** | **3,232** | **896** | **5,922** | **3,586** | **4,128** | **6,818** |

Ethanol yield from fermented juice, bagasse, and sorghum seeds differs between each variety and genotype. Of the 11 genotypes and three sweet sorghum local varieties in Table 2, Watar Hammu Putih and 15021A genotypes have the highest ethanol yields: 7,139 l/ha and 7,819 l/ha respectively, while the average ethanol yield is only 5,381 l/ha. The yield of stem biomass has a range of 12.99–33.47 t/ha. Watar Hammu Putih and 15021A genotypes have higher stem yields than other genotypes: 29.98 t/ha and 33.47 t/ha, respectively. Besides having a high stem biomass yield, Watar Hammu Putih and 15021A have higher sugar content in their stem juice than other genotypes: 12.33% and 12.67%, respectively. The high stem yield and sugar content in the Watar Hammu Putih and 15021A genotypes result in both genotypes having high ethanol yield [6,12].

Watar Hammu Putih, 15011B, and 15021A genotypes have the highest bagasse yields of 5.43 t/ha, 5.55 t/ha, and 7.10 t/ha respectively. Ethanol yield from fermented bagasse ranges from 4,0561 l/ha to 9,1661 l/ha with an average of 6,463 l/ha. Genotypes 5-193C, 15011A, and 15021A has the highest ethanol yields of 7,406 l/ha, 9,166 l/ha, and 8,349 l/ha respectively. The high ethanol yield in genotype 15021A is due to the high yield of bagasse per unit area. However, the high ethanol yield in genotypes 5-193C and 15011A is due to the high volume of ethanol obtained per kg of bagasse. Genotype 5-193C has an ethanol yield of 11.9% and ethanol volume per kg bagasse of 1.79 ml, while genotype 15011A has an ethanol yield of 14.2% and ethanol volume per kg bagasse of 2.13 ml. These figures are higher than the average ethanol yield and the volume of ethanol per kg bagasse, which is 11.2% and 1.68 ml [32].

Sweet sorghum seeds yield ranges from 1.59–2.64 t/ha with an average of 2.05 t/ha, while the yield of ethanol from sweet sorghum seeds ranges from 1,297–2,577 l/ha with an average of 1,792 l/ha.The productivity of ethanol from sweet sorghum seeds is influenced by the tannin and starch content in the seeds. Sorghum seeds with high glucose levels and low tannin levels tend to have high ethanol yield. High tannin level inhibits the process of glucose fermentation; thus seeds with low tannin levels are preferred. Genotypes 15020A is one of the genotypes with a low tannin level (0.04%). Conversely, genotype 15020A has the highest ethanol yield (2,577 l/ha). Nevertheless, the tannin level is not the only deciding factor. Despite having higher tannin content (0.82%), 15011A has high seed yield and glucose levels in its seeds, so the fermentation of the seeds still produces a relatively high ethanol volume [32].

The fermentation substrate may consist of a mixture of raw materials. In all possible combinations of three parts of sorghum crop, except bagasse-and-seed mixture, genotype 15021A has the highest ethanol yield among all genotypes listed in Table 2. This is due to genotype 15021A having high stem biomass yield, juice sugar content, bagasse yield, and seed starch content altogether. Meanwhile, for the mixture of bagasse and seeds, genotype 15011A has the highest ethanol yield. This is due to genotype 15011A having high seed yield, seed starch level, and ethanol volume per kg of bagasse [32].

**Table 3.** Sweet sorghum products price comparison [33,34,35,36,37,38].

| Product | Price/Unit | Raw material price | Value-add |
|---|---|---|---|
| Fuel grade bioethanol | \$0.89/kg | \$0.16/kg sorghum seed; \$0.24/kg sorghum stem | 445% |
| Sorghum seed flour | \$0.51/kg | \$0.16/kg sorghum seed | 318% |
| Sorghum seed flake | \$1.05/kg | \$0.16/kg sorghum seed | 656% |
| Sweet sorghum syrup | \$2.46/kg | \$0.24/kg sorghum stem | 1025% |

[b] 2020 wholesale price.

### *3.2. Economic added-value comparison between sweet sorghum as food and sweet sorghum as bioethanol feedstock*

Before sweet sorghum was considered an energy crop, sweet sorghum had been cultivated as a food and feed crop. Sorghum seeds can be flaked into cereal and also processed into flour, which is an ingredient of various foods. The nutritional content of sorghum seeds is superior compared to other cereals such as wheat or rice. High minerals and protein content is a big advantage of eating sorghum seeds. Sweet sorghum juice contains a lot of sugar, so it has great potentials to be made into syrup, sugar, and alcoholic drinks. Additionally, sorghum seeds are commonly added to animal feed as a protein supplement. However the dose is limited because sorghum seeds contain lots of tannins that can interfere with nutrient absorption. Air-dried sorghum leaves and sorghum bagasse have high cellulose content, which makes them popular animal feeds. The nutrient content of sorghum leaves is equivalent to that of napier grass and sugarcane shoots, two types of plants commonly used as animal feed [39]. The potential of sweet sorghum as an energy crop is changing market demand for sweet sorghum. Table 3 presents the comparison of sweet sorghum economic values as food and as an energy crop. Table 3 shows that the sorghum economic value as a feedstock for bioethanol is small, compared to the value of sweet sorghum as food. This is one of the challenges of utilizing sorghum as a renewable energy source. The value-add of sorghum as an energy source can be increased if bioethanol's selling price is increased, but this is not a simple task since the government controls the energy price.

## 4. Conclusion

As one of the bioethanol-producing plants utilized as a renewable energy source, sweet sorghum has enormous potential for further development. Bioethanol can be produced by processing sweet sorghum juice, seeds, and bagasse from the same plant. In this review, two alternative routes have been compared to see the profitability of sweet sorghum as a source of bioethanol. The comparison is also made between sweet sorghum genotypes to see which variety produces the highest bioethanol. Although value-add comparison gives a rough estimate of economic viability, more accurate assumptions in economic aspects such as capital cost, utility cost, and labor cost are needed to make a better estimate of development feasibility. Besides, the development of sweet sorghum as a feedstock for bioethanol must be supported by the Indonesian government, especially the development of a superior variety of sweet sorghum.

## References

[1] Hunwartzen I 1982 *Modification of CFR Test Engine Unit to Determine Octane Numbers of Pure Alcohols and Gasoline-Alcohol Blends* SAE Technical Paper Series
[2] Lennartsson P R, Erlandsson P, Taherzadeh M J 2014 Integration of the first and second generation bioethanol processes and the importance of by-products *Bioresour Technol* **165**(C) 3-8
[3] Subagio H and Aqil M 2013 Pengembangan Produksi Sorgum Di Indonesia *Pros. Semin. Nas. Inov. Teknol. Pertan.* 199–214
[4] Balai Penelitian dan Pengembangan Pertanian *Sorghum* http://inaagrimap.litbang.pertanian.go.id/index.php/sentra-produksi/tanaman-pangan/sorghum
[5] Direktorat Budidaya Serealia 2012 *Kebijakan direktorat jenderal tanaman pangan dalam pengembangan komoditas jagung, sorghum dan gandum* (Jakarta: Direktorat Jenderal Tanaman Pangan–Kementan RI)
[6] Yulita R and Risda 2006 *Pengembangan sorghum di Indonesia* (Jakarta: Direktorat Budi Daya Serealia. Ditjen Tanaman Pangan)
[7] Kim M and Day D F 2011 Composition of sugar cane, energy cane, and sweet sorghum suitable for ethanol production at Louisiana sugar mills *J. Ind. Microbiol. Biotechnol.* **38** 803–7
[8] Ray R C, Uppuluri K B, Trilokesh C and Lareo C 2019 *Sweet Sorghum for Bioethanol Production: Scope, Technology, and Economics* (Elsevier Inc.)
[9] Almodares A and Hadi M R 2009 Production of bioethanol from sweet sorghum: A review *African J. Agric. Res.* **4** 772–80
[10] Davila-Gomez F J, Chuck-Hernandez C, Perez-Carrillo E, Rooney W L and Serna-Saldivar S O 2011 Evaluation of bioethanol production from five different varieties of sweet and forage sorghums (*Sorghum bicolor* (L) Moench). *Ind Crops Prod.* **33** 611-6
[11] Quintero J A, Montoya M I, Sánchez O J, Giraldo O H and Cardona C A 2007 Fuel ethanol production from sugarcane and corn: Comparative analysis for a Colombian case *Energy* **33** 385-99
[12] Ebrahimiaqda E and Ogden K L 2017 Simulation and Cost Analysis of Distillation and Purification Step in Production of Anhydrous Ethanol from Sweet Sorghum *ACS Sustain Chem Eng.* **5** 6854-62
[13] Ratnavathi C V., Patil J V. and Chavan U D 2016 *Sorghum Biochemistry: An Industrial Perspective*
[14] Bai F W, Anderson W A and Moo-Young M 2008 Ethanol fermentation technologies from sugar and starch feedstocks *Biotechnol Adv.* **26** 89-105
[15] Castro E, Nieves I U, Rondón V, et al 2017 Potential for ethanol production from different sorghum cultivars *Ind Crops Prod.* **109(September)** 367-73
[16] Watson S A 1984 Corn and sorghum starches: Production. in: Whistler R L, Bemiller J N, and Paschall E F, ed. Starch: Chemistry and technology. 2nd ed. (New York, NY: Academic Press)
[17] Barcelos C A, Maeda R N, Santa Anna L M M and Pereira N 2016 Sweet sorghum as a whole-crop feedstock for ethanol production *Biomass and Bioenergy* **94** 46–56
[18] Aggarwal N K, Nigam P, Singh D and Yadav B S 2001 Process optimization for the production of sugar for the bioethanol industry from sorghum, a non-conventional source of starch. World *J Microb Biot.* **17** 411-5
[19] Du Preez J C, de Jong F, Botes P J and Lategan P M 1985 Fermentation alcohol from grain

sorghum starch *Biomass* **8** 101-17

[20] Hamelinck C N, Van Hooijdonk G and Faaij A P C 2005 Ethanol from lignocellulosic biomass: Techno-economic performance in short-, middle- and long-term *Biomass and Bioenergy* **28** 384–410

[21] Sun Y and Cheng J 2002 Hydrolysis of lignocellulosic materials for ethanol production: A review *Bioresour. Technol.* **83** 1–11

[22] Taylor J, Zhang K and Wang D 2019 *Chapter 13 - Industrial and Nonfood Applications* (AACCI)

[23] Sipos B, Réczey J and Somorai Z 2009 Sweet Sorghum as Feedstock for Ethanol Production : Enzymatic Hydrolysis of Steam-Pretreated Bagasse *Appl Biochem Biotechnol.* **151** 151–62

[24] McIntosh S and Vancov T 2010 Enhanced enzyme saccharification of Sorghum bicolor straw using dilute alkali pretreatment *Bioresour Technol.* **101** 6718-27

[25] McMillan J D 1993 Xylose Fermenta to Ethanol : A R. *Natl Renew Energy Lab.* **NREL/TP-42(January)** 51

[26] Meirelles A, Weiss S and Herfurth H 1992 Ethanol dehydration by extractive distillation. *J Chem Technol Biotechnol.* **53** 181-8

[27] Carmo M J and Gubulin J C 1997 Ethanol-water adsorption on commercial 3A zeolites: kinetic and thermodynamic data *Brazilian J Chem Eng.* **14** 217-24

[28] Susilowati S H and Saliem H P 2013 Perdagangan Sorghum di Pasar Dunia dan Asia serta Prospek Pengembangannya di Indonesia (Pusat Sosial Ekonomi dan Kebijakan Pertanian)

[29] Khalil S R A, Abdelhafez A A and Amer E A M 2015 Evaluation of bioethanol production from juice and bagasse of some sweet sorghum varieties *Ann. Agric. Sci.* **60** 317–24

[30] Prasad S, Singh A and Joshi H C 2007 Ethanol production from sweet sorghum syrup for utilization as automotive fuel in India *Energy Fuels* **21** 2415-20

[31] P. Nghiem N, Montanti J and B. Johnston D 2016 Sorghum as a renewable feedstock for production of fuels and industrial chemicals *AIMS Bioeng.* **3** 75–91

[32] Pabendon M B and Sarungallo R S 2015 Pemanfaatan Nira Batang, Bagas, dan Biji Sorgum Manis sebagai Bahan Baku Bioetanol **31** 180–7

[33] Index Mundi. Sorghum Monthly Price Sep 2019 – Mar 2020. https://www.indexmundi.com/commodities/?commodity=sorghum.

[34] Diver S and Sampson J 2003 Sorghum Syrup (ATTRA)

[35] Harga Indeks Pasar (HIP) Bahan Bakar Nabati (BBN) Bulan April 2020. Direktorat Jenderal EBTKE. http://ebtke.esdm.go.id/post/2020/03/17/2512/harga.indeks.pasar.hip.bahan.bakar.nabati.bbn.bulan.april.2020. Published 17 March 2020.

[36] Shi-Zhong L Meeting the Demands of Food, Feed, and Energy by Sweet Sorghum. Presented on: FAO Technical Consultation on-How to Make Integrated Food Energy Systems Work and Benefit Small-Scale Farmers and Rural Communities in a Climate-Friendly Way; 14-15 July 2010; Roma. http://www.fao.org/bioenergy/26369-0887fe6c2880f4aa44c69d0a48457cb6e.pdf.

[37] Sorghum Flour. Indiamart. https://www.indiamart.com/proddetail/sorghum-flour-4634341491.html

[38] Jowar Flakes, Packaging Type: HDPE Bag. Indiamart. https://www.indiamart.com/proddetail/jowar-flakes-18580843988.html

[39] Biba M A 2011 Prospek Pengembangan Sorgum untuk Ketahanan Pangan dan Energi *Iptek Tanam. Pangan* **6** 257–69

# Mini solar water heating biodiesel plant by homogeneous catalyzing

**Rinjani Ratih Rakasiwi[1*], Syaifurrahman[2], and Usman A Gani[2]**

[1] Department of Chemical Engineering, Faculty of Engineering, Universitas Tanjungpura, Jalan Prof. Dr. H. Hadari Nawawi, Pontianak, 78124, Indonesia
[2]Department of Electrical Engineering, Faculty of Engineering, Universitas Tanjungpura, Prof. Dr. H. Hadari Nawawi, Pontianak, 78124, Indonesia

*Corresponding Author's E-mail: rinjani_s@yahoo.com

**Abstract.** Biodiesel can be generated by the catalytic vegetable oil- transesterification at 45-60°C. The heat requirement for a reaction can be obtained from solar thermal energy. West Kalimantan is crossed by the Equator where the sun shines for 10-15 hours /day. Palm Oil Production reaches 2.2 million tons making biodiesel has great potential to be produced. Implementation from the ideas is mini plant technology with solar water heaters for biodiesel production. Solar water heaters as a source of thermal energy are used as heating fluid in reactor plant. This technology is simple, without the need for electrical energy, low operating costs and environmentally friendly. The purpose of this research is to produce biodiesel by using a homogeneous catalyst of 2% KOH in a mini biodiesel plant with solar water heaters. Solar heaters are designed flat which is capable of producing hot water at 60°C, consists of a heat collector and a water storage tank made of stainless steel and polystyrene as an insulator. Transesterification is carried out for 1 hour at 65°C using homogenous catalyst of 2% - KOH mass. The biodiesel yield obtained 83.42%, fulfills the requirements according to SNI in the Acid Number which is 0.42mg KOH/g and the Methyl Ester content is 99,07%.

## 1. Introduction

Scarcity of fossil fuel encourages people to using new renewable energy from natural resources. Indonesian government is now concerning on implementing usage of Biodiesel as renewable fuel. B20 Program is a regulation program to require the mixing of 20% biodiesel and 80% diesel which have been implemented well in the transportation sector (Public Service Obligation) starting 2018. The mandatory has carried out massively in all sectors and will be continues gradually up to B100 [1]. Biodiesel is environmentally energy which produced from vegetable oil that has a high heating value close to the diesel (37-42 MJ/Kg) [2]. Burning biodiesel is more perfect and efficient than diesel and gasoline. Biodiesel does provide a reduction in harmful emissions, such as SOx, CO, hydrocarbons, particle matter, and soot as well as NOx in optimized diesel engines, together with net CO2 emissions. [3,4].

Biodiesel can be produce from crude palm oil by methanolysis reaction (transesterification with methanol). Indonesia is well-known as the fifth biggest country who producer of palm oil in the world in 2019. The amount of palm oil is able to reach 43 million metric ton. Production growth has also increased reaching 3.61% [5].

Transesterification process takes place at 32-65°C for 1-2 hours [6]. Thermal energy from sun can be absorbed directly by the solar water heater as a hot fluid for fulfill the heat of transesterification.

IOP Conf. Series: Materials Science and Engineering 1143 (2021) 012028 doi:10.1088/1757-899X/1143/1/012028

Indonesia is on the Equator, causes the climate to be tropical where sunlight shines around 10-15 hours/day and the energy reach 4.80 kWh/m$^2$/day [7]. Around 60°– 80°C temperature can be attained depending on solar radiation, weather conditions and solar collector system efficiency [8]. Solar water heater replaces an electric water heater, offer long-term benefits that go beyond simple Economics and 50 tons of avoided carbon dioxide emissions alone.

Solar water heating systems include storage tanks and solar collectors. Most solar water heaters require a well-insulated storage tank and it have an additional outlet and inlet connected to and from the solar collector. Solar collectors capture the sun's electromagnetic energy and convert it to heat energy. The efficiency of a solar collector depends not only on its materials and design but also on its size, orientation and tilt. Solar water heating system has been design can be reaches 62°C that can be used for biodiesel production [9].

Transesterification take place in jacket heating reactor with agitation. Insulated jacket to reduce heat loss. Biodiesel is currently synthesized using homogeneous alkaline catalysts because the transesterification reaction by an acid catalyst is much slower than the base-catalyzed reaction. The most common basic catalyst are potassium hydroxide (KOH), sodium hydroxide (NaOH), sodium methoxide ($NaOCH_3$), and sodium ethoxide ($NaOCH_2CH_3$). ethoxide (NaOCH) [10].

The implementation of this idea is a unit operation to produce biodiesel using solar water heater. Design and construction must be simple, easy to operate and efficient to reduce production cost. This tool can be implemented in some removed areas where there is not fuels and electric stock. The unit operation is mobile in order to get optimum thermal energy from the sun. It's called "Mini Solar Water Heating Biodiesel Plant". It's expected that the people can be fulfill the fuel needs independently. This research investigated of biodiesel production in Mini Solar Water Heating Biodiesel Plant using homogenous-base catalyst. The reaction taken place at 65°C for ±1 hour, using 2% - mass of KOH as catalyst. Biodiesel analyzed to check whether it meets the standard requirements.

## 2. Experimental

### *2.1 Materials*

Mini Solar Water Heating Biodiesel Plant is consists of two parts. They are mini biodiesel plant unit and solar water heater unit. Besides that, this unit is added by Solar Cell as source energy for stirrers in the transesterification reactor and washing tank.

Refined palm oil had 0.3 mg KOH/g, methanol (99,8%) pro analysis was produced from J.T. Baker, KOH-base catalyst was technical grade

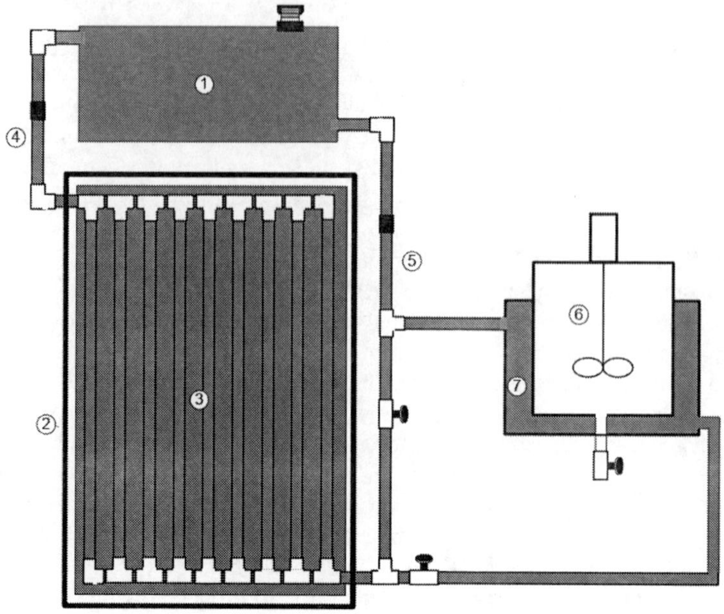


**Figure 1.** Mini Solar Water Heating Biodiesel Plant parts: (1) Storage tank; (2) Solar collection; (3) Tubes; (4) Up-stream; (5) Down-stream; (6) Transesterification reactor; (7) jacket heating.

*2.2 Produce Biodiesel from Palm Oil using homogenous-base*

Solar water heater filled with water with a volume of 20 Liters. Direct the mini plant to an area that is exposed to solar radiation optimally. Temperature in transesterification reactor, atmosphere temperature and sun illumination during process measured. Refined palm oil first heated in reactor at ± 60°C for 1 hour. Catalyst was fixed at 2%-m and methanol was 30%-m at ± 65°C to ± 1 hour in agitated reactor. Biodiesel separated from glycerin and washed to remove impurities (catalyst, methanol, water and glycerin). Water and biodiesel products were stirred in the washing tank until the pH was neutral. Biodiesel distillated to increase the purity for later analyzed.

*2.3 Analysis of Biodiesel Properties*

Parameters tasted for biodiesel were Total Acid Number, Methyl Ester Content, Total Glycerin and Yield. Total acid number of biodiesel was calculated according to the ASTM D 664 standard. The test can be done by adding, a suitable amount of KOH to neutralized the acids in the biodiesel sample in a titration method. The acid value is the mg of KOH required to neutralize one gram of biodiesel sample. Ester Content was obtained from the SNI 7182:2015.

The fuel's ester content indicates the amount of FAME in the fuel, and therefore, it is an indicator of th e fuel's quality. Total Glycerin was calculated using SNI 7182:2015.

Glycerin is a byproduct of the chemical reaction that produces biodiesel, may remain in the fel if th e ester is inadequately separated or washed. Glycerin also may separate out of the liquid during storag e after any methanol, which acts as a solvent, has evaporated [11].

Biodiesel Yield was calculated with Eq. (1)

$$\text{Biodiesel Yield (wt.\%)} = \frac{W_b}{Wo} x\ 100 \quad (1)$$

Where $W_b$ is the mass of biodiesel and $Wo$ is the mass of refined palm oil.

## 3. Result and Discussion

### *3.1 Specification of Mini Solar Water Heating Biodiesel Plant*

The material used to make solar water heaters is lightweight, economical and easily obtained. Specifications of solar water heaters are presented in Table 1. The designed solar water heater consists of a solar collector and a water storage tank. Solar collectors consist of several parts, namely:

1. The heat transfer pipe is made of stainless steel so that it can reduce the rate of corrosion and heat can be absorbed well by the pipe.
2. The outer cover is made of acrylic glass, so it can optimize the absorbed solar energy.
3. A black coating made of polystyrene material painted black is intended to increase the absorption of heat energy by water in pipes in solar water heaters (black layer emissivity = 1)[12].

**Figure 2**. Mini Solar Water Heating Biodiesel Plant.

The water storage tank is made of stainless steel and an insulating layer consisting of foam and aluminum foil. The insulator layer is expected to reduce the heat of the water thereby reducing heat energy lost to the environment. The water in the pipe absorbs heat from the sun so that the temperature of the water rises. The difference in density between hot water and cold water, so hot water flows upward and cold water flows downward. Hot water is then flowed into the biodiesel reactor. This is sustainable so as to form circulation. So without the help of a pump power water can flow [13].

**Table 1**. Specification of water solar heater.

| Parts | Materials | Dimension (length x width) cm |
|---|---|---|
| Storage tank | Shell: Stainless steel<br>Isolator : Polystyrene | 70 x 7,62 |
| Collector pipes | Stainless steel | 70 x 2,5 |
| Cover | Acrylic glass | 70 x 100 |
| Absorber | Black Polystyrene | 70 x 100 |
| Stream pipes | Stainless steel | 150 x 2,5 |

The maximum temperature of the transesterification reactor reached 65°C and the maximum temperature of atmosphere at that time reached 37.8°C with the range of sun illumination were 1020-1280 Lux. Some things that can affect the optimum temperature of the water obtained are:

1. Sunlight shining on the earth is unstable during the experiment time
2. The heat absorbed and stored in the reactor is partially lost to the environment (Q loss).

**Figure 3**. Solar water heating system.

In additional solar cell installed could produced 100W of power (Dimension: 1085 x 675 x 25 mm) which could turned on two direct current mixers for reaction and for washing process.

*3.2 Analysis of Biodiesel Properties*

Based on the results of the parameter test, the biodiesel produced using a 2% KOH catalyst at the Mini Solar Water Heating Biodiesel Plant has met the requirements. Ester content was 99.07%. A low amount of ester may indicate that un-reacted compounds, such as triglycerides, or process related compounds such as catalyst (KOH/NaOH) or methanol remain in the fuel. Low levels also may indicate contamination with non-methyl ester compounds. These impurities may cause fuel filter plugging, engine deposits or other problems. Local low may require measuring that ester content, ester measurement also can help prevent or minimize fraud under incentive programs [11].

**Table 2**. Biodiesel Properties

| Property | Units | Biodiesel Product | Limit | Test Method |
|---|---|---|---|---|
| Ester content | %- mass | 99.07 | 96.5 min. | SNI 7182:2015 |
| Total Acid Number | mg KOH/g | 0.42 | 0.5 max. | ASTM D 664 |
| Total Glycerin | %- mass | 0.09 | 0.24 max. | SNI 7182:2015 |

Acid number is a measure of the acids in the fuel. Biodiesel result had acid number of 0.42 mg KOH/g which was suitable for standard requirement. These acids came from two sources: acid used in the production of the biodiesel that are not completely removed in the production process and acid as a byproduct from degeneration by oxidation. The acids were measured in terms of amount of KOH required to neutralize a gram of biodiesel. The acid number would change as a result of the normal oxidation process over time and recent research has shown that this change is a good indicator of B100 stability. If the fuel is not used immediately after purchase, buyers should monitor their biodiesel fuel blends for change in acid number, as an indicator of fuel degradation. The presence of acid in the fuel can harm injection system and other metallic components.

Total glycerin of 0.09 %-mass was also suitable with the biodiesel standard. Once separated, the glycerin will fall to the tank bottom and attract such polar compounds as water, monoglycerides and soap that can block filter, damage injection, cause injection coking and other engine deposits and otherwise make the fuel incompatible with vehicle material and reduce engine durability. Glycerin can be controlled through the use of good operating practices during production.

Biodiesel with a 2%-mass KOH base-catalyst had gave the yield of 83.42%. It was a low number. When the separation process accrued, it was seen that there was still an oil phase and soap phase. It was indicated that conversion reaction was not optimum yet. This could be caused by the temperature reaction which is not optimum and fluctuated. Other than that, separation process has gone quite hard. Saponification had taken placed due to the excessive use of the catalyst.

## Conclusion

The maximum temperature of hot water in reactor from solar water heater was reached 65°C. Biodiesel has been produced in mini plants using 2%-m KOH catalyst give a yield of 83.42% , a high methyl ester content of 99.07%, a acid number of 0.42 mg KOH/g oil and a total glycerin of 0.09%-mass. And therefore it could be concluded that mini biodiesel plant-solar water heater produced biodiesel that have fulfilled the requirements according to biodiesel standard.

## Acknowledgement

This research was funded by the Ministry of Education and Culture, Republic of Indonesia.

## References

[1] Ministry of Energy and Mineral Resources 2018 *Journal of Energy Potential of Solar* **2** 20
[2] Wasley 2015 *Design of a Small-Scale Biodiesel Plant* Bellairs Reasearch Institute: McGrill University
[3] Katabathini Narasimharao, Adam F Lee, Keren Wilson 2007 Catalysts in Production of Biodiesel: A Review *J. Biodiesel Materials and Energy* **1** 1-12
[4] Sarina Sulaiman, Nur Syakirah 2016 Overview of Catalysts In Biodiesel Production *J. Engineering and Applied Sciences* **11** 11
[5] Indonesian Forestry statistic 2014 *Kelapa Sawit Palm Oil* Direktorat Jenderal Perkebunan
[6] Clifford, Caroline 2017 *The Reaction of Biodiesel: Trans-eterification* The Pennsylvania State University.
[7] Philibert, Cedric 2014 *Solar Energy Perspectives, Renewable Energy Technologies* French: International Energy Agency.
[8] Homola, Christopher 2014 *Solar Domestic Hot Water Heating Systems Design, Installation and Maintenance* Solar Rating and Certification Corporation.
[9] Weiss W 2011 *Solar Thermal, State of the art, Application Areas and Preview of Technology Development* presentation at the IEA Solar Heating and Cooling Roadmap Workshop, 28-29 April, Paris.
[10] Wendi, Valentinoh Cuaca, Taslim 2014 Effect of Reaction Temperature and Catalyst Concentration for Producing Biodiesel from Waste Beef Tallow Using Heterogeneous Catalyst CaO from waste Eggshell *Pap-Am Chem. Soc. Div Fuel Chem* **55** (2), 410
[11] Biodiesel Guidelines March 2009 From the Worldwide Fuel Charter Committee, Ivan Hodac, European Automobile Manufactures Association, Belgium
[12] India Gov 2014 *Solar Water Heating Systems*. Ministry of New and Renewable Energy: India.

# Process Modeling of Drying and Torrefaction of Oil Palm Trunk (OPT)

**Dendy Adityawarman, Vika Fujiyama, Hyung Woo Lee, Retno Gumilang Dewi, and Johnner P. Sitompul***

Department of Chemical Engineering, Faculty of Industrial Technology, Institute of Technology Bandung, Jalan Ganesha 10 Bandung 40132 Indonesia

*E-mail: sitompul@che.itb.ac.id

**Abstract.** This paper concerns on process modeling of drying and torrefaction of oil palm trunk (OPT), abundant biomass resources coming from replantation of oil palm. The processes are very crucial in treating of OPT before utilization for energy source due to high level of water content and volatile matters in the biomass. Modeling with Aspen Plus is performed to predict the composition of product in mass in fastest way without doing experiment. The study uses Flash Dryer Model and Reactor Model available in Aspen as tool box for modeling the drying process and torrefaction process. Objective of the paper is to validate the process model with experimental results from that of Prasetyo et al.(2016) in identifying final mass and calorific value of torrefied OPT. The process modeling of OPT drying and torrefaction gives the charcoal yield of 0.2972 compared to experiment yield of 0.24. The calorific value (HCV) of torrefied OPT with RGibbs is 6,901.0 kcal/kg that is similar with experiment result of 6,939.2 kcal/kg.
**Keywords:** calorific value, drying, oil palm trunk, simulation, torrefaction

## 1. Introduction

The energy demand, reported 505 quadrillion BTU in 2008, is expected to increase by 53% in 2035 [1]. Energy security and environmental sustainability are the major emerging issues in the world that can only be addressed through diversification in the energy resources and clean fuels. The promotion of indigenous renewable energy sources and the low carbon fuels could be good solution while addressing issues of the global warming and climate change. Kyoto Protocol, legally binding EU 20-20-20 targets, and volatility of oil prices have encouraged the global community to reduce dependence on oil and replace it with clean and renewable energy resources [2]. Biomass energy, one of a renewable energy, could be a good candidate for replacement of fossil fuels with prospective and emerging technologies for converting biomass through pyrolysis, gasification and torrefaction. It can be converted into three valuable products, solid (briquette, pellet, char), liquid (ethanol, biodiesel, biofuel), and gaseous (synthesis gas, biogas) [3].

Due to their availability in Indonesia, oil palm wastes such as empty fruit bunch (EFB) and oil palm trunk (OPT) are considered to be the best options for fuel generation and biochar production. Indonesia is the world's biggest producer and consumer of palm oil, currently providing about half of world the supply. Based on USDA data, Indonesia produced 33,500,000 MT of palm oil in 2014. It gave an increase of about 8.06% from the previous year. [7]. In general, OPT are collected during replanting and can be utilized as resource for renewable chemicals as well as for renewable energy. OPT are available in significant amount before replantation is carried out and thus the utilization of OPT will increase its added value.

The commercial utilization of biomass for energy requires integration of intensive logistic system [5]. Torrefaction is regarded as an emerging thermal biomass pretreatment method aiming to produce high quality solid biomass products. It has an ability to reduce the major limitations of biomass such as heterogeneity, lower bulk density, lower energy density, hygroscopic behavior, and fibrous nature. Torrefied biomass can be efficiently used in the co-firing power plant and, in the future, is expected to replace coal in the metallurgical process and promoted as an alternative for charcoal [6]. Basically, torrefaction can be divided into two consecutive processes: drying and isothermal heating or roasting for removing volatiles through different decomposition reactions at 200-300°C in an inert environment at an atmospheric pressure.

This study is aimed to modeling the drying and torrefaction of Oil Palm Trunk (OPT). Process modeling using Aspen Plus V10 could represent laboratory experiment and ease the process to identify the torrefaction products. In addition, prior to torrefaction process, drying process was simulated to eliminate $H_2O$ content in OPT. Modeling analysis conducted to obtain the highest calorific value of OPT. The result of this study are expected to the further research development.

## 2. Process Modeling

Prasetyo et al. (2016) have conducted research on drying and torrefaction of OPT. The operating temperature condition for OPT drying was 110-225°C while torrefaction temperature was 245°C. Yield of torrefied OPT from previous experiment was 24,01%. These experiment data were used as reference for process modeling.

The drying and torrefaction processes of Oil Palm Trunk (OPT) were simulated in Aspen Plus. As shown in Fig. 1, the overall process contains a drying and torrefaction processes. The simulation was designed to validate the yield of torrefaction process of OPT and calorific value of torrefied OPT. The general simulation assumptions were as follows:

1. Oil Palm Trunk had a high content of cellulose therefore it was defined and assumed as cellulose.
2. Feedstock basis was defined 10 kilograms with ash content of 5.85% and moisture content ($H_2O$) of 5.89%.
3. Ash in OPT was defined as a non-conventional solid with weight percentage of 5.85%.
4. The stream class used in the model was MCINCPSD that defines the particle size distribution for non-conventional and conventional solids.
5. The PENG-ROB property model was used to estimate the properties of gas phase while the IDEAL property model was used for solid and liquid phase.

Non-conventional solids in Aspen Plus was assumed that do not participate in phase and chemical equilibrium calculations [21].

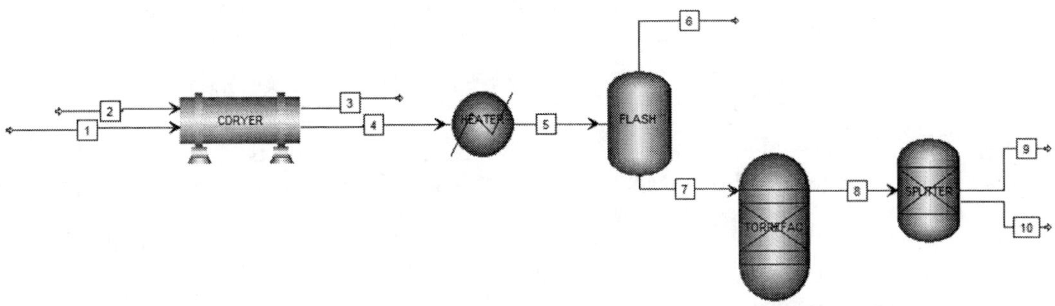


**Figure 1.** A process flow diagram for drying and torrefaction of Oil Palm Trunk (OPT).

### 2.1. *Overview of Process Modeling*

Figure 1 shows the general layout of a OPT drying and torrefaction process. The process configuration consists of OPT drying using contact dryer to remove moisture content ($H_2O$) and heater to increase the temperature of OPT. A flash separator is required to separate gases and dried OPT. Torrefaction reactor

IOP Conf. Series: Materials Science and Engineering 1143 (2021) 012029 doi:10.1088/1757-899X/1143/1/012029

is a reactor to convert dried OPT to charcoal, carbon dioxide, water, and hydrogen gas. Carbon monoxide is assumed not present in torrefaction product. Torrefaction product enters a splitter to separate $H_2O$ as vapor phase and torrefied OPT as solid component.

The reaction of OPT torrefaction is given as:

$$OPT \rightarrow Charcoal_{(s)} + CO_2 + H_2O + H_2 \quad (1)$$

### 2.2. *Contact Dryer*

Moisture content in Oil Palm Trunk was removed as vapor phase using dryer. The contact dryer was selected from the available solid unit operations in Aspen Plus. Air was used as an inlet gas in dryer with mass flowrate of 15 kg/hr and temperature of 132°C.

**Table 1.** Dryer specification for OPT drying.

| Parameter | Unit | Value |
|---|---|---|
| Length | meter | 4 |
| Heat transfer coefficient between wall and solids | Watt/sqm-K | 400 |
| Heat transfer area wall/solids | sqm | 10 |
| Wall temperature | °C | 170 |
| Heat transfer coefficient between gas and solids | Watt/sqm-K | 10 |
| Heat transfer area gas/solids | sqm | 3 |

### 2.3. *Heater and Flash Separator*

Before entering the torrefaction reactor, heater model was used to increase the temperature of dried OPT while flash separator was used to separate vapor phase in solid phase. The operation conditions for heater and flash separator are 225°C and 0.5 atm.

### 2.4. *Torrefaction Reactor and Splitter*

Torrefaction reactor are modeled by Reactor Yield and Reactor Gibbs, shown in the Fig. 1. as RYield and RGibbs. In the RYield, data of product yield of OPT torrefaction process is specified and showed on Table 2. Yield value of charcoal is obtained from previous experiments [19] while for $CO_2$ and water is obtained from literature [20]. The process parameter that needs to be studied in this research is yield of torrefaction product to obtain calorific value.

**Table 2.** RYield specification for Torrefaction of OPT.

| Component | Basis | Yield | Reference |
|---|---|---|---|
| Water ($H_2O$) | Mass | 0.1397 | Haryadi et al (2009) |
| Charcoal (C) | Mass | 0.24 | Prasetyo et al (2016) |
| $CO_2$ | Mass | 0.042 | Haryadi et al (2009) |

RGibbs works based on excess Gibbs value in order to predict the products of torrefaction. Moreover, RGibbs is a reactor that performs reaction by minimalizing Gibbs energy and calculate the product composition based on the properties of chemicals which involved in chemical reaction. The chemical reaction that occurred in RGibbs as torrefaction reactor referred to Eq. (1). The principle of RGibbs depends on chemical and phase equilibrium by specifying temperature in chemical system. The temperature conditions for RGibbs are 245°C according to previous experiment of OPT torrefaction process [19].

Splitter is used to separate solids products from gases or liquids. In this simulation, split fraction of $H_2O$ component as gases is 1. The products of splitter model are torrefied OPT and torrefied gas. Validation are performed by comparing yield and calorific value of torrefaction product from simulation result with previous experiment.

## 3. Results and Discussion

### *3.1. Comparison of Simulation and Experimental Results of OPT Drying using Contact Dryer and Torrefaction*

In this simulation, Oil Palm Trunk was assumed as cellulose and charcoal was used as torrefied product. Contact dryer was able to remove water ($H_2O$) and air as exhaust gases while the dried OPT entered the heater and flash separator. Table 3 and Table 4 showed the simulation results of OPT Drying and Torrefaction Process using RYield and RGibbs. The operation temperature of reactor was set to 245°C. From the simulation, Gibbs Reactor produce $H_2O$ and $CO_2$ mass flowrate higher than Yield Reactor produced. In contrast, it yields lower mass flowrate of charcoal as torrefied product. Moreover, Gibbs Reactor produce $H_2$ in torrefaction reaction.

**Table 3.** Simulation result of OPT Drying and Torrefaction Process with RYield.

| Stream | Units | 1 | 2 | 3 | 4 | 5 | 6 | 7 | 8 | 9 | 10 |
|---|---|---|---|---|---|---|---|---|---|---|---|
| Mass Flows | kg/hr | 10.000 | 15.000 | 15.589 | 9.411 | 9.411 | 0.000 | 9.411 | 9.411 | 2.462 | 6.652 |
| $H_2O$ | kg/hr | 0.589 | 0.000 | 0.589 | 0.000 | 0.000 | 0.000 | 0.000 | 2.462 | 2.462 | 0.000 |
| Charcoal | kg/hr | 0.000 | 0.000 | 0.000 | 0.000 | 0.000 | 0.000 | 0.000 | 4.230 | 0.000 | 4.230 |
| $CO_2$ | kg/hr | 0.000 | 0.000 | 0.000 | 0.000 | 0.000 | 0.000 | 0.000 | 1.837 | 0.000 | 1.837 |
| Ash | kg/hr | 0.585 | 0.000 | 0.000 | 0.585 | 0.585 | 0.000 | 0.585 | 0.585 | 0.000 | 0.585 |
| Cellulose | kg/hr | 8.826 | 0.000 | 0.000 | 8.826 | 8.826 | 0.000 | 8.826 | 0.000 | 0.000 | 0.000 |
| Air | kg/hr | 0.000 | 15.000 | 15.000 | 0.000 | 0.000 | 0.000 | 0.000 | 0.000 | 0.000 | 0.000 |

**Table 4.** Simulation result of OPT Drying and Torrefaction Process with RGibbs.

| Stream | Units | 1 | 2 | 3 | 4 | 5 | 6 | 7 | 8 | 9 | 10 |
|---|---|---|---|---|---|---|---|---|---|---|---|
| Mass Flows | kg/hr | 10.000 | 15.000 | 15.589 | 9.411 | 9.411 | 0.000 | 9.411 | 9.411 | 3.096 | 6.314 |
| $H_2O$ | kg/hr | 0.589 | 0.000 | 0.589 | 0.000 | 0.000 | 0.000 | 0.000 | 3.096 | 3.096 | 0.000 |
| Charcoal | kg/hr | 0.000 | 0.000 | 0.000 | 0.000 | 0.000 | 0.000 | 0.000 | 2.797 | 0.000 | 2.797 |
| $CO_2$ | kg/hr | 0.000 | 0.000 | 0.000 | 0.000 | 0.000 | 0.000 | 0.000 | 2.686 | 0.000 | 2.686 |
| Ash | kg/hr | 0.585 | 0.000 | 0.000 | 0.585 | 0.585 | 0.000 | 0.585 | 0.585 | 0.000 | 0.585 |
| Cellulose | kg/hr | 8.826 | 0.000 | 0.000 | 8.826 | 8.826 | 0.000 | 8.826 | 0.000 | 0.000 | 0.000 |
| Air | kg/hr | 0.000 | 15.000 | 15.000 | 0.000 | 0.000 | 0.000 | 0.000 | 0.000 | 0.000 | 0.000 |
| $H_2$ | kg/hr | 0.000 | 0.000 | 0.000 | 0.000 | 0.000 | 0.000 | 0.000 | 0.246 | 0.000 | 0.246 |

Table 5 shows the yield comparison of OPT torrefaction process from the simulation and experiment. Based on the simulation result, yield of charcoal at end of torrefaction process is 0.2972 compared with experiment yield of 0.24 (Prasetyo et al, 2016). It could be concluded that the simulation results represent the real behavior and relatively close to the experimental results.

**Table 5.** Yield Comparison and Validation of OPT Torrefaction process.

| Mass Flows | Units | Torrefied Gas | Torrefied OPT | Yield Simulation (RGibbs) | Yield Experiment (Prasetyo et al, 2016) | OF |
|---|---|---|---|---|---|---|
| $H_2O$ | kg/hr | 3.0960 | 0.0000 | 0.3290 | | |
| Charcoal | kg/hr | 0.0000 | 2.7970 | 0.2972 | 0.2400 | $5{,}68.10^{-2}$ |

| | | | | |
|---|---|---|---|---|
| $CO_2$ | kg/hr | 0.0000 | 2.6860 | 0.2854 |
| Ash | kg/hr | 0.0000 | 0.5850 | 0.0622 |
| $H_2$ | kg/hr | 0.0000 | 0.2460 | 0.0261 |
| **Total** | **kg/hr** | **3.0960** | **6.314** | **1.0000** |

$$OF = \left(\frac{Yield\ simulation - yield\ experiment}{yield\ experiment}\right)^2$$

*3.2. Comparison of OPT and Torrefied OPT Calorific Value from Simulation and Experiment*
Dulong's formula for calculating the calorific value is given as [22]:

$$\text{Gross calorific value (HCV)} = \frac{1}{100}\left[8080C + 34500\left(H - \frac{O}{8}\right) + 2240S\right]\frac{kcal}{kg} \quad (2)$$

$$\text{Net calorific value (LCV)} = [\text{HCV-0.09H x 587}]\frac{kcal}{kg} \quad (3)$$

Table 6. shows the comparison of OPT and Torrefied OPT Calorific Value in kcal/kg. In this simulation, cellulose was assumed as OPT and gave the calorific value of 2,920.0 kcal/kg (HCV) and 2,570.3 kcal/kg (LCV) compared with the calorific value of non-processed OPT of 5,232.1 kcal/kg. The calorific value of experiment was higher than the simulation because non processed OPT consists of cellulose, hemicellulose, lignin, and extractives.

The calorific values of torrefied OPT from RYield and RGibbs were similar with calorific value of experiment results which means that the process modeling of drying and torrefaction of Oil Palm Trunk could be the reference for further research development.

**Table 6.** Comparison of OPT and Torrefied OPT Calorific Value.

| Component | Simulation | | Experiment | Reference of Experiment |
|---|---|---|---|---|
| | HCV (kcal/kg) | LCV (kcal/kg) | Calorific Value (kcal/kg) | |
| Cellulose | 2,920.0 | 2,570.3 | n/a | Prasetyo et al, 2016 |
| Non processed OPT | n/a | | 5,232.1 | |
| Torrefied OPT at 245°C (RYield) | 6,547.0 | | 6,939.2 | |
| Torrefied OPT at 245°C (RGibbs) | 6,901.0 | 6,627.4 | | |
| Bagasse | | | 4,034.2 | Sahito et al, 2013 |
| Banana plant trash | | | 3,401.2 | |
| Barley straw | | | 4,048.5 | |
| Canola straw | | | 3,874.2 | |
| Coconut coir | | | 3,907.6 | |
| Corn cob | | | 3,972.1 | |
| Corn stover | | | 4,225.3 | |
| Cotton gin waste | | | 3,422.7 | |
| Cotton stalks | | | 3,833.5 | |
| Groundnut shell | | | 4,390.1 | |
| Mango wood | | | 4,318.4 | |
| Millet stalks | | | 3,976.9 | |
| Neem wood | | | 4,781.8 | |
| Rice husk | | | 3,587.5 | |
| Rice straw | | | 3,463.3 | |

| | |
|---|---|
| Saw dust | 3,960.1 |
| Soybean straw | 3,876.5 |
| Sugarcane straw | 3,726.1 |
| Sunflower stover | 3,747.6 |
| Sunflower seed shell | 4,079.6 |
| Wheat straw | 3,597.1 |
| Wood chips | 4,757.9 |

## 4. Conclusions

The drying process using contact dryer helps to remove $H_2O$ content in OPT prior to torrefaction process. Torrefaction is a promising process for processing OPT as it leads to higher heating value and improved energy quality. The process modeling of OPT drying and torrefaction gives the charcoal yield of 0.2972 compared to experiment yield of 0.24. The calorific value (HCV) of torrefied OPT with RGibbs is 6,901.0 kcal/kg that is similar with experiment result of 6,939.2 kcal/kg.

## Acknowledgements

The research was funded by research grant *P3MI from Bandung Institute of Technology*

## References

[1] IEO. International energy outlook 2011. U. S. Energy Information Administration 2011. Retrieved at www.eia.gov/ieo/pdf/0484 (2011).pdf, 24th October 2011
[2] Deutmeyer M, Bradley D, Hektor B, Tumuluru J S, Hess R, Nikolaisen L and Wild M 2012 *Possible Effect of Torrefaction on Biomass* Trade IEA Bioenergy, Task:40 Sustainable International Bioenergy Trade
[3] Koh M P and Hoi W K 2003 Sustainable Biomass Production for Energy in Malaysia *Biomass and Bioenergy* **25** 517-529
[4] Ogawa M, Okimori Y, Takahashi F 2006 Carbon sequestration by carbonization of biomass and forestation: three case studies *Mitigation and Adaptation Strategy for Global Change* **11** 421–36
[5] Atkinson C J, Fitzgerald J D, Hipps N A 2010 Potential mechanisms for achieving agricultural benefits from biochar application to temperate soils: a review *Plant Soil* **337** 1–18
[6] Nhuchhen D R, Basu P, Acharya B 2014 A Comprehensive Review on Biomass Torrefaction *International Journal of Renewable Energy & Biofuels*
[7] http://www.indexmundi.com/agriculture/?country=id&commodity=palm-oil&graph=production
[8] Chan K W 2005 Best-developed practices and sustainable development of the oil palm industry *Journal of Oil Palm Research* **17** 124–35
[9] Puah C W, Balasundram N, Choo Y M, Lim W S 2011 Malaysian oil palm industry: responding to the sustainability criteria of greenhouse gas emission savings under the European Union Renewable Energy Directive *Palm Oil Development* **54** 5–7
[10] Goh J, Leong K H 2008 Development of pulp and paper products from oil palm fiber *Proceedings of the International Conference on Oil Palm Biomass*
[11] Hilstrøm T, Mikkelsen M J, Ahring B K 2005 Ethanol potential for empty fruit bunches pre-treated by wet-explosion *Proceedings of the international palm oil congress* 348–56
[12] Yamada H, Tanaka R, Sulaiman O, Hashim R, Hamid Z A A, Yahya M K A, et al 2010 Old oil palm trunk: a promising source of sugars for bioethanol production *Biomass Bioenergy* **34** 1608–13
[13] Mokhtar A, Hassan K, Abdul Aziz A, Wahid M B 2011 Plywood from oil palm trunks *Journal of Oil Palm Research* **23** 1159–65
[14] Singh R P, Embrandiri A, Ibrahim M H, Esa N 2011 Management of biomass residues generated from palm oil mill: vermicomposting a sustainable option *Resoure Conservation and Recycling* **55** 423–34

[15] Loh S K, James S, Ngatiman M, Cheong K Y, Choo Y M, Lim W S 2013 Enhancement of palm oil refinery waste – spent bleaching earth (SBE) into bio organic fertilizer and their effects on crop biomass growth *Industrial Crops Production* **49** 775–81
[16] Sumathi S, Chai S P, Mohamed A R 2008 Utilization of oil palm as a source of renewable energy in Malaysia *Renewable Sustainable Energy* **12** 2404–21
[17] Deris R R R, Sulaiman M R, Darus F M, Mahmud M S, Bakar N A 2006 Pyrolysis of oil palm trunk (OPT) *Proceedings of the 20th Symposium of Malaysian Chemical Engineers* **19-21th of Dec** 245-250
[18] Mori Y 2007 Oil oil palm trunks as promising feedstock for biofuel and bioplastics *Presented at JIRCAS 4th BMWS*
[19] Prasetyo D, Pramudita D, Lee H W, Sitompul J P 2016 Effect of Vacuum and Atmospheric Drying on Torrefaction of Oil Palm Trunk (OPT) *International Seminar on Chemical Engineering In conjunction with Seminar Soehadi Reksowardojo*
[20] Haryadi Hardianto T, Pasek A D, Suwono A, Azhari R, Ardiansyah W 2009 The Aspen TM Software Simulation of a Peat Torrefaction System Using RYield and SSplit Block as Reactor Model *Proceedings of International Symposium on Sustainable Energy and Environmental Protection (ISSEEP)*
[21] Haydary 2004 J. Processes with Nonconventional Solids *Wiley Online Library Chapter 14*
[22] Demirbas A, Demirbas A H 2004 Estimating the Calorific Values of Lignocellulosic Fuels *Energy Exploration & Exploitation Volume 22 Number 2*
[23] Sahito A R, Mahar R B, Siddiqui Z, Brohi K M 2013 Estimating Calorific Value of Lignocellulosic Biomass from Volatile and Fixed Solids *International Journal of Biomass & Renewables*

# Novel Approach of Biodiesel Production Waste Utilization to Support Circular Economy in Biodiesel Industry

**Aghietyas Choirun Az Zahra[1], Ilya Arina Rusyda[1], Andini Hizbiyati[1], Felix Giovani[2], Nabila Zahara[2], Bramantha Jiwandaru[1], David Gunawan[1], Giovani Andre Halim[1], Meiti Pratiwi[1], Astri Nur Istyami[1], Atmy Verani Rouly Sihombing[3], Ardiyan Harimawan[1], Dwiwahju Sasongko[1], Jenny Rizkiana[1,2,4*]**

[1]Department of Chemical Engineering, Institut Teknologi Bandung
[2]Department of Bioenergy Engineering & Chemurgy, Institut Teknologi Bandung
[3]Department of Civil Engineering, Politeknik Negeri Bandung
[4]Center for Catalysis and Reaction Engineering, Institut Teknologi Bandung

*corresponding author: jr@che.itb.ac.id

**Abstract.** Current development of industry nowadays is more towards sustainability and greener production that minimize the production of waste, for example the production of biodiesel energy from Refined, Bleached, and Deodorized Palm Oil (RBDPO). However, biodiesel production is still could not eliminate the hazardous waste that is glycerin pitch. Glycerin pitch is a waste found in the glycerin purification unit which is a unit used to produce the high-grade glycerin, a side product of biodiesel production. Up until now, the existing glycerin pitch is still not well managed and remains as a hazardous waste. Glycerin pitch contains a sodium salt that can be recovered to produce biodiesel transesterification catalyst, sodium methoxide. Applying a concept of circular economy, a catalyst recovery building block is attached to integrated biodiesel process with glycerol purification. The process consists of three units, solid-liquid extraction to separate inorganic salt contains in the pitch, Chlor-Alkali process to convert inorganic sodium salt to sodium hydroxide, and catalyst making reactor to produce a sodium methoxide. Using a simulation process, the amount of catalyst recovered in the process is 43.78%.
Keywords: *circular economy, glycerin pitch, novel approach, process recovery, sodium methoxide*

## 1. Introduction

Indonesia has a rapid growth of crude palm oil production in the last five years, with the number of 47 million tons in 2019 [1]. Domestic consumption only takes 32,3% of its total cumulative production (crude palm oil and palm kernel oil), with the highest of utilization after food industry is for biodiesel making [1]. Biodiesel is an alternative energy that has similar chemical and physical properties to petroleum diesel so that it can be used directly in diesel engines or mixed with petroleum diesel [2]. In Indonesia, biodiesel is mixed with petroleum diesel in certain composition and widely used as a transportation fuel. As an alternative fuel, biodiesel has many advantages: it is derived from a renewable, domestic resource, it is biodegradable and non-toxic, also by utilizing biodiesel, we slowly relieving reliance on petroleum fuel imports [3]. Compared to petroleum fuel, biodiesel has more preferable combustion emission profile, such as low emission of carbon monoxide, particulate matter, and unburned hydrocarbons [4]. Carbon dioxide produced by combustion of biodiesel can be recycled by photosynthesis, thereby minimizing the greenhouse effect [5].

With B-30 policy implemented in January 2020, which requires 30% biodiesel composition in fuel, the need for biodiesel in Indonesia is increasing [6]. Biodiesel is produced using crude palm oil as raw material through transesterification reaction with methanol catalyzed with acid or base catalyst [7]. Crude glycerol, the by-product in biodiesel production, has percentage about 10%wt of the biodiesel product and still contains about 20% of impurity in it [8]. Using distillation, crude glycerol can be purified to produce glycerin, which commonly has more than 95% purity [9]. Process of glycerol purification generate waste called glycerin pitch, which is found at the bottom product in distillation unit [10]. By mid of 2019, it is found that glycerin production capacity is 0.88 million tons in Indonesia [11], with the current production of 150,000 tons/year. Anticipating the B-30 policy, glycerin production can be increased as high as 500,000 tons/year, thus resulting much higher waste.

Glycerin pitch is a viscous alkaline gel that has brown to dark brown appearance containing non-glycerol organic compound and has a high ash/mineral content [12]. Those compounds exist in various length of carbon chains, from short chain such as leftover methanol until a long fatty acid chain. Soaps and residual oil also found in glycerin pitch. Metal catalyst during the biofuel making also being carried over through the whole process of glycerin purification and is contained in the pitch.

Government regulation in Indonesia categorized this waste as a hazardous waste that requires special treatment to remove. Industrial practice in dealing glycerin pitch is by sending it to a third party for treatment, which requires fee to pay this service. Considering the possibility of increasing glycerin production, the amount of fund required to treat glycerin pitch waste will be much higher. A better alternative is needed as a solution to this matter. As stated in Government Regulation no. 101 of 2014, the first act of hazardous waste management is to reduce the waste by using material substitution, process modification, or the application of environmentally friendly technology. The next step is to reuse and recycle the waste, which in the case of hazardous waste can be tricky and needs a more pre-treatment process.

Waste management is a sustainable process which will be beneficial to the environment. However, the acts described above is not enough, since there is also a possibility to utilize the waste. Hence comes the circular economy. Circular economy is a further concept beyond recycling, which is to treat the waste produced as a new resource in a particular process [13]. This concept will drive people forward to create a new innovation that can utilize unwanted waste becomes a new product that is beneficial and has an added value. Circular economy concept should be applied within the whole process, in this case is from biodiesel making to glycerin purification. The present study elaborates alternative pathway to utilize the glycerine pitch. It is expected that the proposed process pathway may reduce the amount of waste produced and promote the circular economy in biodiesel industries.

## 2. Biodiesel process

Biodiesel is an alternative energy source that is now widely used to replace fossil diesel. There are currently four generation of biodiesel that are distinguished by raw materials: (a) first generation biodiesel, which use edible oils; (b) second generation biodiesel, produced from non-edible oils; (c) third generation biodiesel which utilize waste oils as raw material; and (d) fourth generation, which is a newest development of biodiesel making that the feedstocks are advance solar oils, including photobiological solar fuels and electro-fuels, using a synthetic biology technologies [14]. Biodiesel that is currently available in the market are the first-generation biodiesel. In Indonesia, biodiesel making commonly needs a Refined, Bleached, Deodorized Palm Oil (RBDPO), which is categorized as an edible oil. RBDPO, which is a product of palm oil production, is more preferrable rather than using raw palm oil, because of its low free fatty acid (FFA) which can produce more biodiesel [15]. Raw palm oil also contains 1% of useful materials that becomes impurity in biodiesel production, such as carotenoids, vitamin E, sterols, phospholipids, glycolipids, thus the process of refining, bleaching, and deodorizing is carried out [16].

IOP Conf. Series: Materials Science and Engineering 1143 (2021) 012030 doi:10.1088/1757-899X/1143/1/012030

The main process in biodiesel making is transesterification, the reaction between fatty acids or fatty oils with alcohol in the presence of catalyst [17]. During the reaction, triglycerides contains in fatty acids convert into diglycerides then monoglycerides and ended in the form of glycerol while one molecule of ester is produced at each step of the reaction [18]. Based on the reaction, the stoichiometric ratio of alcohol and triglycerides of the reaction is 3:1. However, in industrial practice, this ratio is usually increased to 6:1 with excess alcohol so that the equilibrium of the reaction is always toward the product. The high content of free fatty acids and water from raw oil material is a problem faced by the transesterification process. Water reacts with catalyst therefore the yield of the reaction will decrease.

Although there are various catalyst in biodiesel making (homogenous, heterogenous, acid, basic, or enzyme catalysts), homogenous basic catalysts are the most prevalent, such as sodium methoxide, potassium methoxide, sodium hydroxide and potassium hydroxide [19]. Between those four, sodium methoxide ($CH_3ONa$) catalyst is the most active one that can induced a better process separation after the reaction, thus the use of sodium methoxide is more commonly found in industry [20]. About 70% of biodiesel industry in North America used this catalyst. All major technology providers including Lurgi PSI, Desmet Ballestra and Crown Iron Works offer biodiesel process technologies and plant designs that utilize sodium methoxide [21]. This type of catalyst is also widely used in Indonesian biodiesel industries. Another parameter that plays an important role in the yield of biodiesel is temperature of transesterification [15]. A temperature cannot be too low nor too high: should be optimum where the time needed to convert triglycerides is short with a high amount of conversion. Optimal temperature falls in the range of 60-70°C in the use of homogenous basic catalyst with 6:1 ratio of alcohol and triglycerides [22].

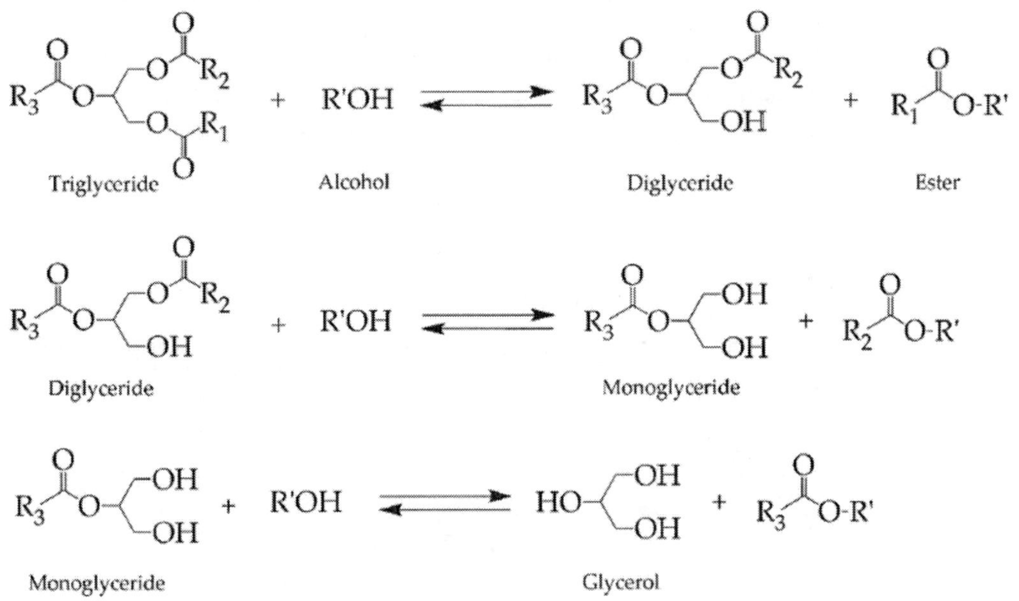


**Figure 1.** Transesterification reaction of triglycerides and alcohol [23].

The process of biodiesel production in industry starts from fresh RBDPO that is rich in triglycerides entering transesterification reactor to conduct the transesterification reactor in the presence of homogenous catalyst and methanol. Once transesterification finished at the specified time, two different liquid phases will form in the reactor, which the oil phase mostly contains fatty acid methyl ester (FAME) and water phase is rich in glycerol and excess methanol. Due to the mixing, those liquid phases will be seen as one phase, thus, separation is needed. This mixture is transferred into a separating funnel. Separation is carried out until two distinct layers of the product is achieved. Glycerol as a by-product is formed in the bottom layer, mixed with the excess methanol, and is associated with a higher density than fatty acid methyl esters (FAMEs) in the upper layer [24].

Through some stage of purification, FAMEs, methanol, and crude glycerol is separated [25]. To purify the upper layer that contain methyl ester from the separator, it is usually washed by water and then distilled to produce high purity biodiesel. The glycerol produced from the transesterification process is known as crude glycerol, which still contains components other than glycerol such as methanol, fatty acid compounds, soap, and others. So that the refining process need to be carried out to get pure glycerol, which is valuable, that can be used in industry. The purification process of crude glycerol or known as sweet water begin with adding solution of acid into crude glycerol to decompose the soap and form free fatty acids [26].This process is also called neutralization or acidification because there is an addition of acid to the crude glycerol compound. Methanol that has been purified can be re-used in the process, and crude glycerol is distilled to produce a pure glycerin.

Before carrying out the acidification process, the crude glycerol stream introduced into the evaporation column to separate methanol from crude glycerol. After it separated, the output methanol stream was fed back to the feed stream, making use of the recovered methanol. During the glycerol formation process, triglycerides react with alcohol with the help of an alkaline catalyst. So that the glycerol produced as a by-product of this transesterification reaction is alkalic. The acidification process which has been mentioned above is then took place. The purpose of this acidification process is to neutralize the glycerol from the alkaline catalyst that is still contained therein. In this process, acid solution such as hydrochloric acid, carbonic acid or sulfuric acid with certain concentration and amount is added to crude glycerol solution. In this study, hydrochloric acid which acted as neutralizing solution was used. Crude glycerol that has been acidified still contains water, so it is necessary to separate it to obtain pure glycerol. The products obtained after glycerol purification are pharma grade glycerol and glycerin pitch. The glycerin pitch that obtained from biodiesel production contains several components such as water, matter organic non-glycerol (MONG), and salt.

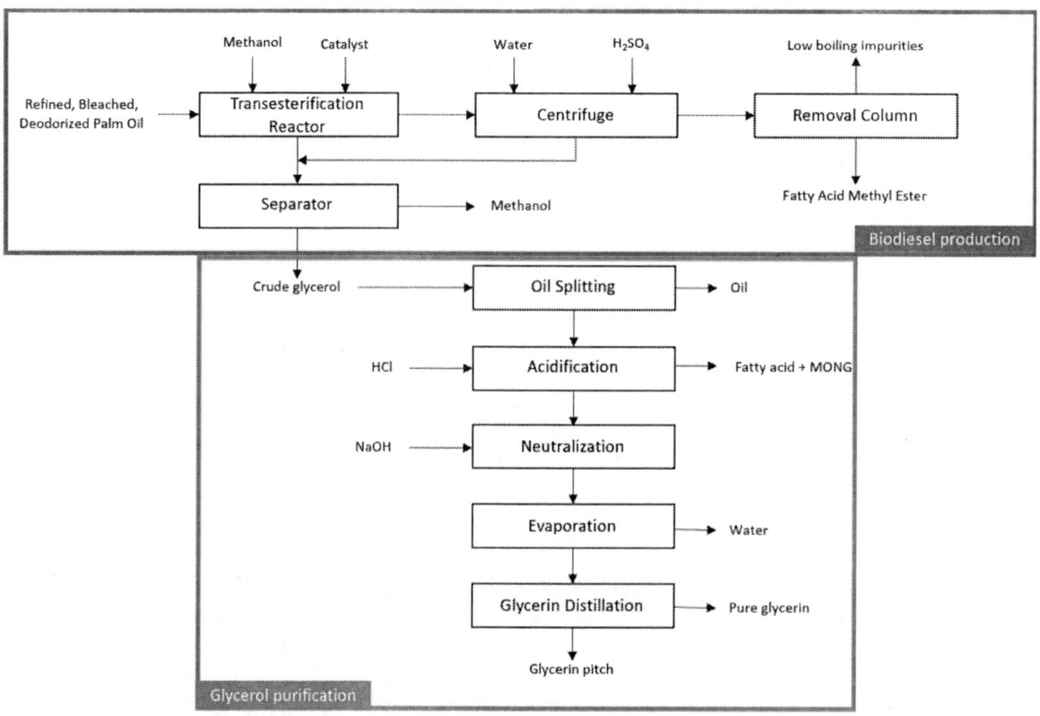


**Figure 2.** Block flow diagram of biodiesel production and glycerol purification.

The mass balance calculation of the biodiesel production process is carried out using a simulator, Aspen Plus v.10. The design basis is to produce 1 ton/h biodiesel from the transesterification reaction of palm oil and methanol which is catalyzed by sodium methoxide. Fresh methanol which has been mixed with recycled methanol, fed to the reactor with a molar to oil ratio of 6. The catalyst used in the transesterification reaction is added and mixed with fresh methanol in a mixing tank with a quantity of 1%-wt. from the amount of oil treated [27]. The residual methanol from the reactor fed into the methanol purification tower, then the pure methanol product recycled to be mixed with fresh methanol as feed for the reaction. Based on the simulation and assumption, to produce 1 ton/hour of biodiesel, feed of 1042.5 kg/h RBDPO and 119.97 kg/h fresh methanol is required. In addition, to obtain pure glycerol by-products (107 kg/h), 7.2 kg/h solution of hydrochloric acid fed in neutralization reactor. The waste produced from the whole process is glycerin pitch with a rate of 61.1 kg/h which contained matter organic non-glycerol (MONG) and sodium chloride salt.

**Table 1.** Fatty acid composition of RBDPO used in simulation [28].

| Fatty acid | Composition (%) |
|---|---|
| Saturated | |
| Myristic | 0.92 |
| Palmitic | 46.3 |
| Stearic | 3.52 |
| Total | 50.74 |
| Unsaturated | |
| Olein | 39.58 |
| Linoleic | 9.68 |
| Total | 49.26 |

## 3. Glycerin Pitch Treatment

Glycerin pitch from the process contains some materials, such as fatty acid, glycerol, methanol, and inorganic salts. Sodium methoxide and sodium hydroxide are feed to the process in the step of transesterification of biodiesel and neutralization of glycerol, respectively. Thus, inorganic matter in glycerin pitch is none other than sodium salt. It is proven in the simulation that glycerin pitch contains 19% of inorganic salt in the form of sodium chloride. Sodium chloride is formed after the sodium from neutralization phase reacted with chlorine atom from the acidification. In the industry, however, the percentage of each component differs based on the process parameters carried out. Typically, inorganic salt contains in glycerin pitch is about 10% or under 5% in crude glycerol after neutralization phase [8, 29]. This value might seem insignificant but anticipating the increase in glycerin production of 500,000 tons/year, the amount of inorganic salts that can be extracted will also as large as 40,000 tons/year.

**Table 2.** Glycerin pitch composition.

| Component | Typical range** | Simulation output |
|---|---|---|
| Glycerol | 0-30 | 5 |
| Water | $<2$ | $<1$ |
| Ash | 20-60 | 19 |
| MONG* | 40-60 | 75 |

*MONG = matter organic non glycerol
**typical range obtained from Apolin, 2019

Liquid extraction process can be carried out to separate inorganic salt from glycerin pitch. A study proved that liquid-solid extraction can be used to recover 87.59% of inorganic salt with methanol as solvent with solute mass: solvent volume ratio of 1:5 in the operation temperature of 70°C [8]. As the continuation of the process in Fig 2, an extraction vessel is installed to extract inorganic salts in form on sodium chloride from glycerin pitch using methanol as a solvent. The amount of inorganic salt that can be recovered from the process with a basis of 1 ton/h biodiesel is 9.87 kg/hr.

**Table 3.** Mass balance of solid-liquid extraction.

| Component | Glycerin pitch | Solvent | Raffinate | Extract |
|---|---|---|---|---|
| Mass flowrate (kg/hr) | 61.10 | 9787.60 | 9834.1 | 11.28 |
| organic fraction | 0.815 | 1 | 0.995 | 0.876 |
| inorganic fraction | 0.185 | | 0.005 | 0.124 |

The continuation of solid-liquid extraction is how to convert sodium chloride (NaCl) into sodium hydroxide (NaOH). Sodium hydroxide is the material that is needed to produce catalyst that is used in biodiesel transesterification process. Reaction between sodium hydroxide and methanol will be resulting in sodium methoxide and water. One of the process that can be used to convert sodium chloride into sodium hydroxide is using Chlor-Alkali process [30]. This process has been established and widely used in industry, since caustic soda or sodium hydroxide is one of the main components in FMCG industry. Chlor-alkali process carried out using electrolysis process in either mercury cell, diaphragm cell, or membrane cell [31]. Industry nowadays tend to use the membrane cell because of safety consideration and power consumption [30]. A brine with concentration of 26% wt. will enter the cell and electrolysis process will be carried out. Reaction resulting in sodium hydroxide occurs in the anode side of the reactor, alongside with the production of chlorine gases. Hydrogen gases also formed in the cathode side of the reactor. The product stream, sodium hydroxide, comes out from the reactor having a concentration of 32%. The overall reaction has 50% of conversion [32]. An evaporator is installed after the reactor to eliminate the water content in the NaOH until 50% wt.

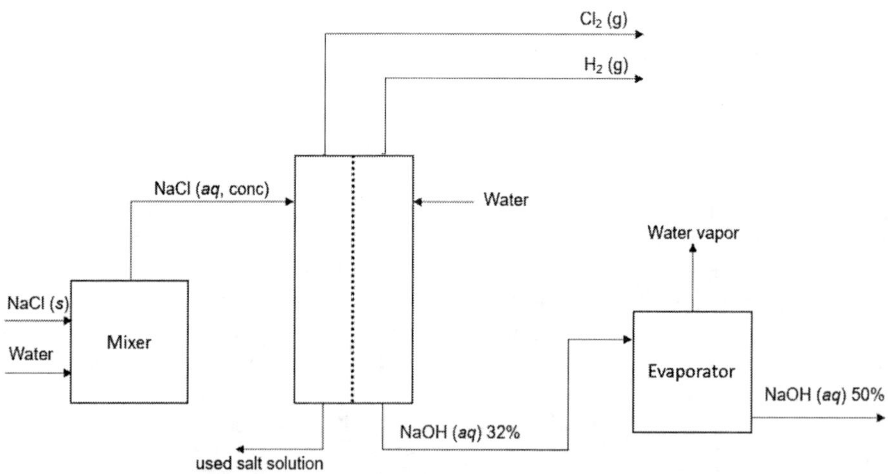


**Figure 3.** Chlor-alkali process for NaOH making.

**Anode:**

$$2\ NaCl\ (aq) + 2\ H_2O\ (l) \rightarrow 2\ NaOH\ (aq)\ + Cl_2\ (g) + 2\ H^+\ + 2\ e^-$$

**Cathode:**

$$2\ H^+ + 2\ e^- \rightarrow H_2\ (g)$$

**Overall reaction:**

$$2NaCl\ (aq)\ + 2H_2O\ (l) \rightarrow Cl_2\ (g) + H_2\ (g) + 2NaOH\ (aq)$$

**Figure 4.** Reactions take place in electrolysis reactor.

**Table 4.** Mass balance of Chlor-Alkali process.

| stream/ component | NaCl (*aq*, conc) | used salt solution | NaOH (*aq*) 32% | $Cl_2$ | $H_2$ | NaOH (*aq*) 50% |
|---|---|---|---|---|---|---|
| Mass flow (kg/hr) | 37.99 | 36.99 | 13.93 | 2.99 | 0.08 | 6.75 |
| NaCl (% wt.) | 26.0 | 24.0 | 32.0 | | | 50 |
| $H_2O$ (% wt.) | 70.3 | 72.2 | 68.0 | | | 50 |
| NaOH (% wt.) | | | | | | |
| $Cl_2$ (% wt.) | | | | 100 | | |
| $H_2$ (% wt.) | | | | | 100 | |
| impurities | 3.7 | 3.8 | | | | |

The last process that is conducted is the catalyst making. The reaction takes place in a reactor with a process temperature of 160°C and pressure 3 bar with 98.3% reaction conversion [33]. Feed used are methanol ($CH_3OH$) and sodium hydroxide obtained from the previous process. The aim of this process is to produce a 20% wt. sodium methoxide catalyst in methanol that used in the biodiesel transesterification reaction. The simplified process is shown in the Figure 5 while the mass balance is in Table 5. Later on, a purification process can be conducted to separate unconverted NaOH and water.

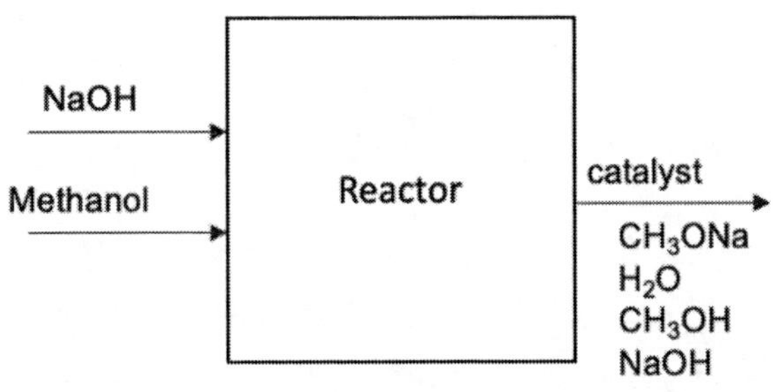


**Figure 5.** Transesterification catalyst making process.

**Table 5.** Mass balance of catalyst making.

| stream/ component | NaOH | Methanol | Catalyst |
|---|---|---|---|
| Mass flow (kg/hr) | 6.75 | 22.41 | 27.64 |
| NaOH (% wt.) | 50.00 | | 10.62 |
| $CH_3OH$ (% wt.) | | 95.00 | 61.58 |
| $CH_3ONa$ (% wt.) | 50.00 | | 12.42 |
| $H_2O$ (% wt.) | | 5.00 | 15.38 |

## 4. Circular Economy Concept

Circular economy is a further concept beyond recycling, which in to treat the waste produced in the process as a new resource in particular process [14]. This concept will drive people forward to create an innovation that can utilize unwanted waste becomes a new product that is beneficial and has an added value. In biodiesel making, the process consists of two different main blocks, that are methyl-ester production from RBD Palm Oil and glycerol purification to obtain a high purity glycerin. This process generates a glycerin pitch waste that is contains a sodium salt, component of transesterification catalyst. Attaching a recovery process of sodium salt, a new and fresh sodium methoxide can be produced. Some of the methanol that recovered from biodiesel production also can be used as a feed to the downstream process. Sodium methoxide recovered can be re-used in the transesterification thus reducing the needs of fresh sodium methoxide.

However, sodium catalyst recovery process can only recover about 43.78% of the catalyst. This low recovery is mainly because the Chlor-Alkali process that only has 50% conversion. Another thing that is become concern is the amount of methanol used in this system is quite high, 9813 kg/h. This due to the fact that solid-liquid extraction process requires a high ratio of methanol as solvent vs solute. Another thing to consider is that the process of Chlor-Alkali requires a high amount of energy, 3 MWh per ton caustic produced [34]. Comparing the cost of production in catalyst recovery to the case if there is no catalyst recovery in Fig. 2, the latter one has a higher profit. Thus, to apply a circular economy concept in biodiesel production, a better approach should be proposed rather than recycling a sodium-based salt that is abundant in the market with an affordable price.

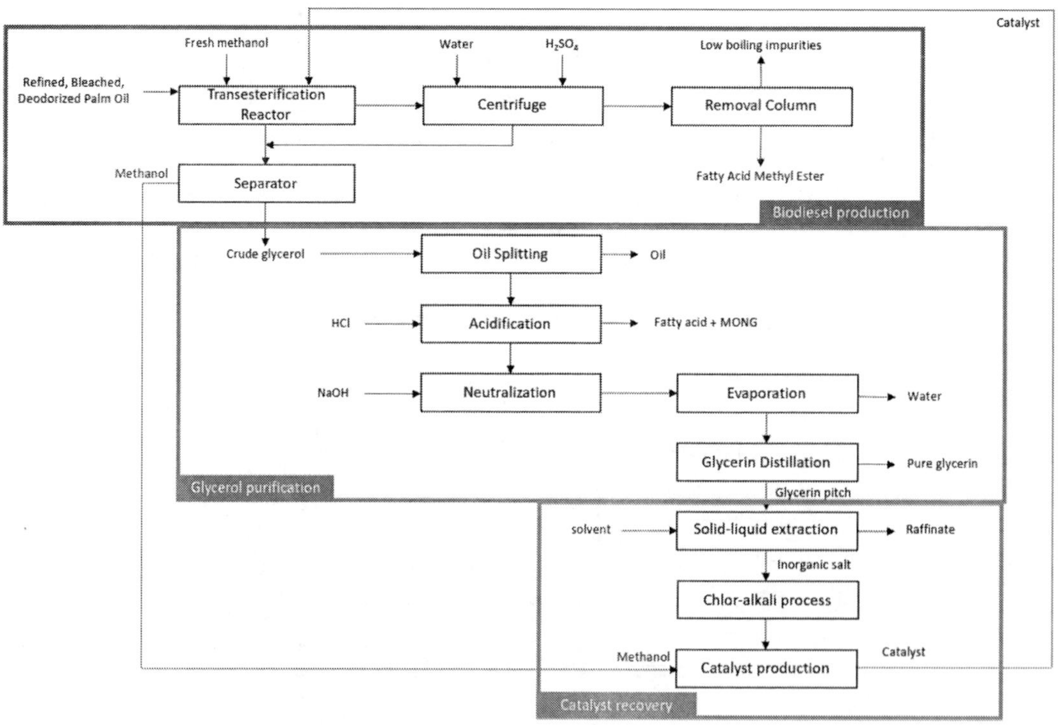


**Figure 6.** Circular economy concept.

## 5. Conclusion

Using the concept of circular economy, a catalyst recovery process can be attached to biodiesel and glycerin purification process to recycle glycerin pitch to recover the sodium salt. However, this concept has some concerns that need to take into an account before integrating in the real life.

**Acknowledgement**
This work is funded by Indonesian Palm Oil Fund Management Agency (Badan Pengelola Perkebunan Kelapa Sawit BPDPKS) year one of two.

**References**

[1] I P O Association 2020 Performance of Indonesian Palm Oil Industry *GAPKI*
[2] Mahmudul H M, Hagos F Y, Mamat R, Adam A A, Ishak W F W and Alenezi R 2017 Production, characterization and performance of biodiesel as an alternative fuel in diesel engines – A review *Renew. Sustain. Energy Rev.* **72** 497–509
[3] Luque R, Lovett J C, Datta B, Clancy J, Campelo J M and Romero A A 2010 Biodiesel as feasible petrol fuel replacement: A multidisciplinary overview *Energy Environ. Sci.* **3** 1706–21
[4] Chincholkar, Srivastava S., Rehman A., Dixit S. and Lanjewar A. 2005 Biodiesel as an Alternative Fuel for Pollution Control in Diesel Engine *Asian J. Exp. Sci.* **19** 13–22
[5] Rao T V, Rao G P and Chandra K H 2008 Experimental Investigation of Pongamia, Jatropha and Neem Methyl Esters as Biodiesel on C.I. Engine *Jordan J. Mech. Ind. Eng.* **2** 117–22
[6] Directorate General of New, Renewable Energy, and Energy Conservation 2019 Mandatori B30 Mulai Diimplementasikan Awal Januari 2020, Harga Biosolar di Pasaran Tetap *Directorate General of New, Renewable Energy, and Energy Conservation*
[7] Luque R and Melero J A 2012 *Introduction to advanced biodiesel production* (Woodhead Publishing Limited)
[8] Demaman Oro C E, Bonato M, Oliveira J V, Tres M V, Mignoni M L and Dallago R M 2019 A new approach for salts removal from crude glycerin coming from industrial biodiesel production unit *J. Environ. Chem. Eng.* **7** 102883
[9] Appleby D B 2005 *Glycerol* (Urbana, IL:AOCS Press) 255-267
[10] Hidawati E N and Sakinah a M M 2011 Treatment of Glycerin Pitch from Biodiesel Production *Int. J. Chem. Environ. Eng.* **2** 309–13
[11] APOLIN 2019 Industri Oleokimia Semakin Menarik *APOLIN*
[12] Hazimah a H, Ooi T L and Salmiah A 2003 Recovery of glycerol and diglycerol from glycerol pitch *J. Oil Palm Res.* **15** 1–5
[13] Ghosh S K 2020 Introduction to Circular Economy and Summary Analysis of Chapters *Circular Economy: Global Perspective* ed S K Ghosh (Singapore: Springer Singapore) pp 1–23
[14] Singh D, Sharma D, Soni S L, Sharma S, Kumar Sharma P and Jhalani A 2020 A review on feedstocks, production processes, and yield for different generations of biodiesel *Fuel* **262** 116553
[15] Shahbazi M R, Khoshandam B, Nasiri M and Ghazvini M 2012 Biodiesel production via alkali-catalyzed transesterification of Malaysian RBD palm oil – Characterization, kinetics model *J. Taiwan Inst. Chem. Eng.* **43** 504–10
[16] Song E-S, Lim J, Lee H-S and Lee Y-W 2008 Transesterification of RBD palm oil using supercritical methanol *J. Supercrit. Fluids* **44** 356–63
[17] Singh S P and Singh D 2010 Biodiesel production through the use of different sources and characterization of oils and their esters as the substitute of diesel: A review *Renew. Sustain. Energy Rev.* **14** 200–16
[18] Ali B, Yusup S, Quitain A T, Bokhari A, Kida T and Chuah L F 2019 Heterogeneous Catalytic Conversion of Rapeseed Oil to Methyl Esters: Optimization and Kinetic Study *Advances in Feedstock Conversion Technologies for Alternative Fuels and Bioproducts* ed M B T-A in F C T for A F and B Hosseini (Elsevier) pp 221–38
[19] Vicente G, Martínez M and Aracil J 2004 Integrated biodiesel production: a comparison of different homogeneous catalysts systems *Bioresour. Technol.* **92** 297–305

[20] KoohiKamali S, Tan C P and Ling T C 2012 Optimization of sunflower oil transesterification process using sodium methoxide *Sci. World J.* **2012** 1–8
[21] Jackson S 2006 Standard-for good reason *Biodiesel Magazine*
[22] Athar M and Zaidi S 2020 A review of the feedstocks, catalysts, and intensification techniques for sustainable biodiesel production *J. Environ. Chem. Eng.* **8** 104523
[23] Haro J C de, Rodríguez J F, Carmona M and Pérez Á 2018 Revalorization of Grape Seed Oil for Innovative Non-Food Applications *Grapes and Wines - Advances in Production, Processing, Analysis and Valorization* vol 205 (InTech) pp 1373–97
[24] Sanli H and Canakci M 2008 Effects of Different Alcohol and Catalyst Usage on Biodiesel Production from Different Vegetable Oils *Energy & Fuels* **22** 2713–9
[25] Lam M K, Jamalluddin N A and Lee K T 2019 Production of Biodiesel Using Palm Oil *Biofuels: Alternative Feedstocks and Conversion Processes for the Production of Liquid and Gaseous Biofuels* (Elsevier) pp 539–74
[26] Hidayati R, Allah A H and Arita S 2012 Pengaruh Penambahan H3PO4 dan Resin Kation – Anion Terhadap Persen Total Gliserol Hasil Samping Pembuatan Biodiesel *J. Tek. Kim.* **18** 31–8
[27] Narváez P C, Rincón S M and Sánchez F J 2007 Kinetics of Palm Oil Methanolysis *J. Am. Oil Chem. Soc.* **84** 971–7
[28] Man Y B C, Haryati T, Ghazali H M and Asbi B A 1999 Composition and thermal profile of crude palm oil and its products *J. Am. Oil Chem. Soc.* **76** 237–42
[29] Darnoko D and Cheryan M 2000 Kinetics of palm oil transesterification in a batch reactor *J. Am. Oil Chem. Soc.* **77** 1263–7
[30] Lakshmanan S and Murugesan T 2013 The chlor-alkali process: Work in Progress *Clean Technol. Environ. Policy* **16** 225–34
[31] Crook J and Mousavi A 2016 The chlor-alkali process: A review of history and pollution *Environ. Forensics* **17** 211–7
[32] Millet P 2013 Chlor-alkali technology: fundamentals, processes and materials for diaphragms and membranes *Handbook of Membrane Reactors* (Elsevier) pp 384–415
[33] Granjo J F O and Oliveira N M C 2015 Process Simulation and Techno-Economic Analysis of the Production of Sodium Methoxide *Ind. Eng. Chem. Res.* **55** 156–67
[34] Euro Chlor 2010 The European Chlor-Alkali industry : an electricity intensive sector exposed to carbon leakage *Production* 2–6

# The potential of Biogas in Energy Transition in Indonesia

**Elisabeth Rianawati[1*], Saut Sagala[1,2], Ichsan Hafiz[1], Johannes Anhorn[3], Sinshaw Alemu[4], Jorge Hilbert[5], Dwight Rosslee[6], Mutala Mohammed[7], Yaseen Salie[8], Dominik Rutz[9], Michael Rohrer[10], Angela Sainz[11], Franz Kirchmeyr[12], Aleksejs Zacepins[13], Frank Hofmann[14]**

[1]Resilience Developement, Bandung 40135 Indonesia
[2]Bandung Institute of Technology (ITB), Bandung 40132 Indonesia
[3]Deutsche Gesellschaft für Internationale Zusammenarbeit, Germany
[4]Iceaddis, Ethiopia
[5]Instituto Nacional de Tecnologia Agropecuaria, Argentina
[6]Selectra, South Africa
[7]Institute for Sustainable Energy and Environmental Solutions, Ghana
[8]Green Cape, South Africa
[9]WIP Renewable Energies, Germany
[10]Austrian Energy Agency, Austria
[11]European Biogas Association, Belgium
[12]Austrian Compost & Biogas Association – Kompost und Biogasverband Österreich, Austria
[13]Latvijas Lauksaimniecibas Universitate, Latvia
[14]Fachverband Biogas, Germany

*Corresponding Author's E-mail: elisabeth.rianawati@rdi.or.id

**Abstract.** Indonesia is an agrarian country that has a rich bioenergy potency in liquid (biodiesel, bioethanol). The Government of Indonesia (GoI) has set the target to achieve 23% of renewable energy utilization into the national energy mix by 2025. In addition, the GoI also aims to increase the production of biofuel to 7.21 million kilolitres by 2019. Theoretically, biogas technology will be a strategic measure in achieving the target, however, at the moment the biogas technology market in Indonesia is still in a nascent state, especially for the direct utilization of biogas for electricity production. Alternatively, biogas provides Indonesia with a promising source of energy, which can be injected directly into natural gas grids and hitchhike existing distribution infrastructure, resulting in reduced costs along the production-distribution pipeline. For this reason, biomethane has been the focus of some developing countries (e.g Argentina, Republic of South Africa) in moving toward energy transition. This paper examines the state of the biogas market in Indonesia using literature review. The status of natural gas is mapped out through its available potential and the existing initiation of national programs related to biogas. Finally, the study provides recommendations on how biogas technology could accelerate the energy transition in Indonesia.
**Keywords:** Energy Transition, Biogas, Indonesia

## 1. Introduction

Since 1978, Indonesia has been included in the list of the largest natural gas producing countries with a total production of 11.2 million $m^3$, after which the figure has increased following the opening of several oil and gas block points. At its peak in 2000, Indonesia became the highest ranking (ranked $6^{th}$) Asian country in the world in producing natural gas. Total production at that time reached 70.5 million $m^3$/year. However, in the last 10 years, Indonesia's natural gas production has decreased by 5% each year. In 2018, Indonesia's natural gas production reached 73.7 million $m^3$. If natural gas is produced at the same rate as in 2016, the natural gas will only last for up to 47 years in the future. Eventually, Indonesia will become a net importer of gas in 2028. Natural gas production is predicted to be unable to meet the national gas demand in the future, so it is necessary to import gas in the form of LNG [1]. On the other hand, the use of LPG for household and commercial use faces many obstacles, as more than 65% of the current LPG supply comes from imports. Since this program was implemented in 2007, the price of 3 kg LPG is Rp. 4,250/kg and has never been increased ever since, even though the import price of LPG has reached Rp. 10,000 per kg. This can cause the amount of LPG subsidies to increase and be burdensome to the national economy as it will increase the trade balance deficit [2]. Therefore, the use of LPG should be gradually reduced to as small as possible [3].

Due to economic and population growth, it is predicted that the need of energy will continuously grow, reaching up to 450 × $10^9$ kWh by 2026. The irresponsible use of energy, especially fossil fuels, is producing emissions that pollute the environment and increase the earth's temperature. The use of fossil fuels contributes to greenhouse gas ($CO_2$) emissions by 76%, which reached through usage in industrial processes (65%) and other land uses (11%). Regarding the energy problem and the greenhouse gas effect, bioenergy development can be an alternative solution. Biogas is biomass-derived fuels produced by anaerobic digestion of organic feedstock through a process through a four-step process that is mediated microbially: hydrolysis, acidogenesis, acetogenesis, and methanogenesis [33]. Biogas is one of the alternative energies that can be used to fulfill some of the domestic gas needs. The use of biogas is an effort to increase economic value from liquid waste to renewable energy and simultaneously reduce greenhouse gas (GHG) emissions [4, 5].

Globally, there are various forms of feedstocks for biogas production, these can be obtained from animal farms, municipal solid waste, agricultural residues, sludges from many industries, and wastewater treatment units [34]. Biogas from POME is composed of 50-75% methane (CH4), around 24-25% carbon dioxide (CO2) and H2O, H2S, and other gases [6]. Biogas can be upgraded by removing the CO2 content using technologies such as water or organic physical scrubbing, chemical scrubbings, pressure swing adsorption, membrane separation, and cryogenic technology [35]. The upgrading process yields biomethane with methane content up to 90-95% [36]. This methane in biogas or biomethane then could be used for several purposes, such as producing heat, electricity, and vehicle fuel, after being cleaned of any impurities to satisfy certain limits or natural gas specifications. The mentioned impurities that must be removed are water vapor and H2S. Water vapor should be extracted as it causes corrosion, accumulation of water in pipes, and freezing at high pressure. On the other hand, H2S is toxic, causes corrosion, and would produce SO2 and SO3 compounds that are more toxic than H2S while in combustion [7, 8]. One of the technologies that can be used to purify biogas into biomethane is a water scrubber.

While Indonesia's power generation still comes primarily from fossil fuel, the Indonesian government is targeting to achieve a total of 23% renewable energy in the energy mix by 2025 [37]. Currently, most biogas production case studies in Indonesia are done in small scale using animal waste as its potential feedstock. This is due to the increasing number of cattle population in Indonesia [38]. By 2015, 16,000 biogas plants had been disseminated in mostly rural areas covering ten provinces: Bali, Banten, Central Java, D.I. Yogyakarta, East Java, East Nusa Tenggara, Lampung, South Sulawesi, West Java, and West Nusa Tenggara [39]. Other case studies have also been conducted to assess the feasibility of biogas production in Indonesia using other feedstocks, e.g. Arabica coffee pulp [40], green seaweeds [41], human excreta [42], and municipal solid waste (MSW) [43]. Since most biogas plants are disseminated in rural areas, the biogas produced are focused to be used directly for household purposes

IOP Conf. Series: Materials Science and Engineering 1143 (2021) 012031 doi:10.1088/1757-899X/1143/1/012031

e.g. cooking, water heating, etc. However, there are attempts to increase the quality of the biogas produced from these plants through purification methods, namely pressurized gas scrubbing [44].

## 2. Literature Review

### *2.1. Biogas upgrading*

CO2 gas from biogas and purify it to increase the heating value and reach the expected standard specifications. In general, commercial upgrading technologies are membranes, pressure swing adsorption (PSA), water scrubbers, organic physical scrubbers, and amine scrubbers [9,10]. However, water scrubber is the common choice to reduce the CO2 content in biogas since it is a commonly used technology with the average yield of the biomethane (CH4) upgrade reaching up to 96.78% [8, 11, 12]. This technology also does not require chemicals in the process even though it requires a lot of water [13]. The process of cleaning $CO_2$ in biogas depends on the principle that $CO_2$ is more soluble in water than $CH_4$. Biogas with high pressure enters the absorption tank and meets the flow of water being sprayed from the top of the tank. CO2 will dissolve into the water and will flow into the flash tank. The gas that comes out of the top of the absorption tank already has a high concentration of biomethane ($CH_4$) (> 96%) yet still needs to be dried to reduce water content. In the flash tank, the dissolved $CO_2$ in the water still carries the remaining $CH_4$ and other impurities. The remaining CH4 is then flown back into the biogas inlet, while CO2 and other contaminants are flown into the desorption tank for disposal. Lastly, the low-pressure air is flown to the desorption tank, while the water is mostly recycled. In case the water is lacking, extra water is added (makeup water) [8,14].

### *2.2. Gas supply in Indonesia*

Initially, the development of Indonesia's natural gas industry was more for export. The construction of LNG (liquefied natural gas) refineries in Badak (East Kalimantan) and Arun (Aceh) in the 1970s made Indonesia the world's largest LNG exporter. In the same era, natural gas was exploited in Cilamaya (West Java), channeled through transmission pipes to be used as raw material for fertilizer (Cikampek), fuel/support for the iron smelting plant (Cilegon), and as city gas in some of the routes it passes. In the 1980s several "Pemjadig" (gas distribution network development) projects were carried out in a number of cities in Java and Sumatra. Natural gas reserves in the north-west sea of Java and in the Madura Strait are exploited, piped to a combined cycle power plant in Java (PLTGU Muara Karang and Tanjung Priok in West Java, PLTGU Gresik in East Java, etc.) and used as fertilizer and raw materials for petrochemicals (in Gresik, East Java). Natural gas transmission Grissik (Sumbagsel) - Duri, Riau (544 km) was built, starting its operation in 1998. This development has been followed by the Grissik -Batam-Singapore and Sumbagsel-West Java transmission lines. However, these efforts have not been able to shift the share of petroleum in the national primary energy mix. As national LPG production was inadequate, LPG was imported as much as 3.3 million tons in 2013 and 3.6 million tons in 2014 due to the increasing demand for LPG. Indonesia's natural gas content is generally in the form of methane and ethane, while LPG is produced from the content of propane and butane natural gas [2].

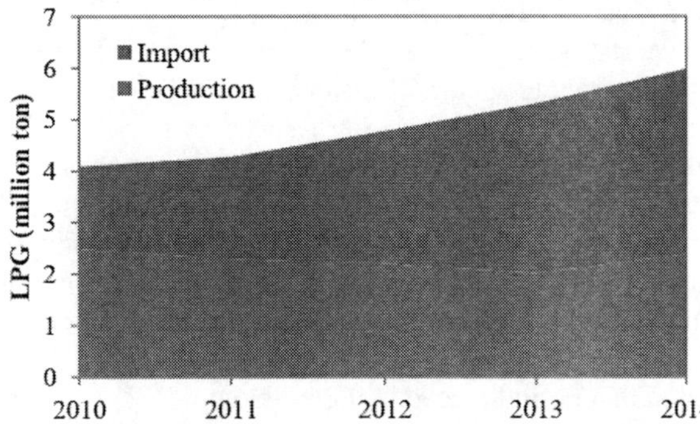


**Figure 1.** Production and import of LPG [2].

## 3. Methodology

This article is written based on the results of studies or literature related to biomethane. The research method includes secondary data collection from the Central Bureau of Statistics and the Ministry of Energy and Mineral Resources in the form of official websites, online books, review of journals related to the conversion of bioethanol from agricultural waste and organic waste, data processing (calculation of bioethanol potential that can be produced) and analysis, and discussions.

## 4. Result and Discussion

### *4.1. Energy transition in Indonesia*

One effort in energy transition in Indonesia can be seen through the energy mix policy. Under the Collective Regulations, these policies will put Indonesia on a path to decarbonization. Government Regulation Number 79/2014, concerning National Energy Policy, sets out the ambition to transform, in 2025 and 2050, that the main energy supply mix is as follows: a) new and renewable energy at least 23% in 2025 and at least 31% in 2050; b) oil must be less than 25% in 2025 and less than 20% in 2050; c) coal at least 30% in 2025 and at least 25% in 2050; d) gas at least 22% in 2025 and at least 24% in 2050. Currently, the contribution of renewable energy (RE) has only reached around 9% (9.15%) of the total national energy mix. Achieving the target of RE contribution in 2025 of 23% is a formidable challenge although it is not impossible to achieve [15].

In the industrial sector, energy diversification from fuel oil to coal, gas, and biomass has taken place so that fuel consumption is no longer dominant. Natural gas has the third-largest reserves in the world after coal and oil. Natural gas was not initially consumed as an energy source due to transportation difficulties, so it was always burned when it was co-generated with petroleum. The products of natural gas used are LPG (Liquid Petroleum Gas), CNG (Compressed Natural Gas), LNG (Liquid Natural Gas), and Coal Bed Methane (CBM) which are non-conventional energy sources that are being developed in Indonesia. CNG is a natural gas that is compressed without going through a refining process and stored in metal cylinders. CNG is relatively cheaper because it does not go through a refining process and is more environmentally friendly. LPG and LNG are natural gas produced from the refining and separation of petroleum. Butane and propane will turn into LPG and methane will turn into LNG [16].

Gas energy is believed to be clean energy which price is relatively cheap compared to other energies. The Downstream Oil and Gas Regulatory Agency (BPH Migas) notes that the average pipeline gas price in Java is US$9-US$11 per MMBtu. The price of other gases such as compressed natural gas (CNG) is US$11 - US$15 per MMBtu and retail liquefied natural gas (LNG) is US$17 - US$19 per MMBtu. Whereas, the average price of fuel oil (BBM) and liquefied petroleum gas (LPG) is US$20 - US$23 per MMBtu [17]. As such, to increase its domestic utilization, in 2008, the government implemented the construction of city gas networks (*jargas*) through APBN funding, starting with the implementation of the FEED and DEDC. In 2014, 16,949 household gas networks were built in 5 locations, namely Semarang City, Bulungan, Sidoarjo, Bekasi District, and Lhoksumawe. From 2009 s.d. 2014, the cumulative development of the city gas network through APBN funding reached 25 locations and was allocated to 86,460 house connections. Apart from funding from the state budget, up to 2014, PGN has also built jargas for 92,858 house connections in 10 cities [2].

Meanwhile, in the transportation sector, the role of BBM is still difficult to replace with other types of fossil energy. As for the household sector, electricity and LPG are still the most consumed types of energy, following the usage of LPG to replace kerosene. In line with this, the decline in fuelwood consumption was also followed by an increase in LPG consumption from 5.3 million tonnes (14%) in 2013 to 5.8 million tonnes (17%) in 2014. Conversely, kerosene consumption decreased from 6.4 million BOE (1.9%) in 2013 to 4.9 million BOE (1.7%) in 2014. The decline in kerosene consumption was caused by the substitution policy for kerosene with LPG for cooking in households and small businesses. This program has been implemented since 2007 and succeeded in reducing kerosene consumption by around 8.28 million kiloliters by 2014, saving around 40 trillion rupiahs. The number of 3 kg LPG units that have been distributed in that period reached 49,269,666 units [2, 14].

### *4.2. Potential of biomethane*

Indonesia has quite a large potential for renewable energy resources, one of which is from waste. Waste in the form of urban waste, agricultural sector, industrial sector, and others can be used to be converted into energy, either in the form of fuel/heating or electricity. According to the Ministry of Energy and Mineral Resources [15], Indonesia has a biomass potential of 885.2 million Gigajoules (GJ) per year. Potential calories obtained include the type of rubber plantation rejuvenation waste (496.0 million GJ per year), remaining lodging (11.0 million GJ per year), sawmill industry waste (10.6 million GJ per year), empty palm oil bunches (15 , 4 million GJ per year), palm oil residue (35.3 million GJ per year), palm fruit shells (17.2 million GJ per year), sugarcane bagasse (78.0 million GJ per year), rice husks (179.0 million GJ per year), coconut shell (18.7 million GJ per year) and coconut husk (24.0 million GJ per year) (Figure 2). The potential source of electricity from the waste can reach 50 thousand MW, which is the second largest potential energy resource after hydro on a large scale. Utilization of this waste is still around 1600 MW or about 3.25 percent of the existing potential [2].

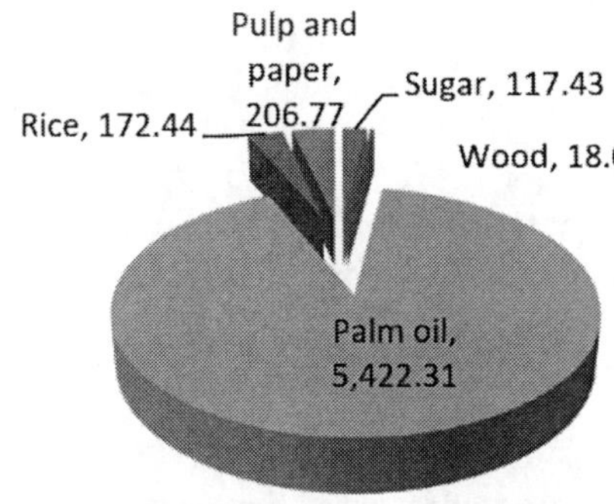


**Figure 2.** Potential of bioenergy (MW) [15].

From the Bioenergy Based Power Plant Development Program conducted by National Electricity Company (PLN-Persero), the existing capacities of the existing power plants originating from Biomass (oil palm-based), biogas, and urban waste connected to the electricity grid PLN was only 61 MW in February 2012, which will be increased later by 197 MW in 2013 and further increased by 544 MW in 2014 (so that it becomes 741 MW in 2013/2014). The utilization of biogas is one of the targets for bioenergy-based renewable energy development as stipulated in the National Energy General Plan (RUEN). However, the achievement of biogas utilization is still far from the target of RUEN in 2025. From the electricity sector, for example, the Biogas Power Plant (PLTBg) is targeted to reach a capacity of 5.5 GW in 2025, yet, the realization is only around 1.33% [15].

Non-electricity-based bioenergy development can also be implemented in the biogas development program. In this biogas development, there is the development of household-scale biogas and communal biogas. The biogas is developed by various parties including the State Budget of the Directorate General of EBTKE, the BIRU Program in collaboration with DJEBTKE and HIVOS, other ministries/ institutions, Regional Revenues and Expenditures Budget (APBD), special allocation funds, and other private sectors. Most of this biogas development utilizes livestock manure as raw material, but communal biogas has also been developed by utilizing human waste which is implemented in the communal biogas boarding school. The Ministry of Energy and Mineral Resources has built communal biogas in 20 Islamic boarding schools in 10 provinces since 2016. Based on the data collection on the implementation of direct biogas development that has been carried out by the Directorate General of EBTKE, as of May 29, 2020, installed household biogas has reached 47,505 units throughout Indonesia, producing 75,044.2 m3/day of biogas or around 26.72 million m3/year. In connection with the commercial biogas PLTBg, currently, there are nearly 96.21 MW of data.

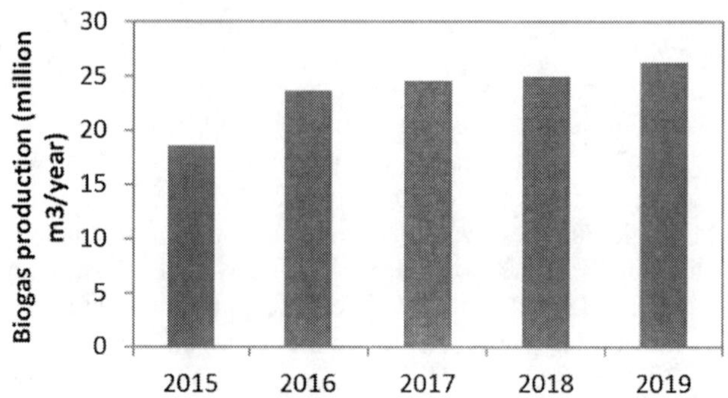


**Figure 3.** Progress of biogas production [15].

*4.3.1. Livestock manure.* Indonesia has quite a lot of livestock potential, including large ruminants such as beef cattle and dairy cows. Based on the implementation of the 2018 inter-census agricultural survey (SUTAS), it is known that the estimated population of large ruminants in 2018 was 17.9 million. From this figure, there are 16.43 million beef cattle, 581,822 dairy cows, and 894,278 buffalos. In 2019, the population is projected to grow to 18.12 million large ruminants [18]. Considering that a single livestock can produce up to 10 kg of manure per day, it has the potential to become an alternative energy source (biogas) and reduce the dependency of the community, especially family farmers, on fuel, oil and electricity. If an average of 10 kg of manure is produced per cow per day, Indonesia can produce 179,000 tons of cow feces every day.

*4.3.2. Sewage Sludge.* Based on the last recorded data in 2015, the total population in Indonesia is 238,518,000. This number is projected to increase in 2020 to 271,066,000 people. However, not all members of the population are supported by adequate sanitation facilities. Based on the 2006 National Socio-Economic Survey, [18] only 40.67% of the total number of households were equipped with a septic tank; the number is worse in the village, which is around 24.37%. This means that most Indonesian people dispose of their black water directly into the environment without any treatment. According to Nagamani and Ramasamy [19], human feces can produce up to 28 L/kg of biogas. With 1 m3 of biogas, we can turn on a 60-100 Watt lamp for 6 hours or cook 3 meals for 5-6 people, as it produces the equivalent of 1.25 kWh of electricity [20].

*4.3.3. POME.* In 2017, Palm oil production in Indonesia reached 37.8 million tons with large palm oil mills (POM) spread across the regions of Sumatra, Kalimantan, and Sulawesi [22, 23]. The by-products of the POM include liquid waste known as palm oil mill effluent (POME). POME can produce biogas which can be further processed into biomethane. The Winrock study [24] shows that the production capacity of fresh fruit bunches (FFB) from POM in Indonesia reaches 34,280 tons per hour with 23,996 m3 of POME produced per hour. This production capacity will be able to produce biomethane of 54.5 million standard cubic feet (SCF) or the equivalent of 11.5 million BOE (equivalent to barrels of oil) annually if all these by-products are being maximized. Meanwhile, the current biomethane production is equivalent to 10% of the current natural gas consumption of 564.3 million SCF or equivalent to 20% of LPG consumption (57 million BOE) [25].

*4.3.4. Municipal solid waste.* With a high number of the population living in the urban area, the amount of waste produced reaches 175,000 tons/day. Most of this waste is transported and dumped in a landfill (69%) and only a small percentage is composted and recycled (7%) [26]. Furthermore, only 10.28% of the population sort and reuse household waste in 2013 [27]. Based on its characteristics, most of the waste in Indonesia is classified as organic waste. In fact, Indonesia is one of the countries with the highest content of organic waste in Asia (70.2%), followed by China (67.3%) [28]. In urban areas, the volume of organic waste can reach 70% of the total waste [29]. Based on a survey conducted in 2005-2006, the composition of organic waste in Indonesian cities could reach up to 52% of all urban waste [30].

Vegetable waste dominates the total waste with an average of 2 tons/day. The composition of vegetable waste that is mostly found, among others, are spinach, chicory, mustard greens, cabbage, and a small portion of other vegetables. This vegetable waste will usually be transported to landfills, while the rest will be left behind. Vegetable waste that is left behind has not received special handling and will cause environmental pollution [31]. However, vegetable waste could provide good nutrition for microbes and does not contain pathogens, as such it is suitable for anaerobic digestion or composting [32].

The Ministry of Environment aims to develop a waste-to-energy (WtE) investment financing program for several projects, including (a) collecting methane gas in cattle farm waste (biogas); (b) increasing the efficiency and collects methane gas in tofu industrial waste (biogas); (c) agricultural waste (biomass); (d) processing industry (heat to energy); (e) general WtE (biomass from oil palm fronds, biogas from palm oil mills, and domestic waste (human waste)); [5]. Of the various types of WtE, four types of WtE are proposed to be supported through financial investment, namely (a) biogas reactor from cattle farm waste; (b) biogas reactor from tofu industrial waste; (c) developing power generator in the oil palm plantation and industry; and (d) utilization of agricultural biomass for heating/drying, particularly for rice husk biomass.

## 5. Conclusion

In line with the direction of energy conversion, LPG has become a strategic commodity that is widely needed by the community as a substitute for kerosene. However, this leads to several problems, mainly related to scarcity and likely price surges. This study aims to reexamine the concept of energy conversion by considering and utilizing optimally the already available domestic energy sources. The prospect of biomethane as an environmentally friendly renewable energy has considerable potential in meeting energy needs due to limited fossil fuel reserves and concerns about increasing GHG emissions.

## Acknowledgment

The authors would like to thank the European Commission and the INEA for the support of the DiBiCoo project. DiBiCoo has received funding from the European Union's Horizon 2020 research and innovation programme under grant agreement No 857804.

Disclaimer: The sole responsibility of the content of this paper lies with the authors. It does not necessarily reflect the opinion of the European Union. Neither the INEA nor the European Commission is responsible for any use that may be made of the information contained therein.

## Reference

[1] BPPT 2018 *Indonesia's Energy Outlook 2018: Sustainable Energy for Land Transportation* Jakarta

[2] Ministry E S D M 2015 *ESDM Strategic Plan 2015-2016* Jakarta: Ministry of Energy and Mineral Resources

[3] Priyanto U 2018 *Perspectives, Indonesian Energy Potential and Resilience* Jakarta: Tempo

[4] Hosseini S E and Wahid M A 2013 Feasibility study of biogas production and utilization as a source of renewable energy in Malaysia *Renewable and Sustainable Energy Reviews* **19** 454-462

[5] BKF 2014 *Analysis of Costs and Benefits of Financing Waste to Energy Investment through Credit Programs* Fiscal Policy Agency

[6] Rahayu A S, Karsiwulan D, Yuwono H, Trisnawati I, Mulyasari S, Rahardjo S, Hokermin S, Paramita V 2015 *Converting POME to Biogas: Project Development in Indonesia* Winrock International

[7] Ryckebosch E, Drouillon M, Vervaeren H 2011 Technique for Transformation of Biogas to Biomethane *Biomass and Bioenergy* **35** 1633-1645
[8] Khan I U, Othman M H D, Hashim H, Matsuura T, Ismail A F, Arzhandi M R, Azelee I W 2017 Biogas as a renewable energy fuel - A review of biogas upgrading, utilization and storage *Energy Conversion and Management* **150** 277-294
[9] Wien T U 2012 *Biogas to Biomethane Technology Review* Vienna University of Technology
[10] Bauer F, Hulteberg C, Persson T, Tamm D 2013 *Biogas upgrading - Review of commercial technologies* Malmö: SGC Rapport
[11] Sun Q, Li H, Yan J, Liu L, Yu Z, Yu X 2015 Selection of appropriate biogas upgrading technology-a review of biogas cleaning, upgrading and utilisation *Renewable and Sustainable Energy Reviews* **51** 521-532
[12] Huseby H H 2015 *Biogas Upgrading: Technoeconomic Evaluation of Different Technologies Based on Norwegian Potential of Raw Materials* Master Thesis, Norwegian University of Life Sciences
[13] Yang L, Ge X, Wan C, Yu F, and Li Y 2014 Progress and perspectives in converting biogas to transportation fuels *Renewable and Sustainable Energy Reviews* **40** 1133-1152
[14] DJMIGAS-1 2015 *Realization and Plan of Conversion and Savings in the Conversion of Kerosene to 3 KG LPG*
[15] EBTKE 2020 *Strategic Plan (Renstra) of the Directorate General of EBTKE 2020-2024* http://ebtke.esdm.go.id/post/2020/05/18/2540/rencana.strategis.renstra.ditjen.ebtke.2020-2024
[16] Secretariat General Of The National Energy Council 2019 *Indonesia's Energy Outlook* https://www.esdm.go.id/assets/media/content/content-outlook-energi-indonesia-2019-bahasa-indonesia.pdf
[17] Syukur H 2016 Potential of Natural Gas in Indonesia *Swara Patra* **6** (1)
[18] Ministry of Agriculture Republic of Indonesia 2019 *Livestock and Animal Health Statistics* Directorate General of Animal Husbandry and Animal Health
[19] Nagamani B and Rasamany K 1999 Biogas Technology –an Indian Perspective *Current Science* **77** 44-55.
[20] 2006 National Socio-Economic Survey - Basic Information on Households and Household Members (Vsen2006-K) https://sirusa.bps.go.id/sirusa/index.php/kuesioner/103
[21] Gladstone N 2006 *Biogas* http://www.paceproject.net
[22] BPS, Statistics Indonesia 2018 Central Bureau of Statistics
[23] Directorate General of Plantation 2016 *Indonesian Plantation Statistics 2015-2017: Palm Oil Commodities* Ministry of Agriculture
[24] Rahayu A S, Karsiwulan D, Yuwono H, Trisnawati I, Mulyasari S, Rahardjo S, Hokermin S, Paramita V 2015 Converting POME to Biogas: Project Development in Indonesia *Winrock International*
[25] MEMR 2017 *Handbook of Energy and Economic Statistics of Indonesia 2017* Jakarta: Ministry of Energy and Mineral Resources
[26] MENLH I M o E 2015 *Hlh 2015 Series - Plastic Waste Handling Dialogue* (10 June 2015). Available: http://www.menlh.go.id/r circuit-hlh-2015-dialog-panganan-sampah-plastik/
[27] BPS B P S I 2015 *Population Projection by Province, 2010-2035 (thousands)*. Available: https://www.bps.go.id/linkTabelStatis/view/id/1274
[28] Idris A, Inanc B, and Hassan M N 2004 Overview of waste disposal and landfills/dumps in Asian countries *Journal of material cycles and waste management* **6** 104-110
[29] Terazono A, Sakai S, Moriguchi Y, Yang J, Inanc B, and S. Y 2003 *Structural analysis of material cycles and waste management in Asia* National Institute for Environmental Studies Japan, Tsukuba
[30] Damanhuri E 2008 A Future prospect of municipal solid waste management in Indonesia *Keynote Lecture in the 5th Asian-Pacific Landfill Symposium, Sapporo, Japan*

[31] Febriyantiningrum Kuntum, Nurfitria N, Rahmawati A 2018 *Study on the Potential of Vegetable Waste in Pasar Baru Tuban as Liquid Organic Fertilizer* National Seminar on Community Research and Development

[32] Kalpana P, Sai Bramari G, Anitha L 2011 Formulation Of Potential Vegetable Waste Compost in Association With Microorganisms and Spirulina platensis *Asian Journal of Plant Science and Research* **1** (3) 49-57

[33] Kapoor R, Ghosh P, Tyagi B, Vijay V K, Vijay V, Thakur I S, Kamyab H, Nguyen DD, Kumar A 2020 Advances in biogas valorization and utilization systems: A comprehensive review. *Journal of Cleaner Production* **273** 1–15

[34] Thiruselvi D, Kumar P S, Kumar M A, Lay C H, Aathika S, Mani Y, Jagadiswary D, Dhanasekaran A, Shanmugam P, Sivanesan S, Show P-L 2020 A critical review on global trends in biogas scenario with its up-graduation techniques for fuel cell and future perspectives *International Journal of Hydrogen Energy* 1–17.

[35] Nguyen L N, Kumar J, Vu M T, Mohammed J A H, Pathak N, Commault A S, Sutherland D, Zdarta J, Tyagi V K, Nghiem L D 2020 Biomethane production from anaerobic co-digestion at wastewater treatment plants: A critical review on development and innovations in biogas upgrading techniques *Science of the Total Environment* **765** 1–13

[36] Feiz R and Ammenberg J 2017 Assessment of feedstocks for biogas production, part I—A multi-criteria approach *Resources, Conservation and Recycling* **122** 373–387

[37] Khalil M, Berawi M A, Heryanto R, Rizalie A 2019 Waste to energy technology: The potential of sustainable biogas production from animal waste in Indonesia *Renewable and Sustainable Energy Reviews* **105**, 323–331

[38] Central Bureau of Statistics 2020 *Peternakan dalam Angka 2020* Retrieved from https://www.bps.go.id/publication/2020/06/10/93c6d3265760176e2a87c8cf/peternakan-dalam-angka-2020.html on February 11, 2021.

[39] Roubík H and Mazancova J 2020 Suitability of small-scale biogas systems based on livestock manure for the rural areas of Sumatra *Environmental Development* 100505

[40] Setyobudi R H, Wahono S K, Adinurani P G, Wahyudi A, Widodo W, Mel M, Nugroho Y A, Prabowo B, Liwang T 2018 Characterisation of Arabica coffee pulp–hay from Kintamani, Bali–as prospective biogas feedstocks *MATEC Web of Conference* **164**

[41] Sitompul J, Bayu A, Soerawidjaja T H, Lee H W 2012 Studies of biogas production from green seaweeds *International Journal of Environment and Bioenergy* **3** (3) 132–144

[42] Andriani D, Wresta A, Saepudin A, Prawara B 2015 A review of recycling of human excreta to energy through biogas generation: Indonesia case *Energy Procedia* **68** 219–225

[43] Anshar M, Ani F N, Kader A S 2014 The energy potential of municipal solid waste for power generation in Indonesia *Jurnal Mekanikal* **37** (2) 42–54

[44] Sinaga N, Nasution S B, Mel M 2018 Process optimization of biogas production from palm oil mill effluent: a case study of a crude palm oil factory in Muaro Jambi, Indonesia *Journal of Advanced Research in Fluid Mechanics and Thermal Science* **49** (2) 155–169

IOP Conf. Series: Materials Science and Engineering 1143 (2021) 012032 doi:10.1088/1757-899X/1143/1/012032

# Biogas Utilization in KPBS Pangalengan: History and Challenges

**Pramujo Widiatmoko[1*], Jenny Rizkiana[2], Susilo Yuwono[3], Mohammad Taufiq[3] and Candra Purnama Hadi[3]**

[1] Department of Chemical Engineering, Institut Teknologi Bandung, Bandung 40132 Indonesia
[2] Department of Bioenergy and Chemurgy Engineering, Institut Teknologi Bandung, Bandung 40132 Indonesia
[3] Cendikia Salim Mulia Foundation, Bandung, Indonesia

*pramujo@che.itb.ac.id

**Abstract.** Pangalengan has been known as business centre of dairy farming in southern Bandung area since Dutch colonization. The Dairy Farmer Cooperative of South Bandung (KPBS) Pangalengan is organizing the collection of raw milk from dairy farmer and sent them to further production system. There are three districts i.e. Pangalengan, Kertasari and Pacet. In this paper, we highlight the history and challenges of biogas utilization in the Pangalengan working field. Although utilization of biogas from dairy manure was long time ago initiated in the area, coverage of biogas application is still as low as 14%. A breakthrough is required to tackle the challenge. We introduced fostered village program in a collaboration between university and the farmer cooperative to ensure continuity of the biogas program.

## 1. Introduction

Utilization of biogas from waste is important for providing a renewable local-source energy in a dairy farming area. It also environmentally friendly methods to reduce organic content in the dairy farming waste [1]. Biogas is produced during anaerobic fermentation of manure. Composition of biogas depend on percentage of fat, protein, and carbohydrate in the substrate used. It usually consists of 50-75% methane and 25-45% carbon dioxide [2]. With energy content of 4800-6700 kcal/$m^3$ [3], the biogas is potential to replace subsidized gas for cooking, even further for electricity generation in an energy self-sustained farming.

International Renewable Energy Agency (2020) showed that the biggest producer of biogas in the world is United States with capacity of 2368 MW [4]. In Asia, China becomes the top ones with biogas capacity of 799 MW. In Indonesia, biogas has been introduced since the 1970s. The utilization was accelerated in 1981 during biogas development projects funded by Food and Agriculture Organization (FAO) [5]. In 2019, the biogas capacity of Indonesia reached 22 MW. The capacity was same since 2015, which indicate deceleration on biogas installation. Meanwhile, National Energy Masterplan (Rencana Umum Energi Nasional, RUEN) mandates 5.5 GW bioenergy contribution on new and renewable energy mix and specifically 489.8 billion metrics cubic as biogas by 2025 [6]. Data show that realization of the target on 2019 is only 25.67 billion metrics cubic. Therefore, it is mandatory to re-accelerate the biogas utilization.

As centre of dairy farming, Pangalengan and Lembang in Bandung area, West Java, Indonesia has high potential for biogas utilization. Pangalengan area is known as the earliest dairy farming in the West Java region. Therefore, this work highlights history of biogas utilization in Pangalengan area and its challenges.

## 2. History of Biogas Utilization in Pangalengan Area

Utilization of biogas in Indonesia can be traced since 1970s. Here, we describe the biogas application in Pangalengan in three milestone periods, i.e. before 1980s, 1980s - 2010, and 2010 up to now. Dairy farming in Pangalengan can be classified into three periods i.e. before 1945, 1945 to 1975, and after 1975, based on number of cows, land occupancy, and technology used [7]. However, biogas technology in the world just been developed in early 1900s and intensive biogas utilization in Indonesia began since 1980. Therefore, we set 1980 as a milestone. *Biogas Rumah* (BIRU) project was started on 2009 as a second wave of biogas project in Indonesia. The program entered Pangalengan area on late of 2010, which then set as the second milestone of biogas in the area.

### *2.1. Before 1980s*

Since 1860s, Fries Holland dairy cows have been imported to Pangalengan, West Java, Indonesia and managed by Dutch companies i.e. De Friesche Terp, Almanak, Van Der Est, and Bigman [8]. The dairy product from those companies was then collected and marketed in Bandoengche Melk Centrale (BMC) in Bandung city. Biogas was not generated in the farming at that time because the technology itself was developed in Europe on early 1900s.

In 1911, Dutch quickly adopted the biogas technology by installing digester in Bandung city (well-known as Inhoftank in Tegallega) which produce gas, electricity, and fertilizer from domestic waste [9]. The name of Inhoftank came from Germany engineer, Karl Imhoff, where his digester design was used in the installation. He had experience on installing waste treatment system in German. When Japan entered Bandung, the plant mainly produced biogas for military vehicles. After 1945, the plant was abandoned.

In 1950s, biogas became a trend in India and China. Widespread dissemination of biogas digesters in other developing countries stems from 1970s. In Indonesia, the manure-based biogas technology development was leads by universities, notably the Bandung Institute of Technology (ITB) [9]. Using two metal oil drums of 200 L as a digester and a gas holder, the system became susceptible for corrosion and leads to a shorter lifetime of the biogas plants. However, to our knowledge, there are no on-line reports available on biogas application in Pangalengan area until this period.

### *2.2. Between 1980s to 2010s*

In 1981, the biogas project from Ministry of Agriculture supported by Food and Agriculture Organization of The United Nations in a Technical Cooperation Programme was initiated in Indonesia. It was considered as the first major wave of biogas project. Objectives of the program was setting up demonstration units, providing training, and preparing biogas program in the integrated rural development. Bali was set as focal development center. A demonstration of 200 fixed dome plants were installed [10]. However, high construction cost of the plant and high availability of subsidized kerosene hinder wide adoption of the technology.

**Figure 1.** Concrete fixed dome Biogas reactor was installed in Unit Usaha Swa Karya Kartika, Pangalengan (Adisasmito & Sukandar, 1994).

Following the major project, sporadic biogas reactor was installed in Pangalengan Area. In 1994, Adisasmito & Sukandar installed concrete fixed dome biogas reactor in *Unit Usaha Swa Karya Kartika*, Pangalengan (Figure 1) [11]. With volume of 33 $m^3$, the reactor handled manure from 50 dairy cows. The produced biogas was used to substitute cooking woods, kerosene, and lighting in the farming with potential saving up to 2 million rupiah per year.

Until 2000s, there are no significant improvement on biogas utilization due to wood availability and subsidized kerosene. However, increasing price of kerosene due to economic crisis made biogas become interesting alternative. Between 2002 to 2004, various organization developed and installed biogas plant. Institut Teknologi Bandung and Padjajaran University developed and installed plastic bag digesters in Pangangalengan. The digester consists of 4 $m^3$ plastic digester and 2.5 $m^3$ plastic gas holder [10].

In 2003, a business start-up under Business Incubator Center of Institut Teknologi Bandung, Cipta Tani Lestari, tried to accelerate plastic-based biogas reactor installation in Pangalengan area. Plastic bag digester was chosen due to its low capital cost which is suitable for farmer. From 2003 to 2005, they have installed about 35 reactors. Despite low installation cost, the plastic-based reactor easily broken. Therefore, The Cipta tani Lestari is developing a notable fiberglass-based reactor known as TENARI Biogas Reactor.

It was noted that from 2006 to 2011, there are 796 reactors were installed in Kabupaten Bandung through Government funding, including in Kecamatan Arjasari, Cilengkrang, Pasirjambu, Ciwidey, Kertasari and Pangalengan [12]. The West Java Province through Energy and Mining Office has been installing 750 biogas plants till the end of 2007. By the middle of 2008, another 369 biogas plants were installed. Almost all plants are plastic bag digesters, with some fixed domes as well [10]. Easily-broken plastic bag digester seems discourage the farmer in Pangalengan to further adopt the biogas technology [13].

### *2.3. Between 2010s to 2020s*

"*Biogas Rumah*" (Household biogas) program can be considered as the second major wave of biogas utilization in Indonesia. The program, also known as its acronym BIRU program, started from May 2009 as collaboration between Dutch and Indonesia governments. The Humanist Institute for Development Cooperation (HIVOS) was appointed by the Dutch government as program manager, with technical assistance from the Netherlands Development Organization (SNV). *Yayasan Rumah Energi* and the Ministry of Energy and Mineral Resources Republic Indonesia became the coalition partners from Indonesia side. Target of the program was installing 10,000 bio-digester by the end of 2013 [14]. Typical concrete fixed-dome reactor they built is shown in Figure 2. By September 2020, they already installed 24,945 biogas reactors in 12 provinces.

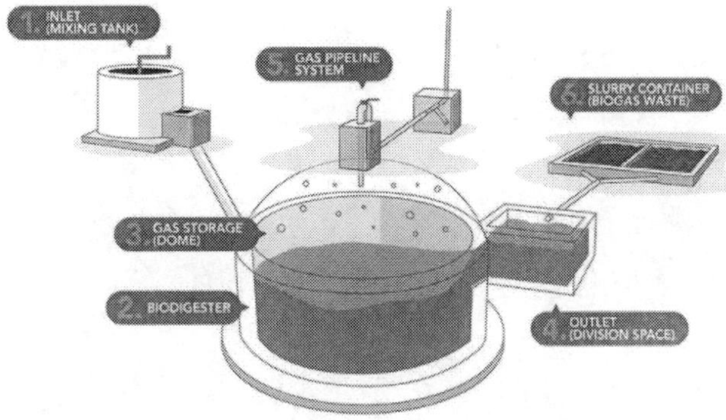


**Figure 2.** Schematic arrangement of BIRU Fixed Dome Biogas Plant (https://www.biru.or.id/en/about-fixed-dome).

West Java Province is one of target for the BIRU program. From 24 October 2009 to February 2015, they installed 1,283 reactors in the West Java Province (out of total 14,173 installation) [15, 16]. The distribution of reactor included dairy farmers from KPSBU Lembang Bandung Barat, KPGS Cikajang Garut Selatan, KUD Tanjungsari Sumedang, KPS Gunung Gede Pangrango Sukabumi, and other area such as Cianjur, Sukabumi, Bogor, Depok, Serang Banten, Cariu, and Bekasi [17]. In first semester of 2017, the BIRU program reported three cooperation which actively gave financial support for the biogas reactor installation, i.e. *Koperasi Peternak Sapi Bandung Utara* (KPSBU), *Koperasi Peternak Sapi Cianjur Utara* (KPSCU), and *Koperasi Peternak Garut Selatan* (KPGS).

In Pangalengan area, the BIRU program significantly increases the biogas adoption. In 2009, it was reported that 7 farmers in Pangalengan have received biogas reactor from government [18]. The BIRU program in 2017 also cooperated with Perum Jasa Tirta II, Pangalengan to install the fixed dome concrete biogas reactors.

## 3. Recent condition and Challenges

Distribution of dairy cows in Pangalengan area is shown in Figure 3, which is potential as source of biogas [19]. Margamekar is the most populated village by 4647 dairy cows, followed by Margamukti village with 2589 cows. In average, each farmer has 3 cows with a limited area for digester installation. The limited area of farm often restricts installation of biogas reactor within its comfortable operation zone. Some of the farmer cease to use the reactor due to filling process of manure using bucket was considered as a bothersome work.

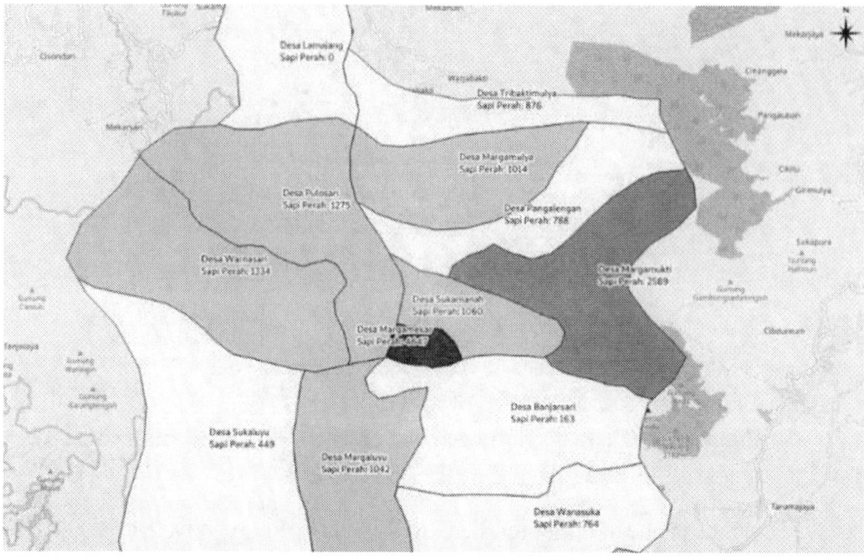


**Figure 3.** Potency of Dairy Cows for Biogas Utilization in Pangalengan Area (Virgana & Hamdani, 2020).

To increase the utilization of biogas, we are conducting community service activities in Margamulya, Pangalengan, cooperated with KPBS Pangalengan from 2019 to 2020. In the initial activity, a TENARI biogas reactor 4 $m^3$ in volume was installed in a dairy farm with 20 cows [20]. The biogas then distributed to the farm housing and three closest neighbour household as substitution of liquified petroleum gas (LPG) for cooking fuel. It was estimated that the substitution of LPG using the biogas save IDR 69,000 to 295,000 a month.

Figure 4 shows the typical consumption rate of biogas in the location. We found that low usage of biogas in the location mainly due to lack of produced biogas. Only one cooker can be used at a time. Also, there is an objection with odour released by the gas. To overcome the problems, we did second activity of community service by installed two additional 4 $m^3$ biogas reactors equipped by $H_2S$ absorber (Figure 5.a and 5.b). Therefore, total volume of reactor became 12 $m^3$. Those reactors are supplying biogas to the farm housing and five neighbour households with potential saving up to IDR 885,000 a month.

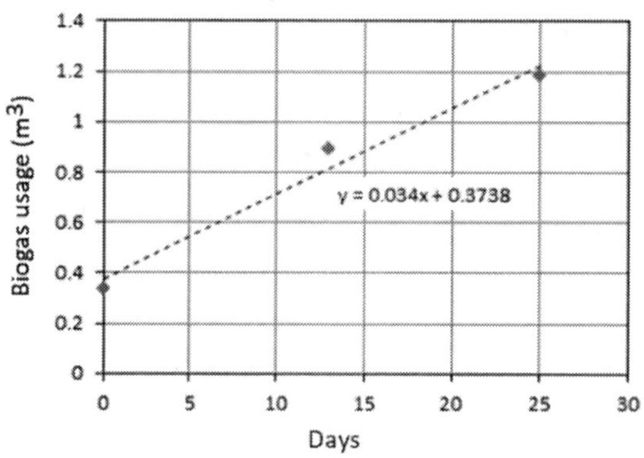


**Figure 4.** Typical biogas consumption rate from the installed biogas reactor.

Adoption of biogas technology to farmer is challenging [18, 21, 22]. Dyah and Sriharti (2019) present the challenges of biogas development in West Java area, especially Subang, Sumedang, and Bandung area [23]. There are technical and social barriers. Damaged biogas stoves, far location of manure source from reactor, and low quantity of produced biogas often cause the farmer stop using biogas. Previous projects mostly left the biogas adopter farmers without continuous training, inspection and maintenance program. Having difficulty due to leaking reactor and damaged stove lead the farmer to abandon their biogas system. A continuous program is required to overcome this condition. It also needs to tackle social barriers such as availability of commercial LPG, reluctance to collect manure, and no available space to build the reactor. The BIRU project also identify barriers on the development of biogas in West Java province e.g. limited number of certified constructors of biogas, financial support mainly come from dairy cooperative only, limited area for installation, and difficult terrain condition. Socially, there is high dependency for free installation support from government [17].

In our case in Pangalengan, challenges for adoption of biogas technology seems to be installation cost and operability of the reactor. Small-scale animal husbandry businesses having 1-3 cows with dependency on milk income left the farmer in poverty line [24, 25]. To get decent life, they need to do another business e.g. sale of calves, heifers, bulls, or rejected cattle. Therefore, the small holder farmers hardly afford expensive biogas reactor like concrete fixed-dome or fiberglass. Alternatively, the small holder farmers can join a communal biogas reactor. It will reduce installation cost and overcome limited area restriction. Technically, operation of such communal reactor is handled by a competence operator. This approach in the past however was hindered economically by operational cost. Biogas users are reluctant to pay for the cost [26].

In term of operability, previous farmer's self-installed fixed dome reactor was abandoned after several month operation. Collecting manure by bucket and filling the reactor consume farmer's precious time to maintain the cows. To overcome the problem, our biogas plant was designed to collect the manure directly from manure wasting system. The bucket filling is not needed anymore, therefore reduce the burden of the farmer (Figure 5.c).

Further, biogas technology dissemination should emphasize on how to maximize its economic potential. Despite significant saving when using biogas, the usage does not increase the farmer income. Their income still depends on the price of dairy milk, which does not increase for years. By developing the biogas and its derivative products as secondary competitive source of income, adoption of the technology will increase significantly. The further utilization of biogas in Pangalengan area needs to consider the limited cattle and area of small holder farmer. Last important thing, we introduced fostered village program in a collaboration between university and the farmer cooperative for continuous monitoring and technology dissemination.

(a)

(b)

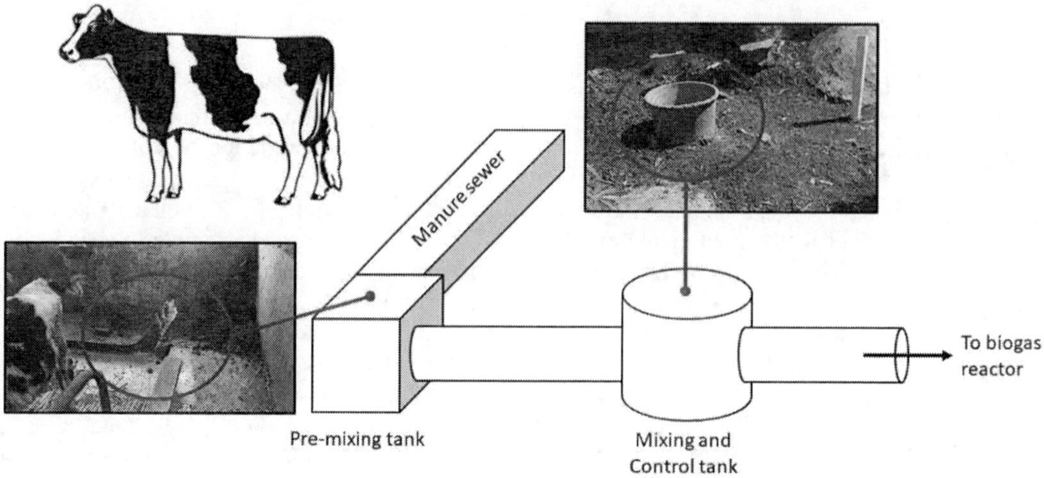


(c)

**Figure 5.** Photograph of (a) location preparation and (b) installed additional biogas reactor (c) schematic diagram of modified reactor feeding system.

## 4. Conclusion and Future Works

Biogas is potential for renewable fuel especially in Pangalengan area which populated by dairy farms. Despite long history on the dairy farming, utilization of biogas in the area just begins in 1980s. Many technical and social challenges cause low adoption of the technology. With the funding from both enterprise and government, the biogas technology become more familiar to the farmers. It will be great advantage to develop the biogas technology and its derivative product as secondary source income for the farmer. Further work needs to improve quality of the biogas to comply with a commercial standard.

## Acknowledgments

The biogas reactor installation in this work was financially supported by the Research and Community Service 2020 program, Faculty of Industrial Technology, Institut Teknologi Bandung. Authors gratefully acknowledge the support from KPBS Pangalengan as partner of this project. We deeply thank Mr. Adang Shalahudin for providing the location for installation of biogas reactor.

## References

[1] Tasmaganbetov A B, Ataniyazov Z, Basshieva Z, Muhammedov A U and Yessengeldina A 2020 *IJEEP* **10** 5
[2] Al Seadi T, Rutz D, Prassl H, Kottner M, Finsterwalder T, Volk S and Janssen R 2008 Biogas Handbook (Esbjerg: University of Southern Denmark) p 41
[3] Widodo T W and Hendriadi A 2005 Proc. Int. Sem. on Biogas Technology for Poverty Reduction and Sustainable Development 1
[4] Whiteman A, Rueda S, Akande D, Elhassan N, Escamilla G and Arkhipova I 2020 Renewable Energy Statistics 2020 (Abu Dhabi: The International Renewable Energy Agency) p 79
[5] Alberdi H A, Sagala S A H, Wulandari Y, Srajar S L and Nugraha D 2018 IOP Conf. Series: Earth and Environmental Science **158** 012031
[6] Ministry of Energy and Mineral Resources Republic of Indonesia 2017 Handbook Energy & Economy Statistic of Indonesia 2017 (Jakarta: Center for Data and Information Technology on Energy and Mineral Resources)
[7] Mauludin M A, Alim S and Sari V P 2017 Sosiohumaniora **19** (1) 37
[8] Kunto H 1984 Wajah Bandoeng tempo doeloe (Bandung: Granesia) p 90

[9] Hutagalung R 1995 Kisah Perjuangan Unsur Ganesa 10 Kurun Waktu 1942-1950 (Bandung: ITB Publisher)
[10] van Nes W J, Tumiwa F, Setyadi I 2009 Feasibility of a national programme on domestic biogas in Indonesia (Jakarta: SNV Netherlands Development Organisation)
[11] Adisasmito S and Sukandar U 1994 Pembangkitan biogas dari limbah peternakan sapi perah di Unit Usaha Swa Karya Kartika Pangalengan
[12] https://jabarprov.go.id/index.php/news/2170/2011/04/27/Kab-Bandung-Pasok-Susu-Nasional-Hingga-2386, accessed on 20 October 2020
[13] Dianawati M, Mulijanti S L, Teddy S, Yulyatin A, Nurnayetti, Histifarina D and Riswita 2014 Laporan Akhir m-P3MI pada Agroekosistem Lahan Kering Dataran Tinggi di Kabupaten Bandung (Bandung: BPTP Jawa Barat)
[14] HIVOS 2020 Biogas Indonesia https://www.hivos.org/program/biogas-indonesia/ accessed on 20 October 2020
[15] BIRU 2015 Laporan Provinsi Semester Kedua 2014 Program Biogas Rumah (BIRU), p 11
[16] DJEBTKE 2015 Lima Tahun, Empat Belas Ribu Biogas BIRU Mengolah Limbah Jadi Berkah
[17] BIRU 2014 Laporan Provinsi Semester Pertama 2014 Program Biogas Rumah (BIRU), p 11
[18] Dianawati M and Mulijanti S L 2015 J. Lit. bang. Pert. **34** 125
[19] Virgana R A E and Hamdani D 2020 JAMIKA **10** 96
[20] Widiatmoko P, Rizkiana J, Yuwono S, Taufiq M and Hadi C P 2020 IOP Conf. Ser.: Mater. Sci. Eng. **823** 012020
[21] Budiman I, Muthahhari R, Kaynak C, Reichwein F and Zhang W 2018 IJE **1** (2) 46
[22] Wahyudi J, Kurnani T B A and Clancy J 2015 *Int. J. Renew. Energy Dev.* **4** 219
[23] Dyah S and Sriharti 2019 IOP Conf. Ser.: Earth Environ. Sci. **277** 012018
[24] Sulistyati M, Hermawan and Firman A 2019 IOP Conf. Ser.: Earth Environ. Sci. **334** 012058
[25] Agusta Q T M, Lestari D A H, Situmorang S 2014 JIIA **2** 109
[26] Ruhiyat R and Siami L, IOP Conf. Ser.: Earth Environ. Sci.**106** 012115

# Economic and Life Cycle Assessments of Integrating Gasification Unit into Production Unit of Pellet from Fallen Leaves and Twig

**Fadil Abdul Rahman[1], Gendewa Utomo[1], Indra Purwadi[1] and Herri Susanto[1*]**

[1] Department of Chemical Engineering, Institut Teknologi Bandung, Bandung 40132 Indonesia

[*]Corresponding Author's E-mail: herri@che.itb.ac.id

**Abstract.** Conversion of fallen leaves and twigs from urban waste into pellets may be one of several options to reduce the amount of waste transported to the landfill. However, the production of pellet needs very high mechanical energy or electricity. A case study has been conducted to evaluate the advantages and disadvantages of integrating a gasification technique into the production unit. In this integrated unit, the source of electricity for the pellet production was based on the conversion of a portion of pellet product to gaseous fuel which was then used as substitute of diesel fuel for running an electricity generator set. Three scenarios for supplying the electricity were compared: (i) scenario-A: from PLN (national grid, coal fired power plant), (ii) scenario-B: diesel-generator set; and (iii) scenario-C: diesel-genset and gasification unit. Total $CO_2$ emissions in the production of pellet were as follows (kg $CO_2$ eq /ton pellet): 81.8, 47.2 and 16.3 for scenario-A, B and C respectively. Due to the addition cost of investment, the integrated pellet production and gasification unit needed a Payback Period of about 5.6 years, while those of diesel based electricity and PLN were 5.1 and 4.8 years respectively. Data for this study was obtained from experiments on a pellet production unit with a design capacity of about 9 ton/week. The diesel engine was run in dual fuel mode using producer gas from the gasification of pellet. The producer gas heating values were in the range of 3300-4600 kJ/Nm$^3$. Oil saving in dual fuel operation was up to 70%.

## 1. Introduction

Municipal solid waste (MSW) is a problem in many big cities in Indonesia, including Depok with population of 2.3 million and generation of 1300 ton/day MSW [1]. With this high waste generation, the disposal landfill for City of Depok in Cipayung is already over capacity in this year 2020. A program for sorting waste into several types have been applied at several temporary collection unit (*TPS, tempat penampungan sementara*). A part of sorted waste is woody or other hard biomass that may be utilized as a raw material for pellets.

Biomass pellets become an attractive commodity and grow to international market [2]. Wood pellets may be used as solid fuel for cooking in households and small-medium enterprises. Wood pellets, MSW pellets and also RDF (refuse derived fuel) are also adaptable to cofiring with coal as an option toward a carbon-free energy intensive industries [3].

Gasification is one of three main thermochemical conversions of biomass to gaseous fuel [4, 5, 6]. Producer gas (gaseous fuel from gasification using air as a gasifying medium) contains $H_2$, CO, $CO_2$, and $CH_4$. Having a heat combustion of about 4000 kJ/Nm$^3$, this producer gas can be used as a substitute for liquid fuel an internal combustion engine [7].

Integrating gasification technology in wood pellet production plant may save electricity bill, and in addition $CO_2$ emission may also be reduced. In this concept, the gasification system is fueled with a portion of pellets product. Evaluation of operating cost and $CO_2$ emission in the integrated gasification and pellets production plant are presented in this paper. Comparisons were made for three schemes in supplying electricity (see Figure 1).

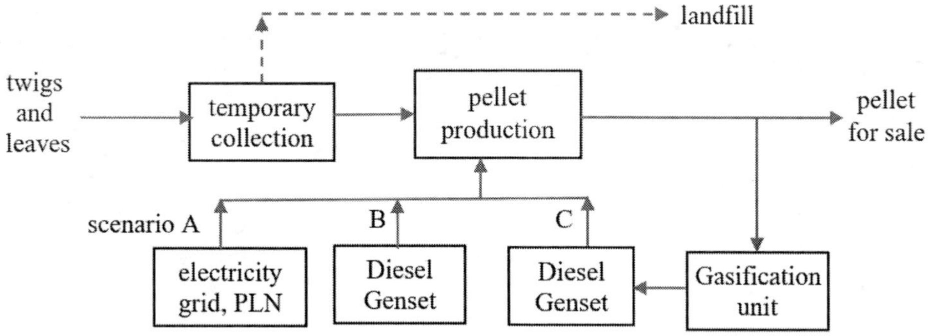


**Figure 1.** Integrated gasification and wood pellet production.

## 2. Methodology

Experiments were carried out in the existing pellet production plant in City of Depok. This pellet production plant consists of a pellet production line and a diesel engine-generator set (Figure 1). Power for this plant is supplied from electricity grid (PLN) with a capacity of about 40 kW. This study dealt with the following aspects:

a. experiments on gasification of pellets to evaluate the performance of the gasification unit
b. operation of the diesel engine with dual fuel (mixture of petroleum fuel and producer gas)
c. testing the real capacity of pellets machineries
d. simple economic evaluation to see the feasibility of self supplying electricity in a pellets production plant
e. study on the potential reduction of $CO_2$ emission.

## 3. Results and Discussion

### *3.1. Gasification and Diesel-Generator*

The gasifier was an open core down draft type with a design capacity of 20 kg/h fuel with palm kernel shell. The producer gas (product of air-gasification) was firstly cooled and then filtered for tar removal. A typical analysis of wood pellets used in this study are presented in Table 1. The proximate and ultimate analysis of wood pellet used in this research was comparable to those reported in [8]. As expected, the gasification unit could be operated smoothly for 6 hours at a capacity up to 20 kg/h. However, the pellet mass conversion in the gasification process was only 73%, as indicated by the difference of ash content in original pellets to that in the char residue after gasification.

The producer gas composition and heating value were satisfactory (Table 2). This producer gas could be fed into the diesel engine for dual fuel operation. At a power output of about 22 kW, the oil saving in the dual fuel operation was up to 70%, meaning that 70% energy intake of the engine came from the producer gas. Of course, the oil saving depended on the producer gas composition and on the adjustment of gas to air ratio by the operator.

**Table 1.** Proximate and ultimate analysis of wood pellet.

| No. | Characteristics | wood pellet, [8] | this work: wood pellet | this work: char residue after gasification |
|---|---|---|---|---|
| | | **Proximate Analysis** (wet basis) | | |
| 1 | Moisture Content | 12.00% | 12.10% | 3.63% |
| 2 | Ash | 1.00% | 4.36% | 14.20% |
| 3 | Volatile matter | 79.19% | 67.68% | 17.60% |
| 4 | Fixed Carbon | 7.83% | 15.86% | 64.57% |
| | | **Proximate Analysis** (dry basis) | | |
| 1 | Ash | 1.14% | 4.96% | 18.27% |
| 2 | Volatile matter | 89.97% | 77.00% | 14.74% |
| 3 | Fixed Carbon | 8.90% | 18.04% | 67.00% |
| | | **Ultimate Analysis** (dry basis) | | |
| 5 | Carbon | 50.09% | 39.29% | 65.46% |
| 6 | Hydrogen | 6.11% | 6.01% | 1.23% |
| 7 | Oxygen (by different) | 40.50% | 48.79% | 13.91% |
| 8 | Nitrogen | 1.98% | 0.85% | 1.14% |
| 9 | Sulfur | 0.18% | 0.10% | 0.00% |
| 10 | Ash | 1.14% | 4.96% | 18.27% |
| 11 | Total | 100.00% | 100.00% | 100.00% |
| 12 | HHV, MJ/kg (dry basis, as measured) | | 16.32 | 24.47 |
| 13 | LHV, MJ/kg (dry basis) | 18.69 | 15,09 | 24,20 |

**Table 2.** Typcial composition of producer gas form gasification of wood pellet at UPS Merdeka 3 Depok.

| No. | Components | wood pellet, [8] | this work: sample-1 | this work: sample-2 |
|---|---|---|---|---|
| 1 | $H_2$ | 21.62% | 20.97% | 14.64% |
| 2 | $CH_4$ | 2.30% | 2.74% | 1.73% |
| 3 | CO | 27.74% | 13.11% | 10.62% |
| 4 | $CO_2$ | 9.43% | 18.75% | 10.35% |
| 5 | $N_2$ | 38.91% | 44.48% | 62.60% |
| 6 | LHV, kJ/Nm$^3$ | 6333.5 | 4655.4 | 3364.7 |

*3.2. Pellet Production Unit*

The pellet production line consists of a chipping machine, a rotary dryer, a sieving machine, powdering machine, two pelletizing machines. Design capacity and power consumption of these machineries are presented in Table 3. Weekly production rate must be scheduled carefully since the limited electricity from the PLN grid was only 40 kW. Assuming that all machineries could be operated at their design capacity, a weekly production rate of 9 ton might achieved (Table 3). The total electricity for running the machineries were 33 kW in day-1, -3 and -5. For day-2, -4 and -6, power consumption were 26 kW.

Unfortunately, the real capacity of both pellet machines were far from their design one, so this was a bottleneck to achieve the production rate of 9 ton/week. In the pelletization, woody biomass powder must be recycled in the pellet machine twice or three times to obtain a proper pelletization temperature of about 70°C in order to get the satisfied mechanical strength of pellet. Thus the real capacity of the pellet machines might only be half or one third of their design. Pellet with a good physical properties was also depended on the moisture content of the woody biomass powder, i.e. about 15 - 20%.

**Table 3.** Specification of machineries and production schedule for handling woody biomass of 9000 kg/week.

| No. | Combination equipment | Power (kW) | Design capacity (kg/h) | Operation time (h) | Day - 1 | 2 | 3 | 4 | 5 | 6 |
|---|---|---|---|---|---|---|---|---|---|---|
| | Batch-1, woody biomass input 3 ton | | | | | | | | | |
| 1. | chipping machine | 5.5 | 750 | 4.0 | 5.5 | | | | | |
| 2. | rotary dryer | 11.0 | 500 | 5.9 | 11.0 | | | | | |
| 3. | sieving machine | 1.5 | 700 | 3.8 | 1.5 | | | | | |
| 4. | powdering machine | 7.5 | 400 | 6.6 | 7.5 | 7.5 | | | | |
| 5. | pellet machine, 6 mm | 7.5 | 120 | 10.9 | 7.5 | 7.5 | | | | |
| 6. | pellet machine, 8 mm | 11.0 | 300 | 4.4 | | 11.0 | | | | |
| | Total power, kW | 44.0 | | | 33.0 | 26.0 | | | | |
| | Batch-2, woody biomass input 3 ton | | | | | | | | | |
| 1. | chipping machine | 5.5 | 750 | 4.0 | | | 5.5 | | | |
| 2. | rotary dryer | 11.0 | 500 | 5.9 | | | 11.0 | | | |
| 3. | sieving machine | 1.5 | 700 | 3.8 | | | 1.5 | | | |
| 4. | powdering machine | 7.5 | 400 | 6.6 | | | 7.5 | 7.5 | | |
| 5. | pellet machine, 6 mm | 7.5 | 120 | 10.9 | | | 7.5 | 7.5 | | |
| 6. | pellet machine, 8 mm | 11.0 | 300 | 4.4 | | | | 11.0 | | |
| | Total power, kW | | | | | | 33.0 | 26.0 | | |
| | Batch-3, woody biomass input 3 ton | | | | | | | | | |
| 1. | chipping machine | 5.5 | 750 | 4.0 | | | | | 5.5 | |
| 2. | rotary dryer | 11.0 | 500 | 5.9 | | | | | 11.0 | |
| 3. | sieving machine | 1.5 | 700 | 3.8 | | | | | 1.5 | |
| 4. | powdering machine | 7.5 | 400 | 6.6 | | | | | 7.5 | 7.5 |
| 5. | pellet machine, 6 mm | 7.5 | 120 | 10.9 | | | | | 7.5 | 7.5 |
| 6. | pellet machine, 8 mm | 11.0 | 300 | 4.4 | | | | | | 11.0 |
| | Total power, kW | | | | | | | | 33.0 | 26.0 |

### *3.3. Economic Evaluation*

The pellet production plant was installed in 2018, with a total investment cost of IDR 2,063,451,600. Using the price index of 711 in 2018 and 735 in 2020, an estimated investment cost in 2020 of IDR 2,132,989,995 was obtained (break down of cost are presented in Table 4). It is understandable that the self supporting electricity required more investment for the diesel generator (half of investment cost of pellet production machineries), even much higher for diesel generator and gasification system. Grants for diesel generators and gasification units are expected from the government or community services from companies.

There three scenarios concerning with supply of electricity, i.e (see figure 1): scenario-A, electricity grid (PLN), scenario-B, self supply using PLTD (diesel genset), and scenario-C, self supply using PLTD and gasification of pellet (integrated system). Parameters for estimating the manufacturing cost are presented in Table 5.

**Table 4.** Capital cost of wood pellet production unit at Depok

| No. | Parameter | estimated investment in 2020, IDR |
|---|---|---|
| 1. | wood pellet production machineries | 465,165,017 |
| 2. | gasification set unit | 237,751,009 |
| 3. | diesel engine and generator | 236,717,308 |
| 4. | cooling Water system | 15,505,501 |
| 5. | power house | 25,842,501 |
| 6. | other utilities | 31,011,001 |
| 7. | Total Installed Cost (TIC) | 1,011,992,336 |
| 8. | factory construction | 640,377,173 |
| 9. | land development (4% TIC) | 40,479,693 |
| 10. | Total Direct Cost (TDC) | 1,692,849,203 |
| 11. | indirect cost (20% TDC) | 338,569,840 |
| 12. | Capital Investment (FCI) | 2,031,419,043 |
| 13. | working capital (5% FCI) | 101,570,952 |
| 14. | Total Capital Cost | 2,132,989,995 |

**Table 5.** Assumptions for estimating direct manufacturing cost

| No | Parameters | Unit | PLN grid | PLTD, diesel genset | PLTD and gasification |
|---|---|---|---|---|---|
| 1 | woody biomass | kg/week | 9,000 | 9,000 | 9,000 |
| 2 | woody biomass transportation cost | IDR/kg | 100 | 100 | 100 |
| 3 | pellet net products | kg/week | 8,212 | 8,212 | 7,456 |
| 4 | electricity consumption | kWh/week | 675.5 | 675.5 | 675.5 |
| 5 | electricity cost | IDR/kWh | 1,467.3 | | |
| 6 | diesel oil consumption | L/week | | 152.3 | 45.7 |
| 7 | diesel oil price | IDR/L | | 9,400 | 9,400 |
| 8 | pellets consumption | kg/week | | | 756.3 |
| 9 | cooling water price | IDR/kWh | | 23 | 38 |
| 10 | working time | h/day | 6 | 6 | 6 |

**Table 6.** Direct manufacturing cost

| No. | Parameter | Manufacturing cost for 9000 kg/week of woody biomass as raw material, IDR | | |
|---|---|---|---|---|
| | | PLN grid | PLTD | PLTD and gasification |
| 1. | electricity cost | 991,161 | | |
| 2. | diesel oil cost | | 1,431,620 | 429,580 |
| 3. | cooling water cost | | 15,536 | 25,669 |
| 4. | Total, weekly production | 1,891,161 | 2,347,156 | 1,335,249 |
| 5. | Annual direct cost (52 weeks) | 98,340,372 | 122,052,112 | 69,432,948 |

It is not surprising, that the direct manufacturing cost in the production using electricity from PLN grid was lower than that of using PLTD (see Table 6). This is understandable because electricities in PLN grid are mostly generated in coal fired power plant. On the otherhand, self supporting electricity using the integration of PLTD and gasification (scenario-C) required a lowest direct manufacturing cost, since the cost of pellet for gasification feed was not taken into account.

**Table 7.** Fixed manufacturing cost.

| No. | Component | Amount | Total cost IDR/year |
|---|---|---|---|
| 1. | salary of manager (*based on Depok plant*) | 1 person | 36,000,000 |
| 2. | salary of technicians (*based on Depok plant*) | 1 person | 36,000,000 |
| 3. | salary of operator (*based on Depok plant*) | 6 person | 201,600,000 |
| 4. | insurance and local taxes (7%FCI) | | 145,537,731 |
| 5. | maintenance cost (2,5 % FCI) | | 51,977,761 |
| 6. | Total fixed manufacturing cost | | 471,115,492 |

Based on the direct cost of production (Table 6), the self supporting electricity using PLTD-gasification (scenario-C) indicated the lowest costs. But the fixed manufacturing cost for all scenarios were the same (Table 7). The fixed manufacturing cost was much higher than those of direct manufacturing costs, and it took a large fraction of the total manufacturing cost. To reduce the fixed manufacturing cost, manpowers should be rearranged, for instance: a total of 4 persons in scenario-B and a total of 6 persons in scenario-C. Insurance and local taxes should be minimized or even excluded from the manufacturing cost since this project dealing with city waste management. Considering only direct and fixed manufacturing costs (Tables 6 and 7, respectively), calculated cost of pellets production (IDR/kg) would be 1333, 1389, and 1394 for scenarios-A, -B and -C respectively.

For economic analysis using *Payback Period* and *Return on Investment*, several assumptions were applied (Table 8). Calculated results for profitability of pellet production are presented in Table 9. In line with the aboved mentioned cost of production, these results of calculated profitability also indicated disadvantages of scenario-C compared to the other two scenarios.

**Table 8.** Assumptions for economic analysis.

| No. | Parameter | Unit | Value | Notes |
|---|---|---|---|---|
| 1. | pellets price | IDR/kg | 2,500 | market price in Depok region |
| 2. | depreciation | IDR/year | 142,199,333 | linear |
| 3. | equity capital structure | | 100% | government of Depok |
| 4. | income tax | | 5% | tax for income less 4.8 billion IDR |
| 5. | project period | year | 15 | |
| 6. | distribution cost | IDR/ton | 100,000 | |
| 7. | factory construction | year | 1 | 100% investment in year -1 |

**Table 9.** Results of calculation for profitability of pellet production.

| No. | Parameter | Value | | |
|---|---|---|---|---|
| | | Scenario-A PLN grid | Scenario-B PLTD | Scenario-C PLTD and gasf. |
| 1. | gross profit, IDR | 455,401,736 | 431,689,996 | 389,960,360 |
| 2. | net profit, IDR | 439,908,535 | 417,382,382 | 377,739,228 |
| 3. | payback period, year | 4.8 | 5.1 | 5.6 |
| 4. | return of Investment | 20.6% | 14.6% | 13.2% |

### *3.4. Life Cycle Analysis*

Life cycle analysis was made based on ISO 14040. Assumptions for this study are presented in Table 10. The system boundary in this study is cradle to gate, as described in Figure 3. The goal in this study was to develop LCA model for wood pellet production using three different electricity sources and compare their impact to the environment.

**Table 10.** Assumptions for calculating $CO_2$ emission.

| No. | Parameter | Unit | Value | Notes |
|---|---|---|---|---|
| 1. | emission factor of diesel generator | $kgCO_2/L$ | 2.697 | [9] |
| 2. | emission factor of coal power plant | $kgCO_2/kWh$ | 1.05 | [10] |
| 3. | emission factor of transportation | $kgCO_2/(ton.km)$ | 0.09 | [11] |
| 4. | subtitution of diesel oil with producer gas | | 70% | [12] |
| 5. | carbon clossure | | 95% | [13] |
| 6. | distance to land fill | km | 10 | assumption |
| 7. | radius of woody biomass collection | km | 2 | assumption |

The $CO_2$ emissions for three scenarios of electrical energy sources are presented in Table 11 and Figure 2. For scenario A (electricity from PLN), the calculated $CO_2$ emission was the highest, i.e. 81.8 $kgCO_2$/ton woody biomass. The use coal as fuel in power plant has been known as a source of $CO_2$ emission [3]. Pellets production using scenario-B generated $CO_2$ emission of 45.6 kg $CO_2$/ton biomass. But it was unexpected that the use of integrated gasification and diesel generator set produced $CO_2$ higher than that of the scenario-B, due to addition process for converting biomass to gaseous fuels.

**Table 11.** $CO_2$ emission at capacity of 9000 kg/week.

| No. | Parameters | Scenario-A PLN grid | Scenario-B PLTD | Scenario-C PLTD-gasf. |
|---|---|---|---|---|
| 1. | type of fuel | coal based | diesel oil | diesel fuel oil, producer gas |
| 2. | woody biomass (kg) | 9,000 | 9,000 | 9,000 |
| 3. | pellet product (kg) | 8,212 | 8,212 | 7,454 |
| 4. | power electricity (kWh) | 675.5 | 675.5 | 675.5 |
| 5. | fuel consumption | | | |
| | a. diesel oil (L) | | 152.3 | 45.7 |
| | b. wood pellet (kg) | | | 756.3 |
| 6. | $CO_2$ emission in transportation ($kgCO_2$/ton wood) | 1.6 | 1.6 | 1.6 |
| 7. | $CO_2$ emission from process, ($kgCO_2$/ton wood) | 81.8 | 45.6 | 66.5 |
| 8. | $CO_2$ emission from process taking into account the carbon closure of 95% ($kgCO_2$/ton wood) | | | 16.3 |

Since the production of gaseous fuel via gasification process consumed a type of biomass, the $CO_2$ emission might be re-evaluated with considering the carbon cycle [13]. Hence by taking a carbon closure of 95% into account, the calculated net $CO_2$ emission from the pellets production using scenario-C dropped to only 16.3 $kgCO_2$/ton of woody biomass.

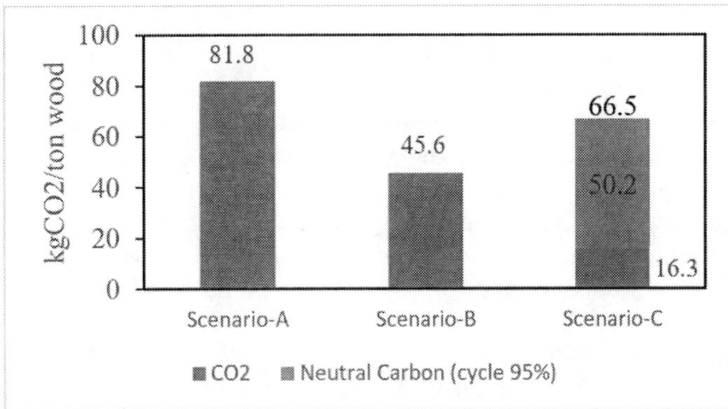


**Figure 2.** $CO_2$ emmision wood pellet production at capacity of 9000 kg/week.

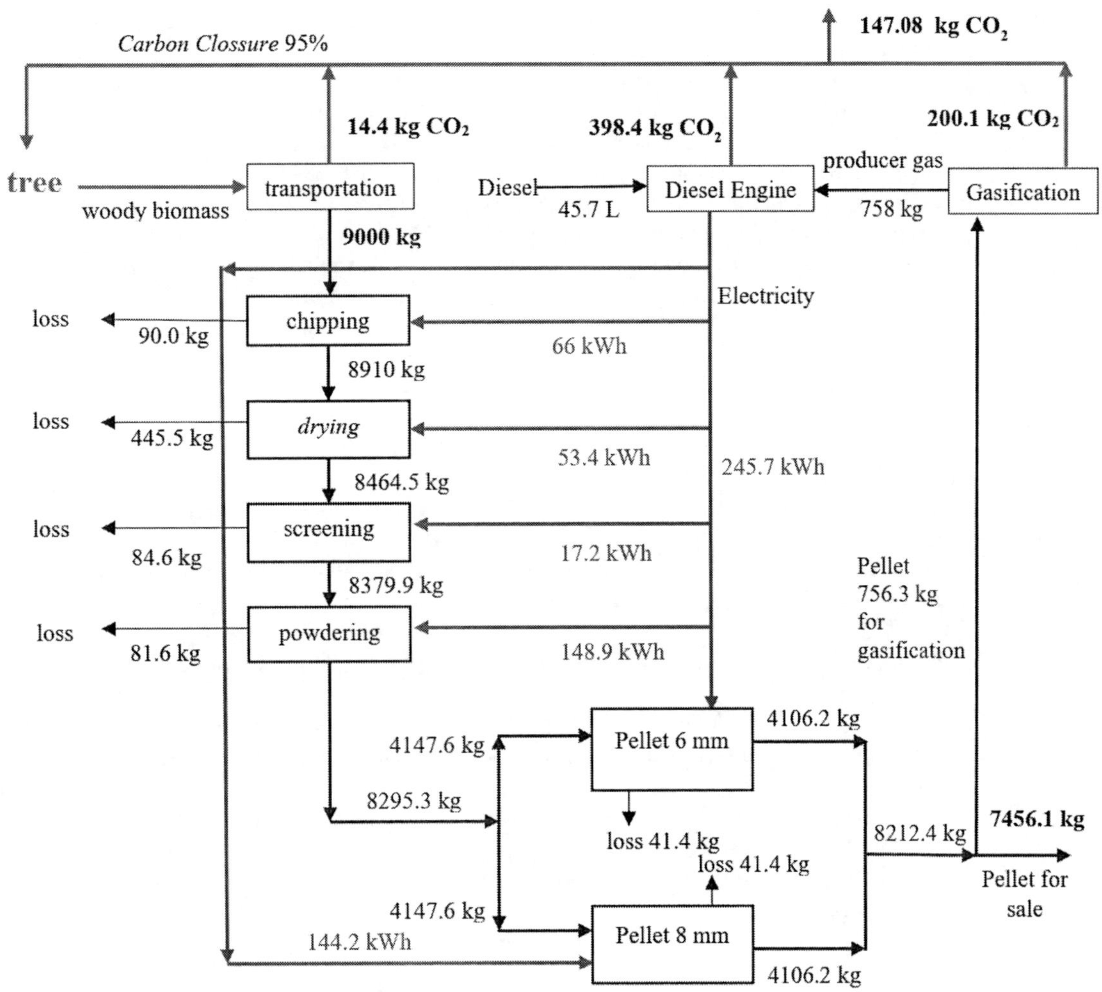


**Figure 3.** $CO_2$ emission in wood pellet production with scenario-C at capacity of 9000 kg/week.

## 4. Conclusions

With a full capacity of 9 ton/week of woody biomass as a raw material, the most economically attractive for pellets production was using electricity from the PLN grid (scenario-A). In this case the calculated payback period was 4.8 year and RoI was 20.6%. However, this scenario of supplying electricity had a highest $CO_2$ emission of 81.8 kg$CO_2$/ton woody biomass. Although the use of integrated gasification process with diesel-genset (scenario-C) was not attrative economically, the $CO_2$ emission was very low, i.e. 16.3 kg$CO_2$/ton woody biomass. The following actions may improve the economic attractiveness of the pellets production with self supporting electricity by scenario-C: (i) lowering investment cost by using locally made gasification unit and building a low cost power house, (ii) employing right manpowers including managements, and (iii) reducing taxes and other administration cost.

## Acknowledgements

This study was funded by LPPM-ITB in PKM-ITB 2020 (Program for Community Services). The authors like to thanks to *Dinas Lingkungan Hidup dan Kebersihan Depok* for facilities and manpowers in the pellets production plant at UPS Merdeka 3, Depok.

## References

[1] Badan Pusat Statistik 2019 *Statistik Daerah Kota Depok 2019* Badan Pusat Statistik Kota Depok
[2] Spelter H and Toth D 2009 *North America's Wood Pellet Sector* USDA Forest Service, Forest Products Laboratory Madison WI p. 21
[3] Mei B and Wetztein M 2017 Burning Wood Pellets for US Electricity generation, A Regime switching analysis *Journal of Energy Economics* Volume **65** 434-441
[4] Puig-Arnavat M, Bruno J C and Coronas A 2010 Review and analysis of biomass gasification models *Renewable and Sustainable Energy Reviews* **14**(9) 2841-2851
[5] Basu Prabir 2013 *Biomass Gasification, Pyrolisis, and Torrefaction Practical Design and Theory,* Second Edition *Elsevier* USA 63-84, 183-198, 199-246
[6] Anca-Couce A 2016 Reaction mechanisms and multi-scale modelling of lignocellulosic biomass pyrolysis *Progress in Energy and Combustion Science* Volume **53** 41-79
[7] Tekin K, Karagoz S and Bektas S 2014 A review of Hydrothermal Biomass Processing *Renewable and Sustainable Energy Reviews* **40** 673-687
[8] Aydin S E, Yucel O and Sadikoqlu H 2019 Experimental Study on Hydrogen-Rich Syngas Production Via Gasification of Pine Cone Particles and Wood Pellets in Fixed Bed Downdraft Gasifier *Journal of Hydrogen Energy*Volume **44** 17389-17396
[9] United States Enveronmental Protection Agency (EPA) 2010 Available and Emerging Technologies for Reducing Greenhouse Gas Emissions From Coal-Fired Electric Generating unit
[10] Feng K, Hubacek K and Siu Y L 2014 The Energy and Water Nexus in Chinese Electricity Production, A Hybrid Life Cycle Analysis *Renewable and Sustainable Energy Reviews* **39** 342-355
[11] United States Enveronmental Protection Agency (EPA) 2015 Emission Factors for Greenhouse Gas Inventories
[12] Susanto H, Suria T and Pranolo S H 2018 Economic Analysis of Biomass Gasification for Generating Electricity in Rural Area in Indonesia *IOP Conference Series: Material Science and Engineering* **334**
[13] Mann MK and PL Spath 1997 *Life Cycle Assessment of a Biomass Gasification Combined Cycle System* Midwest Research Institute
[14] Turton Richard and Bailie Richard C 2008 *Analysis, Synthesis, and Design of Chemical Processes* 3rd Ed. Prentice-Hall, Inc

# Food Safety Analysis and Improvement Concept of β – Carotene Extraction from Fungal Fermented Palm Oil Empty Fruit Bunches (EFB); Extraction Method and Solvent Selection

**Syahdan Amir Muhammad[a*], Clara Novia[b], Achmad Qodim Syafaatullah[c]**

[a] Department of Chemical Engineering
Institut Teknologi Bandung, Bandung 40132 Indonesia
[b] Department of Chemical Engineering
Universitas Indonesia, Depok 16424 Indonesia
[c] Department of Chemical Engineering
Institut Teknologi Sepuluh Nopember, Surabaya 60111 Indonesia

*Corresponding Author's E-mail: samchemicalengineer@gmail.com

**Abstract.** Palm oil is one of the main agricultural commodities in Indonesia. Beside CPO and PKO as the main downstream product, the palm oil industry also produces solid wastes such as shells, fibers, and palm oil empty fruit bunches (EFB). Palm oil EFBs are often dumped nearby the palm oil plantations and being left to be decomposed, but several literature studies indicated that palm oil EFB still contained oil residue rich in β – carotene as well as other lipids components dissolved on it. β – carotene is among the major products of the food industry which have been widely employed as nutrients, food colorants, and additives. It serves as antioxidants and so-called pro-vitamin A. The previous research used n-Hexane as a solvent in β – carotene extraction from fungal – fermented palm oil EFB, whereas n-Hexane has a moderate toxicity level and low solubility of β – carotene which is only 600 ppm. Quality and food safety aspects of β – carotene extract haven't been reviewed or analyzed yet. This paper will be focused on the food safety analysis and improvement concept that can be applied in the extraction method and appropriate solvent selection to obtain high-quality extracts of food-grade β – carotene from fungal – fermented palm oil EFB. Toxicity level and the Hansen Solubility Parameter (HSP) simulation results are the main criteria for solvent selection, while thermal stability, operational cost, and practical aspects are being considered to choose a better extraction method. Based on the HSP simulation results and all criteria mentioned before, n-Hexane can be substituted with 3 recommended solvents; D-Limonene, Tetrahydrofuran (THF), or Tetrahydrofurfuryl Alcohol. According to the economic analysis, the maceration method using THF at room temperature is being preferred to the soxhletation method. Moreover, food safety analysis is being reviewed based on HACCP principles.

## 1. Introduction

Indonesia is one of the top palm oil producers in the world with an average production of 45.861.121 tonnes in 2019 and will be predicted to 49.117.260 tonnes in 2020. A ton of palm fruit will produce about 230 kilograms of palm oil EFB (Empty Fruit Bunches) solid waste [1]. Commonly, a large portion of palm oil EFB has dumped nearby the palm oil plantations and left to be decomposed. Several palm oil industry uses them as mulch, but the abundance of palm oil EFB offers the potential for being a raw material of fine chemicals production. One of them is the carotenoid compound.

Carotenoids contained in palm oil residue are dominated by β – carotene (60%) and α – carotene (34%). β – carotene is among the major products of the food industry which have been widely employed as nutrients, food colorants, and additives [2]. It serves as antioxidants and so-called pro-vitamin A. Several kinds of research have been conducted for carotenoids extraction from palm oil biomass. In his paper, reviewed carotenoids content in several kinds of palm oil biomass represented in Table. 1 [3]

**Table 1.** Carotenoid Composition in Several Kinds of Palm Oil Biomass.

| Oil Palm Biomass | Carotenoid Content (ppm) | References |
|---|---|---|
| Palm Oil | 500 – 1600 | Choo et al. (1992) |
| | 800 – 1000 | Lau et al. (2006) |
| | 800 – 1000 | Chiu et al. (2009) |
| | 330 – 780 | Mustapa et al. (2011) |
| Palm Pressed Fiber (PPF) | 6000 – 7000 | Choo et al. (1996) |
| | 2992 | Lau et al. (2008) |
| Palm Oil Mill Effluent (POME) | 287 – 1665 | Ahmad et al. (2009) |
| | > 1160 | Hudiyono and Septian. (2012) |
| Palm Kernel Cake (PKC) | 1.6 | Ogbuagu. (2008) |
| Oil Palm Leaves (OPL) | 1900 | Ng and Choo. (2010) |

Several literature studies indicated that palm oil EFB still contained oil residue rich in carotenoids as well as other lipid components dissolved on it. Carotenoids can be synthesized biochemically by plants, bacteria, fungi, and algae [4]. They have conducted re-extraction of oil residue from palm oil EFB by maceration using n-hexane at room temperature, which yields carotenoids content of 915,25 ± 300,28 ppm [5]. Improving the previous work studied the effect of natural fungal growth on the carotenoid content of palm oil EFB. They performed carotenoid extraction by using the soxhlet apparatus using n-hexane at its boiling point [6]. According to the obtained results, the extraction process yields 1,87 – 3,08 ppm of β – carotene, and the presence of fungi can improve β – carotene content up to 36% after 3 – 8 days of growth. However, quality and food safety aspects of obtained β – carotene extract haven't been reviewed or analyzed yet. In this work, the explanation will be focused on the food safety analysis and improvement concept that can be applied in the extraction method and appropriate solvent selection to obtain high-quality extracts of food-grade β – carotene from fungal – fermented palm oil EFB.

## 2. Food Safety Analysis

Food safety refers to all hazards that can cause several drawbacks to consumer health. A lack of safety can result in serious injury or even death for the consumer. Safety differs from many other quality attributes since it is almost difficult to be observed. A product can appear to be of high quality (i.e. well colored, flavourful, appetizing, etc). However, unsafe food consumption causes many acute and life-long diseases, ranging from diarrhoeal diseases to various forms of cancer [11]. Moreover, complete guidelines for the Food Safety Management System (FSMS) have been listed in ISO 22000:2018 Standard.

Technical process control of all possible risks and the critical points monitoring provide a solution to prevent the quality or safety defects. Such a procedure for hazard determination and avoidance is presented by the Hazard Analysis and Critical Control Point (HACCP) method [12]. There are seven main principles of HACCP will be explained in this following table:

**Table 2.** Seven Principles of HACCP [9], [12].

| No. | Principle |
|---|---|
| 1 | Perform a hazard analysis / Risk Assessment |
| 2 | Determine the Critical Control Point (CCP) |
| 3 | Determine One or Several Critical Limits |
| 4 | Establish a CCP Monitoring System |
| 5 | Establish Corrective Action |
| 6 | Establish Verification Procedure |
| 7 | Establish Documentation |

*2.1 Risk Assessment*

In hazard analysis, the main point was being objected to incidents, events, or situations that could lead to hazards being introduced to the whole process. Risk assessment is the key to identify entire process risks, assessing their significance, and the controls in the system which systematically manage those risks starting from the whole production process [12].

Almost all of the previous work represented in section 1 uses n-hexane for the extraction solvent. Five potential hazards have been identified in the whole β-Carotene extraction and distillation process using n-hexane; Direct exposure of the solvent, bacterial contamination in raw material, fire and explosion hazard, health hazard, and the excess of solvent usage.

Nowadays, petroleum-based solvent as n-hexane is now strictly regulated by European Directives and "Registration, Evaluation, Authorisation, and Restriction of Chemicals / REACH" [15]. This is because n-hexane can cause several health hazards to the human. In 1975, Spencer et al. (1975) showed that 2,5-hexanedione, which was considered to be one of the n-hexane metabolites could cause polyneuropathy. The previous paper showed the biological monitoring of n-hexane exposure on vinyl-sandals factory workers. The result was 2,5-dimethylfuran, 2,5-hexanedione, and γ-valerolactone, which are considered to be n-hexane metabolites, were detected in the urine of workers by gas chromatography [16]. It has estimated the carcinogenic risk of some organic vapors indoors based on the value of carcinogenic potency [17]. According to his work, n-hexane has a carcinogenic potency of $2{,}6 \, x \, 10^{-5}$ $(mg/kg.day)^{-1}$. In the extraction process, health hazards can be controlled by considering solvent IDLH (Immediately Dangerous to Life or Health) and PDE (Permitted Daily Exposure) value before choosing it as the solvent. n-Hexane has the IDLH and PDE value of 5000 ppm and 0,1 mg/day respectively [14], [18].

Excessive solvent usage is a potential hazard that affects the extraction efficiency and operational cost directly. Moreover, it can rises the carcinogenic potency value due to increasing solvent exposure concentration in the air. This hazard can be controlled by choosing the appropriate green solvent. Hence, "green" is used for describing different types of solvents including the ones which are produced from biomass feedstock and eco-friendly petrochemical-based solvents that are non-toxic and/or biodegradable. The potential of alternative green solvents instead of petrochemical-based solvent has been widely studied. In this work, the selection of the most appropriate solvent among several green solvents for β-Carotene extraction is carried out using Hansen Solubility Parameters (HSP) modeling.

HSP modeling provides a convenient and efficient way to characterize solute-solvent interactions according to the classical "like dissolves like" rule. HSP is based on the concept that the total cohesive energy density ($\delta^2_{total}$) is approximated by the sum of the energy densities required to overcome atomic dispersion forces (London Parameter, $\delta^2_D$), molecular polar forces arising from dipole moments (Polar Parameter, $\delta^2_P$) and hydrogen-bonds between molecules (Electron Transfer Parameter, $\delta^2_H$).

$$\delta^2_{total} = \delta^2_D + \delta^2_P + \delta^2_H \quad (1)$$

A simple affinity parameter, the relative energy difference (RED) number, has been calculated using Eq. (2) to determine the solubility between solute and solvent,

$$RED = R_A / R_o \quad (2)$$

where $R_o$ is the radius of a Hansen solubility sphere, and $R_A$ is the distance of a solvent from the center of the Hansen solubility sphere, which can be calculated using Eqn. (3),

$$R_A^2 = 4(\delta_D A - \delta_D B)^2 + (\delta_P A - \delta_P B)^2 + (\delta_H A - \delta_H B)^2 \quad (3)$$

where A refers to the solute and B refers to the solvent. The lower the $R_A$ value, the greater the affinity between solute and solvent because of their similar properties. It means that good alternative solvents have a RED of less than 1, while medium and poor solvents have RED values of from 1 to 3 and more than 3, respectively [15].

*2.2 Critical Control Points (CCPs)*

The CCPs are being determined by the decision tree method. An application of the CCP decision tree on the extraction and distillation process is depicted in Table 3.

**Table 3.** Application of CCP Decision Tree.

| Process Step | Hazard Category | Q1 | Q2 | Q3 | Q4 | CCP |
|---|---|---|---|---|---|---|
| Extraction | Physical | Yes | No | No | - | NOT CCP |
| | Chemical | Yes | Yes | - | - | CCP |
| | Microbiological | Yes | Yes | - | - | CCP |
| Distilation | Physical | Yes | No | Yes | Yes | CCP |
| | Chemical | Yes | No | No | - | NOT CCP |

The HACCP application plan of previous plan, works has been described in Table 2 [5]. It explains the β-carotene extract production process step, identified hazard category, preventive measures, CCP parameter, monitoring procedure, and corrective actions. In the extraction process, it should be designed to eliminate or reduce the chemical, microbiological, and physical hazard, because preventive action probably can't be managed well on the next process. Before the extraction process, wash water quality for washing the raw palm oil EFB must be specified based on WHO guidelines of WFI (Water For Injection) quality standards. It is a critical control point (CCP) because it can affect the microbiological content of β-carotene, so that preventive action can be taken to reduce microbiological contamination of wash water. The CCP can be monitored from microbiological analysis and inspection results. When the monitoring result deviates extremely from the setpoint or CCP, corrective action must be taken by a comprehensive evaluation of the water treatment system for the wash water, an inspection of wash water analysis results, and find another supplier of wash water if necessary. To overcome the chemical hazard, the operator should maintain the solute-solvent ratio at 1:20 (w/v). The proper solute-solvent ratio must be maintained because this affects the residual solvent level at the end of the product. This can be monitored by doing the frequent inspection and chemical analysis of the obtained β-Carotene extract. When the monitoring result deviates extremely from the setpoint or too much solvent being used, corrective action can be taken by choosing 2 methods; changing back the solute-solvent ratio to the setpoint, or replacing the current solvent (n-hexane) with the alternative green solvent. However, considering the restriction of n-hexane usage for food products by European Directives and REACH, low n-hexane PDE value of 0,1 mg/day from FDA standard, and several carcinogenic risks study of n-hexane, replacing n-hexane with the alternative green solvent is more favorable, because according to the HSP simulation results, there are several potential green solvents can be used to replace n-hexane[15]–[17]. Then, the distillation process uses two temperature settings, the first one by normal distillation is 70°C and the second one by vacuum distillation is 50-55°C. In this process, the temperature level is a CCP and must be controlled properly. Otherwise, it may affect the quality of β-carotene which allows degradation if the temperature is too high. This CCP can be monitored by regular inspection of temperature indicators during the distillation process. When the temperature deviates extremely from the setpoint, corrective action can be taken by checking the temperature controller, and then reset the temperature controller if necessary. If a serious error happens to the temperature controller or indicator, a replacement can be done to fix the problem.

**Table 4.** Principle of HACCP Applied to Extraction and Distillation Process of β-carotene Extract.

| Process Step | Hazards Category | Preventive measures | CCP parameter | Monitoring procedure | Corrective actions |
|---|---|---|---|---|---|
| Extraction | Microbiological | Wash water certificate of analysis (CoA) | WHO guidelines of WFI (Water for Injection) quality standards | Microbiological analysis result and inspection | • A comprehensive evaluation of water treatment system for the wash water<br>• Inspection of wash water analysis result<br>• Replace wash water supplier if necessary |
| | Chemical | Maintain the optimum solute-solvent ratio | The solute-solvent ratio of 1:20 (w/v) in kg and lt | Frequent chemical analysis and inspection | • Change the solute-solvent ratio back to the setpoint<br>• Replace the current solvent with the alternative green solvent |
| Distillation | Physical | Controlling the temperature level | The temperature of 70°C for normal distillation and 50-55 °C for vacuum distillation | Regular inspection of temperature indicator | • Check the temperature controller<br>• Reset the temperature controller<br>• Replace the temperature controller/indicator if a serious error happens |

## 3. Improvement Concept

### 3.1. Solvent Selection

Improvement is the key to achieve operational excellence, while it aims to improve product quality and process efficiency, maintain process safety, and minimize waste, hazards, and contamination. As have been mentioned before, n-hexane must be replaced by new alternative solvents. The appropriate solvent must have a better affinity with the solute, which represented by RED value. Using the obtained β-carotene solubility parameter ($\delta_D$, $\delta_P$, $\delta_H$, $R_o$) on section 2.1, 6 solvents are being tested by HSP simulation: Benzene, Methanol, Ethanol, d-Limonene, Tetrahydrofuran, and Tetrahydrofurfuryl alcohol.

**Table 5.** New Solvent Testing by HSP Simulation Method.

| Solvent | $\delta_D$ | $\delta_P$ | $\delta_H$ | RED | Recommended / Not |
|---|---|---|---|---|---|
| Hexane* | 14.90 | 0.00 | 0.00 | 1.126 | Not |
| Benzene | 18.40 | 0.00 | 2.00 | 0.964 | Not |
| Methanol | 14.70 | 12.30 | 22.30 | 1.230 | Not |
| Ethanol | 15.80 | 8.80 | 19.40 | 1.148 | Not |
| d-Limonene | 17.20 | 1.80 | 4.30 | 1.017 | Yes |
| Tetrahydrofuran (THF) | 16.80 | 5.70 | 8.00 | 1.043 | Yes |
| Tetrahydrofurfuryl Alcohol | 17.80 | 8.20 | 12.90 | 1.020 | Yes |

*: *Reference Value*

According to the simulation results, benzene has the greatest affinity with β-carotene, However, benzene is a petroleum-based solvent which is not environmentally friendly and has bigger carcinogenic potential compared to n-hexane. Benzene has the carcinogenic potential value of $1 x 10^{-3}$ $(mg/kg.day)^{-1}$, which is bigger than n-hexane [17]. It can be inferred that benzene is not a green solvent and can't be used to replace n-hexane. Methanol has the largest value of RED, which means it has the lowest affinity with β-carotene and can be eliminated from the choice. Ethanol, which is categorized as the green and food-grade solvent has the RED value not far from n-hexane. However, ethanol RED value is higher than n-hexane and this will lower the extraction efficiency.

D-Limonene, a monocyclic mono-terpene listed as a flavoring agent in the Code of Federal Regulation, is a major constituent of several kinds of citrus peel essential oil, such as lemon, orange, mandarin, lime, and grapefruit [19]. It has been categorized as a green biodegradable and non-toxic solvent instead of hazardous petroleum based-solvents, because it can effectively be used for lipid extraction from microalgae or plant matrices [20], [21]. Moreover, D-Limonene has the antioxidant potential, anti-diabetic effect and other clinical applications, so that it is recommended to be used for n-hexane substitute [22]–[25].

Tetrahydrofuran (THF) is a multi-purpose polar solvent which has been used for decades in a wide range of applications. It can be synthesized from furfural by hydrogenation reaction using a catalyst consisting of 5% palladium (Pd) on microporous carbon and carried out at 100°C and 20 bar (Godawa et al, 1986). Furfural itself synthesized by dehydrating pentose sugars; xylose and arabinose, or by acid hydrolysis of lignocellulosic biomass obtained from cornstalks, corncobs, wood chips, and other biomass waste products [26]. Toxicological and environmental hazards of THF have been studied comprehensively. Fowles, in his review paper, explained that the acute toxicity of THF in the human body is low to moderate by all routes. The *in vitro* and *in vivo* studies have shown that THF is non-mutagenic [27]. THF has been categorized as "Class-3 Solvent" with the PDE value of $\leq$ 50 mg/day, which has low toxicity and safe to be employed in the manufacturing of drug or food products compared to n-hexane. Nevertheless, it should be limited by GMP or other quality-based requirements [18]. Because of this advantage, THF can be chosen as the alternative solvent replacing n-hexane.

Tetrahydrofurfuryl alcohol (THFAL) is a green industrial solvent because it is less toxic, degradable, and more stable compared to unsaturated furan compounds [29]–[31]. It can be derived from furfural which produced from renewable biomass resources by acid-catalyzed pentose dehydration. Similar to THF, THFAL can be synthesized by furfural catalytic hydrogenation. Nakagawa et al. (2014) has synthesized THFAL from furfural using Ni-Pd/$SiO_2$ alloy catalyst and yielded 94% of THFAL under 8 MPa hydrogen pressure and 40°C for 2 hours with acetic acid solvent. Improving the previous work, have successfully synthesized THFAL from biomass-derived furfural with 100% conversion and yield over a hydroxyapatite-supported Pd catalyst (Pd-HAP) under relatively mild condition (40°C, 1 MPa $H_2$, and 3 hours) in 2-propanol [32]. However, the research of THFAL utilization as the extraction solvent for food and pharmaceutical product processing hasn't been explored yet. The toxicological and carcinogenic effects of THFAL in human still need to be studied comprehensively, for making sure that it will not be harmful to the human body.

### *3.2. Extraction Method*

Several common extraction methods are being used for phytochemical and bioactive compound extraction from plants; Maceration, Soxhletation, Supercritical Fluid Extraction (SFE), and Subcritical Water Extraction (SWE). Thermal stability of the extract, operational cost, and practical aspects are the main considerations for choosing the best method of extraction. Thermal stability can be defined as how difficult the chemical compound changing from its initial molecular form due to the stimulus of heat. The insight about β-carotene thermal stability has been created and applied using multiresponse modeling because its unsaturated structure is very fragile and can degrade via two reaction pathways: isomerization and oxidation [38], [39]. Isomerization of β-carotene involves a configuration change from its natural *trans*-form to various *cis*-isomer species. Although the resulting isomers still able to be converted into vitamin A, the yield will be lower than before isomerization [40].

*3.2.1 Maceration.* Maceration is the method while the coarsely powdered raw material is kept in contact with the solvent in a stoppered container for a defined period with frequent stirring at a certain speed and temperature until the soluble matter is dissolved. This method is best suitable for the extraction of several compounds that have low thermal stability. It is the best method to be used in β-carotene extraction.

**Table 6.** Economic Analysis of Maceration Method using Several Extraction Solvent.

| Extraction Solvent | Fixed Cost | | | | Variable Cost | Total Cost | |
|---|---|---|---|---|---|---|---|
| | Electrical-Stirring (0.408 kWh) | Electrical-Heating | Electrical-Pumping | Cooling Water | Solvent (1 L) | Cost per 1 L of Crude Extract | Cost per mg of β-Carotene |
| n-Hexane* | Rp 423 | - | - | - | Rp 23,000 | Rp 23,423 | **Rp 925** |
| D-Limonene | Rp 423 | - | - | - | Rp 150,000 | Rp 150,423 | **Rp 3,701** |
| THF | Rp 423 | - | - | - | Rp 66,675 | Rp 67,098 | **Rp 1,880** |
| THFAL | Rp 423 | - | - | - | Rp 1,183,000 | Rp 1,183,423 | **Rp 26,275** |

**: reference*

The maceration process only needs electrical power about 8,5 W for stirring purposes, because this process is done in ambient temperature without solvent recirculation (no heating, cooling, and pumping needed). Because the extraction time takes 48 hours, electrical power requirements for stirring purposes will be 0,408 kWh. Variable cost depends directly on the extraction yield. According to the economic analysis presented in Table 6, the total cost of the maceration method is ranging from Rp 1.880 to Rp 26.275 per mg of β-carotene. The maceration using tetrahydrofuran (THF) has the lowest total operational cost, i.e. Rp 1.880 / mg of β-carotene, and considered to be the best extraction method.

*3.2.2 Soxhletation.* Soxhlet extraction method can be chosen while the desired compound has limited solubility in the used solvent and the impurities are insoluble in that solvent. The advantage of this system is that instead of many portions of the warm solvent being passed through the sample, only one batch of solvent is recycled. Nevertheless, this method cannot be used for the compounds having low thermal stability, which degradation or isomerization will easily happen at the high processing temperature [42]. Because of the low β-carotene thermal stability, this extraction method is not suitable to be used.

**Table 7.** Economic Analysis of Maceration Method using Several Extraction Solvent.

| Extraction Solvent | Fixed Cost | | | | Variable Cost | Total Cost | |
|---|---|---|---|---|---|---|---|
| | Electrical-Stirring | Electrical-Heating (1.65 kWh) | Electrical-Pumping | Cooling Water (360 L) | Solvent (0.25 L) | Cost per 0.25 L of Crude Extract | Cost per mg of β-Carotene |
| n-Hexane* | Rp - | Rp 1,709 | - | Rp 1,764 | Rp 5,750 | Rp 9,223 | **Rp 3,976** |
| D-Limonene | Rp - | Rp 1,709 | - | Rp 1,764 | Rp 37,500 | Rp 40,973 | **Rp 6,751** |
| THF | Rp - | Rp 1,709 | - | Rp 1,764 | Rp 16,669 | Rp 20,142 | **Rp 4,930** |
| THFAL | Rp - | Rp 1,709 | - | Rp 1,764 | Rp 295,750 | Rp 299,223 | **Rp 29,325** |

**: reference*

Different from the maceration method, the soxhletation method needs electrical power about 550 W for heating purposes, because this process is done in boiling point with solvent recirculation. Because the extraction time takes 3 hours, electrical power requirements for heating purposes will be 1,65 kWh. Cooling water supply from tap water is needed to condense the solvent which partially vaporizes during extraction. Assuming tap water flow rate is about 2 L/min, the total cooling water being used during extraction is 360 L. The cooling water cost calculation refers to the PDAM (Perusahaan Daerah Air Minum) tariff for 4A Industrial group, i.e. Rp 4.900 / $m^3$. According to the economic analysis presented in Table 7, the total cost of the soxhletation method is more expensive than maceration and ranging from Rp 3.976 to Rp 29.325 per mg of β-carotene.

## 4. Conclusion

The food safety analysis of β-carotene extraction has been reviewed using HACCP principles. Food safety of β-carotene extract can be improved by replacing n-Hexane with green solvents and using the low-cost extraction method. Based on the HSP simulation results, n-Hexane can be substituted with 3 recommended solvents: D-Limonene, Tetrahydrofuran (THF), or Tetrahydrofurfuryl Alcohol. According to the economic analysis, maceration using THF at room temperature is being preferred to the soxhletation method.

## 5. Acknowledgment

Further experimental research of β-carotene extraction using proposed green solvents (D-Limonene, THF, THFAL), solubility measurement, and extract purification method is needed for testing the validity of HSP simulation results and the effectiveness of the proposed improvement concept.

## References

[1] Mandiri 2012 *Manual Pelatihan Teknologi Energi Terbarukan* DANIDA 61
[2] Schroeder M T, Becker E M, and Skibsted L H 2006 Molecular mechanism of antioxidant synergism of tocotrienols and carotenoids in palm oil *Journal of agricultural and food chemistry* **54** (9) 3445-3453
[3] Ofori-Boateng C, and Lee K T 2013 Sustainable utilization of oil palm wastes for bioactive phytochemicals for the benefit of the oil palm and nutraceutical industries *Phytochemistry reviews* **12** (1) 173-190
[4] Stahl W and Sies H 2004 *Antioxidant activity of carotenoids* **2997** May 2018 (doi: 10.1016/S0098-2997(03)00030-X)
[5] Manurung H, Silalahi J, Siahaan D, and Juliani E 2017 The re-extraction of oil from oil palm empty fruit brunch residues and oil palm mesocarp fibers and measures in reducing greenhouse gas emission *Asian J. Agric. Biol.* **5** (4) 346–355
[6] Kresnowati M T A P, Lestari D, Anshori M, Jafar R M 2020 Production of carotenoids from oil palm empty fruit bunches *In IOP Conference Series: Earth and Environmental Science* **460** (1) 012025 IOP Publishing
[7] ISO 2018 *Food safety management systems-Requirements for any organization in the food chain*

En Iso 220002018 **3** 32
[8] Peri C 2006 The universe of food quality *Food quality and preference* **17** (1-2) 3-8.
[9] Training A 2020 Good Manufacturing Practice Hazard Analysis Critical Control Point, no. June
[10] Nations U 2007 Safety And Quality Of Fresh Fruit And Vegetables : A Training Manual For Trainers
[11] Aung M M, and Chang Y S 2014 Traceability in a food supply chain: Safety and quality perspectives *Food control* **39** 172-184
[12] Damikouka I, Katsiri A, and Tzia C 2007 Application of HACCP principles in drinking water treatment *Desalination* **210** (1-3) 138-145
[13] Se U 2003 *General Principles Of Food Hygiene* 1–31
[14] Smallwood I M 2012 *Handbook of Organic Solvent Properties* Handb. Org. Solvent Prop 1–306
[15] Yara-Varon E, Fabiano-Tixier A S, Balcells M, Canela-Garayoa R, Bily A, and Chemat F 2016 Is it possible to substitute hexane with green solvents for extraction of carotenoids? A theoretical versus experimental solubility study *RSC Advances* **6** (33) 27750-27759
[16] Iwata M, Takeuchi Y, Hisanaga N, and Ono Y 1983 A study on biological monitoring of n-hexane exposure *International archives of occupational and environmental health* **51** (3) 253-260
[17] Tancrede M, Wilson R, Zeise L, and Crouch E A C 1987 The carcinogenic risk of some organic vapors indoors: a theoretical survey *Atmospheric Environment (1967)* **21** (10) 2187-2205
[18] 16. FDA Docket No. 97D-0148, "International Conference on Harmonisation, Draft Guideline on Impurities, Residual Solvents, Availability," USA Food Drug Adm. HHS, vol. 38, p. 24303, 1975.
[19] Lopresto C G, Petrillo F, Casazza A A, Aliakbarian B, Perego P, and Calabrò V 2014 A non-conventional method to extract D-limonene from waste lemon peels and comparison with traditional Soxhlet extraction *Separation and Purification Technology* **137** 13-20
[20] Virot M, Tomao V, Ginies C, and Chemat F 2008 Total lipid extraction of food using d-limonene as an alternative to n-hexane *Chromatographia* **68**(3) 311-313
[21] Tanzi C D, Vian M A, and Chemat F 2013 Bioresource Technology New procedure for extraction of algal lipids from wet biomass : A green clean and scalable process *Bioresour. Technol* **134** 271–275
[22] Edris A E 2007 Pharmaceutical and Therapeutic Potentials of Essential Oils and Their Individual Volatile Constituents : A Review **April 2007**
[23] Murali R, Karthikeyan A, and Saravanan R 2013 Protective effects of D-limonene on lipid peroxidation and antioxidant enzymes in streptozotocin-induced diabetic rats *Basic & clinical pharmacology & toxicology* **112** (3) 175-181
[24] Sun J 2007 D-limonene: Safety and clinical applications *Altern. Med. Rev.* **12** (3) 259–264
[25] Murali R and Saravanan R 2012 Antidiabetic effect of d -limonene , a monoterpene in streptozotocin-induced diabetic rats *Biomed. Prev. Nutr.* **2** (4) 269–275
[26] Anthonia E E, and Philip H S 2015 An overview of the applications of furfural and its derivatives *International Journal of Advanced Chemistry* **3** (2) 42-47
[27] Fowles J et al 2013 A review of the toxicological and environmental hazards and risks of tetrahydrofuran *Crit. Rev. Toxicol.* **43** (10) 811–828
[28] Khachik F et al. 2001 (12) United States Patent
[29] Chheda J N, Huber G W, and Dumesic J A 2007 Liquid-phase catalytic processing of biomass-derived oxygenated hydrocarbons to fuels and chemicals. *Angew. Chem. Int. Ed. Engl.* **46** (38) 7164–7183
[30] Song Y, Li W, Zhang M, and Tao K 2007 Hydrogenation of furfuryl alcohol to tetrahydrofurfuryl alcohol on NiB/SiO2 amorphous alloy catalyst *Front. Chem. Eng. China* **1** (2) 151–154
[31] Chen X, Sun W, Xiao N, Yan Y, and Liu S 2007 Experimental study for liquid phase selective hydrogenation of furfuryl alcohol to tetrahydrofurfuryl alcohol on supported Ni catalysts *Chem. Eng. J.* **126** 5–11
[32] Li C, Xu G, Liu X, Zhang Y, and Fu Y 2017 Hydrogenation of Biomass-Derived Furfural to

Tetrahydrofurfuryl Alcohol over Hydroxyapatite-Supported Pd Catalyst under Mild Conditions *Ind. Eng. Chem. Res.* **56** (31) 8843–8849

[33] Chia M et al. 2011 Selective Hydrogenolysis of Polyols and Cyclic Ethers over Bifunctional Surface Sites on Rhodium À Rhenium Catalysts *J. Am. Chem. Soc.* **133** 2675–12689

[34] Liu S, Tamura M, and Nakagawa Y 2014 One-pot selective conversion of furfural into 1,5-pentanediol over a Pd-added Ir–ReOx/SiO2 bifunctional catalyst *Green Chem.* **16** January 617–626

[35] Koso S, Ueda N, Shinmi Y, Okumura K, Kizuka T, and Tomishige K 2009 Promoting effect of Mo on the hydrogenolysis of tetrahydrofurfuryl alcohol *J. Catal.* **267** (1) 89–92

[36] Nakagawa Y, Tamura M, and Tomishige K 2013 Catalytic Reduction of Biomass-Derived Furanic Compounds with Hydrogen *ACS Catal.* **3** (12) 2655–2668

[37] Hirata-Koizumi M, Noda A, Hirose A, Kamata E, and Ema M 2008 Reproductive and developmental toxicity screening test of tetrahydrofurfuryl alcohol in rats *Reprod. Toxicol.* **25** (2) 231–238

[38] Achir N, Pénicaud C, Avallone S, and Bohuon P 2011 Insight into β-Carotene Thermal Degradation in Oils with Multiresponse Modeling *Journal of the American Oil Chemists' Society* **88** (12) 2035-2045

[39] Rodriguez-Amaya D B 2010 Quantitative analysis, in vitro assessment of bioavailability and antioxidant activity of food carotenoids—A review *Journal of Food Composition and Analysis* **23**(7) 726-740

[40] Thurnham D I 2007 Bioequivalence of β-carotene and retinol *Journal of the Science of Food and Agriculture* **87** (1) 13-39

[41] Cheok C Y, Mohd Adzahan N, Abdul Rahman R, Zainal Abedin N H, Hussain N, Sulaiman R, and Chong G H 2017 Current trends of tropical fruit waste utilization *Critical reviews in food science and nutrition* **0** 1-27

[42] Sutar N, Garai R, Sharma U S, Sharma U K, and Jaiswal A 2010 Anthelmintic activity of Platycladus orientalisleaves extract *International Journal of Parasitology Research* **2** (2) 1

[43] Zizovic A I, Senerovic L, Moric I, Jovanovic M, and Krusic M K 2018 SC *J. Supercrit. Fluids* (doi: 10.1016/j.supflu.2018.05.025)

[44] Shi J, Yi C, Ye X, Xue S, Jiang Y, Ma Y, and Liu D 2010 Effects of supercritical CO2 fluid parameters on chemical composition and yield of carotenoids extracted from pumpkin *LWT-Food Science and Technology* **43** (1) 39-44

[45] Essien S O, Young B, and Baroutian S 2020 Recent advances in subcritical water and supercritical carbon dioxide extraction of bioactive compounds from plant materials *Trends in Food Science & Technology* **97** 156-169

# Interesterification of Indonesian Vegetables Oil for Cocoa Butter Alternatives: Its Effect on Slip Melting Point Changes

**Dianika Lestari[1,2], Nathania[1], Oktalia Putri Pratama[1], and Jenny Rizkiana*[2,3]**

[1] Department of Food Engineering, Institut Teknologi Bandung, Bandung 40132 Indonesia
[2] Department of Bioenergy and Chemurgy Engineering, Institut Teknologi Bandung, Bandung 40132 Indonesia
[3] Department of Chemical Engineering, Institut Teknologi Bandung, Bandung 40132 Indonesia

*Corresponding author: jr@che.itb.

**Abstract.** Palm stearin (PS), palm olein (PO), and coconut oil (CO) were used as feed for chemical interesterification to produce fat product with similar slip melting point (SMP) to cocoa butter alternatives. Chemical interesterification was conducted at 75°C for 8 hours using sodium ethoxide as catalyst. The objective of the research was to determine effect of feed composition on slip melting point profile changes during interesterification reaction. The feed composition was determined using Mixture Design method. The experimental data showed that SMP of interesterification product was changed throughout the reaction time. The SMP of chemical interesterification product was affected by the reactivity of the dominant fatty acid in the feed blends. The reactivity of fatty acid in the feed blend directed fatty acid positioning on the triglyserides at the end of reaction. Feed composition determined predominated portion of unsaturated fatty acids and saturated fatty acids in the mixture, and therefore, affected the gradient changes of SMP. Vegetable oil feed mixture which produced interesterification product with the most similar SMP values to cocoa butter (33-34°C) were pure coconut oil feed and palm stearin - palm olein mixture feed (50:50).

## 1. Introduction

Cocoa butter is vegetable fats with unique physical properties since about 80% of cocoa butter is triacylglycerols (TAGs) dominated by symmetrical TAG. The molecule is described as symmetrical TAG and often refer to as a saturated-unsaturated-saturated triglyceride SOS TAGs. The SOS TAGs are palmitic acid- oleic acid- stearic acid (POS, 36-42%), stearic acid- oleic acid- stearic acid (SOS, 23-29%), and palmitic acid- oleic acid- palmitic acid (POP, 13-19%) (Ministry of Industry, 2007). Therefore, cocoa butter has slip melting point at body temperature, 35°C (Beckett, 2008). Other than that, the cocoa butter's production decreases cause relativity high economic value of cocoa butter. This become cocoa butter use's problem in food industry therefore, cocoa butter alternatives (CBA) is more prefered. Various attempts have been made to find an alternative use of cocoa butter, one of them by modifying specialty fats from the cheaper fats/oil to cocoa butter alternatives. Cocoa butter alternatives can be produced by fats blending, hydrogenation reaction and interesterification reaction. Interesterification is one of the most important processes for modifying the pshysicochemical characteristics of oils and fat.

IOP Conf. Series: Materials Science and Engineering 1143 (2021) 012035 doi:10.1088/1757-899X/1143/1/012035

In Industry, chemical interesterification is the most applied interesterification modification process of oil and fats as it is simpler, cheaper, and easier to carry out compared to enzymatic interesterification (Amir, et.al., 2012). Interesterification of solid fats and vegetables oils can produce a fat blend with optimum characteristics. Interesterification product depends on feed composition and interesterification time each blend. Different blends have different triglycerides composition and slip melting point (SMP). Slip melting point is the temperature at which the column of the solid begins to rise in the tube due to buoyancy and because the outside surface of the solid is molten or solid fat content lower than 5%. Fats and oil properties characterization is determined through melting temperature parameter. The objectives of this study were to determine 1) the effect of interesterification time on slip melting point interesterification product; 2) the effect of feed composition on slip melting point interesterification product.

## 2. Materials and Method

### *2.1. Materials*

Palm stearin, palm olein and coconut oil were the raw materials for chemical interesterification. Palm stearin was obtained from PT. Astra Argo Lestari Tbk. Palm olein and coconut oil were purchased from Lion Super Indo. The raw materials were stored at room temperature. Sodium ethoxide was purchased from Sigma Aldrich.

### *2.2. Blend Preparation*

Fat blends, formulated with palm stearin, palm olein and coconut oil were mixed at different ratios. Based on mixture design- second degree lattice, simplex lattice design, fat blends were prepared as indicated in Table 1.

**Table 1**. compositional design of the blends.

| Blends | Mass Compositions (w/w %) of 100grams feed | | |
|---|---|---|---|
| | Palm Stearin (PS) | Palm Olein (PO) | Coconut Oil (CO) |
| 1 | 0 | 100 | 0 |
| 2 | 50 | 0 | 50 |
| 3 | 16,67 | 66,67 | 16,67 |
| 4 | 33,33 | 33,33 | 33,33 |
| 5 | 0 | 0 | 100 |
| 6 | 66,67 | 16,67 | 16,67 |
| 7 | 100 | 0 | 0 |
| 8 | 50 | 50 | 0 |
| 9 | 0 | 50 | 50 |
| 10 | 16,67 | 16,67 | 66,67 |

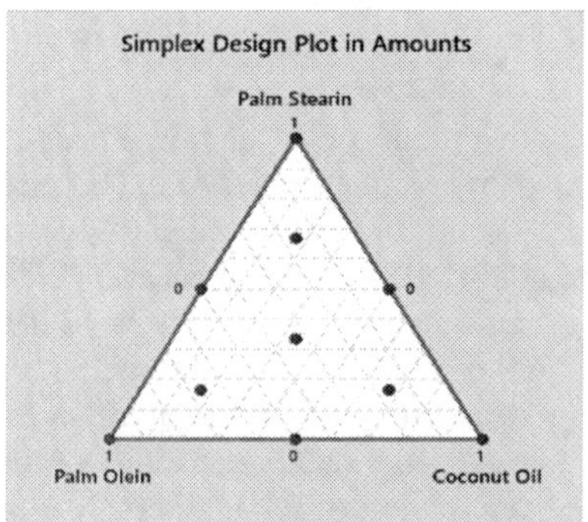


**Figure 1.** simplex design plot.

### 2.3. Water content

The water content of raw material could be determined by heating the sample. Several grams sample was placed in porcelain cup which it was mass known. Then, the sample was dried for 3 hours at 105°C. After that, the sample were cooled in desiccator for 15 minutes and mass was pondered by analytical balance. Water content could be determined using equation below.

$$content = \frac{W2-W1}{W} x\ 100\% \quad (1)$$

where: W2: The wight of porcelain and sample before dried (g), W1: The wight of porcelain and sample after dried (g), W: the sample's weight (g)

### 2.4. Batch Chemical Interesterification

The chemical reaction was started by the addition of 0.4% w/w sodium ethoxide (Sigma Aldrich) as the catalyst into fat blends prepared before according to table 1. Chemical interesterification was conducted at 75°C using sodium ethoxide as catalyst. The chemical interesterification was carried out in laboratory-scale batch reactor with agitator. The sample was agitated with average speed 300-400 rpm for 8 hours. To terminate the reaction, 20 mL 20% (w/w) citric acid solution was added. By water addition, it would non-activate catalyst by changing it into ethanol (Sreenivasan, 1978). Citric acid would react with sodium into sodium salt. Interesterification fat blends was dried again to reduce water content of sample.

### 2.5. Slip Melting Point

The slip melting point of the interesterified blends was determined according to the AOCS Method No. Cc 3-25 and slip melting point digital. A capillary tube filled with 1 cm high column of sample was chilled at 4 ± 1°C for 16 hours before being immersed in a beaker with cold water. The water was stirred and heated slowly. The temperature was recorded when the blend in the tube started to rise due to hydrostatic pressure. The temperature at the point was taken as the SMP.

### 2.6. Iodine value

The iodine number was determined by the volumetric titration method. This method referred to AOCS (1998). Iodine numbers are expressed as grams of iodine absorbed per 100gram sample. The samples were melted at 60-70°C and stirred until smooth. A sample of 0,13-0,15gram was put into the erlenmeyer and then added with chloroform p.a. 25 ml and 25 ml wijs solution, then covered and stirred until homogeneous. The mixture was putted down in a dark room for an hour. After that, the mixture was added with 20ml KI 10% and 150mL aqua dm. The solution was mixed and titrated with 0.1N thiosulfate solution until a purple- purplish colour changes exactly disappear. Next, 10% 2 mL starch indicator is added. The iodine number can be calculated as follows.

$$Iodine\ Value\ (g\ iod\ /100\ g\ sample) = \frac{12.69\ x\ N\ x\ (V2-V1)}{W} \quad (2)$$

where:

N is the normality of thiosulphuric acid, V2 volume of thiosulfate used in blank determination (mL), V1 volume of thiosulfate used in sample determination, W is the test weight (gram). The following is a workflow diagram for analysis of fat iodine numbers.

### 2.7. Experimental Design and Statistical Analysis

Statistical analysis was done using Response Surface Methodology (RSM) and ANOVA. Type of RSM that used was Mixture Design method, simplex lattice design. Experimental results for SMP were applied to obtain the regression models, as a function of the proportions of each ingredient ($x_1$ = palm stearin, $x_2$ = palm olein and $x_3$ =coconut oil) present in blends: SMP (°C) = $\beta_1X_1 + \beta_2X_2 + \beta_3X_3 + \beta_{12}X_1X_2 + \beta_{13}X_1X_3 + \beta_{23}X_2X_3 + \beta_{123}X_1X_2X_3$, where SMP was estimated response, β was coefficient estimated by least-squares method, x was a dependent variable. The quality of models was evaluated by ANOVA and adjusted coefficient of determination ($r^2$). ANOVA that used was Uni-variative Multi Way Analysis of Variance using Microsoft Excel.

IOP Conf. Series: Materials Science and Engineering 1143 (2021) 012035 doi:10.1088/1757-899X/1143/1/012035

## 3. Results and Discussion

### *3.1. Moisture Content*

Moisture content is one of many parameter that indicates raw material quality due to chemical interesterification reaction. Materials that used in this research were palm stearin, palm olein and coconut oil. Moisture content analysis is used to ensure catalyst does not deactivated during reaction, therefore chemical interesterification can occur effectively. Water content describes how much water in the material but does not describe biological activity. The high moisture content in oil can make oil to demage more quickly. That is because water is hydrolysis accelerator that will produce ALB (Raharja & Gunandi, 2000). In general, chemical interesterification requires water content 0,04-11% (w/w). But, water content in chemical interesterification using sodium ethoxide catalyst will be optimum with water content lower than 0,01% (w/w). The low water content is required to support the performance of the catalyst that can be deactivated due to water content (Journal Interesterification of Oil and Fats- A Review, 2012). The water content of PS, PO and CO are shown in Table 2.

**Table 2** Moisture content of Palm Stearin, Palm Olein and Coconut Oil.

| Parameter | Palm Stearin | Palm Olein | Coconut Oil |
|---|---|---|---|
| **Water content** | 0,0143% ± 0,0035 | 0,0179% ± 0,0045 | 0,0006% ± 0,00015 |

According to the Table 2, palm stearin and palm olein have higher moisture content, so they are necessary for drying process as pre-treatment, but coconut oil not.

### *3.2. Effect of Chemical Interesterification Reaction Time on Slip Melting Point (SMP)*

Cocoa butter alternatives can be produced from fat/oil blends, such as palm stearin, palm olein and coconut oil through chemical interesterification. This process was conducted for 8 hours due to overall equilibrium reaction is reached before 8 hours. The sample of interesterification product was taken every hour, then slip melting point was analyzed to evaluate physical properties changes as effect of interesterification reaction. Slip melting point changes indicates triglycerides composition/ configuration changes. To determine effect of reaction time on significance of re-arrangement of fatty acids on TAG, observation of slip melting point changes was done from before reaction to 8 hours reaction. SMP at 8 hours are indicator of fat structure changes massively due to fatty acid re-arrangement to new form of triglycerides.

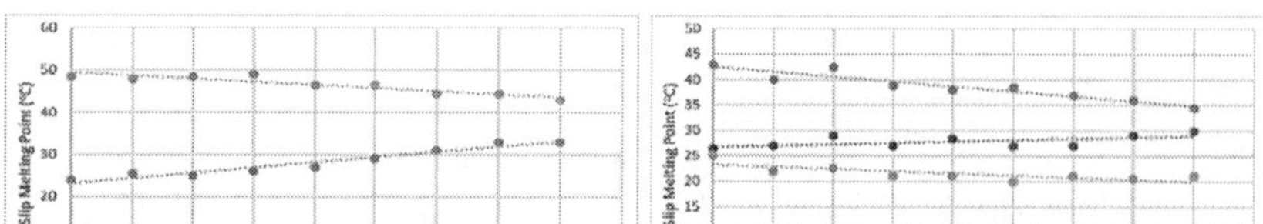


(a) (b)

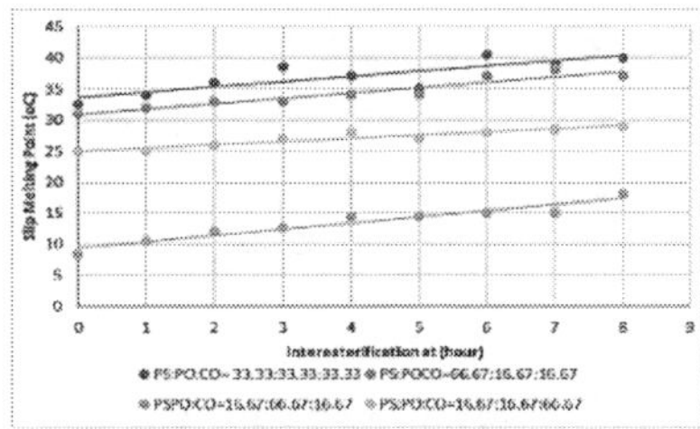


(c)

**Figure 1**. Slip melting point profile changes each product interesterification throughout reaction time on Palm Stearin (PS), Palm Olein (PO), dan Coconut oil (CO) for (a) pure feed, (b) binary blends, and (c) ternary blends.

Figure 2 showed that each variation of feed blends has different characteristic and produce a pattern of slip melting point changes during the reaction. The difference between melting point patterns are caused by fatty acids proportion in feed and also caused by the formation of new TAG with varying slip melting point, such as oleic-oleic-oleic (OOO) with SMP 5.5°C, palmitic-oleic-palmitic (POP) with SMP 37°C, and palmitic-palmitic-palmitic (PPP) with SMP 66°C. Triglycerides configuration changes due to fatty acid rearrangement on any positions of the TAG molecules during reaction, will change SMP of interesterification product. During interesterification, fatty acids (FAs) are exchanged within (interesterification) and among (interesterification triacylglycerols (TAGs)) until a thermodynamic equilibrium is reached. The other hand, interaction between the TAG molecules in mixture and rearrangement TAG for 8 hours will affect fat's polymorphism structure that formed (Hernqvist,1984). Polymorphism changes affects melting properties and fat crystallization behavior. This structure is specific and variative depending on TAG which make interaction. Melting temperature changes of interesterification product through interesterification indicate structure polymorphism changes. So, it can be concluded SMP product's changes throughout reaction time indicate how massive the fatty acids exchange in feed to form new TAG.

### *3.3. Effect of Feed Composition of Chemical Interesterification on Slip Melting Point (SMP)*

Feed composition of chemical interesterification affected slip melting point. As mentioned before, different composition will produce different characteristics. As observed in Figure 2, PS had highest SMP (48°C) fit with Teng Kim-Tiu (2014) (40-46°C), followed by CO (24°C) and PO (8°C) which fit with Siew (2002) (8-25°C). The highest fat's SMP, the longer chain saturated fatty acid it is. According to the figure 2, slip melting point changes in many variations is classified based on slope of SMP changes throughout reaction time into 3 main pattern, scilicet insignificant pattern, pattern increased significantly, pattern decreased significantly. Insignificant pattern means the patterns are insignificant random changing. Insignificant pattern was obtained with slope lower than $\pm$ 0.5, pattern increased significantly was obtained with gradient was obtained with gradient higher than +0.5, and pattern decreased significantly was obtained with gradient higher than -0.5.

**Table 3** Comparison of initial and final SMP products of various composition.

| Feed Composition (%-w) | SMP of Feed (°C) | SMP of Product (°C) | Slope (°C/hour) | Category |
|---|---|---|---|---|
| **Pure Blends** | | | | |
| Pure PO | 8 | 10 | 0.27 | Insignificant ($\leq \pm 0.5$) |
| Pure CO | 24 | 33 | 1.25 | Increased significantly ($> \pm 0.5$) |
| Pure PS | 48 | 43 | -0.72 | Decreased significantly ($> \pm 0.5$) |
| **Binary Blends** | | | | |
| PS:PO = 50:50 | 43 | 34 | -0.98 | Decreased significantly ($> \pm 0.5$) |

| | | | | |
|---|---|---|---|---|
| PO:CO = 50:50 | 22 | 21 | -0.21 | Insignificant (≤ ±0.5) |
| PS:CO = 50:50 | 26 | 30 | 0.3 | Insignificant (≤ ±0.5) |
| **Ternary Blends** | | | | |
| PS:PO:CO = 33.33:33.33:33.33 | 32 | 40 | 0.87 | Increased significantly (> ±0.5) |
| PS:PO:CO = 66.67:16.67:16.67 | 31 | 37 | 1.25 | Increased significantly (> ±0.5) |
| PS:PO:CO = 16.67:66.67:16.67 | 8 | 18 | 1.05 | Increased significantly (> ±0.5) |
| PS:PO:CO = 16.67:16.67:66.67 | 25 | 29 | 0.50 | Insignificant (> ±0.5) |

Insignificant pattern can be observed in variation of pure palm olein (PO 100), PS:CO (50:50), PO:CO (50:50), and PS:PO:CO (16.67:16.67:66.67). Overall, Insignificant pattern on SMP changes can occur due to non-identical fatty acids or the phenomena of the exchange of identical fatty acids (Mat Dian, 2007). These phenomena can occur due to feed contains a high proportion of saturated fatty acids but not balanced with the proportion of unsaturated fatty acids sufficient to trigger a massive exchange reaction. Other conditions, the phenomena are also caused by a high proportion of reactive fatty acids that cause the exchange of identical fatty acids, such as oleic acid which is quite reactive exchanging with each other because when the reaction is triggered, and oleic acid becomes free ions that produce empty spaces in the triglyceride branches so that eventually oleic free fatty acids occupy these empty branches again (Mat Dian, 2007).

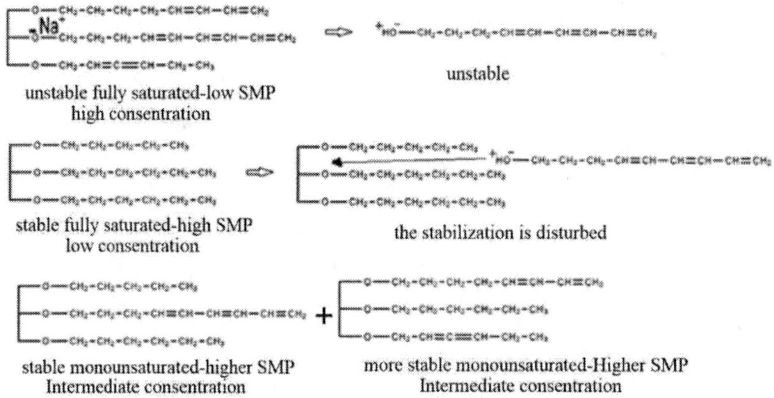


**Figure 2** Illustration of rearrangement TAGs Phenomena.

Pattern increased significantly can be observed in variation of pure coconut oil (CO 100), PS:PO:CO (16.67: 66.67: 16.67), and PS:PO (33.33: 33.33: 33.33). Pattern increased significantly in SMP changes can occur due to fatty acids exchange within (interesterification) and among (interesterification triacylglycerols (TAGs)) into di/tri-saturated triglycerides which have higher SMP values (Fitri.et al, 2009). The saturated TAG configuration is caused by the tendency of reactive unsaturated fatty acids to leave the triglyceride branch of full-unsaturated and attack stable saturated fatty acids (not reactive), so saturated fatty acids which is released will occupy the empty position which is left by reactive fatty acids and form triglycerides monounsaturated or monosaturated, illustration of the phenomena can be observed in Figure 3 (Fitri.et al, 2009). This reactivity phenomena can also occur in short chain fatty acids (reactive fatty acids) which are widely available in coconut oil, that every reactive fatty acid tries to find a stable position therefore make a new form (Soekopitojo,2011 & Ratnayake, 2009). Stable branches are generally composed of long-chain saturated fatty acids.

The third pattern of melting point changes is the pattern decreased significantly which can observed in variations of pure palm stearin (PS 100), PS: PO (50:50), and feed PS: PO: CO (66.67: 16.67: 16.67). These phenomena occurred due to the exchange of fatty acid chains on triglycerides from saturated TAG into mono-saturated TAG (Fitri.et al, 2009). In all variations of blends, the feed is dominated by full-saturated triglycerides (PPP) and saturated TAGs such as POLa / POS / POP which are classified as monounsaturated TAGs as products so that patterns of temperature changes occur. Monounsaturated triglycerides have lower melting temperatures than saturated TAGs. Thus, the mixture has a decreased melting point due to a decrease in the concentration of saturated triglycerides in the mixture which is replaced by an increase in the number of unsaturated TAGs (Ratnayake, 2009).

As mentioned before, different proportion of feed will produce different characteristic, specifically in slip melting point and iodine value due to proportion of fatty acids in feed to form new TAGs. Slip melting point analysis in Figure 2 indicate that SMP changes during reaction depending on compositions and configuration of fatty acids on TAG. The compositions and configurations are produced from three types of fatty acids with different ratio. Therefore, the different proportions of each fatty acids in each feed composition affects SMP value of interesterification product, so comparison of SMP value for each variation needs to be done. The comparison is done on interesterification product for 8 hours. The data was taken at 8 hours due to overall, equilibrium reaction is reached before 8 hours. To determine effect of proportion of fatty acids in SMP, iodine value is observed to identify the proportion of saturated fatty acids in each feed.

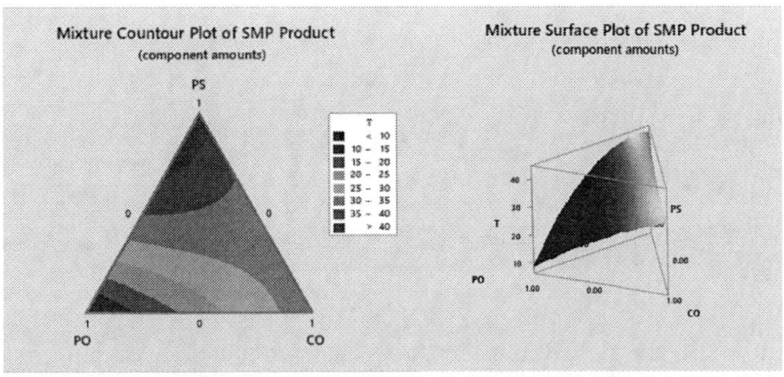


**Figure 3** contour plot and SMP response plot of the 8th hour interesterification product from various fat blends.

Each raw material has a different response, in this case slip melting point as a response. Statistical analysis using Response Surface Method (RSM) show that the addition proportion of palm stearin in the mixture, it will increase SMP, while the reduction proportion of palm stearin in the mixture, it will give an antagonist effect; the addition proportion of palm olein in the mixture, it will decrease SMP exponentially, while the reduction proportion of palm olein in the mixture, it will increased SMP; the addition proportion of coconut oil in the mixture, it will decrease SMP exponentially, while the reduction proportion of palm olein in the mixture, it will increased SMP. Illustration is shown in Figure 4. Then, effect of feed composition on iodine value is determined due to link the correlation with the effect of PS, PO, CO on iodine value. It can be concluded that palm stearin works antagonistic with palm olein and coconut oil.

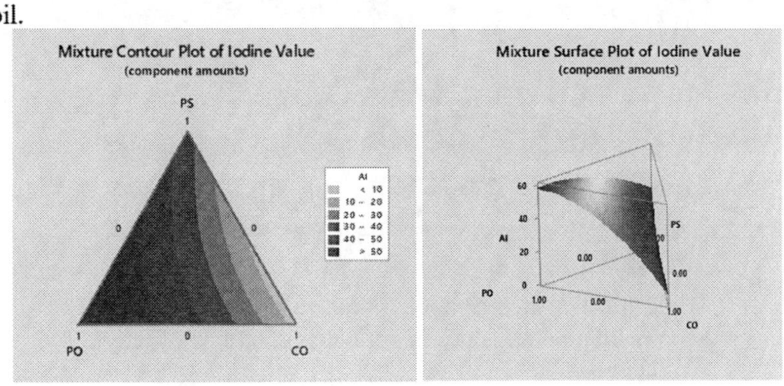


**Figure 4** contour plot and IV response plot of the $8^{th}$ hour

interesterification product from various fat blends

According to Figure 5, the addition proportion of palm stearin and coconut oil in the mixture, it will decrease iodine value, while the addition of palm olein in the mixture, it will increased iodine value. illustration is shown in Figure 5. It can be concluded that palm stearin and coconut oil work antagonistic with palm olein.

This phenomena is clarified by observing nature fat characteristic. Palm stearin is dominated by long chain saturated fatty acids, palm olein is dominated by long unsaturated fatty acids, and coconut oil is dominated by medium chain saturated fatty acids. The addition of fatty acids in TAGs form in the mixture will change the fatty acids proportions and affect on slip melting point, polimorfism and solid fat content. The fatty acids proportions is indicated through iodine value of mixture.

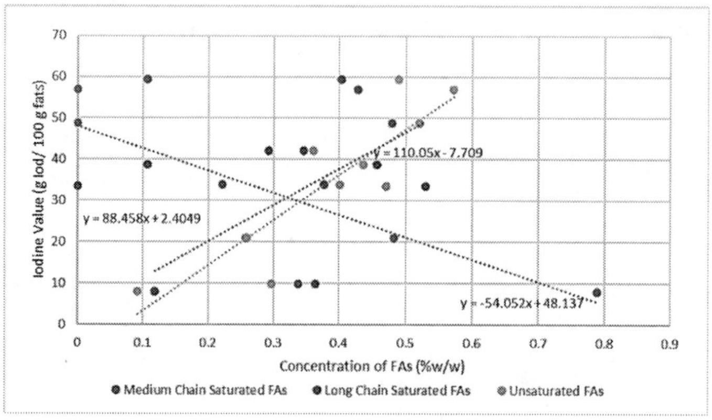


**Figure 5** Effect of medium chain saturated FAs, long chain saturated FAs, and long chain unsaturated FAs composition on iodine value.

Based on Figure 6, the correlation between fatty acids and iodine value match with Silvia Oliver (2019) that unsaturated fatty acids and iodine value have linier correlation. Fatty acids also affect slip melting point blends. Figure 7 illustrates mixture contour plot of temperature.

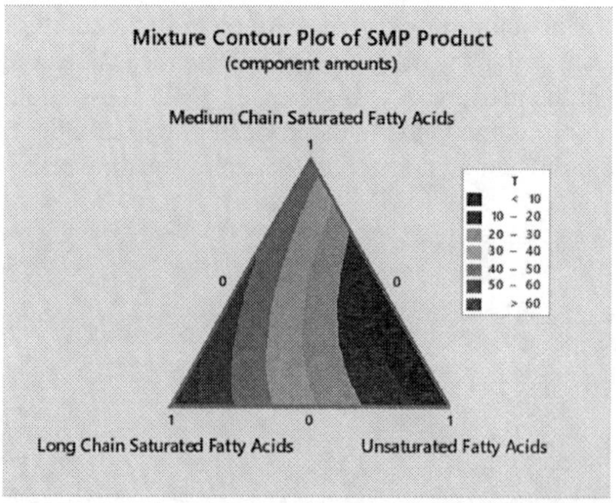


**Figure 6** plot contour of the distribution of slip melting point values in various proportions of fatty acids.

Figure 7 shows that unsaturated fatty acids have the effect of producing fat with a low melting temperature which is below 10°C at high concentrations and below 30°C at low concentrations. While medium chain saturated chain fatty acids can produce fat at medium temperatures in the range 30-40°C at high concentrations. In long-chain saturated fatty acids, the resulting melting temperature can reach more than 60°C at high concentrations. Observation of the value of iodine number and melting temperature (SMP) can form an interconnected pattern where the melting temperature value in general will be inversely proportional to the value of iodine number. High melting temperatures will be obtained from saturated fatty acids which have low IV, while low melting temperatures will be obtained from unsaturated fatty acids which have high IV. However, the values obtained cannot be connected in a linear line or a simple polynomial equation.

$$T\ (^{o}C) = -3{,}1 \times 10^{-5} IV^4 + 3{,}36 \times 10^{-4} IV^3 - 0{,}145\ IV^2 + 3{,}39 \quad (3)$$

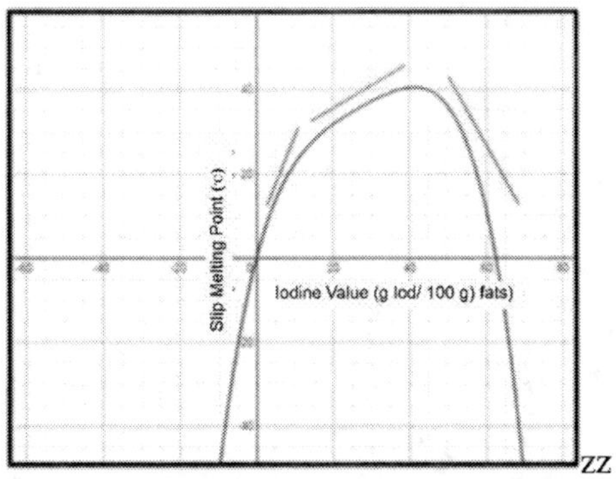


**Figure 7.** The correlation between slip melting point and iodine value.

Based on the equation of the correlation between IV and SMP, it can be indicated that the iodine value varies in melting temperature values. The variation in temperature values is based on the range of iodine values. This range of variation causes the relationship between iodine value and SMP not to cause a constant rise. Figure 8 shows that the range of variations in the iodine value (abscise axis) value with the slip melting point (ordinate axis) is divided into 3 parts, iodine range 0-12 which is the identity of fat with the predominance of saturated fatty acids, range 12-40 which is an identity for fat with a predominance of saturated fatty acids, and a range of 40-63 which is an identity for fats with a predominance of unsaturated fatty acids.

## 4. Conclusions

The interesterification reaction successfully changes the slip melting point value throughout the reaction time with insignificant pattern, the increased significantly pattern and the decreased significantly pattern. The insignificant pattern occurred on variations of pure palm olein, PS:CO (50:50), PO:CO (50:50), and PS: PO: CO (16.67: 16.67: 66.67), the increased significantly pattern occurred in variations of pure coconut oil, PS: PO: CO (16.67: 66.67: 16.67), PS: PO: CO (33.33: 33.33: 33.33 ), the decreased significantly pattern occurred in pure palm stearin, and PS: PO (50:50), PS: PO: CO (66.67: 16.67: 16.67).

Variation of composition of fat mixture determines the portion of unsaturated fatty acids and saturated fatty acids. The portion of fatty acids in the mixture caused variations in the feed's and product's slip melting point and the iodine value. Feed composition which produced 33-36°C as slip melting point are PS: PO (50:50)-34°C and CO (100)-33°C.

Iodine value changes is directly proportional to the addition of saturated fatty acids and /or unsaturated fatty acids in the mixture. The more percentage of unsaturated fatty acids in the mixture, the higher the value obtained.

**Acknowledgments**

This research was funded by P3MI Research Grant provided by Institut Teknologi Bandung.

## References

[1] Berry S E E 2014 *Rich fats : An overview and implications for cardiovascular disease Triacylglycerol structure and interesterification of palmitic and stearic acid-rich fats : an overview and implications for cardiovascular disease* (London: King's College London)

[2] Chapman D 1962 The Polymorphism of Glycerides *Chemical Reviews* **62** (5) 433-456.

[3] da Silva Oliveira et all 2019 Mature chemical analysis methods for food chemical properties evaluation. In *Evaluation Technologies for Food Quality* (pp. 63-90). Woodhead Publishing.

[4] Dian N L H M et all 2007 Effect of chemical interesterification on triacylglycerol and solid fat contents of palm stearin, sunflower oil and palm kernel olein blends *European Journal of Lipid Science and Technology* **109** (2) 147-156

[5] Hernqvist L and Herslöf M 1984 Polymorphism of interesterified triglycerides and triglyceride mixtures *Fette, Seifen, Anstrichmittel* **86** (10) 393-397

[6] Hazirah S 2012 *Effects of Chemical Interesterification On The Physicochemical Properties Of Palm Stearin, Palm Kernel Oil And Soybean Oil Blends*. (Malaysia: Universiti Teknologi MARA.*16*(3), 297–308.

[7] Karupaiah T & Sundram K 2007 Effects of stereospecific positioning of fatty acids in triacylglycerol structures in native and randomized fats: a review of their nutritional implications *Nutrition & metabolism* **4** (1) 16

[8] Kumar P K P & Krishna A G G 2015 Physicochemical characteristics of commercial coconut oils produced in India *Grasas y Aceites* **66** (1)

[9] Neff et all 2001 Triacylglycerol structures of food fats high in saturated acids by HPLC and mass spectrometry (*Journal of liquid chromatography & related technologies* (USA: Food Quality and Safety Research, NCAUR, ARS, USDA) **24** (6) 837-854

[10] Omar K A et all 2017 Triacylglycerol composition, melting and crystallization profiles of lipase catalyzed anhydrous milk fats hydrolyzed *International journal of food properties* (Jiangsu: Jiangnan University) *20*(sup2) 1230-1245

[11] Palm I E (n.d.) *Oil Palm: Fractions & Derivatives Palm Oil process Oil Palm : Fractions & Derivatives Palm Kernel Oil process*

[12] Karabulut I 2003 Effects of chemical interesterification on solid fat content and slip melting point of fat / oil blends Turkey *Hacettepe University* 224–229

[13] Soares D M et al. 2012 *Chemical Interesterification of Blends of Palm Stearin, Coconut Oil, and Canola Oil: Physicochemical Properties*. (Brazil: Sao Paulo)

[14] Soekopitojo S 2012 Asidolisis enzimatik fraksi tengah minyak sawit dengan asam stearat untuk sintesis cocoa butter equivalents (Bogor: Institut Teknologi Bogor) **34** (2)

# Statistical Mixture Design for Modelling and Optimization of Feed Mixture in the Chemical Interesterification to produce Cocoa Butter Alternatives

**Jenny Rizikiana [2,3], Nathania[1], Oktalia Putri Pratama[1], and Dianika Lestari[3]**

[1] Department of Food Engineering, Institut Teknologi Bandung, Bandung 40132 Indonesia

[2] Department of Bioenergy Engineering and Chemurgy, Institut Teknologi Bandung, Bandung 40132 Indonesia

[3] Department of Chemical Engineering, Institut Teknologi Bandung,Bandung 40132 Indonesia

*Corresponding author: dianika@che.itb.ac.id

**Abstract.** This study aimed to find the optimum feed mixture consist of palm stearin (PS), palm olein (PO), and coconut oil (CO), for chemical interesterification to produce cocoa butteralternatives (CBA) with the most similar melting profile with cocoa butter. The feed mixture was chemically interesterified using sodium ethylate as the catalyst at a temperature of 75°C for 8 hours.Based on the melting profile analysis of the products, the pure coconut oil (CO-100) and pure palm stearin (PS-100) gave the most similar melting profile of cocoa butter alternatives. The CO-100 melting profile gave the least value of Sum of Square Errorcompared to Cocoa Butter Substitute while the PS-100 melting profile gave a similar result compared to Cocoa Butter Equivalent (CBE) and Cocoa Butter Replacer (CBR). However, PS-100 product does not meet the other specifications, such as Iodine value and slip melting point. Therefore, it can be concluded that pure coconut oil was the best feed for chemical interesterification to produce cocoa butteralternatives (CBA).

**Keywords**: Chemical interesterification, Melting profile, Mixture Design, Optimization

## 1. Introduction

Cocoa butter is a unique fat as it is the only type of fat consumed directly and is widely producedin the market. Cocoa butter has a melting temperature of around 35 °C or close to the human body temperature [1]. Chocolate fat (cocoa butter) is fat with unique physicochemical characteristics because its 80% triacylglycerol composition is dominated by three symmetric, saturated- monounsaturated-saturated triglycerides (TAGs), namely palmitate-oleate-stearate (POS, 36-42%), stearate- oleate-stearate (SOS, 23-29%) and palmitate-oleate-palmitate (POP, 13-19%) [1]. Aside from the declining production, the high price of cocoa butter, which reachedRp 45,000 per 100 grams, constrains chocolate fat for the food industry. Various attempts have been made to find alternative cocoa butter use, one of them by developing specialty fats fromother oils with lower prices to produce alternative cocoa butter. Cocoa butter alternative can be made using various vegetable oil sources through methods such as mixing fat, hydrogenation reaction, and interesterification reaction. In this research, coconut and palm oil were used due to the abundant availability in Indonesia. In industry, chemical interesterification is widely used because it requires a relatively faster time than enzymatic interesterification [2]. The results of interesterification were dependent on the composition of the feed mixture and each fat mixture's interesterification time with different characteristics of triglyceride content andmelting characteristics of the fats. Slip melting point (SMP) is the temperature when fat is observed tomelt entirely or visually change its shape and color due to solid fat content has reached below 10% [3]. Solid fat content control is the key to fat character modification. Solid fat content responsible for controlling the hardness of the fatin various temperatures. Therefore, the effect of the feed mixture's composition on the content of solid fat produced on the interesterification reaction product needs to be investigated.

## 2. Materials and Method

*2.1. Materials*

Palm stearin, palm olein, and coconut oil were the raw materials for chemical interesterification. Palm stearin was obtained from PT. Astra Argo Lestari Tbk. Palm olein and coconut oil were purchased from Lion Super Indo. The raw materials were stored at room temperature. Sodium ethoxide was purchased from Sigma Aldrich.

*2.2. Blend Preparation*

Fat blends formulated with palm stearin, palm olein, and coconut oil were mixed at different ratios. Based on mixture design- second-degree lattice, simplex lattice design, fat blends were prepared as indicated in Table 1.

**Table 1**. compositional design of the blends.

| Blends | Mass Compositions (w/w %) of 100grams feed | | |
|---|---|---|---|
| | Palm Stearin (PS) | Palm Olein (PO) | Coconut Oil (CO) |
| 1 | 0 | 100 | 0 |
| 2 | 50 | 0 | 50 |
| 3 | 16,67 | 66,67 | 16,67 |
| 4 | 33,33 | 33,33 | 33,33 |
| 5 | 0 | 0 | 100 |
| 6 | 66,67 | 16,67 | 16,67 |
| 7 | 100 | 0 | 0 |
| 8 | 50 | 50 | 0 |
| 9 | 0 | 50 | 50 |
| 10 | 16,67 | 16,67 | 66,67 |

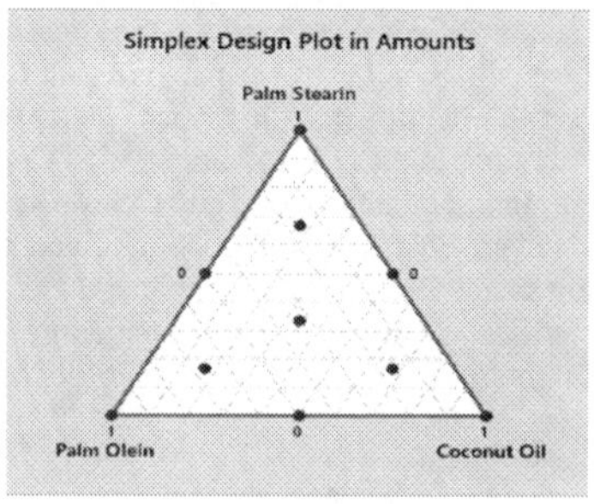


**Figure 1**. simplex design plot.

*2.3. Water content*

The water content of raw material could be determined by heating the sample. About 1 g of sample was placed in a porcelain cup which was mass known. Then, the sample was dried for 3 hours at 105°C. After that, the sample was cooled in a desiccator for 15 minutes, and mass was pondered byanalytical balance. Water content could be determined using the equation below.

$$content = \frac{W2-W1}{W} \times 100\% \quad (1)$$

where: W2: The wight of porcelain and sample before dried (g), W1: The wight of porcelain and sample after dried (g), W: the sample's weight (g)

*2.4. Batch Chemical Interesterification*
The chemical reaction was started by adding 0.4% w/w sodium ethoxide (Sigma Aldrich) as thecatalyst into fat blends prepared before, according to table 1. Chemical interesterification was conductedat 75°C using sodium ethoxide as the catalyst. The chemical interesterification was carried out in a laboratory-scale batch reactor with an agitator. The sample was agitated with an average speed of 300-400 rpm for 8 hours.To terminate the reaction, 20 ml 20% (w/w) citric acid solution was added. By water addition, it wouldnon-activate catalyst by changing it into ethanol (Sreenivasan, 1978). Citric acid would react with sodium into sodium salt. Interesterified fat blends was dried again to reduce the water content of a sample.

*2.5. Slip Melting Point*
The slip melting point of the interesterified blends was determined according to the AOCS Method No. Cc 3-25 and slip melting point digital. A capillary tube filled with a 1 cm high column of the sample was chilled at 4 ± 1°C for 16 hours before being immersed in a beaker with cold water. The water was stirred and heated slowly. The temperature was recorded when the blend in the tube started to rise due to hydrostatic pressure. The temperature at the moment was taken as the SMP.

*2.6. Iodine value*
The iodine number was determined by the volumetric titration method. This method referred to AOCS (1998). Iodine numbers are expressed as grams of iodine absorbed per 100gram sample. The samples were melted at 60-70°C and stirred until smooth. A sample of 0,13-0,15gram was put into the Erlenmeyer and then added with chloroform p.a. 25 ml and 25 ml Wijs solution, then covered and stirred until homogeneous. The mixture was put down in a dark room for an hour. After that, the mixture was added with 20ml KI 10% and 150mL aqua dm. The solution was mixed and titrated with 0.1N thiosulfate solution until a purple-purplish color changes precisely disappear. Next, a 10% 2 mL starch indicator is added. The iodine number can be calculated as follows.

$$Iodine\ Value\ (g\ iod\ /\ 100\ g\ sample) = \frac{12.69 \times N \times (V2-V1)}{W} \quad (2)$$

Where N is the normality of thiosulfuric acid, V2 volume of thiosulfate used in the blank determination (mL), V1 volume of thiosulfate used in sample determination, W is the test weight (gram). The following is a workflow diagram for the analysis of fat iodine numbers.

*2.7. Experimental Design and Statistical Analysis*
Statistical analysis was done using Response Surface Methodology (RSM) and ANOVA. The type of RSMthat was used was the Mixture Design method, simplex lattice design. Experimental results for SMP was appliedto obtain the regression models, as a function of the proportions of each ingredient ($x_1$ = palm stearin, $x_2$ = palm olein and $x_3$ =coconut oil) present in blends: SMP (°C) = $\beta_1X_1 + \beta_2X_2 + \beta_3X_3 + \beta_{12}X_1X_2 + \beta_{13}X_1X_3 + \beta_{23}X_2X_3 + \beta_{123}X_1X_2X_3$, where SMP was estimated response, β was coefficient estimated by least-squares method, x was a dependent variables. The quality of the models was evaluated by ANOVA and adjusted coefficient of determination ($r^2$). ANOVA that used was Univariative Multi-Way Analysisof Variance using Microsoft Excel

## 3. Results and Discussion

*3.1. Effect of Variation in Chemical Interesterification Raw Material Composition onSolid Fat Content (SFC)*
In cocoa butter, solid fat content is responsible for the characteristics of fat hardness and quickmelting. SFC is also essential for the implications of fat stability at room temperature. High storage temperatures at room conditions above 20 °C determine product stability and resistance to oil exudation,which requires SFC to survive above 10% (Graef, 2012). This standard specification is also needed to produce cocoa butter equivalent and cocoa butter replacer. Cocoa butter equivalent and cocoa butter replacer are alternatives to

cocoa butter, similar to the original cocoa butter seen from the resulting SFC profile. CBS's SFC value does not maintain solid fat content far above 10% at 25 °C. The CBE, CBR, and CBS standard specifications are presented in the following table. The addition of fatty acids in the form of triglyceride alloys to the mixture will changethe ratio between saturated and unsaturated fatty acids that affect the content of solidfat in the mix, the polymorphic structure and, melting temperature as described previously in Figure 4. The gentle melting pattern results in melting points (SMP) visually identified appear to be lower dueto decreased solid fat content below 10% (Hernqvist, 1990). Under conditions of 10% SFC, colorless fat appears to melt visually completely. This melting profile is undesirable in alternative chocolate fat products. Profile of melting of fat resulting from interesterification of variations 1 to 10 is presented in Figure 2.

**Table 2.** Differences of CBE, CBS, and CBR.

| Characteristic | CBE | CBS | CBR |
|---|---|---|---|
| Main Fatty Acid | Palmitic, stearic, oleic, linolec, arachidic | Laurac, myristic | Elaidic, stearic, palmitic, linoleic |
| *% Solid Fat Content* | 10 °C = 90-95 %<br>20 °C = 85-92 %<br>25 °C = 80-88 %<br>30 °C = 65-89 %<br>35 °C = 10-25 % | 16 °C = 70-72%<br>18°C = 60-68%<br>22 °C = 31-40 %<br>26 °C = 9-18%<br>34 °C = 5-7% | 10 °C = 91-93 %<br>20 °C = 65-70 %<br>25 °C = 40-43 %<br>30 °C = 28-30 %<br>35 °C = 15-20 % |
| Iodine Value | 30-34 (g iod/ 100 g) | 4-17 (g iod/ 100 g) | 30-34 (g iod/ 100 g) |
| SMP | 33- 42 °C | 33- 42 °C | 33- 40 °C |

Sources: SNI Document and Jurnal *Cocoa Butter and Its Alternatives: A Review*, 2014 dan DeltaWilmar

The profile of solid fat content is specific to the content of fatty acids in oil/fat. The melting profile in Figure 2 was obtained from the results of reprocessing the literature data from the research of Andreia, F eet al. (2012) using multiple regression methods to get the coefficient of determination of the SFC value.

$$SFC(T) = B_1X_1 + B_2X_2 + B_3X_3 + B_{1,2}X_1X_2 + B_{1,3}X_1X_3 + B_{2,3}X_2X_3 + B_{1,2,3}X_1X_2X_3$$

Prediction of solid fat content percentage at the measurement temperature using slip melting point experimental results of each feed variation. Slip melting point experimental results at the 8th hour presented in Table 3.

**Table 3.** Experimental Slip Melting Point Result Feed Mixture Variations.

| Variations | Palm Stearin | Palm Olein | Coconut Oil | Slip Melting Point (°C) |
|---|---|---|---|---|
| 1 | 0% | 100% | 0% | 10 |
| 2 | 50% | 0% | 50% | 30 |
| 3 | 17% | 67% | 17% | 18 |
| 4 | 33% | 33% | 33% | 40 |
| 5 | 0% | 0% | 100% | 33 |

| | | | | |
|---|---|---|---|---|
| 6 | 67% | 17% | 17% | 32 |
| 7 | 100% | 0% | 0% | 43 |
| 8 | 50% | 50% | 0% | 34 |
| 9 | 0% | 50% | 50% | 21 |
| 10 | 17% | 17% | 67% | 29 |

Based on Figure 2, it can be observed that the profile of fat melting has a very gentle melting pattern, gentle melting pattern ,and steep melting pattern. However, based on the CBR triglyceride component's basic specifications, no alternative fat types will be produced from this mixture so that only alternative CBE and CBS fats will be identified from the melting profile variations.

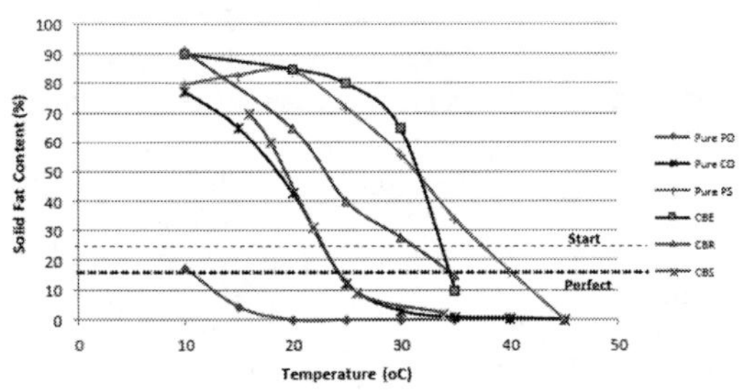


(a)

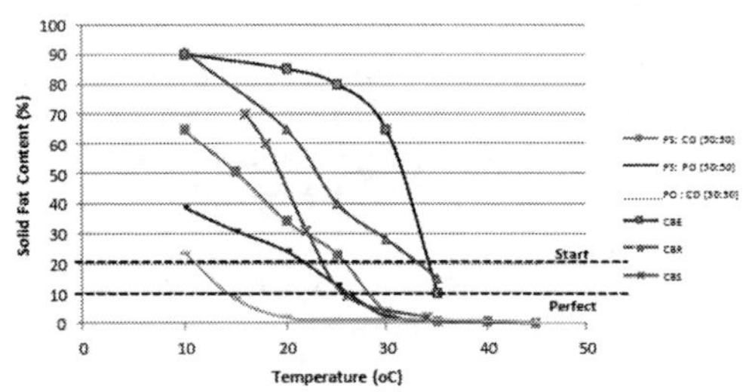


(b)

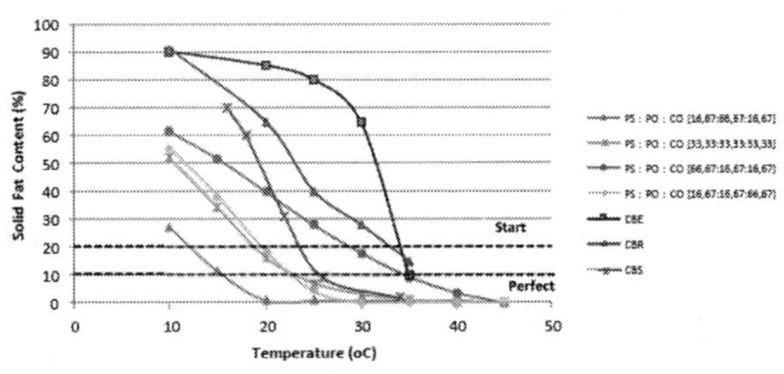


(c)

**Figure 2.** Modeling the profile of solid fat content of interesterification products to temperature on variations in the composition of Palm Stearin (PS), Palm Olein (PO), and Coconut Oil (CO) for (a) single feed, (b) mixture of2 oils, and (c) mixture of 3 oil.

A slight slope of melting profile is produced by variations in a composition dominated by a mixture of palm olein such as pure palm olein, PS: PO: CO (16.67: 66.67: 16.67), and PO: CO (50:50).The melting profile produced is very gentle, with 10 °C indicating that the percentage of solid fat is in the range of 20%. Thus from this mixture, it can be seen that the characteristics of the melting characteristics of palm olein significantly affect the characteristics of the fat variety. The dominant saturatedfatty acid content makes it difficult to form solid fat structures from polyunsaturated or monounsaturated triglycerides so that the fat content at low temperatures is very low.

The mixture shows different things with more solid fat content due to the mixing process. In this condition, a slightly steeper melting profile is formed, which is included in variations of the composition of PS: PO: CO (50: 0: 50), PS: PO (50:50), PS: PO: CO (16.67: 16, 67: 66.67), PS: PO: CO (66.67:

16.67: 16.67), and PS: PO: CO (1: 1: 1). In this composition, the gentle melting pattern is obtained. This better melting pattern results from adding the amount of saturated fatty acids to the mixture. This melting pattern is expected to occur in margarine functional fat (Soekopitojo, 2011) but is undesirable in the production of CBE and CBS fat. This is because the pattern of gentle melting produces less comfortable melting in the mouth (melting behavior).

Furthermore, the steep melting profile from the data analysis results resulted from variations in the composition of 100% palm stearin and 100% coconut oil. This composition found that coconut oil produces SMP between 25-33°C (SFC <10%) and palm stearin spanning 42-45°C. Of these two variations, only 100% coconut oil has a melting profile similar to cocoa butter alternatives. The value of 100% palm stearin SMP is very far above the human body temperature causing this variation to have no potential to produce cocoa butter. Based on the graph, it can be seen that the100% coconut oil melting curve coincides with the CBS line crew starting at a measurement temperatureof 20°C. As such, 100% coconut oil is the only variation with a similar melting profile to CBS-type butter alternatives.

### *3.2. Optimization Feed Mixture for Solid Fat Content Profile by Response SurfaceMethod*

Cocoa butter alternative based on the literature consists of 3 types, namely cocoa butter equivalent (CBE), cocoa butter substitution (CBS), and cocoa butter replacer (CBR). The three alternative types of chocolate have differences from varieties of fatty acids, enabling properties, physicochemical properties, solid fat content, iodine numbers, and melting profiles (Naik & Kumar, 2014). Based on these characteristics, the desired interesterification product has a melting profile similarto alternative cocoa butters and has the appropriate melting temperature of the product and iodine number. Thus, the melting profile is the main factor used to determine the best variation in producing alternative cocoa butter, verified by melting temperature and iodine number. Slip melting point data for each variation is presented in table 3, while iodine number data for experimental results for each variation is shown in table 4.

**Table 4.** Experimental Iodine Value Result Feed Mixture Variations.

| Variations | Palm Stearin | Palm Olein | Coconut Oil | Iodine Value |
|---|---|---|---|---|
| 1 | 0% | 100% | 0% | 56.86 |
| 2 | 50% | 0% | 50% | 9.79 |
| 3 | 17% | 67% | 17% | 59.35 |
| 4 | 33% | 33% | 33% | 34.04 |
| 5 | 0% | 0% | 100% | 7.98 |
| 6 | 67% | 17% | 17% | 38.72 |
| 7 | 100% | 0% | 0% | 33.56 |
| 8 | 50% | 50% | 0% | 48.7 |
| 9 | 0% | 50% | 50% | 42.05 |
| 10 | 17% | 17% | 67% | 20.97 |

The statistical mixture design is used to determine the mixture's optimal composition to be adjusted to the response factor. All variations of the experimental design presentedin Table 1 observed the closeness of the response factor's actual value. This optimization technique is called the Response Surface Method. The response factor used to determine the optimal composition is the melting fat profile against the Cocoa Butter (CB) referencefat. Tracing the optimal value of all variations of the mixture is done by minimizing the difference in profile (Sum of Square Error / SSE) the results of various composition experiments on the trend of Melting reference fat. The least value of SSE means that the melting profile of the sample is closer to the profile of CB melting profile. The results of the SSE calculations for each variation of the composition are presented in Figure 3.

IOP Conf. Series: Materials Science and Engineering 1143 (2021) 012036 doi:10.1088/1757-899X/1143/1/012036

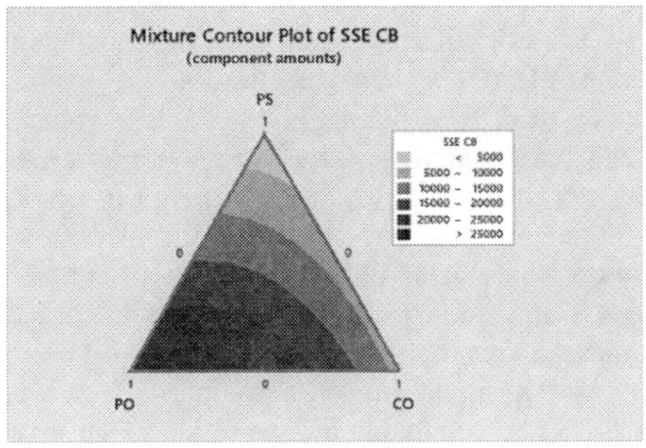


**Figure 3.** Distribution of SSE values to Cocoa butter Profile (CB) in all variations of a mixture of PalmOlein (PO), Palm Stearin (PS), and Coconut Oil (CO).

Figure 3 shows that the acquisition of a minimum SSE value will be obtained from 100%palm stearin variation. This statement is verified from the optimizer's response which shows that an SSE value with a target response close to zero will be obtained from this variation. A low SSE value indicates thatpure palm stearin produces a melting profile similar to cocoa butter. However, from observing the melting profile in Figure 2 (a), the 100% palm stearin curve line is still too right, or the temperature dropvalue is still too high compared to cocoa butter. To explore the optimum potential variation for cocoa butter, the SSE value calculation was performed against alternative profiles of CBE, CBR, and CBS cocoa butter types whose plot contour distribution values are presented in Figure 4.

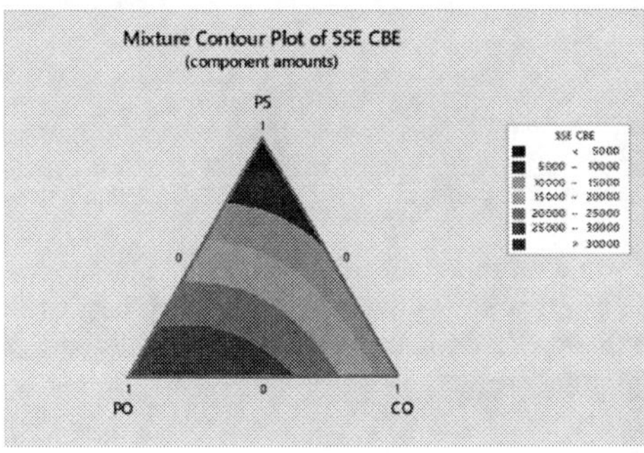


(a)

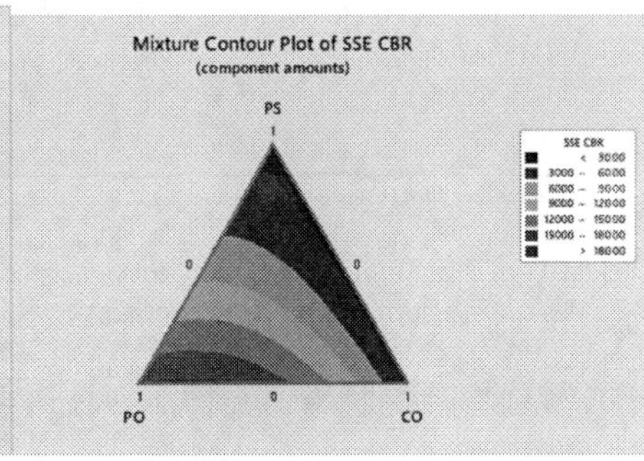


(b)

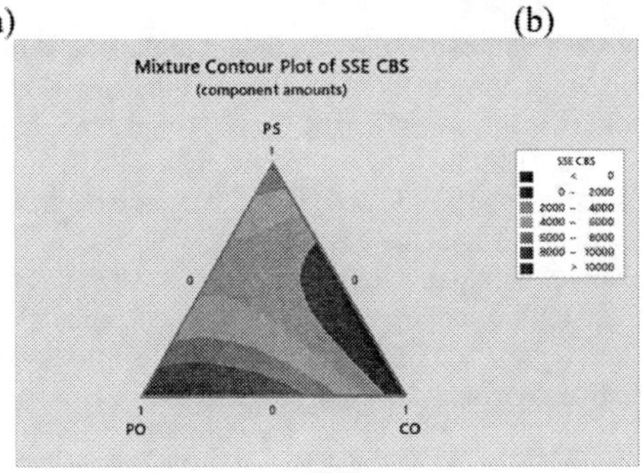


(c)

**Figure 4.** Distribution of SSE values to Equivalent Cocoa butter Profile (a) Exchanging Cocoa butter Profile (b) Substitution Chocolate Fat Profile (c) for all variations of Palm Olein (PO), Palm Stearin (PS), and Coconut Oil (CO) mixtures

Based on the analysis results of Figure 4, it was observed that the equivalent cocoa butter profile was approximated by variations of pure palm stearin as the cocoa butter profile (CB). In contrast, the exchanging cocoa butter profile (CBR) and substitution fat profile (CBS) was approximated by variations in pure coconut oil feed. The optimization of the SSE values on the reference profile showed that pure palm stearin and pure coconut oil are the optimum variations resulting from the surface response (RSM). However, coconut oil variations 1 is superior in advanced selection aspects other than near-zero SSE values. Based on the basic specifications of commercial cocoa butter alternatives regarding SMP values in the range of 33-36°C, it is obtained that coconut oil (100) is following the characteristics of the target SMP value with a melting point value of 33°C. Then, based on observations of the product iodine number, it was found that coconut oil (100) produced an iodine number that was within the range of alternative specifications for lauric fat (4-17 g iodine / 100g fat), which was 7.98 g iodine / 100 g fat. Thus, it can be concluded that variation of virgin coconut oil has amelting profile similar to cocoa butter substitute, the melting temperature of the product close to the target, and the iodine number that meets the standard. Table 2 is the melting profile of the results of interesterification of virgin coconut oil and CBS melting profile literature.

**Table 5.** Comparison of SFC CBS with CO Optimization Fat (100).

| Temperature | SFC CBS | SFC Coconut Oil (100) | % SFC Diffenrences CBS dan Coconut Oil |
|---|---|---|---|
| 10 °C | 84 % | 77.15 % | 6,85% |
| 20 °C | 42 % | 42.85 % | 8,5% |
| 25 °C | 10 % | 12.27 % | 2,27% |
| 30 °C | 4 % | 2.76 % | 1,24% |
| 35 °C | 2 % | 0 % | 2% |

Sources: SNI Document and Jurnal *Cocoa Butter and Its Alternatives: A Review*, 2014 dan DeltaWilmar

Based on Table 5, it can be concluded that the value of important solid fat content at a temperature of 25 °C is very close to the CBS percentage value of literature with a reasonably steep decrease when reachingthe target melting temperature of 35°C so that the fat from the optimization results sufficiently meets the characteristics of the commercial CBS melting profile.

## 4. Conclusion

Fat melting profile from the observation of solid fat content in the mix showed that from 10 variationsin the composition of the feed, only one profile was obtained that matched the characteristics of cocoa butter alternative that variation of pure coconut oil-CO (100) with a melting profile that resembled cocoa butter substitute. The optimization of statistical analysis using the RSM-simplex latticedesign factor with the melting cocoa butter (CB) response profile and alternative cocoa butter CBE,CBR, and CBS produce pure palm stearin and pure coconut oil as the best variation with minimum Sum Square of Error value with the target profile. However, pure palm stearin does not meet the advanced specifications so that it can be categorized as a cocoa butter alternative so that purecoconut oil is a variation that meets the commercial criteria for alternative cocoa butter in terms of slipmelting point, iodine number, and melting profile as an alternative chocolate substitution product.

**Acknowledgments**
This research was funded by P3MI Program 2020 from Institut Teknologi Bandung.

**References**

[1] Becket S 2008 The Science of Chocolate 2nd Ed.

[2] Dian N L H M, Sundram K, Idris N A 2007 Effect of chemical interesterification on triacylglycerol and solid fat contents of palm stearin, sunflower oil and palm kernel olein blends*European Journal of Lipid Science and Technology* **109** (2) 147-156

[3] Paper O 2004 *Effects of chemical interesterification on solid fat content and slip the melting pointof fat/oil blends*. 224–229. https://doi.org/10.1007/s00217-003-0847-4

[4] Berry S E E 2014 Rich fats: An overview and implications for cardiovascular disease Triacylglycerol structure and interesterification of palmitic and stearic acid (June 2009)

[5] Chapman D 1962 The Polymorphism of Glycerides *Chemical Reviews* **62** (5) 433-456.

[6] da Silva Oliveira W, Neves D A, and Ballus C A 2019 Mature chemical analysis methods for food chemical properties evaluation. In *Evaluation Technologies for Food Quality* (pp. 63-90). Woodhead Publishing.

[7] Dian N L H M, Sundram K, and Idris N A 2007 Effect of chemical interesterification on triacylglycerol and solid fat contents of palm stearin, sunflower oil and palm kernel olein blends*European Journal of Lipid Science and Technology* **109** (2) 147-156

[8] Hernqvist L, Herslöf B, and Herslöf M 1984 Polymorphism of interesterified triglycerides and triglyceride mixtures *Fette, Seifen, Anstrichmittel* **86** (10) 393-397

[9] Interesterifikasi K, Stearin M, Sawit M I, Soya K, and Hazirah S 2012 *Effects Of Chemical Interesterification On The Physicochemical Properties Of Palm Stearin, Palm Kernel Oil, And Soybean Oil Blends* **16** (3) 297–308

[10] Karupaiah T, and Sundram K 2007 Effects of stereospecific positioning of fatty acids in triacylglycerol structures in native and randomized fats: a review of their nutritional implications *Nutrition & metabolism* **4** (1) 16

[11] Kumar P K P, and Krishna A G G 2015 Physicochemical characteristics of commercial coconut oils produced in India *Grasas y Aceites* **66** (1)

[12] Neff W E, Byrdwell W C, and List G R 2001 Triacylglycerol structures of food fats high insaturated acids by HPLC and mass spectrometry *Journal of liquid chromatography & related technologies* **24** (6) 837-854

[13] Omar K A, Gounga M E, Liu R, Mwinyi W, Aboshora W, Ramadhan A H and Wang X 2017 Triacylglycerol composition, melting, and crystallization profiles of lipase-catalyzed anhydrous milk fats hydrolyzed *International journal of food properties* **20** (sup2) 1230-1245

[14] Palm I E (n.d.) *Oil Palm : Fractions & Derivatives Palm Oil process Oil Palm : Fractions & Derivatives Palm Kernel Oil process*.

[15] Soares D M, Claro R, Capacla I R, Viccola E R, Maruyama J M, Gioielli L A, and Prestes L 2012 *Chemical Interesterification of Blends of Palm Stearin, Coconut Oil, and Canola Oil: Physicochemical Properties*

[16] Soekopitojo S 2012 Asidolisis enzimatik fraksi tengah minyak sawit dengan asam stearat untuk sintesis cocoa butter equivalents *Teknologi Kejuruan: Jurnal Teknologi, Kejuruan, dan Pengajarannya* **34** (2)

IOP Conf. Series: Materials Science and Engineering 1143 (2021) 012037 doi:10.1088/1757-899X/1143/1/012037

# Pediocin and Grape Seed Extract as Antimicrobial Agents in Nanocellulose Biobased Food Packaging: A Review

**Timotius Weslie*, Vincent Felixius, Zulfah Amala, Dian Shofinita**

Department of Chemical Engineering
Institut Teknologi Bandung, Bandung 40132 Indonesia

*E-mail: timotiusweslie@students.itb.ac.id

**Abstract**. The food industry is a life-long and highly demanded industry as it is an essential human need. Food safety has become an urgent discussion related to this industry since the foodborne disease has gradually increased. One of the causes of the disease is *L. monocytogenes,* usually discovered in fresh meat, ready-to-eat meat, vegetables, and milk. *L. monocytogenes causes* an infection known as listeriosis. Food packaging plays an essential role as it protects the food from external contaminants to increase the shelf life. The high usage of conventional food packaging derived from fossil fuel contributes to the environmental issue as it creates long-term wastes. Therefore, biobased food packaging has been in favor as it is biodegradable. However, it lacks antimicrobial properties, so the development of biobased material as the antimicrobial food packaging is a potent solution in the food safety scope. This review paper intends to summarize current advancements in incorporating antimicrobial agents with nanocellulose biobased food packaging to increase the packaging's functional value. Pediocin is the antimicrobial agent produced by *Pediococcus sp.* integrated with Grape Seed Extract (GSE), which gives an antioxidant property that boosts the food packaging's antimicrobial effect. Observations show that incorporating these antimicrobials agents obstructs the growth of *L. monocytogenes* in biobased food packaging. Incorporating antimicrobial agents into nanocellulose matrix shifts the tendency to make biobased packaging that gives better mechanical strength and longer shelf life. Overall, this greener antimicrobial food packaging could be a solution to environmental waste as well as foodborne pathogens.

## 1. Introduction

Food is an essential part of our lives. A complex workflow such as supply chain, food production, consumption, and services are integrated to provide this primary human need. A significant concern is food safety issues regarding product quality, sanitation, and health [1]. By 2017, the global food and beverage industry has seen substantial growth over the last five years, and this trend is expected to continue as the industry is projected to be $5776 billion worth [2]. However, this positive industrial growth is not balanced with the global foodborne disease trend that still gradually increases. The World Health Organization (WHO) stated that nearly one in 10 people worldwide fall ill every year after eating contaminated food, leading to over 420,000 deaths [3]. Foodborne diseases are caused by microbial contamination of food and occur at many stages of the food lines. In the United States, *L. monocytogenes* is one of the top 5 pathogens causing foodborne illnesses resulting in death by 2011 [4]. Infection of *L. monocytogenes* causes listeriosis, which has a mortality rate of 30%[5].

Food packaging plays a significant role in protecting food from pathogens and preserving its condition [6]. On the other hand, food packaging waste has been a significant problem. The packaging industry still heavily relies on products derived from fossil fuels. Packaging contributes to 42% of plastic waste [7]. A new sustainable approach is needed to solve these problems. One solution is by incorporating natural antimicrobial and antioxidant into bio-based food packaging, which promotes a longer food shelf life. Two common antimicrobial agents that have been used in food packaging are grape seed extract (GSE) and pediocin. Grape seed is a by-product derived from wine processing and grape juice industries [8]. Phenolic compounds in grape seeds have been known to have antimicrobial and antioxidant effects [9]. GSE has been proven to act as antimicrobial both *in vivo* and *in vitro* [10], [11]. Pediocin is a unique bacteriocin with anti-listeria activity. According to Acuña et al. [12], Pediocin is the most studied bacteriocins and commercially used as bio preservatives. However, commercial pediocin has already existed and is widely used. The antimicrobial property comes from its complex peptide structure, and its efficacy to reduce *L. monocytogenes* pathogen in food has been proven by many researchers.

Bio-based packaging uses materials derived from primarily annual renewable resources, thus excluding the paper-based material [13]. One of the renewable resources that are being developed is cellulose. Cellulose is the most abundant organic polymer on Earth [14]. Many biomass contains a high concentration of cellulose. The role of cellulose in biomass is a structural component in the cell wall of green plants, algae, and oomycetes. To enhance the mechanical properties of cellulose, cellulose is modified into a nanocellulose structure. This review paper aims to summarize the potentials of incorporating antimicrobial and antioxidant agents from GSE and pediocin into the nanocellulose matrix. This incorporation gives advantages to the better mechanical strength of packaging and longer shelf life of food.

## 2. Antimicrobial Agents

### *2.1 GSE*

*2.1.1 GSE Characteristic.* GSE contains 74 to 78 % polyphenols, which are oligomeric proanthocyanidins ($B_1$, $B_2$, $B_3$, $B_4$, $B_1$-G, $B_2$-G, $B_2$-G', $C_1$) [15], [16]. These compounds' activity is believed to be the reason for GSE antimicrobial, antioxidant, and health benefit effects [17]. Studies have reported some health benefits of consuming GSE, particularly in treating cardiovascular diseases, obesity, type 2 diabetes, inflammatory bowel disease, neurogenerative disorders, asthma, eye disease, and osteoarthritis [18]. Another research also reported that GSE could be used as nutraceuticals or new drugs for treating Malignant Mesothelioma (MM) cancer [19]. Giving its health benefit, GSE is not reported to exhibit toxicity or adverse health effects on daily consumption [20], [21].

#### *2.1.2 Extraction of GSE and its Recent Advancement*

Research showed that grape seed contains the highest amount of polyphenols, making it the primary extraction [22], [23]. Many methods have been developed to increase the yield of polyphenolic compounds in GSE. The conventional method, such as soxhlet extraction, involves organic solvents such as *n*-hexane and ethanol in extraction, separation, and GSE purification [24]. Dimić et al. [25] conducted the extraction by using *n*-hexane to extract grape seeds sample for 6 hours with 15 exchanges of extracts and filtration of the solvent. The solvent was then evaporated under vacuum conditions at 40°C. A novel extraction of GSE was conducted by using supercritical fluid to reduce the usage of toxic solvents, such as *n*-hexane [26]. Coelho et al. [27] conducted the process using supercritical $CO_2$ at various temperatures, pressures, and flow rates. Grape seed was dried for 48 h at 343 K before being extracted with a continuous flow rate of supercritical $CO_2$. They concluded that extraction using supercritical fluid yield the same amount of extract as the conventional *n*-hexane extractions while using a less harmful solvent.

### *2.2 Pediocin*

*2.2.1 Pediocin Characteristic*

Bacteriocin refers to the bacteria that can inhibit or kill other bacteria because of the ribosomal synthesized antimicrobial peptides. Produced bacteriocins vary depending on the action against the targeted bacteria [28]. Lactic Acid Bacteria (LAB) is the most widely used bacteriocin producer in a food fermentation context. It includes the genera *Pediococcus, Lactococcus, Lactobacillus,* and *Streptococcus*. LAB converts sugar into lactic acid and other substances that act as bio-preservatives, such as diacetyl, hydrogen peroxide, acetone, and other organic acids. Pediocin is the bacteriocin that is classified as Class IIA in the product of LAB classification. The class is unique for its anti-listeria activity that has drawn interest in research and application [29].

Aromatic and aliphatic amino acids arrange pediocin to be a structure of the 44-residue peptide. N-terminal is a highly conserved, hydrophilic, and cationic region. It poses a three-desolate β-sheet reinforced by a disulfide bond, which consists of two cysteine residues. A consensus motif -YGNGV- was found in the N-region. This motif is responsible for the antilisterial activity of pediocin. The C-terminal region is a less conserved region located at the structure tip like a tail. According to Drider et al., it is essential to determine the target cell specificity of pediocin [30]. The C-terminal consists of two cysteine residues. It marks a hairpin domain as the disulfide bond bends the cysteine residues. In their study, Johnsen et al. proposed that the overall pediocin structure is amphiphilic, centered, and α-helix corresponding to the N-terminal [31].

Pediocin is heat stable both in high temperatures like the sterilization process and cold temperature. Pediocin SA-1, the pediocin isolated from *P. acidilactici,* remains stable at 121 °C for more than 60 minutes. Moreover, the antimicrobial activity was not affected even by storage at -80, -20, 4, and 30°C for four weeks. It was also not interrupted even under a week of incubation at pH values from 3.0 to 12.0, which indicates pediocin activity is retained at a wide pH range. [32]. Furthermore, pediocin antimicrobial activity is still working subsequent to lipase, lysozyme, and D/R-Nase treatments, but it could be damaged by proteolytic enzymes [33].

*2.2.2 Antimicrobial mechanism of Pediocin.* Pediocins are sensitive to gram-positive bacteria like *L. monocytogenes* [34]. Before the pore formation in the cytoplasmic membrane, pediocin interacts with the targeted bacteria in prior, which occurs by the electrostatic effect of the cationic N-terminal with the lipoteichoic acid from the bacteria. [35]. After that, the native divalent cations are removed from the cell surface, causing outer membrane destabilization for the peptide to enter and contact the cytoplasmic membrane. This phenomenon is referred to as auto-promoted uptake [36]. After the contact occurred, the C-terminal of pediocin establishes the pore to penetrate the targeted cell's hydrophobic site [36]. The pores effluence the intracellular substances, reduce the adenosine triphosphate in the cytoplasm, vanish the proton force, and, eventually, cell death [37].

*2.2.3. Synthesis of Pediocin.* Pediocins are synthesized successfully in rich media [38]. Process variables such as oxygen, pH, temperature, and NaCl are different in every production strains. They are homofermentative as glucose is fermented to DL-lactate in every case via the Embden-Meyerhof-Parnas pathway, but patterns of assimilation fermentation carbohydrate may differ between strains. [39]. *P. acidilactici* is the most widely used producer strain. The strains mostly ferment glucose, fructose, galactose, ribose, and xylose to DL-lactate. However, it is also reported that a few strains can ferment lactose, sucrose, and maltose [34]. Amado et al. produced pediocin in the culture of *P. acidilactici* NRRL B-5627 with Man, Rogosa, Sharpe (MRS) medium [40]. The synthesis was held in a fermenter under 200 rpm agitation and 30 °C. It took 92 hours of cultivation, and the pH medium was adjusted to 3,5 at the end of the culture to promote the pediocin release from the cell surface. In order to increase the bacteriocinogenic activity, an ultrafiltration process was added. Papagianni and Anastasiadou [41] concluded that the optimum temperature for growth is 40°C, but it can grow until 50°C. A pH of 6.0 is regarded as the optimum condition for starting cultivation, and the pH could fall to levels as low as 3.6 during bacterial growth.

## 3. Potential Usage of GSE and Pediocin in Food Industries

### *3.1 GSE Usage as Antioxidant*

Recent studies have shown that GSE effectively increases the shelf-life of cold storage meat and fruit while used in food film packaging [20]–[22]. This likely happens due to GSE's capability to decrease aerobic mesophilic bacteria's growth rate, Pseudomonas spp., mold, and yeast in food, thus reducing the rate of oxidation and maintaining the color stability of the product [42], [45]. However, a high GSE concentration could lead to a higher oxidation rate as it can reduce metals to a lower valence state [46]. Table 2 shows the GSE concentration used in different methods and their effects on different food products.

**Table 1.** The concentration of GSE and their effects on different types of food products.

| Food Products | GSE Concentration (w/w) | Methods | Effect | References |
|---|---|---|---|---|
| Fresh strawberries | 1% | Incorporation of GSE into chitosan film | Higher shelf-life of strawberries up to 30 day | [42] |
| Roasted chicken | 0.2- 1% | Solution spraying on the meat surface | Lower bacterial count on treated meat than control on 0.5% of the concentration | [45] |
| Snakehead fillets | 0.052% | Direct immersion in GSE solution | Chilled fillet shelf-life increase three days than control | [47] |
| Refrigerated salmon | 0.8% | Incorporation of GSE into chitosan film via microcapsules | Lower *Pseudomonas* spp. on the seventh day of storage than chitosan-alone film | [48] |

Based on **Table 1,** it was observed that 0.5% - 1% w/w concentration of GSE shows a reduction of the bacterial count, ranging from 4 – 8 log CFU/g, which promotes longer shelf life. The direct method of using GSE to the food product and indirect method by using film still shows the usage of GSE as an antioxidant. However, previous literature suggested that direct usage of GSE could affect the taste of food [47]. The downside of incorporating GSE into food packaging is that it lowers the tensile strength of packaging and thickens the film, increasing its opacity [44], [49].

### *3.2 Pediocin Application as the active biopackaging to control L. monocytogenes*

Biopreservation implies using hostile microorganisms or their metabolic products to inhibit or wipe out undesired microbes [50]. High consumer demand for good quality and natural foods and strict government provisions make the food manufacturers struggle to warrant safety aspects [51]. The adverse side effects of artificial chemical additives also promote natural antimicrobial substances, like pediocin, to be widely researched and tested for their performance [52].

Active packaging is defined as packaging which able to extend food shelf life and safety by changing the package's property [53]. Antimicrobial packaging is an active packaging that is reducing, inhibiting, or retarding microbial growth, particularly on the food surface, in order to extend food shelf life. Espitia et al. stated three main methods to incorporate antimicrobial peptides into the active packaging [54]. First, direct peptide incorporation into the polymer. This method was successfully done by Espita et al. [55], who developed the pediocin antimicrobial films to control *L. monocytogenes* in sliced bologna. The polymer matrix was solubilized by mixing it with acetone to make a film-forming solution. The powdered commercial Pediocin ALTA$^{TM}$ 2341 was incorporated in the polymeric solution and mixed homogeneously.

The film solution was distributed on glass slides and dried at 24°C. This method is the simplest and suitable for a polymer that requires mild processing conditions. Second, peptide coating on the polymeric surface. This method is pointed out to be an alternative when it must process the polymer in high pressure and temperature to make the film. The polymer is immersed in a peptide solution To produce the antimicrobial coating. The polymeric material and peptide are processed separately because extreme conditions are possibly inactivated the antimicrobial activity [56]. Those two methods are resulting in the bioactive components' migration from packaging toward the food. In the opposite, the third method, peptide immobilization within the polymer, the bioactive components are tied up with the packaging material so that no bioactive components migrate [57]. However, this method has not been widely used because of several drawbacks, such as limitations on diffusion, bioactivity, and molecular mobility losses. Table 2 summarizes some pediocin applications in the active packaging.

**Table 2.** Application of Pediocin in The Active Packaging

| Producer strain | Packaging material | Subject | Result | References |
|---|---|---|---|---|
| *Pediococcus acidilactici (*in the commercial ALTA™ 2341*)* | Cellulose | Sliced bologna | *L. monocytogenes* growth reduced by 1.2 log cycles after nine days | [58] |
| *Pediococcus acidilactici (*in the commercial ALTA™ 2351*)* | Cellulose | Sliced ham | *L.innocua* growth reduced for 2 log cycles after 15 days of storage | [59] |
| *Pediococcus acidilactici JBL1095* | Nylon | Beef wieners | An average reduction of 2.7 log CFU/ g during storage at 25°C for eight days | [60] |
| *Pediococcus pentosaceus BCC3772* | PLA polymer 4042D and wood sawdust | Raw sliced pork | *L. monocytogenes* was reduced by 2 log cycles or a 99% reduction of listeria activity | [61] |

Researches have shown that pediocin incorporated with polymer or biopolymer significantly reduced *L. monocytogenes* activity in the food. Most of the application is on processed meat that requires longer shelf life. However, the application with P*ediococcus acidilactici* is more widely used compared with other strains. The commercial pediocin strains like ALTATM probably influence this as it does not have to isolate the strain from natural sources anymore.

## 4. Nanocellulose potential as antimicrobial bio packaging

### *4.1 Nanocellulose properties as biopackaging from lignocellulosic biomass*

Cellulose is a polymer of carbohydrates with D-glucose as the monomer. It has a linear structure due to β(1→4) linked to each monomer [62]. The linear structure of cellulose makes this polymer has a good mechanical property. Recent studies have developed a process to increase cellulose's mechanical properties to broaden the benefit of cellulose. One of the strategies that have been developed is transforming cellulose into nanocellulose. Table 3 shows the recent studies which reconstruct cellulose to nanocellulose and the result of its properties.

**Table 3.** Natural resources of nanocellulose and the result properties.

| Nanocellulose sources | Isolation Technique | Yield (%) | Properties | References |
|---|---|---|---|---|
| Oil palm empty fruit bunch, wheat starch, chitosan | Acid hydrolysis | 90 | The addition of nanocellulose could enhance the mechanical strength of the film. | [63] |
| Industrial waste cotton | Acid hydrolysis | 90 | The CMC's from industrial waste cotton has high thermal stability of 97.13 % | [64] |
| Wood | Ethanol and peroxide with ultrasonication | 77 | The value of tensile strength and Young's modulus after the addition of CNC were more than 50 MPa and 5 GPa. | [65] |
| *L. usitissimum* | Acid hydrolysis | 82 | Cellulose nanocrystals (CNC) showed a good mechanical strength by the value of modulus young achieved 52.35 MPa after the loading 20% of CNC. | [66] |
| Wheat straw | Acid hydrolysis | 75 | Blending the films with cellulose nanocrystals (CNC) increased the mechanical properties. | [67] |

Research has shown that the addition of cellulose nanocrystals from a diverse source of carbohydrates produces stronger mechanical properties to the film. The process of synthesizing nanocellullose as the pretreatment in biodegradable packaging is unique to the carbohydrate source. This pretreatment process disrupted and depolymerized cellulose structure to isolate nanocellulose [68]. Most of the carbohydrate source is utilized by using the acid hydrolysis process to synthesis nanocellulose. The yield of nanocellulose from natural resources varies between 75-90%.

### *4.2 Impact of nanocellulose in sustainable food packaging*

The primary fundamental role of food packaging is to preserve food quality and safety. If all food packaging did its role correctly, food packaging could reduce foodborne disease and food chemical contamination. Meanwhile, commercial packaging focus on practical and easy-to-use aspect rather than the role of reducing food loss and improving food shelf-life. This condition is supported because uneaten food's carbon footprint is estimated to be equivalent to 495 million tons of $CO_2$ and occupies almost 210 million hectares of land [69]. If there is no real action to resolve this problem, food waste is estimated to rise to over 200 million tons by 2050 [70]. The food packaging issue is not just misunderstood about the primary role of food packaging but also the amount of waste. If there is no real action to resolve this problem, food waste is estimated to rise to over 200 million tons by 2050 [71]. The plastic packaging problem is the chemical leaching in plastics is hormonally active and can be dangerous to the food. The problem of plastic packaging continues when this packaging becomes a waste. The plastic material was hard to achieve sustainability. Only up to 2% of the waste can be utilized in another sector [72]. The rest of them will contaminate their surrounding environment.

Nanocellulose has a high abundance, low weight, high strength, stiffness, non-toxicity, low cost, and biodegradable [73]. This characteristic makes nanocellulose become a promising material that has been utilized in diverse fields. It has been used in medical, packaging, prepared coating, electronics, and membrane [74]–[77]. In the packaging sector, nanocellulose will support the goal to make sustainable and green packaging. Studies have demonstrated that the use of nanocellulosic based material can enhance the mechanical and functional properties [78]. These properties include biodegradability, transparency, gas barrier properties specific surface area, and heat stability [79], [80].

## 5. Conclusions

Pediocin and GSE show an effective antimicrobial and antioxidant effect while being used as bioactive food packaging agents. However, most experiments are conducted using low mechanical strength film. Nanocellulose has an immense potential to be used as a substitute for conventional packaging as it is the most abundant polymer on Earth. The modification of cellulose structure into diverse kinds of nanocellulose showed increasing mechanical properties to the packaging. Biobased food packaging has advantages in creating less waste and easier processing than conventional. Many researches has\ve been done to utilize natural resources for replacing conventional fossil packaging. However, sustainability is about using biosource and needs to fulfill the role of packaging to preserve food quality and safety. The procedure for making food packaging does not only utilize a single process. It needs integration of several operations to achieve sustainability in packaging that will be used commercially. Therefore, this review suggests that research needs to be done to scale up, integrate active agents with bio-polymer, and optimize antimicrobial biobased packaging production. Therefore, antimicrobial biobased food packaging can be economically feasible in the future.

## References

[1] Sadiku M N O, Musa S M, and Ashaolu T J 2019 Int. J. Trend Sci. Res. Dev 3(4) 128–130.
[2] Lucintel 2012 https://www.lucintel.com/press/global-food-beverage-retail-industry-2012-2017.aspx
[3] WHO 2020 https://www.who.int/news-room/fact-sheets/detail/food-safety#:~:text=An estimated 600 million – almost,healthy life years (DALYs)
[4] CDC 2014 *US Department of Health and Human Services, CDC*, Atlanta, GA
[5] Ryser E T 2007 *Listeria,listeriosis, and food safety* Florida: CRC Press
[6] Marsh K, Bugusu B 2007 *J. Food Sci* **72** (3) R39–R55
[7] Geyer R, Jambeck J R, Law K L 2017 *Sci. Adv.* **3** (7) e1700782
[8] Lau D W, King A J 2003 pp. 1602–1607
[9] Shahidi F, Janitha P K, Wanasundara P D 1992 *Crit. Rev. Food Sci. Nutr.* **32** (1) 67–103
[10] Yilmaz Y, Toledo R T, Agric J 2004 *Food Chem* **52** (2) 255–260
[11] Kong F, Su Z, Zhang L, Qin Y, Zhang K 2019 *LWT* **101** 819–826
[12] Acuña L, Morero R D, Bellomio A 2011 *Food Bioprocess Technol.* **4** (6) 1029–1049
[13] Robertson G L 2012 *Food Packaging : Principles and Practice, Third Edition* USA: Taylor & Francis Group
[14] Klemm D, Heublein B, Fink H P, Bohn A 2005 Angew. Chemie - Int. Ed. **44** (22) 3358–3393
[15] Yali S, Xiaochun L, Xinyu W, Junli W, Xin L, Dayong S 2018 *J. Chifeng Univ. (Natural Sci. Ed.*

**34** (09) 40–42
[16] Burdock G A 2004 *Fenaroli's Handbook of Flavor Ingredients* CRC Press
[17] Moniharapon E, Hashinaga F 2004 *Pakistan J. Biol. Sci.* **7** (6) 1057–1061
[18] Unusan N 2020 *J. Funct. Foods* **67** (no. November 2019) 103861
[19] Di Meo F et al. 2019 *J. Funct. Foods* **61** (no. March) 103515
[20] Wren A F, Cleary M, Frantz C, Melton S, Norris L 2002 *J. Agric. Food Chem.* **50** (7) 2180–2192
[21] Chen Y, Wen J, Deng Z, Pan X, Xie X, Peng C 2020 *J. Funct. Foods* **73** (no. July) 104113
[22] Perumalla A V S and Hettiarachchy N S 2011 *Food Res. Int.* **44** (4) 827–839
[23] Rhodes P L, Mitchell J W, Wilson M W, Melton L D 2006 *Int. J. Food Microbiol.* **107** (3) 281–286
[24] Torres-Acosta M A, Mayolo-Deloisa K, González-Valdez J, Rito-Palomares M 2019 *Biotechnol. J.* **14** (1) 1–31
[25] Dimić I et al. 2020 *Antioxidants* **9** (7) 1–19
[26] Fiori L, Lavelli V, Duba K S, Sri Harsha P S C, Ben Mohamed H, Guella G 2014 *J. Supercrit. Fluids* **94** 71–80
[27] Coelho J P, Filipe R M, Robalo M P, Stateva R P 2018 *J. Supercrit. Fluids* **141** 68–77
[28] Chen H and Hoover D 2003 *Compr. Rev. Food Sci. Food Saf.* **2** 82–100
[29] Ennahar S, Sashihara T, Sonomoto K, Ishizaki A 2000 *FEMS Microbiol. Rev.* **24** (1) 85–106
[30] Drider D, Fimland G, Héchard Y, McMullen L M, Prévost H 2006 *Microbiol. Mol. Biol. Rev.* **70** (2) 564–582
[31] Johnsen L, Fimland G, Nissen-Meyer J 2005 *J. Biol. Chem.* **280** (10) 9243–9250
[32] Anastasiadou S, Papagianni M, Filiousis G, Ambrosiadis I, Koidis P *2008* **99** 5384–5390
[33] Rodríguez J M, Martínez M I, Kok J 2002 *Crit. Rev. Food Sci. Nutr.* **42** (2) 91–121
[34] Ray B 1995 *Food Biotechnol. Microorg.* 745–795
[35] Papagianni M 2003 *Biotechnol. Adv.* **21** (6) 465–499
[36] Powers J P S, Hancock R E W 2003 *Peptides* **24** (11) 1681–1691
[37] Montville T J, Chen Y 1998 *Appl. Microbiol. Biotechnol.* **50** (5) 511–519
[38] Atlas R 2004 *Handbook of microbiological media* USA: CRC Press
[39] Gasson M 1994 *Genetics and biotechnology of lactic acid bacteria* Glasgow, UK: Blackie Academic & Professional
[40] Amado I R, Fuciños C, Fajardo P, Pastrana L 2016 *J. Dairy Sci.* **99** (10) 8070–8080
[41] Chen Y, Ludescher R D, Montville T J 1997 *Appl. Environ. Microbiol.* **63** (12) 4770–4777
[42] Duran M, Seckin M, Demirel N N, Temizkan R, Burak M, Caner C 2016 *Food Bioprod. Process.* **98** 354–363
[43] Xiong Y, Chen M, Warner R D, Fang Z 2019 *Food Control*
[44] Sogut E. and Seydim A C 2018 **18** (no. July) 13–20
[45] Guo Y, Huang J, Chen Y, Hou Q, and Huang M 2012 *Poult. Sci.*
[46] Frankel E N 1998 *Lipid Oxidation* Dundee, Scotland: Oily Press
[47] Li Y et al. 2020 *J. Food Microbiol.* **91** (no. November 2019) 103492
[48] Vieira M C 2017 *LWT - Food Sci. Technol.*
[49] Álison M et al. 2020 *Int. J. Biol. Macromol.* **160** 769–779
[50] Schillinger U, Geisen R, and Holzapfel W H 1996 *Trends Food Sci. Technol.* **7** (5) 158–164
[51] Franz C M A P, Cho G S, Holzapfel W H, Alvez A 2010 in Biotechnology of Lactic acid Bacteria: Novel Applications, F. Mozzi, R. . Raya, and G. . Vignolo, Eds 341–359.
[52] Woraprayote W, Malila Y, Sorapukdee S, Swetwiwathana A, Benjakul S, Visessanguan W 2016 *Meat Sci.* **120** 118–132
[53] Vermeiren L, Devlieghere F, Van Beest M, De Kruijf N, Debevere J 1999 *Trends Food Sci. Technol.* **10** (3) 77–86
[54] Perez Espitia P J, de Fátima Ferreira Soares N, dos Reis Coimbra J S, de Andrade N J, Souza Cruz R, Alves Medeiros E A 2012 *Compr. Rev. Food Sci. Food Saf.* **11** (2) 187–204
[55] Espitia P J P, Otoni C G, Soares N F F 2016 Pediocin Applications in Antimicrobial Food

Packaging Systems *Elsevier Inc.*
[56] Appendini P and Hotchkiss J 2002 *Innov. Food Sci* **3** 113–126
[57] Barish J A and Goddard J M 2011 *J. Food Sci.* **76** (9)
[58] Espitia P J P, Pacheco J J R, de Melo N R, Soares N de F F, Durango A M 2013 *Brazilian J. Food Technol.* **16** (3) 226–235
[59] Santiago-Silva P et al. 2009 *Food Control* **20** (1) 85–89
[60] Degnan A J, Yousef A E, Luchansky J B 1992 *J. Food Prot.* **55** (2) 98–103
[61] Woraprayote W et al. 2013 *Int. J. Food Microbiol.* **167** (2) 229–235
[62] Updegraff D M 1969 *Anal. Biochem.* **32** (3) 420–424
[63] Salehudin M H, Salleh E, Mamat S N H, Muhamad I I 2014 *Procedia Chem.* **9** 23–33
[64] Thambiraj S and Ravi Shankaran D 2017 *Appl. Surf. Sci.* **412** 405–416
[65] Li Y et al. 2016 *Green Chem.* **18** (4) 1010–1018
[66] Mujtaba M, Salaberria A M, Andres M A, Kaya M, Gunyakti A, Labidi J 2017 *Int. J. Biol. Macromol.* **104** 944–952
[67] Oun A A and Rhim J W 2016 *Carbohydr. Polym.***150** 187–200
[68] Lee J A, Yoon M J, Lee E S, Lim D Y, Kim K Y 2014 *Macromol. Res.* **22** (7) 738–745
[69] Elbehri A, Elliott J, Wheeler T 2015 *Assessments of climate change impacts on global-food productivity and implications for food security and trade: Key messages and recommendations*
[70] Hall K D, Guo J, Dore M, Chow C C 2009 *PLoS One* **4** (11) 9–14
[71] Guillard V, Gaucel S, Fornaciari C, Angellier-Coussy H, Buche P, Gontard N 2018 *Front. Nutr.* **5** (December) 1–13
[72] Luttenberger L R 2019 (June)
[73] Abdul Khalil H P S et al. 2014 *Carbohydr. Polym.* **99** 649–665
[74] Lavoine N, Givord C, Tabary N, Desloges I, Martel B, Bras J 2014 *Innov. Food Sci. Emerg. Technol.* **26** 330–340
[75] Cherian B M, Leão A L, de Souza S F, Thomas S, Pothan L A, Kottaisamy M 2010 *Carbohydr. Polym.* **81** (3) 720–725
[76] Nogi M and Yano H 2008 *Adv. Mater.* **20** (10) 1849–1852
[77] Rezayati Charani P, Dehghani-Firouzabadi M, Afra E, Blademo Å, Naderi A, and Lindström T 2013 *Cellulose* **20** (5) 2559–2567
[78] Abdul Khalil H P S et al. 2016 *Renew. Sustain. Energy Rev.* **64** 823–836
[79] Henriksson M, Berglund L A, Isaksson P, Lindström T, and Nishino T 2008 *Biomacromolecules* **9** (6) 1579–1585
[80] Li J et al. 2014 *Carbohydr. Polym.* **113** 388–393

IOP Conf. Series: Materials Science and Engineering 1143 (2021) 012038 doi:10.1088/1757-899X/1143/1/012038

# Taro Ice Cream: Addition of *Colocasia esculenta* Stem to Improve Antioxidant Activity in Ice Cream

**A H Asaduddin[1*], U N Maulani[1], A Y Sari[2], K Hawari[1], A A Ayusari[1,3]**

[1]Faculty of Medicine, Universitas Sebelas Maret, Jl. Kolonel Sutarto, Jebres, Kec. Jebres, Kota Surakarta, Jawa Tengah, Indonesia 57126
[2]Vocational School, Universitas Sebelas Maret, Jl. Ir Sutami No.36 A, Pucangsawit, Kec. Jebres, Kota Surakarta, Jawa Tengah, Indonesia 57126
[3]Nutrition Departement, Faculty of Medicine, Universitas Sebelas Maret, Jl. Kolonel Sutarto, Jebres, Kec. Jebres, Kota Surakarta, Jawa Tengah, Indonesia 57126

*Corresponding Author's E-mail: aimanhilmi02@student.uns.ac.id

**Abstract.** *Colocasia esculenta* (*C. esculenta*) is a widely cultivated plant for consumption and traditional medicine. Recent evidences showed lack studies about the stem of *C. esculenta* as medicinal agent. Besides, ice cream is a food, which contains milk that play beneficial role as antioxidant. Addition of *C. esculenta* in ice cream may improve antioxidant activity by its phytochemical compounds. Thus, the aim of this nutritional food study is to determine the antioxidant activity, phytochemical compounds and nutritional minerals. The methods used were done by thin layer chromatography (TLC) test and antioxidant assays characterized using 2,2-diphenyl-1-picrylhydrazyl (DPPH). The nutritional minerals was also measured on the study. The results showed the extract had antioxidant activity that measured at 675.283pg/ml. The stem extract of *C. esculenta* positively contained Flavonoid, Terpenoid, Saponin, and Steroid based on TLC assays. Based on nutritional minerals test, *C. esculenta* contained vitamin C, vitamin E, Potassium, and % Calcium. Flavonoids, Terpenoid, Saponin, and steroid had been proved as potent treatment for oxidative stress. Vitamin C and E also had evidence as antioxidant properties. Activity of antioxidant enzymes were improved in Potassium and Calcium treatment studies. Therefore, the addition of *C. esculenta* in ice cream could improve the activity of antioxidant.

## 1. Introduction

Antioxidant is natural or man-made substance that can prevent or delay cell damaged caused by free radicals [1]. Antioxidants have an important role as protective agents against oxidative damage and scavenging reactive oxygen species [2]. Antioxidants can be found from natural source, such as fruits and vegetables that contains certain phytochemical compounds, such as lutein, beta-carotene, vitamin C and E, and the others [3]. Based on previous research, spices, herbs and supplements are included to the most antioxidants rich product [4].

In the other hand, Taro (*Colocasia esculenta* (L.) Schott), also known as eddoes or dasheen, which originated in the Bay of Bengal region of South-east Asia [5], had been known as one of antioxidant sources [6]. Taro is a vegetative propagated tropical root and important source of carbohydrates as and energy source [6]. The corm of taro is relatively low in protein (1.5%) and fat (0.2%) and this is similar to many other tuber crops. It is a good source of starch (70–80 g/100 g dry taro), fiber (0.8%), and ash

(1.2%). Taro is also a good source of thiamine, riboflavin, iron, phosphorus, and zinc and a very good source of vitamin B6, vitamin C, niacin, potassium, copper, and manganese [6].

**Figure 1.** Stem of *Colocasia esculenta* (L.) Schott.

Besides, ice cream is a dairy product that is much loved by the public, especially children, because of its sweet taste, soft texture, and high nutritional value [7]. The nutritional content of ice cream in every 100 g contains 12.5 g of fat, 4 g of protein, 20.6 g of carbohydrates, and 0.1 mg of iron [8,9]. The use of milk is essential in the ice cream mixture, where its fat content has a function to enhance the taste, soften the texture, help shape the ice cream, and give good melting properties to the ice cream [10]. Also, milk as an essential ingredient makes ice cream potential to be developed as a probiotic food [11].

Probiotic is a living bacteria given as a food supplement for humans and animals by improving their balance of the gut bacteria [12]. However, in recent years there has been researched on probiotics that have benefits for preventing gastrointestinal diseases and have other potential benefits such as antioxidant activity [13]. Antioxidants are indispensable to the body to cope and prevent oxidative stress [14]. The new formulation of ice cream with *C. esculenta* may improve the antioxidant needs in human body. Thus, the aim of this nutritional food study is to determine the antioxidant activity, phytochemical compounds and nutritional minerals of taro ice cream, a new approach of antioxidant improvement in ice cream.

## 2. Methods

This research was conducted by analyzing the antioxidant activity of *C. esculenta* stalk extract using the free radical scavenging method in the Chemistry laboratory of The Indonesian Institute of Sciences (LIPI). This method was characterized by 2,2-diphenyl-1-picrylhydrazyl (DPPH) and used a standard compound in the form of Quercetin. The scavenging characterization of *C. esculenta* stalk extracts was calculated using present equation below, where $Abs_{control}$ was the absorbance of DPPH + ethanol; $Abs_{sample}$ ws the absorbance of DPPH radical + sample.

$$Percentage\ of\ DPPH\ scavenging = \frac{Abs_{control} - Abs_{sample}}{Abs_{control}}\ x\ 100 \quad (1)$$

The phytochemical compound test on the stalk extract of *C. esculenta* was carried out using the thin layer chromatography (TLC) method at the Faculty of Pharmacy, Gadjah Mada University. Furthermore, ice cream was made with the formulation of 600 grams of *C. esculenta*, 30 grams of skim milk powder, 30 grams of vegetable creamer, 100 grams of granulated sugar, 150 ml of liquid milk, and

750 ml of water. Furthermore, we analysed the mineral nutrients, which were vitamin C, vitamin E, potassium and calcium, using spectrophotometry analysis and repeated twice.

## 3. Results

### *3.1. Antioxidant Activity*

The result of antioxidant activity was shown in table 1 and figure 1. The data showed that the antioxidant activity of the extract was $IC_{50}$ = 675.283. It meant that *C. esculenta* stem extract takes ± 675.283pg / mL to show the antioxidant activity (with the DPPH test as free radical). When it was compared to antioxidant activity of Quercetin, which had $IC_{50}$= 2.4, the antioxidant activity of *C. esculenta* stem extract was still included to low intensity. The difference in the antioxidant activity of these compounds could be caused by the differences in the characteristics of the tested compounds. Quercetin is a single compound and the *C. esculenta* stem extract is still a common extract.

**Table 1.** Regression Analysis of *C. esculenta* Stem Extract Antioxidant Inhibition.

| No | Sample | Absorbance | Consentration (ppm) | Inhibition (%) | $IC_{50}$ (pg/mL) |
|---|---|---|---|---|---|
| 1 | Blanco | 1.8326 | | | |
| 2 | Extract | 1.1661 | 400 | 36.369 | 675.283 |
| | | 1.3053 | 200 | 28.773 | |
| | | 1.437 | 100 | 21.587 | |
| | | 1.5503 | 10 | 15.404 | |

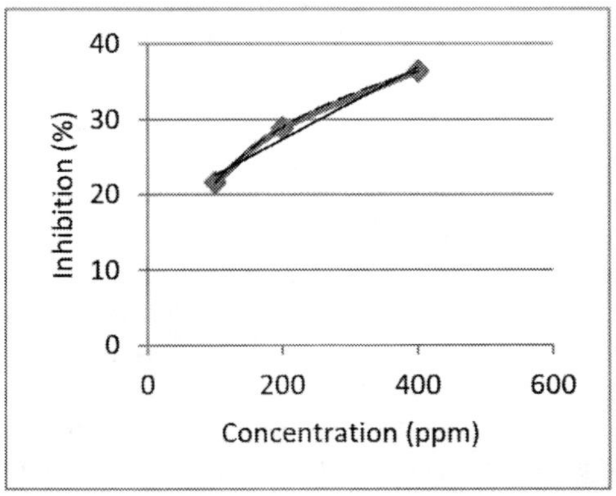


**Figure 1.** Linear Regression Graphic of *C. esculenta* Stem Extract Antioxidant Inhibition. Solid blue line showed linear regression result, solid black line showed standard compound and dotted black line showed log of the standard compound.

### *3.2. Phytochemical Compounds*

Phytochemical compounds of *C. esculenta* stem were analysed by Thin Layer Chromatography (TLC) assay. According to table 2, *C. esculenta* stem extract contained flavonoid, saponin, steroid, and terpenoid. Saponin had proportion in the extract at the most based on the observed spot and highest restriction factor (Rf) and steroid have highest hRx value compared to the comparison.

**Table 2.** Thin Layer Chromatography Result of *C. esculenta.*

| No. | Phytochemical Compund | Comparison | Result | Rf | hRx |
|---|---|---|---|---|---|
| 1. | Alkaloid | Quinin | - | - | - |
| 2. | Flavonoid | Quercetin | + | - | - |
| | | Rutin | + | - | - |
| 3. | Saponin | Saponin | ++ | 0.594 | 71 |
| 4. | Tannin | Tannin | - | - | - |
| 5. | Steroid | Stigmasterol | + | 0.45 | 83.72 |
| 6. | Phenol | Gallic Acid | - | - | - |
| 7. | Terpenoid | Tymol | + | 0.125 | 28.6 |

### 3.3. *Nutrient Compounds*

Taro stem ice cream were conducted within formulation ratio 6:1 of *C. esculenta* powder and sugar. Then, we analyzed the potential antioxidant compound of its mineral nutrients, which were vitamin C, vitamin E, potassium and calcium, using spectrophotometry analysis and repeated twice. The results of antioxidant mineral nutrients could be seen on table 3.

**Table 3.** Antioxidant Mineral Nutrients of Taro Ice Cream.

| Nutrient | 1$^{st}$ test | 2$^{nd}$ test | Average |
|---|---|---|---|
| Vitamin C | 658.62 mg/100 gr | 614.7 mg/100 gr | 636.66 mg/100 gr |
| Vitamin E | 0.5834 % | 0.5946 % | 0.589 % |
| Potassium | 0.2374 % | 0.3391 % | 0.288 % |
| Calcium | 3.1393 % | 3.2241 % | 0.318 % |

Based on nutritional minerals test that was done at Center for Food and Nutrition Studies, Universitas Gadjah Mada, *C. esculenta* contained vitamin C as much as 636.66 mg/100 gram samples. It also contained 0.589 % vitamin E, 0.288 % Potassium, and 0.318 % Calcium. These data showed that addition of taro stem could increase nutritional mineral on ice cream.

## 4. Discussion

*C. esculenta* is widely cultivated plant worldwide, especially in Indonesia. It has been reported that the leaf, tuber, and root of *C. esculenta* had potential role in various pharmaceutical activities [15]. But, there were still lack of information about *C. esculenta* stem nutrition and pharmaceutical activities. Recent studies showed that *C. esculenta* stem extract had an antioxidant activity in certain concentration. Antioxidants are molecules that could neutralize and limiting the process of free radical oxidation by terminating the chain reactions [16,17]. These free radicals are the cause of various health problems, such as aging, inflammatory diseases, cancers, and many others [18,19]. In this study, we characterized the antioxidant compounds by the phytochemical compounds of *C. esculenta* stem extract and mineral nutrients of ice cream, which added by *C. esculenta* stem formulation.

The first results showed that *C. esculenta* stem extract positively contained Flavonoid, Saponin, Steroid, and Terpenoid. Previous studies reported that Flavonoid acted as antioxidant [20], anti-inflammatory [21], anti-mutagenic [22], and anti-carcinogenic [23] properties to improve human health [24]. Saponin were also known as antioxidant properties that plays a role as hepato-protective, anti-tumor, anti-microbial, and anti-inflammatory activities [25]. Mooradian [26] also stated that some steroids had been related to significant antioxidant activities, such as 17β-estradiol, 17α-estradiol and estriol. Furthermore, other study also revealed the antioxidant activities of terpenoid [27]. These phytochemical compounds may represent the antioxidant properties of *C. esculenta* stem extract.

The result of mineral nutrient assays showed that the formulation of *C. esculenta* stem in ice cream contained potential antioxidant compound, which were vitamin C, vitamin E, potassium and calcium.

Several studies revealed the role of vitamin C as antioxidant agent that could prevent human diseases such as cardiovascular disease and cancer [28,29,30]. Vitamin E was also reported to plays a significant role as antioxidant agent [31]. It had a function to protect long chain poly-unsaturated fatty acids and thus maintain their concentrations for important signaling events against free radicals process [31]. Other previous study also showed the result of potassium and calcium roles as antioxidant properties. The research showed that potassium and calcium could enhanced the activity of antioxidant enzymes, which activates more than 50 enzymes [32]. These facts represent the potential role of *C. esculenta* stem formulation in ice cream to improve its antioxidant activity based on its mineral nutrients.

## 5. Conclusion

This study revealed the antioxidant activity *C. esculenta* stem extract ($IC_{50}$ = 675.283) but still included to low intensity. The antioxidant activity of *C. esculenta* stem extract may be regulated by its phytochemical compounds that were Flavonoid, Saponin, Steroid, and Terpenoid. The addition of *C. esculenta* stem formulation in ice cream also could improve the activity of antioxidant by mineral compounds, which were vitamin C, vitamin E, potassium and calcium.

## References


[1] Lobo V, Patil A, Phatak A, Chandra N 2010 Free radicals, antioxidants and functional foods: Impact on human health *Pharmacogn Rev* 4(8):118–26.

[2] Colville L. Seed Storage. In: Thomas B, Murray BG, Murphy DJBT-E of APS (Second E, editors. Oxford: Academic Press; 2017. p. 335–9. Available from: http://www.sciencedirect.com/science/article/pii/B9780123948076000800

[3] Chun O, Frei B, Gardner C, Alekel D, Killen J. Antioxidants in depth [Internet]. NCCIH. 2013. Available from: https://www.nccih.nih.gov/health/antioxidants-in-depth.

[4] Carlsen M H, Halvorsen B L, Holte K, et al 2010 The total antioxidant content of more than 3100 foods, beverages, spices, herbs and supplements used worldwide *Nutr J.* **9**:3. Published 2010 Jan 22. doi:10.1186/1475-2891-9-3

[5] O'Hair S K, Maynard D N VEGETABLES OF TROPICAL CLIMATES | Edible Aroids. In: Caballero BBT-E of FS and N (Second E, editor. Oxford: Academic Press; 2003. p. 5970–3. Available from: http://www.sciencedirect.com/science/article/pii/B012227055X012451

[6] Rashmi, Raghu, Gopenath, Palanisamy P, Bakthavatchalam P, Karthikeyan M, et al 2018 Taro (Colocasia esculenta): An overview *J Med Plants Stud.* **6**(4):156–61.

[7] Simanungkalit H, Indriyani, Ulyarti 2016 Kajian Pembuatan Es Krim dengan Penambahan Kacang Merah (Phaseolus vulgaris L) *Jurnal Penelitian Universitas Jambi Seri Sains* **18**(1):20-26.

[8] Astawan, M 2010 *Teknologi pengolahan pangan dan gizi* IPB (Bogor)

[9] Sandjaja A 2009 *Kamus Ilmu Gizi* Jakarta : PT Kompas Media Nusantara

[10] Octadayani E 2011 *Pengaruh Subtitusi Ubi Jalar (Ipomea batatas) dengan Susu Skim terhadap Pembuatan Es Krim* Faperta Universitas Hasanuddin (Makassar). Skripsi

[11] Saputri E 2015 *Pembuatan Es Krim Fungsional Dengan Bahan Baku Soyghurt dan Susu Rendah Lemak* Jom Faperta 3(1)

[12] Fuller R 1999 *Probiotics a Critical Review*. Wymondham, UK: Horizon Scientific, Probiotics for farm animals pp. 15–22.

[13] Daliri E B-M, Lee B H. New perspectives on probiotics in health and disease. Food Sci Hum Wellness [Internet]. 2015;4(2):56–65. Available from: http://www.sciencedirect.com/science/article/pii/S2213453015000300

[14] Kemenkes RI 2014 *Peran Antioksidan Bagi Kesehatan* Pusat Biomedis dan Teknologi Dasar Kesehatan Balitbangkes. Kemenkes RI

[15] Pawar H A, Choudhary P D, Kamat S R 2018 An Overview of Traditionally Used Herb, Colocasia

esculenta, as a Phytomedicine. *Med Aromat Plants* 07(02):1–7.

[16] Lakhtakia R, Ramji MT, Lavanya K, Rajesh K, Jayakumar K, Sneha C, et al 2011 The role of antioxidant in human health maintenance: Small molecules with infinite function *Int J Pharm Sci Res* **2**(6):1395–402.

[17] Kattappagari K K, Teja C S R, Kommalapati R K, Poosarla C, Gontu S R, Reddy B V R 2015 Role of antioxidants in facilitating the body functions: A review *J Orofac Sci*. **7**(2):71–5.

[18] Harman, D 1956 Aging: a theory based on free radical and radiation chemistry *Journal of Gerontology* **11** 298–300.

[19] Dreher D, Junoda A F 1996 Role of oxygen free radicals in cancer development *European Journal of Cancer* **32** 30-38.

[20] Srivastava N & Bezwada R 2015 Flavonoids: The Health Boosters. White Paper. Hillsborough, NJ: Indofine Chemical Company

[21] Hossain H, Jahan I A, Nimmi I, Hasan K 2012 Evaluation of Antinociceptive and Antioxidant Potential from the Leaves of Spilanthes paniculata Growing in Bangladesh *Int J Pharm Phytopharm Res* **1**(178):178–86.

[22] Snijman P, Swanevelder S, Joubert S, et al 2007 The antimutagenic activity of the major flavonoids of rooibos (*Aspalathus linearis*): some dose–response effects on mutagen activation–flavonoid interactions *Mutat Res* 631: 111–123.

[23] LeJeune T M, Tsui H Y, Parsons L B, et al. 2015 Mechanism of action of two flavone isomers targeting cancer cells with varying cell differentiation status *PLOS ONE* 10: e0142928.

[24] Panche A N, Diwan A D, Chandra S R 2016 Flavonoids: an overview *J Nutr Sci* **5**:e47. Published 2016 Dec 29. doi:10.1017/jns.2016.41

[25] Barky A El, Hussein S A 2017 Saponins-and-Their-Potential-Role-in-Diabetes-Mellitus *Diabetes Manag* 7:148–58.

[26] Mooradian A D 1993 Antioxidant properties of steroids *J Steroid Biochem Mol Biol* **45**(6):509–11.

[27] Graßmann J 2005 *Terpenoids as Plant Antioxidants* Vitam Horm 72(05):505–35.

[28] Buettner G R 1993 The pecking order of free radicals and antioxidants: lipid peroxidation, alpha-tocopherol, and ascorbate *Arch Biochem Biophys* 300:535–543.

[29] Halliwell B 1999 Vitamin C: poison, prophylactic or panacea? *Trends Biochem Sci* **24**:255–259.

[30] Padayatty S J, Katz A, Wang Y, Eck P, Kwon O, Lee J H, Chen S, Corpe C, Dutta A, Dutta S K, Levine M 2003 Vitamin C as an antioxidant: evaluation of its role in disease prevention *J Am Coll Nutr* Feb;22(1):18-35. doi: 10.1080/07315724.2003.10719272. PMID: 12569111.

[31] Traber M G, Atkinson J 2007 Vitamin E, antioxidant and nothing more *Free Radic Biol Med* 43(1):4-15. doi:10.1016/j.freeradbiomed.2007.03.024

[32] Siddiqui M H, Al-Whaibi M H, Sakran A M, Basalah M O, Ali H M 2012 Effect of calcium and potassium on antioxidant system of Vicia faba L. Under cadmium stress *Int J Mol Sci* 13(6):6604-6619. doi:10.3390/ijms13066604

# Production of Coconut Oil and Protein for Food and Cosmetic Ingredients

**Dianika Lestari*[1,2], Amilah Ridho Rahmani[2], and Danu Ariono[2]**

[1]Department of Food Engineering, Faculty of Industrial Technology, ITB, Indonesia
[2]Department of Chemical Engineering, Faculty of Industrial Technology, ITB, Indonesia

*Corresponding author's e-mail: dianika@che.itb.ac.id

**Abstract**. Coconut (Cocos nucifera) is a tropical plant that is almost spread throughout the archipelago. Coconut is a plant of the genus Cocos widely used in various fields ranging from food to cosmetics. Most coconut parts can be utilized so that the coconut is dubbed a versatile plant and has high economic value. This tree has a fruit which is also called coconut. Coconut fruit is the most widely used and processed. Coconut meat can be directly consumed or processed into coconut milk, coconut flour (desiccated coconut), and cooking oil. Coconut oil was obtained from dried or fresh coconut flour using the Soxhlet extraction method with organic solvents. Defatted coconut flour contained a high protein of approx. 19-20%. Coconut protein has a good quality compared to other vegetable protein sources. This research investigated solvent's effect on solid ratio and oil content on coconut protein extraction yield and recovery from coconut flour. Based on experiments, the highest value of extracted protein and yield was obtained after coconut flour extraction at a solvent to flour ratio of 30 g/g. Coconut protein isolate and coconut whey protein were expected to have different functional properties which need to be further evaluated to determine their application as cosmetics or food ingredients.
**Keywords:** coconut oil, coconut protein, protein extraction, food ingredients, cosmetic ingredients

## 1. Introduction

Proteins are macromolecules composed of 20 types of amino acids linked to each other by peptide bonds to form protein molecules. Amino acids can be classified as polar, nonpolar, hydrophobic, hydrophilic, acidic, alkaline, aliphatic, and aromatic. Proteins are the main structure in cells and tissues in the body. They have essential roles in growth, maintaining, repairing damaged tissue, producing metabolite and digestive enzymes, and certain primary hormone constituents. Protein sources can be obtained from animals and plants, vegetable protein [1]. As the largest archipelago country, Indonesia has the most prominent coconut (*Cocos nucifera*) plantation in the world of approx. 3,745,000 ha of a smallholder plantation and income for two and a half million farmer families. The coconut plantation in Indonesia constitutes 31.2% of the total world coconut area.

Coconut has a pulp with high potential to be developed as a food raw material with high economic value. Coconut meat protein has good quality due to its high amino acid score and low content of anti-nutritional compounds [2]. Processing of coconut into coconut flour was one of the alternatives to using coconut meat to increase economic value, maintain its nutritional value while increasing consumer convenience [3-6]. Coconut meat can be directly consumed or processed into coconut milk, coconut flour (desiccated coconut), and cooking oil. Coconut oil can be obtained from dried or fresh coconut flour using the Soxhlet extraction method with organic solvents. Defatted coconut meat will have higher protein content approx. 19-20%-wt [7]. Coconut flour is coconut meat dried, milled, and processed under hygienic conditions for human consumption [5]. The use of desiccated coconut is mainly for the confectionery and bread industry. Defatted coconut flour may contains 21.6% protein, 10.4% fibre, 59.8% carbohydrates, and 8.4% fat [4]. The high protein and fiber content in coconut flour allows this flour as a raw material for food products, especially for consumers at high risk of obesity, cardiovascular disease, and diabetes [4-6]. While coconut oil is already a common ingredient in food and cosmetics formulations, Patil and Benjakul [2] also showed that coconut protein was a natural emulsion stabilizer in coconut milk. Therefore, coconut protein can be used as healthy plant-based proteins and may be potential as an emulsifier or emulsion stabilizer in food and cosmetics products.

Based on the solubility of the protein, there are four types; albumin (soluble in water), globulin (soluble in salt), prolamin (soluble in alcohol), and glutelin (soluble in dilute NaOH). The type of protein that is mainly found in defatted coconut flour/meat is 40.1% globulin, 21% albumin, 14.4% glutelin-1, 4.8% glutelin-2, and 2.2% prolamin. Each type of protein is composed of amino acids at different levels. The amino acids present in the globulin are 14.9% arginine, 19.7% glutelin / glutamic acid, 7.2% leucine, and others. Meanwhile, the amino acids in albumin are 16.5% arginine, 24.1% glutamine, 6.3% linin, and others [2]. The high content of indigestible carbohydrates can inhibit protein digestibility; therefore, the protein was extracted and isolated from coconut flour to improve its digestibility. The extraction of the protein in coconut flour can be carried out using the solid-liquid extraction method.

Solid extraction (leaching) separates a solute in a solid by bringing the solid into contact with the solvent (solvent) to dissolves solutes into the solvent. In solid-liquid extraction, there are two phases: the overflow phase (extract) and the underflow phase (raffinate/dregs) [8]. Solid-liquid extraction involves three processes: 1) solute desorption and solute dissolution in the solvent, 2) internal diffusion of the solute from the matrix to the solid matrix surface, and 3) the external diffusion of the solute from the surface of the solid matrix into the bulk phase of the solvent [9]. Protein extraction was carried out at alkaline conditions far from its isoelectric point to increase protein solubility. Proteins that have been extracted in the solution can be separated by precipitation to obtain protein isolates. Protein was precipitated from crude extract at coconut protein isoelectric pH of about 4 [10]. At isoelectric pH, the protein will form a neutral charge. The attractive forces between protein molecules in the solution are more influential than the electrostatic repulsion so that the protein molecules will aggregate, then settle, and precipitate [11].

This study aimed to investigate the effect of solvent ratios and oil content of ingredients on the extractability of protein and recovery of protein isolates from full-fat and defatted coconut flour. The high content of indigestible carbohydrates in coconut flour can inhibit protein digestibility. Therefore, protein extraction from coconut flour can improve its digestibility for better utilization to fulfill the increased protein demand and other nutrients, which will benefit food security in Indonesia.

## 2. Materials and methods

### *2.1. Materials*

In this research, the materials used were commercial coconut flour, NaOH solutions, HCl; Aqua dm; hexane; Bradford assay kit reagent, and Bovine Serum Albumin protein standard.

### *2.2. Oil extraction from dried coconut flour*

A total of 50 g of dried coconut flour was placed in the thimble, and then the oil was extracted using a hexane solvent of 300 mL in a 500 mL round flask. Oil extraction was carried out by Soxhlet extraction method for four cycles to produce oil-free coconut flour. The result of the extraction process was solids containing hexane and hexane which were rich in oil. Coconut flour containing hexane was put into an aluminum foil cup, weighed, and dried in an oven at a temperature of $80^0$C for 1 hour. The mass of the container and coconut flour after drying were weighed to calculate the mass of hexane-free dry flour.

The oil-rich hexane phase was evaporated using a rotary evaporator at 30-40 °C until all hexane was removed from coconut oil.

### 2.3. *Extraction of protein from dried coconut flour*
About 1 g of oil-free dried coconut flour and oil-rich dried coconut flour were weighed and put into a 50 ml Falcon tube. Then alkali solvent (0.05 M NaOH) was added with a ratio of solvent to teas of 10, 20, and 30 g of solvent/gram of coconut flour. The mass of the container and coconut flour were weighed, and then the protein extraction was done using an orbital shaker for 30 minutes at a speed of 150 rpm. After that, the mixture was centrifuged at 6000 rpm for 20 minutes. The crude extract (supernatant 1) was poured into another pre-weighed 50 mL Falcon tube. The total mass of the mixture was calculated and measured in volume. The resulting residue was transferred to a pre-weighed aluminum foil cup and dried at 105°C overnight. The mass of the mixture after drying was weighed and used to calculate solution retention. Protein content from crude extracts was measured by the Bradford method [12].

### 2.4. *Isoelectric precipitation from crude extracts*
Protein crude extract was added with 0.1 M NaOH or 0.5 M HCL to adjust the pH into its isoelectric pH of 4 [10]. Then the mixture was left overnight at 4°C. The next day, the mixture was removed from the refrigerator and centrifuged at 6000 rpm for 20 minutes at room temperature to obtain a clearer supernatant. The supernatant was then filtered to another 50 mL pre-weighed Falcon tube, then the mass of the protein precipitate was weighed. The non-precipitated protein of the supernatant was measured by the Bradford method [12].

## 3. Results and discussion

### 3.1. *Drying and oil extraction from coconut flour*
The coconut flour was dried to reduce its moisture content before the oil and protein extraction process. Moisture might lessen the effectivity of the nonpolar hexane solvent to extract oil. Hexane is a nonpolar solvent, and the presence of water will interfere with the interaction between the hexane solvent and oil [4]. The drying process was carried out at a temperature of 40°C to prevent physical damage to the flour, such as the binding of lipids to protein and carbohydrates, decomposition of carbon, and oil entry into protein. In general, the moisture content in the oil can cause rancidity and is therefore disliked. Oil extraction was carried out with dried coconut flour using Soxhlet extraction to facilitate repeated contact between the solvent and coconut flour. We obtained 30.5 g oil from 50 g coconut flour or about 61% recovery in this experiment. Oil content in defatted coconut flour obtained from the experiments was comparable to Ohtora [13]. The moisture content and oil content of coconut flour are shown in Table 1.

**Table 1.** Moisture and oil content of coconut flour.

| Coconut flour | Moisture content (%-wt) | Oil content (%-wt) | Protein content (%-wt) |
|---|---|---|---|
| Full-fat-coconut flour (FCF) | 5.3 | 64.5 | 6.9 |
| Defatted-coconut flour (DCF) | 3.6 | 5.3 | 18.1 |
| Defatted-coconut flour from the cold press [13] | 4.2 | 8.4 | 21.6 |

### 3.2. *Production of protein isolate from coconut flour*
Protein isolate (90-95% purity) is produced by dissolving proteins from plant material or cells, followed by precipitation at its isoelectric pH. The protein extraction process was carried out on defatted-coconut flour and full-fat-coconut flour at alkaline conditions using 0.05 M NaOH solution (pH of slurry 11.0-11.4). In alkaline conditions, the OH- ion will bind H+ ions to amino acids as protein monomers, causing

interactions [14]. In opposite, protein then is recovered from the crude extract by precipitation at isoelectric pH of proteins, where the proteins have the lowest solubility. Block flow diagram of protein extraction from coconut flour is shown in Figure 1, while the mass balance calculation is presented in Table 2.

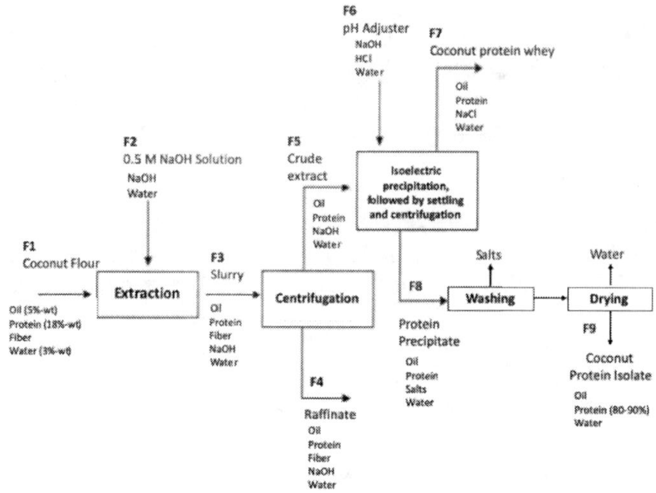


**Figure 1.** Block Flow Diagram of Protein extraction from coconut flour.

**Table 2.** Mass Balance of protein extraction from coconut flour.

| Coconut Flour | Solvent to flour ratio (g/g) | Coconut flour (g) | NaOH solution (g) | Slurry (g) | Raffinate (g) | Crude Extract (g) | HCL (g) | Coconut whey protein (g) | Wet Protein Isolate (g) | Coconut protein isolate (g) |
|---|---|---|---|---|---|---|---|---|---|---|
| | | F1 | F2 | F3 | F4 | F5 | F6 | F7 | F8 | F9 |
| Full fat-coconut flour (FCF) | 10 | 1 | 10 | 10.9 | 6.3 | 3.5 | 0.83 | 5.3 | 0.1 | 0.017 |
| | 20 | 1 | 20 | 21.1 | 6.3 | 14.7 | 3.19 | 18 | 0.9 | 0.150 |
| | 30 | 1 | 30 | 31 | 7.1 | 23.2 | 4.84 | 25.9 | 0.2 | 0.033 |
| Defatted Coconut flour (DCF) | 10 | 1 | 10 | 10.9 | 7.8 | 2.74 | 0.59 | 4.3 | 0.24 | 0.024 |
| | 20 | 1 | 20 | 20.8 | 8 | 10.08 | 0.53 | 7.4 | 1.07 | 0.107 |
| | 30 | 1 | 30 | 30.7 | 10.9 | 20.18 | 11.80 | 30.1 | 1.08 | 0.108 |

### *3.3. Protein extraction from coconut flour*

The effect of solvent to flour ratio on protein extractability in crude extract flow (F5 in Figure 1) is shown in Figure 2. Solvent to flour ratio of 30 g/g resulted in the highest protein extractability and yield. The higher amount of solvent increased the contact with the solid matrix of flour and increased protein dissolution from the solid matrix. Figure 2a showed that protein extractability from full fat-coconut flour (17 to 53 g protein extracted/ 100 g protein in flour) was about two times higher than the protein extractability from defatted-coconut flour (5 to 22 g protein extracted/ 100 g protein in coconut flour). The increased solvent to flour ratio from 10 to 30 g/g increased protein extractability up to three-times for both full fat or defatted coconut flour. Table 1 presented the protein content of full-fat coconut flour based on the information in the packaging (6.9 %-wt), which was well corresponded to the result of Samuel et al. [3], and the protein content of defatted-coconut flour based on calculation (18.1 %-wt.). High protein content in flour was occurred due to the removal of oil from the flour, according to the experiment of Gunathialake et al. [4]. After oil removal, coconut flour contained higher protein content and higher fiber content, which may increase the difficulties of protein extraction. Based on the solubility, protein can be classified into four types; albumin (water-soluble), globulin (salt soluble), a prolamin (soluble in alcohol), and glutelin (soluble in dilute NaOH). The type of protein that is mostly found in defatted coconut flour/ meat is 40.1% globulin, 21% albumin, 14.4% glutenin-1, 4.8% glutelin-2, and 2.2% prolamin [2]. In this experiment, the dilute alkaline solvent was expected to extract most albumin, globulin, and glutenin fractions.

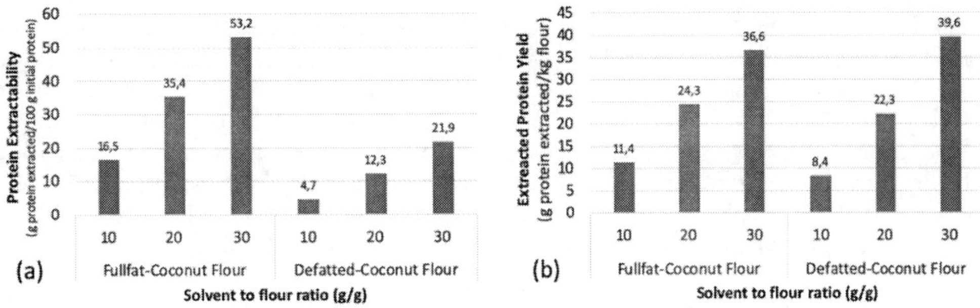


**Figure 2.** The effect of solvent to feed ratio on (a) protein extractability and (b) protein yield from coconut flour in crude extract (stream F5 in Figure 1).

Figure 2b showed that the yield of protein extracted from full-fat coconut flour at a solvent to flour ratio of 10 and 20 g/g was slightly higher than that from defatted coconut flour, while the yield of protein extracted from defatted coconut flour at 30 g solvent/g coconut flour (40 g protein extracted/ kg flour) was slightly higher than full fat-coconut flour (37 g protein extracted/ kg flour). Although protein extractability was much lower from the defatted coconut flour, the protein yield from flour for both full-fat and defatted coconut flour was comparable. The higher the solvent ratio to the solid, the greater the concentration gradient and caused an increase in the diffusivity of the compound from the solid material into the solvent. The contact surface area between the solvent and the material also increased with the amount of the solvent so that the amount of extracted protein also increased. These results corresponded to the research conducted by Cui *et al.* [15], which stated that solvent's ratio to solids affected the amount of protein extracted. Protein interaction with other compounds, such as oil and fiber, seemed to affect protein extractability. Residual oil content in coconut flour increased protein dissolution to solvent, while fiber interaction with proteins reduced protein solubility in the solvents. The higher protein yield from defatted coconut flour was caused by the higher initial amount of proteins (18 %-wt.) compare to the protein content in full-fat coconut flour (7 %-wt.).

### *3.4. Retention of solution*

The solution's retention is defined as the amount of (extract) solution carried to the extraction residue/ underflow (extracted coconut flour). The solvent moved to the residue contains dissolved protein so that high retention in underflow reduced the amount of recovered protein in the extract bulk phase. Solution retention analysis data are presented in Table 3. Solution retention in defatted coconut flour underflow (7-10 g crude extract/ g inert solid) was higher than full-fat coconut flour (5-6 g crude extract/ g inert solid). Residual oil in coconut flour might facilitate protein diffusion into the solvent, while fiber binds protein and reduces protein diffusivity into the solvent.

**Table 3.** Protein solution retention in underflow/raffinate flow (coconut flour residue).

| Coconut flour | Solvent to flour ratio (g/g) | Crude Extract Retention (g crude extract/g inert solid) |
|---|---|---|
| Full-fat-coconut flour (FFCF) | 10 | 5.5 |
| | 20 | 6.1 |
| | 30 | 5.6 |

| | | |
|---|---|---|
| Defatted-coconut flour (PDCF) | 10 | 6.8 |
| | 20 | 10.4 |
| | 30 | 9.8 |

### *3.5. Protein Precipitation from crude extract at isoelectric pH*

Coconut flour crude extract was precipitated at coconut protein isoelectric point of pH 4.00 [10] with the addition of HCl or NaOH solutions at a specific concentration to adjust the solution's pH containing coconut protein. The isoelectric point of protein is the pH value where the protein molecule has zero net charged so that protein coagulated and precipitated [11]. Table 4 showed that protein precipitation rate was higher in defatted coconut flour crude extracts (44-68%) than in full-fat coconut flour crude extracts (27-44%). As a result, the protein isolates yield from defatted coconut flour (4-27 g protein/ kg flour) was higher than the protein isolate from full-fat coconut flour (5-11 g protein/kg flour). Protein in the crude extract was not completely precipitated at the experimental range of isoelectric pH (3.98 to 4.23). The fraction of unprecipitated proteins remained soluble in the whey fraction (stream F8 in Figure 1).

**Table 4.** Isoelectric precipitation of coconut protein from crude extract at pH 4 to produce coconut protein isolate (stream F7 in Figure 1).

| Coconut flour | Solvent to flour ratio (g/g) | Crude Extract Protein concentration (mg /ml) | Whey Protein concentration (mg/ml) | Protein precipitation rate (%) | Protein isolation rate (g protein/ 100 g protein in flour) | The yield of Protein Isolate (g protein/ kg flour) |
|---|---|---|---|---|---|---|
| Full-fat-coconut flour (FCF) | 10 | 2.53 | 1.42 | 43.9 | 7.3 | 5.0 |
| | 20 | 1.75 | 1.28 | 26.9 | 9.5 | 6.5 |
| | 30 | 1.5 | 1.05 | 30.0 | 16.0 | 11.0 |
| | | | | | | |
| Defatted-coconut flour (DCF) | 10 | 2.64 | 1.47 | 44.3 | 2.1 | 3.7 |
| | 20 | 2.32 | 0.36 | 84.5 | 10.4 | 18.8 |
| | 30 | 1.96 | 0.62 | 68.4 | 15.0 | 27.1 |

Due to the higher protein precipitation rate from defatted coconut flour crude extract, protein isolation rate from full-fat coconut flour (7-16 g protein/100 g protein in flour) was similar to that from defatted coconut flour (2-15 g protein/100 g protein in flour). Therefore, the yield of coconut protein isolates from full-fat coconut flour of approx. 5 - 11 g protein/kg coconut flour was lower the yield of coconut protein isolates from defatted coconut flour of approx. 4 - 27 g protein/ kg coconut flour.

## 4. Conclusion

Experimental data showed that the increase of solvent to coconut flour ratio increased protein extractability and yield of protein extracted. The highest protein recovery was obtained from defatted flour at a solvent ratio of 30 g solvent/ g coconut flour. The yield of coconut protein isolates from defatted coconut flour were within the range 4 - 27 g protein/ kg coconut flour. Both coconut protein isolate and coconut whey protein were expected to have different functional properties, which need to be further evaluated to determine their application as cosmetics or food ingredients.

**Acknowledgment**
This research was funded by P3MI Research Grant 2020 provided by Institut Teknologi Bandung.

## References

[1] Khan R H, Siddiqi M K, Salahuddin P 2017 Protein Structure and Function *Interdisciplinary Biotechnology* Unit 1–39
[2] Patil U, Benjakul S 2017 Characteristics of albumin and globulin from coconut meat and their role in emulsion stability without and with proteolysis *Food Hydrocolloids* **69** 220–228

[3] Samuel A O 2018 *Valorization Strategies of Coconut Flour* Faculty of Bioscience Engineering
[4] Gunathilake K D, Prasanna A A N, Kumara C Yalegama 2009 Use coconut flour as a source of protein and dietary fibre in wheat bread *Asian Journal of Food and Agro-Industry 3* **2** (3) 386-395
[5] Trinidad P T, Aida C M, Divinagracia H V, Anacleta S L, Faridah C A, Joan C C, Rosario R E, Dina B M, Angelica S M and Modesto T C 2006 Dietary fiber from coconut flour: a functional food *Innovative Food Science and Emerging Technologies* **7** (4) 309–317
[6] Kumolontang Nova 2014 *Tepung Kelapa sebagai Substituen Parsial dalam Pembuatan White Bread* Manado: Balai Riset dan Standardisasi Industri Manado.
[7] Kwon K, Park K H, and Rhee K C 1996 Fractionation and characterization of proteins from coconut (Cocos nucifera L.) *Journal of Agricultural and Food Chemistry* **44** (7) 1741– 1745
[8] S J, Samson A S, Khaund R N, Cater C M, and Mattil K F 1971 Extractability of Coconut Proteins *Journal of Food Science* **36** 725-728
[9] Albert Ibarz and Gustavo V. Barbosa-Cánovas 2003 Chapter 21 Solid-Liquid Extraction, *Unit Operations in Food Engineering*, CRC Press.
[10] Rodsamran P and Sothornvit R 2018 Physicochemical and functional properties of protein concentrate from byproduct of coconut processing *Food Chemistry* **241** 364–371
[11] Audain et al. 2015 *Accurate estimation of isoelectric point of protein and peptide-based on amino acid sequences* Oxford: Oxford University Press
[12] Bradford M M 1976 A rapid and sensitive method for the quantitation of microgram quantities of protein utilizing the principle of protein-dye binding *Analyst. Biochem* **72** 248–254
[13] Okhtora Angelia 2016 *Analisis Kadar Lemak pada Tepung Ampas Kelapa* Unit Penelitian dan Pengabdian Masyarakat, Politeknik Gorontalo
[14] Zhang et al. 2011 *An improved method for whole protein extraction from yeast Saccharomyces cerevisiae* New York: John Wiley and Sons, Inc
[15] Cui Qiaoyun, Ni X, Zeng L, Tu Z, Li J, Sun K, Chen X, and Li X 2017 Optimization of Protein Extraction and Decoloration Conditions for Tea Residues *Horticultural Plant Journal* 172-176

# Preliminary Evaluation of Halal Protein Hydrolysate Production in Indonesia

**Made Tri Ari Penia Kresnowati*, Cantika Rahayu Affandi, Cindi Pratiwi**

Food and Biomass Processing Technology Research Group, Chemical Engineering, Faculty of Industrial Technology, Institut Teknologi Bandung

*kresnowati@che.itb.ac.id

**Abstract**. Protein hydrolysate is widely used in industry, for example as a substrate for microbial fermentation. With respect to the halal certification of the final fermentation products, the halal status of all media components, including protein hydrolysate, needs to be clarified as well. Indonesia has abundant protein-rich natural resources as well as protein-rich industrial by-products that have not been utilized optimally. Industrial production of halal protein hydrolysate has been overlooked. This research explored the potential of using protein-rich industrial/agricultural by-products, such as cassava leaves, soybean waste (tofu and soy sauce dregs), cow waste (bones and cow skin), fish waste (fish bones and skin), chicken waste (chicken feet and skin), cheese whey, and corn steep liquor, to be processed using halal and green processes to produce halal protein hydrolysate. The best combination of raw material and protease was obtained by simulating the breaking of the peptide bond of the raw material by a protease to determine the effectiveness of the protein hydrolysis process of each combination. Further simulations were carried out using the Analytical Hierarchy Process (AHP) method, to consider the availability/accessibility of raw materials, the protein content of the raw materials, the cleavage of peptide bonds by the enzyme, the price of the enzymes, and the ease of processing which included pre-treatment of raw materials.

## 1. Introduction
Protein hydrolysate is one of the essential raw materials in the development of fermentation products and biotechnology. Currently, protein hydrolysate is widely used in the food and pharmaceutical industries to produce flavorings, drugs, and supplements. Protein hydrolysate is also used in the microbial growth media for Nitrogen source [1]. In its application, the halalness of a protein hydrolysate is important to be considered, especially in Indonesia, where the majority of the population is Muslim. The critical points for the halalness of protein hydrolysate are in the raw material and the hydrolyzing enzyme (protease) used, as well as the potential of cross-contamination with non-halal materials during the production process. To ensure that, the whole production process, from raw materials to the packaging of the end products, needs to be carefully considered. Despite the big market for protein hydrolysate in Indonesia, a commercial scale protein hydrolysate industry has not been established in Indonesia until now. According to data from Statistics Indonesia, in 2018 Indonesia imported 4,014,254 kg of protein hydrolysate with a value of 20,025,992 US $ [2].

On the other hand, Indonesia has abundant protein-rich industrial/agricultural by-products which have not been utilized optimally including cassava leaves, the waste from soybean derivate products (such as tofu and soy sauce), the waste from slaughtering house, the waste from fish canning industry, cheese whey, and corn steep liquor. Food processing mostly leads to waste or by-products such as industrial cheese (85-90%), corn starch (41-43%), beef slaughter (40-52%), poultry slaughter (31-38%), and fish' filleting, salting, preserving, and smoking (50-75%) [3].

One of the methods used to produce protein hydrolysate is through the enzymatic hydrolysis process using proteases. Proteases that can be used to produce protein hydrolysate include papain, bromelain, ficin, rennin, pepsin, and thermolysin. Literature review shows that papain can be used to hydrolyze yellowtail fish (*Caesio cuning*) waste [4], bromelain can be used to hydrolyze nyamplung seeds (*Calophyllum inophyllum*) [5], pepsin can be used to hydrolyze cow and pork skin [6].

This study aimed to evaluate the potential of protein hydrolysate production in Indonesia by determining the best combination of protein-rich industrial/agricultural by-products as the raw materials and proteases for the development of industrial-scale protein hydrolysate and further evaluate the potential of this process to be industrialized commercially.

## 2. Methodology

This research was generally divided into several stages including analyzing the amino acid composition of raw materials, determining the peptide bond cleavage site by proteases, simulating the protein hydrolysis process, determining the optimum protein hydrolysate production process, that was the combination of raw material and protease, using the Analytical Hierarchy Process (AHP) method; and analyzing of the economics. Secondary data from various literature were used throughout the research.

Information regarding the amino acid composition of each raw material was obtained from various literature by adjusting the data of amino acid composition units that have been collected to grams of amino acids per 100 grams of protein. When more than one piece of literature was found, the average value was used.

The determination of the specific cleavage site of protease was carried out by gathering information on the preferred amino acids for cleaving the peptide bond by a protease from various references. Due to the broad specificity of the action of some proteases, the preferential amino acid used in the study was the amino acid mentioned in all literature.

The protein hydrolysis process was simulated by utilizing the raw material amino acid composition data and the specific cleavage site of each protease, to estimate the efficiency of the hydrolysis process. The degree of hydrolysis was calculated by summing up the amino acid content of the raw material which was the site for cleaving the peptide bonds by proteases for each variation of the raw material.

Determination of optimum protein hydrolysate production process, which was the combination of raw material and protease used in the process, was further conducted using the AHP method. Information related to the required data was obtained from various literature such as books, journal articles, proceedings, publications from institutions, news media, and personal information from reliable sources.

The economic added value of the selected raw material-protease combination from the AHP method was then analyzed by determining the difference between the commercial protein hydrolysate price and the raw material price used to produce the protein hydrolysate in the same amount.

## 3. Results and Discussion

### 3.1. *Amino Acid Composition of Various Protein-Rich Industrial/Agricultural Byproducts*

Eleven types of protein-rich industrial/agricultural byproducts were evaluated as the potential raw materials for the protein hydrolysate production. They were cassava leaves, tofu waste, soy sauce solid waste, the waste from fish canning industry: fish bones and fish skin; the waste from beef and poultry slaughtering house: beef bones, cow skin, chicken feet, chicken skin; cheese whey, and corn steep liquor. Each of these raw materials has a unique amino acid composition which can affect the effectiveness of the protein hydrolysis process in the production of protein hydrolysate. The amino acid compositions of these materials are summarized in Table 1.

**Table 1.** Amino acid composition of raw materials (grams of amino acids per 100 grams of protein).

| Amino acid | Cassava leaves[a] | Tofu waste[b] | Soy sauce solid waste[c] | Fish bones[d] | Fish skin[e] | Beef bones[f] | Cow skin[g] | Chicken feet[h] | Chicken skin[i] | Cheese whey[j] | Corn steep liquor[k] |
|---|---|---|---|---|---|---|---|---|---|---|---|
| Alanine (Ala) | 5.39 | 4.28 | 0.68 | 7.28 | 10.38 | 8.52 | 8.18 | 2.39 | 8.32 | 4.20 | 10.17 |
| Arginine (Arg) | 5.01 | 7.62 | 0.40 | 7.80 | 7.54 | 7.67 | 7.46 | 3.89 | 7.57 | 2.05 | 12.62 |
| Asparagine (Asn) | N/A | N/A | N/A | N/A | 6.64 | 7.12 | N/A | N/A | N/A | N/A | N/A |
| Aspartic Acid (Asp) | 9.27 | 10.42 | 2.54 | 8.09 | 5.76 | 4.78 | 4.87 | 8.14 | 5.91 | 9.31 | 4.16 |
| Cysteine (Cys) | 1.04 | 1.41 | N/A | 1.57 | N/A | 0.13 | N/A | N/A | N/A | 1.51 | 3.22 |
| Glutamic Acid (Glu) | 11.69 | 18.97 | 3.23 | 11.56 | 10.14 | 9.77 | 8.93 | 10.63 | 9.59 | 17.92 | 11.97 |
| Glutamine (Gln) | N/A | N/A | N/A | N/A | N/A | N/A | N/A | N/A | N/A | N/A | N/A |
| Glycine (Gly) | 7.09 | 4.17 | 0.32 | 15.15 | 21.52 | 22.23 | 20.60 | 15.20 | 20.26 | 1.64 | 4.56 |
| Histidine (His) | 2.22 | 2.76 | N/A | 2.05 | 0.88 | 0.98 | 0.72 | 4.97 | 0.74 | 1.27 | 19.37 |
| Hydroxyproline (Hyp) | N/A | N/A | N/A | 4.81 | 5.68 | 11.51 | 9.80 | 11.10 | 11.36 | N/A | 0.23 |
| Isoleucine (Ile) | 4.73 | 4.99 | 4.03 | 2.72 | 1.40 | 1.46 | 1.38 | 5.56 | 1.48 | 2.89 | 2.11 |
| Leucine (Leu) | 8.49 | 7.96 | 7.08 | 4.88 | 2.58 | 3.08 | 2.89 | 9.46 | 3.25 | 8.90 | 6.21 |
| Lysine (Lys) | 5.93 | 6.54 | 6.16 | 5.25 | 3.82 | 3.38 | 2.76 | 5.84 | 3.21 | 6.52 | 1.54 |
| Methionine (Met) | 1.90 | 1.26 | 0.53 | 2.46 | 2.25 | 0.89 | 0.47 | 4.27 | 1.12 | 1.95 | 3.39 |
| Phenylalanine (Phe) | 5.59 | 5.30 | 5.00 | 2.92 | 2.51 | 2.25 | 1.96 | 4.99 | 2.76 | 3.35 | 1.91 |
| Proline (Pro) | 4.83 | 4.83 | 0.82 | 7.65 | 12.04 | 12.03 | 21.23 | 3.32 | 15.12 | 6.76 | 7.86 |
| Serin (Ser) | 4.62 | 5.19 | N/A | 5.00 | 3.06 | 3.22 | 3.11 | N/A | 2.67 | 5.52 | 1.42 |
| Treonine (Thr) | 4.43 | 3.99 | 4.50 | 3.55 | 2.84 | 2.10 | 1.84 | N/A | 2.70 | 5.28 | 2.17 |
| Tryptophan (Trp) | 2.31 | 1.37 | N/A | 0.47 | N/A | N/A | N/A | N/A | N/A | 2.05 | 0.11 |
| Tyrosine (Tyr) | 4.06 | 3.93 | 3.35 | 2.22 | 0.90 | 0.75 | 0.60 | 5.53 | 0.82 | 2.97 | 1.65 |
| Valine (Val) | 5.65 | 4.79 | 6.13 | 3.50 | 2.06 | 2.32 | 1.90 | 4.69 | 2.07 | 3.32 | 4.44 |

Resource: [a]processed [7, 8]; [b]processed [9, 10, 11, 12, 13]; [c][14]; [d]processed [15, 16]; [e]processed [17, 18]; [f]processed [17, 19, 20]; [g][21]; [h][22]; [i][21]; [j][23, 24]; [k]processed [25, 26].

Table 1 shows that the amino acid composition varies with the sources (Table 1). Cheese whey is dominated by glutamic acid. Corn steep liquor is dominated by the amino acid histidine. The amino acid composition of cassava leaves is the average of amino acid compositions of the very young, young, and mature cassava leaves [7,8]. It is dominated by glutamic acid, aspartic acid, leucine, and glycine (Table 1).

The amino acid composition of tofu waste is very similar to those of soybeans, dominated by glutamic acid, aspartic acid, and leucine (Table 1). Although soy sauce is produced from soybeans, the different amino acid composition is found in soy sauce solid waste. The latter is dominated by leucine, lysine, and valine (Table 1). According to Fishanda [27], the lysine and methionine content of the soybean decrease during the fermentation process for soy sauce production. Tryptophan is also denatured in this process.

Wastes from animal body parts including fish bones and skins, beef bones and cow skins, as well as chicken feet and skin are dominated by collagen fiber so that they contain lots of glycine, proline, and hydroxyproline. Fish bones and skin are both dominated by the amino acid glycine, but the glycine content in fish skin is higher than fish bones. Fish skin also contains high proline, glutamic acid, and alanine, while in fish bones the highest amino acids after glycine are glutamic acid, aspartic acid, and arginine. Beef bones and cow skin are both dominated by proline and glycine, but the proline content of cow skin is higher than that of beef bones, while the glycine content of both is relatively similar. Chicken feet and skin are dominated by the amino acid glycine, but the content in chicken skin is higher, as is the case with fish bones and fish skin.

### 3.2. *Specific Cleavage Site of Protease*

Each protease has its own specificity, as well as different cleavage strengths. For example, papain cleaves strongly in phenylalanine while weak in cysteine [28]. Therefore, in this study, an analysis of specific cleavage of proteases was carried out from several data to obtain strong cleavage for each protease. The specific cleavage sites of each protease are summarized in Table 2.

**Table 2.** The specificity of proteases (based on cleavage site).

| Protease | Cleavage site |
|---|---|
| Papain | arginine, phenylalanine, lysine[a] |
| Bromelain | alanine, glycine, lysine, tyrosin[b] |
| Ficin | glycine, lysine, serine, tyrosin[c] |
| Rennin | alanine, phenylalanine, tyrosin[d] |
| Pepsin | phenylalanine, leucine, tyrosin[e] |
| Thermolysin | alanine, phenylalanine, isoleucine, leucine, methionine, valin[f] |

Resource: [a]processed [28, 29, 30, 31]; [b]processed [31, 32, 33]; [c]processed [31, 34, 35]; [d]processed [28, 35, 36, 37]; [e]processed [28, 38, 39, 40]; [f]processed [40, 41, 42, 43].

Plant-derived proteases have very broad specificity compared to other sources. Papain has the tendency to cleave peptide bonds in alkaline amino acids such as arginine, lysine, and phenylalanine. Bromelain typically cleaves peptide bonds in alanine, lysine, glycine, and tyrosine. Ficin has several preferential amino acid similarities with bromelain, those are lysine, glycine, tyrosine, and serine. On the other hand, animal protease has a narrow specificity. Rennin cleaves peptide bonds in alanine, phenylalanine, and tyrosine. Pepsin, another animal protease, cleaves peptide bonds with a specificity similar to rennins in hydrophobic amino acids such as leucine, phenylalanine, and tyrosine. Thermolysin is a microbial protease, produced by *Bacillus sp.* Thermolysin cleaves peptide bonds at isoleucine, leucine, valine, alanine, methionine, and phenylalanine. The information regarding the cleavage sites of each protease is used to simulate the hydrolysis process.

### 3.3. *Hydrolysis Process Simulation*

The efficiency of the hydrolysis process is indicated by the degree of hydrolysis, which is the proportion of cleaved peptide bonds in a protein hydrolysate [44]. A high degree of hydrolysis indicates a more proportion of cleaved peptide bonds, thus the higher efficiency of the hydrolysis process. Based on the amino acid composition of the raw materials (Table 1) and the specificity of the protease (Table 2), the maximum/theoretical degree of hydrolysis was simulated. The results are presented in Table 3.

**Table 3.** The estimated theoretical degree of hydrolysis of various raw materials hydrolyzed by specific protease (%)

| Protease | Cassava leaves | Tofu waste | Soy sauce solid waste | Fish bones | Fish skin | Beef bones | Cow skin | Chicken feet | Chicken skin | Cheese whey | Corn steep liquor |
|---|---|---|---|---|---|---|---|---|---|---|---|
| Papain | 16.5 | 19.5 | 11.6 | 16.0 | 13.9 | 13.3 | 12.2 | 14.7 | 13.5 | 11.9 | 16.1 |
| Bromelain | 22.5 | 18.9 | 10.5 | 29.9 | 36.6 | 34.9 | 32.1 | 29.0 | 32.6 | 15.3 | 17.9 |
| Ficin | 21.7 | 19.8 | 9.8 | 27.6 | 29.3 | 29.6 | 27.1 | 26.6 | 27.0 | 16.7 | 9.2 |
| Rennin | 15.0 | 13.5 | 9.0 | 12.4 | 13.8 | 11.5 | 10.7 | 12.9 | 11.9 | 10.5 | 13.7 |
| Pepsin | 18.1 | 17.2 | 15.4 | 10.0 | 6.0 | 6.1 | 5.5 | 20.0 | 6.8 | 15.2 | 9.8 |
| Thermolysin | 31.8 | 28.6 | 23.5 | 23.8 | 21.2 | 18.5 | 16.8 | 31.4 | 19.0 | 24.6 | 28.2 |

Table 3 shows that all raw materials can be hydrolyzed by all types of proteases in spite of the different efficiency. Each of the raw material-protease combinations results in a unique degree of hydrolysis due to differences in the amino acid composition and the specificity of the protease.

Cassava leaves, tofu waste, and soy sauce solid waste were most efficiently hydrolyzed by thermolysin giving the degree of hydrolysate of 31.7, 28.6, and 23.5% correspondingly. The cleavage of peptide bonds in these three raw materials occurs mostly in leucine.

Fish bones, fish skins, beef bones, cow skin, and chicken skin are most efficiently hydrolyzed by using bromelain or ficin. This is due to their high content of glycine, which is one of the cleavage sites of bromelain and ficin. Hydrolysis of chicken feet by thermolysin, bromelain, and ficin proteases were relatively high giving the degree of hydrolysis of 31.4, 29.0, and 26.6%, respectively.

The amino acid composition of leucine, phenylalanine, isoleucine, valine, and methionine in chicken feet is as such that it provides high efficiency of thermolysin hydrolysis (Table 2). The alanine content in chicken feet is much smaller than in other animal-derived waste (Table 2) thereby its degree of hydrolysis by bromelain is also lower than the five other animal-derived waste discussed previously.

Thermolysin also gives the highest degree of hydrolysis for cheese whey and corn steep liquor (Table 3). This is related to the high content of leucine in cheese whey and the high content of alanine in corn steep liquor (Table 1), whereas leucine and alanine are the specific cleaving site of thermolysis (Table 2).

### *3.4. Determination of the Optimum Protein Hydrolysate Production Process*

The AHP method was used to obtain the optimum protein hydrolysate production process, which was the combination of raw material and protease used in the process, was further conducted using the AHP method. Between the five criteria used in the decision-making process, the protein content of the raw materials (Table 4) and the protein hydrolysis process were considered to be the most important as they directly influenced the yield and the quality of the product. High protein content of the raw material would give high product yield. The efficient hydrolysis process, which is the optimum combination of raw material and type of peptone used, would lead to high degree of hydrolysis (Table 3) or good protein hydrolysate quality.

The availability of raw materials is an important factor for the sustainability of the protein hydrolysate industry. Two sets of important data were considered here: the productivity of the raw materials (Table 4), that was the availability of this raw material in Indonesia, and the resource capacity (Table 4), which indicated the accessibility of the raw materials. As an example, the national scale availability of tofu waste was 1,062,284 ton/year however, a protein hydrolysate industry can only process the waste from the nearby tofu industries, which was around 3 ton/month. The availability (productivity and accessibility) of raw materials has a lower level of importance than the two previously described criteria because basically the protein hydrolysate industry can be set up on a small scale. Nonetheless, the high availability of raw materials is important for scaling up the industry.

The ease of processing criterion had a lower level of importance. Even though the ease of processing, for example, the types and number of related pretreatment processes, would affect the overall production costs, it impacts less than the accessibility of the raw materials and the sustainability of the industry.

Of least important was the price of protease. Despite the small amount of protease that was used in the process, the price of enzyme/protease commonly significantly contribute to the production cost. The price of protease varies a lot: bromelain, papain, and ficin are relatively cheap, whereas rennin and thermolysin may reach 367,000 US $ (per 100 grams of protease).

The price of raw materials was not considered here because all evaluated raw materials were industrial by-products and the development process would provide an added value, supporting the circular economy. Based on these considerations, the weight of importance of all the criteria is presented in Table 5.

**Table 4.** Protein composition, productivity, accessibility, and the pretreatment process of various type of raw materials.

| Raw material | Protein content of raw | Productivity (ton/year) | Resource capacity (ton/month) | *Required Pre-treatment Process* |
|---|---|---|---|---|

| | material (%) | | | |
|---|---|---|---|---|
| Cassava leaves | 29.3-38.6[s] | 4,794,580[a] | 5[k] | Chopping and boiling to reduce cyanide levels[ad] |
| Tofu waste | 3.6-4.6[t] | 1,062,284[b] | 3[l] | Drying to prevent decomposition of the dregs[t] |
| Soy sauce solid waste | 14.8-20.8[u] | 155,518[c] | 475[m] | No special pre-treatment is required[u] |
| Fish bone | 26.02[v] | 497,091[d] | 474[n] | Washing and soaking to remove the remaining fat content found in the bones[ae] |
| Fish skin | 23.13[w] | 117,494[d] | 112[n] | Storage in a chiller/freezer to avoid rot and boiling to remove fat content[ae] |
| Beef bone | 21.9[x] | 4,480,670[e] | 1,220[o] | Size reducing, washing and soaking to remove the remaining fat content found in the bones[ae] |
| Cow skin | 33[y] | 2,327,965[f] | 633[o] | Storage in a chiller/freezer to avoid rot, size reducing and boiling to remove fat content[ae] |
| Chicken feet | 41.21[z] | 270,535[g] | 177[p] | Washing and soaking to remove the remaining fat content found in the bones[ae] |
| Chicken skin | 8.9-14.2[aa] | 785,424[h] | 514[q] | Storage in a chiller/freezer to avoid rot and boiling to remove fat content[ae] |
| Cheese whey | 6-10[ab] | 1,185[i] | 27[r] | Storage at 5°C to prevent microbial growth[af] |
| Corn steep liquor | 16.4[ac] | 9,900[j] | 900[j] | No special pre-treatment is required[ag] |

Resource: [a]processed [2,44]; [b]processed [45, 46, 47]; [c]processed [48, 49]; [d]processed [45, 50] ; [e]processed [2, 51, 52]; [f]processed [2, 52]; [g]processed [2, 53, 54] ; [h]processed [2, 53]; [i]processed [45]; [j]processed [55]; [k]processed [44]; [l]processed [47, 56]; [m]processed [48]; [n]processed [50, 57]; [o]processed [52, 58]; [p]processed [54, 59]; [q]processed [53, 59]; [r]processed [60]; [s][61]; [t][13]; [u][62]; [v][16, 63]; [w][50, 64]; [x][65]; [y][66]; [z][67]; [aa][68]; [ab][23, 69]; [ac][26, 70]; [ad][71]; [ae][17]; [af][72]; [ag][73].

**Table 5.** Weight of importance between criteria.

| Criterion | Weight |
|---|---|
| Raw materials' availability | 0.183 |
| Raw materials' protein content | 0.319 |
| Degree of hydrolysis | 0.319 |
| Protease's price | 0.069 |
| Processing's ease level | 0.110 |

Prior elimination of some raw materials and protease was carried out to minimize the alternatives evaluated using the AHP process. Cassava leaves, tofu waste, and cheese whey were eliminated due to their limited accessibility. Soy sauce solid waste, corn steep liquor, and chicken skin were also eliminated because these three raw materials have the lowest initial protein content compared to other raw materials. For the protease, pepsin was eliminated because it was mostly obtained from a nonhalal source. Further, rennin and thermolysin were also eliminated due to their extremely high price. This prior elimination resulted in 5 potential raw materials: fish bones, fish skins, beef bones, cow skin and chicken feet, and 3 proteases: papain, bromelain, and ficin. In total, there were 15 alternatives that were considered for the AHP process.

**Table 6.** The value of alternative process' final weight.

| Alternatives | Raw material's availability | Raw material's protein content | Degree of hydrolysis | Protease's price | Ease of processing | **Final weight** |
|---|---|---|---|---|---|---|
| Fish bone-Papain | 0.032 | 0.041 | 0.015 | 0.059 | 0.033 | **0.032** |
| Fish bone-Bromelain | 0.032 | 0.041 | 0.082 | 0.108 | 0.033 | **0.056** |
| Fish bone-Ficin | 0.032 | 0.041 | 0.062 | 0.033 | 0.033 | **0.044** |
| Fish skin-Papain | 0.013 | 0.041 | 0.015 | 0.059 | 0.139 | **0.040** |
| Fish skin-Bromelain | 0.013 | 0.041 | 0.185 | 0.108 | 0.139 | **0.097** |
| Fish skin-Ficin | 0.013 | 0.041 | 0.080 | 0.033 | 0.139 | **0.059** |
| Beef bone-Papain | 0.186 | 0.041 | 0.014 | 0.059 | 0.021 | **0.058** |
| Beef bone-Bromelain | 0.186 | 0.041 | 0.140 | 0.108 | 0.021 | **0.102** |
| Beef bone-Ficin | 0.186 | 0.041 | 0.080 | 0.033 | 0.021 | **0.077** |
| Cow skin-Papain | 0.082 | 0.078 | 0.011 | 0.059 | 0.083 | **0.057** |
| Cow skin-Bromelain | 0.082 | 0.078 | 0.131 | 0.108 | 0.083 | **0.098** |
| Cow skin-Ficin | 0.082 | 0.078 | 0.054 | 0.033 | 0.083 | **0.069** |
| Chicken feet-Papain | 0.020 | 0.131 | 0.015 | 0.059 | 0.059 | **0.061** |
| Chicken feet-Bromelain | 0.020 | 0.131 | 0.060 | 0.108 | 0.059 | **0.079** |
| Chicken feet-Ficin | 0.020 | 0.131 | 0.056 | 0.033 | 0.059 | **0.072** |
| | | TOTAL | | | | **1.000** |

The final results of all alternative weightings for each criterion are shown in Table 6. The highest score was obtained by a combination of beef bone-bromelain, indicating that the hydrolysis of beef bones using bromelain is the most potent combination for protein hydrolysate production in Indonesia. An alternative process that also potential to be developed is the hydrolysis of cow skin and fish skin by bromelain.

### 3.5. *Simple Economic Analysis*

A simple economic analysis was conducted to calculate the added value of the raw materials. As the baseline the price of protein hydrolysates was set to be 45.8 US $ per kilogram (alibaba.com), this is the average price of protein hydrolysate used for nutrition for microbial growth.

Table 7 shows that the processing of these food industrial waste, in particular waste of slaughtering house and fisk canning industry: beef bones, cow skin, and fish skin, into protein hydrolysate would increase the added value of the materials by 1.8 - 17.7 times.

**Table 7.** The raw material average price and added value after being processed as protein hydrolysate.

| Raw materials | Price (Rp/kg) | Raw material's added value (Rp/kg) | Increase in economic value (fold) |
|---|---|---|---|
| Beef bone | 80,000[a] | 308,400 | 1.8 |
| Cow skin | 12,000[b] | 638,000 | 17.7 |
| Fish skin | 9,000[c] | 635,100 | 17.3 |

Resource: [a,b][74]; [c][75]

## 4. Concluding Remarks

The use of different proteases affects the protein hydrolysis efficiency, thus careful consideration should be given in choosing the raw materials-protease combination. By considering the availability/accessibility of raw materials, the protein content of raw materials, the hydrolysis efficiency, the enzyme price, and the ease of processing which included pretreatment of raw materials; beef bone and bromelain protease was chosen to be the optimum raw material and protease combination for the production of protein hydrolysate in Indonesia. Other potential alternatives are cow skin-bromelain and fish skin-bromelain. These processes would increase the added value of the raw materials, food industrial waste, by 1.8 – 17.7 times and would provide an interesting example for the circular economy.

## References

[1] Naegeli C 1880 *Klasse Akad. Wiss Meunchen* **10** 277
[2] Statistics Indonesia (Accessed at http://www.bps.go.id/)
[3] de las Fuentes L, Sanders B, Lorenzo A and Alber S 2004 *Awarenet: Agro-Food Wastes Minimisation and Reduction Network* In: *Protein Byproducts: Transformation from Environmental Burden into Value-added Product* ed. Dhillon G (Canada: Elsevier Inc) 2016 p 5
[4] Bernadeta P A and Silalahi I H 2012 *J. Kim. Khatulistiwa* **1** 26–30
[5] Restiani R 2016 *J. Biota* **1** 103–10
[6] Vidal A R, Cechin C F, Cansian R L, Mello R O, Schmidt M M, Demiate I M, Kempka A P and Dornelles R C P 2018 *Ciênc. Rural* **48** 1–9
[7] Fasuyi A O and Aletor V A 2005 *Pak. J. Nutr.* **4** 43–9
[8] Ravindran G and Ravindran V 1988 *Food Chem.* **27** 299–309
[9] Kuiken K A, Lyman C M, Bradford M, Trant M and Dieterich S 1949 *J. Biol. Chem.* **177** 29–36
[10] Cavins J F, Kwolek W F, Inglett G E and Cowan J C 1972 *JAOAC* **55** 686–91
[11] Yamazaki K I, Takao S and Yamamoto K 1990 *U.S. Patent No. 4,940,662*
[12] Liu K S 1997 *Chemistry and nutritional value of soybean components*. In: Mateos-Aparicio I, Redondo Cuenca A, Villanueva-Suárez M J and Zapata-Revilla M A 2008 *Nutr. Hosp.* **23** 305–312
[13] Li S, Zhu D, Li K, Yang Y, Lei Z and Zhang Z 2013 *ISRN Ind. Eng.* **2013** 1–6
[14] Susanti S 2006 *Buana Sains* **6** 59–66
[15] Bechtel P J, Bland J M, Watson M A, Lea J M and Bett K L 2019 *Food Sci. Nutr.* **7** 1396–1405
[16] Toppe J, Albrektsen S, Hope B and Aksnes 2007 *Comp. Biochem. Phys.* **146** 395–401
[17] Gauza-włodarczyk M, Kubisz L and Włodarczyk D 2017 *Int. J. Biol. Macromol.* **104** 987–91
[18] Gunawan F, Suptijah P and Uju 2017 *JPHPI* **20** 568–581
[19] Eastoe J E 1955 *Biochem. J.* **61** 589–600
[20] Masirah 2018 *Proc. Seminar Nasional Kelautan Dan Perikanan (Swiss-Berlin, Tunjungan-Surabaya 05 September 2018) vol IV* p 285–292
[21] Aykın-Dinçer E, Koç A and Erbaş M 2017 *Poult. Sci.* **96** 4124–31
[22] Araujo Í B S, Bezerra T K A, Nascimento E S, Gadelha C A A, Santi-Gadelha T and Madruga M S 2018 *Food Sci. Tech.* **2061** 167–73
[23] Yasmin A, Butt M S, Sameen A and Shahid M 2013 *Pak. J. Nutr.* **12** 455–9
[24] Paskaš S, Miočinović J, Savić M, Ješić G, Rašeta M and Becskei Z 2019 *Mac. Vet. Rev.* **42** 151–61
[25] Hofer A A, Hauer S, Kroll P and Herwig C 2018 *Process Biochem.* **70** 20–8
[26] Nebraska Corn Board 2005 *Corn Processing Co-Products Manual* (Lincoln: University of Nebraska-Lincoln) p 6
[27] Fishanda R R 2019 Thesis, Universitas Pasundan (Accessed at http://repository.unpas.ac.id/)
[28] Belitz H, Grosch W and Schieberle P 2009 *Food Chem.* 4th revised (Heidelberg: Springer) p 78
[29] Tapal A and Tiku P K 2019 *Enzymes in Food Biotechnology chapter 27* (Myysuru: Elsevier Inc.) p 471–81
[30] Menard R, Khouri H E, Plouffe C, Dupras R, Ripoll D, Vernet T, Ressier D, Laliberte F, Thomas D Y and Storer A C 1990 *Biochemistry* **29** 6706–13
[31] Hou Y, Wu Z, Dai Z, Wang G and Wu G 2017 *J. Anim. Sci. Biotechnol.* **8** 24
[32] Biocon 2016 *Product data sheet: Bromelain purified enzymatic preparation* (Accessed on 22 July 2020 at http://biocon.es/)
[33] Buyukyavuz A 2014 Thesis, Clemson University (Acccessed at https://tigerprints.clemson.edu/)

[34] Yang Y, Shen D, Long Y, Xie Z and Zheng H 2017 *Sci. Rep.* **7** 1–8
[35] Rawlings N D, Barrett A J, Thomas P D, Huang X, Bateman A and Finn R D 2018 *Nucleic Acids Res.* **46** D624–32
[36] Jiang T, Chen L J, Xue L and Chen L S 2007 *J. Dairy Sci.* **90**, 3126–33
[37] Simpson B K 2000 *Digestive proteinases from marine animals. In Seafood Enzymes: Utilization and Influence on Postharvest Seafood Quality,* ed N. F. Haard and B.K. Simpson (New York: Marcel Dekker) pp 531–40
[38] Palashoff M H 2008 Doctoral dissertation, Northeastern University (Accessed at https://repository.library.northeastern.edu/)
[39] Dunn B M 2001 *Curr. Protoc. Protein Sci* **25** 21.3.1–6
[40] Keil B 1992 *Specificity of proteolysis* (Heidelberg: Springer) pp 43–228
[41] Dua A and Desai S S 2013 *Journal of Histology* **2013** 1–5
[42] Promega 2018 (Accessed on 18 July 2020 at https://worldwide.promega.com/)
[43] Cold Spring Harb Protoc 2007 (Accessed on 31 July 2020 at http://cshprotocols.cshlp.org/)
[44] Rutherfurd 2010 *J. AOAC Int.* **93** 1515–22.Almasyhuri, Yuniarti H, Luciasari E, and Muhilal1996J. *Nutr and Food Res.* **19** p 1-8
[45] [Kementerian Perindustrian Republik Indonesia(Accessed at www.kemenperin.go.id)
[46] Bisnis.com (Accessed on 5 August 2020 at http://m.bisnis.com/)
[47] Puger A W, Suasta I M, Astawa P A and Budaarsa K 2015 *MIP Universitas Udayana* **18** 22–5
[48] Rohmayanti T, Novidahlia N and Damayanti I *2019 J. Agroindustri Halal* **5** p 113–21
[49] Beritagar.id (Accessed on 5 August 2020 at http://beritagar.id/)
[50] Hadinoto S and Idrus S 2018 *Majalah BIAM Kemenperin* 51–57 (Accessed at ejournal.kemenperin.go.id/bpbiam)
[51] Rahaldo P 2012 Thesis, Institut Pertanian Bogor (Accessed at https://repository.ipb.ac.id/)
[52] Dewantara B F, Hamdani M D I, Sulastri S and Adhianto K 2017 *JIPT* **15** 35–40
[53] Ulupi N, Nuraini H, Parulian J and Kusuma S Q 2018 *JIPTHP* **6** 1–5
[54] Fathoni R M 2017 *JTHP* **6** 5-8 (Accessed at http://jurnal.unpad.ac.id/ejournal/)
[55] P.T Tereos Indonesia 2020 *Private communication*
[56] Sukmawati D 2011 *Japerti Universitas Winaya Mukti* **1** 1–10
[57] Tribun News Medan (Accessed on 18 August 2020 at https://medan.tribunnews.com/)
[58] PD Dharma Jaya (Accessed on 18 August 2020 at https://dharmajaya.co.id/)
[59] Republika (Accessed on 18 August 2020 at https://republika.co.id/)
[60] Dhillon G S (ed) 2016 *Protein Byproducts: Transformation from Environmental Burden into Value-added Product.* (Canada: Elsevier Inc.) chapter 9.3 p 171
[61] Yeoh C 1976 *Phytochemistry* **15** 1597–99
[62] Takeshi N, Yasuhiro T, Norihisa K and Nobutaka S 2012 *Food and Nutr. Sci.* **03** 1121
[63] Nurilmala M, Wahyuni M and Wiratmaja H 2006 *JTHP* **9** 22–33
[64] Astiana I, Nurjanah and Nurhayati T 2016 *JPHPI* **19** 170–81
[65] Septimus S 1961 *Anatomy of Domestic Animal. Mc.* (New York: Graw Hill)
[66] Sabtu B, Djojowidagdo S and Triatmojo S 2000 *J. Agric. Sci.* **13** 211–24
[67] Mayasaroh I, Rusmana D and Wiradimadja R 2012 *JTHP* **1** 3–4
[68] Kafri I, Cherry J A, Jones D E and Siegel P B 1985 *Poult. Sci.* **64** 2143–49
[69] Djuric M, Caric M, Milanovic S, Tekic M and Panic M 2004 *Eur. Food Res. Technol.* **219** 321–28
[70] Nisa M, Khan M A, Sarwar M, Lee W S, Ki K S, Ahn B S and Kim H S 2006 *Asian-Australas J. Anim. Sci.* **19** 1610–16
[71] Murdiana A and Saidin S 2001 *J. Food Nutr. Res.* **24** 33–7
[72] Malaka R 2018 Thesis Universitas Hassanudin (Accessed at https://digilib.unhas.ac.id/)
[73] Russo J M, Watson S A and Heiman V 1960 *Poult. Sci.* **39** 1408–12
[74] Liputan 6 News (Accessed on 18 August 2020 at https://liputan6.com/)
[75] Agromaret (Accessed on 18 August 2020 at https://agromaret.com/)

# Techno-economic Analysis: Antioxidant-Rich Natural Colorant Production from Dragon Fruit Peel

**D Shofinita***, **Y Bindar, R U Dinnita, and F Rizqi**

Department of Food Engineering, Faculty of Industrial Technology, Bandung Institut of Technology, Bandung, Indonesia

*shofi@che.itb.ac.id

**Abstract**. In this study, the economics of the various way production of natural colorant has been evaluated, specifically related to the effect of the drying methods: spray drying and freeze-drying. It has been found in this study that, the manufacturing costs of the food colorant in the form of extract, spray-dried powder, and freeze-dried powder were USD 6.01/kg, USD 29.89/kg, and USD 56.6/kg total solid content, respectively. The economic parameters consist of NPV, DPBP, PVR, and IRR of powder production using spray drying method were USD 1,989,412.68, 4.3 years, 3.31 and 20.6%, respectively, while that of using freeze-drying were USD 2,156,866.29, 7.27 years 1.78, and 8%. It is also known that drying using spray drying shows better economic parameters than freeze-drying. The sensitivity analysis of powder production using the spray drying method shows that the selling price parameter is a parameter that has a significant effect on the economics of production activity, followed by the cost of peel and cost of labor.
Keywords: Extraction, spray dryer, antioxidant, pigment, betacyanin

## 1. Introduction

Dragon fruit (*Hylocereus undatus*) is a fruit that belongs to the Cactaceae family. Dragon fruit plant is originally from Mexico, Central America, North America and is now widely cultivated in Indonesia. Indonesia has great potential for the dragon fruit business [1]. Dragon fruit flesh is mostly consumed and utilized in fresh form or juice, while dragon fruit peel (30-35% of the total weight) is often disposed of as waste. Dragon fruit peel contains antioxidants that can reduce the body's cholesterol levels [2]. Red dragon fruit peel (*Hylocereus polyrhizus*) contains betacyanin, which is widely used as an antioxidant and natural colorant [3]. A previous study reported that dragon fruit peel has greater antioxidant potential when compared to the antioxidant content contained in the flesh [4]. This antioxidant content in dragon fruit peel may protect human bodies from the dangers of free radicals.

Betacyanin can be produced from dragon fruit peels by extraction. This extract can be consumed or used directly in the food processing process. However, betacyanin extract in liquid form has a limitation in transportation and storage flexibility, and also shelf life. Processing of liquid extracts into powder can increase its shelf life, minimize transportation cost, and also making it easier to apply as a mixture of food products. One of the methods for making extract powder is drying using a spray dryer. Spray drying is a method of drying liquid extracts into powder which is widely applied in the industry [5]. Freeze drying is another drying method for turning liquid extracts into powder. The advantage of freeze-drying is that the sample is in a frozen condition so that it may avoid degradation of the temperature-sensitive components (e.g. protein, flavor, and color).

The production of extract powder from dragon fruit peel has economic potential. To determine the production process to be carried out, it is necessary to assess which process is the most economical to avoid the risk of loss and to achieve optimum profit. Achieving these objectives can be done through analytical studies using methods of economic approaches that are commonly used. To answer this problem, this study was conducted to determine the feasibility of the production process with different drying methods: spray drying and freeze-drying. The cost of manufacturing of the powder will also be compared with that in the form of extract.

## 2. Methods

The economics that is evaluated in this study is influenced by production costs and revenue from the selling of natural colorant powder dragon fruit peel extract. The economic evaluation was carried out using experimental data obtained from previous studies included yield, powder moisture content, optimal extraction time, the solvent mass fraction (water) remaining in the residue, and ambient air humidity [14].

The production cost of dragon fruit peel extract powder was calculated using the approach existing method proposed [10]. Cost of Manufacturing (COM) was calculated as a function of Fixed Capital Investment (FCI), labor wages (COL), utility costs (CUT), water treatment costs (CWT), and raw material costs (CRW). FCI is calculated using the cost correlation method approach [11, 10] for each equipment. The process flow used in powder production using the spray drying method consists of three extractors, one reservoir, 3 extraction tanks, one mixing tank, 5 centrifugal pumps, one compressor, one indirect heater, one filter press, and one spray dryer and for powder production. For the freeze-drying method, a process production consists of three extractors, one reservoir, 3 extraction tanks, one mixing tank, 5 centrifugal pumps, one filter press, one freezer, and one freeze dryer are used. The PFD (Process Flow Diagram) process is shown in Figure 2.1.

### 2.1. *Production cost*

The production costs are calculated using the methodology proposed [10]. Production costs (COM) are a function of investment costs (FCI), operational labor costs (COL), utility costs (CUT), water treatment costs (CWT), and raw material costs (CRM), which can be determined using the following eq.

$$COM\left(\frac{USD}{year}\right) = 0{,}280\ x\ FCI + 2{,}37\ x\ COL + 1{,}23\ x(CUT + CWT + CRM) \quad (1)$$

Industrial-scale process units and laboratory-scale operating units with constant solvent, solids ratio value, temperature, pressure, and maceration extraction time are assumed to have the same performance. The efficiency values for pumps, fans, indirect heaters are assumed to be 75%, 70%, and 85%, respectively. Efficiency values are obtained from the literature of [10], [12]. CWT is neglected because the process produces the main residue which is non-toxic and can be used for fertilizing crops, put into the soil, or commercialized as a by-product.

### 2.2. *Cost of labor (COL).*

Operational labor costs in this study covering the total costs incurred to pay the salaries of all factory operators in one year are determined using the existing approach proposed [10]. The calculation of the number of operators required to run and control the process activities per shift is done using the following equation.

$$Nol = (6{,}23 + 31{,}7P^2 + 0{,}23N_{np})^{0{,}5} \quad (2)$$

P is the number of processing steps that involve handling particulate solids, Nnp is the number of non-particulate processing steps.

The maximum labor cost using the maximum labor wage in Indonesia, namely in Karawang Regency, is IDR 4,594,325 / month or 327 USD / month. The lowest minimum wage in Indonesia is in the city of Banjar, which is IDR 1,831,884.83 / month or USD 130.42 / month which is used as the minimum value in this study.

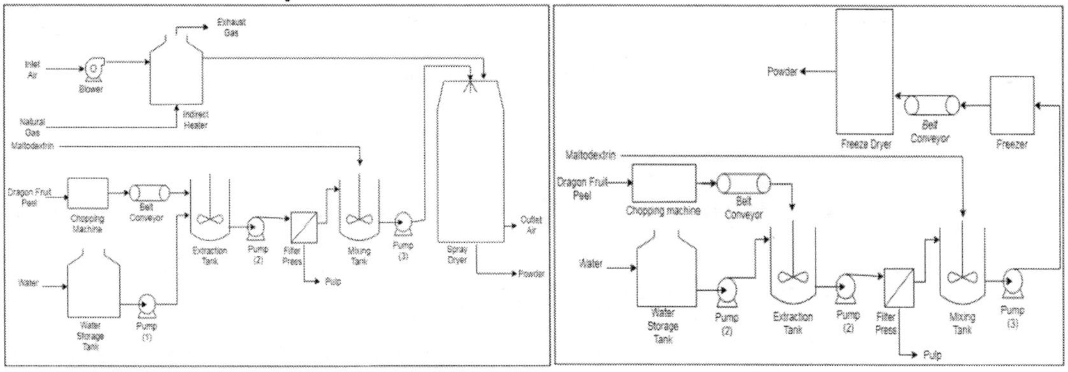


(a) (b)

**Figure 2.1** Process Flow Diagram of the production (a) spray-dried (b) freeze-dried extract

### *3.1. Cost of raw material (CRM)*

The raw material costs in the production of extract powder consist of components of the cost of purchasing dragon fruit peel and water. The maximum price used for dragon fruit peel is Rp 3,000 per kg, and the minimum price for dragon fruit peel is Rp. 0, with the possibility that the peel is a waste from other industrial processes. The cost of purchasing water is obtained from the market price of reverse osmosis water in certain areas. The minimum price for reverse osmosis water in Indonesia is Rp. 210,526.3 / kL or 8.58 USD / kL, and the maximum price for reverse osmosis water is Rp. 263,157,895 / kL or 18.74 USD / kL.

### *3.2. Cost of utility (CUT)*

The utility costs for powder production in this study consist of electricity and natural gas costs. Electricity costs include all costs that need to be paid to run all process units that require electrical energy. The cost of electricity is determined through the basic electricity tariff by PLN at the industrial use scale set by the Indonesian Ministry of Energy and Mineral Resources. The minimum tariff used is Rp. 997 per kWh which is the basic tariff for industrial-scale electricity in Indonesia in 2020, and taking into account inflation of 5%, the maximum value is set at Rp. 1,046 per kWh. Natural gas in Indonesia is supplied by PGN (Perusahaan Gas Negara). The minimum price is set at USD 6 per mmbtu based on the 2020 SKK Migas, and the maximum price is set at USD 10 based on the price of natural gas in Medan.

### 3.3. *Fixed Capital Investment.*

The Fixed Capital of Investment in this study includes the costs incurred to purchase all process equipment for the manufacture of dragon fruit peel powder. The calculation of the cost of purchasing equipment to be purchased is carried out using the correlation cost method with references to [10], [13], [11].

### *3.4. Revenue*

Revenue is obtained from selling powders. The price of powders on the market can vary due to several factors, such as purity, antioxidant content, market trends, and so on. In this study, the selling price used is 36.3 USD/kg.

### *3.5. Calculation of Economic Evaluation Parameters*

2.5.1 *NPV (Net Present Value).* NPV is calculated using Microsoft Excel 2013. Input data are FCI and WC for the first year and revenue, COM, gross profit, tax, net income, and cash each year.

$$NPV = \sum_{t=1}^{n} \frac{Rt}{(1+i)^t} \quad (3)$$

2.5.2 *DPBP (Discounted Pay Back Period).* DPBP in this study is calculated mathematically using the following equation

$$DPBP = \frac{Initial\ Investment}{Annual\ Cash\ Flow} \quad (4)$$

2.5.3 *PVR (Present Value Ratio).* PVR in this study is calculated mathematically using the following equation

$$\mathrm{VR} = \frac{\sum_{t=1}^{T} PV}{FCI+WC} \quad (5)$$

*2.5.4 IRR (Internal Return Ratio).* IRR is calculated using the IRR function in Microsoft Excel 2013. IRR in this study is calculated mathematically using the following equation.

$$0 = NPV = \sum_{t=1}^{T} \frac{Ct}{(1+IRR)^t} - Co \quad (6)$$

## 3. Result and discussion

### *3.1. Production cost*

*3.1.1. Liquid extract production cost.* The cost of producing the liquid extract in this study is calculated from the cost of the extraction and filtering processes. The extraction operating conditions used are the optimum operating conditions of previous research in the form of extraction time of 60 minutes, the solvent in the form of desalinated water, and the ratio of solvent to solids ratio is 3 [14].

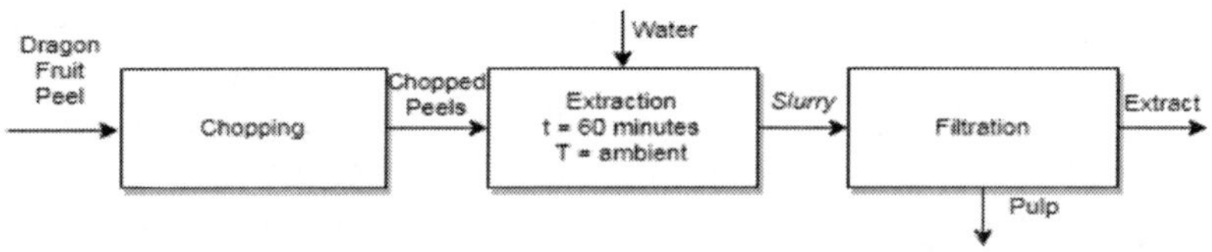


**Figure 3.1** Process flow diagram production liquid extract.

**Table 3.1** Liquid extract cost of production.

| | FCI | COL | CRM | CUT | COM |
|---|---|---|---|---|---|
| Annual COM (USD/Year) | 16,643.69 | 42,065.48 | 77,582.06 | 5,859.88 | 115,494.24 |
| COM (USD/kg solid content) | 0.87 | 2.19 | 4.04 | 0.30 | 6.01 |
| Percentage to COM (%) | 14.48 | 36.60 | 67.50 | 5.09 | 100 |

Based on **Table 3.1**, the required manufacturing cost (COM) per kg of total solid content dragon fruit peel extract solids is 6.01 USD. CRM in this study gave the most significant proportion of COM with a percentage of 67.50%. The next largest proportion is COL with a percentage of 36.6%. COL and CRM which have the largest contribution to COM are in line with the study reported [15] in a study of Jussara pulp extraction using ultrasound-assisted and agitated bed-extraction technology and also reported [16] on the production of orange peel extract powder using the microwave-assisted extraction method and the spray drying method. FCI in the production of liquid extract ranks third with a percentage of 14.48%. This condition occurs because the process of making liquid extracts is simple and is operated in-room operating conditions so that high technology is not required. CUT has the smallest percentage of 5.09% because the tools used for the extraction process are relatively simple and require relatively low energy.

*3.1.2. Production cost of spray-dried dragon fruit peel extract.* The powder production cost for the spray drying method in this study is calculated from the total cost of processing the peel until the powder is obtained. **Table 3.2** presents the COM value along with the percentage of FCI, COL, CRM, CUT against COM for powder production with operating conditions of 170°C and a total powder production of 20.9 tons of powder/year.

**Table 3.2** COM composition of each powder production cost of spray drying method.

| | FCI | COL | CRM | CUT | COM |
|---|---|---|---|---|---|
| Annual COM (USD/Year) | 209,781.51 | 42,065.48 | 270,471.09 | 102,175.98 | 624,494.06 |
| COM (USD/kg powder) | 10.04 | 2.01 | 12.94 | 4.89 | 29.89 |
| Percentage to COM (%) | 31.21 | 6.26 | 47.32 | 15.21 | 100 |

Based on **Table 3.2**, the production cost of spray-dried powder is 29.89 USD/kg powder. CRM in this study gave the most significant proportion of COM with a percentage of 47.32%. CRM that has the greatest contribution to COM is in line with the study reported [15] in a study of Jussara pulp extraction using ultrasound-assisted and agitated bed-extraction technology and also reported on the research [16] on the production of orange peel extract powder using the microwave-assisted extraction method and the spray drying method. The next largest proportion is FCI with a percentage of 31.21%. The second-largest proportion is FCI (31.21%) where this trend is in line with the economic evaluation of orange peel extract spray drying. It was reported that the proportion of FCI to COM for the production of orange peel extract powder was 34.6% for the variation of the solvent to orange peel ratio of 14, which is the largest proportion to COM [16]. This condition is due to the high costs incurred for the procurement of a spray dryer. CUT ranks third with a percentage of 15.21%. CUT is dominated by the need for the drying process in the form of electricity and natural gas, which is large in the spray drying process as well as the consumption of electricity by the main and other supporting equipment in the production process. The large energy consumption of the spray dryer was reported [12] in a study on the energy consumption of industrial spray dryers. In his report, he wrote, based on a survey of 32 spray dryers with an average evaporation capacity of 67.8 tons of water per hour, an average heat input of 333.3 GJ per hour was required. The smallest percentage is owned by COL, which is 6.26% due to the low wages of labor in Indonesia when compared to the wages of workers in the US as in [15].

*3.1.3. Production cost of freeze-dried dragon fruit peel extract.* The cost of powder production using the freeze-drying method in this study is calculated from the total cost of processing the peel to obtain the powder. The processes involved include extraction, filtering, and drying using freeze-drying. Table 3.3 presents the COM values along with the percentage of FCI, COL, CRM, CUT against COM powder production with total powder production of 20,889 tons of powder/year.

**Table 3.3** COM composition of each powder production cost of the freeze-drying method.

| | FCI | COL | CRM | CUT | COM |
|---|---|---|---|---|---|
| Annual COM (USD/Year) | 601,825.35 | 42,065.48 | 77,582.06 | 438,172.25 | 1,159,645.13 |
| COM (USD/kg powder) | 28.81 | 2.01 | 3.71 | 21.07 | 55.6 |
| Percentage to COM (%) | 52.7 | 3.68 | 5.78 | 37.85 | 100 |

Based on **Table 3.3,** the COM per kg of powder is 55.60 USD. The FCI in this study provided the most significant proportion of COM with a percentage of 52.70%. This trend is in line with the economic evaluation of orange peel extract spray drying. It was reported that the proportion of FCI to COM for the production of orange peel extract powder was 34.6% for the variation of the solvent to orange peel ratio of 14, which is the largest proportion to COM [16]. A similar condition was also reported by [22] in an economic evaluation study on the reduction of fat content in annatto fruit seeds using supercritical carbon dioxide. The large FCI value is due to the purchase value of the expensive freeze dryer. The investment value of the freeze dryer was also reported [23] on the research of the investment value of various drying methods. The second-largest proportion is CUT (37.85%). CUT is dominated by the need for the drying process in the form of freeze-drying and refrigerator electricity. Freeze-drying requires a lot of energy, which is an obstacle to the application of this drying method to many materials [24]. CRM ranks third with a percentage of 5.78%. The small percentage of CRM is because the raw materials for production are relatively cheap. Dragon fruit peel is a waste from the production of other factories and the extraction process in this study uses water solvents that are widely available in Indonesia at relatively cheap prices. The smallest percentage is owned by COL, which is 3.68%, this is due to the low labor wages in Indonesia when compared to the wages of workers in the US as in research [15].

*3.1.4. Comparison of production costs (COM/kg solid content) liquid extract, powder from the spray-dried, and freeze-dried extract.*

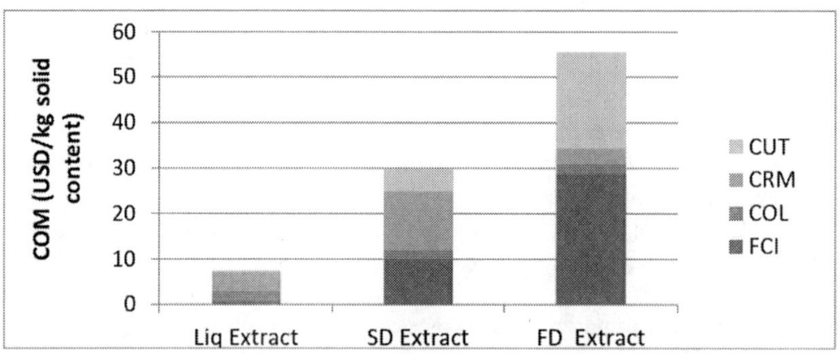


**Figure 3.2**, The COM cost graph for liquid extract, spray dried, and freeze-drying extract.

The graph presented in **Figure 3.2** shows the COM for powder is greater than liquid extract. There is an increase in COM from liquid extract production to powder production from spray drying 4 times. Meanwhile, the increase in COM from liquid extract production to freeze-drying production is 7.5 times. Spray drying and freeze-drying provide substantial COM additions. The process increases the values of FCI, CUT, and COL which contribute to COM computation. However, the conversion to powder provides significant mechanical and economic advantages, where the powder has better stability, longer shelf life, and lower storage requirements and distribution costs [25].

In the production process using a spray dryer and freeze dryer, the FCI value increased significantly due to the cost of purchasing the main equipment in the form of a spray dryer and other companion tools such as the inlet air heater and pump set. CUT also increases due to the need for electricity as the main energy for the spray dryer, and the need for gas supply to heat the inlet spray dryer air supply. COL increases along with the addition of processes requiring an increase in the number of workers.

The spray drying method is the cheapest and most effective method among other microencapsulation methods because it is suitable for continuous operation, provides convenience in the aspects of transportation and distribution, thereby reducing storage and transportation costs [26]. The resulting product stability is the same as it had when it is made. The drying process to powder reduces the moisture content significantly where the extract powder from spray drying has a moisture content of 0.08. A very large reduction in moisture content due to the drying process inhibits microbial growth and results in longer shelf life. Besides, the range of distribution of powders is wider than that of liquid extracts. The product in the form of a liquid extract has a low shelf life and stability, which hinders the distribution process over long distances.

### *3.2. Economic Analysis and Sensitivity Analysis of Drying Powder Production (Spray Drying and Freeze Drying) Dragon Fruit Peel Extract*

Assumptions and the basis for the economic feasibility of operating a factory are evaluated using the net present value [20]. In this study, the feasibility of economic analysis methods is sought including NPV (Net Present Value), DPBP (Discounted Payback Period), IRR (Internal Return Ratio), PVR (Present Value Ratio). Making cash flow refers to the book Analysis, Synthesis, and Design of Chemical Process [10]. The assumptions and bases used are summarized in **Table 3.4.**

**Table 3.4** Assumptions and basis for economic evaluation.

| No | Basis | Assumption |
|---|---|---|
| 1 | Plant operation period | 15 years |
| 2 | Construction period | 1 year |
| 3. | Interest, the tax rate | 25, 6% per year |
| 4. | Exchange rate USD to Rp | 14,046.18 |
| 5. | Annual operation period | 24 hours for 330 days/year |

*3.2.1. Spray-dried and Freeze-dried extract production.* This study aims to analyze the production economics of both powder manufacturing methods. The results of a further analysis can be used to compare and get an overview of the two most economical methods.

**Table 3.5** Economic analysis result of the production.

| Economic Parameters | Spray Drying | Freeze drying |
|---|---|---|
| DPBP (Discounted Pay Back Period) | 4.30 | 7.27 |
| NPV (Net Present Value) (USD) | 1,989,412.68 | 2,156,866.29 |
| PVR (Present Value Ratio) | 3.31 | 1.78 |
| IRR (Internal Rate of Return) | 20.6% | 8.7% |

The economic analysis of the spray-dried extract production gives a Discounted Payback Period (DPBP) value of 4.3 years, which shows that it takes 4.3 years after start-up to return fixed capital investment. A positive Net Present Value (NPV) indicates that the factory is profitable [28]. The Present Value Ratio (PVR) shows that the amount of incoming or outgoing costs is 3.3 times the costs incurred and shows that the factory operation provides a profit of 3.3 times the investment value. The Internal Rate of Return (IRR) value shows a positive value above the bank interest discount rate of 6%. In the investment criteria, the IRR must be greater than the OCC or opportunity cost of capital so that the investment plan is feasible to implement [28]. So that with an IRR that is far above the discount rate, investment in production is feasible. This can be caused by the relatively low COM, which is 29.89 USD, and because the high purity causes the selling price in the big market. The selling price is a component that greatly influences the factory's annual revenue. Revenue is known to be one of the main contributors affecting the economy of production activity.

The economic analysis on Freeze-dried extract production gives a DPBP value of 7.27 years, NPV is positive, PVR is 1.78, and IRR is 8.7%. The Internal Rate of Return (IRR) value shows a positive value above the bank interest discount rate of 6%. In the investment criteria, the IRR must be greater than the OCC or opportunity cost of capital so that the investment plan is feasible to implement [28].

*3.2.2. Sensitivity Analysis Spray Dried and Freeze-Dried Extract.* Sensitivity analysis is carried out to observe the impact of changes in several economic parameters on NPV so that it can be found that the most significant parameters affect the profitability of a plant [21]. Therefore, in this case, the sensitivity analysis is determined by projecting COL, CW (Cost of Water), COE (Cost of Electricity), COP (Cost of Peel), and selling price. The value of each factor will be varied with a relative change range of 5%. The results of the analysis of the impact of parameter changes on the NPV on spray-dried and freeze-dried extract shown in **Figure 3.3.**

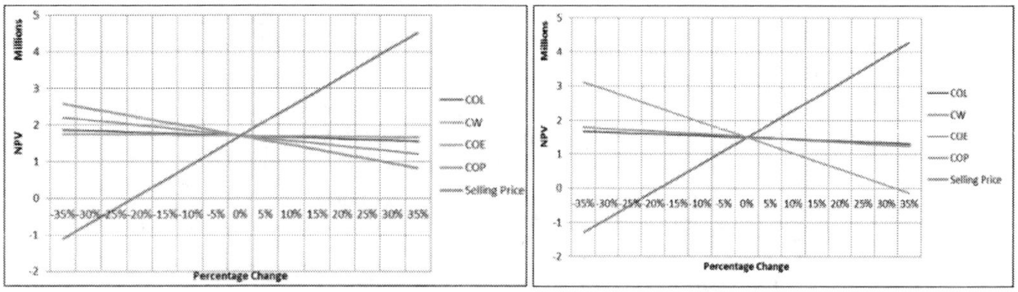


(a) (b)

**Figure 3.3**. The sensitivity of each parameter to NPV operation in (a) spray-dried extract (b) freeze-dried extract.

Sensitivity analysis for the spray-dried extract production provides that the selling price aspect is the aspect that has the most significant influence on NPV. Therefore, this uncertainty needs to be minimized, among others, by negotiating with the buyer. The trend found is in line with a research report on the economic analysis of citrus peel extract powder using the spray drying method [17]. It was reported that selling price is the most influencing aspect of the economics of powder production. Sensitivity analysis for the freeze-dried extract production also provides that the selling price aspect is the aspect that has the most significant influence on NPV. Therefore, this uncertainty needs to be minimized, by negotiating with the buyer.

## 4. Conclusion

The production cost of dragon fruit peel liquid extract (COM/kg dry solids) is USD 6.01, the production cost of dragon fruit extract powder using spray drying and freeze-drying methods is USD 29.89 and USD 55.6, respectively. The increase in the production cost of liquid extract for powder is 4 times using the spray drying method and 7.5 times using the freeze-drying method. The economic parameters of powder production using spray drying method are obtained NPV, DPBP, PVR, IRR respectively USD 1,989,412.68, 4.3 years, 3.31, and 20.6%. Whereas economic parameters of powder production using freeze-drying are obtained NPV, DPBP, PVR, IRR respectively 7.27 years, USD 2,156,866.29, 1.78, and 8%. It is also known that drying using spray drying shows better economic parameters than freeze-drying. The sensitivity analysis of powder production using the spray drying method shows that the selling price parameter is a parameter that has a significant effect on the economics of production activity followed by COP and COL.

## References

[1] Kristanto D 2010 *Buah Naga : Pembudidayaan Pot dan Kebun* Jakarta: Penebar Swadaya
[2] Kanner J, Harel S, and Granit R 2001 Betalains-A new class of dietary cationized antioxidants *J. Agric. Food Chem.* **49** 5178-5185
[3] Cao Jun Ming, Yan Jing, Huang Yan Hua, Wang Guo Xia, Zhang Rong Bin, Chen Xiao Ying, Wen Yuan Hong, Zhou Ting Ting 2012 Effects of replacement of fish meal with housefly maggot meal on growth performance, antioxidant and non-specific immune indexes of juvenile Litopenaeus vannamei *J. Fish China* **36** (4) 529-537
[4] Darmawi A W 2011 Optimasi proses ekstraksi, pengaruh pH dan jenis cahaya pada aktivitas antioksidan dari kulit buah naga (H. polyrhizus)
[5] Shofinita D, and Langrish T A 2016 Redox (pro-oxidant/antioxidant) balance in the spray drying of orange peel extracts *Drying Technology* **34** (14) 1719-1725
[6] Kha T C, Nguyen M N, Roach P D 2010 Effects of Spray Drying Conditions on the Physicochemical and Antioxidant Properties of the Gac (Momordicacochinchinensis) Fruit Aril Powder *Journal of Food Engineering* **98** 385–392
[7] Shishir M R I, Taip F S, Aziz N A, and Talib R A 2014 Physical properties of spray-dried pink guava (Psidium guajava) powder *Agriculture and Agricultural Science Procedia* **2** 74e81
[8] Lee K C, Yoon Y S, Li F Z, and Eun J B 2017 Effects of inlet air temperature and concentration of carrier agents on physicochemical properties, sensory evaluation of spray-dried mandarin (Citrus unshiu) beverage powder *Applied Biological Chemistry* **60** (1) 33–40
[9] A Nafiunisa et al 2017 *J. Phys.*: Conf. Ser. 909 012084
[10] Turton R, Bailie R C, Whiting W B, Shaeiwitz J A 2009 *Analysis, Synthesis, and Design of Chemical Processes, third ed* Prentice-Hall, New Jersey.
[11] Towler G, Sinnott R K 2013 *Chemical Engineering Design: Principles, Practice, and Economics of Plant and Process Design seconded* Butterworth- Heinemann, Boston.
[12] Baker C G J, McKenzie K A 2005 Energy consumption of industrial spray dryers *Drying Technol* **23** (1–2) 365–386
[13] Perry R H, and Green D W 2008 *Perry's chemical engineers' handbook* New York: McGraw-Hill
[14] Shofinita Dian, Bindar Yazid, Samadhi Tjokorde, Jaelawijaya Arwinda, Fawwaz Mifta 2020 Reducing the stickiness of dragon fruit peel extract powder as food colorant by addition of maltodextrin during freeze-drying *AIP Conference Proceedings*
[15] Vieira G S, Cavalcanti R N, Meireles M A A, Hubinger M D 2013 A chemical and economic evaluation of natural antioxidant extracts by ultrasound-assisted and agitated bed extraction from jussara pulp (Euterpe edulis) *J. Food Eng* **119** (2) 196–204,
[16] Shofinita Dian, Langrish Timothy A G, Huang Guo-Lin and Lu Xiuyang 2016 *Techno-economic analysis of spray-dried orange peel extract powders compared with liquid extracts* In: Chemeca 2016: Chemical Engineering - Regeneration, Recovery, and Reinvention. Melbourne, Vic.: Engineers Australia, 2016: 536-546
[17] Shofinita D, TAG Langrish *Journal of Food Engineering* 139, 31-42, 2014. 24, 2014.
[18] Simanjuntak L & Sinaga C 2014 Ekstraksi Pigmen Antosianindari Kulit Buah Naga Merah (*Hylocereus polyrhizus*) *Jurnal Teknik Kimia USU* **3** (2 Juni) 25–29
[19] Lee Woun Tan, and friend 2011 Empirical Modeling for Spray Drying Process of Sticky and Non-sticky Product *Procedia Food Science* **1** 690-697

[20] Helfert E R 2001 *Financial Analysis: Tools and Techniques, A Guide for Managers* McGraw-Hill 221–296

[21] Saltelli A, Ratto M, Tarantola S, and Campolongo F 2012 (Perennial Review of the 2005 paper), Sensitivity Analysis for Chemical Models, Chemical Reviews, 112 (5), pp PR1–PR21.

[22] Albuquerque U P, Ramos M A, Melo J G 2012 New strategies for drug discovery in tropical forests based on ethnobotanical and chemical ecological studies *J. Ethnopharmacol.* **140** 197-201

[23] Nijhuis H H 1998 Approaches to improving the quality of dried fruit and vegetables *Trends in Food Science & Technology* **9** 13–20

[24] Huang M, et al. 2014 Choosing the right metal: case studies of class I ribonucleotide reductases *J Biol Chem* **289** (41) 28104-11

[25] Vehring R, Foss W R, Lechuga-Ballesteros D 2007 Particle formation in spray drying *J. Aerosol Sci.* **38** (7) 728-746

[26] Holsinger K E 2000 Reproductive systems and evolution in vascular plants *Proceedings of the National Academy of Sciences*, USA **97** 7037 – 7042

[27] Vadas E 2000 Stability of Pharmaceutical Products *Gennaro,. A.R.Remington: The Science and Practice of Pharmacy.20 th edition* **I** 972-973

[28] Harahap Sofyan Syafri 2002 *Analisa Kritis Atas Laporan Keuangan* Jakarta: PT Raja Grfindo Persada

IOP Conf. Series: Materials Science and Engineering 1143 (2021) 012042 doi:10.1088/1757-899X/1143/1/012042

# Fermented cassava as an alternative flour for pasta noodle

**R Purwadi*, C F Teguh and D Mazaya**

Food Engineering Study Program, Faculty of Industrial Technology, Institut Teknologi Bandung – Jl. Let. Jend. Purn. Dr (HC) Mashudi no. 1 Jatinangor, Sumedang 45363 – Indonesia.

*Email: ronny.purwadi@che.itb.ac.id

**Abstract.** Cassava is widely cultivated in South East Asia countries like Indonesia. However, ordinary cassava flour cannot substitute wheat flour due to lack of gluten. Fermented cassava flour (fercaf) is a protein enriched cassava flour which is produced from a fermentation process of cassava flakes by lactic acid bacteria. Fercaf may serve as substitution to wheat flour, especially in pasta production. Nevertheless, a number of modifications should be exercise in product formulation to meet the ordinary 'wheat base' pasta product specifications. The current study was focus on these modifications i.e., addition of other flour and addition of hydrocolloids. The properties of pasta such as elasticity, yield, cooking weight, hardness, stickiness as well as sensory property were examined to determine the best pasta formulation. The results show that pasta made from fercaf with the addition of rice and corn flour gave strong and elastic texture while hydrocolloid addition increased water-binding of the pasta. The final pasta formulation included ratio of fercaf:rice:corn flour by 70:10:20 and addition of 2,5% xanthan gum.

## 1. Introduction

Pasta and noodle are loved by many people around the world. The consumption of noodle increases due to its variety and readiness. Even though noodle is not a main food for Indonesian people, but noodle consumption significantly increases every year. However, pasta and noodle are made from wheat which is mostly grown in sub-tropical countries. Tropical country like Indonesia must import wheat from other countries. In the other side, there are many food carbohydrate sources grown in tropical area such as rice, corn, sweet potato, cassava etc. The difference between these carbohydrate sources and wheat is the gluten protein content. This wheat protein may build a structure thus yield a good texture of the food.

A number of study have been carried out in replacing wheat [1–6]. Charoenthaikij [2] has developed a method of making noodle from brown rice with addition of pre-gelatinized cassava starch and xanthan gum. Abidin [1] has successfully developed a method of noodle making using a mixture of cassava flour and wheat flour with a portion of 5:1.

Cassava is a tropical crop which is usually used as food carbohydrate source. Unlike wheat, cassava tuber contains quite small protein so for some extent it cannot substitute wheat due to the absence of gluten protein. However, cassava can be partly fermented to increase its protein content [7,8]. This product is named as fermented cassava or "fercaf". Fercaf has been successfully used as flour for bread and some cake making. However, it is very limited publication that mentions the use of fercaf in noodle making.

The purpose of this study is to develop the method of noodle making using fercaf as main flour. Some noodle properties have been improved by addition of other flours such as maize and rice flours. The noodle properties such as elasticity; water content as cooking weight; yield; and texture as hardness and stickiness were evaluated. In addition, a number of hydrocolloids such as xanthan gum, guar gum, arabic gum and carboxymethyl cellulose were also employed for property improvements.

## 2. Materials and Methods

### *2.1. Materials*

Fermented cassava (fercaf) flour was prepared from local cassava tubers according to the method from Kresnowati [7,8]. Multi-purpose wheat flour, rice fluor and maize flour were purchased in local markets. All hydrocolloid agents such as xanthan gum, guar gum, arabic gum, and carboxy-methylcellulose (CMC) were food grade and purchased from local supplier. Whole fresh eggs were used in all experiments.

### *2.2 Methods*

Determination of the best composition of flour mixture were carried out according to the pre-set composition design which is shown in table 1. An automatic noodle maker, Renoodle RN-77 was used in all noodle preparation to maintain consistent product. The noodle recipe supplied by Renoodle RN-77 was used in all experiment: 150 g flour was entered to the machine followed by 60 g water and 5 g scrambled egg (hydrocolloid was added at last if applicable). Renoodle RN-77 kneaded the dough automatically for 9-12 minutes. The noodle was then extruded from the noodle maker and flour dusting was done to prevent the noodle stick one to each other. A couple of raw noodles were withdrawn for texture and water content analysis while the rest was cooked in boiled water for 2-3 minutes. The cooked product was drained and used for analysis of elasticity, weight and texture. The analyses were carried out directly after noodle making and cooking process.

**Table 1.** Composition of flours in every experiment runs.

| Sample Code | Flour (%) | | | |
|---|---|---|---|---|
| | Wheat | Fercaf | Maize | Rice |
| CTRL | 100 | - | - | - |
| F1 | - | 100 | - | - |
| M1 | - | 70 | 30 | - |
| M2 | - | 70 | - | 30 |
| M3 | - | 60 | 40 | - |
| M4 | - | 60 | - | 40 |
| M5 | - | 70 | 15 | 15 |
| M6 | - | 70 | 10 | 20 |
| M7 | - | 70 | 20 | 10 |

The effect of hydrocolloid in noodle recipe was evaluated as follow. The best flour composition which had been formerly determined was used afterward. Four type of hydrocolloid were applied in the experiments i.e., xanthan gum (XG), guar gum (GG), arabic gum (AG) and carboxymethylcellulose (CMC). The quantities of each hydrocolloid were 0.5, 1.0, 1.5 and 2.0% of flour weight basis.

### *2.3. Analyses*

The yield of noodle preparation was determined by the ratio of the weight of noodle product and the weight of raw dough. Cooking weight and water content expressed how the noodle absorbed water during cooking process. The cooking weight is the difference between the weight of cooked noodle and the weight of raw noodle. The water content was determined as follow: the sample of cooked noodle was weighed and then dried in the oven at temperature of 105°C for 4 hours. The dried sample was also weighed. The difference between cooked noodle weight and dried cooked noodle weight was considered as the weight of water contained in the sample. The elasticity of cooked noodle was determined as follow: first, the length of a piece of noodle was measured using a ruler. The piece of noodle was then drawn out until its broke up. The elasticity was calculated using those data according to Li et.al [9]. The hardness, chewiness and gumminess were determined using TA.XTplus Texture Analyzer. About 4 pieces of noodle were used in every measurement. Different probes were used for each texture analysis measurement.

## 3. Results and Discussions

Noodles were made using different combination of flour (table 1) and the characteristics of those products are shown in table 2. Noodle made from a whole fercaf (F1) was failed and seemed to be soft, slimy, watery and had no elasticity at all. It indicates that additional protein developed during casava fermentation could not imitate the function of gluten protein contained in wheat flour in noodle making process. Other flour such as maize and rice flour were used and mixed with fercaf to improve the characteristic of noodle products. The quantities of additional flour were set to be less than that of fercaf as main flour. Maize flour addition improved the yields, the cooking weight and the elasticity. The maize also gave better noodle appearance (compare with that of wheat noodle). In the other hand, rice flour addition improved the texture (hardness and stickiness) of the noodles. Base on these results, the mixtures of fercaf, maize and rice flour were used for subsequent experiments.

**Table 2.** The characteristics of noodle from different source of flour.

| Sample Code | Hardness | Stickiness | Cooking Weight (%) | Strain (%) | Yields (%) |
|---|---|---|---|---|---|
| CTRL | 16.25 | -0.16 | 61.83 | 51.64 | 70.00 |
| F1 | 11.01 | -0.04 | 101.25 | - | 47.05 |
| M1 | 12.58 | -0.21 | 103.99 | 39.02 | 52.50 |
| M2 | 18.31 | -0.03 | 133.08 | 32.43 | 45.91 |
| M3 | 17.33 | -0.09 | 47.81 | 42.92 | 42.73 |
| M4 | 25.30 | -0.06 | 54.17 | 30.82 | 45.68 |

The characteristics of noodles made from mixtures of fercaf, maize and rice flours are shown in tabel 3. The noodle from M7 experiment resulted a lower elasticity (19.34) and bad texture (with stickiness of -0.17), while other experiments (M5 and M6) yielded noodles with better characteristics. Since these noodles had similar results, it was hard to distinguish and choose the better one among other. Therefore, an analytic hierarchy process (AHP) technique was employed [10–12], and every characteristic item were weighted through comparison matrix (not shown). The AHP vector was calculated from eigenvector of comparison matrix and the parameter values reflected the importance of particular characteristics (tabel 3). The highest AHP score was 27.11 for sample M6 (the composition of fercaf:maize:rice flour was 70:10:20). This composition then used for subsequent experiments.

**Table 3.** The characteristic of noodles from fercaf:maise: rise flour at different composition; and analytic hierarchy process (AHP) scoring.

| Parameters | AHP vector parameters | Sampel Code | | |
|---|---|---|---|---|
| | | M5 | M6 | M7 |
| Strain (%) | 0.4684 | 33.25 | 33.39 | 19.34 |
| Yield (%) | 0.0441 | 60.91 | 50.68 | 55.91 |
| Cooking Weight (%) | 0.0759 | 86.49 | 84.14 | 71.16 |
| Hardness | 0.1436 | 14.69 | 19.75 | 18.17 |
| Stickiness | 0.2681 | -0.05 | -0.06 | -0.17 |
| AHP score | | 26.94 | 27.11 | 19.58 |

The noodle made from fercaf had no water binding capability due to lack of gluten protein. The addition of hydrocolloid might alter the gluten function thus improve the water binding. A number of hydrocolloids i.e.: xanthan gum (XG), guar gum (GG), arabic gum (AG) and carboxymethyl cellulose

(CMC) with the composition of 1, 1.5, 2 and 2.5% to the flour weight were applied to the flour mixture (M6). The elasticity, cooking weight, yield and texture (hardness and stickiness) of the resulted noodles were analyzed. The results are shown in table 4. The selection of the best hydrocolloid addition was also carried out using analytic hierarchy process (AHP) technique. The AHP vector parameter used in the selection of flour composition was also used in this selection.

**Table 4.** The characteristic of noodles from mix flour (fercaf:maize:rice of 70:20:10) with addition of hydrocolloid of xanthan gum (XG), guar gum (GG), arabic gum (AG) and carboxymethyl cellulose (CMC).

| Hydrocolloid | Amount (%) | Strain (%) | Yield (%) | Cooking Weight (%) | Hardness | Stickiness | AHP Score |
|---|---|---|---|---|---|---|---|
| XG | 1 | 34.22 | 52.61 | 51.15 | 24.29 | -0.19 | 25.77 |
| | 1.5 | 36.15 | 39.27 | 49.12 | 22.64 | -0.12 | 25.67 |
| | 2 | 48.40 | 49.23 | 61.50 | 20.07 | -0.09 | 32.42 |
| | 2.5 | 49.52 | 51.74 | 61.46 | 21.83 | -0.05 | 33.29 |
| GG | 1 | 27.62 | 60.05 | 57.83 | 15.18 | -0.10 | 22.18 |
| | 1.5 | 39.13 | 52.43 | 60.94 | 12.03 | -0.11 | 27.02 |
| | 2 | 38.64 | 51.24 | 65.55 | 8.41 | -0.10 | 26.57 |
| | 2.5 | 27.50 | 51.41 | 47.35 | 10.19 | -0.11 | 20.23 |
| AG | 1 | 20.02 | 57.68 | 63.91 | 12.51 | -0.05 | 18.58 |
| | 1.5 | 24.42 | 47.03 | 90.29 | 9.26 | -0.07 | 21.71 |
| | 2 | 18.26 | 49.56 | 81.68 | 19.32 | -0.10 | 19.74 |
| | 2.5 | 20.77 | 58.78 | 71.11 | 7.85 | -0.07 | 18.86 |
| CMC | 1 | 25.91 | 64.12 | 77.05 | 16.66 | -0.11 | 23.23 |
| | 1.5 | 20.98 | 53.44 | 103.91 | 15.75 | -0.16 | 22.37 |
| | 2 | 20.26 | 49.23 | 89.69 | 11.28 | -0.08 | 20.11 |
| | 2.5 | 15.49 | 44.37 | 88.61 | 24.25 | -0.15 | 19.46 |

The elasticity of noodles is an important property where it serves the strength and strain of the noodle. Table 4 show that XG resulted a better elasticity with the average strain value of 42.07% compare to other hydrocolloid such as GG with the average strain value of 33.20%. In addition, XG was also shown better hardness property (with value of 24.29). The experimental results show that the hardness decreased with the increasing amount of XG. The hardness is an important noodle property since the noodle with a high hardness value has a firm character.

Cooking weight property indicates the capability of cooked noodle to hold water. CMC in noodle could hold more water inside the cooked noodle among others up to 48%. It seem that CMC made a better structure in holding water [13]. However, noodle with high moisture content (i.e., high cooking weight) could be gummy.

Stickiness property of the noodle show how easy a noodle sticks one to other. The high negative value of stickiness property indicates the easiness of noodles to lose each other. Since less stickiness noodle are preferable, thus a lower stickiness value is better. AG (1%) and XG (2.5%) resulted the better stickiness properties.

The addition of hydrocolloid did not alter the yield significantly because the amount of hydrocolloid added into the flour mixture was relatively small (up to 2.5%). The experimental results show this phenomenon. The difference values among results might be caused by measurement random error.

By means of AHP technique, the addition of 2.5% XG has the highest score. It is obvious since XG improved elasticity and hardness of the noodles. In addition, 2.5% XG also gave a better stickiness.

In conclusion, the current study has shown that sole fercaf could not be used for noodle making. However, a mixture of fercaf along with maize and rice flours was successfully used for noodle making and gave good noodle properties. In addition, hydrocolloid such as xanthan gum can be used to improve the noodle properties.

**References**

[1] Abidin A Z 2009 Mie Basah Berbahan Dasar Tepung Singkong: Pengembangan Formulasi, Proses Produksi Dan Karakterisasi Produk *Seminar Nasional Teknik Kimia Indonesia* 19–20

[2] Charoenthaikij P, Jangchud K, Jangchud A, Piyachomkwan K, Tungtrakul P and Prinyawiwatkul W 2009 Germination conditions affect physicochemical properties of germinated brown rice flour *J. Food Sci.* **74** 658–65

[3] Raungrusmee S, Shrestha S, Sadiq M B and Anal A K 2020 Influence of resistant starch, xanthan gum, inulin and defatted rice bran on the physicochemical, functional and sensory properties of low glycemic gluten-free noodles *Lwt - Food Sci. Technol.* **126** 109279

[4] Milde L B, Chigal P S, Olivera J E and González K G 2020 Incorporation of xanthan gum to gluten-free pasta with cassava starch. Physical, textural and sensory attributes *Lwt - Food Sci. Technol.* **131** 109674

[5] Wangtueai S, Phimolsiripol Y, Vichasilp C, Regenstein J M and Schöenlechner R 2020 Optimization of gluten-free functional noodles formulation enriched with fish gelatin hydrolysates *Lwt - Food Sci. Technol.* **133**

[6] Charles A L, Huang T C, Lai P Y, Chen C C, Lee P P and Chang Y H 2007 Study of wheat flour-cassava starch composite mix and the function of cassava mucilage in Chinese noodles *Food Hydrocoll.* **21** 368–78

[7] Kresnowati M T A P, Listianingrum, Zaenudin A and Trihatmoko K 2015 The effect of microbial starter composition on cassava chips fermentation for the production of fermented cassava flour *AIP Conf. Proc.* **1699**

[8] Kresnowati M T A P, Bindar Y and Rahmina F 2018 Development of instant microbial starter for production of fermented cassava flour: Effect of vacuum drying temperature, carrier media, and storage temperature *J. Eng. Technol. Sci.* **50** 832–40

[9] Li Q, Obadi M, Qi Y, Liu S, Jiang Y, Zhang Q, Sun J, Jiang S and Xu B 2020 Softness, elasticity, and smoothness characteristics of cooked udon noodles based on texture analysis *J. Texture Stud.* **51** 444–52

[10] Saaty T L 1988 What is the Analytical Hierarchy Process? *NATO ASI Ser.* **48** 109–21

[11] Leal J E 2020 AHP-express: A simplified version of the analytical hierarchy process method *MethodsX* **7**

[12] Chai J, Liu J N K and Ngai E W T 2013 Application of decision-making techniques in supplier selection: A systematic review of literature *Expert Syst. Appl.* **40** 3872–85

[13] Boruvkova K and Wiener J 2011 Water absorption in carboxymethyl cellulose *Autex Res. J.* **11** 110–3

IOP Conf. Series: Materials Science and Engineering 1143 (2021) 012043 doi:10.1088/1757-899X/1143/1/012043

# The effect of size and solid content in hydrolysis of sweet potato starch using endogenous beta-amylase enzyme

**R Purwadi*, D Lestari, C A Lohoo, and J L Tirtaadji**

Food Engineering Study Program, Faculty of Industrial Technology, Institut Teknologi Bandung – Jl. Let. Jend. Purn. Dr. (HC) Mashudi no. 1 Jatinangor, Sumedang 45363 - Indonesia.

*Email: ronny.purwadi@che.itb.ac.id

**Abstract.** Sweet potato contains significant amount of endogenous amylase enzymes. The endogenous enzymes can be used in maltose production from sweet potatoes to carry out pre-hydrolysis of the starch to some extent, follow by complete hydrolysis using external enzymes. The main objective of the current study is to investigate the effect of particle size and solid to liquid ratio in hydrolysis process of sweet potato starch using endogenous enzymes. The extent of hydrolysis process is measured by dextrose equivalent of the crude maltose solution. Hydrolysis process consists of liquefaction and saccharification process. The liquefaction process is carried out at 71.5°C and pH 6 for 25 minutes while saccharification process is carried out at 53°C and pH 5,5 for 72 hours. Dextrose equivalent is measured using Lane-Eynon method. The results show that the access of enzyme to the potato tubers is the key to hydrolysis process. The smaller size of tubers gives bigger surface area while lower solid-to-liquid ratio increases enzyme mobility but decreases sugar concentration. Base on the results, a simulation shows that the pre-hydrolysis of sweet potatoes using endogenous enzymes can save almost a half of enzymes required for traditional hydrolysis process.

## 1. Introduction

Maltose is a disaccharide sugar formed from two units of glucose molecules. Even though maltose is less sweet than table sugar and fructose, it has been commonly used as ingredients in production of hard candies and frozen desserts due to its unique tolerance to heat and cold conditions. Nowadays, maltose is frequently used as sweetener replacing sucrose and fructose. Food companies have attempted to avoid an increasing negative public perception of fructose by replacing high-fructose syrup with high-maltose syrup. Maltose syrup can be produced by starch hydrolysis using β-amylase enzymes [1–3].

Sweet potato (*Ipomoea batatas*) is a tropical dicotyledonous plant and has large, starchy, sweet-tasting tuberous roots. Sweet potato starch is widely used in manufacture of food starch and sweetener; sizing of paper; as well as in textile industry [4]. The pulp is used for animal feed while the juice and spent process water is used for fertilizer or as a feedstock for biogas production. A number of study were also carried out in the use of sweet potato as a feedstock of the production of chemical precursors such as bioethanol and biobutanol [5–8]. In Indonesia, sweet potato is widely grown and the annual production exceed around 2 million tones and about 90% is used as food and its derivatives [9].

A sweet potato tuber contains around 22-28% of starch or about 70-80% in dry-basis [9]. The protein content of sweet potato is not more than 1% but a part of it forms as endogenous α-amylase and dominantly as β-amylase enzymes [1]. These enzymes can readily hydrolyze the starch into simple sugars in favorable conditions. A local famous snack called "ubi-cilembu" is made from a baked sweet potato where the potato is heated at 50-60°C prior to completely baked. The sweet taste arises from starch hydrolysis products. Since β-amylase enzyme is dominant in sweet potato, the hydrolysis product is dominated by maltose sugar.

As sweet potato is rich of starch as well as its endogenous amylase enzymes, the production of maltose syrup from sweet potato may be more suitable. Nevertheless, almost no publication reports the uses of sweet potato endogenous enzymes for carry out auto-hydrolysis in maltose syrup production.

In the current study, the utilization of sweet potato endogenous enzymes in carrying out starch auto-hydrolysis was investigated. Since hydrolysis reaction is heterogenous, thus contact area affects the reaction rate. Furthermore, enzyme mobility during hydrolysis reaction is also affect the reaction rate. Therefore, two process variables i.e., sweet potato particle size and solid to water ratio were examined to determine the effect of those variables on hydrolysis process.

## 2. Materials & Methods

### *2.1. Materials*
Sweet potatoes (SPs) were bought from local market. The tuberous roots were washed thoroughly. About 3 portions of 300 g of SPs were prepared differently i.e.: chopped into 0.5 cm dices, grated using a kitchen grater, and milled into slurry using a kitchen blender. A group of sample contains 5 individual samples that every sample was taken out along the saccharification process for analysis. Thus, every portion was divided into 20 g samples and transferred into 15 of 15 ml falcon tubes. Every 5 samples were diluted with water with ratio of 1:1; 1:3 and 1:5. There were 9 sample groups for every SPs preparation and solid-to-liquid ratio. The samples were directly used to avoid an extensive bacterial deterioration.

### *2.2. Methods*
The pH of all samples was adjusted to value of 6.0 by addition of 1 N HCl or 1 N NaOH. The samples were then stirred for about 2-3 minutes [10] using stirring rod. Furthermore, all samples were incubated at temperature of 71.5°C for 30 minutes to carried out liquefaction process.

After liquefaction process, all samples were readjustment to pH of 5.5 by addition of 1 N HCl. During pH readjustment, the samples were stirred for about 2-3 minutes using stirring rod. The saccharification process was carried out in temperature of 53°C for 72 hours. Every certain period of time, one falcon tube from every sample group was taken out for analysis. The analyses were carried out directly afterward.

### *2.3. Analyses*
In every sample, the liquid was separated from the solid by filtration using ordinary filter paper. The solid was dried at temperature of 60°C for a night, while the liquid was centrifuged at 3000 rpm for 5 minutes to separate the starch from hydrolysate. The hydrolysate was then directly analyzed using Lane-Eynon method [11,12].

## 3. Results and Discussions
Sweet potato contains number of endogenous enzymes such as α–amylase and β–amylase, while the β–amylase enzyme is about 5% of total protein contained in sweet potato [1]. These enzymes carry out hydrolysis reaction to break down its starch polymer into simple sugars such as monosaccharides (e.g. glucose) and disaccharides (e.g. maltose). The hydrolysis reaction is a heterogenous reaction where the enzymes must diffuse from bulk of solution into a polymer surface to carry out a hydrolysis reaction. Certainly, the hydrolysis process is affected by the mobility of enzymes as well as the surface area that enzyme can be attached [13].

The mobility of enzymes usually is affected by its dilution in water. The more dilute the enzyme, the more mobile the enzyme. On the contrary, more water is unfavorable to sugar solution as the main

IOP Conf. Series: Materials Science and Engineering 1143 (2021) 012043 doi:10.1088/1757-899X/1143/1/012043

product. The product concentration is preferred to be high enough to minimize the further separation process. In addition, the large surface area is usually provided by smaller particle size. However, preparation of smaller particles needs more energy, thus increase production operating cost. In this study, the effect of dilution and particle size of sweet potato to the hydrolysis process using endogenous enzymes is investigated.

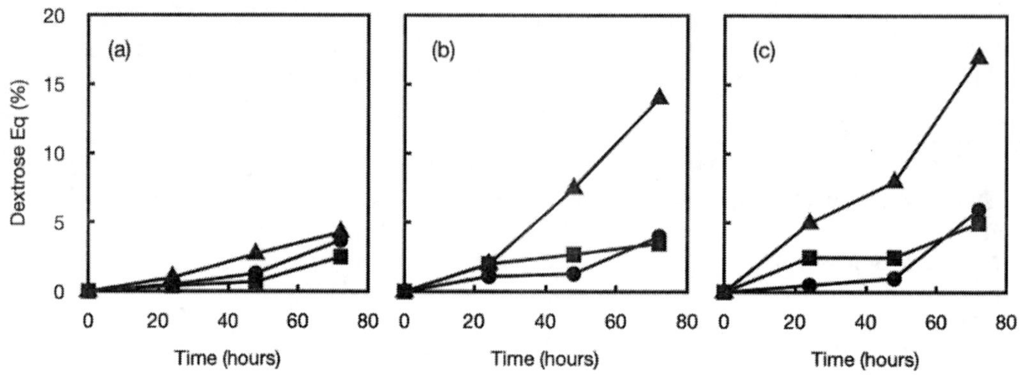


**Figure 1.** The profile of dextrose equivalent (%) during saccharification process of (▲) milled; (●) gratted; and (■) chopped sweet potatoes with solid-to-water ratio of (a) 1:1; (b) 1:3; and (c) 1:5.

This study carried out hydrolysis processes of sweet potatoes in different particle sizes and dilution ratios. The results of these hydrolysis processes are shown in figure 1. The dextrose equivalent (DE) indicates sugar contain in the solution. Along the time, the hydrolysis reaction was carried out in every samples and resulted increasing DE values. However, the hydrolysis of chopped and grated sweet potatoes resulted low dextrose equivalent (DE) values in every dilution ratio, while hydrolysis of sweet potato slurry in 1:3 and 1:5 dilution ratio results more than 3 times higher.

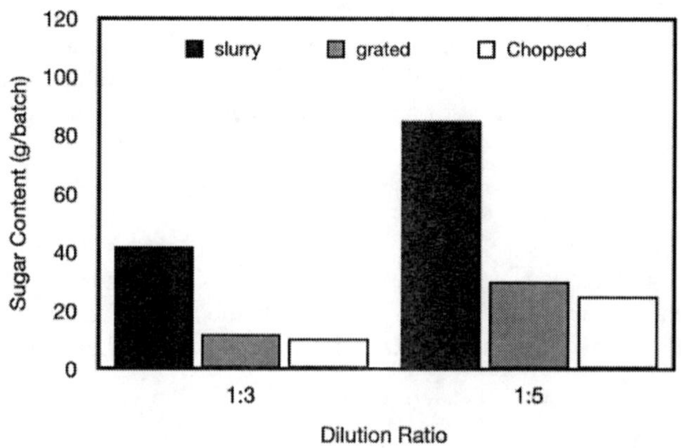


**Figure 2.** Comparison of sugar production at different dilution ratio from hydrolysis of slurry, grated and chopped sweet potato using its endogenous amylase enzymes.

Diced and grated sweet potatoes have less surface area thus hinder the transportation of enzymes into the starch surface even though in dilute solutions. Figure 1 shows that the particle sizes of diced and grated sweet potato were still big enough thus hinder the enzymes to access the inner starch polymer. In the hydrolysis of sweet potato slurry especially in high dilution ratio, the enzymes could easily access the starch polymer and yielded higher hydrolysis product with high sugar concentrations [14]. Nevertheless, the sweet potato slurry could not be hydrolysis successfully in low dilution ratio such as 1:1. Low dilution ratio leaded to low mobility of sugars produced from hydrolysis reactions thus stimulated an accumulation of sugars around enzymes. The high sugar concentration around enzymes might inhibit the hydrolysis reactions [15]. The product inhibition is usually occurred in enzymes mechanism due to its allosteric property.

The hydrolysis of sweet potato slurry in higher dilution ratio yielded higher DE values compare with that of hydrolysis of chopped and grated sweet potatoes. Even though the DE difference was trivial, the product volume made them different. This means that hydrolysis of potato slurry in dilution ratio of 1:5 produced more sugars in the solution (figure 2) and degree of hydrolysis is much higher.

According to International Starch Institute [4], maltose syrup has DE value of 50-55%. If this DE value is a target, endogenous enzymes in sweet potato can serve an adequate preliminary hydrolysis prior to ordinary enzymatic hydrolysis using external enzymes. Lin [16] shows that 0.033 mg amylase enzymes with activity of 750 IU/mg protein could increase the DE value of 87 point. These data were then used to develop a simple proportional calculation to compare the enzyme requirement of an ordinary hydrolysis process using sole external enzymes with that of a hydrolysis process with a help from endogenous enzymes. The simulation show that the use of endogenous enzymes can save about half of external enzymes (figure 3).

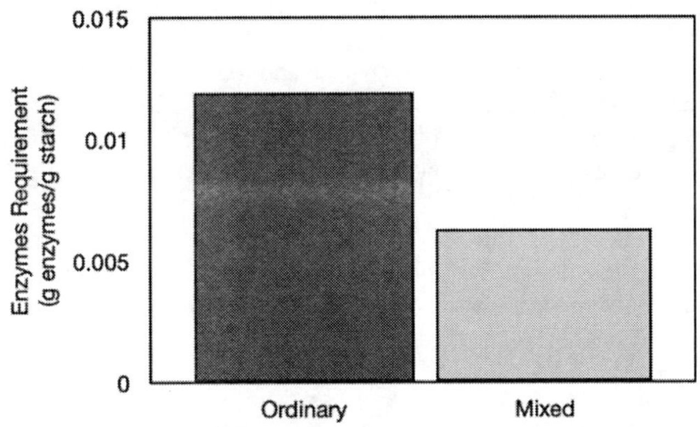


**Figure 3.** Comparison of external enzymes requirement in starch hydrolysis using sole external enzymes with that using endogenous enzymes followed with external enzymes addition

In conclusion, this study has shown that endogenous enzymes contained in sweet potato is very useful for preliminary starch hydrolysis with significant results. Furthermore, it also shows that endogenous enzymes in sweet potato may reduce the requirement of external enzymes needed to complete the hydrolysis to certain extend. These results are very useful for sweet potato hydrolysis process design which includes preliminary hydrolysis using endogenous enzymes. In addition, these results also contribute data and knowledge for further research and development.

## References

[1] Nakamura K, Ohto M A, Yoshida N and Nakamura K 1991 Sucrose-induced accumulation of β-amylase occurs concomitant with the accumulation of starch and sporamin in leaf-petiole cuttings of sweet potato *Plant Physiol.* **96** 902–9

[2] Das R and Kayastha A M 2019 Enzymatic hydrolysis of native granular starches by a new β-

amylase from peanut (Arachis hypogaea) *Food Chem.* **276** 583–90

[3] Pereira C R, Resende J T V, Guerra E P, Lima V A, Martins M D and Knob A 2017 Enzymatic conversion of sweet potato granular starch into fermentable sugars: Feasibility of sweet potato peel as alternative substrate for α-amylase production *Biocatal. Agric. Biotechnol.* **11** 231–8

[4] International Starch Institute 2006 *Starch Technology: Sweet potato starch production* vol 1

[5] Zhang L, Chen Q, Jin Y, Xue H, Guan J, Wang Z and Zhao H 2010 Energy-saving direct ethanol production from viscosity reduction mash of sweet potato at very high gravity (VHG) *Fuel Process. Technol.* **91** 1845–50

[6] Weber C T, Trierweiler L F and Trierweiler J O 2020 Food waste biorefinery advocating circular economy: Bioethanol and distilled beverage from sweet potato *J. Clean. Prod.* **268**

[7] Virgínio e Silva J O, Almeida M F, da Conceição Alvim-Ferraz M and Dias J M 2018 Integrated production of biodiesel and bioethanol from sweet potato *Renew. Energy* **124** 114–20

[8] He C R, Huang C L, Lai Y C and Li S Y 2017 The utilization of sweet potato vines as carbon sources for fermenting bio-butanol *J. Taiwan Inst. Chem. Eng.* **79** 7–13

[9] Ginting E, Utomo J S and Yulifianti R 2015 Potensi Ubijalar Ungu sebagai Pangan Fungsional *Iptek Tanam. Pangan* **6**

[10] Pontoh J and Low N H 1995 Glucose syrup production from Indonesian palm and cassava starch *Food Res. Int.* **28** 379–85

[11] H.C.S. de Whalley 1964 *ICUMSA Method of Sugar Analysis* (Amsterdam)

[12] Hildreth A . and Brown G B 1942 Modification of Lane-Eynon Method for sugar determination *J. AOAC Int.* **25** 775–8

[13] Snow P and O'Dea K 1981 Factors affecting the rate of hydrolysis of starch in food *Am. J. Clin. Nutr.* **34** 2721–7

[14] Bjorck I, Granfeldt Y, Liljeberg H, Tovar J and Asp N 2018 Food properties affecting the digestion and absorption of carbohydrates *Am. J. Clin. Nutr.* **59** 699–705

[15] Yankov D, Dobreva E, Beschkov V and Emanuilova E 1986 Study of optimum conditions and kinetics of starch hydrolysis by means of thermostable α-amylase *Enzyme Microb. Technol.* **8** 665–7

[16] Lin H J, Xian L, Zhang Q J, Luo X M, Xu Q S, Yang Q, Duan C J, Liu J L, Tang J L and Feng J X 2011 Production of raw cassava starch-degrading enzyme by Penicillium and its use in conversion of raw cassava flour to ethanol *J. Ind. Microbiol. Biotechnol.* **38** 733–42

IOP Conf. Series: Materials Science and Engineering 1143 (2021) 012044 doi:10.1088/1757-899X/1143/1/012044

# Cradle to Gate Life Cycle Assessment of Palm Oil Industry

**Ahmadi[1], Mahidin[2]*, M Faisal[2], Hamdani[3], K Siregar[4], Erdiwansyah[5]*, R Masturah[6], Nasrullah[6]**

[1] Graduate School of Chemical Engineering, Universitas Syiah Kuala, Banda Aceh 23111, Indonesia
[2] Department of Chemical Engineering, Universitas Syiah Kuala, Banda Aceh 23111, Indonesia
[3] Department of Mechanical Engineering, Syiah Kuala University, Banda Aceh 23111, Indonesia
[4] Department of Agricultural Engineering, Universitas Syiah Kuala, Banda Aceh 23111, Indonesia
[5] Faculty of Engineering, Universitas Serambi Mekkah, Banda Aceh 23245, Indonesia [6]Graduate School of Environmental Management, Universitas Syiah Kuala, Banda Aceh 23111, Indonesia

*Corresponding Author's Email: mahidin@unsyiah.ac.id, erdi.wansyah@yahoo.co.id

**Abstract.** Crude palm oil production Indonesia especially in Aceh every year continues to increase because the consumption of its derivative products such as cosmetics, food and fuel increases. The increasing demand for CPO has caused palm oil companies expanding their plantations. Based on data from the Ministry of Agriculture in 2018, the total area of Indonesian palm oil reaches 14.3 million hectares. The focus of this paper aims to evaluate the cradle to gate impact produced at land preparation, the nursery stage and the process of producing palm oil. Life cycle assessment is used to evaluate the impact of product for its entire life cycle. This research was conducted with simulation using SimaPro 9.0 software and impact 2002+ method. The results of the data processing with the simulation showed that fertilization stage was the stage with the largest impact with persentage 83% of the category, namely human health, ecosystem quality, climate change and resources 0.0000567 DALY, 9.44 PDF.m$^2$.year, 148 kg $CO_2$ eq and 1650 main MJ, respectively. Meanwhile, the processing of palm oil into CPO the largest environmental impact was in the category of non-renewable energy and global warming of 90 MJ and 9.6 kg $CO_2$ eq.

## 1. Introduction

Palm oil is one of Indonesia's mainstay commodities whose development is very fast and occupies important position in the plantation and agriculture sector because it is considered a good investment opportunity for the country's economic growth. Aceh is a province with fairly rapid development of the plantation sector, especially palm oil. At the end of 2018, palm oil plantations in Aceh reached 26,660 hectares with volume of production capacity of 95,234 metric ton [1]. Every year, the development of the Aceh palm oil industry continues to increase, as well as world market demand for CPO. The estimated world demand for CPO in 2025 is 218,300 million tons [2]. This shows that the future the Indonesian palm oil industry, including in Aceh, has the potential to continue to grow in order to meet the world's CPO needs.

Rapid development in the palm oil industry cannot be separated from various problems, negative issues and adverse environmental impacts among others as the cause of climate change [3], increasing greenhouse gases [4], deforestation, conversion of peatland and decreasing of biodiversity [5]. This led the European Union boycotted Indonesian palm oil products so that the impact on Indonesian palm oil exports was rejected by the European Union arguing that Indonesian palm oil is not environmentally friendly.

The palm oil industry consists of several stages, including land clearing, nursery, planting, fertilization, protection, harvesting and processing of fresh fruit bunch (FFB). These stages have their own environmental impact because in the process they use chemicals that have negative impact on environment, including human health, ecosystem quality, climate change and resources. Al Hakim et al reported that the largest source of emissions in palm oil nurseries is the use of diesel, which is 65.5%, where the use of diesel is used as fuel to operate a water pump twice a day followed by the use of fertilizer of 33.6% [6]. Research conducted by Aziz and Hanafiah showed that changes in fertilizer input, transportation and use of diesel in the palm oil industry had an effect on characterization results when compared to the basic scenario. The impact contribution of the alternative scenarios decreased by 2% - 5% [7].

One of the popular methods used to calculate energy, environmental impacts and costs arising from the product life cycle stages from the production cycle to waste handling is the Life Cycle Assessment (LCA). Life Cycle Assessment can assess and ensure the sustainability of production through comprehensive assessment and holistic approach so that it is able to provide accurate information that can be applied to environmental management, especially palm oil industry in Aceh [8]. In the data processing stage in this study, the SimaPro software was used. This software is chosen because its flexible and can be used in multi-user-versions [9].

## 2. Materials and Methods

### *2.1. Goal and scope, process unit, functional unit and boundary system*

In this paper, the focus chosen an approach cradle to gate on the processing of FFB, including land clearing, seedling, planting, fertilizing, protection, harvesting and processing of FFB at palm oil mills. The goal of this paper focuses on evaluating the environmental impacts of the Aceh palm oil industry. The functional unit used is one ton of FFB. In Figure 1 shows the boundary system in the study.

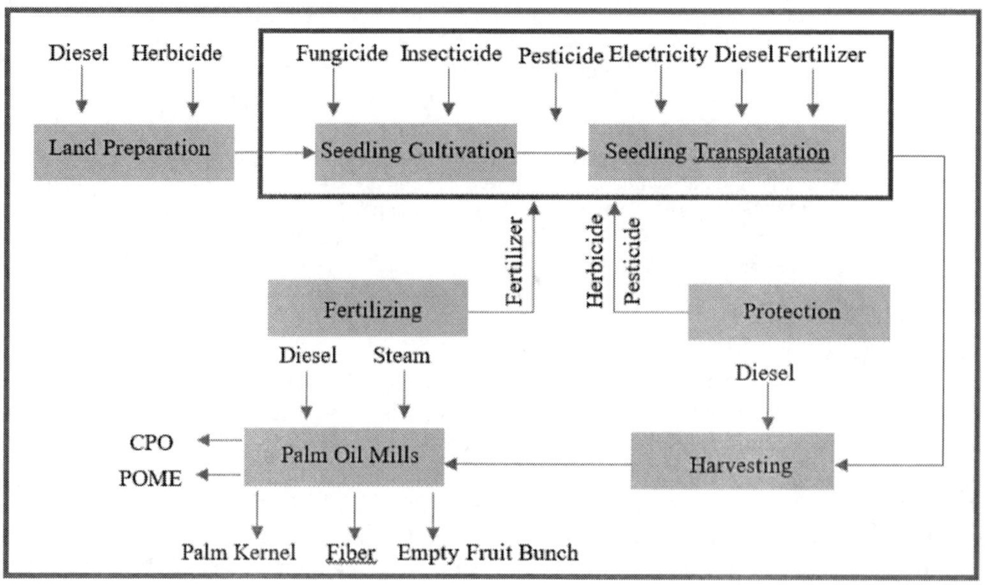


**Figure 1.** Boundary system of the study.

*2.2 Analysis method*

SimaPro is a software that can determine the sustainability of a product by evaluating the environmental impact of the product's inventory data. In this paper, SimaPro is used to analyse inventory data in the Aceh palm oil industry. Life cycle impact category indicators were calculated using Impact 2002+. Impact 2002+ evaluates 15 different impact categories at the midpoint level namely carcinogens, non-carcinogens, respiratory inorganics, ionizing radiation, ozone layer depletion, respiratory organics, aquatic eco-toxicity, terrestrial eco-toxicity, terrestrial acidification / nutrification, land occupation, aquatic acidification, aquatic eutrophication, global warming, non-renewable energy, mineral extraction. The impact category is then divided into four damage assessment categories, namely human health, ecosystem quality, climate change and resources. The division of the 15 impact categories at the midpoint level based on the damage assessment categories can be seen in Table 1.

**Table 1.** Impact categories in Impact 2002+.

| Impact Category | Midpoint Category | Damage Unit |
|---|---|---|
| Human Health | Carcinogens | DALY |
| | Non-Carcinogens | |
| | Respiratory Inorganics | |
| | Ionizing Radiation | |
| | Ozone Layer Depletion | |
| | Respiratory Organics | |
| Ecosystem Quality | Aquatic Ecotoxicity | PDF.m$^2$.year |
| | Terrestrial Ecotoxicity | |
| | Terrestrial Acidification/Nutrification | |
| | Land Occupation | |
| | Aquatic Acidification | |
| | Aquatic Eutrophication | |
| Climate Change | Global Warming | kg $CO_2$ eq |
| Resources | Non-Renewable Energy | MJ Primary |
| | Mineral Extraction | |

Several damage units to estimate the damage of which is DALY, PDF.m$^2$.year, kg $CO_2$ eq and MJ Primary. DALY (Disability-Adjusted Life Years) is person received measure of the overall burden of disease expressed as the number of years lost due to premature death or health problems. PDF.m$^2$.year (Potentially Disappeared Fraction) is the part of a species / ecosystem that potential to be lost per m$^2$ .per year. Kg $CO_2$ eq is a unit to determine the effect resulting from an activity that has an impact on global climate. MJ Primary is the basic amount of energy needed to extract natural resources [10].

2.3 Life cycle inventory

Inventory data comes from the Aceh palm oil industry. The amount of material and energy is important part of the inventory data that will be analysed its impact on the environment. The data obtained came from questionnaires, scientific journals, and SimaPro 9.0 database. Primary data were obtained from several palm oil plantations and mills that represented the Aceh palm oil industry. General information such as description of the process, the input and output of the boiler emissions and generator and the amount of POME generated from palm oil obtained based on the questionnaire that was developed based on ISO 14041 [11]. Interview directly with the head of the mill was also performed to validate the data obtained by agricultural practices and mills operations.

IOP Conf. Series: Materials Science and Engineering 1143 (2021) 012044 doi:10.1088/1757-899X/1143/1/012044

**Table 2.** Life cycle inventory in Aceh palm oil industry per ton of FFB.

| No | Process | Mass and Energy | Amount |
|---|---|---|---|
| 1 | Land Clearing | Herbicide (kg) | 0.16359 |
| | | Diesel (L) | 0.13357 |
| 2 | Seedling | Fungicide (kg) | 0.14706 |
| | | Insecticide (kg) | 0.01007 |
| | | Pesticide (kg) | 0.03477 |
| | | Meister of Fertilizer (kg) | 0.01539 |
| | | Urea/Sulfonylurea (L) | 0.21337 |
| | | $P_2O_5$ (kg) | 0.02033 |
| | | Muriate Photash (kg) | 0.00019 |
| | | Dolomite (kg) | 0.00038 |
| | | NPK (kg) | 0.11742 |
| | | Electricity (kWh) | 5.07301 |
| | | Diesel (L) | 0.93024 |
| 3 | Planting | TSP/SP36 (kg) | 1.8316 |
| | | Rock Phosphate (kg) | 0.23123 |
| 4 | Fertilizing | ZA (kg) | 26.2709 |
| | | ZA-Edta (kg) | 0.04754 |
| | | HGF Borate (kg) | 0.01636 |
| | | Rock Phosphate (kg) | 24.2835 |
| | | Cu-Edta (kg) | 0.05371 |
| | | Kieserit (kg) | 20.5037 |
| | | NPK (kg) | 86.4924 |
| | | HGFB (kg) | 0.44204 |
| 5 | Protection | Isopropyl Amine Glicosphate (L) | 0.16052 |
| | | Paraquat Dichloride (L) | 0.20589 |
| | | Troclopir (L) | 0.01451 |
| | | Florosipir (L) | 0.03025 |
| | | MethylMetsulfuron (kg) | 0.00633 |
| | | Fipronil (L) | 0.00247 |
| | | Polydor 25EC (L) | 0.00062 |
| | | Dipel (Bacillus thuringiensis) (L) | 0.0247 |
| | | Thuricider HO (kg) | 0.00154 |
| | | Pesticide (kg) | 0.36404 |
| 6 | Palm Oil Mill | Diesel (L) | 1.63584 |
| | | CPO (kg) | 202.7972 |
| | | Steam Consumption (kg) | 11.3563 |
| | | Palm Kernel (kg) | 97.9021 |
| | | Empty Fruit Bunch (kg) | 160.8392 |
| | | POME (kg) | 699.3 |
| | | Fibre (kg) | 97.9021 |

Land clearing is carried out with manual and semi-mechanical systems so that soil fertility is not damaged and is well preserved. Land clearing requires a lot of worker in short time and incidental. Therefore, the implementation of this work is given to contractors who are experienced in preparing land clearing until the land is ready for planting. The Seedling stage begins when the sprouts are planted in small polybags. After 90 days, the grown seedlings are transferred to large polybags and waited for

225 days to become mature palms. This stage requires fertilizer, herbicide and fungicide as its application. After the seeds are ready, they will enter the planting stage. At the planting stage, a planting hole is made a few days before planting with size of 50×40 cm and depth of 40 cm. The excavated top soil with thickness of 20 cm is separated from the bottom soil. The time chosen for planting is during the rainy season. This is intended to provide sufficient water for seeds to grow well. Palm oil cultivation only hopes for rainfall because there is no irrigation in Aceh's oil palm plantations.

Fertilization is done according recommendation of the plantation RND team. The types of fertilizers used are inorganic fertilizers namely NPK, ZA, HGF Borate, Rock Phosphate, Cu-Edta, Kieserit and HGFB. Fertilization is not done when the rain intensity is high because it reduces water pollution by fertilizers. Meanwhile, protection is carried out manually and chemically according predetermined rotation of 6-12 rotations/year. After three years, the palm oil is ready to be harvested and delivered to the palm oil mill to process for CPO. In palm oil mills, FFB is weighed and sorted according to ripeness of the fruit. The selected fruit is ripe right. The sorted FFB then goes into the sterilization and pressing stages to extract crude palm oil. At the oil pressing stage, fibre is formed which can be used as boiler fuel. Nut that is generated from the pressing stage has to be separated in order to be cracked so that palm kernel in it obtained. Emissions generated from palm oil mills are $SO_2$, $NO_2$, $NH_3$, HCl, HF and particulates from boilers and generators used in palm oil mills and POME is wastewater that can be used as methane gas producer.

## 3. Result and Discussion

Life cycle impact assessment aims to evaluate the impact of an activity/process caried out. Potential impacts were determined using the results from inventory data analysis. The inventory data obtained were analysed by translating all relevant outputs and emissions into environmental impact score. In this paper, evaluation of the environmental impact of plantation activities and palm oil mills was carried out. The results of data processing using SimaPro software in analysis of the impact of palm plantation activities can be seen in Figure 2 Figure 3 and Table 3. Meanwhile, the analysis of the impact of palm oil mill activities can be seen in Table 4.

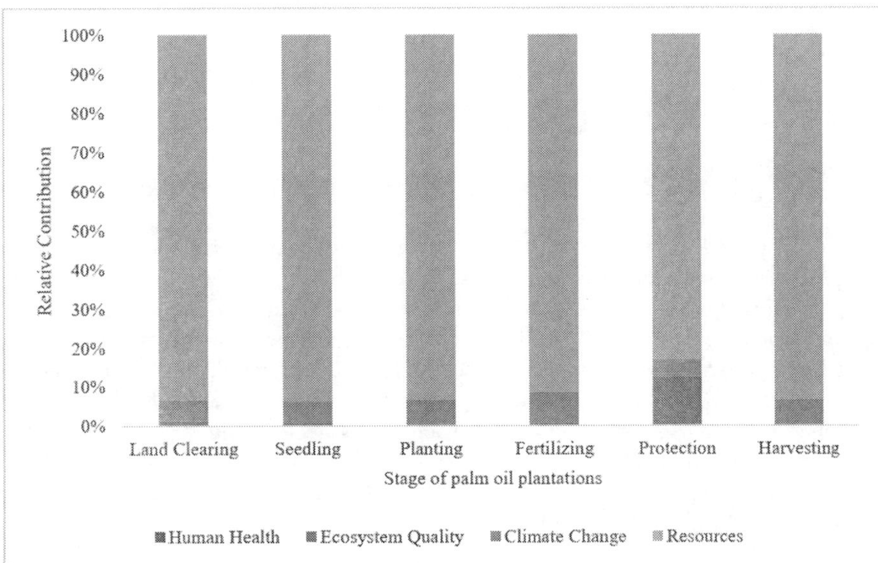


**Figure 2.** Damage assessment of six stages in palm oil plantations.

Figure 2 shows that the largest percentage is generated on the impact of resources, namely greater than 80% at each stage. This is due to the use of non-renewable energy sources such as natural gas, coal and petroleum in the manufacturing process of chemicals for palm plantation applications [12]. While the smallest impact was obtained in the category of impacts on human health and ecosystem quality, but at the protection stage, percentage of impact on ecosystem quality was the highest among other stages. This is because the use of pesticides and herbicides at the protection stage not only kills the pests that exist in palm oil plants but can also disrupt the ecosystem in the soil and water [13]. Pests can reduce the productivity of oil palm plants by 20-22%. But the use of pesticides and herbicides to certain extent can increase soil erosion, reducing organic matter, increasing pest resistance to pesticides and herbicides. The impact of ecosystem quality and climate change is very influential on the impact of resources [14]. The use of chemical fertilizers in the fertilization stage provides the highest impact resources with value of 1650 MJ primary.

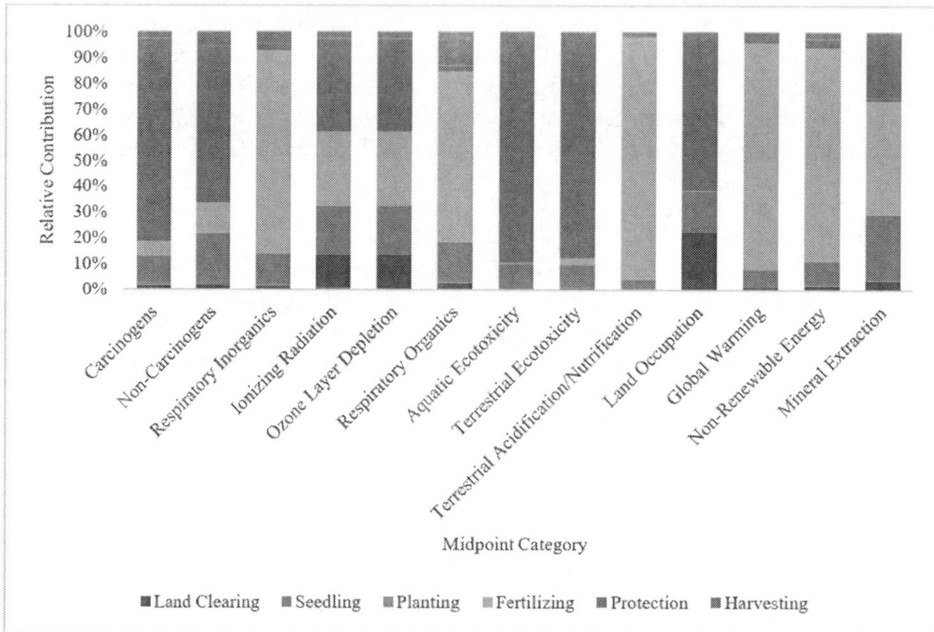


**Figure 3.** Midpoint damage assessment in palm oil plantations.

At the midpoint level, the impact results are given in different units. Materials entered in ton are converted into emission-based impact categories. Figure 3 shows the relative contribution of each stage for each impact category. From this figure, it can be seen that the most dominant impacts are the stages of fertilization, protection and seedling and land clearing. This is because the use of inorganic fertilizers from N fertilizers, pesticides, herbicides and insecticides is the biggest contributor to the impact of the plantation stage [15] [16]. Research conducted by Hidayatno found that land clearing using non-burning techniques can reduce environmental impacts and selecting forest as land for palm oil plantations is also better than using peatlands [17]. Globally, peatlands store around 329-525 giga tons of carbon and around 14% are in the tropics. This large carbon stock also causes the high amount of carbon released into the atmosphere when peatlands are burned during line clearing [18].

**Table 3.** Damage assessment of six stages in oil palm plantations.

| Kategori Dampak | Unit | Land Clearing | Seedling | Planting | Fertilizing | Protection | Harvesting |
|---|---|---|---|---|---|---|---|
| Human Health | DALY | $1.16\times10^{-6}$ | $9.17\times10^{-6}$ | $1.93\times10^{-8}$ | $5.67\times10^{-5}$ | $2.93\times10^{-6}$ | $3.6\times10^{-6}$ |
| Ecosystem Quality | PDF.m$^2$.year | 0.457 | 1.57 | $9.75\times10^{-4}$ | 9.44 | 10.3 | 0.183 |
| Climate Change | kg $CO_2$ eq | 1.83 | 11.4 | 0.0187 | 148 | 3.67 | 3.50 |
| Resources | MJ Primary | 32.5 | 192 | 0.277 | 1650 | 69 | 51.9 |

Based on Table 3, it is known that the stage that has the greatest impact from each impact category is the fertilization stage. This stage is the main contributor to the impact category on oil palm plantations, accounting for 83% of the total impact. This fertilization stage has impact categories, namely human health, ecosystem quality, climate change and resources with respective values of 0.0000567 DALY, 9.44 PDF.m$^2$.year, 148 kg $CO_2$ eq and 1650 MJ primary. Meanwhile, the other five stages only have value of up to $9.17\times10^{-6}$ DALY, 10.3 PDF.m$^2$.year, 11.4 $CO_2$ eq, and 192 MJ primary.

For human health impact units, the scale of damage for DALY has range of values from $0 \leq$ DALY $\leq 1$, where the closer to 0, the smaller the impact on humans. The resulting human health impact is very small and can be ignored. Research conducted by Siregar et al. found that the fertilization stage produced the highest contribution when compared to other plantation stages [19]. Similar results were also reported by Zulkifli et al., the use of N fertilizer and N fertilizer production were the biggest contributors to the impact in the production of fresh fruit bunches for processing CPO in palm oil mills [20].

**Table 4.** Damage assessment of palm oil mills.

| Midpoint Category | Unit | Amount |
|---|---|---|
| Carcinogens | DALY | $2.75\times10^{-8}$ |
| Non-Carcinogens | | $1.12\times10^{-7}$ |
| Respiratory Inorganics | | $2.97\times10^{-5}$ |
| Ionizing Radiation | | $1.97\times10^{-9}$ |
| Ozone Layer Depletion | | $8.96\times10^{-11}$ |
| Respiratory Organics | | $6.59\times10^{-9}$ |
| Aquatic Ecotoxicity | PDF.m$^2$.year | $2.51\times10^{-4}$ |
| Terrestrial Ecotoxicity | | $3.80\times10^{-2}$ |
| Terrestrial Acidification/Nutrification | | 1.78 |
| Global Warming | kg $CO_2$ eq | 9.6 |
| Non-Renewable Energy | MJ primary | 90 |
| Mineral Extraction | | 0.0242 |

Based on Table 4, it can be seen that the most influential impact of palm oil mills is on non-renewable energy and global warming with the respective value of 90 MJ and 9.6 kg of $CO_2$ eq. The use of diesel fuel in generators and boilers in generating electricity for palm oil mills is the cause of the high impact of these two categories. In the research conducted by Shafie et al., the use of non-renewable fuels such as diesel is the key variable in the high impact of global warming and non-renewable energy [21]. Similar results were also shown in the research of Kaewmai et al., the use of fossil fuels in energy fulfilment palm oil mills is the highest contributor to global warming [22]. Midpoint categories that are quite influential on palm oil mills are aquatic eco-toxicity and terrestrial eco-toxicity, the values were $2.51\times10^{-4}$ dan $3.80\times10^{-2}$ respectively. PDF units show potentially lost species / ecosystems calculated in $m^2$year units. This means that the higher the value of PDF, the greater the potential for species / ecosystems to be lost.

## 4. Conclusion

In oil palm plantations, fertilization stage was the stage with the largest impact with percentage of 83% of the category, namely human health, ecosystem quality, climate change and resources 0.0000567 DALY, 9.44 PDF.m2.year, 148 kg $CO_2$ eq and 1650 main MJ respectively. Meanwhile, the processing of palm oil into CPO, the largest environmental impact was in the category of non-renewable energy and global warming of 90 MJ and 9.6 kg CO2 eq. The use of diesel fuel in generators and boilers in generating electricity for palm oil mills is the cause of the high impact of these two categories.

**Acknowledgments**

This research was supported by grant (No: 292 / UN11 / SPK / PNBP / 2020) from Universitas Syiah Kuala, Darussalam, Banda Aceh, Republic of Indonesia

## References

[1] PTPN 2019 *Kelapa Sawit Dominasi Sektor Perkebunan di Aceh* Available: http://www.bumn.go.id/ptpn5/berita/1-Kelapa-Sawit-Dominasi-Sektor-Perkebunan-di-Aceh. [Accessed: 13-Feb-2020].

[2] Gapki 2018 *Proyeksi Vegetable Oil Dunia Tahun 2025: Bagaimana Posisi Indonesia di Masa Mendatang* Available: https://gapki.id/news/3124/proyeksi-vegetable-oil-dunia-tahun-2025-bagaimana-posisi-indonesia-di-masa-mendatang. [Accessed: 13-Feb-2020].

[3] Oosterveer P 2015 Promoting sustainable palm oil: Viewed from a global networks and flows perspective *J. Clean. Prod* **107**

[4] Jamaludin N F, Muis Z A, and Hashim H 2019 An integrated carbon footprint accounting and sustainability index for palm oil mills *J. Clean. Prod.*

[5] Teoh C H 2010 *Key sustainability issues in the palm oil sector: A discussion paper for multi-stakeholders consultations (commissioned by the World Bank Group)* IFC (International Financ. Corp. World Bank Group)

[6] Al Hakim H, Supartono W, and Suryandono A 2014 Life Cycle Assessment Pada Pembibitan Kelapa Sawit Untuk Menghitung Emisi Gas Rumah Kaca *Ziraa'ah Maj. Ilm. Pertan.* **39** (2) 72–80

[7] Aziz N I H A and Hanafiah M M 2020 Life cycle analysis of biogas production from anaerobic digestion of palm oil mill effluent *Renew. Energy* **145** 847–857

[8] Finnveden G et al. 2009 Recent developments in life cycle assessment *J. Environ. Manag.* **91** (1) 1–21

[9] Putri H P and Lingkungan D T 2017 Life Cycle Assessment ( LCA ) Emisi Pada Proses Produksi Bahan Bakar Minyak ( BBM ) Jenis Bensin Dengan Pendekatan Analytical Hierarchy Process (AHP )

[10] Jolliet O et al. 2003 IMPACT 2002+: A New Life Cycle Impact Assessment Methodology *International Journal of Life Cycle Assessment*

[11] Ekvall T and Finnveden G 2001 Allocation in ISO 14041 - a critical review *J. Clean. Prod.*

[12] Andrian D and Irawati D Y 2019 Dampak Proses Pengolahan Air Bersih Terhadap Lingkungan *Heuristic* 16 (1)

[13] Sari Y K, Niswati A, Arif M A S, and Yusnaini S 2015 Pengaruh Sistem Olah Tanah dan Aplikasi Herbisida Terhadap Populasi dan Biomassa Cacing Tanah Pada Pertanaman Ubi Kayu ( Manihot utilissima ) *Agrotek Trop.*

[14] Darwia S, Ichwana, and Mustafril 2017 Jurnal Ilmiah Mahasiswa Pertanian Unsyiah Volume 2,

Nomor 1, Februari 2017 www.jim.unsyiah.ac.id/JFP, J. Ilm. Mhs. Pertan. Unsyiah, vol. 2 (1) 320–330

[15] Choo Y M, Muhamad H, Hashim Z, Subramaniam V, Puah C W, and Tan Y 2011 Determination of GHG contributions by subsystems in the oil palm supply chain using the LCA approach *Int. J. Life Cycle Assess.*

[16] Stichnothe H and Schuchardt F 2011 Life cycle assessment of two palm oil production systems *Biomass and Bioenergy*

[17] Hidayatno A, Zagloel T Y M, Purwanto W W, and Anggraini L 2011 Cradle To Gate Simple Life Cycle Assessment of Biodiesel Production in Indonesia *Makara Technol. Ser.*

[18] Nurhayati A D, Aryanti E, and Saharjo B H 2010 Kandungan Emisi Gas Rumah Kaca Pada Kebakaran Hutan Rawa *J. Ilmu Pertan. Indones.*

[19] Siregar K, Tambunan A H, Irwanto A K, Wirawan S S, and Araki T 2015 A Comparison of Life Cycle Assessment on Oil Palm (Elaeis guineensis Jacq.) and Physic Nut (Jatropha curcas Linn.) as Feedstock for Biodiesel Production in Indonesia *in Energy Procedia*

[20] Zulkifli H, Halimah M, Chan K W, Choo Y M, and Mohd Basri W 2010 Life cycle assessment for oil palm fresh fruit bunch production from continued land use for oil palm planted on mineral soil (part 2) *J. Oil Palm Res.*

[21] Shafie S M, Mahlia T M I, Masjuki H H, and Andriyana A 2011 Current energy usage and sustainable energy in Malaysia: A review *Renewable and Sustainable Energy Reviews.*

[22] Kaewmai R, H-Kittikun A, and Musikavong C 2012 Greenhouse gas emissions of palm oil mills in Thailand *Int. J. Greenh. Gas Control*

IOP Conf. Series: Materials Science and Engineering 1143 (2021) 012045 doi:10.1088/1757-899X/1143/1/012045

# Techno-Enviro-Economic Study of Hydrogenated Vegetable Oil Production from Crude Palm Oil and Renewable Hydrogen

**A H Vivadinar[1], W W Purwanto[1,*]**

Sustainable Energy Systems and Policy Research Cluster, Department of Chemical Engineering, Faculty of Engineering, Universitas Indonesia, Depok, 16424, Indonesia

*widodo@che.ui.ac.id

**Abstract.** From the perspective of technical, environment, and economics, hydrogenated vegetable oil (HVO) production with renewable hydrogen from biomass gasification (BG), geothermal electrolysis (GEO-E), and solar photovoltaic electrolysis (PV-E) is compared to steam methane reforming (SMR). The purpose of this study is to evaluate the energy efficiency, greenhouse gases emission factor, and production cost of HVO from various hydrogen production technologies. Production technologies are simulated using Aspen Plus ®. HVO is produced by hydrotreating and hydroisomerization reactions. The processes produce three main products: HVO, green naphtha, and bio-jet fuel. The feedstock used to produce hydrogen from BG is empty fruit bunch (EFB). Renewable electricity produced by geothermal combination of flash system and organic Rankine cycle (ORC) and solar photovoltaic (PV) with battery. Technical analysis is done by calculating the energy efficiency of overall system. Production cost of HVO is calculated by levelized cost of energy (LCOE). Life cycle analysis (LCA) is carried out to calculate $CO_2$-eq intensity. The result shows that HVO production with SMR has the highest energy efficiency, 55.67%, which then followed by BG, 31.47%, PV-E, 9.34%, and GEO-E, 7.89%. The order of LCOE obtained from lowest to highest is HVO production from hydrogen produced by SMR (15.78/GJ-HVO), BG ($16.37/GJ-HVO), GEO ($22.83/GJ-HVO), and PV ($27.29/GJ-HVO). However, for $CO_2$-eq intensity, the order is HVO productions with the hydrogen produced by GEO-E, PV-E, SMR, and BG are 1.63 kg$CO_2$-eq/kg HVO, 1.86 kg$CO_2$-eq/kg HVO, 5.57 kg$CO_2$-eq/kg HVO, and 16.52 kg$CO_2$-eq/kg HVO, respectively.

## 1. Introduction

Through the biofuel blending mandate (Regulation of MEMR No. 12/2015), the government of Indonesia has decided to shift transportation fuel into an environmentally friendlier option. This mandate rules the government's target of the biodiesel uses for non-public service obligation (PSO) transportation by 2020 and 2025 to be 30% from total need [1]. It should be noted that 30% is also the highest blending target possible to be reached by the fatty acid methyl ester (FAME), due to the nature of the fuel in an unmodified diesel engine complying to the blended fuel standard ASTM D7467 [2]. While government has expressed its plan to blend 50% of biodiesel, such plan will not be plausible if FAME is still being used, unless a substitute for FAME is proposed. One of the potential alternatives is hydrogenated vegetable oil from crude palm oil (CPO).

Commercial production of HVO has been demonstrated from edible vegetable oils with claims of usage in unmodified diesel engine. Neste, one of the producer of HVO, has their production plant widely spread in Europe and Asia [3], [4]. Eni with Honeywell-UOP have also converted their conventional refineries into bio-refineries to produce HVO [5]. These production technology of HVO through

IOP Conf. Series: Materials Science and Engineering 1143 (2021) 012045 doi:10.1088/1757-899X/1143/1/012045

hydrogenation reactions followed by isomerization [5], [6]. The former involves treating triglyceride with hydrogen to produce long chain paraffin, while the latter restructure paraffin to improve its cold properties.

Technical and economic analysis had also been discussed widely. Technical analysis simulates the production of HVO mainly through a series of reaction of hydrogenation, decarboxylation, and decarbonylation to attain high levels of n-$C_{15}$ to n-$C_{18}$ [7], [8]. Although most process simulation studies only simulated the hydrogenation process and hydrogenation followed by isomerization corresponding to available commercial production with co-products of green naphtha and bio-jet fuel [9], [10], [11]. A similar study analysed the sensitivity of HVO production from rubber seed oil to evaluate the economics and it showed that the production cost is driven by feedstock and hydrogen prices [11]. A techno-economic of HVO production using crude palm oil (CPO) is also performed by Martinez-Hernandez [9] where it revealed that production cost for HVO is a more viable investment than for bio-jet fuel due to current product price trend. It also explained that the most affecting cost for HVO production is feedstock price, followed by hydrogen price and capital cost [9]. These previous studies have shown sensitivity of pure hydrogen price, but none of them has ever counted the cost of hydrogen from several supply paths.

Based on a report published by IEA [12] , the supply of hydrogen is one of the challenges to produce HVO. HVO production must have an economically and environmentally viable hydrogen supply. An ERIA study [13] reported that various hydrogen production technologies have differ in cost trends and $CO_2$ footprints, making these two factors as determining acceptable decision for HVO production path. Neither any of the previous studies has considered these two factors. Therefore, this work aims to assess the technical, economical, and environmental aspects of HVO production with different hydrogen supply sources. These hydrogen production technologies include steam methane reforming (SMR), biomass gasification (BG), solar photovoltaic (PV) power with electrolysis (PV-E), and geothermal power with electrolysis (GEO-E).

## 2. Methods

The technical, economic, and environmental perspective is carried out in this study. Overall system efficiency is performed to evaluate the technical performance. Life cycle analysis (LCA) is measured to evaluate $CO_2$-eq intensity. Finally, an economic analysis is done by levelized cost of production for each configuration.

### 2.1. Process Simulation

Process simulation is conducted using Aspen Plus® v.11 with a flow diagram as shown on Figure 1. The capacity of HVO production is set to one million tons per year corresponding to the typical commercial HVO production. An assumption of hydrogen being directly fed to the HVO production is taken.

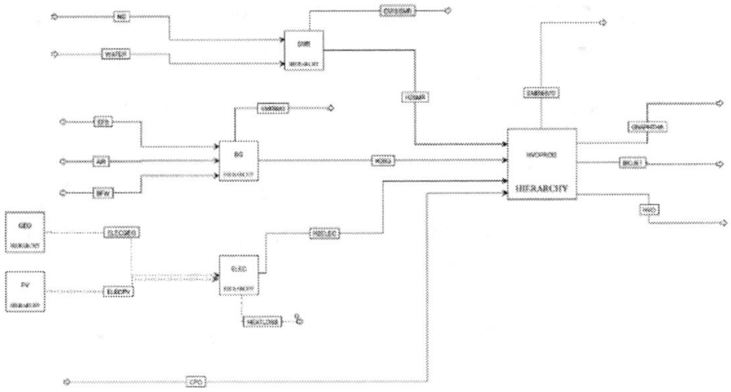


**Figure 1.** Process flowsheet of hydrogen and HVO production.

*2.1.1. Hydrogen Production.* Reactions of SMR are simulated by the reactor blocks in Aspen Plus. The first reactor is to simulate the "pre-reforming" section which is consequent to the reforming reactor. This reforming reactor is modelled using the tool block RPLUG, with the condition operation and reaction kinetics adopted from [14]. In order to maximize the yields of hydrogen production, water gas shift reactor is incorporated after the reforming reactor [15]. Several studies has stated an efficient performance to purify the hydrogen from reforming reactor using pressure swing adsorption [16], [17]. Thus, a pressure swing adsorption is modelled to purify the hydrogen to 99% using the SEPARATOR block with condition operation adopted from the experiment study of Ribeiro, et. al. [17].

BG system is simulated the gasifier followed by water gas shift and pressure swing adsorption. The biomass used is palm oil empty fruit bunches (EFB). Modelling the EFB in the simulator is done by inputting the proximate and ultimate analysis adopted from [18]. A series of reaction occurs in the reactors modelled using the kinetics that is adopted from [19]–[22] with operation condition of the gasifier adopted from [18], while the steam to biomass ratio is set to be 2, complying to the range set in [23].

Electrolysis used for the production of hydrogen is alkaline electrolysis (AEL) with solar energy and geothermal energy as its energy source. Due to the unavailability to model AEL reaction in Aspen Plus, CALCULATOR block is used alongside RSTOIC. CALCULATOR block has a built-in Fortran which enables to model the reaction using the equation adopted from [24]. Physical characteristics of AEL module, such as active area and current density are based on operating data taken from [25].

PV array is located in South Sumatra with daily GHI of 5.21 kWh/m$^2$ and equipped with electricity storage because electricity supply must be continuous to support the production of HVO for intermittent electricity from Solar PV [26]. The efficiency of the Solar PV is 17.54% based on the World Bank Report [26], while the efficiency of the Li-ion battery is 90% [27]. PV array is designed using the equation 1 with $A_{PV}$ being area of PV array needed (m$^2$), $L_{el}$ being the electricity needed, $H_{avg}$ being the GHI (kWh/m$^2$). Battery size is designed using the equation 2 where $B_{sc}$ is battery capacity (MWh), N is hours of no sun (h), and DOD is Depth of Discharge (%).

$$A_{PV} = \frac{L_{el}}{H_{avg} \times \eta_{PV} \times \eta_B} \tag{1}$$

$$B_{sc} = \frac{N \times L_{el}}{\eta_B \times DOD} \tag{2}$$

Similar to PV array, geothermal power plant is set to be at South Sumatra. A combination of Organic Rankine Cycle (ORC) and flash type geothermal power plant is assumed to be used and the efficiency of power plant is 13.37% [28].

*2.1.2. HVO Production.* The production process is assumed to have a capacity factor of 91%. Products resulted from the plant are HVO, green naphtha, and bio-jet fuel. The simulation is ensured to produce those three products complying to the standard and commercial product specification, such as EN 15940:2016, EN 590:2013, ASTM D975:15b, and the commercially available products as seen on [29]–[31]. CPO, as the feedstock, is modelled as refined, bleached, deodorized, palm oil (RBDPO), where the composition of this feedstock is adopted from [10], due to higher HVO yield achieved by reacting RBDPO with hydrogen instead of untreated CPO according to [32]. The processes of HVO production consists of three section: (1) hydrotreating, (2) hydroisomerization, and (3) product separation.

The reactors used in the processes are for hydrotreating reaction and hydroisomerization reaction. Hydrotreating reactor is modelled using the tool REQUIL in Aspen Plus where the stoichiometric reaction adopted from Manco et al[10]. Hydroisomerization reactor is modelled using the RYIELD tool and based on the experiment in [33]. Purifying the paraffin from unreacted hydrogen and water is done by using two phase separators.

Product separation uses two distillation columns and are modelled using the RADFRAC tool. The top product of the first distillation column separates C3 to C9, which consequently must be passed into a decanter to separate the C3 from the product. C3 produced is used as fuel for the plant, while the rest C4 – C9 is sold as green naphtha. Bottom product of the first distillation column is passed into the second distillation column to produce bio-jet fuel (C10-C14) and HVO (C15-C18).

### 2.2. *System Efficiency*

Overall system efficiency for SMR, BG, and HVO production are defined by the equation 3.

$$\eta_j = \frac{\sum m_i HHV_i}{\sum m_k HHV_k} \quad (3)$$

Where $\eta_j$ is energy efficiency for each technology, $m$ is mass flow (kg/h), $HHV$ is Heating Value (MJ/kg), $i$ is the type of input flow, $k$ is the type of output flow.

Equation 4 is used to calculate the system efficiency in electrolysis is as followed.

$$\eta_{electrolysis} = \frac{m_{H_2} HHV_{H_2}}{P} \quad (4)$$

Where $\eta_{electrolysis}$ is system efficiency for electrolysis, $m_{H2}$ is mass flow of hydrogen (kg/s), $P$ is power needed for electrolysis (W), $HHV_{H_2}$ is hydrogen heating value (J/kg).

### 2.3. *Cost Evaluation*

Cost evaluation is done using levelised cost of production of hydrogen and biofuels. The methodology used is levelised cost of energy (LCOE) which considers capital expenditures (CAPEX), fixed operation and maintenance expenditures (OPEX), and feedstock cost of expenditure (FS). The equation used for this methodology is expressed in Equation 5 with $C$ being the CAPEX ($), $O$ being the OPEX ($/year), $F$ being the FS ($/unit feedstock), $CRF$ being the cost recovery factor as a function of lifetime and discount factor, where the plant lifetime is assumed to be 30 year and the discount factor is 10%, $Cf$ being the capacity factor (%), and $q$ being the production capacity (ton/year). CAPEX is estimated based on Table 1 and using the sixth-tenth rule for adjustment of the plant scale.

$$sLCOE_{it} = \left( \frac{[C_{it} CRF_i] + O_{fixed,it} + F_{it}}{8760 \times Cf_{it} \times q_{it}} \right) \quad (5)$$

**Table 1.** CAPEX and OPEX data.

| Technology | CAPEX | | OPEX (USD/y) (percentage of CAPEX) | Ref. |
|---|---|---|---|---|
| | Value | Unit | | |
| Solar PV | 1,158.00 | USD/kW | 1% | [34] |
| Battery | 298.00 | USD/MWh | 2.50% | [35] |
| Geothermal Power Plant | 5.67 | M USD/MW | 2% | [36] |
| Electrolyser | 840.00 | USD/kW | 2% | [37] |
| SMR | 1.02 | M USD/ton/day | 5% | [38] |
| BG | $360 \times \left(\frac{capacity}{150}\right)^{0.75}$ | M USD/ton/day | 5% | [39] |
| HVO Plant | 1,200 | M USD/MT/year | 6% | [40] |

**Table 2.** Feedstock cost of expenditure.

| Feed | Price |
|---|---|
| CPO | $0.60 per kg CPO |
| EFB | $0.045 per kg EFB |
| Natural Gas | $6 per MMBtu |

The CAPEX projection is predicted to evaluate cost reduction in the future for HVO production with renewable hydrogen, especially from GEO-E and PV-E with high learning rate. Learning rate of PV, electrolyser, and battery are 37%, 55.8%, 18%, respectively [41], [42], [43].

### 2.4. *Life Cycle Analysis*

$CO_2$ intensity is calculated $CO_2$ footprint every configuration. The methodology used is life cycle analysis (LCA) of cradle-to-gate with the system boundary covering the production of hydrogen to the end-product of HVO (Figure 2). The functional unit in this paper is set as kg-HVO.

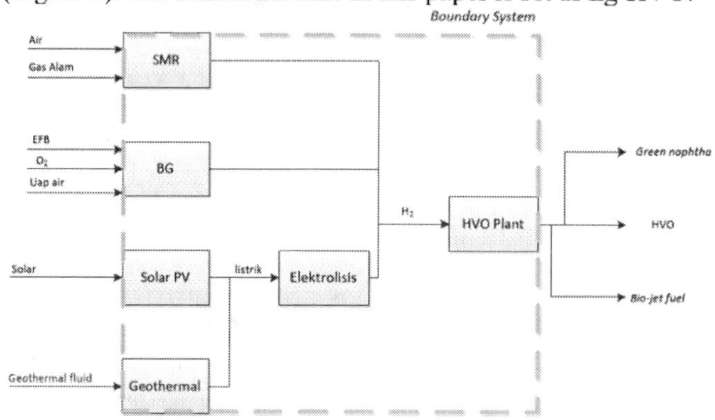


**Figure 2.** LCA system boundary.

Sources of life cycle inventories for this study are based on process simulations for hydrogen productions and HVO productions, and the others (CPO and EFB) are taken from [43], natural gas [44], PV electricity [42], and geothermal power plant [28].

## 3. Results and Discussion

### 3.1. *System Efficiency*

The comparison of energy efficiency can be represented by the Sankey diagram (Figure 3). All HVO based renewable hydrogen technologies have lower energy efficiency than compared to SMR. The efficiency of SMR-HVO, BG-HVO, PV-E – HVO, and GEO-E – HVO are 55.76%, 31.47%, 9.34%, and 7.89%, respectively.

It is noted that renewable based electricity productions have poor efficiencies (7.89% and 9.34%) compared to its non-renewable competitor, SMR – HVO (55.76%). The fall of efficiencies are rooted from electricity productions as additional process which required more energy consumptions. For continuous production of HVO, PV-E – HVO requires Solar PV to generate electricity and the battery to handle intermittently production of electricity, while GEO-E – HVO requires electricity generation through geothermal power plant. Although PV-E – HVO requires battery and GEO-E – HVO does not, GEO-E – HVO still has worse efficiency due to low efficiency of electricity generation (13.37%), while electrical generation using solar PV alone has better efficiency of 15.82% and a good efficiency of battery (90%).

BG – HVO efficiency falls behind SMR – HVO due to higher loss of energy during hydrogen production (Figure 3b and 3d). Higher loss of energy happens due to $CO_2$ emission. BG – HVO losses

3017.6 MWh due to $CO_2$ emission in hydrogen production, a significantly higher number compared to SMR – HVO (880.6 MWh). The wide differences were due to the C/H ratio in the feedstock. SMR uses natural gas as feedstock, which has lower C/H ratio compared to EFB, the feedstock of BG.

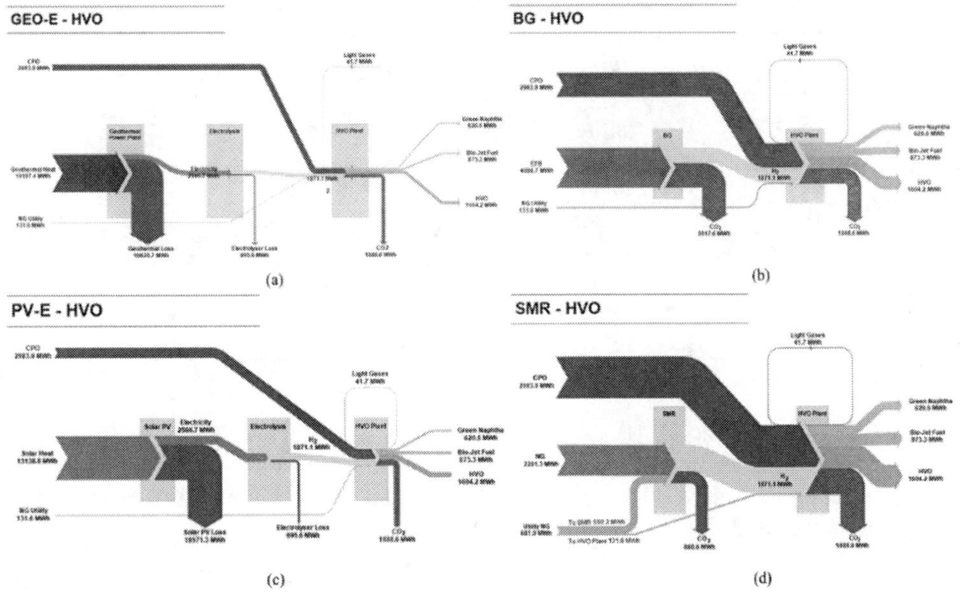


**Figure 3.** Energy flow analysis of (a) GEO-E - HVO; (b) BG - HVO; (c) PV-E - HVO; (d) SMR – HVO.

### *3.2. $CO_2$-eq Intensity*

The $CO_2$-eq intensity for all systems without weighting are shown at Figure 7. SMR – HVO and BG – HVO have higher $CO_2$-eq emission than HVO produced from hydrogen-based renewable electricity. SMR-HVO presents 5.57 kg $CO_2$-eq/kg HVO, BG – HVO produces 16.52 kg $CO_2$-eq/kg HVO, while GEO-E-HVO and PV-E-HVO are 1.63 kg CO2eq/kg HVO and 1.86 kg CO2eq/kg HVO, respectively. The higher $CO_2$-e emissions mainly due to the hydrogen production for BG and SMR – HVO. BG uses empty fruit bunch (EFB) as feedstock with higher C/H ratio than SMR. GEO-E – HVO has lowest emission due to highest electricity generation efficiency with 0.02 kg $CO_2$-e/kWh [28] compare to PV electricity of .0.04 kg $CO_2$-eq/kWh [44].

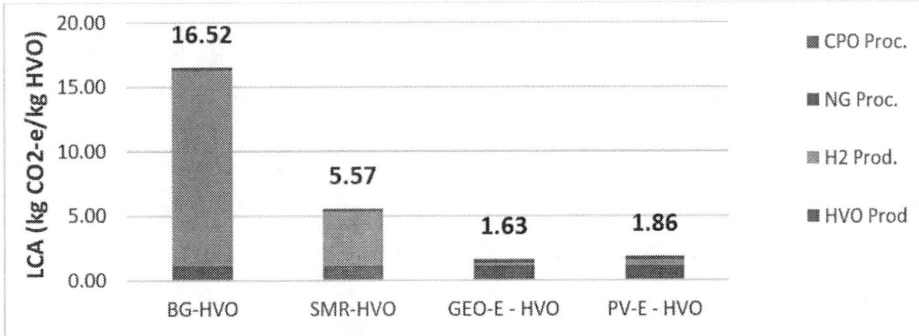


**Figure 4.** LCA of HVO production.

Although EFB has highest $CO_2$-eq emission, the biomass is carbon neutral by considering the $CO_2$ absorption during photosynthesis. Khoon, et. al. (2011) calculated the net fixation of $CO_2$ from palm oil plantation is 25.7 kg-$CO_2$/ha/hr [45]. Additionally, by using carbon capture, bioenergy carbon capture and storage (BECCS) that is a reliable path for negative emissions as shown in Figure 5. HVO production with BECCS produces -44.4 kg $CO_2$-e/kg of all products.

### *3.3. Economic Performance*

CAPEX breakdown of each configuration is presented in Table 3. Hydrogen production-related CAPEX dominates share of HVO investment. This is affected by the developing technologies for all renewable hydrogen technologies, whereas SMR, as non-renewable hydrogen technology, has already in its lowest cost.

**Table 3.** CAPEX value and share.

| Technology | $H_2$ Prod. Plant | | HVO Prod. Plant | | Electricity Generator | |
|---|---|---|---|---|---|---|
| | Total Inv.[a] | Share | Total Inv.[a] | Share | Total Inv.[a] | Share |
| **SMR – HVO** | 139.3 | 47% | 123.2 | 53% | - | - |
| **BG – HVO** | 234.4 | 66% | 123.2 | 34% | - | - |
| **GEO-E – HVO** | 279.4 | 48% | 123.2 | 21% | 180.3 | 31% |
| **PV-E – HVO** | 285.1 | 12% | 123.2 | 5% | 1973.9 | 83% |

[a]Total Inv. (total investment) is in M USD

Figure 7 revealed that SMR – HVO configuration is the cheapest option driven by cost of hydrogen productions. HVO production from renewable hydrogen with the lowest production cost is BG-HVO. While hydrogen productions via electrolysis give high production cost around \$44.48/GJ and \$53.16/GJ of total products for GEO-E – HVO and PV-E – HVO, respectively.

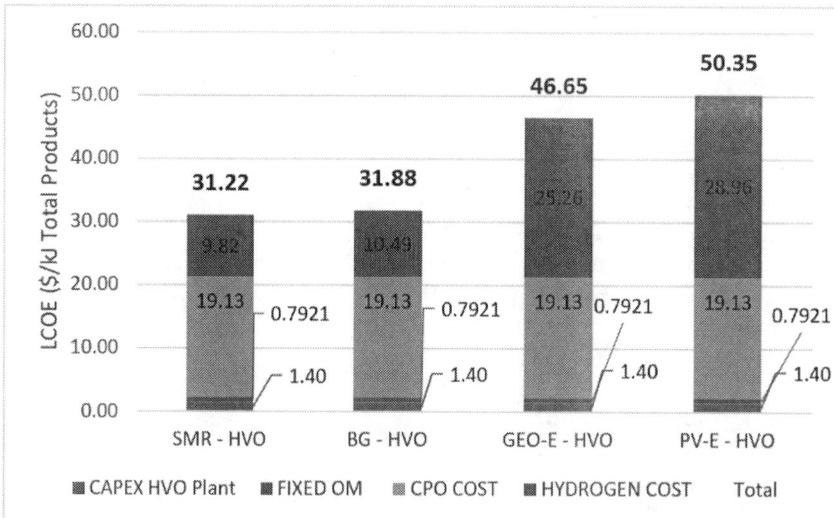


**Figure 5.** Levelised Cost of Products.

The levelised cost of each product is also estimated based on joint cost (Figure 8). For every technological configuration, HVO production cost based on PV electrolysis has the highest cost.

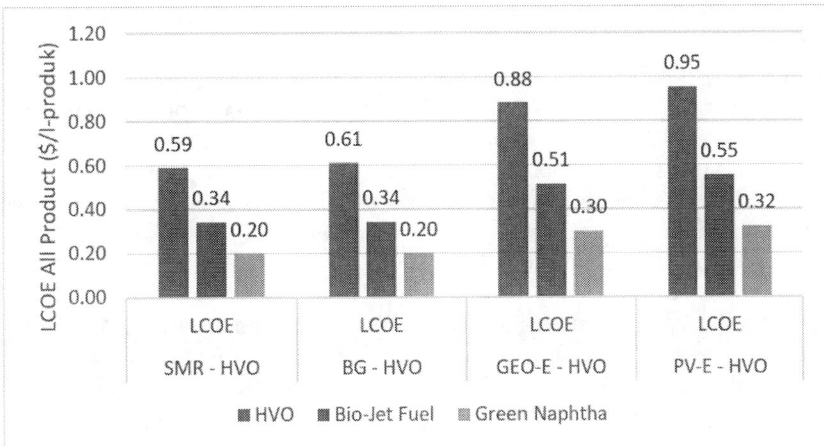


**Figure 6.** Levelised Cost of Energy for each product.

Based on higher costs of renewable electricity, battery and electrolyser, the projection of HVO production cost is carried out to evaluate when these new technologies can compete to conventional one PV-E – HVO will surpass SMR – HVO in 2050, while GEO-E – HVO will surpass SMR-HVO in 2070. PV-E – HVO can surpass SMR – HVO faster than GEO-E – HVO due to greater learning rate.

## 4. Conclusion

According to the study, the best overall energy efficiency is achieved by SMR – HVO. SMR – HVO has the highest efficiency of 55.67%, while the renewable options' efficiencies are 31.47%, 9.34%, 7.89%, for BG – HVO, PV-E – HVO, and GEO-E – HVO, respectively. Configuration with the lowest to highest $CO_2$ emission are GEO-E – HVO (1.63 kg$CO_2$-e/kg HVO), PV-E – HVO (1.86 kg$CO_2$-e/kg HVO), SMR – HVO (5.57 kg$CO_2$-e/kg HVO), BG – HVO (16.52 kg$CO_2$-e/kg HVO). However, if BECCS is used, BG – HVO can have a negative emission with -22.79 kg$CO_2$-eq/kg HVO.The lowest to highest LCOE are SMR – HVO ($0.59/l-HVO), BG – HVO (0.61/l-HVO), GEO-E – HVO ($0.88/l-HVO), and PV-e – HVO ($0.95/l-HVO). A projection is carried out and concluded that PV-E – HVO will be the first to surpass SMR – HVO in 2050.

## Acknowledgment

Th study is funded by Hibah Publikasi Terindeks Internasional (PUTI) Prosiding Universitas Indonesia, Contract Number: NKB-1105/UN2.RST/HKP.05.00/2020.

## References

[1] Kementrian Energi dan Sumber Daya Alam 2014 *Regulation of MEMR No. 12 Year 2015* Kementrian Energi dan Sumber Daya Alam
[2] Ali O M , Mamat R, Abdullah N R, and Abdullah A A 2015 Analysis of blended fuel properties and engine performance with palm biodiesel-diesel blended fuel *Renew. Energy* **86** 59–67
[3] Neste Oil, "Neste Oil starts up its new renewable diesel plant in Singapore," 2010. https://www.neste.com/neste-oil-starts-its-new-renewable-diesel-plant-singapore (accessed Sep. 24, 2020).
[4] "News: Neste Oil to build a NExBTL renewable diesel plant in Singapore," *Oil Gas European Magazine*, vol. 34, no. 1. p. 50, 2008, Accessed: Sep. 24, 2020. [Online]. Available: https://www.neste.com/neste-oil-build-nexbtl-renewable-diesel-plant-rotterdam.
[5] Eni, "EcofiningTM," *Eni*, 2019. https://www.eni.com/en-IT/operations/biofuels-ecofining.html (accessed Sep. 24, 2020).
[6] Neste, "NexBTL technology," 2019. https://www.neste.com/about-neste/innovation/nexbtl-technology (accessed Sep. 24, 2020).

[7] Kittisupakorn P, Sae-ueng S, and Suwatthikul A 2016 Optimization of Energy Consumption in a Hydrotreating Process for Green Diesel Production from Palm Oil *Comput. Aided Chem. Eng.* **38** 751–756

[8] Kiatkittipong W, Phimsen S, Kiatkittipong K, Wongsakulphasatch S, Laosiripojana N, and Assabumrungrat S 2013 Diesel-like hydrocarbon production from hydroprocessing of relevant refining palm oil *Fuel Process. Technol.* **116** 16–26

[9] Martinez-Hernandez E, Ramírez-Verduzco L F, Amezcua-Allieri M A, and Aburto J 2019 Process simulation and techno-economic analysis of bio-jet fuel and green diesel production — Minimum selling prices *Chem. Eng. Res. Des.* **146** 60–70

[10] Vélez Manco J F 2014 Conceptual design of a palm oil hydrotreatment reactor for commercial diesel production 122

[11] Cheah K W, Yusup S, Gurdeep Singh H K, Uemura Y, and Lam H L 2017 Process simulation and techno economic analysis of renewable diesel production via catalytic decarboxylation of rubber seed oil – A case study in Malaysia *J. Environ. Manage.* **203** 950–961

[12] Karatzos S, Mcmillan J, and Saddler J 2014 *IEA Bioenergy Task 39 - The potential and challenges of "drop in" biofuels*, July

[13] ERIA 2019 The Potential and Costs of Hydrogen Supply *Demand and Supply Potential of Hydrogen Energy in East Asia, ERIA Research Project Report FY 2018* **01**140–183

[14] Xu J and Froment G F 1989 Methane steam reforming, methanation and water-gas shift: I. Intrinsic kinetics *AIChE J.* **35** (1) 88–96

[15] Muritala I K, Guban D, Roeb M, and Sattler C 2019 High temperature production of hydrogen: Assessment of non-renewable resources technologies and emerging trends *Int. J. Hydrogen Energy*

[16] Nikolaidis P and Poullikkas A 2017 A comparative overview of hydrogen production processes *Renew. Sustain. Energy Rev.* **67** 597–611

[17] Ribeiro A M, Grande C A, Lopes F V, Loureiro J M, and Rodrigues A E 2012 Four Beds Pressure Swing Adsorption for Hydrogen Purification: Case of Humid Feed and Activated Carbon Beds *AIChE J.* **59** (4) 215–228

[18] Mohammed M A A, Salmiaton A, Wan Azlina W A K G, and Mohamad Amran M S 2012 Gasification of oil palm empty fruit bunches: A characterization and kinetic study *Bioresour. Technol.* **110** 628–636

[19] Hejazi B, Grace J R, Bi X, and Mahecha-Botero A 2017 Kinetic Model of Steam Gasification of Biomass in a Bubbling Fluidized Bed Reactor *Energy and Fuels* **31** (2) 1702–1711

[20] Abbas S Z, Dupont V, and Mahmud T 2017 Kinetics study and modelling of steam methane reforming process over a NiO/Al2O3 catalyst in an adiabatic packed bed reactor *Int. J. Hydrogen Energy* **42** (5) 2889–2903

[21] Yu J and Smith J D 2018 Validation and application of a kinetic model for biomass gasification simulation and optimization in updraft gasifiers *Chem. Eng. Process. - Process Intensif.* **125** February 214–226

[22] Olsbye U, Wurzel T, and Mleczko L 1997 Kinetic and Reaction Engineering Studies of Dry Reforming of Methane over a Ni/La/Al2O3 Catalyst *Ind. Eng. Chem. Res.* **36** (12) 5180–5188

[23] Inayat A, Ahmad M M, Mutalib M I A, and Yusup S 2012 Process modeling for parametric study on oil palm empty fruit bunch steam gasification for hydrogen production *Fuel Process. Technol.* **93** (1) 26–34

[24] Sánchez M, Amores E, Abad D, Rodríguez L, and Clemente-Jul C 2020 Aspen Plus model of an alkaline electrolysis system for hydrogen production *Int. J. Hydrogen Energy* **45** (7) 3916–3929

[25] Ruth M, Mayyas A, and Mann M 2017 Manufacturing Competitiveness Analysis for PEM and Alkaline Water Electrolysis Systems *Fuel Cell Seminar and Energy Exposition* 5–11

[26] World Bank Group 2017 *Solar Resource and Photovoltaic Potential of Indonesia* May 86, [Online]. Available:

http://documents.worldbank.org/curated/en/729411496240730378/Solar-resource-and-photovoltaic-potential-of-Indonesia.

[27] Tesla "Powerwall Datasheet." https://www.tesla.com/sites/default/files/pdfs/powerwall/Powerwall 2_AC_Datasheet_en_northamerica.pdf (accessed Sep. 27, 2020).

[28] Wang Y, Du Y, Wang J, Zhao J, Deng S, and Yin H 2020 Comparative life cycle assessment of geothermal power generation systems in China *Resour. Conserv. Recycl.* **155** November 2019

[29] Kalnes T N, Marker T, Shonnard D R, and Koers K P 2008 *Green diesel production by hydrorefining renewable feedstocks*

[30] Neste Corporation 2015 *Neste Renewable Diesel Handbook*

[31] P S & Global P 2017 *Methodology and Specifications Guide Asia Pacific & Middle East Refined Oil Products*

[32] Kiatkittipong W, Phimsen S, Kiatkittipong K, Wongsakulphasatch S, Laosiripojana N, and Assabumrungrat S 2013 Diesel-like hydrocarbon production from hydroprocessing of relevant refining palm oil *Fuel Process. Technol.* **116** 16–26

[33] Hengsawad T, Srimingkwanchai C, Butnark S, Resasco D E, and Jongpatiwut S 2018 Effect of Metal-Acid Balance on Hydroprocessed Renewable Jet Fuel Synthesis from Hydrocracking and Hydroisomerization of Biohydrogenated Diesel over Pt-Supported Catalysts *Ind. Eng. Chem. Res.* **57** (5) 1429–1440

[34] International Renewable Energy Agency (IRENA) 2019 *Renewable Power Generation Costs in 2018*

[35] IRENA 2019 *Utility-Scale Batteries*

[36] Sarulla Operations Ltd 2020 *No Title* https://sarullaoperations.com/id/company/overview.

[37] IRENA *Hydrogen From Renewable Power: Technology outlook for the energy transition*, no. September. 2018.

[38] Jakobsen D and Åtland V 2016 Concepts for Large Scale Hydrogen Production *Master Thesis* 145 [Online]. Available: https://brage.bibsys.no/xmlui/handle/11250/2402554.

[39] Acar C and Dincer I 2013 Comparative environmental impact evaluation of hydrogen production methods from renewable and nonrenewable sources *Causes, Impacts Solut. to Glob. Warm.* 493–514

[40] "Neste invests 1.4b euros, builds new Singapore plant, Companies & Markets - THE BUSINESS TIMES." https://www.businesstimes.com.sg/companies-markets/neste-invests-14b-euros-builds-new-singapore-plant (accessed Sep. 26, 2020).

[41] International Renewable Energy Agency - IRENA 2019 Future of solar photovoltaic:Deployment, investment, technology, grid integration and socio-economic aspects International Renewable Energy Agency, Abu Dhabi

[42] Gielen D, Taibi E, and Miranda R 2019 *Hydrogen: a Renewable Energy Perspective* September. 2019

[43] Cole W and Frazier A W 2019 *Cost Projections for Utility- Scale Battery Storage Cost Projections for Utility- Scale Battery Storage* doi: NREL/TP-6A20-73222.

[44] IRENA 2017 *Technology Brief: Geothermal Power*

[45] Khoon K L, Cobb A, and Harun M H 2011 The Potential of Oil Palm in the Global Carbon Cycle *Palm Oil Dev.* **54** 8–13

IOP Conf. Series: Materials Science and Engineering 1143 (2021) 012046 doi:10.1088/1757-899X/1143/1/012046

# Multi-Objective Optimization of Blending Strategy of FAME, HVO, and Petroleum Diesel

**Shanti Mustika, Widodo Wahyu Purwanto***

Sustainable Energy and Policy Research Cluster, Department of Chemical Engineering, University of Indonesia, Depok, 16424

*widodo@che.ui.ac.id

**Abstract.** The utilization of biofuels based on palm oil could decrease greenhouse gasses (GHG) emitted by fossil fuels. The aim of this research is on the optimization of blending strategy of Fatty Acid Methyl Esters (FAME), Hydrogenated Vegetable Oil (HVO), Diesel CN48 (DCN48), and Diesel CN53 (DCN53) to meet the fuel quality standard, Euro2 and Euro4 by minimizing levelized cost of fuel supply and GHG emission. The process simulation is performed by Aspen Plus software, proceed by Life Cycle Analysis (LCA) calculation. The optimization is carried out by General Algebraic Modelling System (GAMS) with Cplex solver. The result showed the percentage of the optimal blending composition of fuel quality Euro2 were FAME 43.9-51.1%, HVO 2.6-40.1%, DCN48 15.3-17.6%, and DCN53 46.3-100% with LCOE was 0.55-0.864 USD/Liter and GHG intensity 599.46-3000.78 $gCO_{2eq}$/Liter. For Euro4 specification consists of FAME 32.5%, HVO 28.6%, and DCN53 38.8% with LCOE were 0.637-0.786 USD/Liter and GHG intensity of 902.69-2863.03 $gCO_{2eq}$/Liter.

## 1. Introduction

The demand for Indonesian's diesel in 2018 reached 30.36 million kiloliters (kL) along with the domestic production of petroleum diesel and biodiesel was 22.72 and 3.04 million kL, respectively [1]. Thus, increasing the import of diesel reached 4.61 million kL. The country has to face a rise in its current account deficit due to this import. To reduce petroleum fuel imports, starting from 2015, the Ministry of Energy and Mineral Resources issued Regulation No. 12/2015 regarding the mandate of mixing diesel fuel with FAME. In 2015, the government targets biodiesel blending for 20% later on, in 2020, the percentage of FAME will be increased to reach 30%. Using 30% of FAME, Indonesia's diesel imports are predicted to reduce by 3.82 million kL.

Biodiesel has potential to improve engine performance and reduce particulate exhaust emissions in compression-ignition (CI) engines. However, the blending fuel is required to meet the standard specifications set by the World Fuel Charter. Indonesia's emission target is an adaptation of the European emission target called the Euro in which the emission target in 2021 is Euro4.

Regarding the standard specification mentioned above, the blending diesel fuel no longer stems from petroleum diesel and FAME only, as paraffinic fuel such as HVO has the characteristic similar to petroleum diesel. Thus adding HVO into consideration has become an approach to fulfill the requirement of the standards. In the previous study, optimization is carried out by obtaining the least GHG intensity, as previously conducted by Abdulrazik [2], and obtaining the least LCOE with an optimal ternary blending HVO, FAME, and alcohol [3]. However, the two research which mentioned above does not focus on multi-objective optimization. The aim of this study is the optimization of biofuel blending by minimizing production costs and GHG intensity. The result of the optimization could potentially be a guidance for the development of blended fuels in Indonesia.

## 2. Methods

This study is divided into five steps, first, build scenarios of the blending presented in Figure 1

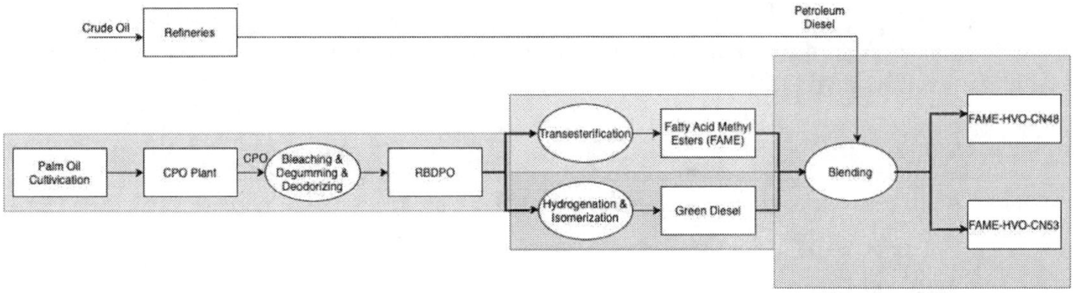


**Figure 1.** Blending fuels scenarios.

There are 8 scenarios in total in which each scenario consists of three fuels, the two of the fuels are biofuels (FAME and HVO) with different combination on the petroleum diesel cetane number (CN48 and CN53). Besides from the blending fuels, the standards of the blending fuels are based on the World Fuel Standard from Europe which called Euro and following Indonesia's existing condition which follow Euro2 and the Indonesia's goal toward Euro 4 specification. However, considering the petroleum diesel specification from Indonesia there are adjustments formed in order to achieve the Euro standard, which later be called Indo-Euro2 and Indo-Euro4.

The difference between the Indo-Euro2 and Indo-Euro4 scenarios is due to the condition of Indonesian refineries which are unable yet to produce diesel fuel with sulphur content less than 10 ppm compare to refineries in Europe which able to produce Ultra Low Sulphur Diesel (ULSD). Therefore, additional scenario is needed as an adaptation of Indonesia with the provisions of the Euro2 and Euro4 specifications standard. The scenarios can be seen in Table 1.

**Table 1.** Blending scenarios.

| Scenarios | Standard | Blending Fuels |
|---|---|---|
| **Scenario 1 (S1)** | EURO-2 | FAME - HVO - DCN 48 |
| **Scenario 2 (S2)** | | FAME - HVO - DCN 53 |
| **Scenario 3 (S3)** | INDO-EURO2 | FAME - HVO - DCN 48 |
| **Scenario 4 (S4)** | | FAME - HVO - DCN 53 |
| **Scenario 5 (S5)** | EURO-4 | FAME - HVO - DCN 48 |
| **Scenario 6 (S6)** | | FAME - HVO - DCN 53 |
| **Scenario 7 (S7)** | INDO-EURO4 | FAME - HVO - DCN 48 |
| **Scenario 8 (S8)** | | FAME - HVO - DCN 53 |

After completing the scenarios design, the first step is to simulate the process production of FAME and HVO to attain the diesel product properties, technical performance, sizing, along with the cost production through process simulation using Aspen Plus V11. However, for certain specifications cite, viscosity, sulphur content, lubricity are extracted from [4] and [5].

Second, LCA analysis calculated with system boundary of Well-to-Wheel (WTW). The LCA starts with cultivation of palm oil until the extraction, FAME and HVO distribution and storage are extracted from GREET based calculation [6], the inventory process taken from the simulation data, and the petroleum diesel data along with the combustion earned from Toyota Motor Corporation [7].

Result of the process simulation and calculation of the GHG intensity are being optimized using General Algebraic Modelling System (GAMS) to earn the minimum LCOE and minimum GHG intensity. Therefore, two different values are obtained for each objective function. From the two values, one objective function is determined to be the limit because the multi-objective optimization method chosen is a *ε-Constraint* Method. The objective function being constrained is the GHG intensity. Afterwards, the points of minimum and maximum emission can be determined and then, pareto curves can be formed. From the Pareto curve, then the compromised solution can be determined using the Multi-Criteria Decision Making (MCDM) with the TOPSIS method.

### *2.1. Process Simulation*

A process flowsheet is designed using Aspen Plus V11 with base capacity of HVO production is set to 1 million tons per year similar to the commercial HVO production and the base capacity of FAME production is 300,000 tons per year.

*2.1.1. FAME.* The manufacture of FAME in this simulation uses Refined Bleached Degummed Palm Oil (RBDPO) as raw material, which is a derivative product of CPO that has gone through a previous refining process. With production capacity mentioned above, the required CPO in the process is 7031.02 kg of CPO/hour. The process of making FAME itself consists of several steps, namely the transesterification reaction, methanol recovery, water washing, FAME purification, and alkaline removal.

*2.1.2. Glycerol Steam Reforming (GSR) – HVO.* GSR is being processed to produce hydrogen which will later on be utilized in the HVO manufacturing process. With the HVO production capacity mentioned above, 47,704 kg of hydrogen/hour is needed. The process of GSR described further in the research of steam reforming glycerol [8]. In this simulation, the HVO plant not only produces HVO but also bio-jet fuel and naphtha with HVO specifications referring to the EN 15940: 2016 standard. The process of HVO production described further in the preliminary design of the HVO production [9]. The hydroisomerization reactor uses a yield reactor based on experimental data [10]. Separation was carried out for hydrogen recovery from both reactors. Then, the C3 and C4 hydrocarbons are separated to be burned so that the steam can be reused for production. The resulting product is also purified using distillation to separate green naphtha, HVO, and bio-jet fuel.

### *2.2. Life Cycle Analysis*

Well-to-Wheel (WTW) is more applicable to crude oil resources and their derivatives. WTT includes the process of producing raw materials up to the car's fuel tank. Tank-to-Wheel (TTW) includes the process of burning fuel in the engine so that WTW is a total of WTT and TTW. However, for the vegetable oil such as FAME and HVO the calculation starting from cultivation, fuel production process, until the distribution of the fuels which assumed as WTT.

### *2.3. Multi-Objective Optimization*

After LCOE is obtained, the optimization could be proceeded to get the minimum LCOE and minimum GHG intensity with GAMS (Cplex solver) which can be plot into pareto curves so that the results of the compromised solution can be obtained. The objective function equations are stated the following equations:

$$\text{Objective function 1} \equiv OF_1 = \sum x_i C_i \quad (1)$$

$$\text{Objective function 2} \equiv OF_2 = \sum x_i FE_i \quad (2)$$

The $x_i$ is the volume fraction of the fuel, $C_i$ is the production cost of the diesel (USD/Liter), $FE_i$ is the GHG intensity to produce the diesel ($gCO_2eq$/Liter). In this study, there are constraints on the fuel property specifications as follows:

$$CN_{min} \leq CN_{blend} \quad (3)$$

$$\rho_{min} \leq \rho_{blend} \leq \rho_{max} \quad (4)$$

$$\mu_{min} \leq \mu_{blend} \leq \mu_{max} \quad (5)$$

$$SC_{blend} \leq SC_{max} \quad (6)$$

$$WC_{blend} \leq WC_{max} \quad (7)$$

$$L_{blend} \leq L_{max} \quad (8)$$

CN is cetane number, $\rho$ is density ($kg/m^3$), $\mu$ is viscosity ($mm^2/s$), SC is sulphur content (ppm), WC is water content (mg/kg), and L stands for lubricity (micron). The minimum and maximum value mentioned above use the standardized upper and lower limit shown in Table 2 [11].

**Table 2.** Specification properties Euro-2 and Euro-4.

| Properties | EURO-2 | | EURO-4 | |
|---|---|---|---|---|
| | Lower Limit | Upper Limit | Lower Limit | Upper Limit |
| **Cetane number** | 51 | - | 53 | - |
| **Density (kg/m³)** | 815 | 850 | 815 | 840 |
| **Viscosity (mm²/s)** | 2 | 4 | 2 | 4 |
| **Sulphur content (ppm)** | - | 500 | - | 50 |
| **Lubricity (micron)** | - | 460 | - | 460 |
| **Water content (mg/kg)** | - | 200 | - | 200 |

There are mixing rules to blend the diesel, specifically for each property such as density, viscosity, sulphur content, lubricity, and water content [12] and [13]. For $x_i$ is the volume fraction and $y_i$ is the mass fraction. However as stated earlier in this study, the condition of refineries in Indonesia has not yet fully comply with the standard therefore, Indonesia has its own standard which refer to the worldwide fuel charter.

**Table 3.** Indo-Euro2 and Indo-Euro4.

| *Properties* | INDO-EURO2 | | INDO-EURO4 | |
|---|---|---|---|---|
| | Lower limit | Upper limit | Lower limit | Upper limit |
| **Cetane number** | 48 | - | 51 | - |
| **Density (kg/m³)** | 815 | 860 | 815 | 860 |
| **Viscosity (mm²/s)** | 2 | 4.5 | 2 | 4.5 |
| **Sulphur content (ppm)** | - | 2500 | - | 300 |
| **Lubricity (micron)** | - | 460 | - | 460 |
| **Water content (mg/kg)** | - | 500 | - | 500 |

## 3. Result and discussion

### *3.1. Overall system efficiency*

The system efficiency for the simulated fuels is shown in Table 4.

**Table 4.** Process simulation result.

| Parameter | Units | FAME | [14] | HVO | [9] |
|---|---|---|---|---|---|
| **Product/Feed Ratio** | - | - | - | 3 | 1,74 |
| **Yield Conversion** | % | 91.65 | 90-95 | 52 | 93 |
| **Energy Efficiency** | % | 53.11 | - | 76 | 79 |
| **Purity** | %w/w | 99.82 | 99.7 | 52.17 | 77.23 |
| | | **Properties** | | | |
| **Cetane Number** | - | 66.66 | 63 | 78.49 | 99.00 |
| **Density** | kg/m$^3$ | 877.64 | 874 | 823 | 783 |
| **Viscosity [4]** | mm$^2$/s | 4.5 | 4.372 | 3 | 3.704 |
| **Sulphur Content [5]** | ppm | 0 | 1 | 1 | 9 |
| **Lubricity [5]** | micron | 186 | 203 | 650 | 607 |
| **Water content** | mg/kg | 270.87 | 62 | 42 | 39 |

Based on the simulation result, the mass purity of FAME reached 99.82% and has been validated with the standard product of ASTM 6751 with the minimum mass purity of 99.65%. The deviation data for the ration of FAME to CPO is 4.71% as the result of the greater the feed compare to the study [5]. For HVO simulation result, the yield conversion is different because the product in the study [4] are not producing bio-jet fuel as in this study the products are HVO, bio-jet fuel, and naphtha. Another reason is the hydrogenation reaction used in this study is using equilibrium equation therefore ratio of product/feed greater which then affected the greater yield production efficiency.

**Table 5.** GSR plant simulation technical performance evaluation.

| Parameter | Unit | Simulation |
|---|---|---|
| **Glycerol conversion** | % | 100.00 |
| **$H_2$/Glycerol Mole Ratio** | - | 4.01 |
| **$H_2$Yield** | % | 30.86 |
| **$H_2$ Selectivity** | % | 100.00 |
| **Energy Efficiency** | % | 50.42 |
| **PSA Recovery** | % | 71.30 |
| **Glycerol/Hydrogen Ratio** | w / w | 8.13 |

Yield of the hydrogen has increased for 0.3%. Actually, the conversion results from Water Gas Shift (WGS) reactor is large, which is equal to 40.52% but after entering the Pressure Swing Adsorption (PSA) the final results obtained reduced because of the impurities with $H_2$ 2%, $H_2O$ 16%, CO 28%, and

$CO_2$ 52% as the result of impurities. Those impurities are being recirculated again in natural gas burning, the $CO_2$ content that is recirculated in the combustion of natural gas is equal to 6% and still follow into the limit of $CO_2$ content for burning natural gas, which is 12% [15].

### *3.2. Cost Evaluation*

Costs are calculated by using Aspen Process Economic Analyzer. Additional costs for the catalyst and the reactor are estimated separately the added to the capital cost result of Aspen. LCOE of FAME is 0.69 USD/Liter however, this calculation is not a final calculation because there is a joint cost effect between productions FAME with glycerol as a by-product. The percentage for FAME is 95% and glycerol is 5%. Therefore, the LCOE of FAME is 0.66 USD/Liter and LCOE of glycerol is 0.03USD/Liter.

The LCOE GSR is 4.69 USD/kg hydrogen not much different from reference data [16] which stated 4.86 USD/kg of hydrogen for the base case *scenario*. When compared with the LCOE Steam Methane Reforming (SMR) which was 1.27 USD/kg hydrogen, LCOE GSR is much more expensive than LCOE of SMR. This is due to the small hydrogen conversion of 4 moles hydrogen per 1 mole of glycerol so that there is little hydrogen produced but the need for raw materials is large.

The LCOE for HVO is 2.44USD/Liter which shows the total product produced in the simulation, namely HVO, naphtha and bio-jet fuel therefore by using weighted joint cost, the HVO percentage is 51.33% so that the LCOE earned for HVO is 1.25 USD/Liter. The cost breakdown can be generated for FAME and HVO, in which has been integrated with the LCOE GSR in Figure 2.

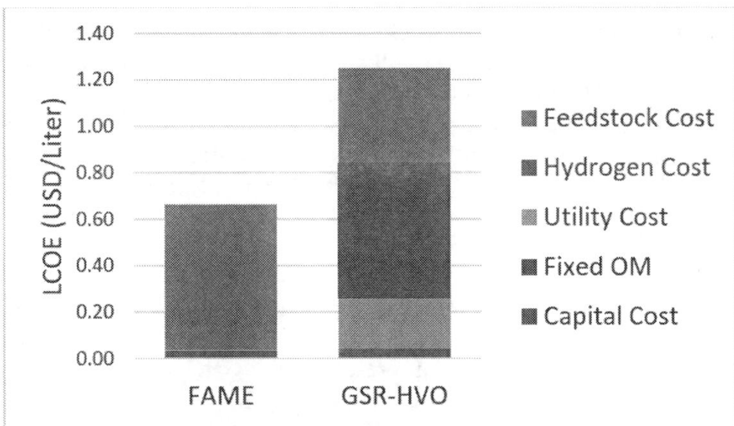


**Figure 2.** LCOE of FAME and GSR-HVO.

It can be seen in Figure 2 above, that the percentage of (CPO) as FAME raw material taken up to 94.8% while according to APEC (2013) the percentage of raw material costs is 83%. This difference is due to the conversion generated by the process simulation 4.71% greater. Thus, for FAME, the LCOE is heavily influenced by the feedstock cost.

For HVO production itself, there is no literature using GSR so the cost breakdown cannot be validated. However, it can be seen from Figure 2 that the cost of producing hydrogen is very large, which is 58%, this is because the cost of producing 1 kg of hydrogen costs 4.69 USD whereas when compared to hydrogen production using other technologies, the cost of producing hydrogen through GSR technology is more expensive.

The fundamental technological difference between FAME and HVO production is one of the factors that causes the production cost between FAME and HVO to differ up to 0.59USD/Liter. The hydrogenation process, which requires more expensive costs by using two reactors, namely hydrogenation and hydroisomerization, are certainly bigger when compared to the production of FAME which only uses a transesterification reactor.

### *3.3. GHG Intensity*

The GHG intensity of $CO_2$ of the fuels produced and burned are presented in the Figure 2.

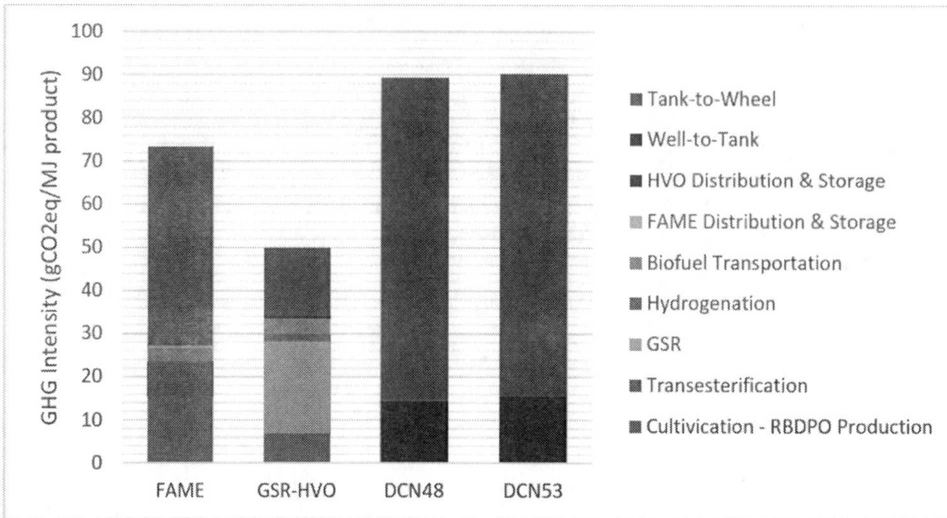


**Figure 3.** LCA FAME, HVO, and Petroleum diesel (Well-to-Wheel).

The GHG intensity of GSR is 78.19 $gCO_2eq/MJ$. The difference in the reference study [17] with amount of 105.42 gCO2eq /MJ is because of the allocation factor with FAME is 0.96 and glycerol is 0.04. In addition, the operating conditions in the study use a large pressure of 25 bar and a temperature of 800°C while the GSR simulation in this study uses a temperature of 4 bar and a temperature of 575°C because of these differences, the resulting emissions are different. When compared with the SMR emission factor of 10.99 $kgCO_2eq/kg$ hydrogen. Not much different from SMR, although GSR from biomass. Steam reforming using glycerol produces 3 moles of $CO_2$ stoichiometry. This $CO_2$ release makes the GHG value even greater.

As in the Figure 3, the WTT of the petroleum diesel also presented and through this figure, it can be concluded that the production of petroleum diesel is producing lower emission compare to FAME and HVO. However, the diesel emission factor is much greater during the internal combustion process with the amount of 74.75 $gCO_2eq/MJ$. Thus, the result of the GHG intensity is 89.3-90.28 $gCO_2eq/MJ$ petroleum diesel and HVO has a GHG value of 50 $gCO_2eq/MJ$ HVO.

The composition of HVO with diesel is different so that the diesel engine, which is why HVO cannot yet fully utilize the properties of HVO. Adjustment of the machine parameters need to be made. One of them is the cetane number so that it can compensate the fuel ignition when the engine is started. The fuel consumption using HVO will be around 2-8% more than diesel because it has a lower HVO density of 6% due to the natural properties of paraffinic fuels. Previously it was stated that with a lower density diesel consumption would increase, but with a higher HHV HVO value, HVO consumption would remain lower than diesel.

For the Tank-to-Wheel of HVO, there is not enough data on the use of HVO in light-duty vehicles (vehicles with feeds weighing less than 3 tons), so far the HVO has only been tested under steady operating conditions, while most light-duty vehicles operate in a transient state. The emission reduction by HVO actually occurs the most in the Particulate Matter (PM), CO, and unburned hydrocarbon. $CO_2$ reduction is only about 6% with $CO_2$ emissions of 1146g with diesel emissions of 1194g, this lower emission is due to the carbon content of HVO as much as 84% while the carbon content of diesel is 86.2%.

### *3.4. Optimization of the blends*

IOP Conf. Series: Materials Science and Engineering 1143 (2021) 012046 doi:10.1088/1757-899X/1143/1/012046

Initially the date of WTW with two solutions on each scenario are being plot into the pareto curve to get the compromised solution presented in Table 6.

**Table 6.** Compromised solution of WTW.

| **Scenario** | **Compromised Solution** | | | | |
|---|---|---|---|---|---|
| | Emissions (gCO2eq / Liter) | LCOE (USD / Liter) | FAME (%) | HVO (%) | Solar (%) |
| **1 (DCN48)** | 2443.94 | 0.864 | 44.7 | 40.1 | 15.3 |
| **2 (DCN53)** | 3000.78 | 0.624 | 51.1 | 2,6 | 46.3 |
| **3 (DCN48)** | 3045.09 | 0.552 | 48.6 | 0 | 51.4 |
| **4 (DCN53)** | 2863.03 | 0.637 | 68.8 | 1.6 | 29.6 |
| **6 (DCN53)** | 2739.12 | 0.786 | 32.5 | 28.6 | 38.8 |
| **7 (DCN48)** | 2521.48 | 0.784 | 64.4 | 24.9 | 10.8 |
| **8 (DCN53)** | 2863.03 | 0.637 | 68.8 | 1.6 | 29.6 |

In a compromised solution, when MCDM is carried out with the ideal point at point (x, y) with (x min $OF_2$, y-axis $OF_1$) for scenarios 1, 6, and 7, they have the same characteristics as the optimal point chosen (low LCOE but high GHG intensity). This is because the difference between maximum and minimum emissions is not much different, namely 15.86-23.14 compared to other scenarios with an average difference of 47.94. Since the GHG intensity value for each point is not much different, the optimal point is chosen with the trend of the minimum LCOE.

Scenarios 2, 4, and 8 have the same characteristics in choosing a compromised solution, when the minimum HVO percentage is blended. Compared to emissions, the optimal points taken from scenarios 2, 4, and 8 are looking for the lowest LCOE possible but still considering the GHG intensity produced.

Scenario 3 chooses the optimal point right when the HVO component has not been added to the blending diesel. This shows that the optimal point is chosen to consider LCOE because after the HVO component is added to the mixed fuel, the LCOE value will increase. When the HVO component has not being blended, the difference between each LCOE value is 0.17 but when the HVO is blended, the difference in LCOE value increases to 0.23 and will continue to increase as the HVO component increases.

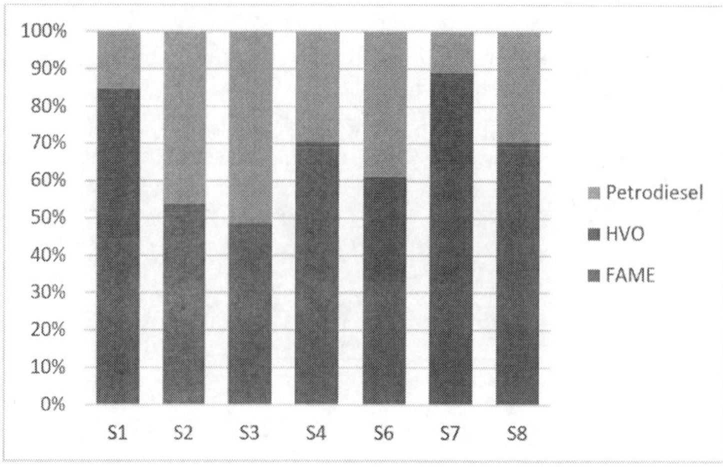


Figure 4. Portfolio blending.

In the portfolio blending in Figure 4 above, scenario 3 only contains two mixed fuels, FAME and diesel, so this scenario has a lower selling price compared to other scenarios that contain HVO. Scenarios 2,4 and 8 have three fuel components, but the HVO component is blended in small amount compared to other fuel components so that the selling price of the 2,4,8 scenario is lower than the scenario 1,6, and 7 which has a greater percentage of HVO.

## 4. Conclusion

Blending portfolio for Euro2 diesel specifications contains FAME 43.9-51.1%, HVO 2.6-40.1%, DCN48 15.3-17.6%, and DCN53 46.3-100% along with the LCOE of the blending is 0.55-0.864 USD/Liter and the GHG intensity reached 599.46-3000.78 g$CO_2$eq/Liter. Then, blending portfolio for Euro4 diesel specifications contains FAME 32.5%, HVO 28.6%, and DCN53 38.8% along with the LCOE of the blending is 0.637-0.786 USD/Liter and the GHG intensity reached 902.69-2863.03 g$CO_2$eq/Liter. Further research should be conducted for the HVO combustion in diesel engines, especially in light duty vehicles, specifically in transient operating conditions.

## Acknowledgment

Th study is funded by Hibah Publikasi Terindeks Internasional (PUTI) Prosiding Universitas Indonesia, Contract Number: NKB-1105/UN2.RST/HKP.05.00/2020.

## References

[1] Mayasari F, Dalimi R, and Purwanto W 2019 Projection of Biodiesel Production in Indonesia to Achieve National Mandatory Blending in 2025 using System Dynamics Modeling *International Journal of Energy Economics and Policy* **9** 421-429

[2] Abdulrazik A, et al. 2017 Multi-products productions from Malaysian oil palm empty fruit bunch (EFB): Analyzing economic potentials from the optimal biomass supply chain *Journal of Cleaner Production* **168**.

[3] Razak N H, et al. 2019 Optimization of ternary green diesel blends for diesel/palm methyl ester/alcohol using product design optimization *Chemical engineering transactions* **72** 391-396

[4] Ari Engman Q G 2020 *Tuuka Hartikka, Markku Honkanen, Ulla Kiiski, Terhi Kolehmainen, Anna Kunnas, Markku Kuronen, Sari Kuusisto, Kalle Lehto, Seppo Mikkonen, Jenni Nortion, Jukka Nuittimaki, Reetu Sallinen, Pirjo Saikkonen, Supali Tripathi, and Eerika Vuorio* Neste Renewable Diesel Handbook, Neste Corporation: Espoo.

[5] Setiapraja H, et al. 2019 *An Effect of Utilization B30 from Various Blends of B0:FAME and HVO on Emissions, Fuel Consumption and Power of Euro4 Vehicle Technology* SAE International.

[6] Bonomi A, et al. 2018 *Comparison of Biofuel Life Cycle Analysis Tools Phase 2, Part 1: FAME and HVO/HEFA*

[7] Corporation T M. 2004 *Well-to-Wheel Analysis of Greenhouse Gas Emissions of Automotive Fuels in the Japanese Context* Japan

[8] Cheng C K, Foo S, and Adesina A 2011 *Steam reforming of glycerol over Ni/Al2O3 catalyst.* Catalysis Today **178**.

[9] Paniagua J M F -V.a.A.H.d.M. 2018 *Preliminary design of the green diesel production process by hydrotreatment of vegetable oils* EURECHA

[10] Hengsawad T, et al. 2018 Effect of Metal–Acid Balance on Hydroprocessed Renewable Jet Fuel Synthesis from Hydrocracking and Hydroisomerization of Biohydrogenated Diesel over Pt-Supported Catalysts *Industrial & Engineering Chemistry Research* **57**(5) 1429-1440

[11] Charters W F 2019 *Gasoline and Diesel Fuel (Sixth Edition)*

[12] Jiang H, Yanhui H, and Huixin W 2011 A Model for Blending Properties of Diesel Oils and Its Optimization Application *Petroleum Science and Technology* **29** 201-207

[13] Sivaramakrishnan K and Ravikumar P 2012 Determination of cetane number of biodiesel and it's influence on physical properties *Asian Journal of Applied Sciences* 7 205-211

[14] Zhang L, et al. 2013 Upgrading of bio-oil from biomass fast pyrolysis in China: A review *Renewable and Sustainable Energy Reviews* **24** 66-72

[15] Slefarski R, et al. 2018 Analysis of Combustion Process in Industrial Gas Engine with Prechamber-Based Ignition System *Energies* **11** 336

[16] Ahmed S, D P 2010 *Hydrogen from glycerol : A Feasibility Study* U.S. Department of Energy: Argonne National Laboratory.

[17] Susmozas A, Iribarren D, and Dufour J 2015 Assessing the Life-Cycle Performance of Hydrogen Production via Biofuel Reforming in Europe *Resources* **4**

# Simulation of $CO_2$ Capture Process for Coal based Power Plant in South Sumatra Indonesia

**M. Eviani[1], H. Devianto[2], P. Widiatmoko[2], I. F. Sukmana[2], H R. Fitri[2], F. Yusupandi[2]**

[1]Pusat Penelitian dan Pengembangan Teknologi Minyak dan Gas Bumi LEMIGAS
[2]Department of Chemical Engineering, Institut Teknologi Bandung, Jl. Ganesha 10, Bandung, 40132, Indonesia

mitra_eviani@lemigas.esdm.go.id, hardev@che.itb.ac.id

**Abstract**. Indonesia committed to reduce greenhouse gas emission by 29% in 2030. Carbon dioxide ($CO_2$) emissions from flue gas in coal power plant must be captured to reduce greenhouse gas emissions. Technology to decrease greenhouse gases on a large scale and in a relatively short period is Carbon Capture and Storage (CCS). Chemical absorption method is a more advantageous CCS technology than other methods owing to high efficiency, low cost, and mature technology. Solvents in chemical absorption include amine-based solvent such as monoethanolamine (MEA) was frequently utilized to absorb $CO_2$ from low-pressure flue gas streams, especially fossil fuel-based power plant, owing to rapid reaction rate with $CO_2$ and low cost of raw amines compared to other amines. In this study, Aspen Hysys was used to simulate the $CO_2$ capture process and 30 wt% of MEA was selected as a solvent which is mainly handy for $CO_2$ capture from flue gas. The results show that lean amine temperature, flue gas temperature, and regenerator feed temperature has effects on the $CO_2$ capture and energy consumption. The rich-split stream configuration was developed to examine energy reduction in various split fractions. The minimum in energy consumption and reboiler duty occurs when 30% of the solvent is split to the top of the column.
**Keywords:** $CO_2$, coal-based power plant, absorption, MEA, Aspen Hysys.

## 1. Introduction

For the past two decades the greenhouse gas (GHG) emissions have increased, and threated to the earth's ecological system and human race [1-2]. Combustion of fossil fuels such as natural gas, coal for power generation, industrial processes and transportation has led to the huge increase of $CO_2$ concentration in the atmosphere [3]. Indonesia has joined a global wave of countries that submitted their post-2020 climate pledges to the United Nations Framework Convention on Climate Change (UNFCCC) called intended nationally determined contributions (INDCs) in 2015. Since then, Indonesia has signed the Paris Agreement, and formally submitted the first nationally determined contribution (NDC) in 2016. Indonesia committed to decrease unconditionally GHG emissions by 29% against a 2030 business-as-usual (BAU) scenario and GHG emissions by up to 41% below the 2030 BAU level which subjects to international assistance for finance, technology transfer, and capacity building [1].

Carbon dioxide ($CO_2$) capture and storage is a process consisting of the separation of $CO_2$ from industrial and energy-related sources, transport to a storage location and long-term isolation from the atmosphere. There are three major types for CCS: post-combustion capture, pre-combustion capture and oxyfuel process [4]. Post-combustion capture provides many advantages as existing combustion methods such as easier to implement as a retrofit option to existing plants compared to the other two types [5]. There are several separation methods that could be employed with post-combustion capture such as adsorption, chemical absorption, physical absorption, cryogenics separation and membranes [4]. Among various carbon capture technologies, chemical absorption method is more advantageous CCS technology than other methods due to high efficiency, low cost, and mature technology [6]. Solvents in chemical absorption include amine-based solvent such as monoethanolamine (MEA), diethanolamine (DEA), hindered amine (KS-1) and alkaline solvents such as NaOH, $K_2CO_3$, and $Ca(OH)_2$ and Ionic liquids [7]. The utilization of amine solution is the most commonly methods for gas sweetening. Among the amines, MEA was frequently utilized to absorb $CO_2$ from low-pressure flue gas streams, especially fossil fuel-based power plant, owing to rapid reaction rate with $CO_2$ and low cost of raw amines compared to other amines [8-10].

Knudsen, et.al [11] informed that $CO_2$ removal efficiency of $CO_2$ capture pilot plan from coal-fired power plant at Esbjergværket, Denmark as part of EU CASTOR project achieved 90% with MEA 30 wt% during 4,000 h. In coal-fired power station at Queensland, Australia, amine-based $CO_2$ absorption pilot plant has been operated to result in 84% of $CO_2$ capture and regeneration energy of 3.6 GJ/ton $CO_2$ with MEA 30wt% [12,13]. Meanwhile, Gjernes, et.al [15] reported that $CO_2$ Technology Centre Mongstad (TCM), located at Statoil Refinery and one of largest post-combustion $CO_2$ capture, treated flue gas from natural gas-based combined heat and power (CHP) and a residual fluid catalytic cracker (RFCC) which contain 3.5 and 13-14 vol% $CO_2$, respectively. 83.4% of $CO_2$ capture was attained using MEA 30 wt% at TCM [14-16]. On the other hand, Øi, et.al [17-19] has modelled three configurations namely standard, split-stream and vapor compression model of $CO_2$ capture from natural gas-based power plant using Aspen Hysys. In standard configuration, heat consumption in regenerator was 3.26 GJ/ton $CO_2$ while it could be decreased about 8% and 11% with split-stream and vapor compression scheme, respectively. Additionally, Aspen Hysys was also applied to simulate $CO_2$ absorption of flue gas from cement industries which has $CO_2$ content of 23.75 mole%. With $CO_2$ capture of 85%, the energy consumption for MEA regeneration was 3.71 GJ/ton $CO_2$ [20,21].

In this work, we demonstrated a model of $CO_2$ absorption of flue gas from coal-fired power plant which has installed capacity of 4 x 65 MW (260 MW) in South Sumatra, Indonesia. Aspen Hysys was used to simulate the $CO_2$ capture process and 30 wt% of MEA was selected as a solvent which is mainly handy for $CO_2$ capture from flue gas. The paramaters observed were the effect of lean amine, flue gas and regenerator feed mix temperature to $CO_2$ capture, energy consumption and MEA losses. Moreover, the rich-split stream configuration was developed to examine energy reduction in various split fractions.

## 2. Methodology

### *2.1. Location*
A coal-based power plant in South Sumatra, Indonesia was chosen for the location of $CO_2$ capture simulation.

### *2.2. Data Collection*
The data was obtained from a confidential third-party report that surveyed properties of flue gas in coal-based power plant, South Sumatra.

**Table 1.** Coal Power Plant Exhaust Emissions Characteristics.

| Unit | Flue Gas Emission Flow Rate ($Nm^3/h$) | Flue Gas Compositions (%vol) | Flue Gas Temperature (°C) | Flue Gas Pressure (Bar) |
|---|---|---|---|---|
| 1 | 172,500 | $CO_2$: 13-14<br>$O_2$: 4-5 | 361.20 | 1.5 |
| 2 | 142,500 | $CO_2$: 13-14<br>$O_2$: 4.5-5 | 332.00 | 1.5 |

IOP Conf. Series: Materials Science and Engineering 1143 (2021) 012047 doi:10.1088/1757-899X/1143/1/012047

| Unit | Flue Gas Emission Flow Rate ($Nm^3/h$) | Flue Gas Compositions (%vol) | Flue Gas Temperature (°C) | Flue Gas Pressure (Bar) |
|---|---|---|---|---|
| 3 | 158,600 | $CO_2$: 14-15<br>$O_2$: 4.5-5 | 369.04 | 1.6 |
| 4 | 158,150 | $CO_2$: 13-14<br>$O_2$: 4.5-5 | 378.54 | 1.6 |

*2.3. $CO_2$ Capture Simulation using Aspen Hysys V.10.0*

$CO_2$ capture process was simulated using Aspen Hysys V10.0 software using the Acid Gas fluid package that is suitable to remove acid gases ($H_2S$ and $CO_2$). Several parameters for simulation were taken from other studies [17,20,22]. The baseline flowsheet and process parameter of $CO_2$ capture is shown in Figure 1.

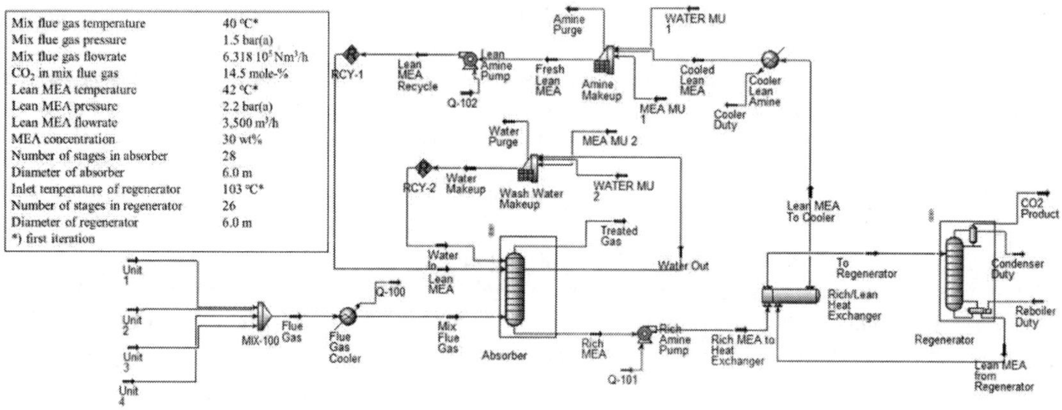


**Figure 1.** Process flow diagram and modelling parameter of standard $CO_2$ capture from flue gas of coal-based power plant.

The paper focused on $CO_2$ absorption-desorption process, excluding the pre-treatment process such as electrostatic precipitation, flue gas desulfurization (FGD), and selective catalytic reduction (SCR). Initially, the flue gas from four units was mixed and cooled to 40 °C before entering absorber column. MEA was chosen as amine solvent and contacted counter-current with flue gas in absorber. Thus, the rich MEA solution, containing $CO_2$ and trace of $N_2$ and $O_2$, was pumped to rich/lean amine heat exchanger for preheating to 103 °C. The gases in rich MEA were stripped at regenerator column and $CO_2$ was recovered in the top of column.

Hot lean MEA which still contains $CO_2$ from regenerator was cooled down to 40 °C in rich/lean amine heat exchanger and lean amine cooler. A makeup unit was added to ensure the desired flowrate and concentration of lean MEA as there was possibility of MEA and water losses at outlet of the absorber and regenerator column. Then, fresh lean MEA solution was pumped to absorber to start new absorption-desorption cycles. Due to high vapor pressure, the water wash unit was installed at the top of absorber in order to minimize MEA losses through vaporization [23]. Influence of lean amine,flue gas and regenerator feed mix temperature to $CO_2$ capture, energy consumption and MEA losses were studied. $CO_2$ capture was calculated using equation (1) from Qing, et. al work [25] which can be expressed as

$$CO_2 \text{ capture} = \frac{P}{S} \times 100\% \quad (1)$$

Where P and S show the mass flow of the $CO_2$ product and supply, respectively.

Moreover, rich-split model proposed by Cousins, et.al [13] and Eisenberg and Johnson [24] was built to analyze the reduction potential of regeneration energy of MEA. The system split unheated rich amine entering to the top of regenerator column and the heated rich amine was still flowed to stage 8. The split percentage of cold rich amine was varied. Figure 2 shows the rich-split flowsheet of $CO_2$ absorption.

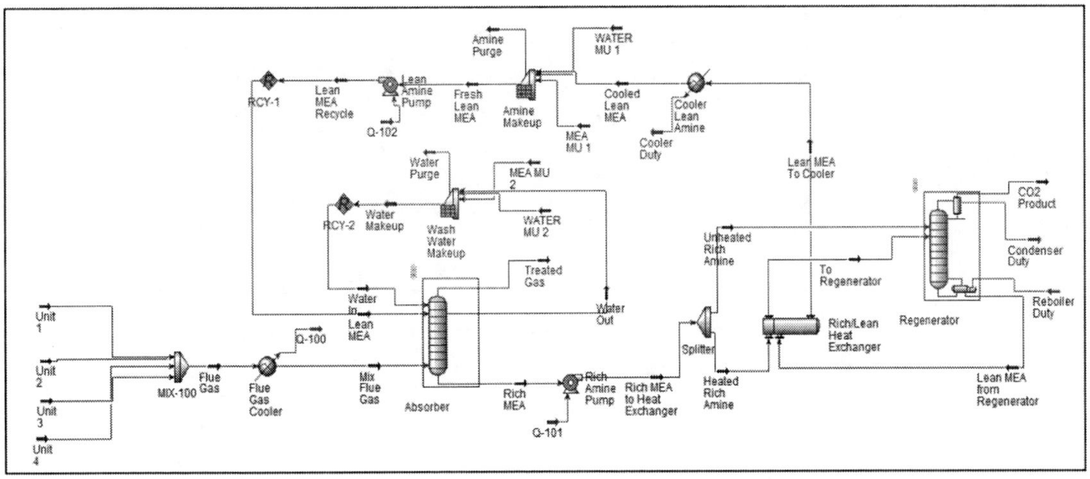


**Figure 2.** Rich-split configuration in $CO_2$ capture.

## 3. Results and Discussion

Based on standard configuration, the $CO_2$ capture using MEA 30 wt% reached 85% with energy consumption of 5.42 GJ/ton $CO_2$. The percentage of $CO_2$ capture was similar with the work of Gervasi, et.al [20]. For validation, the regeneration energy of MEA 30 wt% in this study was in good agreement with industrial $CO_2$ absorption data reported by Wilson, et.al [26] and very closely with the experiment of Sakwattanapong, et.al [27]. Compared to another study, Badea and Dinca [28] stated that when MEA 30 wt% was applied, the heat consumption of 85% $CO_2$ capture was about 6.40 GJ/ton $CO_2$ which is greatly higher than that of this work. Moreover, installation of wash water was proven to decrease the MEA losses. Table 2 shows the comparison of $CO_2$ absorption process between this work and others.

**Table 2.** Comparison of $CO_2$ capture process result.

| Parameter | This study | Gervasi, et.al [20] | Sakwattanapong, et.al [27] | Badea and Dinca [28] |
|---|---|---|---|---|
| $CO_2$ Capture (%) | 85 | 85 | - | - |
| Lean $CO_2$ loading (mol $CO_2$/mol MEA) | 0.23 | 0.31 | 0.23 | 0.26 |
| MEA losses (ppm mole) | 1 | 197 | - | - |
| Regeneration Energy (GJ/ton $CO_2$) | 5.42 | 3.71 | 5.20 | 6.40 |
| Purity of $CO_2$ (%) | 97 | 98 | - | - |

### *3.1. Effect of Lean Amine Temperature*

The result of changing lean amine temperature is shown in Figure 3. The plot shows that the increase in temperature decreases the energy consumption and the $CO_2$ capture was stable over the simulation range. Usually, the absorption of acid gases with primary and secondary amines performance can be enhanced

IOP Conf. Series: Materials Science and Engineering 1143 (2021) 012047 doi:10.1088/1757-899X/1143/1/012047

by lowering the operating temperature of the absorber [29,30]. The lean amine temperature effect on the MEA losses is shown in Figure 3b. Based on the results it is clear that the increase in temperature increases the MEA losses. At a higher lean amine temperature, the absorber as a whole will be operated at a higher temperature. This higher operating temperature will increase the evaporation rate of MEA from the top of the absorber [31]. Compared to Gervasi's study, the MEA losses in this study is relatively low as shown in Table 2.

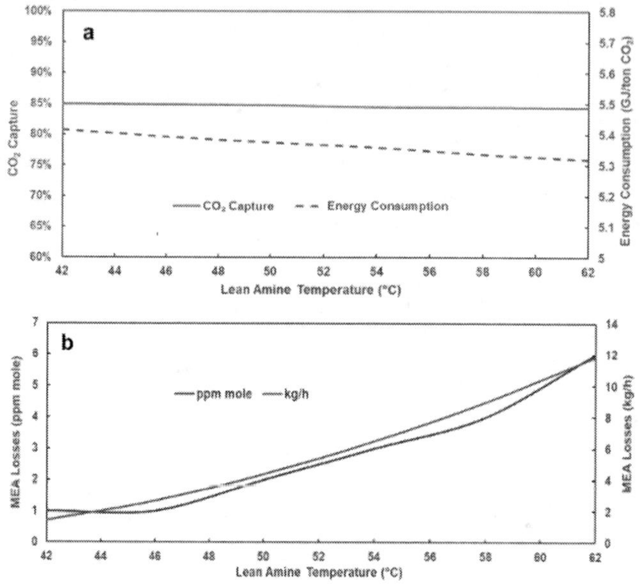


**Figure 3.** Effect of lean amine temperature on (a) $CO_2$ capture and energy consumption (b) MEA losses.

## 3.2. *Effect of Flue Gas Temperature*

The effect of flue gas temperature is studied along the temperature range of 40–60 °C in plant's performance condition and the response is plotted in Figure 4. The result shows that flue gas temperature variation affects the energy consumption and $CO_2$ capture. The energy consumption slightly increases as the temperature increases. Meanwhile, the $CO_2$ capture slightly decreases as the temperature increases. This is due to the fact that the $CO_2$ solubility decreases as the temperature increases [32].

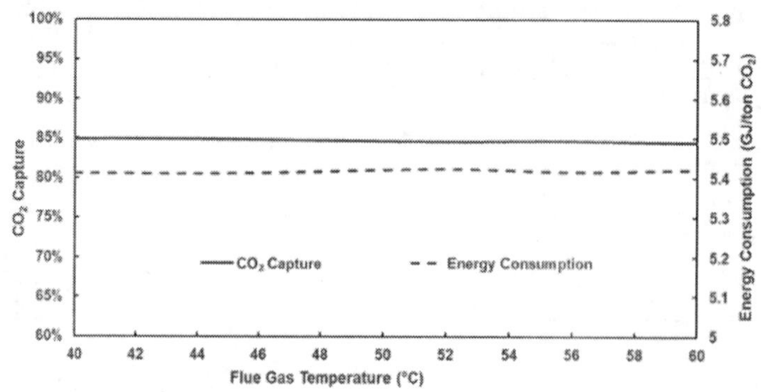


**Figure 4.** Effect of flue gas temperature on $CO_2$ capture and energy consumption.

### *3.3. Effect of Regenerator Feed Temperature*

In order to investigate the response regenerator feed temperature on the process, the simulation was performed at temperatures range 96-110 °C. The simulation results are shown in Figure 5.

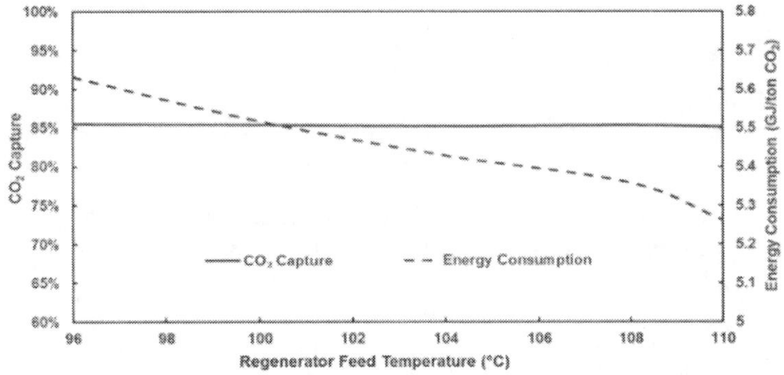


**Figure 5.** The effect of regenerator feed temperature on $CO_2$ capture and energy consumption.

The plot shows that the increase in regenerator feed temperature decreases the energy consumption. At the higher temperatures, the amount of carbon dioxide vapor in the feed stream increases, so regeneration energy needed was decrease [32]. The result shows that regenerator feed temperature variation does not affect the $CO_2$ capture and energy consumption.

### *3.4. Rich-Split Model*

Rich-split model proposed by Cousins, et.al [13] and Eisenberg and Johnson [24] was built to analyze the reduction potential of regeneration energy of MEA. $CO_2$ capture efficiency of 85% and lean $CO_2$ loading of 0.23 were achieved for these runs. The effect of the fraction of solvent split to the top of the column on energy consumptions and reboiler duty is outlined in Figure 6. The minimum in energy consumption and reboiler duty occurs when 30% of the solvent is split to the top of the column. The minimum in reboiler duty should occur when the energy of vapor entering in the hot rich solvent stream matches the additional sensible heat requirement of the cold rich solvent entering at the top of the stripping column [13]. This minimum energy consumption was 5.08 GJ/ton $CO_2$ together with 759.43 GJ/h reboiler duty. This is a 35% reduction in energy consumption compared to the reference case without rich-split process. To investigate why this saving occurs, the $CO_2$ vapor mass flow rate and temperature along the stages in regenerator column for the baseline case was compared to the case when 30% of the cold rich solvent is split to the top of the stripping column.

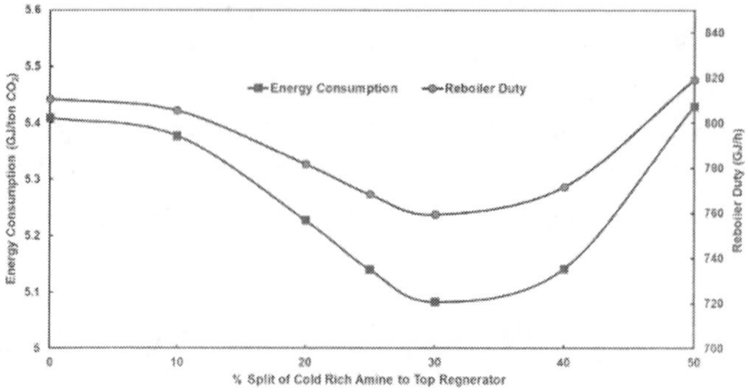


**Figure 6.** The effect of split fraction on total energy consumption with 85% of $CO_2$ capture and 0.23 of lean $CO_2$ loading

As shown in Figure 7a, the $CO_2$ vapor mass flow rate in regenerator column was lower than baseline. This indicated that the driving force of the $CO_2$ desorption process was increased, which accelerates the $CO_2$ desorption rate and reduce the energy consumption of solvent regeneration. The split cold rich solvent decreased the temperature at the top stage of the stripper (stage 1) in the rich-split process, which was favorable because it could recover the steam and reduce the reboiler duty. Meanwhile, the temperatures along column after stage 2 were elevated. The temperature rise came from the heat exchanger where the unsplit rich solvent was heated to a higher temperature as a result of the decreasing solvent flow rate. The effectiveness of this rich split will be significantly influenced by the efficiency of the lean/rich heat exchanger [33].

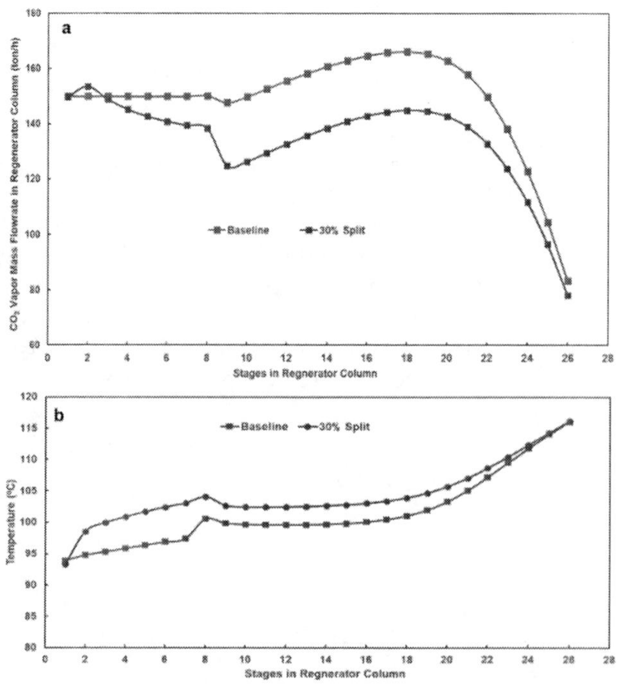


**Figure 7.** Profile of (a) $CO_2$ vapor mass flowrate and (b) temperature at baseline and 30% split in regenerator column

IOP Conf. Series: Materials Science and Engineering 1143 (2021) 012047 doi:10.1088/1757-899X/1143/1/012047

## 4. Conclusion

The $CO_2$ removal of flue gas from coal-fired power plant model was developed in Aspen Hysys. The lean amine temperature, flue gas temperature, and regenerator feed temperature are varied to test its effect on the $CO_2$ capture and energy consumption. The increase in lean amine temperature decreases the energy consumption and the $CO_2$ capture was stable over the simulation range, besides that the increase in temperature increases the MEA losses. Flue gas temperature affects the energy consumption and $CO_2$ capture. The energy consumption slightly increases as the temperature increases, meanwhile the $CO_2$ capture slightly decreases as the temperature increases. The regenerator feed temperature on $CO_2$ capture and energy consumption has been studied. The result shows that regenerator feed temperature variation does not affect the $CO_2$ capture and the energy consumption. The Rich-Split model was simulated to analyze the reduction potential of regeneration energy of MEA. The simulation result shows that the minimum in energy consumption and reboiler duty occurs when 30% of the solvent is split to the top of the column. The effectiveness of rich split will be significantly influenced by the efficiency of the lean/rich heat exchanger

## References


[1] Wijaja A, Chrysolite H, Ge M, Wibowo C K, Pradana A, Utami A F and Austin K 2017 *How Can Indonesia Achieve Its Climate Change Mitigation Goal? An Analysis of Potential Emissions Reductions from Energy and Land-Use Policies* (Jakarta: World Resources Institute) 1-35

[2] Modak N M, Ghosh D K, Panda S and Sana S S 2018 *CIRP J. Manuf. Sci. Tech* **20** 1–11

[3] Edenhofer O, Pichs-Madruga R, Sokona Y, Farahani E, Kadner S, Seyboth K, Adler A, Baum I, Brunner S, Eickemeier P, Kriemann B, Savolainen J, Sclömer S, Stechow C.V, Zwickel T and Minx J.C 2014 : *Summary for Policymakers. In: Climate Change 2014: Mitigation of Climate Change* (Cambridge: Cambridge University Press) 1-30

[4] Wang Y, Zhao L, Otto A, Robinius M and Stolten D 2017 *Energy Procedia* **114** 650-665

[5] Wang X and Song C 2020 *Front. Energy Res* **8** 1-24

[6] Zhao X, Smith K H, Simioni M A, Tao W, Kentish S E, Fei W and Stevens G W 2011 *Int. J. Greenh. Gas Con* **5** 1163–1169

[7] Cuèllar-Franca R M and Azapagic A 2015 *Journal of $CO_2$ Utilization* **9** 82–102

[8] Borhani T N and Wang M 2019 *Renew. Sus. Energy Rev* 114 109299

[9] Wu X, Yu Y, Qin Z and Zhang Z 2014 *Energy Procedia* **63** 1339-1346

[10] Gomes J, Santos S and Bordado J 2014 *Environ. Technol* **36** 19-25

[11] Kndusen J N, Jensen J N, Vilhelmsen P-J and Biede O 2009 *Energy Procedia* **1** 783-790

[12] Cousins A, Cottrell A, Lawson A, Huang S and Feron P H M 2012 *Greenhouse Gas Sci. Technol* **5** 329-345

[13] Cousins A, Wardhaugh L T and Feron P H M 2011 *Chem. Eng. Res. Des* **89** 1237–1251

[14] Thimsen D, Maxson A, Smith V, Cents T, Falk-Pedersen O, Gorset O and Hamborg E S 2014 *Energy Procedia*; 2014.

[15] Hamborg E S, Smith V, Cents T, Brigman N, Falk-Pedersen O, de Cazanove T, Chhagnlal M, Feste J K, Ullestad Ø, Ulvatn H, Gorset O, Askestad I, Gram L K, Fostås B F, Shah M I, Maxson A and Thimsen D 2014 *Energy Procedia* **63** 5994-6011

[16] Gjernes E, Pedersen S, Cents T, Watson G, Fostås B F, Shah M I, Lombardo G, Desvignes C, Flø N E, Morken A K, de Cazenove T, Faramarzi L, Hamborg E S 2017 *Energy Procedia* **114** 1146-1157

[17] Øi L E 2007 *The 48th Scandinavian Conference on Simulation and Modeling (SIMS)* (Gøteborg: Telemark University College)

[18] Øi L E and Vozniuk I.O 2010 *Processes and Technologies for a Sustainable Energy (PTSE)* (Ischia: Telemark University College)

[19] Øi L E, Bråthen T, Berg C, Brekne S K, Flatin M, Johnsen R, Moen I G and Thomassen E 2014 *Energy Procedia* **51** 224-233

[20] Gervasi J, Dubois L and Thomas D 2014 *Energy Procedia* **63** 1018-1028

[21] Dubois L and Thomas D 2017 *Energy Procedia* **114** 1409-1423
[22] Hasan N S M 2005 *Techno-Economic Study of $CO_2$ Capture Process for Cement Plants* Master Thesis (Chemical Engineering: University of Waterloo)
[23] GPSA 2004 *Engineering data book* (Oklahoma: Gas Processors Suppliers Association)
[24] Eisenberg B and Johnson R R 1979 Patent No. US 4152217 A1 Exxon Research and Engineering Company
[25] Faramarzi L, Thimsen D, Hume S, Maxon A, Watson G, S P, Gjernes E, Fostås B F, Lombardo G, Cents T, Morken A.K, Shah M.I, Cazenove T.D and Hamborg E.S 2017 *Energy Procedia* **114** 1128-1145
[26] Wilson M, Tontiwachwuthikul P, Chakma A, Idem R, Veawab A, Aroonwilas A, Gelowitz D and Stobbs R 2005 *The 7th International Conference on Greenhouse Gas Control Technologies 5* (Vancouver: Elsevier)
[27] Sakwattanapong R, Aroonwilas A and Veawab A 2005 *Ind. Eng. Chem. Res* **44** 4465-4473
[28] Badea A A and Dinca C F 2012 *U.P.B. Sci. Bull. Ser. D* **74** 21–32
[29] Umer Z 2020 *Chem. Prod. Process. Model* **44** 1-17
[30] Dashti S S and Nikou M R K 2015. Asia-Pac. J. Chem. Eng.
[31] Muhammad R M, Abu Z, Leon H J, John P M, Paul H M, Geert F V 2006 *Int. J. Greenh. Gas. Con* **1** 37-46
[32] Naji S Z and Abd A A 2019 Front. Energy **13** 317–324
[33] Kangkang L, Ashleigh C, Hai Y, Paul F, Moses T, Weiliang L, Jian C 2016 *Energy. Sci. Eng* **4** 23-29

IOP Conf. Series: Materials Science and Engineering 1143 (2021) 012048 doi:10.1088/1757-899X/1143/1/012048

# Plastic marine debris study based on Indonesian plastic national balance (IPNB) and seashore approach

**A Z Abidin[1] and S Steven[1*]**

[1] Department of Chemical Engineering, Institut Teknologi Bandung, Jl. Ganesha 10, Bandung 40132, Indonesia

* Email: soen_steven@students.itb.ac.id / soensteven@yahoo.com

**Abstract.** A clear and accurate plastic marine debris amount to clarify Indonesian marine debris issues still does not exist. Only Jambeck's study estimated that millions of tonnes of plastic enter the Indonesian ocean every year, but the result is often disputed. This study, hence, was conducted to calculate the Indonesian plastic marine debris amount using Indonesian plastic national balance (IPNB) and seashore approach. The study was based on Indonesia's 2019 data of plastic production, plastic raw material and finished-product balance, plastic recycling industry, plastic consumption, and plastic distribution. Indonesian plastic industries have a total capacity of 2.66 MT/y, production of 2.31 MT/y, import of 1.67 MT/y, and recycle of 1.655 MT/y. The results from Indonesian plastic national balance (IPNB) imply that after-used plastic of 5.755 MT/y is categorized into material still-consumed, material recycled, and plastic waste. About 90% of the plastic waste is managed which further recycled (0.654 MT/y) and ended up in the landfill (0.868 MT/y), while the rest (about 10%) is mismanaged and become marine debris (0.17 MT/y). For the results from the seashore approach, after-used plastic is burned and buried (3.228 MT/y), recycled (1.655 MT/y), ended up in the landfill (0.868 MT/y), flowed to the river (0.009 MT/y), picked up in the river (0.005 MT/y), and the rest become marine debris (0.004 MT/y). Both results differ from Jambeck's calculation (0.48-1.29 MT/y) and therefore, clarify that Indonesia is not the second-largest contributor country for plastic marine debris.

## 1. Introduction

Plastic marine debris is a global environmental problem and challenging to Indonesia as a maritime country [1–5]. Due to rapid population growth in Indonesia, waste management infrastructure becomes unbalance and poor [1,2,4,6–8]. Plastic is used in many aspects of daily life resulting in the accumulation of mismanaged plastic waste [9–13]. In many places, plastic waste is commonly seen along a beach and floating in the ocean. This condition makes plastic waste become a visible part of the marine debris problem [3,4,14]. In the ocean, plastic will continue to degrade into small pieces that difficult to be seen anymore, usually called micro-plastic if less than 5 mm long. Due to the low temperature of the ocean, the degradation process occurs slower than on the land [14–17]. It makes plastic marine debris truly threaten underwater living beings, such as fish, microorganisms, and can damage the marine ecosystem [3,4,18,5–8,11,14,16,17].

An accurate Indonesian plastic marine debris amount, however, still does not exist. Only Jambeck et al. (2015) estimated that millions of tonnes of plastic enter the Indonesian ocean every year. Nevertheless, the result from Jambeck's study is still unclear and often disputed if compared to the ministry and recycling association data. Thus, it is necessary to clarify Indonesian marine debris issues with a more accurate approach. This study was conducted to calculate the plastic marine debris using

Indonesian plastic national balance (IPNB) and seashore approach. IPNB calculation refers to the data of ministry of industry, Indonesian trade map, and Indonesia's Olefin, Aromatic and Plastics Industry (INAPLAS), whereas the seashore approach is assisted with data from the ministry of industry and Indonesian Statistics 2019 (BPS).

## 2. Methods

There were three calculations in this study. Jambeck's calculation as a baseline or control, then IPNB and seashore calculations for verifying Jambeck's calculation. These calculations had several different parameters such as coastal population definition and determination, waste generation rate, percentage of plastic waste, and percentage of mismanaged waste. Jambeck assumed that Indonesian coastal population lived within 50 km from the ocean and the plastic marine debris calculation according to Jambeck's calculation follows equation 1 [3].

$$Indonesian\ Plastic\ Marine\ Debris = F * CP * PWG * MPW = F * CP * WGR * PPW * MPW \quad (1)$$

where $F$ is marine debris factor (15-40%), $CP$ is coastal population within 50 km from the ocean (person), $PWG$ is plastic waste generation rate (T/person/y), $WGR$ is waste generation rate (T/person/y), $PPW$ is percentage of plastic waste (11% for Jambeck's calculation), and $MPW$ is percentage of mismanaged plastic waste (%).

For IPNB calculation, the flow diagram should be generated in advance using the data adapted from the ministry of industry, Indonesian trade map, and INAPLAS. From this flow diagram, the after-used plastic will be recycled, still-consumed, and ended up as plastic waste. The plastic waste is then categorized as managed and mismanaged with the percentage referring to the IPNB flow diagram. According to this approach, the value of mismanaged plastic waste was equal to the Indonesian plastic marine debris. Subsequently, the determination of plastic waste generation rate ($PWG$) could be seen in equation 2.

$$PWG = \frac{Plastic\ Waste\ Amount}{Indonesian\ Population} \rightarrow Indonesian\ 2019\ pop. = 268\ Millions \quad (2)$$

For seashore approach, the Indonesian coastal population should be first determined by the data from Indonesian statistics 2019. Actually, the coastal population from Jambeck's assumption is too far and absurd so this approach justified that coastal population lived in more than 10 km distance from the ocean. The seashore apporach utilized coastal sub-district population which was calculated from the coastal district population in Java Island ($JCDP$), which follows the justification that Indonesian population is concentrated in Java Island [9]. The Indonesian coastal sub-district population calculation procedure was served in equation 3. The plastic waste generation rate calculation from seashore approach was identical to IPNB. Moreover, plastic flow diagram was also generated and the after-used plastic will be recycled, burned, buried, ended up in the landfill, thrown into the river, and picked up in the river [19]. Plastic waste entering ocean as plastic marine debris was calculated from the difference between the plastic waste entering river and the picked-up plastic waste.

$$ICsDP = \frac{JCDP * \%sD}{J} \quad (3)$$

where $ICsDP$ is Indonesian coastal sub-district population, $\%sD$ is percentage of sub-district population in Java, and $J$ is Java population proportion.

## 3. Results and discussion

### *3.1. Indonesian plastic marine debris amount using IPNB*

The Indonesian plastic capacity is 2.66 MT/y, the country production is 2.31 MT/y, imports 1.67 MT/y of plastic, and recycles 1.655 MT/y of plastic. The annual plastic production as well as import and recycle will then be consumed. Average plastic consumption is 21.47 kg/person/y which is much lower than neighboring countries Singapore, Malaysia, and Thailand which consume more than 60 kg/person/y [1,2]. The consumed plastic products are then distributed into seven categories: household, packaging, building, shopping bag, Styrofoam, others, and finished goods plastic import [1,2,9,10]. After-used plastic is generated after plastic consumption and becomes material still-consumed (53.2%), material recycled (17.4%), and plastic waste (29.4%). The plastic waste generation is 1.692 MT/y or equal to 0.017 kg/person/day. Afterward, about 90% of plastic waste is managed which is then thrown into the landfill and recycled again, while about 10% of plastic waste, 0.17 MT/y, is categorized as mismanaged plastic waste and entered the ocean. From the exposition above, a detailed plastic flow diagram based on IPNB is depicted in Figure 1.

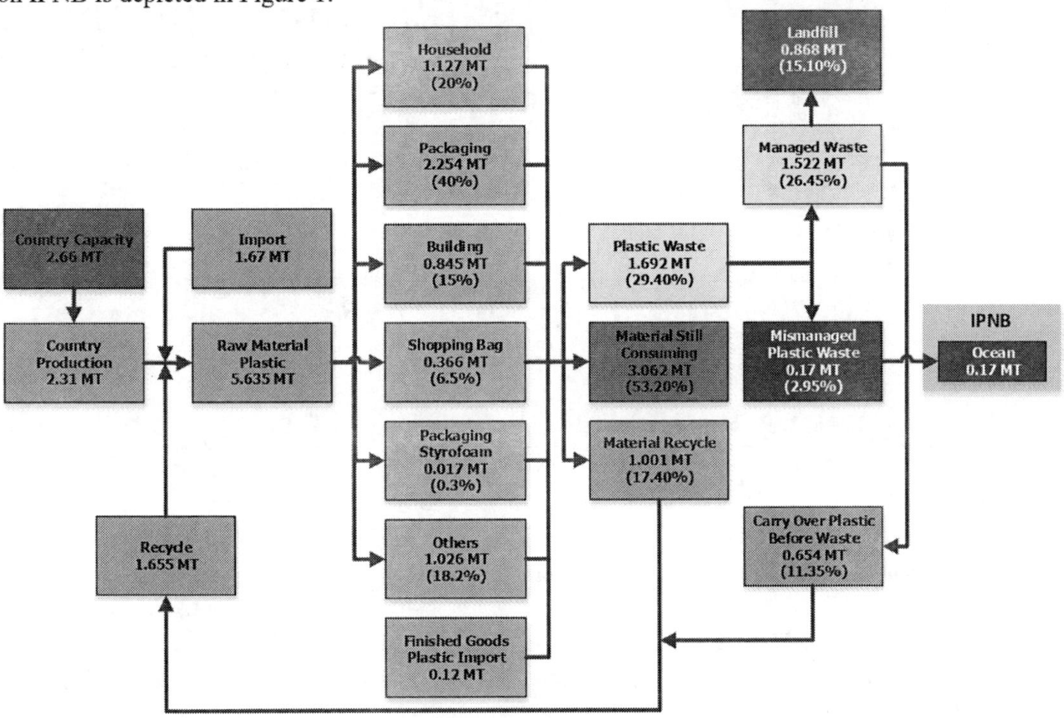


**Figure 1.** Indonesian plastic balance based on IPNB, adapted from: [1,2,19].

### *3.2. Indonesian plastic marine debris amount using seashore approach*

The seashore approach presumes that plastic marine debris comes from the river stream and due to the fishermen who throw the waste in the ocean. Plastic waste generated in the river initiates from the riverbank population which is concentrated in the sub-district area. The population is first determined from the coastal district population in Java Island ($JCDP$) which can be seen in Figure 2. The average % sub-district population in Java ($\%sD$) is also calculated at 34.7% (Table 1) and approximately 56.3% ($J$) of the Indonesian population is concentrated in Java, as described in Figure 3. Therefore, the Indonesian coastal sub-district population occupies about 44.75 million people, as shown in Table 2.

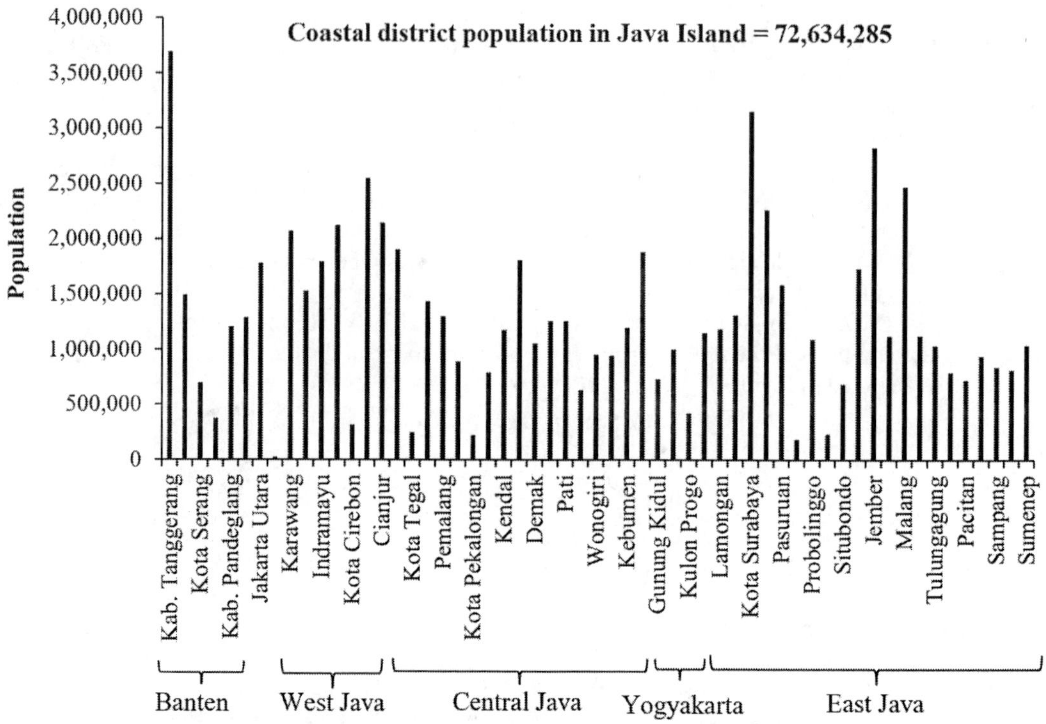


**Figure 2.** Coastal district population in Java 2019 [9,20].

**Table 1.** Percentage of coastal sub-district population in Java [9,20].

| Indramayu | | Pandeglang | | Surabaya | |
|---|---|---|---|---|---|
| Sukra | 43884 | Carita | 33139 | Benowo | 29506 |
| Kandanghaur | 86603 | Labuan | 56455 | Asemrowo | 23058 |
| Losarang | 54033 | Pagelaran | 35113 | Krembangan | 59805 |
| Cantigi | 31879 | Sukaresmi | 34657 | Pabeancantikan | 41595 |
| Sindang | 50563 | Panimbang | 51142 | Semampir | 96054 |
| Indramayu | 111009 | Cigeulis | 35182 | Kenjeran | 78385 |
| Balongan | 38675 | Cimanggu | 38409 | Bulak | 21192 |
| Juntinyuat | 78644 | Sumur | 23876 | Mulyorejo | 42343 |
| Karangampel | 63134 | Cibitung | 22190 | Sukolilo | 54022 |
| Krangkeng | 63944 | Cikeusik | 52866 | Rungkut | 54256 |
| | | | | Gununganyar | 27144 |
| Sub-district | 622368 | Sub-district | 383029 | Sub-district | 527360 |
| Total pop. Indramayu | 1718495 | Total pop. Pandeglang | 1194911 | Total pop. Surabaya | 1473640 |
| % sub-district | 36.22 | % sub-district | 32.06 | % sub-district | 35.79 |
| **Average % sub-district → 34.70%** | | | | | |

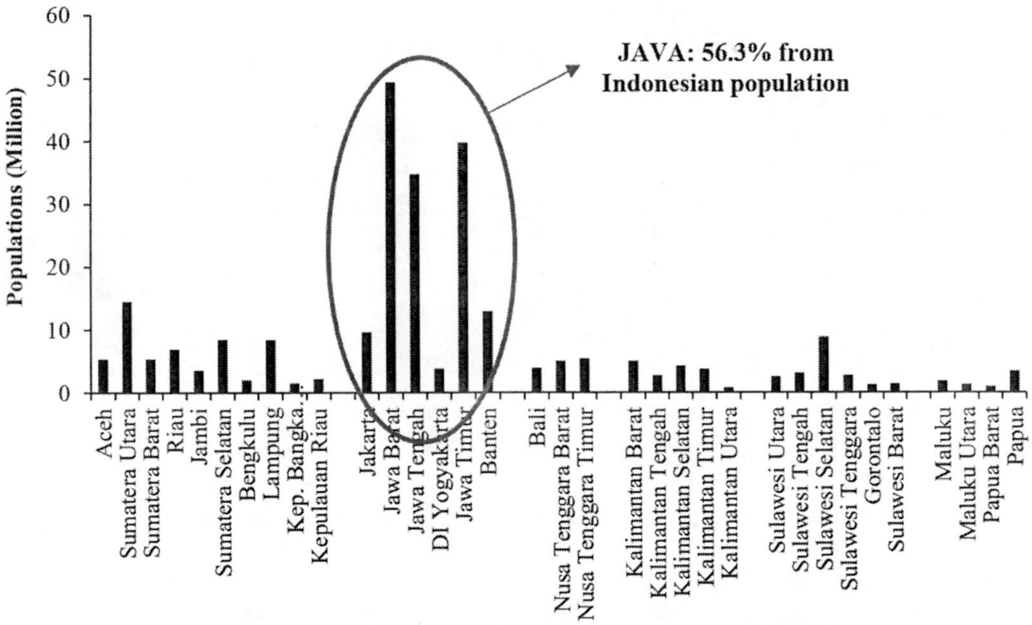


**Figure 3.** Indonesian province population in 2019 [9,20].

**Table 2.** Indonesian coastal sub-district population.

| Parameters | Amount |
|---|---|
| Coastal district population in Java | 72,634,285 |
| Coastal sub-district population in Java | 25,193,714 |
| Indonesian coastal sub-district population | 44,749,048 |

Slightly differ to IPNB, the after-used plastic from seashore approach tends to be burned, except for shopping bag which is mostly recycled [9,10,12,13,19]. The rest will then be recycled, buried, ended up in the landfill, and thrown into the river with the percentage given in Table 3. The percentage of after-used plastic which is burned and recycled is quite similar for building sector whereas the percentage of after-used plastic which is buried and recycled is quite similar for automotive and electronics sectors. From Table 3, about 71.08% of the after-used plastic will be thrown in the landfill (0.868 MT/y) as well as burned and buried (3.223 MT/y), nearly 28.76% is recycled (1.655 MT/y), and the rest (0.16% or 0.009 MT/y) is thrown into the river. The river plastic waste of 0.005 MT/y is picked up and 0.004 MT/y enters the ocean. The detailed plastic flow diagram based on seashore approach is served in Figure 4.

**Table 3.** Percentage of after-used plastic calculation [9,10,12,13].

| Plastic Distribution Sector | Proportion (%) | Burned (%) | Recycled (%) | Buried (%) | To River (%) | Total Burned (%) | Total Recycled (%) | Total Buried (%) | Total to River (%) |
|---|---|---|---|---|---|---|---|---|---|
| Household (Aneka) | 31% | 61.74 | 27.20 | 10.90 | 0.16 | 19.14 | 8.43 | 3.38 | 0.05 |
| Packaging | 33% | 62.35 | 27.20 | 10.30 | 0.15 | 20.58 | 8.98 | 3.40 | 0.05 |
| Building | 6% | 39.74 | 38.20 | 21.90 | 0.16 | 2.38 | 2.29 | 1.31 | 0.01 |
| Shopping bag | 4% | 19.35 | 69.20 | 11.30 | 0.15 | 0.77 | 2.77 | 0.45 | 0.01 |
| Automotive | 22% | 49.74 | 24.20 | 25.90 | 0.16 | 10.94 | 5.32 | 5.70 | 0.04 |
| Electronics | 4% | 49.74 | 24.20 | 25.90 | 0.16 | 1.99 | 0.97 | 1.04 | 0.01 |
| | | | **Percentage of after-used plastic** | | | **55.80** | **28.76** | **15.28** | **0.16** |

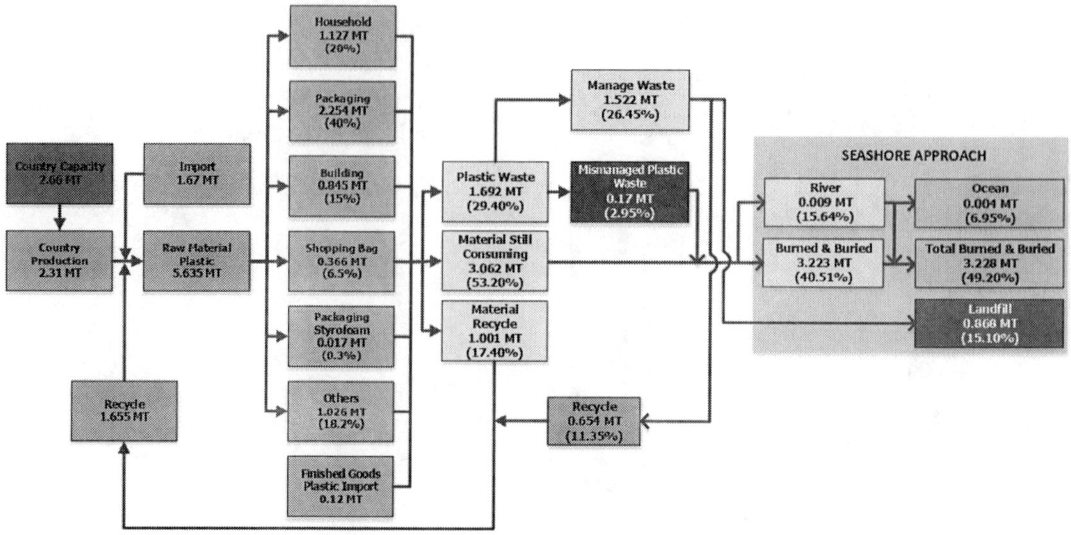


**Figure 4.** Indonesian plastic flow diagram based on seashore approach.

### *3.3. Comparison of Jambeck, IPNB, and seashore results*

The results discrepancies between these three calculations lie in the different justifications for coastal population, the amount of waste generation, the percentage of plastic waste, and the percentage of mismanaged waste. Jambeck identifies that Indonesian coastal population is 187.2 million due to the population within 50 km from the ocean [3]. Also, the assumption for 11% of total waste is categorized as plastic waste and about 83% of total plastic waste is mismanaged [3,21]. Jambeck's calculation on waste generation rate is 0.52 kg/person/day and the amount of Indonesian plastic marine debris is calculated 0.48-1.29 MT/y [3].

Contrary to Jambeck's, IPNB and seashore presumed 44.75 million of coastal population. That is the population who lived within 10 km from the ocean with the justification that people are very rare coming to the ocean within 50 km of distance. According to Indonesian plastic product distribution data, plastic waste generation rate is only 0.017 kg/person/day [9,10,12,13]. Moreover, the percentage of plastic waste is only 9% [12,13]. The waste generation rate is then equal to 0.19 kg/person/day which is lower than Jambeck's. IPNB calculation has a result of 0.17 MT/y of mismanaged plastic waste becomes plastic marine debris [1,2,9,10,12,13].

Also, seashore approach strengthens and complements IPNB results. The after-used plastic is thrown into the landfill, burned and buried, recycled, thrown into the river, and picked up. About 0.004 MT/y of plastic waste flow to the ocean as plastic marine debris. Both calculations prove and clarify that only a small part of mismanaged plastic waste contributes to plastic marine debris. The results comparison is outlined in Table 4.

## 4. Conclusions and future outlook

The plastic marine debris calculation using IPNB and seashore approach clarifies Jambeck's calculation and claim. Based on this study, there is only a partial amount of mismanaged plastic waste entered the ocean. Both calculations give a lower amount of plastic marine debris than Jambeck's. Jambeck's calculation on Indonesian plastic marine debris is 0.48-1.29 MT/y, where the IPNB and seashore calculation resulting in 0.17 MT/y and 0.004 MT/y, respectively. Although Indonesia is not the second-largest contributor country for plastic marine debris, some activities either zero waste management education program; plastic recycling performance improvement; clean river, beach, and ocean maintenance, or trash trap installation on the river are also interesting to be offered in order to solve the plastic waste problem.

**Table 4.** Plastic marine debris calculation results comparison.

| Parameters | Units | Jambeck | IPNB | Seashore |
|---|---|---|---|---|
| Indonesian coastal population | million | 187.2 | 44.75 | 44.75 |
| Waste generation rate | kg/person/day | 0.52 | 0.19 | 0.19 |
| Percentage of plastic waste | % | 11 | 9 | 9 |
| Plastic waste generation rate | kg/person/day | 0.057 | 0.017 | 0.017 |
| Plastic waste amount | MT/y | 3.89 | 1.69 | 1.69 |
| Mismanaged waste | % | 83 | 10 | 10 |
| Mismanaged plastic waste | MT/y | 3.22 | 0.17 | 0.17 |
| Total plastic marine debris | MT/y | 0.48-1.29 | 0.17 | 0.004 |

**Acknowledgments**
Authors wishing to acknowledge assistance and data support from Polymer and Membrane Technology Laboratory and INAPLAS. The authors would also thank the STKSR 2020 committee's assistance and the financial support provided by DOW Chemicals.

**References**

[1] Rivai E 2017 *Plastics and lifecycle* (Jakarta: Forum Group Discussion Marine Debris Waste Management)
[2] Rivai E 2017 *Indonesia national plan of action on marine plastic debris* (Yangon, Myanmar: AFPI Meet)
[3] Jambeck J, Geyer R, Wilcox C, Theodore S, Perryman M, Andrady A, Narayan R and Law K L 2015 Marine pollution: Plastic waste inputs from land into the ocean *Science* **347** 768–71
[4] Purba N P, Pangestu I F and Nurrahman Y 2016 Marine debris in Indonesia: Finding debris *Indonesia Marine Plastic Debris Summit*
[5] Purba N P, Padjadjaran U, Syamsuddin M, Padjadjaran U and Sandro R 2017 Distribution of marine debris in Biawak Island, West Java, Indonesia *World Sci. News* **66** 281–92
[6] Hartman A, Loven E and Olsson M 2014 *Plastic as marine debris and its potential for economic value: A practical and analytical estimation of the marine debris characteristics and a comparative evaluation of possible treatment procedures* (Stockholm: Bachelor Thesis Energy Environment)
[7] Sur C, Abbott J M, Ambo-rappe R, Asriani N, Hameed S O, Jellison B M, Lestari H A, Limbong S R, Mandasari M, Ng G, Satterthwaite E V and Syahid S 2018 Marine debris on small islands: Insights from an educational outreach program in the spermonde archipelago, Indonesia *Perspect. Front. Mar. Sci.* **5** 1–5
[8] Putri M R 2017 ASEAN trans-boundary marine debris and consortium for Indonesia oceanic research *Regional Workshop in support of the second cycle of the Regular Process for Global Reporting and Assessment of the State of the Marine Environment, including Socioeconomic Aspects*
[9] Statistics Indonesia 2019 *Indonesian coastal population*. https://www.bps.go.id/. (Central Bureau of Statistics)
[10] Indonesian Trade Map 2019 *Indonesian plastic production, import, and export*. http://www.trademap.org/Country_SelProductCountry
[11] Dili 2007 *Comparison of existing life cycle analysis of shopping bag alternatives* (Melbourne: Sustainability Victoria)
[12] Indonesian Ministry of Industry 2015 *Indonesian petrochemical industry review (Tela'ah industri petrokimia Indonesia)* (Indonesia: Ministry of Industry)

[13] Indonesian Ministry of Industry 2015 *Natural resources final report (Final report SDA)* (Indonesia: Ministry of Industry)
[14] Kershaw P J 2015 *Biodegradable plastics & marine litter: Misconceptions, concerns, and impacts on marine environments* (Nairobi: United Nations Environmental Programe)
[15] Masura J, Baker J, Foster G and Arthur C 2015 Laboratory methods for the analysis of microplastics in the marine environment: Recommendations for quantifying synthetic particles in waters and sediments *Natl. Ocean. Atmos. Adm. Mar. Debris Program-U.S. Dep. Commer.* 1–30
[16] Mclauchlin A, Thomas N L, Patrick S G and Clarke J 2012 Oxo-degradable plastics: Degradation, environmental impact, and recycling *Waste Resour. Manag.* **165**
[17] Hann S, Ettlinger S, Gibbs A and Hogg D 2016 *The impact of the use of "oxo-degradable" plastic on the environment, final report for the European commission DG environment* (Luxembourg: Publications Office of the European Union)
[18] UN Environment 2018 *Single-use plastics: A roadmap for sustainability*
[19] World-Economic-Forum 2020 *Radically reducing plastic pollution in Indonesia: A multistakeholder action plan (National Plastic Action Partnership)*
[20] Statistics Indonesia 2019 *Statistical Yearbook of Indonesia* (Central Bureau of Statistics)
[21] Shuker I G and Cadman C A 2018 *Indonesia marine debris hotspot rapid assessment: Synthesis report* (Washington D.C.: World Bank Group)

# Study on Indonesian plastic marine debris based on river survey

**A Z Abidin[1] and S Steven[1*]**

[1] Department of Chemical Engineering, Institut Teknologi Bandung, Jl. Ganesha 10, Bandung 40132, Indonesia

[*] Email: soen_steven@students.itb.ac.id / soensteven@yahoo.com

**Abstract.** In response to Jambeck's claim on Indonesian plastic marine debris, some academicians, scientists, as well as associations such as The Indonesian Olefin and Plastic Industry Association (INAPLAS), Indonesian Plastic Recycle Association (ADUPI), and Indonesian Plastic Recycle Industry Association (APDUPI) conducted a deeper plastic balance examination to clarify and enlight the Indonesian plastic marine debris amount. This study continued the previous work, which employed the river survey method to conduct a more accurate calculation in the actual condition. From the river survey results, Indonesian plastic marine debris amount is about 0.038 MT/y and it will reduce to 0.004 MT/y if the trash trap is strictly-applied. This result is not much different compare to the seashore approach but again also clarifies that Indonesia is not the second-largest contributor country for plastic marine debris. Moreover, asides from conserving the marine and river environments, installing a trash trap on the river could increase local people's income at once.

## 1. Introduction

Based on the research conducted by Jambeck et al. (2015), Indonesia is the second-largest country for plastic marine debris after China with the amount of 0.48-1.29 MT/y and this happened due to Jambeck's misleading assumptions such as Indonesian coastal population within 50 km from the ocean, 0.52 kg/person/day of waste generation rate, 11% of total waste is categorized as plastic waste, and 83% of total plastic waste is mismanaged [1]. This article raises some issues and speculations that plastic has become an enemy of human life. The results of Jambeck's calculation have forced our government to implement plastic packaging tax according to Circular Letter of the Ministry of Environment and Forestry No. S.1230/PSLB3-PS /2016 about mechanism and price of paid plastic bags [2]. To follow up this matter, some academicians, scientists, as well as associations such as The Indonesian Olefin and Plastic Industry Association (INAPLAS), Indonesian Plastic Recycle Association (ADUPI), and Indonesian Plastic Recycle Industry Association (APDUPI) conducted a deeper Indonesian plastic balance examination [3,4].

The previous study on Indonesian plastic marine debris using national balance (0.17 MT/y) and seashore approach (0.004 MT/y) have different results with Jambeck's due to the improved justifications in terms of coastal population definition which is 10 km from the ocean, 0.19 kg/person/day of waste generation rate, 9% of waste is categorized as plastic waste, and 10% of plastic waste is mismanaged [5]. To assign a more accurate Indonesian plastic marine debris amount, a direct river survey was conducted on several rivers in Java Island which lead into the ocean. It was realized that Java Island contained 6% of the Indonesian Area and contributed to 56.3% of the total population [5–8]. As consequence, a river survey in Java Island could be applied to represent the Indonesian river [9]. The purpose of this study is to determine the actual Indonesian plastic marine debris through river survey and compared the results with Jambeck's. A trash trap was installed in every surveyed river and the amount of plastic waste trapped was calculated annually. The plastic flow diagram from river survey was also generated.

## 2. Methods

The method in this study encompassed direct rivers survey, trash trap installation, trapped plastic waste calculation, as well as local people interview to justify the riverbank width. The interview was held for 200 respondents. This study only focused on the plastic waste that flows in the river, because has a high potential to entering the ocean. As aforementioned, river survey was conducted in Java Island's rivers because it could represent the Indonesian river.

Eleven dominant rivers in Java Island were surveyed, namely Ciliwung-Cisadane, Citarum, Cidanau-Ciujung-Cidurian, Cisadea-Cibareno, Ciwulan-Cilaki Jratun-Seluna, Wiso-Gelis, Pemali-Comal, Serayu, Bengawan Solo, and Brantas. These rivers were chosen because of their length and wide anatomy, which were distributed in five major provinces in Java: Banten, DKI Jakarta, West Java, Central Java, and East Java which directly connect to the ocean. The map of each river is shown in Figure 1.

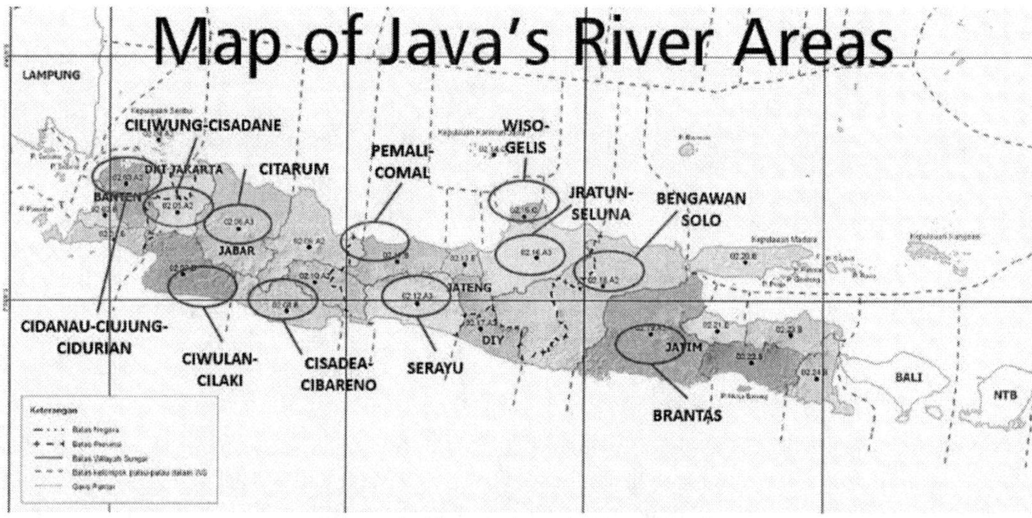


**Figure 1.** Eleven rivers in Java Island.

The trash trap installation was performed in every surveyed location and the trapped plastic waste was then weighted every day. Subsequently, Indonesian plastic marine debris calculation according to river survey was first derived from the monthly trapped plastic waste ($\boldsymbol{RPW_m}$) in every river. The annual trapped plastic waste in every river ($\boldsymbol{RPW_A}$) was calculated based on equation 1. To provide a plastic flow diagram, the riverbank area ($\boldsymbol{RBA}$), the riverbank population density ($\boldsymbol{RPD}$), and specific plastic waste generation ($\boldsymbol{SPWG}$) should be determined. The riverbank area ($\boldsymbol{RBA}$) calculation was presumed as a rectangle with the length ($\boldsymbol{RBL}$) which was measured using google map tracing distance and the width ($\boldsymbol{RBW}$) which was obtained from the interview. The riverbank population ($\boldsymbol{RP}$) was determined through equation 2. The calculation for plastic waste generation ($\boldsymbol{PWG}$) in every river was served in equation 3. Finally, the Indonesian plastic marine debris was determined as seen in equation 4 if no trash trap was installed and follows equation 5 if there is trash trap was installed.

$$\boldsymbol{RPW_A = 12 \times RPW_m} \quad (1)$$

$$\boldsymbol{RP = RPD \times RBA = RPD \times RBL \times RBW} \quad (2)$$

$$\boldsymbol{PWG = RP \times SPWG} \tag{3}$$

$$Indonesian\ Plastic\ Marine\ Debris = \sum_{i=1}^{n} \frac{SF \times PWG}{J}\ , n = total\ river \tag{4}$$

$$Indonesian\ Plastic\ Marine\ Debris\ = \sum_{i=1}^{n} \frac{SF \times (PWG - RPW_A)}{J}\ , n = total\ river \tag{5}$$

where ***SF*** is season factor (the escalation of river waste amount due to 6 months rainy season and 6 months dry season in Indonesia) and ***J*** is Java population proportion.

## 3. Results and discussion

### *3.1. Riverbank length and width*

The interview is intended to determine the riverbank width and to convince the plastic marine debris contributor. According to 200 respondents, interestingly, 85.6% of them (172 respondents) were not throwing garbage into the river. Some of them fear heavy flooding due to the water drainage plugged; some of them said that it is a bad habit; some of them argued that the river and marine ecosystem and esthetical value will be damaged. Nearly 61% of respondents who ever throw garbage into the river (18 respondents, 9% of all respondents), stayed near the riverbank with a distance of less than 1 km. The people who have a house more than 1 km from the river do not throw garbage into the river due to the long-distance, so they like to be buried, burned, or disposed of the waste in the landfill.

The interview also convinced the argument that plastic marine debris is not generated by people who throw waste on the ocean because it will be dragged back to the beach due to the ocean waves. This condition is supported in Cianjur Coast survey that the twig and tree waste were dominantly found along the coastal line and not found in the ocean. Furthermore, there is a low amount of plastic waste found on Cianjur Coast. The supporting picture is served in Figure 2.

(a) (b) (c) (d)

**Figure 2.** Domination of wood and twig trash in Ciwulan River downstream: Surveyed on November 22nd, 2017 (a,b) and November 24st, 2017 (c,d).

The plastic waste generation in the river is then strongly coming from the riverbank population who throw waste into the river, with 1 km distance from the river. The riverbank width should be multiplied by 2 due to river's left side and right side. Meanwhile, the riverbank length measurement procedure using google map tracing distance is illustrated in Figure 3.

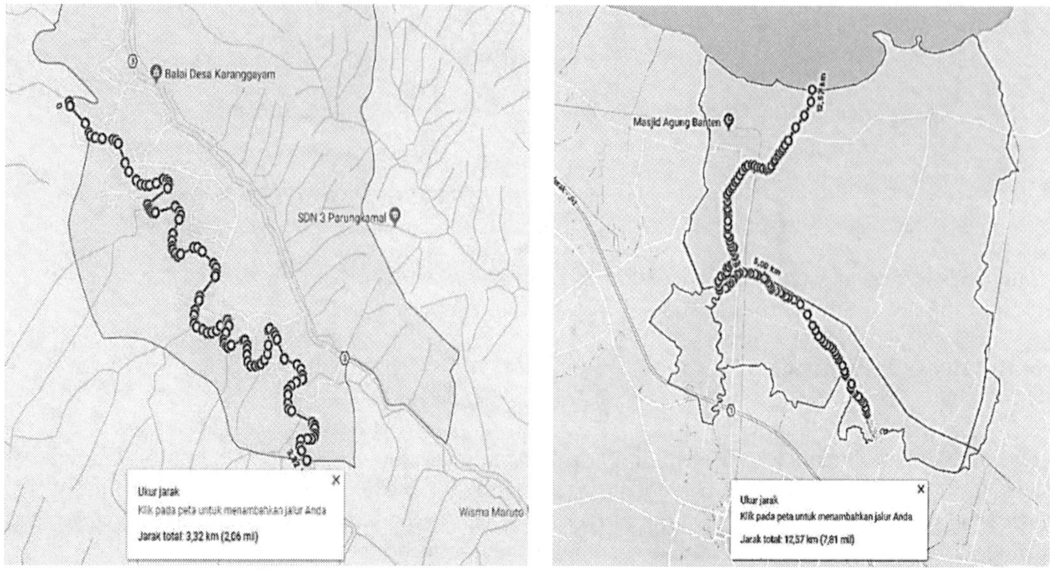


(a) (b)

**Figure 3.** River length measurement procedure: Serayu River in Karanggayam sub-district, Central Java (a); Cidanau-Ciujung-Cidurian River in Serang sub-district, Banten (b).

### *3.2. Plastic waste generation from riverbank population*

The river waste generation inevitably depends on the riverbank population density and riverbank area. The longer river length and denser riverbank population increase river waste generation, indeed. It is also observed that every Indonesian people throw waste theoretically at about 0.75 kg/day (Table 1), although it is lower than Malaysia and Singapore which are 1.5 kg/day [10].

**Table 1.** Theoretical waste generation per person per day.

| No. | Amount | Unit | References |
|---|---|---|---|
| 1. | 0.70-0.80 | kg/person/day | [11,12] |
| 2. | 0.70-0.85 | | [13,14] |
| 3. | 0.75 | | [10] |
| 4. | 0.70 | | [15] |
| **Average** | **0.75** | **kg/person/day** | |
| | **0.27** | **tonnes/person/y** | |

According to Tempo, 9% of total waste is categorized as plastic waste [15] and the updated one is 4.65% taken from INAPLAS, Unilever, and Bali province data. INAPLAS also complements that 17.4% of total plastic waste is recycled and 40.2% of the plastic waste after-recycled is thrown into the landfill [3,4,7]. The rest is entering the river. Therefore, the plastic waste generation per person entering the river ($SPWG$) is 0.0062 tonnes/person/y.

### *3.3. Indonesian plastic marine debris from river survey*

Every year, Indonesia has 6 months of rainy season and 6 months of dry season. During rainy season with a high wind occurs, the river flow rate becomes much higher, the waste amount in the river is 10 times higher, and waste transportation has more frequently occurred, reaching 10-30 times daily. The season factor is 1 when dry season and is 10 when heavy rain and windy. Thus, the annual marine plastic waste is generated from cumulative trapped plastic waste in the river multiplied by 5 as an average season factor. It is found that there is a huge of wood and trunks (Figure 4a) and also 15 tonnes of trunk (Figure 4b) were dragged by the river stream during rainy season and this could increase the probability of trash trap breakage. Frequent maintenance of the trash trap is imperative to hinder plastic waste enter the ocean.

(a) (b)

**Figure 4.** Wood and trunks waste in the Manggarai River during rainy season (a); 15 tonnes of trunk found in Manggarai river (red area).

Plastic marine debris calculated in every province is outlined in Table 2. From the survey, Indonesian plastic marine debris result is about 0.038 MT/y and if the trash trap is strictly-applied, the value is reduced to 0.004 MT/y. The comparison between several methods on Indonesian plastic marine debris amount is served briefly in Table 3. The number of plastic marine debris counted from the river survey lower than the IPNB, yet almost the same with the seashore approach and the three are still lower than Jambeck's calculation. The activity of daily river waste trapping is depicted in Figure 5.

**Table 2.** Indonesian plastic marine debris result based on river survey.

| Province | No Trash Trap (MT/y) | With Trash Trap (MT/y) |
|---|---|---|
| Banten | 0.00264 | 0.00017 |
| DKI Jakarta | 0.00922 | 0.00111 |
| West Java | 0.00337 | 0.00040 |
| Central Java | 0.00329 | 0.00045 |
| East Java | 0.00297 | 0.00039 |
| Total | 0.02149 | 0.00253 |
| **Indonesian Plastic Marine Debris** | **0.038** | **0.004** |

**Table 3.** Comparison of several methods on Indonesian plastic marine debris amount.

| Calculation Methods | Indonesian Plastic Marine Debris | Reference |
|---|---|---|
| Jambeck | 0.48 – 1.29 MT/y | [1] |
| IPNB | 0.17 MT/y | [5] |
| Seashore Approach | 0.004 MT/y | |
| River Survey + Trash Trap | 0.004 MT/y | This study |

(a) (b)

**Figure 5.** Daily river waste trapping by riverbank community (a); River waste trapped (b).

Indonesian population is about 268 million people [6] and coastal population is 44.75 million people. From the river survey, every person generates 0.27 tonnes/person/y of waste and the percentage of plastic waste is 4.65%. The plastic waste generates in the coastal area, hence, is 0.57 MT/y while in the non-coastal area is 2.84 MT/y. The plastic waste is mostly recycled, thrown into the landfill, burned, buried, and the rest is thrown into the river. About 0.034 MT/y of river plastic waste was picked up and together with recycled plastic waste from non-coastal area to form 1.655 MT/y of annual recycled plastic [3,4,7]. After picking up, 0.004 MT/y of plastic enters the ocean as marine debris. The Indonesian plastic material balance from river survey is described in Figure 6. The plastic marine debris amount from river survey is not far different from IPNB and seashore approach [5] as shown in Figures 7 and 8. Again, the river survey calculation does not support Jambeck's claim. The difference is might caused by misleading assumptions and inaccurate methods of Jambeck's calculation.

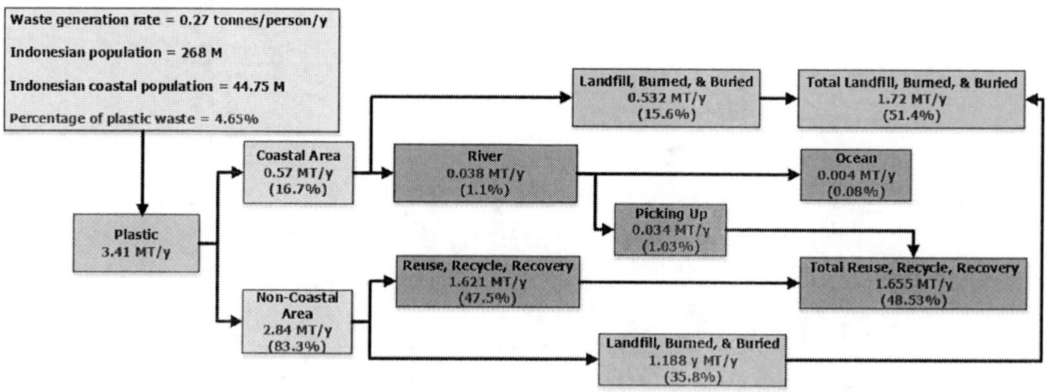


**Figure 6.** Indonesian plastic flow diagram based on river survey.

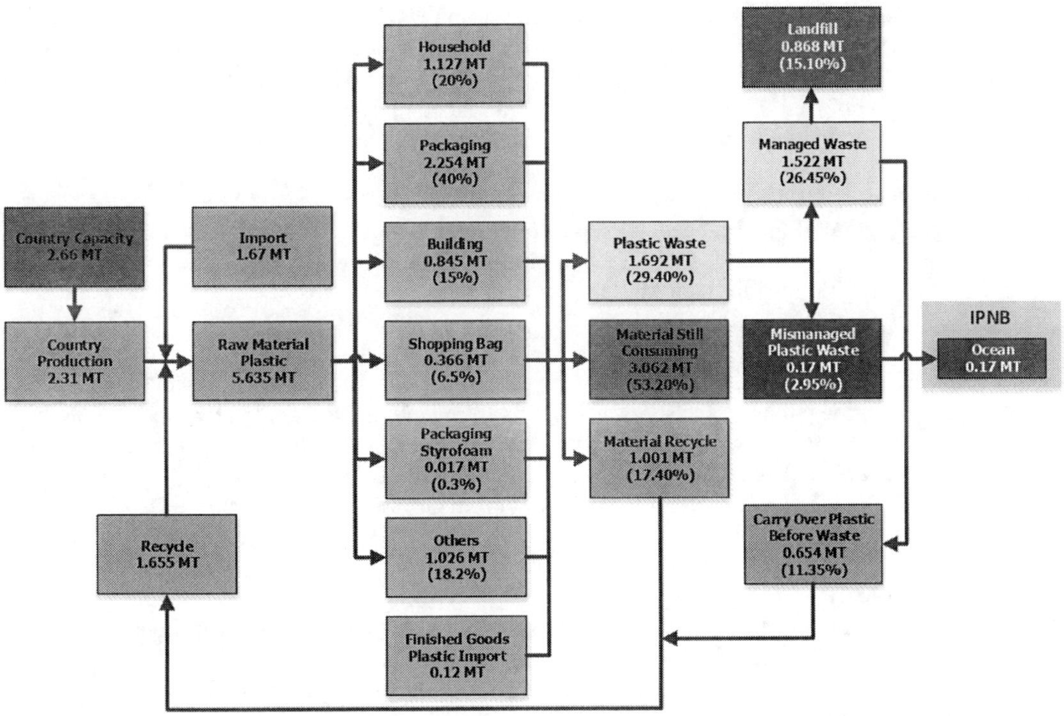


**Figure 7.** Indonesian plastic balance based on IPNB [5].

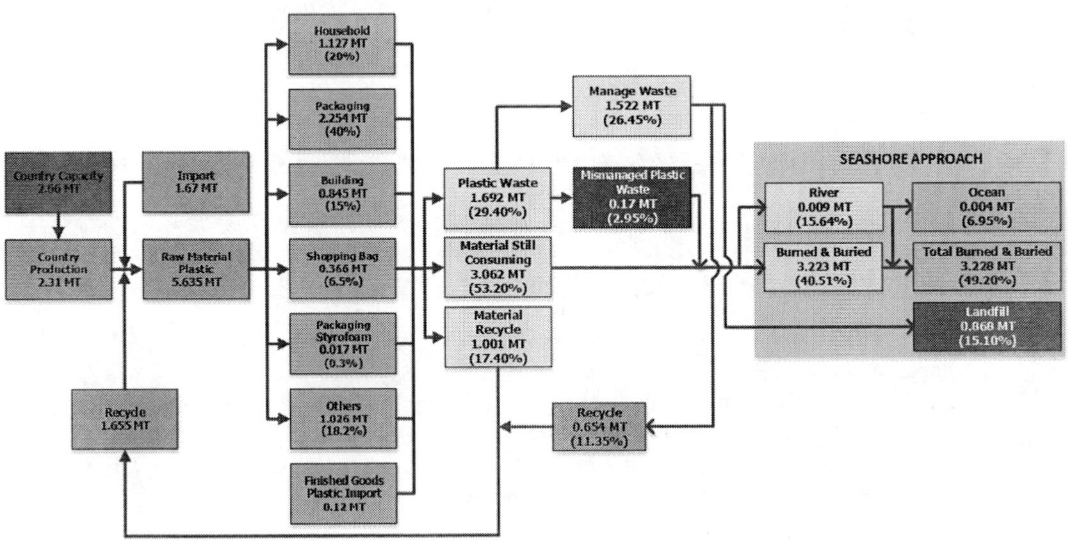


**Figure 8.** Indonesian plastic flow diagram based on seashore approach [5].

## 4. Conclusions and closing remarks

From this study, it is agreed that Indonesia has a problem with plastic marine debris although the number is much lower than Jambeck's calculation. The river survey with the season factor correction gives a more accurate plastic marine debris amount with similar results to the previous study, 0.004 MT/y. However, it could increase to 0.038 MT/y if no trash trap is installed on the river. The river survey study also confirmed that Indonesia is not the second-largest contributor country for plastic marine debris.

Installing a trash trap on the river protects the river waste flow to the ocean, it also facilitates local people to clean the river as well as increasing their income from recycled plastic waste selling activity. From the river survey activities, many local people show their big thanks and hoped that this could be a long-term program. The government should show the synergy and would be more appreciated if each village and district are equipped with trash trap, plastic shredder machine, incinerator, and further plastic pyrolysis unit, as an implementation part of integrated waste management.

**Acknowledgments**

The authors are grateful for the financial support provided by DOW Chemicals and the research data support from INAPLAS, ADUPI, APDUPI, and Ministry of Industry. Also, the authors would also thank the STKSR 2020 committee for their kind assistance.

**References**

[1] Jambeck J, Geyer R, Wilcox C, Theodore S, Perryman M, Andrady A, Narayan R and Law K L 2015 Marine pollution: Plastic waste inputs from land into the ocean *Science* **347** 768–71
[2] KemenLHK 2016 *Ministry of Environment and Forestry (Kementrian Lingkungan Hidup dan Kehutanan)*, website http://www.menlhk.go.id/KemenLHK
[3] Rivai E 2017 *Plastics and lifecycle* (Jakarta: Forum Group Discussion Marine Debris Waste Management)
[4] Rivai E 2017 *Indonesia national plan of action on marine plastic debris* (Yangon, Myanmar: AFPI Meet)
[5] Abidin A Z and Steven S 2020 Plastic marine debris study based on Indonesian plastic national balance (IPNB) and seashore approach *International Seminar on Chemical Engineering Soehadi Reksowardojo (STKSR) in conjunction with Symphosium on Photocatalyst and Photocatalysis*
[6] Statistics Indonesia 2019 *Statistical Yearbook of Indonesia* (Central Bureau of Statistics)
[7] World-Economic-Forum 2020 *Radically reducing plastic pollution in Indonesia: A multistakeholder action plan (National Plastic Action Partnership)*
[8] Statistics Indonesia 2019 *Jakarta in Figures* (Central Bureau of Statistics)
[9] Statistics Indonesia 2019 *Indonesian coastal population*. https://www.bps.go.id/. (Central Bureau of Statistics)
[10] Kompas 2017 *Jakarta produced 7.700 tonnes of waste per day (Jakarta hasilkan 7.700 ton sampah per hari)*. https://sains.kompas.com/
[11] Beritasatu 2015 *Waste production reaches 0.8 kg per person per day (Produksi sampah capai 0,8 kg per orang per hari)*. https://www.beritasatu.com/nasional/
[12] Beritasatu 2019 *It is time for waste to be analogized as energy (Saatnya sampah dianalogikan sebagai energi)*. https://www.beritasatu.com/nasional/
[13] Geotimes 2017 *The waste represent our identity (Sampah adalah wajah kita)*. https://geotimes.co.id/
[14] Geotimes 2019 *Indonesian waste production is 67.1 million of tonnes per year (Produksi sampah di Indonesia 67,1 juta ton sampah per tahun)*. https://geotimes.co.id/arsip
[15] Tempo 2019 KLHK: *32% of Indonesian waste has not been managed (32% sampah di Indonesia belum terkelola)*. https://bisnis.tempo.co/

# Energy and Exergy Analysis on the Rotary Kiln Unit of RKC-2 PT. Semen Gresik – Tuban Plant

**Mala Hayati Nasution[1*], Irwan Rasyid Syahputra[1], and Darul Rahman[1]**

[1] Chemical Engineering Department
Universitas Internasional Semen Indonesia,
Kompleks PT. Semen Indonesia (Persero) Tbk, Jl. Veteran, Kebon Dalem, Sidomoro, Kebomas, Gresik, East Java 61122 Indonesia.

*Corresponding Author's e-mail: mala.nasution@uisi.ac.id

**Abstract.** Cement industry is an industry that intensively uses energy in the process. One of the main units in a cement plant that consumes the most energy is the rotary kiln. Rotary kiln is used for calcination process of cement raw material into clinker. It operates at 900-1450°C. It obtains energy from coal combustion process. Energy conservation in rotary kiln can be used to optimize production costs. Energy conservation can be carried out through energy identification. One method of energy identification is through energy and exergy analysis. In this study, energy and exergy analysis was carried out at the rotary kiln unit of RKC-2 PT. Semen Gresik - Tuban Plant. The analysis was carried out based on the calculation of the mass balance, energy balance, enthalpy balance, entropy balance and exergy balance. These calculations are used to obtain energy efficiency, exergy efficiency and irreversibility. Based on the research results, the energy efficiency, exergy efficiency, and irreversibility were 69,2%; 50,48% and 49,52% respectively.

**Keywords:** Rotary kiln; Energy; Exergy; Irreversibility

## 1. Introduction

The cement industry is one of the industries that is energy intensive, because it absorbs a large amount of energy. The costs incurred for energy consumption in a cement plant range between 20-30% of the total cement production cost (UNIDO, 1994). If the costs for energy consumption can be reduced, the company profits can be increased. Research conducted on several cement factories in Japan in 1992 showed that energy use for the clinker combustion process reached 91.90% of the total energy use in a cement factory, the remaining energy was used for electricity by 7.6%, drying fuel and coal, as well as other processes by 0.5% (UNIDO, 1994).

Theoretically, to produce one ton of clinker a minimum of 1.8 GJ of heat is required. Based on research in several cement factories with dry type cement production processes, the average energy consumption is 3.5 GJ to produce one ton of clinker. The kiln system efficiency is 50% and the rotary kiln efficiency is 96% (UNIDO, 1994). According to Engin and Tahsin, 2005, a cement factory in Turkey with a production capacity of 600 tons of clinker per day requires an energy consumption of 3.6 GJ to produce one ton of clinker products.

One of the units in the cement production process that has high energy requirements is a rotary kiln. Rotary kilns are the main equipment in cement manufacturing which requires combustion heat for the clinker formation process. The main source of combustion heat comes from coal. In the combustion process there is heat loss. Energy conservation in the rotary kiln will have a significant effect on reducing the cost of energy requirements in the cement production process and increasing the performance of the rotary kiln.

One of the methods used in energy conservation is energy and exergy analysis. The analysis is carried out to (i) measure the minimum work required to produce a product; (ii) evaluate energy conversion; and (iii) identify energy and exergy losses. This analytical method not only shows the flow of energy in the process but also provides an overview of the decline in energy quality during the process due to the formation of entropy and irreversibility. The research "Energy Analysis and Exergy at Rotary Kiln Unit RKC-2 PT. Semen Gresik - Tuban Plant " was carried out through the process of calculating the energy balance, enthalpy balance, entropy balance, exergy balance and process irreversibility.

## 2. Theoretical Analysis

The process is assumed to be at steady state conditions. The following balance equations are used to determine the energy and exergy efficiencies and the rate of irreversibility (Ari, 2011).

### *2.1. Mass Balance*

The mass balance in the process is stated in equation (1). Based on the equation, the inlet mass entering is the same as the outlet massa. Mass balance data in the rotary kiln unit were obtained from operating data at the Tuban Cement Plant.

$$\sum m_{in} = \sum m_{out} \quad (1)$$

Mass balance calculations are based on the law of conservation of mass passing through a rotary kiln. These calculations are presented in equations (2) and (3) (Duda, 1985).

$$\sum m_{in} = m_{coal} + m_{kiln\ feed} + m_{air} \quad (2)$$

$$\sum m_{out} = m_{clinker} + m_{exhaust\ gas} \quad (3)$$

*2.1.1. Determination of the residual mass of $CaCO_3$ and $MgCO_3$.* The inlet mass of $CaCO_3$ and $MgCO_3$ to the rotary kiln will undergo a 100% calcination reaction in the kiln. The calcination reaction is the reaction to release $CO_2$ from the compounds $CaCO_3$ and $MgCO_3$, this reaction uses the most energy. Based on Duda, 1985, the equation for the calcination reaction are as follows:

$$CaCO_{3(s)} \rightarrow CaO_{(s)} + CO_{2(g)} \quad (179\ kJ/mol) \quad (4)$$

$$MgCO_{3(s)} \rightarrow MgO_{(s)} + CO_{2(g)} \quad (118\ kJ/mol) \quad (5)$$

To calculate the mass of $CaCO_3$ and $MgCO_3$ the results of calcination are presented as equations (6) and (7).

$$m_{CaCO3\ (calcined)} = degree\ of\ calcination \times m_{CaCO3} \quad (6)$$

$$m_{MgCO3\ (calcined)} = degree\ of\ calcination \times m_{MgCO3} \quad (7)$$

To determine the mass of CaO and MgO the results of calcination are presented as equations (8) and (9).

$$m_{CaO} = m_{CaCO3\ (calcined)} \times \frac{Mr\ CaO}{Mr\ CaCO_3} \quad (8)$$

$$m_{MgO} = m_{MgCO3\ (calcined)} \times \frac{Mr\ MgO}{Mr\ MgCO_3} \quad (9)$$

Meanwhile, to determine the mass of calcination $CO_2$ is presented in equation (10).

$$m_{CO_2} = m_{CaCO_3\ (calcined)} \times \frac{Mr\ CO_2}{Mr\ CaCO_3} \quad (10)$$

IOP Conf. Series: Materials Science and Engineering 1143 (2021) 012050 doi:10.1088/1757-899X/1143/1/012050

*2.1.2. Determination of the coal mass.* The amount of mass of coal entering the rotary kiln is obtained from the daily report of PT. Semen Gresik – Tuban Plant.

*2.1.3. Determination of the mass of combustion air.* Combustion air is the required air for the combustion process in a rotary kiln. The mass of combustion air can be determined by calculating the theoretical coal combustion. The chemical reactions that occur in the coal burning process are shown as follows:

$$C_{(s)} + O_{2(g)} \rightarrow CO_{2\,(g)} \quad (11)$$

$$2H + \frac{1}{2} O_{2(g)} \rightarrow H_2O_{(l)} \quad (12)$$

$$S_{(s)} + O_{2(g)} \rightarrow SO_{2(g)} \quad (13)$$

The chemical reaction above can be written in the form of a general equation for the theoretical combustion reaction of coal as in equation (14) to (19).

$$C_xH_yN_zS_pO_q + \alpha(O_2 + 3{,}762\ N_2) \rightarrow xCO_2 + \frac{y}{2}H_2O + pSO_2 + \left(3{,}762\ \alpha + \frac{z}{2}\right)N_2 \quad (14)$$

$$x = \frac{\%\,C}{Ar\ C} \quad (15)$$

$$y = \frac{\%\,H}{Ar\ H} \quad (16)$$

$$z = \frac{\%\,Z}{Ar\ Z} \quad (17)$$

$$p = \frac{\%\,S}{Ar\ S} \quad (18)$$

$$q = \frac{\%\,O}{Ar\ O} \quad (19)$$

The number of moles of air for theoretical combustion ($\alpha$) is obtained by equalizing the number of moles of chemical elements on the product and reactant sides as in equation (20).

$$\alpha = x + \frac{y}{4} + p - \frac{q}{2} \quad (20)$$

Meanwhile, the number of moles of coal for the total mass fed to the rotary kiln can be determined using equation (21).

$$Mole\ of\ coal = \frac{Mass\ of\ coal}{Mr\ coal} \quad (21)$$

The mass of flue gas formed for the total mass unit of coal fed to the rotary kiln can be determined by equation (22) to (25).

$$mass\ of\ CO_2 = mole\ CO_2 \times Mr\ CO_2 \quad ton/h \quad (22)$$

$$mass\ of\ H_2O = mole\ H_2O \times Mr\ H_2O \quad ton/h \quad (23)$$

$$mass\ of\ H_2O = mole\ H_2O \times Mr\ H_2O \quad ton/h \quad (24)$$

$$mass\ of\ N_2 = mole\ N_2 \times Mr\ N_2 \quad ton/h \quad (25)$$

*2.1.4. Determination of the mass of clinker.* Combustion in a rotary kiln causes a chemical reaction, this reaction is also called the clinker formation reaction. The following are the reactions for clinker formation as follows:

$$CaCO_3 \rightarrow CaO + CO_2 \quad (26)$$

$$2CaO + SiO_2 \rightarrow (CaO)_2.SiO_2 \quad (27)$$

$$CaO.Al_2O_3 + 2CaO \rightarrow (CaO)_3.Al_2O_3 \quad (28)$$

$$CaO.Al_2O_3 + 3CaO + Fe_2O_3 \rightarrow (CaO)_4.Al_2O_3.Fe_2O_3 \quad (29)$$

$$(CaO)_2.SiO_2 + CaO \rightarrow (CaO)_3.SiO_2 \tag{30}$$

The clinker-forming compounds have a special name which is used in the cement production process. These compounds are shown in table 1.

**Table 1.** Clinker forming compounds.

| Chemical Formula | Compound Name | Special Name |
|---|---|---|
| $2CaO.SiO_2$ | *Dicalcium Silicate* | $C_2S$ |
| $3CaO.SiO_2$ | *Tricalcium Silicate* | $C_3S$ |
| $3CaO.Al_2O_3$ | *Tricalcium Aluminate* | $C_3A$ |
| $4CaO.Al_2O_3.Fe_2O_3$ | *Tricalcium Alumino Phase Ferrite* | $C_4AF$ |

The calculation of the clinker compounds formed is carried out using equation (31) to (35).

$$C_3S = 48\%\ CaO + 17\%\ SiO_2 + 0{,}7\%\ Al_2O_3 + 0{,}5\%\ Fe_2O_3 + 0{,}1\%\ SO_3 + 0{,}1\%\ K_2O + 0{,}1\%\ Na_2O + 0{,}7\%\ MgO \tag{31}$$

$$C_2S = 5{,}1\%\ SiO_2 + 0{,}1\%\ MgO + 0{,}3\%\ Al_2O_3 + 0{,}2\%\ K_2O + 8{,}2\%\ CaO + 0{,}1\%\ Fe_2O_3 \tag{32}$$

$$C_3A = 0{,}1\%\ Na_2O + 0{,}1\%\ MgO + 3{,}2\%\ Al_2O_3 + 0{,}3\%\ SiO_2 + 4\%\ CaO + 0{,}3\%\ Fe_2O_3 \tag{33}$$

$$C_4AF = 0{,}2\%\ MgO + 2{,}2\%\ Al_2O_3 + 0{,}4\%\ SiO_2 + 5\%\ CaO + 2{,}1\%\ Fe_2O_3 \tag{34}$$

$$Free\ lime = 0{,}9\%\ CaO \tag{35}$$

### 2.2. *Energy Balance*

The energy balance is a basic method of investigating a process. Through the energy balance it can be seen the requirement of the development in the process,. In general, the energy balance equation is expressed in equations (36) and (37).

$$\sum E_{in} = \sum E_{out} \tag{36}$$

$$Q + \sum m_{in}h_{in} = W + \sum m_{out}h_{out} \tag{37}$$

The assumption in the energy balance calculation is that there is no change in kinetic and potential energy during heat transfer or work so that the energy balance equation is as presented in equation (38).

$$\sum m_{in}h_{in} = W + \sum m_{out}h_{out} \tag{38}$$

*2.2.1. Determination of the coal thermal energy.* The calculation of thermal energy that can be produced by coal is presented in equation (39) to (41).

$$Q_{coal} = (H_{coal} \times m) \tag{39}$$

$$HHV = 33950\,C + 144200\left(H_2 - \left(\frac{O_2}{8}\right)\right) + 9400\,S \tag{40}$$

$$LHV = HHV - 2400\,(M + 9H_2) \tag{41}$$

*2.2.2. Determination of the sensible heat of the coal.* The calculation of the sensible heat of the coal is presented in equation 42.

$$Q_c = c_c \times m_c \times T_c \tag{42}$$

*2.2.3. Determination of the kiln feed heat.* The kiln feed heat calculation is presented in equation 43.

$$Q_{kf} = m_{kf} \times c_{kf} \times T_{kf} \tag{43}$$

*2.2.4. Determination of the heat of CaO.* The calculation of the heat of Cao is presented in equation 44.

$$Q_{Cao} = m_{Cao} \times c_{Cao} \times T_{Cao} \tag{44}$$

*2.2.5. Determination of the heat of MgO.* The calculation of the heat of Cao is presented in equation 45.

$$Q_{Mgo} = m_{Mgo} \times c_{Mgo} \times T_{Mgo} \tag{45}$$

*2.2.6. Determination of the heat of combustion air.* The calculation of the heat of combustion air is presented in equation 46.

$$Q_{air} = m_{air} \times c_{air} \times T_{air} \tag{46}$$

*2.2.7. Determination of the heat of clinker.* The calculation of the heat of clinker is presented in equation 47.

$$Q_{clinker} = m_{clinker} \times c_{clinker} \times T_{clinker} \tag{47}$$

*2.2.8. Determination of the heat of kiln exhaust gas.* The calculation of the heat of kiln exhaust gas is presented in equation 48.

$$Q = m \int_{T\,ref}^{T} Cp\ (T)dT \tag{48}$$

*2.2.9. Determination of the heat loss.* The heat loss that occurs in the kiln shell walls is caused by two factors, namely radiation and convection with the formulas presented in equations (49) to (51).

$$Q_{radiation} = \sigma \varepsilon A_{kiln\ shell}(T_s^4 - T_\infty^4) \tag{49}$$

$$Q_{convection} = h_{con} A_{pk}(T_s - T_\sim) \tag{50}$$

$$hc = \frac{k\ Nu}{D} \tag{51}$$

### 2.3. Exergy Balance

The exergy balance are presented in equations (52) to (54).

$$\sum Ex_{in} - \sum Ex_{out} = \sum Ex_{dest} \tag{52}$$

$$\sum Ex_{dest} = \sum (1 - \frac{T_0}{T_k})Q_k - W + \sum m_{in}\Psi_{in} - \sum m_{out}\Psi_{out} \tag{53}$$

$$\Psi = (h - h_o) - T_o(s - s_o) \tag{54}$$

*2.3.1. Determination of the physical exergy.* Physical exergy is exergy related to changes in pressure and temperature in the flow. According to Bejan, 1996, physical exergy can be calculated using the following equation

$$E^{ph} = m(h - h_o) - T_o(s - s_o) \tag{55}$$

*2.3.2. Determination of the chemical exergy.* Chemical exergy is exergy related to changes in chemical compounds. According to Bejan 1996, chemical exergy can be calculated using the following equation:

$$E^{ch} = m\ x\ e^{ch} \tag{56}$$

## 3. Energy Analysis

The results of the calculation of the heat energy balance in the rotary kiln are shown in table 2. Based on the table 2, there is a difference between the amount of inlet and outlet energy. The inlet energy is much greater than the outlet energy. The difference between the amount of inlet and outlet energy shows that there is heat energy loss or heat loss of 375,644,605.4 kJ / hour.

**Table 2**. Energy balance in the rotary kiln.

| Inlet | | Outlet | |
|---|---|---|---|
| Components | Flow (kJ/h) | Components | Flow (kJ/h) |
| *Kiln feed* | 364,584,546.3 | Exhaust gas | 150,527,377.0 |
| Secondary air | 119,761,060.3 | Clinker | 694,572,383.2 |
| Primary air | 48,758.5 | Heat loss | 375,644,605.4 |
| Coal | 1,101,593.1 | | |
| Clinker reaction | 417,030,673.3 | | |
| Sensible heat of coal | 318,217,734.0 | | |
| Total | 1,220,744,365.6 | Total | 1,220,744,365.6 |

Based on research, heat loss occurs due to three factors, namely radiation, convection and conduction. The amount of heat lost due to radiation, convection and conduction are 9,894,424.44 kJ/hour; 326,429,550.96 kJ/hour and 39,320,629.98 kJ/hour respectively. These three factors are largely determined by the type of the material in the rotary kiln. The conduction factor usually occurs on the inner and outer wall surfaces of the rotary kiln which is limited by refractory materials (refractory bricks) and steel due to the difference in temperature inside and outside the rotary kiln. The convection factor is due to heat transfer through an intermediate substance in the form of fluid, namely liquid and gas, in this case there is direct contact between the combustion hot air and the kiln feed material in the rotary kiln. This causes an increase in temperature at each point (zone) in the rotary kiln. The radiation factor is caused by the heat emitted from an open furnace or fireplace, so that the heat energy escapes to the environment. The percentage of heat loss due to radiation, convection and conduction factors are 0.81%; 26.74% and 3.22% respectively. Based on these results, it is known that the heat loss factor by convection is more significant than by radiation and conduction.

A Sankey diagram showing the energy flow in a rotary kiln is shown in figure 1. Based on this figure, it can be seen that the greatest inlet heat energy comes from the energy resulting from the clinker process of forming C4AF, C3A, C2S and C3S compounds in the clinker, which is 34.16%. The energy generated is quite large compared to the others because the reaction is exothermic so it releases large amounts of heat. The most widely used thermal energy is the energy for clinker formation, which is 56.90% of the total heat energy that enters the rotary kiln. Based on the research results, the energy efficiency value of the rotary kiln was 69.2% when the coal flow rate was 16.875 tons / hour and the mass flow rate of clinker products that came out of the rotary kiln was 351.919 tons / hour. Based on research by Engin, 2002, the optimal kiln system efficiency is 50%. The results of energy efficiency calculations in this study are higher than those of Engin and Vedat, 2005 by ignoring the excess heat which is fed back to the raw mill and coal mill.

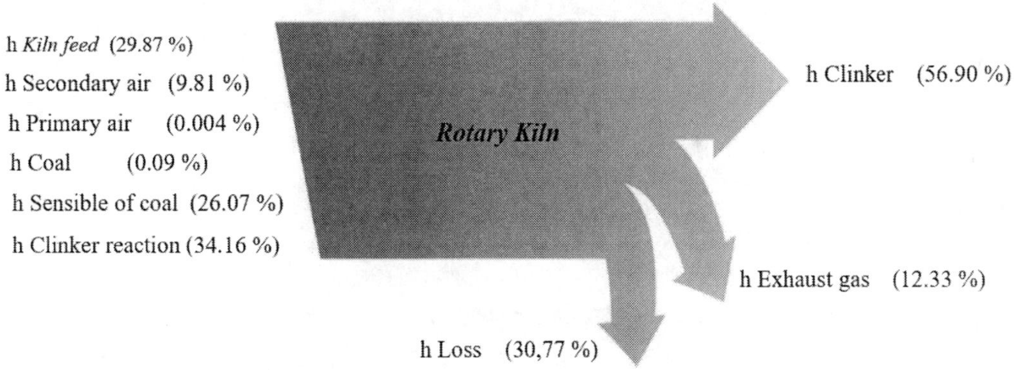


**Figure 1.** Sankey diagram of energy balance.

## 4. Exergy Analysis

The results of exergy calculations which include physical and chemical exergy are shown in table 3. Based on the table 3, the total inlet exergy is 1,419,934,737.1 kJ / hour and the outlet exergy is 716,847,932.0 kJ / hour. Based on these results, there is a difference between the inlet and outlet exergy. The total inlet exergy is much greater than the outlet exergy. The difference between the amount of inlet and outlet exergy shows a damaged exergy or commonly referred to as destructive exergy, which is 703,086,805.1 kJ / hour or equivalent to 49.52% of the total incoming exergy. Destructive exergy is an irreversible amount of heat energy. Destruction exergy also shows the value of irreversibility. The irreversibility value is the value of energy that cannot be reused. The higher the irreversibility value the more irreversible the process will be.

**Table 3.** Exergy balance in the rotary kiln.

| Inlet | | Outlet | |
|---|---|---|---|
| Components | Flow (kJ/h) | Components | Flow (kJ/h) |
| *Kiln feed* | 746,835,463.7 | Exhaust gas | 114,196,099.9 |
| Secondary air | 73,862,852.0 | Clinker | 602,651,832.1 |
| Primary air | 467,300.0 | Destructed Exergy | 703,086,805.1 |
| Coal | 598,769,121.3 | | |
| Total | 1,419,934,737.1 | Total | 1,419,934,737,1 |

A Sankey diagram showing the exergy flow in a rotary kiln is shown in figure 2. Based on figure 2, the largest inlet exergy comes from the kiln feed exergy namely 52.60% from total inlet exergy. This value is the largest when compared to other inlet flows. This is because the kiln feed plays an important role as the main raw material in the clinker formation process in the rotary kiln. The most widely used exergy is exergy for clinker formation, which is 42.44% of the total physical and chemical heat exergy that goes into the rotary kiln. Based on the results of the calculation of the value of exergy efficiency in the rotary kiln is 50.48%. The value of exergy efficiency is smaller than the value of energy efficiency. Energy efficiency shows the quantity of energy while exergy shows the quality of energy. Based on the results of the research, the amount of energy that can be utilized in a rotary kiln is 50.48%.

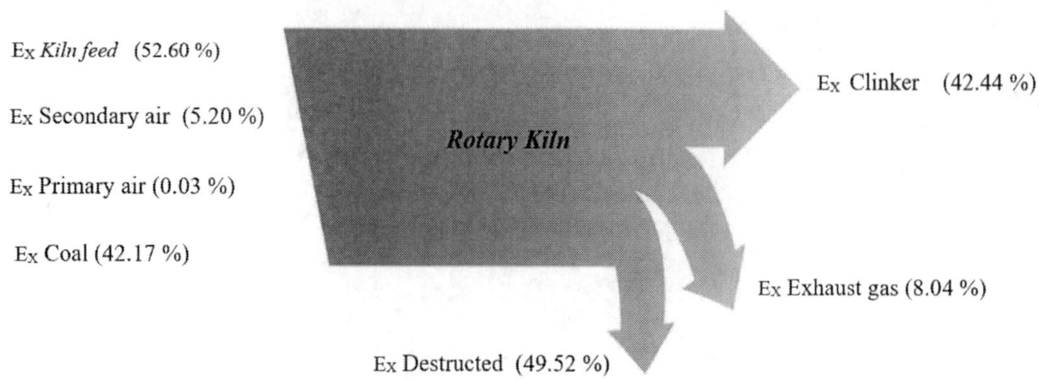


**Figure 2.** Sankey diagram of exergy balance.

## 5. Conclusion

In the rotary kiln RKC-2 PT. Semen Gresik (Persero), Tbk Tuban plant, the total inlet energy is 1,220,744,365.6 kJ / hour. The total outlet energy output is 845,099,760.22 kJ / hour and the heat loss is 375,644,605.4 kJ / hour. Heat loss due to conduction factor is 39,320,629.98 kJ / hour or equivalent to 3.22% of inlet energy. Heat loss due to convection factor is 326,429,550.96 kJ / hour or equivalent to 26.74% of inlet energy. Heat loss due to radiation factor is 9,894,424.44 kJ / hour or equal to 0.81% of inlet energy. Energy and exergy efficiency of the rotary kiln RKC-2 PT. Semen Gresik (Persero), Tbk Tuban plant are 69.2% and 50.48% with excess hot air that is fed back to the rawmill and coal mill is assumed to be negligible.

**Symbols :** **A,** surface area; **c,** specific heat; **Cp,** heat capacity; **D**, characteristic length; **E**, energy; **Ex**, exergy; **Ψ**, exergy; **h**, specific enthalpy; **hc**, convection heat transfer coefficient; **HHV**, higher heating value; **LHV**, lower heating value; **m**, mass; **Mr**, molecular weight; **Nu**, Nusselt number; **Q**, thermal energy; s, entropy; **T**, temperature; **σ**, Stefan-Boltzmann constant; **W**, power;

**Subscription :** **c,** coal; **ch**, chemical; **dest**, destruction; **e**, molar exergy; **gen**, generated; **in,** inlet; **kf**, kiln feed; **o**, reference; **out**, outlet; **ph**, physical; **ref**, reference; **s**, surface;

## 6. References

[1] United Nations Industrial Development Organization (UNIDO) 1994 *Output of a Seminar on Energy Conservation in Cement Industry.*
[2] Engin T and Vedat A 2005 Energy Auditing and Recovery for Dry Type Cement Rotary Kiln Systems – A Case Study *Energy Conversion and Management* **46**, Issue 4 551-562.
[3] Ari V 2011 Energetic and exergetic assessments of a cement rotary kiln system *Scientific Research and Essays* **6(6)** 1428-1438.
[4] Duda WH 1985 *Cement Data Book* (London : Macdonald & Even).
[5] Bejan A 1996 *Advanced Engineering Thermodynamics* (New York : Wiley)

## 7. Acknowledgments

We wish express appreciation to Mr. Adam Wijatmiko, Head of RKC – 2 PT. Semen Gresik (Persero), Tbk Tuban plant for his permission and support to carry out the research.

# Circular Economy on Organic Waste Management with MASARO Technology

**A Z Abidin[1], H Bramantyo[2], M K Baroroh[3], and C Egiyawati[4]**

[1]Department of Chemical Engineering, Institute of Bandung Technology, Bandung 40132, Indonesia

*E-mail: zainal@che.itb.ac.id

**Abstract.** MASARO technology presents a solution in solving the waste problem through an organic waste processing unit, known as IPPO (Organic Fertilizer and Feed Industry). The Masaro system characterises solid waste into five catagories including organic waste, low value plastic waste, incenerated waste, recycled waste, and hazardous waste. The organic waste will be separated into an organic fast-decaying waste to produce POCI (Special Liquid Organic Fertilizer) and KOCI (Special Liquid Organic Concentrate), and an organic slow-decaying waste to produce a compost. POCI is used as an organic fertilizer and its farming produces a higher production number, higher quality product, faster production rate while its production cost is lower. KOCI and compost is applied in the farm industry where the organic waste with low economic value is used as a raw material to produce new materials with higher economic value. These products can produce an environmentally friendly system and reflect the circular economy principle application. The organic waste is used as raw material for IPPO and its products form a circular economy through aplication in the farm industry that create lower cost production and sustainable development.

## 1. Introduction

Today's domestic waste problem is serious to consider that the volume and rate of a landfill are increasing along with the increase in handling costs. In line with the enhancement in population and economic activity, the waste generated an increase in number every day, when the handling faces many obstacles due to limited costs and existing facilities.

National waste accumulation from year to year continues to increase. The collect - transport - waste management paradigm causes the diversion of problems from urban activity sources to a problem at the final landfill location. According to data from the Ministry of Home Affairs Research and Development Agency and the National Waste Management Information System, Indonesia's waste production is 65.2 million tons per year, when only 36% is transported to the landfill, and the rest is burned, dumped, thrown away into rivers, and others. Sources of waste include households, industry, and traditional markets. So far, there are still many illegal piles of garbage on the side of the road. Even when the waste is moved to the landfill, the waste is only stored up because no processing is conducted so that the amount is getting bigger and even creates new problems.

Waste handling that is still done conventionally cannot control the waste produced by the community. Handling municipal waste in Indonesia has until now been a challenge for the local government. Waste management is an environmental problem with a complex level of resolution, so it requires the involvement of all related components, namely the community, government, and the industrial sector, including academics and experts (professionals) in the environmental field, especially in the waste utilization sector and utilization. The management approach that has been done so far is ineffective and sustainable, so a paradigm shift is needed in managing waste from collect-transport-waste to sort-transport-process-sell and from cost centre to profit centre. In this paper, a solution to dealing with waste is with MASARO (Zero Waste Management) directly from the source that is the community.

On the other hand, the problem of Indonesia is that national agriculture is stuck in inorganic chemical fertilization that has an impact on accelerating the degradation of agricultural land fertility. This matter can occur due to various problems, especially the culture of using chemical fertilizers that have been going on for too long. As a result, agricultural land is very critical and slight in soil nutrients that at the end have an impact on low plant productivity and reduced immunization power of plants that results in a large number of plant pests that attack plants (the cost of production increases). Eventually, farmers will always depend on the need for insecticides (to increase the cost of the crop) and destroy the qualifications of the production of residue-free crops. To be able to improve agricultural land that is already critical and poor in nutrients, it takes courage from all parties to go out of the box, to leave inorganic chemical fertilization and return to using more natural organic fertilizers.

Apart from the fertilization aspect, another thing that can be improved in the agricultural industry is the sustainability aspect. Currently, the high dependence of agriculture compared to other fields, such as the fertilizer industry, has resulted in a lack of control and sustainability of the industry. MASARO technology can form a circular economy in the agriculture industry, thus helping the sustainability of the industry by processing organic waste into POCI (*Pupuk Organik Cair Istimewa*/special liquid organic fertilizer) that has the potential to eliminate problems caused by landfills and while producing fertilizers at the same time.

## 2. Method

The principles of MASARO Technology in Waste Processing are (1) Waste sorting on source, (2) waste treatment nearby the source of waste, (3) Community, government, and industries involvement, (4) educate the community, (5) environmentally friendly technology implementation, and (6) create sustainability management.

MASARO technology principles are intended to reduce waste pile to 0% by waste sorting on source to classify waste in to 5 categories, in which are: fast-decaying organic waste, slow-decaying organic waste, waste-to-energy, recyclable waste, and plastic/film waste. The sorted waste is going to be transferred to the waste treatment facility to eliminate the need for landfill, waste will be kept on houses or waste sources and transferred to the waste treatment facility on a scheduled period based on the type of waste that is produced. This system will increase the awareness towards waste piles. Therefore, developing waste reduction behavior within the community. To increase the effectiveness of waste reduction, educational programs must be paired with adequate waste treatment facilities located in areas such as houses, markets, offices and other public facilities. Hence, the need for communal, governmental and industrial involvement. The purpose of this system is to reduce waste by shifting the paradigm towards waste itself that used to be seen as cost-center material. With the implementation of MASARO technology, the paradigm towards waste will be shifted from cost-center material to profit-center material.

MASARO technology is designed to process all waste and produce high value products as pictured in figure 1. The overall products are differentiated into two categories based on the type of waste being used. The products produce from organic waste are POCI (*Pupuk Organik Cair Istimewa*/Special Liquid Organic Fertilizer), KOCI (*Konsentrat Organik Cair Istimewa*/Special Liquid Organic Concentrate), Growth Medium, and MASARO Compost. This paper will focus on organic waste treatment using IPPO (*Industri Pupuk dan Pakan Organik*/Organic Fertilizer and Animal Food Industry).

IOP Conf. Series: Materials Science and Engineering 1143 (2021) 012051 doi:10.1088/1757-899X/1143/1/012051

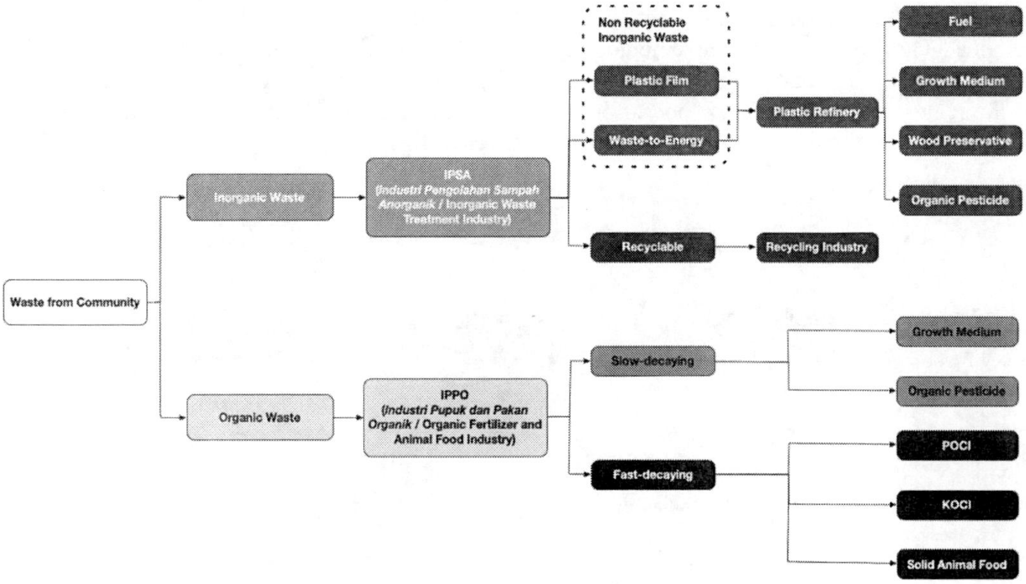


**Figure 1.** MASARO's waste classification and treatment.

MASARO's organic waste treatment is divided into two programs, which are social programs and industrial program. These two programs are aiming for waste pile reduction but with different scales. MASARO's social program for organic waste treatment is scoping on community scale organic waste management. The program itself is held to increase the awareness towards waste management while also creating a space for social interaction in the community. MASARO's social program is consists of slow-decaying organic waste collection from local sources and process them through MASARO compost process with MASARO Bio-composter. The compost later be used as fertilizer for the polybag farming system that is implemented in the community. The polybag farming is using the 4:3:2:1 volumetric ratio, with 4 parts growth medium, 3 parts animal feces, 2 parts MASARO compost, and 1-part charcoal. The development of this system can create RPL (*Rumah Pangan Lestari*) with integrated polybag farming. Crops from RPL is used as one of the food sources of the community. Waste created from the food production process is also the source of organic waste thus creating a circular economy. The whole social program system is illustrated in figure 2.

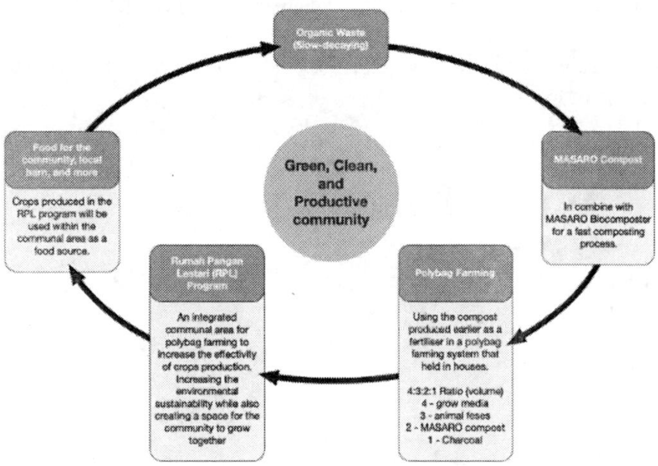


**Figure 2.** Circular economy system for the social program

The Industrial program is aiming for the agricultural, farming, and fishery industries. Organic waste produced within the industry are processed in IPPO to produce POCI/KOCI/growth medium/MASARO compost. These products are implemented directly to the industry by using POCI as fertilizer, KOCI as animal food additive, growth medium and MASARO compost in polybag farming system. The product of the industries is allocated as food and waste produces in the food industries are transferred back to IPPO creating a circular economic system as illustrated in figure 3.

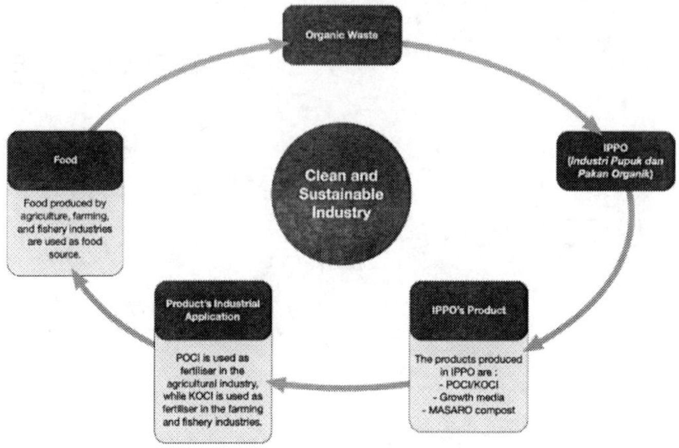


**Figure 3.** Circular economy system for the social program.

## 3. Result and Discussion

### *3.1 Social Program*

One of the case studies in the MASARO's social program implementation is in the Pasawahan region, Bandung. One of the critical points in this system is the local market. This market is acting as the place for crops selling and waste source. The waste is then transported back to the community where it will be shredded. The shredding process is held by 2 methods, using a shredding machine or with the help from the community. This step is also acted as a vessel for social interaction and movement towards a green, clean, and productive Pasawahan region community. The shredded slow-decaying organic waste will be processed for composting. The composting process is done using MASARO bio-composter. This process is more effective compared to a typical composting process by only taking 7 days of composting process. Product of this process is used as fertilizer for the polybag farming system with 4:3:2:1 volumetric ratio, 4-parts growth medium, 3-parts animal feces, 2-parts MASARO compost, and 1-part charcoal. This polybag farming is held from houses in the community. Social program implementation in Pasawahan region can be seen in figure 4.

**Figure 4.** MASARO's social program implementation in Pesawahan Region (a) Slow-decaying organic waste collection, (b) communal waste shredding, (c) shredder machine, (d) polybag farming, (e) local community working with MASARO.

### *3.2 Industrial Program*

*3.2.1 Agricultural.* The agriculture industry using MASARO technology will improve the saving of farming costs and decrease the amount of waste from agriculture. Waste generated from agricultural products, both fast-decaying organic waste and slow-decaying organic waste will be processed using MASARO technology. Fast-decaying organic waste is processed using IPPO to produce POCI, while slow-decaying organic waste will be processed using MASARO compost technology to conduce MASARO compost. Then the two products will be reused in the agriculture industry to produce organic agriculture that has better quality agricultural products. The leftover of the agricultural product will then be reprocessed in IPPO and form a circular economy system in the agriculture industry. The circular economy diagram in the agriculture industry can be seen in figure 5.

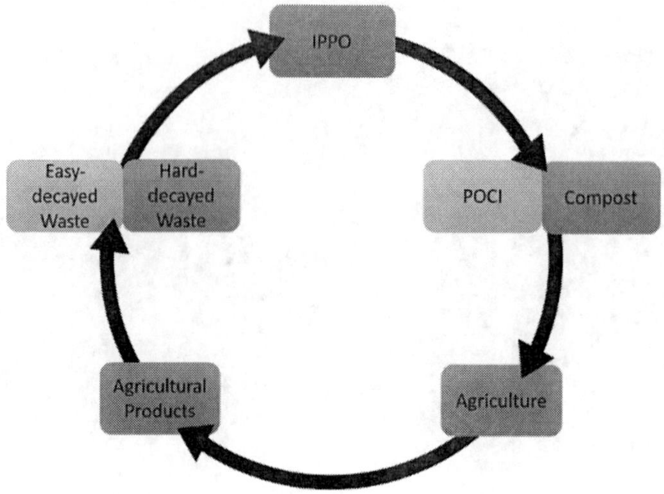


**Figure 5.** Circular economic system in agriculture industry.

With the use of MASARO as fertilizer, several advantages will be obtained, including: improve agricultural land and fertilize the soil, accelerate plant growth, accelerate the cropping period, accelerate, and improve the growth of fruit, prolong plant life, improve flower quality, color, a taste of fruits and vegetables, prevent pests and grass growth, stabilize soil pH and grow NPK from nature.

*3.2.2 Farming.* MASARO technology on organic waste treatment can also be the key for the circular economy in farming industry. As fast-decaying organic waste is used as material to produce POCI/KOCI in IPPO, residual solid material is also produced. This solid material can be used as animal food. This food is still contains a residual amount of KOCI but more KOCI can be added if needed. Meat produced by the farm is then consumed. The leftover food waste and animal feces are classified as fast-decaying organic waste and processed again with IPPO, creating a circular economy system in farming industry as illustrated in figure 6.

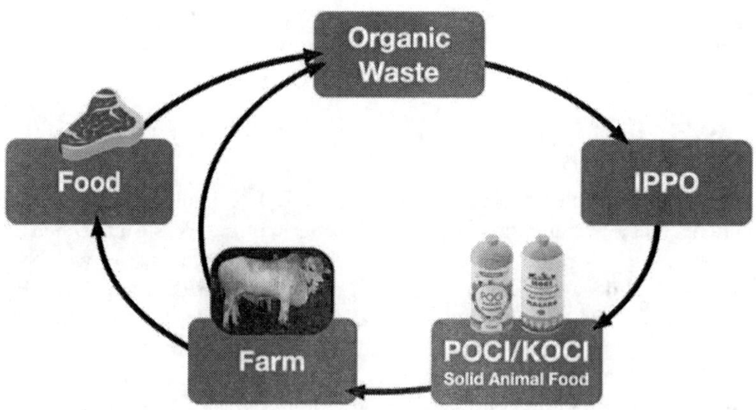


**Figure 6.** Circular economic system in farming industry.

Another advantage of using MASARO technology in creating a circular economic system in farming industry is the impact towards the animal growth, immune, or fertility. In the case of cow farm in Padalarang, KOCI usage had led to increase of cow's growth from the typical 30-45 kg monthly to 57 kg of weight gain within a month with cows that are more immune towards diseases.

*3.2.3 Fishery.* KOCI MASARO is a special liquid organic feed concentrate made from perishable organic waste which has no sale value, but after processing it in IPPO it becomes a special, highly valuable feed concentrate, and becomes a milestone in circular economies in the field of organic fisheries. Circular economy in fishery industry has a similar concept to the circular economy in farming industry. The fast-decaying organic waste will be processed in IPPO and produces KOCI that will be used in the fishery industry. After the fish is consumed, the leftover food will be processed again using IPPO. The circular economy diagram in fishery industry can be seen in figure 7.

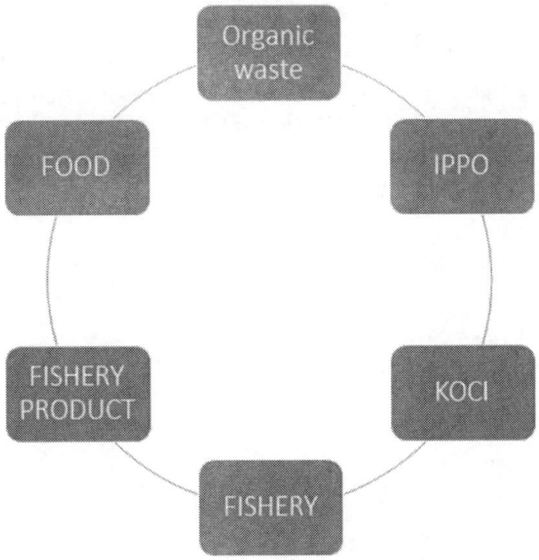


**Figure 7.** Circular economic system in the fishery industry.

KOCI MASARO applications in the field of fisheries have been proven to produce fish that are healthier, shiny, longer and have more significant changes than fisheries without using KOCI MASARO. Industrial Program implementation of fisheries using KOCI MASARO is in the Gorontalo area in catfish fisheries this skin which indicates disease susceptibility, higher appetite, 2 months the size reaches 4 cm ang there are catfish that are ready to harvest.

## 4. Conclusion

Organic waste processing with MASARO technology and "sort-collect-transport-though-sell" system can save lots of costs and operational land when compared to the current waste processing system with some impacts such as:

1. Circular economy from organic waste treatment with IPPO MASARO. Based on organic waste characterisation into 2 groups, fast-decaying and slow-decaying organic waste. These wastes will be processed accordingly with fast-decaying organic waste will be treated in IPPO to produce POCI/KOCI while slow-decaying organic waste will be treated with MASARO Bio-composter to produce MASARO compost.
2. MASARO technology implementation on social program has been applied in Cileunyi Kulon village and Pasawahan region with highly satisfactory result and ability to process 100% of organic waste while also creating a social movement in the area. This process is integrating in house planting in the neighborhood/community association and increasing the social harmony of the area. Hence, MASARO technology implementation is solving environmental problem and increasing social environment quality.
3. MASARO technology implementation on industrial scale can be applied in agricultural, farming, and fishery industries with IPPO as the key for the creation of circular economy. Food waste and

IOP Conf. Series: Materials Science and Engineering 1143 (2021) 012051 doi:10.1088/1757-899X/1143/1/012051

animal feces are some of the material used in IPPO. POCI/KOCI produced in IPPO can be used in the three industries and has been showing a positive result in the crops growth. The crops will be used as food source and the food sector will produce food-waste used as material for IPPO. Thus, creating a circular economy.

4. MASARO technology implementation with both social and industrial programs in applicable to Indonesian villages to create a circular economy waste management system that eliminate the reliance on landfill.

## References

[1] Patent P00201704782.
[2] Vos M, Wullink F, Lange M, Acoleyen M, Staveren D, Meijenfeldt V 2016 The Circular Economy-What is It and What Does It Mean For You? *Arcadis Briefing Paper* 2-3.
[3] Ratriyanto A, Widyanti S D, Suprayogi W P S, Prastowo S, Widyas N 2019 *Jurnal Semar* **8** (1) 9-13

IOP Conf. Series: Materials Science and Engineering 1143 (2021) 012052 doi:10.1088/1757-899X/1143/1/012052

# Circular Economy on Non-Biodegradable Waste Management with MASARO Technology

**A Z Abidin[1*], E V Yemensia[1], K W Wijaya[1], and A P Rahardjo[1]**

[1] Department of Chemical Engineering, Bandung Institute of Technology, Jl. Ganesha 10, Bandung 40132, Indonesia

*Email: zainal@che.itb.ac.id / mitraiqro@yahoo.com

**Abstract.** The study was based on indonesia and global waste managament issue especially on non-biodegradable waste handling and the circular economy of the method to solve the problem. Non-biodegradable waste is a problem in the world and its management especially for non-recycled waste does not reflect the circular economy principle and not environmentally friendly. This needs a new method that can process this type of waste through an environmentally friendly recovery process. Zero Waste Management/*Manajemen Sampah Zero* (MASARO) technology presents a solution to this problem by processing non-recycled non-biodegradable waste through a plastic refinery unit in Non-Biodegradable Waste Management Installation/*Instalasi Pengolahan Sampah Anorganik* (IPSA). The application of the circular economy principle has been carried out in this MASARO plastic refinery where All non-biodegradable waste is processed into other useful products, such as: fuel, planting media, wood preservatives and organic pesticides. There is no non-biodegradable waste ended up in Temporary Waste Site (TPS) and Final Landfill (TPA). Therefore, MASARO plastic refinery is an environmentally friendly solotion for non-biodegradable waste and suitable for sustainable development.

## 1. The first section in your paper

Indonesia has serious problem regarding Municipal Solid Waste (MSW) that keep increasing from year to year. World Bank estimates that the amounts of MSW generated by urban area is 85.000 tons daily. Among the top polutter, Indonesia is ranked second behind China when it comes polutting ocean with plastics [9]. The current waste problem is quite serious considering the increasing volume waste produced along with handling cost.

The problem of non-biodegradable waste management now is not only the absence or not updated technology of waste management facilities but also the attitude and the behavior of the citizen that do not reflect waste management itself. Most of the garbage is often mixed up between organic waste and non-biodegradable waste when throwing to the trash can. Non-biodegradable waste is divided by two group, recycled waste and un-recycled waste. Recycled waste often sold to recycling industry to be converted into new product while un-recycled waste is transported every week to TPS/TPA area. The garbage is then only stockpiled in TPS and TPA so that this activity does not deal with the problem of waste because the waste is only moved and not processed.

Non-biodegradable waste can be divided by two: (1) un-recycled plastic waste and (2) other un-recycled non-biodegradable waste. Unrecycled plastic waste has been a concern in the world for many decades, as it is non-biodegradable polymers mostly containing carbon, hydrogen, and few other elements such as chlorine, nitrogen, etc [1]. Un-recycled plastic waste is includ plastic film Recycling plastic has proven difficult and can be costly because of the constraints on water contamination and inadequate separation prior to recycle that is labor intensive [2]. Other type of un-recycled waste can be considered as waste to energy. The waste included dirty paper woods, tissue, fabric, leather, diaper. This type of waste is processed by burning it in incenerator. The waste cannot be recycled again but contain a lot of energy. By burning this waste as energy sources for other waste processing such as pyrolysis.

Plastics come in different types, and not all of them can be recycled. Hence, separation is needed since plastics are made of different resin compound, transparency, and color [3]. Clearly transparent plastics are often desirable bu the manufacturers since they can be dyed to transform into new products, thus have greater flexibility. Pigmented or dyed plastics have lower market value, and sometimes cannot be recycled [4]. Plastic waste management is needed to overcome this problem. Plastic pyrolysis is one of the solution to overcome plastic waste problem. Pyrolysis is the process of thermally degrading long chain polymer molecules into smaller, less complex molecules through heat and pressure [5]. This process is a part of de-polymerization, which is in contrast to the production of plastics, which uses polymerization of syngas and crude oil [6]. The process requires intense heat with shorter duration and in absence of oxygen. The major products from pyrolysis process are oil, gas, and char which are valuable for industries [3]. The liquid oil produced can be used as fuel for furnaces, boilers, turbines, and diesel engines without the needs of upgrading or treatment [7]. Pyrolysis does not cause water contamination and is considered as green technology [8].

MASARO (Management Sampah Zero) is a zero waste management system that offers the handling of waste until it is fully reduced to zero. MASARO change the old paradigm "collect and transport waste" or "reduce-reuse-recycle" in waste management system into a new paradigm of "sort-collect-process-sell" without producing waste that pollutes the environment. 100% of waste is processed and becomes valueable products, and hence MASARO needs no more TPS/TPA area that currently are big problems. Certainly this reflects the real circular economy application that is now promoted all over the world. The purpose of this paper is to show the system and technology MASARO for non-biodegradable waste is one of the best solution for handling waste because it has provide environementally friendly technology and also represent the circular economy.

## 2. Method

The circular economic of the non-biodegradable waste processing with MASARO Plastic Refinery Technology is carried out by reviewing several aspects of the waste processing process according to circular economic principles [10] including:

1. Renewable sources are continually being regenerated over time. [10]
2. Fast trade enables optimal use and access of products and assets. [10]
3. Increase product life by designing products to be durable. [10]
4. Associated value chains that produce no waste. [10]

Considering that waste is an environmental issue, it is necessary to consider environmentally friendly aspects in the waste management process. To produce a circular economic process of environmentally friendly waste processing technology, it is necessary to analyze the MASARO plastic refinery technology in the form of :

1. The reduction of environmental and health quality caused by incineration waste, including the emission of CO2, SOx, NOx, and ash (fly ash and precipitation) [11]
2. Global warming effect caused by CO2 production. [11]
3. Acid rain potential or acid deposition by SOx and NOx oxidation that can cause the reduce the life and population of water ecosystem, as well as reduce soil nutrient. [11]
4. Nearby air pollution caused by dust particullate, smoke mist, and the formation of bad ozone caused by photochemical reaction of NOx with volatile organic compound that can harm health [11]
5. Dioxins content in flue gas that releases to the environment which can cause cancer, damage the immune system and growth.

MASARO plastic refinery analyst data was analyzed by the third party comprise: (1) flue gas of environmental standard test was tested by Sucofindo (Table 1 and Table 2), (2) dioxins content was tested by CRL Calabarquez Corporation (Table 3), (3) waste water environmental standard test was tested by Sucofindo (Table 4)

Basically, non-biodegradable waste management in MASARO consists of five principles as follows

1. At source separation;
2. At source treatment;
3. Involving Society, Government, and Industry: (society to do separation at sources, government to provide facility and treat the hazardous waste, and industries to recycle the waste);
4. Educating the people to separate dan treat the waste at source;
5. Applying the appropriate environmentally friendly technology.

The first step in non-biodegradable waste processing with MASARO technology is waste sorting on source to classify waste in to 3 categories, in which are: recycled waste, un-recycled plastic waste, and other un-recyled waste. Non-biodegradable Waste that has been sorted is completely processed in IPSA by MASARO plastic refinery without any waste dumped into TPS/TPA. Non-biodegradable waste processing in MASARO involve the participation of the community, government and industry where MASARO changes the habits and behavior of the community in waste processing, the government as Masaro's technology provider in waste processing and industry in further processing recycle waste. MASARO technology convert non-biodegradable wsste into a valuable products such as fuel, polybag farming growing medium, organic pesticide, and wood preservative. This MASARO system will transform waste processing that was originally cost center into a profit center waste management. The non-biodegradable waste management by MASARO technology is presented in Figure 1.

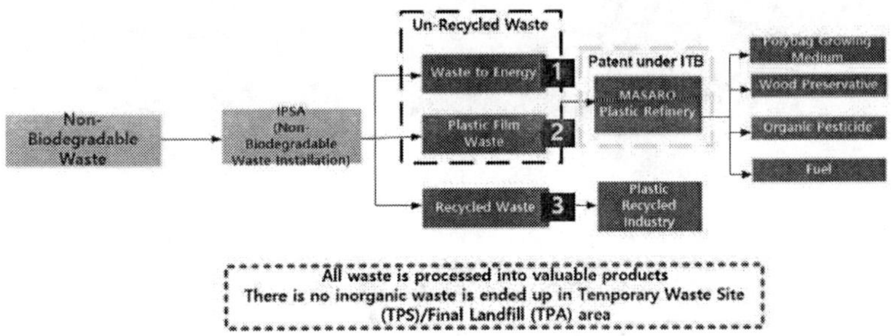


**Figure 1**. MASARO non-biodegradable waste management.

Circular economy of MASARO Non-biodegradable Waste Mangement is also carried out by studying the case of the implementation of MASARO non-biodegradable waste processing unit in MAN 2 Cirebon and Babakan Ciwaringin Village.

## 3. Results and Discussion

### 3.1. *Masaro Plastic Refinery in Non-biodegradable waste processing*

MASARO technology presents a solution to non-biodegradable waste problem by processing non-recycled non-biodegradable waste through a plastic refinery unit. MASARO plastic refinery divides non-recycled non-biodegradable waste into two groups, such as low value plastic waste (plastic film, thermosetting plastic, biodegradable plastic, and other plastics which are of poor quality) and waste to energy. The plastic refinery consists of a incinerator, pyrolyzer, and wet scrubber.

The incinerator burn waste such as woods, diapers, unrecycled papers, and etc that clasified as waste to energy (WTE) garbage. The feed is fed into the incinerator chamber by a conveyor to prevent the smoke from garbage burning release into the environment and to ensure the safety of the production process. The incinerator is designed with utilizes the buoyancy force of air that make oxygent flow into the furnace chamber naturally. that make the operation process does not require any external energy source, only requires a match. The flue gas temperature of this incineration process is about 800-1000o C and will be contacted with pyrolizer chamber directly. The low value plastic waste will be process in pyrolyzer that installed vertically with the incinerator to maximizing exhaust gas heat for the pyrolysis process and preventing heat loss during the process. The fuel vapor flows into condenser by the top of pyrolizer and being cooled along the spiral condenser pipe until the fuel is produced. The exhaust hot gas flowing through the output pipe that connected the pyrolizer and wet scrubber. Water injection is carried out along the pipe to reduce the temperature of the exhaust gas and to dissolve the solid particles contained in the exhaust gas. The exhaust gas cleaning process is carried out on the wet scrubber unit. In the wet scrubber room, the gas will be contacted with water so that the toxic gas and dirt particles in the exhaust gas are absorbed by water and the resulting clean air can be discharged into the environment through the chimney pipe.

MASARO plastic refinery has four configurations based on waste processing capacity. The unit configuration consists of one to four incinerators and pyrolyzers with each incinerator processing capacity of $4m^3$/hour. MASARO plastic refinery configuration with two and four incinerator and pyrolyzer unit configurations is presented in Figure 2

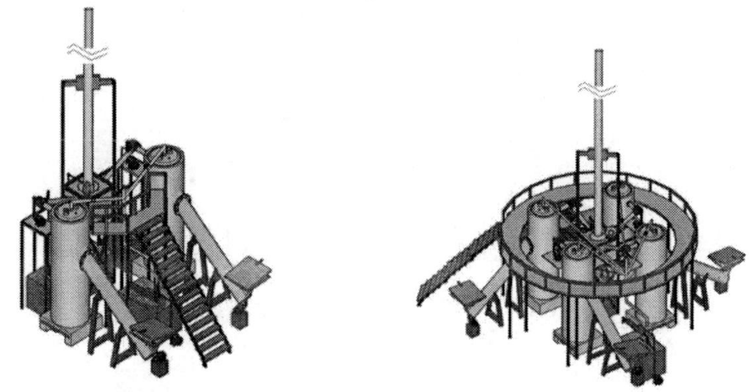

(a) (b)

**Figure 2**. MASARO plastic refinery configuratios (a) two incinerator and pyrolyzer unit, (b) four incinerator and pyrolyzer unit.

MASARO plastic refinery processes waste into a product without any residue. All non-biodegradable waste can be processed into useful products in the form of fuel, planting media, wood preservatives and organic pesticides. The fuel comes from processing low-value plastic films in pyrolyzers, planting media is the ash from incinerator processing, and wood preservative and large organic pesticides from wet scrubber wastewater treatment.

MASARO plastic refineries have been tested to be safe and environmentally friendly. Tests have been carried out to ensure that the exhaust gas and waste water from the MASARO plastic refinery are safe to be released to environment. The combustion process which is carried out at high temperatures also does not produce dioxins which has also been tested based on laboratory results. The test results are shown below in tables 1-4.

**Table 1.** Result of flue gas test in MAN 2 Cirebon by Sucofindo (2019) [11].

*Kep-03/BAPEDAL/09/1995*

| Parameter | Unit | Result | Quality Standard | Method |
|---|---|---|---|---|
| Particulate | mg/Nm$^3$ | 49.6 | 50 | Gravimetric |
| Sulfur Dioxide ($SO_2$) | mg/Nm$^3$ | 6.2 | 250 | Electrometric |
| Nitrogen Oxide ($NO_2$) | mg/Nm$^3$ | 4.1 | 300 | Electrometric |
| Hydrogen Flouride (HF) | mg/Nm$^3$ | <0.1 | 10 | SNI 19-7117.9-2005 |
| Carbon Monoxide (CO) | mg/Nm$^3$ | 102 | 100 | Electrometric |
| Hydrogen Chloride (HCl) | mg/Nm$^3$ | 0.4 | 70 | SNI 19-7117.8-2005 |
| Total Hydrocarbon ($CH_4$) | mg/Nm$^3$ | 11.3 | 35 | Electrometric |
| | | Metals: | | |
| Arsen (As) | mg/Nm$^3$ | <0.01 | 1 | SNI 7117.20-2009 |
| Cadmium (Cd) | mg/Nm$^3$ | <0.02 | 0.2 | SNI 7117.20-2009 |
| Chromium (Cr) | mg/Nm$^3$ | <0.2 | 1 | SNI 7117.20-2009 |
| Plumbum (Pb) | mg/Nm$^3$ | <0.1 | 5 | SNI 7117.20-2009 |
| Mercury (Hg) | mg/Nm$^3$ | <0.004 | 0.2 | SNI 7117.20-2009 |
| Thallium (Th) | mg/Nm$^3$ | <0.02 | 0.2 | SNI 7117.20-2009 |
| Opacity | % | 20 | 10 | SNI 19-7117.11-2005 |

*< : Less than the detection limit indicated*

**Table 2.** Result of flue gas test in MAN 2 Cirebon by Sucofindo (2019) [11].

| Parameter | Unit | Result |
|---|---|---|
| $O_2$ | % | 10.3 |
| $CO_2$ | % | 1.78 |
| Gas Temperature | °C | 60.0 |
| Barometer Pressure | mmHg | 757.6 |
| Gas Velocity | m/sec | 9.76 |
| Pressure | hPa | 0.52 |

**Table 3.** Result of flue gas dioxins content in MAN 2 Cirebon by CRL Calabarquez Corporation. (2019)

| Parameter | Unit | Result | | | Average result | NESSAP standard |
|---|---|---|---|---|---|---|
| | | Run 1 | Run 2 | Run 3 | | |
| PCDDs/ PCDFs | μg/kgm$^3$ | 0,216 | 0,04 | 0,04 | 0,099 | 0,1 |

**Table 4.** Result of waste water test in MAN 2 Cirebon by Sucofindo (2019) [11].

| Parameter | Units | Result | Quality Standard | Method |
|---|---|---|---|---|
| Temperature | °C | 34.45 | 38 | 2550B |
| Total Dissolved Solid | mg/L | 402 | 2000 | 2540C |
| Total Suspended Solid | mg/L | 136 | 200 | 2540D |
| Iron | mg/L | 3.11 | 5 | 3111B |
| Manganese | mg/L | 0.1 | 2 | 3111B |
| Barium | mg/L | < 0.12 | 2 | 3111D |
| Copper | mg/L | < 0.02 | 2 | 3111B |
| Zinc | mg/L | 0.16 | 5 | 3111B |
| Chrom Hexavalent | mg/L | < 0.02 | 0.1 | 3500 Cr-B |
| Chrom Total | mg/L | < 0.04 | 0.5 | 3111B |

| Parameter | Units | Result | Quality Standard | Method |
|---|---|---|---|---|
| Cadmium | mg/L | < 0.003 | 0.05 | 3111B |
| Mercury | mg/L | < 0.0008 | 0.002 | 3112B |
| Lead | mg/L | < 0.02 | 0.1 | 3111B |
| Tin | mg/L | < 0.3 | 2 | 3111B |
| Arsen | mg/L | < 0.002 | 0.1 | 3114C |
| Selenium | mg/L | < 0.001 | 0.05 | 3114C |
| Nickel | mg/L | < 0.04 | 0.2 | 3111B |
| Cobalt | mg/L | < 0.04 | 0.4 | 3111B |
| Cyanide | mg/L | < 0.01 | 0.05 | 4500-CN-E |
| Sulfide | mg/L | < 0.01 | 0.2 | 4500-$S^{2}$-D |
| Fluoride | mg/L | 0.19 | 0.4 | 4500-F-D |
| Free Chlorine | mg/L | < 0.01 | 0.05 | 4500-$Cl_2$-G |
| Ammonia | mg/L | < 0.05 | 0.5 | 4500-$NH_3$-F |
| Nitrate | mg/L | 11.03 | 2 | 4500-$NO_3$-B |
| Nitrite | mg/L | 0.28 | 1 | 4500-$NO_2$-B |
| BOD5 20°C | mg/L | 362 | 50 | 5210B |
| COD by Pottasium Dichromate | mg/L | 1208 | 100 | 5220B |
| Surfactant | mg/L | 0.53 | 5 | 5540C |
| Phenol | mg/L | 1.07 | 0.5 | 5530C |
| Oil and Grease | mg/L | 10 | 10 | 5520B |

The flue gas analysis were done by Sucofindo as shown in Table I and II above. All aspects from gas meet the environmental standard, which means that inceneration unit and wet scrubber unit are working well. The analysis results also showed all aspects from waste water in Table IV meet the environmental standard, except the values of nitrate, phenol, and COD by potassium dichromate. These parameters need to bring down by using waste water treatment [11]. Another test that were done in MASARO plastic refinery in MAN 2 Cirebon is dioxins content by CRL Calabarquez Corporation as shown in Table III. The analysis results showed dioxin content in flue gas lower than the standard by NESSAP which means the waste is burned in high temperature in incinerator chamber.

### 3.2. *Masaro Non-biodegradable Waste Management in MAN 2 Cirebon and Babakan Ciwaringin Village*

The waste problem in MAN 2 Cirebon and Babakan village is not only because of the absence of waste management facilities but also attitude and behavior that do not reflect the waste management itself. There is still a lot of waste scattered around the school environment and school residents are still throwing garbage mixed up/no waste sorting. MAN 2 Cirebon has a temporary trash shelter located at the back of the school. Non-biodegradable waste is transported every week using a garbage truck to the landfill area located in Indramayu, Arjawinangun, and Majalengka. The garbage is then only stockpiled so that this activity does not deal with the problem of waste because the waste is only moved and not processed. MAN 2 Cirebon spends around 12-15 million rupiahs every month just to dispose of garbage. The similar problem also occurred in Babakan village where the community disposed of waste without any sorting it beforehand. The local people just simply dump trash openly everywhere. Once there is enough trash, they will burn them all in an open fire without sorting out any hazardous waste, such as electronic waste. After the rainy season, this burned trash will be washed away into a nearby river, which is also a key source for drinking and living water. Babakan village government has provided a waste bank as a waste processing facility for the community. However, the waste bank only functions as a recycled waste collector without processing other types of non-biodegradable waste. Non-biodegradable waste handling in MAN 2 Cirebon and Babakan Village are presented in Figure 3 and Figure 4.

**Figure 3**. Non-biodegradable waste handling in MAN 2 Cirebon.

**Figure 4**. Non-biodegradable waste handling in Babakan Ciwaringin Village.

MASARO technology presents a solution of non-biodegradable waste problem in MAN 2 Cirebon and Babakan Village with processing non-biodegradable waste in IPSA (non-biodegradable waste installation) by plastic refinery unit which is represented in Figure 5. MASARO plastic refinery processing un-recycled waste with clean and environmentally friendly technology. The plastic refinery consists of a incinerator, pyrolyzer, and wet scrubber unit. The incinerator burn waste such as woods, diapers, unrecycled papers, and etc that clasified as other un-recycled waste. The incinerator is designed by utilizing the buoyancy force of air that make oxygent flow into the furnace chamber naturally. The flue gas of this incineration process will be contacted with pyrolizer chamber directly. The low value plastic waste will be processed in the pyrolyzer installed vertically with the incinerator to maximizing exhaust gas heat for the pyrolysis process and preventing heat loss during the process. The fuel vapor flows into condenser by the top of pyrolizer and being cooled along the spiral condenser pipe until the fuel is produced.The exhaust gas cleaning process is carried out on the wet scrubber unit clean air can be discharged into the environment through the chimney pipe.[11]

MASARO plastic refinery processes waste into a product without any residue. All non-biodegradable waste can be processed into useful products in the form of fuel, planting media, wood preservatives and organic pesticides. The fuel comes from processing low-value plastic films in pyrolyzers, planting media is the ash from incinerator processing, and wood preservative and large organic pesticides from wet scrubber wastewater treatment.

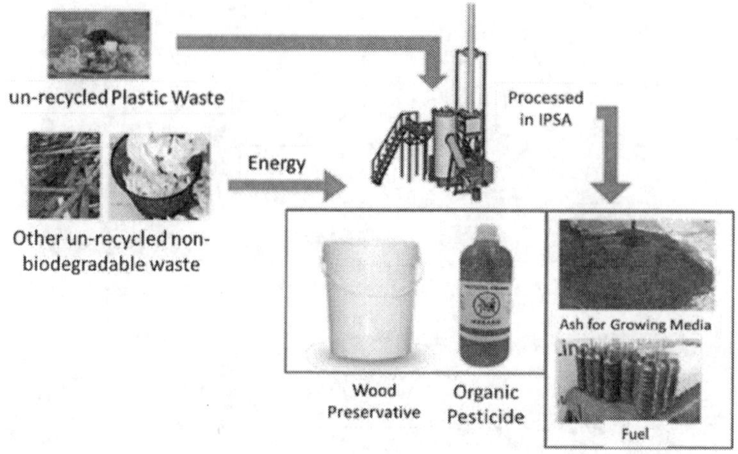


**Figure 5.** Non-biodegradable wastemanagement in IPSA MASARO.

MASARO technology is applied in MAN 2 Cirebon by using one of the classrooms in the Man 2 Cirebon as IPSA. While, in Babakan village MASARO plastic refinery is placed in the Babakan village waste bank. To support the MASARO technology implementation, it is necessary to educate all school academics and villagers through MASARO Workshop so that they are more aware of clean living behavior, waste sorting according to its type, and the benefits of waste sorting. The purposes of this event are educating the importance of maintaining cleanliness, introducing MASARO, educating the waste sorting according to MASARO, and educating the benefits of waste sorting.

The implementation of MASARO technology in MAN 2 Cirebon and Babakan Village has a positive impact where students and villagers sorting waste into segregated trash bins that sorted according to MASARO. All of the non-biodegradable waste is processed at IPSA in MAN 2 Cirebon and Babakan Village waste bank. IPSA products is used by students and villagers for greening and farming by polybag farming and fulfillment of residents' stove fuel. The implementation of MASARO non-biodegradable waste managemnt are represented in Figure 6 and Figure 7.

**Figure 6.** MASARO Non-biodegradable waste management in MAN 2 Cirebon.

**Figure 7.** MASARO Non-biodegradable waste management in Babakan Ciwaringin Village.

### 3.3. *Circular Economy of MASARO Plastic Refinery*

Circular economy of MASARO technology for non-biodegradable waste can be divided by 2 path: recycle path and recovery path. Figure 8 represent MASARO circular economy for non-biodegradable waste management. In recycle path, valuable waste will be recycled by recycle industry to become another product for consumption the it will become non-biodegradable waste and going through recycle cycle again. Meanwhile, non-valuable waste in recovery path processed to MASARO plastic refinery and convert it into MASARO product such as fuel, ash for growing media, organic pesticide, wood preservative. Output from MASARO plastic refinery will be used to agriculture and machinery that use fuel. Product from farming will go through manufacturer and going through cycle again. Recycle path and recovery path is represent by Figure 9. No waste is generated by consumption of MASARO product.

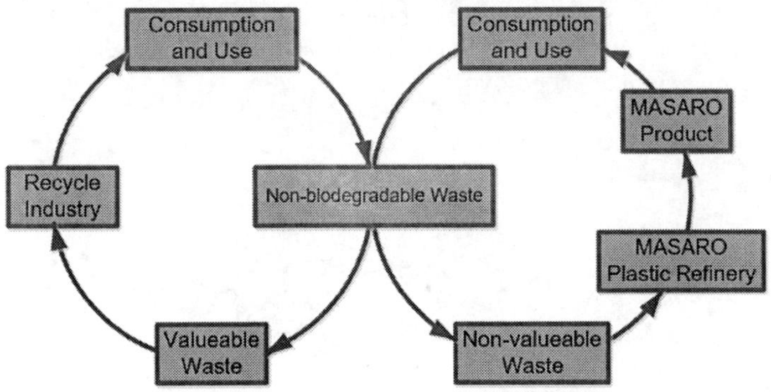


**Figure 8**. MASARO circular economy diagram.

MASARO non-biodegradable waste treatment is divided by two programs, which are social and industrial program. Figure 10 represent the MASARO program with (a) is MASARO's social program and (b) is MASARO's industrial program. MASARO's social program is started with collection of non-biodegradable waste especially other un-recycled waste type. The waste the burned in the incenerator and produced ash. The ash later be used as growth medium for polybag farming, the system that implemented in the community. The polybag content is using the 4:3:2:1 volumetric ratio of, with 4 parts growth medium, 3 parts animal feces, 2 parts MASARO compost, and 1-part charcoal. The program is to use ash (product of MASARO plastic refinery) for polybag farming. The polybag farming movement by community is developed and form Sustainable Food House/Rumah Pangan Lestari (RPL). Crop produced from RPL is used as food sources for the community. Residue of polybag farming create plastic film waste and processing crops into food also create non-biodegradable waste like food packaging. The waste created from the process become the source of non-biodegradable waste thus creating circular economy.

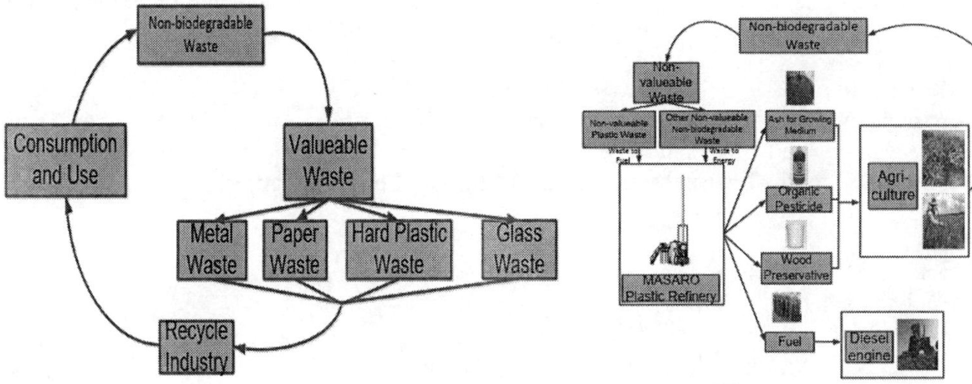


(a) (b)

**Figure 9**. (a) MASARO recycle path, (b) MASARO recovery path.

MASARO's industrial program represent MASARO plastic refinery as Non-biodegradable Waste Management Industry (IPSA). Non-biodegradable waste is processed in IPSA to produce 4 type of product. Non-biodegradable waste that entered IPSA is divided by two type which is un-recycled plastic waste and other un-recycled non-biodegradable waste. Plastic waste is converted into fuel that has the same quality as diesel oil. Other un-recycled waste is burned produced to produce hot gas as energy source for pyrolize process and the recidue can be used as growth medium. Meanwhile, the condensed water used for cooling the hot gas can be used as wood preservative and organic pesticide. The product of IPSA include fuel, growing media, organic pesticide, wood preservative. Fuel from IPSA is used for machinery such as diesel engine. Meanwhile, growing media, organic pesticide, and wood preservative is used mainly in agriculture sector. The food processed from the harvested crops will generate non-biodegradable waste (from polybag and food packaging) and transferred back again into MASARO plastic refinery creating circular economy.

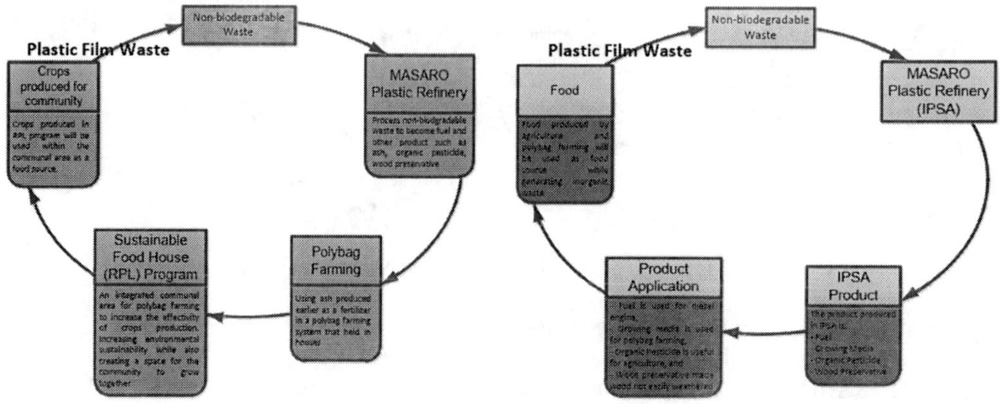


(a) (b)

**Figure 10**. (a) MASARO social program, (b) MASARO industrial program.

## 4. Conclusion and Future Outlook

The circular economy of non-biodegradable waste processing by MASARO technology can solve the whole problem of non-biodegradable waste in the community as a source of community income with the MASARO plastic Refinery. MASARO technology in handling non-biodegradable waste can save lots of costs production and energy consumption when compared to the current waste processing system. MASARO Plastic refinery is a very profitable and environmentally friendly to be apllied in small scale or industrial scale in the future that can regenerate and explore the maximum end-of-life value of non-biodegradable material that can be used as long as possible in its life cycle.

**Acknowledgements**

Authors wishing to acknowledge assistance and data support from MASARO MAN 2 Cirebon, Babakan Ciwaringin Village, and Babakan Ciwaringin Waste Bank. Special thank to Dow Chemical International, USA, for the financial sponshorship of Masaro Showcase Project.

**References**

[1] Grigore M E 2017 Methods of recycling, properties and applications of recycled thermoplastic polymers *Recycling* **2**(4) 24
[2] Kukreja R 2009 Advantages and disadvantages of recycling *Conserve Energy Future*.
[3] Sharuddin S D A, Abnisa F, Daud W M A W, Aroua M K 2016 A review on pyrolysis of plastic wastes *Energy Convers. Manag.* **115** 308-326.

[4] Masanet E, Auer R, Tsuda D, Barillot T, Baynes A 2002 An assessment and prioritization of "design for recycling" guidelines for plastic components *IEEE* 5-10.

[5] Martynis M, Mulyazami, Winanda E, Harahap A N 2019 Thermal pyrolysis of propylene plastic waste into liquid fuel: reactor performance evaluation, *IOP conf. ser., Mater. sci. eng.* **543**.

[6] Olufemi A S, Olagboye S A 2017 Thermal conversion of waste plastics into fuel *Int. j. petrochem. sci. eng.* **2** (8).

[7] Bridgwater A V, Review of fast pyrolysis of biomass and product upgrading 2012 *Biomass Bioenergy* **38** 68-94.

[8] Abnisa F, Daud W M A W, A review on co-pyrolysis of biomass:nan optional technique to obtain a high-grade pyrolysis oil 2014 *Energy Convers Manage* **87** 883-900.

[9] Kwakwa V, Chaves R A, Kemper K E, Glauber A J, Warden F V 2019 Improvement of solid waste management to support regional and metropolitan cities *World Bank Project Appraisal Document*.

[10] Vos M, Wullink F, Lange M, Acoleyen M, Staveren D, Meijenfeldt V. The circular economy-what is it and what does it mean for you? *Arcadis Briefing Paper* 2-3

[11] Abidin A Z, *Manajemen sampah zero (MASARO)* Indonesian Patent P00201704782

[12] Abidin A Z, Yemensia E V, Choliq N S, Hastuti R, Study on Environmental Health Aspect of Plastic Refinery in MASARO Cirebon Unit in Indonesia 2020 $4^{th}$ *ICGEA*.

IOP Conf. Series: Materials Science and Engineering 1143 (2021) 012053 doi:10.1088/1757-899X/1143/1/012053

# Integrated Biorefinery Technology: Monetization of Oil Palm Empty Fruit Bunch to Biofuel & Bio-based Chemicals, and Beyond

**Rina Mariyana[1]*, Azka Azkiya Choliq[1], Kharis Adi Rahmanto[1], Yusuf Nugroho[1], Ronny Purwadi[2], MTAP Kresnowati[2], Krisna Septiningrum[3], Frans B.M. Dabukke[4]**

[1]Research & Development Department, PT Rekayasa Industri, Jakarta, Indonesia
[2]Department of Chemical Engineering, Institut Teknologi Bandung, Bandung, Indonesia
[3]Center for Agro-Based Industry, Ministry of Industry, Bogor, Indonesia
[4]Indonesian Centre for Agricultural Socio Economic Policy Studies, Ministry of Agriculture, Jakarta, Indonesia

*Corresponding Author's E-mail: rina_mariyana@rekayasa.co.id

**Abstract.** As the largest producer of palm oil, Indonesia has abundant oil palm empty fruit bunch (EFB), as one of the solid wastes produced from palm oil mill, which utilization has not yet been optimal although it has a potency to be used as feedstock for biofuel and bio-based chemicals. However, due to the recalcitrant characteristic of EFB as lignocellulosic material, EFB conversion into biofuel and bio-based chemicals has been a challenge, especially in regard to its economic viability. To obtain the economically viable of the conversion of EFB, the integrated EFB biorefinery technology concept was proposed to produce multiple products which are precursors for biofuel and bio-based chemicals. This paper presents the integrated EFB biorefinery technology concept, economic analysis of the technology, the role of the technology for circular bioeconomy in Indonesia, and the biorefinery industrialization concept in Indonesia as part of an effort for "Making Indonesia 4.0".

## 1. Introduction

Indonesia has the largest palm oil plantation in the world with palm oil production reaching 43 million tonnes in 2019 [1]. Oil Palm Empty Fruit Bunch (EFB) is the largest solid waste generated from palm oil mills, and as much as 1.1 tonnes of EFB is generated for every tonne of palm oil produced [2]. The production of palm oil is expected to increase. Based on extrapolation using data obtained from the Ministry of Agriculture [1], EFB production is projected to reach 94 million tonnes by 2030.

As lignocellulose, EFB mainly consists of cellulose, hemicellulose and lignin, which can be used for the production of biofuel and a vast array of bio-based chemicals, such as butanediol, ethanol, ethylene, acetic acid, xylitol, levulinic acid, BTX (benzene, toluene, xylene) [3]. Due to its abundance, low price, and its great utilization potential, EFB is considered as one of the most promising biomasses for biorefinery feedstocks, particularly in Indonesia.

Various research related to EFB conversion into bio-based chemicals have been reported [4–17]. However, none of the research is developed until the commercialization stage. There are some constraints encountered during the implementation of this technology, such as: (1) low density of EFB which results in the high transportation and storage cost; (2) current utilization of EFB as mulch or compost feedstock which may affect the EFB supply for biorefinery; and (3) technical barriers which affect the economic viability of the technology, such as the inconsistency of biomass composition which affects the product yield, low concentration of sugar in the hydrolysate, high consumption of enzyme/chemicals accompanied by their high cost, and difficulties in high solid loading operations [18]. Some methods have been reported to overcome these problems, such as the installation of feed handling facility in the palm oil mills to increase the density of EFB prior to transporting [18], the use of solid residue-by products obtained from biorefinery as a fertilizer to replace mulch [18], fed-batch operation [7], enzyme recycling [19], and membrane separation [20] to increase the product yields and to reduce the enzyme/chemical consumptions.

In Indonesia, many studies regarding the conversion of EFB into biofuel and bio-based chemicals have been carried out. Most researchers, however, focus on single constituent of EFB while neglecting the others despite their potential utilization [4–15, 21–24]. Only few studies reported the optimization of the utilization of more than one component of EFB as in biorefinery [25]. The concept of biorefinery is one of the solutions to increase the economic viability of EFB monetization as well as to maintain the sustainability of the process and helps to move towards a circular bioeconomy [26–28]. A biorefinery technology is needed as a bridge which connects all EFB-based research and promotes in accelerating the development of biofuel and bio-based chemicals production technology, particularly in Indonesia.

PT Rekayasa Industri and researchers from Institut Teknologi Bandung, have succeeded in developing the technology for processing EFB into glucose on a lab scale with satisfactory results, where the consumption of chemicals and enzyme was minimized, while maintaining high product yields and concentration which is described in our patent pending (Indonesia Patent No.P00201803638; International PCT patent No.PCT/IB 2019/054125). Currently, in collaboration with researchers from Institut Teknologi Bandung and Centre for Agro Based Industry-Ministry of Industry, we are developing EFB biorefinery technology on a pilot scale which is funded by The Indonesian Palm Oil Plantation Fund Management Agency.

This paper presents the integrated EFB biorefinery technology concept, economic analysis of the technology, the role of the technology for circular bioeconomy in Indonesia, and its position in the industrialization concept as part of an effort for "Making Indonesia 4.0".

## 2. Methodology

### *2.1. Process modelling*

The process model used to simulate the biorefinery technology to convert EFB into glucose, xylose and lignin is based on our patent pending, publications and non-confidential inputs from Institut Teknologi Bandung as our research partner. The techno-economic calculation model is prepared in Excel Spreadsheet. The design basis for pretreatment and hydrolysis reaction models for glucose production are obtained based on our patent pending (Indonesia Patent No.P00201803638; International PCT patent No.PCT/IB 2019/054125), while pretreatment and hydrolysis reaction models for xylose production as well as pretreatment reaction model for lignin production are prepared based on data from Institut Teknologi Bandung as our collaborator [11]. The heat exchange and evaporation processes were simulated using Aspen HYSYS software.

The composition of cellulose, hemicellulose and lignin, the main constituent of EFB, varies depending on plant's age, growth condition, the climate and soil conditions [29, 30]. Table 1 shows the composition of the EFB used in this study.

**Table 1.** Composition of EFB in dry basis (%-wt).

| Component | %-wt |
|---|---|
| Cellulose | 34 |
| Hemicellulose | 31 |
| Lignin | 22 |
| Others | 13 |

### 2.2. *Total Capital Investment calculation method*

Equipment sizing was done manually based on capacity. The purchase price of equipment is estimated by means of scaling-down or scaling-up from a literature data such as National Renewable Energy Laboratory (NREL) Report [33] or our database of a known size as depicted in Eqn (1):

$$Purchace\ cost = Purchase\ cost\ of\ a\ known\ size \times \left(\frac{size}{known\ size}\right)^{n} \quad (1)$$

Where n is the power coefficient with a value of 0.5–0.8, depending on the type of equipment. The factors used to estimate other direct cost and indirect cost to calculate total capital investment cost are obtained from NREL Report [33].

### 2.3. *Unit cost of main raw materials and utilities*

Table 2 shows the unit cost of main raw materials and utilities used to calculate operating cost and economic analysis. The costs are obtained from our database and National Renewable Energy Laboratory (NREL) Report [33], and has been corrected using inflation rate.

**Table 2.** The unit cost of main raw materials and utilities.

| **Materials** | **Price** |
|---|---|
| Butanol | $1,000.00 ton$^{-1}$ |
| Chemical Cooling Tower | $2,716.00 ton$^{-1}$ |
| Chemical for Waste Water Treatment | $410.00 ton$^{-1}$ |
| Cooling Water | $0.14 ton$^{-1}$ |
| Electricity | $0.08 kWh$^{-1}$ |
| Cellulase Enzyme | $4,000.00 ton$^{-1}$ |
| Oil Palm Empty Fruit Bunch | $6.90 ton$^{-1}$ |
| Process Water | $0.55 ton$^{-1}$ |
| Steam | $1.53 ton$^{-1}$ |
| Xylanase Enzyme | $4,000.00 ton$^{-1}$ |

### 2.4. *Other assumptions*

Table 3 summarizes other main assumptions used to calculate the economic analysis. The plant operates for 20 years. The depreciation used is straight line depreciation. The income tax rate is 25%. The inflation rate is 6%. The construction was carried out for 2 years. In the first year of construction, the costs incurred for construction are as much as 40% of the total investment cost, while the remaining costs are used in the second year. In the first year of the operation, only 80% of maximum working capacity is used.

**Table 3**. Economic analysis assumptions.

| | |
|---|---|
| Plant lifetime (years) | 20 |
| Operating hours per year | 7920 |
| Equity | 100% |
| General plant depreciation | straight line depreciation |
| Depreciation period (years) | 10 |
| Construction period (years) | 2 |
| % spent in year -1 | 40% |
| % spent in year 0 | 60% |
| Revenues in year 1 | 80% |
| Revenues in year 2 | 100% |
| Income tax rate | 25% |
| Working Capital | 5% of TCI |

## 3. Result & Discussion

### *3.1. The integrated EFB biorefinery technology concept: EFB Fractionation*

The concept of the integrated EFB biorefinery technology is to fractionate the three main constituents of EFB and convert them into glucose, xylose and lignin through sequential processes as described in Figure 1. The processes are divided into three main stages, namely: (1) Xylose production: Hemicellulose fractionation through hydrothermal pretreatment followed by an enzymatic hydrolysis reaction to convert hemicellulose to xylose, (2) Lignin production: lignin fractionation through organosolv pretreatment, and (3) Glucose production: cellulose fractionation using steam explosion pretreatment followed by an enzymatic hydrolysis reaction to convert cellulose to glucose. The process sequence is determined based on the consideration that hemicellulose is the most easily degraded part among all EFB constituents, thus requires the mildest operating condition for recovery [15, 31]. Hydrothermal was considered as the most suitable technique for hemicellulose recovery due to its ability to selectively remove hemicellulose while leaving the other constituents as essentially intact [31, 32].

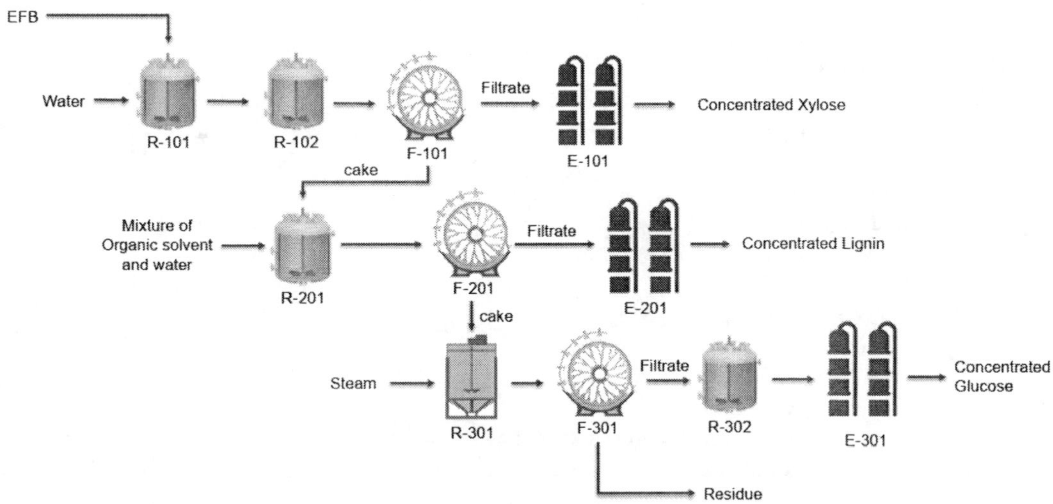


**Figure 1.** EFB fractionation process scheme.

*3.1.1. Xylose production.* Shredded EFB is pretreated hydrothermally in R-101 at temperature 140 - 200°C under saturated pressure. The reaction was held for 20 - 40 minutes. Both the pretreated solid and liquid are sent to hydrolysis reactor R-102. The hydrolysis reaction is carried out at 50°C and pH 5.0 for 24 - 48 hours using commercial xylanase. The broth is separated from the pretreated solid prior to purification. Purification process is conducted using evaporator E-101 to produce concentrated xylose (60%-wt).

*3.1.2. Lignin recovery.* The remaining solids from R-102 are pretreated using organosolv method in R-201 at temperature 140 – 200°C under saturated pressure. A mixture of butanol and water is used as the solvent. The reaction was held for 20 - 40 minutes. The broth is separated from the remaining solid prior to purification. Purification process is conducted using evaporator E-201 to produce concentrated lignin (60%-wt).

*3.1.3. Glucose production.* The remaining solid from R-201 is pretreated using steam explosion method in R-301 at temperature 140 – 200°C under saturated pressure. The reaction was held for 10 - 20 minutes. The liquid is recovered prior to be sent to hydrolysis reactor R-302. The hydrolysis reaction is carried out at 50°C and pH 5.0 for 24 - 48 hours using commercial cellulase. The broth is then purified in evaporator E-301 to produce concentrated glucose (60%-wt).

The yields for concentrated xylose, concentrated lignin and concentrated glucose are 0.8, 0.46, and 0.38 ton/ton dry EFB, respectively.

### *3.2. Economic analysis*

The EFB biorefinery capacity used in this study is 60 k-ton EFB wet basis per annum. This capacity selection is based on the average EFB production rate from one palm oil mill with the capacity of 240 k-ton fresh fruit bunch per annum. This is in accordance with the concept of "*one palm oil mill, one EFB biorefinery plant*", which helps in reducing the cost for transportation as well as minimizing the need for storage. EFB produced from palm oil mill is washed and dried before being shredded.

Table 4 shows the detail of total investment cost. The total capital investment cost for EFB biorefinery with capacity of 60 k-ton EFB wet-basis per annum is 40.98 million USD.

**Table 4.** Total capital investment cost breakdown.

| | | |
|---|---|---|
| **Direct costs** | | |
| Total installed equipment cost | 1.00 | 29.27 million USD |
| **Other direct costs** | | |
| Warehouse | 0.04 | 1.17 million USD |
| Site development | 0.04 | 1.17 million USD |
| Additional piping | 0.04 | 1.17 million USD |
| **Total direct costs** | **1.12** | **32.78 million USD** |
| **Indirect costs** | | |
| Field Expense | 0.056 | 1.64 million USD |
| Home Office and Construction Fee | 0.056 | 1.64 million USD |
| Project Contingency | 0.056 | 1.64 million USD |
| Other Cost (start up, permits, insurance, etc) | 0.112 | 3.28 million USD |
| **Total Capital Investment (TCI)** | **1.40** | **40.98 million USD** |

Figure 2 shows the share of each components constituting direct cost of the total capital investment in a pie chart. Xylose production area accounts for the largest portion of the direct cost, 38%, as the equipment capacity in this area is bigger than in other areas. Despite its bigger capacity as the second process area, the lignin production area accounts for roughly the same cost, 21%, as the glucose production area, 19%. This is because lignin production area comprises only pretreatment reactor, filter and evaporator as its main equipment, while glucose production area includes hydrolysis reactor.

Figure 3 shows the share of components constituting the operating cost in a pie diagram. Butanol contributes the largest portion of the operating cost, which is about 52%, followed by enzyme 20% and utilities 11%. The high consumption of solvent due to moderate solid loading operation (S/L 15–25%) in organosolv pretreatment as well as the high price are the main reasons for the major contribution of butanol. On the other hand, despite its high price, the moderate contribution of enzyme is attributed to the high solid loading operation (S/L 25–35%) in hydrolysis.

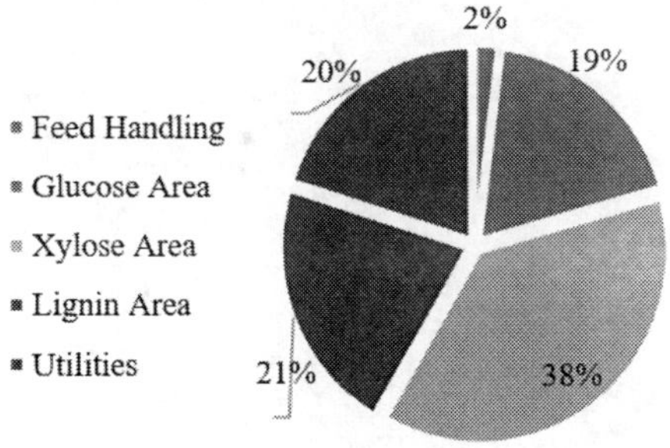


Figure 2. The direct cost of TCI categorized by process area.

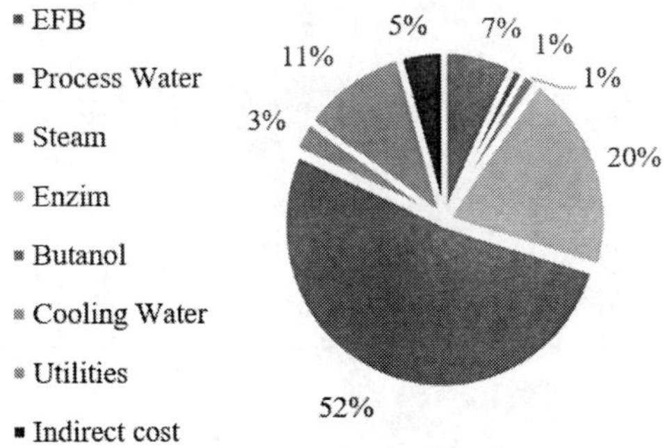


**Figure 3.** Operating cost component share.

Table 5 summarizes the results of the economic analysis. The economic analysis was calculated at selling price of concentrated xylose, concentrated lignin and concentrated glucose 308.70, 594.01[34] and 113.00 USD/ton, respectively. The internal rate of return (IRR), the payback period (PBP), and the net present value (NPV) are 12.4%, 7 years and 6 million USD, respectively. The result shows that the IRR value is greater than bank interest which indicates that the EFB Biorefinery has good potential in terms of economic viability. The selling price of glucose used in this economic analysis is selected based on the consideration that glucose will be used as raw material for bioethanol production so that the price of bioethanol produced can compete with market price. The demand for bioethanol in Indonesia is quite high, especially after the government decree the mandatory use of bioethanol fuel into conventional fuel mixtures, as stated in MEMR (The Ministry of Energy and Mineral Resources) Regulation No. 12/2015. This also indicates that the biorefinery technology where multiple products are produced plays a crucial role in enabling the production of cheap glucose, and consequently cheap bioethanol fuel or other glucose derivatives where the price can compete with fossil derived products.

**Table 5.** Economic analysis result.

| | |
|---|---|
| IRR | 12.4% |
| Payback period | 7 years |
| Net Present Value | $ 6 million |

*3.3. Palm oil circular bioeconomy in Indonesia and the role of EFB biorefinery technology*

Circular bioeconomy (CBE) concept, which is the combination of circular economy and bioeconomy, is introduced to achieve a resource-efficient biomass utilization by focusing on the use of biomass residues and wastes as resources and integrating several bio-based processes to achieve high value application of biomass. Moreover, it also focuses on extending product lifespan through improved design and servicing, and using resources more efficiently by using them over and over [28, 35].

Utilization of EFB into high-value added products has manifold roles in the circular bioeconomy in Indonesia. Firstly, it relocates waste from the end of supply chain to the feedstock. Secondly, it adds value from a relatively low-value bioproducts into high-value bioproducts. Thirdly, it replaces some products derived from fossil resources which consequently reduces Indonesia dependency on external sources [27, 28].

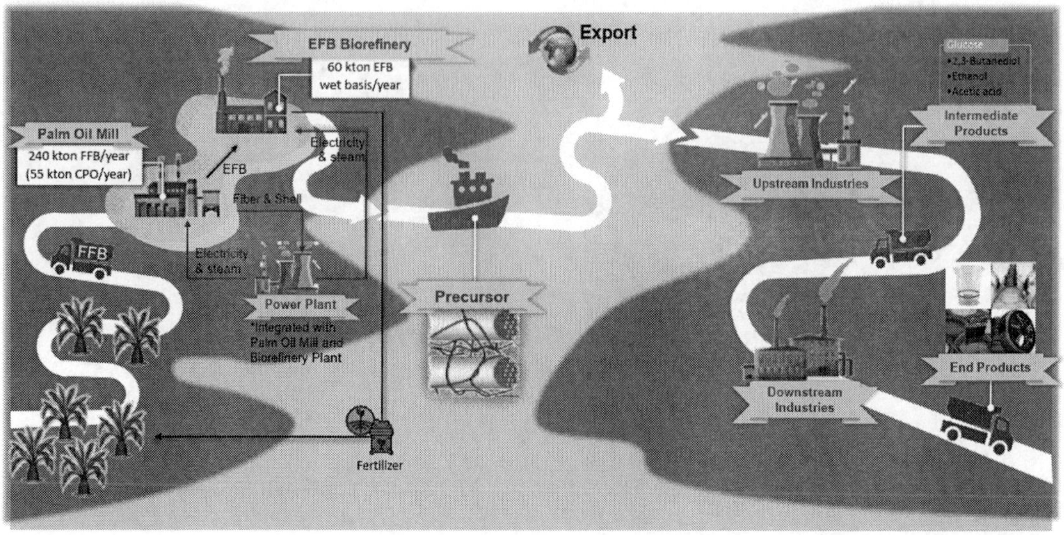


**Figure 4**. EFB industrialization scheme: EFB biorefinery coupled with upstream/downstream industries.

Figure 4 shows the industrial concept of EFB utilization. EFB biorefinery plant will be built near the palm oil mill. EFBs will either be washed, dried and shredded in palm oil mill prior to transfer to biorefinery plant or will be transferred directly to biorefinery plant to be processed there. Since the location of the mill and biorefinery plant are close to each other, some facilities such as the feed handling, storage and utilities can be combined, thus minimizing the capital investment cost. Additionally, the cost for feedstock transportation, which gives relatively high contribution to the operational cost due to the low density of EFB [18], can also be reduced. Shredded EFBs are then fractionated and converted into glucose, xylose, and lignin in EFB biorefinery plant. These precursors will then be transported and distributed to related upstream or downstream industries for further processing into intermediates or end products. The solid residue as by-product of biorefinery plant contains some nutrients which are beneficial for the soil, thus can be used as a complement to fertilizer. Since the location of biorefinery plant is close to palm oil plantation, the transportation cost for the solid residue application as fertilizer is negligible.

*3.3.1. Benefit.* There are several positive impacts of EFB circular bioeconomy in Indonesia in terms of economic, environmental and social impacts. Economically, EFB biorefinery plant could trigger the development of bio-based downstream industries in Indonesia to meet the increasing demand for bio-based products, thereby increasing the production volume and the country's Gross Domestic Product. As the awareness of investors and consumers about sustainable and environmentally benign products increases, the demand for bio-based products will increase.

EFB derivative products, such as 2,3-butanediol, lactic acid, ethanol, sorbitol, lysine, succinic acid, lactic acid, acetone, gluconic acid which are glucose derivative products; xylitol, furfural, and levulinic acid which are xylose derivative products; as well as benzene, toluene, and xylene, which are lignin derivative products, are classified as the top value-added bio-based chemicals based on its market demand, market size, and commercial application [3]. Currently, the fulfillment of these products in

Indonesia depends on foreign supply with total import value reached USD 500 million [36]. The development of EFB biorefineries in Indonesia will accelerate the development of its downstream industries, thereby strengthening Indonesia's sovereignty in energy, food and chemical production. In addition, as a country having the second largest peatland area amounting to 22.5 million ha [37], Indonesia has the potential to supply bio-based products not only domestically, but also globally.

Environmentally, the use of EFB as a replacement of non-renewable material could mitigate the climate change. EFB derivative products have the potential to replace the conventional products produced from non-renewable raw materials, which will significantly contribute to the reduction of greenhouse gasses emissions. Furthermore, it also reduces Indonesia dependency on non-renewable raw materials, particularly on food, chemical, and energy sectors.

Socially, the use of EFB as biorefinery feedstock could ensure the food security. Unlike first-generation ethanol which is produced from food, EFB is an agricultural waste therefore its utilization does not compete with food. Additionally, the development of EFB biorefineries will trigger the growth of the downstream industries, thus creating more jobs for Indonesian people. The biorefinery plant itself is proposed to be built near an established palm oil mill and plantations and will therefore absorb local workers. This will create jobs for thousands of people and consequently increasing the well-being of the people in Indonesia.

### *3.4. EFB biorefinery as part of Making Indonesia 4.0*

In order to strengthen the manufacturing sector in Indonesia for making Indonesia as the top 5 global economy, the government plans to implement I4.0 in Indonesia as outlined in the "Making Indonesia 4.0" program. Several commitments are made by the government to successfully implement this program which includes doubling the productivity to cost ratio, increasing exports to 10% of GDP, and increasing the allocation of R&D funds to 2% of GDP [30]. There are five main industrial sectors prioritized in the "Making Indonesia 4.0" program which includes food and beverages, chemical, textile, automotive, and electronic [31]. EFB biorefineries play an important role in supporting the successful and long-term implementation of "Making Indonesia 4.0" because the biorefineries can act as a trigger for the development of the downstream industries in Indonesia, particularly in food, energy and chemical sectors, replacing the non-renewable with bio-based products, thereby reducing Indonesia dependency on non-renewable source and strengthening Indonesia's sovereignty in energy, food and chemical sectors. Further, some long-term benefits of the implementation of "Making Indonesia 4.0", which are expected to happen, are (1) the increase in the production volume thereby increasing the country's GDP, (2) higher investment on technology research & development which can further support the growth of the industry to achieve higher productivity and efficiency, and move towards cleaner energy, and (3) higher living quality and education standard of the people in Indonesia [30].

## 4. Conclusion

The integrated EFB biorefinery technology concept through sequence fractionation of main components of EFB shows a good potency to be applied in industrial scheme, for the production of various bio-based chemicals and biofuel as an alternative to conventional ones. The economic analysis shows that EFB biorefinery with capacity as much as 60 k-ton/year has an IRR of 12.4%, payback period of 7 years, and NPV of 6 million USD at selling price of concentrated glucose, xylose and lignin 113.00, 308.70, and 594.01 USD per ton, respectively. Furthermore, the advances of this technology in Indonesia will provide several benefits, such as connecting various oil palm-based research in Indonesia, thus accelerating the commercialization of these research, as well as strengthening Indonesia's sovereignty in energy, food and chemical sectors, helping to move toward circular bioeconomy, increasing the country's GDP, and elevating the well-being of Indonesian people.

## References

[1] Widayanto R K P 2019 *Buku Outlook Komoditas Perkebunan Kelapa Sawit* (Jakarta: Ministry of Agriculture)

[2] Elbersen W, Meesters K and Bakker R 2013 *Valorization of palm oil (mill) residues. Identifying and solving the challenges* (The Netherlands: Wageningen UR, Food & Biobased Research)
[3] Werpy T and Petersen G 1992 *Top Value Added Chemicals from Biomass: Volume I -- Results of Screening for Potential Candidates from Sugars and Synthesis Gas* (United States)
[4] Sitompul J P *et al.* 2014 *Procedia Chemistry* **9** 88
[5] Kresnowati M, Mardawati E and Setiadi T 2015 *Mod. Appl. Sci.* **9** 206
[6] Triwahyuni E *et al.* 2015 *Energy Procedia* **68** 138
[7] Sugiharto Y E C *et al.* 2016 *Bioresour. Technol.* **207** 175
[8] Panjaitan J R H, Monica S and Gozan M 2017 *IOP Conf Ser: Earth Environ. Sci.* **65** 012042.
[9] Gozan M *et al.* 2018 *IOP Conf Ser: Mater. Sci. Eng.* **334** 012058
[10] Gozan M *et al.* 2018 *Int. J. Chem. Eng.* **2018** 1
[11] Harahap B M and Kresnowati M T A P 2018 *Biomass Conv. Biorefin.* **8** 255
[12] Isroi I and Cifriadi A 2018 *J. selulosa* **8** 51
[13] Ahda Y *et al.* 2019 *IOP Conf Ser: Mater. Sci. Eng.* **543** 012061
[14] Burhan K H, Kresnowati M T A P and Setiadi T **2019** *Bull. Chem. React. Eng. Catal.* **14** 559
[15] Meilany D *et al.* 2020 *Appli. Sci.* **10** 1391
[16] Medina J D C *et al.* 2016 *Bioresour. Technol.* **199** 173
[17] Pangsang N *et al.* 2019 *Energy Rep.* **5** 337
[18] How B S *et al.* 2019 *Renew. Sust. Energ. Rev.* **113** 109277
[19] Kim J K *et al.* 2019 *Biotechnol. Biofuel* **12** 138
[20] Wijaya H *et al.* 2020 *Processes* **8** 619
[21] Setyaningsih D *et al.* 2018 *IOP Conf Ser: Earth Environ. Sci.* **141** 012027
[22] Hendriansyah R *et al.* 2018 *MATEC Web Conf* **156** 03018
[23] Hendriansyah R *et al.* 2017 *4th Int. Conf. on Electric Vehicular Technology* (Bali: IEEE) p 13
[24] Sukmana I F *et al.* 2020 *IOP Conf Ser: Mater. Sci. Eng.* **778** 012011
[25] Burhani D and Triwahyuni E 2019 *AIP Conf. Proc.* **2175** 020034
[26] Ali A A M *et al.* 2015 *J. Clean. Prod.* **91** 96
[27] Ubando A T, Felix C B and Chen W H 2020 *Bioresour. Technol.* **299** 122585
[28] Stegmann P, Londo M and Junginger M 2020 *Resour. Conserv. Recycl.* **6** 100029
[29] Kumar P *et al.* 2009 *Ind. Eng. Chem. Res.* **48** 3713
[30] Ching Y C and Ng T S 2014 *BioResources* **9** 6373
[31] Ahmad N and Zakaria M R 2019 *Oligosaccharide from Hemicellulose (Lignocellulose for Future Bioeconomy)* ed Arrifin H, Sapuan S M and Hassan M A (Elsevier) p 135
[32] Garrote G, Domınguez H and Parajo J C 1999 *J. Chem. Technol. Biotechnol.* 9
[33] Humbird D *et al.* 2011 *Process Design and Economics for Biochemical Conversion of Lignocellulosic Biomass to Ethanol: Dilute-Acid Pretreatment and Enzymatic Hydrolysis of Corn Stover* (Colorado: National Renewable Energy Laboratory)
[34] Leaf Resources 2018 *FEL 2 Feasibility Review Report – Executive Summary* (Mackay: Aurecon australasia Pty Ltd)
[35] Jeroen G *et al.* 2018 *The circular economy and the bioeconomy: partners in sustainability* (LU: Publications Office)
[36] Statistik Ekspor Impor Industri Indonesia https://www.kemenperin.go.id/statistik/exim.php
[37] Luas Gambut Indonesia Terbesar Kedua di Dunia - Infografik Katadata.co.id, https://katadata.co.id/timpublikasikatadata/infografik/5e9a519433cb1/luas-gambut-indonesia-terbesar-kedua-di-dunia (2019, accessed 9 October 2020)

# Microplastic pollution in Surabaya River Water and Aquatic Biota, Indonesia

**P Lestari, Y Trihadiningrum*, M Firdaus, and I D A A Warmadewanthi**

Department of Environmental Engineering
Faculty of Civil, Planning, and Geo Engineering
Institut Teknologi Sepuluh Nopember, Kampus ITS Sukolilo, Surabaya 60111, Indonesia

*Corresponding author's e-mail address: trihadiningrum@gmail.com

**Abstract**. Microplastics (MPs/MP) have been considered as an emerging environmental threat worldwide. It occurs due to its persistent, ubiquitous presence, and potential ecotoxicological risks to almost all aquatic environments. The MPs could release chemical additives and adsorb persistent organic pollutants. It could also be ingested mistakenly by aquatic biota due to its similarity with their original prey's size and colour. This situation could impact water resource quality, aquatic biota biodiversity, even human health. Additionally, rivers have been estimated as the main pathway of plastic transport from the land to the ocean. Surabaya River, is the main lower part of the Brantas, which is one of the top 20 plastic polluted rivers globally. The Surabaya River takes essential role in clean water supply for Surabaya City, aquatic biota habitat, and irrigation. This study aimed to investigate the distribution and characteristics of the MPs in water and aquatic biota in Surabaya River. Fish and bivalve were selected as the representative aquatic biota due to their specific habitat, movement, and feeding behavior. Water and biota samples were collected from five sampling sites in the river. The MP abundance in the water ranged from 9.66 ± 8.87 to 21.16 ± 19.35 particles/m$^3$. The highest MP abundance (21.16 ± 19.35 particles/m$^3$) was found at Joyoboyo. The MP abundance in the fishes of *Oreochromis niloticus, Barbonymus gonionotus,* and in the bivalves of *Elongaria orientalis* were 105.25 ± 45.07 – 155.50 ± 61.96; 62.13 ± 20.33 – 155.00 ± 81.71; and 36.00 ± 13.67 – 76.17 ± 29.46 particles/individual, respectively. The MPs in the water were dominated by film shaped, transparent coloured particles. Meanwhile, the MPs in the fishes and the bivalves were generally found in transparent and black fiber particles. Only small amounts of film particles were discovered in the biota. Moreover, most of the MPs in the water and the biota were in large sized (1 – 5 mm) particles.

## 1. Introduction

Plastics are synthetic or semi-synthetic organic polymers that are persistent, durable, lightweight, versatile, resistant to corrosion, and low-cost production [1-3]. These characteristics make plastics become commonly used materials for daily products [4]. Annual global plastic production has reached up to 322 million tonnes in 2016 [4]. It is estimated that the global plastic production will be increased in double by 2025 due to plastic versatility in various daily products [3, 5]. About 75 – 90% of global plastic demand is limited to several types of plastic: high-density polyethylene (HDPE), low-density polyethylene (LDPE), polypropylene (PP), polyethylene terephthalate (PET), polyvinyl chloride (PVC), polystyrene (PS), and polyurethane (PU) [1, 6]. Meanwhile, plastic waste could be released and polluted the environment due to improper solid waste management and insufficient service coverage [6 – 11].

Plastics can be fragmented and degraded in the environment due to photodegradation, weathering, mechanical-physical, water current, and biodegradation mechanisms [12 – 13]. One of the emerging environmental concerns of plastic degradation results is microplastics (MPs/MP), plastic particles smaller than 5 mm in size [12 - 15]. The MPs can be originated from primary (household product application and personal care products) or secondary sources (breakdown of larger plastic particles) [15 - 18]. The MPs can be identified based on size, shape, colour, and polymer density [15]. These characteristics could affect its distribution in different environmental compartments and make the particle susceptible to be ingested by the biota [19 – 23].

The MPs have been found in rivers, coastal area, marine, even in aquatic biota in the global [24 – 31]. This condition could give various potential impacts to the environment and the biota [32]. The MPs also has an ability to adsorb environmental pollutants from their surroundings [32 – 33]. Furthermore, biofilm growth on the surface of the MP particles may become a source of microbial pathogens [34]. However, the MPs have been found in commercial edible fishes and mussels, tap and bottled drinking water, table salt, honey, and beer [35 – 37]. This condition could give potential ecotoxicological effects not only to the aquatic biota food web, but also a risk to food security and human health [38 – 40].

Rivers are predicted as the main pathway of plastic waste transport from the land to the marine [41 – 42]. Regrettably, Brantas River, the largest river in East Java Province, Indonesia, was included as one of the top 20 plastic polluted rivers in the global [25, 43]. It has been estimated that about 3.23 – 6.37 x $10^4$ tonnes per year of mismanaged plastic waste originating from the Brantas flowing into the marine [25]. However, this river has an essential function as the primary source of clean water supply for 15 densely populated cities in East Java Province [44]. One of the main lower parts of the Brantas, Surabaya River, is also used for clean water supply for Surabaya City, aquatic biota habitat, flood control, and recreation [43, 45]. However, the Surabaya River has suffered from water quality deterioration due to 215 point sources of domestic effluent and ten industrial outlets that discharged the wastewater into the Surabaya River [46 - 48]. Former study investigated 92 – 590 MP particles per kg dry weight sediment contained in the Surabaya river mouth and the Jagir Estuary [49]. This evidence showed that MP had polluted the Surabaya River. Therefore, this study aimed to investigate the distribution of the MPs in the water and biota of the Surabaya River. Fish and bivalve were selected as representatives of edible biota due to their high potential to be contaminated by the MPs [21]. Fish is highly potential to ingest the MPs due to its feeding behavior and active movement in the water [22]. Meanwhile, the bivalve as a sedentary organism could adsorb and accumulate the MPs due to its feeding behavior as a filter feeder [50]. Therefore, this study could contribute for further ecological risk evaluation of MP pollution [26]. More importantly, a better understanding of the MP pollution study in the freshwater, especially in the urban river, is urgently needed for effective prevention of the MP pollution in aquatic environment [51].

## 2. Materials and Methods

### *2.1 Study area and sampling sites*

The sampling sites were: S1 in Driyorejo, S2 in Bambe, S3 in Karang Pilang, S4 in Joyoboyo, and S5 in Jagir Districts (Figure 1). The water samples were collected during the rainy season from February to March 2019. Meanwhile, the biota samples collection was conducted from the end of September to early October 2019. Both sampling activities were conducted in two replications. The water and biota sampling activities were conducted at different times as this study did not consider time variation, which might influence the MP distribution. The particular reason for this situation was also related to technical challenges (limited sampling accessibility, the availability of supported equipment, and human resources) and hydraulic conditions (river discharge fluctuation). However, both sampling activities were still carried out within the same range of the rainy season in Indonesia.

Water samples collection was done by using a modified Manta trawl with pore size of 333 μm [52, 53]. The dimension of the manta trawl net was 3 m, 61 cm, and 16 cm in length, width, and height, respectively [53]. A cod-end at the end of the Manta trawl net was placed for collecting the MP particles from the filtered water with an average volume of 55.34 $m^3$. The Manta trawl net was placed in three stratified depths to collect the samples from the surface, middle, and bottom of the river water column. This method was meant for obtaining average MP abundance and avoiding under- or over-estimation of the actual MP abundance [54]. The trawl was towed for approximately 20 minutes in the opposite direction of the river current alongside the boat [55, 56]. Afterward, the cod end, which retained and collected the MP particles, was carefully released from the trawl and preserved in aluminum foil zip lock containing 70% alcohol.

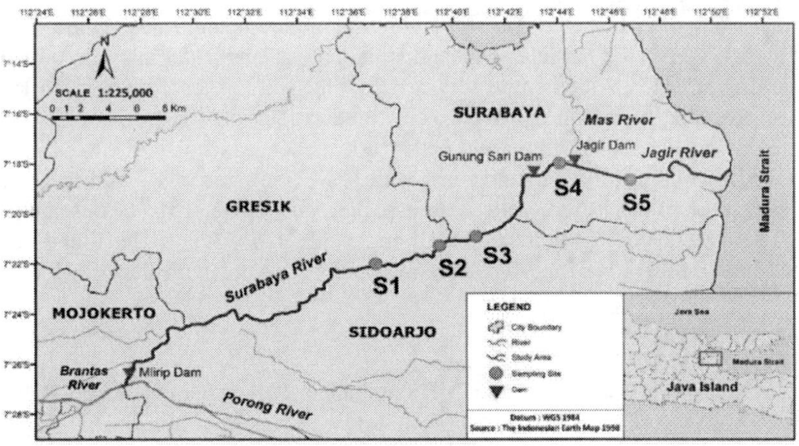


**Figure 1.** Map of sampling sites

The fishes of *Oreochromis niloticus* and *Barbonymus gonionotus,* and the bivalves of *E. orientalis* samples were collected by manual random sampling in each site. The *O. niloticus* was found in all sites, while *B. gonionotus* was only found at S1, S4, and S5. Meanwhile, the bivalves were only found at S1, S2, and S4. Only 15 individuals of *O. niloticus,* 6 individuals of *B. gonionotus,* and 16 individuals of *E. orientalis* were selected for their similarities in size. Each sample was preserved in an aluminum zip lock containing 70% alcohol and placed in storage box.

### *2.2 Sample preparation and extraction*

*2.2.1 Water samples.* The MP extraction from water samples in this study was done based on National Oceanic and Atmospheric Administration (NOAA) method [57]. First, the cod end of the trawl was rinsed by using distilled water. Then, it was wet filtered by two stratified stainless steel sieves with 5.6 mm and 0.3 mm pore sizes. Afterward, the samples were dried at 90°C for 24 h. The next step was wet peroxide oxidation (WPO). The WPO step aimed to remove organic material in the samples. In WPO, the samples were oxidized by adding 20 mL $H_2O_2$ 30% and 20 mL Fe(II) solutions, then heated until 75°C for 30 minutes. Hereafter, the samples were treated by density separation. About 6 g NaCl per 20 mL solution was added into the samples in this process to increase the solution density to 1.15 g/mL, so that the MP particles could be buoyant and easily separated [57]. After 24 h settling process in density separator, the solution was filtered by 0.3 mm filter. The particles resembling the MPs were manually separated from the filter into a clean petri dish under dissecting microscope (Sunshine SZM45T-B1).

*2.2.2 Biota samples.* First of all, characteristics of each individual biota sample were measured for obtaining similar size of the samples. Then, MP extraction steps for the biota samples followed by dissection, WPO, density separation, and visual sortation [57 – 59]. For fish samples, the dissection was done by separating the gill and the gastrointestinal tract [58]. Afterward, the wet weight of the gill and the gastrointestinal tract of each individual were measured [58]. Meanwhile, for the bivalve samples, the whole soft tissue (ST) of each individual bivalve was taken from the shell. Then, the wet weight of whole ST of each individual bivalves was measured and placed in a 500 mL beaker glass [59]. Hereafter, all dissection results were measured in wet weight (ww) [58].

Afterward, all the samples were treated with the same WPO step above using 200 mL $H_2O_2$ 30% for removing and digesting the organic material. After the WPO step, the biota samples were treated with the same density separation process for 24 h. Then, the solution was filtered by 0.3 mm filter until potential MP particles were obtained. These particles were separated from the filter, identified, and characterized under dissecting microscope.

### *2.3 Identification and characterization of the MPs*

Identification of MP particles can be done based on three fundamental rules: 1) the particles can not be ripped apart by tweezer, 2) the particles have a distinct colour, and 3) the particles are cleared from the bound organic structure [15]. All potential MP particles then were observed and identified visually under a dissecting microscope based on the shape, size, and colour [15, 60]. The MP shapes commonly observed as fragment, film, foam, fiber, and pellet [60]. Based on its size, the MPs can be divided into small (SMP) in sized of 1 μm - 1 mm and a large one (LMP) in sized of 1 – 5 mm. The MP colours can be categorized into blue, black, red, yellow, transparent, and white [15]. Representative MP particles were photographed by dissecting microscope (Olympus SZX16). All particles found in this study could be categorized and considered as the MPs according to the fundamental rules of MP identification [15].

## 3. Results and Discussion

### *3.1 MP abundance*

*3.1.1. MP abundance in the water.* Average MP abundance at all sites ranged from 9.66 ± 8.87 to 21.16 ± 19.35 particles/m$^3$ (Figure 2). The highest MP abundance was found in 21.16 ± 19.35 particles/m$^3$ at S4 (Joyoboyo), which located before the Surabaya River branches, namely the Jagir and the Mas Rivers. The average MP abundance at other sites (S1, S2, S3, S5) were obtained as 20.62 ± 12.32; 9.66 ± 8.87; 12.32 ± 7.90; and 16.03 ± 5.29 particles/m$^3$, respectively. This situation showed a fluctuating trend of MP abundance along the river (Figure 2). These results were in accordance to several previous studies, which stated that the MP abundance was spatially and temporally heterogeneous in the urban river [28, 61]. This condition might be affected by several causes, such as population density, domestic and industrial discharges, wastewater effluent, land use, hydrodynamic condition, water velocity, turbulence, drainage system, dams, and stream geomorphology [27, 28, 51, 62, 63]. The population density from the upper to the lower part of the Surabaya River increased up to 150,608 inhabitants/km$^2$ [64]. In addition, the water velocity condition might become a crucial factor that contributes to MP distribution in the river [42].

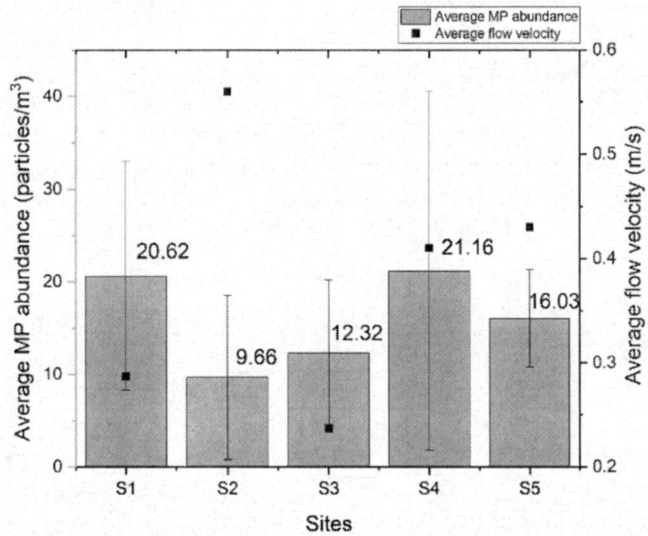


**Figure 2.** Average MP abundance in the Surabaya River

*3.1.2 MP abundance in the fishes.* All of the fish samples contained MP particles (Table 1). The highest MP abundance of *O. niloticus* was found at S3 (155.50 ± 61.96 particles/individual). Meanwhile, the highest MP abundance of *B. gonionotus* was found at S1 (155.00 ± 81.71 particles/individual). These results showed that the MP abundance in *O. niloticus* was higher than in *B. gonionotus* based on average MP particles/individual (Table 1). This condition might be affected by the different feeding behavior of each fish species [65]. Based on their feeding behavior, *O. niloticus* is categorized as omnivorous with carnivorous tendency [68], while *B. gonionotus* is categorized as herbivorous [66, 67]. This condition could explain the higher MP abundance in *O. niloticus* than in *B. gonionotus.*

**Table 1.** MP abundance in the fishes.

| Site | MP abundance in | | | | | |
|---|---|---|---|---|---|---|
| | *O. niloticus* | | | *B. gonionotus* | | |
| | MP particles/g ww | | Average MP particles/individual | MP particles/g ww | | Average MP particles/individual |
| | gill | gastrointestinal | | gill | gastrointestinal | |
| S1 | 53.13 ± 53.13 | 8.10 ± 8.10 | 123.25 ± 123.25 | 16.88 ± 16.88 | 10.29 ± 10.29 | 155.00 ± 155.00 |
| S2 | 303.59 ± 293.09 | 265.71 ± 133.34 | 138.83 ± 86.03 | ND | ND | ND |
| S3 | 43.78 ± 43.78 | 53.21 ± 53.21 | 155.50 ± 155.50 | ND | ND | ND |
| S4 | 21.42 ± 21.42 | 34.64 ± 34.64 | 105.25 ± 105.25 | 35.16 ± 35.16 | 7.95 ± 7.95 | 99.00 ± 99.00 |
| S5 | 8.05 ± 2.63 | 15.12 ± 11.01 | 108.50 ± 56.21 | 32.70 ± 9.83 | 17.32 ± 12.59 | 62.13 ± 20.33 |

*ND (no data)

The MP abundance was generally found higher in the gill than in the gastrointestinal tract of the fishes (Table 1). These results were in accordance to the previous research results [69, 70], in which this condition might be affected by fish interaction and MP pollution levels in their aquatic environments. Furthermore, investigation of the MPs in specific organs of the fishes is essential due to its potential translocation and effects within the body, and further risk of indirect transfer to human [31]. The MPs in the gill could occur as the result of the respiratory process of the fish [69]. Meanwhile, the MPs in the gastrointestinal could origin from ingestion in the food chain, in which large MPs might retain and accumulate in the gastrointestinal [31]. The average MP abundance per individual fish was determined by averaging the amount of the MPs in the gill and the gastrointestinal, so that it can represent the MP abundance of both organs per individual fish.

*3.1.3 MP abundance in the bivalves.* The MP abundance in *E. orientalis* varied from 20.25 ± 9.60 to 47.28 ± 8.65 particles/g ST, or from 36.00 ± 14.15 to 76.17 ± particles/individual (Table 2). The average MP abundance in *E. orientalis* was 31.33 ± 10.75 particles/g ST or 61.14 ± 21.91 particles/individual. The highest MP abundance of the bivalves was found at S2 (26.47 ± 17.91 particles/g ST or 71.25 ± 59.75 particles/individual). These results showed that MP particles have been found and contaminated the bivalves of *E. orientalis* in the Surabaya River. Furthermore, the bivalves are potential to become bioindicator for the MPs in aquatic environments due to their sedentary movement [59, 71, 72].

**Table 2.** MP abundance in the bivalves.

| Site | MP abundance | |
|---|---|---|
| | MP particles/g ST | MP particles/individual |
| S1 | 7.28 ± 8.65 | 76.17 ± 29.46 |
| S2 | 26.47 ± 17.91 | 71.25 ± 59.75 |
| S4 | 20.25 ± 9.60 | 36.00 ± 13.67 |
| Average | 31.33 ± 10.75 | 61.14 ± 21.91 |

*ND (no data)

*3.2 MP characteristics*

*3.2.1 MP characteristics in the water.* The MP shapes found at all sampling sites comprised film (69.17 – 88.50 %), fragment (3.63 – 19.57%), fiber (0.67 – 7.20%), foam (0.67 – 8.33%), and pellet (0.20 – 1.00%) (Figure 3a). The film particles were the most dominant found of MP shapes in this study. These results were consistent with research findings in the Qing and the Pearl Rivers, China, which found film as dominant MP particles [74, 75]. The MP shapes also might predict its potential sources. In this study, the film particles morphology has similarly resembled used plastic packaging ruins. Former studies at Singapore's coastal mangrove and in the Gulf of Mexico, Texas, estimated that the MP particles in film-shaped were originated from the breakdown of used plastic packaging and bags [76 - 78]. This condition might relate to the high demand for plastic packaging [3].

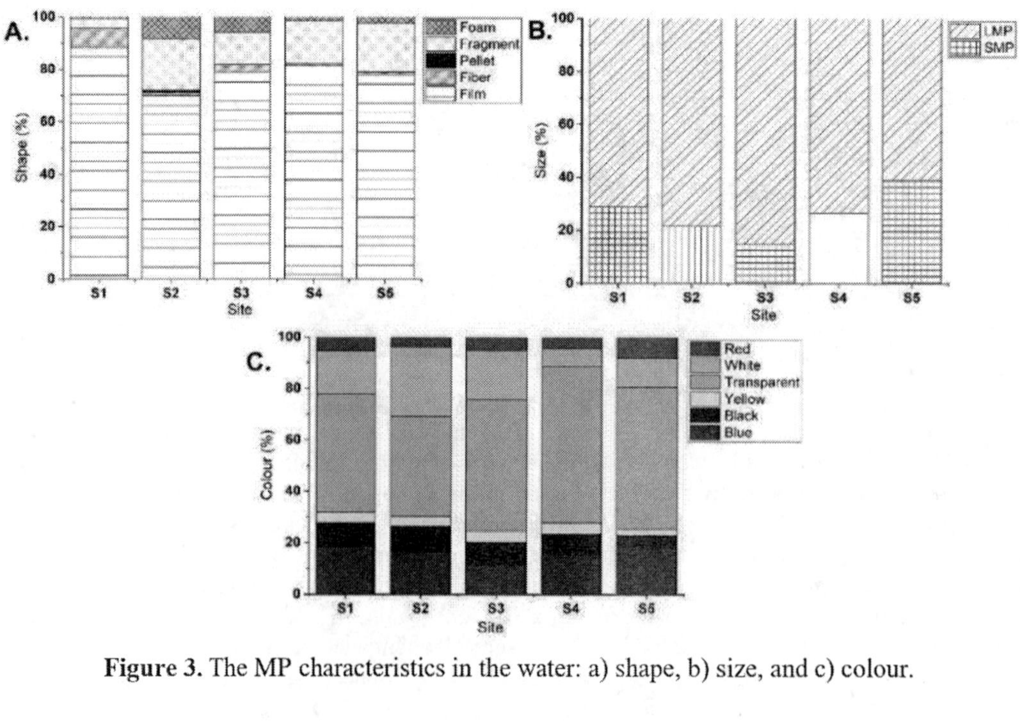


**Figure 3.** The MP characteristics in the water: a) shape, b) size, and c) colour.

Based on the size, the MP particles were dominantly found as LMP (61.07 – 78.63%). Only a small proportion of SMP was found (15.07 – 38.93%) (Figure 3b). The particle size could affect the vertical distribution of the MPs in the aquatic environment and potentially trigger the particles to be more susceptible ingested by the biota [79 - 81]. The transparent particles were the most abundant MP color in the water (Figure 3c). The MP colour comprised transparent (38.79 – 60.86%), white (7.02 – 27.16%), blue (11.05 – 17.95%), black (5.23 – 10.02%), red (3.75 – 8.18%), and yellow (2.31 – 4.34%) (Figure 3c). Besides, the colour, as plastic additives substances in the MPs could cause potential ecotoxicological risk to the biota as it could leached out to the environment [82].

*3.2.2 MP characteristics in the fishes.* Fibers were the most dominant MP shape found in both fish species. Fibers were found about 73.13 – 100% in the gill and 94.7 – 100% in the gastrointestinal of *O. niloticus* (Figure 4a - b). In *B. gonionotus,* fibers were also found sdominant in the gill (98.17 – 100%) and in the gastrointestinal (73.79 – 99.74%). These results were similar to the finding research from commercial fishes in Hangzhou Bay and Yangtze Eztuary, China, which found fibers as dominant MPs [81]. Other shapes of MP particles were discovered only in a small amount (Figure 4a - b). It might be caused by the fiber size, in which the smaller the particles, the more easily to be uptaken or ingested by the biota [81, 83].

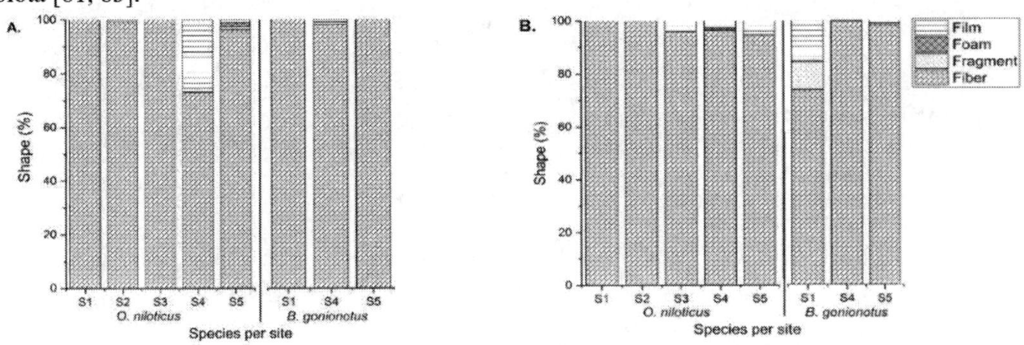


**Figure 4.** The MP shapes in: a) the gill and b) the gastrointestinal of the fishes.

Based on the size, the LMP was the most commonly found MP particles in the gill and the gastrointestinal of both fish species (Figure 5a - b). The proportion of the LMP was 64.81 – 76.87% in the gill and 59.73 – 82.73% in the gastrointestinal of *O. niloticus*. Meanwhile, the LMP proportion in *B. gonionotus* was found about 57.80 – 70.34% in the gill and 67.02 – 71.47% in the gastrointestinal. Only small amount of the SMP was found in the gill and the gastrointestinal of both fishes (Figure 5a – b). Furthermore, the transparent MPs were the most commonly discovered particles in the gill (29.94 – 59.34%) and the gastrointestinal (27.52 – 58.80%) of *O. niloticus* (Figure 6a - b). However, black MP particles were observed as the most dominant ones in the gill (28.52 – 45.19%) and the gastrointestinal (28.72 – 47.93%) of *B. gonionotus* (Figure 6a - b). Different feeding behavior of the fishes might influence this situation, in which *O. niloticus* is omnivorous, while *B. gonionotus* is herbivorous [65]. Representative MP particles in the fishes were photographed and presented in Figure 7.

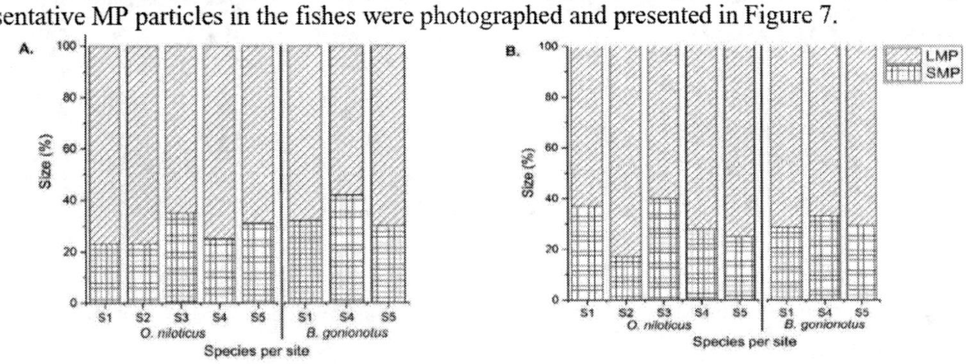

**Figure 5.** The MP size in: a) the gill and b) the gastrointestinal of the fishes.

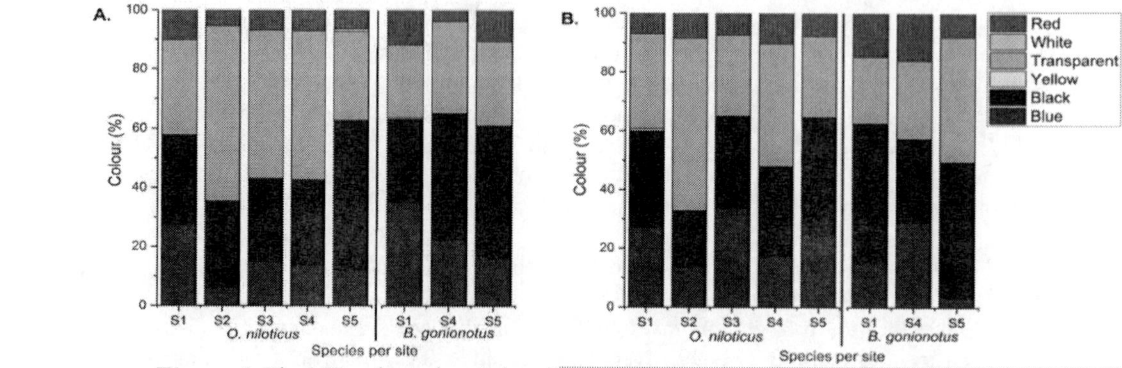


**Figure 6.** The MP colours in: a) the gill and b) the gastrointestinal of the fishes.

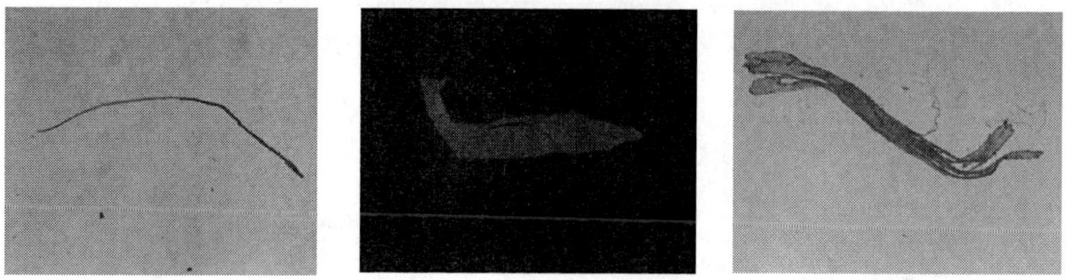

a) b) c)

**Figure 7.** The MP shapes in: a) black fiber, b) transparent film, c) blue film particles
(Figure 7b was affected by microscope light)

*3.2.3 MP characteristics in the bivalves.* The MPs in the bivalves were mostly discovered as black fibers in the sized of LMP (Figure 8a - c). Fiber was found about 98.54 – 99.31% of the MPs in the bivalves at all sites, and only small amount of film particles (0.69 – 1.46%) (Figure 8a). These particles were categorized as the LMP (56.53 – 78.61%) and the SMP (16.04 – 43.37%) (Figure 8b). The MP colour in the bivalves were dominated by black particles (Figure 8c). This condition were similar to the finding results in commercial bivalves from Qingdao, China [84]. The deposition process and settling velocity of the fibers made the particles more easily to be settled than other shape particles in nature [85].

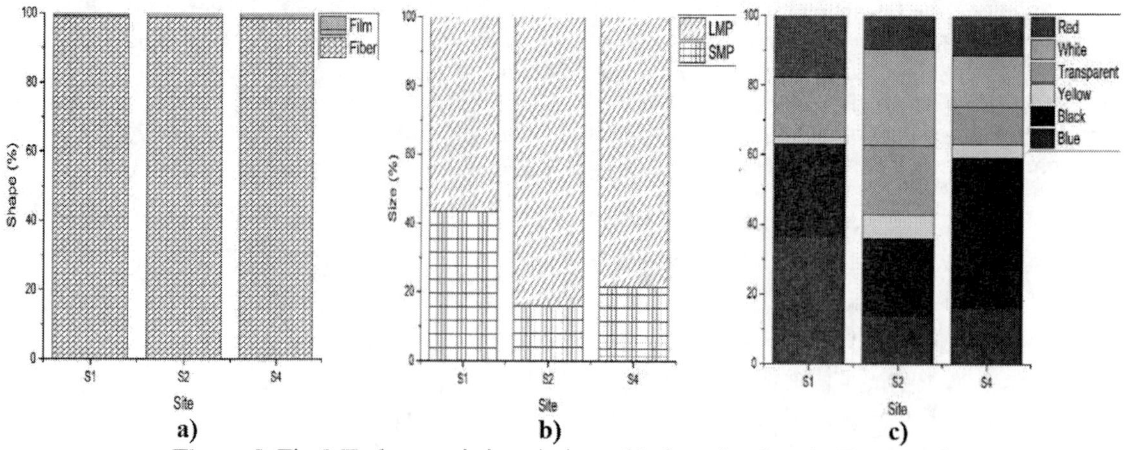


a) b) c)

**Figure 8.** The MP characteristics: a) shape, b) size, c) colour in *E. orientalis*.

The MP particles were found in the water and the aquatic biota of the Surabaya River (Figure 2, Table 1 - Table 2). These results showed that MP had polluted not only the water, but also contained in the fishes and the benthic bivalves of the Surabaya River. The highest MP abundance in the water was found at S4 (Figure 2). Meanwhile, the highest MP abundance of *O. niloticus* and *B. gonionotus* was found at S1 and S3, respectively (Table 1). Besides, the highest MP abundance in the bivalves was found at S2 (Table 2). These results indicated that the relationship of MP distribution in the water and the biota has remained unclear [73]. This situation could be affected by several environmental conditions, even MP characteristics itself [19 – 21]. The dominant characteristics of MP particles in the water and the biota were dissimilar (Table 3). In the water, the MPs were dominated by transparent film particles. Besides, the MP particles in the gill and the gastrointestinal of *O. niloticus* were transparent fibers. Meanwhile, black fibers were dominant found in *B. gonionotus* and *E. orientalis*. The only similarity of MP characteristics in the water and the biota was particle sized of LMP. These results have confirmed that MP distribution in the environment can be affected by their characteristics itself, which is also supported by other environmental conditions [20, 21].

**Table 3.** Comparison of dominant MP characteristics in the water and the biota

| Sample | Dominant MP | | |
|---|---|---|---|
| | Shape | Size | Colour |
| Water | film | LMP | transparent |
| *O. niloticus* (gill) | fiber | LMP | transparent |
| *O. niloticus* (gastrointestinal) | fiber | LMP | transparent |
| *B. gonionotus* (gill) | fiber | LMP | black |
| *B. gonionotus* (gastrointestinal) | fiber | LMP | black |
| *E. orientalis* | fiber | LMP | black |

## 4. Conclusion

Microplatics have polluted the Surabaya River water and the aquatic biota. The MP pollution in the water showed a fluctuating trend along the river. The highest MP abundance in the river water was found at Joyoboyo (S4). Meanwhile, the highest MP content in the fishes was found at Driyorejo (S1) and Karang Pilang (S3). Besides, the highest MP content in the bivalves was found at Bambe (S2). The MP abundance was also found higher in the gill than in the gastrointestinal of the fishes. Based on their characteristics, the MPs were mostly discovered as transparent film particles in the water, while transparent and black fibers were dominant in the biota. The film particles in the biota were only found in small amount. Polymer identification of MP particles is needed for plastic type identification.

## Acknowledgement

The authors gratefully acknowledged the Kurita Water and Environment Foundation (KWEF) who funded this study through Kurita Overseas Research Grants [Grant Number 19Pid041]. The authors are also grateful for the financial support of the Ministry of Education and Culture of the Republic of Indonesia through the Master toward Doctoral Education Program for Excellent Graduate (PMDSU) research grant [Contract Number 1273/PKS/ITS/2020]. The authors also thank Perum Jasa Tirta I for the research permit.

## References

[1] Li WC, Tse HF, Fok L. 2016a. Plastic waste in the marine environment: A review of sources, occurrence and effects. *Sci Total Environ.* **566** 333-49.

[2] Leslie HA, Van der Meulen MD, Kleissen FM, Vethaak AD. 2011. *Microplastic litter in the Dutch marine environment: Providing facts and analysis for Dutch policymakers concerned with marine microplastic litter*. Deltares: Amsterdam, The Netherlands.

[3] Thompson RC, Moore CJ, Vom Saal FS, Swan SH. 2009. Plastics, the environment and human health: current consensus and future trends. *Philos. Trans. R. Soc. B* **364** (1526) 2153-66.
[4] Lusher A, Hollman P, Mendoza-Hill J. 2017. *Microplastics in fisheries and aquaculture: status of knowledge on their occurrence and implications for aquatic organisms and food safety*. FAO.
[5] Bellasi A, Binda G, Pozzi A, Galafassi S, Volta P, Bettinetti R. 2020. Microplastic contamination in freshwater environments: a review, focusing on interactions with sediments and benthic organisms. *Environments* **7**(4) 30.
[6] Thushari GG, Senevirathna JD. 2020. Plastic pollution in the marine environment. *Heliyon* **6**(8).
[7] Dobaradaran S, Schmidt TC, Nabipour I, Khajeahmadi N, Tajbakhsh S, Saeedi R, Mohammadi MJ, Keshtkar M, Khorsand M, Ghasemi FF. 2018. Characterization of plastic debris and association of metals with microplastics in coastline sediment along the Persian Gulf. *Waste Manage.* **78** 649-58.
[8] Lestari P, Trihadiningrum Y. 2019. The impact of improper solid waste management to plastic pollution in Indonesian coast and marine environment. *Mar. Poll.Bull.* **149** 110505.
[9] Saleem J, Riaz MA, Gordon M. 2018. Oil sorbents from plastic wastes and polymers: A review. *J. Hazard. Mater.* **341** 424-37.
[10] Andrady AL. 2011. Microplastics in the marine environment. *Mar. Pollut. Bull.*. **62**(8) 1596-605.
[11] Law KL. 2017. Plastics in the marine environment. *Annu. Rev. Mar Sci.* **9** 205-29.
[12] Lambert *et al.* 2018. Microplastics are contaminants of emerging concern in freshwater environments: an overview. In *Freshwater microplastics.* Springer, Cham.
[13] Klein *et al.*. 2018. Analysis, occurrence, and degradation of microplastics in the aqueous environment. In *Freshwater microplastics* Springer, Cham.
[14] Lippiatt *et al.* 2013. *Marine debris monitoring and assessment: recommendations for monitoring debris trends in the marine environment.* National Oceanic and Athmospheric Administration.
[15] Peng *et al.* 2017. Microplastics in sediments of the Changjiang Estuary, China. *Environ. Pollut.* **225** 283-90.
[16] Horton *et al.* 2017. Large microplastic particles in sediments of tributaries of the River Thames, UK – abundance, sources and methods for effective quantification. *Mar. Pollut. Bull.* **114**(1) 218-26.
[17] Duis *et al.* 2016. Microplastics in the aquatic and terrestrial environment: sources (with a specific focus on personal care products), fate and effects. *Environ. Sci. Eur.* **28**(1).
[18] Kooi *et al.* 2018. Modeling the fate and transport of plastic debris in freshwaters: review and guidance. In *Freshwater microplastics*. Springer, Cham.
[19] De Sá *et al.* 2018. Studies of the effects of microplastics on aquatic organisms: what do we know and where should we focus our efforts in the future? *Sci. Total Environ.* **645** 1029-39.
[20] Anderson *et al.* 2016. Microplastics in aquatic environments: implications for Canadian ecosystems. *Environ. Pollut.* **218** 269-80.
[21] Cole *et al.* 2011. Microplastics as contaminants in the marine environment: a review. *Mar. Pollut. Bull.* **62**(12) 2588-97.
[22] Lusher *et al.* 2016. Microplastic interactions with North Atlantic mesopelagic fish. *ICES J.Mar. Sci.* **73**(4) 1214-25.
[23] Van Cauwenberghe *et al.* 2015. Microplastics in sediments: a review of techniques, occurrence and effects. *Mar. Environ. Res.* **111** 115-17.
[24] Lahens *et al.* 2018. Macroplastic and microplastic contamination assessment of a tropical river (Saigon River, Vietnam) transversed by a developing megacity. *Environ. Pollut.* **236** 661-71.
[25] Lebreton *et al.* 2017. River plastic emissions to the world's oceans. *Nat. Commun.* **8** 15611.
[26] Ding *et al.* 2019. Microplastics in surface waters and sediments of the Wei River, in the northwest of China. *Sci. Total Environ.* **667** 427-34.
[27] Mani *et al.* 2015. Microplastics profile along the Rhine River. *Sci. Rep.* **5**(1) 1-7.
[28] McCormick *et al.* 2016. Microplastic in surface waters of urban rivers: concentration, sources, and associated bacterial assemblages. *Ecosphere* **7**(11).

[29] Zhang *et al.* 2017. Microplastic pollution in the surface waters of the Bohai Sea, China. *Environ. Pollut.* **231** 541-8.
[30] Digka *et al.* 2018. Microplastics in mussels and fish from the Northern Ionian Sea. *Mar. Pollut. Bull.* **135** 30-40.
[31] Jabeen *et al.* 2017. Microplastics and mesoplastics in fish from coastal and fresh waters of China. *Environ. Pollut.* **221** 141-9.
[32] Menéndez-Pedriza *et al.* 2020. Interaction of environmental pollutants with microplastics: a critical review of sorption factors, bioaccumulation and ecotoxicological effects. *Toxics* **8**(2) 40.
[33] Bakir *et al.* 2012. Competitive sorption of persistent organic pollutants onto microplastics in the marine environment. *Mar. Pollut. Bull.* **64**(12) 2782-9.
[34] Kershaw *et al.* 2019. Guidelines for the Monitoring and Assessment of Plastic Litter in the Ocean. GESAMP reports and studies.
[35] European Food Safety Authority (EFSA) Contam Panel 2016. Statement of the presence of microplastics and nanoplastics in food, with particular focus on seafood. *EFSA Journal* **14**(6) 4501.
[36] Barboza *et al.* 2018. Marine microplastic debris: An emerging issue for food security, food safety and human health. *Mar. Pollut. Bull.* **133** 336-48.
[37] Koelmans *et al.* 2019. Microplastics in freshwaters and drinking water: critical review and assessment of data quality. *Water R.* **155** 410-22.
[38] Wang *et al.* 2019. The ecotoxicological effects of microplastics on aquatic food web, from primary producer to human: a review. *Ecotoxicol. Environ. Saf.* **173** 110-7.
[39] De-la-Torre GE. 2020. Microplastics: an emerging threat to food security and human health. *J. Food Sci. Technol.* **57**(5) 1601-8.
[40] Neves *et al.* 2015. Ingestion of microplastics by commercial fish off the Portuguese coast. *Mar. Pollut. Bull.* **101**(1) 119-26.
[41] Lestari *et al.* 2020. Distribution of microplastics in Surabaya River, Indonesia. *Sci. Total Environ.*138560.
[42] Van Emmerik *et al.* 2019. Seasonality of riverine macroplastic transport. *Sci. Rep.* **9**(1) 1-9.
[43] Mariyanto *et al.* 2019. Heavy metal contents and magnetic properties of surface sediments in volcanic and tropical environment from Brantas River, Jawa Timur Province, Indonesia. *Sci. Total Environ.* **675** 632-41.
[44] Regulation of the Minister of Public Works No 268/2010. Management of the Brantas River Resources. Ministry of Public Works of the Republic of Indonesia (In Indonesian).
[45] Hayati *et al.* 2017. Water quality and fish diversity in the Brantas River, East Java, Indonesia. *J. Biol. Res.* **22**(2) 43-9.
[46] Sornlertlamvanich *et al.* 2018. *Information Modelling and Knowledge Bases* XXIX. IOS Press.
[47] Suwari *et al.* 2011. Dynamic model of water pollution control in Surabaya River. *Jurnal Bumi Lestari* **11**(2) 234-248. (In Indonesian).
[48] Novitasari A.K. 2015. *Analysis of identification and inventory of pollutant sources in Surabaya River.* Thesis. Institut Teknologi Sepuluh Nopember. (In Indonesian).
[49] Firdaus *et al.* 2020. Microplastic pollution in the sediment of Jagir Estuary, Surabaya City, Indonesia. *Mar. Pollut. Bull.* **150** 110790.
[50] Jara-Marini *et al.* 2013. Comparative bioaccumulation of trace metals using six filter feeder organisms in a coastal lagoon ecosystem (of the central-east Gulf of California). *Environ. Monit. Asses.* **185**(2) 1071-85
[51] Kataoka *et al.* 2015. Assessment of the sources and inflow processes of microplastic in the river environments of Japan. *Environ. Pollut.* **244** 958-65.
[52] Silva *et al.* 2018. Microplastics in the environment: challenges in analytical chemistry: a review. *Anal. Chim. Acta* **1017** 1 – 19.
[53] Eriksen *et al.* 2013. Microplastic pollution in the surface waters of the Laurentian Great Lakes. *Mar. Pollut. Bull.* **77** 177-82.

[54] Lenaker *et al.* 2019. Vertical distribution of microplastics in the water column and surficial sediment from the Milwaukee River Basin to Lake Michigan. *Environ. Sci.Technol* **53**(21) 12227-37.
[55] Michida *et al.* 2019. *Guidelines for harmonizing ocean surface microplastic monitoring methods version 1.1.* Ministry of the Environment Japan.
[56] Schmidt *et al.* 2018. Occurrence of microplastics in surface waters of the Gulf of Lion (NW Mediterranean Sea). *Prog. Oceanogr.* **163** 214-20.
[57] Masura *et al.* 2015. *Laboratory methods for the analysis of microplastics in the marine environment: recommendations for quantifying synthetic particles in waters and sediments.* National Oceanic and Atmospheric Administration Technical Memorandum NOS-OR&R-48.
[58] Barboza *et al.* 2020. Microplastics in wild fish from North East Atlantic Ocean and its potential for causing neurotoxic effects, lipid oxidative damage, and human health risks associated with ingestion exposure. *Sci. Total Environ.* **717** 134625.
[59] Li *et al.* 2016b. Microplastics in mussels along the coastal waters of China. *Environ. Pollut.* **214** 177-184.
[60] Free *et al.* 2014. High-levels of microplastic pollution in a large, remote, mountain lake. *Mar. Pollut. Bull.* **85**(1) 156-63.
[61] De Troyer, N., 2015. *Occurrence and distribution of microplastics in the Scheldt River*. Thesis. Ghent University, Ghent.
[62] Van Emmerik *et al.* 2019. Riverine plastic emission from Jakarta into the ocean. *Enviro. Res. Lett.* **14**(8).
[63] Watkins *et al.* 2019. The effect of dams on river transport of microplastic pollution. *Sci. Total Environ.* **664** 834-40.
[64] BPS of Surabaya City. 2018. *Jambangan district in Figure*. BPS Surabaya City (In Indonesian).
[65] Hastuti *et al.* 2019. The presence of microplastics in the digestive tract of commercial fishes of Pantai Indah Kapuk coast, Jakarta, Indonesia. *Biodiversitas* **20**(5) 1233 – 42.
[66] Nugroho, A.A.K. 2013. *Study of feeding and ecological niche of Mystus planiceps, Hemibragus nemurus (Bagridae family), Pangasius micronemus (Pangasidae family) in Surabaya River*. Thesis. University of Airlangga. (In Indonesian).
[67] Buwono *et al.* 2019. Analysis of the stomach contents of Tawes fish (*Barbonymus gonionotus)* in the lower part of the Bengawan Solo, Lamongan Regency. *JAFH* **8**(1) 8-14. (In Indonesian).
[68] Tresna, L.K., Dhahiyat, Y., Herawati, T. 2012. Feeding behavior and fish niches area in the upstream of Cimanuk River, Garut Regency, West Java Province. *Jurnal Perikanan dan Kelautan* **3**(3) 163 – 73. (In Indonesian).
[69] Yona *et al.* 2020. Microplastic analysis in gills and gastrointestinal of coral fish in three outside islands, Papua, Indonesia: preliminary study. *Jurnal ITKT* **12**(2) 497-507. (In Indonesian).
[70] Abbasi *et al.* 2018. Microplastics in different tissues of fish and prawn from the Musa Estuary, Persian Gulf. *Chemosphere* **205** 80-87.
[71] Qiu *et al.* 2015. Occurrence of microplastics in the coastal marine environment: first observation on sediment of China. *Mar. Pollut. Bull.* **98** 274 – 280.
[72] Su *et al.* 2018. Using the Asian Clam as an indicator of microplastic pollution in freshwater ecosystem. *Environ. Pollut.* **234** 347 – 55.
[73] Scherer *et al.* 2018. Interactions of microplastics with freshwater biota. In *Freshwater microplastics* 153-80. Springer, Cham.
[74] Wang *et al.* 2019. The ecotoxicological effects of microplastics on aquatic food web, from primary producer to human: a review. *Ecotoxicol. Environ. Saf.* **173** 110-17.
[75] Yan *et al.* 2019. Microplastic abundance, distribution and composition in the Pearl River along Guangzhou city and Pearl River estuary, China. *Chemosphere* **217** 879-86.
[76] Nor and Obbard. 2014. Microplastics in Singapore's coastal mangrove ecosystems. *Mar. Pollut. Bull.* **79**(1-2) 278-83.

[77] Browne *et al.* 2010a. Ingested microscopic plastic translocates to the circulatory system of the mussel, *Mytilus edulis*. *Environ. Sci. Technol.* **42** 5026–31.
[78] Browne *et al.* 2010b. Spatial patterns of plastic debris along estuarine shorelines. *Environ. Sci. Technol.* **44** 3404–09.
[79] Schwarz *et al.* 2019. Sources, transport and accumulation of different types of plastic litter in aquatic environments: a review study. *Mar. Pollut. Bull.* **143** 92-100.
[80] Cole *et al.* 2013. Microplastic ingestion by zooplankton. *Environ. Sci. Technol.* **47**(12) 6646-55.
[81] Su *et al.* 2019. Occurrence of microplastic in specific organs in commercially caught fishes from coast and estuary area of east China. *J. Hazard. Mater.* **365** 716-24.
[82] Gewert *et al.* 2015. Pathways for degradation of plastic polymers floating in the marine environment. *Environmental Sci. Processes Impacts* **17** 1513-1521.
[83] Walkinshaw *et al.* 2020. Microplastics and seafood: lower trophic organisms at highest risk of contamination. *Ecotoxicol. Environ. Saf.* **190** 110066.
[84] Jin-Feng *et al.* 2018. Separation and identification of microplastics in digestive system of bivalves. *Chinese J. Anal. Chem.* **46**(5) 690-7.
[85] Kane and Clare. 2019. Dispersion, accumulation, and the ultimate fate of microplastics in deep-marine environments: A review and future directions. *Front. Earth. Sci.* **7** 80.

# Microwave-Assisted Synthesis of $TiO_2$/GO Composite and Its Adsorption-Photocatalysis Property under Visible Light

**S Setiawan[1], A Hardiansyah[2], C W Kartikowati[3], A F Arif[4], S Priatmoko[1], and O Arutanti[5*]**

[1]Department of Chemistry, Universitas Negeri Semarang, Semarang, Indonesia
[2]Research Center for Physics, Indonesian Institute of Sciences, Kawasan Puspiptek, Serpong, Indonesia
[3]Departement of Chemical Engineering,Universitas Brawijaya, Malang, Indonesia
[4]Department of New Investment PT. Rekayasa Industri, Jakarta, Indonesia
[5]Research Center for Chemistry, Indonesian Institute of Sciences, Kawasan Puspiptek, Serpong, Indonesia

*osiarutanti@yahoo.com

**Abstract**. The photocatalytic performance of low-grade Titanium Dioxide ($TiO_2$) anatase was successfully improved by compositing with graphene oxide (GO) via microwave-assisted, activated by UV and Xe light. The various experimental conditions were conducted. X-ray diffraction (XRD), Particle size analyzer (PSA), Prepared $TiO_2$/GO photocatalyst was characterized by several instruments such as Specific surface area (BET) and Scanning electron microscopy (SEM). Effect of various annealing times (30 to 120 minutes) and temperatures (200-1000°C) were studied to reach the optimum pure $TiO_2$ as a further based material in composite with GO. The result showed that both annealing time and temperature influenced crystallinity and particle size only, not the phase structure of $TiO_2$. A similar phenomenon was founded by adding GO in a small concentration. A comparison of photocatalytic activity under UV and Xe irradiation has been proposed to show the GO effect independently. Under UV irradiation, the presence of GO enhancing the active sites of the catalyst. 1.0 wt% of GO succeeds in decomposing 8ppm of RhB more than 90%. Meanwhile, under Xe irradiation, the efficiency of the degradation of RhB was only 30%, resulted in using 1.2wt% of GO. Interestingly, the photodecomposition rate ($k$) of $TiO_2$/GO composite under Xe irradiation was 3 times over pure $TiO_2$ higher than that of UV irradiation (1 time). The study present that the small concentration of GO affected not only the active site of the $TiO_2$/GO composite but also the electronic properties, which give different effects under different energy activated.
**Keyword**: Photoadsorption, Rhodamine B, Organic dye, Water treatment

## 1. Introduction

Rhodamine B is a commercial azo dye for textile industries with high sunlight energy to decompose[1]. Since their toxicity, the high effort regarding the degradation of this compound from the environment is proposed. Unfortunately, the conventional method to remove Rhodamine B from water, such as precipitation, separation, and adsorption, is inefficient [2]. Almost the technology only converts the non-biodegradable dye into sludge, resulting in further post treatment[3]. Furthermore, green technology to solve this problem is needed.

For decades, photocatalytic technology attracts environmental remediation applications research. Semiconductor material as a photocatalyst material has been known effectively to decompose organic pollutants [4]. One of the common photocatalyst material is Titanium dioxide ($TiO_2$). As well as their advantages, their disadvantages, such as fast recombination of the electron (e-) and hole (h+) during the process and broader bandgap energy making their only active under UV irradiation, are the problems[5]. Accordingly, many research works have been proposed. Composite with another semiconductor, metal-doped, and inorganic doped are the several techniques to answer the disadvantageous[5,6].

Among the co-catalyst material, Graphene Oxide (GO) is a promising material that can be considered. GO known as an oxide of grapheme which has excellent adsorption properties[7]. Their reactive oxygen functional groups make this material suitable for forming Van der Waals force with $TiO_2$[8,9]. Moreover, GO could enhance the adsorption process due to their high specific surface area, while, $TiO_2$-GO bonding resulted in decreasing bandgap energy and electron-hole recombination[9]. Composite of $TiO_2$ and GO for decompose organic pollutants has been proposed by several researchers [10][11][12][13]. Kumaran et al. proposed the synthesis of nanocomposite GO-$TiO_2$ for dye wastewater treatment [14]. The studies showed that GO succeeded in the improvement of photoadsorption organic dyes. Unfortunately, almost the reports used $TiO_2$ P25, which has been known as a highly photocatalytic activity. Some of them synthesize $TiO_2$ from the $TiO_2$ salt. Importantly, they only used UV or visible light. Based on those of literature studies, GO's effect on the photocatalytic activity of low-grade $TiO_2$and their comparison study under different energy sources to deeply understanding is hard to find.

Based on that rationale, the purpose of this study was to prepare the $TiO_2$/GO composite by microwave-assisted. Here, low-grade $TiO_2$ particles have been used. The low price of the material making this research feasible to be developed on a large scale. Photocatalytic activity of the prepared composite $TiO_2$/GO was studied under UV and Xe light irradiation independently. An interesting result was founded when the GO present affected the catalyst's active site dominantly under UV light and affected the electronic properties dominantly under Xe light irradiation. It is shown by different efficiency and photodecomposition rate results between UV and Xe irradiation.

## 2. Experimental

### *2.1. Material*

Low-grade Titanium Dioxide anatase ($TiO_2$, Bratachem) powder has been used as a $TiO_2$ source. The materials used to prepare Graphene Oxide (GO) in this study were as follows: Graphite powder (<20 nm, Aldrich), potassium permanganate ($KMnO_2$, Baker), fuming nitric acid ($HNO_3$ ≥99.5%, sigma Aldrich), hydrogen peroxide ($H_2O_2$ (aq) 35%, Acros), Nitric acid (HNO(aq), 69%, Panreac), and hydrochloric acid (HCl(aq) 37%), scharlau). Rhodamine B (RhB; Wako pure chemical industries Ltd, Japan) was used as a waste model organic dye.

### *2.2.* Preparation of $TiO_2$/GO composite

Graphene oxide has been prepared by a modified Hummers method described in detail in our previous work [15]. The annealing process has optimized $TiO_2$. The annealing condition, such as various temperatures (from 200 to 900°C) and various times (from 30 to 120 minutes), was conducted. The optimum generated annealed $TiO_2$ was further composited with prepared GO by microwave-assisted synthesis at 300 watts for 10 minutes. A small concentration of GO from 0.6 to 1.2 wt% was directed.

### *2.3.* Characterizations

Phase analysis of the prepared particles has been done by X-ray diffraction (XRD, PANALITICAL, with Cu K$\alpha$ radiation; angular domain 20-80° ($2\theta$)), where the crystallite size was measured by Scherrer's equation[16]. The morphology and surface area were analyzed by Scanning Electron Microscopy (SEM, JEOL JIB 4610F), Tunneling Electron Microscopy (TEM, Tecnai G2 20 S-TWIN), Particle Size Analyser (PSA), and Brunauer-Emmett-Teller (BET, respectively.

To investigate the photocatalytic performance, 8 ppm Rhodamine B dissolved in 250 ml for one minute. Further, the wastewater model was mixed with the prepared photocatalyst (1.6 g/L). The solution was left in the dark for 30 minutes under continuous agitation to established equilibrium (adsorption/desorption) between RhB and the prepared photocatalysts and continuously irradiated by UV-A (365 nm) for 90 minutes. Xenon lamp was used as a visible light source with the wavelength peak at 450 and 540 nm as depicted in Figure 1. Further, photodegradation of RhB was monitored for 90 minutes. The degradation of RhB was measured by UV-Vis spectrophotometer (UV-3150, Shimadzu, Japan).

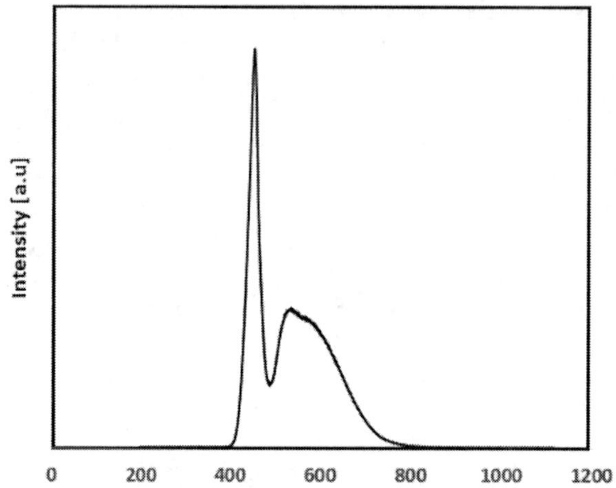


**Figure 1.** Wavelength characteristic of Xenon lamp.

## 3. Results and Discussion

### *3.1 Effect of annealing time*

Figure 2 depicted the physical properties of annealed-$TiO_2$ at 600°C for 30 to 120 minutes. The XRD patterns all prepared annealed $TiO_2$ has a similar peak with raw $TiO_2$. The peaks generated at 25.27° [101] and 37.78° [004] represents the anatase phase, confirmed that annealing time did not influence the crystal structure. The annealing times were confirmed affected the crystallite and particle sizes, as shown in Figure 2b. Raw $TiO_2$, as a reference, has the smallest at 34.39 nm for crystallite size and 800 nm in particle size. By increasing the annealing time from 30 to 120 minutes, the crystallinity increased from 45.34 up to 47.78 nm. Unlike crystallite size, the particle size of $TiO_2$ at 30 minutes shows the biggest size at 1054 nm. The further increase of annealing time resulted in a decrease of particle size down to 777 nm.

The photodegradation of RhB under UV irradiation over the times is shown in Figure 3. Annealing temperature from 30 to 90 minutes resulted in the remaining of RhB was at 30%. The highest photocatalytic activity was shown by using annealed $TiO_2$ at 120 minutes, with the remaining RhB was almost 20%. The photodegradation activity of RhB was significantly influenced by the size of crystallinity and particle of the prepared annealed $TiO_2$. The smallest particle size and the biggest crystallite size enhanced the photocatalytic performance. A similar phenomenon was also proposed in detail by Arutanti et al. [17]. Based on this condition, 120 minutes was the optimum annealing time.

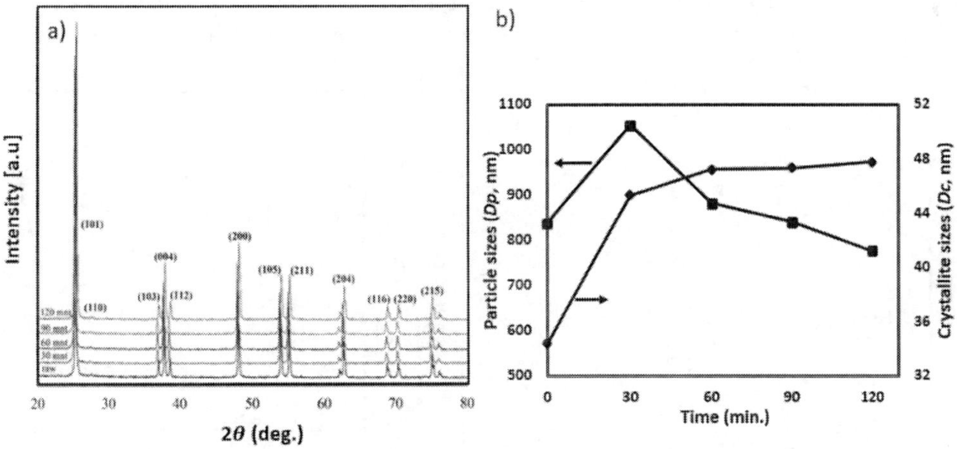

**Figure 2.** a) XRD patterns of the annealed low-grade $TiO_2$ for 600°C at different times, and b) correlation the particle and crystallite size of the prepared particles.

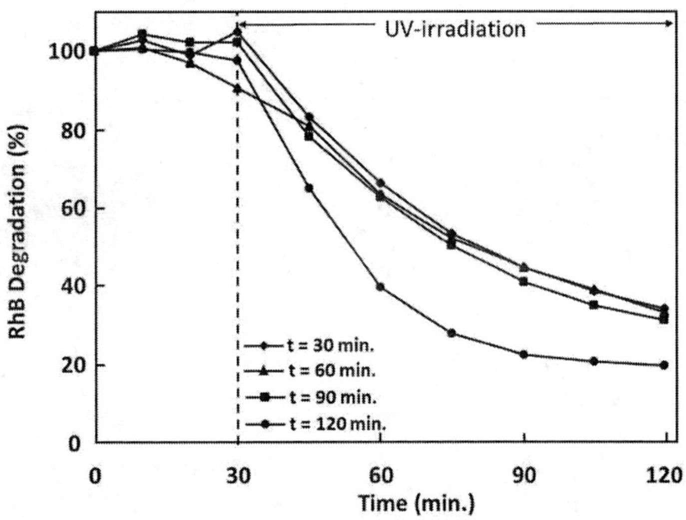

**Fig. 3.** Effect of annealing time on the photodegradation of 8 ppm RhB under UV-irradiation.

*3.2 Effect of annealing temperature*

The XRD peaks of the annealed $TiO_2$ from 200 to 1000°C are shown in Figure 4a. The results showed that various annealing temperatures did not affect the phase structure of $TiO_2$, except the peak intensity. All prepared $TiO_2$ peaks refered to anatase structure. Correlation between particle and crystallite size as the effect of annealing temperature is shown in Figure 4b. Due to the sintering effect, particle size increased by increasing the annealing temperature. Over 600°C, generated $TiO_2$ over 1000 nm.

The photodegradation of RhB under UV irradiation continuously decreased over time by adding annealed $TiO_2$ (Figure 5a). Annealing temperature influencing the increase of photodegradation efficiency higher than that of raw $TiO_2$. The remaining RhB decreased from 45 to 20% by increasing the annealing temperature from 200 to 1000°C. Photocatalytic activity of the annealed $TiO_2$ from 400 to 1000°C showing a similar efficiency. Even though the crystallinity of annealed $TiO_2$ increased proportionally with annealing temperature, the particle size also increased. The bigger particle size quite possibly affected decreasing the specific surface area resulted in the lower photocatalytic efficiency due to the surface-active site. This correlation phenomenon has been proven in our previous study [17].

IOP Conf. Series: Materials Science and Engineering 1143 (2021) 012055 doi:10.1088/1757-899X/1143/1/012055

Figure 5b and 5c show the morphology of raw $TiO_2$ and annealed $TiO_2$ at 600°C. The SEM images depicted that morphology change was hardly founded. This result inline with Figure 4b. Under different annealing temperatures, the optimum condition to prepare the best $TiO_2$ was 600°C for 120 minutes.

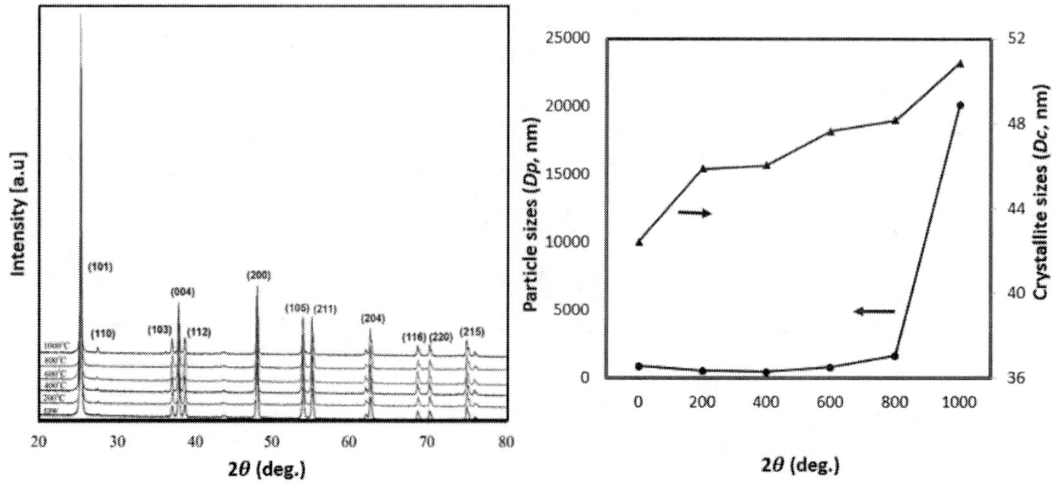


**Figure 4.** a) Effect of anneling temperature on the XRD patterns of the prepared $TiO_2$, and b) correation between crystallite and particle size.

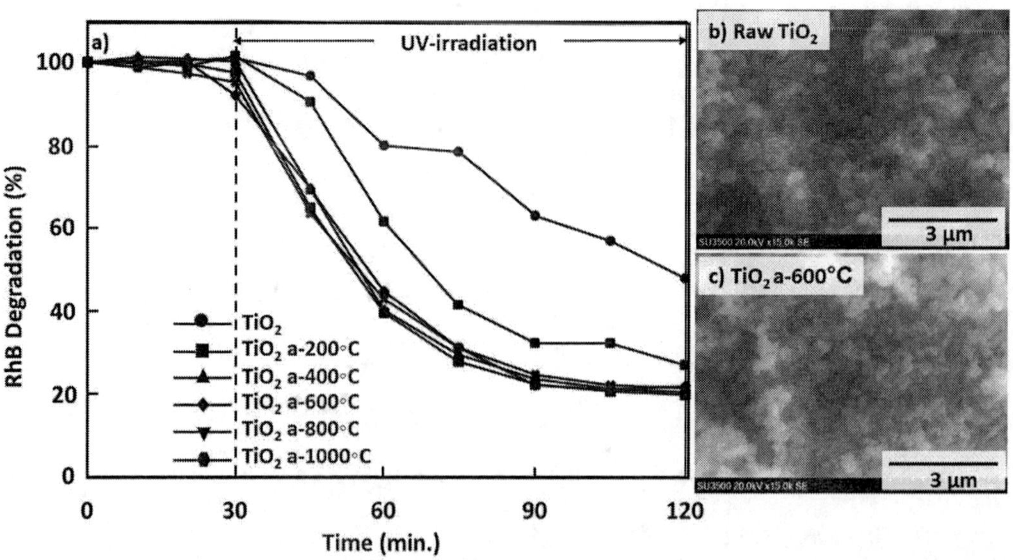


**Figure 5.** Effect of annealing temperature on the photodegradation of 8 ppm RhB under UV-irradiation with the SEM images of the morphology of low-grade $TiO_2$ b) before and c) after annealing process at 600°C and 120 minutes.

*3.3 Composite $TiO_2$/GO*

XRD patterns of the prepared $TiO_2$/GO composite under various GO ratio from 0.6 to 1.2 wt% was performed. The peak results obtained for the composite are presented in Figure 6. All the diffraction peaks observed correspond to the characteristics $TiO_2$ anatase. The expected GO peak (black line) was not found, probably due to their low concentration.

Figure 7 shows the Nitrogen adsorption-desorption isotherm of raw $TiO_2$ and composite $TiO_2$/GO. Porosity was also identified and presented on the table insert the figure. The result shows that the $S_{BET}$

of composite $TiO_2$/GO was lower than raw $TiO_2$, while all prepared composite $TiO_2$/GO has a similar surface area. Based on BJH analysis, volume pores of raw $TiO_2$ was higher, possibly due to the cavity between aggregate $TiO_2$. Composite of $TiO_2$ and small concentration GO block the porosity of composite particles.

The investigation of the photocatalytic activity of composite $TiO_2$/GO through the degradation of RhB under UV and Xe irradiation is plotted in Figure 8a and Figure 8b, respectively. A systematic study was performed on the adsorption process, confirming the prepared photocatalyst's complete saturation in the first 30 minutes under dark condition. Compared to pure $TiO_2$, $TiO_2$/GO showed better adsorption of RhB. Since the specific surface area of composite $TiO_2$/GO and raw $TiO_2$ was almost comparable, the enhancement of the adsorption process by adding GO was due to electrostatic interaction between RhB and the catalyst surface[8]. Under UV irradiation, composite $TiO_2$/GO shows a greater photocatalytic performance than that pure $TiO_2$ (81%). Regarding the remaining or RhB, 1.0 wt% of GO contributing to the highest efficiency of RhB degradation was 93% after 120 minutes. The presence of sufficient GO concentration increase the life charge carriers resulted in the improvement of interface charge transfer to the photocatalyst surface. Here, additional a higher GO concentration at 1.2 wt% showed the lowest photodegradation or similar to pure $TiO_2$ at 80%. Heterojunction $TiO_2$-GO promotes the charge separation avoiding the recombination between electron and hole. However, the excessive GO concentration generates over states influencing the faster recombination process[18]. Under Xe irradiation, the efficiency of RhB degradation proportional to the increase of GO concentration. 1.2 wt% of GO resulted in the highest efficiency was of 31%, three times higher than that of pure $TiO_2$. Additional of GO was effectively improved the photodegradation of RhB under Xe-light. A comparison of photodecomposition rate ($k$) under UV and Xe irradiation is shown in Figure 9. Even though the degradation efficiency of RhB under UV irradiation showed better efficiency than that of under Xe irradiation, the $k$ rate of composite $TiO_2$/GO over pure $TiO_2$ slightly increased. The $k$ rate of composite $TiO_2$/GO over pure $TiO_2$ was of 1 and 3 times under UV and Xe irradiation, respectively. Since $TiO_2$ has an energy band gap of around 3.2 eV, which can be activated under UV irradiation, the GO influences the composite catalyst surface activity. The presence of GO enhanced interaction between RhB and the photocatalyst. The presence of GO responsible for introducing new C 2p and O 2p energy states below the conduction band of $TiO_2$ resulted in the transition of O 2p and Ti 3d as a new state[10,19,20]. Additional energy states influenced the $TiO_2$/GO composite photoexcitation of electron and the visible adsorb ability of light. In simply, the presence of sufficient number of GO accommodate facilitates the transportation of electron due to the conductive structure of composite photocatalyst [21]. The summary of comparison of RhB photodegradation by composite $TiO_2$/GO during 2020 is proposed in Table 1. A view of paper is just for overview of the previous works related to the composite $TiO_2$ and GO to decompose RhB. The comparison showed that current result has the lowest photodegradation activity. The different of presentation degradation was due to the different of characteristic from the produced photocatalysts which influenced by material source, experimental method, kind of co-catalyst used. Different from the previous works, here, $TiO_2$ technical grade priced at 0.6 USD per kG has been used. To enhance the photocatalytic performance, more optimization is still required, but still focus on the utilization of a cheaper material resources.

This research proposed the idea of improving the photocatalytic activity of low-grade $TiO_2$ under visible light. Photoabsorption activity was influenced by physical and electronic properties. The physical properties can be determined by the specific surface area, morphology, particle and crystallite size [17]. Meanwhile, the electronical properties can be verified by electronic band structure and valence state which can be analysed by DR-UV and XPS, respectively [10]. Further optimization is inevitable to reach the optimum photodegradation efficiency.

IOP Conf. Series: Materials Science and Engineering 1143 (2021) 012055 doi:10.1088/1757-899X/1143/1/012055

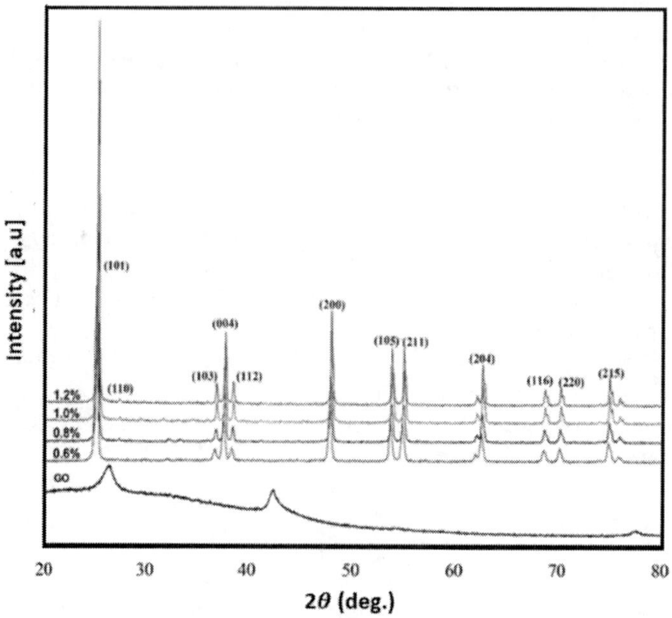


**Figure 6.** XRD patterns of the prepared photocatalyst $TiO_2$/GO composite with different GO ratios.

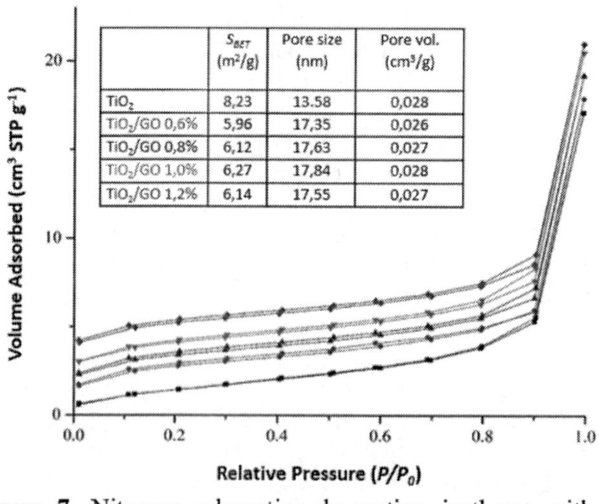


| | $S_{BET}$ ($m^2$/g) | Pore size (nm) | Pore vol. ($cm^3$/g) |
|---|---|---|---|
| $TiO_2$ | 8,23 | 13.58 | 0,028 |
| $TiO_2$/GO 0,6% | 5,96 | 17,35 | 0,026 |
| $TiO_2$/GO 0,8% | 6,12 | 17,63 | 0,027 |
| $TiO_2$/GO 1,0% | 6,27 | 17,84 | 0,028 |
| $TiO_2$/GO 1,2% | 6,14 | 17,55 | 0,027 |

**Figure 7.** Nitrogen adsorption-desorption isotherm with the insert table of porosity characteristic based on BJH analysis result.

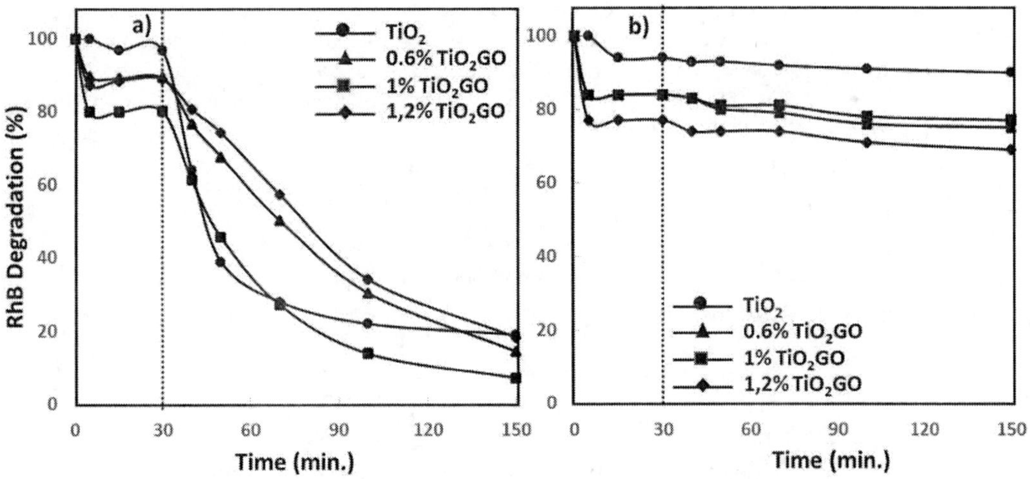


**Figure 8.** Photocatalytic activity of the prepared $TiO_2$/GO composite under a) Xe-light, and b) UV-light irradiation.

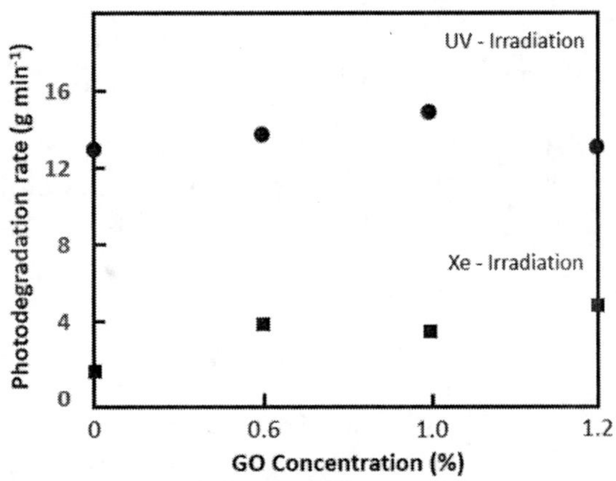


**Figure 9.** Photodecompotition rate under UV- and Xe- irradiation.

**Table 1**. Summary of photodegradation of RhB.

| No. | Material | Light source | C, V, catalyst | %, t | Ref. |
|---|---|---|---|---|---|
| 1. | $TiO_2$/GO/$H_2O_2$ | UV-light | 10 ppm, 250 ml, | 100 %, 80 min | [22] |
| 2. | $TiO_2$/GA | Xe-lamp | 20 mg $L^{-1}$, 25, 5 mg | 90%, 60 min | [23] |
| 3. | AgNw/$TiO_2$/GO | Xe-lamp | 12 mg $L^{-1}$, 30 mg | 90%/, 100 min | [24] |
| 4. | $TiO_2$/GO | Xe-Lamp | 8 ppm, 250 mL, 40 mg | 30%, 120 min | current |

## 4. Conclusion

A comprehensive study related to the improvement of photocatalytic activity of low-grade $TiO_2$ under UV and Visible light has been proposed. Additional GO in small concentration successfully enhanced the photodegradation efficiency and photodecomposition rate. The existence of GO under UV irradiation influencing the active surface sites of photocatalyst $TiO_2$/GO. Where 1.0 wt% is the optimum condition to increased efficiency one time over pure $TiO_2$, almost 100% of Rhb succeed to decompose. In comparison, GO's presence could enhance composite $TiO_2$/GO three times over than that of pure $TiO_2$. Under Xe irradiation, the presence of GO dominantly influenced the electronic properties of the prepared composite $TiO_2$/GO. Unfortunately, the efficiency of photodegradation under visible light is not optimum. Therefore, further optimization and study related to surface active sites of the composite $TiO_2$/GO is required.

## 5. Acknowledgment

This research was financially supported by KURITA Water and Environment Foundation No. 19Pid007, and ELSA LIPI – Indonesian Institute of Sciences.

## References

[1] Anon 1978 Rhodamine B. IARC Monogr. Eval. Carcinog. Risk Chem. Man

[2] Kandisa R V and Saibaba KV N 2016 Dye Removal by Adsorption: A Review *J. Bioremediation Biodegrad* **7(6)** 371

[3] Katheresan V, Kansedo J and Lau S Y 2018 Efficiency of various recent wastewater dye removal methods: A review *J. Environ. Chem. Eng.* **6**: 4676-4697

[4] Ameta R, Solanki M S, Benjamin S and Ameta S C 2018 *Photocatalysis Adv. Oxidation Proc. Wastewater Treat.: Emerging Green Chem.* Tech.135-175

[5] Lee S Y and Park S J 2013 TiO2 photocatalyst for water treatment applications *J. Ind. Eng. Chem.***19(6)**: 1761-1769

[6] Hernández-Alonso M D, Fresno F, Suárez S and Coronado J M 2009 Development of alternative photocatalysts to TiO2: Challenges and opportunities *Energy Environ. Sci.***2**:1231-1257

[7] Dreyer D R, Park S, Bielawski C W and Ruoff R S 2010 The chemistry of graphene oxide *Chem. Soc. Rev.*61-95

[8] Krishnamoorthy K, Mohan R and Kim S J 2011 Graphene oxide as a photocatalytic material *Appl. Phys. Lett.* **98**, 244101

[9] Julkapli N M and Bagheri S 2015 Graphene supported heterogeneous catalysts: An overview *Int. J. Hydrogen Energy* **40(2)**:948-979

[10] Vallejo W, Rueda A, Díaz-Uribe C, Grande C and Quintana P 2019 Photocatalytic activity of graphene oxide-TiO2 thin films sensitized by natural dyes extracted from Bactris guineensis R. Soc. Open Sci.6:181824

[11] Pérez-Molina Á, Morales-Torres S, Maldonado-Hódar F J and Pastrana-Martínez L M 2020 *Functionalized graphene derivatives and TiO2 for high visible light photodegradation of azo dyes Nanomaterials* 10:1106

[12] Fu Z, Zhang S and Fu Z 2019 Preparation of multicycle GO/TiO2 composite photocatalyst and study on degradation of methylene blue synthetic wastewater *Appl. Sci.* **9(16)**:3282

[13] Trapalis A, Todorova N, Giannakopoulou T, Boukos N, Speliotis T, Dimotikali D and Yu J 2016 TiO2/graphene composite photocatalysts for NOx removal: A comparison of surfactant-stabilized graphene and reduced graphene oxide *Appl. Catal. B Environ.* 180:637-647.

[14] Kumaran V, Sudhagar P, Konga A K and Ponniah G 2020 Photocatalytic degradation of synthetic organic reactive dye wastewater using GO-TiO2 nanocomposite Polish *J. Environ. Stud.* 29(2):1683-1690

[15] Mevold A H H, Hsu W W, Hardiansyah A, Huang L Y, Yang M C, Liu T Y, Chan T Y, Wang K S, Su Y A, Jeng R J, Wang J K and Wang Y L 2015 Fabrication of Gold Nanoparticles/Graphene-PDDA Nanohybrids for Bio-detection by SERS Nanotechnology *Nanoscale Res. Lett.* 10:397

[16] He K, Chen N, Wang C, Wei L and Chen J 2018 Method for Determining Crystal Grain Size by

X-Ray Diffraction *Cryst. Res. Technol.* 52(2).

[17] Arutanti O, Ogi T, Nandiyanto A B D, Iskandar F and Okuyama K 2014 Controllable crystallite and particle sizes of WO3 particles prepared by a spray-pyrolysis method and their photocatalytic activity *AIChE J.* **60**(1) 41-49

[18] Hsu K C, Fang T H, Lee C I, Chen T H and Hsieh T H 2020 Mechanistic Insights and Photodegradation of Heterostructure Graphene Oxide/Titanium Dioxide *Top. Catal.* **63** :956–963.

[19] Mahima S, Kannikka B, Subhasha N, Moniska J 2018 TiO2-GO Nanocomposite for Energy and Environmental Applications: A green Syntheis Approach *Vacuum.* **156**: 434-439.

[20] Arham S A, Tanveer A, Nafees A, Mohammad Z K 2019 Removal Enhancement of Acid navy Blue Dye by GO-TiO2 Nanocomposites Synthesized using Sonication Method *Mater. Chem. Phys.***238**:121906

[21] Mai F E, Ghada A M, Amanay I R, Amr E A, El said A S 2020 Fabrication of TiO2 Reduced Graphene Oxide Based Nanocomposites for Effective of Photocatalytic Decolorization of Dye Effluent *J. Inorg. Organometal. Poly. Mater.* **30**:2720-2735.

[22] Munikrishnappa C, Kumar S, Shivakumara S, ao G M, Munichandraiah N 2019 The Tio2-graphene oxide-hemin Ternary Hybrid composite Material as an Efficient heterogeneous catalyst for the Degradation of Organic Contaminants *J. Sci. Adv. Mater. Dev.* **4 (1)** 80-88.

[23] Zhang J, Fang S, Mei J, Zheng G-P, Zheng X, Guan, X 2018 High efficiency removal of rhodamine B dye in water using g-C3N4 and TiO2 co-hybridized 3D graphene aerogel composites *Separ. Purif. Technol.* **194**: 96-103

[24] Pejman H, Abbas B, Maryam Y M, Willem D V D, Kouchi Z 2020 Facile Synthesis of Ag Nanowire/TiO2 and Ag Nanowire/TiO2/GO Nanocomposites for Photocatalytic Degradation od RhodamineB *Mater.* **14**:763

IOP Conf. Series: Materials Science and Engineering 1143 (2021) 012056 doi:10.1088/1757-899X/1143/1/012056

# Non-Solvent Selection for Cellulose Acetate/Polyethylene Glycol/Polyethylene Glycol-grafting-Graphene Oxide Membranes

**A Ramadhani, R D Nyamiati, I Berin, N A Murtadho, Y Rahmawati and S Nurkhamidah[*]**

Department of Chemical Engineering, Institut Teknologi Sepuluh Nopember (ITS), Kampus ITS Sukolilo, Surabaya 60111 Indonesia

[*]Corresponding Author's E-mail: nurkhamidah@chem-eng.its.ac.id

**Abstract.** Fabrication process of membrane is key factor for producing reverse osmosis membrane with a good performance for desalination process. One of the methods to fabricate membrane is phase inversion where casted membrane was immersed in the non-solvent. Non-solvent is important parameter to determine that phase inversion is success or not in resulting the morphology of the membrane. As known that morphology of the membrane contributes to the performance of the membrane. In this study, several non-solvents have been used to fabricate CA/PEG/PEG-g-GO membrane. Non-solvent used in this study were water, isopropanol, methanol, isopropanol-water and methanol-water. The morphology of the membrane was analyzed using a Scanning Electron Microscope (SEM). The thermodynamic and kinetic properties of non-solvent/solvent/polymer CA/PEG systems are studied and correlated with membrane morphology. Membrane performance was determined by salt rejection, permeate flux, and permeability. The experiment results show that the best non-solvent is water followed by isopropanol, isopropanol-water, methanol, and methanol-water. CA/PEG/PEG-g-GO membrane with water as non-solvent has a salt rejection (% R) of 75%, permeate flux (F) of 1,985 $L/m^2.h$, permeability of 0.0005 $L/m^2.h.kPa$ and the morphology is sponge-like with the pore size diameter in average is 0.471 μm.

## 1. Introduction

Providing a hygiene and affordable clean water is one of an urgent global necessity. One of the urges is desalination technology which is potentially able to meet the global demand on a clean water. Reverse osmosis is one of the most promising and important methods for produce clean water from seawater sources and another wastewater [1]. In the reverse osmosis method, the reverse osmosis membrane acts as a barrier semipermeable which retains salt concentration resulting in clean water. Terms of the membrane that is good to use is a small pore size and pore distribution evenly, the smaller the pore, the more impurity will be left. Membrane research currently being developed aims to determine performance membrane such as flux, salt rejection and surface morphology, and hydrophilicity.

In this desalination process using the phase inversion method, which is a method in which the polymer solution is cast on a glass plate and immersed in a non-solvent coagulation bath after a given evaporation time, there will be an exchange of solvent and non-solvent to produce a porous membrane structure. The structure of the membrane is highly dependent on the thermodynamic and kinetic properties of the non-solvent, solvent and polymer ternary systems [2,3-5]. Of the several types of polymers that exist, Cellulose Acetate (CA) was chosen to be used in the desalination process because it has the potential to produce high flux, moderate salt rejection properties, and chlorine resistance.

Graphene oxide (GO), is one of the nanoparticles that shows a very promising potential to improve the performance of RO membranes, due to its high surface area, mechanical strength, and high hydrophilicity. In addition, the oxygen functional groups of GO, including the epoxide, hydroxyl, and carboxyl functional groups on the GO surface provide good dispersion properties in a wide variety of polymers [6]. In the study of Ghaeminezhad et al. [6] GO with a size of 2.7 nm was dissolved in the CA polymer matrix to form a GO/CA nanocomposite membrane. The resulting CA membrane character is influenced by differences in GO concentrations. The addition of GO to the CA polymer matrix changed the pore structure of the membrane from finger-like to sponge-like. The CA/GO membrane produced in this study with the addition of 1 %wt GO showed a salt rejection rate of up to 90% (an increase of 80% compared to membranes made of pure CA); but, decreased in permeability flux by 30%. However, GO still has drawbacks because the dispersion is not optimal. So, in the study of Y-B. Li et al. [7] grafting PEG onto the GO surface (PEG-g-GO), PEG-g-GO is easily dispersed and the distribution of PEG-g-GO is better than GO, so that the uniform dispersion of PEG-g-GO in the fiber matrix composites improve mechanical properties as well as composites.

In this study, the CA/PEG/PEG-g-GO membrane, DMF as a solvent, was prepared using the phase inversion method. Water, isopropanol, methanol, isopropanol-water, and methanol-water as non-solvent are used in the CA/PEG/PEG-g-GO membrane immersion process. We added PEG-g-GO by 0.005 % wt, because of the best variable from our previous study [8]. The resulting asymmetric membrane will be analysed for its characteristics and performance to find the best non-solvent.

## 2. Materials and Methods

### *2.1. Materials*

Cellulose acetate (CA, 30.000 Da, with acetyl content 39.8%) and polyethylene glycol-400 (PEG, 400 Da) were purchased from Sigma Aldrich. Graphene oxide (GO) with bulk density of 1,8 g/cm$^3$, was purchased from Sigma Aldrich. *N,N*-Dimethylformamide (DMF) was used as solvent. The non-solvent were water, isopropanol (IPA), and methanol. NaOH was purchased from Sigma Aldrich. $ClCH_2COONa$ was purchased from Sigma Aldrich. HCl was purchased from Sigma Aldrich. *N,N*-dicyclohexylcarbodiimide (DCC), dicloromethane, and dimethylaminopyridine were purchased from Sigma Aldrich.

### *2.2. Grafting-Graphene Oxide (PEG-g-GO)*

An amount of 3.75 g of NaOH and 3.75 g of $ClCH_2COONa$ were dissolved in 300 mL GO suspension (1 mg/mL), then sonicated for 2.5 h, followed by neutralization with 0.1 M HCl. The resulting GO (GO-COOH) suspension was purified by repeated rinsing and centrifugation (15000 rpm for 7 min) then heated at 100 $^0$C. 1.5 g PEG dissolved in 30 mL dichloromethane, add 0.06 g of N,N-Dimethylaminopyridine and GO-COOH dissolved in 20 mL of DMF then stirred for 5 min. 0.5 g of N,N dicyclohexylcarbodiimide (DCC) was dissolved in 5 mL of dicloromethane and added to the GO-COOH solution after that stirred for 24 h at mild temperatures, repeated rinsing and centrifugation then heated to 100 $^0$C, produced Grafting GO (PEG-g-GO) [9].

### *2.3. Preparation of membranes*

The CA/PEG polymer ratio 80/20 was prepared by mixing 2 g of CA and 0.5 g of PEG in 17 mL DMF. PEG-g-GO was added to the CA/PEG membrane solution with PEG-g-GO composition 0.005 wt.% of the solvent. The solution was then mixed using a hotplate at 70 °C for 12 h, then cast on glass and immeresed into a non-solvent water, isopropanol (IPA), methanol, isopropanol-water and methanol-water for 15 minutes to produce a flat sheet membrane.

### 2.4. *Solubility and interaction parameters*

Solubility parameters are used to determine the solubility of a particular polymer. The representation of the solubility parameters developed by Hansen are dispersion (δd), polar (δp) and hydrogen bonding (δh) [10]. To determine the interaction between non-solvent-solvent and polymer-non-solvent, it can be seen from the results of the difference in solubility parameters between non-solvent-solvent (Δδ (ns-s)) and polymer-non-solvent (Δδ (p-ns)). The difference in solubility parameters between non-solvent-solvent (Δδ (ns-s)) were calculated using Eq. (1) [11].

$$\Delta\delta_{ns-s} = \sqrt{\left(\delta_{d,ns} - \delta_{d,s}\right)^2 + \left(\delta_{p,ns} - \delta_{p,s}\right)^2 + \left(\delta_{h,ns} - \delta_{h,s}\right)^2} \quad (1)$$

The difference in solubility parameters between polymer-non-solvent (Δδ (p-ns)) were calculated using Eq. (2) [10].

$$\Delta\delta_{p-ns} = \sqrt{\left(\delta_{d,p} - \delta_{d,ns}\right)^2 + \left(\delta_{p,p} - \delta_{p,ns}\right)^2 + \left(\delta_{h,p} - \delta_{h,ns}\right)^2} \quad (2)$$

**Table 1.** Molar volume, density and boiling point non-solvent, solvent and polymer [2, 10].

| | V ($cm^3$/mol) | Density (g/$cm^3$) | Bp ($^0$C) |
|---|---|---|---|
| **Water** | 18 | 0.997 | 100 |
| **IPA** | 76.8 | 0.781 | 82 |
| **Methanol** | 40.7 | 0.7918 | 65 |
| **DMF** | 77 | 0.95 | 153 |
| **CA** | - | 1.3 | 210 |
| **PEG** | - | 1.12 | 290 |

**Table 2.** Solubility parameter non-solvent, solvent and polymer [12, 13, 14].

| | $\delta_d$ | $\delta_p$ | $\delta_h$ | $\delta_t$ |
|---|---|---|---|---|
| **Water** | 15.5 | 16 | 42.3 | 47.8 |
| **IPA** | 15.8 | 6.1 | 16.4 | 23.5 |
| **Methanol** | 15.1 | 12.3 | 22.3 | 29.6 |
| **DMF** | 17.4 | 13.7 | 11.3 | 24.8 |
| **CA** | 16.9 | 16.3 | 3.7 | 23.8 |
| **PEG** | 16.8 | 10.2 | 8.6 | 21.5 |

Table 1 shows the molar volume, density and boiling point data for non-solvent, solvent and polymer. Table 2 shows the dispersion (δd), polar (δp), hydrogen bonding (δh), and the total solubility parameters of each non-solvent, solvent and polymer. Total solubility parameter (δt) was calculated using Eq. (3) [10, 15].

$$\delta_t = \sqrt{\delta_d^2 + \delta_p^2 + \delta_h^2} \quad (3)$$

### 2.5. *Membrane Characterization*

Analysis of membrane characteristics among others analysis of Scanning Electron Microscopy (SEM), analysis of water content, and porosity. Scanning Electron Microscopy (SEM) is an analysis used to determine the morphology of the fracture side of the membrane. Water content analysis was calculated using Eq. (4) [16, 17].

$$WC\% = \frac{Ww - Wd}{Ww} \times 100\% \quad (4)$$

The porosity ε of the resulting membrane was calculated using Eq. (5) [2].

$$\varepsilon = \frac{(W_w - W_d)/\rho_w}{\frac{W_w - W_d}{\rho_w} + \frac{W_d}{\rho_p}} \tag{5}$$

Where, Ww is the weight of the wet membrane, Wd is the weight of the dry membrane, $\rho w$ and $\rho d$ are densities of water and CA/PEG polymer, respectively.

*2.6. Membrane Performance*

Salt rejection is the efficiency of membrane and its ability to remove contaminates. Salt rejection was calculated using Eq. (6) [18, 19].

$$R = \left(1 - \frac{C_p}{C_f}\right) \times 100\% \tag{6}$$

Where, R is the percentage of salt rejection, $C_p$ (in ppm) is the salt concentration of the permeate, $C_f$ (in ppm) is the salt concentration in the feed water.

The permeate flux (J) represents the amount of pure water collected per unit time and per unit area at variable pressures. Permeate flux was calculated using Eq. (7) [20].

$$J = \frac{Q}{t} \times A \tag{7}$$

Where, J is the permeate flux ($mL/h.m^2$), Q is the amount of permeate (mL), t is the time and A is the area ($m^2$).

## 3. Results and discussion

The non-solvent type affects the morphological structure of the membrane. The exchange rate of solvent and non-solvent at the polymer solution/coagulation bath interfaces was theoretically estimated by the diffusion coefficient calculated from Tyn-Calus equation [21, 2]. The results of the diffusion coefficient can be seen in the Table 3.

$$D_{AB} = 8{,}93 \times 10^{-8} \left(\frac{V_B^{0.267}}{V_A^{0.433}}\right)\left(\frac{T}{\eta_B}\right)\left(\frac{\sigma_B}{\sigma_A}\right)^{0.15} \tag{8}$$

Where, $D_{AB}$($cm^2/s$) is the diffusion coefficient of species A in B, $V_A$ dan $V_B$ ($cm^3/mol$) are the molar volumes of species A and B, T (K) is the absolute temperature, $\eta_B$ (mPa/s) is the viscosity from species B, $\sigma_A$ and $\sigma_B$ (mN/m) is the surface tension of species A and B [2].

**Table 3.** Surface tension, viscosity and diffusion coefficient of solvent and non-solvent [2, 10].

| Solvent or non-solvent | Surface Tension σ (mN/m) | Viscosity η (mPa/s) | $D_{DMF-NS}$ ($10^{-6}$ $cm^2/s$) | $D_{NS-DMF}$ ($10^{-6}$ $cm^2/s$) |
|---|---|---|---|---|
| **Water** | 72.8 | 1 | 9.7442 | 27.2972 |
| **IPA** | 20.8 | 2.4 | 4.9563 | 17.5752 |
| **Methanol** | 22 | 0.6 | 16.8750 | 22.9431 |
| **IPA- water** | 46.8 | 1.7 | 6.9469 | 19.1788 |
| **Methanol- water** | 47.4 | 0.8 | 21.2545 | 23.5575 |
| **DMF** | 36.4 | 0.802 | - | - |

In Table 3, which shows the highest $D_{DMF-NS}$ values are methanol-water and methanol because it has a low viscosity, this indicates a faster exchange rate between solvent and non-solvent, in other words it only takes a very short time for DMF solvent to diffuse in the coagulation bath. non-solvent isopropanol. As can be seen in Table 3 the order of the $D_{DMF-NS}$ values is methanol-water > methanol > water >

isopropanol-water > isopropanol. Having a low viscosity and high diffusion coefficient causes the non-solvent methanol membrane to become thin, brittle and the SEM results show no pores.

### *3.1. Solubility parameters for non-solvent-solvent and polymer-non-solvent*

Solubility parameters is used to determine the solubility of a particular polymer. The representation of the solubility parameters developed by Hansen is dispersion (δd), polar (δp) and hydrogen bonding (δh) [10]. To determine the interaction between non-solvent-solvent and polymer-non-solvent, it can be seen from the results of the difference in solubility parameters between non-solvent-solvent (Δδ (ns-s)) from Eq. 1 and polymer-non-solvent (Δδ (p -ns)) from Eq. 2 in Table 4.

**Table 4.** The difference in solubility parameters for non-solvent-solvent and polymer-non-solvent.

| Polymer | Solvent | Non-Solvent | $\Delta\delta_{(ns-s)}$ | $\Delta\delta_{(p-ns)}$ |
|---|---|---|---|---|
| **CA-PEG** | DMF | Water | 31.1432 | 37.6565 |
| **CA-PEG** | DMF | IPA | 9.2914 | 14.8042 |
| **CA-PEG** | DMF | Methanol | 11.3248 | 17.9266 |
| **CA-PEG** | DMF | IPA- Water | 18.3272 | 25.0272 |
| **CA-PEG** | DMF | Methanol- Water | 21.1095 | 27.6808 |

The values Δδ (ns-s) and Δδ (p-ns) play an important role in the formation of asymmetric membranes. In this study, the non-solvent sequence of Δδ (ns-s) and Δδ (p-ns) values is the same, namely water> methanol-water> isopropanol-water> methanol> isopropanol. The lowest value of the difference in solubility parameters shows high affinity and a high value of the difference in solubility parameters shows low affinity [2, 10, 11]. The non-solvent with the lowest affinity for the difference in the solubility parameter of the non-solvent-solvent (Δδ (ns-s)) is water, this shows that there is a short interaction between water and DMF so that when the phase inversion process occurs instant demixing. Non-solvent with the highest affinity is indicated by the value of the difference in the solubility parameter of the polymer-non-solvent (Δδ (p-ns)) and the difference in the solubility of the non-solvent-solvent parameter (Δδ (ns-s)) which is low, namely the non-solvent -solvent isopropanol, this shows that there is an interaction between the CA-PEG polymer and non-solvent isopropanol during the phase inversion process so that there is a demixing delay. In Saeed Mazinani's research, states that non-solvent strength is in the order of water> glycerol> methanol. Non-solvent water has the highest δh value, hence higher hydrogen bonding ability and better non-solvent polymer compatibility, thus indicating lower non-solvent tolerance of polymer/solvent systems. Water is considered a strong non-solvent for membrane formation while methanol can be considered a weak non-solvent [10].

### *3.2. Microscopy Scanning Electron (SEM) analysis of CA/PEG/PEG-g-GO membranes with non-solvent Isopropanol, water, and methanol*

**Table 5.** Results Scanning Electroscopy Microscope Analysis of membrane CA/PEG/PEG-g-GO with non-solvent variations.

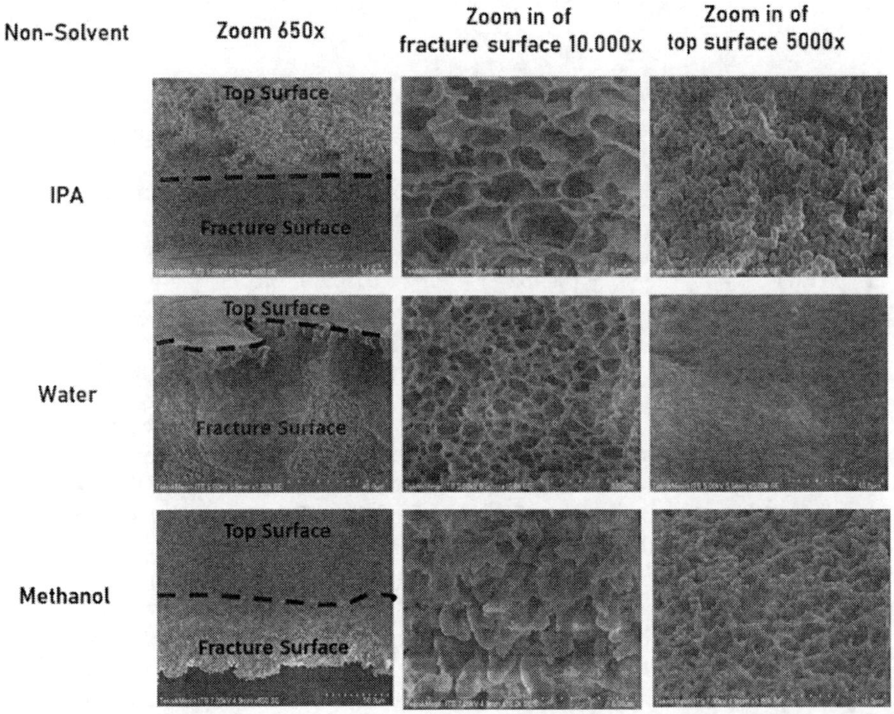


The mechanism for the formation of asymmetric membrane structures is determined by the thermodynamics and kinetics of the non-solvent, solvent and polymer. The SEM analysis results of CA/PEG/PEG-g-GO membranes for non-solvent isopropanol, water and methanol can be seen in Table 5. The difference in non-solvent has an effect on membrane morphology, the value of Δδ (ns-s) between nonsolvent water and solvent DMF is high, this indicates that nonsolvent water has the lowest affinity with DMF solvent, so when the phase inversion process occurs instant demixing occurs, so it is indicated by the presence of a finger-like morphological structure, while the Δδ (ns-s) value between non-solvent isopropanol and solvent DMF is low, this indicates that non-solvent isopropanol has a high affinity for DMF solvent, so it occurs demixing delays during the phase inversion process, which causes the non-solvent isopropanol morphological structure of the membrane to be spong-like. Seen in Table 5, the asymmetric membrane made using non-solvent water has the most dense top layer and has finger-like and sponge-like sub-layer characteristics, whereas the isopropanol and methanol non-solvent membranes showed a thin top layer and a sponge-like sublayer [15]. In the SEM results, PEG-g-GO nanoparticles can be dispersed well, because it does not show any agglomeration of PEG-g-GO nanoparticles, on the top surface and fracture surface there is no showing lumps or black spots of PEG-g-GO nanoparticles. The CA/PEG/PEG-g-GO membrane which is non solvent isopropanol and methanol has an uneven top surface, this is due to the influence of non-solvent on the polymer.

**Table 6**. Average pore diameter of CA/PEG/PEG-g-GO membrane

| Non Solvent | Average pore diameter *sponge-like* (μm) |
|---|---|
| **IPA** | 1.011 |
| **Water** | 0.471 |
| **Methanol** | N. A |

It can be seen in Table 6 CA/PEG/PEG-g-GO membrane with non-solvent isopropanol resulting a pore diameter larger than that with non-solvent water. In the research of Ghorani et al. [15] the same results were also obtained, the pore size of non-solvent isopropanol membranes was larger than non-solvent water. This membrane is classified as Microfiltration (MF) membrane, it is quite difficult to get 99% salt rejection.

### *3.3. Membrane Hydrophilicity Analysis*

**Table 7.** Results of water content and porosity analysis CA/PEG/PEG-g-GO membrane with non-solvent variables.

| | Water Content | | | Porosity |
|---|---|---|---|---|
| Non solvent | Wet Weight | Dry Weight | %X | (ε) |
| **Water** | 0.0482 | 0.0096 | 80% | 0.8360 |
| **Methanol-water** | 0.0212 | 0.0057 | 73% | 0.7752 |
| **IPA-water** | 0.0371 | 0.0106 | 71% | 0.7602 |
| **Methanol** | 0.0176 | 0.0054 | 69% | 0.7412 |
| **IPA** | 0.0101 | 0.0052 | 49% | 0.5443 |

The results of water content analysis are one of the method used to determine how hydrophilicity a membrane is. In Table 7 shows the results of water content and porosity of the CA/PEG/PEG-g-GO membrane with different non-solvents. The values of water content and porosity are obtained from equations 4 and 5. The non-solvent water membrane has the highest value in the water content analysis followed by methanol-water, isopropanol-water, methanol and isopropanol. In this study, the results of water content analysis are comparable to the results of membrane porosity. The non-solvent water membrane had the highest porosity value followed by methanol-water, isopropanol-water, methanol and isopropanol. In the CA/PEG/PEG-g-GO membrane, the non-solvent isopropanol and methanol have hydrophobic properties. In this study, the non-solvent isopropanol membrane has a thin structure like the plastic and the non-methanol solvent membrane has a structure that is easily broken. Other than that, the factors that cause water content and porosity of the non-solvent membrane isopropanol and methanol are small because they both have a sponge-like membrane structure and do not have finger-like, so that the surface area for water absorption in the non-solvent membrane structure isopropanol and methanol is small.

### *3.4. Analysis of Salt Rejection Performance and Membrane Flux*

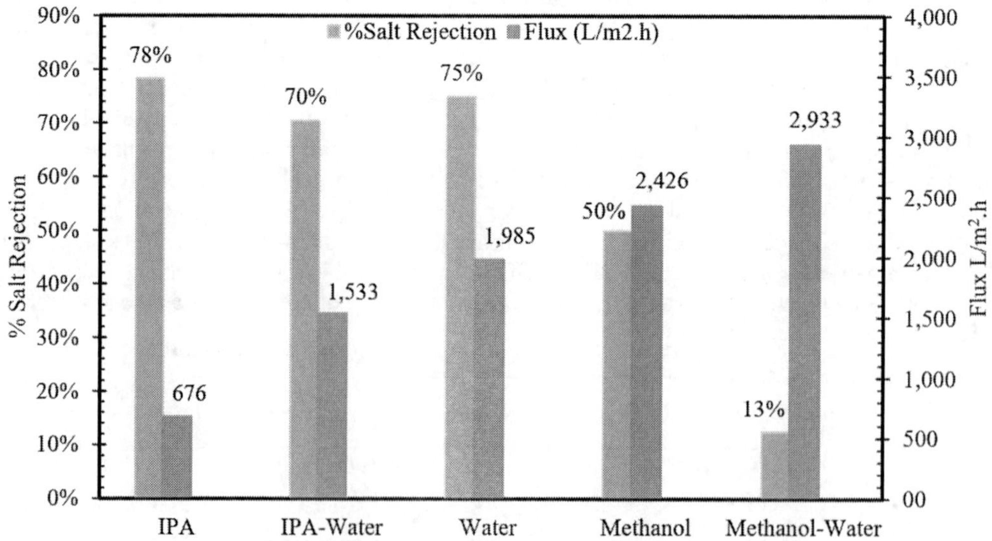


**Figure 1.** Salt rejection and permeate flux of CA/PEG/PEG-g-GO membrane with non-solvent variables.

Salt rejection analysis and permeate flux were performed to determine the performance of different non-solvent CA/PEG/PEG-g-GO membranes. This research produces membrane pores with a pore range of 0.4-1 μm which is included in the membrane microfiltration (MF) category, but this membrane can reject salt because the salt solution used has a very small concentration of 1000 ppm. When the salt solution is passed through the CA/PEG/PEG-g-GO membrane, there are NaCl ions that stick to the membrane walls and partially into the product. So that a relatively small rejection is generated. Figure 1 shows the results of the highest salt rejection analysis are isopropanol followed by water, isopropanol-water, methanol, and methanol-water. The average pore size of non-solvent isopropanol membranes is larger than that of non-solvent water membranes, but the high rejection is thought to be due to the dense or sponge-like membrane structure. So that the surface area to reject salt is wider. The non-solvent isopropanol membrane produced the smallest permeate flux followed by isopropanol-water, water, methanol, and methanol-water. This is due to the texture of the non-solvent isopropanol which resembles plastic, has low hydrophilicity, has a sponge-like sublayer and is supported by low porosity. This can also result in a blockage. The non-solvent water membrane has the smallest pores compared to isopropanol and methanol, has a sponge-like and finger-like structure, this membrane also has the highest hydrophilicity resulting in high flux and relatively small blockage. On the surface of the non-solvent water membrane is very dense, it is suspected that NaCl ions are stuck on the surface of the membrane, so that it produces high rejection. The non-solvent methanol membrane has high flux but low rejection, this is due to the leakage of the membrane by the texture of the membrane which is brittle and breaks easily. In this study, the optimum membrane was the CA/PEG/PEG-g-GO membrane with non-water solvent because it had high salt rejection and permeate flux. From the results of the flux did not show a decrease in flux due to agglomeration of PEG-g-GO nanoparticles.

### *3.5. Membrane Permeability Performance Analysis*

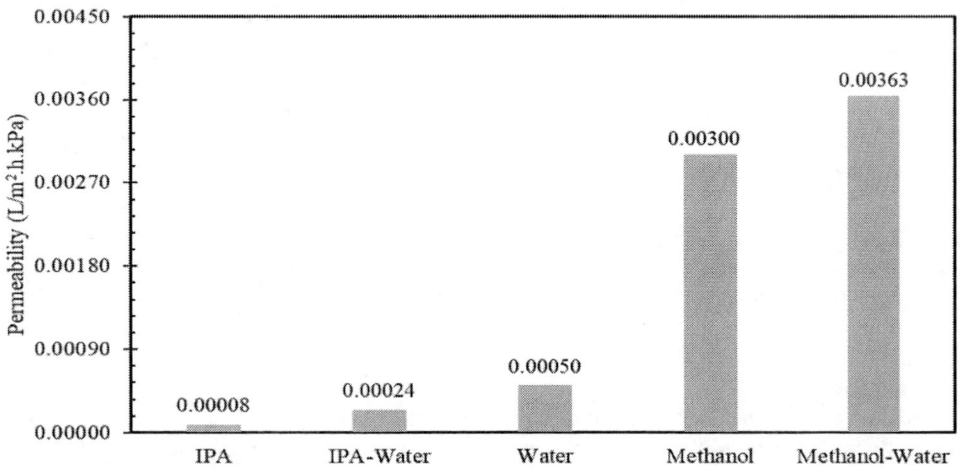


**Figure 2.** Permeability of CA/PEG/PEG-g-GO membrane with non-solvent variables.

The results of membrane permeability CA/PEG/PEG-g-GO with different non-solvent can be seen in Figure 2 the permeability results of the CA/PEG/PEG-g-GO membrane with non-solvent variables were proportional to the flux results. The non-solvent isopropanol membrane produced the lowest permeability followed by isopropanol-water, water, methanol, and methanol-water.

### *3.6. Effect of non-solvent type on membrane pressure*

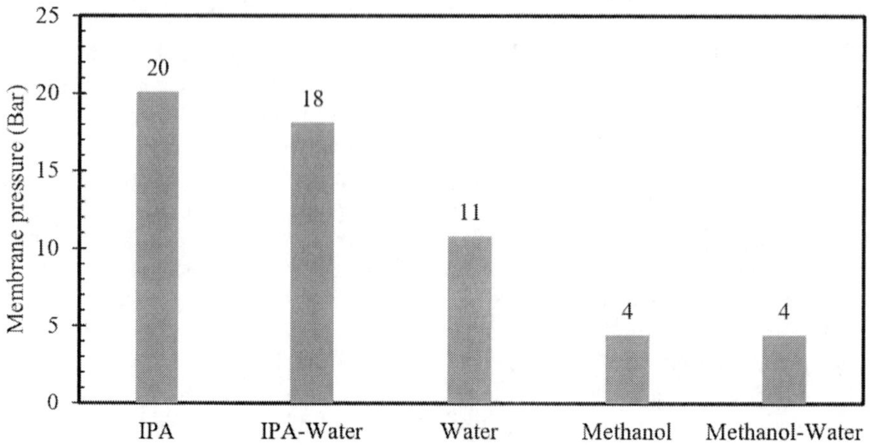


**Figure 3.** Membrane pressure of CA/PEG/PEG-g-GO membrane with non-solvent variables.

The non-solvent type affects the membrane strength level, the membrane strength will affect the membrane pressure. In this study, the CA/PEG/PEG-g-GO membrane using non solvent isopropanol has a strong texture like plastic so that it has a large membrane pressure. The membrane that uses non-methanol solvent has a fragile texture so it has a small membrane pressure. The order of membrane pressure based on the non-solvent type is isopropanol > isopropanol-water > water > methanol > methanol-water. From Figure 3, the optimum membrane pressure is non-solvent water because the texture of the membrane is not easily brittle and the membrane pressure in water is below isopropanol so it is more economical.

## 4. Conclusion

In this study, the CA/PEG/PEG-g-GO membrane using a non-water solvent had high hydrophilicity and porosity. The smallest pore size is owned by membranes that use water as non-solvents. The highest salt rejection analysis results were isopropanol followed by water, isopropanol-water, methanol, and methanol-water, but non-solvent isopropanol membranes produced the smallest permeate flux followed by isopropanol-water, water, methanol and methanol-water. The non-solvent isopropanol membrane produced the lowest permeability followed by isopropanol-water, water, methanol, and methanol-water. The optimum membrane pressure is a membrane that uses water as non-solvents. From the above results it can be concluded that the most optimal non-solvent is water, able to produce high salt rejection and permeate flux with medium pressure compared to another non-solvent.

**Acknowledgements**

This work has been financially supported by ministry of Research, Technology and Higher Education of the Republic of Indonesia with contract's number 3/AMD/E1/KP.PTNBH/2020 and 1189/PKS/ITS/2020 to which the authors express their sincerest gratitude.

**References**

[1] Chae H R, Lee J, Lee C H, Kim I C and Park P K 2015 *J. Membr. Sci* **483** 128–135.
[2] Lua A C and Shen Y 2013 *J. Membr. Sci* **429** 155-167.
[3] Reuvers A J, van den Berg J W A and Smolders C A 1987 *J. Membr. Sci* **34** 45.
[4] Reuvers A J and Smolders C A 1987 *J. Membr. Sci* **34** 67.
[5] Yilmaz L and McHugh A J 1986 *J. Membr. Sci* **31** 997.
[6] Ghaseminezhad S M, Barikani M and Salehirad M 2019 *Composites Part B: Engineering* **161** 320–327.

[7] Li Y B, Liu H H, Wang X C and Zang X X 2019 *Materials Today Sustainability* **3-4** 100006
[8] Nurkhamidah S, Ramadhani A, Devi B C, Febriansyah B A, Nyamiati R D and Rahmawati Y 2019 *Indonesian Polymer Journal* **22(2)** 16-20.
[9] Wang C Y, Wang W, Li G L, Li W, Tian W H and Li X G 2013 *Advanced Materials Research* **800** 459-463.
[10] Mazinani S, Darvishmanesh S, Ehsanzadeh A and der Bruggen B V 2017 *J. Membr. Sci* **526** 301-304.
[11] Bottino A, Camera-Roda G, Capannelli G and Munari S 1991 *J. Membr. Sci* **57** 1-20.
[12] Brandrup J, Immergut E H and Grulke E A 1999 *Polymer Handbook (4th ed.). New Jersey: John Wiley & Sons, Inc.*
[13] Hansen C M 1967 Three Dimensional Solubility Parameter and Solvent Diffusion Coefficient. Their Importance in Surface Coating Formulation *J. Paint Technology* **104**.
[14] Hansen C 2007 Hansen Solubility Parameters: *A User's Handbook 2nd ed* Vol. **53**.
[15] Ghorani B, Kadkhodaee R, Rajabzadeh G and Tucker N 2019 *Reactive and Functional Polymers* **134** 121-132.
[16] Sabir A, Shafiq M, Islam A and Sarwar A 2015 *Carbohydrate Polymers* **132** 589-597.
[17] Ahumada E A, Delgado D R and Martínez F 2012 *Fluid Phase Equilibria* **332** 120–127
[18] Badawi N E, Ramadan A R, Esawi A M K and El-Morsi M 2014 *Desalination* **344** 79-85.
[19] Li Z, Ren J, Fane A G, Li D F and Wong F -S 2006 *J. Membr. Sci* **279** 601-607.
[20] Gebru K A and Das C 2017 *Chinese Journal of Chemical Engineering* **25** 911-923.
[21] Van't Hof J A, Reuvers A J, Boom R M, Rolevink H H M and Smolders C A 1992 *J. Membr. Sci* **70** 17.

IOP Conf. Series: Materials Science and Engineering 1143 (2021) 012057 doi:10.1088/1757-899X/1143/1/012057

# Effect of Additive on Microstructure, Hydrophilicity and Ultrafiltration Performance of Polyethylene Terephthalate Membranes

**Samuel P. Kusumocahyo*, Syarifa K. Ambani, Shelly Marceline, Franzesca Michelle**

Department of Chemical Engineering, Faculty of Life Sciences and Technology, Swiss German University, Tangerang 15143, Indonesia

*Corresponding author's e-mail address: samuel.kusumocahyo@sgu.ac.id

**Abstract**. Ultrafiltration is a pressure-driven separation process through a porous membrane that can separate particles or macromolecules from a solution. Ultrafiltration is mostly applied for water treatment processes in various industries such as pharmaceutical, chemical, food and beverage industries. Commercial ultrafiltration membranes are mostly fabricated from polymer materials such as polysulfone, polyethersulfone and cellulose acetate. These polymers are expensive, and for the time being they are not produced in Indonesia. In this work, polyethylene terephthalate (PET) bottle waste was used as the polymer material to prepare ultrafiltration membranes. PET is commonly used as a packaging material for foods and beverages due to its low price and excellent mechanical properties. The PET membranes were prepared via a phase-inversion technique using polyethylene glycol (PEG) as the additive. It was observed that the addition of PEG improved the flexibility and the hydrophilicity of the membranes. The water contact angle decreased to 61.5° by the addition of PEG. The microstructures of the membranes could be controlled by the molecular weight of the PEG. The result of the ultrafiltration experiment showed that the membrane with a high porosity, a large pore size and a high hydrophilicity exhibited a high water permeate flux. In the ultrafiltration experiment using a model feed solution of an aqueous solution containing Bovine Serum Albumin with a molecular weight of 66,000 Da, the membranes showed high rejection values of up to 90% with water permeate fluxes of 14.7 - 23.0 kg/($m^2$h).

## 1. Introduction

Membrane separation technology have attracted many attentions for the last few decades because of its high selectivity and low-use energy [1,2,3,4]. Reverse osmosis, nanofiltration, ultrafiltration and microfiltration membranes have been applied for various separation processes in many industries such as pharmaceutical, chemical, food, beverages, dairy, drinking water, and other industries [4]. The separation using these membranes is basically a pressure-driven separation process through a selective layer, where particles, macromolecules, ions or other solutes are retained, while smaller substances and solvents can permeate through the membranes. Among these types of membranes, ultrafiltration membranes have medium pore sizes that range from 0.001 μm to 0.1 μm, and they can retain molecules with molecular weights ranging from 500 Da to 500,000 Da. Nowadays, ultrafiltration membranes are one of the most important separation technologies for water treatment to remove microorganism, bacteria, viruses, colloidal substances, and suspended particles [4]. Commercial ultrafiltration membranes are usually fabricated from polymers such cellulose acetate, polysulfone, or polyethersulfone. For the time being, these polymers are not produced in Indonesia, and must be imported. This condition hinders the development of cheap ultrafiltration membranes in Indonesia, meanwhile the demand of cheap ultrafiltration membranes is high.

In this study, polyethylene terephthalate (PET) bottle waste was utilized as the material to prepare ultrafiltration membranes. The aims of this work are to contribute in the efforts of plastic recycling for the environment conservation, and to develop low-cost ultrafiltration membranes. PET is abundantly produced in Indonesia as it is widely used as a packaging material mostly for foods and beverages. PET is a good packaging material because of its good transparency, good mechanical and chemical properties, as well as good gas-barrier resistance [5]. However, this single-use packaging resulted in a serious environmental problem because it produced a lot of plastic waste that is non-biodegradable. Recently, it has been reported that only 9% of plastic waste is recycled, while the remaining 91% of the plastic waste is just discarded or combusted resulting in water, soil and air pollution [6]. In spite of its excellent mechanical and chemical properties, there are only few reports about the utilization of PET for the development of separation membranes. Hao et al reported the use of PET to develop fibrous membranes by the electrospinning method, and applied the membranes as separator in lithium ion battery [7]. Other studies reported the development of track-etched PET membranes for membrane distillation [8, 9]. To the best of our knowledge, at the time being there is only one study on the use of PET for the development of ultrafiltration membranes as reported by Rajesh et al [10]. It was reported that the ultrafiltration membranes showed rejection values of about 70-90% for solute molecules with molecular weights of more than 500,000 Da (molecular weight cut off/ MWCO > 500,000 Da). Since the application of ultrafiltration membranes with MWCO larger than 500,000 Da is very limited, it is desired to develop PET ultrafiltration membranes that have high rejection values for solute molecules with lower molecular weights. To develop such kind of membranes, the source of the PET polymer and the type of additive are crucial factors in the membrane fabrication. In this study, waste of PET bottles that are commonly used as a packaging for drinking water were utilized as a polymer source to prepare ultrafiltration membranes using the phase inversion technique. The effect of the type of additive on the membrane morphology and the hydrophilicity was studied. The ultrafiltration performance of the membranes was studied by measuring the permeate flux and the rejection through an ultrafiltration experiment using pure water and an aqueous solution containing solute molecules of Bovine Serum Albumin with a molecular weight of 66,000 Da as a feed model.

## 2. Experimental

### *2.1. Materials*
PET bottles that were previously used for mineral water packaging were used. Phenol, ethanol (96%) polyvinyl pyrrolidone (PVP), polyethylene glycols with molecular weights of 400, 4000, and 6000 Da (PEG 400, PEG 4000, and PEG 6000), Bovine Serum Albumin (BSA, MW: 66,000 Da) were all supplied by Merck, Germany, and were used as received. Distilled water was used.

### *2.2. Membrane preparation*
First, the used PET bottles were washed, dried and cut to obtain small pieces of PET with a size of approximately 3 mm x 3 mm. Phenol was used as a solvent to dissolve the PET bottles. First, phenol was heated at 40°C for 20 min to liquify the solid phenol prior to be used as a solvent. Then, 4 g of the PET shards was dissolved in 15 g of phenol at 100°C until a clear homogeneous polymer solution was obtained. Meanwhile, solutions containing polyvinyl pyrrolidone (PVP) or polyethylene glycol (PEG) as the additives were prepared by dissolving 1.5 g of the additive in 5 g of phenol at 100°C. The additive solution was then blended with the PET solution at 100°C for 60 min to obtain a casting solution. Further, the casting solution was poured out onto a glass plate, and was cast using a casting knife to obtain a liquid thin film. After the casting, the glass plate was immediately immersed in a coagulation bath containing water and ethanol (1:12 v/v) as the non-solvent. The membrane was rinsed with distilled

water several times to remove the remaining residue, and then stored in a plastic container containing distilled water for further characterization.

### *2.3. Membrane characterization*

Analysis using Scanning Electron Microscopy was carried out to study the microstructure of the PET membranes using SEM Quanta 650 equipment. The membrane sample for the SEM analysis was obtained by dipping the membrane in a liquid nitrogen and then fracturing the membrane. The membrane porosity was measured by gravimetric method through measuring the weight of the wet membrane dan that of the dry membrane [11, 12]. The wet membrane was obtained by immersing the membrane in distilled water at room temperature for 24 h, whereas the dry membrane was obtained by drying the wet membrane in an oven at 110°C for 3 h. The following equation was used to determine the porosity $\varepsilon$ [8,9]:

$$\varepsilon = \frac{(w_1 - w_2)/d_w}{((w_1 - w_2)/d_w) + w_2/d_p} \quad (1)$$

where $w_1$ and $w_2$ are the weight of the wet membrane and that of the dry membrane, respectively, whereas $d_w$ and $d_p$ are the density of the water and that of the polymer, respectively.

The hydrophilicity of the membranes was analyzed by measuring the water contact angle. The measurement was carried out using a water contact angle meter with the equipment type of Face CA-D (Kyowa Kaimengaku Co. Ltd, Japan).

The ultrafiltration performances of the membranes were studied by conducting ultrafiltration experiments using the experimental set up as shown in Figure 1. The membrane was placed in a membrane module having an effective membrane area of 51.8 cm2. The feed solution was pure water or a phosphate-buffered saline solution containing 1000 ppm BSA as the solute model. The trans-membrane pressure was 1 bar and the temperature was room temperature. The collected permeate was weighed, and the permeate flux $F$ was calculated from the mass of the collected permeate $m$, the effective membrane area $A$, and the time interval $\Delta t$, by using the equation below:

$$F = \frac{m}{A\,\Delta t} \quad (2)$$

The rejection $R$ of the membrane was determined by measuring the BSA concentration in the permeate $C_P$ and that in the feed solution $C_F$, using the following equation:

$$R = 1 - \frac{C_P}{C_F} \times 100\% \quad (3)$$

A UV-vis spectrophotometer was used to measure the BSA concentration. The measurement was carried out at a wavelength of 280 nm using a PG instrument T-60 (UK).

IOP Conf. Series: Materials Science and Engineering 1143 (2021) 012057 doi:10.1088/1757-899X/1143/1/012057

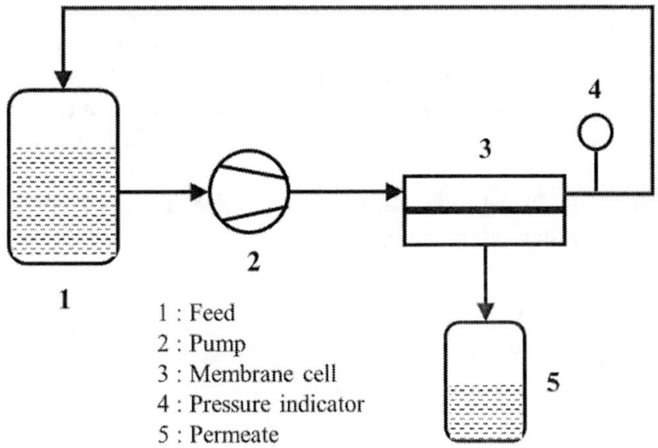


**Figure 1.** Ultrafiltration experimental set up.

## 3. Results and Discussion

The PET bottle shards and the additive (polyvinyl pyrrolidone or polyethylene glycol) were completely soluble in phenol, and the blending of both solutions resulted in clear polymer solutions. The polymer solution was then cast onto a glass plate, and was immediately submerged in the water-ethanol bath for coagulation to obtain the membranes. The average thickness of the membranes was 149 ± 13 μm. Figure 2 shows the appearance of the surfaces of the membranes. It can be seen that the surface of the PET membrane without additive (a) and that of the PET membrane with PEG 4000 additive (c) were smooth, however the surface of the PET membrane with PVP as the additive (b) contained agglomerates that were formed after the immersion of the membrane in the coagulation bath. Although PVP was completely soluble in phenol and could be blended with the PET in phenol, PVP and PET formed agglomerates during the immersion process in the coagulation bath containing water-ethanol as the non-solvent, since PVP could not easily diffused from the PET polymer matrix into the coagulation bath. Since the use of PVP additive resulted in defects on the membrane surface, PEG was used as the additive for further experiment. Furthermore, it was observed that the PET membrane without additive was stiff, while the PET membranes with the addition of PVP or PEG showed a better flexibility. It has been known that the additives such as PVP and PEG are plasticizing agents for many polymer films that increase the flexibility of polymer films [13, 14].

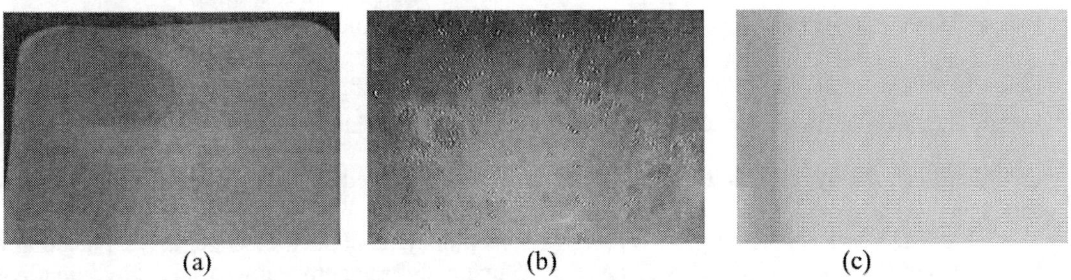

(a) (b) (c)

**Figure 2.** Photographs of the membranes (a: PET membrane without additive, b: PET membrane with PVP as the additive, c: PET membrane with PEG 4000 as the additive).

To study the influence of the additive molecular weight on the membrane performance, PET membranes were prepared using PEG 400, PEG 4000 and PEG 6000. All casting solutions were clear, indicating that PET and PEGs were completely soluble in phenol as the solvent. Figure 3 shows the photograph of the PET membranes that were prepared using PEGs with various molecular weights. The PET membranes with PEG 400 additive and PEG 4000 additive exhibited smooth surfaces without defects, however the PET membrane with PEG 6000 showed defects in form of holes that could be clearly observed visually using bare eyes. The formation of defects in the PET-PEG 6000 membrane might be caused by the rapid diffusion of PEG 6000 from the PET polymer matrix into the coagulation bath containing the non-solvent during the immersion process. This result showed that the choice of appropriate molecular weight of the PEG additive is very critical to obtain membranes without defects. Table 1 summarizes the effect of the additive on the appearance of the casting solutions and the membranes.

**Figure 3.** Photographs of the surfaces of the PET membranes using PEG additives with various molecular weights (a: PET-PEG 400, b: PET-PEG 4000, c: PET-PEG 6000).

**Table 1.** Summary of the appearances of the casting solutions and the membranes with various additives.

| Membrane | Additive | Appearance of casting solution | Membrane appearance |
|---|---|---|---|
| PET without additive | No additive | Clear solution | The membrane was stiff |
| PET-PVP | PVP | Clear solution | Membrane defects; The membrane contained agglomerates on the surface |
| PET-PEG 400 | PEG 400 | Clear solution | The membrane showed no defect, and had a good flexibility |
| PET-PEG 4000 | PEG 4000 | Clear solution | The membrane showed no defect, and had a good flexibility |
| PET-PEG 6000 | PEG 6000 | Clear solution | Membrane defects; Holes were found in the membrane |

Furthermore, the hydrophilicity of the membranes was analyzed by the measurement of the water contact angle. The result of the measurement of the water contact angle is listed in Table 2. There was a decline of the water contact angle when the PET membrane contained PEG additive, indicating an increment of the hydrophilicity. The addition of PEG in the PET membrane improved the hydrophilicity since PEG is known as a hydrophilic polymer. The increase in the hydrophilicity is important in the development of ultrafiltration membranes that are applied for water treatment. Similar phenomenon of increasing hydrophilicity by the addition of PEG was also reported for other polymer membranes such as polysulfone (PS) and polyether sulfone (PES) membranes [15].

**Table 2.** Water contact angle of the PET membranes without and with various additives

| Membrane | Additive | Additive concentration (wt%) | Water contact angle (°) | Remarks |
|---|---|---|---|---|
| PET without additive | No additive | 0 | 65.5 ± 1.4 | |
| PET-PVP | PVP | 5.9 | - | Could not be measured due to membrane defects |
| PET-PEG 400 | PEG 400 | 5.9 | 64.1 ± 1.6 | |
| PET-PEG 4000 | PEG 4000 | 5.9 | 61.5 ± 0.5 | |
| PET-PEG 6000 | PEG 6000 | 5.9 | - | Could not be measured due to membrane defects |

The microstructures of the membranes were then analyzed by using SEM. Figure 4 shows the SEM photographs of the cross section and the surface of the PET-PEG 400 membrane. The membrane showed an asymmetric structure consisting of a macro-porous cross section as a supporting layer and a thin smooth layer as the active surface of the membrane. Further, Figure 5 shows the SEM photographs of the PET-PEG 4000 membrane. This membrane also showed an asymmetric structure, however it can be obviously seen that the pores of the cross section of the PET-PEG 4000 membrane were larger than that of the PET-PEG 400 membrane. Since PEG has been known as a pore former [10,11], the presence of PEG in the polymer solution enhanced the formation of pores during the immersion process in the non-solvent. The pore size of the membrane was strongly affected by the molecular weight of the PEG. The membrane pore size increased with increasing molecular weight of the PEG, however it is important to be noted here that the use of PEG with too large molecular weight such as PEG 6000 resulted in membrane defect. Furthermore, the result of the determination of the membrane porosity using gravimetric method showed that the membrane PET-PEG 400 had a porosity of 76.8%, whereas the PET-PEG 4000 membrane had a higher porosity of 78.1%. Thus, increasing the molecular weight of the PEG resulted in an increment of both pore size and porosity of the membrane. Similar finding was also observed in other membranes such as polyether sulfone (PES) membrane and polyvinylidene fluoride-co-hexafluropropylene (PVDF-HFP) membrane [16, 17]

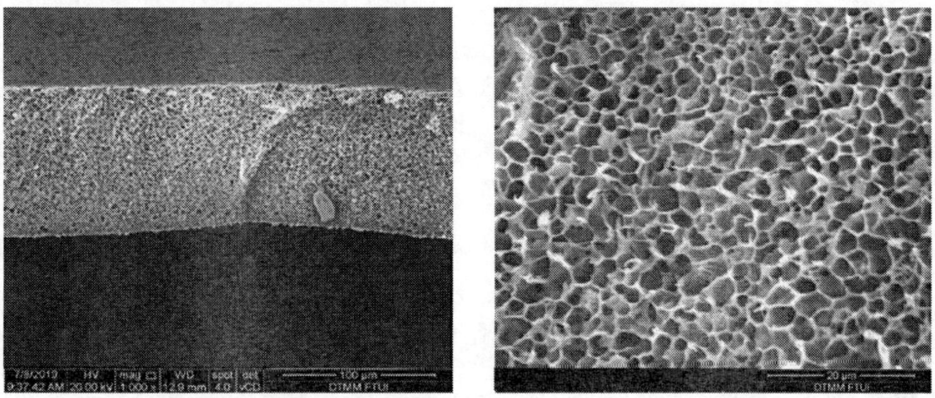

(a) (b)

**Figure 4.** SEM photographs of the PET-PEG 400 membrane (a: cross section and surface, b: cross section with magnification).

IOP Conf. Series: Materials Science and Engineering 1143 (2021) 012057 doi:10.1088/1757-899X/1143/1/012057

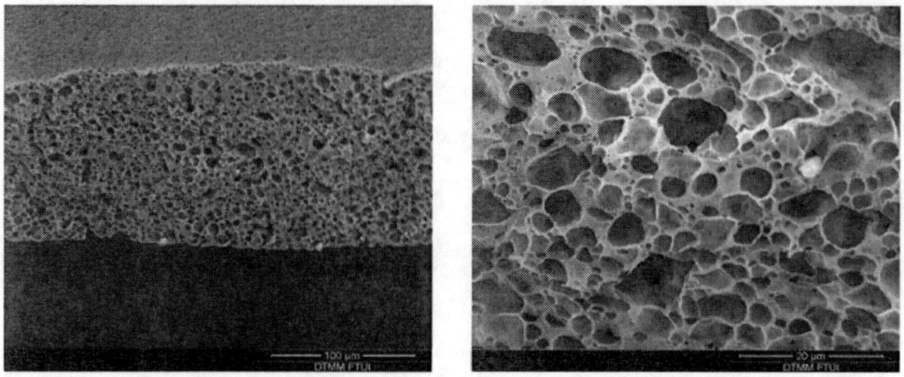

**Figure 5.** SEM photographs of the PET-PEG 4000 membrane (a: cross section and surface, b: cross section with magnification).

Furthermore, ultrafiltration experiments were conducted using the membranes that were successfully prepared without defects, namely the PET-PEG 400 membrane and the PET-PEG 4000 membrane. The other membranes, namely the PET membrane without additive, the PET-PVP membrane and the PET-PEG 6000 membrane could not be used for ultrafiltration experiment since the PET membrane without additive had a low flexibility and could not be fitted in the membrane module, while the PET-PVP membrane and the PET-PEG 6000 showed defects as mentioned previously. Table 3 depicts the permeate flux and the BSA rejection as a result of the ultrafiltration experiment. The values of the permeate flux and the rejection were taken after the ultrafiltration experiment reached a steady state condition. In comparison with the PET-PEG 400 membrane, the PET-PEG 4000 membrane exhibited a higher water permeate flux, because the PET-PEG 4000 membrane had larger pores as revealed by the SEM analysis and also higher porosity as mentioned above. Moreover, the PET-PEG 4000 membrane exhibited a higher hydrophilicity than the PET-PEG 400 membrane as measured using the water contact angle, thus resulting in a higher water permeate flux. On the other hand, the rejection of the PET-PEG 400 membrane against BSA molecules was 90.6%, which was higher than that of the PET-PEG 4000. This phenomenon is also in agreement with the result of the SEM analysis that showed smaller pores of the PET-PEG 400 membrane. Since the molecular weight of the BSA molecule is known to be 66,000 Da, the rejection value of 90.6% indicated that the PET-PEG 400 membrane had a molecular weight cut off (MWCO) of 66,000 Da. This showed that the rejection of this membrane is higher than that of the PET membrane reported in other study [10].

**Table 3.** Result of ultrafiltration experiment.

| Membrane | Additive | Permeate flux ($kg/m^2h$) | Rejection (%) | Remarks |
|---|---|---|---|---|
| PET without additive | No additive | - | - | Could not be used for UF because too stiff |
| PET-PVP | PVP | - | - | Could not be used for UF due to defect |
| PET-PEG 400 | PEG 400 | 14.7 | 90.6 | No defect |
| PET-PEG 4000 | PEG 4000 | 23.0 | 68.0 | No defect |
| PET-PEG 6000 | PEG 6000 | - | - | Could not be used for UF due to defect |

## 4. Conclusion

Ultrafiltration membranes were fabricated by utilizing PET bottle waste as the polymer material. Various additives were used for the membrane preparation. Polyethylene glycol (PEG) was found as a good additive for the PET membrane. It was revealed that the use of polyethylene glycol (PEG) additives such as PEG 400 and PEG 4000 improved the flexibility, the hydrophilicity and the porosity of the membranes, however PEG with a too large molecular weight (6000 Da) resulted in membrane defects. The use of PEG with a molecular weight of 400 Da or 4000 Da resulted in membranes without defects that showed good ultrafiltration performances as these membranes were able to retain Bovine Serum Albumin molecules as the solute model. The water permeate flux and the rejection of the membranes were strongly affected by the molecular weight of the PEG. The result of this study revealed that the utilization of PET bottle waste for the fabrication of ultrafiltration membranes can contribute in the plastic recycling and also in the efforts of the development of cheap ultrafiltration membranes for water treatment.

## Acknowledgement

The authors are very grateful to the Ministry of Research, Technology, and Higher Education of Republic Indonesia for their financial support for this work through the World Class Research (WCR) grant program 2019-2020.

## References

[1] Kusumocahyo S P, Kanamori T, Sumaru K, Aomatsu S, Matsuyama H, Teramoto M, Shinbo T 2004 Development of polymer inclusion membranes based on cellulose triacetate: carrier-mediated transport of cerium(III) *Journal of Membrane Science* **244** (1-2) 251-257

[2] Kusumocahyo S P, Ichikawa T, Shinbo T, Iwatsubo T, Kameda M, Ohi K, Yoshimi Y, Kanamori T 2005 Pervaporative separation of organic mixtures using dinitrophenyl group-containing cellulose acetate membrane *Journal of Membrane Science* **253** (1-2) 43-48

[3] Kusumocahyo S P & Sudoh M 2016 Pervaporation of Acetic Acid-Water Mixture Using Silica Membrane Prepared by Sol-Gel Method *Journal of Engineering and Technological Sciences* **48** (1) 99-110

[4] Baker R W 2004 *Membrane Technology and Applications, Second Edition* John Wiley & Son Ltd.

[5] Sanchez-Garcia M D, Gimenez E, Lagaron J M 2007 Novel PET nanocomposites of interest in food packaging applications and comparative barrier performance with biopolyester nanocomposites *Journal of Plastic Film and Sheeting* **23** (2) 133-148

[6] Nace T, Forbes 2017 Jul 26 *We're now at a million plastic bottles per minute - 91% of which are not recycled* https://www.forbes.com/sites/trevornace/2017/07/26/million-plastic-bottles-minute-91-not-recycled

[7] Hao J, Lei G, Li Z, Wu L, Xiao Q, Wang L 2013 A novel polyethylene terephthalate nonwoven separator based on electrospinning technique for lithium ion battery *Journal of Membrane Science* **428** 11-16

[8] Korolkov I V, Yeszhanov A B, Zdorovets M V, Gorin Y G, Olgun Güven, Dosmagambetova S S, Khlebnikov N A, Serkov K V, Krasnopyorova M V, Milts O S, Zheltov D A 2019 Modification of PET ion track membranes for membrane distillation of low-level liquid radioactive wastes and salt solutions *Separation and Purification Technology* **227** 115694

[9] Zdorovets M V, Yeszhanov A B, Korolkov I V, Güven O, Dosmagambetova S S, Shlimasa D I, Zhatkanbayeva Zh.K, Zhidkov I S, Kharkin P V, Gluchshenko V N, Zheltov D A, Khlebnikov N A, Kuklin I E 2020 Liquid low-level radioactive wastes treatment by using hydrophobized track-etched membranes *Progress in Nuclear Energy* **118** 103128, https://doi.org/10.1016/j.pnucene.2019.103128

[10] Rajesh S, Murthy Z V P 2014 Ultrafiltration membranes from waste polyethylene terepththalate

and additives: Synthesis and characterization *Quimica Nova* **37** (4), 653-657

[11] Hamid N A A, Ismail A F, Matsuura T, Zularisam A W, Lau W J, Yuliwati E, Abdullah M S 2011 Morphological and separation performance study of polysulfone/titanium dioxide (PSF/$TiO_2$) ultrafiltration membranes for humic acid removal *Desalination* **273** 85–92

[12] Li J H, Xu Y Y, Zhu L P, Wang J H, Du C H 2009 Fabrication and characterization of a novel $TiO_2$ nanoparticles self-assembly membrane with improved fouling resistance *Journal of Membrane Science* **326** (2) 659-666

[13] Cao N, Yang X, Fu Y 2009 Effects of various plasticizers on mechanical and water vapor barrier properties of gelatin films *Food Hydrocolloids* **23** (3) 729-735

[14] Oh H J, Freeman B D, McGrath J E, Ellison C J, Mecham S, Lee K-S, Paul D R 2014 Rheological studies of disulfonated poly (arylene ether sulfone) plasticized with poly (ethylene glycol) for membrane formation *Polymer* **55** (6) 1574-1582

[15] Liu Y, Koops G H, Strathmann H 2003 Characterization of morphology controlled polyethersulfone hollow fiber membranes by the addition of polyethylene glycol to the dope and bore liquid solution *J. Membr. Sci.* **223** 187–199

[16] Wongchitphimon S, Wang R, Jiraratananon R, Shi L, Loh C H 2011 Effect of polyethylene glycol (PEG) as an additive on the fabrication of polyvinylidene fluoride-co-hexafluropropylene (PVDF-HFP) asymmetric microporous hollow fiber membranes *Journal of Membrane Science* **369** (1-2) 329-338

[17] Idris A, Zaina N M, Noordin M Y 2007 Synthesis, characterization and performance of asymmetric polyethersulfone (PES) ultrafiltration membranes with polyethylene glycol of different molecular weights as additives *Desalination* **207** (1–3) 324–339

# Separation of Potassium from The Model Solution

**Ratna Puspita[1] and Herri Susanto[1,*]**

[1] Department of Chemical Engineering, Institut Teknologi Bandung, Bandung 40132 Indonesia

*herri@che.itb.ac.id

**Abstract.** Potassium content in biomass ashes, especially empty fruit bunches was quite high and had not been utilized maximally. This research dealt with the development of a technique for recovering potassium salt from a solution of ash extraction. Experiments were carried out usung a solution model containing KCl, NaCl, $MgCl_2$, and $CaCl_2$ with concentrations of 164.19, 8.39, 7.44, dan 39.60 mg/L respectively. Results of a preliminary experiment indicated that the solubilities of KCl and NaCl were reversed: 280.4 and 280.4 (g/L in water) to 6.9 and 14.2 (g/L in absolute ethanol), while the solubilities of $MgCl_2$ and $CaCl_2$ were still high (68.7 and 383.5 g/L in ethanol). Using these values of solubility, a combined evaporation and adding ethanol was trialed firstly to increase the precipitation of KCl. Evaporation of KCl solution 164.19 g/L until an evaporation leves of 15% did not give any precipitation yet, while adding ethanol to the remaining solution (at solution ratios of 4/1 and 3/2 ethanol) yielded precipitations with the potassium recoveries of 60% and 48% respectively. At an evaporation level of 30%, the evaporation yielded a 5% potassium recovery, while the combined technique gave 60% and 50% recoveries for adding ethanol with ratios of 4/1 and 3/2 respectively. The application of the combined technique on the mixed salts solution was modified slightly. The precipitate from the evaporation (containing KCl and NaCl) was redissolved into water and followed by adding ethanol to precipitate only KCl. Thus, two precipitates were finally obtained: (i) precipitate from adding ethanol to filtrate of evaporation, and (ii) precipitate from dissolving precipitate from evaporation. At an evaporation level of 30%, the evaporation only gave a precipitate recovery of about 10%, while the application of combined technique gave bout 80% and 60% recoveries by adding ethanol with ratios of 4/1 and 3/2 respectively. Hopefully this technique could separate KCl from its mixture with NaCl, $MgCl_2$, and $CaCl_2$.
**Keywords**: selective precipitation of potassium salt, combined evaporation and adding ethanol, solubilities of salt in water-ethanol

## 1. Introduction

Palm oil is a commodity produced in large quantities in Indonesia, which used in the food and renewable energy industries. Production of crude palm oil (CPO) is accompanied with generation of biomass approximately 58% palm oil mill effluent (POME), 21% oil palm empty fruit bunches (EFB), 5% palm kernel shell (PKS) and 14% fiber [1]. These by-products can be used as energy resources through a thermal conversion process. Combustion of fiber and PKS in the boiler of CPO mills produces ash containing various mineral such as the sufficiently high content of K2O, $Na_2O$, Al2O3, Fe2O3, CaO, and $SiO_2$ [2]. Many researches also indicate that the ash of EFB has a high potassium content [3]. However, this potential potassium for fertilizer has not been utilized maximally.

Besides potassium, ash from EFB, PKS, and fiber contain various salts of sodium, magnesium, and calcium in the form of carbonate, chloride, phosphate, and sulfate [4]. Extraction of salts from ash is a well-known separation of those minerals from ash or event directly from biomass. A problem in separating potassium from the solution is due to the close solubilities of potassium and sodium salts in water. Evaporation of salts solution and followed addition of organic compounds to the solution has been proposed in the literature. Organic compounds may reduce and change the order of solubility of sodium and potassium, as well as other salts from ash [5, 6]. Evaporation followed by ethanol addition has been investigated to separate potassium from fly ash of municipal waste incineration [7].

Results of our preliminary experiment are reported in this paper. The objective of the research is to find a procedure for precipitating KCl from its solution. In the preliminary step of research, a model solution containing KCl, NaCl, MgCl, and $CaCl_2$ was used with a composition based on our experiments on the hydrothermal process of EFB (oil palm empty fruit bunches).

## 2. Materials and Methods

### 2.1. *Materials*

All reagents used in this research were of analytical reagent grade, i.e.: KCl, NaCl, MgCl2, and CaCl2. Distilled water was used as the solvent. Analytical ethanol p.a. was used for precipitating salts from its aqueous solution.

### 2.2. *Methods*

The first series of experiments was the precipitation of KCl from its solution with a concentration of 164.19 g/L, below its solubility of about 280 g/L (measured at 30°C). Solution of KCl (80 mL) was firstly evaporated at a temperature of 50°C to evaporation levels of 15%, up to 80%. The precipitate obtained from this stage was filtered, dried and weighed (as precipitate-1, in Figure 1a). Each filtrate was treated further by adding ethanol:

(a) 4 mL ethanol to 1 mL filtrate-1, designated as evaporation and ethanol 80%

(b) 3 mL ethanol to 2 mL filtrate-1, designated as evaporation and ethanol 60%.

These additions of ethanol might give more precipitation (as precipitate-2).

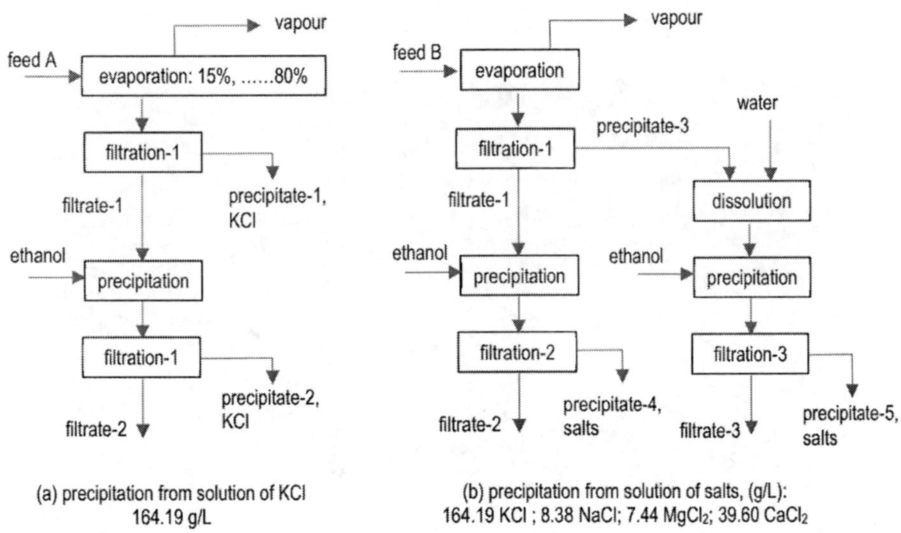


**Figure 1.** Procedure of experiment s of evaporation combined with ethanol addition.

In the second series of experiments, the feed was the model solution containing (mg/L): KCl 164.19, NaCl 8.39, $MgCl_2$ 7.44 and $CaCl_2$ 39.60. The proportion of these concentrations were about the same as those in the filtrate obtained from the hydrothermal treatment of EFB.

The experimental procedure (Figure 1b) was slightly different from that in the first series of experiments. Since the feed was the solution of four salts, precipitate-1 might contain all salts. Therefore, in order to get purer of KCl, the precipitate-3 was redissolved in water and followed by adding ethanol to precipitate KCl (precipitate-5). Meanwhile, ethanol was added to filtrate-1 to get precipitate-4. As in the first series of experiments, the ratio of ethanol to filtrate were 4/1 and 3/2 by volume. This method was based on [7].

Performance of precipitation was expressed as *potassium recovery*:

$$recovery = \frac{mass\ of\ potassium\ in\ precipitates}{mass\ of potassium\ in\ the\ initial\ solution} \tag{1}$$

## 3. Results and Discussion

Evaporations were carried out up to the evaporation levels of 15% up to 80%. Theoretically, if the evaporation level achieved 40%, the solution with the initial KCl concentration of 164.19 g/L would become a saturated solution with a concentration of 280 g/L (see Table 1). So the precipitation became more and more with increasing the evaporation level as presented visually in Figure 2.

**Table 1.** Solubility of salts, g/L (measured at 30°C).

| | KCl | NaCl | $MgCl_2$ | $CaCl_2$ |
|---|---|---|---|---|
| in water, 30°C | 280.4 | 258.6 | 336.2 | 484.2 |
| in ethanol, 30°C | 6.9 | 14.2 | 68.7 | 383.5 |

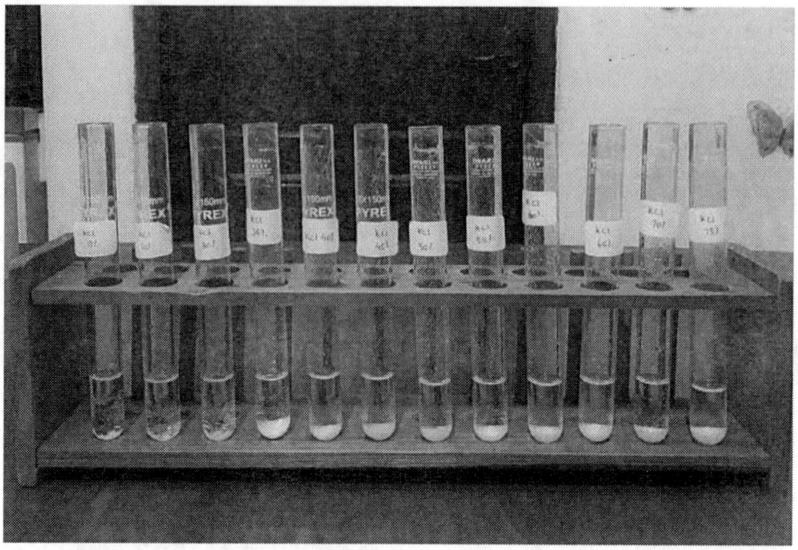

**Figure 2**. Precipitation due to addition of ethanol to filtrate after evaporation and filtration evaporation levels.

15% 20% 30% 40% 45% 50% 55% 60% 65% 70% 75% 80%

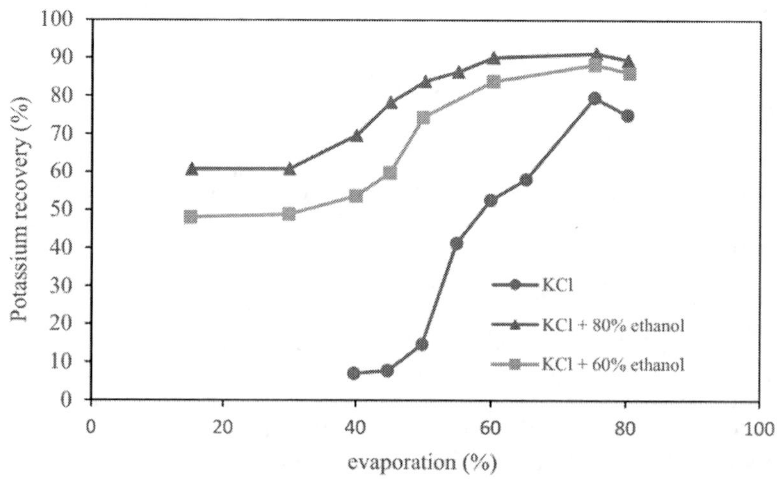


**Figure 3.** Potassium precipitation by evaporation and adding ethanol.

| | |
|---|---|
| KCl | potassium recovery from evaporation only, precipitate-1 (*see Figure 1a*) |
| KCl+ 80% ethanol | evaporation and adding 4 mL ethanol to 1 mL filtrate-1 → precipitate-1 and -2 |
| KCl+ 60% ethanol | evaporation and adding 3 mL ethanol to 2 mL filtrate-1 → precipitate-1 and -2 |

By evaporation only, the precipitation was not obtained at evaporation levels of 15%, 20%, and 30% (see Figures 2 and 3). Precipitate-1 (see diagram in Figure 1a) was obtained at the evaporation levels of 40% and more. Since the evaporations were conducted at a fixed temperature, the results of the experiment indicated that the potassium recoveries increased more or less linear with the evaporation level. The potassium recovery as a function of evaporation level is presented as the blue curve in Figure 3.

By adding ethanol, more precipitate could be obtained from filtrate-1 (see diagram in Figure 1a). Even the adding ethanol to filtrate-1 from the evaporation levels of 15%, significant precipitations were obtained, i.e. 50%, and 60% from adding of 60%, and 80% ethanol (as presented respectively with green and red dots in Figure 3). More precipitates were obtained by adding ethanol, since the solubility of KCl in ethanol was much lower than that in water (measured at 30°C, see Table 1).

Unlike the results from evaporation only, the combined effects of evaporation and adding ethanol to the potassium recovery were clearly in the form of S-curves for both adding ethanol of 60% and 80%. This phenomenon is challenging for further study with respect to the solubility of salts in an ethanol-water mixture.

From the second series of experiments, the precipitation started already at the evaporation level of about 15% (Figure 4). This is understandable because the feed solution was a solution of KCl with the concentration similar to that in the first series of experiments, plus three other salts (see Figure 1). In this solution, the solubility product of $[K^+]$ and $[Cl^-]$ became very high that caused precipitation.

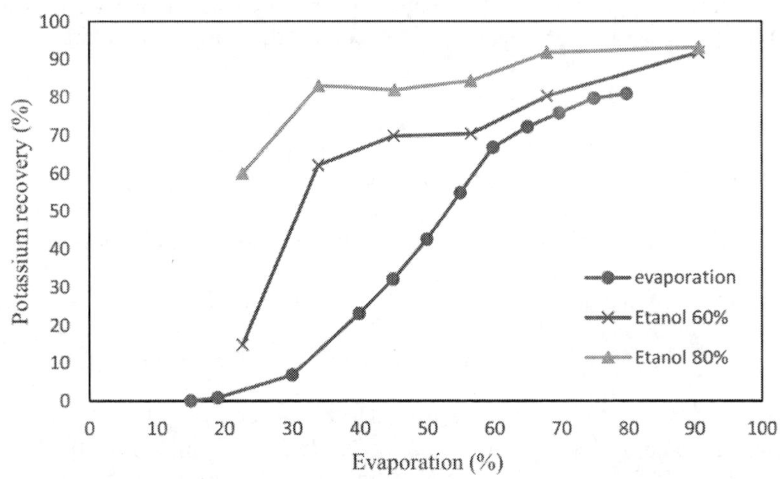


**Figure 4.** Potassium precipitation by evaporation and adding ethanol.
solution model (mg/L): KCl 164.19, NaCl 8.39, $MgCl_2$ 7.44 and $CaCl_2$ 39.60

| | |
|---|---|
| evaporation | potassium recovery from evaporation only, precipitate-3 (*see Figure 1b*) |
| ethanol 60% | evaporation and adding 4 mL ethanol to 1 mL filtrate-1 → precipitate-4 and -5 |
| ethanol 80% | evaporation and adding 3 mL ethanol to 2 mL filtrate-1 → precipitate-4 and -5 |

Based on their solubilities (Table 1), the precipitate-3 might contain KCl, NaCl, $MgCl_2$, and $CaCl_2$. Therefore, the precipitate-3 was further treated by dissolving into water and then adding ethanol, so the precipitate-4 might contain mainly KCl, a small amount of NaCl, and expectantly no-$MgCl_2$ and no-$CaCl_2$. Again, adding ethanol to the filtrate gave more precipitates. The curves green and red in Figure 4 are the sum of precipitate-4 and precipitate-5. Although the concentration of KCl in these precipitates were not measured yet, potassium content in these precipitates presumably very high based on the different solubilities of salts in ethanol.

The evaporation gave a higher potassium recovery from the mixed salts solution than that from the KCl solution (compare blue curves in Figure 4 and Figure 3). This phenomenon is understandable because of the effect of solubility product in the mixed salts solution in water. However, the potassium recoveries using evaporation and adding ethanol were the other way around. More significant precipitates were observed in the precipitation from the single salt solution than those from the mixed salt solutions. This was probably due to the solubilties of $CaCl_2$ and $MgCl_2$ in ethanol were still relatively high, while those of KCl dan NaCl dropped. These values of solubility were advantageous in keeping $CaCl_2$ and $MgCl_2$ in the water-ethanol solution instead of precipitation together with KCl and NaCl.

## 4. Conclusions

The solubilities of KCl, NaCl, $MgCl_2$, and $CaCl_2$ in ethanol were much lower than those in water. The solubility of NaCl in water was higher than that of KCl, but it changed in reverse in ethanol. This behavior was used for separating KCl from the solution of mixed chloride salts.

Compared to the method of evaporation only, the method of combined evaporation and adding ethanol could increase the recovery of KCl and advancing the incipient precipitation from the single KCl solution. Application of this combined method also indicated the increase in the precipitate recovery assuming that the precipitate contained mainly potassium.

## References

[1] Hambali E, Rivai M 2017 The potential of palm oil waste biomass in Indonesia in 2020 and 2030 *IOP Conf. Ser. Earth Environ. Sci.* **65** (1) 012050

[2] Vassilev S V, Baxter D, Andersen L K, Vassileva C G 2013 An overview of the composition and application of biomass ash, Part 1. Phase-mineral and chemical composition and classification *Fuel* **105** 40-76.

[3] Samadhi T W, Wulandari W, Amalia R A, Khairunnisah R. 2019 Potassium recovery from tropical biomass ash *AIP Conf Proc* **2085** 020003

[4] Udoetok I A 2012 Characterization of ash made from oil palm empty fruit bunches (OEFB), Int. *J. Environ. Sci.* **3** 519-524

[5] Pinho S P, Macedo E A 1996 Representation of salt solubility in mixed solvents: A comparison of thermodynamic models *Fluid Phase Equilib.* **116** 209-216

[6] Pinho S P, and Macedo E A 2005 Solubility of NaCl, NaBr, and KCl in water, methanol, ethanol, and their mixed solvent *J. Chem. Eng. Data* **50** 29-32

[7] Tang H, Erzat A, and Liu Y 2014 Recovery of soluble chloride salts from the wastewater generated during the washing process of municipal solid wastes incineration fly ash *Environ. Technol* **35** (22) 2863-2869

# Effect of Graphene Oxide on the Performance of Cellulose Acetate/ Polyethylene Glycol Membrane with Blending Method for Desalination of Brackish Water

**R D Nyamiati[1], B C Devi[1], B A Febriansyah[1], A Ramadhani[1], Y Rahmawati[1], S Nurkhamidah[1]***

[1]Department of Chemical Engineering, Institut Teknologi Sepuluh Nopember, Kampus ITS Sukolilo, Surabaya 60111 Indonesia

*Corresponding Author's E-mail : nurkhamidah@chem-eng.its.ac.id

**Abstract.** Polymeric membrane must be strong during the reverse osmosis process. Carbon-based materials like carbon nanotubes (CNTs), graphene, and its derivative graphene oxide (GO) have strong mechanical strength, high resistance to strong acids/ alkaline and organic solvents, and easy in accessibility. In this study, GO were blendingd into Cellulose Acetate/ Polyethylene Glycol (CA/PEG) membranes prepared by phase inversion method. Performance of the membrane was evaluated from the water flux, salt rejection, and permeability. Morphology, hydrofilicity, tensile strength were isvestigated by using Scanning Electron Microscopy (SEM), Fourier-Transformed Infra-Red (FTIR) and contact angle analysis, and Dynamic Mechanical Analysis (DMA), respectively. The experiment results show that hydrophilicity and porosity increase with the addition of GO with the addition of 0.005% GO exhibits the highest value of them resulting membrane with better flux and salt rejection values as compared to the GO-free membrane. With the addition of 0.005 wt% GO resulted in salt rejection (%R) 83.3% and flux permeate (F) 2194 L/m2.h. and tensile strength of 0.0148 Mpa. Membrane with CA/PEG/GO has more advantages, where GO affects the porosity of the membrane by increasing the pore size such as fingers and reducing the thickness of the wall which will result in increased membrane performance ability.

## 1. Introduction

The increasing human population combined with exploitation of water resources for domestic purposes, industry, and irrigation has resulted in a shortage of fresh water supply in many parts of the world. Only 0.5% of the 1.4 billion cubic kilometers of water in the world is accessible fresh water, which is furthermore poorly distributed across the globe [1]. Development of synthetic membranes opened up a new avenue to membrane separation tech- nology in the 1960s thus during the past 40 years, membrane separation technology has been widely applied in various fields of industries worldwide Nowadays membrane processes are needed for a wide spectrum of separations including supply of high-quality water for communities and industries (to remove surfactants, microparticles and macromolecules, organic colloidal, dissolved organic matters (DOM) etc. Food and pharmaceutical industries to obtain high-grade products, and removal or recovery of toxic or valuable components from various industrial effluents [2]. The phase inversion process induced by immersion precipitation is a well-known technique to prepare asymmetric membranes. The phase inversion is the most widely used method in the process of membranes where phase inversion has asymmetrical membrane [3], both flat sheet and hollow fiber membrane type [4][5].

Recently, carbon-based materials like carbon nanotubes (CNTs) [6], graphene, and its derivative graphene oxide (GO) [7], have shown notable potential in membrane based separation fields because of their strong mechanical strength, high resistance to strong acids/ alkaline and organic solvents, and easy accessibility. Among them, GO was served as one of the emerging nano-building materials for the fabrication of novel separation membrane owing to its distinct two-dimensional (2D) and single-atomic-thick structure, high mechanical strength, high chemical inertness, nearly frictionless surface, and good flexibility combined with large-scale and cost-effective production in solution. Based on these advantages as well as high surface-to-volume ratio structure of GO nanosheets, various GO-based membranes have been widely developed and exhibited great promise in many membrane separation fields such as gas separation, water purification, desalination, and pervaporation (PV) [8]. In recent years, rapid advances have been made in the study of nanoadditives in CA/PEG. These nanoparticles commonly include silica [9], titanium dioxide (TiO2)[5][10], $Al_2O_3$ [8][9], graphene oxide (GO) [13][14], and carbon nanotubes (CNTs) [15]–[17]. Among them, graphene oxide is a low-dimensional carbon nanomaterial and is widely used in traditional organic film preparations because of its large surface area, surface functional groups, and other characteristics. GO as another promising type of nanomaterial, has been considered as a novel high efficiency membrane material [18]. Dai et al. [19] studied permeation and ion rejection in GO nanochannels layered by layer using molecular dynamics (MD) simulation and found that laminated GO membranes with reasonable interlayer spacing exhibited high ion rejection capabilities. Meanwhile, GO has also increased research hotspots in hybrid membranes [20]. Wang et al. [21] prepared GO-blended PVDF membranes, and the contact angel decreased from 79.2° to 60.7°, which implied an improved antifouling ability. Kabiri and Namazi [22] synthesized the CA–GO nanocomposite films with 63.88 and 61.92% increase in Young's modulus and tensile strength, respectively. In a study conducted by Shi 2017 [23], the addition of GO could increase the % rejection value but the permeate flux value was still low.

In the present study, GO were introduced into Cellulose Acetate/ Polyethylene Glycol (CA/PEG) membranes for increased value of performance and characteristic of membrane. The water flux, salt rejection, Scanning Electron Microscopy (SEM), Fourier-Transformed Infra-Red (FTIR), contact angle analysis and Dynamic Mechanical Analysis (DMA) of membranes were investigated to characterize the effect of GO on the characterization and performance of CA/PEG membrane in the reverse osmosis.

## 2. Experimental

### *2.1 Material*
Cellulose acetate (CA, 30.000 Da, with acetyl content 39,8 %), polyethylene glycol-400 (PEG, 400 Da) are purchased from Sigma Aldrich. Graphene oxide (GO) with bulk density of 1.8 g/cm$^3$, was purchased from Sigma Aldrich. Aceton and Dimethylformamide (DMF) was used as solvent and water as nonsolvent

### *2.2 Membrane Preparation*
Preparation of CA/PEG polymeric membrane was started by mixing 2 gram of CA and 0.5 gram of PEG in 17 mL of acetone-DMF. GO solution was made by dissolving GO in distilled CA/PEG membrane solution with variable GO composition, which are 0; 0.0025; 0.0050; 0.0075; 0.0100; 0.0125 wt% of membrane solvent. Then, the solution mixed using hotplate at 70°C for 12 hours. After that, membrane was casted and immersed in water resulting flat sheet membrane.

### *2.3 Membrane Characterization*
Cross-sectional and surface morphologies of the membranes were observed by using scanning electron microscopy (SEM) observed by using scanning electron microscopy (SEM) (Hitachi, FlexSEM 1000, Japan) with 5 kVA. The membranes were frozen in liquid nitrogen, broken, and sputtered with gold before SEM analysis.

Fourier Transform Infrared (FTIR) spectra was performed at the wave number of 500 - 4,000 cm$^{-1}$ by using Thermo Scientific Nicolet iS10. The water contact angle between the water and the membranes was directly measured using a contact angle measuring by Camera [Sony a6000] and measuring by

IOP Conf. Series: Materials Science and Engineering 1143 (2021) 012059 doi:10.1088/1757-899X/1143/1/012059

Image J Application for evaluation of their hydrophilicity. De-ionized water was used as the probe liquid in all the measurements.

In terms of membrane's performance, water permeate flux and salt rejection are measured. Water flux and salt rejection were characterized using cross-flow system. This experimental set-up consists of a reservoir, a pump, valves, pressure gauges, and flat sheet membrane module using Sterlitech CF042A membrane cell with an effective membrane area of 42 cm$^2$. The flat sheet membrane with a size of 13 x 15.5 cm was used to analyze membrane's performance by using synthetic salt water with concentration of NaCl is 1000 ppm. Membrane flux was calculated using the following equations:

$$\textbf{\textit{Permeate FLux}} = \frac{Q}{A.\Delta t} \tag{1}$$

Where Q is the quantity of permeate (L), A is the effective membrane area (m$^2$) and $\Delta$t is the sampling time (h) [24]. Salt Rejection was calculated using the following equations [25]:

$$\%\boldsymbol{R} = \left(\frac{C_f - C_p}{C_f}\right) \boldsymbol{x}\ \mathbf{100\%} \tag{2}$$

Where R is the salt rejection, and $C_f$ and $C_p$ are the concentration of the feed and permeate (ppm).

## 3. Result and Disscusion

### *3.1. Morphology of Membrane*

The results of the FTIR analysis show the spectrum of the groups contained in the CA/PEG/GO membrane. The hydrophilicity character of a membrane can be analyzed from wavenumber at 3500 cm$^{-1}$ as the -OH group in FTIR in **Figure 1** and shows the difference in the character of CA/PEG membrane hydrophilicity by the addition of GO which varies from 0 wt% polymer to 0.0125 wt% polymer. The existence of groups in the FTIR results can be seen from the peak that appears at several wavenumber points.

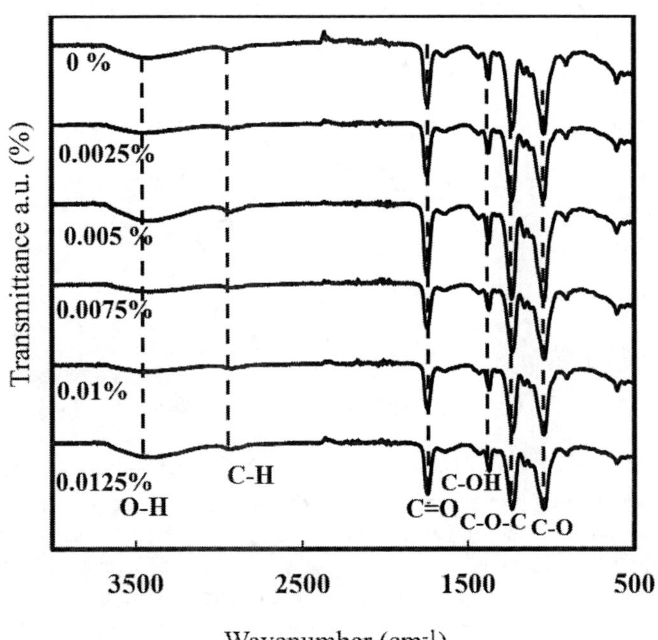


**Figure 1**. Results of FTIR spectroscopy membrane CA/ PEG/ GO with blending method.

Peak area of -OH of CA/PEG/GO membranes have been calculated from FTIR spectra as shown in Table 1. With the addition of 0.005 wt% GO, it shows the most significant increase in the peak area of the -OH group compared to other variables. Based on the peak area of the FTIR analysis graph, it is known that the addition of GO by using the blending method can increase the CA/PEG membrane hydrophilicity. This is indicated by an increase in peak area of -OH group with the addition of 0.005 %wt GO. Other evidence that can indicate changes due to the addition of GO can be explained by changes in functional groups that occur in the FTIR results, such as -OH and -COC groups [26].

**Table 1**. Peak -OH area on the FTIR graph CA / PEG / GO blending method.

| GO Content (%wt) | Area of -OH peak (area unit) |
|---|---|
| 0 | 2335.33 |
| 0.0025 | 1621.67 |
| 0.005 | 2921.00 |
| 0.0075 | 1197.33 |
| 0.01 | 2607.33 |
| 0.0125 | 1346.33 |

The results of hydrophilicity analysis for the CA/ PEG membrane with the addition of GO using the blending method can be seen in **Table 2**. In Table 2, the addition of 0.005 %wt GO exhibit the lowest contact angle values. Lower contact angle leads to the conclusion that, the membrane surface is more hydrophilic in nature. The low contact angle value indicates that the membrane has good surface hydrophilicity because GO which is hydrophilic is spread on the upper surface of the membrane during the phase inversion. This indicates that the surfaces of GO doped membranes are relatively more hydrophilic in nature. This behavior can be explained by the mechanism of action involved during the phase inversion process. During membrane formation, the hydrophilic GO migrates towards the top surface of the membrane as the top layer is more exposed to water (non-solvent). This is also clear from the IR spectra of the membranes which confirmed the presence of -OH group [27].

**Table 2.** Contact angle values on the membrane CA/ PEG/ GO with Blending method.

| GO Content (%wt) | Contact Angle (°) |
|---|---|
| 0 | 43.567 |
| 0.0025 | 44.095 |
| 0.005 | 39.517 |
| 0.0075 | 43.095 |
| 0.01 | 43.665 |
| 0.0125 | 43.895 |

SEM analysis was carried out with magnification of 2,000 times and 10,000 times to show the morphological structure of the membrane in the top surface and cross-section area. From the results of the SEM analysis, the morphological parameters observed were pore distribution, pore type, and pore diameter. Pore distribution in the blending method shows asymmetrical tendencies, where the membrane has finger-like structure and sponge-like pores. The existence of these pore variations due to the addition of GO affects the porosity of the membrane by increasing the finger-like pore size and reducing the thickness of the wall which will result in increased membrane permeability capability. The increase in porosity is also due to the nature of GO which is hydrophilic that helps form larger and wider pores. The addition of GO to the CA/ PEG membrane increases the viscosity of the membrane, causing a slower demixing process. The slow demixing process makes changes to the pore structure of the CA/ PEG membrane added with GO [28]. The results of SEM analysis for all CA/PEG/GO membrane variables by the blending method can be seen in Figure 2. The membrane is composed of a dense sponge-like pore in the outermost layer, then forms a finger-like pore in the middle layer separated by sponge-like pore walls.

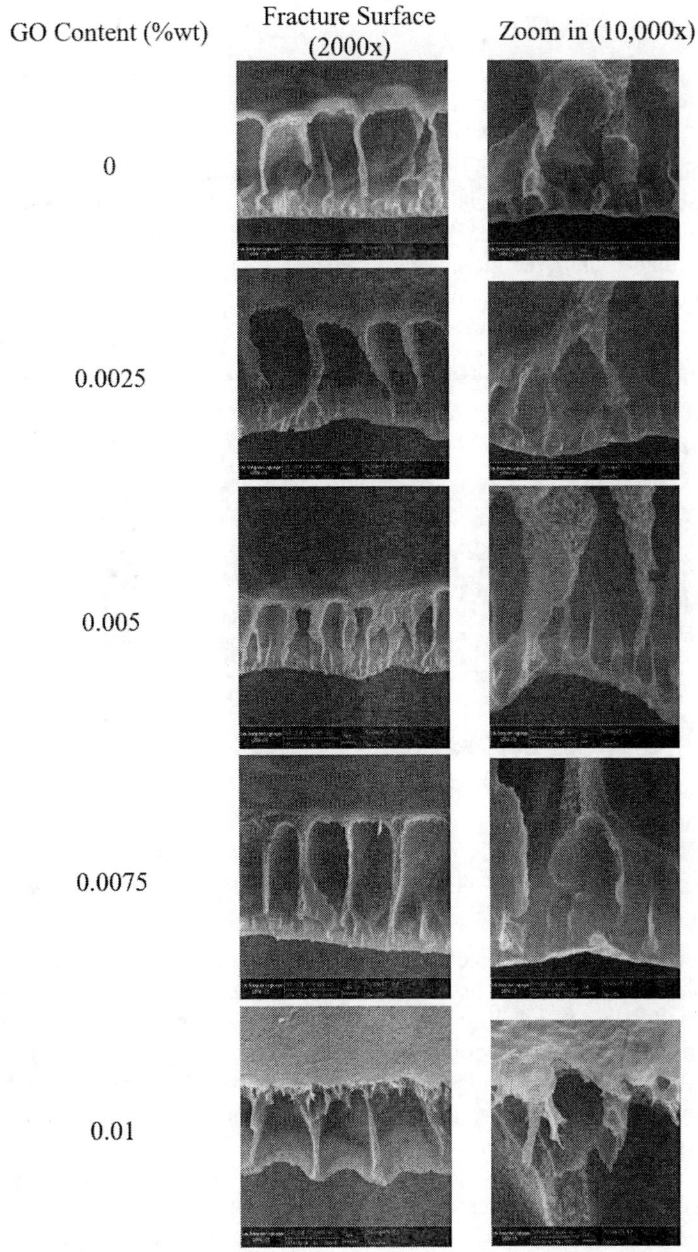

0.0125 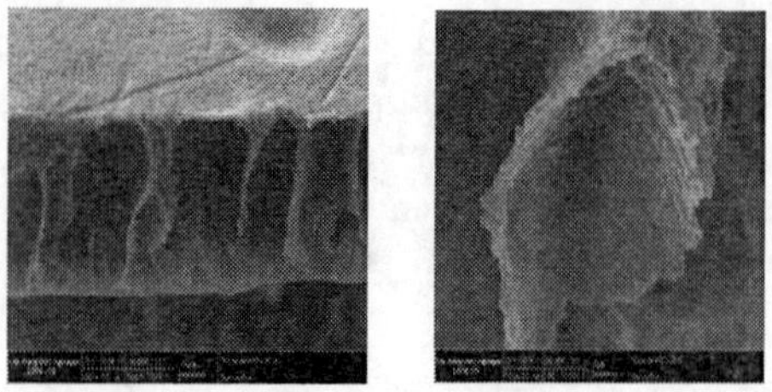

**Figure 2**. SEM morphology for CA/ PEG/ GO membrane prepared by the blending method.

*3.2. Performance of Membrane*

The DMA anaylsis shows that the CA/ PEG/ GO membrane has the higher tensile strength. It can be seen from **Figure 3** that the 0.0075 wt% GO membrane was able to handle more tensile stress before permanent deformation occurred. It could be assumed that a small concentration of GO perphaps did enhance the stress placed upon the membrane. However, a higher concentration had adverse effect on the membrane, thus causing it to fail under less stress [28].

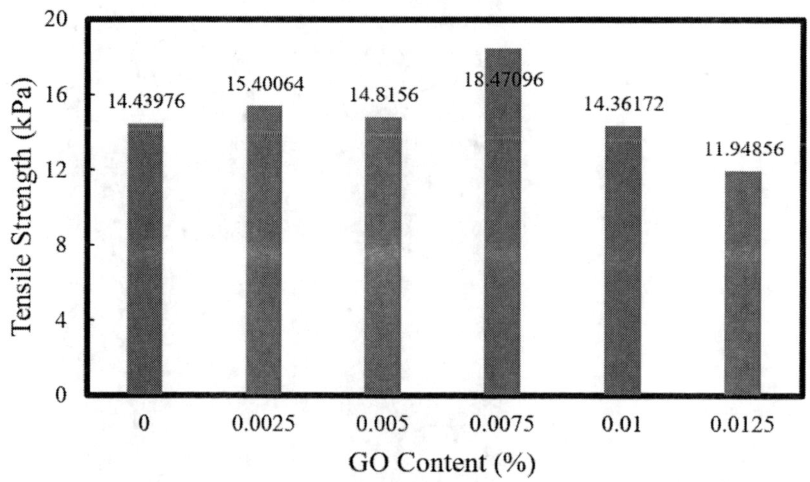


**Figure 3**. Tensile strength results of CA/PEG/GO membranes with blending method.

The results of the analysis of salt rejection performance for the CA / PEG membrane with the addition of GO using the blending method can be seen in **Figure 4**. The best salt rejection performance is 83.3% with addition GO 0.005% wt. While the lowest salt rejection performance is in the addition GO 0.0025% and 0.01 %wt which have a salt rejection value of 33%. On Figure 4, we can see a decrease in salt rejection performance, which is due to the addition of GO causing an increase in pore size which causes more salt to escape. That is because the presence of GO can cause hydrophilic properties of non-solvent which can increase the unstable thermodynamic nature of the membrane, causing nucleation which will increase pore size [29]. The mass transport in Reverse Osmosis (RO) occurs due to the solution-diffusion phenomena. One of the characteristics of RO is dense structure [30]. Reverse osmosis (RO) is a filtration method able to set aside many types molecules and large ions from solution by giving pressure on the solution that is on one side of the membrane selective. External pressure applied on the solution to fight its osmotic pressure. So that the result is water displacement from hypertonic solution to solution hypotonic [31]. The resistance to mass transfer is determined by the thickness of the membrane, the membrane with a high thickness, the permeation will decrease, while if the thickness is thin, the permeation will increase. This asymmetric membrane consists of two layers, namely a pore support layer with a thickness of 50 μm - 150 μm and a thin and dense layer on the membrane surface and has a thickness of 0.1 μm -0.5 μm this layer is often called the active layer [31].

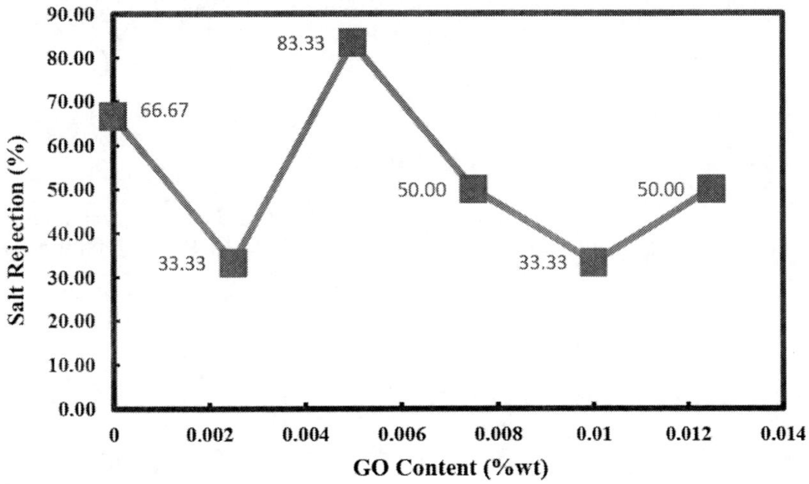


**Figure 4**. Salt rejection CA/PEG/GO membranes with blending method.

In the results of the analysis of permeate flux performance for CA/PEG/GO membranes using the blending method shows in **Figure 5**, the highest flux results obtained in the addition of GO 0.005 %wt where membrane variable has a larger macro-void size compared to other membranes. Therefore, the rate of water through a variable membrane with addition of GO 0.005 %wt is greater than that of other membranes. Meanwhile, the amount of GO addition to the CA/PEG membrane with blending method influences the tendency of permeating flux to increase on CA/PEG/GO membrane with blending method as shown on Figure 5. The best variable is the addition of GO 0.005% with Salt rejection is 82%, this is in accordance with research conducted by Shi, 2017 [23], where performance increases with added GO content 0.005%. In this study, it has an increasing of permeate flux value.

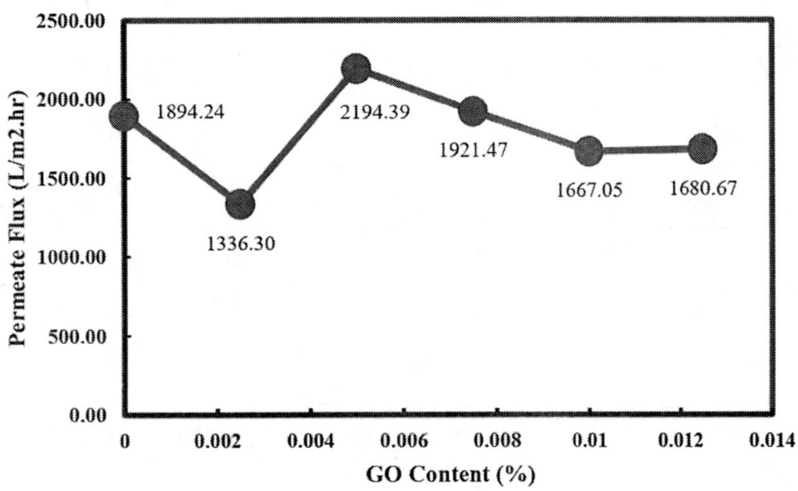


**Figure 5.** Permeat flux CA/PEG/GO membranes with blending method.

## Conclusion

GO-blended with CA/PEG membranes were successfully prepared using Aceton-DMF as solvent. The prepared membranes displayed enhanced hydrophilicity and porosity, which improved their overall flux and salt rejection as compared to the GO-free membrane. The addition of GO by the blending method contributed to improving the performance of CA/ PEG membrane with the best results on the addition of 0.005 wt% GO resulted in salt rejection (%R) 83.3% and flux permeate (F) 2194 L/m$^2$.h The results from this study highlight the role of GO and Acetone-DMF in obtaining a highly porous membrane that provides improved flux and enhanced rejection of salt without compromising on membrane performance.

## Acknowledgements

This work has been financially supported by ministry of Research, Technology and Higher Education of the Republic of Indonesia with contract's number 882/PKS/ITS/2019 to which the authors express their sincerest gratitude.

## References

[1] Humplik T *et al* 2011 *Nanotechnology* **22** 29
[2] Mohammadi T and Saljoughi E 2009 *Desalination* **243** pp 1–7
[3] Nurkhamidah S, Dwi R , Ramadhani A, and Cintya B 2019 *Indonesian Polymer Journal Article* **22** pp 10–15
[4] Saljoughi E, Amirilargani M, and Mohammadi T 2010 *Desalination* **262** pp 72–78
[5] Abedini R, Mahmoud S, and Aminzadeh R 2011 *Desalination* **277** pp 40–45
[6] El Badawi N, Ramadan A R, Esawi A M K, and El-Morsi M 2014 *Desalination* **344** pp 79–85
[7] Chae H, Lee J, Lee C, Kim I, and Park P 2015 *J. Memb. Sci* **483** pp 128–135
[8] Ma J, Ping D, and Dong X 2017 *Materials* **3** 127-149
[9] Ahmad A *et al* 2015 *Desalination* **355** pp 1–10
[10] Saleem H and Zaidi S J 2019 *Desalination* **475** p 114171
[11] Yan L, Li Y S, Xiang C B, and Xianda S 2006 *J. Memb. Sci* **276** pp 162–167
[12] Ebrahimpour M, Akbar A, and Mahmoud S 2017 *Desalination and Water Treatment* **79** 19-29
[13] Feng B, Xu K, and Huang A 2016 *Desalination* **394** pp 123–130
[14] Zhu Z, Wang L, Xu Y, Li Q, Jiang J, and Wang X 2017 *J. Colloid Interface Sci* **504** pp 429–439
[15] Kim J and Lung M, 2007 *Nanoscience and Technology* **123** pp 1407–1410
[16] El-Din L A N, El-Gendi A, Ismail N, Abed K A, and Ahmed, A I **2015** *J. Ind. Eng. Chem* **26** pp 259–264
[17] Osorio A G, Silveira I C L, Bueno V L, and Bergmann C P 2008 *Appl. Surf. Sci* **255** pp 2485–2489
[18] Nurkhamidah S, Ramadhani A, Cintya B, and Arief B 2019 *Indonesian Polymer Journal* **22** pp 16–20
[19] Yang X, Dai H, and Xu Z 2016 *The Journal of Physical Chemistry* **120** pp 1-40
[20] Sun P *et al* 2013 *ACS Nano* **1** pp 428–437
[21] Wang Z *et al* 2012 *Desalination* **299** pp 50–54
[22] Kabiri R and Namazi H 2014 *Cellulose pp 1-13*
[23] Shi Y, Li C, He D, Shen L, and Bao N 2017 *J. Mater. Sci* **52** pp 13296–13306
[24] Saljoughi E, Sadrzadeh M, and Mohammadi T 2008 *J. Memb. Sci* **326** pp 627–634
[25] Celik E, Park H, and Choi H 2011 *Water Res* **45** pp 274–282
[26] Ionita M *et al* 2015 *Compos. Part B* **72** pp 108–115
[27] Ganesh B M, Isloor A M, and Ismail A F 2013 *Desalination* **313** pp 199–207
[28] Ravishankar H, Christy J, and Jegatheesan V 2018 *Membranes* **8** 77
[29] Samu R, Moulee A, and Kumar V G 1999 *Journal of Colloid and Interface Science* **268**, pp 260–268
[30] Paul D R 2004 *Journal of Membrane Science* **241** pp 371–386
[31] Mulder 1996 Basic *Principles of Membrane Technology* 2nd Edition

# Progress, Challenges, and Prospects of Forward Osmosis (FO) and Pressure Retarded Osmosis (PRO) as An Alternative solution for Water and Energy Crisis

**Graecia Lugito*, Danu Ariono, Mochamad Rizqy Trihutama Putra, and Zoealya Nabilla Zafra**

Department of Chemical Engineering, Institut Teknologi Bandung, Bandung 40132 Indonesia.

*graecia@che.itb.ac.id

**Abstract**. Blue energy is found fascinating to be implemented in Indonesia, the largest archipelago country of which 70% territory is covered with water. The utilization of osmotic-driven membranes in addressing water and energy scarcity has received much attention. Forward osmosis (FO) and pressure-retarded osmosis (PRO) are two osmotic driven membrane processes that utilize draw solution with higher osmotic pressure than the feed solution to drive the water flux. These processes are less energy-intensive compared to other pressure-driven membranes. However, the applications of each of these processes are still limited due to three main challenges, which are the production of high-performance membrane materials for high water flux and selectivity, the selection of draw solutions, and the need for post-treatment to recover the draw solution. In this study, recent developments in FO and PRO membrane processes are reviewed, then the potentials of the utilization of osmotic-driven membrane in addressing the water and energy crisis are discussed and evaluated. The review is based on asymmetric polyamide (PA) membranes with strong porous support performing in FO and PRO processes. Draw solution plays a significant role in attaining good performance in osmotic-driven membrane processes. Integrated FO/PRO/RO hybrid processes have been proposed and evaluated in terms of their energy consumption and carbon footprint. The results indicate positive prospects of these hybrid processes pushing forward the research on continuous and self-sustaining osmotic-driven water and energy productions.

## 1. Introduction

Indonesia, of which 70% territory is covered by water, is apparently vulnerable to clean water and energy scarcity. The constant decline of the amount of available groundwater drives the government to promote seawater reverse osmosis (SWRO) desalination projects; yet these projects require large amounts of energy, investment costs, and operating costs [1]. The current status of energy use for water desalination plant is approximately 2.5 – 4 kWh/m$^3$ [2] and produce anthropogenic waste of concentrated brine [3]. As an alternative, forward osmosis membranes (FO) based on osmotic pressure have been developed to lower the energy requirements. Furthermore, FO membranes also have a lower fouling propensity due to hydraulic pressure than RO membranes [4]. However, to reclaim the drawn water molecules and regenerate the draw solution, the FO has to be integrated with another separation post-treatment [5]. Integrated FO with nanofiltration (NF) in a hybrid closed-loop system has proven to be the best in terms of economic and environmental aspects for water reclamation from wastewater. However, for seawater desalination, the FO-RO hybrid system is superior in both economic and environmental aspects. The diluted draw solution from FO can be directly processed in less energy-intensive compared to SWRO desalination [6]

Energy requirements from desalination of seawater and reclamation of wastewater are also important to note, considering that excessive energy consumption has also contributed to the climate crisis. Therefore, we need a sustainable integrated process that can supply renewable energy to produce clean water. One potential untapped renewable energy is water salination or blue energy, which comes from mixing river water and seawater [7]. There are two emerging membrane technologies in harvesting the blue energy, known under the names of pressure-retarded osmosis (PRO) and reverse electrodialysis (RED). Yip and Elimelech [3] have encompassed a comparative study between PRO and RED regarding their power densities and feasibilities for salinity energy extraction. They imply that PRO is more suitable for blue energy harvesting. Furthermore, integrating PRO on RO site as brine effluent treatment facility as well as power generation can partially offset the desalination energy cost [3].

Although there is a growing interest in the FO and PRO processes, there are still some challenges that hamper their commercial applications. Those challenges include the need for a membrane material that could mitigate concentration polarization and fouling providing high membrane performance and efficiency, the selection of suitable draw solution, and the need for post-treatment to regenerate the draw solution for the recycling process. In this study, the recent developments and technological advances of FO and PRO membranes for water and energy reclamation are reviewed and evaluated to propose their prospects in integration with SWRO membranes to overcome the water and energy crisis.

## 2. Osmotic-driven Membrane

In osmotic-driven processes, water molecules move from a feed solution with a lower salt concentration to a draw solution with a higher salt concentration. As such, this prolonged separation process will end up with a diluted draw solution. The osmotic-driven membrane has to have as minimum water transport resistance and as high solute rejection as possible yet strong enough to retain its form under the operating pressure. Both FO and PRO processes use selective FO membranes promoting the water flux from FS to DS while blocking the reverse solute flux from DS. The main challenges in FO membranes include a trade-off between permeability (water flux) and selectivity (reverse solute flux), concentration polarization, and fouling. Asymmetric membrane with an ultra-thin dense active layer (AL) and a porous support layer (SL) has been introduced to achieve a high-performance and selective separation with low hydrodynamic resistance and potentially high water flux [8]. **Figure 1** displays the typical solute concentration profiles in FO and PRO processes.

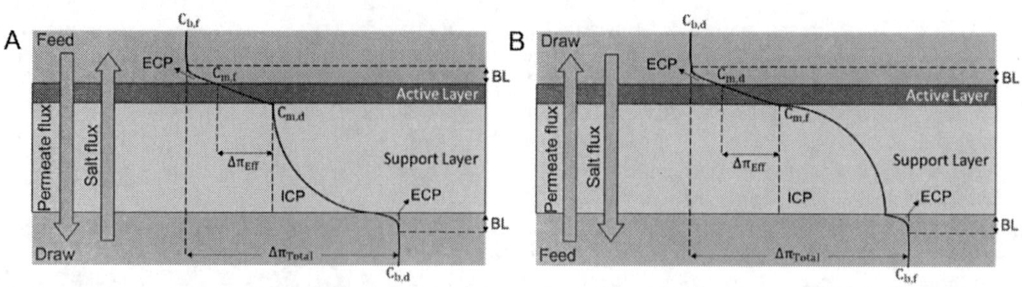


**Figure 1.** Typical solute concentration profiles in (A) AL-FS FO and (B) AL-DS PRO processes. Adapted from ref. [Bogler, 2017].

### 2.1. Membrane configurations

In FO processes, the feed solution typically flows in contact with the active layer (AL-FS mode) [4]. Ideally, the available osmotic pressure difference is shown as $\Delta\pi_{total}$ that drives the movement of water molecules from feed solution to draw solution in a determined rate ($J_w$). Under certain hydrodynamic conditions, a boundary layer (BL) can be formed on top of the dense active layer which causes additional resistance to mass transport. As such, accumulation of solutes in this layer is prone to happen, resulting in a higher solute concentration on the membrane surface ($C_{m,f}$) compared to solute concentration on the bulk phase of the feed solution ($C_{b,f}$) [9]. In contrast on the other side of the membrane, the accumulation of water molecules on the surface of the support layer results in the reduction of solute concentration from the initial concentration of draw solution on bulk phase ($C_{b,d}$). This phenomenon, called concentrative and dilutive external concentration polarization (ECP) respectively for AL and SL, creates a decline in osmotic pressure difference. On the other hand, internal concentration polarization (ICP) happens when water flux through the relatively thick support causes a lower concentration of solute on the interface between AL and SL ($C_{m,d}$) compared to $C_{b,d}$. As a result of both ICP and ECP, the initially available driving force reduces tremendously from $\Delta\pi_{total}$ to $\Delta\pi_{eff}$ [10].

While ECP has been proven to be easily overcome by manipulating hydrodynamic parameters, ICP still proves as the major drawback in FO applications. ICP in FO processes is highly related to the membrane's structural parameter (S) which is defined as a product of membrane thickness (t) and ratio of tortuosity ($\tau$) to porosity ($\varepsilon$) :

$$S = t\frac{\tau}{\varepsilon} \tag{1}$$

The higher the structural parameter of a support layer, the more severe the ICP is [4]. As such, to achieve a membrane with higher performance a thin, highly porous, and not tortuous support layer is required [11]. Besides low structural parameters, a support layer of FO membranes has to have high hydrophilicity, stability, and mechanical strength to accommodate water flux [12].

As a result of the asymmetrical configuration, the FO membrane is still prone to fouling even without additional hydraulic pressure [4]. The foulants deposit on the active layer surface foulants in the form of a loose matrix which is easy to wash by physical cleaning [13]. Even so, fouling still contributes to a decline of water flux which requires periodical cleaning hence increasing maintenance cost. One way to suppress fouling is to incorporate inorganic nanomaterials in the membrane matrix, forming a nanocomposite with enhanced properties [14]. Ultrasonication can also be done to improve the active layer's performance and provide anti-fouling properties on the active layer.

An apparent difference between the PRO and FO process is the presence of external hydraulic pressure applied to the draw solution of the PRO process to offer higher conversion of mechanical energy to electricity [10]. This additional external pressure is maintained not to be higher than osmotic pressure difference as the main driving force [15], causing water molecules to permeate from feed to draw solution. This permeate will be pressurized upon mixing with the draw solution and provide net potential energy in the draw solution flow [10,15]. Theoretically, the maximum energy density of the RO process can be achieved at hydraulic pressure equals to half of the available osmotic pressure difference [16].

In contrast to FO, PRO processes commonly operate in AL-DS configuration in which the active layer directly in contact with the draw solution, to reduce the impact of ICP and increase the specific power generation of the process [10]. Similar to that of FO processes, ICP, along with ECP, also poses a major threat in PRO processes by causing a significant drop in water flux and power density. While the ECP phenomenon is exactly the same as what happens in FO processes, instead of dilutive, concentrative ICP takes place at the interface of AL and SL of the membrane. Accumulation of solutes in the matrix of the support layer increases $C_{m,f}$ compared to $C_{b,f}$ which reduces the initial osmotic pressure difference.

Overall, membrane characteristics to attain good-performance PRO processes are similar to FO membrane characteristics. It is required for the support layer to have high hydrophilicity, stability, chemical and thermal durability, resistance to oxidizing agents, and low structural parameter [11,12]. In addition to the appliance of external pressure in the PRO process, the support layer must offer adequate mechanical strength while still exhibiting good transport properties [16].

AL-DS configuration used in the PRO process, while offering lower ICP [17] causes more severe fouling problems compared to fouling in FO processes, resulting in unstable water flux. Pores of the support layer which directly in contact with feed solution are really prone to plugging by foulants [18].

The foulants can deposit even inside the matrix of the support [19]. As such, a more complicated and difficult fouling cleaning system is required compared to FO processes [20]. To overcome this fouling problem, research has been conducted such as grafting, support pre-treatment, and feed solution pH adjustment. As for now, hydraulic backwash and flushing with alkaline cleaning agents have been used to wash off the foulants from the support matrix [21].

## 2.2. Membrane materials

The active layer of an FO membrane has to be thin, dense, and hydrophilic. Cellulose acetate (CA), cellulose triacetate (TCA), polyamide (PA), and even the biomimetic aquaporin have been considered as the candidate. However, in this review, the evaluation is made based on the polyamide active layer due to its ability to form an ultrathin layer. Depending on its monomer, PA can be either aliphatic, aromatic, or even the combination of the two. In general, aromatic PA has shown to exhibit superior properties compared to aliphatic PA ranging from a stronger polymeric chain and higher chemical stability [22].

The support layer of an asymmetric membrane mainly functions as a base to offer sufficient mechanical strength so the membrane will not deform under certain operating conditions [4]. The intrinsic properties of a membrane are closely related to the material and fabrication method used. In this regard, several studies have been conducted to fabricate supports that can provide good mechanical strength by keeping polarization concentrations low, as shown in **Figure 2**.

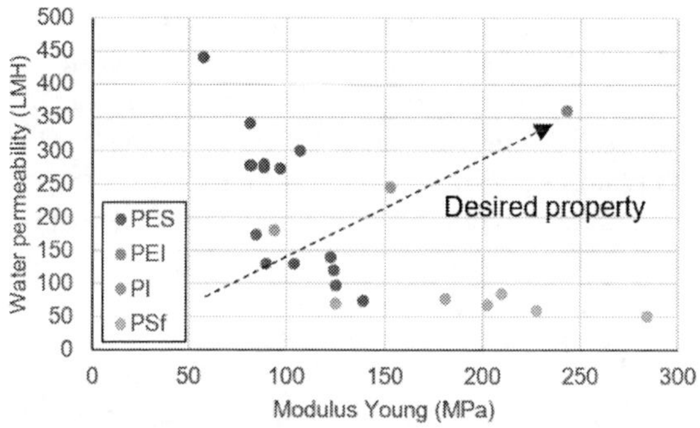


**Figure 2.** Permeability-mechanical strength diagram of various supports.

From Figure 2, it can be seen that recent studies have focused on using PES and PSf as support due to their ease of fabrication. From the few studies that have been carried out for PI and PEI, it can be seen that PI and PEI can provide support with higher mechanical strength accompanied by fairly good flux. The hydrophilic nature of PI and PEI also supports the development of these two materials for the use of PRO and FO for desalination. On the other hand, PP appears as an alternative material for the support layer due to the availability and stability of the material. The use of PP as a support material can expand the use of FO towards separation that involves organic compounds such as obtaining clean water from produced water. In the future, the authors suggest further studies on PI, PEI, or other hydrophilic materials, as well as PP as support for FO and PRO. Especially for PROs, studies on highly hydrophilic supports should also be developed to solve the fouling problem in the support layer.

## 2.3. Membrane fabrication methods

There have been various methods introduced to coat support layer with a polymer, such as dip-coating, spray-coating, spin-coating, interfacial polymerization (IP), in-situ polymerization, plasma polymerization, and grafting [23]. Among them, IP emerges as the most common route to fabricate an active layer of a membrane due to its capability of forming an ultrathin layer and simple reaction mechanism [24]. In general, IP is conducted in five steps: pre-treatment of support, the immersion of support in aqueous monomer, removal of excess solution, the immersion of support in organic monomer, and post-treatment [10]. The precursor solutions, both aquatic and organic, contain two different reactive monomers and are immiscible [25]. In some cases, only one of the precursors is required to contain monomers while the other only keep the reaction to only occurs on the interface of the two precursors [26].

Properties of the active layer are significantly affected by monomer concentration and composition used in the IP process [24]. Petersen, on the other hand, has reported that diffusivity and solubility of the polymer in solvent precursors are among the most important factors that control membrane properties [23]. In the most recent study, ultrasonic power could enhance the PA coverage on the support surface leading to a more hydrophilic membrane surface and higher water flux while simultaneously forming a thick layer with excellent rejection as well as lower fouling propensity [27].

## 3. Draw solutions

The need for a draw solution with high osmotic pressure and low salt return flux is still very much needed to realize good FO and PRO performance. Previous studies have mostly explored inorganic and organic draw solutions based on their water flux and their salt return flux. A graph comparing several types of inorganic and organic draw solutions is presented in **Figure 3** as follows.

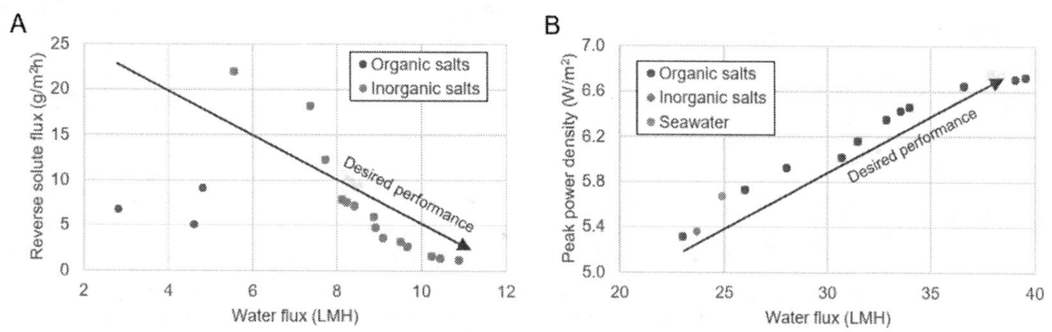


**Figure 3.** (A) Reverse solute-water flux diagram of various draw solution for FO processes (B) Peak power density-water flux diagram of various draw solution for PRO processes.

Based on **Figure 3**, most of the inorganic draw solutions are still superior to organic draw solutions because of their high-water flux and low salt return flux. Therefore, studies regarding the use of inorganic draw solutions for the FO process are considered feasible to be developed. The use of seawater as a draw solution is also a good option, considering that seawater is a widely available source of inorganic draw solutions.

For the PRO process, the draw solution with organic salts has a high-water flux which is also accompanied by a large energy density. Based on this study, several types of organic draw solutions turned out to have high osmotic pressures accompanied by lower salt fluxes than some types of inorganic draw solutions that are often studied. This shows that some organic salts are also interesting for further research as an alternative to lab-scale PRO process draw solutions.

However, apart from having high osmotic pressure and low salt return flux, draw solutions for a PRO process should also be widely available, particularly if they are used to produce large amounts of energy. Therefore, the PRO membrane currently being developed generally uses seawater as a draw solution and river water as a feed solution. However, this method has not been proven to be economically sound because the difference in salinity between seawater and river water is not too large so that the energy generated is not comparable to the operational costs of the PRO. In addition, the use of seawater as a draw solution requires a pre-treatment step to maintain the sustainability of the PRO membrane.

Several studies were conducted to find an alternative to seawater as a PRO membrane draw solution. One of them is the use of RO retentate as a draw solution. Retentate RO which has high salinity, high pressure, and has gone through a pre-treatment process. RO brine has an osmotic pressure of 1.6 to 2 times higher than seawater so that the energy density generated from the PRO process can increase up to 4 times as high as the use of seawater. In addition, the use of RO brine as a draw solution for the PRO process can mitigate desalination waste that can damage marine ecosystems. The development of a suitable PRO membrane support layer material also allows the use of wastewater from various oil and gas processing processes as a feed solution for the PRO process. The large salinity difference between RO brine and wastewater can be exploited to produce large amounts of energy. Therefore, future PRO studies can focus on the use of RO desalination retentate as a draw solution and river water as a feed solution.

## 4. Post-treatment and case study

Regeneration of draw solutes and reclamation of the drawn water molecules from diluted draw solutions to produce clean water is still the main challenge in the FO process. The concentration of draw solutes for the FO process is important because replenishment costs for draw solutes are a significant component of operational costs. The draw solutes reconcentration method is chosen based on the application of the FO process itself. Some of the post-treatments carried out to concentrate draw solutes include heating, distillation, the NF process, the RO process, and so on. As explained in the background, several ways to overcome the water crisis through the FO process are wastewater reclamation and seawater desalination. For FO applications as wastewater reclamation, the FO-NF hybrid closed-loop system has proven to be the best in terms of economic and environmental aspects compared to other hybrid closed-loop systems. However, for the application of FO in the seawater desalination process, the FO-RO hybrid system is superior in both economic and environmental aspects, because the energy consumption of RO desalination is drastically reduced with FO as the pre-treatment of the RO process. That way, the product diluted draw solution from FO can be further processed in the form of a feed solution from the RO process.

In the PRO process, post-treatment that needs to be done is the conversion of hydrostatic energy into electrical energy. This can be achieved by flowing the diluted draw solution to turn the turbine. Furthermore, the PRO process can also be combined with the RO process to supply the RO process energy into a PRO-RO hybrid system. Therefore, the RO process acts as a post-treatment of the PRO process by utilizing the salinity energy of the PRO process as a power source for the high-pressure pump used to pump the RO process feed. For PRO processes that use lab-scale inorganic or organic salts, the draw solutes reconcentration process still needs to be done to obtain draw solutes so that they can be reused. However, for PRO membranes that use seawater or RO brine as a draw solution, the reconcentration process is no longer a problem, but a new problem arises, namely fouling. Therefore, the next development of PRO membrane post-treatment lies in innovative and effective membrane cleaning techniques.

Based on a comprehensive review of this study, the water and energy crisis still needs resolution. One source of water and energy that can be utilized simultaneously is seawater, through a process of desalination of seawater using various membrane processes. Indonesia is a coastal country with a wide sea line so that it has the potential to develop desalination of seawater as a source of water and new renewable energy. For further research, the FO-RO-PRO hybrid closed-loop system is a good system to develop because of its potential to be able to answer the three main challenges of FO and PRO membranes, such as the research conducted by Cheng et al. in 2018 [28]. A schematic of this combined process is shown in **Figure 4**.

Reconcentration and recovery of draw solution in osmotic-driven membranes is done to significantly reduce the cost of draw solution replenishment by allowing reuse of the draw solution. In the case of FO, this additional process is a crucial part; without a post-treatment, clean water is not obtainable and still in form of a diluted draw solution. Hybrid systems in certain applications have been previously proposed to negate the need for complicated post-treatment processes. In impaired water reclamation using FO, a hybrid system of FO-NF has been proposed and proven to have a sustainable and economically viable continuous closed-loop of osmotic-driven process. In seawater desalination, however, integration of FO as RO feed pretreatment can result in energy reduction. Another example is the hybrid system of RO-PRO, in which the diluted draw solution of the PRO process can be directly flowed to move water turbines and safely discharged while continuously supplied from RO as concentrated brine.

The use of the hybrid RO-PRO system has been previously mentioned as a way to simultaneously desalinate seawater and reclaim wastewater instead of using river water, addressing the water crisis. This system, unfortunately, faces a major problem in which PRO membranes are very susceptible to foulants in the feed solution. As such, Cheng et al. in 2018 proposed a hybrid system of FO-PRO-RO for simultaneous seawater desalination and wastewater reclamation as shown in **Figure 4**. In this system, FO acts as a bridge to prevent direct contact of wastewater feed to PRO support, removing potential severe fouling in the PRO system using an interloop cycle of additional draw solution. The diluted draw solution from the FO system is then flowed to the PRO system where it is reconcentrated and used to produce energy by direct mixing with RO brine from the RO system. Theoretically, this proposed hybrid system potentially constructs self-sustaining and continuous osmotic energy and water production.

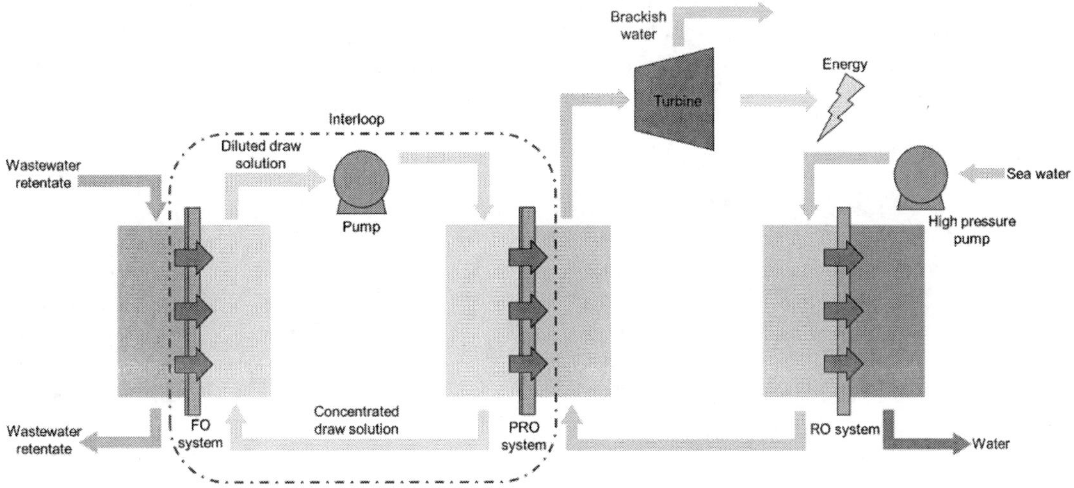


**Figure 4.** Simulation case study of FO-PRO-RO hybrid system. Adapted from ref. [29].

In this system, the interloop concentration of the draw solution acts as an important parameter to be determined as it affects both FO and PRO performance contrarily. A high interloop concentration will increase the driving force in the FO system but simultaneously decrease the power density generated by PRO as a result of water flux decline. To determine an optimum interloop concentration, a simplistic simulation is conducted to see the effect of interloop concentration variation, ranging from 0 to 0.6 M, to process parameters of another system. In this simulation, RO is operated under 50% recovery and 60 bar pressure referring to the commercial application of seawater desalination RO. Dilution ratio, defined as the ratio of diluted solution TDS to that of concentrated solution, of 0.5 is applied at the PRO and FO system to produce diluted draw solution with the same salinity as seawater, avoiding environmental damage of discharge [28]. Applied retarded pressure in the PRO system is assumed to be half of the available osmotic pressure [29] Lastly, membrane intrinsic properties, water permeability coefficient (A), salt permeability coefficient (B), and structural parameter (S) are listed in **Table 6.1** from previous studies. Water-salt diffusivity (D) of 1.48 x 10-9 m/s2 is obtained from ref. [28].

**Table 1.** Intrinsic properties of the membrane used in the simulation.

| System | Parameters | | | Ref. |
|---|---|---|---|---|
| | A (μm/s Pa) | B (μm/s) | S (mm) | |
| FO | 6.20 | 0.056 | 0.59 | [30] |
| PRO | 4.22 | 0.067 | 0.61 | [31] |
| RO | 3.33 | 0.069 | 0.14 | [32] |

Four system parameters are evaluated in this simulation: FO and PRO water flux, PRO power density, net specific energy consumption (Net SEC) of the overall hybrid system, and required membrane are of FO and PRO system as shown in **Figures 5** Net SEC is estimated by subtracting total energy produced by PRO system from RO high-pressure pump energy requirement.

**Figure 5A** confirmed that an increase in interloop concentration will result in higher water flux in the FO process while a decline in PRO water flux, which in turn decreases the power density of the PRO system as shown in **Figure 5C.** In **Figure 5D**, it can be seen that the use of lower interloop concentration will result in a lower net SEC which is desirable towards self-sustaining osmotic energy and water production. While energy-wise it looks better to operate with lower interloop concentration, the very low water flux in FO operation results in an infinite required membrane is to keep the amount of draw solution in the interloop constant as shown in **Figure 5B.**

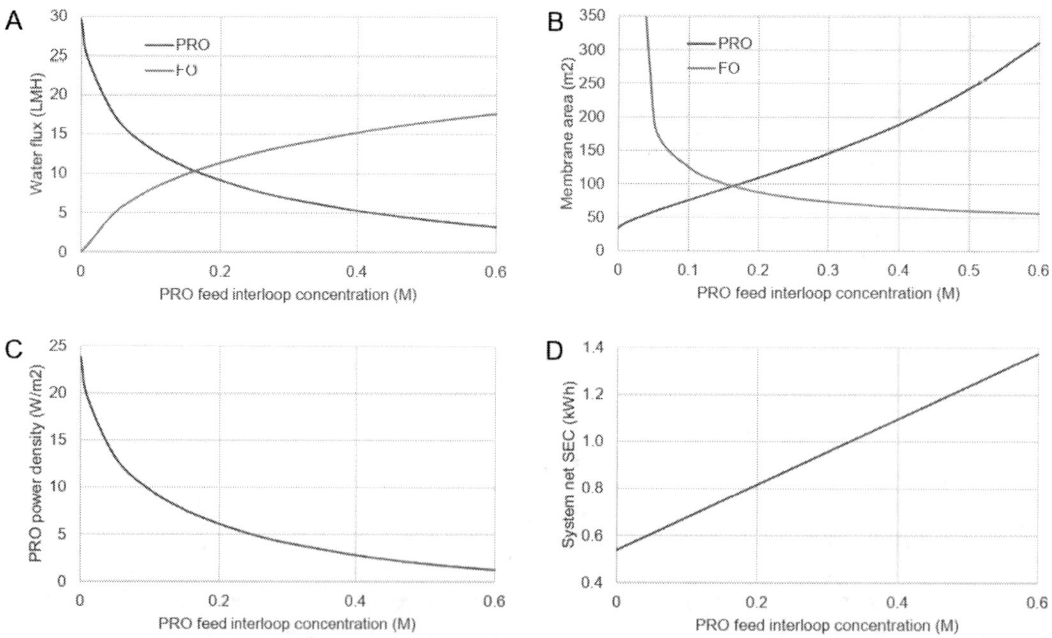


**Figure 5.** Effect of interloop concentration to: (A) FO and PRO water flux and (B) required membrane area as well as to (C) PRO power density and (D) net SEC of the overall hybrid system.

In this simulation, a trade-off point of 0.163 M interloop concentration is chosen as a base of further analysis to operate this hybrid FO/PRO/RO system with equal FO and RO membrane area. At 0.163 M interloop concentration, the system is simulated to be capable of producing 7.15 W/m2 while requiring a net SEC of 0.76 kW electricity for every 1 m3/hour clean water produced. Compared to the stand-alone RO process with the same parameters which costs 2.15 kW, simulation-wise this hybrid system can reduce energy consumption by up to 64% (1.39 kW). This electrical energy reduction, according to the United States Environmental Protection Agency, is equal to a carbon emission reduction of 0.98 kg/hour. Results of this simulation are summarized in **Table 2**.

**Table 2.** Simulation result of hybrid FO/PRO/RO system.

| Parameter | Value | Unit |
|---|---|---|
| Concentration | | |
| RO feed (seawater) | 0.600 | M |
| RO brine | 1.200 | M |
| PRO discharge | 0.600 | M |
| Interloop (PRO feed) | 0.163 | M |
| Interloop (FO draw) | 0.325 | M |
| RO Recovery | 50 | % |
| PRO Dilution Ratio | 0.5 | |
| FO Dilution Ratio | 0.5 | |
| Applied Pressure | | |
| RO system | 60.0 | bar |
| PRO system | 25.9 | bar |
| Membrane area | | |
| FO | 96.6 | $m^2$ |
| PRO | 96.6 | $m^2$ |
| RO | 28.8 | $m^2$ |
| Water flux | | |
| FO | 10.35 | LMH |
| PRO | 10.35 | LMH |
| RO | 34.73 | LMH |
| Efficiency | | |
| Pump | 80 | % |
| Pressure Exchanger | 95 | % |
| Water Turbine | 90 | % |
| PRO power density | 7.15 | $W/m^2$ |
| Net SEC | 0.76 | kWh |
| Global warming impact | | |
| Electricity reduction | 1.39 | kWh |
| $CO_2$ emission reduction | 0.98 | kg $CO_2$/jam |

Although simulation-wise this hybrid system has shown a lot of potentials both by performance and environmental impacts, as this simulation is still based on an ideal operating condition, FO and PRO with susceptible mechanical strength and low reverse solute flux are still necessary to carry out operating under high pressure with high water flux. Economic analysis and thorough life cycle assessment are still required to be conducted to further study the feasibility of this system as a solution towards self-sustaining and continuous osmotic energy and water production.

## 5. Conclusion

Osmotic-driven membranes have been garnering a lot of interest in membrane processes due to their low energy consumption. Although showing excellent potentials theoretically, these membranes still face concentration polarization and other major problems that prohibit their commercial use due to their poor performance. To achieve better performances, materials play an important role to tune membrane properties which in turn affect separation performance. An idea of using RO brine as PRO draw solution has also been proposed to obtain a high-osmotic pressure draw solution while simultaneously reducing the environmental effect of brine discharge. As for post-treatment necessity, application of hybrid systems such as FO/RO, PRO/RO, or even FO/PRO/RO have been previously suggested to remove the need for a post-treatment system. The latest invention of FO/PRO/RO hybrid system simulation-wise is capable of saving 0.42 kWh of electricity per hour for each 1 m3/hour water produced. This reduction of electricity is equivalent to 0.99 kg CO2 emission reduction per hour. However, as this S is still based on ideal operating conditions, further studies in the economical and life-cycle aspects of this system are required.

**References**

[1] Le N L, Nunes S P 2016 Materials and membrane technologies for water and energy sustainability *Sustainable Materials and Technologies* **7** 1–28.

[2] Voutchkov N 2018 Energy use for membrane seawater desalination – current status and trends *Desalination* **431** 2-14.

[3] Yip N Y, Elimelech M 2014 Comparison of Energy Efficiency and Power Density in Pressure Retarded Osmosis and Reverse Electrodialysis *Environmental Science and Technology* 48 11002–12

[4] Eyvaz M, Arslan S, Imer D, Yüksel E, Koyuncu I 2018 Forward Osmosis Membranes – A Review : Part I *Osmotically Driven Membrane Processes - Approach, Development and Current Status* 11–40.

[5] Nagy E 2019 Forward Osmosis *Basic Equations of Mass Transport Through a Membrane Layer* 447–56.

[6] Chekli L, Phuntsho S, Vigneswaran S, Shon H K, Kandasamy J, Chanan A 2012 A review of draw solutes in forward osmosis process and their use in modern applications *Desalination and Water Treatment* **43** 167–84

[7] Aaberg R J 2003 Osmotic power: A new and powerful renewable energy source? *Refocus* **4** 48–50

[8] Scott K 1998 Introduction to membrane separation *Handbook of Industrial Membranes* 3–185.

[9] Bogler A, Lin S, Bar-Zeev E 2015 Biofouling of membrane distillation, forward osmosis and pressure retarded osmosis: Principles, impacts and future directions *Journal of Membrane Science* **542** 378-98.

[10] Alsvik I L, Hägg M B 2013 Pressure retarded osmosis and forward osmosis membranes: Materials and methods *Polymers* **5** 303–27

[11] Tiraferri A, Yip N Y, Phillip W A, Schiffman J D, Elimelech M 2011 Relating performance of thin-film composite forward osmosis membranes to support layer formation and structure *Journal of Membrane Science* **367** 340–52

[12] Han G, ChungT S, Toriida M, Tamai S 2012 Thin-film composite forward osmosis membranes with novel hydrophilic supports for desalination *Journal of Membrane Science* 423–24:543–55

[13] Lee S, Boo C, Elimelech M, Hong S 2010 Comparison of fouling behavior in forward osmosis (FO) and reverse osmosis (RO) *Journal of Membrane Science* **365** 34–9

[14] Qin D, Liu Z, Sun D D, Song X, Bai H 2015 A new nanocomposite forward osmosis membrane custom-designed for treating shale gas wastewater *Scientific Reports* **5** 14530

[15] Achilli A, Cath T Y, and Childress A E 2010 Selection of inorganic-based draw solutions for forward osmosis applications *Journal of Membrane Science* **364** 233–41

[16] Han G, Zhang S, Li X, Chung T S 2015 Progress in pressure retarded osmosis (PRO) membranes for osmotic power generation *Progress in Polymer Science* **51** 1–27

[17] Klaysom C, Cath T Y, Depuydt T, Vankelecom I F J 2013 Forward and pressure retarded osmosis: potential solutions for global challenges in energy and water supply *Chemical Society Reviews* **42** 6959–89

[18] Tang N, Jia Q, Zhang H, Li J, Cao S 2010 Preparation and morphological characterization of narrow pore size distributed polypropylene hydrophobic membranes for vacuum membrane distillation via thermally induced phase separation *Desalination* **256** 27–36

[19] Mi B, Elimelech M 2008 Chemical and physical aspects of organic fouling of forward osmosis membranes *Journal of Membrane Science* **320** 292–302
[20] Yip N Y, Elimelech M 2013 Influence of natural organic matter fouling and osmotic backwash on pressure retarded osmosis energy production from Natural Salinity Gradients *Environmental Science and Technology* **47** 12607–16
[21] Han G, Zhou J, Wan C, Yang T, Chung T S 2016 Investigations of inorganic and organic fouling behaviors, antifouling and cleaning strategies for pressure retarded osmosis (PRO) membrane using seawater desalination brine and wastewater *Water Research* **103** 264–75
[22] Raaijmakers M J T, Benes N E 2016 Current trends in interfacial polymerization chemistry *Progress in Polymer Science* **63** 86–142
[23] Petersen R J 1993 Composite reverse osmosis and nanofiltration membranes *Journal of Membrane Science* **83** 81–150
[24] Gohil J M, Ray P 2017 A Review on Semi-Aromatic Polyamide TFC Membranes Prepared by Interfacial Polymerization : Potential for Water Laboratory for Advanced Research in Polymeric Materials ( LARPM ), Central Institute of Plastics Engineering & Technology Reverse Osmosis Membrane *Separation and Purification Technology* **181** 159–82
[25] Morgan P W 2011 Interfacial Polymerization *Encyclopedia of Polymer Science and Technology*
[26] Bean K, Black C F, Govan N, Reynolds P, Sambrook M R 2012 Preparation of aqueous core/silica shell microcapsules *Journal of Colloid and Interface Science* **366** 16–22
[27] Shen L, Hung W S, Zuo J, Tian L, Yi M, Ding C, Wang Y 2020 Effect of ultrasonication parameters on forward osmosis performance of thin film composite polyamide membranes prepared with ultrasound-assisted interfacial polymerization *Journal of Membrane Science* **599** 117834.
[28] Cheng Z L, Li X, Chung T S 2018 The forward osmosis-pressure retarded osmosis (FO-PRO) hybrid system: A new process to mitigate membrane fouling for sustainable osmotic power generation *Journal of Membrane Science* **559** 63–74
[29] Luo L, Wang P, Zhang S, Han G, Chung T S 2014 Novel thin-film composite tri-bore hollow fiber membrane fabrication for forward osmosis *Journal of Membrane Science* **461** 28–38
[30] Wang R, Shi L, Tang C Y, Chou S, Qiu C, Fane A G 2010 Characterization of novel forward osmosis hollow fiber membranes *Journal of Membrane Science* **355** 158–67
[31] Chou S, Wang R, Fane A G 2013 Robust and High performance hollow fiber membranes for energy harvesting from salinity gradients by pressure retarded osmosis *Journal of Membrane Science* **448** 44–54
[32] Xiaoxiao S, Prince J A, Sun D D 2016 Relating Water/Solute Permeability Coefficients to the Performance of Thin-Film Nanofiber Composite Forward Osmosis Membrane *Journal of Membrane Science and Technology* **6** 4

**Acknowledgement**
This research has been financially supported by Research Programs, Community Services, and Innovation, Institut Teknologi Bandung (P3MI ITB).

# Regeneration of Spent Bleaching Earth for PLA Nanocomposite Filler

**Hizkia M V Gultom[1], Tika Paramitha[2] and Johnner P Sitompul[1]**

[1]Department of Chemical Engineering, Institute of Technology Bandung, Bandung 40132, Indonesia
[2]Department of Chemical Engineering, Universitas Sebelas Maret, Surakarta, 57126, Indonesia

sitompul@che.itb.ac.id

**Abstract**. This paper concerns on regeneration of spent bleaching earth (SBE) as a filler in biodegradable nanocomposite. The nanocomposite were produced from two different production methods, solvent casting and extrusion. SBE can be used as filler after being regenerated by removing oil and impurities. A regeneration method for SBE was conducted using chemical treatment. Regeneration process was proved to be succesful as shown in FTIR with the absence of peak at 2850 $cm^{-1}$ – 2930 $cm^{-1}$ and 1730 $cm^{-1}$ indicating the disappearance of free fatty acids and ester bonds from regenerated bleaching earth (RBE). The RBE was then applied as filler for PLA-Nanocomposite, biodegradable plastic, a suitable substitute for conventional plastic. The production of nanocomposite used two different surfactants, namely octadecyl amine (ODA) and trimethyl stearyl ammonium chloride (TSC) at two different concentration (20 mmol and 40 mmol). The mechanical property of PLA-Bentonite nanocomposite was then analyzed for tensile strength and permeability. The highest tensile strength and lowest gas permeability was obtained by nanocomposite that used 40 mmol TSC as surfactant, with 12.48 MPa and 0.017 g/day, respectively. Moreover, addition of regenerated bleaching earth to PLA-Nanocomposite during production using extrusion and solvent casting had slight different effect. XRD pattern of all extruded PLA-nanocomposite samples indicated the formation of exfoliated structure, as shown in XRD pattern with very low intensity peak around 2 nm at $2\theta = 5$, while only a few of samples of PLA-nanocomposite created by solvent casting indicating the same structure.

## 1. Introduction

PLA is an environmental-friendly polymer because it can easily degrade and can be produced from biomass material. Apart from that, PLA has the advantage of simple processing and biocompatible [1]. However, PLA has several weaknesses such as fast physical aging, sensitivity to water, limited impact resistance, and high rigidity [2]. Therefore, PLA modification really needs to be done to correct its weaknesses, such as modification with plasticization [3] or modification with addition of fillers. Commonly used fillers in PLA modification are nanoclay [4] and cellulosic fibers [5]. In Indonesia, bentonite is one of the most abundant clay.

Bentonite is used during the bleaching step of palm oil. After use, the bleaching earth loses its adsorption properties and becomes a waste dumped in landfills without any treatment. This waste, known as spent bleaching earth (SBE), contains up to 30%-wt residual oil that produces unpleasant odors and classified as hazardous waste [6]. Reactivation and reuse of SBE have been proposed as a way to reduce the pollution arising from the inappropriate disposal of this waste. Some studies have examined the possibility of using reactivated SBE as a sorbent for different pollutants [7][8][9].

The activation process for SBE aims to create a distance between bentonite surfaces (distance between layers, so-called basal spacing), in order that regenerated bentonite can form a variety of nanocomposites shapes with PLA [10]. The basal spacing is then improved by surfactants [11], octadecylamine (ODA), and trimethyl stearylammonium chloride (TSC), in the activation process. A comparison of both surfactants need to be conducted.

Oil removal from SBE can be done by the extraction process. The main advantages of extraction methods are a short time and the use of minimum solvents. Extraction is carried out within a short time to minimize the potential degradation of the active components of bentonite. There are several methods of oil removal from SBE such as soxhlet extraction [12], membrane technology [13], subcritical water technology [14], and supercritical liquid extraction [15].

In this study, we used the soxhlet extraction method. Soxhlet extraction has traditionally been used for a solid sample with limited solubility in a solvent in the presence of insoluble impurities. A porous thimble loaded with a solid sample is placed inside the main chamber of the soxhlet extractor. By refluxing the solvent through the thimble using a condenser and a siphon sidearm, the extraction cycle is typically repeated many times. Soxhlet extraction is a rugged, well-established technique and permits unattended extraction [16].

Concerning nanocomposite, solvent casting is mostly used for the preparation of films containing nanocomposite scaffolds. This method is widely used because of low costs, shorter preparation time, and most importantly easy variation in reaction conditions [17]. On the other hand, the extrusion method can operate a continuous operation and can obtain high production volumes [18].

This study focuses on the use of regenerated SBE as fillers for producing PLA nanocomposite in order to improve the mechanical properties of PLA. Regenerated SBE as filler was compared with commercial bentonite and cloisite with regard to mechanical properties testing and water vapor permeability for PLA-Clay-Nanocomposite. In addition, the nanocomposite resulted from two methods are also compared with regards to intercalated and exfoliated structures.

## 2. SBR Regeneration Method and Nanocomposite Synthesis Method

### *2.1. Materials and Equipment*

Bentonite, hexane and NaOCl was the main materials of SBE regeneration process. Used bentonite is obtained from the oil refining industry (Sinarmas Group) while hexane and NaOCl are obtained from Brataco Chemica, a local chemical company. To extract bentonite, soxhlet extractor was applied, equipped with heating mantle, two-neck round-bottom flask, timble, condensor, siphon tube and thermometer.

Previous research have been made to create a PLA-bentonite nanocomposite. In this study, PLA-nanocomposite that obtained from solvent casting method and extrusion method are analysed with several methods such as Universal Testing Machine (UTM), X-ray Powder Diffraction (XRD), Fourirer-Transform Infrared Spectroscopy (FTIR), Water Vapor Permeability (WVP) methods.

XRD patterns were obtained from a Rigaku Smartlab diffractometer equipped with a Cu Ka generator (k = 1.5406 Å) operating at 45 kV, 30 mA, and at room temperature. On the XRD test, the basal spacing ($d_{001}$) of regenerated bleaching earth compared to the basal spacing of commercial bentonite.

The morphology of PLA-clay nanocomposites was characterized using scanning electron microscopy (SEM). Surfaces of nanocomposites were investigated in a Jeol, JSM-6510A (Japan) with the acceleration voltage was 5 kV. The specimens for the PLA-clay nanocomposites was carried out according to ASTM D683 type IV. The tensile test were performed using a universal testing machine RTF-1310, with a load cell of 10 kN and with crosshead speed of 2mm/min.

### *2.2. Working Procedure*
SBE regeneration process started with soxhlet extraction. After the extraction, it was bleached using NaOCl before modified with surfactant to produce organobentonite. The resulting organobentonite is sized under 325 mesh or 0.044 mm. Organobentonite was then mixed with PLA by two different method, solvent casting and extrusion. Organobentonite was dispersed to PLA to create a nanoscale interlayer distance. The details for both methods can be found elsewhere [19, 26].

In brief, a solvent casting method, PLA and bentonite was mixed using chloroform as a solvent. The mixture then molded in a film shape. The extrusion method mixed PLA and bentonite. The mixture of PLA granule and bentonite was heated at high temperature in extrusion process until it reach its melting point. Extrusion result molded using compression molding to obtain a film shaped nanocomposite.

## 3. Results and Discussion

### *3.1. Characterization of Bentonite*
Figure 1 shows XRD result of RBE. The addition of 20 mmol of ODA surfactant to regenerated bentonite increase the distance between layers. Increasing distance between layers of bentonite with surfactants was occured due to inorganic ions Na/Ca/Mg available between the bentonite layers swapped with positive ions possessed by organic surfactants [20]. The absorption of these ions resulted the surfactant entering the space between bentonite layers. The addition of surfactant by doubling concentration it also increased the distance between layers. Increasing amount of surfactant would increase the number of ions that can be exchanged, so that the repulsion force will be bigger and result in a greater distance [21].

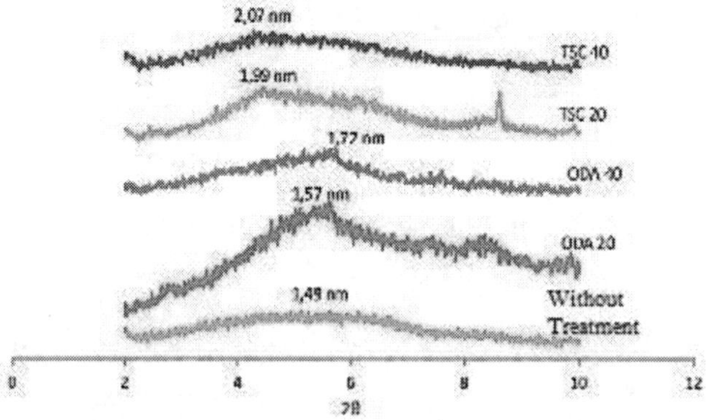


**Figure 1.** XRD spectrum and space between bentonite layers before and after treatment [19].

Another type of surfactant TSC also gives similar result. The bentonite was treated with more surfactant portion showed a larger distance between layers. Therefore, it can be shown that the additional distance between layers is not only affected by the amount of surfactant but also influenced by the type of surfactant.

In general, bentonite contains montmorillonite and other minerals in the form of biedellite, nontronite, iron impurities, caolinite, feldspar, cristobalite, crystalline and amorf silica, and calcite. The mineral composition of bentonite can be analyzed using XRD to confirm the minerals.

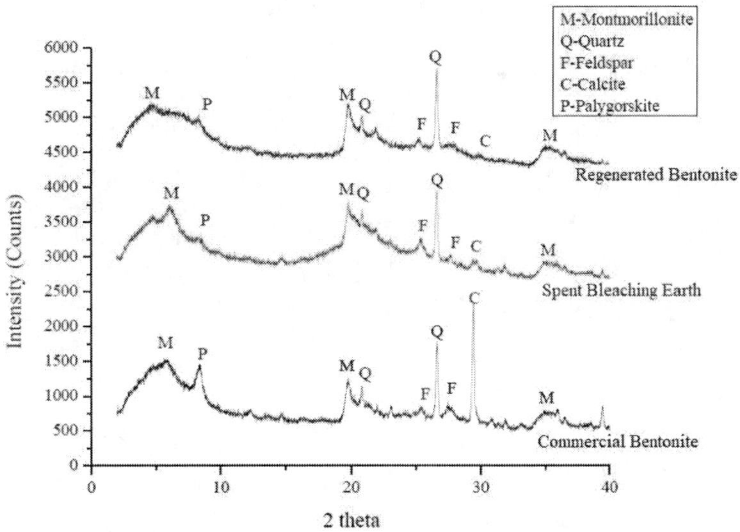


**Figure 2.** XRD spectrum of commercial bentonite, SBE and regenerated SBE.

XRD pattern of commercial bentonite, SBE and regenerated SBE visualized in Figure 2. XRD pattern showed that commercial bentonite mainly contains montmorillonite. Commercial bentonite, SBE and regenerated SBE have similar diffraction peak. Therefore, the removal of oil in SBE by adsorption process and regeneration of SBE do not affect the main structure of bentonite in RBE.

FTIR analysis was also conducted to find the effect of modification with surfactants on bentonite in RBE.

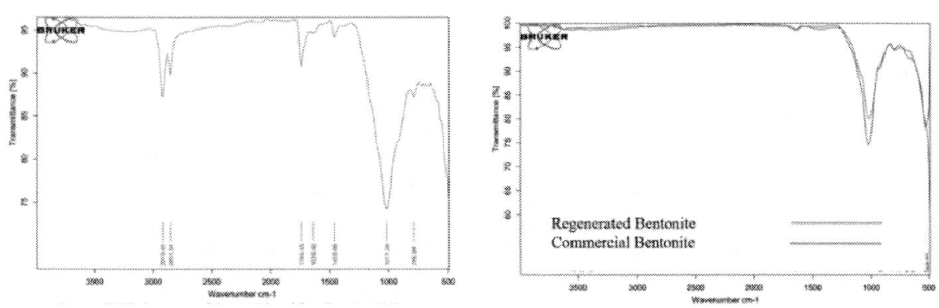


**Figure 3.** FTIR analysis on bentonite.

Figure 3 shows the regeneration process of spent bleaching earth with n-hexane and NaOCl solutions caused some of the absorption peaks at 2850-2930 cm-1 and 1730 cm-1 to disappear, indicating oil elimination.

### *3.2. Surface Structure of PLA-Bentonite Nanocomposite*

The structures of the nanocomposites were analysed using X-ray diffraction (XRD). XRD was operated at 45 kV and 30 mA. The interlayer distance is calculated using Bragg's equation [22]. This interlayer distance determines the structure of the nanocomposite whether it exfoliated or intercalated following Equation 1 below.

$$n \times \lambda = 2 \times \sin\theta \tag{1}$$

IOP Conf. Series: Materials Science and Engineering 1143 (2021) 012061 doi:10.1088/1757-899X/1143/1/012061

Table 1 summarized the nanocomposites produced using solvent casting. Only TSC 40 mmol results in exfoliation structure compared with other concentration and surfactant, only intercalation.

**Table 1.** Surface structure of nanocomposite film with various surfactant by solvent casting [19].

| Variation | Concentration (mmol) | Morphology |
|---|---|---|
| ODA | 20 | Intercalated |
| | 40 | Intercalated |
| TSC | 20 | Intercalated |
| | 40 | Exfoliated |

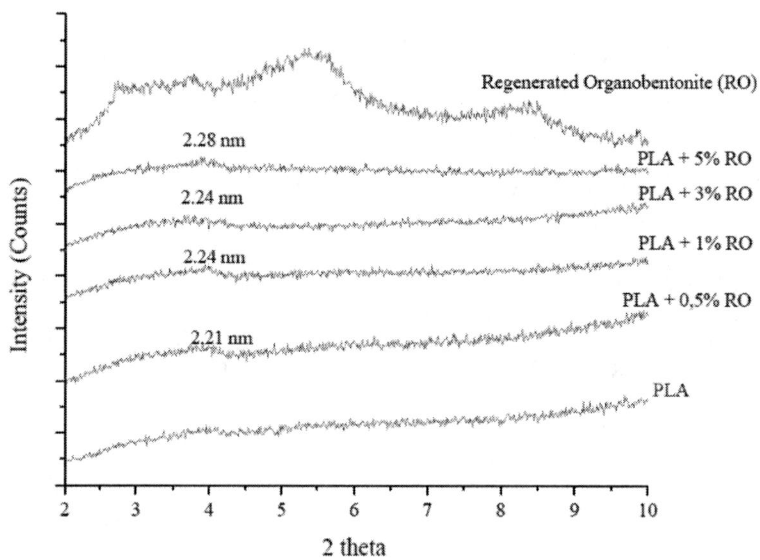


**Figure 4.** XRD Pattern for PLA-Regenerated Organobentonite with various compositions of regenerated organobentonite without surfactant [23].

Nanocomposite structure from the extrusion method was estimated by XRD pattern of PLA-regenerated organobentonite that shown in Figure 4. XRD pattern for nanocomposites have a very low intensity peak at the same angle with that of regenerated organobentonite It indicates the formation of partially exfoliated structure of all nanocomposites.

### *3.3. Mechanical Strength of PLA-Bentonite Nanocomposite*

In order to test the mechanical properties, pull test was held using Universal Testing Machine. Mechanical properties testing was carried out at all PLA-Bentonite film. The addition of bentonite which is used as filler in the PLA matrix nanocomposites decrease the elongation at break value, but it increases Young's modulus. Use of activated bentonite has a different effect on film's mechanical strength. For example, the use of bentonite that have been activated with TSC have a a higher tensile strength than film with ODA bentonite. The same can be done observed at the elongation at break of the film with TSC bentonite having a higher value than films with ODA bentonite.

**Table 2.** Mechanical properties of PLA-Bentonite nanocomposite film with various surfactant [19].

| Variation | Concentration (mmol) | Tensile Strength (MPa) | Young's Modulus (MPa) | Elongation at break (%) | PLA Molecular Weight (g/mol) |
|---|---|---|---|---|---|
| Without bentonite | - | 1.55 | 59.72 | 19.27 | 111.174 |
| Without surfactant | - | 1.59 | 82.21 | 5.03 | 111.174 |
| ODA | 20 | 7.87 | 329.72 | 2.66 | 111.174 |
| | 40 | 2.17 | 194.62 | 3.00 | 111.174 |
| TSC | 20 | 10.31 | 148.85 | 14.90 | 111.174 |
| | 40 | 12.68 | 194.58 | 8.02 | 111.174 |

To find the better surfactant type, we need to compare the result from both surfactant type. The highest tensile strength value obtained by activated bentonite with using 40 mmol of TSC, which also have the longest distance between layers. Longer distance between layers can increase the strength and stiffness of the nanocomposite film due to bentonite which has been activated can be intercalated inside the PLA matrix. Enhancement of mechanical properties are also affected by the increase of clay dispersion degree that can promote increased contact area and surface interaction between clay and PLA.

In general, it can be seen that bentonite activation using TSC surfactant result in higher tensile strength than the film activated using ODA surfactant. An increase in tensile strength can be affected by the presence of the evenly dispersed bentonite in the PLA matrix. This form of bentonite dispersion can be represented from the distance between the layers produced from the film. This large distance between layers can increase the area of contact and interaction between the surface of bentonite. The mechanical test results also showed that the film those using ODA surfactants have high value of Young's modulus but low elongation at break value and TSC surfactant delivers the opposite result. However, because of bentonite use is intended to improve PLA brittleness, it can be concluded that the TSC 40 film produces the best tensile strength value.

In this study, there are also some relation between SBE composition and nanocomposite mechanical properties. Mechanical properties increase can be caused by stiffness, ratio and interaction area between filler and polymer matrix [24]. If filler composition exceed a certain value, it can create a local tension and decreased tensile strength [25]. Mechanical properties characterization of PLA-nanocomposites in various composition showed at Table 3.

**Table 3.** Mechanical properties of nanocomposite film (various filler composition).

| Filler | Composition (%) | Elongation at Break (%) | Tensile Strength (MPa) | Young's Modulus (MPa) | PLA Molecular Weight (g/mol) | Referemce |
|---|---|---|---|---|---|---|
| - | - | 3.02 | 39.59 | 1328..40 | 120000 | [26] |
| SBE | 5 | 1.69 | 18.76 | 1230.60 | 120000 | [26] |
| Modified regenerated SBE | 3 | 3.20 | 41.10 | 1327.00 | 120000 | [26] |
| Modified regenerated SBE | 5 | 3.26 | 42.22 | 1508.90 | 120000 | [26] |
| - | - | 19.27 | 1.55 | 59.72 | 120000 | [26] |
| Regenerated SBE | 4 | 5.03 | 1.59 | 82.21 | 120000 | [26] |
| Modified SBE | 4 | 2.66 | 7.87 | 329.72 | 120000 | [26] |
| PLA/PBAT Cellulose Nanocomposite | 13.05 | 4.90 | 27.00 | 1328.00 | 155000 | [27] |
| Commercial Bentonite | 0.5 | 4.2 | 49.00 | 1280 | 120000 | [28] |
| Cloisite 30B | 1 | 3.75 | 47.00 | 1380.00 | 120000 | [28] |
| Commercial Organobentonite | 0.5 | 3.90 | 51.00 | 1420.00 | 120000 | [23] |
| Regenerated Organobentonite | 0.5 | 4.00 | 52.00 | 1400.00 | 120000 | [23] |

As showed above, 5 % addition of modified regenerated SBE have a better mechanical properties compared to both cellulose nanocomposite and PLA-cloisite nanocomposite. The highest tensile strength obtained from the PLA with filler addition. This result corresponds with another research about filler addition in HDPE [29]. However, addition of higher concentration of filler gives a drawback in elongation at break value as compared with no filler which have 19.27 % elongation at break value. The highest elongation at break was obtained at low clay compositions. The decrease in the elongation at break of nanocomposite can be caused by decreasing of PLA chain mobility due to the addition of clays [30].

Table 2 and table 3 are nanocomposites produced by solvent casting method and extrusion, respectively. By comparing the elongation at break value, tensile strength value and Young's modulus value, it can be concluded that mechanical properties of nanocomposite from extrusion process are better than solvent casting method.

### *3.4. Water Vapor Permeability Test*

In this study, the WVP of the samples was determined by calculating the change in mass of the silica gel which is covered by the sample. Mass data from silica gel was taken every 24 hours for 5 days. The largest mass change, at 1.1 g, was seen on films treated with 40 mmol ODA. Films formed in the 40 mmol ODA variation has a non-smooth surface structure as well as porous. This may contribute to higher water vapor absorption by silica gel compared to other samples. This may also contribute to higher water vapor absorption by silica gel compared to other samples. Lowest WVP is shown by the film with 40 mmol TSC with only 0.122 g after 7 days. The result confirms that films with bentonite TSC 40 has a better dispersion structure than other films. Evenly dispersed bentonite across the film can block water vapor to enter through the film that result in low value of WVP [30]. As distance between layers gets bigger, bentonite tends to form random structures in the film so that it blocks the entry of moisture. In addition, the presence of bentonite which contains a silicate coating can extends the path that water vapor has to travel water to penetrate the film [31].

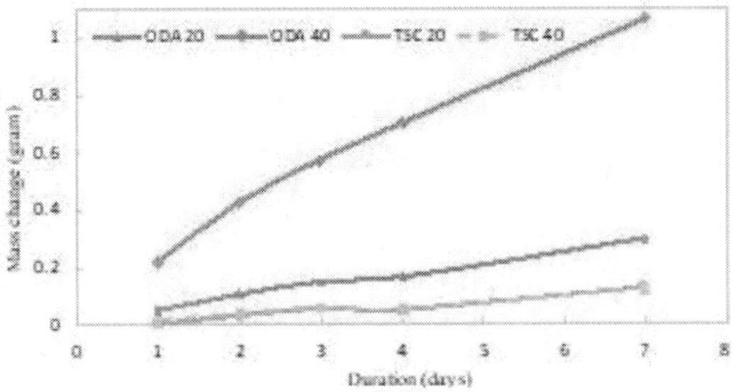


**Figure 5.** Water vapor permeability curve at various PLA-bentonite nanocomposite film [19].

Water vapor permeability of nanocomposite using commercial Cloisite 30 B [32] and bentotite can be seen in Table 4.

**Table 4.** Water vapor transmission rates of PLA-Cloisite 30 B nanocomposites.

| Mass fraction (%) | WVTR (g/m$^2$.day) | |
|---|---|---|
| | Bentonite | Cloisite [32] |
| 0 | - | 177 |
| 1 | 60 | 156 |
| 2 | 25 | 139 |
| 3 | 27 | 126 |
| 4 | 39 | 114 |
| 5 | 58 | 107 |
| 6 | - | 101 |

Table 4 showed WVTR comparison between PLA-Cloisite 30 B and PLA-Bentonite nancomposite. PLA-bentonite nanocomposite showed better WVTR performance due to their low transmission rate. It indicated the ability to be a good food packaging materials especially in high humidity environment.

## 4. Conclusion

The regeneration method for SBE has been successfully conducted by soxhlet extraction method continued by modification using surfactants. Those method was proved to eliminate oil from bentonite indicated by the absence of hydrocarbon peaks at the FTIR result. The use of surfactants on various concentration increases the distance between layers of bentonite. Furthermore, activated bentonite TSC 40 has the largest distance between layers. The increase in the distance between layers may be caused by the entry of surfactant chains into the bentonite layers, as shown by the FTIR results.

When a mixture of PLA and bentonite was sonicated and molded, the distance between the layers of bentonite larger. This indicates that the incoming PLA chain between the layers of bentonite in forming nanocomposite film. Most of the samples produce an intercalation structure between the bentonite with PLA, except the PLA-Bentonite TSC 40 producing exfoliated structures. Hence, samples containing TSC 40 have the best mechanical properties with regards to tensile strength result.

The film containing TSC 40 is the best film to withstand the entry of moisture into the sample. Compared to Cloisite 30 B, bentonite as a filler resulted in a better water vapor permeability of nanocomposite. Mechanical properties characterization also showed that PLA that added with regenerated bentonite have a better mechanical properties compared to PLA that added with cellulose and cloisite. When nanocomposite production from solvent casting and extrusion compared, their mechanical strength showed that the extruded nanocomposite have a better properties. It may be caused by a better mixing condition which reachs its melting point. Therefore, it can be concluded that extrusion result in better nanocomposite quality.

**Acknowledgements**
We acknowledge the funding from Korea Institute of Chemical Technology (KRICT), Republic of Korea.

## 5. References

[1] Rapa M, Mitelut A C, Tanase E E, Grosu E, Popescu P, Popa M E, Vasile C, et al 2016 Influence of chitosan on mechanical, thermal, barrier and antimicrobial properties of PLA-biocomposites for food packaging *Composites Part B: Engineering* **102** 112-21

[2] Csizmadia R, Faludi G, Renner K, Moczo J and Pukanszky B 2013 PLA/wood biocomposites: Improving composite strength by chemical treatment of the fibers *Composites Part A: Applied Science and Manufacturing* **53** 46-53

[3] Kumar R, Ofosu O and Anandjiwala R D 2013 Macadamia Nutshell Powder Filled Poly Lactic Acid Composites with Triacetin as a Plasticizer *Journal of Biobased Materials and Bioenergy* **7(5)** 541-8

[4] Piekarska K, Sowinski P, Piorkowska E, Haque M M U and Pracella M 2016 Structure and properties of hybrid PLA nanocomposites with inorganic nanofillers and cellulose fibers *Composites Part A: Applied Science and Manufacturing* **82** 34-41

[5] Song Z, Xiao H, Zhao Y 2014 Hydrophobic-modified nano-cellulose fiber/PLA biodegradable composites for lowering water vapor transmission rate (WVTR) of paper *Carbohydrate Polymers* **111** 442-8

[6] Sapawe N and Hanafi M F 2018 Analysis of the pyrolysis product from spent bleaching clay *Materials Today; Proceedings* **5** 21940-7

[7] Wambu EW, Muthakia GK, Joseph K, Shiundu PM 2009 Regeneration of spent bleaching earth and its adsorption of copper (II) ions form aqueous solutions *Applied Clay Science* **46(2)** 176-80

[8] Malakootian M, Fatehizadeh A, Youse N, Ahmadian M and Moosazadeh M 2011 Fluoride removal using Regenerated Spent Bleaching Earth RSBE from groundwater : Case study on Kuhbonan water *Desalination* **277** 244-9

[9] Shehab A, Basir S, Khalek M, et al 2019 Dye Removal from Aqueous Solution by Regenerated Spent Bleaching Earth *International Journal of Chemical and Materials Engineering* **13(9)** 452-61

[10] Lagaron J M and Cabedo L 2015 *Poly(lactic acid) Science and Technology: Processing, Properties, Additives and Applications* vol 1, ed Jimenez A, Peltzer M A, et al (Cambridge: The Royal Society of Chemistry) pp 215-24

[11] Ouellet-Plamondon C M, Stasiak J and Al-Tabbaa A 2014 The effect of cationic, non-ionoc and amphiphilic surfactants on the intercalation of bentonite *Colloids and Surfaces A: Physicochemical and Engineering Aspects* **444** 330-7

[12] Mat R, Ling O S, Johari A, and Mohamed M 2011 In situ biodiesel production from residual oil recovered from spent bleaching Earth *Bulletin of Chemical Reaction Engineerng and Catalysis* **6** 53-7

[13] Gonçalves L A G, Viotto L A, Ribeiro A P B, Chiu M C, Basso R C and de Morais C C 2009 State of the art of the application of membrane technology to vegetable oils: A review *Food Research International* **42** 536-50

[14] Abdelmoez W, Ashour E and Naguib S M 2015 A Review on Green Trend for Oil Extraction Using Subcritical Water Technolog and Biodiesel Production *Journal Of Oleo Science* **64** 467-78

[15] Herrero M, Mendiola J A, Cifuentes A and Ibáñez E 2010 Supercritical fluid extraction: Recent advances and applications *Journal of Chromatography* **1217** 2495-511
[16] Kim J, Choi K and Chung D S 2012 Comprehensive Sampling and Sample Preparation: Analytical Techniques for Scientists *Elsevier Reference Collection on Chemistry, Molecular Sciences and Chemical Engineering* **3** 701-21
[17] Wahid F, Khan T, Hussain Z and Ullah H 2018 *Applications of Nanocomposite Materials in Drug Delivery* vol 1, ed Inamuddin, Asiri A M, et al (Sawston: Woodhead Publishing) pp 701-35
[18] Hietala M, Mathew A P and Oksman K 2013 Bionanocomposites of thermoplastic starch and cellulose nanofibers manufactured using twin-screw extrusion *European Polymer Joirnal* **49** 950-6
[19] Wonoputri V, Emanuella N, Angelica E and Sitompul J 2020 Variasi Jarak Antar *Layer* Bentonit Pada Pembuatan Nanokomposit Pla-Bentonit Sebagai Kemasan Makanan *Indo. J. Chem. Res* **8(1)** 57-65
[20] Patel H A, Somani R S, Bajaj H C and Jasra RV 2010 Synthesis of Organoclays with Controlled Particle Size and Whiteness from Chemically Treated Indian Bentonite *Ind. Eng. Chem. Res.* **49(4)** 1677-83
[21] De Paiva L B, Morales A R and Diaz F R V 2008 Organoclays: Properties, Preparations and Applications *Appl. Clay. Sci.* **42** 8-24
[22] Le Pevelen D D 2010 *Encyclopedia of Spectroscopy and Spectrometry* vol 2, ed Lindon J C (Cambridge: Academic Press) pp 2559-76
[23] Paramitha T, Sitompul J P and Lee H W 2018 The effect of organobentonites from spent bleaching earth (SBE) and commercial bentonite on nanocomposite properties *International Journal of Engineering & Technology* **7(4)** 2000-5
[24] Rhim J W 2013 Effect of PLA lamination on performance characteristics of agar/κ-carrageenan/clay bio-nanocomposite film *Food Research International* **51(2)** 714-22
[25] Lai J, Rahkam M R, and Hamdan S 2018 Comparative studies of thermo-mechanical and morphological properties of polylactic acid/fumed silica/clay (1.28E) and polylactic acid/fumed silica/clay (1.34TCN) nanocomposites *Polymer Buletin* **75** 135-47
[26] Paramitha T and Sitompul J P 2020 Karakterisasi Nanokomposit Poly(Lactic Acid)-*Spent Bleaching Earth* Regenerasi Termodifikasi *KOVALEN: Jurnal Riset Kimia* **6(2)** 90-8
[27] Ludvik C N, Glenn G M, Klamczynski A P, et al 2007 Cellulose Fiber/Bentonite Clay/Biodegradable Thermoplastic Composites *J Polym Environ* **15** 251-7
[28] Paramitha T, Wonoputri V, Sitompul D S D, Lee H W and Sitompul J P 2020 Properties of clays reinforecd PLA nanocomposites by melt extrusion technique *Malaysian Journal of Fundamental and Applied Sciences* **16(4)** 453-7
[29] Ma F, Lu X, Wang Z, Sun Z, Zhang F, Zheng Y 2011 Nanocomposites of poly(L-lactide) and surface modified magnesia nanoparticles: Fabrication mechanical property and biodegradability *Journal of Physics and Chemistry of Solids* **72** 111-6
[30] Buzarovska A, Bogoeva-Gaceva G and Fajgar R 2016 Effect of the Talc Filler on Structural Water Vapor Barrier and Mechanical Properties of Poly (Lactic Acid) Composistes *J. Polym. Eng.* **36(2)** 181-8
[31] Oritz-Zarama M A, Jimenez-Aparicio A R and Solorza-Feria J 2016 Obtainment and Partial Characterization of Biodegradable Gelatin Films with Tannic Acid, Bentonite and Glycerol *J. Sci. Food Agric.* **96(10)** 3424-31
[32] Duan Z, Thomas N and Huang W 2013 Water vapour permeability of poly(lactic acid) nanocomposites *Journal of Membrane Science* **445** 112-8

IOP Conf. Series: Materials Science and Engineering 1143 (2021) 012062 doi:10.1088/1757-899X/1143/1/012062

# Purification of vitamin E from palm fatty acid distillate through neutralization, extraction, and adsorption methods

**Anggita Veningtia Sari, Ardiyan Harimawan, and Dianika Lestari***

Department of Chemical Engineering, Institut Teknologi Bandung, Bandung 40132 Indonesia

*corresponding author's e-mail: dianika@che.itb.ac.id

**Abstract.** Vitamin E is one of the compounds in palm fatty acid distillate (PFAD) which was carried away due to thermal treatment in the deodorization unit in the refining of crude palm oil (CPO). The process of purifying vitamin E can be done by combining several common processes such as neutralization, extraction, and adsorption. The neutralization process is carried out to remove the content of free fatty acids (FFA), which are the saponifiable compounds and the largest component in PFAD. FFA in PFAD were first neutralized using magnesium oxide and most of FFA removed during process. Vitamin E and the remaining FFA are carried away during the extraction process with hexane due to similar polarity. Vitamin E extracted from neutralized PFAD was then subjected to batch adsorption process using incubator shaker with silica as the adsorbent. The adsorption process in this study resulted in a vitamin E uptake percentage of 98%. Isopropanol is used as solvent in the desorption process and produces a vitamin E recovery percentage of 96,9%. The concentration of final concentrate was obtained with vitamin E purity of 5,6% (12282 ppm) and the amount increased 9.2 times compared to concentrate from neutralization followed by extraction process. The concentrate has strong antioxidant activity as evidenced by the IC50 value of 23,3 ppm.
Keywords: PFAD, vitamin E, neutralization, extraction, adsorption

## 1. Introduction

Palm fatty acid distillate (PFAD) is a by-product from refining process of crude palm oil (CPO) and widely used for oleochemical industry and a raw material for animal feed. PFAD consists of 81,7% free fatty acids (FFA), 14,4% glycerides, 0,8% squalene, 0,5% vitamin E, 0,4% sterols, and 2,2% other substances [1]. Among these components, vitamin E has been reported to have powerful antioxidant properties and the highest economic value [2]. Vitamin E is a fat-soluble natural antioxidant and has two main structures namely tocopherol and tocotrienol. As the essential vitamin, vitamin E cannot be self-produced by the human body and needs to be obtained from the food supplement [3]. Thus, turning PFAD into value-added products such as vitamin E would seem beneficial.

Since vitamin E in PFAD mixed with various other components, separation of vitamin E from PFAD is difficult task due to relatively low concentration. However, the availability of abundant stocks in Indonesia makes PFAD a potential raw material for producing vitamin E. The method of separating vitamin E from PFAD consists of two stages i.e. the separation of saponifiable components (FFA and glycerides) and unsaponifiable component (squalene, sterols, etc.). Many attempts are made to separate vitamin E from PFAD such as molecular distillation [4], esterification and crystallization [5], and neutralization and adsorption chromatography [6]. Understanding the separation processes is crucial to have products with high vitamin E purity and recovery.

In this study, vitamin E in PFAD were concentrated by removing the extraneous matters, especially FFA. Better removal of the fatty components resulted in better concentration of vitamin E in the final product [6]. There are three stages for vitamin E purification, namely neutralization, extraction, and batch adsorption. FFA in PFAD were first neutralized with alkaline saponification using magnesium oxide. Vitamin E is then extracted using hexane and subjected to batch adsorption using incubator shaker and silica as the adsorbent. To produce the final concentrate, desorption was carried out using isopropanol followed by evaporation. The antioxidant activity of vitamin E is generally evaluated using DPPH spectrophotometry method. The result from this study provide information on the separation of vitamin E in PFAD by combining neutralization, extraction, and adsorption methods and their effect on the purity of vitamin E produced.

## 2. Materials and methods

### *2.1. Materials*
PFAD with an iodine value of 52,2 g of $I_2$/100 g and slip melting point of 45-50°C was obtained from PT. Tunas Baru Lampung Tbk., Lampung, Indonesia. Column chromatography grade silica gel 60 (diameter size of 0,063-0,200 mm; specific pore volume of 0,74-0,84 ml/g) was purchased from Merck. Magnesium oxide and hexane solvent used were technical grades and other chemicals used for analysis were either of analytical grades.

### *2.2. Neutralization of PFAD*
Neutralization of PFAD was carried out using a modified fusion method (Patent US 2,890,232). PFAD is preliminary heated in neutralization reactor at temperature 60-65°C. After completely melted, magnesium oxide (MgO) is added with a mole ratio PFAD:MgO of 1:1,1. After mixing evenly, then water added as a catalyst until the mixture expands. This reaction lasts for 5-7 minutes with stirring. The metal salt formed is Mg-PFAD salt.

### *2.3. Extraction*
The Mg-PFAD salt produced from neutralization reaction was then extracted using hexane as an organic solvent. The mass ratio Mg-PFAD:hexane of 1:4. The extraction process is carried out in three steps with the same solvent at temperature 60-65°C for 45 minutes for each step. The solids and solvents in the mixture are then followed by filtration. The resulting extract then used to the next stage.

### *2.4. Batch Adsorption*
Batch mode adsorption studies were carried out using a modified version of the method of Chu et al. [7-8]. Weighed amount of silica with silica mass ratio of 0,5 g/50 ml extract of known initial vitamin E concentration in covered Erlenmeyer flasks. The samples were agitated in an incubator shaker at 180 rpm and 25°C for 60 minutes. The contents of the flask were then filtered through 400-micron nylon mesh filter. Adsorption is carried out in two steps on the same extract. Silica saturated with vitamin E then mixed and shaken with isopropanol with silica mass ratio 1 g/50 ml solvent at 180 rpm and 40°C for 60 minutes. After desorption process, the content of the flask was filtered. The solution was collected and then solvents removed by rotary evaporation.

### *2.5. Analysis*

*2.5.1. Determination of FFA.* The content of FFA was determined by titration method. A sample of 0,1 g was weighed and then mixed with solution containing with 25 ml of ethanol, 25 ml of chloroform, and three drops of phenolphthalein pH indicator in 250 ml Erlenmeyer flask and previously neutralized to pink color. This mixture is titrated with 0,1 N KOH to the same color using 25 ml burette. The resulting titrant volume is used to calculate acid number and FFA content in the sample.

*2.5.2. Determination of total tocopherol.* The content of tocopherol compounds was determined according to the colorimetric method described by Wong et al. [9]. In volumetric flask, the samples (0,2 g) was dissolved in 5 ml of toluene, then added 3,5 ml of 2,2-bipyridine (0,07%-w/v in ethanol) and 0,5 ml of $FeCl_3.6H_2O$ (0,2%-w/v in ethanol). The solution was made up to 10 ml with ethanol. After standing for one minute, the absorbance at 520 nm was determined by using spectrophotometer (Genesis 10 VIS). A calibration curve of α-tocopherol in toluene was performed in a concentration range of 0-1000 ppm.

*2.5.3. DPPH radical scavenging activity.* The antioxidant activity in the sample was determined using 2,2-diphenyl-1,1-picrylhydrazyl (DPPH) by the decolorization of methanol solution. DPPH (0,1 mM in methanol) was prepared and then 2 ml of this solution was mixed with 8 ml of sample with different concentration. The solution was then shaken and incubated at 30°C for 30 minutes in the dark. The absorbance was read at 517 against a blank. The radical scavenging activity was calculated as inhibition concentration percentage (%IC) according to the following equation:

$$\%IC = 100(Abs_{blank}-Abs_{sampel})/Abs_{blank.} \quad (1)$$

The IC50 values (the concentration of sample required for 50% inhibition of DPPH radicals) were obtained through extrapolation from regression analysis. The antioxidant was evaluated based on IC50 value.

## 3. Results and discussion

### *3.1. Effect of neutralization process*

The physicochemical characteristic of PFAD and Mg-PFAD are shown in Table 1. Yield of Mg-PFAD produced from neutralization process was 1,16 g Mg-PFAD/g PFAD. Changes in the fatty components were reflected in acid number and FFA content, both of which lower in Mg-PFAD compared to PFAD due to the removal of FFA [6]. Neutralization of PFAD decreased acid number and FFA content in Mg-PFAD was 87,71% and 87,66%, respectively. No degradation of vitamin E during the process. However, there is still 12,34% of FFA content which had not been saponified in Mg-PFAD salt.

Table 1. Characteristic of PFAD and Mg-PFAD.

| Characteristic | PFAD | Mg-PFAD |
|---|---|---|
| Acid number (mg KOH/g sample) | 209,87 ± 1,52 | 25,84 ± 0,03 |
| FFA content (%-w/w) | 95,61 ± 0,69 | 11,77 ± 0,02 |
| Vitamin E content (%-w/w) | 0,57 ± 0,01 | 0,48 ± 0,01 |

### *3.2. Effect of extraction process*

The Mg-PFAD salt produced from neutralization was further extracted using hexane as an organic solvent. In this process, the percentage of vitamin E and FFA content in the extract in the process stages will be reviewed. Figure 2 shows the vitamin E and FFA extracted with three-stages process. The amount of extracted vitamin E and FFA increased with the addition of the extraction stage. Hexane solvent has the ability to extract vitamin E and remaining FFA in Mg-PFAD with three extraction steps up to 84,4% and 94,2%, respectively. Due to vitamin E and fatty acids are nonpolar compounds and hexane has low polarity [8], these compounds will be extracted with the solvent easily. Vitamin E is relatively more polar than FFA, which tends to make FFA extracted more than vitamin E [10]. However, the condition of Mg-PFAD which was homogeneously dispersed in hexane form favorable interaction between solvent and material and most of vitamin E extracted from Mg-PFAD salt. Neutralization followed by extraction process could concentrate vitamin E about 0,62%(w/w).

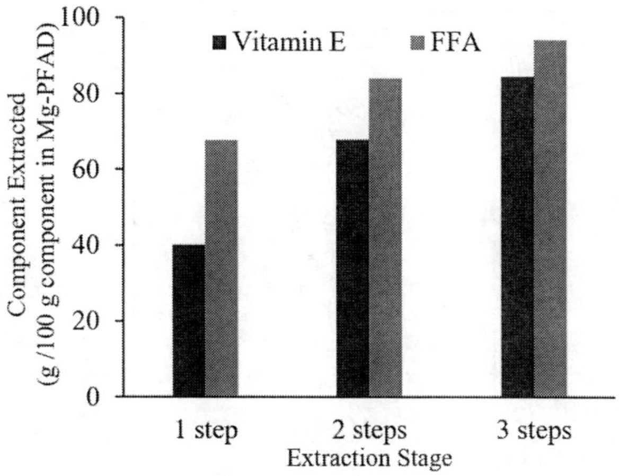


**Figure 1.** Vitamin E and FFA extracted with three-stage process .

### *3.3. Effect of batch adsorption process*

The extract produced from extraction process was further concentrated using batch adsorption with silica as adsorbent. Silica was polar adsorbent with large surface areas [10]. Table 2 shows uptake percentages (adsorption) and recovery percentages (desorption) of vitamin E and FFA at equilibrium. Since vitamin E is relatively more polar than hexane and FFA, it gets adsorbed to a polar adsorbent. Although FFA can be adsorbed into surface of silica, the resulting uptake of vitamin E was higher than FFA. Adsorption was an important step for purify vitamin E due to could adsorbed more vitamin E than FFA in the extract.

Table 2. Percentage uptake and recovery of vitamin E and FFA from batch adsorption process.

| **Component** | **%uptake (mg component adsorbed/mg total component)** | **%recovery (mg component desorbed/ mg component adsorbed)** |
|---|---|---|
| Vitamin E | 98,0 ± 1,11 | 96,9 ± 0,14 |
| FFA | 51,2 ± 0,90 | 13,7 ± 0,06 |

Same as adsorption process, recovery percentage of vitamin E was higher than FFA. FFA recovery percentage was lower than the uptake percentage indicating that lots of remaining FFA was still in surface of the adsorbent. High polarity isopropanol desorbed vitamin E molecules better from the adsorbent. Thus, Isopropanol solvent is more selective to release vitamin E than FFA from the adsorbent. This seems beneficial considering that FFA is another adsorbate that can be adsorbed besides vitamin E and its presence can affect the level of purity of vitamin E.

### *3.4. Final product overview*

Vitamin E was mainly found in isopropanol solvent with high degree of recovery (>95%). This indicates that batch adsorption was successful in concentrating vitamin E in the samples. To produce the final concentrate, batch adsorption process followed by evaporation. Table 3 shows characteristic of pre-concentrate (concentrate from neutralization followed by extraction) and final product concentrate from a combination of neutralization, extraction, and adsorption methods.

Table 3. Characteristic of pre-concentrate and final product concentrate.

| Characteristic | Pre-concentrate | Final product concentrate |
|---|---|---|
| Visual concentrate |  | 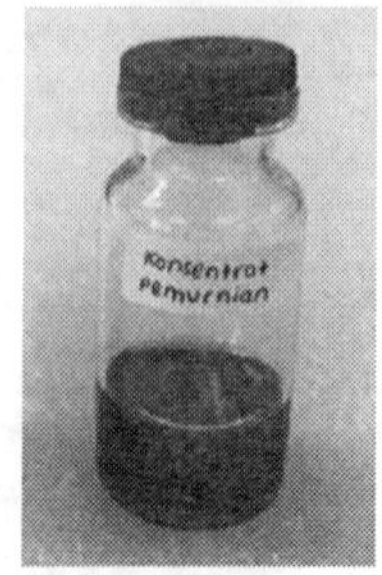 |
| Concentrate yield (mg/g PFAD) | 395,9 | 51,7 |
| Vitamin E content (ppm) | 1350 (0,62%-w/w) | 12282 (5,6%-w/w) |
| FFA content (mg FFA/mg concentrate) | 0,20 | 0,09 |
| IC50 (ppm) | 247,9 (weak activity) | 23,3 (strong activity) |

Removal of FFA in the neutralization process followed by extraction and batch adsorption greatly concentrated vitamin E in the sample compared with concentrate without adsorption process. There was a visual change from a solid yellow suspension to a thick brown concentrate. The concentration of final concentrate was obtained with vitamin E purity of 5,6% (12282 ppm) and the amount increased 9.2 times compared to pre-concentrate (0,62%). Moreover, there was a decrease in the FFA concentration 2,3 times compared to pre-concentrate. In terms of the antioxidant activity, IC50 value is a measure of the effectiveness of a substance inhibiting the oxidation. The lowest value of IC50 means the highest antioxidant activity due to lots of antioxidant molecules in the sample [11]. The IC50 value of <50 ppm indicates the sample has a strong antioxidant activity, whereas >150 ppm indicates has a weak antioxidant activity [12]. Thus, pre-concentrate has a weak antioxidant activity and final concentrate has a strong antioxidant activity due to the IC50 value of 23,3 ppm.

## 4. Conclusion

Neutralization process followed by extraction and batch adsorption process successfully concentrated vitamin E from PFAD. There was a change in antioxidant activity from weak (concentrate from neutralization followed by extraction) to strong activity (neutralization-extraction-batch adsorption method). The methods described in this study used low temperature (60-65°C) and in mild alkali conditions (neutralization process) which minimize degradation of vitamin E during separation. The methods were relatively simple and not energy-intensive compared to the existing methods such as molecular distillation. This combination method could be easily adapted to the existing process in industry for vitamin E purification with some modifications.

**Acknowledgments**
This research was funded by Badan Pengelola Dana Perkebunan Kelapa Sawit (BPDPKS) Indonesia. Another support by PT Tunas Baru Lampung Tbk for providing PFAD.

## References

[1] Top A G M 2010 Production and utilization of palm fatty acid distillate (PFAD) *Journal of Lipid Technology* **22** (1) 11-13
[2] Damrongwattanakool N and Raviyan P 2018 Enrichment of vitamin E in palm fatty acid distillate using sequential-cooling urea-fatty acid complexation *Songklanakarin J. Sci. Technol.* **40** (5) 1175-1180

[3] Maarasyid C, Muhamad I I, and Suproyanto E 2014 Potential Source and Extraction of Vitamin E from Palm-Based Oil: A Review *Jurnal Teknologi* **69** (4) 43-50

[4] Posada L R, Shi J, Kakuda Y, and Xue S J 2007 Extraction of tocotrienols from palm fatty acid distillate using molecular distillation *Journal of Separation and Purification Technology* **57** 218-227

[5] Top A G M, Leong L W, Ong A S H, Kawada T, Watanabe H, and Tsuchiya N 1993 Production of high concentration tocopherols and tocotrienols from palm oil by-product *United States Patent* **5** 190 618.

[6] Chu B S, Baharin B S, Che Man Y B, and Quek S Y 2003 Separation of tocopherols and tocotrienols from palm fatty acid distillate using hydrolysis-neutralization-adsorption chromatography method *Journal of Food Lipids* **10** (2) 141-152

[7] Chu B S, Baharin B S, Che Man Y B, and Quek S Y 2004 Separation of vitamin E from palm fatty acid distillate using silica: I. Equilibrium of batch adsorption *Journal of Food Engineering* **62** 97-103

[8] Chu B S, Baharin B S, Che Man Y B, and Quek S Y 2004 Separation of vitamin E from palm fatty acid distillate using silica: III. Batch desorption study *Journal of Food Engineering* **64** 1-7

[9] Wong M, Trimms R, and Goh E 1988 Colorimetric determination of total tocopherols in palm oil, olein, and stearin *Journal of the American Oil Chemists' Society* **65** 258-261

[10] Chu B S, Baharin B S, Che Man Y B, and Quek S Y 2005 Comparison of selected adsorbents for adsorption and desorption of vitamin E from palm fatty acid distillate *Journal of Food Lipids* **12** 23-33

[11] Jadid N, Hidayati D, Hartanti S R, Arraniry B A, Rachman R Y, and Wikanta W 2017 Antioxidant activities of different solvent extract of *Piper retrofractum* Vahl. using DPPH assay *AIP Conference Proceedings* ***1854*** *020019*

[12] Tristantini D, Ismawati A, Pradana B T, and Jonathan J G 2016 Pengujian aktivitas antioksidan meggunakan metode DPPH pada daun tanjung (*Mimusops elengi* L) *Prosiding Seminar Nasional Teknik Kimia "Kejuangan"* ISSN 1693-4393

# Effect of Dimethyl Sulfoxide (DMSO) as a Green Solvent and the Addition of Polyethylene Glycol (PEG) in Cellulose Acetate/Polybutylene Succinate (CA/PBS) Membrane's Performance

**R D Nyamiati[1], Y Rahmawati[1], A Altway[1] and S Nurkhamidah*[1]**

[1]Department of Chemical Engineering, Institut Teknologi Sepuluh Nopember, Kampus ITS Sukolilo, Surabaya 60111 Indonesia

*Corresponding Author's E-mail : nurkhamidah@chem-eng.its.ac.id

**Abstract.** In this study, performance of CA/PBS membranes, prepared with the non-toxic solvent dimethyl sulfoxide (DMSO) compared to Dimethylformamide (DMF), was investigated. The membranes were prepared using immersion precipitation technique at coagulation bath at temperatures of 25°C. Ternary diagram was used to show the interaction between polymer (CA/PBS), solvents (DMF or DMSO) and isopropanol as non-solvent by measuring the cloud point instead of Hansen solubility parameter differences. The effect of Polyethylene Gycol (PEG) to the membrane performance was also investigated. Morphology and tensile strength were analyzed by using Scanning Electron Microscopy (SEM) and Dynamic Mechanical Analysis (DMA), respectively. Experimental results proved that DMSO could be used to replace (DMF) as solvent, as the membranes had a higher value of salt rejection (84.9%). With the addition of PEG, salt rejection, mechanical property, and porosity of CA/PBS membrane also increases. The best result shows that PEG with concentration of 7% wt of polymer, the salt rejection is 88.89%; Permeate Flux is 29.036 L/m$^2$.hr; tensile strength is 68.366 kPa; porosity is 75.16 % and the morphology is sponge-like with the pore size diameter in average is 0.262 μm

## 1. Introduction

The phase inversion is the most widely used method in the fabrication of membranes resulting asymmetrical membrane, both flat sheet or hollow fiber membrane types [1]–[3]. The phase inversion method involves transformation from a homogeneous polymer membrane solution to a final membrane structure (solid) through a phase separation of polymer solution [4],[5]. In the phase inversion method, there are two basic techniques, dry technique and wet technique. In the dry technique the solvent is evaporated from the membrane solution into the atmosphere and or absorbs the vapor phase (non-solvent) from humidity. However, the membrane structure is formed by immersing the solution membrane into a coagulation bath containing non-solvents in the wet technique. Both of these methods are known as the immersion precipitation method where it has been widely used in the manufacture to fabricate membranes commercially [6]. The solidification process is very frequently started by the change from one liquid phase into two liquid phases (liquid- liquid demixing). At certain period during demixing, the polymer rich phase solidifies; thus, a solid membrane matrix is formed. Membrane morphologies, the pore size distributions can be controlled by selecting the different solvent, non-solvent, polymer, pore former, and fabrication parameters depending on the particular application [5]–[9]. Liquid-liquid demixing (L-L demixing) is strongly influenced by the thermodynamic and kinetic properties of the membrane system. The formation of membrane structure related to the phase equilibrium which is in the thermodynamic property can be described by Flory-Huggins theory and in the kinetics of transport process can be described by the mutual diffusion and transport between components. A ternary diagram is commonly used as a tool to analyze the thermodynamic and kinetic behavior of the membrane system.

Petroleum-based non-biodegradable polymers are widely used in industry due to cheaper and affordable production costs, but the materials are recalcitrant to these microorganisms and result in the formation of solid waste which is a problem [10]. Alternative to reduce sources of environmental pollution is to select and replace membrane-forming materials with materials that have a high biodegradation rate and are more environmentally friendly [11]. Cellulose Acetate is an economically valuable polymer of particular concern in the field of green and sustainable chemistry and stimulates the development of environmentally friendly technologies. Cellulose Acetate is a polymer which has a high biodegradation rate, is low cost, is non-toxic and is a renewable source [12], but he first generation cellulose acetate (CA) membranes yield low flux and are susceptible to chemical and bacteriological agents. CA performance can be improved by mixing it with appropriate additives to meet new requirements and related membrane properties [4]. Among the different polymeric materials that are used for the membrane preparation Polly Butylene Succinate is high biodegradable, such as softness, tensile and gas barrier properties, melt viscosity for further processing [13]. However, the solvents used for membrane preparation are largely toxic to human health and the environment, typical examples are chloroform, dichloromethane, N-dimethylformamide (DMF), dichloromethane and dimethylacetamide (DMAc) with more environmentally friendly solvents for the manufacture of polymeric membranes. DMSO, which can be extracted from lignin or synthesized by oxidation of dimethyl sulfide, is an aprotic organic solvent for membrane synthesis. DMSO not only has relatively low intrinsic toxicity, but is also biodegradable, forming a non-toxic product. Therefore, DMSO is considered a green solvent for the manufacture of CA/PBS membranes [14], [15].

Therefore, many researchers wished to further modify it to obtain a high-performance membrane. In this study, Polyethylene Glycol (PEG) was added to the CA/PBS membran. PEG, commonly used as additive in membrane manufacture, because it is known as pore former and can act as plasticizer [16]. It is expected that the pore size of membrane decreases, and its mechanical property increases. Thus, the effect of using DMSO as a green solvent and the addition of PEG to the performance of CA/PBS membrane was investigated.

## 2. Experimental

### *2.1. Materials*

Cellulose acetate (CA, Mn = 30.000 Da, density = 1.3 g/mL) and poly(1,4-butylene succinate) (PBS, density = 1.3 g/mL at 25°C) were purchased from Sigma Aldrich. Polyethylene Glycol (PEG, 400 Da) was purchased from Merck and used as additive. Dimethylformamide (DMF) and Dimetyhl Sulfoxide (DMSO) were purchased from Merck and used as solvent. Isopropanol with a purity of 96 wt% were purchased from Merck and used as non-solvents. NaCl were purchased from Merck and used to make saline solutions for desalination tests.

### *2.2. Membrane Preoaration*

CA/PBS membranes with a composition of 80/20 with total polymer 2.5 g were prepared by phase inversion method. CA/PBS were dissolved in DMF or DMSO with volume of 17 mL. PEG was added by dissolving it into CA-PBS solution with composition: 0; 2; 7 and 8 wt% of total polymer until the solution is homogeneous. The homogeneous solution was cast on a glass plate by a film applicator then immersed into coagulation bath containing non-solvents for 15 minutes at room temperature. Then, CA/PBS membranes were dried and then before analyzed, defects in the membrane are visually observed put the membrane under the light. To make sure that the membrane has no defects, all the membranes are checked using this visual observation before further use.

### *2.3. Phase Diagram*

The phase diagram of the CA-PBS/Solvent/Isopropanol ternary systems was determined by cloud point measurement. Ternary phase diagram was obtained from the turbid points. To reach the turbid point, the non-solvent was added slowly into the polymer solution under stirring. During titration, the solution temperature was controlled at 25°C with a water bath and the addition of isopropanol was continued until the clear polymer solution visually turned to look milky-like, the observation of the first sign of turbidity, the addition of non-solvent was stopped and the solution was stirred for an additional 20-40 min to see whether the turbid solution changes to a clear solution or not. If the solution turned to a clear solution, more non-solvent was added; otherwise, the determined point was considered as the onset of real turbid point. The ternary composition of the turbid point was then calculated from the amount of non-solvent, solvent, and polymer present in the glassware.

### *2.4. Hansen Solubility Parameter*

The affinity between the solvent, polymer and nonsolvent can be predicted by solubility parameter ($\boldsymbol{\delta}$), which is described as the square root of the cohesive energy density and express the strength of attractives force among the molecules. Solvent - polymer; solvent – non-solvent; and polymer - non-solvent interaction in a polymer solution has been estimated based on the difference between polymer, solvent and nonsolvent as already clearly mentioned by Hansen Solubility [17]. According to Hansen, the permanent dipole-dipole interaction ($\boldsymbol{\delta_p}$), dispersive ($\boldsymbol{\delta_d}$), hydrogen bonding forces ($\boldsymbol{\delta_h}$) should be taken into consideration. Consequently, the solubility parameters can be calculated by using Equation 1 [18].

$$\delta^2 = (\delta_d^2 + \delta_h^2 + \delta_p^2) \qquad (1)$$

The Hansen Solubility parameter differences (Δ) between solvents and polymers was calculated by using Equation 2 [18].

$$\Delta = \sqrt{[(\delta_{p,d} - \delta_{s,d})^2 + (\delta_{p,p} - \delta_{s,p})^2 + (\delta_{p,h} - \delta_{s,h})^2} \qquad (2)$$

where, S and P denotes as solvent and polymer, respectively. *p, d* and *h* represent for polar, dispersive, and hydrogen bonding elements of Hansen Solubility Parameter, respectively.

### *2.5. Membrane Characterization*

*2.5.1. Morphology of Membrane. Morphology of the membranes were observed by using scanning electron microscopy (SEM) (Hitachi, FlexSEM 1000, Japan) with 5 kVA. To observe cross sectional area of the membrane, it must be broken during dipping in liquid nitrogen; then, sputtered with gold. The thickness of the membranes was measured using SEM Photo and Image J. The membrane porosity ε was determined by gravimetric method [13], [19],[20]. The membranes were immersed in water for 24 hours, and the wet membranes are dried in an oven (Oven Memmered UN55, Germany) at 70 °C for 48 h, the weights of wet and dry membrane using analytical balance (Ohaus PA214, USA). The membrane porosity ε was determined using the Equation (3).*

$$\varepsilon = \frac{(w_1 - w_2)/d_w}{\frac{w_1 - w_2}{d_w} + \frac{w_2}{d_p}} \qquad (3)$$

Where $w_1$ is the weight of the wet membranes (g), $w_2$ is the weight of the dry membranes (g), $d_w$ is the density of water (0.998 g/cm$^3$) and $d_p$ is the density of the polymer (1.3 g/cm$^3$). The pore membrane was measured using by combining the threshold in the Image J application assisted by Photoshop and by using Guerout–Elford–Ferry equation based on the data of porosity and permeation rate of pure water [19]–[22].

$$r_m = \sqrt{\frac{(2.901.75\varepsilon)x\ 8\eta l Q}{\varepsilon \text{ x A x } \Delta P}} \quad (4)$$

Where η is the viscosity of water (8.9 x $10^{-4}$ Pa.s), *l* is the thickness of membrane (m), A is the effective of membrane area ($m^2$), Q is the permeation rate if water ($m^3/s$) and ΔP is the pressure membrane (Pa).

*2.5.2. Hydrophilicity of Membrane. Water content analysis is carried out to determine the ability of the membrane to absorb water. It indicates hydrophilicity of membrane. Water content analysis is done by calculating the difference between the wet mass and the dry mass. The wet mass is the mass of the membrane after immersion into water; while, the dry mass is the mass of the membrane that has been oven-dried up to a constant weight. Water content (WC) analysis was calculated using Equation 5 [23] where Ww is wet sample (g) and Wd is dry sample (g).*

$$WC\% = \frac{Ww - Wd}{Ww} \times 100\% \quad (5)$$

*2.5.3. Mechanical Property.* Tensile strength was analysed using Dynamic Mechanical Analysis (DMA/SDTA 861$^e$, Mettler Toledo, Singapore). All tests were done according to ASTM D5026 standard.

*2.5.4. Membrane Performance.* Membrane performance was analysed in terms of salt rejection with NaCl for make saline solutions for desalination testsand permeate flux in cross flow system using Sterlitech CF042A membrane cell with an effective membrane area of 42 $cm^2$. The flat sheet membrane with a size of 13 x 15.5 cm was used to analyze membrane's performance by using synthetic salt water with concentration of NaCl is 1000 ppm. Salt rejection (R) and Permeate flux was calculated by using Equation 6 and 7, respectively.

$$\%\boldsymbol{R} = \left(\frac{C_f - C_p}{C_f}\right) \boldsymbol{x}\ \mathbf{100\%} \quad (6)$$

where R is the salt rejection (%), and $C_f$ and $C_p$ are the concentration of the feed and permeate (ppm), respectively [24].

$$\boldsymbol{Permeate\ FLux} = \frac{Q}{A.\Delta t} \quad (7)$$

where Q is the quantity of permeate (L), A is the effective membrane area ($m^2$) and Δt is the sampling time (h) [25].

## 3. Result and Disscusion

*3.1 Effect of Dimetyhl Sulfoxide as a Green Solvent*

*3.1.1. Phase Diagram CA-PBS/ Solvent/ Isopropanol.* Phase diagrams are useful for predicting the phase transitions occurred when the phase separation is induced according to the immersion precipitation method. The membrane formation mechanism is based on the demixing process. This process occurs when the addition of non-solvents to the compaction process so that the polymer solution becomes thermodynamically unstable, the demixing process is one sign that the binodal process in the solution has been achieved [9], [15], [26], [27].

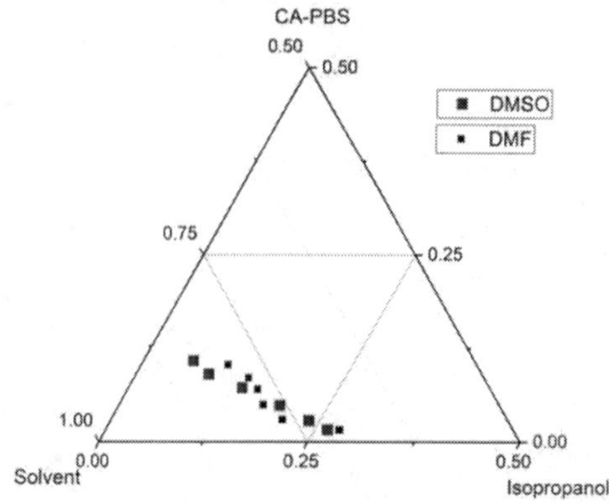


**Figure 1**. Ternary Diagram CA-PBS/Solvent/Isopropanol Membrane.

In **Figure 1**, the binodal curve with DMSO and DMF solvents shows that the DMSO solvent has a curve closer to polymer (CA-PBS) and isopropanol as non-solvent which means that liquid-liquid and solid-liquid demixing processes will occur compared to DMF. It means the polymer will diffuse more stably and form a membrane matrix. Thus, the membrane made using DMSO solvent resulting more porous membrane because the non-solvent - solvent exchange rate will be higher [15].

*3.1.2. Solubility Parameter CA-PBS/ Solvent/ Isopropanol.* The polymer-solvent-nonsolvent concentration parameter is inversely proportional with Hansen solubility parameters ($\Delta$)[18]. The solubility parameter difference between polymer- solvent ($\Delta_{p,s}$) and solvent – non-solvent ($\Delta_{s,ns}$) shown in the **Table 1.** The results show that $\Delta_{p,s}$ for DMSO is smaller than that for DMF. It means that interaction between CA/PBS-DMSO is higher than that of CA/PBS-DMF. However, the $\Delta_{s,ns}$ value show that the interaction between DMF – isopropanol is higher than that of DMSO – isopropanol. It indicates that DMF as a solvent can accelerate the process of exchange rate from solvent to non-solvent which causes instantaneous demixing processes. Because the process is very fast, resulting in the formation of large pore size, large cavities and macrovoid on the membrane. However, with DMSO as a solvent, the exchange process from solvent to non-solvent is slower which causes a liquid-liquid demixing process followed by solid-liquid demixing. Because of this, CA/PBS with DMSO membrane resulting in smaller pores and can remove macrovoid in the pores that is in agreement with the previous study [7], [11].

**Table 1.** Hansen solubility parameters for CA/PBS membrane with variation of solvents.

| **Component** | $\Delta_{p,s}$ | $\Delta_{s,ns}$ |
|---|---|---|
| CA/PBS-DMF-Isopropanol | 6.4765 | 9.2914 |
| CA/PBS-DMSO-Isopropanol | 5.7361 | 12.3000 |

*3.1.3. Morphology of Membrane CA-PBS/ Solvent/ Isopropanol*

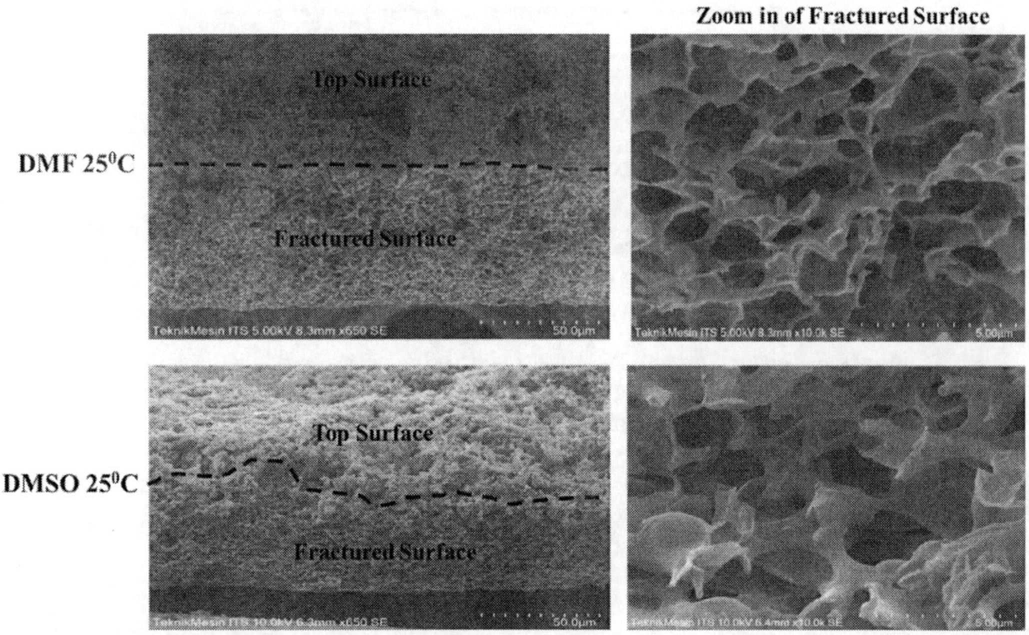


**Figure 2.** SEM morphology for CA/PBS membrane with variation of solvents, DMF and DMSO.

SEM analysis was carried out with magnification of 5000x and 10,000x to show the morphological of the membrane in the top surface and cross-section area, respectively in **Figure 2**. Morphology of membrane CA/PBS either with DMF or DMSO as solvents exhibit dense membrane with sponge like-structure. Top surface in the both membranes shows rough surface; however, the top surface of membrane with DMSO as solvent is denser than that of membrane with DMF as solvent. Porosity of CA/PEG membrane with DMF and DMSO as solvent is 58.258% and 58.843%, respectively. These porosity values are almost the same, because the pore size of both of them is also similar, about 0.3 μm.

*3.1.4. Hydrophilicity of Membrane CA-PBS/ Solvent/ Isopropanol.* Hydrophilicity plays an important role in liquid filtration [15]. The membrane hydrophilicity was determined by a water content measurement. **Table 2** shows the measured water content of CA/PBS membrane with variation of solvent. The result shows that hydrophilicity of CA/PBS membrane with DMSO as a solvent is higher than that of CA/PBS membrane with DMF as a solvent.

**Table 2.** Water Content of CA/PBS membrane with variation of solvents.

| Component | Water Content (%) |
|---|---|
| CA/PBS-DMF-Isopropanol | 51.724 |
| CA/PBS-DMSO-Isopropanol | 52.326 |

*3.1.5. Mechanical Property of Membrane CA/PBS Membrane.* **Figure 3** shows the mechanical property of CA/PBS membranes with variation of solvent. In general, the tensile strength of membrane with DMSO as solvent is higher compared to membrane with DMF as a solvent. Membranes with DMSO solvent variations have a relatively small pore size compared to the DMF solvent (0.331 vs 0.351 μm), resulting in an increased of mechanical strength of membrane with DMSO as a solvent [15].

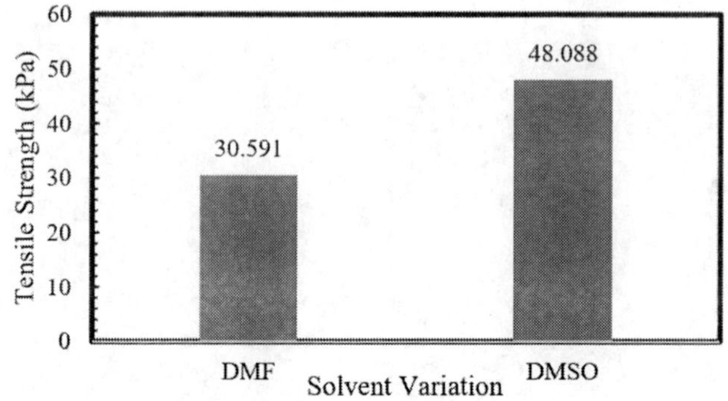


**Figure 3.** Tensile Strength Results CA-PBS/ Solvent/ Isopropanol Membrane.

*3.1.6. Hydrophilicity of Membrane CA-PBS/ Solvent/ Isopropanol.* Based on **Figure 4**. the membrane whose DMSO solvent has increased the value of salt rejection, this means that the DMSO solvent is proven to be in accordance with the prediction of phase separation analysis where by using the DMSO solvent the resulting line is closer to the polymer-non-solvent line which results in a liquid-liquid demixing process, resulting in a more uniform pore and more able to hold the solute that passes through the membrane and is a solvent that is more environmentally friendly can also increase the performance of the membrane [14], [15], [28].

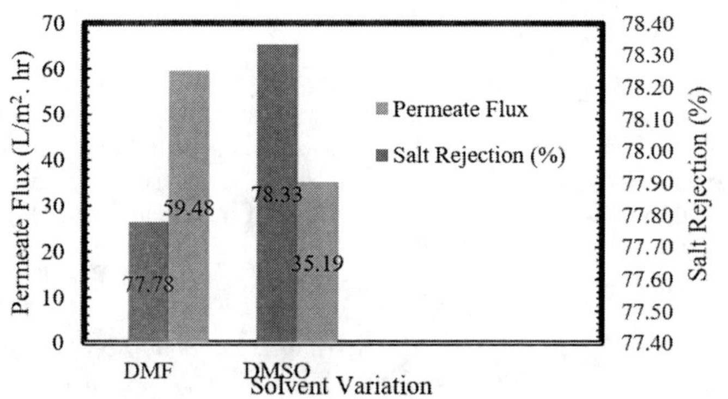


**Figure 4.** Performance Test for Variation Solvent in CA/PBS Membrane.

From the membrane performance as shown in Figure 4, it can be concluded that DMSO is better than DMF. Salt rejection of CA/PBS membrane with DMSO as a solvent is 78.33 % and permeate flux is 35.19 L/m$^2$.hr.

*3.2 Effect of Additive Polyethylene Glycol in CA-PBS/DMSo/Isopropanol Membrane*

*3.2.1 Morphology of Membrane CA-PBS/ DMSO/ Isopropanol with additive Polyethylene Glycol.* SEM analysis was carried out with magnification of 650x, 5000x and 10,000x to show the morphological structure of the membrane in the top surface and cross-section area. From the results of the SEM analysis, the morphological parameters observed were pore distribution, pore type,pore diameter and porosity. Pore distribution in the additive PEG can be seen in **Figure 5,** shows asymmetrical tendencies where the membrane has sponge-like pores. When the addition of PEG is more, the smaller the pore size occurs. It is because PEG is a pore-forming and acts as a plasticizer agent that makes the membrane stronger [2], [29], [30].However, when the addition of PEG is 8% wt of polymer resulted in a more dense, symmetrical membrane and a larger pore size.

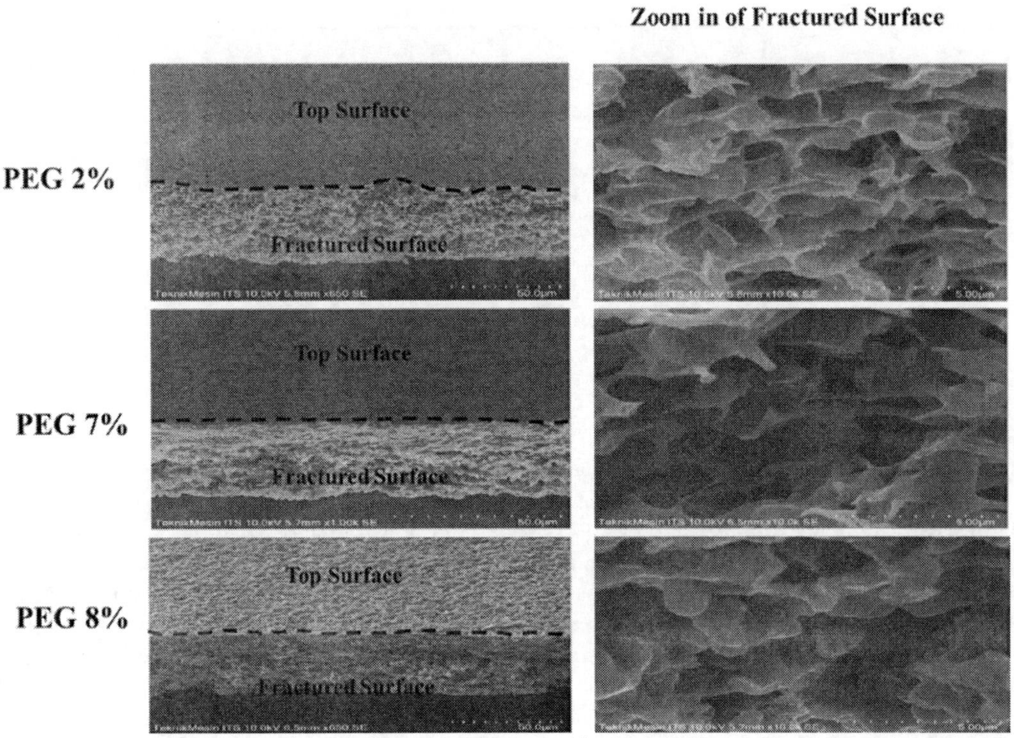


**Figure 5**. SEM morphology for CA/PBS membrane with the addition of PEG by using DMSO as solvent.

**Table 3** shows pore size and porosity of CA/PBS membrane with the addition of PEG. It can be seen that with the addition of PEG at 7 wt% resulting the smallest pore size diameter in the fractured surface. This result is an agreement with the value of the porosity. It exhibits the highest value in the porosity. The mass transport in Reverse Osmosis (RO) occurs due to the solution-diffusion phenomena, not by the molecular sieving. One of the characteristics of RO is dense structure [31]. The resulting pore has a dense pore structure, it's just that the pore size is still relatively large, this is what causes the use of the sample in the rejection salt test using a salt concentration only 1000 ppm, when the concentration is increased to above 1000 ppm the results obtained are not maximal. Therefore, the membrane in this research work is still limited to a salt concentration of 1000 ppm.

**Tabel 3.** Pore size and Porosity of CA/PBS Membrane with the addition of PEG.

| PEG Content % wt | Pore size (μm) | Porosity (%) |
|---|---|---|
| 2 | 0.441 | 59.00 |
| 7 | 0.262 | 75.16 |
| 8 | 0.596 | 67.47 |

*3.2.2 Mechanical Evaluation of CA-PBS/ DMSO Membrane with additive PEG.* Tensile strength of the CA-PBS/DMSO with additive of PEG membrane is depicted in **Figure 6.** The result show that for membrane prepared with addition of 7%wt PEG increased the tensile strength until 68.366 kPa. At higher PEG concentration, some undesired big pores were formed at the surface. It was known that high porosity and large cavities in the membrane structure reduced the tensile strength [32], And then tensile strength of PEG 8% wt membrane was decreased.

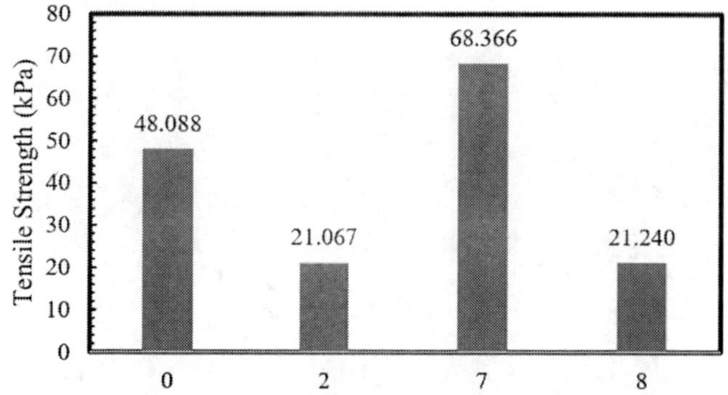


**Figure 6.** Tensille Strength Result Membrane with additive PEG.

*3.2.3 Performance of CA-PBS/ DMSO Membrane with the Addition of PEG.* In **Figure 7,** it can be seen that salt rejection increases with the addition of PEG. It also increases with the increasing of PEG content up to 7 wt%. With the addition of 7 wt% PEG resulting salt rejection of 88.89 %. When CA-PBS/PEG membrane is immersed in the coagulation bath, the phase separation process begins. The PEG has an almost low molecular chain length and consequently good mobility which results in a smaller downward movement of the polymer on the membrane surface along with the outflow of the solvent to the coagulation bath. Therefore, the aggregated PEG molecules are distributed uniformly on the membrane surface causing the formation of a more porous surface [33], and the presence of PEG molecules on the surface makes the diffusion process of non-solvent molecules onto the films created during the compaction stage in the coagulation bath to occur which resulting in pore formation [34].

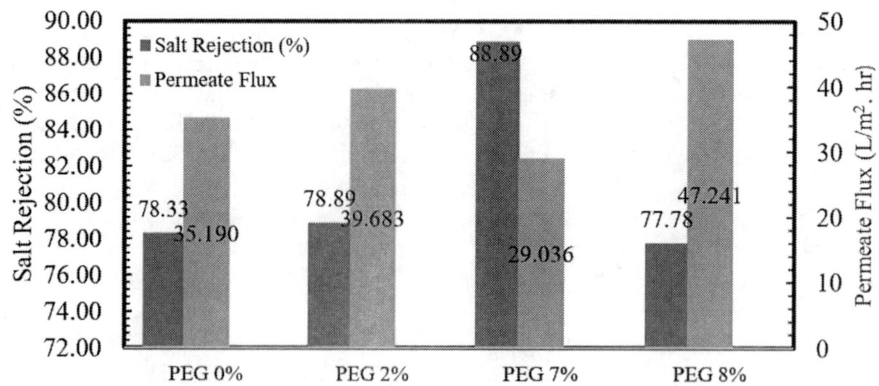


**Figure 7.** Membrane's Performance with the addition of PEG.

## Conclusion

DMSO is non-toxic solvent that is suitable for an alternative solvent in the casting solution for the preparation of CA/PBS membranes, instead of DMF. The ternary phase diagrams showed that using

DMSO instead of DMF in the same concentration leads to more permeable membranes. Both the salt rejection and permeate flux of the membranes with DMSO was higher than with DMF as a solvent. This difference in performance was caused by a denser morphology on the top surface and a more porous sublayer (fractured surface) of asymmetric membranes with DMSO as solvent. The effect of the addition of PEG in CA/PBS-DMSO membrane increases the membrane's performance, mechanical strength, hydrophilicity and porosity. With the addition of 7 wt % PEG, salt rejection of CA/PBS membrane is 88.89 % and the tensile strength is 68.366 kPa.

**Acknowledgements**
This work has been funded by ministry of Research, Technology and Higher Education of the Republic of Indonesia with contract's number 2117/PKS/ITS/2020 to which the authors express their sincerest gratitude.

**References**
[1] Abedini R, Mahmoud S, and Aminzadeh R 2011 *Desalination* **277** p 1-40
[2] Kim J.H and Lee K H 1998 *J. Memb. Sci.***138** p 153–163
[3] Waheed S 2014 *Desalination* **351** p 59–69
[4] Nurkhamidah S, Dwi R D, Ramadhani A, and Cintya B 2019 *Indonesian Polym Journal Article* **22** p 10–15
[5] Zhang Z, An Q, Ji Y, Qian J, and Gao C 2010 *Desalination* **260** p 43–50
[6] Albrecht W, Weigel T, Kneifel K, Peinemann K, and Paul D 2001 *Journal of Membrane Science* **192** p 217–230
[7] Boom R M , van den Boomgaard T, van den Berg J W A and Smolders C A 1993 *Polymer* **34**, p 2348–2356
[8] Tian Y, Booth J, Meehan E, Jones D S , Li S, and Andrews G P 2013 *Mol. Pharm* 10 p 236–248
[9] Ebrahimpour M, Safekordi A A, Mousavi S M, and Nasab A H 2017 *Bulg. Chem. Commun. Spec. Issue J* **49** pp 389–395.
[10] Tanaka T 2010 *Desalin. Water Treat* **17** 176–182
[11] Ishii N, Inoue Y, Tagaya T, and Mitomo H 2008 *Polymer Degradation and Stability* **93** pp 883–888
[12] Cantor P A, Mechalas B J, and Monte E 1969 *J Polym Sci* **241** pp 225–241
[13] Chen L 2018 *Chinese Journal of Polym Sci* **10** pp. 1–7.
[14] Soroko L, Bhole Y, and Livingston A G 2011 *Green Chem* **13** pp 162–168
[15] Evenepoel N, Wen S, Tsehaye M T, and Van Der Bruggen B 2018 *Journal of Membrane Sci* **2** pp. 1–10
[16] Sabir A. 2015 *Carbohydr. Polym.* **132** pp 589–597
[17] Hansen C M 1967. *Copenhagen Danish Technical Press*
[18] Gebru K A and Das C 2017 *Chinese J. Chem* **25** pp 911–923
[19] Kusumocahyo S P, Ambani S K, Kusumadewi S, Sutanto H, Widiputri D I, and Kartawiria I S 2020 *J. Environ. Chem. Eng* **8** p 104381
[20] Hamid N A A 2011 *Desalination* 273 pp 85–92
[21] Zambare R S, Dhopte K B, Patwardhan A V and Nemade P R 2017 *Desalination* **403** pp 24–35
[22] Zhu Z, Wang L, Xu Y, Li Q, Jiang J, and Wang X 2017 *J. Colloid Interface Sci* **504** pp 429–439
[23] Zhao J L 2013 *J. Tianjin Polytech. Univ* **32** pp. 1–7
[24] Celik E, Park H, and Choi H 2011 *Water Res* **45** pp 274–282
[25] Saljoughi E, Sadrzadeh M, and Mohammadi T 2008 *J. Memb. Sci* **326** pp 627–634
[26] Aryanti P T P, Ariono D, Hakim A N and Wenten I G 2018 *J. Phys. Conf. Ser* **1090** p 1-13
[27] Altena F W and Smolders C A 1982 *Macromolecules* **15** pp 1491–1497
[28] Nu D T T, Hung N P, an Hoang C V, and Van der Bruggen B 2019 *Appl. Sci* **9** p 16
[29] Mohammadi T and Saljoughi E 2009 *Desalination* **243** pp 1–7
[30] Yoo J E, Kim J H, Kim Y, and Kim C K 2003 *J. Memb. Sci* **216** pp 95–106

[31] Paul D R 2004 *Journal of Membrane Sci* **241** pp 371–386
[32] Vahideh Q, Mahmoud S, Bahreini M, and Jalaei H 2013 *J. Ind. Eng. Chem* **1460** pp 1–8
[33] Zuo D, Xu Y, Xu W and Zou H 2008 *Chinese of Polym Science* **26** pp 405–414
[34] Taylor P, Ghaffarian V, Mousavi S M, Bahreini M, and Chamani H 2015 *Desalination and Water Treatment* **1944** pp 1-10

# Natural Gas Network Design using Superstructure Method in East Java Indonesia

**Rendra Panca Anugraha*, Renanto, and Juwari**

Department of Chemical Engineering, Faculty of Industrial Technology and Systems Engineering, Institut Teknologi Sepuluh Nopember (ITS), Kampus ITS Sukolilo, Surabaya 60111 Indonesia

*rendra@its.ac.id

**Abstract.** East Java is one of the provinces in Indonesia which has total natural gas reserves of 4.66 trillion standard cubic feet (TSCF) spread across several locations. Natural gas in this province is used by a variety of consumers, such as petrochemical industries, power plants, industrial fuel, transportation and household needs. However, there are obstacles in the utilization of natural gas due to differences in operating time and differences capacity between suppliers and consumers. Therefore, to optimize the utilization of natural gas in East Java, a natural gas network design is required which is considering the operating time and the capacity of suppliers (source point) and consumers (sinks point). In this study, a natural gas network design of East Java area was developed by modelling superstructure methods which consider the operating time and capacity. The superstructure natural gas network model developed in this study was optimized using GAMS software. From 5 source points and 6 sink points, an optimum natural gas network design has been obtained with a total gas distribution of 4832.8 billon standard cubic feet (BSCF) in a period of 30 years. Due to mismatch of operating time, it is also known that the amount of excess gas supply from this area (export gas) is 1364.1 BSCF and the demand for gas supply from other areas (import gas) is 1105.7 BSCF. With this superstructure method, it is possible to know the optimum network configuration and the natural gas balance in an area which has different operating time, flowrate and capacity between sources and sinks.

## 1. Introduction

Indonesia has large natural gas reserve of 135.55 trillion standard cubic feet (TSCF) which are spread over several regions. Especially in East Java region, it has natural gas reserve of 4.66 TSCF and predicted still be available for the next 30 years [1]. The large availability of natural gas in this region is spread over several locations with different capacities, flow rates and time operations. Natural gas in this province is used by a variety of consumers, such as petrochemical industries, power plants, industrial fuel, transportation and household needs [1]. However, there are obstacles in the utilization of natural gas due to differences in operating time and differences capacity between suppliers and consumers. Therefore, to optimize the utilization of natural gas in East Java, a natural gas network design is required which is considering the operating time and the capacity of gas suppliers (source point) and consumers (sinks point).

Several studies are discussing about the natural gas network design and optimization. Anugraha et al (2020) [2] were developed a systematical method to design a natural gas network system in a single region using pinch analysis and introduced the heuristics of natural gas pairing between source and sink streams in a grid diagram based on the cascade natural gas calculation. Su et al (2019) [3] were developed a multi-objective optimization methods for natural gas pipeline networks by considering trade-off reliability and power demand in the systems to obtain the minimum gas supply shortage risk and minimum power demand. Mikolajková et al (2017) [4] were developed a pipeline network design for gas distribution which considering different supply of gas, such as external gas networks, injected biogas, or gasified liquefied natural gas (LNG) at terminals. Sidarto et al (2017) [5] were developed an approachment to determine the natural gas pressure distribution in natural gas pipeline networks system using Broyden method. Ríos-Mercado at al (2015) [6] reported a review in optimization of natural gas transportation systems which focused on specific categories such as short-term basis storage, gas quality satisfaction and compressor station modelling. Rodríguez et al. (2014) [7] were used an sequential quadratic programming (SQP) algorithm to obtain optimum natural gas networks with minimum fuel gas consumption of the compressors. However, as far as authors know, there are no research in open literature discussing about the natural gas network design using superstructure method which considering different capacity, flowrate and time operation between gas suppliers and consumers.

Therefore, in this study, natural gas network design of East Java area was developed by modelling superstructure methods [8] which consider the suitable operating time and capacity. The developed superstructure model was optimized by GAMS software to obtain maximum natural gas transferred from source points to sink point with suitable operating time, flowrate and capacity.

## 2. Research Methodology

### *2.1. Supply and Demand Data*

The supply and demand data for this study were obtained from the annual report on national gas balance issued by Energy & Mineral Resources Ministry of Republic Indonesia. as shown in Table 1 and Table 2. The supply and demand data listed in Table 1 and 2 were the simplified data from the annual report on national gas balance in East Java region by consideration of the nearest location.

**Table 1** Supply data of natural gas in East Java [1].

| Code | Source Place | Start Time (year) | Duration (year) | End Time (year) | Natural Gas Production Rate (BSCF/year) | Natural Gas Capacity (BSCF) |
|---|---|---|---|---|---|---|
| SRM1 | Madura Field 1 | 2 | 16 | 18 | 142.4 | 2278.4 |
| SRM2 | Madura Field 2 | 5 | 18 | 23 | 87.6 | 1576.8 |
| SRG1 | Gresik Field | 3 | 13 | 16 | 34.7 | 451.1 |
| SRB1 | Bojonegoro Field | 6 | 23 | 29 | 56.6 | 1301.8 |
| SRL1 | Lengo Field | 7 | 23 | 30 | 25.6 | 588.8 |
| | | **Total Source** | | | | **6196.9** |

BSCF = billion standard cubic feet

**Table 2** Demand data of natural gas in East Java [1].

| Code | Sink Place | Start Time (year) | Duration (year) | End Time (year) | Natural Gas Consumption Rate (BSCF/year) | Consumed Natural Gas (BSCF) |
|---|---|---|---|---|---|---|
| SKP1 | PT. Petrokimia | 2 | 28 | 30 | 54.8 | 1534.4 |
| SKG1 | Gas Power Plants | 3 | 22 | 25 | 113.2 | 2490.4 |
| SKE1 | PT. Pertamina Tuban | 10 | 20 | 30 | 45.6 | 912.0 |
| SKI1 | Industrial Complex | 4 | 16 | 20 | 54.8 | 876.8 |
| SKR1 | Domestic Gas | 8 | 22 | 30 | 0.7 | 15.4 |
| SKT1 | Transportation Gas | 5 | 15 | 20 | 7.3 | 109.5 |
| | | **Total Sink** | | | | 5938.5 |

BSCF = billion standard cubic feet

The "start time" column in Table 1 and 2 means the year from the supply points to start delivering the natural gas and the year from demand points to start receiving the transferred natural gas, respectively. The "end time" column in Table 1 and 2 shows the year when the supply points no longer produce the natural gas and the year when demand points had run out of the gas contracts, respectively.

### *2.2. Modelling and Optimization*

The natural gas network modelling was approached using superstructure pairing concept which shown in Figure 1. Figure 1 shows the possible scenario of the natural gas network of the supply and demand data from Table 1&2 which can be obtained by superstructure pairing network.

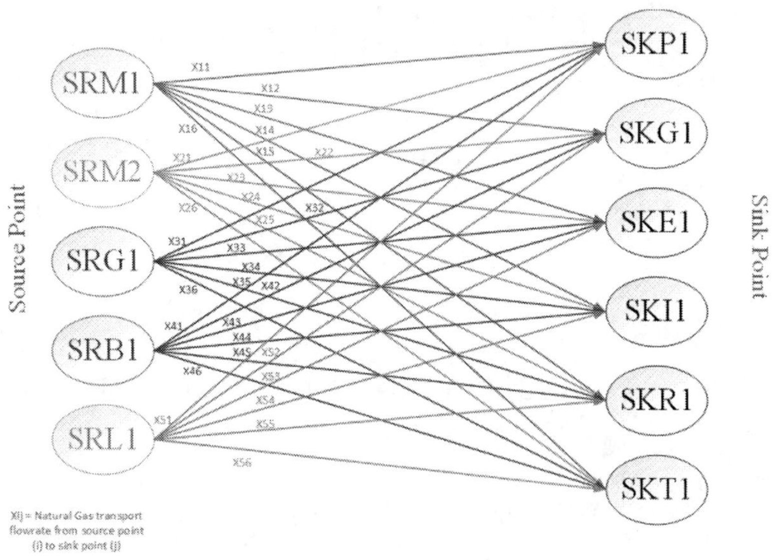


**Figure 1** The possible scenario of the natural gas network by superstructure pairing network.

Each source points were connected to all sinks points and created multiple possible connections which is equal to the number of source points times the number of sink points. The start time and end time of each connection were determined by choosing the larger start time between paired source and sink point and choosing the smaller end time between paired source and sink point, respectively. For each connection, from a source point ($i$) to a sink point ($j$), delivers a certain flowrate natural gas transported ($x_{i,j}$) in BSCF/year. Reminder, the value of flowrate natural gas transported ($x_{i,j}$) must be less than the annual flowrate of source point and sink point.

From this possible scenario of superstructure modelling, linear programming (LP) formulation can be obtained [9] and also the constrains (explained in subsection 2.3). The LP mathematic model was optimized using GAMS software [10] to obtain the maximum total natural gas transferred from source to sink. Then, the optimum transfer flowrate and configuration was drawn in a grid diagram to visualize the natural gas network and to calculate the alternative supply and excess supply. All steps of the modelling and optimization procedures were described in a flowchart as shown in Figure 2.

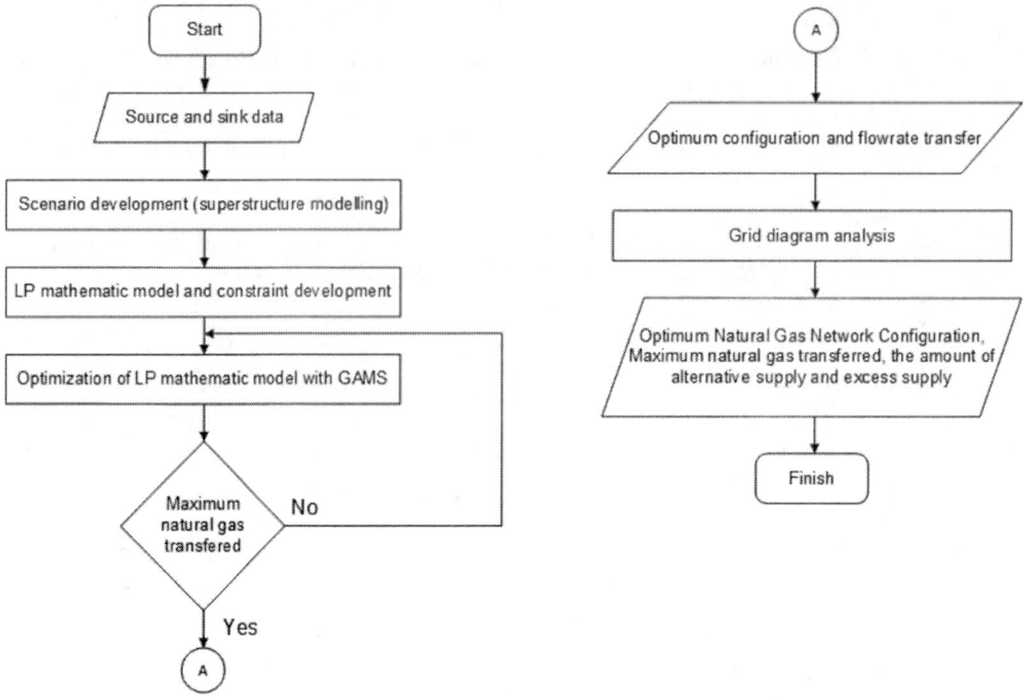


**Figure 2** The flowchart of natural gas network modelling and optimization.

### *2.3. Parameters and Equations*

As described in previous sub-section, the objective function ($z$) in this study is maximizing total natural gas transferred from source to sink which described in following equation:

$$z = NG\ transfered = \sum_{i=1}^{5}\sum_{j=1}^{6} t_{i,j}\, x_{i,j} \tag{1}$$

where, $i$ = numbering of source points (1 – 5)
$j$ = numbering of sink points (1 – 6)
$t_{i,j}$ = duration transfer of natural gas connection from source ($i$) to sink ($j$) in years
$x_{i,j}$ = flowrate transfer of natural gas connection from source ($i$) to sink ($j$) in BSCF/years.

The value of flowrate natural gas transported ($x_{i,j}$) cannot be negative value and must be less than the annual flowrate of source ($S_i$) and sink ($D_i$) point. This statement could be defined as the constraints which described in Equation 2 – 4:

for each source point,

$$\sum_{j=1}^{6} x_{i,j} \leq S_i \tag{2}$$

for each sink point,

$$\sum_{i=1}^{5} x_{i,j} \leq D_j \tag{3}$$

for all connections,

$$X_{i,j} \geq 0 \tag{4}$$

where, $S_i$ = the maximum annual flowrate of source point ($i$)
$D_j$ = the maximum annual flowrate of sink point ($j$)

This means that all transferred gas to all sink points from source point ($i$) could not be higher than the maximum annual flowrate of source point ($S_i$) and all received gas from all source points to sink point ($j$) could not be higher than the maximum annual flowrate of sink point ($D_j$).

After optimizing the value of the flowrate transfer of natural gas connection ($x_{i,j}$), the amount of excess supply and excess demand of the system could be obtained with following equation:

excess supply ($ES$) :

$$ES = \left[\sum_{i=1}^{5} S_i(t_{end,i} - t_{start,i})\right] - NG\ transfered \quad (5)$$

excess demand ($ED$) :

$$ED = \left[\sum_{j=1}^{6} D_j(t_{end,j} - t_{start,j})\right] - NG\ transfered \quad (6)$$

where, $t_{start,i}$ = start time of source point ($i$)
$t_{end,i}$ = end time of source point ($i$)
$t_{start,j}$ = start time of sink point ($j$)
$t_{end,j}$ = end time of sink point ($j$)

## 3. Results and Discussion

### 3.1. Optimization Result

In this study, the flowrate natural gas transported ($x_{i,j}$) for each connection between source point ($i$) to sink point ($j$) were optimized using GAMS Software. Table 3 shows the optimized value of the flowrate natural gas transported ($x_{i,j}$) for each connection.

**Table 3.** The optimized value of the flowrate natural gas transported ($x_{i,j}$) for each connection between source point ($i$) to sink point ($j$) in BSCF/year.

| $x_{i,j}$ value | SKP1 $D_1$ = 54.8 | SKG1 $D_2$ = 113.2 | SKE1 $D_3$ = 45.6 | SKI1 $D_4$ = 54.8 | SKR1 $D_5$ = 0.7 | SKT1 $D_6$ = 7.3 | $\sum_{j=1}^{6} x_{i,j}$ |
|---|---|---|---|---|---|---|---|
| SRM1 $S_1$ = 142.4 | 18.9 | 25.6 | – | 54.8 | – | 7.3 | **106.6** |
| SRM2 $S_2$ = 87.6 | – | 87.6 | – | – | – | – | **87.6** |
| SRG1 $S_3$ = 34.7 | – | – | – | – | – | – | **0** |
| SRB1 $S_4$ = 56.6 | 11 | – | 45.6 | – | – | – | **56.6** |
| SRL1 $S_5$ = 25.6 | 24.9 | – | – | – | 0.7 | – | **25.6** |
| $\sum_{i=1}^{5} x_{i,j}$ | **54.8** | **113.2** | **45.6** | **54.8** | **0.7** | **7.3** | |

$x_{i,j}$ = flowrate transfer of natural gas connection from source ($i$) to sink ($j$) in BSCF/years.
$S_i$ = the maximum annual flowrate of source point ($i$) in BSCF/years.
$D_j$ = the maximum annual flowrate of sink point ($j$) in BSCF/years.

As shown in Table 3, the total flowrate natural gas transported from a source point to all sink points is less than the annual flowrate of the source point ($S_i$). And also, the total flowrate natural gas transported to a sink point from all source points is less than the annual flowrate of the sink point ($D_j$). There are no negative value of the flowrate natural gas transported ($x_{i,j}$). Therefore, by this optimum pairing configuration, the maximum total natural gas transferred from sources to sinks is 4832.8 BSCF.

### 3.2. *Grid Diagram of Optimized Natural Gas Network*

The optimum pairing configuration can be drawn in a grid diagram to visualize the natural gas network and to calculate the alternative supply and excess supply. Figure 3 shows the grid diagram of the optimized natural gas network by superstructure method.

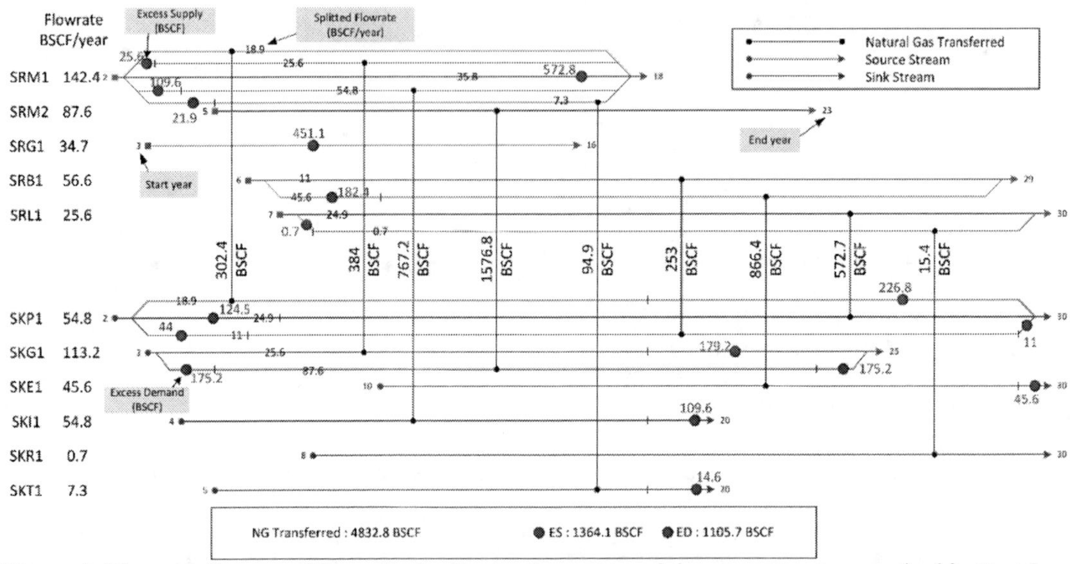


**Figure 3.** The grid diagram of the optimized natural gas network by superstructure method in East Java Region.

As shown in Figure 3, all pairing between source and sink point were connected in suitable time, no time is overturned or delayed. From the grid diagram, the excess supply and excess demand were obtained with value of 1364.1 BSCF and 1105.7 BSCF, respectively. The excess supply means the amount of natural gas that must be exported from this region to other regions, meanwhile the excess demand means the amount of natural gas that must be imported from other regions to this region. Therefore, from the grid diagram, it is known that how much natural gas can be utilized directly domestically and the amount of gas that must be exported and imported in a certain region. With this superstructure method, it is possible to know the optimum configuration and the natural gas balance in an area which has different operating time, flowrate and capacity between sources and sinks.

## 4. Conclusion

In this study, natural gas network design of East Java area was developed by modelling superstructure methods which consider the suitable operating time, flowrate and capacity. The developed superstructure model was optimized by GAMS software to obtain maximum natural gas transferred from source points to sink point with suitable operating time, flowrate and capacity. The maximum total natural gas transferred from sources to sinks is 4832.8 BSCF. The excess supply and excess demand were obtained with value of 1364.1 BSCF and 1105.7 BSCF, respectively. With this superstructure method, it is possible to know the optimum network configuration and the natural gas balance in an area which has different operating time, flowrate and capacity between sources and sinks.

**Acknowledgements**
This work was supported by a research local grant from Institut Teknologi Sepuluh Nopember under Contract No. 862/PKS/ITS/2020 with *Penelitian Doktor Baru* research scheme.

**References**

[1] Ministry of Energy and Mineral Resource of Indonesia 2018 *Natural Gas Balance in Indonesia: 2018-2027* (Jakarta)
[2] Anugraha R P, Annasit, Altway A, Juwari and Renanto 2020 The optimization of natural gas utilization network in single region using pinch analysis method *ASEAN J. Chem. Eng.* **20** 57–66
[3] Su H, Zio E, Zhang J, Li X, Chi L, Fan L and Zhang Z 2019 A method for the multi-objective optimization of the operation of natural gas pipeline networks considering supply reliability and operation efficiency *Comput. Chem. Eng.* **131** 106584
[4] Mikolajková M, Haikarainen C, Saxén H and Pettersson F 2017 Optimization of a natural gas distribution network with potential future extensions *Energy* **125** 848–59
[5] Sidarto K A, Kania A, Mucharam L, Darmadi D and Widhymarmanto R A 2017 Determination of gas pressure distribution in a pipeline network using the broyden method *J. Eng. Technol. Sci.* **49** 750–69
[6] Ríos-Mercado R Z and Borraz-Sánchez C 2015 Optimization problems in natural gas transportation systems: A state-of-the-art review *Appl. Energy* **147** 536–55
[7] Rodríguez T, Sarabia D, Valbuena M and de Prada C 2014 Modelling and Optimization of Natural Gas Networks *Comput. Aided Chem. Eng.* **33** 367–72
[8] Smith R 2005 *Chemical Process Design and Integration* (New Jersey: John Wiley & Sons)
[9] Konukman A E ş., Çamurdan M C and Akman U 2002 Simultaneous flexibility targeting and synthesis of minimum-utility heat-exchanger networks with superstructure-based MILP formulation *Chem. Eng. Process.* **41** 501–18
[10] Kalvelagen E 2002 *Solving Systems of Linear Equations With GAMS* (GAMS Technical Report, 12 Nov 2002)

IOP Conf. Series: Materials Science and Engineering 1143 (2021) 012065 doi:10.1088/1757-899X/1143/1/012065

# Isomerization of Raw Turpentine Using Various Combination of Strong and Weak Acid Catalyst for Eucalyptol Production

**Nicolaus Elka Yudhatama[1], Diva Almira Chairany[1], Muhammad Mufti Azis*[1], Antonius Indarto[2]**

[1]Department of Chemical Engineering, Faculty of Engineering,Universitas Gadjah Mada, Yogyakarta 55281 Indonesia
[2]Department of Chemical Engineering, Institut Teknologi Bandung, Kampus ITB Ganesha, Indonesia

*Corresponding Author's E-mail: muhammad.azis@ugm.ac.id

**Abstract**. There has been large interest to isomerize raw turpentine to produce a number of derivative products that has higher economic value. This process is often conducted using homogenous acid catalysts. Catalyst screening is often challenging to target a certain derivative product, such as cineole. The aim of present work is investigate various combination of strong and weak acid catalyst to obtain cineole. Strong acids used in this research were PTSA, hydrochloric acid, sulfuric acid. In addition, the weak acids were oxalic acid dihydrate, citric acid, and formic acid. Commercial turpentine was mixed with strong and weak acid with 4:1:6 ratio respectively, heated up to 85°C for 6 hours for each combination of strong and weak acid. The result from GC-MS analysis shows that combination of hydrochloric acid and formic acid gave the highest eucalyptol yield around 9.67% along with other valuable product such as camphene, limonene, and α-terpinolene.

## 1. Introduction

Turpentine is a compound with molecular formula of $C_{10}H_{16}$ which is produced from the hydrodistillation process of pine sap as a top product, and this product has distinctive odor that is widely used in fragrance, pharmaceutical, and food industries [1]. Turpentine can be used raw as an organic solvent in paints or coatings. However, with further processing, turpentine can produce numerous compounds with wider application with higher values in market [2]. The main components contained in turpentine compound are α- and ß- pinene. In this process, α-pinene compound can be isomerized into various derivative product using an acid catalyst to produce cyclic and bicyclic derivative compounds. One of the promising product is eucalyptol or 1,8-cineole which is a valueable derivative compound that has the most expensive market price. This oil shows antibacterial, antiseptic, antioxidant, anti-inflammatory, and anticancer activities as at its wide application on perfumery, cosmetics, aromatherapy, phytotherapy, food beverages, and pharmaceutical industries [3]

Cineole produced from α-pinene isomerization has 2 isomers, namely 1,8-cineole (eucalyptol) and also 1,4-cineole and can be isomerized from turpentine using a combination of strong and weak acid catalyst to increase the selectivity of α-pinene to cineole compounds [4]. Previous work by Putra et al.[5] reported that the use of this acid catalyst can reduce the activation energy so that even low temperatures can facilitate the reaction. The acid catalyst in this reaction provides $H^+$ ions that attack the double bonds in the α-pinene ring and produce pinanyl cation. This carbocation component is a key intermediate and a precursor for the next two reaction steps. The first is the ring expansion step which gives rise to bicyclic and tricyclic products such as camphene and tricyclene. The second path yields monocyclic products such as limonene and cineole [6]. The reaction mechanism is as illustrated in **Figure 1**.

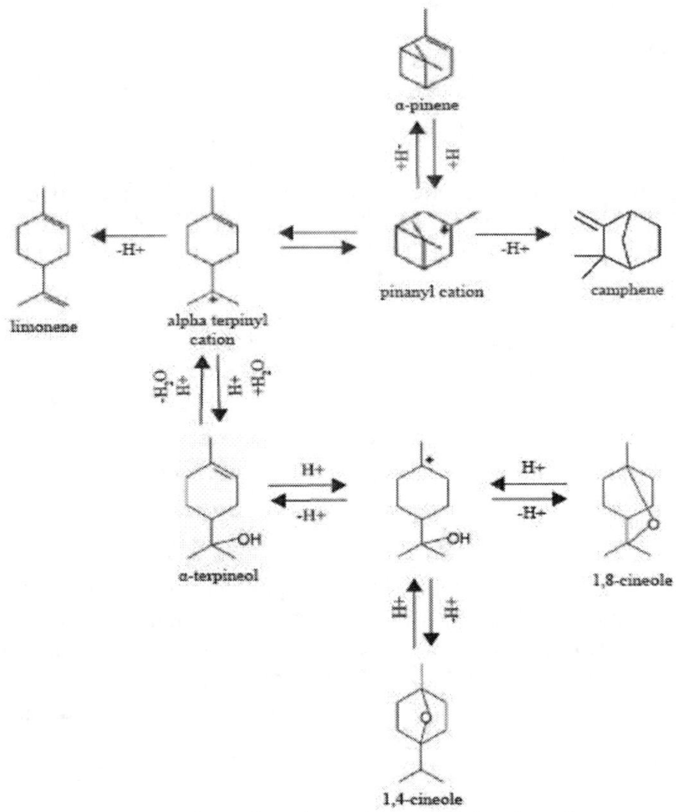


**Figure 1.** Isomerization mechanism of α-pinene into its derivatives from several studies[2,7].

Several studies showed that the isomerization of turpentine using catalyst combination of PTSA and oxalic acid was able to produce 21.7% wt of cineole [8]. Other study conducted by Lana et. al. showed that the isomerization of a-terpineol with $H_3PW_{12}O_{40}$ gave 50-90% conversion of a-terpineol with selectivity of 25% for 1,8-cineole and 23-27% for 1,4-cineole [9]. The use of temperature of 85°C was chosen into this research because this condition is the optimization result based on previous research [5]. The purpose of turpentine isomerization to cineole is to change the composition of a-pinene into eucalyptol. In general, eucalyptol has been obtained as byproduct from the hydration process of α-pinene into terpineol. Meanwhile, this process is considered very difficult and costly because it still requires further distillation to get the pure eucalyptol. Therefore it is necessary to have an optimal synthesis pathway in the production of cineole, one of which is the isomerization process of turpentine to cineol using various acid catalysts. The objective of present work is to conduct acid catalyst screening to obtain high 1,8-cineole yield using combination of strong and weak acids. The strong acids used in this work are PTSA, HCl and $H_2SO_4$. In addition, a number of weak acids such as citric, formic and oxalic acid have been used.

## 2. Material and Methods

### *2.1. Material*

The material used was commercial raw turpentine which contained 73.21% α-pinene along with other isomer compound such as α-terpinolene (1,156%), Camphene (1,243%), o-cymene (1,676%), and limonene (1,597%). The acid used in this research were sulphuric acid, oxalic acid, PTSA, as strong acid component, and formic acid, oxalic acid, and citric acid as the weak acid component. All materials (commercial grades) were obtained from Coal, Gas and Petroleum Technology Laboratory, Chemical Engineering Department, Universitas Gadjah Mada.

### 2.2. *Methods*

*2.2.1. Sample Preparation.* 6.4 mL of turpentine oil was taken from turpentine flask and 0.01 mol HCl solution was prepared by dissolving 3.1 mL of 37% HCl in 100 mL of aquadest. Subsequently, 25 mL of the HCl solution was used in the reactor. For the weak acid, 0.06 mol formic acid was taken directly from the 98-100% formic acid as much as 2.3 mL and later was placed into a three neck flask. The experiment was repeated with 0.01 mol of sulfuric acid, 0.01 mol of PTSA acid and 0.06 mol of oxalic acid and citric acid with different combinations.

*2.2.2. Batch Reactor Experiment.* A mixture of strong acid, weak acid, turpentine, and a magnetic stirrer were placed into the three-neck flask reactor. The three neck flask equipped with condenser was submerged in an oil heater using clamps. The reaction initiated by adjusting the hotplate stirrer magnetic controller and temperature control to mix and heat the mixture until the temperature reading in the alcohol thermometer shows 85°C. This process was maintained for 6 hours duration. The samples that contained the oil layer which was located on the top was taken using a pipette and put into a vial bottle for storage.

*2.2.3. Sample Analysis.* The process of analyzing each product from using the combination of acids was carried out using Shimadzu GC-MS QP-2010 with Rtx-5MS capillary column at Laboratory of Analysis and Instrumentation, Chemical Engineering Department, Faculty of Engineering, Universitas Gadjah Mada. For data processing, the α-pinene and eucalyptol normalization area ($N_{pinene}$ & $N_{eucalyptol}$) were evaluated by using total normalization method from α-pinene and eucalyptol peak to give fair comparison result between initial and after reaction[10]. The eucalyptol yield ($Y_{Eucalyptol}$) were also evaluated to give precise number of eucalyptol produced by isomerization reaction. Below are the equations used.

$$N_{\alpha-pinene} = \frac{A_{A_{\alpha-pinene}}}{A_{\alpha-pinene} + A_{eucalyptol}} x100\% \quad (1)$$

$$N_{eucalyptol} = \frac{A_{eucalyptol}}{A_{\alpha-pinene} + A_{eucalyptol}} x100\% \quad (2)$$

$$Y_{Eucalyptol} = \frac{A_{Eucalyptol\ product} - A_{Eucalyptol\ initial}}{A_{\alpha-pineneinitial}} x100\% \quad (3)$$

## 3. Result and Discussion

As advantageous essentials oil, eucalyptol can shows various beneficial biological activities such as antimicrobial agent against many pathogens and exhibit antioxidant properties as well as good odor in fragrance industries. Eucalyptol which is a monoterpenoid compound derivative α-pinene from turpentie was isomerized to produce eucalyptol using various acid catalyst combination and analyzed using GC-MS to identify the component present in 9 sample combination. Based on GC-MS results, the isomerization reaction produces many peaks which indicated the presence of numerous components in the final reaction product as shown in the chromatogram of **Figure 2.**

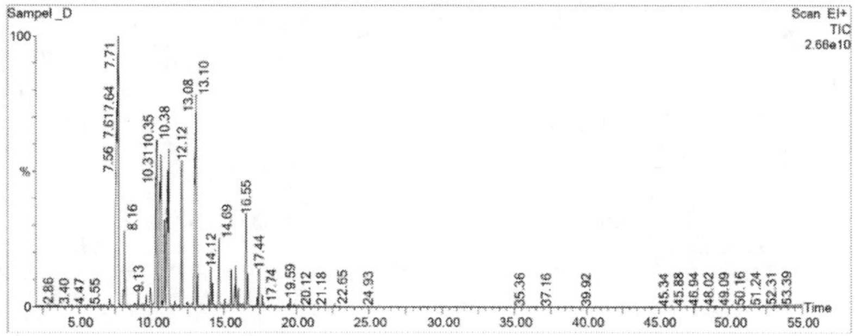


**Figure 2.** Chromatogram from GC-MS Analysis of HCl and Formic Acid Turpentine Isomerization.

As seen in **Figure 2**, it is shown that there are several peaks. It also indicated that the resulting product cannot solely lead to the formation of eucalyptol. From **Table 1,** it can also be seen that in addition to the desired main product in the form of eucalyptol there are also new by-products in the form of other essential compounds that were not previously found in turpentine such as α-terpinolene, γ-terpinene, and 1,4-cineole or existing ones such as camphene, cymene, and limonene. These compounds are the side products of this isomerization reaction but fortunately can have higher price then pine oil. Small amount of it can be further challenge in the products purification.

This reaction product distribution is due to the poor selectivity of the acid combination to direct the conversion of α-pinene into the desired main product. This product distribution does not apply only to the combination of hydrochloric acid and formic acid but applies to all acid combinations used in this study.

**Table 1.** Major Product Distribution of Turpentine Isomerization Using Various Acid Combination.

| Acid Combination | %GC-MS Chromatogram Area | | | | | | | |
|---|---|---|---|---|---|---|---|---|
| | A | B | C | D | E | F | G | H |
| HCl + Citric | 80.706 | 1.708 | 3.519 | 2,137 | 2.221 | 1.357 | 0.637 | 0.656 |
| HCl + Oxalic | 13.926 | 1.726 | 3.778 | 2.958 | 10.774 | 4.467 | 2.518 | 2.985 |
| HCl + Formic | 34.167 | 2.013 | - | - | 19.74 | 6.668 | 5.542 | 4.542 |
| $H_2SO_4$ + Citric | 73.842 | 1.381 | 1.986 | 1.381 | 1.388 | 0.815 | 0.992 | 0.558 |
| $H_2SO_4$ + Oxalic | 55.27 | 1.824 | 3.153 | - | 8.432 | 4.916 | 3.639 | 4.273 |
| $H_2SO_4$ + Formic | 76.171 | 1.381 | 2.122 | 1.441 | 3.07 | 0.962 | 1.77 | 0.992 |
| PTSA + Citric | 18.212 | 1.467 | 2.848 | 1.892 | 1.905 | 1.036 | 0.47 | 0.413 |
| PTSA + Oxalic | 14.143 | 1.425 | 3.151 | 2.005 | 2.832 | 0.975 | 0.975 | 1.005 |
| PTSA + Formic | 65.622 | 1.74 | 3.032 | 3.163 | - | - | 2.479 | 3.163 |
| Raw Turpentine | 73.21 | 1.243 | 1.676 | 1.597 | - | - | - | - |

*Note :* A = α-pinene, B = Camphene, C = O-Cymene, D = Limonene, E = α-terpinolene
F = γ-terpinene, G = 1,4-Cineole, H = Eucalyptol (1,8-Cineole)

The application of a combination of hydrochloric acid and formic acid catalyst for the turpentine isomerization reaction gives quite good product results consistently where the entire composition of the components consisting of the main product and by-products has the highest percentage compared to the use of other catalyst combination. Each combination of strong and weak acids has different interaction where there is no single type of strong acid catalyst or weak acid catalyst that consistently provides high eucalyptol yield. For PTSA and HCl, good result of eucalyptol composition achieved, if combined with formic acid (HCOOH) in the isomerization reaction, while for sulfuric acid interacts well with oxalic acid. In this study, citric acid gave the smallest yield in all of its combinations with strong acids. All of the catalysts used is also classified as not selective for the formation of cineole. The low selectivity of these acid combinations in isomerizing α-pinene is probably due to the ratio of the amount of the strong acid catalyst with the turpentine used [4]. This unbalanced ratio could reduce the ability of the weak acid catalyst in the reaction to produce the desired compound, but leads to uncontrolled reactions that occurred for all types of catalyst combinations used.

**Table 2.** The α-pinene and Eucalyptol Area on GC-MS Chromatogram.

| Acid Combination | α-pinene Area | Eucalyptol Area |
|---|---|---|
| HCl + Citric | 10,240,145,790.90 | 83,213,544.00 |
| HCl + Oxalic | 1,975,730,624.00 | 423,468,896.00 |
| HCl + Formic | 5,633,158,144.00 | 748,833,856.00 |
| $H_2SO_4$ + Citric | 8,228,521,936.00 | 57,640,040.00 |
| $H_2SO_4$ + Oxalic | 8,269,096,039.80 | 639,247,616.00 |
| $H_2SO_4$ + Formic | 8,298,675,584.00 | 108,055,664.00 |
| PTSA + Citric | 2,172,861,408.00 | 49,278,836.00 |
| PTSA + Oxalic | 1,717,492,000.00 | 122,083,128.00 |
| PTSA + Formic | 8,430,150,144.00 | 406,333,280.00 |
| Raw Turpentine | 7,742,696,512.00 | 0 |

The analysis of GC-MS result can also be conducted from the change of α-pinene area as displayed in **Table 2.** As seen, there are several acids that have increased chromatogram area. The addition of the chromatogram area is possible because this isomerization reaction not only undergoes a one-way reaction, but also undergoes a reversible reaction. Hence, probably some derivative compounds previously present in the turpentine feed were isomerized to α-pinene, since most of α-pinene derivatives formed by alpha terpinyl cation intermediates so it needed further study to identify whether the derivative compound present in raw turpentine could be isomerized into α-pinene by this isomerization process (Figure 3).

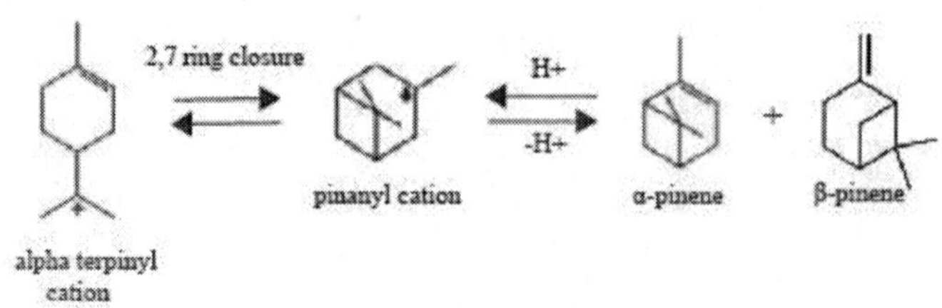


**Figure 3.** Reversible Reaction of Turpentine Isomerization [11].

By using the chromatogram area data in **Table 2.**, we can determine the normalized of α-pinene and eucalyptol area. In addition, we can also estimate the eucalyptol production yield. An illustration of α-pinene conversion and eucalyptol yield calculation from the chromatogram area based on the HCl + Citric in **Table 2** is as follows.

IOP Conf. Series: Materials Science and Engineering 1143 (2021) 012065 doi:10.1088/1757-899X/1143/1/012065

$$N_{pinene} = \frac{10{,}240{,}145{,}790.90}{10{,}240{,}145{,}790.90 + 83{,}213{,}544.00} x100\% = 99{,}19\%$$

$$N_{eucalyptol} = \frac{83{,}213{,}544.00}{10{,}240{,}145{,}790.90 + 83{,}213{,}544.00} x100\% = 0{,}81\%$$

$$Y_{Eucalyptol} = \frac{83{,}213{,}544.00 - 0}{7{,}742{,}696{,}512.00} x100\% = 1.07\%$$

The calculations were carried out for other acid combinations to obtain the normalized area of α-pinene and eucalyptol area as presented in **Table 3.** Further, the eucalyptol yield is presented in **Figure 4.**

**Table 3.** The Calculation of Normalized Area of α-pinene & Eucalyptol.

| Acid Combination | α-pinene Area Normalization (%) | Eucalyptol Area Normalization (%) |
|---|---|---|
| HCl + Citric | 99,19 | 0,81 |
| HCl + Oxalic | 82,35 | 17,65 |
| HCl + Formic | 88,27 | 11,73 |
| $H_2SO_4$ + Citric | 99,30 | 0,70 |
| $H_2SO_4$ + Oxalic | 92,82 | 7,18 |
| $H_2SO_4$ + Formic | 98,71 | 1,29 |
| PTSA + Citric | 97,78 | 2,22 |
| PTSA + Oxalic | 93,36 | 6,64 |
| PTSA + Formic | 95,40 | 4,60 |
| Raw turpentine | 100 | 0 |

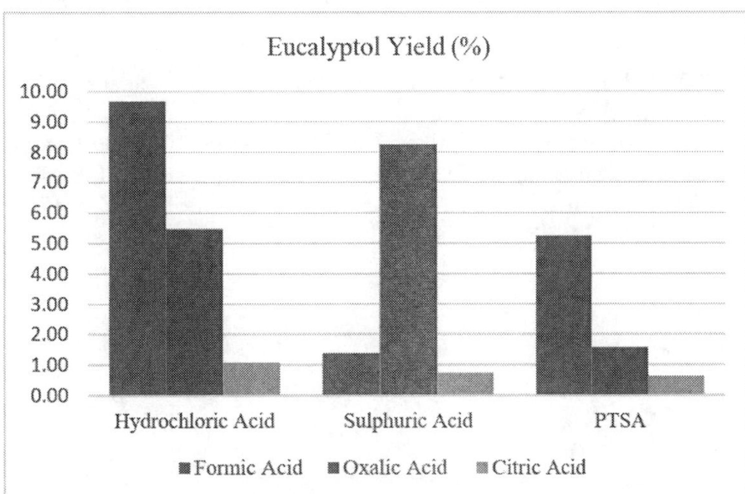


**Figure 4.** Result of Calculation based on Chromatogram Area.

Based on **Table 3**, it is known that the use of HCl with oxalic acid and formic acid are the only combination which are able to achieve an α-pinene composition below 90% after normalization value. These results indicate that HCl is effective to be used as a catalyst to produce 1,8-cineole with high yield compared to other strong acids when combined with provided weak acid. To obtain more accurate information in determining the optimal catalyst type, normalization comparisons are not used but using the highest eucalyptol yield calculation in **Figure 4**. From the calculation, it was obtained that the highest yields produced from the combination of HCl and HCOOH. This type of acid combination gives the best yield due to HCOOH is a weak organic acid with good solubility in both polar and non-polar compounds [4]. In addition, as inorganic acid, it is good if we could pair the strong acid such as hydrochloric acid as shown in **Figure 4** which gives the highest eucalyptol yield.

## 4. Conclusion

The present work has conducted acid catalyst screening for turpentine isomerization using combination of strong and weak acid catalysts. Our results showed that all combinations of strong and weak acid catalysts could produce eucalyptol compounds with various selectivity. It was also found that there are many side products other than eucalyptol. The combination of hydrochloric acid with oxalic acid or formic acid catalyst in this study has shown a promising result as it provides good results of lesser α-pinene composition to eucalyptol composition compared to the other acid combinations. The highest eucalyptol product was obtained at 9.67% with the combination of hydrochloric acid and formic acid. This research was carried out a reaction with types of catalyst that are still not selective in the isomerization of α-pinene to produce cineole, so further research is needed to increase the selectivity of α-pinene into eucalyptol.

## References

[1] Mercier B, Prost J, and Prost M 2009 *International Journal of Occupational Medicine and Environmental Health* **22**(4) 331–42.
[2] Handojo L, Putra I A, Azis M M, Prakoso T, Soerawidjaja T H, and Indarto A 2019 *AIP Conference Proceedings* 2085 020056
[3] Vecchio M G, Minto C 2016 *The Open Agriculture Journal* **10** (53) 52-57
[4] Prakoso T, Putra I A, Handojo L, Soerawidjaja T H, Winoto H P, and Indarto A 2020 *Heliyon 6* e04984
[5] Putra I A, Azis M M, Soerawidjaja T H, and Indarto A 2019 *AIP Conference Proceedings* 2085 020057
[6] Unveren E, Gunduz G, and Cakicioglu-Ozkan F 2006 *Chemical Engineering Communication* **192** (3) 386-404
[7] Ringgani R, Budhijanto, and Budiman A 2014 *Eksergi* **13** 6-12
[8] Azis M M, Qani'a F R, Pratiwi S I, Kristanto J, Purwono S, and Indarto A 2020 *IOP Conf. Ser. : Mater. Sci. Eng. 823 012042*
[9] Lana E J L, Rocha K A d S, Kozhevnikov I, and Gusevskaya EV 2006 J.Mol.Catal. **259** 99-102
[10] Noonan M J, Tinnesand H V, and Buesching C D 2018 *BioEssays* 1700210
[11] Piechulla B, Bartelt R, Brosemann A, Effmert U, Bouwmeester H, Hippauf F, and Brandt W 2016 *Plant Physiology* **172** 2120–31

# Depolymerization Kinetics of Aqueous Cassava Starch under Sonication Process using Free-Radical Depolymerization Model and its Correlation with Radical Production from Acoustic Cavitation

**Bramantyo Airlangga, Dewangga Widyanindra A, Ahmad Adnan Billah A, Prida Novarita Trisanti, Juwari, Sumarno***

Chemical Engineering Department, Institut Teknologi Sepuluh Nopember (ITS), Kampus ITS Sukolilo, Surabaya 60111, Indonesia

*onramus@chem-eng.its.ac.id

**Abstract.** The technology of biopolymer degradation to produce its derivative compounds is an important topic nowadays. The use of environmentally friendly methods, such as sonication, is a promising method for obtaining biopolymer derivative compounds. Sonication creates acoustic cavitation in the liquid body to produce radicals and microjets that can substitute acid substances in the degradation process. Through the sonication process, starch-based compounds can be degraded to oligosaccharides and reducing sugars, which have a variety of benefits in the food, pharmaceutical, and fermentation industries. Many experiments have been conducted to investigate the role of sonication in starch degradation. However, there is much process model that can describe the role of sonication in depolymerization. This work aims to evaluate the ability of sonication in producing reducing sugars and lower-molecular-weight starch by using the Free-radical depolymerization model and correlating it with the production of radicals from acoustic cavitation. Experiments were carried out using horn-type sonication with a power of 500W with various amplitudes. The reducing sugar analysis was carried out by the DNS method and the average molecular weight was estimated with Ubbelohde viscosimetry. From the simulation results, sonication with 30% of the maximum amplitude gives the largest radical production, which well correlates with the production of reducing sugar. The experimental results show a maximum yield of reducing sugar of 0.570 g / L in the process using S30 for 60 minutes.

## 1. Introduction

The utilization of biomass derivative products has increased attention in the food and pharmaceutical industry [1]. Cassava starch is one of the most abundant polysaccharide resources derived by depolymerization or degradation into various valuable products such as soluble starch, modified starch, oligosaccharide, and reducing sugars [2]. The acid and enzyme method becomes the most common method to produce the starch derivative product [3]. Long processing time and chemical solvent waste become problems that can affect the efficiency of the process. Water-insoluble granule form of starch is often addressed as the main factor that caused these drawbacks. To overcome these problems, many methods have been proposed, including sonication [4], high shear mixing [5], and hydrothermal [6].

Sonication of starch has become an exciting topic as an attractive reaction medium and environmentally-friendly solvent for numerous applications. It utilizes an ultrasonic wave to give acoustic pressure to create cavitation on the liquid body [7]. By high amount intensity energy transferred through this processed media, the cavitation produces a local hotspot to introduce microjets and radicals [8]. Microjets create high velocity of the liquid particle, so that erodes the material surface. Radicals can trigger various chemical reactions, including starch depolymerization. Several works have been focused on starch degradation using the sonication process, which successfully affects reducing sugar production, granule defect, microstructural properties change, and granule size reduction.

Although many sonication applications on starch degradation have been researched, it is essential to enhance the process controllability by developing the depolymerization kinetics model. This article concentrates on the depolymerization kinetics of cassava starch. We develop the free-radical induced depolymerization and the n-order depolymerization reaction. The model includes the number-average DP profile that is gained from intrinsic viscosity measurement. The depolymerization rate constant was obtained by fitting the model parameter by reducing sugar production results. Radical production was also presented to correlate the concentration of the radical to the sonication degradation ability. The amplitude of sonication is a controllable equipment parameter strongly related to how much energy is delivered. Hence, the variation of amplitude in this works is obliged to study the effect of energy delivered to starch degradation.

## 2. Process Model

Generally, a typical polymeric compound depolymerization model may be applied to this case: randomized chain scission and end-chain scission [9]. Fully randomized chain scission may be modeled as a probability term and provide fast chain length reduction [10]. End chain scission enables the process to produce a high yield of the monomeric compound but often resulted in a slower reaction [2].

Using the nth-order reaction, the number average molecular weight profile was obtained. This function will be correlated with classical depolymerization induced by radicals to predict the total monomer product. The equations are time dependent on the function of Number Average Molecular Weight ($M_n$, g/L) and monomeric material weight concentration ($C_s$, g/L). Number-average molecular weight is an important parameter that depicts the average chain length and easy to be measured. Much monomeric material can be produced throughout this process, such as glucose, fructose, maltose, and other reducing sugars can be estimated using the DNS method.

### 2.1. *Number Average Molecular Weight from Ubbelohde Viscosimetry Data.*

We developed the equation based n$^{th}$-order reaction, which has been applied to other polysaccharide degradation using sonication [11]. The change in polymer molar concentration ($[M_P]$) was determined by the constant rate ($k'$), as shown in **Eq. 1**.

$$\frac{d\,[M_P]}{dt} = k'[M_P]^n \quad (1)$$

Constant total weight of ploymer solution ($W_P$) and time depended-number average molecular weight ($M_n$) are inscribed in **Eq. 2**:

$$[M_P] = \frac{W_P}{M_n} \quad (2)$$

Inserting **Eq. 2** to **Eq. 1** to obtain:

$$\frac{d\,M_n}{dt} = k' W_P^{(n-1)}\, M_n^{(2-n)} \quad (3)$$

These parameters ($k', n$) was obtained by parameter fitting using following intrinsic viscosities function that combined Mark Houwink Sakurada equation (**Eq. 4**) and n-order depolymerization reaction according to Li et al. [11] as depicted in **Eq. 5-6**.

$$[\eta] = K\, M_V^{\alpha} \quad (4)$$

$$[\eta]_t = \left( (1-n)k''\, t + [\eta]_0^{\left( {(n-1)}/{\alpha} \right)} \right)^{\alpha/(n-1)} \tag{5}$$

$$k' = k''(K\,[(1+\alpha)\Gamma(1+\alpha)]\, W_P^{\alpha})^{(1-n)/\alpha} \tag{6}$$

### 2.2. Glucose Production Prediction using the Free-radical Degradation Model.

The equation was generated from initiation, propagation, transfer, and termination of radicals, as shown in **Fig. 1**. The equation was first proposed by Robert Simha and successfully simulated the radical degradation of several polymeric compounds in pyrolysis. Two parameters (σ and ε) were introduced to measure the radicals transfer effectivity, as shown in **Eq. 7-8**.

$$\left(\frac{1}{\varepsilon} - 1\right) = \frac{propagation}{transfer+termination} = number\ of\ zip\ length \tag{7}$$

$$\sigma = \frac{transfer}{initiation} = transfer\ ratio \tag{8}$$

Near zero value of $\sigma$ means, the mechanism approximated chain-end only because of only a small radical transfer. Term "zip length" describes radicals' propagation compared to the sum of transfer and termination. Small value zip length will need ε value near one that implied a random scission point mechanism. The general expression was written in **Eq. 9**.

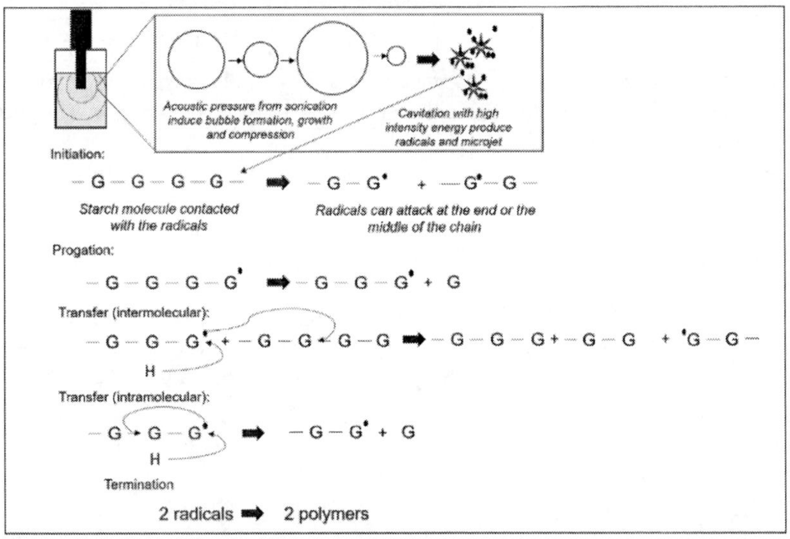


**Figure 1** Radicals role mechanism on the degradation process adapted from other research [10].

$$\frac{d\,C_S}{dt} = \frac{A-B}{E} \tag{9}$$

$$A = 2k_{dp}\left[\frac{1}{\varepsilon} + \sigma + \frac{\left(\frac{3}{\varepsilon}-1\right)}{\bar{P}} + \frac{\left(\frac{\sigma}{2}\right)\left(\frac{3}{\varepsilon}-7\right)}{\bar{P}}\right](1-C_S) \tag{9a}$$

$$B = 2\left[\left(\frac{1}{\varepsilon} - 2\right)\left(\frac{1-C_S}{\bar{P}}\right)\frac{d\,\bar{P}}{dt}\right] \tag{9b}$$

$$E = \left[1 + \frac{2}{\bar{P}}\left(\frac{1}{\varepsilon} - 2\right)\right] \tag{9c}$$

The average degree of depolymerization ($\bar{P}$) as the number average molecular weight ($M_n$) divided D-glucose Molecular Weight ($MW_{DG}$)

$$\bar{P} = \frac{M_n}{MW_{DG}} \quad \textbf{(10)}.$$

### 2.3. *Bubble Dynamics with Radical Production Model in Starch-Water Suspension*

2.3.1. *Bubble Dynamics.* The single bubble sonochemistry model adopted the Keller-miskis equation (**Eq. 11-13**) with the first-order term in Mach number ($M = \dot{R}/c$) that have been fully described in other research [12]. The ideal gas rule follows temperature and pressure. The equation correlated single bubble compression cycle through the bubble pressure ($p_w$) balance that induced by acoustic pressure. The expansion of the bubble is considered isothermal, and its total compression is followed by an adiabatic term (**Eq. 4**).

$$\left(1-\frac{\dot{R}}{c}\right) R\, \ddot{R} + \left(\frac{3}{2}-\frac{\dot{R}}{2c}\right) \dot{R}^2 = \frac{1}{\rho_0}\left(1+\frac{\dot{R}}{c}\right) p_w + \frac{R}{\rho_0 c}\frac{dp_w}{dt} \quad \textbf{(11)}$$

$$p_w = \left(p_0 + \frac{2\sigma}{R_0}\right)\left(\frac{R_0}{R}\right)^{3\gamma} - \frac{2\sigma}{R} - \frac{4\mu\dot{R}}{R} - p_0 - P_A sin(\omega t) \quad \textbf{(12)}$$

$$\frac{dp_w}{dt} = \left(p_0 + \frac{2\sigma}{R_0}\right) \gamma \left(\frac{{R_0}^3}{R^3}\right)^{(\gamma-1)} \left(-3\frac{\dot{R}\,{R_0}^3}{R^4}\right) - \frac{-2\sigma\dot{R}}{R^2} - \frac{-4\mu\dot{R}^2}{R^2} - \left(-P_A\,(2\pi f)cos((2\pi f)t)\right) \quad \textbf{(13)}$$

In this equation (**Eq. 11-13**), dots showed time derivative (d/dt), R is the radius of a bubble, c is the speed of sound, $\rho_0$ is bulk density, $\sigma$ is surface tension, $\gamma$ is Laplace constant, $P_A$is acoustic amplitude, $f$ is frequency velocity. Acoustic pressure depended on acoustic intensity as **Eq. 14**. The acoustic intensity can be measured as the energy dissipated in a cross-sectional area ($I_a = \frac{1}{A_{cs}}\frac{dE}{dt}$). An alternative way to measure the actual acoustic intensity was depicted by Sathiskumar and Madras [13], which use calorimetry. By measuring bulk temperature, the intensity can be written as **Eq. 15**. The cross-sectional area $A_{cs}$ was sonotrode surface area.

$$P_A = (2\, I_a \rho_L c)^{1/2} \quad \textbf{(14)}$$

$$I_a = \frac{m\, C_p\, d\, T_L}{A_{cs}\, dt} \quad \textbf{(15)}$$

2.3.1. *Radical Reaction.* Chemical kinetics model and reaction constant was simulated followed the series radical formation reaction by Merouani et al. [14]. The reaction equation is considered a reversible elementary reaction in bubble thin film with bubble temperature ($T_b$) followed **Eq. 16-19**. Reaction rates ($r_i$) was the function

$$\boldsymbol{T_b = T_L\left(\frac{R_0}{R}\right)^{3(\gamma-1)}} \quad \textbf{(16)}$$

$$\sum_{k=1}^{K} v'_{ki}\, X_k \leftrightarrow \sum_{k=1}^{K} v''_{ki}\, X_k \quad \textbf{(17)}$$

$$\dot{w}_k = \sum_{i=1}^{l} (v''_{ki} - v'_{ki})\; r_i \quad \textbf{(18)}$$

$$r_i = k_{fi} \prod_{k=1}^{K} [X_k]^{v'_{ki}} - k_{ri} \prod_{k=1}^{K} [X_k]^{v''_{ki}} \quad \textbf{(19)}$$

## 3. Method

### 3.1. *Experimental*

Cassava starch was a commercial-grade from Budi Jaya (Jakarta, Indonesia). Potassium hydroxide (1310-58-3), sodium hydroxide (1310-73-2), potassium iodide (7681-11-0), and perchloric acid 70% (7601-90-3) were obtained from SAP Chemical (Jakarta, Indonesia). Potassium sodium tartrate (6381-

59-5/ 108087) and anthrone reagent (90-44-8/ 101468) and were purchased from Merck (Darmstadt, Germany).

Sonication was conducted as previous work [15]. The aqueous starch mixture was made by mixing 5 g starch in 100 mL distilled water. Sonication was carried out under various amplitudes for 60 min, and the bulk liquid temperature was recorded. The variable was listed in **Table 1**.

**Table 1** List of variables and amplitude used.

| Name | The percentage from the maximum amplitude (%) | Ideal Equipment Amplitude (μm) |
|---|---|---|
| S30 | 30 | 34.2 |
| S40 | 40 | 45.6 |
| S50 | 50 | 57.0 |
| S60 | 60 | 68.4 |
| S70 | 70 | 79.8 |

### *3.2. Chemical Analysis*

3.2.1. *Ubbelohde Viscosimetry.* The intrinsic viscosities of the measurement method was based on another experiment [16]. $M_v$ was calculated by Mark Houwink equation for amylose of starch (**Eq. 4**), taking 'K' as 0.00183 and 'a' as 0.89 [17]. Solutions were prepared with a sample concentration of 0.1%-0.4% in 1 M potassium hydroxide. It was filtered, and the viscosity was measured using Schott-Gerate Ubbelohde Capillary Viscometer Model 525 (Darmstadt, Germany). The measurement was repeated at least three times.

3.2.2. *Total Reducing Sugar for DNS Method.* The liquid product was analyzed using the DNS method. The method was adopted from elsewhere [18]. The standard solutions were made from the D-glucose solution. The spectrophotometry was conducted at 502 nm wavelength (Genesys 10S UV-Vis spectrophotometer).

### *3.3. Simulation*

The constant was adapted from other experiments, as listed in **Table 2**. The constant density of 1050 kg/m³ applied in this work was obtained from the experiment. The bubble dynamic and chemical reaction equations solved simultaneously using the ODE package of MATLAB 2020 (academic version). The fitting parameter was done by evaluating the parameter value to non-linear least square regression.

$$s^2 = \sum_i^n \frac{(y_i - y_{exp})^2}{N-K}$$
**(20)**

**Table 2** List of constant used in the bubble dynamic equation.

| Symbol | Unit | Value | Ref. |
|---|---|---|---|
| $\gamma$ | | $4/3$ | [19] |
| M | Pa s | $124.68 - 0.36\,T_L$ | [20] |
| $p_0$ | Pa | $1.013250 \times 10^5$ | [21] |
| $c_p$ | J/g C | $4.70518 + (-3.71017)X_s + 0.00343\,(T_L - 273) + 3.24167\,(X_s^2) + (-2 \times 10^{-4})(T_L - 273)^2$ | [22] |
| $\sigma$ | N/m | $0.078$ | [23] |
| $p_v$ | Pa | $133.32 \exp\left(18.3036 - \frac{3816.44}{T_L - 46.13}\right)$ | [21] |

| | | | |
|---|---|---|---|
| c | m/s | $1402.7 + 4.9045\,(T_L - 273) - (4.72 \times 10^2)(T_L - 273)^2 + (1.242 \times 10^{-4}) - 4(T_L - 273)^3$ | [21] |

## 4. Results and Discussion

The Sonication process induces starch degradation by microjet and radicals. The microjet eroded the granule and detached the starch molecule. The radicals contacted a free starch molecule, therefore made charge destabilization of an oxygen atom at glycosidic bond [7]. This effect is shown by depolymerization kinetics, which combined the n-th order reaction and free radical degradation model.

### *4.1. Depolymerization Kinetics*

The number average molecular weight and total reducing sugar data were utilized to obtained 5 equation constant listed in **Table 3**. Firstly, the determination of $k''$ and $n$ were derived from the correlation between the intrinsic viscosities measurement and the n-order reaction kinetics equation. The value of $k_{dp}$, $\sigma$, and $\varepsilon$ were obtained from the fitting between total reducing sugar (monomer) concentration profile and free radical depolymerization equation.

**Table 3** Equation constant obtained through parameter fitting.

| Variable | $k_{dp}(\times10^6)$ | $\sigma$ | $\varepsilon$ | $k''$ $(\times10^{13})$ | $n$ |
|---|---|---|---|---|---|
| S30 | 61 | 0.9 | 1.0 | 7.1 | -3.2 |
| S40 | 85 | 0.1 | 1.0 | 6 | -3.2 |
| S50 | 8.5 | 1.0 | 0.6 | 30.1 | -3.2 |
| S60 | 12 | 0.6 | 0.5 | 7.1 | -3.2 |
| S70 | 8.5 | 0.1 | 0.1 | 48 | -3.2 |

The intrinsic viscosities profile and average molecular weight were presented in **Fig. 2**. The sonication reduced a high amount of starch molecule chain early. Sonication delivered immense energy intensity, inducing cavitation that made the sonication have the ability to degrade chemically by radicals and physically by microjet [3,24]. The result shows that the process has negative value of reaction order (n), which can be classified as physical process control. This negative value of reaction order also means that increase of polymer concentration made the reaction rate decrease [25]. This also implied that the chemical degradation was much faster than the physical degradation so that the average process can be considered physical process control.

Besides, the higher bulk temperature will trigger the starch granule to be gelatinized. The amount of water penetrated inside the granule matrices as the intermolecular bond of starch weakened. It decreased the total amount of free water in bulk liquid; therefore, the viscosity increase. A more viscous solution may retard the microjet power that hit starch granules.

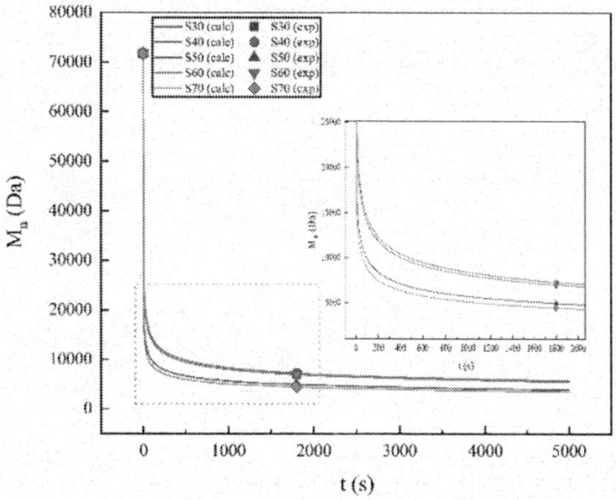


**Figure 2** Number-average molecular weight profile of the starch sonication process.

The monomeric product of the starch depolymerization process can be predicted as reducing sugar production depicted in **Fig. 3**. It is found that the calculated concentration of reducing sugar tends to increase over the processing time. However, the experimental results show a fluctuating trend over time with a maximum yield of reducing sugar of 0.570 g/ L in the process using S30 for 60 minutes. From the graph, the value of apparent reaction rates may appear in sequence from S30>S40>S70>S60>S50. Practically, it is agreed that the fluctuating monomer production may come from radicals based degradation. Radicals tend to have a random scissoring point, which hard to be stable. This semi-empirical model may help the rough prediction of monomer production from the starch sonication process.

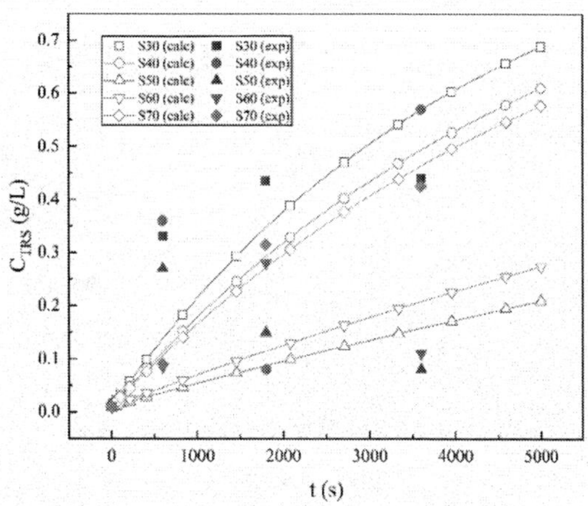


**Figure 3** Monomer production profile as Total Reducing Sugar concentration.

### *4.2. Radical Production from Acoustic Cavitation in Starch-Water Suspension*

Bulk liquid temperature profile from experimental data was converted to calculated dissipated energy used in radical production prediction. The cavitation comes from high-intensity acoustic pressure caused by an extreme compression-rarefaction cycle to generate a local hotspot. The cavitation bubble's

IOP Conf. Series: Materials Science and Engineering 1143 (2021) 012066 doi:10.1088/1757-899X/1143/1/012066

implosion made the liquid contacted the local hotspot, then raise the bulk liquid temperature. From **Fig. 4**, the higher amplitude achieved higher stability at about 12 minutes process that may occur by a steady-state between heat generation by cavitation and heat dissipation. Higher amplitude means that higher accoustic pressure was applied to the bulk liquid, resulting in a high compression ratio from the cavitation bubble. It generates a local hotspot with high local pressure and temperature, resulting in higher heat generation from the process.

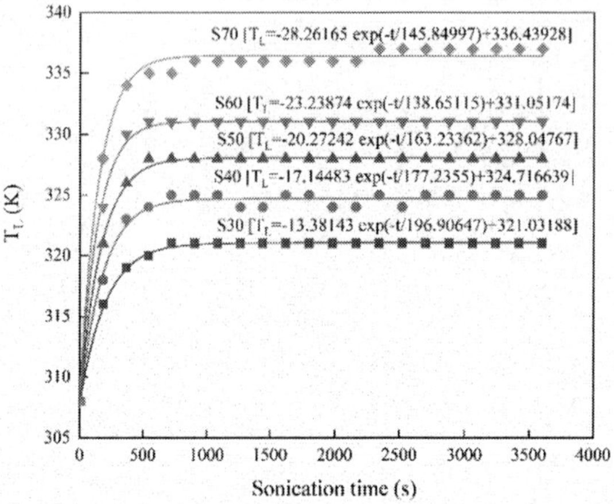


**Figure 4** Temperature profile of the starch sonication process.

The bubble dynamics of sonication is presented in **Fig. 5**. This result was different from sonication in constant bulk temperature pure water [14], which demonstrated a high bubble maximum compression ratio. It was expected that cavitation in aqueous starch solution showed milder cavitation than it is in pure water due to the bubble growth strongly inhibition by high density and surface tension. It is also found that a higher amplitude corresponded to a higher maximum bubble radius. The sonication wave's oscillatory pattern made a high amplitude to have high positive and high negative acoustic pressure; therefore, it can result in a higher bubble radius. As the bulk temperature rises at a more prolonged process time, the speed of sound was reduced. Lower speed of sound makes a lower the sound intensity to decrease the maximum acoustic pressure.

IOP Conf. Series: Materials Science and Engineering 1143 (2021) 012066 doi:10.1088/1757-899X/1143/1/012066

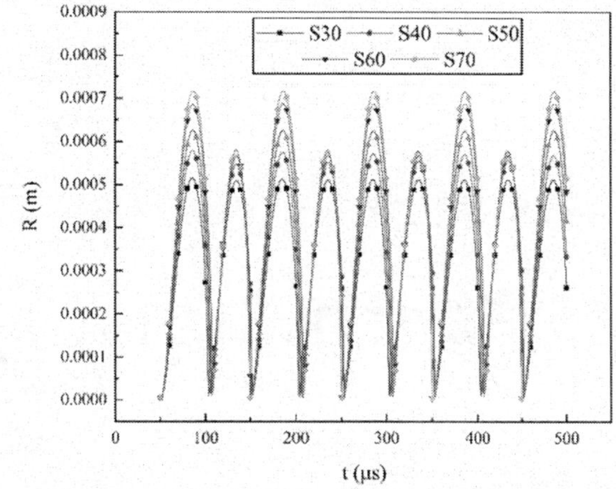


**Figure 5** Bubble radius profile of the starch sonication process.

Radicals may be produced because the water components (water molecule and dissolved gases ($H_2$ and $O_2$)) create contact with the local hotspot [21]. The local hotspot's high temperature opens the possibilities to the water components take a sudden molecular cleavage that can occur at the thin bubble film layer. The predicted maximum of three radical products (H*, OH*, and $HO_2$*) is presented in **Fig. 6**. The maximum value of H* concentration was 0.00278 M, which was obtained from S30. It is also found that the OH* concentration was reached its maximum value at S70 (0.01051 M). S50 produced the most $HO_2$* with a value of 2.29957 x $10^{-13}$ M. Generally, H* showed values more than 0.00012 M. Meanwhile, OH* have concentration below 1.2 x $10^{-5}$ M, except S70 which show the highest value among the radicals. It is also showed that the $HO_2$* value was below 2.5 x $10^{-13}$ M.

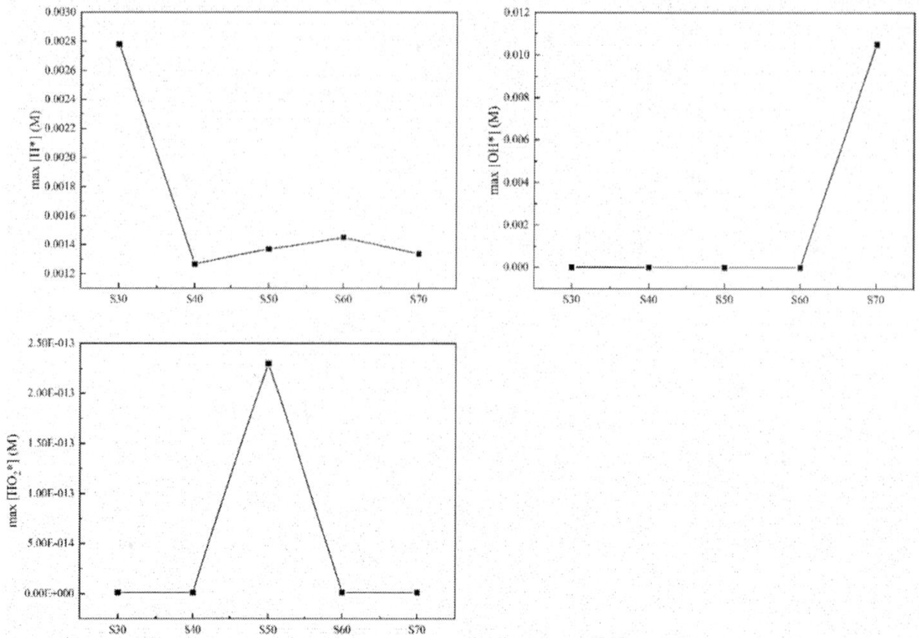


**Figure 6** Maximum radicals (H*, OH* and $HO_2$*) concentration formed in the bubble wall from starch sonication process.

The contribution of the product of each radical to the degradation of starch is still much debate. From the value of radical concentration, H* and OH* may be positioned to have the most significant effect among the radicals. Further evidence showed that the highest apparent reaction rates correlate with the highest H* concentration (S30), and the second-highest apparent reaction rates correlate with the highest OH* concentration (S70). From this result, further work may be focused on the total of H* and OH* to be applied in the radical-depended reaction rate for the depolymerization model.

## 5. Conclusion

The experimental and simulation of the starch sonication process has been done in this work. The experimental result shows fluctuating trend monomer production over time with a maximum yield of reducing sugar of 0.570 g / L in the process using S30 for 60 minutes. The free-radical degradation model can be a rough prediction of monomer production from the starch sonication process. The bubble dynamics simulation has been done with the Keller-Miskis equation. The highest value of H*, OH*, and $HO_2$* is 0.00278 M, 0.01051 M, and $2.29957 \times 10^{-13}$ M, respectively. The highest apparent reaction rate of the free-radical degradation model correlates well with the highest H* concentration depicted by the S30 process.

## Acknowledgment

The authors thank the Ministry of Research, Technology, and Higher Education of Indonesia. This study was financially supported by the Program Magister Doktor Untuk Sarjana Unggul (PMDSU) Scholarship Program.

## References

[1] Trisant P N, Gunardi I and Sumarno 2020 The Influence of Hydrolysis Time in Hydrothermal Process of Cellulose from Sengon Wood Sawdust *Macromol. Symp.* **391** 1–5

[2] BeMiller J and Whistler R 2009 *Starch Chemistry and Technology. 3rd Edition* vol 1

(Academic Press)

[3] Sumarno, Bramantyo A, Febriyati P, Gunardi I and Trisanti P N 2019 Ultrasound Pre-treatment for Intensification of Hydrothermal Process in Reducing Sugar Production from Cassava Starch *IOP Conf. Ser. Mater. Sci. Eng.* **543**

[4] Monroy Y, Rivero S and García M A 2018 Microstructural and techno-functional properties of cassava starch modified by ultrasound *Ultrason. Sonochem.* **42** 795–804

[5] Airlangga B, Puspasari F and Trisanti P N 2020 Structural Properties Change of Cassava Starch Granule Induced by High Shear Mixer **2000004** 1–8

[6] Nagamori M and Funazukuri T 2004 Glucose production by hydrolysis of starch under hydrothermal conditions *J. Chem. Technol. Biotechnol.* **79** 229–33

[7] Airlangga B, Sugianto A M, Parahita G, Puspasari F, Mayangsari, Novi Eka, Trisanti P N, Sutikno J P and Sumarno 2020 Study of Cassava Starch Degradation using Sonication Process in Aqueous Sodium Chloride *J. Sci. Food Agric.*

[8] Adam F F, Istiqomah R A, Budianto M A, Trisanti P N and Sumarno S 2020 Study of microcellular foaming of polystyrene aided with 45 kHz of ultrasound waves energy *Polym. Technol. Mater.* **00** 1–7

[9] Simha R 1941 Kinetics of degradation and size distribution of long chain polymers *J. Appl. Phys.* **12** 569–78

[10] Simha R 1962 Degradation of Polymers. *Adv. Chem.*

[11] Li R and Feke D L 2015 Rheological and kinetic study of the ultrasonic degradation of locust bean gum in aqueous saline and salt-free solutions *Ultrason. Sonochem.* **27** 334–8

[12] Johansen K 2018 *Stable-inertial Cavitation* (University of Glasgow)

[13] Sathiskumar P S and Madras G 2012 Ultrasonics Sonochemistry Ultrasonic degradation of butadiene , styrene and their copolymers *Ultrason. - Sonochemistry* **19** 503–8

[14] Merouani S, Hamdaoui O, Rezgui Y and Guemini M 2014 Theoretical estimation of the temperature and pressure within collapsing acoustical bubbles *Ultrason. Sonochem.* **21** 53–9

[15] Airlangga B, Erlangga D, Solikhah L K M, Puspasari F, Gunardi I, Novarita P T and Sumarno 2019 Effect of amplitude on ultrasound treatment for reducing sugar production from cassava starch *AIP Conf. Proc.* **2085**

[16] Ahmad F B, Williams P A, Doublier J L, Durand S and Buleon A 1999 Physico-chemical characterisation of sago starch *Carbohydr. Polym.* **38** 361–70

[17] Cowie M G 1962 Studies on Amylose and its Derivatives **53** 189–200

[18] Miller G L 1959 Use of Dinitrosalicylic Acid Reagent for Determination of Reducing Sugar *Anal. Chem.* **31** 426–8

[19] Johansen K, Song J H and Prentice P 2017 Validity of the Keller-Miksis equation for "non-stable" cavitation and the acoustic emissions generated *IEEE Int. Ultrason. Symp. IUS* 1–4

[20] Iida Y, Tuziuti T, Yasui K, Towata A and Kozuka T 2008 Control of viscosity in starch and polysaccharide solutions with ultrasound after gelatinization *Innov. Food Sci. Emerg. Technol.* **9** 140–6

[21] Merouani S, Hamdaoui O, Rezgui Y and Guemini M 2015 Computer simulation of chemical reactions occurring in collapsing acoustical bubble: Dependence of free radicals production on operational conditions *Res. Chem. Intermed.* **41** 881–97

[22] Cansee S, Watyotha C, Thivavarnvongs T, Uriyapongson J and Varith J 2008 Effects of temperature and concentration on thermal properties of cassava starch solutions *Songklanakarin J. Sci. Technol.* **30** 405–11

[23] Lawton J W 1995 Surface Energy of Extruded and Jet Cooked Starch *Starch - Stärke* **47** 62–7

[24] Savitri E, Juliastuti S R, Handaratri A, Sumarno and Roesyadi A 2014 Degradation of chitosan by sonication in very-low-concentration acetic acid *Polym. Degrad. Stab.* **110** 344–52

[25] Li R and Feke D L 2015 Rheological and kinetic study of the ultrasonic degradation of xanthan gum in aqueous solutions *FOOD Chem.* **172** 808–13

IOP Conf. Series: Materials Science and Engineering 1143 (2021) 012067 doi:10.1088/1757-899X/1143/1/012067

# Kinetic studies and performance analysis of Indonesian rice husk pyrolysis

**L D Kasmiarno[1], S Steven[1], J Rizkiana[1,2,3], E Restiawaty[1,2,4], and Y Bindar[1,2*]**

[1] Department of Chemical Engineering, Faculty of Industrial Technology, Institut Teknologi Bandung, Bandung 40132, Indonesia
[2] Department of Bioenergy Engineering and Chemurgy, Faculty of Industrial Technology, Institut Teknologi Bandung, Bandung 40132, Indonesia
[3] Center for Catalysis and Reaction Engineering, Institut Teknologi Bandung, Bandung 40132, Indonesia
[4] Biosciences and Biotechnology Research Center, Institut Teknologi Bandung, Bandung 40132, Indonesia

[*] Email: ybybyb@fti.itb.ac.id

**Abstract**. Rice husk is abundantly available biomass in Indonesia and it can be a potential source of high-value energy source. In the present work, pyrolysis of rice husk was subjected to various temperature and used as a method to produce bio-oil. The decomposition profile of rice was examined at different heating rates (10, 20, 30, and 40°C/min) using a thermogravimetric analysis (TGA) method. The pyrolysis kinetics were evaluated using three different kinetic models: Kissinger-Akahira-Sunose (KAS), Ozawa-Flynn-Wall (OFW), and Coats-Redfern (CR). The activation energy for KAS, OFW, and CR was centered at 251.62 kJ/mol, 253.32 kJ/mol, and 33.92 kJ/mol, respectively. Varying activation energy with conversion was observed, which reveals that the pyrolysis of rice husk processes through complex reaction. The production of bio-oil from rice husk using pyrolysis method was studied at different temperature conditions (300, 400, 500, 600°C). Results showed that bio-oil and gas products increase at higher temperature. The highest yield of bio-oil was produced at 600°C with 46.3 grams (37%-wt). The specific test showed that the density, viscosity, and acid value of bio-oil decreased at higher temperatures indicating that different pyrolysis conditions might lead to the different bio-oil quality.

## 1. Introduction

Petroleum-based products such as fuel and chemicals are slackening in energy supplied. Hence, alternative energy such as biomass is not only directed to alternative fuel sources but also more benignant fuels for industrial sectors. Biomass such as rice husk, corn cobs, and cotton stalks are potential for energy sources. Indonesia as an agricultural country has expanded rice production by around 59.2 million tonnes in 2018 [1]. Technologies for biomass utilization are not well-established in Indonesia. Rice husk yield was attained in 20% of rice production [2]. Recently, rice husk utilization is disposed of as agricultural waste, however, rice husk has the potential for producing electricity and can be used in the cement industry. Biomass waste such as rice husk could be converted into high-value energy since it has low nitrogen and sulfur content. Rice husk contains 34.4% cellulose that favorable for liquid fuel production [3].

Biomass can be converted to liquid, gases, or solid fuel products through various methods, which are combustion, gasification, and pyrolysis. Pyrolysis is a thermal degradation process that takes place in an inert atmosphere with the absence of oxygen. The range temperature of pyrolysis is 200-800°C. Large biopolymers (20,000-400,000 a.m.u) break into smaller molecules (less than 200 a.m.u) in pyrolysis process [4]. Biomass was used as feed in pyrolysis process to produce decomposed products including solid char, gas, and bio-oil or a liquid, with low viscosity that contains complex mixture such as aldehydes, ketones, and carboxylic acids. Pyrolysis also can produce advanced materials such as carbon materials including carbon nanotubes and graphene-like materials [5].

Kinetic studies of pyrolysis may help better understanding and planning of industrial process since kinetic modelling of chemical and physical can be beneficial for process design and optimization. Non-isothermal thermogravimetric analysis (TGA) is verified as a method for the understanding of kinetic pyrolysis. Saddawi et al. reported that TGA methods have been used in a wide temperature range and followed with various model fitting kinetics such as iso-conversional methods [6,7]. Iso-conversional methods can provide the estimation of the kinetic parameters with rather accurate results compared to the other methods since it is considered using a single-step process, with the presumption that activation energy as a function of the heating rate and temperature determined as the total reaction [8]. The activation energy, as kinetic parameters did not change during the conversion range in iso-conversional methods.

This work was designed to determine the rice husk pyrolysis kinetic parameters. The rice husk pyrolysis kinetics study is figured out in thermogravimetric analysis which is conducted with TGA-DSC 600, while main pyrolysis experiment as well as for pyrolysis product examination are held using a fixed bed reactor.

## 2. Methods

### *2.1. Materials and characterization*

The rice husk was provided by the local rice milling industry in West Java, Indonesia. It was first dried at 105°C for 24 h. After grinding, the rice husk was sieved to obtain a uniform size particle, 10 mesh by cutting mill. The sample was characterized using elemental analysis, which was performed using NC Technology Organic Elemental Analyzer ECS8020. The moisture content was estimated gravimetrically before and after drying. Ultimate analysis dry ash free basis of rice husk is presented in Table 1.

**Table 1.** Ultimate analysis of rice husk (dry ash free basis).

| Element | Mass Composition (%) |
|---|---|
| C | 32.29 |
| H | 4.31 |
| O | 43.08 |
| N | 0.28 |

### *2.2. Thermogravimetric analysis*

The instrument involved in thermogravimetric analysis (TGA) of rice husk was TGA-DSC 600. The TGA process was referred to ASTM-E43 operational standard. About 12-14 mg of rice husk sample was heated from room temperature to 600°C using various heating rate conditions (10, 20, 30, 40, and 50°C $min^{-1}$) with an inert atmosphere of $N_2$, as a purge gas, was injected into the system with a flowrate of 100 mL/min.

### *2.3. Pyrolysis process*

The pyrolysis observed in this work was slow pyrolysis and was assisted with fixed bed reactor. Pyrolysis was held with 125 g of rice husk under various temperatures 300, 400, 500, and 600°C. The reactor temperature was held isothermally for 50 minutes when reached the end of pyrolysis temperature. The produced char and bio-oil was collected in the flask.

Physicochemical analysis for bio-oil encompassed density and viscosity were identified. Density measurement was performed with pycnometer. Bio-oil viscosity was explored through Ostwald Viscometer. The bio-oil functional group was then identified using Thermo-scientific Fourier Transform Infrared Spectroscopy.

Bio-oil acid value was identified under titrimetric analysis. It was solubilized in ethanol and then was dripped with phenolphthalein (PP) indicator. This solution was titrated with 0.1 mol/L potassium hydroxide (KOH). The indicator color change was carefully monitored and KOH amount for the indicator color change was recorded.

*2.4. Pyrolysis kinetic parameters determination*

Three different kinetic models were occupied in the present study, namely Kissinger-Akahira-Sunose (KAS), Ozawa-Flynn-Wall (OFW), and Coats-Redfern (CR) method [9]. KAS equation follows differential form as served below:

$$\ln\left(\frac{\beta}{T^2}\right) = -\frac{Ea}{R}\left(\frac{1}{T}\right) - \ln\left[\frac{A.Ea}{R.g(a)}\right] \tag{1}$$

where β, Heating rate, °C/min; A, Pre-exponential factor, $s^{-1}$; E, Activation energy, kJ/mol; R, Gas constants, J/kg.K; α, Mass conversion; T ,Temperature, °C.

The linear plot of the curve $\ln\left(\frac{\beta}{T^2}\right)$ versus the inverse temperature produces the activation energy value corresponding to each conversion step.

OFW method is another iso-conversional method which is also widely found in kinetic study. It is determined from global kinetic equation [8]. The formula for OFW method is based on the integral form of the kinetic curve from TG analysis and the correlation is based on Doyle's approximation [10].

$$log\beta = \log\left(\frac{A.Ea}{R.g(a)}\right) - 2.315 - 0.4567\frac{Ea}{R}\left(\frac{1}{T}\right) \tag{2}$$

where β, Heating rate, °C/min; A, Pre-exponential factor, $s^{-1}$; E, Activation energy, kJ/mol; R, Gas constants, J/kg.K; α, Mass conversion; T, Temperature, °C. The slope of the curve log $\beta$ vs $\left(\frac{1}{T}\right)$ obtain the activation energy for the specified conversion.

Another method was developed by Coats-Redfern could predict the activation energy and pre-exponential factor from pyrolysis process [11]. This equation was given below:

$$\ln\left[\frac{g(a)}{T^2}\right] = ln\left[\frac{AR}{\beta E}\left(1 - \frac{2RT}{E}\right)\right] - \frac{E}{RT} \tag{3}$$

where β, Heating rate, °C/min; A, Pre-exponential factor, $s^{-1}$; E, Activation energy, kJ/mol; R, Gas constants, J/kg.K; α, Mass conversion; T, Temperature, °C; n, the order reaction. The slope of the curve $\left[\frac{g(a)}{T^2}\right]$ vs $\left(\frac{1}{T}\right)$ to obtain the activation energy and pre-exponential factor.

## 3. Results and discussion

*3.1. Thermal decomposition characterization at different heating rate condition*

Figure 1 shows the thermal decomposition curves of rice husk pyrolysis at various heating rate conditions (10, 20, 30, and 40°C). The thermal gravimetric curves of rice husk at different heating rates of pyrolysis have the same trend curves and typical appearance of lignocellulosic material decomposition. The peaks of biomass decomposition are detected in DTG curves. The biomass decomposition appeared below temperature 100°C or can clearly observe at first peak in Figure 1(b). It indicated that water content starts to evaporate, followed by a large amount of weight loss in temperature range 200-450°C which can be seen at second or the highest peak in DTG curves. By comparing the DTG peaks at different heating rates, it can be observed that by increasing the heating rate, the DTG curve can be more cuspate and the fraction decomposed was higher.

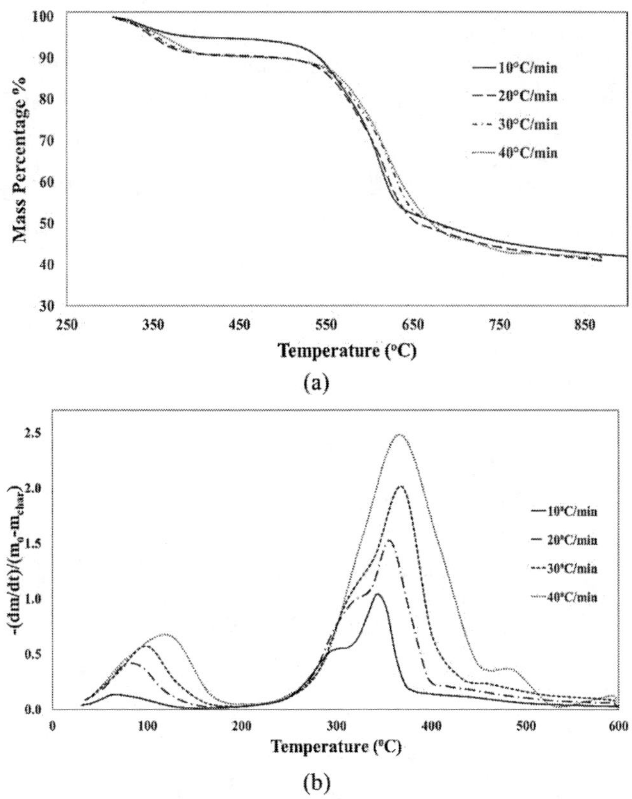


**Figure 1.** Thermal gravimetric curves (a) and DTG curves (b) of rice husk pyrolysis at various heating rates.

According to TG-DTG curves, rice husk pyrolysis process is divided into several stages. Biomass was absorbing heat to decompose and break down the molecules. The first stage occurs before 100°C where the water content in biomass (about 4-7%) is evaporated. TG curves almost stagnant at second stage around 100-240°C when slow decomposing and polymerization to break down the larger molecules started. From temperature 240-450°C, large molecules of rice husk decomposed into small molecules which consist of gas and condensable volatile elements. The percent weight loss was about 50-65%. It is indicated that the main phase of the pyrolysis occurred. The last stage (char production) undergoes at temperature exceeding 450°C and is considered to be C-C and C-H bonds further cracking phase [12].

*3.2. Kinetic parameters*

Biomass pyrolysis follows a complicated process since biomass has various chemical compositions of components. In this study, the pyrolysis kinetic parameters were determined with three different kinetic methods, i.e. Kissinger-Akahira-Sunose (KAS) and Ozawa-Flynn-Wall (OFW) as iso-conversional methods [9] and Coats-Redfern (CR) model. KAS and OFW methods assume that the kinetics of the reactions does not depend on the heating rate and the reaction conversion of biomass into products as one-step process. Iso-conversional method also has the advantages of avoiding a potential error in different assumptions of mechanism reaction function.

Figure 2(a) and (b) show the linear plot of the curve ln $\left(\frac{\beta}{T^2}\right)$ vs $\left(\frac{1}{T}\right)$ and log $\beta$ vs $\left(\frac{1}{T}\right)$ at different conversion to determine the kinetic parameters using KAS and OFW methods respectively. The calculated activation energies and correlation coefficient can be obtained through further calculation as listed in Table 2. Both methods have a high correlation coefficient. Each conversion for both methods has slightly different activation energies. The lowest values of activation energy for KAS and OFW about 211 kJ/mol and the highest value is 283.142 kJ/mol for KAS and 294.749 kJ/mol for OFW. It is evidenced that pyrolysis progresses through multi-step kinetics with various apparent activation energies [8,13].

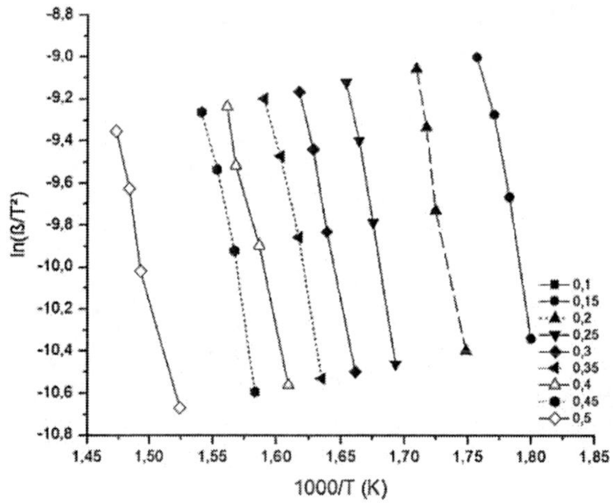


(a)

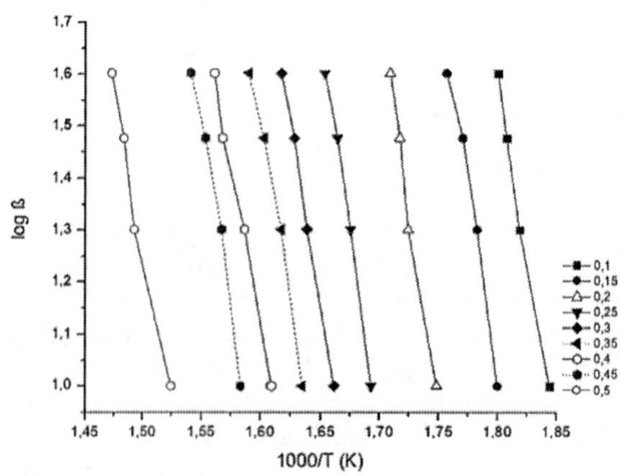


(b)

**Figure 2.** Estimation of activation energy using (a) KAS method and (b) OFW method.

**Table 2.** Estimation of activation energies (Ea), Pre-exponential factor (A), and correlation coefficients for different conversion values using the KAS and OFW methods.

| Conversion | Kissinger-Akahira-Sunose (KAS) | | | Ozawa-Flynn-Wall (OFW) | | |
|---|---|---|---|---|---|---|
| | Ea | A | Correlation coefficient | Ea | A | Correlation coefficient |
| 0.1 | 247.167 | 7.54E+16 | 0.991 | 243.722 | 1.70E+23 | 0.991 |
| 0.15 | 259.580 | 4.53E+17 | 0.982 | 294.749 | 6.36E+27 | 0.997 |
| 0.2 | 281.886 | 1.07E+19 | 0.988 | 277.200 | 2.69E+25 | 0.989 |
| 0.25 | 283.142 | 2.74E+18 | 0.995 | 278.693 | 7.34E+24 | 0.995 |
| 0.3 | 257.751 | 7.35E+15 | 0.997 | 254.754 | 2.04E+22 | 0.997 |
| 0.35 | 246.568 | 4.72E+14 | 0.979 | 245.415 | 1.41E+21 | 0.983 |
| 0.4 | 217.952 | 1.08E+12 | 0.992 | 217.234 | 3.19E+18 | 0.993 |
| 0.45 | 259.106 | 1.48E+15 | 0.981 | 256.520 | 4.25E+21 | 0.982 |
| 0.5 | 211.500 | 1.32E+10 | 0.994 | 211.664 | 1.39E+17 | 0.995 |

The CR plot is showed in Figure 3 and the results for activation energy and Arrhenius pre-exponential factor are summarized in Table 3 below. Coats-Redfern equation was simulated with mampel (first-order) reaction and geometrical contraction models at different heating rates.

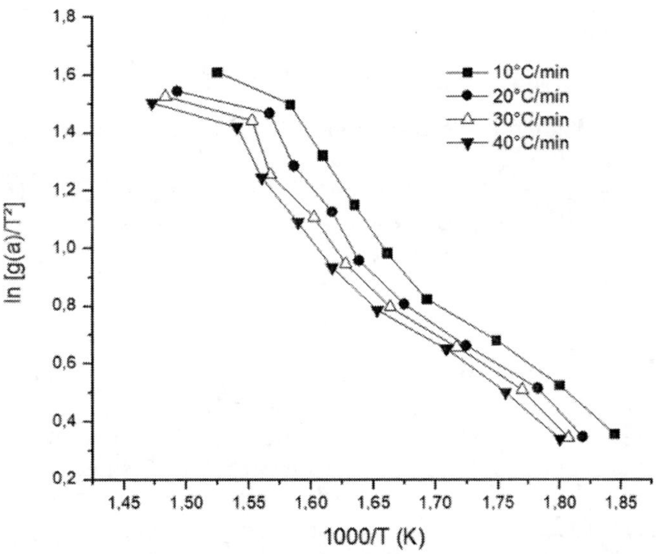


**Figure 3.** Estimation of activation energy using CR method.

**Table 3.** Estimation of activation energies (Ea), Pre-exponential factor (A), and correlation coefficients and correlation coefficients for different conversion values using CR method

| β | Ea | A | Correlation coefficient |
|---|---|---|---|
| 10 | 34.298 | 11.30E+04 | 0.975 |
| 20 | 32.711 | 13.99E+04 | 0.968 |
| 30 | 31.928 | 16.48E+04 | 0.974 |
| 40 | 31.329 | 18.19E+04 | 0.976 |

*3.3. Pyrolysis product*

Figure 4 shows the liquid product or bio-oil increase at higher temperature, reaching 46.3 g (37%-wt) at 600°C. On the other hand, the solid product or charcoal has decreasing trend at higher temperature. The highest amount of charcoal obtained, 82.7 g (66%-wt), is subjected to a temperature of 300°C. The pyrolysis temperature condition will increase the breaking rate of volatile compounds in rice husk.

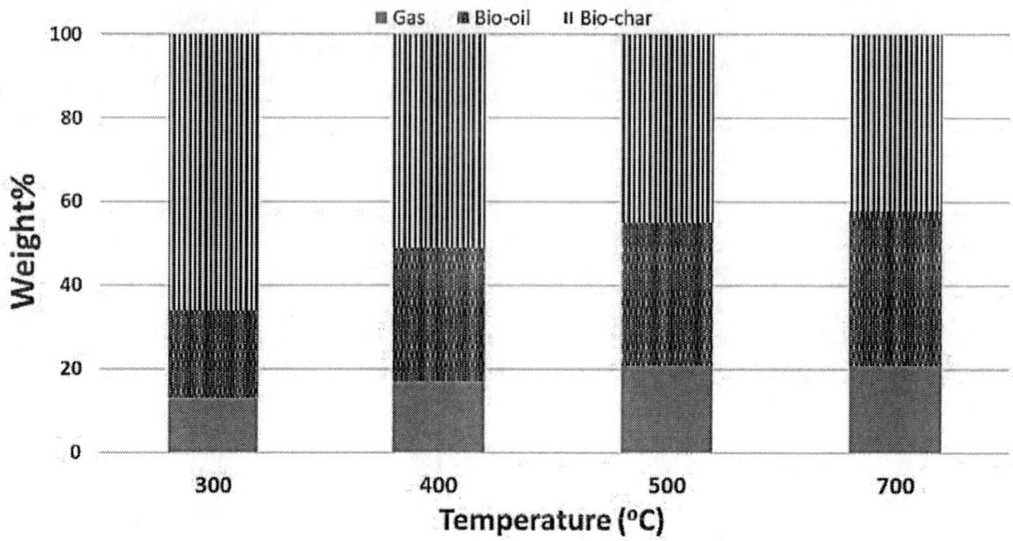


**Figure 4.** Pyrolysis product at different temperatures.

The pyrolysis process also produces gas products consist of $H_2$, CO, $CO_2$, $CH_4$, and a limited amount of hydrocarbon gas which cannot be condensed with a condenser and will exit into gas products [14]. Bio-oil obtained is the result of the condensation of gases formed during the pyrolysis process. According to Atadana et al, liquid product contains more complex compounds such as pyrone, catechol syringol, guaiacol, water, vanillin, furan carboxaldehyde, isoeugenol, acetic acid, formic acid, and carboxylic acid [15].

From figure 5, bio-oil contain functional groups such as alcohol groups with O-H stretch vibration at wavenumber 3550-3200 $cm^{-1}$, C-H stretch vibrations at wavenumber 2990-2850 $cm^{-1}$ indicate the presence of alkane groups, C=O stretch vibrations at wavenumber 1715-1680 $cm^{-1}$ and 1700-1665 $cm^{-1}$ represent the aldehyde and ketone groups, then the vibration C=C stretch at wavenumber 1625-1440 $cm^{-1}$ and C-H bend at wavenumber 900-680 $cm^{-1}$ are attributed to the aromatic groups, C=C stretch vibration at wavenumber 1650-1600 $cm^{-1}$ is attributed to alkene groups, while C=O stretch vibrations at wavenumber 1320-1000 $cm^{-1}$ denote the presence of alcohol, carboxylic acids, esters, and ether. The solid product or charcoal is a component of the remaining biomass. The highest yield of char produces at low temperature or below 500°C.

Figure 6 shows the bio-oil characteristic analysis at different temperatures which includes density, viscosity, total acid number, and water content. The highest acid number at temperature 300°C with 78.4 mg KOH/g sample and the lowest acid number at temperature 600°C, with 58.24 mg KOH/g sample. Various research showed the same results that the acid number will decrease with the heating time and the reaction temperature, the lowest acid number indicates a reduction in the fatty acid content in bio-oil due to an increase in the temperature of the pyrolysis process [16,17].

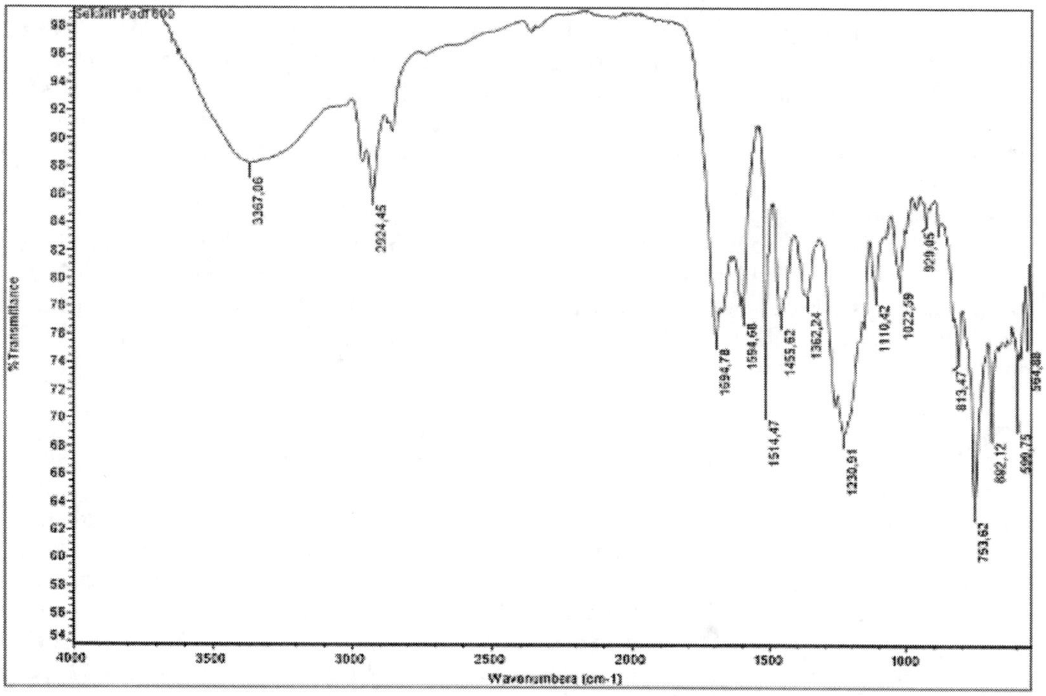


**Figure 5.** FTIR analysis of bio-oil product.

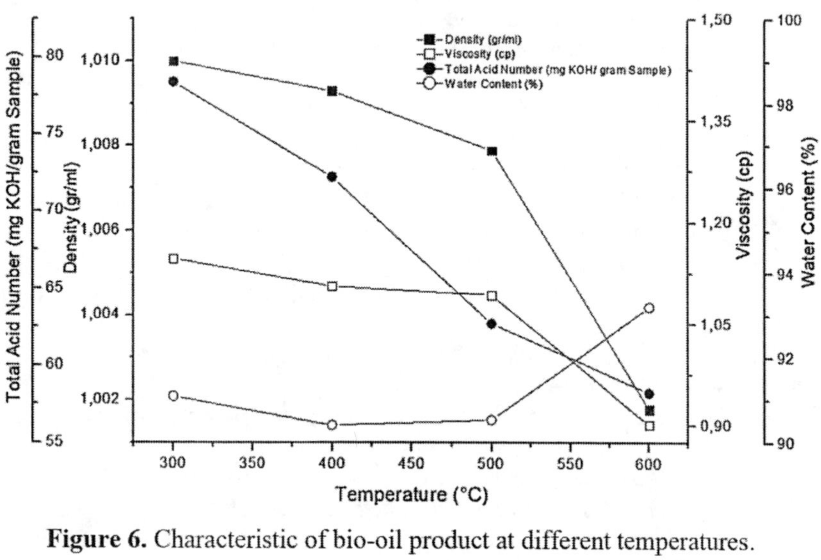


**Figure 6.** Characteristic of bio-oil product at different temperatures.

Furthermore, density is a unit that represents the mass or weight of an object relative to the volume of the solution. Because it relates to the mass of an object, the main factor that affects or the weight of the molecules that make up the substances. The decreasing bio-oil density is found at higher temperature. Figure 6 shows the density of the bio-oil produced is closer to the density of water (1 g/mL) than to the density of biodiesel (0.867 g/mL) since the water content of bio-oil is above 90%-wt.

IOP Conf. Series: Materials Science and Engineering 1143 (2021) 012067 doi:10.1088/1757-899X/1143/1/012067

## 4. Conclusion

The present work revealed different approaches for kinetic study using iso-conversional methods including KAS and OFW model and CR method to estimate the kinetic parameters. The results of thermal decomposition of rice husk showed in the temperature range 100-650°C, followed by the stagnant curves that indicate no more decomposition process occurred above temperature 650°C for all different heating rate conditions. The bio-oil products from rice husk pyrolysis have potential as an alternative for fuel since it contains hydrocarbon functions with low sulphur content.

## Acknowledgments

The authors wishing to acknowledge Mr. Achmad Muhajirin, assistant laboratory of Universitas Pertamina, for his contribution to assist TGA Analysis, Bioenergy and Chemurgy students for helping to collect pyrolysis data as part of laboratory activities (Labtek II). Ms. Laksmi Dewi Kasmiarno would also like to thank the Indonesia Endowment Fund for Education (LPDP) for providing scholarship during her study.

## References

[1] Indonesian Central Bureau of Statistics 2020 *The harvested area and rice production in 2019* (Indonesia: Statistics Indonesia)
[2] Ghodke P and Mandapati R N 2017 Kinetic Modeling of Indian Rice Husk Pyrolysis *International Journal of Chemical Reactor Engineering* **16** 1-15
[3] Naufala W and Pandebesie E 2015 Hydrolysis of Water Hyacinth and Rice Husk for Producing Reducing Sugar as a Beginning Stage of Bioethanol Production *Jurnal Teknik ITS* **4** 109-14
[4] Dauenhauer P J, Colby J L, Balonek C M, Suszynski W J and Schmidt L D 2009 Reactive boiling of cellulose for integrated catalysis through an intermediate liquid *Green Chem.* **11** 1555–61
[5] Arancon R A D, Lin C S K, Chan K M, Kwan T H and Luque R 2013 Advances on waste valorization: new horizons for a more sustainable society *Energy Science & Engineering* **1** 53–71
[6] Saddawi A, Jones J M, Williams A and Wójtowicz M A 2010 Kinetics of the Thermal Decomposition of Biomass *Energy Fuels* **24** 1274–82
[7] Wilson L, Yang W, Blasiak W, John G R and Mhilu C F 2011 Thermal characterization of tropical biomass feedstocks *Energy Conversion and Management* **52** 191–8
[8] Damartzis Th, Vamvuka D, Sfakiotakis S and Zabaniotou A 2011 Thermal degradation studies and kinetic modeling of cardoon (Cynara cardunculus) pyrolysis using thermogravimetric analysis (TGA) *Bioresource Technology* **102** 6230–8
[9] Flynn J H 1983 The isoconversional method for determination of energy of activation at constant heating rates *Journal of Thermal Analysis* **27** 95–102
[10] Doyle C D 1965 Series Approximations to the Equation of Thermogravimetric Data *Nature* **207** 290–1
[11] Coats C and Redfern R 1964 Kinetic Parameters from Thermogravimetric Data *Journal of Nature* **201** 68–9
[12] Liu R and Yuan H 2008 Kinetics of the low-temperature pyrolysis of walnut shell *International Journal of Global Energy Issues* **29** 248–61
[13] Cai J and Bi L 2009 Kinetic analysis of wheat straw pyrolysis using isoconversional methods *Journal of Thermal Analysis and Calorimetry* **98** 325–30
[14] Minh Loy A C 2018 Comparative study of in-situ catalytic pyrolysis of rice husk for syngas production: Kinetics modelling and product gas analysis *Journal of Cleaner Production* **197** 1231–43
[15] Atadana F W 2016 *Catalytic Pyrolysis of Cellulose, Hemicellulose and Lignin Model Compounds* (Virginia: Virginia Tech Electronic Theses and Dissertations)

[16] Likeng-Li-Ngue B C, Ntsomboh-Ntsefong G, Ngando-Ebongue G F, Ngalle-Bille H, Nyouma A and Bell J M 2017 A review of main factors affecting palm oil acidity within the smallholder oil palm (Elaeis guineensis Jacq.) sector in Cameroon *Afr. J. Food Sci.* **11** 296–301
[17] Gerde J, Hardy C, Fehr W and White P J 2007 Frying Performance of No-trans, Low-Linolenic Acid Soybean Oils *J Amer Oil Chem Soc* **84** 557–63

IOP Conf. Series: Materials Science and Engineering 1143 (2021) 012068 doi:10.1088/1757-899X/1143/1/012068

# Isomerization of Raw Turpentine for Cineole using Response Surface Methodology (RSM) : Influence of Acid Ratios and Residence Time

**Diva Almira Chairany[1], Nicolaus Elka Yudhatama[1], Muhammad Mufti Azis[1*], Antonius Indarto[2]**

[1]Department of Chemical Engineering, Faculty of Engineering, Universitas Gadjah Mada,Yogyakarta 55281 Indonesia
[2]Department of Chemical Engineering, Institut Teknologi Bandung, Kampus ITB Ganesha, Indonesia

*Corresponding Author's E-mail : muhammad.azis@ugm.ac.id

**Abstract**. Turpentine is a potential non-wood product from pine tree forest in Indonesia. Turpentine has a wide application in industry mainly as solvent. Isomerization of turpentine is an attractive route to obtain a higher value fine chemical such as cineole. Acid catalyst is often used for this purpose and screening of numerous combination of acid catalysts is often challenging. Therefore, an investigation of isomerization operating condition is important. The aim of present work is to investigate the influence of several acids ratios, and residence time. The raw turpentine was obtained from PT Perhutani Anugerah Kimia (PAK) Trenggalek, East Java. Here, we have investigated the influence of raw turpentine to acid catalyst ratios as well as residence time using central composite design (CCD) of $2^2$ factorial design. In this work, the strong acid catalysts used is hydrochloric acid (HCl) and the weak acid catalysts is formic acid (HCOOH). The result from GC-MS analyses showed that isomerization of raw turpentine using combination of HCl and formic acid can produce cineole compound along with other valuable products such as camphene and limonene. Herein, we have reported the highest yield of cineole as high as 11.5% which is in close agreement with model prediction. It is expected that this work may provide a useful path for cineole production in a larger production scale.

## 1. Introduction

Turpentine oil in Indonesia are generally obtained from pine resin as a top product of the distillation process. This compound has greater potential because it can be processed into various derivative compounds that have higher economic value. One of potential product of turpentine isomerization is cineole or commercially known also as eucalyptol [1,2]. Cineole has a wide range of uses, especially in pharmaceutical industry. Cineole has been obtained as a byproduct from the hydration process of α-pinene to terpineol. In this case, the distillation process is still needed to obtain the cineole. This method is considered less economical and impractical because it still requires further separation method to obtain the pure cineole. Therefore, further research is needed to find suitable methods and formulations in order to obtain a more effective way of producing cineole from raw turpentine.

The composition of turpentine includes beta-pinene (1%-3%), 3-deltacarena (10%-18%), limonene (1%-3%), champhene (<1%), and α-pinene (65-85%) [3]. Alpha-pinene as the main component in turpentine is a key in processing this compound. Alpha-pinene can take part in isomerization reaction

IOP Conf. Series: Materials Science and Engineering 1143 (2021) 012068 doi:10.1088/1757-899X/1143/1/012068

with the help of acid catalyst to produce various isomerization products as shown in **Figure 1.** Isomerization of α-pinene can produce products in the form of monocyclic, bicyclic, etc.. These bicyclic products such as camphene [4] and tricyclene. Meanwhile, monocyclic products are such as limonene [5] and terpineol [6].

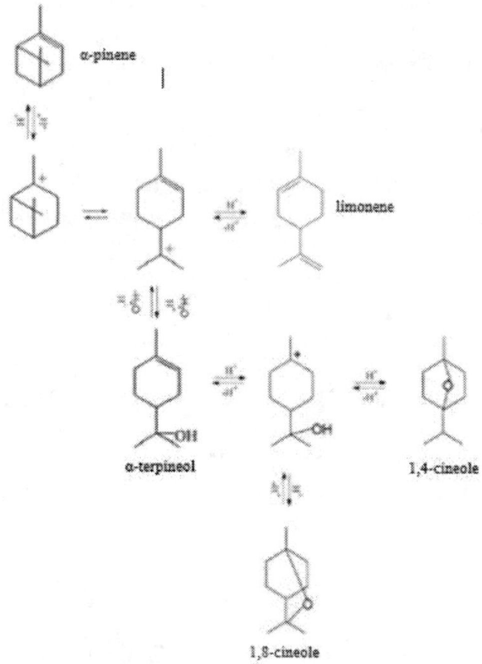


**Figure 1.** Isomerization reaction mechanism of α-pinene.

In the present work, we have investigated the influence of turpentine-acid ratio and residence time using Central Composite Design (CCD) in conjunction with Response Surface Methodology (RSM) [7]. This type of method is useful collection of statistical and mathematical technique to develop and optimize a process which the response of this process is influenced by factors known as independent variables. For this study, the independent variables are residence time and the ratio of turpentine and acid catalyst. This method will provide mathematical as well as graphical modelling of current chemical process to correlate those two factors and the response which is yield of cineole produced from reaction. Furthermore, RSM can also provide the value of independent variables which can give the optimized value of cineole yield.

The use of low cost homogeneous catalyst in the form of a combination of strong and weak acids are commonly used for turpentine isomerization [3,8,9]. Homogeneous catalysts have advantages including not being easily poisoned by impurities and each molecule of the catalyst has an active function as a catalyst. Meanwhile, the disadvantages of this type of catalyst is the difficulty to separate from the mixture and easily decomposed at high temperatures. Here, we have investigated the influence of hydrochloric acid (HCl) and formic acid (HCOOH) for cineole production. This type of catalyst was selected based on our previous research where the turpentine isomerization was carried out at 85°C for 6 hours with variations in the type of catalyst from various combinations of strong and weak acids. It was obtained that the combination of HCl-HCOOH catalyst gave the highest cineole product, which is 9.67 % along with other valuable products such as limonene, camphene, etc.

This study aims to find the optimum setting of the various mixing ratio of the catalyst and residence time for cineole production using central composite design (CCD). We have focused on the use of HCl and HCOOH as a catalyst mixture to maximize cineole production.

## 2. Research Methodology

*2.1 Material.*
Raw turpentine in this research was obtained from PT. Perhutani Anugerah Kimia which is located in Trenggalek, East Java. The raw turpentine contains 85.68% α-pinene. The strong acid used was hydrochloric acid (HCl), meanwhile the weak acid used was formic acid (HCOOH). Water was also used as diluent for the acid solutions. The acids and the water were obtained from Coal, Gas and Petroleum Technology Lab at Department of Chemical Engineering, Universitas Gadjah Mada.

*2.2 Methods*

*2.2.1 Design of Experiment.* Minitab software was used to find the required variation of catalyst ratios and reaction time. The data collected from experiment was analyzed by Response Surface Methodology (RSM) design called Central Composite Design (CCD). In this software, a factorial design method with 2 factors and 2 levels of input were used. The first factor is weak catalyst ratio, which has minimum and maximum value input of 6 and 10 respectively. Meanwhile, the second factor is the residence time, which has minimum value input of 120 and maximum value of 300. The response analyzed in this study is the yield of the cineole. In order to simplify the calculations, the factors were coded to ±1 for the factorial points, ±1α for axial points, and 0 for the center points. The value of α depending on factors used as shown in equation (1) with n as the amount of factor analyzed. Hence the value of α is 1.414. From this method, it was found that 11 experiments with various ratios and residence time variations were needed as shown in **Table 1** including 4 factorial and 4 axial points, as well as 3 replicates of center points to avoid error.

$$\alpha = 2^{\frac{n}{4}} \tag{1}$$

*2.2.2 Sample Preparation.* As much as 6.4 mL of raw turpentine was used to obtain 0.04 mol of raw turpentine. Hydrochloric acid (HCl) solution with the amount of 0.01 mol was prepared by diluting 3.9 mL of HCl 37% into 100 mL-volumetric flask and continued by taking 25 mL of this solution. Further, 0.08 mol of formic acid (HCOOH) was added to the mixture by taking 3.1 mL of HCOOH 98-100%.

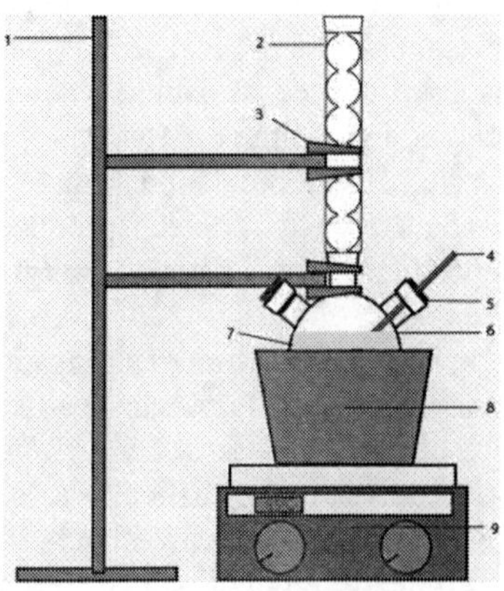


Notes :
1. Stative
2. Condenser
3. Clamp
4. Thermometer
5. Plug
6. Mixture of turpentine and catalyst
7. Three-neck flask
8. Oil container
9. Heater and Magnetic Stirrer

**Figure 2.** Experimental set-up of turpentine isomerization.

*2.2.3 Isomerization Reaction.* Turpentine, weak acid, and strong acid with each required ratio and residence time were poured into a three-neck flask. Plugs, thermometers, condensers, stirrers, and heaters were installed. The reaction in the three-neck flask was run for the appropriate reaction time and catalyst ratio with the temperature of 85°C. After a certain reaction time, the product was then cooled and poured into a beaker glass. The oil layer which is on the top layer of the mixture was then taken using a pipette. Approximately 10 mL of the product was taken and put into a vial bottle. From this experiment, we obtained 11 samples. The experimental set-up of this study is depicted in **Figure 2**.

*2.2.4 Sample Analysis Using GC-MS.* Furthermore, the sample resulting from the reaction was analyzed using the Shimadzu GC-MS QP-2010 with Rtx-5MS at Laboratory of Analysis and Instrumentation, Chemical Engineering Department, Universitas Gadjah Mada. From the analysis, the peak area parameter was used to evaluate the value of cineole yield. For each sample, cineole yield is defined as the percentage of the addition of cineole peak area divided by its initial peak area. Initial cineole peak area was found from the raw turpentine analysis.

$$Yield_{cineole} = \frac{A_{cineole,product} - A_{cineole,initial}}{A_{pinene,initial}} \times 100 \quad (2)$$

*2.2.5 Sample Analysis Using Minitab Software.* The response of cineole yield was then analyzed using Minitab software. Both ANOVA table and Response Surface Methodology (RSM) model were obtained from the software. In addition, we have made contour plot, surface plot, and response optimization to study the experiment results.

## 3. Result and Discussion

Based on the analysis using GC-MS, it was found that all samples with 11 variations in the catalysts ratio and residence time can produce cineole along with other compounds as shown in chromatogram for sample X1 by **Figure 2**.

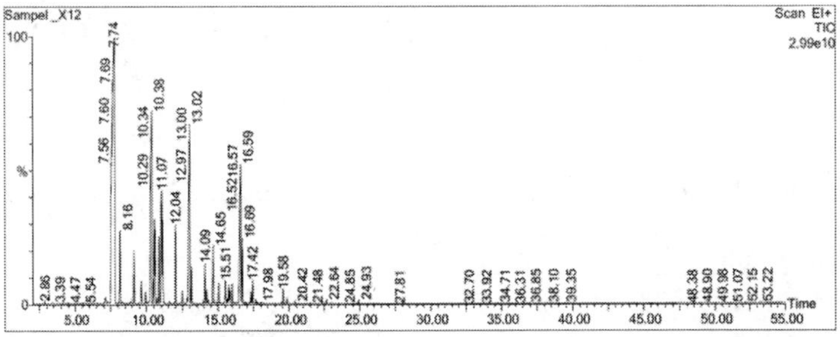


**Figure 3.** Chromatogram from GC-MS analysis of sample X1.

Peak area of each compound is obtained from a chromatogram. Peak area of cineole in sample X1 and raw turpentine are 349,631,168.0 and 3,935,045.5 respectively. Meanwhile, the peak area of α-pinene in the raw turpentine is 10,473,374,880.0. Thus, the yield of the cineole for sample X1 can be calculated using equation (2) as follows.

$$Cineole\ yield_{X1} = \frac{349{,}631{,}168.0 - 3{,}935{,}045}{10{,}473{,}374{,}880.0} \times 100\% = 3.3007\%$$

Based on these data, it can be seen that the highest yield of cineole was obtained in sample X8 by using the turpentine, HCl, HCOOH ratio of 4 : 1 : 8 and the residence time of 337.279 minutes with the yield value of 11.5834 %. Meanwhile, the lowest amount of cineole was produced when the ratio of 4 1 : 10 was used for 120 minutes residence timetime in sample X4.

**Table 1.** Cineole yield of each sample.

| Sample | Coded Variable | | Ratio[c] | Residence Time (min) | Cineole Yield (%) |
|---|---|---|---|---|---|
| | $x_1$[a] | $x_2$[b] | | | |
| X1 | -1 | -1 | 4 : 1 : 6 | 120 | 3.3007 |
| X2 | 1 | -1 | 4 : 1 : 10 | 120 | 1.8407 |
| X3 | -1 | 1 | 4 : 1 : 6 | 300 | 7.8299 |
| X4 | 1 | 1 | 4 : 1 : 10 | 300 | 5.7786 |
| X5 | -1.414 | 0 | 4 : 1 : 5.1716 | 210 | 5.5979 |
| X6 | 1.414 | 0 | 4 : 1 : 10.8284 | 210 | 5.8384 |
| X7 | 0 | -1.414 | 4 : 1 : 8 | 82.72 | 3.9582 |
| X8 | 0 | 1.414 | 4 : 1 : 8 | 337.28 | 11.5834 |
| X9 | 0 | 0 | 4 : 1 : 8 | 210 | 5.9609 |
| X10 | 0 | 0 | 4 : 1 : 8 | 210 | 3.0975 |
| X11 | 0 | 0 | 4 : 1 : 8 | 210 | 5.2634 |

*Note* : [a] $x_1$ represents coded variable for factor 1 which is the amount of weak acid
[b] $x_2$ represents coded variable for factor 2 which is the residence time
[c] Ratio represents mole ratio of turpentine, strong acid, and weak acid

Subsequently, the yield was analyzed using Analysis of Variance (ANOVA) in Minitab software. The result gained by this analysis is listed in the **Table 2**. It can be seen that residence time appears as the single factor which exhibits significant impact with p-value less than the value of $\alpha$=0.05. Other factors such as weak acid catalyst ratio, interaction and quadratic terms were found to be insignificant. Perhaps, this is due to the small range of weak acid ratio that has been selected in this work.

**Table 2.** Analysis of Variance (ANOVA) by Minitab.

| Source | DF | Adj SS | Adj MS | F-Value | P-Value |
|---|---|---|---|---|---|
| Model | 5 | 53,8174 | 10,7635 | 3,48 | 0,098 |
| Linear | 2 | 47,5805 | 23,7903 | 7,70 | 0,030 |
| Weak catalyst ratio | 1 | 1,2572 | 1,2572 | 0,41 | 0,552 |
| Residence time | 1 | 46,3234 | 46,3234 | 15,00 | 0,012 |
| Square | 2 | 6,1495 | 3,0748 | 1,00 | 0,433 |
| Weak catalyst ratio*Weak catalyst ratio | 1 | 0,0100 | 0,0100 | 0,00 | 0,957 |
| Residence time*Residence time | 1 | 5,4698 | 5,4698 | 1,77 | 0,241 |
| 2-Way Interaction | 1 | 0,0874 | 0,0874 | 0,03 | 0,873 |
| Weak catalyst ratio*Residence time | 1 | 0,0874 | 0,0874 | 0,03 | 0,873 |
| Error | 5 | 15,4461 | 3,0892 | | |
| Lack-of-Fit | 3 | 10,9874 | 3,6625 | 1,64 | 0,400 |
| Pure Error | 2 | 4,4587 | 2,2293 | | |
| Total | 10 | 69,2635 | | | |

A second order model was proposed from the data to obtain:

$$\text{Cineole yield} = Y = 4.0 + 0.14A - 0.0177B - 0.011A^2 + 0.000122B^2 - 0.00082AB \quad (2)$$

with A = weak acid catalyst ratio
B = residence time

Equation (2) represent cineole product that could be obtained from the isomerization reaction in terms of weak acid ratio (A) and residence time (B). The negative sign in the equation shows antagonistic effect while the positive sign indicates synergistic effect to cineole formation in the mixture. The benefit

IOP Conf. Series: Materials Science and Engineering 1143 (2021) 012068 doi:10.1088/1757-899X/1143/1/012068

of knowing this equation is that we could predict the cineole formed just by putting the weak acid ratio and residence time factor that we want to implement in our reaction system.

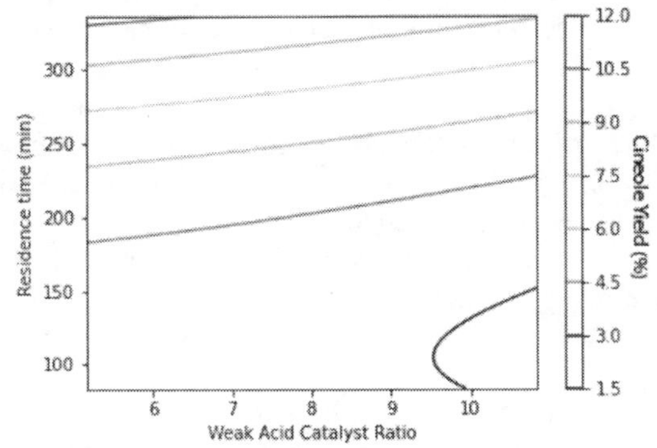


**Figure 4.** Contour plot of yield vs catalyst ratio and residence time.

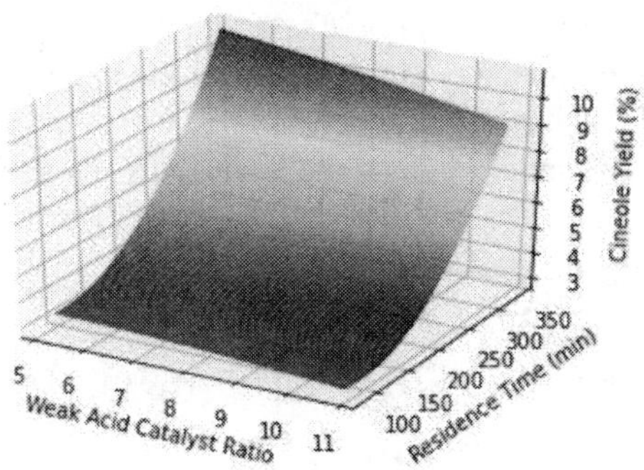


**Figure 5.** Surface plot of yield vs catalyst ratio and residence time.

Next, the data can also be analyzed graphically as shown in **Figure 4** and **Figure 5**. From the graph, it is shown that the area with the red color give the highest yield of cineole with the value of up to 10%. This cineole yield was obtained by using less weak acid catalyst in the mixture and longer reaction time, so that the α-pinene in raw turpentine could be optimally isomerized into cineole.

Furthermore, response optimization was also used in this analysis. From the Minitab, the maximum optimization result the yield of cineole of 10.9173 using the turpentine, HCl, HCOOH ratio of 4: 1 : 5.1716 and the residence time of 337.2792 was obtained from the response optimizer.

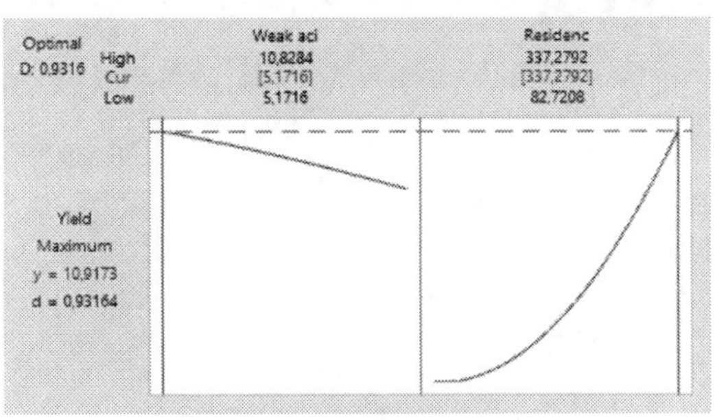


**Figure 6.** Response optimization.

Based on the graph, it is shown that the use of a lower amount of weak acid catalyst is favorable with respect to cineole yield. On the opposite, higher weak acid concentration in the mixture will reduce the cineole present in the product. Our previous work also indicated that the lower weak acid concentration gives higher α-pinene conversion which leads to higher amount of cineole [3].

From this analysis, it can also be seen that residence time has an effect on the yield value of the resulting cineol. The longer the reaction time, the higher the cineole yield. This phenomenon is also in agreement to a previous study by Prakoso et al. where the formation of terpineol from a-pinene isomerization also gave the same trend in the correlation between reaction time and terpineol yield [9]. Therefore, the longer reaction time let a-pinene forms more terpineol so that this compound reacts further to form cineole. A further study is needed to find the optimum reaction time to obtain maximum cineole yield. In addition, there is a need to further investigate this integrate the reaction and separation process in a continuous operating system.

## 4. Conclusion

A study of RSM has been presented to investigate the influence of weak acid and residence time to maximize cineole yield. All 11 combinations of catalyst ratio and residence time in the isomerization of turpentine oil can produce cineole with the highest yield obtained in the turpentine, HCl, HCOOH ratio of 4 : 1 : 8 and the residence time of 337.279 minutes with the yield value of 11.5834 %. The result of experiment is in close agreement with model prediction. From the ANOVA, the only factor which is significance to the production of cineole yield is residence time. The influence of weak acid catalyst here is insignificane that might be due to too small range of weak acid range that has been used in this work. This work provides better understanding and useful path on the screening of acid catalyst for cineole production from raw turpentine. A further study is however still needed to scale up this process in a continuous reactor.

## References

[1] Putra I A, Azis M M, Soerawidjaja T H, and Indarto A. 2019 *AIP Conference Proceedings* **2085** 020057

[2] Lena E J L, Rocha K A, Kozhevnikov I V, Gusevskaya E V. 2006 *Journal of Molecular Catalysis A: Chemical*. **259**. 99-102.

[3] Azis M M, Qani'a F R, Pratiwi S I, Kristanto J, Purwono S, and Indarto A. 2020 *IOP Conf. Ser. : Mater. Sci. Eng.* **823** 012042

[4] Findik S and Gunduz G. 1997 *Journal of the American Oil Chemists' Society* **74**(9) 1145-1151.

[5] Becerra J A and Villa A L. 2020 *Chemical Engineering & Technology* **43** (11) 2295-2306.

[6] Aguirre M R, Torre-Saenz L D, Flores W A, Robau-Sanchez A, Elguezabal A A. 2005 *Catalysis Today* **107-108** 310-314

[7] Montgomery D C 2013 *Design and analysis of experiments* ed 8 (Arizona: Arizona State University) chapter 11 478-522

[8] Prakoso T, Hanley J, Soebianto M N, Soerawidjaja T H, and Indarto A 2018 *Catalysis Letters* **148** 725-731.

[9] Prakoso T, Putra I A, Handojo L, Soerawidjaja T H, Winoto H P, and Indarto A 2020 *Heliyon* **6** e04984

# Effects of Operating Conditions on the Production of Sodium Stearoyl 2- Lactylate

**L Handojo, D Shofinita*, K Yuventia and Lindawaty**

Department of Food Engineering, Faculty of Industrial Technology, Bandung Institut of Technology, Bandung, Indonesia

*shofi@che.itb.ac.id

**Abstract.** Emulsifier is added in food products to maintain the shape of the food and stabilize it in an emulsion system. One of the emulsifiers commonly used in bread is sodium steroil-2-lactylate (SSL). SSL is the reaction product of stearic acid and lactic acid, with the addition of sodium hydroxide. This study aims to determine the effect of temperature, mole ratio, reaction time, and water vapor elimination in the manufacturing operation of SSL. The variations in this study include the composition ratio of stearic acid, lactic acid, and sodium hydroxide (2.2: 1: 1; 2.4: 1: 1; 2.6: 1: 1), as well as reaction conditions that include temperature (140-200 ° C) and reaction time (3-5 hours). The analysis showed that the acid and saponification values of SSL were inversely related to the increase in temperature and reaction time. The acid and saponification values of the products for the reaction with a mole ratio of 2.6: 1: 1 (lactic acid: stearic acid: NaOH) at 180° C and the reaction time of 5 hours show the lowest acid and saponification values, which were 150.96 mg KOH / g and 246.41 mg KOH / g, respectively. The reaction conversion was inversely proportional to the acid number with a conversion range of 61.73% - 67.79%.
Keywords: emulsifier, stearic acid, lactic acid, SSL.

## 1. Introduction

Emulsifiers are one of the additives that are widely used in the chemical industry, one of which is the food industry. The role of emulsifiers is very important in the manufacture of certain food ingredients, for example bread, pasta, and other foods that require baking (bakery products) [1]. Apart from helping to improve the texture of food, this substance can also increase the shelf life of food because the emulsifier is able to maintain stability in an emulsion system. Sodium stearoyl-2 lactylate (SSL) is an anionic emulsifier that is widely used in the bakery industry [1] because of its ability to improve texture and increase the organoleptic value of taste and texture in general. Apart from that, SSL can also be used in the manufacture of cakes, pasta and noodles [2].

SSL can be made by mixing lactic acid, stearic acid, and sodium hydroxide [3]. Lactic acid comes from the fermentation of sugar substrates, such as sugar beet, while the stearic acid used in the manufacture of SSL is saturated fatty acids obtained from the processing of palm oil. The role of emulsifiers is broad and important makes many industrial sectors in Indonesia need emulsifiers in their processing, however this is not balanced with the production of emulsifiers by domestic industries due to various limitations. SSL is an example of an emulsifier that is still being imported today. Along with the increasing consumption of bread which has an impact on the growth of the bread and cake business, SSL manufacturing has high potential and usability if the operating conditions can be known.

## 2. Methods

SSL esterification was carried out using an experiment set up (Figure 1) with a reaction volume of 500 mL. SSL was made by mixing lactic acid in a liquid phase with commercial grade stearic acid. This mixture was stirred without heating, then NaOH is added. The esterification reaction was then carried out in a three-neck flask. Nitrogen was introduced during the reaction using a sparger to avoid oxidation of reactants and products, and to remove water. The variations in this study include the composition ratio of stearic acid, lactic acid, and sodium hydroxide (2.2: 1: 1; 2.4: 1: 1; 2.6: 1: 1), as well as reaction conditions that include temperature (140-200 ° C) and reaction time (3-5 hours). The final product obtained from this esterification was SSL powder. The sample was then analyzed for the acid number and saponification number. According to Franzke and Kroll [4], acid value and saponification value is the most important parameter of SSL.

**Figure 1** Experiment Setup for SSL esterification.

The procedure to determine acid number was started by weighing about 1 gram of the substance (SSL) in Erlenmeyer flask. Note the weight of the sample in gram (c). Heat until the substance has completely melted, cool; 25 mL of ethanol is added to Erlenmeyer flask. The solution will be refluxed for one hour, cool. Add 2-3 drops of phenolphthalein indicator to the flask. Titrate with potassium hydroxide (0.1 M), constantly shaking the contents of the flask until a pink color, which persists for 15 seconds, is obtained. Note the number of mL (a) required. As the next step, do another titration step in blank sample. Add 25 mL of ethanol to Erlenmeyer flask. Reflux for one hour, cool. Add 2-3 drops of phenolphthalein indicator to the flask. Titrate with potassium hydroxide (0.1 M), constantly shaking the contents of the flask until a pink color, which persists for 15 seconds, is obtained. Note the number of mL (b) required for blank sample. Calculate the acid value from the following formula:

$$Acid\ Value = \frac{56.1\ x\ (a-b)\ x\ N\ KOH}{c} \quad (1)$$

The desired acid number for SSL products is within the range of 60-80 mg KOH/gr.

The procedure to determine saponification number is defined as two stages; titration with blank sample and titration with SSL sample. 25 mL of potassium hydroxide in 0.5 M ethanol are added to Erlenmeyer; add 2-3 drops of phenolphthalein. Add 0.5 M HCl is added to the burette. Titrate with hydrochloric acid (0.5 M), constantly shaking the contents of the flask until a clear color, which persists for 15 seconds, is obtained. Note the number of mL hydrochloric acid (a) required for the sample. Secondly, weigh about 10 grams of SSL in Erlenmeyer flask. Note the weight of the sample in gram (c). Add 25 ml of ethanolic potassium hydroxide (0.5 M); reflux for 1 hour; cool; add 2-3 drops of phenolphthalein. Titrate with hydrochloric acid (0.5 M), constantly shaking the contents of the flask until a clear color, which persists for 15 seconds, is obtained. Note the number of mL hydrochloric acid (b) required for the sample. Calculate the acid value from the following formula:

$$Saponification\ Value = \frac{56.1\ x\ (a-b)\ x\ N\ HCl}{c} \quad (2)$$

## 3. Result and discussion

### *3.1 The Effect of Temperature Variation, Mole Ratio, and Time on Acid Number and Conversion*

Based on Figures 2, 3, and 4, an increase in temperature causes a decrease in the acid number. This result occurs because the higher the reaction temperature, the lower the free fatty acid content produced. According to Srinivasan et al. [5], the SSL synthesis reaction takes place in an endothermic reversible state which shows better results at high temperatures. At temperature variations, the acid number values of the three figures with the lowest fixed variable mole ratio are at the reaction temperature of 180°C, while the lower temperature indicates a higher acid number. This is consistent with the theory of endothermic reactions. In an endothermic reaction, the higher the temperature, the reaction will shift towards the product, so that the product acid number is lower.

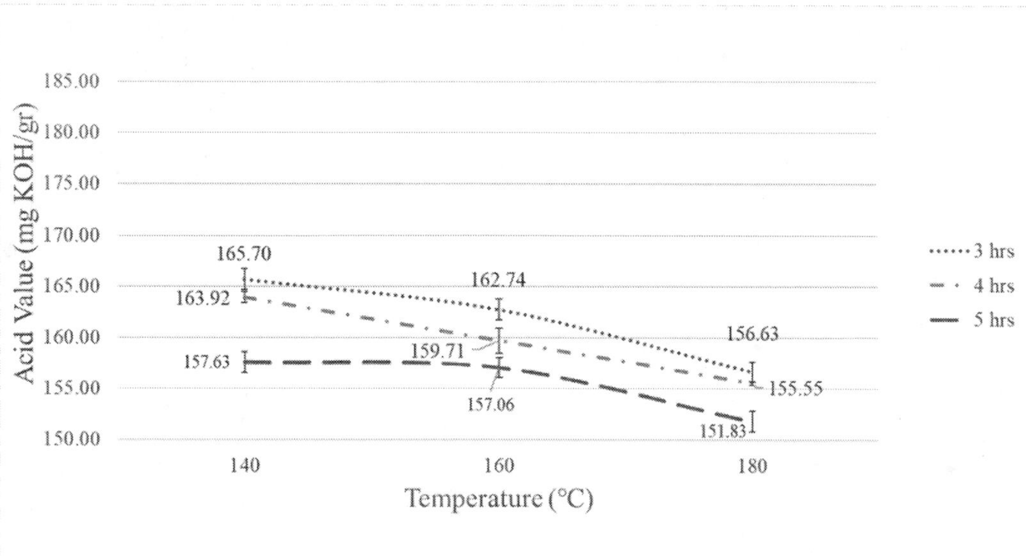


**Figure 2** Acid value of SSL as a function of temperature and reaction time at a reactant mole ratio of 2.2:1:1.

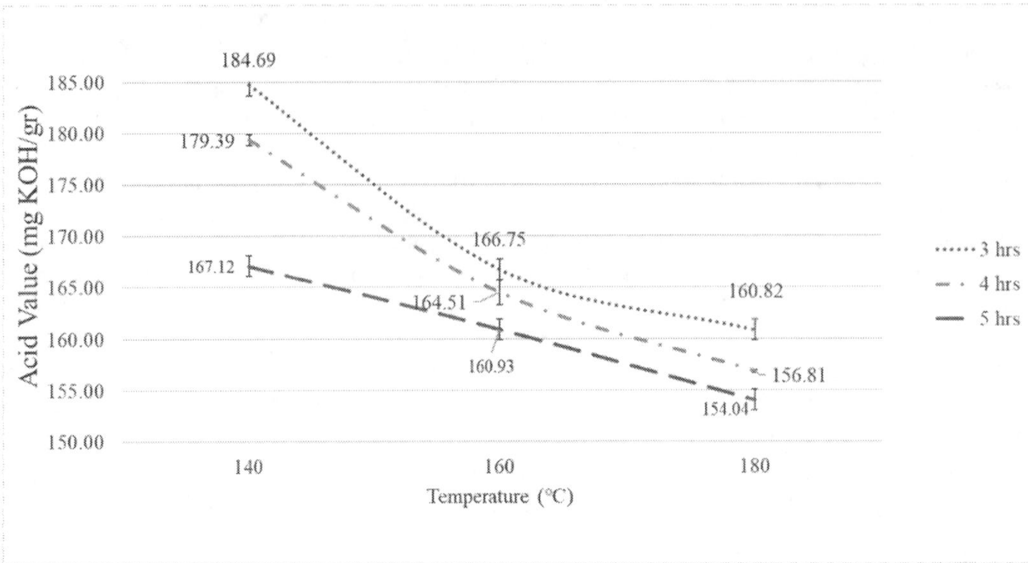


**Figure 3** Acid value of SSL as a function of temperature and reaction time at a reactant mole ratio of 2.4:1:1.

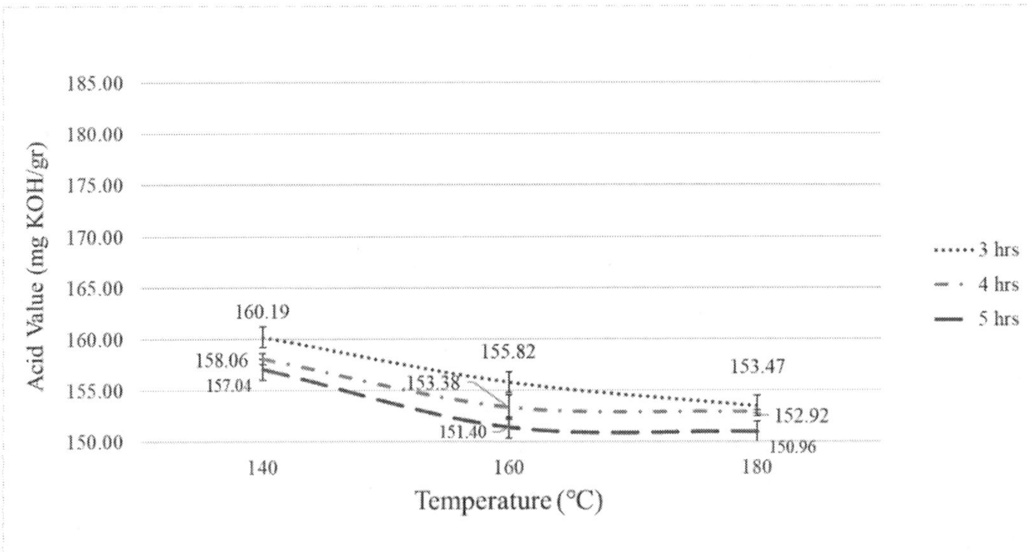


**Figure 4** Acid value of SSL as a function of temperature and reaction time at a reactant mole ratio of 2.6:1:1

The acid value obtained at a temperature of 140-180°C is not in accordance with the standard, namely 60-130 mg KOH / g sample (Commission Regulation (EU) No 231/2012) [6]. This is due to the presence of excess water in the product formation reaction. The water content comes from the initial stage reaction, namely the polymerization reaction of lactic acid into poly-lactides. The polymerization process that occurs is polycondensation, which is a reversible reaction. The presence of water in the reaction product mixture under certain conditions causes the reaction to shift towards the reactants. In

Figures 2, 3, and 4, the longer the reaction time, the lower the acid number. The longer reaction time allows greater contact between substances, so that the acid number obtained is lower.

*3.2 The Effect of Temperature Variation, Mole Ratio, and Time on Saponification Value*

Figures 5 to 7 show that the highest saponification value was achieved at a mole ratio of 2.4:1:1, which is consistent with the trend of acid values. This may occur because water hammering consistently happened in the flask of reactant mole ratio of 2.4:1:1. The hot plate temperature is lowered in that specific flask to avoid water hammering, resulting in anomaly in the data at mole ratio of 2.4:1:1.

Figures 5 to 7 also show that the saponification values of the SSL decreased with an increase in time and temperature. This trend is the same as the trend in the acid number, as shown previously. The saponification value may indicate the average length of the fatty acid chains. The lower the saponification value, the longer the chain the fatty acid has. A higher saponification value indicates that there are more fat and free fatty acids that can be soaped in the sample. The desired saponification value for SSL product is within the range of 190-230 mg KOH/gr. With the acid number, the saponification value is useful in providing information about the quantity of glyceride. The trend of the acid and the saponification values together indicates that in one gram of sample there is approximately the same quantity of glyceride.

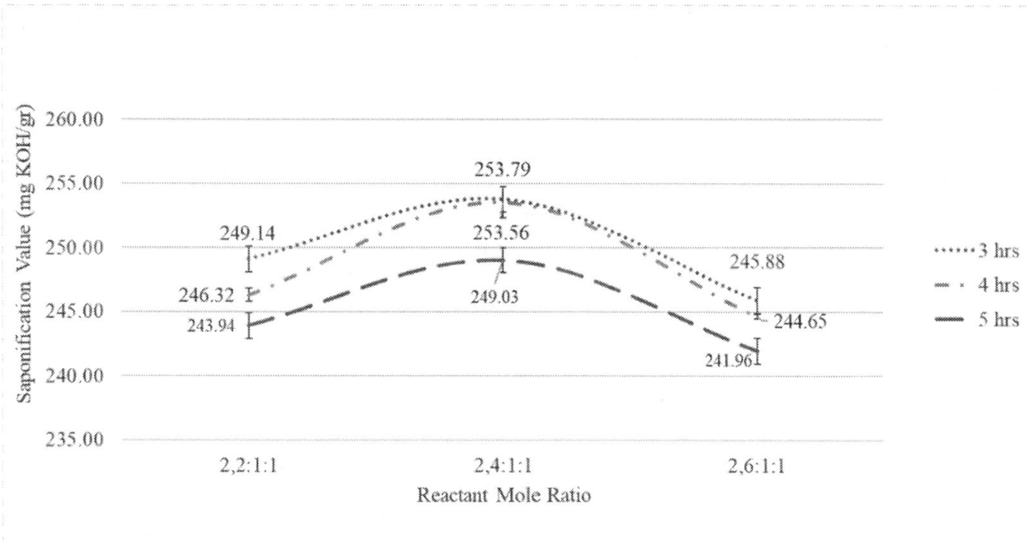


**Figure 5** Saponification value as a function of reactant mole ratio at 140°C.

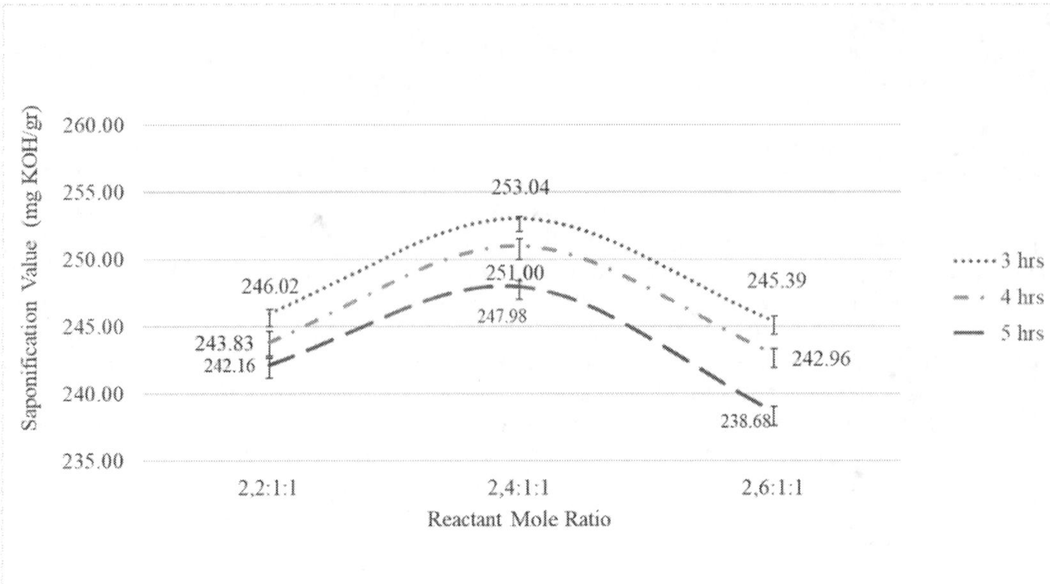


**Figure 6** Saponification value as a function of reactant mole ratio at 160°C.

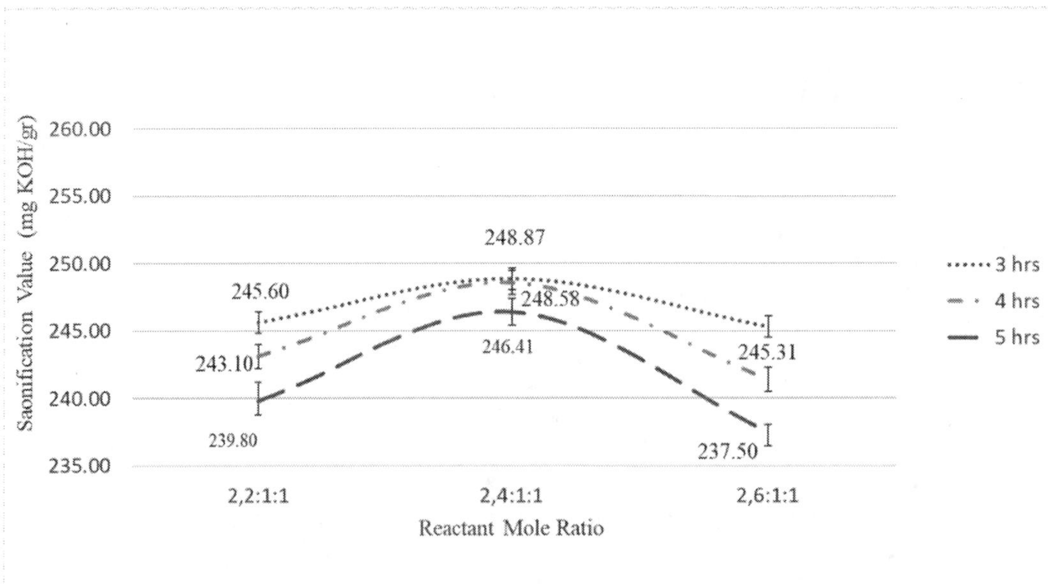


**Figure 7** Saponification value as a function of reactant mole ratio at 180°C.

### *3.3 Order Reaction & Reaction Rate Constant Determination*

In order to know whether the experiment fits certain reaction order, the experiment model is matched with zero, first, second order reaction model. In this case, the experiment model was based on the total lactic acid left in the product at third, fourth, and fifth hour of reaction. Based on the modelling of the reaction on Figure 8, the reaction order is 1, therefore reaction concentration affects reaction rate. The coefficient determination ($R^2$) of this experiment is 0.9969 ($R^2$ value is close to 1).

Table 1 shows that the temperature increases, the rate increases too. In first order reaction, reaction rate will increase as lactic acid concentration is increased. Reactant mole ratio should affect reaction rate, but the data shows that reaction rate lowered at mole ratio 2.4:1:1. The anomaly may happen due to water hammering that was already explained earlier.

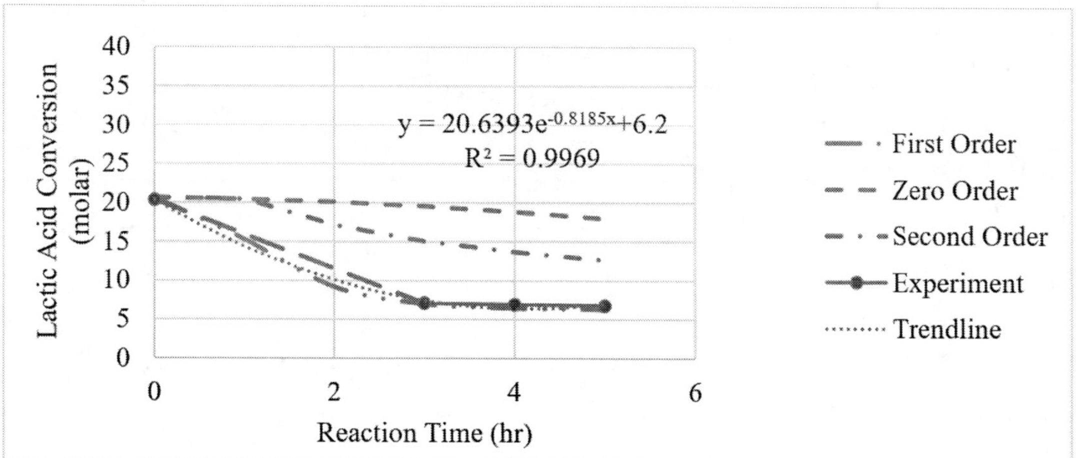


**Figure 8** Modelling Result Comparison (For Reaction at *140°C with reactant mole ratio of 2.2:1:1).*

**Table 1.** Reaction Constant at Every Temperature.

| Run | Temperature (°C) | Mole Reactant Ratio | Reaction rate constant (1/hour) |
|---|---|---|---|
| 1 | 140 | 2.2:1:1 | 0.8092 |
| 2 | 160 | | 0.8169 |
| 3 | 180 | | 0.8364 |
| 4 | 140 | 2.4:1:1 | 0.7136 |
| 5 | 160 | | 0.7935 |
| 6 | 180 | | 0.8195 |
| 7 | 140 | 2.6:1:1 | 0.8364 |
| 8 | 160 | | 0.8417 |
| 9 | 180 | | 0.8543 |

*3.4 The Effect of Temperature on the Colour of the Product*

Table 2 shows the effect of reaction condition on the colour of the SSL. Total colour difference (ΔE*) was obtained from the calculation based on the L*, a*, and b* values of the sample. The higher ΔE*, the higher the total color difference between the sample and the standard. Table 1 shows that the reaction temperature significantly affects the colour of the product. It was found that the ΔE* increases as an increase in the temperature. The product obtained at 140°C was generally has clear white in colour, while the product obtained at 160°C was generally yellowish in colour. In addition, the product obtained at 180°C had a slight brown colour.

**Table 2.** Total color difference for SSL as a function of reaction time, reactant mole ratio, and temperature.

| Mole Ratio | Duration (hours) | ΔE* | | |
|---|---|---|---|---|
| | | 140°C | 160°C | 180°C |
| 2.2:1:1 | 3 | 1.49 | 2.14 | 11.17 |
| | 4 | 1.84 | 2.77 | 12.13 |
| | 5 | 1.06 | 2.33 | 11.15 |
| 2.4:1:1 | 3 | 1.17 | 2.06 | 16.49 |
| | 4 | 0 | 3.25 | 17.34 |
| | 5 | 1.75 | 4.66 | 22.81 |
| 2.6:1:1 | 3 | 1.44 | 4.47 | 15.15 |
| | 4 | 1 | 2.73 | 15.28 |
| | 5 | 1.54 | 2.21 | 14.95 |

## 4. Conclusions

In this study, the operating conditions of the SSL production has been evaluated. It was found that the higher the reaction temperature, the lower the acid and saponification values, which indicates the higher the reaction conversion. The increase in temperature also affects the total colour difference of the product. In addition, the longer the reaction time, the acid and saponification values. It was also found that the mole ratio affects the reaction rate of SSL formation because the SSL synthesis reaction follows a first-order reaction.

## References

[1] Hasenhuettl G L 2008 Food Emusifiers and Their Applications, Blackwell Publishing Ltd, Oxford
[2] Whitehurst Robert J. (Ed.) 2020 *Emulsifiers in Food Technology* Blackwell Publishing Ltd, Oxford
[3] Tsen Cho C, William J Hoover 1975 Method of Increasing the Protein Content of Wheat Flour Based, Low Shortening, low Sugar Baked and Fried Product *US Patent* 3 883 669
[4] Franzke C, Kroll J, Nahrung 1980 **24** (1) 89–90
[5] Srinivasan M, Singh H and Munro P A 2001 Creaming stability of oil-in-water emulsions formed with sodium and calcium caseinates *Food J. Food Sci.* **66** (3) 441-446
[6] European Food Safety Authority (EFSA) 2013 *Scientific Opinion on the Re-Evaluation of Sodium Stearoyl-2-Lactylate (E481) and Calcium Stearoyl-2-Lactylate (E482) as Food Additives* European Food Safety Authority, Parma

# Application of Reverse Flow Reactor for Vent Gas Emission Reduction in Catalytic Oxidation Unit at Purified Terephthalic Acid Plant

**F M Hanafiah and Y W Budhi***

Department of Chemical Engineering, Faculty of Industrial Technology, Institut Teknologi Bandung, Jalan Ganesha 10 Bandung 40132, Indonesia

E-mail: Y.Wibisono@che.itb.ac.id

**Abstract.** PTA (Purified Terephthalic Acid) manufacturing process generates waste gases containing VOC (Volatile Organic Compounds) that have negative health effects, one of it is benzene. Catalytic oxidation method consists of a recuperative (need external energy to reach the reaction temperature) and reverse flow reactor (RFR) which uses the principles of auto thermal (requires no external energy). Fuel consumption for heating the feed gas is a major operating cost in the catalytic oxidation, so that the fuel reduction becomes important. The objective of this study is to assess the technical possibility of RFR to lower the fuel consumption; and to assess the operating conditions for autothermal conditions. The mathematical model consisted of the unsteady state mass and energy balances and was numerically solved using a software package of FlexPDE verion 6.32. The recuperative system was simulated in the steady state condition, while RFR with the same amount of catalyst was simulated in the unsteady state condition. As the key parameter, the switching time was varied to consider the performance of RFR. At various switching times and inlet concentrations, the autothermal was achievable even heat extraction was required to prevent the catalyst overheat. At the feed gas linear velocity of 1.2 m/s and the switching time of 7.5 s or higher, RFR provides energy saving that is equivalent to US$ 0.1208/ton feed gas up to US$ 0.7248/ton of feed gas compared to recuperative system.

## 1. Introduction

Terephthalic acid (TA) is an organic compound with formula $C_6H_4(CO_2H)_2$ and widely used as commodity chemical. Terephthalic acid in a purified condition is well known as Purified Terephthalic Acid (PTA). Almost all PTA productions in the world are used as raw material for the polyethylene terephthalate fibre (PET polyester) manufacturer. From 1999 to 2009, PTA worldwide demand grows 7-8% per year and mostly it is influenced by Asia market which is up to 70% [1].

The exhaust gases of the PTA process typically contain CO, methyl bromide, benzene, and others VOC (Volatile Organic Compounds). According to the US Environmental Protection Agency (EPA), VOC refers to any compounds of carbon, excluding carbon monoxide, carbon dioxide, carbonic acid, metallic carbides or carbonates and ammonium carbonate. The VOC may participate in atmospheric photochemical reactions, except those designated by EPA as having negligible photochemical reactivity. VOC indicate negative health effects, such as respiratory disorders, allergies, immune effects in infants and children, and carcinogenic [2]. Therefore, the VOC must be treated before discharging to the atmosphere. Worldwide, the environmental regulation shows more stringent, especially in USA and

Europe. The PTA plant requires to reduce VOC containing in its exhaust gases. Some PTA plants in Indonesia install a catalytic converter unit to reduce the VOC before venting the exhaust gas.

The application of carbon adsorption may capture the VOC in the exhaust gas up to 97% recovery but is not enough for most PTA plants. Thermal oxidation may reduce VOC and CO containing in the vent gases up to 99% or even more. However, more fuel consumption is required to reach the operating temperature above 800 °C. The catalytic oxidation process destructs VOC and CO simultaneously and offers a fuel consumption saving, due to lower operating temperatures, typically around 275–500 °C [3].

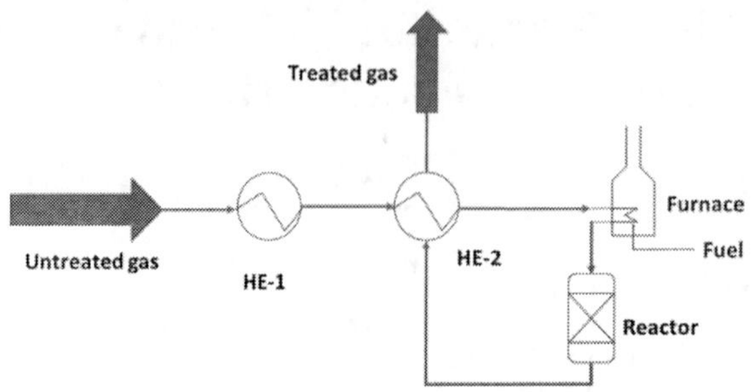


**Figure 1.** Simple Flow Diagram of Catalytic Oxidation

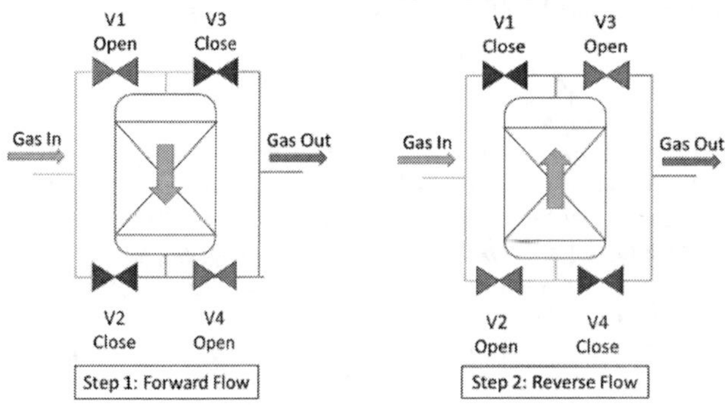


**Figure 2.** Common Flow Diagram for Regenerative System or RFR.

The main operating cost in the catalytic oxidation is furnace fuel. The fuel reduction is always important, meanwhile achieving the environment and health concerns. Various technologies for the abatement of the VOC concerning catalyst and reactor developments and controlling methods have been applied. The selection method is based on operating cost and initial capital cost, VOC type and concentration, as well as gas flow rate. In general, the PTA plant exhaust gas flow rate has capacity above 20 $m^3/s$ and has VOC concentration in the range of 500–1000 ppm, while the expected conversion is at least 98% [3-4]. Thus, the most feasible options for industrial applications are thermal and catalytic oxidations. Given the operating cost consideration, the catalytic oxidation is mostly preferred [5].

The catalytic oxidation for regenerative type by heat storage media (direct heating and cooling) is also known as reverse flow reactor (RFR). Figure 2 shows the reverse flow reactor concept. The heat transfer effectiveness may reach 85% or higher when using the regenerative type of reactor. The most important advantage is no external heat supply (fuel) needed when the autothermal operation is achieved. This kind of operation may reach up to 99% conversion [5].

The working principle of RFR as shown in Figure 2: in the first half cycle, the feed gas flows through valve V1 then enters the reactor to carry out the reaction and generate heat released by exothermic reaction. The released heat is stored in the catalyst bed or inert solid in the reactor. The reaction products leave the reactor through V4. During this time period, the valves V2 and V3 are closed. In the second half cycle, valves V1 and V4 are closed while valve V2 and V3 are open, so that the feed gas enters the reactor from the bottom to the top. When the feed gas enters the reactor, this stream is heated up by the heat that is previously stored in the catalyst bed or inert solid [6-7].

In this study, benzene was considered as representative compound of VOC. Benzene is widely agreed as dangerous compound and must be removed because it can cause carcinogenic and harmful to health

IOP Conf. Series: Materials Science and Engineering 1143 (2021) 012070 doi:10.1088/1757-899X/1143/1/012070

[8]. In addition, when benzene has been converted above 98%, it can be assured other VOCs also have been converted (above 98%) [5]. The exception is for $CH_4$ in very diluted conditions, which is difficult to be oxidized and it requires a relatively higher temperature [10-12].

In general, the catalytic oxidation of VOC uses platinum group catalyst. Several other studies also used 0.5% Pt/γ-$Al_2O_3$ as catalysts for benzene oxidation [5, 9, 13-17]. The oxidation of benzene can be written as follows:

$$C_6H_6 + 7.5\ O_2 \rightarrow 6\ CO_2 + 3\ H_2O \qquad \Delta H = -\ 3259\ \text{kJ/mol}$$

The mechanism of the catalytic oxidation of benzene follows two-stage redox model, also known as the Mars-Van Krevelen Model. It is based on several studies comparing power-law kinetic model, Langmuir-Hishelwood model, Eley-Rideal model, and Mars-Van Krevelen model. It is found that the Mars-Van Krevelen model indicates most proper model in this case [14].

Based on Mars-Van Krevelen model, hydrocarbon oxidation is assumed to occur through a surface redox cycle. The reaction rate of the benzene oxidation is as follows:

$$r = \frac{k_{oi}\ k_i\ C_{O_2} C_{benzena}}{k_{oi}\ C_{O_2} + \ v_i\ k_i\ C_{benzena}} \tag{1}$$

$v_i$ is the oxygen stoichiometric coefficient in the benzene oxidation equation, which is 7.5. $C_{O_2}$ is the oxygen concentration in mol%, and $C_{benzena}$ is the benzene concentration. Based to the research [15] $k_{oi} = 1.069\ x\ 10^{-4} e^{\left(\frac{-3795}{RT}\right)}$ and $k_i = 5.565\ x\ 10^{-2} e^{\left(\frac{-22582}{RT}\right)}$

When using RFR, some of the difficult reaction to react without external heat source becomes feasible. The reaction also sustains even with lower feed temperature and gets higher conversion when compared to the reaction performed in an adiabatic reactor with a one directional flow. During start up, the RFR may need preheating to bring the catalyst bed at the reaction temperature, but after the reaction takes place, it withstands without external heat source (autothermal). This because the solid inert or catalyst bed has three times volumetric heat capacity larger than the gas volumetric heat capacity [18].

After reaching the autothermal, heat extraction may be needed to maintain the reactor temperature. The two most potential and efficient methods: first, partial flow withdrawal for heat extraction by an external heat exchanger, and the flow is not returned to the reactor. The second is full flow withdrawal for heat extraction by an external heat exchanger, and all the flow is returned to the reactor [19].

The switching time of flow reversal is a key performance to maintain the autothermal operation. The exothermic energy must be stored in the solid material inside the reactor. In this case, the reactor is a fixed bed reactor with or without a solid inert section. In general, the inert (either monolith or packed bed) is placed on the catalyst both sides to store the heat and preheat the feed [10, 18-23]. The study purpose is to examine the technical feasibility of RFR for optimizing the process equipment configuration and get lower fuel consumption. Meanwhile assess the autothermal operating conditions.

## 2. Methodology

This research consists of six stages: data collection, modelling, model validation, steady state simulation, unsteady state simulation and result analysis. The flow condition used in this study is shown in Table 1 which is typical flue gas at PTA plants with medium production capacity [3]. The FlexPDE Software v6.32s [24] is used with the following equations:

The mass balance equation [25]:

$$\frac{\partial C_i}{\partial t} = -U_{(t)} \frac{\partial C_i}{\partial z} + D_{eff} \frac{\partial^2 C_i}{\partial z^2} + R_{(x,T_s)} \tag{2}$$

The energy balance equation:

$$\epsilon \rho_g c_{p,g} \frac{\partial T_g}{\partial t} = -U_{(t)} \rho_g^0 c_{p,g} \frac{\partial T_g}{\partial z} - \alpha_0 a_v (T_g - T_s) \tag{3}$$

$$\left[(1-\epsilon)\epsilon_p \rho_s c_{p,s} + \epsilon_g \rho_g c_{p,g}\right] \frac{\partial T_s}{\partial t} = \alpha_0 a_v 0 (T_g - T_s) + (-\Delta H) R_{(x,T_s)} + K_{eff} \frac{\partial^2 T_s}{\partial x^2} \tag{4}$$

$U_{(t)} = K_{(t)}u$ [K has (+) value for right direction flowrate and (-1) for left direction flowrate]

Initial condition:

$C_i = C_{i,input}$

$T_g = T_{g,input}$

$T_{s,\ inert} = T_{s,\ one\ direction\ steady\ state}$

$T_{s,\ katalis} = T_{s,\ one\ direction\ steady\ state}$

Boundary condition:

$C_i = C_{i,input}, T_g = T_{g,input}\ and\ \frac{\partial T_s}{\partial z} = 0\ at\ z = 0$

$\frac{\partial C_i}{\partial z} = 0, \frac{\partial T_s}{\partial z} = 0\ at\ z = L$

The modelling assumptions are:

- The reactor is assumed to be dynamic, pseudo homogeneous and one-dimensional (axial).
- Gas is assumed to be an ideal gas at operating conditions with constant physical properties (to P & T).
- The fixed bed reactor is assumed as a plug flow reactor, axial dispersion is not considered.
- Pressure drop along the reactor is ignored. Constant pressure along the reactor.
- The reactant adsorption and the product desorption are assumed to be unaffected to the gas velocity in the reactor due to dilution with the inert gas.
- $N_2$ and other VOC present in the feed are not involved in the reaction and assumed as inert.

**Table 1.** Off Gas Composition and Condition Data as Feed of the Reactor.

| | Units | Value |
|---|---|---|
| Pressure | $kg/cm^2$ G | 10.7 |
| Temperature | °C | 100 |
| Mass Flow rate | kg/hour | 160,000 |
| Composition | | |
| Nitrogen | % mass | 92.5000% |
| Oxygen | % mass | 4.8000% |
| Water | % mass | 0.3000% |
| Carbon dioxide | % mass | 2.2200% |
| Benzene | % mass | 0.0100% |
| Methanol | % mass | 0.0400% |
| Other VOCs | % mass | 0.1300% |
| Total | % mass | 100.0000% |

Model is validated by comparing the model in this research to a previous research. Initial simulations are carried out in steady conditions and used as initial conditions for unsteady state. The results of steady-state simulation are concentration profile ($C_i$) along the reactor and temperature profile ($T$) along the reactor. The purpose of unsteady conditions simulation is to know the reactor temperature profile during a reverse flow operation. This simulation is performed with and without catalyst length optimization and inert solids presence. This simulation results a temperature profile against switching time, temperature profile against concentration, and heat extracted to maintain the temperature.

## 3. Result and Discussion

### 3.1. Model Validation

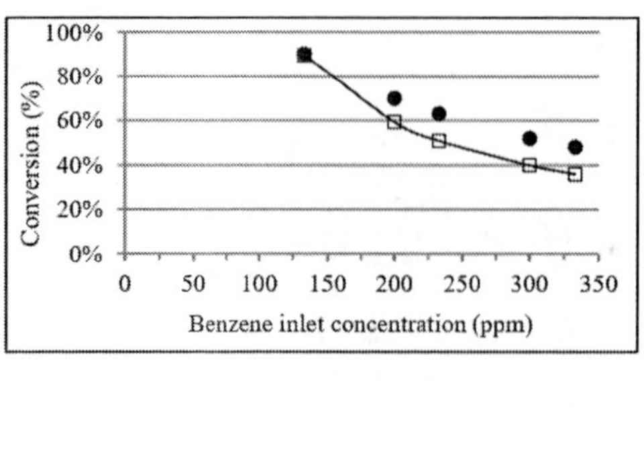


**Figure 3.** The profile of the benzene inlet concentration against the reaction conversion at 300°C temperature and gas superficial velocity = 0.147 m/sec. The symbol (●) refers to experimental results [15]; the symbol (□) refers to the simulation in this study.

The previous study results [15] describes reactor feed inlet concentration against conversion at temperature 300°C in a one directional reactor at steady state. The comparison is under the same operating conditions and reactor dimensions. The result in Figure 3 shows a similarity and has average error of 16.75%. So, it is concluded that the model is usable to simulate the case in this study.

### *3.2. Steady State One Directional Reaction Simulation*

One directional flow reaction simulation in steady state is the base case for other simulations. In this simulation is used the dimension and operating condition of industrial scale reactor. Commonly, the industrial scale reactor is designed with a margin to the theoretical sizing. This is to overcome the process fluctuation and the catalyst deactivation effect so the interval between catalyst replacements is long and economic. In this study the industrial scale reactor is used including the over design. The catalyst bed length is 1.95 m and reactor diameter is 2.11 m. The maximum allowable catalyst bed length is 2.9 m.

The one directional reaction simulation in steady state checks the reaction effect to the catalyst bed effective length. The catalyst bed effective length is the catalyst length where the reaction occurs and achieves 100% conversion on new catalyst conditions. The effective length may be shorter than the catalyst bed overall length. This simulation varies the reactor inlet temperature and intends to know the effective length effect. The reactor inlet concentration is 15 ppm benzene.

The temperature is varied from 80°C up to 400°C. The temperature prior entering the catalytic oxidation system is 80°C. The HE-1 in Figure 1 functions as pre-treatment to heat up the feed up to 100°C (to vaporize the water). Because bromine presence in the feed stream may form HBr upon meet with water (liquid phase). HBr is highly corrosive. Heating up to 400°C can be achieved with the furnace. HE-2 is a recuperative heater to reduce the energy load of the furnace.

Referring to Figure 4, as the inlet reactor temperature increase, shortening the reactor effective length. The effective length is about 24%-37% of the total reactor length, indicating a large over design to encounter the catalyst deactivation. Commonly $Pt/Al_2O_3$ catalysts normal lifetime is 5 years [9].

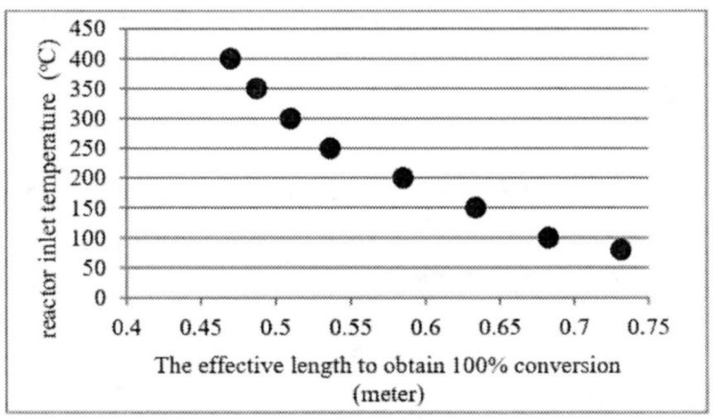


**Figure 4.** The reactor's inlet temperature profile against the effective length to obtain 100% conversion at the initial concentration of 15 ppm and the linear gas velocity of 1.2163 m/sec.

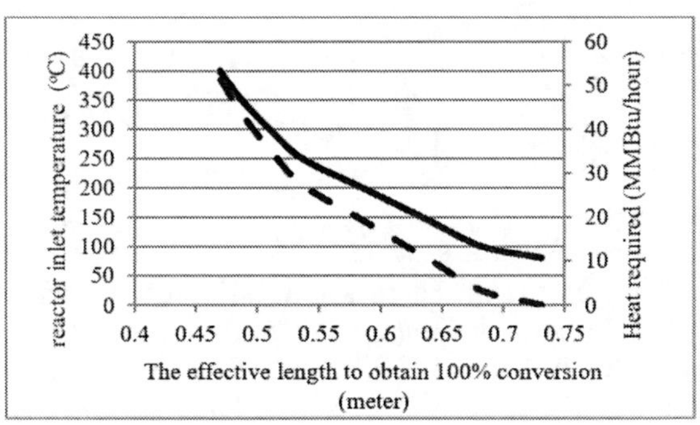


**Figure 5.** The supplied energy profile to meet a suitable inlet temperature with the effective length to get 100% conversion at 15 ppm initial concentration and a linear gas velocity of 1.2163 m/sec. The profile of inlet temperature against the reactor effective length (solid line), the additional energy required to increase the temperature from 80°C to the targeted inlet temperature (dashed line).

By the inlet temperature increases, the reactor effective length is shorter, but it also must be compensated by the higher fuel consumption in the furnace to increase the temperature. Figure 5 shows the energy required to increase the temperature to 400°C. Exothermic reaction is the major source of the supplied energy, and partly supplied from the furnace. The recuperative heater output can reach a temperature of 280°C, while the furnace-supplied heat is to increase from 280°C to the required reaction temperature, and this heat is from the fuel combustion that may be a natural gas, diesel, or others.

Figure 6 shows the required additional energy from the furnace to raise the temperature from 280°C, about 3 to 19 MMBtu/hour, equivalent to a heating cost of US$ 160,000 to US$ 1 million per year (US$ 0.1208/ton feed gas up to US$ 0.7248/ton feed gas). This refers to the natural gas price US$ 6/MMBtu as stated in the Indonesian Ministry of Energy and Mineral Resource Decree Number 8 Year 2020. Hence alternative method for benzene catalytic oxidation may gains saving. So, the RFR is more effective than the recuperative method. Figure 7 shows the inlet benzene concentration increases along with the reactor effective length. The required reactor length is equivalent to 30%-45% of the total typical industrial reactor length, which means the over design is large. This is a design optimization potential and combines it with the RFR concept.

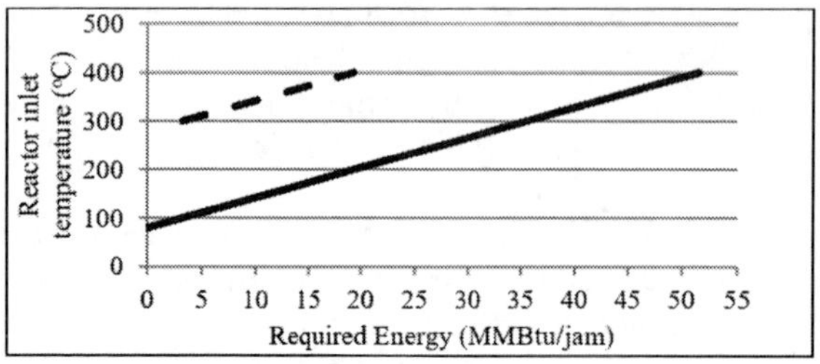


**Figure 6.** The additional energy to get suitable reactor inlet temperature at various reactor inlet temperature (gas linear velocity 1.2163 m/sec). The energy required to increase the temperature from 80°C (solid line), the energy required to increase the temperature from 280°C (dashed line).

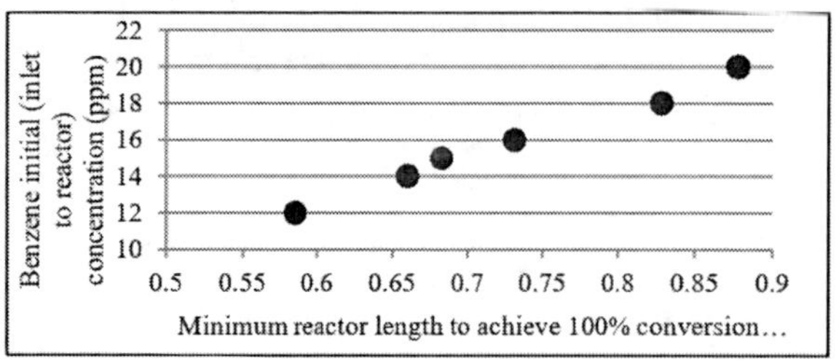


**Figure 7.** The reactor inlet concentration Profile against minimum reactor length to achieve 100% conversion (linier gas velocity 1.2163 m/sec).

### *3.3. Simulation of RFR without Catalyst Length Optimization and Inert Solid Presence.*

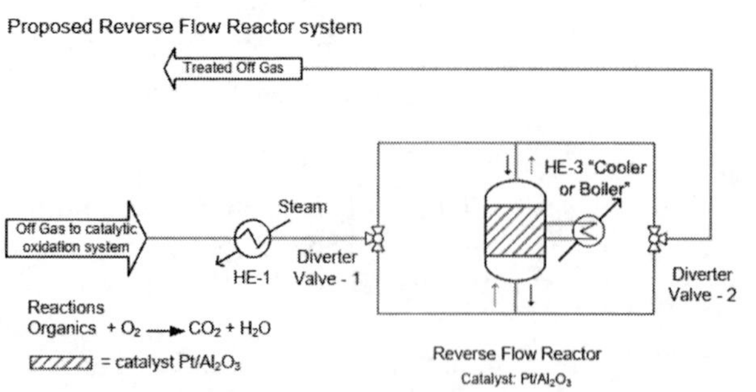


**Figure 8.** Simple Diagram of the proposed RFR.

This part explains the RFR simulation when using industrial scale reactor (includes the over design), with bed catalyst length of 1.95 m and reactor diameter 2.11 m. There is no catalyst length optimization and inert solids as heat transfer medium, so all the solid in the reactor are $Pt/Al_2O_3$ catalyst. The RFR simulation is developed by varying the switching time (ST) and the benzene inlet concentration; besides observing the heat extraction possibility. Variations in flowrate are not conducted because by changing the flowrate for same reactor size, it changes the residence time in the reactor, which is the characteristic of variations in flowrate indirectly reflected in ST variations.

Figure 8 is the proposed RFR system. The reactor feed is also the heating product on HE-1 to keep the temperature at 100°C. So that the whole simulations on RFR is fed at 100°C. The HE-3 is a "Cooler or Boiler" to extract excess heat avoiding reactor overheat. The HE-3 cold side can be a heat transfer fluid, cooling water or boiler feed water. This dynamic simulation determines the effect of Switching Time (ST) for selecting of the right operating parameters to maintain reactor operation without external heat and no overheating. The linear velocity is 1.2163 m/sec with the residence time 2.4336 seconds.

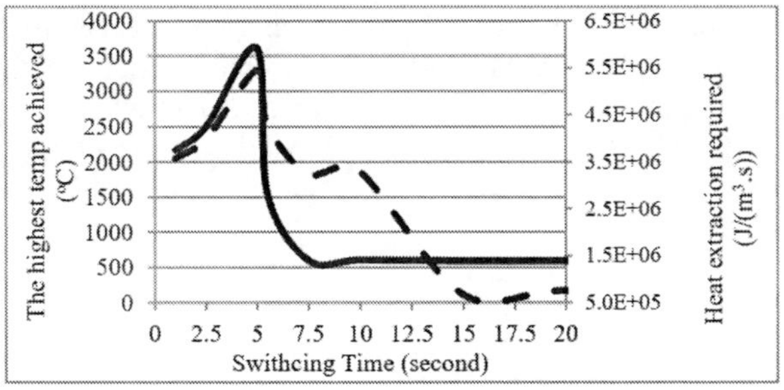


**Figure 9.** The Switching Time profile against the highest temperatures achieved (the solid line) and the Switching Time profile against the heat extraction required to avoid exceeding 400°C (the dashed line).

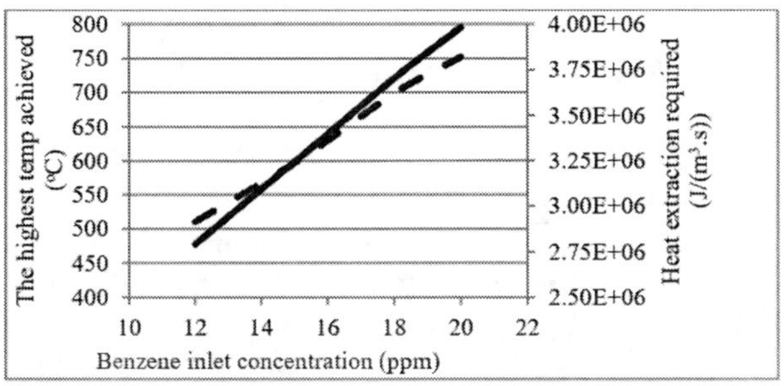


**Figure 10.** Benzene inlet concentration against the highest temperature achieved (the solid line) and Benzene inlet concentration against the required heat extraction to avoid exceeding 400°C (the dashed line).

In Figure 9 shows a significant temperature rise at small ST and peaking at ST about 5 seconds (twice the residence time), then the temperature decreases and stabilize at ST 7.5 seconds (three times the residence time) and above. While the ST profile against the required heat extraction to not exceed 400°C, has similar profile to the ST profile for the highest temperature, the different is the heat extracted profile tend to be flat on ST 15 seconds and above. The temperature limit of 400°C is taken as the reactor operating envelop. When temperatures exceeding 550°C may cause agglomeration or sintering [9].

It is simulated a reverse flow reaction with a variation in the inlet benzene concentration with the same reactor dimension as the previous simulation. The simulation results is in Figure 10 and shows the higher temperature as the inlet benzene concentration increase into the reactor. Because more inlet concentration means more benzene reacts thus produces more exothermic heat. The temperature rise in this simulation is not too extreme, since the chosen ST is a stable ST obtained from the previous sub-chapter on ST variation (7.5 seconds). The heat extraction required also increase with the inlet benzene concentration increased; this is certainly related to the more benzene reaction.

### *3.4. Reverse Flow Reaction Simulation with Catalyst Length Optimization and Inert Solid Presence.*

This section simulates RFR with industrial scale reactor, the differences with previous sub-chapter is the optimization of catalyst length, which only 50% (0.975 m) of catalyst length compare to the previous sub-chapter. The 50% catalyst length is based on the sub-chapter 3.2, where the largest effective length obtained is 45% of the total catalyst length; it is rounded to 50% of the total catalyst length. In addition, on the right and left sides of the catalyst is placed inert solid (25% of the catalyst length in the previous sub-chapter or 0.4875 m). Thus, the overall length of the catalyst and the inert solids in this simulation is equal to the total catalyst length in the simulations at the previous. This optimization aims to reduce the catalyst amount; the 50% catalyst length decrease provides significant saving due to expensive Pt price. The average Pt price in January to August 2020 on the London Bullion Exchange is about \$ 871.07/troy ounce or equal to \$ 28005.61/kg. In Figure 11 shows the proposed RFR system. The system is very similar to the system proposed in Figure 8, the difference is the presence of inert solids in the reactor. The linear velocity is 1.2163 m/sec.

In Figure 12, there is a significant temperature rise at small ST, and peaking at ST about 5 sec. Then the temperature decreases and flats at ST 7.5 seconds and above, which is at the highest temperature about 650°C. The decrease of heat extraction required to avoid exceeding 400°C after ST 5 seconds is because longer ST means more heat stored utilization to heat up the feed. While the cause of the temperature increase in ST less than 5 seconds is the small ST causes a high heat accumulation in the reactor. From the data then it is concluded that the suitable ST is 7.5 second and above, because the highest temperature achieved stable at about 650°C, means more controlled reaction.

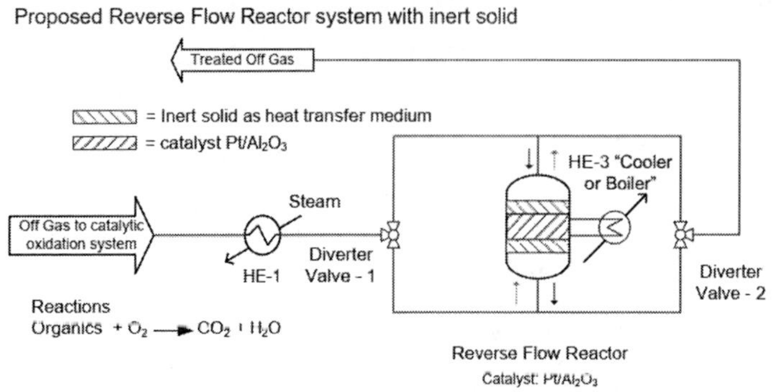


**Figure 11.** The proposed RFR reactor simple diagram with optimize catalyst length and inert solid.

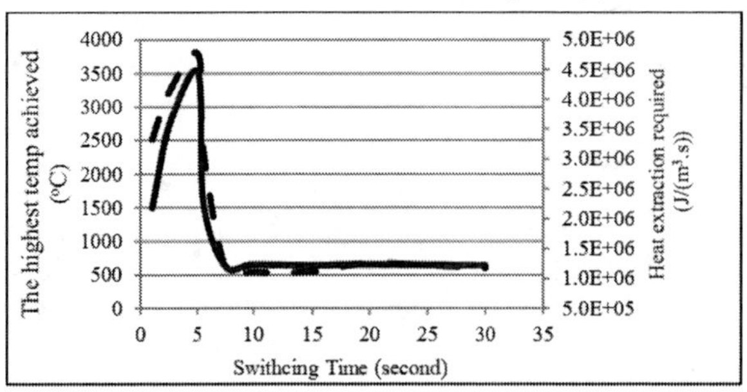


**Figure 12.** The Switching Time profile against the highest temperatures achieved (the solid line) and the Switching Time profile against the heat extraction required to avoid exceeding 400°C (the dashed line).

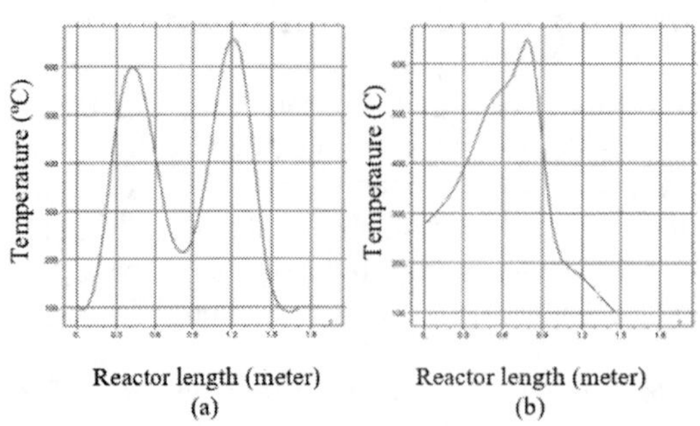


**Figure 13.** Temperature profile along the reactor. (a) 1 second ST, example of ST smaller than the reactor residence time. (b) 7.5 seconds ST, example of ST bigger than the residence time.

Another interesting thing found (as shown in Figure 13) is when ST is smaller than the reactor residence time there are two peak temperatures on the right and left side of the reactor. When ST is greater than or equal to the reactor residence time then only one peak temperature is found, one peak temperature is the same as the profile description proposed by Liu [23]. The comparison of the temperature profile illustrates if ST is smaller than the reactor residence time, then fluid unable to flow from one side to other side. When the direction of fluid flow is reversed, there is a fluid trapped in the centre of the reactor, this is the cause of the two peak temperatures in the profile.

The next is simulation of reverse flow reaction with a benzene inlet concentration variation at the same reactor dimension as the previous simulation. The simulation results are presented in Figure 14, and it shows as the benzene inlet concentration rises, the higher peak temperature can be achieved. Because more inlet concentration means more benzene reacts and produce exothermic heat. The temperature rise in this simulation is not too extreme, since the chosen ST in this simulation is a stable ST obtained from the previous (7.5 seconds). The heat extraction required also increase with more inlet benzene concentration; this is certainly related to the more exothermic heat produced.

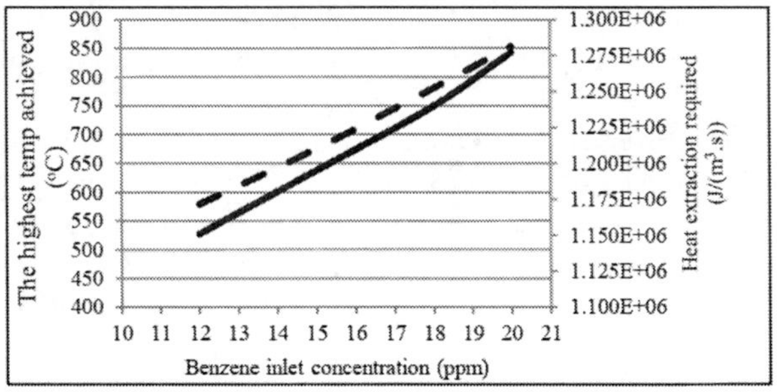


**Figure 14.** Benzene inlet concentration profile against peak temperatures achieved (solid line) and Benzene inlet concentration profile against heat extraction required to avoid exceeding 400°C (dashed line).

From the simulation of ST variations and benzene inlet concentrations it shows the autothermal is happened in benzene oxidation, even heat extraction is required to allow good temperature control in the reactor. Thus, this section main conclusion is by optimizing the length of the catalyst, the reactor operation still takes place with 100% conversion and auto thermally. Additionally, potential capital expenditure reduction is also observed because of optimization of the catalyst length.

## 4. Conclusion

The study focuses on industrial scale reactor. The taken reactor size affect to the conclusion. There is a large over-design factor in industrial scale reactors to counter catalyst deactivation during operation.

In one directional flow operation:

a. As the reactor inlet temperature increase, the catalyst effective length is decreased. The results show by increasing reactor inlet temperature from 80°C to 400°C, the effective length decreased from 0.73 m to 0.47 m or decrease 36%.
b. As the benzene inlet concentration increases there is an increase in effective length. The study shows by increasing benzene inlet concentration from 12 ppm to 20 ppm, then the effective length increased from 0.59 m to 0.88 m or increase 49%.

In reverse flow reactor:

a. ST 7.5 second and above should be used. The heat extracted to stabilize the RFR at 400°C is in the ranges of $5x10^5$ to $5.5x10^6$ J/($m^3$. second).
b. As the benzene inlet concentration increases, the higher peak temperature. At benzene inlet concentration 12 ppm to 20 ppm, the peak temperature ranging from 475°C to 800°C. The heat extracted to stabilize the RFR at 400°C is ranging $2.75x10^6$ to $3.75x10^6$ J/($m^3$. second).

In RFR with Catalyst Length Optimization and Inert Solid Presence:

a. ST 7.5 second and above should be used. The heat extracted to stabilize the RFR at 400°C in the ranges of $1x10^6$ to $4.5x10^6$ J/($m^3$. second).
b. As the benzene inlet concentration increases, the higher peak temperature. At the benzene inlet concentration of 12 ppm to 20 ppm, the peak temperature varies from 525°C to 850°C. The heat extracted to stabilize the RFR at 400°C is ranging $1.15x10^6$ to $1.275x10^6$ J/($m^3$. second).

The benzene catalytic oxidation can operate in autothermal by RFR and may get energy savings of US$ 0.1208/ton feed gas up to US$ 0.7248/ton feed gas compared to recuperative types. Heat extraction may be needed to avoid catalyst overheat. By optimizing the catalyst effective length (50% of the reactor length), the industrial scale reactor still can achieve 98% conversion and auto thermally.

**References**

[1] Ainsworth S 1994 Purified Terephthalic Acid Capacity Takes Off *C&E News* **72** 19–20

[2] Code of Federal Regulations, 40: Chapter 1, Subchapter C, Part 51, Subpart F, 51100.

[3] Belmonte F, Abrams K, and Oppenheim J 2001 *Catalytic Vent Gas Treatment System For Abatement of Volatile Chemical Emissions*, US Patent 6,207,120 B1.

[4] Kirk R and Othmer D 1978 *Encyclopedia of Chemical Technology Volume 17* (New York: John Wiley & Sons)

[5] Hermia J and Vigneron S 1993 Catalytic Inceneration for Odor Abatement and VOC Destruction, *Catalyst Today* **17** 249-258.

[6] Budhi Y W, Jaree A, Hoebink J H B J, Schouten J C 2004 Simulation of reverse flow operation for manipulation of catalyst surface coverage in the selective oxidation of ammonia *Chemical Engineering Science* **59** 5365-5377.

[7] Matros Y and Bunimovich G 1996 Reverse-Flow Operation in Fixed Bed Catalytic Reactors *Catalysis Reviews: Science and Engineering* **38** 1-68

[8] Longo D, Kasper D, Jameson J, Fauci A, Hauser S, Loscalzo J 2004 *Harrison's Principles of Internal Medicine, 16th editon* (New York: McGraw-Hill Profesional) p 618

[9] Heck R, Farrauto R and Gulati S 2009 *Catalytic Air Pollution Control: Commercial Technology* (New Jersey: Wiley and Sons)

[10] Budhi Y W, Effendy M, Bindar Y, Subagjo S 2014 Dynamic Behavior of Reverse Flow Reactor for Lean Methane Combustion *Journal of Engineering and Technological Sciences* **46 (3)** 299-317.

[11] Subagjo S, Budhi Y W, Effendy M, Bindar Y,2014 Homogenity of Continuum Model of an Unsteady State Fixed Bed Reactor for Lean $CH_4$ Oxidation *Journal of Engineering and Technological Sciences* **46 (2)** 195-210

[12] Nuryaman A, Gunawan A Y, Sidarto K A, Budhi Y W 2012 Singular Perturbation Problem for Steady State Conversion of Methane Oxidation in a Reverse Flow Reactor *Journal of Mathematical and Fundamental Sciences* **44 (3)** 275-284

[13] Sawyer J E and Abraham M A 1994 Reaction Pathway During The Oxidation of Ethyl-Acetate on Platinum Alumina Catalyst *Industrial and Engineering Chemistry Research* **33** 2084-2089

[14] Ordonez S, Bello L, Sastre H, Rosal R, and Diez F V 2002 Kinetics of The Deep Oxidation of Benzene, Toluene, n-Hexane, and Their Binary Mixture Over a Platinum on γ-Alumina Catalyst *Applied Catalyst B: Environmental* **38** 139-149

[15] Lee S B, Kang, S L, Lee J D, Hong I K 1999 Catalytic Oxidation Kinetics of Vapor Phase BTX over Pt/$\gamma Al_2O_3$ *Journal of Industrial and Engineering Chemistry* **5 No.3** 170-176.

[16] Rusu A O and Dumitriu E 2003 Destruction of Volatile Organic Compounds by Catalytic Oxidation *Environmental Engineering and Management Journal* **2** 273-302

[17] Santos V P 2010 *Catalytic Oxidation of Volatile Organic Compounds* Doctoral thesis (Porto: University of Porto) p 9 – 12

[18] Salomons S 2003 *Modelling the Behaviour of a Reverse-Flow Catalytic Reactor for the Combustion of Lean Methane* Master degree thesis (Edmonton: University of Alberta) p 6-14.

[19] Marin P, Ordonez S, and Diez F V 2009 Procedures for Heat Recovery in The Catalytic Combustion of Lean Methane–Air Mixtures In A Reverse Flow Reactor *Chemical Engineering Journal* **147** 356–365

[20] Budhi Y W 2020 Reverse Flow Reactor with Side Feeding as a Novel Strategy to Create Dynamic Oxygen Coverage and Heat Propagations in Lean Ammonia Oxidation *Chemical Engineering and Processing: Process Intensification* **144** 108064

[21] Budhi Y W, Jaree A, Hoebink J H B J, Schouten J C 2004 Reverse flow operation with reactor side feeding: analysis, modeling, and simulation *Industrial and Engineering Chemistry Research* **43** 6955-6963.

[22] Budhi Y W 2020 Reverse Flow Reactor with Side Feeding as a Novel Strategy to Create Dynamic Oxygen Coverage and Heat Propagations in Lean Ammonia Oxidation *Chemical Engineering*

*and Processing: Process Intensification* **144** 108064

[23] Liu B 2000 *Experimental and Modelling Study of Reverse Flow Catalytic Converters for Natural Gas/Diesel Dual Fuel Engine Pollution Control* Doctoral thesis (Edmonton: University of Alberta)

[24] PDE Solutions Inc 2014 *FlexPDE 6 Version 6.35 2/17/2014* (Spokane Valley: PDE Solutions Inc)

[25] Suharti P H, Budhi Y W and Bindar Y 2011 Kajian Pemanfaatan Reverse Flow Reactor Untuk Oksidasi Katalitik Uap Bensin *Jurnal Teknik Kimia Indonesia* **10** 26-36

# Synthesis of Maleic-Modified Rosin Ester from Pine Rosin

**Tirto Prakoso[1,2], Ihda Kumalasari[1], Bramantha Jiwandaru[1], Tatang Hernas Soerawidjaja[1,2], Muhammad Mufti Azis[3], Antonius Indarto[1,2*]**

[1]Department of Bioenergy Engineering and Chemurgy, Insitut Teknologi Bandung, 45363 Sumedang, Indonesia
[2]Department of Chemical Engineering, Institut Teknologi Bandung, 40132 Bandung, Indonesia
[3]Department of Chemical Engineering, Universitas Gadjah Mada, 55281 Yogjakarta, Indonesia

*antonius.indarto@che.itb.ac.id

**Abstract**. Rosin is the major component of Pine sap that has a composition of up to 90%. Rosin is an organic compound that consists of diterpene tricyclic compounds with carboxyl functional group. The presence of this functional group makes rosin becoming acidic and corrosive. Furthermore, rosin also has double bonds that are prone to oxidation when exposed to air. To improve the weaknesses, modifications shall be done. Esterification reaction can be done through transforming into a more stable functional group, ie. ester, while Diels-Alder reaction will saturate the double bonds into more stable cyclic compounds. The results indicate that synthesizing maleic modified rosin ester could improve the properties of pine rosin. Esterification was done using a ratio of glycerol to rosin of 1:3 mol. Esterification was done for 6 hours at 250°C at room and vacuum pressure. The Diels-Alder reaction was done using maleic anhydride at 1:1 stoichiometric ratio to rosin, reacted at 150°C for 1 hour at vacuum condition.
**Keywords:** Rosin modification, Esterification, Glycerol, Diels-Alder, Maleic

## 1. Introduction

Pine rosin is one of the most abundant natural resources in Indonesia. From global market data, Indonesia was ranked third in production with a total of 8% global production, after China and Brazil. Indonesia produces around 60 kilotonnes annually where 80% of it was exported while the rest supplies the domestic needs [1].

Pine rosin is commonly used as antibacterial agent [2], paints, ink toners, and also coatings [3]. It is an organic compound that consists of two major groups of rosin acids, ie. abietic acid (consists of abietic, livomiric, polustric, neoabietic, dehydroabietic, and tetrahidroabietic acid) and pimaric acid (pimaric and isopimaric). These rosin acids are diterpene tricyclic ($C_{20}H_{30}O_2$) compound that has double bonds in their molecular structure [4]. The double bonds are easy to be oxidized, making rosin unstable and prone to oxidation such as oxygen [5]. The carboxyl group that is also present in rosin increases the acidity of rosin. By modifying rosin by chemical reactions, these weaknesses will be eliminated and the process produces rosin that is more commercially desirable. Several chemical reactions can be used; for example, hydrogenation, esterification, disproportionation, and polymerization [5,6]. In this study, a preliminary reaction to improve the property of pine rosin was done by esterification using glycerol and Diels-Alder reaction using maleic anhydride.

## 2. Experimental Setup

The experimental work was divided into three major steps. First, evaluation of optimum parameters to react rosin with glycerol in an esterification reaction. The second is determining the parameters to optimally react rosin with maleic anhydride in a Diels-Alder reaction. The last is to examine the resistance of the product of the reaction against oxidation at open-air conditions. The parameters include the reaction time, temperature, stoichiometric ratio of reactants, and also the use of vacuum conditions to improve the conversion of the reaction. The use of vacuum condition could be deduced from the previous results [7,8]. To quantify the conversion of both esterification and Diels-alder reaction, acid and iodine values were done for each reaction respectively while percent clarity was measured directly to determine the oxidation resistance of the product [9].

Reagents used in this study include raw pine rosin, glycerol, and maleic anhydride. Pine rosin was the distilled pine resin obtained from Perhutani Pine Chemical Industry (PPCI) located in Pemalang, Indonesia with quality of X on the Gardner scale. Glycerol used was in technical grade while maleic anhydride was in analytical grade.

The reactor consists of a three-headed round-bottomed flask that is connected to a vacuum pump to adjust the reaction pressure and a thermometer to measure the temperature of the reaction during the experiment. Nitrogen injection was done to purge off all possible oxygen in the overhead of the flask before and during the experiment. The condenser that was connected further to a vacuum pump was used to condense all vaporized rosin.

Solid pine rosin was first broken apart into smaller chunks however the size was maintained just enough to fit into the opening of the three-headed flask. The small size of rosin will have more risk of being oxidized and will turn solid rosin into brown color. At first, 80 grams of rosin was introduced into the flask. Nitrogen was then injected to purge the reactor of any oxygen that remains inside the flask; this was done to further protect rosin from early oxidation [10,11]. Then the heating mantle was turned on to melt the solid rosin. Once the rosin was perfectly melted, maleic anhydride was added to the reactor and the vacuum pump was turned on. In case of room pressure condition was used, the vacuum adaptor was left unconnected to the vacuum pump system. The final step was to melt of product into an appropriate aluminum cast (preferably from metal due to the high temperature of rosin). The procedure for the esterification reaction was identical to the one previously done in the Diels-Alder reaction with maleic anhydride. The difference is maleic anhydride was used instead of glycerol. Furthermore, due to the commercial demand for rosin triesters; the stoichiometric ratio used was 1:3 moles of rosin esters.

For the oxidation resistance test, the product was exposed to conditions that promote oxidation. Modified rosin will be put inside an oven at 150°C for 24 hours. Afterward, the color of the rosin was measured using a light source and a lux meter, mentioned in the previous method [9]. Unmodified rosin will also be subjected to an identical treatment and measurements to be used as a comparison.

## 3. Result and Discussions

### *3.1. Diels-Alder reaction with Maleic Anhydride*

Based on the Iodine value comparison between feed and products, the reaction time required to react maleic anhydride was roughly 1 hour as seen in Figure 1. The reaction conversion of rosin at above 1 hour did not improve or change significantly. This indicates that longer reaction times (more than 1 h) would not yield any better results. Furthermore, the optimal stoichiometric ratio of pine rosin to maleic anhydride was 1:1. The conversion of rosin was very low for a stoichiometric ratio of 1:0.5 that gave ca. 26.6% of conversion. Meanwhile, ratios of 1:1 and 1:2 have a better conversion at 84.7% and 85.65%, respectively. However, the use of 1:2 ratio causes a problem that is the darkening of the rosin product, making 1:1 ratio a more desired option especially since a lesser amount of maleic was used.

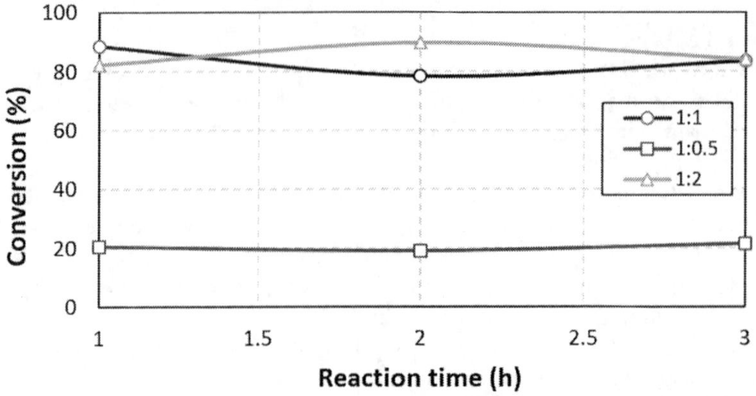


**Figure 1.** Diels-Alder conversion as a function of reaction time and rosin to maleic ratio.

The influence of vacuum condition on the conversion reaction was conducted at rosin to maleic ratio of 1 to 1. Results indicate that higher conversion is achieved when a vacuum condition was used. At vacuum condition, the conversion reached 89% while at atmospheric condition, the conversion was only around 61%.

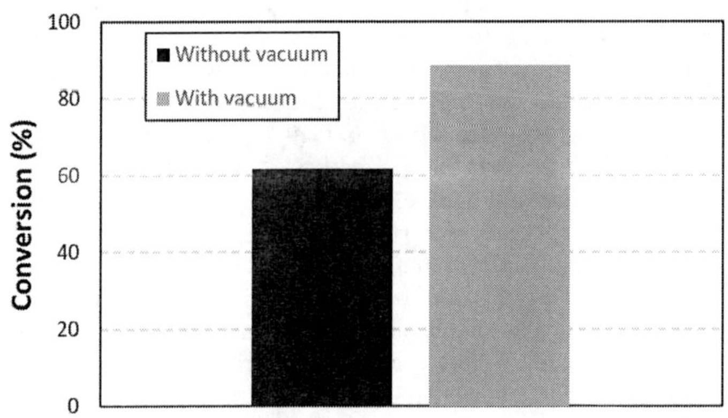


**Figure 2.** Influence of pressure on the conversion reaction at ratio rosin to maleic of 1 to 1.

*3.2. Esterification with Glycerol*

In the case of the esterification reaction, glycerol was used as the reactant. The selection of glycerol was done to result in ester group product with improved properties compared to the properties of the original rosin. The reaction will result in water as a by-product and the use of vacuum condition will be an advantage. The purpose of using vacuum condition was mainly to draw out water vapor that is produced by esterification. Water left in rosin can react with maleic anhydride later on to produce maleic acid that will cause rosin color darkening.

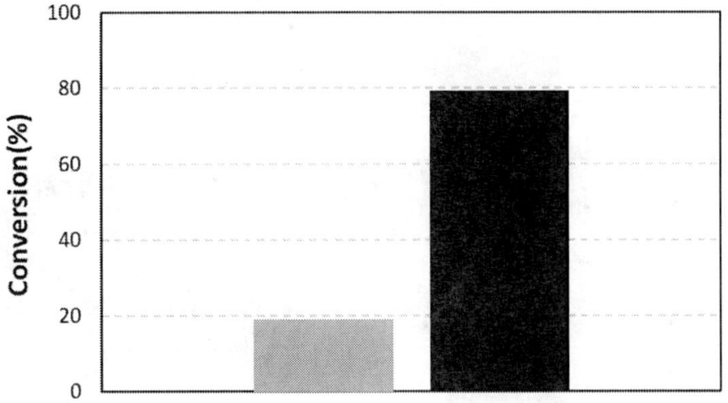


**Figure 3.** Influence of pressure on the conversion at 6h of reaction time.

As shown in Figure 3, the esterification conversion percentage is higher when vacuum condition was applied (black color bar = vacuum condition while grey color bar = room pressure condition). This is following the theoretical evaluation that the presence of a vacuum will help to remove water from the reactor. This will drive the reaction to the product side and increase the conversion of the reaction. Moreover, vacuum conditions will help to melt the rosin at lower temperatures. The reaction will be more reactive and result higher in terms of reaction conversion. Rosin boiling point at atmospheric condition was around 250°C. However, the vacuum condition should be adjusted, as a strong vacuum condition could lead the rosin into vapor condition and reduce the yield of the product. In very low vacuum conditions, vapor rosin will be sucked into the vacuum pump together with vapor water (steam). This condition should be avoided.

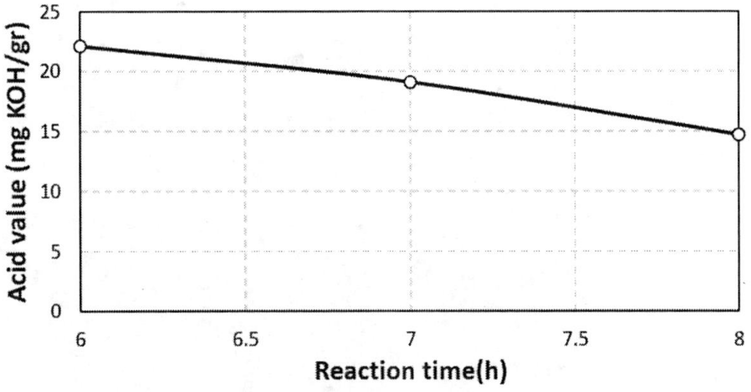


**Figure 4.** Influence of reaction time on the acid value of the product.

In terms of reaction time, the conversion of reaction will increase at a longer reaction time. As shown in Figure 4, the acid value of the product was lower at a longer reaction time. Using a reaction temperature of 250°C, the acid value of the product reached 22 mg KOH/gr sample and reduced to 15 mg KOH/gr sample at 8 hours of reaction time. The possibility of reducing the reaction temperature was also conducted. However, Figure 5 shows clearly that the optimum reaction occurred at 250°C. Some literature has pointed out the optimal reaction for esterification occurred at 250°C. For rosin that is reacted at 200°C, the conversion only reached 40% while the rosin that is reacted at 150°C, the conversion is even lower at less than 10%. From this evaluation, it is proven that the optimal temperature was 250°C.

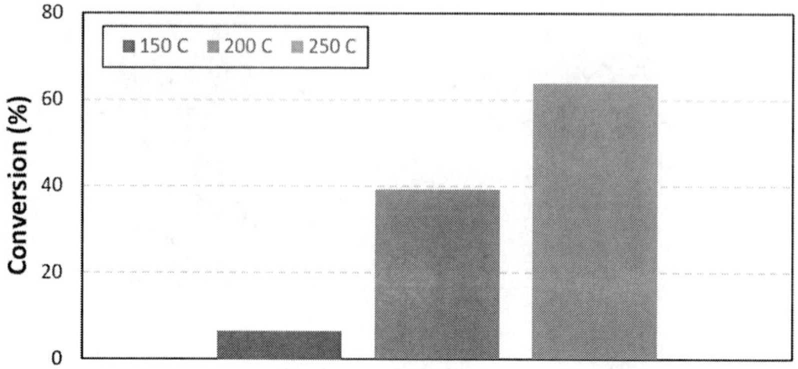


**Figure 5**. Influence of reaction temperature on the esterification conversion at 6 h of reaction time.

*3.3. Effect of Different Additives Thickness on the Evaporation of Crude Oil*

A direct comparison between the tested rosin is shown in Figure 6. It is visually clear that modified rosin (see red arrow in Figure 6b) using esterification and Diels-Alder resists oxidation better than the others. No significant change occurs before and after being oxidized.

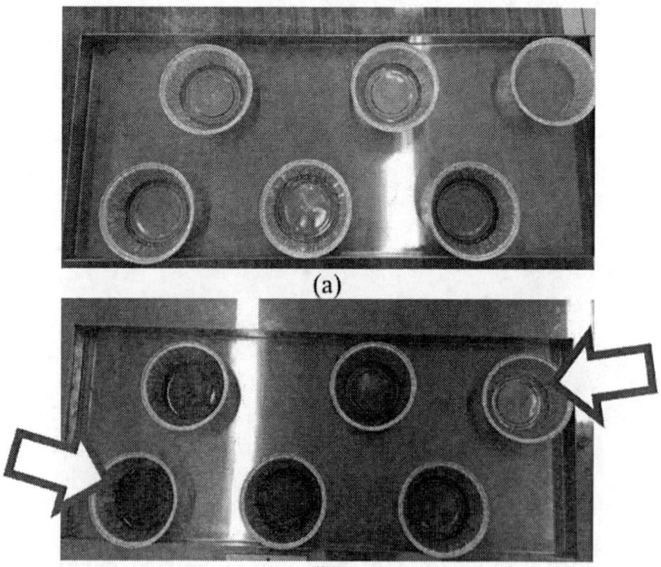

(a)

(b)

*Note for Fig. 6b*: red arrow: maleic modified rosin ester; blue arrow: unmodified rosin

**Figure 6**. Pine rosin samples (a) before being oxidized and (b) after the oxidation process.

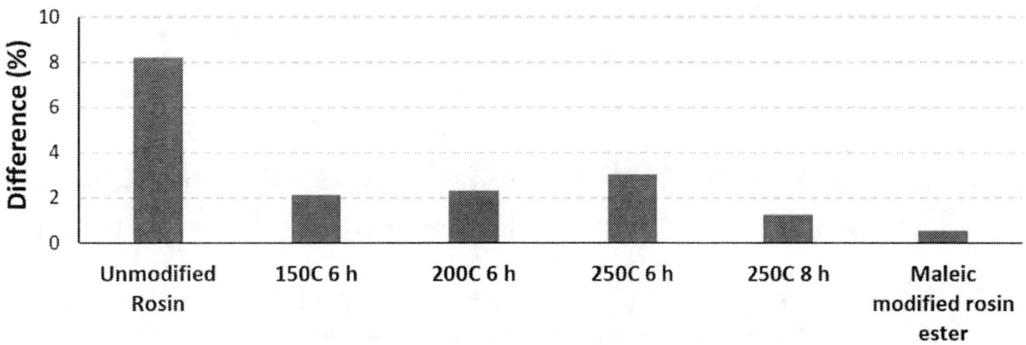


**Figure 7**. The difference in clarity before and after the oxidation test.

As a comparison, the unmodified rosin has the highest value of color change. It means that the product color is the darkest among other products (see blue arrow in Figure 6b). The other samples were obtained from products of esterification with variation in reaction temperature and time. From Figure 7, it can be observed that it is indeed true that the maleic modified rosin ester produces the least amount of change in color as it only changes by 0.5% while the unmodified rosin reaches up to 8%.

## 4. Conclusions

In order to obtain oxidized resistance maleic modified rosin ester, esterification is done first then followed by Diels-Alder reaction. Esterification was done at 250°C for 6 hours and glycerol to rosin ratio used was 1:3. For the Diels-Alder reaction, it was done at 150°C for an hour and rosin to maleic ratio used was 1:1. The modified rosin product was proven to be resistant to oxidation with a color change of approximately only 0.5%.

## References

[1] A.S. Fachrodji, U. Suhendang, Harianto. J. Manajemen dan Agribisnis **6**, 140-151 (2009).
[2] T. Prakoso, I.A. Putra, L. Handojo, T.H. Soerawidjaja, H.P. Winoto, A. Indarto, Heliyon **6**, e04984 (2020).
[3] R.E. Kirk, D.F. Othmer, Rosin and Rosin Derivate. Encyclopedia of Chemical Technology. Vol. 21. New York: The Interscience Encyclopedia. Inc. (2007).
[4] B. Wiyono, S. Tachibana, D. Tinambunan. Pakistan J. Bio. Sci. **9(1)**, 7 – 14 (2006).
[5] R.F.G. Mota, J.C.M. Bordado, J.F.D. Goncalves, Increase of Rosin Ester Shelf Life: Improvement of Oxidative Stability, University of Lisbon. (2016).
[6] B. Wiyono, Chemical Treatment on Indonesia Pine Oleoresin and Rosin in Making Fortified Rosin used for Sizing Agent in Paper Making Process. Master Thesis, Ehime University (2009).
[7] A. Indarto, Res. Chem. Intermed. **38(3-5)**, 1029-1041 (2012)
[8] A. Indarto, H. Lee, J.W. Choi, H.K. Song, Energy Source A **30(17)**, 1628-1636 (2008)
[9] Y. Hartanto, Y. Yaswari, M. Zunita, T. H. Soerawidjaja and A. Indarto, Sep. Sci. Technol. **52**, 967-1972 (2017)
[10] A. Indarto, D.R. Yang, J.W. Choi, H. Lee, H.K. Song, Chem. Eng. Comm. **194(8)**, 1111-1125 (2007).
[11] A. Indarto, J.W. Choi, H. Lee, H.K. Song, Plas. Dev. Oper. **14(1)**, 1-14 (2006).
[12] I.J. Wati, Esterifikasi Gondorukem dengan Penambahan Gliserol/Pentaerithritol, Bachelor Thesis. Bogor: Fakultas Kehutanan, Institut Pertanian Bogor (2005).

# Activity Test of $CuO/\gamma\text{-}Al_2O_3$ as Catalyst of Methanol Dehydration to Dimethyl Ether at Atmospheric Pressure

**Edi Susanto[1], Aisyah Ardy[1], and Herri Susanto[1*]**

[1]Department of Chemical Engineering. Institut Teknologi Bandung, Bandung 40132 Indonesia

*Corresponding Author's E-mail: herri@che.itb.ac.id

**Abstract**. Dimethyl ether (DME) is already used as a propellant and intermediate chemicals. Now, Indonesia will use DME as alternative energy for Liquid Petroleum Gas (LPG), as a result of the continued increase in LPG imports. In this study, DME was synthesized by dehydration of methanol process. The $\gamma\text{-}Al_2O_3$ and modified $CuO/\gamma\text{-}Al_2O_3$ ITB use as catalysts for the process. The objective of this research was to compare of catalytic activity of two catalysts at methanol dehydration to DME. The catalyst activity test was carried out in a tubular fixed-bed reactor (ID 1 cm and length 21 cm). Methanol dehydration was carried out at 1 bar, GHSV 1,100-2,200/h, methanol 1.9-4 mL/h, $N_2$ 10-30 mL/min (27.5°C and 0.93 bar). The downstream gas of the reactor was analyzed using gas chromatography and the condensate was analyzed using a refractometer. The $\gamma\text{-}Al_2O_3$-ITB catalyst gives the highest performance with 89.6% of methanol conversion, 6.6% of DME yield, and 16.6% of DME selectivity. Whereas $CuO/\gamma\text{-}Al_2O_3$ catalyst has a lower performance by 85.5% of methanol conversion, 1.4% of DME yield, and 3.7% of DME selectivity.

## 1. Introduction

Dimethyl Ether (DME) is environmentally friendly energy that is low in emissions of CO, NOX, SOX, and other harmful pollutants. DME can be produced from coal or biomass so that it is following the National Plan on Energy of Indonesia (RUEN 2017) and the National Energy Policy (KEN). The Synthesis of DME could be done by two methods, namely direct synthesis, and indirect synthesis. Direct synthesis of DME was conducted by the conversion of syngas (a mixture of $H_2$ and CO) using a bi-functional catalyst. Meanwhile, the indirect synthesis of DME was conducted by dehydration of methanol using an acid solid catalyst. The reaction of the dehydration of methanol to DME was as follows:

$$2CH_3OH \rightleftharpoons CH_3OH + H_2O \quad (1)$$

Currently, DME was synthesized by dehydration of methanol using solid acid catalysts such as $\gamma\text{-}Al_2O_3$, modified $Al_2O_3$, and zeolites. Dehydration of methanol to DME using the $\gamma\text{-}Al_2O_3$ catalyst was carried out at a temperature of 150-340°C and a pressure of 1-10 bar (1, 2). The results have shown that the increase in temperature could reduce the yield of DME (an exothermic reaction).

The increase of synthesis temperature increased methanol conversion approaching equilibrium conversion, and an increase of the methanol flow rate causes decreased methanol conversion. Meanwhile, the pressure did not change the methanol conversion. The methanol conversion was 84% at a temperature of 290°C and a pressure of 1 bar (2). The HZSM-5 catalyst could be used in the dehydration of methanol to DME, and the activity of this catalyst was 78% of methanol conversion and 100% of selectivity of DME at 190°C (1). However, the HZSM-5 catalyst formed light hydrocarbons as a by-product of methanol dehydration.

Dehydration of methanol to DME also carried out by 3% $CuO/\gamma\text{-}Al_2O_3$, 2% $Fe_2O_3/\gamma\text{-}Al_2O_3$, and $CuO\text{-}Fe_2O_3/\gamma\text{-}Al_2O_3$ at temperatures of 250°C-290°C and pressure atmospheric. The highest methanol conversion was 70% by $CuO\text{-}Fe_2O_3/\gamma\text{-}Al_2O_3$ catalyst at a temperature of 290°C and 100% of DME selectivity (3). The objective of this research was to compare the activity of $\gamma\text{-}Al_2O_3$ and modified the $\gamma\text{-}Al_2O_3$ catalyst ($CuO/\gamma\text{-}Al_2O_3$) produced by the Chemical Reaction and Catalysis Engineering Laboratory-ITB.

## 2. Methodology

This research was carried out by two steps process: (1) Modified of $\gamma\text{-}Al_2O_3$ the catalyst used impregnation method, and (2) activity test of the catalyst before and after modified.

### *2.1. Modification of Catalyst*

The $\gamma\text{-}Al_2O_3$ catalyst was modified with CuO using 0,1 M of $Cu(NO_3)_2{\cdot}3H_2O$ solution. The $CuO/\gamma\text{-}Al_2O_3$ catalyst with 3% -weight of CuO was prepared by mixing 49 ml of 0.1 M $Cu(NO_3)_2{\cdot}3H_2O$ solution and 10 grams of $\gamma\text{-}Al_2O_3$ catalyst that produce by the Chemical Reaction and Catalysis Engineering Laboratory-ITB. Then stirred at 27.5 °C and atmosferic pressure for 4 hours. After the impregnation process, the catalyst was drying at a temperature of 80 °C for 10 hours and calcined at a temperature of 650°C for 6 hours (Figure 1).

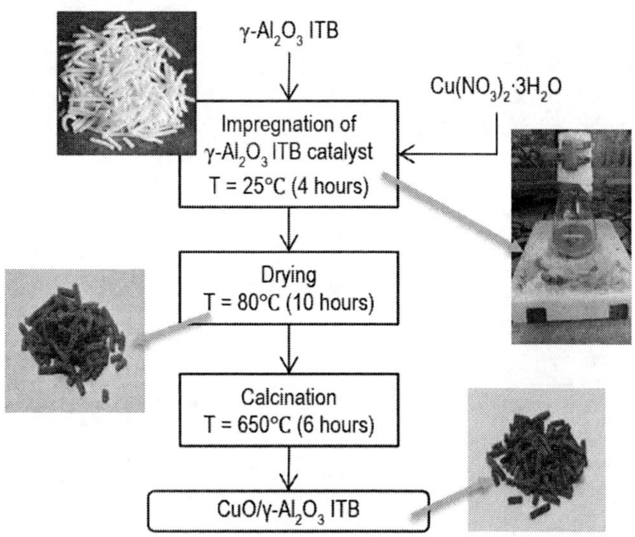


**Figure 1.** Modification of $\gamma\text{-}Al_2O_3$ using the impregnation method.

### *2.2. Activity Test.*

The dehydration of methanol to DME was carried out in a fixed bed tubular reactor with a 15 mm of ID, 50 mm of bed height, and 2 grams of catalyst (Figure 2). In the first step, the reactor was purging using $N_2$ gas. The flow rate of $N_2$ gas was 100 mL/minute at 27.5°C and 1 bar. At the same time, the reactor heated to an operating temperature of 260-300°C. The purging process was carried out for 1 hour.

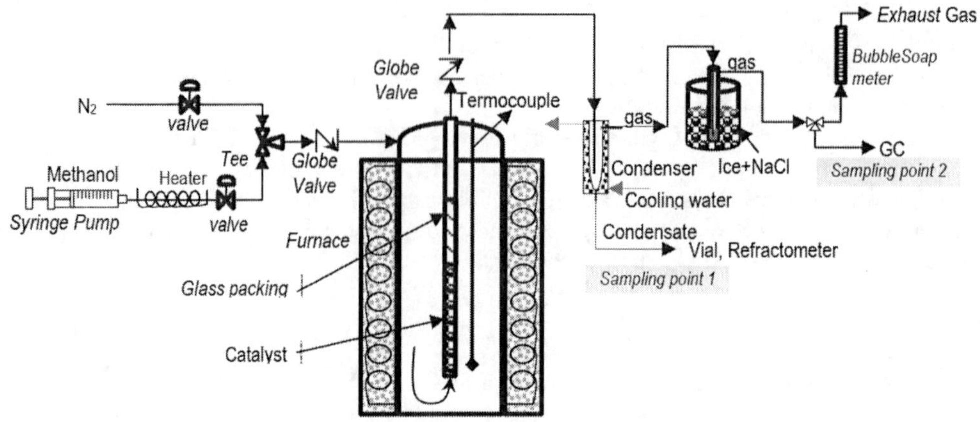


**Figure 2.** Experimental setup.

The next step was the activity test of catalyst. The flow rate of $N_2$ gas was 10-30 mL/min at 27.5°C and 1 bar. Liquid methanol was injected into the reactor at 1.9-4 mL/h, before entering the reactor the methanol is heated to vapor phase. The methanol dehydration process was done at a temperature of 260 -300 °C, a pressure of 1 bar, and Gas Hourly Space Velocity (GHSV) of 1,100-2,200 /h.

A gas sample was taken downstream of the reactor at sampling point 2 every one hour and analyzed using Gas Chromatography. Meanwhile, the condensate sample (sampling point 1) was analyzed using a refractometer (Figure 2).

### 2.3. *Evaluation of Data*

The performance of the γ-Al2O3 catalyst and 3% CuO/$\gamma$-$Al_2O_3$ catalyst for the dehydration reaction of methanol to DME showed by the methanol conversion (X), the yield of DME (Y), and the DME selectivity (S) which can be calculated as follows.

$$X_{MeOH\,(\%)} = \frac{\text{Mole of upstream methanol} - \text{Mole of downstream methanol}}{\text{Mole of upstream methanol}} \times 100\% \qquad (2)$$

$$Y_{DME\,(\%)} = \frac{\text{mole of DME}}{\text{Mole of upstream methanol}} \times 100\% \qquad (3)$$

$$S_{DME\,(\%)} = \frac{\text{mole of DME}}{\frac{1}{2}(\text{Mole of upstream methanol} - \text{Mole of downstream methanol})} \times 100\% \qquad (4)$$

## 3. Results and Discussion

### 3.1. *The comparison of catalyst activity*

Methanol conversion on the $\gamma$-$Al_2O_3$-ITB catalyst was constant at 82% for 4 hours with GHSV 2,142/hour, and 80% of methanol conversion at GHSV 1,100/hour for 12 hours (Figure 3). While the highest activity of CuO/$\gamma$-$Al_2O_3$ catalyst shown 90% of methanol conversion that exceeds the equilibrium conversion. The CuO/$\gamma$-$Al_2O_3$ catalyst decreased its activity after 4 hours at a temperature of 280 °C, but at a temperature of 300 °C the results obtained were quite constant (Figure 4). The uneven CuO dispersion on the surface of -Al2O3 during impregnation made the fluctuate of the catalyst activity.

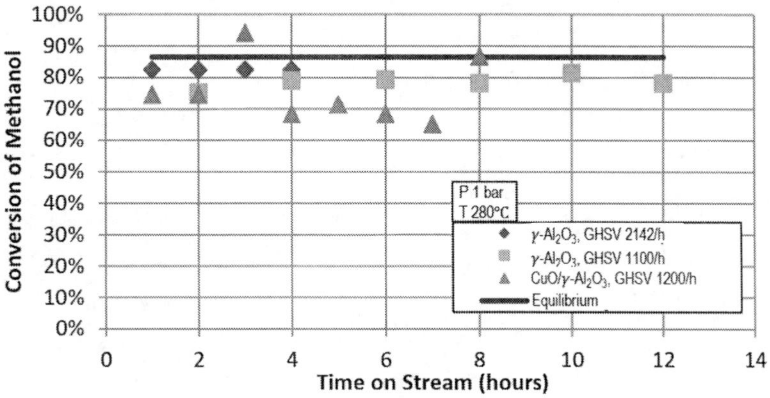


**Figure 3.** Comparison of catalytic activity of the $\gamma$-$Al_2O_3$ catalyst and the $CuO/\gamma$-$Al_2O_3$ catalyst at a temperature of 280°C.

Water was produce during the process of methanol dehydration, which could make water and methanol to compete to adsorb on the catalyst surface (Xu et al., 1997). Cu metal can partially cover the active site of the $\gamma$-$Al_2O_3$ catalyst which decreased the acidity of the catalyst. The experiment of Armenta et al. (2018) shown 45% of methanol conversion using the $CuO/\gamma$-$Al_2O_3$ catalyst. But even so, the addition of CuO to the $\gamma$-$Al_2O_3$ ITB catalyst also showed an increase in the methanol conversion, such as 90% of the methanol conversion synthesis obtained at 3 hours experiment process.

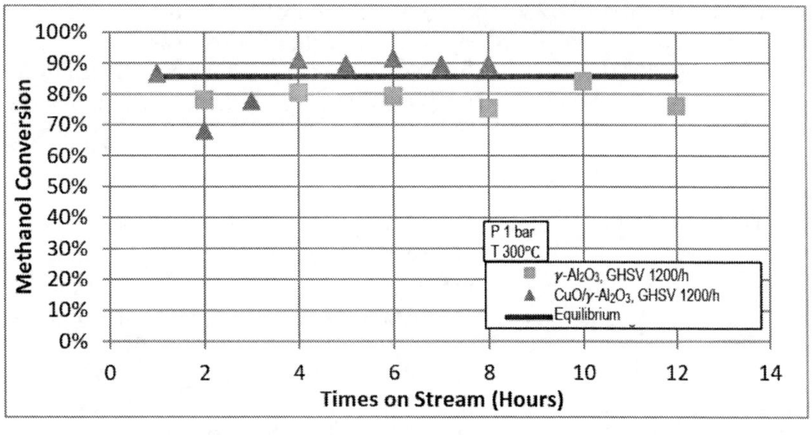


**Figure 4.** Comparison of catalytic activity of the $\gamma$-$Al_2O_3$ catalyst and the $CuO/\gamma$-$Al_2O_3$ catalyst at a temperature of 300°C.

This shows that the addition of CuO given an effect on increasing the methanol conversion which was closer to the equilibrium conversion. These results were in line with the statement from Armenta et al. (2018) that the addition of CuO to a commercial $\gamma$-$Al_2O_3$ catalyst can increase methanol conversion.

### *3.2. DME Yield*

The experiments that have been carried out using $\gamma$-$Al_2O_3$ ITB catalyst and $CuO/\gamma$-$Al_2O_3$ catalyst at the variation of GHSV at a temperature of 280°C and a pressure of 1 bar (Figure 5). The highest yield of DME was 6.6% at GHSV 2142/hour using the $\gamma$-$Al_2O_3$ ITB catalyst, and the lowest was around 0.2%.

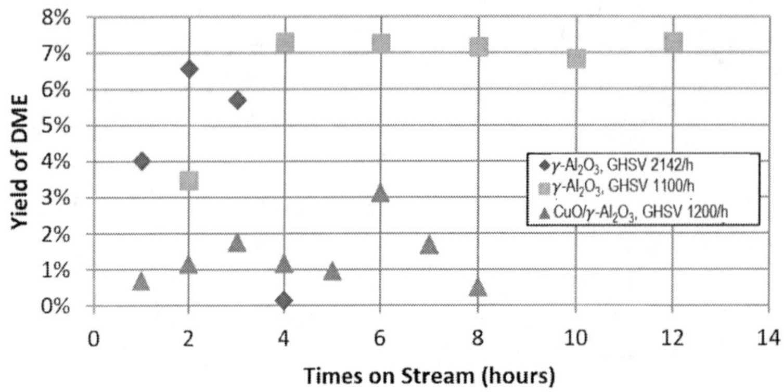


**Figure 5.** Comparation of Yield of DME on catalytic activity of γ-$Al_2O_3$ catalyst and CuO/γ-$Al_2O_3$ catalyst

Dehydration reaction using CuO/γ-$Al_2O_3$ catalyst was given a lower DME yield than γ-$Al_2O_3$. The highest DME yield was 3% at GHSV 1,200/hour which lasted for 8 hours, and the lowest was about 0.5%. This relatively low yield was thought to be due to side reactions such as the formation of $H_2$ and light hydrocarbons, and also because the addition of CuO masks the active site of the catalyst.

## 4. Conclusion

The effect of the addition of CuO increased the methanol conversion. The methanol conversion was 79% on the γ-$Al_2O_3$ ITB catalyst and increased to 86% on the CuO/γ-$Al_2O_3$ catalyst, under the same conditions at 300 °C and GHSV 1,200/hour. But the yield of DME that formed of the γ-$Al_2O_3$ ITB catalyst was better than the CuO/γ-$Al_2O_3$ catalyst. The yield of DME was 6.55% at 280°C and GHSV 1,100/ hour using the γ-$Al_2O_3$ catalyst. Whereas 1.4% of the yield of DME using the CuO/γ-$Al_2O_3$ catalyst at 280°C and GHSV 1200/hour. The low of the yield of DME was thought to be due to side reactions.

## 5. Acknowledgment

This study is a part of Doctoral Research at the Department of Chemical Engineering, ITB. The authors would like to thank the Ministry of Research, Technology, and Higher Education of The Republic of Indonesia for funding this research On Improvement Of γ-Alumina Activity And Selectivity as a Catalysis of Methanol Dehydration to DME: Variations of Cu and Mg concentrations, in the Fiscal Year of 2020.

## 6. References

[1] Xu M, Lunsford J H, Goodman D W, Bhattacharyya A 1997 Synthesis of dimethyl ether (DME) from methanol over solid-acid catalysts *Applied Catalysis A: General* **149** (2) 289-301
[2] Zhang L, Zhang H, Ying W, Fang D 2013 Dehydration of methanol to dimethyl ether over γ-$Al_2O_3$ catalyst: Intrinsic kinetics and effectiveness factor *Can J Chem Eng* **91** (9) 1538-46
[3] Armenta M A, Valdez R, Quintana J M, Silva-Rodrigo R, Cota L, Olivas A 2018 Highly selective CuO/Γ–$Al_2O_3$ catalyst promoted with hematite for efficient methanol dehydration to dimethyl ether *International Journal of Hydrogen Energy* **43** (13) 6551-60

IOP Conf. Series: Materials Science and Engineering 1143 (2021) 012073 doi:10.1088/1757-899X/1143/1/012073

# Titania Modified Silica from Sugarcane Bagasse Waste for Photocatalytic Wastewater Treatment

**Wibawa Hendra Saputera*, Candra Egiyawati, Amellia Setyani Putrie, Aryan Fathoni Amri, Jenny Rizkiana, Dwiwahju Sasongko**

Department of Chemical Engineering, Institut Teknologi Bandung, Jalan Ganesha 10 Bandung, Bandung 40132 Indonesia

*E-mail: hendra@che.itb.ac.id

**Abstract.** Photocatalytic oxidation is one of the technologies to overcome pollution that can be applied for air and water purification. $TiO_2$ has been widely used as a photocatalyst, however, several disadvantages of $TiO_2$ including low absorption of visible or solar radiation, rapid recombination of electron and hole as well as low stability limits its practical applications especially for wastewater treatment. Thus, to overcome this problem, this study aims to develop highly adsorbent photocatalyst using $TiO_2/SiO_2$ composites with sugarcane bagasse waste act as $SiO_2$ source. The experimental results show that the photocatalytic performance of $TiO_2/SiO_2$ composite in the decolorization of methyl orange exhibits three-fold enhancement compared to neat $TiO_2$. Several catalyst characterizations were obtained including X-ray diffraction (XRD), scanning electron microscopy (SEM), and X-ray fluorescence (XRF). Characterization data show that a phase transformation was obtained from amorphous to crystalline phase by increasing $TiO_2$ content. These results proved that the feasibility of $SiO_2$ from sugarcane bagasse waste coupled with $TiO_2$ can be utilized for wastewater degradation.

## 1. Introduction

In today's era, the world is witnessing a social economic revolution from various types of industries. However, the progress of this industry is not balanced between the processes contained in the industry and the waste produced, resulting in environmental pollution which is very dangerous for its sustainability. One of the largest industries in the world is the textile industry. Based on the process taking place in the textile industry, a lot of water, dyes and chemicals are used to produce textile products. Currently, about 10,000 types of dyes are used in the textile industry and more than $7x10^5$ tons of dyes are produced annually worldwide which are widely used for textile dyeing, paper printing, color photography, and as additives in petroleum products[1]. It is estimated that 10–15% of these dyes are contained in industrial waste[1]. Due to the use of these materials, the end result of this process is a lot of waste that cannot be utilized. Consequently, this textile waste will be very dangerous for the environment if it is thrown away without being processed.

Among the existing technologies, photocatalytic oxidation is one of the promising technologies to overcome pollution issue that can be applied to water and air purification due to low-cost, environmentally friendly and sustainable technology. The basic principle of photocatalysis is the excitation of electrons from the valence band to the conduction band after being illuminated by light source (with the energy higher than band gap energy of semiconductor-based photocatalyst used). The electron in the conduction band react with oxygen to generate superoxide radical while the holes in the valence band react with water to form hydroxyl radical. These reactive radicals act as important role in oxidizing organic pollutants, i.e. textile waste (dye). Among the various metal oxide semiconductors, titania ($TiO_2$) is a semiconductor frequently used due to non-toxic and cheap. However, the main drawback of $TiO_2$ is the wide band gap energy (3.0 – 3.2 eV), which limits its light response only in the UV region. In addition, other disadvantages of $TiO_2$ including agglomeration, the low efficiency of the catalyst, high electron-hole recombination rate, and low adsorption of organic pollutants[2]. In recent decades, several attempts have been reported that immobilization of $TiO_2$ into porous materials such as clay[3], zeolite[4], activated carbon[5], and silica[6-11] can increase photocatalytic activity.

In this study, we report on the developing silica-based material to support $TiO_2$based photocatalyst due to its high specific surface area and high stability and good adsorption of organic pollutants capacity. Waste biomass, i.e. sugar cane bagasse is used as a silica source to synthesize $TiO_2/SiO_2$ composites due to its abundant, unutilized waste and high silica content. The effect of $TiO_2/SiO_2$ ratio is also examined for photocatalytic dye degradation and physiochemical properties of $TiO_2/SiO_2$ composites.

## 2. Methodology

### *2.1. Synthesis of Silica from Sugar Cane Bagasse*

10 grams of sugar cane bagasse was cut, dried, cleaned and weighed. Bagasse was calcined at 600°C for 4 hours. Then the calcined sugar cane bagasse was washed using HCl to remove alumunium and iron contents. A total of 5 grams of bagasse ash was stirred and mixed with 50 ml of 1 M HCl for 2 hours at room temperature with a stirring speed of 200 rpm. The solid was then filtered with filter paper and the residual solid was washed with 20 ml of distilled water to remove metal ions. Finally, the residue was dried for 5 hours at 100°C for further processing.

### *2.2. Synthesis of $TiO_2/SiO_2$ composites*

The preparation of $TiO_2/SiO_2$ composites is adopted from Fatimah and Lusi Sopia [12]. $TiO_2$ supported by silica bagasse ash was prepared using the impregnation method. Briefly, the source of $TiO_2$ was obtained from the dilution of $TiO_2$ P25 in isopropanol as a solvent followed by stirring at 750 rpm for 2 hours at room temperature before being dispersed into the silica suspension of bagasse ash in distilled water. The mixture was stirred at 750 rpm for 4 hours at room temperature followed by solvent evaporation until a dry powder was obtained at 90°C for 4 hours. The dry powder obtained was then calcined at 600°C for 4 hours. In this study, the ratio of $TiO_2/SiO_2$ molar ratio used are 1:0, 1:1, 1:3, 3:1, and 0:1.

### *2.3. Catalyst characterizations*

The $SiO_2$ content of sugarcane bagasse was measured using wavelength dispersive X-ray fluorescence (WD-XRF, Rigaku Supermini 200). The crystal structure of the catalyst was determined in air by X-ray Diffraction using a Rigaku SmartLab X-Ray Diffractometer instrument. The samples were ground and sieved to obtain a fine powder of particle sizes <50 μm. Data were collected varying 2θ between 15° and 65° with a step size of 0.01°. The anatase and rutile ratio was estimated using Spurr equation:

$$X_R = \frac{1.26 I_R}{I_A + 1.26 I_R} \tag{1}$$

XR is the Rutile fraction and IR and IA are the strongest intensities in the Rutile ((110)) and Anatase ((101)) diffraction pattern, respectively. The morphologies of the samples were investigated by scanning electron micrsocopy (SEM), using a JEOL 6510 equipped with Energy Dispersive Spectrometry (EDS). Diffuse reflectance UV–Vis spectra were collected at ambient temperature on an Evolution 220 (Thermo Scientific) spectrometer, using $BaSO_4$ as a reference. Spectra were recorded in the wavelength range of 200–1000 nm.

### *2.4. Photocatalytic activity assessment*

The photocatalytic activity of the prepared materials was tested in the decolorization reaction of Methyl Orange (MO) in the home-built reactor. The housing consists of an aluminum box with dimensions of 55 cm in height, 85 cm in width, and 50 cm in depth and four Philips black light tubes (F20 T8 BLB, 18W, 60 cm in length) for illumination. In a typical reaction, 50 mg of the catalyst was introduced into a beaker containing 50 ml of MO (20 mg/L). The suspension was ultra-sonicated for approximately 10 min and then placed into the reactor. Once in the reactor housing, the solutions were stirred for 30 min in the dark at room temperature to allow for complete chemisorption. Once this time had elapsed, a sample of the solution was taken, and the remaining reaction solution irradiated with UV light. Additional samples were taken every 20 min, filtered through a 0.2-mm PTFE Millipore membrane filter to remove suspended catalyst agglomerates, and finally analysed using a UV–Vis spectrometer 220 (Thermo Scientific)) at the wavelength of 456 nm to determine the concentration of the dye.

The reaction rate constant (ka) was then determined assuming quasi first order kinetics using the following equation:

$$\ln\frac{C_0}{C} = k_a t \quad (2)$$

where $C_0$ is the concentration of the dye in solution after chemisorption, $C$ is the concentration at time $t$, and $k_a$ is the reaction rate constant.

## 3. Results and Discussion

### 3.1. *Synthesis of Silica from Sugar Cane Bagasse*

The $SiO_2$ source in this study was sugarcane bagasse ash. The analysis of the compound compositions was carried out by WD-XRF analysis. The XRF spectra and analysis data of sugarcane bagasse ash are shown in **Figure 1** and **Table 1**, respectively. The XRF results indicated that sugarcane bagasse ash was dominated by the $SiO_2$ component as much as 92%. The $SiO_2$ content increased after dealumination with 1 M HCl which helped to remove impurities in sugarcane bagasse ash [13]. The finding of XRF results in this study are in line with the research reported by Norsuraya et al. [13] which found that after acid treatment, the silica composition increased by 1.7 times from the initial composition of 53%. to 88%. This silica content is also in accordance with the research of Worathanakul et al. [14] which reported that the acid treatment method was able to increase the silica composition in ash with the obtained silica composition by 89%. In addition, research reported by Drummond and Drummond [15] that the $SiO_2$ composition obtained was 97% which is higher than the composition obtained in this study.

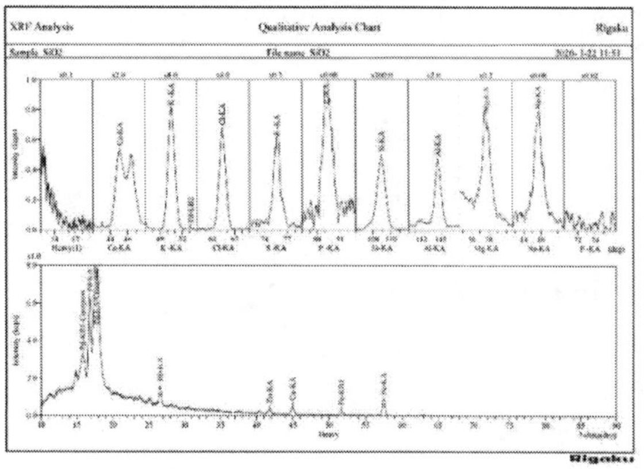


**Figure 1.** XRF spectra of sugar cane bagasse after acid treatment.

IOP Conf. Series: Materials Science and Engineering 1143 (2021) 012073 doi:10.1088/1757-899X/1143/1/012073

**Table 1.** XRF analysis data of sugarcane bagasse after acid treatment.

| Composition | %-weight |
|---|---|
| $Na_2O$ | 0,420 |
| MgO | 0,363 |
| $Al_2O_3$ | 0,618 |
| **$SiO_2$** | **91,7** |
| $P_2O_5$ | 0,171 |
| $SO_3$ | 0,182 |
| Cl | 0,740 |
| $K_2O$ | 4,86 |
| CaO | 0,587 |
| $Fe_2O_3$ | 0,249 |
| CuO | 0,0507 |
| ZnO | 0,0235 |
| $Rb_2O$ | 0,0255 |

### *3.2. Catalyst characterizations*

**Figure 2a** shows the XRD patterns $TiO_2/SiO_2$ composites with varying $TiO_2/SiO_2$ ratio compared with neat $SiO_2$ and $TiO_2$. The XRD pattern of neat $TiO_2$ samples shows peaks at $2\theta = 25.3°$, 37.9°, 48°, 54°, 55.1° and 62.7° which are characteristic of the (101), (004), (200), (105), (211), and (204) anatase phase crystal facets (JCPDS 84-1286), respectively, while peaks at $2\theta = 27.5°$, 36.1°, and 41.3° corresponded to the (110), (101), and (111) rutile phase crystal facets (JCPDS 88-1175). The XRD pattern of the neat $SiO_2$ was characterized by a broad peak at $2\theta = 21.5°$ corresponded to the characteristic of amorphous $SiO_2$ phase. In addition, the peak at $2\theta = 21.9°$, 28,3°, and 40.5° are corresponded to the presence of distributed $K_2O$ on the $SiO_2$ surface (JCPDS 47-1701) which is in agreement with the XRF results (**Table 1**). The diffraction peaks of $TiO_2$ with a main component of anatase phase and a small fraction of rutile phase can be observed for $TiO_2/SiO_2$ composites. The existence of amorphous $SiO_2$ phase was still observed in $1TiO_2$-$3SiO_2$ and $1TiO_2$-$1SiO_2$ composites indicating that there is a transformation of amorphous into crystalline phase by incorporating more $TiO_2$ content. The XRD patterns provide clear evidence that incorporation of Si atoms into the $TiO_2$ lattice, suggesting Ti-O-Si bonding is existed. The fraction of anatase and rutile phase was calculated by using Spurr equation. Based on the **Figure 2b**, it can be seen that the ratio of rutile phase fraction ($X_R$) to anatase phase fraction ($X_A$) increased by increasing $SiO_2$ content.

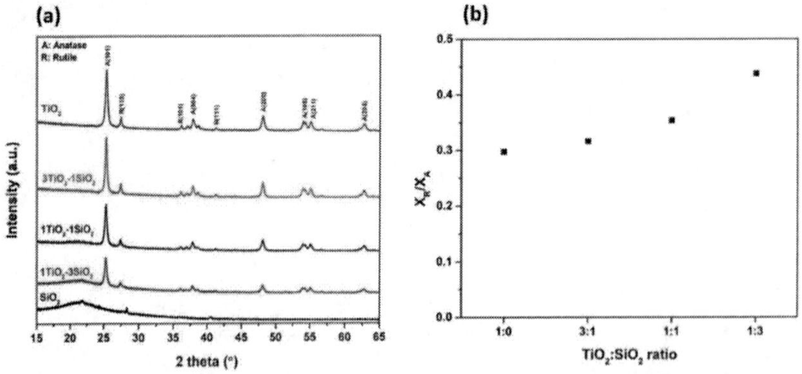


**Figure 2.** (a) XRD patterns and (b) ratio of rutile phase to anatase phase fraction ($X_A/X_R$) for $TiO_2$-$SiO_2$ composites compared to the neat $TiO_2$ and $SiO_2$.

**Figure 3** provide representative SEM images and particle sizes distribution of the $TiO_2/SiO_2$ composites compared with neat $TiO_2$ and $SiO_2$ catalysts. The SEM images of $SiO_2$ from sugarcane bagasse ash (**Figure 3a**) show that the particles were found to be irregular-shaped with pre-dominantly tubular-shaped and rich in porous particles with fibrous morphology. For the neat $TiO_2$ samples (**Figure 3b**), the particles cluster consisting of agglomerated nanoparticles and can be identified as a non-ordered and porous structures. In the case of $3TiO_2$-$1SiO_2$ (Figure 3c), $1TiO_2$-$1SiO_2$ (**Figure 3d**), and $1TiO_2$-$3SiO_2$ (**Figure 3e**) depicts that amorphous $SiO_2$ (in the form of mainly tubular-shaped) is interspersed and surrounded highly crystalline agglomerated spherical $TiO_2$ nanoparticles. It is also show that the majority of the $TiO_2$-$SiO_2$ particles are also well-dispersed nanospheres within a lightly compacted aggregate and with a broad particle size distribution. Larger tubular-like shape amorphous $SiO_2$ particles were present with $TiO_2$ particles are being observed on/in those particles.

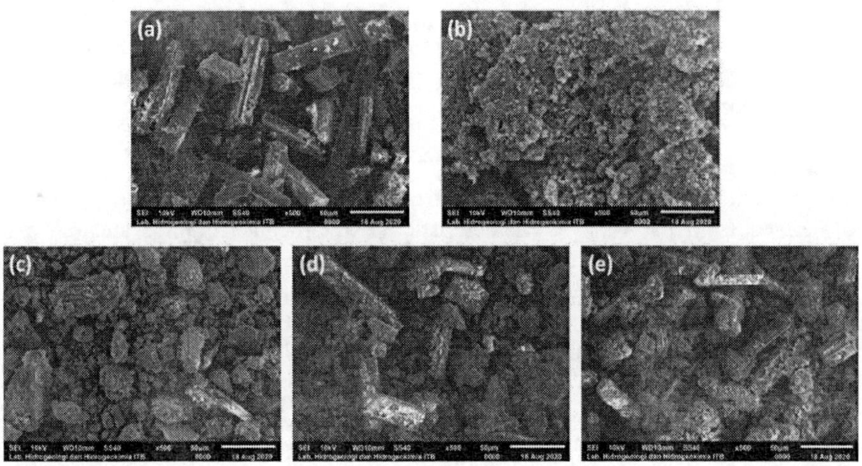


**Figure 3** SEM image of (a) $SiO_2$ (from sugarcane bagasse) (b) $TiO_2$ (c) $3TiO_2$-$1SiO_2$ (d) $1TiO_2$-$1SiO_2$ (e) $1TiO_2$-$3SiO_2$ in magnification of 500 times.

### *3.3. Catalytic activity assessment*

The photocatalytic activity of the $TiO_2/SiO_2$ composites synthesized at different $TiO_2/SiO_2$ ratio was evaluated in the decolorization reaction of Methyl Orange (MO) as depicted in **Figure 4**. **Figure 4a** clearly shows that MO is a very stable dye which is resistant to self-photo degradation. Neat $SiO_2$ exhibited negligible activity as expected given by the insulating properties of the material. Marginal activity was displayed by $TiO_2$. For $TiO_2/SiO_2$ composites, the photocatalytic activity of $1TiO_2$-$3SiO_2$, $1TiO_2$-$1SiO_2$, and $3TiO_2$-$1SiO_2$ enhanced by 1.8, 2.8, and 3-folds compared with neat $TiO_2$. The enhanced of photocatalytic activity might be due to several aspects: (i) $TiO_2/SiO_2$ composites provide higher adsorption capacity and specific surface area compared to neat $TiO_2$ which can promote strong surface adsorption ability to dye molecules and thus reactive radicals can easily react with dye molecules adsorbed on catalyst surface[16]. (ii) $TiO_2/SiO_2$ composite possess more hydroxyl group on the surface than neat $TiO_2$ which can capture photoinduced holes to generate reactive hydroxyl radicals[17]. (iii) $TiO_2/SiO_2$ composite promote generation of $Ti^{3+}$ which can lower band gap energy and trap photoinduced electron to superoxide radicals[9, 18, 19]. In addition, the role of hydroxyl and superoxide anion radicals can be determined and quantified using isopropyl alcohol (IPA) and benzoquinone (BQ) as quenchers/scavengers, respectively[20]. It has been proved that both radicals play significant roles in the photocatalytic degradation of methyl orange[21].

IOP Conf. Series: Materials Science and Engineering 1143 (2021) 012073 doi:10.1088/1757-899X/1143/1/012073

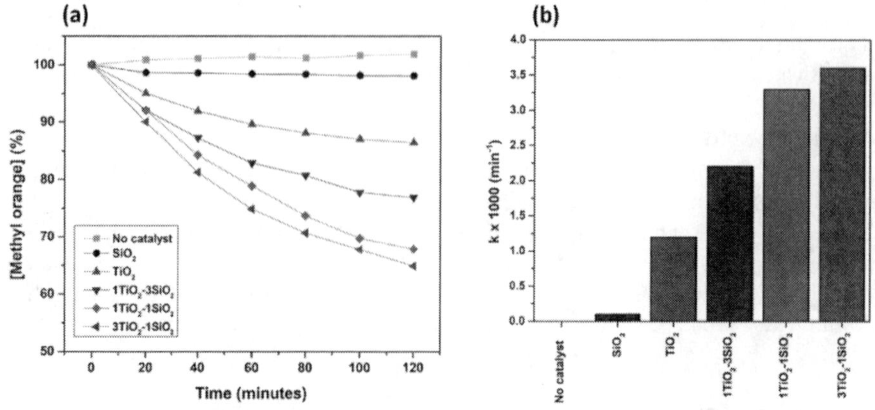


**Figure 4** (a) Photocatalytic degradation of methyl orange and (b) the apparent rate constant without a presence of catalyst (orange), neat $SiO_2$ (black), neat $TiO_2$ (red), $1TiO_2$-$3SiO_2$ (blue), $1TiO_2$-$1SiO_2$ (pink), and $3TiO_2$-$1SiO_2$ (green) under UV light irradiation.

### *3.4. Proposed mechanism*

A schematic depicting photocatalytic degradation of methyl orange over $TiO_2/SiO_2$ composite has been proposed, derived from the catalyst characteristics, experimental findings and relevant literatures. Initially, UV light promote the excitation of electron from valence band to the conduction band. Then, electron in the conduction band react with oxygen to generate superoxide radical ($\bullet O_2^-$) and holes in the valence band react with water to form hydroxyl radicals (•OH). Since the adsorption capacity of $TiO_2/SiO_2$ composites is higher than neat $TiO_2$ and neat $SiO_2$ [16], thus the methyl orange either in the interface of $TiO_2$ and $SiO_2$ particles or $SiO_2$ particles react with both radicals ($\bullet O_2^-$ and •OH). In addition, the $TiO_2/SiO_2$ composite possess more surface hydroxyl groups than the neat $TiO_2$ and neat $SiO_2$ sample so that surface hydroxyl groups can capture the photoinduced holes to produce active hydroxyl radicals and prevent electron–hole recombination, simultaneously[17]. This proposed mechanism clearly highlights that modification of $TiO_2$ photocatalyst with $SiO_2$ (sourced from sugarcane bagasse waste) could enhance photocatalytic efficiency.

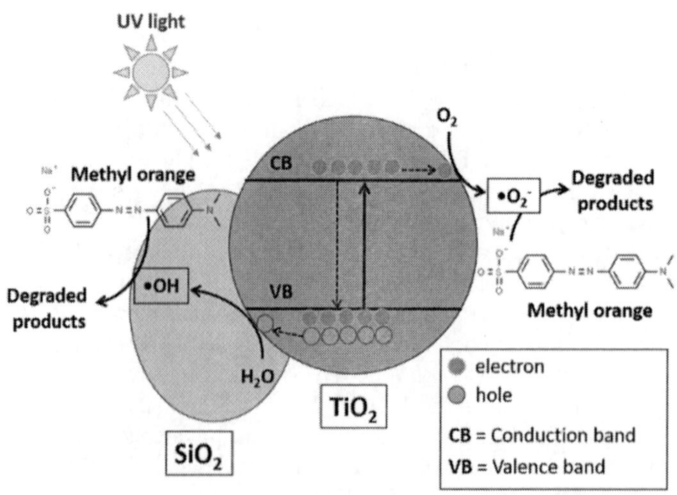


**Figure 5.** Schematic mechanism of photocatalytic degradation of methyl orange on $TiO_2/SiO_2$ composite.

## 4. Conclusion

$TiO_2$-$SiO_2$ composites was successfully synthesized by sol-gel method using sugarcane bagasse ash waste as a silica source. By increasing the addition of $TiO_2$ content, the transformation phase from amorphous to crystalline phase was obtained and amorphous $SiO_2$ is interspersed by highly crystalline $TiO_2$ nanoparticles. The small amount of silica addition, i.e. $3TiO_2$-$1SiO_2$ can enhance photocatalytic efficiency for MO degradation by 3-folds compared with the neat $TiO_2$. The study clearly demonstrates the power of renewable silica source (i.e. from sugarcane bagasse ash waste) and the potential it holds as a useful approach to advancement photocatalyst performance in the future research for efficient, cheap and abundant catalysts for photocatalytic degradation of organic pollutant.

## 5. References

[1] Selvam K, Swaminathan K, and Chae K S 2003 Decolourization of azo dyes and a dye industry effluent by a white rot fungus Thelephora sp. *Bioresour Technol* **88**(2) 115-9.
[2] Pant B, Park M, and Park S J 2019 Recent Advances in TiO2 Films Prepared by Sol-Gel Methods for Photocatalytic Degradation of Organic Pollutants and Antibacterial Activities *Coatings* **9**(10) 613
[3] Bel Hadjltaief H, et al. 2019 TiO2/clay as a heterogeneous catalyst in photocatalytic/photochemical oxidation of anionic reactive blue 19 *Arabian Journal of Chemistry* **12**(7) 1454-1462
[4] Al-Harbi L M, et al. 2015 The Photocatalytic Activity of TiO2-Zeolite Composite for Degradation of Dye Using Synthetic UV and Jeddah Sunlight *Journal of Nanomaterials* **2015** 565849
[5] Singh P, et al. 2016 Photocatalytic degradation of Acid Red dye stuff in the presence of activated carbon-TiO2 composite and its kinetic enumeration *Journal of Water Process Engineering* **12** 20-31
[6] Zaccariello G, et al. 2014 TiO2–mesoporous silica nanocomposites: cooperative effect in the photocatalytic degradation of dyes and drugs *RSC Advances* **4**(71) 37826-37837
[7] Ekka B, et al. 2019 Titania coated silica nanocomposite prepared via encapsulation method for the degradation of Safranin-O dye from aqueous solution: Optimization using statistical design *Water Resources and Industry* **22** 100071
[8] Yang D, et al. Biogenic Hierarchical TiO2/SiO2 Derived from Rice Husk and Enhanced Photocatalytic Properties for Dye Degradation *PLOS ONE* **6**(9) e24788
[9] Saputera W H, et al. 2020 Role of defects on TiO2/SiO2 composites for boosting photocatalytic water splitting *RSC Advances* **10**(46) 27713-27719
[10] Saputera W H, et al. 2019 Cooperative defect-enriched SiO2 for oxygen activation and organic dehydrogenation *Journal of Catalysis* **376** 168-179
[11] Saputera W H, et al. 2019 Light-Induced Synergistic Multidefect Sites on TiO2/SiO2 Composites for Catalytic Dehydrogenation *ACS Catalysis* **9**(3) 2674-2684
[12] Fatimah I, Sopia L 2017 Preparation of TiO2/MCM-41 photocatalyst using rice husk ash as silica source *AIP Conference Proceedings* **1823**(1) 020124
[13] Norsuraya S, Fazlena H, and Norhasyimi R 2016 Sugarcane Bagasse as a Renewable Source of Silica to Synthesize Santa Barbara Amorphous-15 (SBA-15) *Procedia Engineering* **148** 839-846
[14] Patcharin Worathanakul W P, A M 2009 Charcterization for post treatment effect of bagasse ash for silica extraction *World academy of science, Engineering and Technology* **3**(8) 398-400
[15] Drummond A.-R F and Drummond I W 1996 Pyrolysis of Sugar Cane Bagasse in a Wire-Mesh Reactor *Industrial & Engineering Chemistry Research* **35**(4) 1263-1268
[16] Hoffmann M R, et al. 1995 Environmental Applications of Semiconductor Photocatalysis *Chemical Reviews* **95**(1) 69-96
[17] Fu X, et al. 1996 Enhanced Photocatalytic Performance of Titania-Based Binary Metal Oxides: TiO2/SiO2 and TiO2/ZrO2 *Environmental Science & Technology* **30**(2) 647-653

[18] Hamdy M S, et al. 2014 A novel TiO2 composite for photocatalytic wastewater treatment *Journal of Catalysis* **310** 75-83
[19] Saputera W H, Mul G, and Hamdy M S 2015 Ti3+-containing titania: Synthesis tactics and photocatalytic performance *Catalysis Today* **246** 60-66
[20] Nosaka Y and Nosaka A Y 2017 Generation and Detection of Reactive Oxygen Species in Photocatalysis *Chemical Reviews* **117**(17) 11302-11336
[21] Guo Y, et al. 2011 Enhanced Photocatalytic Activity for Degradation of Methyl Orange over Silica-Titania *Journal of Nanomaterials* **2011** 296953

**Acknowledgements**
The authors would like to thank ASAHI Glass Foundation 2020 and Riset ITB Multidisiplin 2020 (FTI-PN-6.022020) for providing financial support for this research.

IOP Conf. Series: Materials Science and Engineering 1143 (2021) 012074 doi:10.1088/1757-899X/1143/1/012074

# Optimization of UV light source conditions for photocatalytic activity of methyl orange using $TiO_2$

**S O Sativa[1*], M A Zulfikar[1] and A Alni[2]**

[1] Analytical Chemistry Research Group, Department of Chemistry, Institut Teknologi Bandung, Bandung 40135, Indonesia
[2] Organic Chemistry Research Group, Department of Chemistry, Institut Teknologi Bandung, Bandung 40135, Indonesia

*sitioryzasativa@gmail.com

**Abstract**. Methyl orange is one of the azo compounds which carcinogenic and damages the water system if untreatead properly. This compound is used in the textile industry as a coloring agent. Dyestuff waste produced from the textile industry is generally a non-biodegradable organic compound, which cause environmental pollution, especially at the aquatic environment. As organic waste, it is necessary to treat this compound to be harmless and safe for the environment. One of the method to remove dye waste is by using photocatalytic reaction using $TiO_2$ as semiconductor. In this study, optimization of UV light source conditions of the photocatalytic reaction was carried out inside a photoreactor system using $TiO_2$ synthesized by hydrothermal method. Optimization of the distance of UV lamps from the surface of methyl orange solution and UV lamps power has been done under the artificial UV light sources with 254 nm wavelength. The percentage of degradation was determined by measuring the absorbance of methyl orange solution before and after degradation using UV-Vis spectrophotometer. It shows that the optimum photocatalytic activity for the distance of UV lamps and UV lamps power is 20 cm and 20 Watt respectively. Physically, there was color changing during the reaction from orange to colorless.

## 1. Introduction

The textile industry has recently been growing rapidly because clothing is a primary need for humans. The development of this industry has a negative impact on pollution water, because the dyeing process waste containing dye from the industry is discharged into rivers thus changing the quality of the water. The textile industry that does not have good waste treatment can have an impact on environmental pollution [1,2], specially the process of dyeing and staining when the dyed waste is just thrown away without any more processing. One of the methods of treating waste from the dyeing process in the textile industry is to use a photocatalyst [3,4,5]. The semiconductor material that can be used as a photocatalyst is $TiO_2$, whose structure and characteristics can be changed. Photocatalyst is defined as a compound whose function is to accelerate a chemical reaction which is activated by photon absorption. The compounds that are used as photocatalysts are semiconductors with various types. Photocatalyst semiconductors have come to the fore over the past three decades as solutions to environmental and energy problems. The semiconductor materials that are usually used as photocatalyst compounds are ZnO [6], $Fe_2O_3$ [7], CdS [8], ZnS [9], $SnO_2$ [10], $WO_3$[11], $Al_2O_3$[12], $Ag_3PO_4$[13] and BiOCl[14]. The ideal photocatalyst is one that is stable to light, does not react chemically, and can be provided at an inexpensive price [15]. The photocatalytic process has great potential as an inexpensive technology that is good for the environment and is compatible with the concept of zero waste. This means that this technology is environmentally friendly and produces reaction products that are not dangerous. One of the semiconductor materials that is most often used as research material as a photocatalyst recently is $TiO_2$ because of its stable nature to treat pollutants and organic waste. In addition, $TiO_2$ has advantages to be photocatalyst because it has high photocatalytic activity, excellent stability on physical and chemical, low cost, non-corrosive, nontoxicity and reusability [16,17]. Reactions involving photocatalytic compounds are called photocatalytic reactions. The basis of photocatalysis is photoexitation as a result of absorption of radiation. Photocatalyst technique is a photochemical process combined with an integrated catalyst to carry out a chemical reaction. This chemical treaction takes place on the surface of the photocatalyst material involving light source which comes from photons with certain energy. In other words the process photocatalysts can decompose compounds with the help of light. Basic mechanism from this process is formation of electron-hole pairs on the surface semiconductor catalyst when induced by the appropriate photon energy [18].

## 2. Research method

This research was conducted at laboratory of Organic Chemistry, Department of Chemistry, Institut Teknologi Bandung. Material $TiO_2$ that used in this research was synthesized by hydrothermal method. $TiO_2$ was synthesized using titanium (IV) etoxide (pro analysis grade, Sigma-Aldrich, USA) as precursor and isopropyl alcohol (pro analysis grade, Sigma Aldrich, USA) as solvent. In addition, methyl orange (Merck, Germany) was used as analyte. All chemicals were purchased and used without further purification.

Optimization of photocatalytic activities in this research was done by the procedure below. For the distance of UV lamps to the surface of analyte optimization, a total of 50 mL methyl orange solution 10 ppm was put into five 250 mL beaker glasses with pH 5. Into each beaker glass containing methyl orange solution were added 0.03 g of $TiO_2$. Each beaker glass was put inside a photoreactor system (Figure 1) where the photodegradation process took place. The mixture was stirred for 20 minutes under dark condition. Then, it was irradiated under 10-40 Watt UV light sources for 3 hours with 20 cm lamp distance from the surface of solution. The results of photodegradation process were centrifuged to separate the precipitate particles. For UV lamps power optimization, a total of 50 mL methyl orange solution 10 ppm and pH 5 was put into four 250 mL beaker glasses. The paramaters were the same as the distance of UV lamps optimization method. To optimize the UV lamps power, it was adjusted to 20-40 cm. Finally, the remaining methyl orange concentration was determined by using UV-Vis spectrophotometer at 467 nm wavelength by calculating the percentage of degradation using this following equation,

$$\%D = \frac{C0-C}{C0} x\, 100\% \quad (1)$$

Where C0 is the initial concentration of methyl orange in ppm and C is the concentration of methyl orange after photo degradation process in ppm.

The sample were characterized by scanning electron microscope (SEM JEOL JSM-6360LA), X-ray diffraction (XRD Shimadzu XRD-7000 Maxima), diffuse reflectance spectroscopy (DR-UV Thermo Scientific Evolution 220) and brunauer emmett teller (BET Quantachrome Quadrasorb SI)

IOP Conf. Series: Materials Science and Engineering 1143 (2021) 012074 doi:10.1088/1757-899X/1143/1/012074

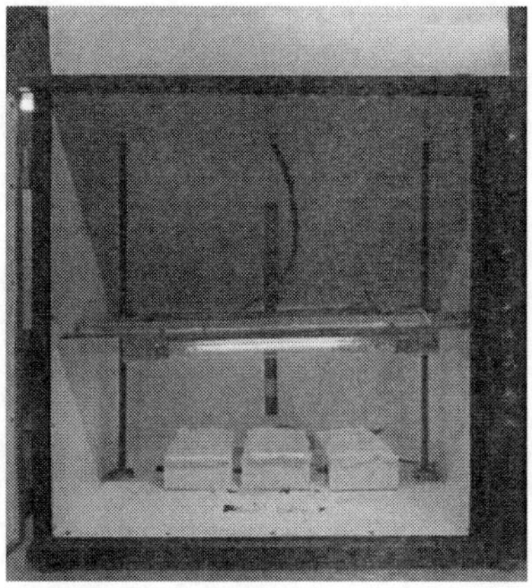

**Figure 1.** Reactor system for photocatalytic process.

## 3. Results and discussion

3.1 Characterization of Synthesized Material.

SEM characterization was done to find out the morphology of synthesized $TiO_2$. Figure 2 shows the SEM images of synthesized $TiO_2$ in 5000x and 25.000x zoom.

**Figure 2**. SEM characterization

The crystallinity degree of synthesized $TiO_2$ is 45,1% by using the X-ray diffractogram characterization. Figure 3 shows the diffractogram of synthesized $TiO_2$. It can be seen that this synthesized $TiO_2$ still has rutile on its structure.

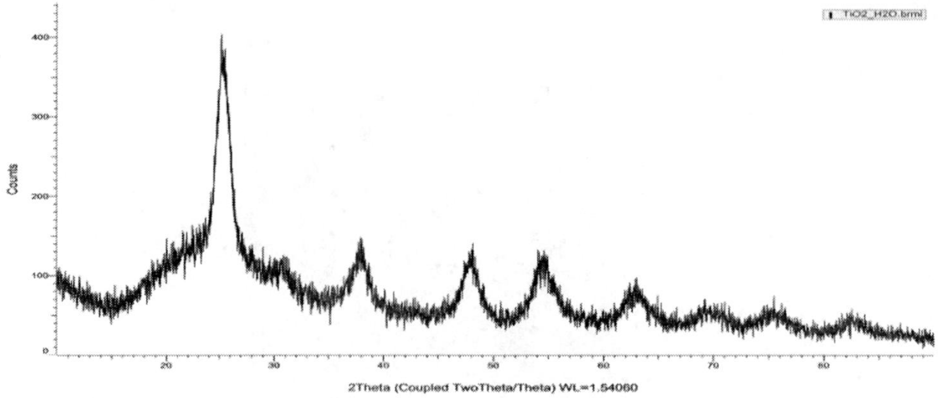


**Figure 3**. XRD diffractogram.

The band gap value of synthesized $TiO_2$ was calculated by Kubelka-Munk formula and DR-UV characterization.

$$F(R) = \frac{(1-R)^2}{2R} \quad (2)$$

with F(R) is Kubelka-Munk factor and R is reflectan.

From the DR-UV data result, the band gap of synthesized $TiO_2$ is 2,98 eV. It means that the range of light source to initiate the photocatalytic process is in the UV light source area. If the band gap value is smaller, the larger the wavelength range of the rays means that photocatalytic reactions can be carried out in visible light [19].

BET characterization was done to find out the surface are, total pore volume, and average pore size. Table 1 shows the result data of BET characterization. Surface area and pore size affect the photocatalytic activity of $TiO_2$. The larger the surface area and the pore size of the photocatalyst, the higher the photocatalytic activity because the larger the active side is used during the photocatalytic process [20].

**Table 1**. BET Characterization.

| $TiO_2$ | Surface Area ($m^2$/gram) | Total Pore Volume (cc/gram) | Average Pore Size (nm) |
|---|---|---|---|
| Synthesized $TiO_2$ | 204, 431 | 0,3184 | 5,96 |

3.2. Optimization of Photocatalytic Activity

Optimization of lamp power is carried out by varying the power of lamps starting from 10 to 40 Watt. From the research result, the optimum lamp power to degrade the methyl orange solution was 20 Watt as shown at Figure 4. The greater the lamp power, the greater percent of degradation of methyl orange solution. UV rays caused electrons in the valence band excited to the conduction band which results in positive holes and free radicals. The greater power of the light that initiates photochemical reaction, the higher photocatalytic activity because the electron pairs are formed quickly.

IOP Conf. Series: Materials Science and Engineering 1143 (2021) 012074 doi:10.1088/1757-899X/1143/1/012074

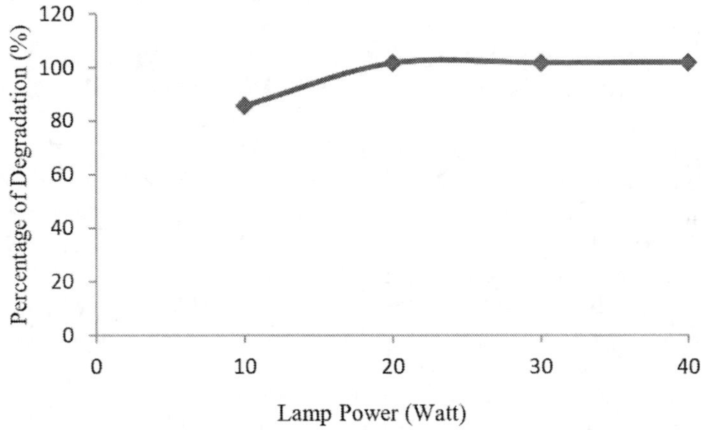


**Figure 4.** Optimization of lamp power on photocatalytic activity of $TiO_2$.

Optimization of the distance of the lamp was done by varying the distance of lamps from the surface analytie starting from 20 to 50 cm. From the research results, the optimum lamp distance to degrade the methyl orange solution was 20 cm as shown at Figure 5. The closer the lamp distance, the greater the degradation percent of the methyl orange solution. UV rays will cause electrons in the valence band to be excited to the conduction band which results in positive holes and free radicals. The closer the distance between the lamps that initiate the photochemical reactions, the higher the photocatalytic activity because the electron pairs are formed quickly.

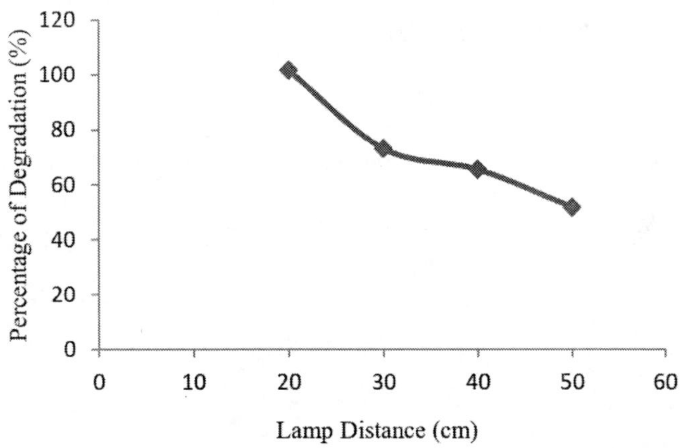


**Figure 5.** Optimization of lamp distance on photocatalytic activity of $TiO_2$.

## 4. Conclusion

Photocatalytic activity of synthesized $TiO_2$ for the photodegradation of methyl orange solution had been carried out. From the results of the study, it was found that the optimum condidition of the UV lamps distance to the surface of analyte was 20 cm and the UV lamps power was 20 Watt.

**Acknowledgments**
The authors are highly thankful to DIKTI, Kementerian Pendidikan dan Kebudayaan Republik Indonesia.

## References

[1] Lellis B, Favaro-Polonio C F, Pamphile J A and Polonio J C 2019 *Biotechnology Research and Innovation* **3(2)** 275-90.
[2] Sandin G and Peters G M 2018 *Journal of Cleaner Production* **184** 353-65.
[3] Al-Mamun M R, Kader S, Islam M S and Khan M Z H 2019 *Journal of Environtmental Chemical Engineering* **7(5)** 103248.
[4] Rajeshwar K, Osugi M E, Chanmanee W, Chenthamarakshan C R, Zanoni M V B, Kajitvichyanukul P and Krishnan-Ayer R 2008 *Journal of Photochemistry and Photobiology C: Photochemistry Reviews* **9(4)** 171-92.
[5] Al-Zahrani F A M, El-Shishtawy R M, Ahmed N S E, Awwad N S, Hamdy M S and Asiri A M 2020 *Arabian Journal of Chemistry* **13(2)** 3633-38.
[6] Nazarkovsky M A, Bogatyrov V M, Czech B, Galaburda M V, Wojcik G, Kolomys O F, Strelchuk V V, Malysheva M L, Oranska O I and Gun'ko V M 2017 *Journal of Photochemistry and Photobiology A: Chemistry* **334** 36-46.
[7] Mishra M and Chun D M 2015 *Applied Catalysis A: General* **498** 126-41.
[8] Zhen W, Ning X, Yang B, Wu Y, Li Z and Lu G 2018 *Applied Catalysis B: Environmental* **221** 243-57.
[9] Ye Z, Kong L, Chen F, Chen Z, Lin Y and Liu C 2018 *Optik* **164** 345-54.
[10] Kim S P, Choi M Y and Choi H C 2016 *Materials Research Bulletin* **74** 85-89.
[11] Dong P, Hou G, Xi X, Shao R and Dong F 2017 *Environmental Science: Nano* **4** 539-557.
[12] Asif S A B, Khan S B and Asiri A M 2015 *Nanoscale Research Letters* **10** 1-10.
[13] Huang G, Ma Z, Huang W, Tian Y, Jiao C, Yang Z M, Wan Z and Pan A 2013 *Journal of Nanomaterials* **2013** 1-8.
[14] Wang Q, Hui J, Huang Y, Ding Y, Cai Y, Yin S, Li Z and Su B 2014 *Materials Science in Semiconductor Processing* **17** 87-93.
[15] Gaya U I and Abdullah A H 2008 *Journal of Photochemistry and Photobiological Chemistry* **9** 1-12.
[16] Chen X and Mao S S 2007 *Chemical Reviews* **107** 2891-2959.
[17] Nakata K and Fujishima A 2012 *Journal of Photochemistry and Photobiology C: Photochemistry Reviews* **13** 169-189.
[18] Regmi C, Joshi B, Ray S K, Gyawali G and Pandey R P 2018 *Frontiers in Chemistry* **6** 1-6
[19] Dette C, Perez-Osorio M A, Kley C S, Punke P, Patrick C E, Jacobson P, Giustino F, Jung S J and Kern K 2014 *Nano Letters* **14** 6533-6238.
[20] Yu J G, Wang, W G and Cheng B 2010 *Chemistry* **5** 2499-2506.

# Improved Visible Light Activity of Copper Oxide/Carbon Nitride Photocatalysts Prepared by Photodeposition for Phenol Degradation

**C P Sagita[1] and L Yuliati[1,2*]**

[1]Ma Chung Research Center for Photosynthetic Pigments, Universitas Ma Chung, Malang, 65151, Indonesia
[2]Department of Chemistry, Faculty of Science and Technology, Universitas Ma Chung, Malang, 65151, Indonesia

*Email: leny.yuliati@machung.ac.id

**Abstract.** A series of copper(II) oxide deposited on carbon nitride (CuO/CN) as visible light-active photocatalyst was successfully prepared via a photodeposition method. The CuO modification was carried out to improve the photocatalytic activity of CN having fast electron-hole recombination. The CuO loading was varied from 0.05 to 0.5 wt% and the photodeposition was carried out at room temperature under UV light illumination. The CuO/CN samples were confirmed to have similar optical properties and functional groups to those of the unmodified CN. Meanwhile, the emission intensity of the CN decreased with the increase of the copper species loading, which could correspond to the suppression of charge recombination on the CN. After a 24-h reaction under visible light irradiation, the best photocatalyst, *i.e.* the CuO(0.1)/CN, gave 41.7% phenol degradation, which was almost two times higher than the unmodified CN (21.3%). However, the photocatalytic activity decreased when the added copper species was more than 0.1 wt%. Therefore, the optimum amount of copper species deposited on the CN surface would have a prominent contribution to improve the photocatalytic activity. It was also confirmed that holes, hydroxyl radicals, and superoxide radicals were important for the photocatalytic degradation of phenol on the CuO(0.1)/CN.

## 1. Introduction

The role of photocatalytic reaction in the toxic organic pollutant degradation has gained much interest in the last decades, as this reaction could be performed under mild conditions by solar light irradiation [1]. It is proven that this safe process allows the organic pollutant to be decomposed into carbon dioxide, water, and benign minerals by involving reactive oxygen species (ROS) [2], thus simplify the further steps in water treatment. As this green technology process has the potential to be applied in practical using sunlight irradiation, the selection of suitable photocatalyst that can work effectively in the wider visible light region becomes a crucial approach. Appearing as the yellow colored-solid, carbon nitride (CN) photocatalyst is considered to be an active photocatalyst under visible light illumination having light absorption ability up to the blue light region (ca. 460 nm) [3,4]. Moreover, the CN also exhibits good resistance both in the various chemical attack conditions and in the heating thermal up to 600°C, implying its high chemical and thermal stability that are raised from its graphite-analog structure consisting of strong C-N bonds [5,6].

Unfortunately, the photocatalytic activity of CN under visible light irradiation is still limited by its rapid recombination rate of electron-hole pairs and narrow light absorption bandwidth [7]. Therefore, the modification of CN is importantly needed. For this reason, the development of a CN-based heterostructure has been widely adopted to repress these shortcomings [8,9]. Among the various CN composite heterostructures, the metal species/CN composites have been frequently reported to suppress the recombination of photogenerated charges, which subsequently improved the photocatalytic activity of the CN [10-12]. Since there is a different position of the conduction band (CB) and valence band (VB) between metal species and CN in the composite structure, the suitable band levels from that difference would produce a driving force for efficient transfer and separation process of electron-hole pairs [11]. In this case, the metal species would be considered as the electron trapper from the conduction band of CN [13], which could reduce the recombination of electron-hole pairs and thus would give a beneficial impact on the photocatalytic performance.

Particularly for photodegradation of organic pollutants under visible light irradiation, it has been reported that introducing the noble metal/CN [13] and other metal species in their oxide forms [14-16] can enhance the photocatalytic activity of CN in comparison to the unmodified CN. Apart from these reported metal species, copper is attractive as the potential metal species owing to its inexpensive, non-corrosive, abundant availability on the earth, and similar properties with gold and silver metals [17]. This makes copper species a promising material for various energy applications [18, 19], including in the photocatalysis field [20-29]. Moreover, the narrow bandgap of CuO (1.3–2.1 eV) indicates their ability to absorb visible light, supporting its use as a favorable semiconductor for the photocatalytic reaction [20]. Therefore, in terms of being composited with the CN, the CuO/CN has been reported as an effective photocatalyst for degradation of dyes [21], thermal decomposition of ammonium perchlorate [22], photoreduction carbon dioxide [23], and degradation of salicylic acid [24].

Common methods to prepare CuO/CN are using hydrothermal [21], facile precipitation [22], and impregnation [30], which all of them need heat treatment. The use of heat treatment would cause aggregation of the metal oxide and if the temperature of the heat treatment is too high, the decomposition of the CN could not be avoided. It has been reported previously that the photodeposition method could produce $Fe_2O_3/TiO_2$ nanocomposite with better catalyst properties and superior photocatalytic activity for the degradation of 2,4-dichlorophenoxyacetic acid as compared to the impregnation method which involved a heat treatment process [31]. The Ag/CN composite prepared by photodeposition method also showed an increased photocatalytic activity under visible light irradiation in comparison to the unmodified CN [13]. Hence, the photodeposition method without employing the heat treatment can be suggested to prepare the CuO/CN catalyst.

When using the photodeposition method, the formed copper species on the titanium dioxide ($TiO_2$) was clarified to be highly dispersed of Cu(II) species [25]. In this study, we studied the effect of depositing CuO to the CN surface by the photodeposition method on the properties and photocatalytic activity of the CN. The photocatalytic activity was evaluated under visible light irradiation for the degradation of phenol, which is known as a toxic compound for human health and the environment [32]. We observed that the optimum amount of CuO deposited on the CN surface resulted in high photocatalytic activity, and thus, the prominent contribution of CuO optimum amount was discussed. Besides, the scavenging experiment was also studied to understand the important role of active species contributing to the photocatalytic activity of the CuO/CN photocatalyst and propose the charge transfer mechanism involved in the reaction.

## 2. Materials and Methods

### *2.1. Materials*

All chemicals used in this study were available commercially and did not receive any purifications before the usage. Urea ($CO(NH_2)_2$, 98%, Sigma-Aldrich) was used as the precursor for the synthesis of the CN, while copper(II) nitrate trihydrate ($Cu(NO_3)_2.3H_2O$, 99–104%, Sigma-Aldrich) was used as the copper precursor for the preparation of CuO/CN samples. Phenol ($C_6H_5OH$, 99.5%, Merck) was used as the pollutant model in the photocatalytic activity test. For the scavenging experiment, ammonium oxalate ($(NH_4)_2C_2O_4.H_2O$, 99.5–101%, Merck) was used as a hole species scavenger, *tert*-butanol ($C_4H_9OH$, 99.5%, Merck) was used as a hydroxyl radical scavenger, and ascorbic acid ($C_6H_8O_6$, 99.5%, Chameleon Reagent, Kishida Chemicals) was used as a superoxide radical scavenger.

### 2.2. *Preparation of CN and CuO (x)/CN photocatalysts*
CN was synthesized by heating urea (25 g) at 550 °C under a limited air atmosphere for 4 h with a heating rate of 2.2 °C $min^{-1}$ to reach the final temperature of 550 °C. The yellow solid was obtained after the calcination and then was ground into fine powder to be used as the photocatalyst. The CuO/CN composites were prepared by the photodeposition method. The CN photocatalyst (0.1 g) was dispersed in 20 ml of water/methanol mixture (1:1 v/v). The required amount of copper species from the copper(II) nitrate trihydrate solution was then added to the mixture. For making a homogeneous suspension, the mixture was sonicated for 1 min, followed by stirring under irradiation of ultraviolet (UV) lamp (UVLS-28 EL Series, 8 W, at 365 nm) from the top at room temperature for 5 h. Afterward, the separated solid of CuO/CN composite was washed a few times by using methanol and distilled water. The solid was finally dried in an oven at 80 °C overnight. The prepared samples were labeled as CuO(x)/CN, where x represents the ratio of added CuO to CN *i.e.*, 0.05, 0.1, 0.2, and 0.5 wt%.

### 2.3. *Characterization of Photocatalysts*
Fourier transform infrared (FTIR) spectra of all CN samples were measured by JASCO FT/IR-6800 spectrometer using attenuated total reflection (ATR) technique in the scan range of 400–4000 $cm^{-1}$, where the air was recorded as the baseline. The diffuse reflectance ultraviolet-visible (DR UV-vis) spectra were obtained by JASCO V-760 UV-vis spectrophotometer in the scan range of 225–600 nm with scan speed 100 nm $min^{-1}$, at room temperature. For the fluorescence properties, both excitation and emission spectra were measured by JASCO FP-8500 spectrofluorometer with very low sensitivity mode. The excitation spectra were monitored at the emission wavelength of 442 nm, while the emission spectra were monitored at three excitation wavelengths of 277, 331 nm, and 369 nm.

### 2.4. *Photocatalytic Activity Test*
Photocatalytic degradation of phenol under ambient conditions was used to evaluate the photocatalytic activity of the CN and the CuO/CN composites. In the 100 ml of a round bottom flask, the photocatalyst (50 mg) was dispersed in 50 ml of aqueous phenol (70 mg $L^{-1}$). To reach the adsorption equilibrium between the photocatalyst and phenol before visible-light illumination, the mixture was stirred under the dark condition for 2 h. While the mixture was stirred constantly, the mixture was then exposed to a visible-light lamp (Dolan-Jenner Fiber Lite MI-157, 150 W halogen lamp) for 24 h. Upon the completion of visible-light lamp exposure, the solution was separated from the photocatalyst by using a filter syringe equipped with a 0.2 μm nylon membrane filter. The remaining concentration of phenol after the reaction was determined using a high-performance liquid chromatography instrument (HPLC Shimadzu LC-20AT, with a photodiode array (PDA) detector and column C-18). The eluent used in this HPLC analysis was acetonitrile 100% and the concentration of phenol was monitored at 272 nm. The percentage of phenol degradation was determined by using the following equation (1), where $C_0$ is the initial concentration of phenol and $C$ is the concentration phenol after reaction under visible light irradiation.

$$\text{Percentage of phenol degradation (\%)} = \frac{C_o - C}{C_0} \times 100 \quad (1)$$

The investigation of the responsible active species for phenol photocatalytic degradation was carried out using a scavenging experiment on the CuO(0.1)/CN composite, which showed the best photocatalytic activity in this work. Scavenger agents used in this experiment were ammonium oxalate, *tert*-butanol, and ascorbic acid for scavenging holes, hydroxyl radicals, and superoxide radicals, respectively. The photocatalytic activity test in the presence of scavenger agents was conducted under irradiation of the visible light lamp (Dolan-Jenner Fiber Lite MI-150, 150 W halogen lamp) for 24 h. It was noted here that the light intensity used for this scavenging experiment was lower than that used for

IOP Conf. Series: Materials Science and Engineering 1143 (2021) 012075 doi:10.1088/1757-899X/1143/1/012075

photocatalytic evaluation. However, it was confirmed that the trend of the activity remained the same and the light intensity difference would not affect the content of the discussion. To ensure that the scavenger agents would react to the active species formed during the photocatalytic reaction, an excess amount of scavenger agent was added into the phenol solution, which was ten times from that of the phenol amount in mol ratio.

## 3. Results and Discussions

### *3.1. Photocatalysts Properties*

Structural properties of CN and CuO/CN samples in this study were investigated by FTIR spectroscopy. From the FTIR spectra shown in Figure 1, the functional groups of the CN and its composites were identified by the appearance of some vibration bands. As the CN structure consists of a triazine ring with single and double bonds between carbon and nitrogen, the bending mode of heterocyclic 1,3,5-triazine ring was observed at 810 $cm^{-1}$, while the C-N heterocycles stretching mode was observed at 1200–1640 $cm^{-1}$ [4,33], where the C=N stretching mode appeared at 1632 $cm^{-1}$. Furthermore, the stretching mode of the uncondensed C=O group from the urea precursor was also detected at 1680 $cm^{-1}$. The observed vibration bands around 3000–3300 $cm^{-1}$ were attributed to the presence of an uncondensed amino group (N–H bond) and the hydroxyl group of adsorbed water (O–H bond) on the CN structure [4,33,34]. Therefore, the FTIR spectra confirmed the successful formation of the CN structure on the CN and CuO/CN samples. While CuO nanoparticle was reported to give a vibration band at 532 $cm^{-1}$ [35], such vibration band could not be observed on the CuO/CN samples. This could be due to the low amount of added copper species on the surface of the CN. The addition of copper species did not give a significant effect on the functional groups of the CN structure, as have been reported in other literature [22,24,36].

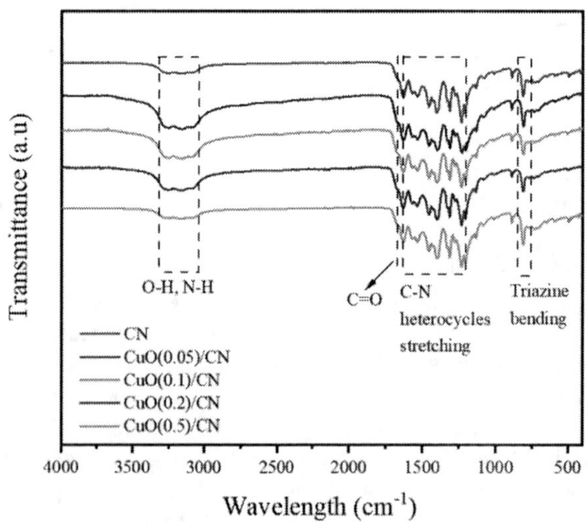


**Figure 1.** FTIR spectra of the unmodified CN and all CuO/CN composites.

The optical absorption properties of unmodified CN and CuO/CN samples were investigated by DR UV-vis spectroscopy and the results are shown in Figure 2(a). All the samples showed light absorption in both UV and visible regions up to the wavelength of 430 nm, which was associated with the blue light absorption [37]. Two strong absorption peaks of the CN sample appeared at 287 and 370 nm, which could be corresponded to the $\pi \rightarrow \pi^*$ charge transfer at the C=N groups in the aromatic ring of triazine and $n \rightarrow \pi^*$ charge transition at C–N terminal of triazine unit, respectively. On the other hand, a weak absorption peak was observed at 330 nm, which could be related to the presence of charge transfer ($\pi \rightarrow \pi^*$ and $n \rightarrow \pi^*$) at the uncondensed C=O group in the triazine unit [33,36,37].

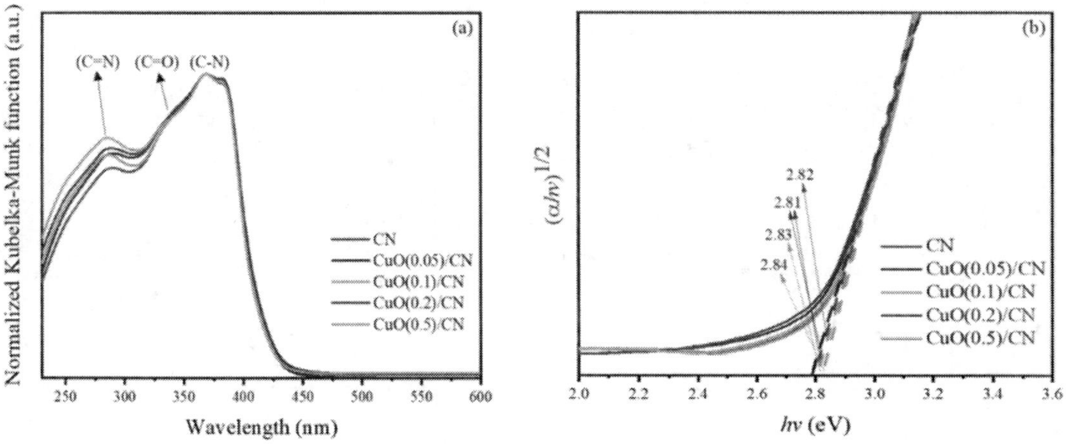


**Figure 2.** (a) DR UV-vis spectra and (b) Tauc plots of the unmodified CN and all CuO/CN composites.

The CuO/CN samples also showed three similar absorption peaks to those of the unmodified CN. It could be observed that the presence of copper species did not extend the absorption edge of the CN to a larger wavelength as the photodeposition method would only deposit a small amount of the copper species on the surface of the photocatalyst to support the photocatalyst activity [38]. However, the absorption peaks below 350 nm increased gradually along with the increase in the copper species loading amount. These increased absorption intensities suggested the presence of copper species on the surface of CN since the absorption peaks below 350 nm could be associated with the charge transfer transition between $Cu^{2+}$ and $O^{2-}$ species [36,39].

As shown in Figure 2(b), the bandgap energies ($E_g$) of CN and CuO/CN samples were evaluated by Tauc plot considering the samples have the indirect bandgap. The linear extrapolation from the $x$-intercept in the plot gave the $E_g$ values. The CN sample gave an $E_g$ value of 2.82 eV, while the $E_g$ values of CuO/CN samples were found to be similar to the CN sample, which was in the range of 2.81–2.84 eV. This result again indicated that the copper species added by the photodeposition method were deposited on the surface of the CN and thus, would not alter the bandgap energy of the CN, similar to the previously reported result on $TiO_2$ [25].

The role of copper species as the electron trapper in suppressing the charge recombination on the CN photocatalyst was studied by fluorescence spectroscopy. As shown in Figure 3(a), the CN and CuO/CN samples exhibited three excitation peaks at 277, 331, and 369 nm, which were associated with the charge transfer at C=N, C=O, and C–N, respectively, in good agreement with their absorption peak from DR UV-vis spectra [33,36]. Meanwhile, the emission peaks from these three excitation peaks were observed only at 442 nm, which were illustrated in Figure 3(b)–(d) for each excitation wavelength of 277, 331, and 369 nm. The intensities of both the excitation and emission spectra of all CuO/CN samples were found to decrease progressively as the loading amount of copper species increased. The lower intensities of CuO/CN samples as compared to the unmodified CN denoted that there were obvious interactions between the copper species and the CN [36]. This phenomenon is well-agreed with the results from the previous literature, which reported that the copper species would act as the electron trap in the photocatalytic reaction system [17,25,40]. The copper species could give a pathway to trap the photogenerated electrons from the conduction band of CN, preventing the recombination with holes and then would lead to the improved phenol photocatalytic degradation.

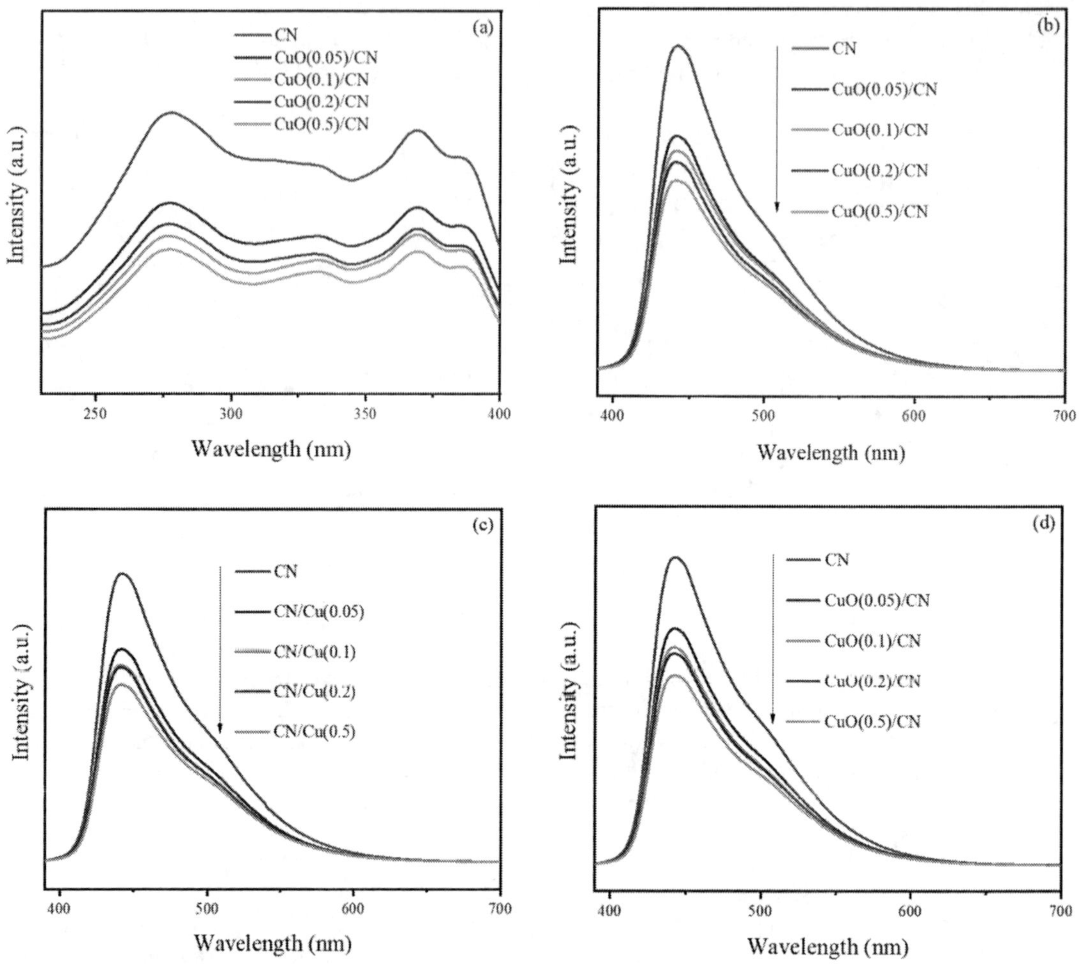


**Figure 3.** (a) Excitation spectra monitored at the emission wavelength of 442 nm and emission spectra monitored at the excitation wavelength of (b) 277 nm, (c) 331 nm, and (d) 369 nm for the unmodified CN and all CuO/CN composites.

Since all the emission sites of the CN were quenched after the addition of CuO, the favorable interaction site between the copper species and the CN structure was further investigated. The relative emission intensity ($I_o/I$) of CN and its composite was evaluated from the ratio of CN emission intensity ($I_o$) to the CuO/CN emission intensity (I) from three excitation wavelengths of 277, 331, and 369 nm for C=N, C=O, and C–N sites. As displayed in Figure 4, the highest $I_o/I$ ratio was observed on CuO(0.5)/CN for all emission sites, while for each CuO/CN sample, the highest $I_o/I$ ratio was observed at the C=N site. This result indicated that the most favorable interactions between copper species and CN occurred at the C=N site, which was consistent with the reported result from other work [36].

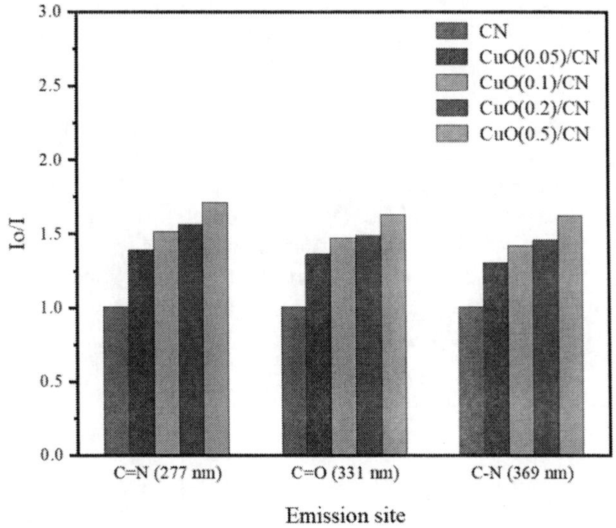


**Figure 4.** Relative emission intensity ($I_o/I$) of the unmodified CN and all CuO/CN composites.

### *3.2. Photocatalytic Activity of Photocatalysts*

The photocatalytic activities of the CN and the CuO/CN samples were evaluated for degradation of phenol under visible light irradiation. The remaining concentration of phenol after the reaction was determined by HPLC and the results are displayed in Figure 5(a) as the percentages of phenol degradation. These results confirmed that all the CN and CuO/CN samples were active for degradation of phenol as they could utilize the visible light of more than 400 nm as the light source, which is in agreement with the medium bandgap energy of CN. The unmodified CN photocatalyst was observed to give 21.30% phenol degradation after a 24-h reaction. A similar phenol degradation (20.9%) was achieved upon the addition of 0.05 wt% copper species on the CN, suggesting that the addition of copper species in a very small amount would not give a significant effect in enhancing the photocatalytic activity of the unmodified CN. On the other hand, the addition of 0.1 wt% copper species on the CN, denoted as CuO(0.1)/CN, was found to give a remarkable increase of photocatalytic activity under visible light (41.70%), which was almost two times higher than the activity of the unmodified CN. This enhancement in photocatalytic activity would be closely related to the role of the copper species in decreasing the electrons-holes recombination as also supported by their fluorescence properties.

Further increases in the CuO loading amount did not further increase the photocatalytic activities of the CN. The CuO(0.2)/CN and the CuO(0.5)/CN gave phenol degradation percentages of 24.1% and 3%, respectively. This could be caused by the formation of CuO clusters along with the increase in the copper species amount added to the CN [17]. The CuO cluster would cover the active sites of CN and prevent the visible light to reach the active sites. This phenomenon is known as the masking effect and has been reported in previous works, either for copper species [24,25,40] and for other metal species [13]. Since the visible light has weaker penetration depth into the active site of photocatalyst as compared to the UV light source, the masking effect in visible light became more prominent which resulted in the decreased activity after the addition of copper species was more than 0.1 wt% [25]. This result confirmed that the copper species deposited to CN by the photodeposition method could enhance the photocatalytic activity of the CN when the optimum amount of the CuO was achieved, which was 0.1 wt% in this work.

To understand the active species that were responsible for the phenol photodegradation reaction, the free radicals and holes scavenging experiments were studied on the best composite *i.e.*, CuO(0.1)/CN by using ammonium oxalate, *tert*-butanol, and ascorbic acid as the hole, hydroxyl radical, and superoxide radical scavenger agents, respectively [13,30,41]. The ascorbic acid has been studied as the superoxide radical scavenger, analogous with *p*-benzoquinone [42]. The results from the scavenging experiments are shown in Figure 5(b). After the addition of ammonium oxalate to the reaction system, the photocatalytic activity of CuO(0.1)/CN was drastically inhibited as there was no phenol degradation occurred. This result strongly suggested that the photogenerated holes were important active species for phenol degradation. Meanwhile, when *tert*-butanol was added to the reaction system, the phenol degradation was reduced from 12.22 to 1.60%, indicating that hydroxyl radicals also played a role as the important active species in the phenol degradation. A dramatic decrease in phenol degradation percentage was also observed after the addition of ascorbic acid to the photocatalytic reaction system, demonstrating that the superoxide radicals have an important role as the active species in the phenol degradation system. From these scavenging experiments, it was revealed that the photogenerated holes, hydroxyl radicals, and superoxide radicals were all the active species in the phenol photocatalytic degradation reaction, with the hydroxyl radicals gave the least inhibition effect to the phenol degradation.

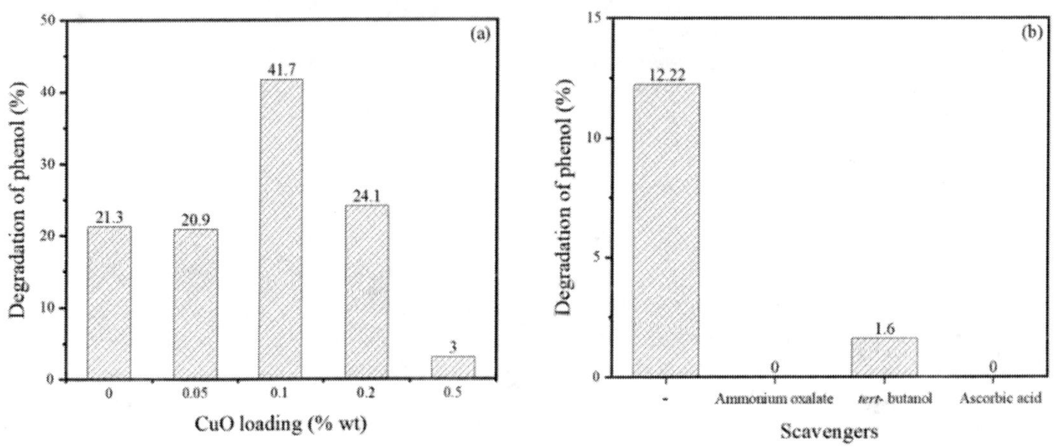


**Figure 5.** (a) Photocatalytic activity of the unmodified CN and all CuO/CN composites under visible light irradiation for 24 hours and (b) Photocatalytic activity of CuO (0.1)/CN in the absence and the presence of various scavengers under visible light irradiation for 24 hours.

Based on the photocatalyst characterizations and scavenging experiments, the schematic mechanism of photocatalytic phenol degradation on the CuO/CN composites is proposed and shown in Figure 6. Under visible light irradiation having larger energy than the bandgap energy of the CN, the electrons would be excited from the VB of the CN to the CB of the CN and/or trapped by the CuO. The photogenerated electrons would further reduce the dissolved oxygen to produce superoxide radicals, which would oxidize phenol to final products of carbon dioxide ($CO_2$) and water ($H_2O$). On the other hand, the photogenerated holes at the VB of the CN would also oxidize phenol. From the scavenging experiment, it could be seen clearly that hydroxyl radicals affected phenol degradation. However, the CN would not be able to directly generate hydroxyl radicals from the water due to the energy restriction and thus, the hydroxyl radicals shall be produced through another pathway, such as multiple-electron reductions of oxygen to produce hydroperoxy radicals, hydrogen peroxide, and finally the hydroxyl radicals [43]. The last process would require a good electron supplier, which could be possibly provided by the CuO.

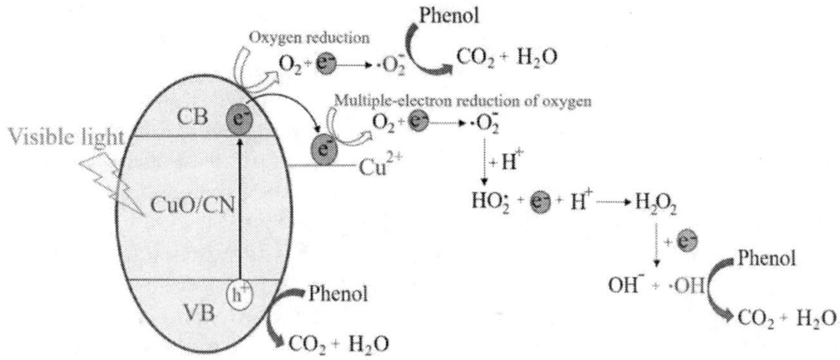


**Figure 6.** The proposed mechanism of photocatalytic phenol degradation on the CuO/CN composite.

The presence of CuO was indeed important to enhance the photocatalytic of the CN. Since most reports used different reaction conditions and CuO amounts, the impact of CuO shall be analyzed from the activity enhancement factor before and after its addition on the photocatalyst. As compared to the reported CuO-based photocatalysts listed in Table 1, the activity enhancement observed on the CuO/CN (1.96 times) was comparable to those of the CuO/$TiO_2$ (2.17 times) [26] and CuO/ZnO (2.05 times) [27], and even better than those of CuO/$Ag_3AsO_4$/GO (1.37 times) [28] and CuO/$BiPO_4$/$Fe_2O_3$ (1.29 times) [29]. This data again supported that the photodeposition is a good method to prepare the CuO/CN photocatalysts.

**Table 1.** Comparison of photocatalytic activity over CuO-based photocatalysts for phenol photodegradation.

| Composite photocatalyst | Preparation method | Reaction conditions | Activity (enhancement factor) | Reference |
|---|---|---|---|---|
| CuO(0.1)/CN | Photo-deposition | Photocatalyst (50 mg), phenol (70 ppm), visible light (halogen, 150 W), reaction time (24 h) | 41.7% (1.96) | This work |
| CuO(2 wt%)/$TiO_2$ | Sol-gel | Photocatalyst (0.5 g/L), phenol (200 ppm), UV lamp, reaction time (2 h), $H_2O_2$ (12.5 mL) | 43.21% (2.17) | [26] |
| CuO(7 mol%)/ZnO | Deposition by wet chemistry method | Photocatalyst (50 mg), phenol (10 ppm), UV lamp (Xe, 300 W), reaction time (3 h) | 78% (2.05) | [27] |
| CuO(11wt%)/$Ag_3AsO_4$/GO | Self-assembly | Photocatalyst (6 mg), phenol (20 ppm), visible light (300 W-halogen lamp), reaction time (20 min) | 85% (1.37) | [28] |
| CuO(0.25)/$BiPO_4$/$Fe_2O_3$ | Impregnation | Photocatalyst (0.1 g), phenol (10 ppm), UV lamp (Xe, 150 W), reaction time (1 h) | 35% (1.29) | [29] |

## 4. Conclusion

The CuO/CN composites were successfully prepared by depositing copper species onto the CN surface via photodeposition method using copper(II) nitrate trihydrate as the copper precursor. This method gave no significant alteration in the CN structure and bandgap energy after the addition of copper species, which were confirmed respectively from the FTIR and DR UV-vis spectra of the CuO/CN composites. The interactions between copper species and CN were revealed by the fluorescence spectra, which favorably occurred at the C=N sites. The CuO/CN composite showed the best photocatalytic activity for phenol degradation under visible light exposure when the copper species loading amount was 0.1 wt%, which gave almost two times higher activity than that of the unmodified CN photocatalyst. The improved photocatalytic activity on the CuO/CN composite was due to the presence of copper species that could suppress the electron-hole recombination of the CN. This study also demonstrated prominent contribution from the optimum amount of copper species loading, considering the importance of low masking effect to achieve high photocatalytic activity of CN under visible light illumination. The active species contributing to the phenol photodegradation reaction were photogenerated holes, hydroxyl radicals, and superoxide radicals as presented by scavenger studies.

**Acknowledgement**

This research was financially supported by the Directorate General of Strengthening Research and Development, Ministry of Research, Technology and Higher Education of the Republic of Indonesia through the World Class Research (WCR 2020, No. 041/SP2H/AMD/LT/MULTI/L7/2020 and No. 008/MACHUNG/LPPM/SP2H-LIT-MULTI/AMD/VI/2020).

**References**

[1] Chen C, Ma W and Zhao J 2010 *Chem. Soc. Rev.* **39** 4206
[2] Ma H Y, Zhao L, Guo L H, Zhang H, Chen F J and Yu W C 2019 *J. Hazard. Mater.* **369** 719
[3] Markushyna Y, Smith C A and Savateev A 2019 *European J. Org. Chem.* **2020** 1294
[4] Zhang W, Zhang Q, Dong F and Zhao Z 2013 *Int. J. Photoenergy* **2013** 1
[5] Yang Z, Zhang Y and Schnepp Z 2015 *J. Mater. Chem. A* **3** 14081
[6] Yang G, Chen T, Feng B, Weng J, Duan K, Wang J and Lu X, 2019 *J. Alloys Compd.* **770** 823
[7] Xu B, Ahmed M B, Zhou J L, Altaee A, Xu G and Wu M 2018 *Sci. Total Environ.* **633** 546
[8] Zhao Z, Sun Y and Dong F 2015 *Nanoscale* **7** 15
[9] Raza W and Ahmad K 2020 *Handbook of Nanomaterials and Nanocomposites for Energy Environmental Applications*, ed O Kharissova *et al* (Cham: Springer International Publishing) p 7
[10] Masih D, Ma Y and Rohani S 2017 *Appl. Catal. B Environ.* **206** 556
[11] Wang L, Wang C, Hu X, Xue H and Pang H 2016 *Chem. - An Asian J.* **11** 3305
[12] Koltsov I, Wojnarowicz J, Nyga P, Koziorowska J S, Stelmakh S, Babyszko A, Morawski A W and Lojkowski W 2019 *Molecules* **24** 1
[13] Yang Y, Guo Y, Liu F, Yuan X, Guo Y, Zhang S, Guo W and Huo M 2013 *Appl. Catal. B Environ.* **142–143** 828
[14] Sridharan K, Jang E and Park T J 2013 *Appl. Catal. B Environ.* **142–143** 718
[15] Sun J X, Yuan Y P, Qiu L G, Jiang X, Xie A J, Shen Y H and Zhu J F 2012 *Dalt. Trans.* **41** 6756
[16] Katsumata K, Motoyoshi R, Matsushita N and Okada K 2013 *J. Hazard. Mater.* **260** 475
[17] Janczarek M and Kowalska E 2017 *Catalysts* **7** 1
[18] Li J, Jin X, Li R, Zhao Y, Wang X, Liu X and Jiao H 2019 *Appl. Catal. B Environ.* **240** 1
[19] Fujimoto K, Oku T, Akiyama T and Suzuki A 2013 *J. Phys. Conf. Ser.* **433** 1
[20] Deng X, Wang C, Shao M, Xu X and Huang J 2017 *RSC Adv.* **7** 4329
[21] Muthukumaran M, Prasath P V, Kulandaivelu R, Sagadevan S, Mohammad F and Oh W C 2020 *J. Mater. Sci. Mater. Electron.* **31** 2257
[22] Xu J, Li S, Tan L and Kou B 2017 *Mater. Res. Bull.* **93** 352
[23] Jiang X X, Hu X D, Tarek M, Saravanan P, Alqadhi R, Chin S Y and Khan M M R 2020 *J. $CO_2$*

*Util.* **40** 101222
[24] Duan Y 2018 *Mater. Res. Bull.* **105** 68
[25] Siah W R, Lintang H O, Shamsuddin M, Yoshida H and Yuliati L 2016 *Catal. Sci. Technol.* **6** 5079
[26] Sohrabi S and Akhlaghian F 2016 *Process Saf. Environ.* **99** 120
[27] Zhu L, Li H, Liu Z, Xia P, Xie Y and Xiong D 2018 *J. Phys. Chem. C* **122** 9531
[28] Rakibuddin M, Mandal S and Ananthakrishnan R 2017 *New J. Chem.* **41** 1380
[29] Chen S, Mu Z, Yan R, Zhang X, Khan I, Li Z, Sun N, Zhang Q and Jing L 2019 *J. Alloy Comp.* **800** 231
[30] Sanni S O and Idemudia O G 2015 *J. Nanomater.* **2015** 1
[31] Lee S C, Lintang H O and Yuliati L 2017 *Beilstein J. Nanotechnol.* **8** 915
[32] Chowdhury P, Nag S and Ray A K 2017 *Phenolic Compounds-Natural Sources, Importance and Applications*, ed M S Hernández *et al* (London: IntechOpen) chapter 16 p 396
[33] Lee S C, Lintang H O and Yuliati L 2014 *J. Exp. Nanosci.* **9** 78
[34] Xia P, Antonietti M, Zhu B, Heil T, Yu J and Cao S 2019 *Adv. Funct. Mater.* **29** 1
[35] El-Trass A, Elshamy H, El-Mehasseb I and El-Kemary M 2012 *Appl. Surf. Sci.* **258** 2997
[36] Jasman S M, Lintang H O and Yuliati L 2017 *Macromol. Symp.* **371** 84
[37] Cui Y, Zhang J, Zhang G, Huang J, Liu P, Antonietti M and Wang X 2011 *J. Mater. Chem.* **21** 13032
[38] Zhang Y, Ligthart D A J M, Quek X Y, Gao L and Hensen E J M 2014 *Int. J. Hydrogen Energy.* **39** 11537
[39] Smith E A, Ambrogi E K, Barnes E and Brame J A 2020 *Colloids Surfaces A Physicochem. Eng. Asp.* **603** 125179
[40] Chiang K, Amal R and Tran T 2002 *Adv. Environ. Res.* **6** 471
[41] Nandi A and Chatterjee I B 1987 *J. Biosci.* **11** 435
[42] Wenzel U, Nickel A, Kuntz S and Daniel H 2004 *Carcinogenesis* **25** 703
[43] Cui Y, Huang J, Fu X and Wang X 2012 *Catal. Sci. Technol.* **2** 1396

# High photocatalytic activity of zinc metatitanate materials for phenol photodegradation

**K T A Priyangga[1], Y S Kurniawan[1] and L Yuliati[1,2,*]**

[1]Ma Chung Research Center for Photosynthetic Pigments, Universitas Ma Chung, Malang 65151, Indonesia
[2]Department of Chemistry, Faculty of Science and Technology, Universitas Ma Chung, Malang 65151, Indonesia

[*]Email: leny.yuliati@machung.ac.id

**Abstract.** In this work, we synthesized zinc metatitanate ($ZnTiO_3$) through a sol-gel method strictly using a 1:1 mol ratio of zinc nitrate and titanium(IV) isopropoxide as the precursors. The calcination temperature was set to 700, 900, and 1100 °C to give ZM-700, ZM-900, and ZM-1100, respectively. These ZM materials were characterized using X-ray diffractometer (XRD), diffuse reflectance ultraviolet-visible (DR UV-vis), Fourier transform infrared (FTIR), and fluorescence spectroscopies. The XRD analyses showed that the ZM-700 contained both cubic-$ZnTiO_3$ and rhombohedral-$ZnTiO_3$ phases with a small part of ZnO, while the ZM-900 contained cubic-$ZnTiO_3$ and cubic-$Zn_2TiO_4$. In contrast to the ZM-700 and the ZM-900, the ZM-1100 contained cubic-$Zn_2TiO_4$ and rutile $TiO_2$ as the main phases with a very small part of the $ZnTiO_3$ phase. The formation of these crystal phases was also supported by their DR UV-vis spectra, FTIR analysis, and fluorescence spectra. Photocatalytic degradation of phenol was carried out under UV light irradiation for 1–24 h. The kinetic study revealed that the reaction followed first-order, in which the reaction rate constants were 0.0351, 0.0323, 0.0298 $h^{-1}$ for the ZM-700, ZM-900, and ZM-1100, respectively. This study demonstrated that the formation of $ZnTiO_3$ was crucial to obtain high photocatalytic activity on the ZM materials.

## 1. Introduction

Water pollution involving organic contaminants has gained significant attention due to its harmful effect on the environment. Phenol is considered as one of the organic pollutants present in the wastewater affecting the aquatic ecosystem [1]. The photocatalysis technique offers a promising way for the removal of this phenolic compound, in which it can degrade this contaminant to the less toxic compound of carbon dioxide ($CO_2$) and water ($H_2O$) [2]. Heterogeneous catalysts such as titanium dioxide ($TiO_2$)-based catalysts have been widely developed for this photocatalysis process [3-5]. Besides, binary metal oxide photocatalysts have also attracted great interest due to their excellent properties. These materials have been reported to exhibit good photocatalytic activity, such as the zinc titanates [6-10] and the zirconium titanates [11,12]. Particular attention is on the use of zinc titanate-based materials, which have been recognized as able to degrade organic pollutants.

Zinc titanates exist in different polymorphs, namely zinc metatitanate ($ZnTiO_3$), zinc orthotitanate ($Zn_2TiO_4$), and meta-stable zinc titanate ($Zn_2Ti_3O_8$). While the $ZnTiO_3$ has two different crystal phases of cubic (c-$ZnTiO_3$) and rhombohedral (r-$ZnTiO_3$) [13,14], the $Zn_2TiO_4$ is found in cubic crystal phase (c-

$Zn_2TiO_4$) or cubic with inverse spinel structure, and $Zn_2Ti_3O_8$ is only available in cubic phase (c-$Zn_2Ti_3O_8$) [15,16]. These polymorphs have been also used to increase the activity of other metal oxide photocatalysts, such as in the $ZnTiO_3/Zn_2Ti_3O_8/ZnO$ [6] and g-$C_3N_4$/c-$ZnTiO_3/TiO_2$ [17]. The $ZnTiO_3/Zn_2Ti_3O_8/ZnO$ was used for the degradation of Rhodamine B and the reduction of Cr(VI) [6], while the g-$C_3N_4$/c-$ZnTiO_3/TiO_2$ was used to degrade methylene blue [17]. In both cases, the $ZnTiO_3$ phase was proposed to be responsible for the photocatalytic activity enhancement of the materials [6,17].

In order to obtain zinc titanates with the desired polymorph and crystal phase, the stoichiometric mol ratio of Zn and Ti as well as the synthesis temperature are the pivotal parameters [18,19]. The $ZnTiO_3$ could be synthesized when the mol ratio of Zn and Ti was 1:1 and at a certain temperature [20,21]. When the $ZnTiO_3$ was prepared by a precipitation method, the temperature of 600 °C was crucial as below the temperature the $Zn_2Ti_3O_8$ was formed as the main crystal phase [20]. Using the hydrothermal method, the $ZnTiO_3$ could be formed at the calcination temperature range of 600–900 °C [21]. However, the preparation of pure $ZnTiO_3$ was quite difficult because the ZnO was also co-existed in the prepared material. In the present study, we employed a sol-gel method to synthesize the $ZnTiO_3$ at various calcination temperatures (700, 900, and 1100 °C). The effects of the calcination temperature on the formation of polymorphs zinc titanates, their properties, and their photocatalytic activity for phenol degradation under UV light irradiation were examined.

## 2. Experimental Section

### *2.1. General*

The materials used in this work were titanium(IV) isopropoxide (TTIP, $Ti(OC_3H_7)_4$, 97%), zinc nitrate hexahydrate ($Zn(NO_3)_2{\cdot}6H_2O$, 98%), sodium hydroxide (NaOH, 99%), ethanol ($C_2H_5OH$, > 99.9%), and phenol ($C_6H_5OH$, 99%). All the materials were purchased from Merck and used without any further purifications.

The instrumentations used in this work were powder X-ray diffractometer (XRD, Rigaku SmartLab, PhotonMax high-flux 9 kW), ultraviolet-visible spectrophotometer (UV-vis, JASCO V-760) attached with an integrating sphere (ISV-922/ISN-923/ISN-901i), Fourier transform infrared spectroscopy (FTIR, JASCO 6800) attached with an attenuated total reflectance (ATR, Pro One), and fluorescence spectrophotometer (JASCO FP-8500). High-performance liquid chromatography (HPLC, Shimadzu LC-20 AT) with a photodiode array detector (PDA, SPD-M20) was used for phenol analysis at 272 nm. The stationary and mobile phase used in the HPLC were a C18 column and 100% acetonitrile, respectively. The column temperature was set to 40 °C and the flow rate was set to 0.8 mL $min^{-1}$.

### *2.2. Procedure*

*2.2.1. Synthesis of ZM materials.* The $ZnTiO_3$ was prepared by a sol-gel method, followed by the calcination process at different temperatures to give zinc metatitanate (ZM) materials. In a typical preparation method, the TTIP (3.50 g, 0.0123 mol) was dissolved in ethanol (10 mL) to produce solution A, while the $Zn(NO_3)_2{\cdot}6H_2O$ (3.65 g, 0.0123 mol) was dissolved with ethanol (10 mL) to produce solution B. Solution B was then poured into solution A, followed by a dropwise addition of NaOH (1 M, 2 mL) into the mixture. The mixture was kept stirred at room temperature for 1 hour to homogenize the mixture. The solvent was evaporated slowly by heating the mixture at 60 °C for 24 hours to allow sol-gel formation. The white solid was ground into a powder and calcined at different temperatures of 700, 900, and 1100 °C for 6 hours to give ZM-700, ZM-900, and ZM-1100, respectively.

*2.2.2. Characterization of ZM materials.* The ZM materials were characterized by XRD to identify their crystal phases. The diffractogram was measured at room temperature in the $2\theta$ range of 20–80 degrees. The diffuse reflectance UV-visible (DR UV-vis) spectrum of each sample was collected on a UV-vis spectrophotometer, in which the measurement was made in the range of 200–800 nm and shown as

Kubelka-Munk (KM) function. The FTIR spectra were recorded by the ATR method at the wavenumber range of 4000–400 $cm^{-1}$. The emission and excitation fluorescence spectra of each sample were recorded at room temperature with an emission wavelength ($\lambda_{emission}$) of 302 nm and an excitation wavelength ($\lambda_{excitation}$) of 274 nm [3].

2.2.3. *Photocatalytic Phenol Degradation on ZM Materials.* The ZM materials were evaluated for photocatalytic degradation of phenol under UV light irradiation at room temperature. An aqueous phenol solution (50 mg $L^{-1}$, 50 mL) was used for such a purpose. The ZM sample (50 mg) was added into the phenol solution and stirred for 2 hours in the dark condition to reach the adsorption equilibrium. The reaction was then started by illuminating the mixture using a UV lamp (UVLS-28 EL Series, 365 nm, 8 W) for 1, 3, 6, 15, and 24 hours. After the illumination, the mixture was filtered and then the phenol concentration in the filtrate was analyzed by the HPLC. The photocatalytic activity was determined from the percentage of phenol degradation, which was calculated from the ratio of decreased phenol concentration after reaction ($C_0$ - $C_t$) to the initial phenol concentration ($C_0$) as shown in equation (1). $C_t$ is phenol concentration after a certain time of reaction.

$$\text{Photodegradation (\%)} = \frac{(C_0 - C_t)}{C_0} \times 100\% \quad (1)$$

## 3. Results and discussion

### *3.1. Physicochemical Properties of ZM Materials*

The XRD pattern was analyzed to identify the crystal phases present on each material calcined at different temperatures of 700, 900, and 1100 °C. As shown in Figure 1, the synthesized ZM materials have various crystal phases, which could be indexed to c-$ZnTiO_3$ (JCPDS 00-039-0190) [22], r-$ZnTiO_3$ (JCPDS 00-026-1500) [13,23], ZnO (JCPDS 00-036-1451) [22], c-$Zn_2TiO_4$ (JCPDS 00-073-0578) [22], c-$Zn_2TiO_4$ inverse-spinel (JCPDS 00-025-1164) [15], and rutile $TiO_2$ or r-$TiO_2$ (JCPDS 00-021-1276) [15].

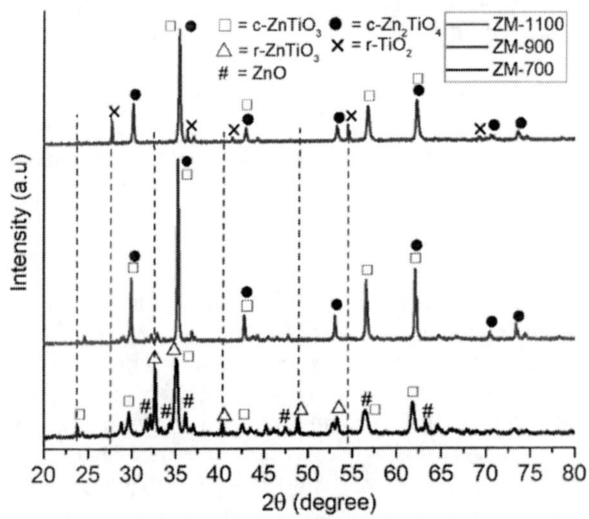


**Figure 1.** XRD patterns of ZM materials.

The ZM-700 consisted of c-$ZnTiO_3$ and r-$ZnTiO_3$, as well as the ZnO phase. The formed crystal phase was affected by the synthesis temperature. When the synthesis was carried out at a higher temperature of

900 °C, the diffraction peaks of the c-$ZnTiO_3$ still could be observed, however, the diffraction peaks of r-$ZnTiO_3$ were diminished and the ZnO peaks were decreased in intensity. In addition to the c-$ZnTiO_3$, the c-$Zn_2TiO_4$ peaks were also observed in the ZM-900. This result showed that at 900 °C the r-$ZnTiO_3$ could be transformed into the c-$ZnTiO_3$ and the r-$ZnTiO_3$ could react with ZnO to produce the c-$Zn_2TiO_4$ as depicted in equation (2). In contrast to the ZM-700 and the ZM-900, the ZM-1100 not only contained c-$ZnTiO_3$ and c-$Zn_2TiO_4$, but also contained r-$TiO_2$. The formation of r-$TiO_2$ was possible since the c-$ZnTiO_3$ could be decomposed to c-$Zn_2TiO_4$ and r-$TiO_2$ at 1100 °C [24,25] as shown in equation (3). It was worthy to note that the formation of the c-$ZnTiO_3$ was further reduced with the increase of the calcination temperature.

$$ZnO\ (s) + r\text{-}ZnTiO_3\ (s) \rightarrow c\text{-}Zn_2TiO_4\ (s) \tag{2}$$

$$2\ c\text{-}ZnTiO_3\ (s) \rightarrow c\text{-}Zn_2TiO_4\ (s) + r\text{-}TiO_2\ (s) \tag{3}$$

The DR UV-vis spectra of the ZM materials were also recorded to study their optical properties. The plots of Kubelka-Munk (KM) function versus wavelength for all the ZM materials are shown as DR UV-vis spectra in Figure 2(a). Different calcination temperatures resulted in the different absorption properties of the ZM materials. The ZM-700 showed several absorption peaks at 219, 257, and 297 nm, while the ZM-900 exhibited absorption peaks at 221, 257, and 311 nm. In addition, the ZM-700 gave the characteristic absorption of ZnO, giving an absorption edge close to 390 nm. The shifted absorption peak from 297 to 311 nm with the increase of the calcination temperature from 700 to 900 °C would be related to the additional formation of c-$Zn_2TiO_4$ as suggested by the XRD pattern. The ZM-1100 gave absorption peaks at 254, 303, and 390 nm. The first and second absorption peaks could be assigned to the c-$ZnTiO_3$ and c-$Zn_2TiO_4$, while the latter would be due to the presence of the r-$TiO_2$, as supported by its XRD pattern.

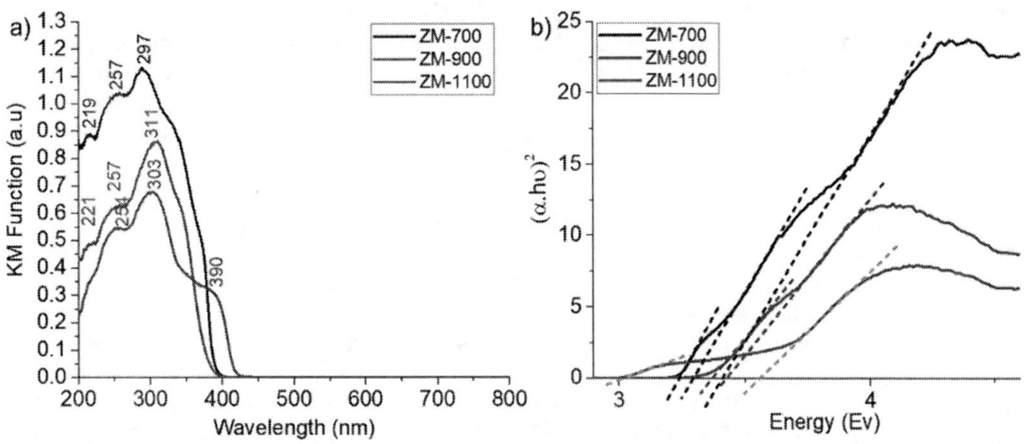


**Figure 2.** (a) DR UV-vis spectra and (b) the Tauc plots for bandgap energy determination of ZM materials.

Bandgap energy is an important parameter related to photocatalytic activity. The bandgap energy value determines the minimum energy required to excite electrons from the valence band to the conduction band of a photocatalyst material. The bandgap energy values of the ZM-materials were determined using the Tauc method with a direct bandgap manner as shown in equation (4) [26]. The bandgap energy is symbolized as *Eg*, while the $\alpha$, $h$, and $\upsilon$ are referring to the absorption coefficient related to the material, Plack constant, and light frequency, respectively.

$$(\alpha h\upsilon)^2 = A(E_g - h\upsilon) \tag{4}$$

The extrapolation of the linear plot between the $(\alpha h\upsilon)^2$ versus $(h\upsilon)$ to the $x$-axis can be used to determine the bandgap energy value of each material. The Tauc plots of the ZM-700, the ZM-900, and the ZM-1100 are shown in Figure 2(b) and the bandgap energy values are summarized in Table 1. It could be observed that ZM materials have multiple bandgap energies. The bandgap energy values of ZM-700 were determined to be 3.36, 3.28, and 3.23 eV, which could be correlated to the presence of the c-ZnTiO, r-$ZnTiO_3$, and ZnO respectively [27,28]. The ZM-900 has two distinguished bandgap energy values of 3.36 and 3.44 eV, which could be assigned to the c-$ZnTiO_3$ and the c-$Zn_2TiO_4$. In contrast to the ZM-700 and ZM-900, the ZM-1100 did not show the characteristic bandgap value of c-$ZnTiO_3$. The ZM-1100 has bandgap energy values of 3.56 eV due to the presence of c-$Zn_2TiO_4$ and 3.02 eV from the r-$TiO_2$ [29,30]. This result was also in agreement with the XRD pattern, in which the ZM-1100 contained the c-$Zn_2TiO_4$ and r-$TiO_2$ as the main crystal phase with c-$ZnTiO_3$ as the minor phase.

**Table 1.** Bandgap values of the ZM materials.

| Materials | Bandgap energy (eV) | Assignment [27-30] |
|---|---|---|
| ZM-700 | 3.36 | c-$ZnTiO_3$ |
| | 3.28 | r-$ZnTiO_3$ |
| | 3.23 | ZnO |
| ZM-900 | 3.36 | c-$ZnTiO_3$ |
| | 3.44 | c-$Zn_2TiO_4$ |
| ZM-1100 | 3.56 | c-$Zn_2TiO_4$ |
| | 3.02 | r-$TiO_2$ |

The FTIR spectra of the ZM materials were analyzed to identify the functional vibrational modes present in each sample. As depicted in Figure 3(a), all the ZM materials exhibited no O–H vibrational stretching mode (3400 $cm^{-1}$) of adsorbed water molecules [31]. This could be caused by the employed high calcination temperature in the synthesis process. On the other hand, all ZM-materials showed weak absorption peaks in the range of 883–970, 693–699, 639–641, 546–548, 490–506, and 400–430 $cm^{-1}$ as shown in Figure 3(b). The strong broad absorption band at 883–970 $cm^{-1}$ could be related to the vibrational mode of Ti–O stretching, involving non-bridging oxygen atoms. In the ZM-900 and the ZM-1100, the intensity of this band was decreased and also shifted to the higher wavenumber. This change could be ascribed as the indication of crystal phase transformation of the ZM materials with the increase in the calcination temperature [32], as also confirmed from the XRD results. The Zn–O–Ti vibrational mode was observed as a shoulder band in the region of 693–699 $cm^{-1}$, which was the characteristic of zinc titanate materials [33]. The other bands in the region of 639–641 $cm^{-1}$ and 400–430 $cm^{-1}$ were related to the vibrational mode of Ti-O bending in $[TiO_6]^{2-}$ octahedral in the ZM materials [6]. In addition, the broad band in the region of 900 $cm^{-1}$ was assigned to be the vibrational mode of O–Ti–O stretching in $[TiO_6]^{2-}$ octahedral, which was in good agreement with the reported literature [34]. On the other side, the presence of $[ZnO_4]^{2-}$ structure in ZM materials was also observed as a shoulder band in the region of 546–548 $cm^{-1}$ and 490–506 $cm^{-1}$, which was related to the vibrational mode of Zn–O bending [35,36]. The more intense peak at 506 $cm^{-1}$ in the ZM-700 as compared to those of the ZM-900 and ZM-1100 would be related to the presence of the ZnO crystal phase in this ZM-700 as confirmed from the XRD result.

Fluorescence properties of the ZM-700, ZM-900, and ZM-1100 materials are depicted in Figure 4. As shown in Figure 4(a), all the ZM materials gave a similar shape of excitation spectrum to each other. The excitation peak was observed at 274 nm when 302 nm was used as the emission wavelength. The increase in the calcination temperature increased the excitation intensity. The excitation intensity of the ZM-700 was still close to that of the ZM-900, but lower than that of the ZM-1100. The emission spectra of the ZM materials were monitored at the excitation wavelength of 274 nm and shown in Figure 4(b). All the ZM

IOP Conf. Series: Materials Science and Engineering 1143 (2021) 012076 doi:10.1088/1757-899X/1143/1/012076

materials gave a similar main emission peak at 302 nm and slightly different shapes of low emission peaks at 360–700 nm. As expected from their excitation intensity, the ZM-700 and the ZM-900 showed similar emission intensity, while the ZM-1100 gave much higher emission intensity. The emission intensity could be ascribed to the recombination of trapped electron-hole [3] at the different energy levels of the ZM-materials. The difference between the ZM-700 and the ZM-900 as compared to the ZM-1100 could be also observed from the emission fine structure at the visible region. Such differences could be correlated to the different main crystal phases in these ZM materials as previously discussed from the XRD patterns and DR UV-vis spectra.

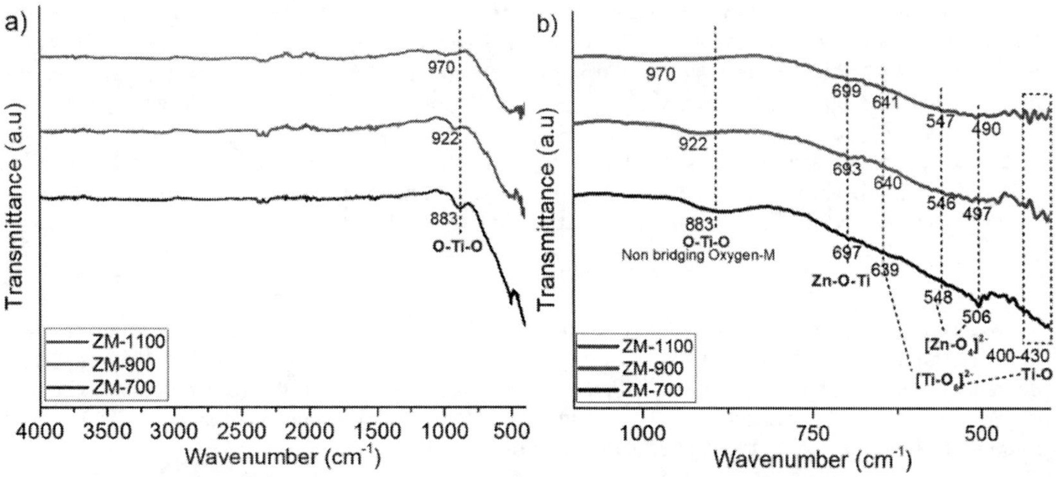


**Figure 3.** FTIR spectra of ZM materials (a) in the region of 400–4000 cm$^{-1}$ and (b) their magnified spectra in the region of 400–1100 cm$^{-1}$.

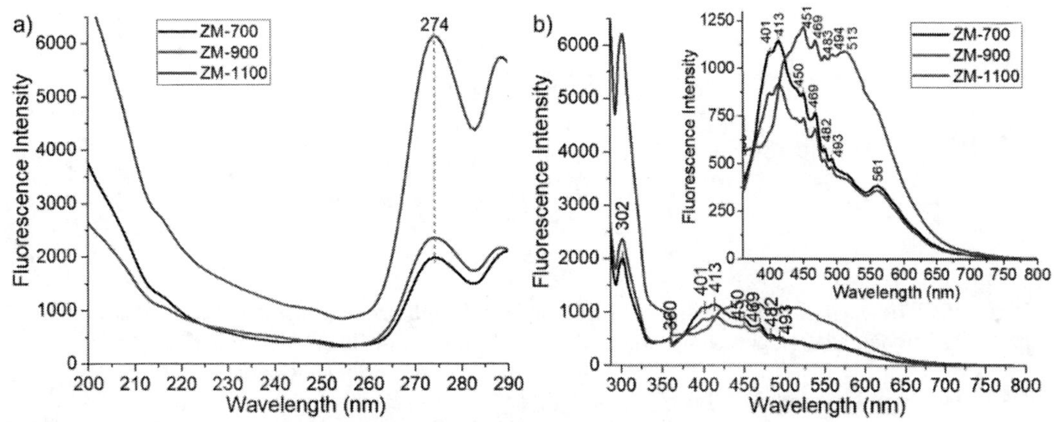


**Figure 4.** (a) Excitation and (b) emission spectra of ZM materials monitored at the emission wavelength of 302 nm and excitation wavelength of 274 nm, respectively. Inset shows the magnified emission spectra at 360–800 nm.

### *3.2. Photocatalytic activity of ZM materials*

The photocatalytic activity of each ZM-700, ZM-900, and ZM-1100 materials was evaluated for the degradation of phenol under UV light. Figure 5(a) shows the plot between the ratio of phenol concentration at a certain reaction time ($C_t$) to the initial phenol concentration ($C_0$) versus the reaction time (t). The ZM-700 gave the highest photocatalytic activity followed by ZM-900, and the ZM-1100. After 24 hours, the phenol degradation percentage for the ZM-700, ZM-900, and ZM-1100 were 57, 56, and 50%, respectively. The difference in the photocatalytic activity could be affected by the different crystal phases present in the ZM materials. The ZM-700 having the c-$ZnTiO_3$ and r-$ZnTiO_3$ as the main crystal phases gave the highest activity. The ZM-900 also has c-$ZnTiO_3$ and thus, a similar level of activity to the ZM-700 was also observed. The slightly decreased activity could be due to the less active c-$Zn_2TiO_4$ in this ZM-900. The decreased activity was more observed on the ZM-1100 having c-$Zn_2TiO_4$ and r-$TiO_2$ as the main crystal phases with the c-$ZnTiO_3$ as the minor phase. Therefore, the presence of r-$TiO_2$ could be also the reason for the decreased activity. Besides, as shown in the fluorescence study, the ZM-1100 gave a high emission intensity, which could be the result of the high electron-hole recombination process. All these results demonstrated that both c-$ZnTiO_3$ and r-$ZnTiO_3$ were the ones corresponding to the high photocatalytic activity of the ZM materials.

The kinetic photocatalytic activity of the ZM materials for the photodegradation of phenol was evaluated using equation (5). The reaction rate constant ($k$) for each ZM material was calculated from the slope of the linear plots between the $-\ln(C_t/C_0)$ and the reaction time (t).

$$-\ln\left(\frac{C_t}{C_0}\right)=kt \qquad (5)$$

As shown in Figure 5(b), it was clarified that all the reactions over the ZM materials followed the first-order reaction kinetic model as usually proposed for the photocatalytic degradation reaction [37]. The degradation rate constants for the ZM-700, the ZM-900, and the ZM-1100 were determined to be 0.0351, 0.0323, and 0.0298 $h^{-1}$, respectively. This result again confirmed that the presence of $ZnTiO_3$ would be important to give high activity for phenol degradation.

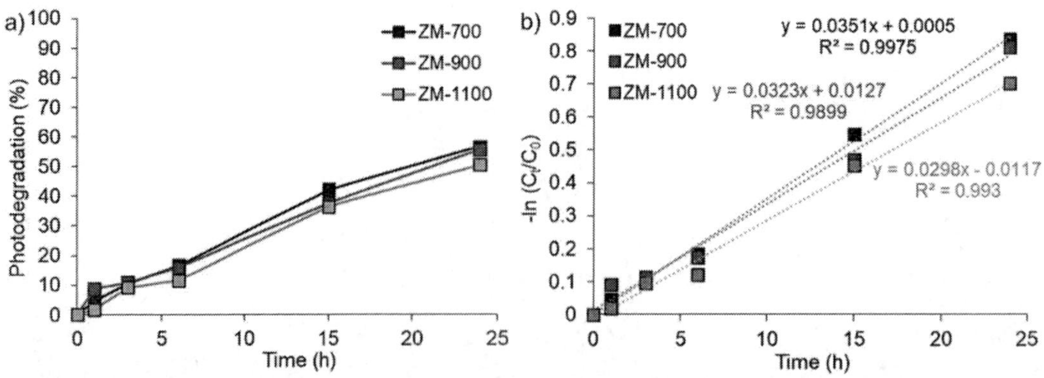


**Figure 5.** Photocatalytic degradation of phenol under UV light irradiation on ZM materials shown as (a) plots of photodegradation percentage versus time and (b) plots of -ln ($C_t/C_0$) versus time.

Since the reported conditions for photocatalytic degradation of phenol involved various parameters, the activity comparison between the ZM-700 and other reported $ZnTiO_3$ photocatalysts [6,31,38] could be determined from the turn over frequency (TOF) values. Considering that the $ZnTiO_3$ was proposed as the main active sites for the phenol degradation, the TOF values could be calculated from equation (6), where the amount of the active sites was determined solely from the amount of photocatalyst.

$$\mathrm{TOF} = \frac{\text{amount of converted phenol}}{\text{amount of active sites} \times \text{reaction time}} \quad (6)$$

Unfortunately, the real amount of $ZnTiO_3$ existing in the ZM-700 was not examined in this study. Judging from the absorbance intensity, the ratio of the $ZnTiO_3$ intensity to the ZnO intensity was close to 70%. Taking into account an assumption if the ZM-700 consisted of $ZnTiO_3$ phase in the range of 70–100%, the TOF value would be in the range of $2.04\times10^{-3}$–$2.91\times10^{-3}$ $h^{-1}$. As listed in Table 2, this value was still at a similar level to the TOF values calculated from the reported data on $ZnTiO_3$ photocatalysts prepared by other methods, which was in the range of $3.29\times10^{-3}$–$3.73\times10^{-3}$ $h^{-1}$. The similar level suggested that the similar active sites, which were the $ZnTiO_3$, were involved in the reaction. The slightly lower TOF value of the ZM-700 would be due to other parameters, such as the use of different light sources which would lead to different light intensity. The reaction in this study was carried out under an 8 W UV lamp, while all the reported reactions were conducted under irradiation of strong UV lamp power (400–500 W) [6,31,38].

**Table 2.** Activity comparison of $ZnTiO_3$-based photocatalysts for phenol degradation.

| Materials | Preparation method (temperature) | Reaction conditions | Photodegradation (%) | TOF ($10^{-3}$ $h^{-1}$) |
|---|---|---|---|---|
| ZM-700 | Sol-gel (700 °C) | Photocatalyst (50 mg), phenol (50 mg $L^{-1}$, 50 mL), irradiation time (24 h), UV lamp (8 W) | 57 | 2.04 – 2.91[a] |
| $ZnTiO_3$ [6] | Solvothermal (700 °C) | Photocatalyst (50 mg), phenol (10 mg $L^{-1}$, 80 mL), irradiation time (2.5 h), UV lamp (500 W) | 30 | 3.29 |
| $ZnTiO_3$ [31] | Solvothermal (700 °C) | Photocatalyst (50 mg), phenol (10 mg $L^{-1}$, 80 mL), irradiation time (2.5 h), UV lamp (400 W) | 34 | 3.73 |
| $ZnTiO_3$ [38] | No information[b] | Photocatalyst (100 mg), phenol (25 mg $L^{-1}$, 200 mL), irradiation time (2.5 h), UV lamp (500 W) | 10 | 3.43 |

[a]Based on the rough assumption of $ZnTiO_3$ content in ZM-700 (70–100%).
[b]Commercially available from Sigma-Aldrich.

The proposed mechanism of photocatalytic phenol degradation on the ZM-700 is displayed in Figure 6. The valence band positions of $ZnTiO_3$ and ZnO were also shown according to the reported values [39]. In general, it is well-known that the metal oxide materials act as a photocatalyst since they can generate the pair of excited electrons and holes during the UV light irradiation process. In this work, the ZM materials were confirmed to have multiple bandgap energy since the ZM materials consisted of several semiconductors. For instance, the ZM-700 was shown to have c-$ZnTiO_3$, r-$ZnTiO_3$, and ZnO phases. Therefore, electron excitations would occur on all these semiconductors if the light source has energy higher than their bandgap energy. Since a UV lamp of 365 nm (3.4 eV) was employed in this work, the electron excitation could occur on c-$ZnTiO_3$ (3.36 eV), r-$ZnTiO_3$ (3.28 eV), and ZnO (3.23 eV) phases. As shown in Figure 6, the excited electrons could reduce the dissolved oxygen to create superoxide radicals that would further oxidize phenol to carbon dioxide and water. On the other hand, the remaining holes would react with water to form hydroxyl radicals which are strong enough to oxidize phenol. In addition, since the conduction band position of the $ZnTiO_3$ is slightly more negative than that of the ZnO, the possibility that electron transfers also occur from the conduction band of the $ZnTiO_3$ to the conduction band of the ZnO could not be neglected. The charge transfer could also proceed between the $ZnTiO_3$ phases. A further study shall be carried out to clarify the detailed charge transfer pathways in the ZM materials.

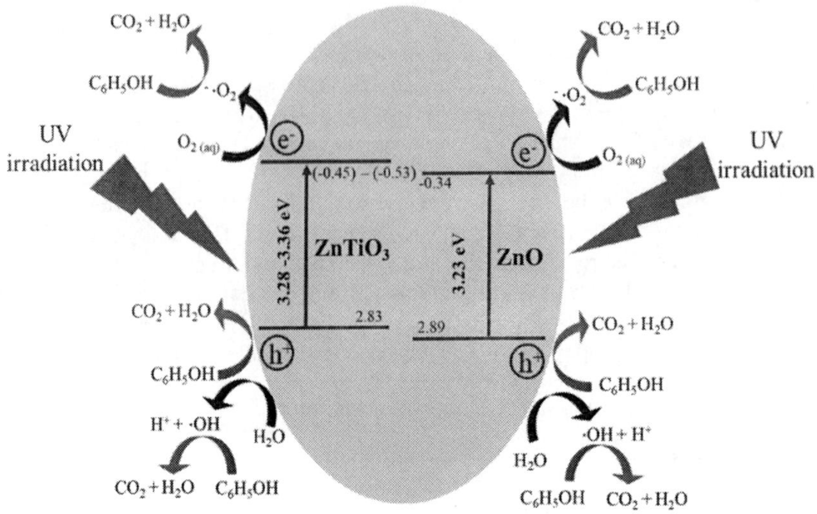


**Figure 6.** Proposed mechanism of photocatalytic phenol degradation on the ZM-700.

## 4. Conclusions

The ZM materials were successfully synthesized by the sol-gel method using titanium(IV) isopropoxide and zinc(II) nitrate as the precursors with 1:1 mol ratio. Different calcination temperatures (700, 900, 1100 °C) were revealed to affect the formation of different crystal phases in the ZM materials. As confirmed from the XRD patterns, the $ZnTiO_3$ crystal phase was mainly formed in the ZM-700 and the ZM-900 but not in the ZM-1100. Similar optical properties of the ZM-700 and the ZM-900 were also confirmed from the DR UV-vis, FTIR, and fluorescence spectroscopies analyses. In contrast, the ZM-1100 contained c-$Zn_2TiO_4$ and r-$TiO_2$ as the main crystal phases as the result of the decomposition of $ZnTiO_3$ at the high temperature. All the photocatalytic degradation phenol over the ZM materials followed the first-order kinetic reaction. The degradation rate constants for the ZM-700, the ZM-900, and the ZM-1100 were 0.0351, 0.0323, and 0.0298 $h^{-1}$, respectively. Both the c-$ZnTiO_3$ and the r-$ZnTiO_3$ crystal phases in ZM materials were proposed as the main active phases for the photocatalytic degradation of phenol.

## References

[1] Michałowicz J and Duda W 2007 *Pol. J. Environ. Stud.* **16** 347.
[2] Ahmed S, Rasul M G, Martens W N, Brown R and Hashib M A 2011 *Water Air Soil Pollut.* **215** 3
[3] Mathew S, Kumar Prasad A, Benoy T, Rakesh P P, Hari M, Libish T M and Vallabhan C P G 2012 *J. Fluoresc.* **22** 1563.
[4] Ahmed S, Rasul M G, Martens W N, Brown R and Hashib M A 2010 *Desalin.* **261** 3.
[5] Choquette-Labbé, M, Shewa W A, Lalman J A and Shanmugam S R 2014 *Water* **6** 1785.
[6] Chen F, Yu C, Wei L, Fan Q, Ma F, Zeng J and Ji H 2019 *Sci. Total Environ.* **706** 136026.
[7] Hong Z, Wei M, Deng Q, Ding X, Jiang L and Wei K 2010 *Chem. Commun.* **46** 740.
[8] Qu Y, Zhou W, Ren Z, Wang G, Jiang B and Fu H 2014 *Chem. Cat. Chem.* **6** 2258.
[9] Zhao M, Bastakoti B P, Li, Y, Xu H, Ye J, Liu Z and Yamauchi Y 2015 *Chem. Commun.* **51** 14582.
[10] Sutanto N, Saharudin K A, Sreekantan S, Kumaravel V and Akil H M 2019 *J. Sol-Gel Sci. Technol.*

**93** 354.
[11] Badli N A, Ali R, Bakar W A W A and Yuliati L 2017 *Arab. J. Chem.* **10** 935.
[12] Andita K R, Kurniawan R and Syoufian A 2019 *Indones. J. Chem.* **19** 761.
[13] Tian H, Wang S, Zhang C, Veder J P, Pan J, Jaroniec M and Liu J 2017 *J. Mater. Chem. A.* **5** 11615.
[14] Wang L, Kang H, Xue D and Liu C 2009 *J. Cryst. Growth* **311** 611.
[15] Lokesh B and Rao N M 2016 *J. Mater. Sci. Mater. Electron.* **27** 4253.
[16] Arin J, Thongtem S, Phuruangrat A and Thongtem T 2017 *Mater. Lett.* **193** 270.
[17] Li X, Xiong J, Huang J, Feng Z and Luo J 2019 *J. Alloys Compd.***774** 768.
[18] Budigi L, Nasina M R, Shaik K and Amaravadi S 2015 *J. Chem. Sci.* **127** 509.
[19] Wang C L, Hwang W S, Chang K M, Co H H, Hsi C S, Huang H H and Wang M C 2011 *Int. J. Mol. Sci.* **12** 935.
[20] Chang Y S, Chang Y H, Chen I G, Chen G J, Chai Y L, Fang T H and Wu S 2004 *Ceramics Int.* **30** 2183.
[21] Jose M, Elakiya M and Dhas S A M B 2017 *J. Mater. Sci. Mater. Electron.* **28** 13649.
[22] Lokesh B, Madhusudhana Rao N, Kaleemulla S and Sivakumar A 2015 *Chemical Papers* **69** 1.
[23] Arin J, Thongtem S, Phuruangrat A and Thongtem T 2016 *Res. Chem. Intermed.* **43** 3183.
[24] Eskandarloo H, Badiei A, Behnajady M A, Tavakoli A and Ziarani G M 2016 *Ultrason. Sonochem.* **29** 258.
[25] Mrázek J, Spanhel L, Chadeyron G and Matějec V 2010 *J. Phys. Chem. C* **114** 2843.
[26] Viezbicke B D, Patel S, Davis B E and Birnie D P 2015 *Phys. Status Solidi B* **252** 1700.
[27] Mullerova J, Sutta P, Medlin R, Netrvalova M, Novak P 2017 *J. Electr. Eng.* **68** 10.
[28] Davis K, Yarbrough R, Froeschle M, White J and Rathnayake H 2019 *RSC Advances* **9** 14638.
[29] Mebrek A, Alleg S, Benayache S and Benabdeslem M 2018 *Ceram. Int.* **44** 10921.
[30] García-Ramírez E, Mondragón-Chaparro M and Zelaya-Angel O 2012 *Appl. Phys. A* **108** 291.
[31] Yu C, Chen F, Zhou W, Xie Y, Zeng D, Liu Z, Wei L, Yang K and Li D 2019 *Nanoscale* **11** 7720.
[32] Babu B C, Naresh V, Prakash B J and Buddhudu S 2011 *Ferroelectr. Lett. Sect.* **38** 114.
[33] Wang J X, Huang J, Xie H L and Qu A L 2014 *Int. J. Hydrog. Energ.* **39** 6354.
[34] Mohammadi M R and Fray D J 2010 *J. Eur. Ceram. Soc.* **30** 947.
[35] Liu G, Li G, Qiu X and Li L 2009 *J. Alloys Compd.* **481** 492.
[36] Musić S, Popović S, Maljković M and Dragčević Đ 2002 *J. Alloys Compd.* **347** 324.
[37] Nickeslat A, Amin M M, Izanloo H, Fatehizadeh A and Mousavi S M 2013 *J. Environ. Public Health* **815310** 1.
[38] Zhou G, Sun H, Wang S, Ang H M and Tadé M O 2011 *Sep. Purif. Technol.* **80** 626.
[39] Lei S, Fan H, Ren X, Fang J, Ma L and Liu Z 2017 *J. Mater. Chem. C* **5** 4040.

**Acknowledgements**
This research was financially supported by the Directorate General of Strengthening Research and Development, Ministry of Research, Technology, and Higher Education of the Republic of Indonesia through the Higher Education Excellent Applied Research Grant (PTUPT 2020, No. 041/SP2H/AMD/LT/MULTI/L7/2020 and No. 002/MACHUNG/LPPM/SP2H-LIT-MULTI/AMD/VI/2020).

# $TiO_2$/AC Composites for Adsorption-Photocatalytic of Methyl Orange

**C W Kartikowati[1*], A L Wulansari[2], B Poerwadi[1], Supriyono[1], A F Arif[3], T Sulistyaningsih[2], and O Arutanti[4*]**

[1]Departement of Chemical Engineering,Universitas Brawijaya, Malang, Indonesia
[2]Department of Chemistry, Universitas Negeri Semarang, Semarang, Indonesia
[3]Dept. New Investment PT. Rekayasa Industri, Jakarta, Indonesia
[4]Research Center for Chemistry, Indonesian Institute of Sciences, Serpong Indonesia

*Corresponding Authors: christinawahyu@ub.ac.id and osiarutanti@yahoo.com

**Abstract**. Herein, the adsorption-photocatalytic performance of composite titanium dioxide ($TiO_2$) and Activated Carbon (AC) to decompose methyl orange was investigated systematically. This work demonstrated the synthesis of $TiO_2$/AC composite via the sol-gel method. After getting the composite $TiO_2$/AC under different ratios of AC, the prepared particles were then annealed at 500°C for 2 h. The presence of AC detected by Fourier transform infrared (FTIR) spectra at 1028 nm. Based on Brunauer–Emmett–Teller (BET) analysis, the result showed that the prepared particles specific surface area increased by increasing the AC ratio. The prepared photocatalyst was used to decompose methyl orange under UV light irradiation for 90 minutes. From the photocatalytic performance, the additional AC did not influence the adsorption process significantly under dark conditions. Interestingly, during the irradiation process, methyl orang could be decomposed until 62.5%, which means 15% higher than that of bare $TiO_2$. The present result showed that an additional AC could enhance photocatalytic performance due to its ability as an electron transfer and avoid the recombination process between electrons and holes.

## 1. Introduction

More than 50% of pollutant water contaminant consists of non-biodegradable organic pollutant produced from human activities. This hazardous pollutant is well thought-out as toxics to humans, including human hepatic dysfunction, carcinogenic, endocrine, etc. [1]. Typical wastewater treatment technologies have been developed during the past decade, including biological and physical treatments to eliminate assorted types of contaminants [2-4] effectively. Unfortunately, these techniques are cost-ineffective and lead to the production of secondary products or pollution. In this regard, non-destructive, green, and sustainable water treatment technology is highly desired.

Among the previous technologies, photocatalytic treatment has been considered one of the appealing options for wastewater treatment due to its ability to effectively decompose organic pollutants and harmful bacteria under appropriate photon energy [5-6]. The abundance of solar energy with a high intensity as a renewable light source making photocatalytic treatment appropriate to be developed and applied. Theoretically, photocatalysis works based on the ability of photocatalyst material in absorbing photon energy to create the pairs of electron (e-) and hole (h+) [6]. Here, the photon energy must be equal or higher than the band-gap energy of the photocatalyst material [7]. The numbers of e- and h+ are crucial in the oxidation-reduction reaction to produce the radical hydroxyl (OH*) or super anion ($O_2^-$) to decompose organic pollutants [7].

Compared to another photocatalyst material, Titanium dioxide ($TiO_2$) is the most effective and can potentially be applied due to its low-cost, simple preparation, and high stability [8,9]. Despite its merits, there are some drawbacks of $TiO_2$, including the wide bandgap and high charge recombination rate [10]. Given the wide bandgap, $TiO_2$ is active under UV radiation, which only accounts for 5% of the total solar light available in nature [11]. Extensive efforts have been devoted to improve the performance of $TiO_2$ as a catalyst [10,12-13]. However, the development of $TiO_2$ photocatalysts in terms of molecular structural variations so that they have high efficiency is still scarce. To increase the efficiency of $TiO_2$ nanostructuring (composite, hollow, porous, etc.) and functionalization (impurity doping, reduction, etc.) [10, 14-18] must be carried out. $TiO_2$ composites with activated carbon (AC) are effective ways to improve the photocatalyst performance.

Activated carbon (AC) has a large surface area with an internal pore structure of 25-100% of the weight of activated carbon [19,20]. AC can improve photocatalytic activity through the adsorption-catalytic process, which increases the contact between pollutants and the catalyst increasing the photodegradation rate [19-23]. Therefore, the research was carried out to synthesize $TiO_2$/AC composites via the sol-gel method to decompose waste organic compounds. To confirm, the fundamental reason for this catalytic improvement was the effect of $TiO_2$/AC ratio on crystallinity, particles structure, and surface area. In this study, the photocatalytic process was conducted under UV light irradiation.

## 2. Experimental method

$TiO_2$/AC composites were prepared by the sol-gel method. Titanium (IV) ethoxide (technical grade, Sigma Aldrich, USA), activated carbon (Sigma Aldrich, USA), and ethanol (absolute, ≥99.8%, Merck, USA) were used as precursors without further purification. The composites were synthesized under different ratios of AC i.e. 1:1, 1:3, 1:5, 1:7. 5.24 mL of titanium (IV) ethoxide was dissolved in 20 ml of ethanol and stirred for 15 minutes. AC was added and stirred for 10 minutes. 10 mL of distilled water was added and followed by stirring for 1 hour until forming a gel. The gel is then oven for 12 hours at 100 °C. The obtained powder was annealed at 500 °C for 2 hours in a furnace.

To examine the crystallinity and crystal size of the $TiO_2$/AC composites, an X-ray diffraction (XRD; Rigaku Denki RINT 2000, Japan; using Cu Kα radiation and a 2θ scanning range of 20−80°). The prepared composites surface area was analyzed by using the Brunauer Emmett−Teller measurement (BET; BELSORP 28SA, Bel, Japan, nitrogen adsorption isotherms at 77.15 K). Fourier-transform infrared spectroscopy was used to determine the functional groups of the $TiO_2$ framework and UV-Vis Diffuse Reflectance spectrophotometer to determine the composites band gap energy.

The photocatalytic activity of the $TiO_2$/AC composites was assessed by mixing 0.2 g of the prepared composites with 4 ppm methyl orange (MO, Merck, USA). The mixed solution was introduced into the batch reactor equipped with a solar simulator system. 4 mL of the mixed solution sampled for several minutes. Prior to analysis, the sampled solution was then centrifuged at 12,000 rpm for 10 min. The MO concentration in the sample was measured using a UV−Vis spectrophotometer (UV3150; Shimadzu Corp., Japan) in the wavelength range of 200-700 nm. The photocatalytic degradation rate (k) was measured using Langmuir-Hinshelwood's pseudo first order kinetic[24]. Where C is the concentration of MO and k is the photocatalytic degradation rate.

$$k = -\frac{ln\left({}^{C}/_{C_0}\right)}{t} \quad (1)$$

## 3. Results and discussion

The structure and crystallite sizes of the $TiO_2$/AC composites were identified by XRD patterns, as depicted in Figure 1. The XRD patterns show that all prepared composites were consistent with the pure $TiO_2$ anatase of standard ICDD no. 04-014-8515. This number was indexed to the tetragonal structure of $TiO_2$ with orientation 101, 004, and 200. The crystallite sizes were determined from the broadening peal area by using the Scherrer method to be decreasing with an increase in AC content from 111.97 to 68.68 nm. This is because the AC has an amorphous structure, which affects the crystallinity of $TiO_2$. The phase and pattern of the $TiO_2$ nanoparticles (NPs) and $TiO_2$/AC composites were similar, suggesting that AC did not induce a change in the structure of $TiO_2$ NPs.

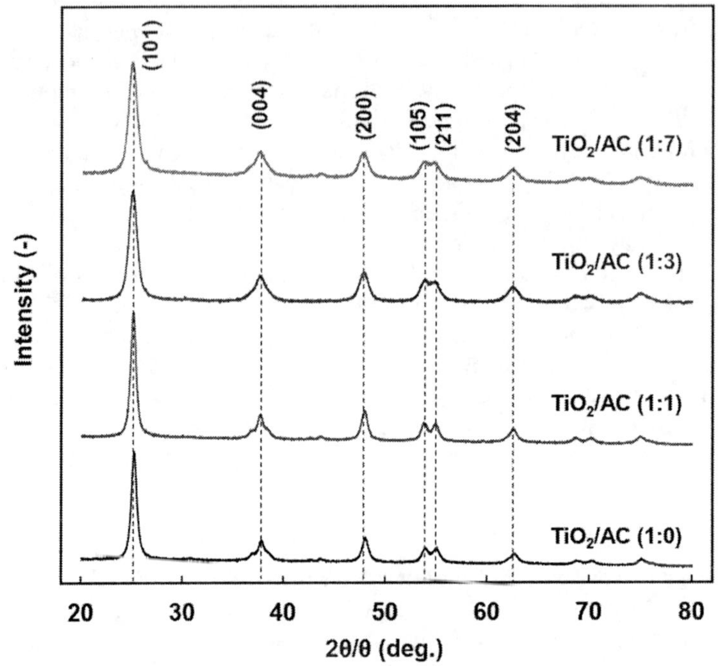


**Figure 1.** XRD paterns of prepared $TiO_2$/AC composites.

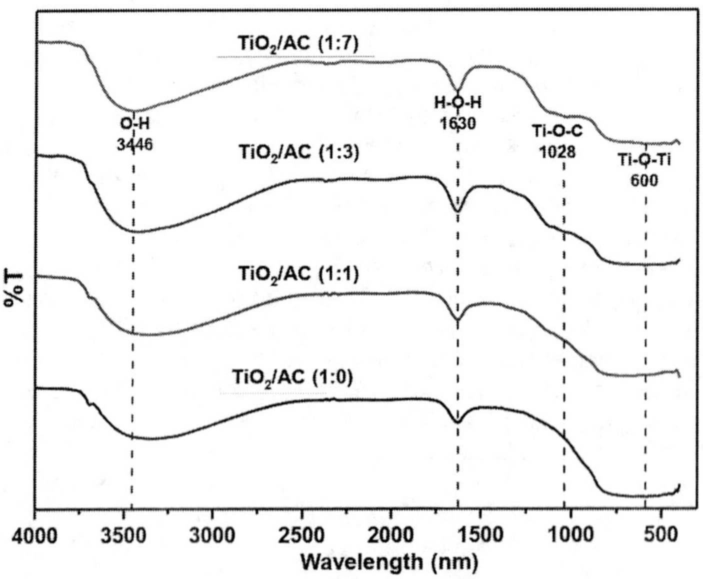


**Figure 2.** FTIR spectra of prepared $TiO_2$/AC composites.

IOP Conf. Series: Materials Science and Engineering 1143 (2021) 012077 doi:10.1088/1757-899X/1143/1/012077

Figure 2 showed the FTIR spectrum of prepared $TiO_2$/AC composites with various compositions. The Ti-O-Ti bond was detected at a wavelength of 600 $cm^{-1,}$ indicating the formation of $TiO_2$. The O-H group at a wavelength of 3369 $cm^{-1}$ and H-O-H group at wavelength 1630 $cm^{-1}$ indicate water absorption during the synthesis process. This result showed that the addition of AC particles only affects the presence of the Ti-O-C group.

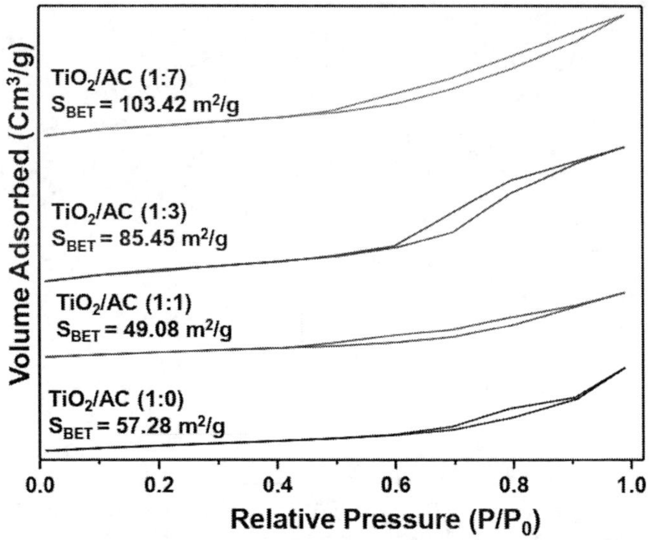


**Figure 3.** Nitrogen adsorption of the prepared particles under various ratio composition.

The surface area of the $TiO_2$/AC composites was determine using BET characterization. Figure 3 shows the nitrogen adsorption analysis of the prepared composites. The result showed that varying AC concentration influenced the specific surface area and porosity from 57.28 $m^2/g$ to 103.42 $m^2/g$ and 0.16 $cm^3/g$ to 0.23 $cm^3/g$, respectively. The high specific surface area possibility produced a better photocatalyst.

The photocatalysis performance of the prepared $TiO_2$/AC composites to degrade 4 ppm methyl orange was examined. Figure 4 shows normalized MO concentrations (concerning the initial concentration) as a function of time. The photocatalytic experiments were conducted under varying the prepared composite $TiO_2$/AC. The solid line represents the pure $TiO_2$ particles, whereas the dotted line shows the $TiO_2$/AC ratio variation. The MO concentrations gradually decreased over time. Increases in the $TiO_2$/AC ratio from 1:1 to 1:7 for the synthesis of the catalyst resulted in an improvement in photocatalytic activity. From the photocatalytic performance, the additional AC did not influence the adsorption process significantly under dark conditions. However, the MO could be decomposed until 62.5%, which means 15% higher than that of bare $TiO_2$. This is showed that an additional AC could enhance photocatalytic performance due to its ability as an electron transfer and avoid the recombination process between electrons and holes. On the $TiO_2$/AC ratio of 1:1, the photocatalytic activity was not better than bare $TiO_2$ particles. This is due to the composite's surface area and porosity that smaller than that of bare $TiO_2$ particles. These results show that the addition of the AC particles, which is an adsorbent, improves the $TiO_2$ photocatalyst materials' photocatalytic activity. Besides increasing the surface area to increase the absorbance ability, the addition of AC also affects the charge transfer electron. Furthermore, the synergistic effect of AC adsorpsion and $TiO_2$ photochatalysis can improve the photocatalytic activity [25-27]. The summary of comparison of MO photodegradation by composite $TiO_2$/AC photocatalyst is proposed in Table 1. A view of paper is just for overview of the previous works related to the composite $TiO_2$ and AC to decompose MO. With the smallest surface area, the current result show the highest photodegradation activity. The different of presentation degradation was due to the different of characteristic from the produced photocatalysts which influenced by material source, experimental method, and $TiO_2$ content. Furthemore, Different from the previous works, here, $TiO_2$ technical grade priced at 0.6 USD per kG has been used. To enhance the photocatalytic performance, more optimization is still required, but still focus on the utilization of a cheaper material resources.

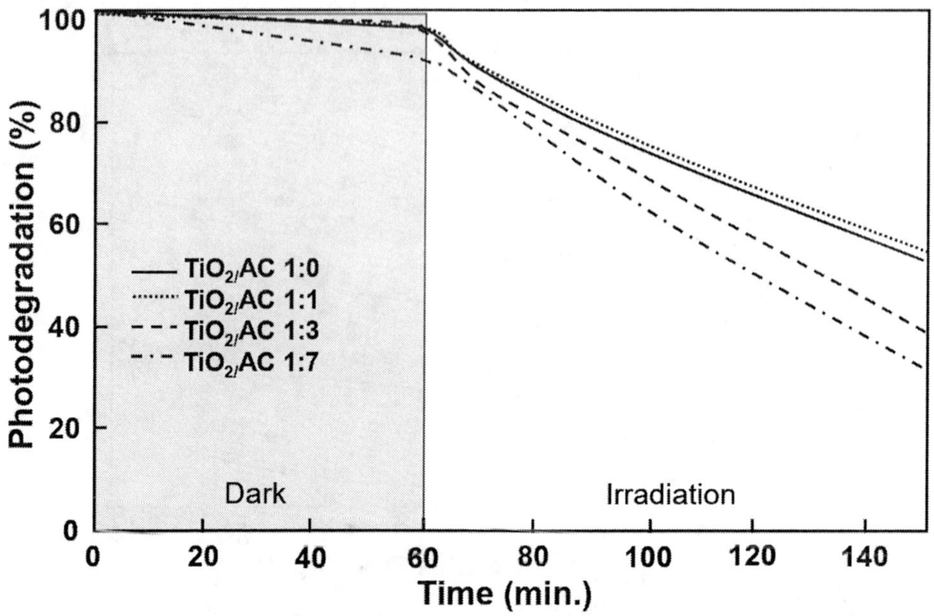


**Figure 4.** Photodecomposition of Methyl Orange by the prepared particles over the time.

**Table 1.** Summary of photodegradation of MO by composite *TiO2/AC photocatalist.*

| No. | Light source | C, V, catalyst | BET surface area ($m^2\ g^{-1}$) | %, t | Ref. |
|---|---|---|---|---|---|
| 1. | Hg-lamp | 20 mg $L^{-1}$, 40 mL, 10 mg | - | 97 %, 150 min | [25] |
| 2. | UV-light | 50 mg $L^{-1}$, 250 mL, 50 mg | 985 | 97%, 60 min | [26] |
| 3. | UV-light | 50 mg/L, 250 mL, 50 mg | 980 | 95%, 60 min | [27] |
| 4 | UV-light | $10^{-3}$ mol $L^{-1}$, 200 mL, 60 m g | 456 | 81 %, 100 min | [28] |
| 5. | UV-light | 4 ppm, 100 mL, 20 mg | 85.45 | 63%, 90 min | current |

The effect of the composite ratio on the physical properties and photocatalytic performance is shown in figure 5. The diameter of the composite increased by increasing AC concentration. To evaluate the photocatalytic degradation rate (k), a simplified Langmuir-Hinshelwood kinetics model was used. As observed, the degradation rate of MO was dependent on the $TiO_2$/AC ratio. The highest photocatalytic rate was obtained using $TiO_2$/AC with a ratio of 1:3, whereas the lowest was obtained using the $TiO_2$/AC ratio with a ratio of 1:7.

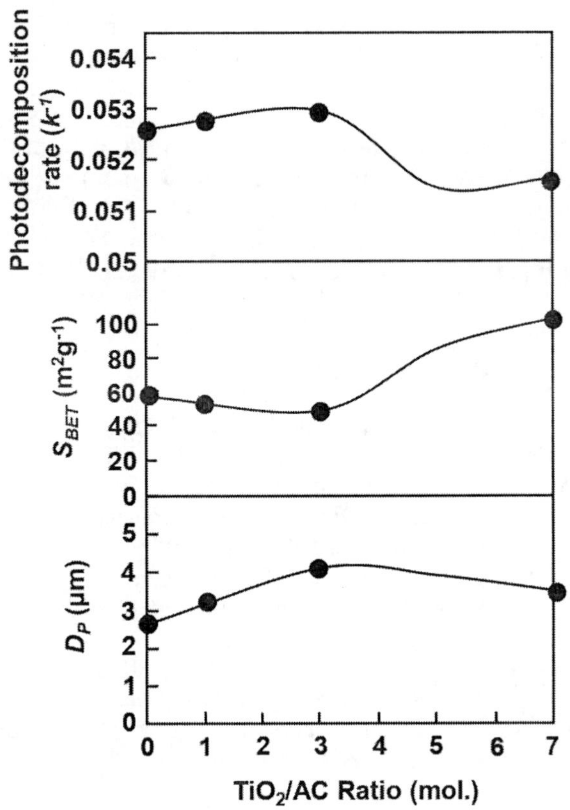


**Figure 5.** Effect of the composite ratio on the physical properties and photocatalytic performance.

## 4. Conclusion

The composites titanium dioxide ($TiO_2$) and Activated Carbon (AC) were successfully synthesized via the sol-gel method by varying the $TiO_2$/AC ratio. The adsorption-photocatalytic performance of prepared composite materials to decompose methyl orange was investigated systematically. After getting the composite $TiO_2$/AC under different ratios of AC, the prepared particles were then annealed at 500°C for 2 h. The result showed that the prepared particles' specific surface area and porosity increased by increasing the AC ratio. The prepared photocatalyst was used to decompose methyl orange under UV light irradiation for 90 minutes. From the photocatalytic performance, the additional AC did not influence the adsorption process significantly under dark conditions.

Interestingly, during the irradiation process, methyl orang could be decomposed until 62.5%, which means 15% higher than that of bare $TiO_2$. The best photocatalytic performance was obtained using $TiO_2$/AC with a ratio of 1:3. The present result showed that an additional AC could enhance photocatalytic performance due to its ability as an electron transfer and avoid the recombination process between electrons and holes. The MO's degradation on the $TiO_2$/AC composite was facilitated by the synergistic relationship between surface adsorption characteristics and photocatalytic potential.

## References

[1] Corominas L I, Foley J, Guest J S, Hospido A, Larsen H F, Morera S, Shaw A 2013 Life cycle assessment applied to wastewater treatment: State of the art *Water Research* **47** 5480-92

[2] Chong M N, Jin B, Chow C W K, and Saint C 2010 Recent Developments in Photocatalytic Water Treatment Technology: A Review *Water Research* **44** 2997-3027

[3] Banerjee P, Das R, Das P, and Mukhopadhyay A 2018 Membrane Technology *Carbon Nanostructures*
[4] Von Sperling M 2015 Wastewater Characteristics, Treatment and Disposal *Water Intell. Online*
[5] Skubi K L, Blum T R, and Yoon T P 2016 Dual Catalysis Strategies in Photochemical Synthesis *Chemical Reviews* **116** 10035–74
[6] Arora A K, Jaswal V S, Singh K, and Singh R 2016 Applications of Metal/Mixed Metal Oxides as Photocatalyst: A Review *Oriental Journal of Chemistry* **32** 2035-42
[7] Hernández-Ramírez, Aracely, Medina-Ramirez, and Iliana 2015 Photocatalytic Semiconductors: Synthesis, Characterization, and Environmental Applications
[8] Fujishima A, Rao T N, and Tryk D A 2000 Titanium Dioxide Photocatalysis. *Journal of photochemistry and Photobiology C: Photochemistry Reviews* **1** 1-21
[9] Xing Z, Zhang J, Cui J, Yin J, Zhao T, Kuang J, Xiu Z, Wan N, and Zhou W 2018 Recent Advances in Floating $TiO_2$-Based Photocatalysts for Environmental Application *Applied Catalysis B: Environmental* **225** 452-467
[10] Zaleska A 2008 Doped-$TiO_2$: A Review *Recent Patents on Engineering* **2** 157-164
[11] Borges M E, Sierra M, Cuevas E, García R D, and Esparza P 2016 Photocatalysis with Solar Energy: Sunlight-Responsive Photocatalyst Based on $TiO_2$ Loaded on a Natural Material for Wastewater Treatment. *Solar Energy* **135** 527-535
[12] Chen X and Mao S S 2007 Titanium Dioxide Nanomaterials: Synthesis, Properties, Modifications and Applications *Chemical Reviews* **107** 2891–2959
[13] Wenderich K and Mul G 2016 Methods, Mechanism, and Applications of Photodeposition in Photocatalysis: A Review *Chemical Reviews* **116** 14587–619
[14] Arutanti O, Nandiyanto A B D, Ogi T, Iskandar F, Kim T O, and K Okuyama 2014 Synthesis of Composite $WO_3/TiO_2$ Nanoparticles by Flame-Assisted Spray Pyrolysis and Their Photocatalytic Activity *Journal of Alloys and Compounds* **591** 121-126
[15] Habisreutinger S N, Schmidt-Mende L, and Stolarczyk J K 2013 Photocatalytic Reduction of $CO_2$ on $TiO_2$ and Other Semiconductors *Angewandte Chemie - International Edition* **52** 7372-7408
[16] Abdullah H, Khan M M R, Ong H R, and Yaakob Z 2017 Modified $TiO_2$ Photocatalyst for $CO_2$ Photocatalytic Reduction: An Overview *Journal of $CO_2$ Utilization* **22** 15-32
[17] Xinghou H, Anzhi W, Pian W, Shibiao T, Yong Z, Lei L, and Ping D 2020 Photocatalytic degradation of microcystin-LR by modified $TiO_2$ photocatalysis: A review *Science of The Total Environment* **743** 140694
[18] Ahmad J J and Mehrdad M 2020 Synthesis of Fe-Doped TiO2 for Photocatalytic Processes under UV-Visible Light: Effect of Preparation Methods on Crystal Size—A Systematic Review Study *Comments on Inorganic Chemistry* **4** 6
[19] Rowan L and Aidan W 2011 Carbonaceous nanomaterials for the enhancement of $TiO_2$ photocatalysis *Carbon* **49** 741-772
[20] Meltem A and Şadiye Ş 2012 $TiO_2$-activated carbon photocatalysts: Preparation, characterization and photocatalytic activities *Chemical Engineering Journal* **180** 354-363
[21] Wendong W, Cláudia G, Silva J, and Luís F 2007 Photocatalytic degradation of Chromotrope 2R using nanocrystalline $TiO_2$/activated-carbon composite catalysts *Applied Catalysis B: Environmental* **70** 470-478
[22] García, E B, Elmouwahidi A, Álvarez M A, Carrasco-marín F, Pérez-cadenas A F, and Maldonado-hódar F J 2017 New Carbon Xerogel-$TiO_2$ Composites With High Performance As Visible-Light Photocatalysts For Dye Mineralization *Applied Catalysis B, Environmental* **201** 29–40
[23] Gang X, Huanhuan L, Quanyuan C, Colin H, Mark T, and Francis I 2011 Synergy between surface adsorption and photocatalysis during degradation of humic acid on $TiO_2$/activated carbon composites *Journal of Hazardous Materials* **186** 765-772

[24] Nandianto A. B. D. N., Iskandar F., and Okuyama K 2009 Macroporous anatase titania particle: Aerosol self-assembly fabrication with photocatalytic performance *Chemical Engineering Journal* **152** 293-296

[25] Lu L, Shan R, Shi Y, Wang S, and Yuan H 2019 A novel $TiO_2$/biochar composite catalysts for photocatalytic degradation of methyl orange *Chemosphere* **222** 391-398

[26] Wang X, Hu Z, Chen Y, Zhao G, Liu Y, and Wen Z 2009 A novel approach towards high-performance composite photocatalyst of $TiO_2$ deposited on activated carbon *Applied Surface Science* **255** 3953–58

[27] Wang X, Liu Y, Hu Z, Chen Y, Liu W, and Zhao G 2009 Degradation of methyl orange by composite photocatalysts nano-$TiO_2$ immobilized on activated carbons of different porosities *Journal of Hazardous Materials* **169** 1061–67

[28] Li Y, Li X, Li J, and Yin J 2006 Photocatalytic degradation of methyl orange by $TiO_2$-coated activated carbon and kinetic study *Water Research* **40** 1119– 26

# AUTHOR INDEX